AF709884

Madan Lal Puri Selected Collected Works
Volume 3

Madan Lal Puri

Dedication

These three volumes of "Selected Collected Works of Madan Lal Puri" will serve to preserve, in a unified and easily accessible form, the knowledge and wisdom conveyed in his many research papers, so as to aid its dissemination to future generations.

The Editors

MADAN LAL PURI
SELECTED COLLECTED WORKS

VOLUME 3

TIME SERIES, FUZZY ANALYSIS AND MISCELLANEOUS TOPICS

Editors

Peter G. Hall, Marc Hallin and George G. Roussas

///VSP///
Utrecht • Boston
2003

VSP
P.O. Box 346
3700 AH Zeist
The Netherlands

Tel: +31 30 692 5790
Fax: +31 30 693 2081
vsppub@compuserve.com
www.vsppub.com

© VSP 2003

First published in 2003

ISBN 90-6764-386-6 (Volume 3)
ISBN 90-6764-387-4 (Set)

All rights reserved. No part of this publication may be reproduced, stored in a retrieval system, or transmitted in any form or by any means, electronic, mechanical, photocopying, recording or otherwise, without the prior permission of the copyright owner.

Printed in The Netherlands by Ridderprint bv, Ridderkerk

CONTENTS

PREFACE

The volumes at hand are the outcome of a concerted effort to make Professor Puri's research works easily available to the research community. The sheer volume of the research output by him and his collaborators, coupled with the broad spectrum of the subject matters investigated, and the great number of outlets where the papers were published, attach special significance in making these works easily accessible.

In compiling these volumes, the Editors are satisfied with the feeling of discharging part of their duty to the profession, and simultaneously expressing their respect and admiration for their colleague, Madan Puri, for his immense contributions to statistics and probability literature. A generous dose of appreciation is due to the publishers of the volumes. The VSP International Science Publishers, for undertaking a less than profitable venture. Also, thanks are due to professional organizations, as well as the private publishers who graciously waived all reprinting copyright fees with a deep sense of service to the profession. An appropriate list of said entities is given at the end of this preface, whereas suitable acknowledgements are cited at the end of each paper in the three volumes.

Professor Puri's published research works number more than 3,000 pages. However, practical publishing considerations had to be taken into account, and expectedly they took their toll. Accordingly, the three volumes would have to comprise a total of about 2,000 pages. The burden of selection was not easy, and works omitted are not to be considered inferior to those selected for inclusion. The need of exclusion also explains the title assigned to this work.

The papers selected for inclusion in this work have been classified into three volumes, each consisting of several parts. Thus, Volume 1 consists of 44 papers distributed into four parts as follows: Part I of 24 papers, falling into the area of *Nonparametric Methods in Univariate Analysis*; Part II of 6 papers from *Nonparametric Methods in Multivariate Analysis*; Part III of 4 papers from *Nonparametric Methods in Design and Analysis of Experiments*; and, finally, Part IV consisting of 10 papers in *Miscellaneous Topics*, 4 of which (#'s [40]–[43]) are also from *Nonparametric Methods in Multivariate Analysis*. Volume 2 consists of 35 papers classified in three parts as follows: Part I of 18 papers under the heading of *Limit Theorems, Rates of Convergence and Related Topics* (*Independent Case*); Part II of 13 papers taken from his contributions in *Limit Theorems* (*Dependent Case*); and Part III consisting of 4 papers from the area of *Extreme Value Theory*. Volume 3 consists of a total of 48 papers distributed into three parts as described below: Part I of 18 papers under the heading of *Time Series and Related Topics*; Part II of 14 papers falling into the area of *Fuzzy Set Theory and Related Topics*; and, finally, Part III comprising 16 papers from *Miscellaneous Topics*. Also, all three volumes carry a final part consisting of the contents of the other two volumes, as well as the complete list of Professor Puri's publications.

It would be appropriate that a brief biographical sketch of Professor Puri be included in this preface. The following few paragraphs are excerpts from the preface of the book *Asymptotics, Nonparametrics, and Time Series: A Tribute to Madan Lal Puri*, edited by Subir Ghosh and published by Marcel Dekker, Inc. in 1999. One of the present Editors was a co-author of that preface.

Madan Lal Puri was born in Sialkot (then in India, now in Pakistan) on February 20, 1929. In 1947, when India gained her independence and Pakistan was created, his family migrated to Delhi as refugees. He received a B. A. degree in 1948 and an M. A. degree in 1950, both in mathematics, from Panjab University in India. From January 1951 to August 1957, he served as a Lecturer in Mathematics in different colleges of Panjab University.

In September 1957, he came to the United States as an instructor and graduate student in mathematics at the University of Colorado in Boulder. In September 1958, he moved to the University of California at Berkeley as a research assistant in the Department of Statistics and received his Ph. D. in statistics in 1962.

In 1962, Dr. Puri joined the renowned Courant Institute of Mathematical Sciences in New York University as an Assistant Professor and became an Associate Professor in 1965. He joined Indiana University at Bloomington in 1968 as a Full Professor of Mathematics and remains there to this day.

Professor Puri is one of the most versatile and prolific researchers in the world in mathematical statistics. His research areas include nonparametric statistics, order statistics, limit theory under mixing, time series, splines, tests of normality, generalized inverses of matrices and related topics, stochastic processes, statistics of directional data, random sets, and fuzzy sets and fuzzy measures. His fundamental contributions in developing new rank-based methods and precise evaluation of the standard procedures, asymptotic expansions of distributions of rank statistics, as well as large deviation results concerning them, span such areas as analysis of variance, analysis of covariance, multivariate analysis, and time series, to mention a few. His in-depth analysis has resulted in pioneering research contributions to prominent journals that have substantial impact on current research.

Professor Puri has done joint work with many researchers of different countries. To date he has collaborated with 89 scholars from 22 countries on 5 continents. He was the Alexander von Humboldt Guest Professor at the University of Göttingen in West Germany in 1974–1975 and Guest Professor at many other universities in Germany, with German National Science Foundation grants. He has been a Distinguished Visitor at the London School of Economics and Political Science, Visiting Professor at the University of Auckland in New Zealand, the Universities of Bern and Basel in Switzerland, the University of New South Wales in Australia, the University of Goteborg and Chalmers University of Technology in Sweden, Université des Sciences et

Technologies de Lille in France, Australian National University, Canberra, and University of Washington, Seattle, among other universities. In 1974, he was invited by the Japanese Society for the Promotion of Sciences to visit Japan under its Visiting Professorship Program to conduct cooperative research with Japanese scientists. He has been an invited speaker as well as a plenary speaker at many international conferences all over the world.

Professor Puri has received numerous honors and awards. He is an elected member of the International Statistical Institute, and a Fellow of the Institute of Mathematical Statistics, a Fellow of the American Statistical Association, and a Fellow of the Royal Statistical Society. In 1975, he was honored with the D. Sc. degree from Panjab University in India. He twice received the Senior U. S. Scientist Award from the Alexander von Humboldt Foundation in 1974 and 1983. In 1974, he was honored by the government of the Federal Republic of Germany, "in recognition of past achievements in research and teaching." In 1984, he received the best paper award from the Seventh European Meeting on Cybernetics and Systems Research, Vienna, Austria. In 1991, he received the Rothrock Faculty Teaching Award in recognition of outstanding teaching in the Department of Mathematics of Indiana University. He was ranked the ninth most prolific author in 1992, and the fourth most prolific author in 1997 in top statistical journals of the world.

Professor Puri has served on various committees of many international conferences in addition to those of the Institute of Mathematical Statistics and the American Statistical Association. He also served as Editor-in-Chief of the *Journal of Statistical Planning and Inference* in 1984–1988.

Professor Puri has directed 16 Ph. D. dissertations. Most of his former Ph. D. students are in research and teaching positions at respectable universities. A few hold responsible positions in industry.

Professor Puri is truly an international academician and a peripatetic scholar who works with missionary zeal. Scientists from all over the world visit him regularly and do research with him while staying at his home. His office and home have always been wide open to bright young scientists from the United States and Overseas, who were more in need of sponsorship and gentle encouragement and guidance in their professional endeavors rather than mere mathematical mentoring. He is a caring colleague with the warmest affection, an international host, a persuasive communicator, a dedicated as well as an outstanding teacher, and a versatile statistician whose work continues to inspire the scientific community.

We are editing these volumes in the hope of facilitating the availability to the research community of a substantial part of Professor Puri's work. We take great pleasure in doing so.

This project has benefited greatly from the generous financial support of Moya Andrews, Vice-Chancellor for Academic Affairs and Dean of the Faculties, Indiana University; Patrick O'Meara, Dean of the International Programs, Indiana University; Curtis R. Simic, President, Indiana University Foundation; Kumble R. Subbaswamy, Dean of the College of Arts and Sciences, Indiana University; and George Walker, Vice-President for Research and Dean of the Graduate School, Indiana University. Special thanks go to Kenneth R. R. Gros Louis, currently Chancellor-Emeritus and

Trustee Professor, Indiana University, for his never-ending enthusiastic encouragement and financial support for several of Professor Puri's research projects during his tenure as Chancellor of Indiana University, Bloomington Campus, and Vice-President for Academic Affairs, Indiana University.

The reprinting copyright fee waivers granted by professional societies and private publishers have made this undertaking financially feasible. Their deep sense of service to our profession is gratefully acknowledged here. They are Akademiai Kiado, Biometrika Trustees (The Oxford University Press), Sankhyā, SIAM, The American Mathematical Society, The American Statistical Association, The Institute of Mathematical Statistics, The Royal Society, and The Statistical Society of Canada. Also, Academic Press, Blackwell Publishing, LTD, Cambridge University Press, Elsevier Science, John Wiley & Sons, Inc., Kluwer Academic / Plenum Publisher, and Springer Verlag GmbH & Co. KG.

July 2002

The Editors
PETER G. HALL
MARC HALLIN
GEORGE G. ROUSSAS

Part I

Time Series and Related Topics

LINEAR SERIAL RANK TESTS FOR RANDOMNESS AGAINST ARMA ALTERNATIVES

MARC HALLIN[1], JEAN-FRANCOIS INGENBLEEK and MADAN L. PURI[2]

Université Libre de Bruxelles and Indiana University

In this paper we introduce a class of linear serial rank statistics for the problem of testing white noise against alternatives of ARMA serial dependence. The asymptotic normality of the proposed statistics is established, both under the null as well as alternative hypotheses, using LeCam's notion of contiguity. The efficiency properties of the proposed statistics are investigated, and an explicit formulation of the asymptotically most efficient score-generating functions is provided. Finally, we study the asymptotic relative efficiency of the proposed procedures with respect to their normal theory counterparts based on sample autocorrelations.

1. Introduction. Nonparametric methods have been developed for the analysis of univariate, and, much later, for the analysis of multivariate observations as a reaction against the distributional assumptions (mostly, that of normality) on which much of the classical inference relies.

The need for nonparametric procedures is even stronger in the area of (univariate and multivariate) time-series analysis: indeed, while there are several quite tractable inference procedures for independently distributed nonnormal observations, almost nothing exists, in time series, except for methods based on the normal theory likelihood ratio approach.

Although many rank-based procedures have been developed in (non–time-series context) for testing randomness against different types of trend alternatives (see, e.g., Dufour et al. (1982)), one of the most significant problems for practical applications, viz. the problem of testing randomness against serial dependence using rank methods has not received much attention—apart from some scattered results here and there. Of course, even if not always specifically time series-oriented, some of the "historical" nonparametric tests, such as the *run test* and the *turning point test* can be used for testing the hypothesis of an independent time series against serial dependence—in fact, the history of the problem can be traced back to Hotelling and Pabst (1936), and Wald and Wolfowitz (1943) who proposed tests based on serial versions of Spearman's correlation coefficient referred to below as *Spearman's autocorrelation coefficient* (see Section 5.3). Jogdeo (1968)—also in a non–time-series context—studied the distribution (under independence) of a class of statistics which is not unrelated with our S_n statistics (to be defined below); however, because of the restrictions he puts on the score functions, his statistics are not adapted to time-series situations.

Received February 1984; revised March 1985.

[1] Research supported by a grant of the Belgium F.N.R.S.

[2] Research supported by National Science Foundation Grant MCS 8301409.

AMS 1980 *subject classifications.* Primary 62M10, 62G10.

Key words and phrases. Linear serial rank statistics, autoregressive, moving average, ARMA models, asymptotic relative efficiency.

Moreover, in none of these early papers is any particular alternative of serial dependence considered, nor any optimality question addressed.

The first attempt to investigate the power of serial rank procedures against specific alternatives of serial dependence is due to Knoke (1977), who studies the asymptotic relative efficiency of several tests based on rank statistics (e.g., *Spearman's first-order autocorrelation coefficient* or the *turning point statistic*) with respect to the first-order sample autocorrelation coefficient, for autoregressive alternatives of order one. This, in some sense, was the first step towards the application of rank methods in time-series analysis problems—but the statistics that he studies are not new, and are not specifically devised for the alternative considered. Recently, in a rather restrictive set-up, Gupta and Govindarajulu (1980) considered the first-order autoregressive and moving average *normal* alternatives, and derived a locally most powerful rank statistic which is a particular case of our *linear serial rank statistic* (cf. (1.1)). Aiyar (1981) proposed, on heuristic considerations, a test (still for first-order alternatives) based on a statistic which is also a particular case of an extended *van der Waerden statistic* which we arrive at in (5.4). Bartels (1982) introduces a rank version of *von Neumann's ratio test*, still for first-order autoregressive alternatives, which he shows to be more efficient than the run test and parametric von Neumann test (under normality).

In a somewhat different direction, a highly systematic and theoretically-based approach is provided in Bell et al. (1970), where the Pitman functions generating the most powerful distribution-free tests are derived for a null hypothesis of randomness against the alternative of a process with stationary independent increments and several other alternatives of dependence.

More recently, letting $Z_t = X_t X_{t-1}$, where $(X_1 \cdots X_n)$ denotes the observed series, Dufour (1982) applied to Z_t some well-known procedures, such as the *sign*, *Wilcoxon*, *signed rank* and *van der Waerden* tests for symmetry about zero. Although the statistics he introduces are new, they still are an adaptation of existing ones: no optimality considerations are made and whether multivariate extensions are possible is unknown. A sign test of the type studied in Dufour (1982) was also considered by Govindarajulu and Dwass (1983). An overall review of some of these procedures is given in Govindarajulu (1983).

In spite of the growing interest in time-series problems, there exists thus no systematic, coherent and unified rank order theory, on the model of the one developed in Hájek and Šidák (1967) and Puri and Sen (1971) for independent observations; the few available results constitute only a very incomplete and piecemeal approach, and the subject remains, to a large extent, unexplored.

The purpose of this paper is to undertake a first systematic time-series oriented study of this important problem of testing for randomness against the types of serial dependence usually considered in time-series analysis (autoregressive, moving average or mixed ARMA dependence). We propose to do so by considering the statistics of the form

$$S_n = (n-p)^{-1} \sum_{t=p+1}^{n} a_n(R_t^{(n)}, R_{t-1}^{(n)}, \cdots, R_{t-p}^{(n)}), \tag{1.1}$$

where $a_n(\cdots)$ is some given *score function* and $R_t^{(n)}$ is the rank of the observation made at time t in an observed series of length n. These statistics are what we call *linear serial rank statistics*. There are many reasons for considering this class of statistics. For example,

(a) it is intuitively a quite natural idea to account for serial dependence (of order 1 through p) by considering the ranks of successive $(p + 1)$-tuples of observations;

(b) special cases of the score functions a_n give traditional test statistics, such as the *run statistic* (with respect to the median)

$$a_n(i_1, i_2) = \begin{cases} 1 & \text{if} \quad (2i_1 - n - 1)(2i_2 - n - 1) < 0 \\ 0 & \text{if} \quad (2i_1 - n - 1)(2i_2 - n - 1) \geq 0, \end{cases}$$

the *turning point* statistic

$$a_n(i_1, i_2, i_3) = \begin{cases} 1 & \text{if} \quad i_1 > i_2 < i_3 \\ 1 & \text{if} \quad i_1 < i_2 > i_3 \\ 0 & \text{elsewhere,} \end{cases}$$

Spearman's rank correlation coefficient of order p (up to additive and multiplicative constants)

$$a_n(i_1, i_2, \cdots, i_{p+1}) = i_1 i_{p+1}/(n + 1)^2, \quad \text{etc.;}$$

(c) it can be shown that the locally most powerful rank statistic for testing randomness against autocorrelation (AR(1)) alternatives belongs to this class of statistics (cf. Ingenbleek (1980));

(d) multivariate versions can be considered (hence, among others, multivariate extensions of the run, turning point, . . . statistics); this latter argument is very important, since the need for results is particularly strong in the multivariate case.

We mainly develop here the asymptotic theory of the class of tests based on linear serial rank statistics, and our approach largely relies on Hájek's projection method and LeCam's results on *contiguous sequences* of hypotheses.

Denote by $H_0^{(n)}$ the null hypothesis under which the observed series (of length n) is *white noise*, and by $H_1^{(n)}$ the alternative under which it is generated by an ARMA(p_1, p_2) model of the form

$$X_t - n^{-1/2} \textstyle\sum_{i=1}^{p_1} a_i X_{t-i} = \varepsilon_t + n^{-1/2} \sum_{i=1}^{p_2} b_i \varepsilon_{t-i}. \tag{1.2}$$

Under fairly mild assumptions (Section 2), we show in Section 3 (Proposition 3.1) that the sequence of hypotheses $H_1^{(n)}$ is contiguous to $H_0^{(n)}$.

This result, together with a lemma by LeCam, allows us to derive the asymptotic distribution of S_n under the alternatives $H_1^{(n)}$ from the knowledge of the asymptotic joint distribution of S_n and the log-likelihood ratio under $H_0^{(n)}$; these distributions are obtained (Section 4) by introducing an adequate *U-statistic*.

The last section (Section 5) is devoted to *efficiency* considerations. We investigate the *asymptotic relative efficiency* (ARE) of linear serial rank statistics and provide (Proposition 5.1) an explicit form of the asymptotically most efficient one—hence, for particular density types (normal, logistic, double exponential, . . .), time-series versions of Wilcoxon's, van der Waerden's, median, . . . tests.

Next, we (Proposition 5.3) compare our nonparametric statistics with their classical (normal theory) competitors, based on autocorrelation coefficients. This comparison seems to be quite favorable to serial rank statistics: the ARE of the "optimal" linear serial rank test with respect to that based on its "optimal" normal theory competitor depends on the Fisher information associated with the white noise density type, taking value one in the case of normal distribution, $\pi^2/9$ in the case of logistic distribution and two in the case of double exponential distribution. Finally, a more detailed study of the first-order case (ARMA(1, 1), AR(1) or MA(1) alternatives) is given, the "optimal" serial rank statistic (in the sense of Proposition 5.1) having then the important additional property that it is *uniformly* "optimal" (over the whole range of ARMA(1, 1), AR(1) and MA(1) alternatives, and within the class of linear serial rank statistics).

2. Notation and basic assumptions. Throughout the paper, we denote by $\mathbb{Z} = \{0, \pm 1, \pm 2, \cdots\}$, $\mathbb{R}$ and $\mathbb{C}$, respectively, the sets of integer, real and complex numbers.

Let $\{\varepsilon_t;\ t \in \mathbb{Z}\}$ be a discrete-time stationary *white noise*, i.e., a sequence of independent and identically distributed random variables with means $E[\varepsilon_t] = 0$, $t \in \mathbb{Z}$; assume that it has a density $f(x)$, and that the following conditions are satisfied.

(i) ε_t has finite moments up to the third order; denote its variance by σ^2.

(ii) $f(x)$ is a.e. derivable, and its derivative $f'(x)$ satisfies $\int_{-\infty}^{+\infty} |f'(x)|\, dx < \infty$.

(iii) $f(x)$ has finite *Fisher's information* $I(f)$, i.e., $f(x)$ is *absolutely continuous on finite intervals*, and

$$0 < I(f) = \int_{-\infty}^{+\infty} \left(\frac{f'(x)}{f(x)}\right)^2 f(x)\, dx < \infty$$

(as usual, $I(f)$ is the Fisher information related to the location parameter family $\{f_\theta(x) = f(x - \theta) \mid \theta \in \mathbb{R}\}$).

(iv) Let $F(x)$ be ε_t's distribution function, and $F^{-1}(u) = \inf\{x \mid F(x) \geq u\}$, $0 < u < 1$. Put

$$\phi(F^{-1}(u)) = -f'(F^{-1}(u))/f(F^{-1}(u)) \quad 0 < u < 1.$$

This function can also be written a.e. as $\phi(x) = -f'(x)/f(x)$, $x \in \mathbb{R}$. Assume $\phi(x)$ is a.e. derivable, and its derivative $\phi'(x)$ satisfies (a.e.) a Lipschitz condition $|\phi'(x) - \phi'(y)| < A\,|x - y|$.

These conditions are satisfied by most of the densities considered in the theory of rank tests. However, they do not include Cauchy and stable distributions.

Under these conditions, we have (Hájek and Šidák (1967) Chapter I)

$$\int_{-\infty}^{+\infty} \phi(x) f(x)\, dx = 0 \tag{2.1}$$

and

$$\int_{-\infty}^{+\infty} \phi^2(x) f(x)\, dx = I(f). \tag{2.2}$$

It is also easy to check that

$$\int_{-\infty}^{+\infty} x\phi(x)f(x)\,dx = 1 \tag{2.3}$$

and

$$\int_{-\infty}^{+\infty} \phi'(x)f(x)\,dx = I(f). \tag{2.4}$$

Note also that $\sigma^2 I(f)$ is independent of the scale transformation (i.e., if we put $f_\sigma(x) = (1/\sigma)f_1(x/\sigma)$, then $\sigma^2 I(f_\sigma) = I(f_1)$).

3. Asymptotic distribution of likelihood ratios.

3.1. *The hypotheses.* Let $a_1, \cdots, a_{p_1}, b_1, \cdots, b_{p_2}$ be an arbitrary $(p_1 + p_2)$-tuple of real numbers, and consider the sequence

$$X_t^{(n)} - n^{-1/2} \textstyle\sum_{i=1}^{p_1} a_i X_{t-i}^{(n)} = \varepsilon_t + n^{-1/2} \sum_{i=1}^{p_2} b_i \varepsilon_{t-i}, \quad t \in \mathbb{Z}, \quad n = 1, 2, \cdots \tag{3.1}$$

of stochastic difference equations. For n sufficiently large, all the roots of the *characteristic equation*

$$z^{p_1} - n^{-1/2} \textstyle\sum_{i=1}^{p_1} a_i z^{p_1-i} = 0, \quad z \in \mathbb{C}$$

lie inside the unit-circle, and (3.1) (see, for example, Wold, 1954, page 99) generates a sequence of stationary processes $\{X_t^{(n)};\ t \in \mathbb{Z}\}$. Denote by $\mathbf{x}^{(n)} = (x_1^{(n)}, \cdots, x_n^{(n)})$ an observed realization of $\mathbf{X}^{(n)} = (X_1^{(n)}, \cdots, X_n^{(n)})$.

If $a_1 = a_2 = \cdots = a_{p_1} = b_1 = b_2 = \cdots b_{p_2} = 0$, the processes $\{X_t^{(n)}\}$ of course all coincide with the generating white noise process $\{\varepsilon_t\}$. The likelihood function of the observation $\mathbf{x}^{(n)}$ is then

$$\ell_n^0(\mathbf{x}^{(n)}) = \textstyle\prod_{t=1}^{n} f(x_t^{(n)}). \tag{3.2}$$

Denote by $H_0^{(n)}$ the sequence of simple (null) hypotheses consisting of these densities.

If (at least) a_{p_1} and b_{p_2} are both different from zero, the processes $\{X_t^{(n)}\}$ are stationary *autoregressive-moving average processes* of order p_1 and p_2 (shortly, ARMA(p_1, p_2) processes). Denote then by $G_{ab}^{(n)}\ (x_{t+1}, \cdots x_{t+p_1})$ the distribution function of p_1 successive values of $\{X_t^{(n)}\}$. The likelihood function of $\mathbf{x}^{(n)}$ is

$$\begin{aligned} \ell_n^1(\mathbf{x}^{(n)}) = \int \textstyle\prod_{t=1}^{n} f(x_t^{(n)} - n^{-1/2} \sum_{i=1}^{p_1} a_i x_{t-i}^{(n)} + \sum_{u=1}^{t-1} g_u^{(n)}(x_{t-u}^{(n)} - n^{-1/2} \sum_{i=1}^{p_1} a_i x_{t-u-i}^{(n)}) \\ + \textstyle\sum_{u=t}^{t+p_2-1} g_u^{(n)} e_{t-u})\, dG_{ab}^{(n)}(x_{-p_1+1}, \cdots, x_0) f(e_{-p_2+1}), \cdots, f(e_0)\, de_{-p_2+1}, \cdots, de_0, \end{aligned} \tag{3.3}$$

where the coefficients $g_u^{(n)}$ are the *Green's functions* associated with the moving average difference operator (L denotes the *lag operator*) $1 + n^{-1/2} \sum_{i=1}^{p_2} b_i L^i$ (cf. Appendix 1). Denote by $H_1^{(n)}$ the corresponding sequence of hypotheses: $H_1^{(n)}$ will be our sequence of alternatives (ARMA dependence); $H_0^{(n)}$ is the sequence of null hypotheses (randomness).

Particular cases of $H_1^{(n)}$ are

(a) the autoregressive case of order p_1 ($b_1 = b_2 = \cdots = b_{p_2} = 0$; $a_{p_1} \neq 0$)

(b) the moving average case of order p_2 ($a_1 = a_2 = \cdots = a_{p_1} = 0$; $b_{p_2} \neq 0$).

The corresponding likelihood functions can be obtained by letting, respectively, $g_u^{(n)} = 0 \; \forall n$, $u = 1, 2, \cdots$ or $a_i = 0$, $i = 1, \cdots, p_1$ in (3.3).

As we shall see, however, the asymptotic behaviour of the likelihood function $\ell_n^1(\mathbf{x}^{(n)})$ is quite similar under the three types of dependence considered (mixed ARMA, pure AR or pure MA).

3.2. *Contiguity.* In what follows, we shall omit the superscripts (n), and write $\mathbf{x}$, x_t, $\mathbf{X}$ and X_t for $\mathbf{x}^{(n)}$, $x_t^{(n)}$, $\mathbf{X}^{(n)}$ and $X_t^{(n)}$ respectively. Consider the likelihood ratio

$$L_n(\mathbf{x}) = \begin{cases} \ell_n^1(\mathbf{x})/\ell_n^0(\mathbf{x}) & \text{if } \ell_n^0(\mathbf{x}) > 0 \\ 1 & \text{if } \ell_n^1(\mathbf{x}) = \ell_n^0(\mathbf{x}) = 0 \\ \infty & \text{if } \ell_n^1(\mathbf{x}) > \ell_n^0(\mathbf{x}) = 0. \end{cases} \tag{3.4}$$

It follows from LeCam's first lemma (we adopt here Hájek and Šidák's (1967) terminology) that, in order to establish that a sequence of hypotheses $H_1^{(n)}$ is contiguous to a sequence $H_0^{(n)}$, it is sufficient to show that $\log L_n(\mathbf{X})$ is asymptotically normal (under $H_0^{(n)}$), with mean $-d^2/2$ and variance d^2.

The following proposition therefore implies that $H_1^{(n)}$ is contiguous to $H_0^{(n)}$.

Proposition 3.1. *Under* $H_0^{(n)}$, $\log L_n(X_1, \cdots, X_n) = \mathcal{L}_n^0(X_1, \cdots, X_n) - d^2/2 + o_p$, *where*

$$\mathcal{L}_n^0(\mathbf{X}) = n^{-1/2} \textstyle\sum_{t=p+1}^n \phi(X_t) \sum_{i=1}^p d_i X_{t-i} \tag{3.5}$$

$$d_i = \begin{cases} a_i + b_i & 1 \leq i \leq \min(p_1, p_2) \\ a_i & p_2 < i \leq p_1 \quad \textit{if} \quad p_2 < p_1 \\ b_i & p_1 < i \leq p_2 \quad \textit{if} \quad p_1 < p_2 \end{cases} \tag{3.6}$$

$$p = \max(p_1, p_2) \quad \textit{and} \quad d^2 = \textstyle\sum_{i=1}^p d_i^2 \sigma^2 I(f). \tag{3.7}$$

Moreover, $\mathcal{L}_n^0(\mathbf{X})$ *is asymptotically normal, with mean zero and variance* d^2.

Proof. See Appendix 2.

The form of this asymptotic distribution shows that, for n sufficiently large, there will be little difference, from a statistical point of view, between AR, MA and ARMA models; as an example, the samples $\mathbf{x}^{(n)}$ generated by the models

$$x_t - n^{-1/2} \textstyle\sum_{i=1}^p a_i X_{t-i} = \varepsilon_t, \quad X_t = \varepsilon_t + n^{-1/2} \sum_{i=1}^p a_i \varepsilon_{t-i},$$

$$X_t - n^{-1/2} \textstyle\sum_{i=1}^p a_i X_{t-i}/2 = \varepsilon_t + n^{-1/2} \sum_{i=1}^p a_i \varepsilon_{t-i}/2$$

have the same asymptotic behaviour.

4. Asymptotic distribution of linear serial rank statistics. In this section, we study the asymptotic distribution under $H_1^{(n)}$ of the linear serial rank statistics S_n defined in (1.1). To this end, we derive the joint asymptotic normality of

$$\begin{pmatrix} n^{1/2}(S_n - m_n) \\ \log L_n \end{pmatrix}$$

under $H_0^{(n)}$, where

$$m_n = E[S_n \mid H_0^{(n)}] = \frac{1}{n(n-1)\cdots(n-p)} \sum_{1 \le i_1 \ne \cdots \ne i_{p+1} \le n} \cdots \sum a_n(i_1, \cdots, i_{p+1}).$$

The asymptotic normality under $H_1^{(n)}$ will then follow from LeCam's third lemma (Proposition 4.2).

Put $U_t = F(X_t)$ and $\underline{U} = (U_1, \cdots, U_n)$. We assume that the score functions $a_n(\cdots)$ are such that there exists a function $J = J(v_{p+1}, v_p, \cdots, v_1)$, defined over $[0, 1]^{p+1}$, such that

$$0 < \int_{[0,1]^{p+1}} J^2(v_{p+1} \cdots v_1)\, dv_{p+1} \cdots dv_1 < \infty \tag{4.1}$$

and

$$\lim_{n\to\infty} E[(J(U_{p+1}, \cdots, U_1) - a_n(R_{p+1}^{(n)}, \cdots, R_1^{(n)}))^2 \mid H_0^{(n)}] = 0 \tag{4.2}$$

(this assumption is satisfied most of the time when a_n is of the form

$$a_n(i_1, i_2, \cdots, i_{p+1}) = J(i_1/(n+1), i_2/(n+1), \cdots, i_{p+1}/(n+1)).$$

Such a function J will be called a *score-generating function* (associated with the serial rank statistic S_n).

We proceed as follows. First (Section 4.1), we establish the asymptotic equivalence of $(n-p)^{1/2}(S_n - m_n)$ with $\mathcal{S}_n - \mathcal{E}_n$, where

$$\mathcal{S}_n(\mathbf{X}) = (n-p)^{-1/2} \textstyle\sum_{t=p+1}^{n} J(F(X_t), F(X_{t-1}), \cdots, F(X_{t-p})) \tag{4.3}$$

and

$$\mathcal{E}_n(\mathbf{X}) = \frac{(n-p)^{1/2}}{n(n-1)\cdots(n-p)} \sum_{1 \le t_1 \ne \cdots \ne t_{p+1} \le n} \cdots \sum J(F(X_{t_1}) \cdots F(X_{t_{p+1}})); \tag{4.4}$$

then (Section 4.2) we show that $n^{-1/2}(\mathcal{S}_n - \mathcal{E}_n)$ and $n^{-1/2}\mathcal{L}_n^0$ are themselves asymptotically equivalent to U-statistics. An appropriate convergence theorem for U-statistics finally establishes (Section 4.3) the desired result.

4.1. *Asymptotic equivalence of $(n-p)^{1/2}(S_n - m_n)$ and $(\mathcal{S}_n - \mathcal{E}_n)$ (under $H_0^{(n)}$).* Let

$$\Delta_n(\mathbf{X}) = (n-p)^{1/2}(S_n(\mathbf{X}) - m_n) - (\mathcal{S}_n(\mathbf{X}) - \mathcal{E}_n(\mathbf{X})). \tag{4.5}$$

Then it suffices to show that $\lim_{n\to\infty} E[\Delta_n^2(\mathbf{X})] = 0$.

First, note that

$$(n-p)^{1/2}S_n - \mathscr{S}_n = (n-p)^{-1/2}\textstyle\sum_{t=p+1}^n [a_n(R_t^{(n)}, \cdots, R_{t-p}^{(n)}) - J(U_{(R_t)}, \cdots, U_{(R_{t-p})})]$$

can be written as

$$(n-p)^{1/2}\textstyle\sum_{t=p+1}^n \alpha_n^{U_{(\cdot)}}(R_t^{(n)}, \cdots, R_{t-p}^{(n)}) = S_n^\alpha(\mathbf{X}; U_{(\cdot)}),$$

where $U_{(\cdot)}$ is the vector $(U_{(1)}, U_{(2)}, \cdots, U_{(n)})$. Conditionally, given the vector $U_{(\cdot)}$, $\alpha_n^{U_{(\cdot)}}(\cdots) = a_n(\cdots) - J(\cdots)$ is indeed a score function, and, consequently, $S^\alpha(\mathbf{X}; U_{(\cdot)})$ is a linear serial rank statistic.

Now

$$E[\Delta_n^2(\mathbf{X})] = E_{U_{(\cdot)}}[E[((n-p)^{1/2}S_n^\alpha(\mathbf{X}; U_{(\cdot)}) - ((n-p)^{1/2}m_n - \mathscr{E}_n(\mathbf{X})))^2 \mid U_{(\cdot)}]],$$

and

$$\begin{aligned} &E[(n-p)^{1/2}S_n^\alpha(\mathbf{X}; U_{(\cdot)}) \mid U_{(\cdot)}] \\ &\quad = (n-p)^{1/2}[Ea_n(R_{p+1}^{(n)}, \cdots, R_1^{(n)}) - EJ(U_{R_{p+1}^{(n)}}, \cdots, U_{R_1^{(n)}})] \\ &\quad = (n-p)^{1/2}\left[m_n - \frac{1}{n(n-1)\cdots(n-p)} \sum_{1\le t_1\ne\cdots\ne t_{p+1}\le n}\cdots\sum J(U_{t_1}\cdots U_{t_{p+1}})\right] \\ &\quad = (n-p)^{1/2}m_n - \mathscr{E}_n. \end{aligned}$$

Hence, denoting by $D^2(S_n^\alpha \mid U_{(\cdot)})$ the conditional variance of $S_n^\alpha(\mathbf{X}; U_{(\cdot)})$, we have

$$E[\Delta_n^2] = (n-p)E_{U_{(\cdot)}}[D^2(S_n^\alpha \mid U_{(\cdot)})].$$

In view of Lemmas 2 and 4 (Appendix 3), we obtain

$$\begin{aligned} E[\Delta_n^2] &\le E_{U_{(\cdot)}}[D^2(\alpha_n^{U_{(\cdot)}} \mid U_{(\cdot)})(2p+1) \\ &\qquad + n\,\mathrm{Cov}(\alpha_n^{U_{(\cdot)}}(R_{p+1}^{(n)}, R_p^{(n)}, \cdots, R_1^{(n)}), \alpha_n^{U_{(\cdot)}}(R_{2p+2}^{(n)}, R_{2p+1}^{(n)}, \cdots, R_{p+2}^{(n)}) \mid U_{(\cdot)})] \\ &\le (2p+1+K)E_{U_{(\cdot)}}[E[(\alpha_n^{U_{(\cdot)}}(R_{p+1}^{(n)}, \cdots, R_1^{(n)}))^2 \mid U_{(\cdot)}]] \\ &= (2p+1+K)E[(\alpha_n^{U_{(\cdot)}}(R_{p+1}^{(n)}, \cdots, R_1^{(n)}))^2], \end{aligned}$$

which by (4.2) converges to 0 (as $n \longrightarrow \infty$).

4.2. We now show that $n^{-1/2}(\mathscr{S}_n - \mathscr{E}_n)$ and $n^{-1/2}\mathscr{S}_n^0$ are asymptotically equivalent to U-statistics.

Define the $(p+1)$-dimensional random variables

$$\mathbf{Y}_t = \begin{pmatrix} Y_{t,1} \\ \vdots \\ Y_{t,p+1} \end{pmatrix} = \begin{pmatrix} U_t \\ \vdots \\ U_{t-p} \end{pmatrix}, \quad p+1 \le t \le n. \tag{4.6}$$

The $\mathbf{Y}_t$'s are identically distributed (uniformly over $[0, 1]^{p+1}$, under $H_0^{(n)}$); of course, they are not independent—but, being p-dependent, they constitute an *absolutely regular process* (Yoshihara, 1976).

Consider the function $G(\mathbf{y})$(of $(p+1)$ arguments $\mathbf{y} = (y_1, \cdots, y_{p+1}) \in [0, 1]^{p+1}$) where

$$G(\mathbf{y}) = G(y_1, \cdots, y_{p+1}) = \phi(F^{-1}(y_1)) \textstyle\sum_{i=1}^{p} d_i F^{-1}(y_{i+1}). \tag{4.7}$$

$G(\mathbf{Y})$ could serve as a kernel (of degree 1) converting the variables $n^{-1/2}\mathscr{L}_n^0$ into a sequence of U-statistics. However, in order to obtain a U-statistic approximation of $n^{-1/2}\mathscr{E}_n$ we shall need a kernel of degree $(p+1)$. Therefore, put

$$\Phi^{\mathscr{L}}(\mathbf{Y}_{t_1}, \cdots, \mathbf{Y}_{t_{p+1}}) = \textstyle\sum_{j=1}^{p+1} G(\mathbf{Y}_{t_j})/(p+1), \tag{4.8}$$

then the corresponding U-statistic is

$$\binom{n-p}{p+1}^{-1} \sum_{p+1 \le t_1 < \cdots < t_{p+1} \le n} \cdots \sum \Phi^{\mathscr{L}}(\mathbf{Y}_{t_1}, \cdots, \mathbf{Y}_{t_{p+1}}) = n^{-1/2}\mathscr{L}_n^0 + o_p(n^{-1/2}).$$

In a similar way, the kernel

$$\begin{aligned} \Phi^{\mathscr{S}}(\mathbf{Y}_{t_1}, \cdots, \mathbf{Y}_{t_{p+1}}) &= \textstyle\sum_{j=1}^{p+1} J(\mathbf{Y}_{t_j})/(p+1) \\ &= \textstyle\sum_{j=1}^{p+1} J(Y_{t_j,1}, \cdots, Y_{t_j,p+1})/(p+1) \end{aligned} \tag{4.9}$$

defines a U-statistic which is asymptotically equivalent to $n^{-1/2}\mathscr{S}_n$:

$$\binom{n-p}{p+1}^{-1} \sum_{p+1 \le t_1 < \cdots < t_{p+1} \le n} \cdots \sum \Phi^{\mathscr{S}}(\mathbf{Y}_{t_1}, \cdots, \mathbf{Y}_{t_{p+1}}) = n^{-1/2}\mathscr{S}_n + o_p(n^{-1/2}).$$

As for $\mathscr{E}_n$, let us consider the kernel

$$\Phi^{\mathscr{E}}(\mathbf{Y}_{t_1}, \cdots, \mathbf{Y}_{t_{p+1}}) = (1/(p+1)!) \textstyle\sum_j J(Y_{j_1,1}, \cdots, Y_{j_{p+1},1}), \tag{4.10}$$

where the summation $\sum_j$ extends over all possible $(p+1)!$ permutations $(j_1, \cdots, j_{p+1})$ of $(t_1, \cdots, t_{p+1})$. The corresponding U-statistic is

$$\begin{aligned} \mathscr{U}_n^{\mathscr{E}} &= \binom{n-p}{p+1}^{-1} \sum_{p+1 \le t_1 < \cdots < t_{p+1} \le n} \cdots \sum \Phi^{\mathscr{E}}(\mathbf{Y}_{t_1}, \cdots, \mathbf{Y}_{t_{p+1}}) \\ &= (1/(n-p) \cdots (n-2p)) \sum_{p+1 \le t_1 \ne \cdots \ne t_{p+1} \le n} \cdots \sum J(U_{t_1} \cdots U_{t_{p+1}}). \end{aligned}$$

From (4.4), we obtain

$$(n-p)^{-1/2}\mathscr{E}_n = ((n-p) \cdots (n-2p))/(n(n-1) \cdots (n-p))\mathscr{U}_n^{\mathscr{E}} + \delta_n,$$

where

$$\delta_n = (1/n(n-1) \cdots (n-p)) \textstyle\sum_{\ell=1}^{p} \displaystyle\sum_{1 \le j_1 \ne \cdots \ne j_\ell \le p} \cdots \sum \; \sum_{1 \le k_1 < \cdots < k_\ell < p+1} \cdots \sum \; \sum_{p+1 \le t_1 \ne \cdots \ne t_{p+1-\ell} \le n} \cdots \sum$$

$$J(U_{t_1}, \cdots, U_{t_{k_1-1}} U_{j_1} U_{t_{k_1}}, \cdots, U_{t_{k_2-2}} U_{j_2} U_{t_{k_2-1}}, \cdots, U_{t_{k_\ell-\ell}} U_{j_\ell} U_{t_{k_\ell-\ell+1}}, \cdots, U_{t_{p+1-\ell}}).$$

The number of terms, in δ_n, is

$$\sum_{\ell=1}^{p} p(p-1)\cdots(p-\ell+1)\frac{(p+1)\cdots(p+2-\ell)}{\ell!}(n-p)\cdots(n-2p+\ell)$$
$$\leq p!(p+1)!(n-p)\cdots(n-2p+1).$$

Thus,

$$E\,|\,\delta_n\,| \leq p!(p+1)!\frac{(n-p)\cdots(n-2p+1)}{n(n-1)\cdots(n-p)}E\,|\,J(U_{p+1}\cdots U_1)\,| = O(n^{-1}),$$

and

$$n^{-1/2}\mathscr{E}_n = \mathscr{U}_n^{\mathscr{E}} + o_p(n^{-1/2}).$$

4.3. *Asymptotic normality.* It follows from the conclusions of the preceding section that, α and β being arbitrary coefficients, $\mathscr{U}_{\alpha\beta}^{n} = n^{-1/2}[\alpha(\mathscr{S}_n - \mathscr{E}_n) + \beta\,\mathscr{L}_n^0]$ is (up to $o_p(n^{-1/2})$ terms) a sequence of U-statistics, with kernel

$$\Phi_{\alpha\beta} = \alpha(\Phi^{\mathscr{S}} - \Phi^{\mathscr{E}}) + \beta\Phi^{\mathscr{L}}. \tag{4.11}$$

A number of results are available concerning the asymptotic behaviour of U-statistics in case the sample observations (here, the $\mathbf{Y}_t$'s) are not independent. The one we are using in Proposition 4.1 below is a theorem established by Yoshihara (1976) for the class of stationary absolutely regular processes—from which the p-dependent processes constitute a subclass.

Let

$$\begin{aligned} J^*(u_{p+1},\cdots,u_1) &= J(u_{p+1},\cdots,u_1) \\ &\quad - \sum_{k=1}^{p+1}\int_{[0,1]^p} J(v_p,\cdots,v_k,u_1,v_{k-1},\cdots,v_1)\,dv_1\cdots dv_p \\ &\quad + p\int_{[0,1]^{p+1}} J(v_{p+1},\cdots,v_1)\,dv_1\cdots dv_{p+1}. \end{aligned} \tag{4.12}$$

Obviously, $E[J^*(U_{p+1},\cdots,U_1)] = 0$. We have then the following results.

Proposition 4.1. *Under* $H_0^{(n)}$,

$$\begin{pmatrix}\sqrt{n}(S_n - m_n)\\ \log L_n\end{pmatrix}$$

is asymptotically normal, with mean

$$\begin{pmatrix}0\\ -\tfrac{1}{2}\cdot\sum_{i=1}^{p} d_i^2\sigma^2 I(f)\end{pmatrix}$$

and covariance matrix

$$\begin{pmatrix}V^2 & \sum_{i=1}^{p} d_iC_i\\ \sum_{i=1}^{p} d_iC_i & \sum_{i=1}^{p} d_i^2\sigma^2 I(f)\end{pmatrix},$$

where

$$V^2 = \int_{[0,1]^{p+1}} J^{*2}(v_{p+1}, \cdots, v_1)\, dv_1 \cdots dv_{p+1} \tag{4.13}$$
$$+ 2 \textstyle\sum_{j=1}^{p} \displaystyle\int_{[0,1]^{p+1+j}} J^*(v_{p+1}, \cdots, v_1) J^*(v_{p+1+j}, \cdots, v_{1+j})\, dv_1 \cdots dv_{p+1+j}$$

and

$$C_i = \int_{[0,1]^{p+1}} J^*(v_{p+1}, \cdots, v_1) \textstyle\sum_{j=0}^{p-i} \phi(F^{-1}(v_{p+1-j}))F^{-1}(v_{p+1-j-i})\, dv_1 \cdots dv_{p+1}. \tag{4.14}$$

PROPOSITION 4.2. *Under $H_1^{(n)}$, $\sqrt{n}(S_n - m_n)$ is asymptotically normal, with mean $\sum_{i=1}^{p} d_i C_i$ and variance V^2.*

The proof of Proposition 4.1 is given in Appendix 4; Proposition 4.2 follows immediately from Proposition 4.1, by application of LeCam's third lemma. Note that the asymptotic variance V^2 depends only on the score-generating function $J(\cdots)$, not on the ARMA model coefficients a_i and b_i (cf. (3.7)), nor on the white noise density f. The mean, on the contrary, depends on both f and the coefficients a_i and b_i; however, it remains invariant under scale transformations because of its dependence on f only through $\phi(F^{-1}(u))F^{-1}(v)$.

5. Asymptotic efficiency of linear serial rank statistics.

5.1. *ARE of two linear serial rank statistics—Optimal scores.* The results of Proposition 4.2 allow for an explicit form of the asymptotic relative efficiency of two linear serial rank statistics.

Let $S_n^{(1)}$ and $S_n^{(2)}$ have asymptotic normal distributions under $H_1^{(n)}$, with means $\sum d_i C_i^{(1)}$ and $\sum d_i C_i^{(2)}$ and with variances $V_{(1)}^2$ and $V_{(2)}^2$. Then, the ARE of $S_n^{(1)}$ with respect to $S_n^{(2)}$ is

$$e(S_n^{(1)}, S_n^{(2)}) = (V_{(2)} \textstyle\sum_{i=1}^{p} d_i C_i^{(1)} / V_{(1)} \sum_{i=1}^{p} d_i C_i^{(2)})^2. \tag{5.1}$$

A test statistic $\bar{S}_n$ such that $e(\bar{S}_n, S_n) \geq 1$ for any linear serial rank statistic S_n will be asymptotically the most efficient statistic (in Pitman's sense) within the class of linear serial rank statistics for testing randomness (i.e., $H_0^{(n)}$) against ARMA dependence (i.e., $H_1^{(n)}$). (A test based on $\bar{S}_n$ will be called the asymptotically optimal test against $H_1^{(n)}$.) We shall denote by $H_d^{(n)}$, a sequence of alternatives characterized by the coefficients $\underset{\sim}{d} = (d_1, \cdots, d_p)$ (cf. (1.2) and (3.6)).

The following result provides an explicit formulation of the optimal test statistic.

PROPOSITION 5.1. *An asymptotically optimal linear serial rank test for $H_0^{(n)}$ against $H_d^{(n)}$ is provided by any statistic S_n^d with score-generating function (up to additive and multiplicative constants) given by*

$$J^d(v_{p+1}, \cdots, v_1) = \textstyle\sum_{i=1}^{p} (d_i/(p+1-i)) \sum_{j=0}^{p-i} \phi(F^{-1}(v_{p+1-j}))F^{-1}(v_{p+1-j-i}). \tag{5.2}$$

Under $H_{\underline{h}}^{(n)}(\underline{h} \in \mathbb{R}^p)$, $n^{1/2}(S_n^{\underline{d}} - m_n^{\underline{d}})$ *is asymptotically normal, with mean* $\sum_{i=1}^p h_i d_i \sigma^2 I(f)$ *and variance* $V_{\underline{d}}^2 = \sum_{i=1}^p d_i^2 \sigma^2 I(f)$.

This optimality result relies on the following lemma whose proof is given in Appendix 5.

LEMMA 5.1. *Let* S_n *be a linear rank statistic with score-generating function* $J^*(v_{p+1}, \cdots, v_1)$(cf. (4.12)), *and let*

$$J_0^*(v_{p+1}, \cdots, v_1) = (\sigma^2 I(f))^{-1} \sum_{i=1}^p (C_i/(p+1-i)) \sum_{j=0}^{p-i} \phi(F^{-1}(v_{p+1-j}))F^{-1}(v_{p+1-j-i}). \tag{5.3}$$

Denote by S_n^0 *a linear serial rank statistic associated with* J_0^*. *Then* $e(S_n, S_n^0) \leq 1$ *for any alternative* $H_n^{\underline{d}}$.

$J_0^*(v_{p+1}, \cdots, v_1)$ is actually the projection of $J^*(v_{p+1}, \cdots, v_1)$ onto the linear L^2-space spanned by $\{\sum_{j=0}^{p-i} \phi(F^{-1}(v_{p+1-j}))F^{-1}(v_{p+1-j-i});\ i = 1, \cdots, p\}$. This immediately follows from (4.14) and

$$\int [\textstyle\sum_{j=0}^{p-i} \phi(F^{-1}(v_{p+1-j}))F^{-1}(v_{p+1-j-i})]^2 \, dv_{p+1} \cdots dv_1 = (p+1-i)\sigma^2 I(f).$$

Similarly, $(S_n^0 - m_n^0)$ is the projection (under $H_0^{(n)}$ and up to $o_p(n^{-1/2})$ terms) of $(S_n - m_n)$ onto the second-order linear space spanned by p linear serial rank statistics of orders 1 through p, associated with the score-generating functions

$$\textstyle\sum_{j=1}^{p-i} \phi(F^{-1}(v_{p+1-j}))F^{-1}(v_{p+1-j-i}), \quad i = 1, \cdots, p.$$

PROOF OF PROPOSITION 5.1. Because of Lemma 5.1, we may restrict our attention to score functions of the form (5.3). The corresponding statistic has an asymptotic variance

$$\begin{aligned} V_0^2 = {} & \textstyle\sum_{i=1}^p \sum_{j=0}^{p-i} c_i^2 \displaystyle\int [\phi(F^{-1}(u))F^{-1}(v)]^2 \, du \, dv \\ & + 2 \textstyle\sum_{s=1}^p \sum_{i=1}^p \sum_{j=0}^{p-i} \sum_{k=1}^p \sum_{\ell=0}^{p-k} c_i c_k \displaystyle\int \phi(F^{-1}(v_{p+1-j}))F^{-1}(v_{p+1-j-i}) \\ & \cdot \phi(F^{-1}(v_{p+1+s-\ell}))F^{-1}(v_{p+1+s-\ell-k}) \, dv \end{aligned}$$

where $c_i = (\sigma^2 I(f))^{-1} C_i/(p+1-i)$.

Omitting routine computations, we obtain

$$\begin{aligned} V_0^2 &= \sigma^2 I(f)[\textstyle\sum_{i=1}^p (p+1-i)c_i^2 + 2 \sum_{i=1}^p c_i^2 \sum_{j=1}^{p-i} (p+1-j-i)] \\ &= \sigma^2 I(f) \textstyle\sum_{i=1}^p c_i^2 (p+1-i)^2 = (\sigma^2 I(f))^{-1} \sum_{i=1}^p C_i^2. \end{aligned} \tag{5.4}$$

The values of the C_i's that maximize

$$(\textstyle\sum_{i=1}^p d_i C_i)^2 / V_0^2 = (\sigma^2 I(f))(\sum_{i=1}^p d_i C_i)^2 / \sum_{i=1}^p C_i^2$$

are thus proportional to the d_i's, and this establishes the desired result. □

Note that the optimal scores (5.2) could also be taken as

$$J^d(v_{p+1}, \cdots, v_1) = \sum_{i=1}^{p} (d_i/(p+1-i)) \sum_{j=0}^{p-i} F^{-1}(v_{p+1-j})\phi(F^{-1}(v_{p+1-j-i})).$$

This is in accordance with the fact that, due to stationarity, if $\{x_t\}$ is generated by an ARMA model of the form (3.1), it can also be seen as generated by an ARMA model of the form

$$X_t - n^{-1/2} \sum_{i=1}^{p_1} a_i X_{t+i} = \varepsilon_t + n^{-1/2} \sum_{i=1}^{p_2} b_i \varepsilon_{t+i}.$$

An additional consequence of Proposition 5.1 is that the "optimal" serial rank statistic against an ARMA(p_1, p_2) alternative is a serial rank statistic of order $p = \max(p_1, p_2)$. Using a statistic of the wrong order—whether too low or too high—always results into a loss of efficiency.

EXAMPLES. The optimal score functions in (5.2) depend only on the *type* of the density of the generating white noise; examples are:

(i) Gaussian white noise (van der Waerden scores)

$$J(v_{p+1}, \cdots, v_1) = \sum_{i=1}^{p} (d_i/(p+1-i)) \sum_{j=0}^{p-i} \Phi^{-1}(v_{p+1-j})\Phi^{-1}(v_{p+1-j-i})$$
$$\left(\text{with } \Phi(x) = (2\pi)^{-1/2} \int_{-\infty}^{x} e^{-u^2/2}\, du\right). \tag{5.5}$$

(ii) Logistic white noise (Wilcoxon scores)

$$J(v_{p+1}, \cdots, v_1) = \sum_{i=1}^{p} (d_i/(p+1-i)) \sum_{j=0}^{p-i} (2\, v_{p+1-j} - 1)\log(v_{p+1-j-i}/(1 - v_{p+1-j-i})). \tag{5.6}$$

(iii) Double exponential white noise (median test scores)

$$J(v_{p+1}, \cdots, v_1) = \sum_{i=1}^{p} d_i/(p+1-i) \sum_{j=0}^{p-i} \operatorname{sgn}(v_{p+1-j} - ½)F_e^{-1}(v_{p+1-j-i})$$
$$\left(\text{with } F_e(x) = \frac{1}{2} \int_{-\infty}^{x} e^{-|u|}\, du\right). \tag{5.7}$$

As an example of an optimal test statistic, let us consider the problem of testing randomness against the sequence of ARMA(2, 1) alternatives

$$X_t + (1.4)n^{-1/2}X_{t-1} + (0.5)n^{-1/2}X_{t-2} = \varepsilon_t + (0.2)n^{-1/2}\varepsilon_{t-1}, \quad t \in \mathbb{Z}$$

where $\{\varepsilon_t\}$ has a logistic density (with unspecified variance). The following serial version of Wilcoxon's test is then optimal:

$$S_n = (n-2)^{-1} \sum_{t=3}^{n} \left\{ \frac{-1.2}{2} \left[\left(2\frac{R_t^{(n)}}{n+1} - 1 \right) \log\left(\frac{R_{t-1}^{(n)}}{n - R_{t-1}^{(n)} + 1} \right) + \left(2\frac{R_{t-1}^{(n)}}{n+1} - 1 \right) \log\left(\frac{R_{t-2}^{(n)}}{n - R_{t-2}^{(n)} + 1} \right) \right] - 0.5 \left(2\frac{R_t^{(n)}}{n+1} - 1 \right) \log\left(\frac{R_{t-2}^{(n)}}{n - R_{t-2}^{(n)} + 1} \right) \right\}$$

$$= -(1.2)(n-1)^{-1} \sum_{t=2}^{n} \left(2\frac{R_t^{(n)}}{n+1} - 1 \right) \log\left(\frac{R_{t-1}^{(n)}}{n - R_{t-1}^{(n)} + 1} \right) - (0.5)(n-2)^{-1} \sum_{t=3}^{n} \left(2\frac{R_t^{(n)}}{n+1} - 1 \right) \log\left(\frac{R_{t-2}^{(n)}}{n - R_{t-2}^{(n)} + 1} \right) + O_p(n^{-1}).$$

5.2. *Comparison with classical parametric procedures.* The parametric version of our problem is treated, in the Gaussian case, by Anderson (1971, Chapter 6). For the problem of testing for randomness against an alternative of AR(p) dependence

$$X_t - \sum_{i=1}^{p} \beta_i X_{t-i} = \varepsilon_t, \quad t \in \mathbb{Z} \tag{5.8}$$

(with $\beta_p \neq 0$), Anderson (1971, page 266) shows that the most powerful similar test is based on a test statistic

$$\sum_{k=1}^{p} \gamma_k \dot{r}_k^{(n)}, \tag{5.9}$$

where

$$\gamma_k = \beta_k + \sum_{i=1}^{p-k} \beta_i \beta_{i+k}, \tag{5.10}$$

$\dot{r}_k^{(n)}$ being *a serial correlation coefficient* of order k. These coefficients are such that

$$\dot{r}_k^{(n)} = r_k^{(n)} + o_p(n^{-1/2}), \tag{5.11}$$

where $r_k^{(n)}$ is the sample autocorrelation coefficient of order k, i.e.,

$$r_k^{(n)} = (n/(n-k)) \sum_{t=k+1}^{n} X_t X_{t-k} / \sum_{t=1}^{n} X_t^2. \tag{5.12}$$

It is therefore natural to compare our linear serial rank statistics (1.1) with linear combinations of the autocorrelation coefficients (also, time-series analysts generally give much attention to the *correlogram* of their series). In order to do this, we need the asymptotic joint distribution of those linear combinations and log L_n under $H_0^{(n)}$.

PROPOSITION 5.2. *Under $H_0^{(n)}$, $(n^{1/2}r_1^{(n)}, \cdots, n^{1/2}r_p^{(n)}, \log L_n)'$ is asymptotically normal, with mean $(0, \cdots, 0, -\frac{1}{2}\sum_{i=1}^p d_i^2\sigma^2 I(f))'$ and covariance matrix*

$$\begin{pmatrix} & & d_1 \\ & I & \vdots \\ & & d_p \\ d_1 & \cdots\ d_p & \sum d_i^2\sigma^2 I(f) \end{pmatrix}$$

(I being the $p \times p$ identity matrix).

COROLLARY 5.1. *Under $H_d^{(n)}$, $n^{1/2}\sum_{k=1}^p \alpha_k r_k^{(n)}$ is asymptotically normal with mean $\sum_{i=1}^p \alpha_i d_i$ and variance $\sum_{i=1}^p \alpha_i^2$.*

COROLLARY 5.2. *The asymptotically most efficient (in Pitman's sense) linear combination of the $r_k^{(n)}$'s against $H_d^{(n)}$ is $\sum_{k=1}^p d_k r_k^{(n)}$.*

PROOF. See Appendix 6. □

Letting $\beta_i = \beta_i^{(n)} = n^{-1/2}a_i$ in (5.10), and by denoting by $\gamma_k^{(n)}$ the corresponding values of γ_k, we obtain in the pure AR(p) case, that

$$\sum_{k=1}^p d_k r_k^{(n)} = n^{1/2}\sum_{k=1}^p \gamma_k^{(n)}\dot{r}_k^{(n)} + o_p.$$

The asymptotically optimal test statistic in Corollary 5.2 is thus asymptotically equivalent to the one providing the most powerful similar test (which is of course not surprising !).

Now, if we denote by e the ARE of the asymptotically optimal serial rank statistic defined in Proposition 5.1 with respect to the asymptotically optimal combination of autocorrelations $\sum_{k=1}^p d_k r_k^{(n)}$, we obtain the following result.

PROPOSITION 5.3.

$$e = \sigma^2 I(f). \tag{5.13}$$

PROOF. On account of Proposition 5.1 and Corollary 5.1, we have

$$e = \frac{(\sum_{i=1}^p d_i^2\sigma^2 I(f))^2/\sum_{i=1}^p d_i^2\sigma^2 I(f)}{(\sum_{i=1}^p d_i^2)^2/\sum_{i=1}^p d_i^2} = \sigma^2 I(f). \quad \square$$

This efficiency thus depends on the density f. For normal, logistic, and double exponential densities, it takes the values 1, $\pi^2/9$, and 2, respectively.

A more detailed study for the case $p = 1$ is given in the next section.

5.3. *The first-order case* (AR(1), MA(1) *or* ARMA(1, 1) *alternatives*). The asymptotic mean of any linear serial rank statistic has the form of an inner product $\sum_{i=1}^p d_i C_i$; any of these statistics has thus the drawback that, when

$p \geq 2$, there exists, for any type of density f, a family of alternatives against which this statistic has a zero efficiency—namely, the alternatives characterized by a vector of coefficients $\underset{\sim}{d} = (d_i)$ (cf. 3.7)) orthogonal to $\underset{\sim}{C} = (C_i)$ (4.14) (recall that C_i depends only on J and f).

This does not happen, however, in the first-order case ($p = 1$); indeed the ARE of a linear serial rank statistic $S_n^{(1)}$ with respect to another $S_n^{(2)}$ is given by

$$e(S_n^{(1)}, S_n^{(2)}) = \left(\frac{C_{(1)}/V_{(1)}}{C_{(2)}/V_{(2)}}\right)^2$$

and hence does not depend on the alternative. As a consequence, the statistic

$$S_n^f = (n-1)^{-1} \textstyle\sum_{t=2}^n \phi\left(F^{-1}\left(\frac{R_t^{(n)}}{n+1}\right)\right) F^{-1}\left(\frac{R_{t-1}^{(n)}}{n+1}\right) \tag{5.14}$$

is *uniformly* asymptotically most efficient against the whole family of ARMA(1, 1) alternatives where f is a *strongly unimodal density* (cf. Hájek and Sidák, 1967).

The classical parametric test statistic in this problem is the first-order autocorrelation coefficient $r_1^{(n)}$. Under the usual gaussian assumptions, this statistic (cf. Anderson, 1971, Chapter 6) provides the (one-sided) uniformly most powerful similar and (two-sided) unbiased tests. It is therefore interesting to investigate the AREs $e(S_n^f)$ of (5.14) with respect to $r_1^{(n)}$.

In Table 1, we give, for various score functions (van der Waerden, Wilcoxon, median test, Spearman) and various density types (normal, logistic and double exponential), the values of $e(S_n^f)$.

Because of its popularity, we also include in this comparative study the Spearman autocorrelation coefficient of order 1, viz.

$$r_S^n = \frac{(n-1)^{-1} \sum_{t=2}^n R_t^{(n)} R_{t-1}^{(n)} - (n+1)/2)^2}{(n^2-1)/12}.$$

r_S^n is equivalent (for strongly unimodal densities $f(\cdot)$) to the linear serial rank statistic S_n^S induced by the score-generating function $J(u, v) = uv$. Indeed, it can be shown that

$$r_S^n = (S_n^S - 1/4)/(1/12(n-1)/(n+1)),$$

with

$$m_n^S = E(S_n^S) = (3\,n^2 - n - 2)/12(n^2-1)$$

and

$$V^2 = \int_{[0,1]^2} \left(uv - \frac{u+v}{2} + \frac{1}{4}\right)^2 du\, dv = (144)^{-1}.$$

Consequently, $n^{1/2}(S_n^S - m_n^S)$ is asymptotically normal, with mean 0 and variance 1/144 (under $H_0^{(n)}$). This confirms a classical result (cf., for example, Kendall and Stuart, 1968) that $n^{1/2} r_S^n$ is asymptotically normal, with mean 0 and variance 1.

TABLE 1

Asymptotic relative efficiencies of some linear serial rank statistics with respect to the first-order sample autocorrelation

Score Generating Functions $J(u, v)$	V^2	Types of Densities		
		Normal	Logistic	Double Exponential
van der Waerden $\Phi^{-1}(u)\Phi^{-1}(v)$	1	1	$\frac{1}{\Pi^3}\left[\int_{-\infty}^{+\infty} e^{-x^2}(\Phi(x))^{-1}\,dx\right]^2 \simeq 1.047$	$\frac{2}{\Pi^3}\left[\int_{-\infty}^{0} e^{-x^2}(\Phi(x))^{-1}\,dx\right]^2 \simeq 1.226$
Wilcoxon $(2u-1)\log(v/1-v)$	$\frac{\Pi^2}{9}$	$\frac{9}{\Pi^6}\left[\int_{-\infty}^{+\infty} e^{-x^2}(\Phi(x))^{-1}\,dx\right]^2 \simeq 0.948$	$\frac{\Pi^2}{9} \simeq 1.097$	$\frac{9}{4\Pi^2}\left[\frac{\Pi^2}{6} - (\log 2)^2 + \log 4\right]^2 \simeq 1.483$
Median Test $\operatorname{sgn}(u-½)F_2^{-1}(v)$	2	$\frac{1}{\Pi^3}\left[\int_{-\infty}^{0} e^{-x^2}(\Phi(x))^{-1}\,dx\right]^2 \simeq 0.613$	$\frac{1}{8}\left[\frac{\Pi^2}{6} - (\log 2)^2 + \log 4\right]^2 \simeq 0.813$	2
Spearman uv	$(144)^{-1}$	$\left(\frac{3}{\Pi}\right)^2 \simeq 0.912$	1	$¼(3/2)^4 \simeq 1.226$
Maximal ARE		1	1.097	2

APPENDIX 1

The Green's functions. Denoting by L the lag operator ($L\psi_t = \psi_{t-1}$), consider the linear difference operator of order p: $C(L) = 1 + \sum_{i=1}^{p} c_i L^i$, where the c_i's are real and $c_p \neq 0$. The Green's function g_u associated with this operator is (cf. Miller, 1968, Chapter 2.2 or Hallin, 1984) the value in $t = u$ of the solution of the homogeneous difference equation $\psi_t + \sum_{i=1}^{p} c_i \psi_{t-i} = 0$, $t \in \mathbb{Z}$, taking on initial values $\psi_0 = 1$, $\psi_{-1} = \cdots = \psi_{-p+1} = 0$.

Let $g_u^{(n)}$ be the Green's function associated with the operator $B^{(n)}(L) = 1 + n^{-1/2} \sum_{i=1}^{p} b_i L^i$. The following lemma will be helpful for the proof of Proposition 3.1.

LEMMA 1.

$$\sum_{u=v}^{w} | g_u^{(n)} | \leq \sum_{u=[(v-1)/p]+1}^{w} (p b_M n^{-1/2})^u \quad \forall w \geq v \geq 0$$

where $b_M = \max_i | b_i |$

COROLLARY 1. *For* $n > 4(pb_M)^2$,

$$\text{(A1)} \qquad \sum_{u=pv+1}^{\infty} | g_u^{(n)} | \leq 2(p b_M n^{-1/2})^{v+1} = o(n^{-v/2}) \quad \forall v \geq 0.$$

PROOF. $g_u^{(n)}$ is a sum of terms of the form

$$\text{(A2)} \qquad (-1)^k (b_{i_{(1)}} b_{i_{(2)}} \cdots b_{i_{(k)}}) n^{-k/2},$$

where the indexes $i_{(1)}, \cdots, i_{(k)} \in \{1, \cdots, p\}$ need not be distinct, and $[(u-1)/p] + 1 \leq k \leq u$ ($[z]$ stands for the largest integer $\leq z$). Each of these terms is absolutely bounded by $(b_M n^{-1/2})^k$.

Denote by ν_u^k the maximal (some might collapse) number of these terms in $g_u^{(n)}$. ν_u^k does not depend on n. Put $\nu_0^k = \delta_{k0}$, and $\nu_u^k = 0$ for $u < 0$ or $k \notin [[(u-1)/p] + 1; u]$. Obviously, $\nu_u^k = \sum_{i=1}^{p} \nu_{u-i}^{k-1}$. Hence, letting $\nu^k = \sum_{u=0}^{\infty} \nu_u^k = \sum_{u=k}^{pk} \nu_u^k$, we get $\nu^k = \sum_{i=1}^{p} \sum_{u=0}^{\infty} \nu_{u-i}^{k-1} = p\nu^{k-1}$ and, since $\nu^0 = 1$, $\nu^k = p^k$; it follows that

$$\sum_{u=v}^{w} | g_u^{(n)} | \leq \sum_{k=[(v-1)/p]+1}^{w} \nu^k (b_M n^{-1/2})^k = \sum_k (p b_M n^{-1/2})^k.$$

The proof of Corollary 1 is straightforward. □

APPENDIX 2

PROOF OF PROPOSITION 3.1. To avoid unnecessary details of computations involved in the proof for the general ARMA(p_1, p_2) case, we successively consider the pure AR(p) and MA(p) cases, since these two cases cover all the problems raised in the general case.

A2.1. *The* AR(p) *case.* Let $\mathbf{X}_0 = (X_{-p+1}, \cdots, X_0)$, and

$$\prod_{t=1}^{n} f(X_t - n^{-1/2} \sum_{i=1}^{p} a_i X_{t-i}) = \lambda^{(n)}(\mathbf{X}; \mathbf{X}_0).$$

Then, we have (a.e.)

$$\log \lambda^{(n)}(\mathbf{X}; \mathbf{X}_0)$$

$$(A3) \qquad = \sum_{t=1}^{n} \log f(X_t) + n^{-1/2} \sum_{t=1}^{n} \phi(X_t) \sum_{i=1}^{p} a_i X_{t-i} - (2n)^{-1} \sum_{t=1}^{n} \phi'(X_t - \theta n^{-1/2} \sum_{i=1}^{p} a_i X_{t-i}) \sum_{i,j=1}^{p} a_i a_j X_{t-i} X_{t-j},$$

with $\theta = \theta(X_n, \cdots, X_{-p+1}) \in [0, 1]$. The first-order term in (A3) is

$$(A4) \qquad \mathscr{L}_n^0(\mathbf{X}) + n^{-1/2} \sum_{t=1}^{p} \phi(X_t) \sum_{i=1}^{p} a_i X_{t-i} = \mathscr{L}_n^0(\mathbf{X}) + o_p.$$

$\mathscr{L}_n^0$ is itself of the form $n^{-1/2} \sum_{t=p+1}^{n} Z_t$, where the Z_t's are p-dependent. It is therefore (cf., for example, Anderson, 1971, page 427) asymptotically normal with mean $E[Z_t] = 0$ and variance

$$E[Z_t^2] + 2 \sum_{i=1}^{p} E[Z_t Z_{t+i}] = (\sum_{i=1}^{p} a_i^2)\sigma^2 I(f) + 0 = d^2.$$

Now consider the second-order term in (A3). Put $|a_M| = \max_i |a_i|$ and $Q_t = \sum_{i,j=1}^{p} a_i a_j X_{t-i} X_{t-j}$. ϕ' being Lipschitzian (assumption (iv)), we have

$$\left| \sum_{t=1}^{n} Q_t \phi'(X_t - \theta n^{-1/2} \sum_{i=1}^{p} a_i X_{t-i}) - \sum_{t=1}^{n} Q_t \phi'(X_t) \right| \le n^{-1/2} A |a_M| \sum_{t=1}^{n} |Q_t| \sum_{i=1}^{p} |X_{t-i}| \le n^{-1/2} A |a_M|^3 \sum_{t=1}^{n} \sum_{i,j,k=1}^{p} |X_{t-i} X_{t-j} X_{t-k}|.$$

Since its summands are p-dependent, and since (assumption (i)) $E|X_r X_s X_t| < \infty\ \forall r, s, t$, this upper bound is $O_p(n^{1/2})$. It follows that the second-order term in (A3) can be written as $-(2n)^{-1} \sum_{t=1}^{n} \phi'(X_t) \sum_{i,j=1}^{p} a_i a_j X_{t-i} X_{t-j} + o_p$, and hence converges in probability to $-\frac{1}{2} \sum_{i=1}^{p} a_i^2 E[\phi'(X_t) X_{t-i}^2] = -d^2/2$. Summing up,

$$\log \lambda^{(n)}(\mathbf{X}; \mathbf{X}_0) = \log \ell_0^{(n)}(\mathbf{X}) + \mathscr{L}_n^0(\mathbf{X}) - (d^2/2) + R^{(n)}(X_p, \cdots, X_0, \cdots, X_{-p+1}),$$

where $R^{(n)}$ is o_p. Hence

$$\log L^{(n)}(\mathbf{X}) = \log E[\exp(\log \lambda^{(n)}(\mathbf{X}; \mathbf{X}_0) - \log \ell_0^{(n)}(\mathbf{X})) \mid \mathbf{X}] = (\mathscr{L}_n^0(\mathbf{X}) - d^2/2) + \log E[\exp(R^{(n)}) \mid \mathbf{X}] = (\mathscr{L}_n^0(\mathbf{X}) - d^2/2) + o_p,$$

where the latter expectation is to be taken with respect to the joint distribution $G_{a,b}^{(n)}$ of $X_{-p+1}, \cdots, X_0$.

Proposition 3.1 now readily follows from the asymptotic normality of $\mathscr{L}_n^0(\mathbf{X})$.

A2.2 *The* MA(p) *case.* Define here $\log \lambda^{(n)}(\mathbf{X}, \varepsilon_0)$ where $\varepsilon_0 = (\varepsilon_{-p+1}, \cdots, \varepsilon_0)$ as

$$(A5) \qquad \begin{aligned} \log \lambda^{(n)}(\mathbf{X}, \varepsilon) &= \sum_{t=1}^{n} \log f(X_t + \sum_{u=1}^{t-1} g_u^{(n)} X_{t-u} + \sum_{u=t}^{t+p-1} g_u^{(n)} \varepsilon_{t-u}) \\ &= \sum_{t=1}^{n} \log f(X_t) - \sum_{t=1}^{n} (\sum_{u=1}^{t-1} g_u^{(n)} X_{t-u} + \sum_{u=t}^{t+p-1} g_u^{(n)} \varepsilon_{t-u}) \phi(X_t) \\ &\quad - \tfrac{1}{2} \sum_{t=1}^{n} (\sum_{u=1}^{t-1} g_u^{(n)} X_{t-u} + \sum_{u=t}^{t+p-1} g_u^{(n)} \varepsilon_{t-u})^2 \\ &\quad \cdot \phi'(X_t + \theta \sum_{u=1}^{t-1} g_u^{(n)} X_{t-u} + \theta \sum_{u=t}^{t+p-1} g_u^{(n)} \varepsilon_{t-u}) \end{aligned}$$

with $\theta = \theta(\mathbf{X}, \varepsilon) \in [0, 1]$.

The first-order term in (A5) is

$$-\sum_{u=1}^{n-1} g_u^{(n)} \sum_{t=u+1}^{n} \phi(X_t) X_{t-u} - \sum_{u=1}^{n+p-1} g_u^{(n)} \sum_{t=u-p+1}^{u} \phi(X_t) \varepsilon_{t-u} = -\sum_{(1)} - \sum_{(2)}.$$

Considering $-\sum_{(1)}$, decompose

$$-\textstyle\sum_{(1)} = n^{-1/2} \sum_{i=1}^{p} b_i \sum_{t=i+1}^{n} \phi(X_t)X_{t-i} - \sum_{u=1}^{p} (g_u^{(n)} + b_u n^{-1/2}) \sum_{t=u+1}^{n} \phi(X_t)X_{t-u}$$

$$- \textstyle\sum_{u=p+1}^{2p} g_u^{(n)} \sum_{t=u+1}^{n} \phi(X_t)X_{t-u} - \sum_{u=2p+1}^{n-1} g_u^{(n)} \sum_{t=u+1}^{n} \phi(X_t)X_{t-u}$$

$$= \textstyle\sum_{(11)} - \sum_{(12)} - \sum_{(13)} - \sum_{(14)}.$$

Up to a finite number of o_p's $\sum_{(11)} = \mathscr{L}_n^0(\mathbf{X})$ (as in (A4), replacing the a_i's with b_i's), and is therefore asymptotically normal, with mean 0 and variance $d^2 = \sum_{i=1}^{p} b_i^2\sigma^2 I(f)$. For $1 \le u \le p$, $g_u^{(n)}$ is a finite sum of terms of the form

$$-b_u n^{-1/2},\ b_{i_{(1)}}b_{i_{(2)}}n^{-2/2},\ \ldots,\ (-1)^u b_{i_{(1)}}b_{i_{(2)}} \cdots b_{i_{(u)}}n^{-u/2}.$$

$(g_u^{(n)} - b_u n^{-1/2})$ is thus $O(n^{-1})$, and, since $E[\phi(X_t)X_{t-u}] = 0$, $\sum_{(12)}$ converges to 0 in probability. Identical arguments apply to the case of $\sum_{(13)}$. As for $\sum_{(14)}$, it follows from Lemma 1 that

$$E\,|\textstyle\sum_{(14)}| \le \sum_{u=2p+1}^{n-1} |g_u^{(n)}|\,(n-2p)E\,|X_t|\,E\,|\phi(X_t)|$$

$$\le 2(qb_M)^3 n^{-3/2}(n-2p)E\,|X_t|\,E\,|\phi(X_t)|.$$

Hence $\sum_{(14)}$ also converges to 0.

Similar arguments can be used to show that $\sum_{(2)}$ is also o_p.

Let $Q_t = |\sum_{u=1}^{t+1} g_u^{(n)}X_{t-u} + \sum_{u=t}^{t+p-1} g_u^{(n)}\varepsilon_{t-u}|$, and consider the second-order term in (A5). Since X_t and ε_t have, under $H_0^{(n)}$, exactly the same distribution, and since our purpose is to prove that this second-order term converges in probability to a constant, we may write Q_t under the more convenient form $Q_t = |\sum_{u=1}^{t+p-1} g_u^{(n)}X_{t-u}|$.

ϕ' being (assumption (iv)) Lipschitzian, this second-order term is of the form

$$-\tfrac{1}{2} \textstyle\sum_{t=1}^{n} Q_t^2\phi'(X_t) + \tfrac{1}{2}T_n(\mathbf{X}, \varepsilon) \quad \text{where} \quad |T_n(\mathbf{X}; \varepsilon)| \le A\theta \sum_{t=1}^{n} Q_t^3.$$

Now

$$\textstyle\sum_{t=1}^{n} EQ_t^3 = \sum_{t=1}^{n} E\,|\sum_{u=1}^{t+p-1} g_u^{(n)}X_{t-u}|^3 \le M \sum_{t=1}^{n} (\sum_{u=1}^{t+p-1} |g_u^{(n)}|)^3$$

$$\le nM(\textstyle\sum_{u=1}^{\infty} |g_u^{(n)}|)^3 \le M(2pb_M)^3 n^{-1/2}.$$

Hence $T_n(\mathbf{X}, \varepsilon)$ is o_p. Finally,

$$-\tfrac{1}{2} \textstyle\sum_{t=1}^{n} Q_t^2\phi'(X_t) = -\tfrac{1}{2} \sum_{u=1}^{p} (g_u^{(n)})^2 \sum_{t=u-p+1}^{n} X_{t-u}^2\phi'(X_t)$$

$$-\tfrac{1}{2} \textstyle\sum_{u=p+1}^{n+p-1} (g_u^{(n)})^2 \sum_{t=u-p+1}^{n} X_{t-u}^2\phi'(X_t)$$

$$-\textstyle\sum_{u=2}^{p+1} \sum_{v=1}^{u-1} g_u^{(n)}g_v^{(n)} \sum_{t=u-p+1}^{n} X_{t-u}X_{t-v}\phi'(X_t)$$

$$-\textstyle\sum_{u=p+2}^{n+p-1} \sum_{v=1}^{u-1} g_u^{(n)}g_v^{(n)} \sum_{t=u-p+1}^{n} X_{t-u}X_{t-v}\phi'(X_t)$$

$$= -\tfrac{1}{2} \textstyle\sum_{(21)} - \tfrac{1}{2} \sum_{(22)} - \sum_{(23)} - \sum_{(24)}.$$

$$\textstyle\sum_{(21)} = \sum_{u=1}^{p} (b_u/\sqrt{n})^2 \sum_{t=u-p+1}^{n} X_{t-u}^2\phi'(X_t)$$

$$+ \textstyle\sum_{u=1}^{p} (g_u^{(n)} - b_u n^{-1/2})^2 \sum_{t=u-p+1}^{n} X_{t-u}^2\phi'(X_t)$$

$$+ 2 \textstyle\sum_{u=1}^{p} (g_u^{(n)} - b_u n^{-1/2})b_u n^{-1/2} \sum_{t=u-p+1}^{n} X_{t-u}^2\phi'(X_t),$$

but $g_u^{(n)^2} - b_u^2 n^{-1}$, as we already know, is $O(n^{-2})$, and thus

$$\Sigma_{(21)} = \sum_{i=1}^{p} b_i^2 E[X_t^2]E[\phi'(X_t)] + o_p = \sum_{i=1}^{p} b_i^2 \sigma^2 I(f) + o_p.$$

Also

$$E\,|\,\Sigma_{(22)}\,| \leq \sum_{u=p+1}^{n+p-1} (g_u^{(n)})^2 (n-u+p)\sigma^2 I(f)$$
$$\leq 2(n-1)\sigma^2 I(f)(pb_M n^{-1/2})^{2(p+1)},$$
$$E\,|\,\Sigma_{(24)}\,| \leq (E\,|\,X_t\,|)^2 I(f)(n+p-2) \sum_{u=2}^{n+p-1} \sum_{v=1}^{u-1} |\,g_u^{(n)} g_v^{(n)}\,|$$
$$\leq (E\,|\,X_t\,|)^2 I(f)(n+p-2)4(pb_M n^{-1/2})^4,$$

and, since $g_u^{(n)} g_v^{(n)}$ is $O(n^{-1})$, $\Sigma_{(23)}$ is o_p.

So far, we have proved that $\log \lambda^{(n)}(\mathbf{X}, \varepsilon) = \log \ell_0^{(n)}(\mathbf{X}) + \mathscr{L}_n^0(\mathbf{X}) - (d^2/2) + R_n(\mathbf{X}, \varepsilon)$ where $R^n(\cdots)$ is o_p. Since $L^{(n)}$ is again a conditional expectation, viz.

$$L^{(n)} = E_\varepsilon[\exp(\log \lambda^{(n)}(\mathbf{X}, \varepsilon) - \log \ell_0^{(n)}(\mathbf{X}))\,|\,\mathbf{X}],$$

the proof follows as in the AR(p) case.

A2.3. *The* ARMA(p_1, p_2) *case.* We briefly sketch here the proof; it is easy to see that it will lead to the same problems as in the AR and MA cases treated above.

Assume $p_2 \leq p_1 = p$. Denoting by $\lambda^{(n)}(\mathbf{X};\, \mathbf{X}_0;\, \varepsilon)$, the integrand in (3.3), we have

$$\log \lambda^{(n)} = \sum_{t=1}^{n} \log f(X_t)$$
$$+ \sum_{t=1}^{n} \phi(X_t)\Big[\sum_{i=1}^{p} X_{t-i}(n^{-1/2}a_i - g_i^{(n)}) + n^{-1/2} \sum_{u=1}^{p} g_u^{(n)} \sum_{i=1}^{p} a_i X_{t-u-i}$$
$$- \sum_{u=p+1}^{t-1} g_u^{(n)}(X_{t-u} - n^{-1/2} \sum_{i=1}^{p} a_i X_{t-u-i}) - \sum_{u=t}^{t+p_2-1} g_u^{(n)} \varepsilon_{t-u}\Big]$$
$$-\tfrac{1}{2} \sum_{t=1}^{n} \Big[n^{-1/2} \sum_{i=1}^{p} a_i X_{t-i} - \sum_{u=1}^{t-1} g_u^{(n)}(X_{t-u} - n^{-1/2} \sum_{i=1}^{p} a_i X_{t-u})$$
$$- \sum_{u=t}^{t+p_2-1} g_u^{(n)} \varepsilon_{t-u}\Big]^2 \phi'(X_t - \theta[\cdots]).$$

The term $\sum_{t=1}^{n} \phi(\mathrm{X}_t) \sum_{i=1}^{p} \mathrm{X}_{t-i}(n^{-1/2}a_i - g_i^{(n)})$ is of the form (3.5), up to o_p terms. The second-order term converges in probability to $-\tfrac{1}{2}[\sum_{i=1}^{p_2} (a_i + b_i)^2 + \sum_{i=p_2+1}^{p} a_i^2]$. The proof follows. □

APPENDIX 3

Some properties of linear serial rank statistics.

LEMMA 2. *The variance of* $(n-p)S_n = \sum_{t=p+1}^{n} a(R_t^{(n)}, R_{t-1}^{(n)}, \cdots, R_{t-p}^{(n)})$ *is*

$$D^2((n-p)S_n) = (n-p)\mathrm{Var}(a(R_{p+1}^{(n)}, \cdots, R_1^{(n)}))$$
$$+ 2\sum_{i=1}^{p} (n-p-i)\mathrm{Cov}(a(R_{p+1+i}^{(n)}, \cdots, R_{1+i}^{(n)}), a(R_{p+1}^{(n)}, \cdots, R_1^{(n)}))$$
$$+ [(n-3p)(n-3p-1) + p(2n-5p-1)]$$
$$\cdot\, \mathrm{Cov}(a(R_{2p+2}^{(n)}, \cdots, R_{p+2}^{(n)}), a(R_{p+1}^{(n)}, \cdots, R_1^{(n)})).$$

Corollary 2.

$$D^2((n-p)S_n) \leq \mathrm{Var}(a(R_{p+1}^{(n)}, \cdots, R_1^{(n)}))(n-p)(2p+1)$$
$$+ n\,\mathrm{Cov}(a(R_{2p+2}^{(n)}, \cdots, R_{p+2}^{(n)}), a(R_{p+1}^{(n)}, \cdots, R_1^{(n)})).$$

Proof. The proof is straightforwardly obtained by expanding $E[\sum a(\cdots)]^2$. □

Lemma 3.

$$E[a(R_{p+1}^{(n)}, \cdots, R_1^{(n)})]$$
$$= \frac{(n-p-1)(n-p-2)\cdots(n-2p-1)}{n(n-1)\cdots(n-p)} E[a(R_{p+1}^{(n)}, \cdots, R_1^{(n)}) \mid R_{2p+2}^{(n)}, \cdots, R_{p+2}^{(n)}]$$
$$+ \sum_{\ell=1}^{p+1} \frac{(n-p-1)(n-p-2)\cdots(n-2p-1+\ell)}{n(n-1)\cdots(n-p)} \sum_{p+2\leq j_1\neq\cdots\neq j_\ell\leq 2p+2}\cdots\sum \;\; \sum_{1\leq k_1<k_2<\cdots<k_\ell\leq p+1}\cdots\sum$$
$$\cdot E[a(R_{p+1}^{(n)}, \cdots, R_{k_\ell+1}^{(n)} R_{j_\ell}^{(n)} R_{k_\ell-1}^{(n)}, \cdots, R_{k_1+1}^{(n)} R_{j_1}^{(n)} R_{k_1-1}^{(n)}, \cdots, R_1^{(n)}) \mid R_{2p+2}^{(n)}, \cdots, R_{p+2}^{(n)}].$$

Proof. Again, the proof follows from a tedious enumeration of the $n(n-1)\cdots(n-p)$ terms of $\sum_{1\leq i_1\neq\cdots\neq i_p\leq n}\cdots\sum a(i_1, \cdots, i_p)$, which contains $(n-p-1)\cdots(n-2p-1)$ terms with $\# \{i_1 \neq \cdots \neq i_p\} \cap \{R_{p+2}^{(n)}, \cdots, R_{2p+2}^{(n)}\} = \emptyset$, $(n-p-1), \cdots, (n-2p)$ terms with $\# \{i_1 \neq \cdots \neq i_p\} \cap \{R_{p+2}^{(n)}, \cdots, R_{2p+2}^{(n)}\} = 1$, $\cdots$, $(n-p-1), \cdots, (n-2p-1+\ell)$ terms with $\# \{i_1 \neq \cdots \neq i_p\} \cap \{R_{p+2}^{(n)}, \cdots, R_{2p+2}^{(n)}\} = \ell$, $\cdots$, and $(n-p+1)$ terms with $\# \{i_1 \neq \cdots \neq i_p\} \cap \{R_{p+2}^{(n)}, \cdots, R_{2p+2}^{(n)}\} = p+1$. □

Lemma 4.

$$n \,|\, \mathrm{Cov}(a(R_{p+1}^{(n)}, \cdots, R_1^{(n)}), a(R_{2p+2}^{(n)}, \cdots, R_{p+2}^{(n)})) \,|\, \leq K E[a^2(R_{p+1}^{(n)}, \cdots, R_1^{(n)})]$$

(*where the constant K does not depend on n*).

Proof. Applying Lemma 3,

$$E[a(R_{p+1}^{(n)}, \cdots, R_1^{(n)}) a(R_{2p+2}^{(n)}, \cdots, R_{p+2}^{(n)})]$$
$$= E[a(R_{2p+2}^{(n)}, \cdots, R_{p+2}^{(n)}) E[a(R_{p+1}^{(n)}, \cdots, R_1^{(n)}) \mid R_{2p+2}^{(n)}, \cdots, R_{p+2}^{(n)}]]$$
$$= \frac{n(n-1)\cdots(n-p)}{(n-p-1)\cdots(n-2p-1)} E^2[a(R_{p+1}^{(n)}, \cdots, R_1^{(n)})]$$
$$- \sum_{\ell=1}^{p+1} [(n-2p-2+\ell)\cdots(n-2p-1)]^{-1} \sum_{p+2\leq j_1\neq\cdots\neq j_\ell\leq 2p+2}\cdots\sum \;\; \sum_{1\leq k_1<\cdots<k_\ell\leq p+1}\cdots\sum$$
$$E[a(R_{p+1}^{(n)}, \cdots, R_{k_\ell+1}^{(n)} R_{j_\ell}^{(n)}, \cdots, R_{j_1}^{(n)} R_{k_1-1}^{(n)}, \cdots, R_1^{(n)}) a(R_{2p+2}^{(n)}, \cdots, R_{p+2}^{(n)})].$$

For every fixed value of ℓ, this latter sum contains

$$((p+1)\cdots(p+2-\ell))^2/(n-2p-2+\ell)\cdots(n-2p-1)\ell!$$

terms and since $|E[a(\cdots)a(\cdots)]| \leq E[a^2(\cdots)]$, this sum is bounded by

$$\sum_{\ell=1}^{p+1} \frac{((p+1)\cdots(p+2-\ell))^2}{(n-2p-2+\ell)\cdots(n-2p-1)\ell!} E[a^2(R_{p+1}^{(n)}, \cdots, R_1^{(n)})],$$
$$= C_n E[a^2(R_{p+1}^{(n)}, \cdots, R_1^{(n)})],$$

where C_n is $O(n^{-1})$. Thus

$$\begin{aligned}\mathrm{Cov}[a(R_{p+1}^{(n)}, \cdots, R_1^{(n)}), a(R_{2p+2}^{(n)}, \cdots, R_{p+1}^{(n)})] &\leq \left(\frac{n(n-1)\cdots(n-p)}{(n-p-1)\cdots(n-2p-1)} - 1\right)(Ea(R_{p+1}^{(n)}, \cdots, R_1^{(n)}))^2 \\ &\quad + C_n E[a^2(R_{p+1}^{(n)}, \cdots, R_1^{(n)})] \\ &\leq C_n' E[a^2(R_{p+1}^{(n)}, \cdots, R_1^{(n)})],\end{aligned}$$

with C_n' another $O(n^{-1})$. □

APPENDIX 4

Proof of Proposition 4.1. The g_1 function in Yoshihara's (1976) Theorem 1 is here

$$g_{\alpha\beta}(\mathbf{Y}_t) = \int_{[0,1]^{p(p+1)}} \Phi_{\alpha\beta}(\mathbf{Y}_t, \mathbf{y}_2, \cdots, \mathbf{y}_{p+1}\, d\mathbf{y}_2 \cdots d\mathbf{y}_{p+1},$$

where $d\mathbf{y}_i$ stands for $dy_{i,1}, dy_{i,2}, \cdots, dy_{i,p+1}$. Since

$$\int_{[0,1]^{p(p+1)}} \Phi^{\mathcal{G}}(\mathbf{Y}_t, \mathbf{y}_2, \cdots, \mathbf{y}_{p+1})\, d\mathbf{y}_2 \cdots d\mathbf{y}_{p+1}$$
$$= (p+1)^{-1}(G(\mathbf{Y}_t) + pEG(\mathbf{Y}_1)) = (p+1)^{-1}\phi(F^{-1}(U_t)) \sum_{i=1}^{p} d_i F^{-1}(U_{t-i}),$$

$$\int_{[0,1]^{p(p+1)}} \Phi^{\mathcal{J}}(\mathbf{Y}_t, \mathbf{y}_2, \cdots, \mathbf{y}_{p+1})\, d\mathbf{y}_2 \cdots d\mathbf{y}_{p+1}$$
$$= (p+1)^{-1}\left(J(\mathbf{Y}_t) + p\int_{[0,1]^{p+1}} J(u_{p+1}, \cdots, u_1)\, du_1 \cdots du_{p+1}\right),$$

and

$$\int \Phi^{\mathcal{E}}(\mathbf{Y}_t, \mathbf{y}_2, \cdots, \mathbf{y}_{p+1})\, d\mathbf{y}_2 \cdots d\mathbf{y}_{p+1}$$
$$= \frac{1}{(p+1)} \sum_{k=1}^{p+1} \int_{[0,1]^p} J(u_p, \cdots, u_k, U_t, u_{k-1}, \cdots, u_1)\, du_1 \cdots du_p,$$

we obtain

$$\begin{aligned}g_{\alpha\beta}(\mathbf{Y}_t) &= g_{\alpha\beta}(U_t, \cdots, U_{t-p}) \\ &= (p+1)^{-1}[\alpha J^*(U_t, \cdots, U_{t-p}) + \beta\phi(F^{-1}(U_t)) \sum_{i=1}^{p} d_i F^{-1}(U_{t-i})].\end{aligned}$$

Clearly, $E[g_{\alpha\beta}(\mathbf{Y}_t)] = 0$; the variance of $g_{\alpha\beta}(\mathbf{Y}_t)$ is

$$D^2(g_{\alpha\beta}(\mathbf{Y}_t))$$

$$= (p+1)^{-2}\Big\{\alpha^2 \int_{[0,1]^{p+1}} J^{*2}(v_{p+1}, \cdots, v_1)\, dv_1 \cdots dv_{p+1} + 2\alpha\beta \textstyle\sum_{i=1}^{p} d_i$$

$$\int_{[0,1]^{p+1}} J^*(v_{p+1}, \cdots, v_1)\phi(F^{-1}(v_{p+1}))F^{-1}(v_{p+1-i})\, dv_1 \cdots dv_{p+1}$$

$$+ \beta^2 \textstyle\sum_{i=1}^{p} d_i^2\sigma^2 I(f)\Big\};$$

the covariances between $g_{\alpha\beta}(\mathbf{Y}_t)$ and $g_{\alpha\beta}(\mathbf{Y}_{t+j})$, $1 \leq j \leq p$, are

$$E[g_{\alpha\beta}(\mathbf{Y}_t)g_{\alpha\beta}(\mathbf{Y}_{t+j})]$$

$$= (p+1)^{-2}\Big[\alpha^2 \int_{[0,1]^{p+1+j}} J^*(v_{p+1}, \cdots, v_1)J^*(v_{p+1+j}, \cdots, v_{1+j})\, dv_1 \cdots dv_{p+1+j}$$

$$+ \alpha\beta \textstyle\sum_{i=1}^{p-j} d_i \displaystyle\int_{[0,1]^{p+1}} F^{-1}(v_{p+1-i})\phi(F^{-1}(v_{p+1}))J^*(v_{j+1+p}, \cdots, v_{j+1})\, dv_{j+1} \cdots dv_{j+1+p}\Big]$$

(with the convention that, for $j = p$, the sum $\sum_{i=1}^{p-j}$ vanishes). On account of assumption (iii) and (4.1),

$$\sigma_{\alpha\beta}^2 = D^2(g_{\alpha\beta}(\mathbf{Y}_t)) + 2 \textstyle\sum_{j=1}^{p} E[g_{\alpha\beta}(\mathbf{Y}_t)g_{\alpha\beta}(\mathbf{Y}_{t+j})]$$

$$= (p+1)^{-2}\{\alpha^2 V^2 + 2\alpha\beta \textstyle\sum_{i=1}^{p} d_i C_i + \beta^2 \sum_{i=1}^{p} d_i^2 \sigma^2 I(f)\}$$

(with V^2 and C_i given in (4.13) and (4.14)) is never zero; the conditions of Yoshihara's (1976) Theorem 1 are therefore satisfied; and $n^{1/2}(\mathscr{U}_{\alpha\beta}^n - E\mathscr{U}_{\alpha\beta}^n)$ is asymptotically normal, with mean zero and variance $(p+1)^2\sigma_{\alpha\beta}^2$. This completes the proof, since

$$n^{1/2}(\mathscr{U}_{\alpha\beta}^n - E\mathscr{U}_{\alpha\beta}^n) = \alpha\sqrt{n}(S_n - m_n) + \beta(\log L_n + \tfrac{1}{2} \textstyle\sum d_i^2\sigma^2 I(f)) + o_P. \quad \square$$

APPENDIX 5

Proof of Lemma 5.1. It is easy to see that

$$\{\textstyle\sum_{j=0}^{p-i} \phi(F^{-1}(v_{p+1-j}))F^{-1}(v_{p+1-j-i});\ i = 1, \cdots, p\}$$

constitutes a p-tuple of L^2-orthogonal functions, with norms

$$[(p+1-i)\sigma^2 I(f)]^{1/2},\ i = 1, \cdots, p.$$

Any square-integrable score-generating function J^* can therefore be decomposed into $J^*(v_{p+1}, \cdots, v_1) = J_0^*(v_{p+1}, \cdots, v_1) + J_1^*(v_{p+1}, \cdots, v_1)$, where J_0^* is given by (5.3) and

$$\text{(A6)} \quad \int J_1^*(v_{p+1}, \cdots, v_1) \textstyle\sum_{j=0}^{p-i} \phi(F^{-1}(v_{p+1-j}))F^{-1}(v_{p+1-j-i})\, dv_{p+1}, \cdots, dv_1 = 0,$$

$$i = 1, \cdots, p.$$

Denote by S_n, S_n^0, and $S_n^\perp$ linear serial rank statistics associated with J^*, J_0^*, and $J_\perp^*$, respectively; denote by V^2, V_0^2, and $V_\perp^2$ the asymptotic variances of $n^{1/2}(S_n - m_n)$, $n^{1/2}(S_n^0 - m_n^0)$ and $n^{1/2}(S_n^\perp - m_n^\perp)$, respectively. Then, it is easy to check that

$$n^{1/2}(S_n - m_n) = n^{1/2}(S_n^0 - m_n^0) + n^{1/2}(S_n^\perp - m_n^\perp) + o_p(1),$$

with $\lim_{n\to\infty} nE[(S_n^0 = m_n^0)(S_n^\perp - m_n^\perp)] = 0$ (under $H_0^{(n)}$).

Thus, we have $V^2 = V_0^2 + V_\perp^2$. Now, since using J^* or J_0^* in (4.14) leads to the same values of the C_i's, we have

$$e(S_n, S_n^0) = \frac{(\sum_{i=1}^p d_iC_i)^2/(V_0^2 + V_\perp^2)}{(\sum_{i=1}^p d_iC_i)^2/V_0^2} \leq 1,$$

which completes the proof. □

APPENDIX 6

Asymptotic normality of $(n^{1/2}r_1^{(n)}, \cdots, n^{1/2}r_p^{(n)}, \log L_n)'$.

Proof of Proposition 5.2. Consider the linear combination (with $\sum_{i=1}^p \alpha_i^2 = 1$)

$$\begin{aligned} &\sqrt{n}\sum_{i=1}^p \alpha_i r_i^{(n)} + \beta \log L^n \\ &= [\sigma^{-2}\sum_{i=1}^p \alpha_i n^{-1/2}(n-k)^{-1}\sum_{t=i+1}^n X_tX_{t-i} + \mathscr{L}_n^0(\mathbf{X})] - \tfrac{1}{2}\beta^2\sigma^2 I(f)\sum_{i=1}^p d_i^2 \\ &\quad - \sum_{i=1}^p \alpha_i(\sigma^{-2} - ((1/n)\sum_{t=1}^n X_t^2)^{-1})(n^{1/2}/(n-i))n^{1/2}\sum_{t=i+1}^n X_tX_{t-i} + o_p \qquad \text{(A7)} \\ &= \textstyle\sum_{(1)} - \tfrac{1}{2}\beta^2\sum_i d_i^2\sigma^2 I(f) - \sum_{(2)}. \end{aligned}$$

Since $n^{1/2}\sum X_tX_{t-k}$ is asymptotically normal, and since $(\sigma^{-2} - ((1/n)\sum_t X_t^2)^{-1})$ converges to 0 in probability, $\sum_{(2)}$ is o_p. Hence

$$\textstyle\sum_{(1)} = n^{-1/2}\sum_{t=1}^{n-p} X_t[\sigma^{-2}\sum_{i=1}^p \alpha_i X_{t+i} + \beta d_i\phi X_{t+i})] + o_p = n^{-1/2}\sum_{t=1}^{n-p} Z_t + o_p.$$

An immediate computation gives

$$E(Z_t^2) = \sum_{i=1}^p \alpha_i^2 + \beta^2\sum_{i=1}^p d_i^2\sigma^2 I(f) + 2\beta\sum_{i=1}^p d_i\alpha_i \qquad \text{(A8)}$$

and

$$E(Z_tZ_{t+k}) = 0.$$

Consequently, (A7) is asymptotically normal, with mean $-\tfrac{1}{2}\beta^2\sum d_i^2\sigma^2 I(f)$ and variance (A8). This proves Proposition 5.2. □

Corollary 5.1 is an immediate consequence of LeCam's third lemma and Corollary 5.2 follows from the Cauchy-Schwarz inequality. □

Acknowledgement. The authors gratefully acknowledge the valuable comments and suggestions of the Editor, Professor Michael Perlman, the Associate Editor and the referee.

REFERENCES

AIYAR, R. J. (1981). Asymptotic efficiency of rank tests of randomness against autocorrelation. *Ann. Inst. Statist. Math.* **33 A** 255–262.

ANDERSON, T. W. (1971). *The Statistical Analysis of Time Series.* Wiley, New York.

BARTELS, R. (1982). The rank version of von Neumann's ratio test for randomness. *J. Amer. Statist. Assoc.* **77** 40–46.

BELL, C. B., WOODROOFE, M. and AVADHANI, T. V. (1970). Some nonparametric tests for stochastic processes. In *Nonparametric Techniques in Statistical Inference.* (M. L. Puri, ed.) 215–258. Cambridge University Press.

DUFOUR, J.-M. (1982). Rank tests for serial dependence. *J. Time Series Anal.* **2** 117–128.

DUFOUR, J.-M., LEPAGE, Y. and ZEIDAN, H. (1982). Nonparametric testing for time series: a bibliography. *Canad. J. Statist.* **10** 1–38.

GOVINDARAJULU, Z. (1983). Rank tests for randomness against autocorrelated alternatives. In *Time Series Analysis: Theory and Practice* 4. (O. D. Anderson, ed.) 65–73. North-Holland, Amsterdam and New York.

GOVINDARAJULU, Z. and DWASS, M. (1983). Simultaneous tests for randomness and location of symmetry. *Proc. Second Prague Conference on Asymptotic Statistics*, 181–197. Hradec Karlove, Czechoslovakia.

GUPTA, G. D. and GOVINDARAJULU, Z. (1980). Nonparametric tests of randomness against autocorrelated normal alternatives. *Biometrika* **67** 375–379.

HÁJEK, J. and ŠIDÁK, Z. (1967). *Theory of Rank Tests.* Academic, New York.

HALLIN, M. (1984). Spectral factorization of nonstationary moving average processes. *Ann. Statist.* **12** 172–192.

HALLIN, M., INGENBLEEK, J.-FR., and PURI, M. L. (1985). Tests de Rangs Linéaires pour une Hypothèse de Bruit Blanc. *C. R. Acad. Sc. Paris* **301** Série 1, 49–52.

HOTELLING, H. and PABST, M. R. (1936). Rank correlation and tests of significance involving no assumption of normality. *Ann. Math. Statist.* **7** 29–43.

INGENBLEEK, J.-FR. (1980). *Statistiques de rangs sérielles univariées et multivariées.* Unpublished D.Sc. dissertation, Université Libre de Bruxelles, Brussels.

JOGDEO, K. (1968). Asymptotic normality in nonparametric methods. *Ann. Math. Statist.* **39** 905–922.

KENDALL, M. and STUART, A. (1968). *The Advanced Theory of Statistics.* Vol. 3. Griffin, London.

KNOKE, D. J. (1977). Testing for randomness against autocorrelation: alternative tests. *Biometrika* **64** 523–529.

LECAM, L. (1960). Locally asymptotically normal families of distributions. *Univ. Calif. Publ. Statist.* **3** 37–98.

MILLER, K. S. (1968). *Linear Difference Equations.* Benjamin, New York.

PURI, M. L. and SEN, P. K. (1971). *Nonparametric Methods in Multivariate Analysis.* Wiley, New York.

YOSHIHARA, K. (1976). Limiting behavior of U-statistics for stationary absolutely regular processes. *Z. Wahrsch. verw. Gebiete* **35** 237–252.

WALD, A. and WOLFOWITZ, J. (1943). On exact tests for randomness in the nonparametric case based on serial correlation. *Ann. Math. Statist.* **14** 378–388.

WOLD, H. (1954). *A Study in the Analysis of Stationary Time Series.* Almqvist and Wiksell, Stockholm.

M. HALLIN
J.-FR. INGENBLEEK
INSTITUT DE STATISTIQUE
C.P. 210 - CAMPUS DE LA PLAINE
UNIVERSITÉ LIBRE DE BRUXELLES
B1050 BRUXELLES
BELGIUM

M. L. PURI
DEPARTMENT OF MATHEMATICS
SWAIN HALL EAST
INDIANA UNIVERSITY
BLOOMINGTON, INDIANA 47405

Reprinted by permission of the publisher,
© 1985 by The Institute of Mathematical Statistics

LINEAR AND QUADRATIC SERIAL RANK TESTS FOR RANDOMNESS AGAINST SERIAL DEPENDENCE

MARC HALLIN[1], JEAN-FRANÇOIS INGENBLEEK, and MADAN L. PURI[2]

Université Libre de Bruxelles and Indiana University

Abstract. The problem of testing randomness against specified and unspecified contiguous ARMA alternatives is considered. In the case of specified alternative, the linear serial rank tests proposed by the authors (Hallin, Ingenbleek and Puri 1985) are shown to be asymptotically most powerful within the class of all possible tests (at the required level). In the case of unspecified alternative, however, any of the above optimal tests is unfortunately completely insensitive against a whole subclass of the alternative. *Quadratic serial rank statistics*, providing tests of the χ^2-type are therefore introduced. Their asymptotic distributions are derived, under the null hypothesis as well as under contiguous ARMA alternatives. The asymptotically maximin most powerful quadratic serial rank tests (for a given density) are then obtained. Because of their close similarity with the Box–Pierce parametric test, we call them *rank portmanteau tests*. The asymptotic relative efficiencies (AREs) of the rank portmanteau tests with respect to one another and their AREs with respect to the corresponding Box–Pierce and quadratic Spearman tests are derived.

Keywords. Rank tests for serial dependence; quadratic rank tests; ARMA models; asymptotically most powerful tests; asymptotically maximin most powerful tests; Box–Pierce portmanteau test.

1. INTRODUCTION

1.1. *Testing randomness against serial dependence*

A large variety of rank procedures including some of the earliest and most classical ones, such as the run test, the turning point test, Spearman's rank autocorrelations tests, is available for testing randomness against serial dependence. Yet, in spite of the crucial importance of this problem and of its long history, very little is known about the power and relative efficiencies of these procedures—and hence about their optimality properties (for a recent review, see Govindarajulu (1983) or Bhattacharyya (1984)).

One of the main reasons for the lack of such results is probably that serial dependence is a particularly vast and diffuse alternative, for which no obvious and natural parametrization exists. Such a parametrization should indeed be altogether rich enough, reasonably tractable, and intuitively appealing. We believe that all these requirements are fairly well satisfied by the class of ARMA dependencies, widely used in time series analysis.

In a previous work on the subject, Hallin, Ingenbleek and Puri (1985) considered the problem of testing randomness against alternatives of ARMA dependence by introducing linear serial rank statistics of the form

$$S^{(n)} = (n-p)^{-1} \sum_{t=p+1}^{n} a_n(R_t^{(n)}, R_{t-1}^{(n)}, \ldots, R_{t-p}^{(n)}), \tag{1.1}$$

where $a_n(\ldots)$ is some *score function* and $R_t^{(n)}$ is the rank of the observation $X_t^{(n)}$ in an observed series $\boldsymbol{X}^{(n)} = (X_1^{(n)}, \ldots, X_n^{(n)})$ of length n. We developed the asymptotic theory of the tests based on such statistics and obtained the asymptotically optimal score-generating functions for specified contiguous ARMA alternatives (a stronger version of this optimality result is given in Proposition 3.2 below). Such an alternative is completely specified (Section 1.2) by a vector $\boldsymbol{d}$ characterizing a family of ARMA models, and the density type f of the input white noise of these models: we denote it by $H_{\boldsymbol{d},f}^{(n)}$.

It turns out, however, that this optimal score-generating function depends on both $\boldsymbol{d}$ and f; moreover, in the case of ARMA alternatives of order $p > 1$, for any test statistic $S^{(n)}$ and any density type f there exists a class of contiguous alternatives $H_{\boldsymbol{d},f}^{(n)}$ under which the asymptotic distribution of $S^{(n)}$ is the same as under the null hypothesis of randomness. This is of course very unpleasant, since it means that any test statistic is completely insensitive to certain types of dependences. (By the way, the same phenomenon also occurs in the parametric approach to the problem, in which the asymptotically most powerful test statistic can be shown to be some linear combination of the sample autocorrelations.)

The usual remedy, in the parametric as well as in the nonparametric approach, consists in considering statistics of the χ^2-type. Examples are the Box–Pierce portmanteau statistic (Box and Pierce, 1970) for testing randomness against ARMA dependence, or the class of quadratic rank tests considered by Beran (1975) in the allied problem of testing randomness against trend alternatives.

We introduce here (Section 2) quadratic serial rank statistics, with the purpose of obtaining good overall performances against alternatives $H_{\boldsymbol{d},f}^{(n)}$ with unspecified $\boldsymbol{d}$. The asymptotic distribution of these statistics is given in Proposition 2.1, under both the null hypothesis $H_0^{(n)}$ of randomness and the alternative $H_{\boldsymbol{d},f}^{(n)}$. In Section 3, the problem of testing $H_0^{(n)}$ against a specified alternative $H_{\boldsymbol{d},f}^{(n)}$ is briefly revisited. Introducing the concept of rank autocorrelation coefficients associated with a (strongly unimodal) density f, we reformulate the asymptotically optimal rank statistic (Hallin *et al.*, 1985) in a particularly simple and intuitively appealing form. Moreover, we establish a stronger result, that this statistic provides asymptotically the most powerful test for $H_0^{(n)}$ against $H_{\boldsymbol{d},f}^{(n)}$, not only within the class of rank tests, but also within the class of all possible tests (at the required probability level). Section 4 is devoted to the problem of testing $H_0^{(n)}$ against $H_{\boldsymbol{d},f}^{(n)}$ with unspecified $\boldsymbol{d}$ by means of quadratic serial rank statistics. We show, in Proposition 4.1, that an asymptotically maximin most powerful rank test can be based on a rank analogue of Box and Pierce's portmanteau statistic. Finally, the mutual AREs of some of these rank portmanteau statistics, which are sums of the squared rank autocorrelation coefficients defined in Section 3, as well as their AREs with respect to a Spearman portmanteau statistic and Box and Pierce's parametric statistic, are investigated and given in Table I.

1.2. *Contiguous ARMA alternatives*

The concept of the contiguity of probability measures, initially developed by LeCam (1960) has been very successfully employed (cf. Hájek (1962) and Roussas (1972) among others) in the study of the local asymptotic power properties and optimality of test procedures.

As stated in Section 1.1, the contiguous ARMA alternatives $H^{(n)}_{\mathbf{d},f}$ that we are considering are characterized by a vector $\mathbf{d} = (d_1, \ldots, d_p)'$ of real coefficients and a density type f. By a density type f we mean a family of densities $\{f_\sigma(x) = (1/\sigma)f_1(x/\sigma);\ \sigma > 0\}$ indexed by scale parameter. We shall denote an unspecified member of this family by $f(x)$. We assume that $\int xf(x)\,dx = 0$ and $\int x^2 f(x)\,dx = \sigma^2$. Furthermore, we assume (as in Hallin *et al.*, 1985, Section 2) that $f(x)$ is a.e. derivable, that its derivative $f'(x)$ satisfies $\int_{-\infty}^{\infty} |f'(x)|\,dx < \infty$, and that $f(x)$ has finite Fisher's information $I(f)$, i.e. $f(x)$ is absolutely continuous on finite intervals, and

$$I(f) = \int_{-\infty}^{\infty} \{f'(x)/f(x)\}^2 f(x)\,dx < \infty.$$

Under these assumptions, letting $\phi(x) = f'_1(x)/f_1(x)$ a.e., $f(x)$ has strictly positive finite Fisher's information $I(f) = \sigma^{-2} \int \phi^2(x) f_1(x)\,dx$. Note that the quantity $\sigma^2 I(f)$ $(= I(f_1))$, which plays an important role in the paper, does not depend on the scale transformation.

Assuming $d_p \neq 0$, consider the set of all vectors $\mathbf{a} = (a_i)$ and $\mathbf{b} = (b_i)$ in $\mathbb{R}^p$ such that $a_i + b_i = d_i$, $i = 1, \ldots, p$. These vectors, together with $f(x)$, characterize a family of ARMA (p_1, p_2) models $(\max(p_1, p_2) = p)$

$$X_t - n^{-1/2} \sum_{i=1}^{p_1} a_i X_{t-i} = \varepsilon_t + n^{-1/2} \sum_{i=1}^{p_2} b_i \varepsilon_{t-i}, \qquad t \in \mathbb{Z}, \tag{1.2}$$

$\{\varepsilon_t;\ t \in \mathbb{Z}\}$ being a white noise with density $f(x)$. Define now the sequence of alternatives $H^{(n)}_{\mathbf{d},f}$ ($\mathbf{d}$ and f specified) as the sequence of (composite) hypotheses under which the vector of observations $\mathbf{X}^{(n)} = (X^{(n)}_1, \ldots, X^{(n)}_n)$ is generated by one of the ARMA models (1.2) associated with $\mathbf{d}$ and f.

The null hypothesis of randomness is easier to characterize. Letting $\mathbf{a} = \mathbf{b} = \mathbf{0}$ in (1.2) yields $X_t = \varepsilon_t$, under which the observations $(X^{(n)}_1, \ldots, X^{(n)}_n)$ would be i.i.d. (independent and identically distributed) with density f. Denote by $H^{(n)}_0$ the family of distributions of all independently and identically distributed n-tuples $\mathbf{X}^{(n)}$. Then, Proposition 3.1 in Hallin *et al.* (1985) establishes the contiguity of $H^{(n)}_{\mathbf{d},f}$ and $H^{(n)}_0$.

Denote by $\mathbf{R}^{(n)} = (R^{(n)}_1, \ldots, R^{(n)}_n)$ the vector of ranks corresponding to $\mathbf{X}^{(n)}$. Assume that there exists some square-integrable function $J(\ldots)$, defined over $[0, 1]^{p+1}$, such that, under $H^{(n)}_0$,

$$\lim_{n\to\infty} E[J(F(X^{(n)}_{p+1}), \ldots, F(X^{(n)}_1)) - a_n(R^{(n)}_{p+1}, \ldots, R^{(n)}_1)]^2 = 0, \tag{1.3}$$

where F is the distribution function with density f. $J(\ldots)$ is called a score-generating function associated with $S^{(n)}$. Equation (1.3), however, is only a limiting condition, and distinct linear serial rank statistics may have the same score-generating function (the converse, as we shall see in Remark 3.4, is also true: distinct score-generating functions can be associated with the same linear serial rank statistic).

Of course, score-generating functions, as well as the corresponding statistics, always can be defined up to additive and multiplicative constants. We therefore may assume that the score-generating functions $J(\ldots)$ to be used in the sequel are such that

$$\int_{[0,\,1]^{p+1}} J(v_{p+1}, \ldots, v_1) \prod_{j \neq i} dv_j = 0 \qquad (i = 1, \ldots, p+1). \tag{1.4}$$

If (1.4) would not be satisfied by some $J(\ldots)$, then replace it by the equivalent function

$$J^*(v_{p+1}, \ldots, v_1) = J(v_{p+1}, \ldots, v_1) - \sum_{i=1}^{p+1} \int_{[0,\,1]^p} J(v_{p+1}, \ldots, v_1) \prod_{j \neq i} dv_j$$
$$+ p \int_{[0,\,1]^{p+1}} J(v_{p+1}, \ldots, v_1)\, dv_{p+1} \ldots dv_1 \,. \tag{1.5}$$

2. QUADRATIC SERIAL RANK STATISTICS

Let $(\boldsymbol{S}^{(n)} - \boldsymbol{m}^{(n)}) = (S_1^{(n)} - m_1^{(n)}, \ldots, S_q^{(n)} - m_q^{(n)})'$ be a q-tuple of linear serial rank statistics centred (under $H_0^{(n)}$) about their means and p_μ be the order of $S_\mu^{(n)}$ ($\mu = 1, \ldots, q$). Let $p = \max_\mu p_\mu$ and denote by $\boldsymbol{V}^2$ the asymptotic covariance matrix of $n^{1/2}(\boldsymbol{S}^{(n)} - \boldsymbol{m}^{(n)})$ under $H_0^{(n)}$. Assume that $\boldsymbol{V}^2$ is of rank q. Then, we define a quadratic serial rank statistic of order p as

$$Q^{(n)} = n(\boldsymbol{S}^{(n)} - \boldsymbol{m}^{(n)})' \boldsymbol{V}^{-2} (\boldsymbol{S}^{(n)} - \boldsymbol{m}^{(n)}). \tag{2.1}$$

Whenever $q = 1$, equation (2.1) reduces to a squared, asymptotically standardized, linear serial statistic. The class of quadratic statistics can thus be viewed as an extension of the previously studied class of linear statistics. As can be expected, quadratic statistics will lead to rank tests of χ^2 types.

We first establish the asymptotic distribution of $n^{1/2}(\boldsymbol{S}^{(n)} - \boldsymbol{m}^{(n)})$, both under $H_0^{(n)}$ and $H_{\boldsymbol{d},f}^{(n)}$. The asymptotic χ^2 distribution of (2.1) immediately follows from the following result:

PROPOSITION 2.1. *Denote by $J_\mu(v_{p_\mu+1}, \ldots, v_1)$ the score-generating functions associated with $S_\mu^{(n)}$ ($\mu = 1, \ldots, q$), and assume that they satisfy (1.4). Then, under $H_0^{(n)}$, $n^{1/2}(\boldsymbol{S}^{(n)} - \boldsymbol{m}^{(n)})$ is asymptotically normal, with mean $\boldsymbol{0}$ and covariance matrix $\boldsymbol{V}^2 = (V_{\mu\nu})$ where (writing $J_\mu(v_{p+1}, \ldots, v_1)$ for $J_\mu(v_{p_\mu+1}, \ldots, v_1)$),*

$$V_{\mu\nu} = \int_{[0,1]^{p+1}} J_\mu(v_{p+1}, \ldots, v_1) J_\nu(v_{p+1}, \ldots, v_1)\, dv_{p+1} \ldots dv_1$$
$$+ \sum_{k=1}^{p} \int_{[0,1]^{p+1+k}} \{J_\mu(v_{p+1}, \ldots, v_1) J_\nu(v_{p+1+k}, \ldots, v_{1+k})$$
$$+ J_\mu(v_{p+1+k}, \ldots, v_{1+k}) J_\nu(v_{p+1}, \ldots, v_1)\}\, dv_{p+1+k} \ldots dv_1$$
$$(\mu, \nu = 1, \ldots, q). \quad (2.2)$$

Under $H^{(n)}_{\mathbf{d},f}$, $n^{1/2}(\mathbf{S}^{(n)} - \mathbf{m}^{(n)})$ is asymptotically normal with covariance matrix $\mathbf{V}^2$, the mean being $\mathbf{C}'\mathbf{d}$ where $\mathbf{C}$ is a $p \times q$ matrix with entries

$$C_{i\mu} = \sum_{j=0}^{p-i} \int_{[0,1]^{p+1}} J_\mu(v_{p+1}, \ldots, v_1)\phi(F_1^{-1}(v_{p+1-j}))F_1^{-1}(v_{p+1-j-i}) dv_{p+1} \ldots dv_1,$$
$$(i = 1, \ldots, p;\ \mu = 1, \ldots, q). \quad (2.3)$$

Considering the particular case in which $q < p$, it immediately follows that whenever $\mathbf{d}$ is orthogonal to the columns of $\mathbf{C}$, the asymptotic distribution of $n^{1/2}(\mathbf{S}^{(n)} - \mathbf{m}^{(n)})$ is the same under $H_0^{(n)}$ as under $H^{(n)}_{\mathbf{d},f}$. As mentioned in the introduction, for any $\mathbf{S}^{(n)}$ and f, such values of $\mathbf{d}$ exist which define a whole family of ARMA models against which $\mathbf{S}^{(n)}$ is totally insensitive. This provides the incentive for our considering quadratic statistics.

PROPOSITION 2.2. *Using the same notation as in Proposition 2.1, and assuming that $\mathbf{V}^2$ is of rank q, the rank statistic (2.1) is, under $H_0^{(n)}$, asymptotically χ^2 with q degrees of freedom and, under $H^{(n)}_{\mathbf{d},f}$, asymptotically noncentral χ^2, still with q degrees of freedom, and with noncentrality parameter*

$$\lambda_f(\mathbf{d}) = \tfrac{1}{2}\mathbf{d}'\mathbf{C}\mathbf{V}^{-2}\mathbf{C}'\mathbf{d}. \quad (2.4)$$

PROOFS OF PROPOSITIONS 2.1 AND 2.2. The proof of Proposition 2.1 is very similar to that of Propositions 4.1 and 4.2 in Hallin *et al.* (1985); we therefore sketch it very briefly. From Section 4.2 in Hallin *et al.* (1985), it follows that the (univariate) statistic

$$n^{1/2}\boldsymbol{\alpha}'(\mathbf{S}^{(n)} - \mathbf{m}^{(n)}) + \beta \log L^{(n)}, \quad (2.5)$$

where $\boldsymbol{\alpha} \in \mathbb{R}^q$ and $\beta \in \mathbb{R}$ are arbitrary coefficients and $L^{(n)}$ is the likelihood ratio for $H_0^{(n)}$ against $H^{(n)}_{\mathbf{d},f}$, is asymptotically equivalent to a sequence of U-statistics with kernel (of degree $p + 1$),

$$\Psi_{\alpha\beta}(\mathbf{u}_{p+1}, \ldots, \mathbf{u}_1) = \sum_{\mu=1}^{q} \alpha_\mu (p+1)^{-1} \Bigg\{ \sum_{j=0}^{p} J_\mu(u_{p+1-j,1}, \ldots, u_{p+1-j,p+1})$$
$$- \sum_{1 \leqslant t_1 \neq \cdots \neq t_{p+1} \leqslant p+1} \cdots \sum J_\mu(u_{t_1,1}, \ldots, u_{t_{p+1},1})/p! \Bigg\} + \beta(p+1)^{-1}$$
$$\times \sum_{j=0}^{p} \phi(F_1^{-1}(u_{p+1-j,1})) \sum_{i=1}^{p} d_i F_1^{-1}(u_{p+1-j,1+i}),$$

where $\boldsymbol{u}_t = (u_{t,1}, \ldots, u_{t,p+1})' \in [0, 1]^{p+1}$ stands for the values of $\boldsymbol{U}_t = (U_t, \ldots, U_{t-p})' = (F(X_t), \ldots, F(X_{t-p}))'$. The asymptotic normality of (2.5), with mean 0 and variance

$$\sigma^2_{\alpha\beta} = \sum_{\mu=1}^{q} \alpha_\mu V_{\mu\mu} + 2 \sum\sum_{1 \leq \mu \neq \nu \leq q} \alpha_\mu \alpha_\nu V_{\mu\nu} + 2\beta \sum_{\mu=1}^{q} \alpha_\mu \sum_{i=1}^{p} C_{i\mu} d_i + \beta^2 \sum_{i=1}^{p} d_i^2 \sigma^2 I(f) \tag{2.6}$$

then follows from a central limit theorem by Yoshihara (1976, Theorem 1) for U-statistics under absolutely regular processes (the $\boldsymbol{U}_t$ here constitute, under $H_0^{(n)}$, a p-dependent process). Equation (2.6), together with Le Cam's third Lemma (Hájek and Šidák (1967), *VI*) then imply the joint asymptotic normality of $n^{1/2}(\boldsymbol{S}^{(n)} - \boldsymbol{m}^{(n)})$, with mean $\boldsymbol{0}$ under $H_0^{(n)}$, $\boldsymbol{C}'\boldsymbol{d}$ under $H_{\boldsymbol{d},f}^{(n)}$, and covariance matrix $\boldsymbol{V}^2$ under both $H_0^{(n)}$ and $H_{\boldsymbol{d},f}^{(n)}$. The asymptotic distribution of $Q^{(n)}$ is then straightforward.

3. TESTING $H_0^{(n)}$ AGAINST $H_{\boldsymbol{d},f}^{(n)}$ ($\boldsymbol{d}$ AND f SPECIFIED)

3.1. *Rank autocorrelation coefficients associated with density f*

Consider the following linear serial rank statistic of order i ($i = 1, 2, \ldots$):

$$r_{(i)f}^{(n)} = \left\{(n-i)^{-1} \sum_{t=i+1}^{n} \phi\left(F_1^{-1}\left(\frac{R_t^{(n)}}{n+1}\right)\right) F_1^{-1}\left(\frac{R_{t-i}^{(n)}}{n+1}\right) - m^{(n)}\right\} \Big/ s^{(n)} \tag{3.1}$$

where

$$m^{(n)} = \frac{1}{n(n-1)} \sum\sum_{1 \leq i_1 \neq i_2 \leq n} \phi_1\left(F_1^{-1}\left(\frac{i_1}{n+1}\right)\right) F_1^{-1}\left(\frac{i_2}{n+1}\right) \tag{3.2}$$

and

$$\begin{aligned}
(s^{(n)})^2 = {} & \{n(n-1)\}^{-1} \sum\sum_{1 \leq i \neq i_2 \leq n} \left\{\phi\left(F_1^{-1}\left(\frac{i_1}{n+1}\right)\right) F_1^{-1}\left(\frac{i_2}{n+1}\right)\right\}^2 \\
& + \tfrac{1}{3}(n-2i)(n-i)^{-1} \binom{n}{3}^{-1} \sum\sum\sum_{1 \leq i_1 \neq i_2 \neq i_3 \leq n} \\
& \times \left\{\phi\left(F_1^{-1}\left(\frac{i_1}{n+1}\right)\right) \phi\left(F_1^{-1}\left(\frac{i_2}{n+1}\right)\right) F_1^{-1}\left(\frac{i_2}{n+1}\right) F_1^{-1}\left(\frac{i_3}{n+1}\right)\right\} \\
& + \tfrac{1}{24}\{n^2 - n(2i+3) + i^2 + 5i\}(n-i)^{-1} \binom{n}{4}^{-1} \sum\sum\sum\sum_{1 \leq i_1 \neq i_2 \neq i_3 \neq i_4 \leq n} \\
& \times \left\{\phi\left(F_1^{-1}\left(\frac{i_1}{n+1}\right)\right) \phi\left(F_1^{-1}\left(\frac{i_2}{n+1}\right)\right) F_1^{-1}\left(\frac{i_3}{n+1}\right) F_1^{-1}\left(\frac{i_4}{n+1}\right)\right\} \\
& - (n-i)(m^{(n)})^2.
\end{aligned}$$

As we shall see in the subsequent sections, $r^{(n)}_{(i)f}$ enjoys most of the distributional and inferential properties of the ordinary sample autocorrelation coefficient

$$r^{(n)}_{(i)} = \left\{n \sum_{t=i+1}^{n} X^{(n)}_t X^{(n)}_{t-i}\right\}\left\{(n-i)\sum_{t=1}^{n}(X^{(n)}_t)^2\right\}^{-1};$$

we therefore call it the rank autocorrelation coefficient of order i associated with the density type f (briefly, the f-rank autocorrelation of order i).

Clearly, $m^{(n)}$ and $s^{(n)}$ are normalizing constants such that $(n-i)^{1/2}r^{(n)}_{(i)f}$ is exactly standardized under $H^{(n)}_0$. Examples are ($m^{(n)}_{\cdot}$ and $s^{(n)}_{\cdot}$ standing for the appropriate means and standard errors):

(i) the normal-rank autocorrelations, or van der Waerden autocorrelations,

$$r^{(n)}_{(i)N} = \left\{(n-i)^{-1}\sum_{t=i+1}^{n}\Phi^{-1}\left(\frac{R^{(n)}_t}{n+1}\right)\Phi^{-1}\left(\frac{R^{(n)}_{t-i}}{n+1}\right) - m^{(n)}_N\right\}\Big/ s^{(n)}_N,$$

where $\Phi(\cdot)$ denotes, as usual, the standard normal distribution function;

(ii) the logistic-rank autocorrelations, or Wilcoxon autocorrelations,

$$r^{(n)}_{(i)W} = \left\{(n-i)^{-1}\sum_{t=i+1}^{n}\left(2\frac{R^{(n)}_t}{n+1}-1\right)\right.$$
$$\left.\times\left(\log(R^{(n)}_{t-i}) - \log(n+1-R^{(n)}_{t-i}) - m^{(n)}_W\right\}\Big/ s^{(n)}_W;$$

(iii) the double exponential-rank autocorrelations, or Laplace rank-autocorrelations,

$$r^{(n)}_{(i)L} = \left\{(n-i)^{-1}\sum_{t=i+1}^{n}\operatorname{sign}\left(R^{(n)}_t - \frac{n+1}{2}\right)F_e^{-1}\left(\frac{R^{(n)}_{t-i}}{n+1}\right) - m^{(n)}_L\right\}\Big/ s^{(n)}_L,$$

where $F_e(x)$ denotes the distribution function $\frac{1}{2}\int_{-\infty}^{x}\exp(-|u|)\,du$, and $F_e^{-1}(u) = \log 2u$ if $0 < u \leqslant \frac{1}{2}$, and $= -\log(2-2u)$ if $\frac{1}{2} < u < 1$.

REMARK 3.1. f-Rank autocorrelations should not be confused with Spearman rank autocorrelations, probably one of the most popular rank-based measures of serial dependence. Actually, it can be shown (Hallin *et al.* (1985, 5.2) that Spearman's autocorrelation coefficient of order i is equivalent to the statistic $S^{(n)}$ with $p = i$ and a score-generating function of the form $J^S(v_{i+1}, \ldots, v_1) = v_{i+1}v_1$. We shall see, on the other hand, that (still under the strong unimodality assumption) $r^{(n)}_{(i)f}$ has the score-generating function given below (3.3). It is easy to check that there exists no density f such that $F^{-1}(v) = v$ and $\phi(F^{-1}(v)) = v$, $0 \leqslant v \leqslant 1$: Spearman's autocorrelations therefore cannot be f-rank autocorrelations.

We now give the asymptotic joint distribution of the vector of f-rank autocorrelations.

PROPOSITION 3.1. *Assume f to be a strongly unimodal density (cf. Hájek and Šidák, (1967), 1.2).*

(i) *Then* $r^{(n)}_{(i)f}$ *admits the score-generating function*

$$J_{(i)f}(v_{i+1}, \ldots, v_1) = \phi(F_1^{-1}(v_{i+1}))F_1^{-1}(v_1)\{\sigma^2 I(f)\}^{-1/2}. \tag{3.3}$$

(ii) *Denote by* $r^{(n)}_f$ *the vector of f-rank autocorrelations of orders 1 through q (q arbitrary—assume q > p). Then,* $n^{1/2}r^{(n)}_f$ *is asymptotically normal, with mean* $(0, \ldots, 0)'$ *(under* $H^{(n)}_0$*), mean* $(d_1, \ldots, d_p, 0, \ldots, 0)'\{\sigma^2 I(f)\}^{1/2}$ *(under* $H^{(n)}_{d,f}$*) and covariance matrix* $I_{q\times q}$ *(under both).*

Notice the strong similarity between the asymptotic behaviour of $n^{1/2}r^{(n)}_f$ and that of the vector of the usual sample autocorrelations $n^{1/2}r^{(n)} = n^{1/2}(r^{(n)}_{(1)}, \ldots, r^{(n)}_{(q)})'$ which, under $H^{(n)}_0$ is (Hallin *et al.* (1985, Section 5.2)) asymptotically normal, with the same mean and the same covariance matrix $I_{q\times q}$. Moreover, f-rank autocorrelations of distinct orders are asymptotically independent, just as the corresponding sample autocorrelations are. The eventual superiority of the f-rank autocorrelations over their parametric counterparts lies in their exact standardized form and in the factor $\{\sigma^2 I(f)\}^{1/2}$ appearing in their means under the alternative.

Proof of Proposition 3.1. (i) Strong unimodality of f implies that $\phi(.)$ is a nondecreasing function (actually, these two properties are equivalent).

Let $U_1, U_2, \ldots$ be a sequence of independent random variables, each uniformly distributed over [0, 1], and let $R^{(n)}_i$ be the rank of U_i, $1 \leqslant i \leqslant n$, in the subsequence $U_1, \ldots, U_n$. Using a two-dimensional version of the reasoning of Hájek and Šidák (1967) in their proof of Lemma a, V.1.6, we obtain

$$\lim_{n\to\infty} E\left\{\phi(F_1^{-1}(U_2))F_1^{-1}(U_1) - \phi\left(F_1^{-1}\left(\frac{R^{(n)}_2}{n+1}\right)\right)F_1^{-1}\left(\frac{R^{(n)}_1}{n+1}\right)\right\}^2 = 0 \tag{3.4}$$

which entails (1.3), with $p = i$, $J(\ldots) = J_{(i)f}(\ldots)$ and

$$a_n(R^{(n)}_{i+1}, \ldots, R^{(n)}_1) = J_{(i)f}\left(\frac{R^{(n)}_{i+1}}{n+1}, \ldots, \frac{R^{(n)}_1}{n+1}\right).$$

Therefore $r^{(n)}_{(i)f}$ admits (3.3) as its score-generating function, and we may apply the asymptotical results of the previous sections.

(ii) The asymptotic normality of $n^{1/2}r^{(n)}_{(i)f}$ follows from Proposition 2.1. The computations of the asymptotic mean and covariance matrix from (2.2) and (2.3) are straightforward (recall that $\sigma^2 I(f) = \iint \{\phi(F_1^{-1}(u))F_1^{-1}(v)\}^2\, du\, dv$).

Remark 3.2. It follows from Proposition 3.1 below that $m^{(n)} \to 0$ and $s^{(n)} \to \{\sigma^2 I(f)\}^{1/2}$ as $n \to \infty$. The asymptotic properties of the f-rank autocorrelations are therefore not affected if these asymptotic values are substituted in (3.1) for the exact ones.

Remark 3.3. It follows from Hallin *et al.* (1985, Section 4.2) that, under normality, the van der Waerden autocorrelations $r^{(n)}_{(i)N}$ are asymptotically equivalent with the usual sample autocorrelations $r_{(i)}^{(n)}$, in the sense that $n^{1/2}(r_{(i)N}^{(n)} - r_{(i)}^{(n)}) \to 0$ in probability, as $n \to \infty$ both under the hypothesis of randomness as well as under contiguous (normal) alternatives.

3.2. *Asymptotically most powerful test against $H_{d,f}^{(n)}$*

The assumption of strong unimodality and the introducing of f-rank autocorrelations allow for a particularly simple and intuitively appealing form of the optimal linear serial rank statistic $S_{d,f}^{(n)*}$ (defined below) for $H_0^{(n)}$ against $H_{d,f}^{(n)}$; moreover we show that $S_{d,f}^{(n)*}$ is asymptotically optimal not only within the class of linear serial rank tests, but also within the class of all possible tests.

PROPOSITION 3.2. *Assume f to be strongly unimodal, and consider the linear serial rank statistic*

$$S_{d,f}^{(n)*} = \sum_{i=1}^{p} \frac{d_i}{\|d\|} r_{(i)f}^{(n)} \qquad \left(\text{with } \|d\| = \left(\sum_{i=1}^{p} d_i^2\right)^{1/2}\right). \tag{3.5}$$

Then

(i) *$n^{1/2} S_{d,f}^{(n)*}$ is asymptotically normal with mean 0 (under $H_0^{(n)}$), mean*

$$\sum_{i=1}^{p} \frac{h_i d_i}{\|d\|} \{\sigma^2 I(f)\}^{1/2}$$

(under $H_{h,f}^{(n)}$, $h = (h_1, \ldots, h_p) \in \mathbb{R}^p$ and variance 1 (under both).

(ii) *The test based on $S_{d,f}^{(n)*}$ (at level α) is asymptotically most powerful, within the class of all tests of level α, for testing $H_0^{(n)}$ against $H_{d,f}^{(n)}$.*

PROOF OF PROPOSITION 3.2. Part (i) is a straightforward consequence of Proposition 3.1. It follows that the asymptotic power of $S_{d,f}^{(n)*}$ against $H_{d,f}^{(n)}$ is $1 - \Phi(k_{1-\alpha} - \|d\|\{\sigma^2 I(f)\}^{1/2})$, where Φ denotes the standard normal distribution function and $k_{1-\alpha}$ the corresponding $(1-\alpha)$-quantile.

Now denote by $l_0^{(n)}$ the particular likelihood function in $H_0^{(n)}$ associated with density f, and by $l_1^{(n)}$ a completely specified likelihood function in $H_{d,f}^{(n)}$, and let $L^{(n)}(X^{(n)})$ be the corresponding likelihood ratio. In order to establish the contiguity of $H_0^{(n)}$ and $H_{d,f}^{(n)}$, we showed in Hallin *et al.* (1985) that log $L^{(n)}(X^{(n)})$ is asymptotically normal with mean $-\|d\|^2\sigma^2 I(f)/2$ and variance $\|d\|^2\sigma^2 I(f)$ under $l_0^{(n)}$, and with mean $\|d\|^2\sigma^2 I(f)/2$ and variance $\|d\|^2\sigma^2 I(f)$ under $l_1^{(n)}$. Using the usual notation, let $\beta(\alpha, H, K)$ be the envelope power function for testing H against K at level α. Then $\beta(\alpha, \{l_0^{(n)}\}, \{l_1^{(n)}\})$ is the power reached by the likelihood ratio test. Hence

$$\lim_{n\to\infty} \beta(\alpha, \{l_0^{(n)}\}, \{l_1^{(n)}\}) = 1 - \Phi(k_{1-\alpha} - \|d\|\{\sigma^2 I(f)\}^{1/2}) = \lim_{n\to\infty} \beta(\alpha, \{l_0^{(n)}\}, H_{d,f}^{(n)}).$$

Obviously,

$$\beta(\alpha, \{l_0^{(n)}\}, H_{d,f}^{(n)}) \geqslant \beta(\alpha, H_0^{(n)}, H_{d,f}^{(n)}) \qquad \text{for every } n,$$

and consequently

$$\lim_{n\to\infty} \beta(\alpha, \{l_0^{(n)}\}, H_{\boldsymbol{d},f}^{(n)}) \geqslant \limsup_{n\to\infty} \beta(\alpha, H_0^{(n)}, H_{\boldsymbol{d},f}^{(n)}).$$

It follows that

$$\limsup_{n\to\infty} \beta(\alpha, H_0^{(n)}, H_{\boldsymbol{d},f}^{(n)}) \geqslant 1 - \Phi(k_{1-\alpha} - \|\boldsymbol{d}\|\{\sigma^2 I(f)\}^{1/2}) = \lim_{n\to\infty} \beta(\alpha, \{l_0^{(n)}\}, H_{\boldsymbol{d},f}^{(n)})$$

whence $\lim_{n\to\infty} \beta(\alpha, H_0^{(n)}, H_{\boldsymbol{d},f}^{(n)})$ exists, and is equal to the asymptotic power of the α-level test based on $S_{\boldsymbol{d},f}^{(n)*}$.

REMARK 3.4. (In what follows, for convenience, we write S^* instead of $S_{\boldsymbol{d},f}^{(n)*}$.) Comparing Proposition 3.2 and Proposition 5.1 in Hallin *et al.* (1985) leads to some apparent contradictions: Proposition 5.1 in Hallin *et al.* (1985) claims that any linear serial rank statistic S^{**} with score-generating function

$$J^{**}(v_{p+1}, \ldots, v_1) = \sum_{i=1}^{p} \frac{d_i}{p+1-i} \sum_{j=0}^{p-i} \phi(F_1^{-1}(v_{p+1-j}))F_1^{-1}(v_{p+1-j-i}) \tag{3.6}$$

is (asymptotically) optimal within the class of linear serial rank statistics.

From Proposition 3.2 above, it is easy to see that S^* is equivalent (up to $o_p(n^{-1/2})$ terms and a multiplicative constant $\{\|\boldsymbol{d}\|^2\sigma^2 I(f)\}^{1/2}$) to any linear serial rank statistic with score-generating function

$$J^*(v_{p+1}, \ldots, v_1) = \sum_{i=1}^{p} d_i \phi(F_1^{-1}(v_{p+1}))F_1^{-1}(v_{p+1-i}) \tag{3.7}$$

which is different from (3.6).

We also establish in Hallin *et al.* (1985, Lemma 5.1) that the asymptotic power of any linear serial rank statistic with score-generating function $J(\ldots)$ is less than or equal to the asymptotic power of the linear serial rank statistic whose score-generating function is the projection $J_0(\ldots)$ of $J(\ldots)$ onto the L^2-subspace spanned by $\{\sum_{j=0}^{p-i} \phi(F_1^{-1}(v_{p+1-j}))F_1^{-1}(v_{p+1-j-i});\ i = 1, \ldots, p\}$. Looking at (3.7), $J^*(\ldots)$ obviously does not belong to this subspace, whereas $J^{**}(\ldots)$ does.

These somewhat contradictory facts can be explained; the two arguments we are giving here are equivalent—except for the fact that the second does not require the strong unimodality property of f (Proposition 3.2 can then be reformulated in terms of score-generating functions only).

(i) Unlike usual linear rank statistics, a linear *serial* rank statistic (of order $p > 1$) generally can be defined by means of several distinct (even asymptotically distinct) score functions $a_n(\ldots)$, characterizing distinct score-generating functions. This is due to the overlapping of the rank arguments in the successive $a_n(\ldots)$ terms (see (1.1)). Equations (3.6) and (3.7) are examples of this phenomenon.

(ii) It is easy to check that J^{**} is actually the projection of J^* onto the linear L^2 space spanned by $\{\sum_{j=0}^{p-i} \phi(F_1^{-1}(v_{p+1-j}))F_1^{-1}(v_{p+1-j-i});\ i = 1, \ldots, p\}$. J^* thus decomposes as $J^* = J^{**} + J^{\perp}$, and according to the above-mentioned Lemma 5.1, the variance V_*^2 of S^* is the sum of V_{**}^2 and $V_{\perp}^2$, where V_{**}^2 and $V_{\perp}^2$ can be obtained by letting $J_\mu(\ldots) = J_\nu(\ldots) = J^{**}$ and $J_\mu(\ldots) = J_\nu(\ldots) = J^* - J^{**}$ respectively in (2.2). A lengthy but straightforward calculation then yields the expected result that $V_\perp^2 = 0$: although using a projection $J_0(\ldots)$ of the score-generating function can only result in an improved test statistic, and notwithstanding the fact that J^* is strictly distinct from its projection J^{**}, the improvement resulting from using J^{**} instead of J^* is here nil.

4. TESTING $H_0^{(n)}$ AGAINST $H_{\boldsymbol{d},f}^{(n)}$ ($\boldsymbol{d}$ UNSPECIFIED, f SPECIFIED)

4.1. *Asymptotically maximin most powerful quadratic serial rank tests*

It is clear from Proposition 3.2 that uniformly asymptotically most powerful tests for $H_0^{(n)}$ against unspecified alternatives of the form $\cup \{H_{\boldsymbol{d},f}^{(n)} \mid \boldsymbol{d} \in \mathbb{R}^p\}$ do not exist. Weaker optimality concepts have thus to be considered. Proposition 4.1 below establishes the asymptotic maximin optimality of a test based on a quadratic rank statistic of the portmanteau type.

Informally speaking, a maximin most powerful test at a given level α is "a test whose worst performance is best" within the class of tests at level α. Formally, a sequence $\phi^{(n)}$ of tests is said to be asymptotically maximin most powerful for the sequence of null hypotheses $H^{(n)}$ against the sequence of alternatives $K^{(n)}$ if its power asymptotically reaches the envelope power function $\beta(\alpha, H^{(n)}, K^{(n)})$, i.e. if

$$\limsup_{n\to\infty} [E_{l^{(n)}} \phi^{(n)} - \alpha] \leqslant 0, \qquad \text{for all } l^{(n)} \in H^{(n)},$$

and

$$\liminf_{n\to\infty} [E_{l^{(n)}} \phi^{(n)} - \beta(\alpha, H^{(n)}, K^{(n)})] \geqslant 0 \qquad \text{for all } l^{(n)} \in K^{(n)}.$$

Now if we do not want the envelope power function $\beta(\alpha, H^{(n)}, K^{(n)})$ to be asymptotically equal to α, the sequence of alternatives $K^{(n)}$ has to be "bounded away" from the null sequence $H^{(n)}$. We therefore consider the sequence of alternatives

$$K_f^{(n)}(d) = \cup \{H_{\boldsymbol{d},f}^{(n)} \mid \boldsymbol{d} \in \mathbb{R}^p, \|\boldsymbol{d}\| \geqslant d\} \tag{4.1}$$

where d is an arbitrary preassigned strictly positive constant. We call (4.1) an unspecified ARMA alternative of order p; indeed, it includes all the sequences of contiguous ARMA (p_1, p_2) dependences of orders p_1 and p_2 less than or equal to p, with unspecified coefficients (but with specified white noise density f).

PROPOSITION 4.1. *Let f be strongly unimodal. Then, the quadratic serial rank statistic*

$$Q_f^{(n)*} = \sum_{i=1}^{p} (n-i)(r_{(i)f}^{(n)})^2 = n \sum_{i=1}^{p} (r_{(i)f}^{(n)})^2 + o_p(1) \tag{4.2}$$

(i) *is asymptotically* χ^2 *with p degrees of freedom (under* $H_0^{(n)}$) *and noncentral* χ^2 *with p degrees of freedom and noncentrality parameter*

$$\lambda^*_{f,g}(\boldsymbol{d}) = \tfrac{1}{2}\|\boldsymbol{d}\|^2\left\{\int \phi(F_1^{-1}(u))\phi_g(G^{-1}(u))\,du \int F_1^{-1}(v)G^{-1}(v)\,dv\right\}^2\{\sigma^2 I(f)\}^{-1} \tag{4.3}$$

(under $H^{(n)}_{\boldsymbol{d},g}$, *with g a strongly unimodal density. Here* $G(x) = \int_{-\infty}^{x} g(u)\,du$ *and* $\phi_g(.) = g'(.)/g(.)$). *Under* $H^{(n)}_{\boldsymbol{d},f}$, *(4.3) reaches, for given* $\|\boldsymbol{d}\| = d$, *its maximal value* $\lambda^*_f(\boldsymbol{d}) = \tfrac{1}{2}\|\boldsymbol{d}\|^2\sigma^2 I(f)$.

(ii) *provides, at any level* $\alpha \in (0, 1)$ *and for any value of* $d > 0$, *an asymptotically maximin most powerful test for* $H_0^{(n)}$ *against* $K^{(n)}(d)$.

Remark 4.1. Being the sum of squared *f*-autocorrelation coefficients of orders 1 through *p*, $Q_f^{(n)*}$ can be considered as a rank-based version of the Box–Pierce (1970) portmanteau statistic

$$Q_{\mathrm{BP}}^{(n)} = n(n+2)\sum_{i=1}^{p}(n-i)^{-1}(r_{(i)}^{(n)})^2 = n\sum_{i=1}^{p}(r_{(i)}^{(n)})^2 + o_p(1). \tag{4.4}$$

We shall henceforth refer to $Q_f^{(n)*}$ as the *f*-rank portmanteau statistic of order *p*, particular cases being (cf. Section 3.1) the van der Waerden, the Wilcoxon and the median-test or Laplace portmanteau statistics.

Remark 4.2. The van der Waerden and Box–Pierce portmanteau statistics are asymptotically equivalent under normality. (The proof follows from Hallin *et al.* (1985, Section 4). Proposition 4.1 above thus also establishes the asymptotic maximin optimality of the well-known Box–Pierce test, bringing some theoretical support to this often-criticized procedure.

Remark 4.3. Since the noncentrality parameter (4.3) depends on $\boldsymbol{d}$ only through $\|d\|$, the surfaces of equal asymptotic power of $Q_f^{(n)*}$, in the *p*-dimensional parametric space $\mathbb{R}^p$, are full-dimensional hyperspheres centred at the origin.

Proof of Proposition 4.1. Part (i) is a straightforward consequence of Propositions 2.2 and 3.1. It follows that the test rejecting $H_0^{(n)}$ whenever $Q_f^{(n)*} > \chi^2_{p;1-\alpha}$ has asymptotic level α and asymptotic power $1 - F(\chi^2_{p;1-\alpha}; \|\boldsymbol{d}\|^2\sigma^2 I(f)/2)$ against $H^{(n)}_{\boldsymbol{d},f}$. (Here $\chi^2_{p;1-\alpha}$ denotes the $(1-\alpha)$-quantile of the χ^2-distribution with *p* degrees of freedom, and $F(.,\lambda)$ denotes the distribution function of the noncentral χ^2 variable with *p* degrees of freedom and noncentrality parameter λ.) This asymptotic power coincides with the asymptotic value of the envelope power function $\beta(\alpha; H_0^{(n)}, K_f^{(n)}(\|\boldsymbol{d}\|))$. (For details, see Hallin *et al.* (1986)). This completes the proof of part (ii).

4.2. *Asymptotic relative efficiencies of rank portmanteau statistics*

Table I gives the noncentrality parameters associated with several particular rank portmanteau statistics, under normal, logistic and double-exponential densities.

TABLE I

VALUES OF $\lambda_f(\boldsymbol{d})/\|\boldsymbol{d}\|^2$ FOR VARIOUS RANK PORTMANTEAU STATISTICS (INCLUDING SPEARMAN'S), AND BOX AND PIERCE'S STATISTIC, UNDER NORMAL, LOGISTIC AND DOUBLE-EXPONENTIAL DENSITIES.

	Density type f		
Statistic	Normal	Logistic	Double-exponential
van der Waerden	0.5	$\frac{1}{2\pi^3}\left\{\int_{-\infty}^{+\infty} e^{-x^2}(\Phi(x))^{-1}\,dx\right\}^2 \simeq 0.524$	$\frac{1}{\pi^3}\left\{\int_{-\infty}^{0} e^{-x^2}(\Phi(x))^{-1}\,dx\right\}^2 \simeq 0.613$
Wilcoxon	$\frac{9}{2\pi^5}\left\{\int_{-\infty}^{+\infty} e^{-x^2}(\Phi(x))^{-1}\,dx\right\}^2 \simeq 0.474$	$\frac{\pi^2}{18} \simeq 0.549$	$\frac{9}{2\pi^2}\left\{\frac{\pi^2}{6} - (\log 2)^2 + \log 4\right\}^2 \simeq 0.741$
Laplace	$\frac{1}{2\pi^3}\left\{\int_{-\infty}^{0} e^{-x^2}(\Phi(x))^{-1}\,dx\right\}^2 \simeq 0.306$	$\frac{1}{16}\left\{\frac{\pi^2}{6} - (\log 2)^2 + \log 4\right\}^2 \simeq 0.406$	1
Spearman	$\frac{9}{2\pi^2} \simeq 0.456$	0.5	$\frac{1}{8}\left(\frac{3}{2}\right)^4 \simeq 0.633$
Box–Pierce	0.5	0.5	0.5

The same information is provided for the Spearman portmanteau statistic (4.10) (which is not an f-rank one, and defines tests which are nonadmissible under any density f) and the parametric (4.3) Box–Pierce one.

It follows from a result of Hallin *et al.* (1985, Section 5.3) that the Spearman autocorrelation coefficient of order i, viz.,

$$r_{(i)S}^{(n)} = \frac{(n-i)^{-1} \sum_{t=i+1}^{n} R_t^{(n)} R_{t-i}^{(n)} - \{(n+1)/2\}^2}{(n^2-1)/12},$$

is equivalent with the serial rank statistic

$$S_{(i)S}^{(n)} = \frac{n-1}{n+1}\, r_{(i)S}^{(n)} + 3$$

with mean $(3n+2)/(n+1)$, score-generating function $J(v_{p+1}, \ldots, v_1) = 12 v_{p+1} v_{p+1-i}$ and asymptotic variance 1. The asymptotic independence of the coefficients $r_{(i)S}^{(n)}$ and $r_{(j)S}^{(n)}$, $i \neq j$ is easy to establish from (2.2). Hence, defining the Spearman portmanteau statistic as

$$Q_S^{(n)} = n \sum_{i=1}^{p} (r_{(i)S}^{(n)})^2 \simeq n \sum_{i=1}^{p} \left(\frac{n-1}{n+1}\right)^2 \left(r_{(i)S}^{(n)} + \frac{1}{n-1}\right)^2 = n \sum_{i=1}^{p} (S_{(i)S}^{(n)} - m_S^{(n)})^2, \tag{4.10}$$

it is easy to see that it is asymptotically χ^2 with p degrees of freedom under $H_0^{(n)}$ and asymptotically noncentral χ^2 with p degrees of freedom and noncentrality parameter $\lambda_{S,f}(\boldsymbol{d}) = 72\|\boldsymbol{d}\|^2\{\iint uv\phi(F_1^{-1}(u))F_1^{-1}(v)\,du\,dv\}^2$ under $H_{\boldsymbol{d},f}^{(n)}$.

As for the Box–Pierce statistic, its noncentrality parameter under $H^{(n)}_{\boldsymbol{d},f}$ is simply $\lambda_{BP}(\boldsymbol{d}) = \frac{1}{2}\|\boldsymbol{d}\|^2$ (this is an immediate consequence of Hallin *et al.* (1985, Proposition 5.2).

It is well known that the ARE of a test statistic Q_1 with respect to a test statistic Q_2, where Q_1 and Q_2 are both asymptotically χ^2-distributed with p degrees of freedom under the null hypothesis, and noncentral χ^2-distributed, still with p degrees of freedom, but with noncentrality parameters λ_1 and λ_2 respectively, under the alternative, is given by

$$e(Q_1, Q_2) = \begin{cases} \lambda_1/\lambda_2 & \text{if } \lambda_2 > 0 \\ \infty & \text{if } \lambda_1 > 0,\ \lambda_2 = 0 \\ 1 & \text{if } \lambda_1 = \lambda_2 = 0. \end{cases}$$

We thus have, with obvious notation, the following general formula for the AREs of rank portmanteau statistics.

PROPOSITION 4.2. *The ARE of an f-rank portmanteau statistic $Q^{(n)*}_g$ with respect to another one, $Q^{(n)*}_h$, when testing against $H^{(n)}_{\boldsymbol{d},f}$, is given by*

$$e_f(Q^{(n)*}_g, Q^{(n)*}_h) = \frac{\sigma^2 I(h)}{\sigma^2 I(g)} \left(\frac{\int \phi_g(G^{-1}(u))\phi_f(F^{-1}(u))\, du \int G^{-1}(v)F^{-1}(v)\, dv}{\int \phi_h(H^{-1}(u))\phi_f(F^{-1}(u))\, du \int H^{-1}(v)F^{-1}(v)\, dv} \right)^2 .$$

The proof of this proposition is trivial and is therefore omitted.

As a consequence of Proposition 4.2, the mutual ARE's of the van der Waerden, Wilcoxon, Laplace, Spearman and Box–Pierce portmanteau statistics can be obtained as the corresponding ratios of noncentrality parameters in the appropriate column in Table I. An inspection of Table I shows immediately that the ARE (with respect to the usual Box–Pierce test) of the van der Waerden portmanteau statistic is equal to 1 in the Gaussian case, and strictly larger than 1 (approximately 1.047 and 1.226 for logistic and double-exponential densities) for any other white noise density (it can be shown, indeed, that $\sigma^2 I(f) \geqslant 1$ for any density f, where equality is achieved iff f is Gaussian). Since the van der Waerden portmanteau statistic also has many additional advantages (it is distribution-free, naturally robust and easy to compute; a tabulation of exact small-sample critical values is possible), it should be extremely attractive for practitioners.

As for the Spearman portmanteau statistic, derived from Wald and Wolfowitz's (1943) serial rank correlations (also studied in a recent paper by Dufour and Roy (1985)), its performances (approximate AREs with respect to Box and Pierce's test: 0.912, 1.000 and 1.226 in the gaussian, logistic and double-exponential cases, respectively) are slightly but uniformly worse than those of Wilcoxon's (approximate AREs of 0.948, 1.097 and 1.483, respectively).

These conclusions are valid independently of the order q of the portmanteau statistics; the choice of q obviously depends on the order of the alternative to be detected, and should be made exactly as in the usual parametric case.

5. CONCLUDING REMARKS

The rank autocorrelation coefficients introduced in Section 3.1 appear to play an essential role in the problem of detecting serial autocorrelation, since they have been shown to provide the asymptotically most powerful and asymptotically maximin most powerful tests for randomness.

Therefore, whenever there is some suspicion that the input white noise of some observed series might have a nonnormal distribution, the f-rank autocorrelation coefficients, being distribution-free under the null hypothesis of randomness, doubtless would constitute a more reliable and robust basis for diagnosis checking than the usual sample autocorrelations. Some of them would certainly figure profitably in the output of standard time series analysis computer programs.

In the case when the departure from randomness also involves a linear trend or seasonal component, it can be shown along the same lines as in Swensen (1985) that the asymptotic distributions and performances of the rank statistics we are considering are not affected.

However, several important questions remain unanswered: what are the small-sample performances of linear and quadratic serial rank tests? How do they behave under nonlocal alternatives? What happens when these tests are performed on residuals from some estimated ARMA model? How can f-rank autocorrelations be used for model identification? Can f-rank autocorrelation coefficients serve as a basis for nonparametric estimation and model-building purposes? These open problems certainly deserve further investigation.

NOTES

[1] Research supported by a grant of the Belgian Ministère de l'Education Nationale et de la Culture Française.

[2] Research supported by the Office of Naval Research Contract N00014-85-K-0648.

REFERENCES

BERAN, R. (1975) Local asymptotic power of quadratic rank tests for trend. *Ann. Statist.* 3, 402–12.

BHATTACHARYYA, G. K. (1984) Tests for randomness against a trend or serial correlations. In *Handbook of Statistics* 4 (eds P. R. Krishnaiah and P. K. Sen). Amsterdam and New York: North-Holland, 89–111.

BOX, G. E. P. and PIERCE, D. A. (1970) Distribution of residual autocorrelations in autoregressive-integrated moving average time series models. *J. Amer. Statist. Ass.* 65, 1509–26.

DUFOUR, J.-M. and ROY, R. (1985) Generalized portmanteau statistics and tests of randomness. *Commun. Statist. Theor. and Meth.* 15, 2953–72.

GOVINDARAJULU, Z. (1983) Rank tests for randomness against autocorrelated alternatives. In *Time Series Analysis: Theory and Practice 4*, (ed. O. D. Anderson). Amsterdam and New York: North-Holland.

HÁJEK, J. (1962) Asymptotically most powerful rank order tests. *Ann. Math. Statist.* 33, 1124–47.

HÁJEK, J. and ŠIDÁK, Z. (1967) *Theory of Rank Tests*. New York: Academic Press.

HALLIN, M., INGENBLEEK, J.-FR. and PURI, M. L. (1985) Linear serial rank tests for randomness against ARMA alternatives. *Ann. Statist.* 13, 1156–81.

—— (1985) Tests de rangs quadratiques pour une hypothèse de bruit blanc. *C. R. Acad. Sci., Paris* 301, Série I, 935–8.

—— (1986) Linear and quadratic serial rank tests for randomness against serial dependence. Technical Report, Department of Mathematics, Indiana University, Bloomington.

LECAM, L. (1960) *Locally asymptotically normal families of distributions. Univ. Calif. Publ. Stat.* 3, 37–98.

ROUSSAS, G. G. (1972) *Contiguity of Probability Measures: Some Applications in Statistics.* London: Cambridge University Press.

SWENSEN, A. R. (1985) The asymptotic distribution of the likelihood ratio for autoregressive times series with a regression trend. *J. Multivariate Anal.* 16, 54–70.

WALD, A. and WOLFOWITZ, J. (1943) An exact test for randomness in the non-parametric case based on serial correlation. *Ann. Math. Statist.* 14, 378–88.

YOSHIHARA, K. (1976) Limiting behavior of U-statistics for stationary absolutely regular processes. *Z. Wahrsch. Verw. Geb.* 35, 237–52.

Reprinted by permission of the publisher,
© 1987 by Blackwell Publishing, LTD

LOCALLY ASYMPTOTICALLY OPTIMAL TESTS FOR RANDOMNESS

MARC HALLIN and MADAN L. PURI

Institut de Statistique, Université Libre de Bruxlles, Brussels, Belgium
Department of Mathematics, Indiana University, Bloomigton, Indiana

SUMMARY

A class of linear serial statistics is introduced for the problem of testing white noise against autocorrelated alternatives. The asymptotic distribution of these statistics is given, both under the null hypothesis as well as under contiguous alternatives of ARMA dependence. It is shown that the asymptotically most powerful test statistic against a specified contiguous ARMA alternative belongs to the class considered. A rank-based version of this optimal test is also provided. Quadratic serial statistics, leading to χ^2-type tests, are also introduced for the problem of testing white noise against unspecified ARMA alternatives. The asymptotically maximin most powerful test statistic is given, and is shown to constitute a generalization of the

---

Research of Madan L. Puri is supported by the Office of Naval Research Contract N00014-85-K-0648.

classical Box–Pierce and Ljung–Box *portmanteau* statistics. A rank-based version of these generalized portmanteau statistics is also provided.

1. INTRODUCTION

The problem of testing for randomness—more specifically, the problem of testing for randomness against serial dependence—is certainly one of the oldest and possibly the most fundamental one in statistical inference, since independence assumption underlies the vast majority of classical inference procedures. Moreover, this independence assumption is often a crucial one, in the sense that, most of the time, whenever it is violated, classical methods cannot be applied safely any longer; much of the development of time-series methods was motivated by the fact that classical linear model methods do not work in the presence of autocorrelated error terms.

Not surprisingly, this problem of testing for randomness has been extensively investigated, from a parametric point of view (procedures based on sample autocorrelations, including, e.g., Durbin–Watson's, von Neumann's ratio, or Box–Pierce's tests) as well as from a nonparametric point of view (the run test, turning point test, serial rank correlation test, etc.).

However, in spite of the recognized importance of this problem, and in spite of its long history [the rank-based approach, for example, can be traced back as far as Hotelling and Pabst (1936) and Wald and Wolfowitz (1943)], very little is known about the power of the many existing procedures—and hence about their optimality properties.

The reason for this lack of power results is that nonindependence is rather a vast and diversified alternative: whereas there exists only one way for identically distributed observations to be independent, they can yield infinitely many distinct serial dependence structures, which makes it particularly difficult to obtain overall efficient testing procedures. Actually, it can be shown [from Hallin and Puri (1985, Proposition 4.5)] that, in the problem of testing for randomness against the whole alternative of contiguous serial dependence, the envelope power function converges to the level α as the sample size n tends to infinity: the trivial test $\phi(\mathbf{X}) \equiv \alpha$ therefore provides an asymptotically (locally) maximin most powerful α-level test for the problem.

Thus it appears hopeless to look for optimal tests against completely unspecified serial dependencies: if efficient testing procedures are to be obtained, more restricted alternatives have to be considered. A convenient way of restricting the alternative consists in introducing some appropriate, finite-dimensional, parametrization of serial dependence; the one we adopt here is provided by the class of ARMA models, the wide applicability of which is now well established.

Our objective is to provide asymptotically locally optimal tests for randomness against two types of ARMA alternatives:

1. *Specified* ARMA *alternatives* (under which the observations are generated by some ARMA model with specified coefficients, driven by a white noise with specified density), against which we shall give asymptotically locally most powerful tests.
2. *Unspecified* ARMA *alternatives* (under which the observations are generated by some ARMA (p_1, p_2) model, with $\max(p_1, p_2) \leq p$, p fixed, unspecified coefficients but specified white-noise density), for which asymptotically locally maximin most powerful tests are given.

In both cases, we shall provide *parametric and nonparametric* (rank-based) versions of the optimal tests; the word *locally* refers to the fact that *contiguous* alternatives are considered.

2. BASIC NOTATION AND TECHNICAL ASSUMPTIONS

Denote by $\{\varepsilon_t | t \in \mathbb{Z}\}$ an absolutely continuous (with respect to the Lebesgue measure) stationary white noise, that is, a sequence of independent and identically distributed (i.i.d.) random variables with density function $f_\sigma(x) = \sigma^{-1} f_1(x/\sigma)$, $\sigma > 0$. Assume that $E(\varepsilon_t) = 0$, $E(\varepsilon_t^2) = \sigma^2$, and $E(\varepsilon_t^4) < \infty$, $t \in \mathbb{Z}$ ($\mathbb{Z} = \{0, \pm 1, \pm 2, \dots\}$). We make the following assumptions on $f_\sigma(x)$:

1. $\dot{f}_\sigma(x) = d f_\sigma(x)/dx$ exists (a.e.); put $\phi_\sigma(x) = -\dot{f}_\sigma(x)/f_\sigma(x) = \sigma^{-1}\phi_1(x/\sigma)$.
2. $f_\sigma(x)$ has finite Fisher's information: $0 < I(f_\sigma) = E[\phi_\sigma^2(\varepsilon_t)] = \sigma^{-2} I(f_1) < \infty$.
3. $\dot{\phi}_\sigma(x) = d\phi_\sigma(x)/dx$ exists (a.e.) and satisfies (a.e.) a Lipschitz condition of the form $|\dot{\phi}_\sigma(x) - \dot{\phi}_\sigma(y)| \leq A_\sigma |x - y|$.
4. $f_\sigma(x)$ is *strongly unimodal* (cf. Hájek and Šidák, 1967).

We say that ε_t has a specified *density type* $f_.(x)$ when its density belongs to the scale parameter family $\{f_\sigma | \sigma > 0\}$ without σ being specified.

Consider a sequence of stochastic difference equations ($n = 1, 2, \ldots$)

$$X_t^{(n)} - n^{-1/2} \sum_{i=1}^{p_1} a_i X_{t-i}^{(n)} = \varepsilon_t + n^{-1/2} \sum_{i=1}^{p_2} b_i \varepsilon_{t-i}, \qquad t \in \mathbf{Z} \tag{2.1}$$

If n is sufficiently large (so that the *stationarity condition* is satisfied), and $a_{p_1} \neq 0 \neq b_{p_2}$, then (2.1) characterizes an ARMA(p_1, p_2) process $\{X_t^{(n)} | t \in \mathbf{Z}\}$, namely, the unique stationary solution of (2.1).

Denote by $\mathbf{X}^{(n)} = (X_1^{(n)}, \ldots, X_t^{(n)}, \ldots, X_n^{(n)})$ an n-tuple of observations, and by $\mathbf{R}^{(n)} = (R_1^{(n)}, \ldots, R_t^{(n)}, \ldots, R_n^{(n)})$ the corresponding n-tuple of ranks.

3. ASYMPTOTICALLY MOST POWERFUL TESTS AGAINST SPECIFIED ALTERNATIVES

3.1. Autocorrelation Coefficients Associated with Density Types

The null hypothesis under which the observation $\mathbf{X}^{(n)} = (X_1^{(n)}, \ldots, X_n^{(n)})$ constitutes a finite realization of a white-noise process with density type $f_{\cdot}(x)$ is denoted by $H_{0;f}^{(n)}$ whenever $f_{\cdot}(x)$ is specified and by $H_0^{(n)}$ otherwise.

The alternative under which $\mathbf{X}^{(n)}$ is a finite realization of some ARMA process characterized by (2.1) with specified parameters

$$\mathbf{a} = \left(a_1, a_2, \ldots, a_{p_1}, \ldots, a_{\max(p_1, p_2)}\right),$$

$$\mathbf{b} = \left(b_1, b_2, \ldots, b_{p_2}, \ldots, b_{\max(p_1, p_2)}\right), \qquad \left(a_i = 0, i > p_1 \text{ and } b_j = 0, j > p_2\right)$$

and specified density type $f_{\cdot}(x)$ is denoted by $H_{\mathbf{a},\mathbf{b};f}^{(n)}$.

Denote by $(S^{(n)})^2 = (n^{-1}\Sigma_{t=1}^n X_t^2)$ the sample variance [note that $E(X_t^{(n)}) = 0$ both under $H_0^{(n)}$ and $H_{\mathbf{a},\mathbf{b};f}^{(n)}$], and let (for $i < n$)

$$r_{i;f}^{(n)} = (n-i)^{-1} \sum_{t=i+1}^{n} \phi_\sigma\left(X_t^{(n)}\right) X_{t-i}^{(n)} \left[\sigma^2 I(f_\sigma)\right]^{-1/2} \tag{3.1}$$

$$\hat{r}_{i;f}^{(n)} = (n-i)^{-1} \sum_{t=i+1}^{n} \phi_1\left(\frac{X_t^{(n)}}{S^{(n)}}\right) \frac{X_{t-i}^{(n)}}{S^{(n)}} \left[I(f_1)\right]^{-1/2} \tag{3.2}$$

and

$$\tilde{r}_{i;f}^{(n)} = \left[(n-i)^{-1} \sum_{t=i+1}^{n} \phi_1\left(F_1^{-1}\left(\frac{R_t^{(n)}}{n+1}\right)\right) F_1^{-1}\left(\frac{R_{t-i}^{(n)}}{n+1}\right) - m^{(n)}\right] \left[\sigma^{(n)}\right]^{-1} \tag{3.3}$$

where

$$m^{(n)} = [n(n-1)]^{-1} \sum\sum_{1 \le t_1 \ne t_2 \le n} \phi_1\left(F_1^{-1}\left(\frac{t_1}{n+1}\right)\right) F_1^{-1}\left(\frac{t_2}{n+1}\right) \qquad (3.4)$$

and

$$\begin{aligned}(\sigma^{(n)})^2 = {} & \frac{1}{n(n-1)} \sum\sum_{1 \le t_1 \ne t_2 \le n} \left[\phi_1\left(F_1^{-1}\left(\frac{t_1}{n+1}\right)\right) F_1^{-1}\left(\frac{t_2}{n+1}\right)\right]^2 \\ & + 2\frac{(n-2i)}{(n-i)} \frac{1}{n(n-1)(n-2)} \\ & \times \sum\sum\sum_{1 \le t_1 \ne t_2 \ne t_3 \le n} \left[\phi_1\left(F_1^{-1}\left(\frac{t_1}{n+1}\right)\right) \phi_1\left(F_1^{-1}\left(\frac{t_2}{n+1}\right)\right)\right. \\ & \qquad \left. \times F_1^{-1}\left(\frac{t_2}{n+1}\right) F_1^{-1}\left(\frac{t_3}{n+1}\right)\right] \\ & + \frac{n^2 - n(2i+3) + i^2 + 5i}{n(n-1)(n-2)(n-3)(n-i)} \\ & \times \sum\sum\sum\sum_{1 \le t_1 \ne t_2 \ne t_3 \ne t_4 \le n} \left[\phi_1\left(F_1^{-1}\left(\frac{t_1}{n+1}\right)\right) \phi_1\left(F_1^{-1}\left(\frac{t_2}{n+1}\right)\right)\right. \\ & \qquad \left. \times F_1^{-1}\left(\frac{t_3}{n+1}\right) F_1^{-1}\left(\frac{t_4}{n+1}\right)\right] \\ & - (n-i)(m^{(n)})^2 \end{aligned}$$

[$F_\sigma(x)$ stands for the distribution function associated with $f_\sigma(x)$.]

As we shall see (Proposition 3.1), (3.1), (3.2), and (3.3) are essentially equivalent measures of dependence (of order i). (3.1), however, is of little practical use, since it depends on σ^2, an unknown quantity in most problems; (3.2) is obtained from (3.1) by substituting $S^{(n)}$ for σ (indeed, $\sigma^2 I(f_\sigma) = I(f_1)$ and $\phi_1(x/\sigma)y/\sigma = \phi_\sigma(x)y$, $\sigma > 0$); as for (3.3), it constitutes a rank-based (hence distribution-free under $H_0^{(n)}$) version of (3.1). Note that $(n-i)^{1/2} r_{i;f}^{(n)}$ and $(n-i)^{1/2} \tilde{r}_{i;f}^{(n)}$ are exactly standardized random variables (under $H_{0;f}^{(n)}$ and $H_0^{(n)}$ respectively) [the exact variance of $\tilde{r}_{i;f}^{(n)}$ follows from Hallin et al. (1985a, Appendix 3)].

In the gaussian case, $\sigma^2 I(f_\sigma) = I(f_1) = 1$, and $\phi_1(x) = x$; $\hat{r}^{(n)}_{i;f}$ (write it $\hat{r}^{(n)}_{i;\mathcal{N}}$) thus reduces to the *sample autocorrelation of order* i: $r_i^{(n)} = \sum_{t=i+1}^{n}(X_t^{(n)}X_{t-i}^{(n)})/\sum(X_t^{(n)})^2$. It is well known that optimal inference procedures for gaussian series can be based [at least asymptotically, cf. Anderson (1971) or Hallin et al. (1985a, Section 5)] on sample autocorrelations $r_i^{(n)} = \hat{r}^{(n)}_{i;\mathcal{N}} \simeq \tilde{r}^{(n)}_{i;\mathcal{N}}$, and remain asymptotically valid for nongaussian densities, in which case, however, their optimality properties are lost. Our objective is to show that optimality can be maintained, provided that $r^{(n)}_{i;f}$, $\hat{r}^{(n)}_{i;f}$, or $\tilde{r}^{(n)}_{i;f}$ are substituted for the usual sample autocorrelations $r_i^{(n)}$.

The prominent role ascribed to sample autocorrelations in testing for randomness (as well as in most of time-series analysis) thus actually belongs to the $\hat{r}^{(n)}_{i;f}$, to which we therefore shall refer as the *autocorrelation coefficient of order i associated with the density type* $f_\cdot(x)$—shortly, the *f-autocorrelation of order i*. Accordingly, we call $\tilde{r}^{(n)}_{i;f}$ the *rank autocorrelation coefficient of order i associated with the density type* $f_\cdot(x)$—shortly, the *f-rank autocorrelation of order i*. Note that f-rank autocorrelations never reduce to rank autocorrelations in the usual (Spearman) sense. Also note that $\hat{r}^{(n)}_{i;f}$ and $\tilde{r}^{(n)}_{i;f}$ are not *stricto sensu* correlation or autocorrelation coefficients (except for gaussian densities). To be precise, they are standardized [up to $(n-i)^{1/2}$ factor] statistics asymptotically equivalent with the *cross*-correlation coefficient of order i between $\phi_\sigma(X_t^{(n)})$ and $X_t^{(n)}$, and they reduce to an autocorrelation coefficient if and only if f is gaussian. The term *f-autocorrelation* coefficient is adopted because of its intuitive appeal.

3.2. Asymptotic Properties of f-Autocorrelations

We say that two random variables X_n and Y_n are asymptotically equivalent (as $n \to \infty$) if their difference converges in probability to zero.

Proposition 3.1. The statistics $n^{1/2}r^{(n)}_{i;f}$, $n^{1/2}\hat{r}^{(n)}_{i;f}$ and $n^{1/2}\tilde{r}^{(n)}_{i;f}$

1. are asymptotically equivalent (under both $H^{(n)}_{0;f}$ and $H^{(n)}_{\mathbf{a},\mathbf{b};f}$),
2. are asymptotically normal, with mean 0 and variance 1 under $H_0^{(n)}$, and with mean

$$\begin{cases} (\mathbf{a}+\mathbf{b})_i\sigma^2[I(f_\sigma)]^{1/2} & \text{for } i \le p \\ 0 & \text{for } i > p \end{cases}$$

   and variance 1 under $H^{(n)}_{\mathbf{a},\mathbf{b};f}$,
3. are asymptotically independent of $n^{1/2}r^{(n)}_{j;f}$, $n^{1/2}\hat{r}^{(n)}_{j;f}$ and $n^{1/2}\tilde{r}^{(n)}_{j;f}$, respectively $j \neq i$ under both $H_0^{(n)}$ and $H^{(n)}_{\mathbf{a},\mathbf{b};f}$.

Proof. 1. *Asymptotic distribution of* $n^{1/2}r_{i;f}^{(n)}$. Let $\mathbf{Y}_t^{(n)} = (X_t^{(n)} \cdots X_{t-q}^{(n)})'$, with $q = \max(p_1, p_2, i)$. Being q-dependent, $\{\mathbf{Y}_t^{(n)} | t \in \mathbb{Z}\}$ constitutes an *absolutely regular process* (cf., e.g., Ibragimov and Rozanov, 1978) under $H_0^{(n)}$. The log-likelihood ratio $\log L^{(n)}$ for $H_{0;f}^{(n)}$ against $H_{\mathbf{a},\mathbf{b};f}^{(n)}$ can be written (cf. Hallin et al., 1985a, Proposition 3.1) as [put $\|\mathbf{a} + \mathbf{b}\|^2 = \sum_{j=1}^{p}(\mathbf{a} + \mathbf{b})_j^2$]

$$\log L^{(n)} = n^{-1/2} \sum_{t=p+1}^{n} \phi_\sigma\left(X_t^{(n)}\right) \sum_{j=1}^{p} (\mathbf{a} + \mathbf{b})_j X_{t-j}^{(n)} - \tfrac{1}{2}\|\mathbf{a} + \mathbf{b}\|^2 + o_p(1)$$

$$= n^{-1/2}\mathcal{L}_{0;f}^{(n)} - \tfrac{1}{2}\|\mathbf{a} + \mathbf{b}\|^2 + o_p(1)$$

An arbitrary linear combination of $r_{i;f}^{(n)}$ and $\mathcal{L}_{0;f}^{(n)}$ therefore takes the form [up to $o_p(n^{-1/2})$ terms] of a differentiable von Mises's statistical functional (with respect to the observed series $(\mathbf{Y}_q^{(n)}, \ldots, \mathbf{Y}_n^{(n)})$. Applying Yoshihara (1976)'s central-limit theorem, we obtain that $n^{1/2}r_{i;f}^{(n)}$ and $\log L^{(n)}$ admit a joint asymptotic normal distribution (under $H_{0;f}^{(n)}$) with mean $(0, -\frac{1}{2}\|\mathbf{a} + \mathbf{b}\|^2\sigma^2 I(f_\sigma))$ and covariance matrix

$$\begin{bmatrix} 1 & \begin{cases} (\mathbf{a} + \mathbf{b})_i\left[\sigma^2 I(f_\sigma)\right]^{1/2} & \text{if } i \le p \\ 0 & \text{if } i > p \end{cases} \\ \cdot & \|\mathbf{a} + \mathbf{b}\|^2\sigma^2 I(f_\sigma) \end{bmatrix} \tag{3.5}$$

The desired result then follows from LeCam's third lemma.

2. *Joint asymptotic normality and asymptotic independence.* The joint asymptotic normality of $n^{1/2}r_{i;f}^{(n)}$ and $n^{1/2}r_{j;f}^{(n)}$, $i \neq j$, can be established as in 1, and their asymptotic covariance matrix can be shown to be diagonal. Note, however, that $r_{i;f}^{(n)}$ and $r_{j;f}^{(n)}$ are exactly uncorrelated under $H_{o;f}^{(n)}$ for any value of n, since $E[\phi_\sigma(X_{t_1}^{(n)})X_{t_1-i}^{(n)}\phi_\sigma(X_{t_2}^{(n)})X_{t_2-j}^{(n)}] = 0$ for every $n, t_1, t_2, 0 \neq i \neq j \neq 0$.

3. *Asymptotic equivalence of* $n^{1/2}r_{i;f}^{(n)}$ *and* $n^{1/2}\tilde{r}_{i;f}^{(n)}$. This equivalence straightforwardly follows from Hallin et al. (1985a, Section 4.1), under $H_{0;f}^{(n)}$.

4. *Asymptotic equivalence of* $n^{1/2}\hat{r}_{i;f}^{(n)}$ *and* $n^{1/2}r_{i;f}^{(n)}$. Put $T^{(n)}(\mathbf{X}^{(n)}; \theta) = (n - i)^{-1}\sum_{t=i+1}^{n}\phi_1(\theta X_t^{(n)})X_{t-i}^{(n)}$. Then $n^{1/2}r_{i;f}^{(n)} = \sigma^{-1}n^{1/2}T^{(n)}(\mathbf{X}^{(n)}; \sigma^{-1})$ and $n^{1/2}\hat{r}_{i;f}^{(n)} = (S^{(n)})^{-1}T^{(n)}(\mathbf{X}^{(n)}; (S^{(n)})^{-1})$. Since $S^{(n)}$ is a consistent estimator of σ, we just have to show that $\Delta^{(n)} = n^{1/2}[T^{(n)}(\mathbf{X}^{(n)}; (S^{(n)})^{-1}) - T^{(n)}(\mathbf{X}^{(n)}; \sigma^{-1})]$ is $o_p(1)$. Since $T^{(n)}(\mathbf{X}^{(n)}; \theta)$ is a differentiably smooth function of θ at $\theta = \sigma^{-1}$, we have $\Delta^{(n)} = n^{1/2}((S^{(n)})^{-1} - \sigma^{-1})\partial_\theta T^{(n)}(\mathbf{X}^{(n)}; \theta^*)$, where $\theta^* = \theta^*(\mathbf{X}^{(n)})$ is some (random) intermediate point be-

tween $(S^{(n)})^{-1}$ and σ^{-1}. Now

$$\partial_\theta T^{(n)}(\mathbf{X}^{(n)};\theta^*) - \partial_\theta T^{(n)}(\mathbf{X}^{(n)};\sigma^{-1})$$

$$= (n-i)^{-1} \sum_{t=i+1}^{n} \left[\left(\dot{\phi}_1\left(\theta^* X_t^{(n)}\right) - \dot{\phi}_1\left(\sigma^{-1} X_t^{(n)}\right)\right) X_t^{(n)} X_{t-i}^{(n)}\right] \quad (3.6)$$

Under $H_0^{(n)}$, the terms of this latter sum are mutually uncorrelated, and the variance of (3.6) is

$$(n-i)^{-1} E\left\{\left[\dot{\phi}_1\left(\theta^* X_2^{(n)}\right) - \dot{\phi}_1\left(\sigma^{-1} X_2^{(n)}\right)\right]^2 \left(X_2^{(n)}\right)^2 \left(X_1^{(n)}\right)^2\right\}$$

$$\leq (n-i)^{-1} A^2 E\left\{\left[(S^{(n)})^{-1} - \sigma^{-1}\right]^2 \cdot \left(X_2^{(n)}\right)^4 \left(X_1^{(n)}\right)^2\right\} = o_p(n^{-1})$$

Hence $\Delta^{(n)} = n^{1/2}((S^{(n)})^{-1} - \sigma^{-1}) \cdot (n-i)^{-1} \Sigma \dot{\phi}_1(\sigma^{-1} X_t^{(n)}) X_t^{(n)} X_{t-i}^{(n)} + o_p(1)$; it then follows from Yoshihara (1976)'s and Slutzky's theorems that $\Delta^{(n)}$ converges to zero in probability, which completes the proof.

3.3. Asymptotically Most Powerful Test

Denote by $\Phi(x)$ and $k_{1-\alpha}$, respectively, the distribution function and $(1-\alpha)$ quantile of the standard normal distribution. It then follows from Proposition 3.1 that the three tests

$$\text{reject } H_0^{(n)} \quad \text{if } \sum_{i=1}^{p} (n-i)^{1/2} \frac{(\mathbf{a}+\mathbf{b})_i}{\|\mathbf{a}+\mathbf{b}\|} r_{i;f}^{(n)} > k_{1-\alpha} \quad (3.7)$$

$$\text{reject } H_0^{(n)} \quad \text{if } \sum_{i=1}^{p} (n-i)^{1/2} \frac{(\mathbf{a}+\mathbf{b})_i}{\|\mathbf{a}+\mathbf{b}\|} \hat{r}_{i;f}^{(n)} > k_{1-\alpha} \quad (3.8)$$

$$\text{reject } H_0^{(n)} \quad \text{if } \sum_{i=1}^{p} (n-i)^{1/2} \frac{(\mathbf{a}+\mathbf{b})_i}{\|\mathbf{a}+\mathbf{b}\|} \tilde{r}_{i;f}^{(n)} > k_{1-\alpha} \quad (3.9)$$

have asymptotic probability level α.

Proposition 3.2. The tests (3.7), (3.8), and (3.9) are asymptotically most powerful against $H_{\mathbf{a},\mathbf{b};f}^{(n)}$ (within the class of all tests of asymptotic level α). Their asymptotic power is $1 - \Phi(k_{1-\alpha} - \|\mathbf{a}+\mathbf{b}\|(\sigma^2 I(f_\sigma))^{1/2})$.

Proof. Since the three test statistics are asymptotically equivalent, let us consider (3.7) only. Since under $H_{0;f}^{(n)}$, $\log L^{(n)}$ is asymptotically normal with mean $-\frac{1}{2}\|\mathbf{a}+\mathbf{b}\|^2\sigma^2 I(f_\sigma)$ and variance $\|\mathbf{a}+\mathbf{b}\|^2\sigma^2 I(f_\sigma)$, an asymptotically most powerful test against $H^{(n)}_{\mathbf{a},\mathbf{b};f}$ is

$$\text{reject } H_0^{(n)} \quad \text{if } \frac{\log L^{(n)}}{\|\mathbf{a}+\mathbf{b}\|\left(\sigma^2 I(f_\sigma)\right)^{1/2}} + \frac{1}{2} > k_{1-\alpha} \tag{3.10}$$

Now it follows from Proposition 3.1 (3) and the form of the autocovariance matrix (3.5) that

$$\sum_{i=1}^{p} (n-i)^{1/2} \frac{(\mathbf{a}+\mathbf{b})_i}{\|\mathbf{a}+\mathbf{b}\|} r_{i;f}^{(n)} - \frac{\log L^{(n)}}{\|\mathbf{a}+\mathbf{b}\|\left(\sigma^2 I(f_\sigma)\right)^{1/2}} - \frac{1}{2} = o_p(1)$$

(3.10), (3.7), (3.8), and (3.9) are thus asymptotically equivalent tests. Their common asymptotic power straightforwardly follows from Proposition 3.1.

REMARK 3.1. The tests (3.7), (3.8), and (3.9) actually are *uniformly* asymptotically most powerful against alternatives of the form

$$U\left\{H^{(n)}_{\boldsymbol{\alpha},\boldsymbol{\beta};f} \,\middle|\, \frac{\boldsymbol{\alpha}+\boldsymbol{\beta}}{\|\boldsymbol{\alpha}+\boldsymbol{\beta}\|} = \frac{\mathbf{a}+\mathbf{b}}{\|\mathbf{a}+\mathbf{b}\|}\right\}$$

Sequences of ARMA models (2.1) associated with coefficients $(\boldsymbol{\alpha},\boldsymbol{\beta})$ and $(\mathbf{a},\mathbf{b})$ such that $\boldsymbol{\alpha}+\boldsymbol{\beta}=\mathbf{a}+\mathbf{b}$ are, indeed, asymptotically observationally equivalent, in the sense that the corresponding likelihood functions are asymptotically equivalent, and the whole problem is scale-invariant. As a consequence, Proposition 3.2 provides uniformly asymptotically locally most powerful tests against unilateral alternatives of positive (negative) first-order dependence [AR(1), MA(1), or ARMA(1, 1)]. As another consequence, the sequences of ARMA models associated with coefficients $(\mathbf{a},\mathbf{b})$ such that $\mathbf{a}+\mathbf{b}=\mathbf{0}$ are asymptotically observationally equivalent with white noise in which case the asymptotic power of the optimal test reduces to its size α.

REMARK 3.2. For n sufficiently large, (2.1) always constitutes a sequence of stationary invertible ARMA models. The tests in Proposition 3.2 thus should be interpreted as optimal against sequences of ARMA processes whose *stationary invertible representations* admit coefficients proportional to $n^{-1/2}(\mathbf{a}+\mathbf{b})$; their corresponding noninvertible or nonstationary representations—if any—*cannot* be of the form (2.1).

4. ASYMPTOTICALLY MAXIMIN MOST POWERFUL TESTS AGAINST UNSPECIFIED ALTERNATIVES

4.1. Asymptotically Maximin Most Powerful Tests and Asymptotically Sufficient Statistics

The asymptotically most powerful tests in Section 3 are perhaps of theoretical rather than practical interest, since the alternative in practice is seldom specified. A uniformly asymptotically most powerful test procedure being impossible (cf. Hallin et al., 1985b, c)—except for the case of unilateral first-order alternatives [such as positive AR(1) or positive MA(1) dependence]—a weaker form of asymptotic optimality has to be considered.

Before establishing the existence of asymptotically locally maximin most powerful tests and providing their explicit form, we first recall some basic definitions.

Denote by $\beta(\alpha; H^{(n)}, K^{(n)})$ the envelope power function for some null hypothesis $H^{(n)} = \{l_0^{(n)}\}$ against some alternative hypothesis $K^{(n)} = \{l_1^{(n)}\}$:

$$\beta(\alpha; H^{(n)}, K^{(n)}) = \sup_{\psi} \inf_{l_1^{(n)} \in K^{(n)}} E_{l_1^{(n)}}[\psi],$$

where the $\sup_\psi$ is taken with respect to all the α-level tests, $\alpha \in (0,1)$. A sequence of tests $\psi^{(n)*}$ is said to be *asymptotically maximin* for (the sequence of tests) $H^{(n)}$ against $K^{(n)}$ if

$$\begin{cases} \liminf_{n\to\infty} \left[E_{l_1^{(n)}}[\psi^{(n)*}] - \beta(\alpha; H^{(n)}, K^{(n)}) \right] \geq 0, & l_1^{(n)} \in K^{(n)} \\ \limsup_{n\to\infty} E_{l_0^{(n)}}[\psi^{(n)*}] \leq \alpha, & l_0^{(n)} \in H^{(n)} \end{cases}$$

Another concept that plays a fundamental role in the asymptotic theory of tests is that of asymptotically sufficient statistics, introduced by LeCam (1960) and studied by Hájek and Šidák (1967, Chapter 7). Consider again the problem of testing $H^{(n)}$ against $K^{(n)}$. Denote by $\mathbf{T}^{(n)}$ a sequence of statistics, and by $H_T^{(n)}$ and $K_T^{(n)}$, respectively, the hypotheses induced by $\mathbf{T}^{(n)}$ (i.e., the families of distributions $\{l_0^{(n)T}\}$ and $\{l_1^{(n)T}\}$ induced from $\{l_0^{(n)}\}$ and $\{l_1^{(n)}\}$ by $\mathbf{T}^{(n)}$). Let $\beta(\alpha; H_T^{(n)}, K_T^{(n)})$ stand for the corresponding envelope power function. If

$$\lim_{n\to\infty} \left[\beta(\alpha; H^{(n)}, K^{(n)}) - \beta(\alpha; H_T^{(n)}, K_T^{(n)})\right] = 0, \qquad \alpha \in (0,1)$$

then $\mathbf{T}^{(n)}$ is said to be *asymptotically sufficient* for $H^{(n)}$ against $K^{(n)}$. This definition immediately implies that any asymptotically maximin most powerful test for $H_T^{(n)}$ against $K_T^{(n)}$ is also asymptotically maximin most powerful for $H^{(n)}$ against $K^{(n)}$. One can thus safely restrict to $\mathbf{T}^{(n)}$-measurable tests in the problem of obtaining asymptotically maximin most powerful tests for $H^{(n)}$ against $K^{(n)}$.

4.2. Unspecified ARMA Alternatives

An asymptotically maximin most powerful test for the null hypothesis of white noise (specified *or* unspecified density type f.) against, say, the whole family of ARMA(p_1, p_2) dependencies (2.1) [specified maximal order p such that $\max(p_1, p_2) \leq p$; specified density type f. but unspecified coefficients a_i and b_i] is clearly the trivial test $\phi \equiv \alpha$. If we allow, indeed, arbitrarily small coefficients a_i and b_i in (2.1), the alternative (for fixed n) contains distributions that are arbitrarily close to the null hypothesis.

If asymptotically maximin most powerful tests are to be obtained, the alternative has thus to be bounded away from the null hypothesis. Let us call unspecified ARMA alternative of order p an alternative of the form

$$K^{(n)}(f.|p, d) = U\left\{ H^{(n)}_{\mathbf{a},\mathbf{b};f} | \mathbf{a}, \mathbf{b} \in \mathbb{R}^p, \ \|\mathbf{a} + \mathbf{b}\| \geq d \right\}$$

($\|\mathbf{a} + \mathbf{b}\|$ denotes the euclidean norm; d is an arbitrary strictly positive constant.) The following result then follows as a particular case of Hallin and Puri (1985, Proposition 3.2).

Proposition 4.1. Any of the three vectors of f-autocorrelations $(r^{(n)}_{1;f}, \ldots, r^{(n)}_{p;f})'$, $(\hat{r}^{(n)}_{1;f}, \ldots, \hat{r}^{(n)}_{p;f})'$ or $(\tilde{r}^{(n)}_{1;f}, \ldots, \tilde{r}^{(n)}_{p;f})$ constitutes an asymptotically sufficient statistic for $H_0^{(n)}$ against $K^{(n)}(f.|p, d)$, for any $d > 0$.

4.3. Optimal Portmanteau Tests

According to the definitions given in Section 4.1, asymptotically (locally) maximin most powerful tests can be based on f-autocorrelations. It follows easily from Proposition 3.1 (2) and (3), an invariance argument, and the Hunt–Stein theorem that this test can be based on either of the quadratic statistics $Q^{(n)*}_{p;f} = \sum_{i=1}^{p}(n-i)(r^{(n)}_{i;f})^2$, $\hat{Q}^{(n)*}_{p;f} = \sum_{i=1}^{p}(n-i)(\hat{r}^{(n)}_{i;f})^2$ or $\tilde{Q}^{(n)*}_{p;f} = \sum_{i=1}^{p}(n-i)(\tilde{r}^{(n)}_{i;f})^2$. Because of their close similarity with the usual Box–Pierce (1970) or Ljung–Box (1978) *portmanteau statistics*, we call

$Q_{p;f}^{(n)*}$ and $\hat{Q}_{p;f}^{(n)*}$ *f-portmanteau statistics*, and $\tilde{Q}_{p;f}^{(n)}$ an *f-rank portmanteau statistic* (of order p).

We then have the following result.

Proposition 4.2. The (asymptotically equivalent under $H_{0;f}^{(n)}$) portmanteau statistics $Q_{p;f}^{(n)*}$, $\hat{Q}_{p;f}^{(n)*}$, and $\tilde{Q}_{p;f}^{(n)}$

1. are asymptotically χ^2 with p degrees of freedom under $H_0^{(n)}$, and asymptotically noncentral χ^2, with p degrees of freedom and noncentrality parameter $\lambda_f^*(\mathbf{a}, \mathbf{b}) = (1/2)\|\mathbf{a} + \mathbf{b}\|^2 I(f_1)$ under $H_{\mathbf{a},\mathbf{b};f}^{(n)}$ (with $\mathbf{a}, \mathbf{b} \in \mathbb{R}^p$).
2. provide an asymptotically locally maximin most powerful test for randomness against the unspecified ARMA alternative $K^{(n)}(f.|p, d)$ (for any positive d):

$$\text{reject } H_0^{(n)} \quad \text{if } Q_{p;f}^{(n)*} \left(\text{or } \hat{Q}_{p;f}^{(n)*} \text{ or } \tilde{Q}_{p;f}^{(n)*}\right) > \chi_{1-\alpha}^2$$

[where $\chi_{1-\alpha}^2$ is the $(1 - \alpha)$-fractile of the χ^2 distribution with p degrees of freedom]. The ARE of these tests with respect to the usual Box–Pierce or Ljung–Box portmanteau tests, based on sample autocorrelations, is $I(f_1) \geq 1$ (cf. Hallin et al., 1985b).

ACKNOWLEDGMENT

We would like to thank the referee for a number of useful comments and suggestions.

REFERENCES

Anderson, T. W. (1971). *The Statistical Analysis of Time Series*. New York: Wiley.

Box, G. E. P. and Pierce, D. A. (1970). Distribution of residual autocorrelations in autoregressive-integrated moving average time series models. *J. Amer. Statist. Assoc.* **65**, 1509–1526.

Hájek, J. and Šidák, Z. (1967). *Theory of Rank Tests*. Orlando, Florida: Academic Press.

Hallin, M., Ingenbleek, J.-Fr. and Puri, M. L. (1985a). Linear serial rank tests for randomness against ARMA alternatives. *Ann. Statist* **13**, 1156–1181.

Hallin, M., Ingenbleek, J.-Fr. and Puri, M. L. (1985b). Tests de rangs quadratiques pour une hypothèse de bruit blanc. *C. R. Acad. Sc. Paris* Série I **301**, 935–938.

Hallin, M., Ingenbleek, J.-Fr. and Puri, M. L. (1985c). Linear and quadratic serial rank tests for randomness. *J. Time Ser. Anal.* (to appear).

Hallin, M. and Puri, M. L. (1985). Optimal rank-based procedures for time series analysis: testing an ARMA model against other ARMA models. Technical Report, Department of Mathematics, Indiana University.

Hotelling, H. and Pabst, M. R. (1936). Rank correlation and tests of significance involving no assumption of normality. *Ann. Math. Statist.* **7**, 29–43.

Ibragimov, I. and Rozanov, Y. (1978). *Gaussian Random Processes*. Berlin and New York: Springer-Verlag.

Le Cam, L. (1960). Locally asymptotically normal families of distributions. *Univ. Calif. Publ. Statist.* **3**, 27–90.

Ljung, G. and Box, G. E. P. (1978). On a measure of lack of fit in time series models. *Biometrika* **65**, 297–303.

Wald, A. and Wolfowitz, J. (1943). An exact test for randomness in the nonparametric case based on serial correlation. *Ann. Math. Statist.* **14**, 378–388.

Yoshihara, K. (1976). Limiting behavior of *U*-statistics for stationary absolutely regular processes. *Z. Wahrsch. Geb.* **35**, 237–252.

Reprinted by permission of the publisher,
© 1987 by John Wiley & Sons, Inc.

OPTIMAL RANK-BASED PROCEDURES FOR TIME SERIES ANALYSIS: TESTING AN ARMA MODEL AGAINST OTHER ARMA MODELS

MARC HALLIN and MADAN L. PURI[1]

Universtié Libre de Bruxelles and Indiana University

The problem of testing a given ARMA model (in which the density of the generating white noise is unspecified) against other ARMA models is considered. A distribution-free asymptotically most powerful test, based on a *generalized linear serial rank statistic*, is provided against contiguous ARMA alternatives with specified coefficients. In the case when the ARMA model in the alternative has unspecified coefficients, the asymptotic sufficiency (in the sense of Le Cam) of a finite-dimensional vector of rank statistics is established. This asymptotic sufficiency is used to derive an asymptotically maximin most powerful test, based on a *generalized quadratic serial rank statistic*. The asymptotically maximin optimal test statistic can be interpreted as a rank-based, weighted version of the classical Box–Pierce portmanteau statistic, to which it reduces, in some particular problems, asymptotically and under Gaussian assumptions.

1. Introduction.

1.1. *Rank tests for time series analysis.* Time series analysis is certainly one of the areas in statistics whose development has been most impressive during the past two decades, and its importance for practical applications, ranging from economics to engineering has been widely recognized. However in spite of the growing interest in the subject and in spite of the fact that the need for robust or rank-based procedures has been emphasized by many authors, not much has been done to introduce such procedures (especially the rank-based ones) in the time series context.

This does not mean, of course, that no attempt was ever made to use rank-based methods beyond the classical problems involving i.i.d. (independent and identically distributed) observations. A very comprehensive bibliography on rank tests for non-i.i.d. observations can be found in a paper entitled "Nonparametric testing for time series analysis: A bibliography," by Dufour, Lepage and Zeidan (1982). This title however is somewhat misleading inasmuch as most of the papers mentioned on rank-based methods deal with such problems as testing for randomness (i.i.d. observations) against *trend alternatives* (trend in location, in dispersion, ordered alternatives,...) under which observations are *nonidentically but still independently distributed*, whereas time series analysis is typically

Received May 1985; revised May 1987.

[1]Research supported by Office of Naval Research Contract N00014-85-K-0648.

AMS 1980 *subject classifications.* Primary 62M10, 62G10.

Key words and phrases. ARMA models, linear serial rank statistics, quadratic rank statistics, rank portmanteau statistics, asymptotic sufficiency, asymptotically most powerful tests, asymptotically maximin most powerful tests.

concerned with nonindependent observed data and their *serial dependence structure*. Thus, it appears from this bibliography along with more recent literature that, apart from some scattered and very piecemeal results on the problem of testing for white noise against serial dependence [cf. Bhattacharyya (1984) or Hallin, Ingenbleek and Puri (1985) for a review], the subject of rank tests in time series analysis remains largely unexplored.

A first step toward a systematic treatment of this subject was taken in two papers by Hallin, Ingenbleek and Puri (1985 and 1987), where locally asymptotically optimal and locally maximin-optimal rank tests (in the Pitman–Noether sense) were derived for the problem of testing randomness against alternatives of ARMA dependence. Two classes of *serial rank statistics* were introduced for that purpose.

The first one is the class of *linear serial rank statistics*, of order p, of the form

$$S^{(n)} = (n-p)^{-1} \sum_{t=p+1}^{n} a^{(n)}\left(R_t^{(n)}, R_{t-1}^{(n)}, \dots, R_{t-p}^{(n)}\right), \tag{1.1}$$

where $R_t^{(n)}$ denotes the rank of the observation $X_t^{(n)}$ in an observed series $\mathbf{X}^{(n)} = (X_1^{(n)}, \dots, X_n^{(n)})$ of length n, and $a^{(n)}(\cdots)$ is some *score function*. Of special interest is the particular linear serial rank statistic, namely, the *rank autocorrelation* $r_{i;f}^{(n)}$ [see (2.5) for definition] *of order* i associated with density $f(\cdot)$ which enjoys most of the asymptotic properties of the usual sample autocorrelation $r_i^{(n)}$ for Gaussian series [cf. Hallin, Ingenbleek and Puri (1987) and Hallin and Puri (1987)].

The second class of rank statistics considered is that of *quadratic serial rank statistics*, which are quadratic forms of linear serial rank statistics [cf. Hallin, Ingenbleek and Puri (1987)]. Of particular interest are the *rank portmanteau statistics* $n\Sigma_{i=1}^{\kappa}(r_{i;f}^{(n)})^2$ which provide a rank version of Box and Pierce's (1970) classical *portmanteau statistic* $n\Sigma_{i=1}^{\kappa}(r_i^{(n)})^2$.

1.2. *Testing randomness against ARMA alternatives.* Consider the sequence of ARMA(p, q) models

$$X_t - n^{-1/2} \sum_{i=1}^{p} a_i X_{t-i} = \varepsilon_t + n^{-1/2} \sum_{i=1}^{q} b_i \varepsilon_{t-i}, \qquad t \in \mathbb{Z}, \tag{1.2}$$

where $\{\varepsilon_t\}$ is an independent white noise, i.e., a family of i.i.d. random variables with mean zero, unspecified variance σ^2 and density function $f(\cdot)$ (f has to satisfy some mild technical assumptions; cf. Section 2). Let $a_i = 0$, $i > p$ and $b_i = 0$, $i > q$; put $\mathbf{a} = (a_1, \dots, a_{\max(p,q)})$ and $\mathbf{b} = (b_1, \dots, b_{\max(p,q)})$ and denote by $\|\cdot\|$ the usual Euclidean norm. The vectors $\mathbf{a}$ and $\mathbf{b}$ and the density f determine what we call here a *specified ARMA alternative* $K^{(n)}(\mathbf{a}, \mathbf{b}; f)$. Denote by $H_0^{(n)}$ the null hypothesis of randomness, under which $\mathbf{X}^{(n)}$ is a vector of i.i.d. random variables with unspecified density function. Then an asymptotically most powerful α-level test $\varphi_\alpha^{(n)*}(\mathbf{a}, \mathbf{b}; f)$ (among all tests of given level α) for testing randomness against $K^{(n)}(\mathbf{a}, \mathbf{b}; f)$ is given by

$$\varphi_\alpha^{(n)*}(\mathbf{a}, \mathbf{b}; f) = 1 \tag{1.3}$$
$$\text{if} \quad \sum_{i=1}^{\max(p,q)} (n-i)^{1/2}(a_i + b_i) r_{i;f}^{(n)} > \|\mathbf{a} + \mathbf{b}\| k_{1-\alpha},$$

where $k_{1-\alpha}$ denotes the $(1-\alpha)$-quantile of the standard normal distribution. In the Gaussian case [assuming $E(X_t) = 0$], (1.3) can also be expressed, under asymptotically equivalent "parametric" form, in terms of the sample autocorrelations $r_i^{(n)} = \sum_{t=i+1}^n X_t^{(n)} X_{t-i}^{(n)} / \sum_{t=1}^n (X_t^{(n)})^2$, as

$$\varphi_\alpha^{(n)*}(\mathbf{a}, \mathbf{b}; f) = 1 \tag{1.4}$$
$$\text{if} \quad \sum_{i=1}^{\max(p,q)} (n-i)^{1/2}(a_i + b_i) r_i^{(n)} > \|\mathbf{a} + \mathbf{b}\| k_{1-\alpha}.$$

These results follow from Proposition 3.2 in Hallin, Ingenbleek and Puri (1985); they can also be obtained as particular cases of Propositions 2.4 and 4.3 by letting $p_1 = q_1 = 0$ [hence $A(L) = B(L) = 1$] in (3.2) and (4.2).

Of primary importance from the practical viewpoint is the development of the theory when the coefficients $\mathbf{a}$ and $\mathbf{b}$ in the ARMA model (1.2) are unspecified. We thus consider the alternative $K_f^{(n)}(m)$ consisting of all ARMA(p, q) dependencies (1.2) such that $\max(p, q) \leq m$: $K_f^{(n)}(m) = U\{K^{(n)}(\mathbf{a}, \mathbf{b}; f) | \mathbf{a} + \mathbf{b} \in \mathbb{R}^m\}$. Also, denote by $K_f^{(n)}(m|d)$, where $d \in \mathbb{R}^+$, the family of all ARMA dependencies in $K_f^{(n)}(m)$ such that $\|\mathbf{a} + \mathbf{b}\| \geq d$.

Then an asymptotically maximin most powerful α-level test $\psi_\alpha^{(n)*}(m; f)$ (among all tests of given level α) for randomness against any alternative $K_f^{(n)}(m|d)$, $d \in \mathbb{R}^+$, can be based on the *rank portmanteau statistic of order m*,

$$\psi_\alpha^{(n)*}(m; f) = 1 \quad \text{if} \quad \sum_{i=1}^{m} (n-i)\left(r_{i;f}^{(n)}\right)^2 > \chi^2_{m;1-\alpha}, \tag{1.5}$$

where $\chi^2_{m;1-\alpha}$ denotes the $(1-\alpha)$-quantile of the χ^2 distribution with m degrees of freedom. In the Gaussian case, (1.5) can also be expressed, under asymptotically equivalent "parametric" form, in terms of the classical *Box–Pierce portmanteau statistic*

$$\psi_\alpha^{(n)*}(m; f) = 1 \quad \text{if} \quad \sum_{i=1}^{m} (n-i)\left(r_i^{(n)}\right)^2 > \chi^2_{m;1-\alpha}. \tag{1.6}$$

These latter results follow, as particular cases, from Proposition 4.3.

1.3. *Testing an ARMA model against another ARMA model.* Testing for white noise (with unspecified density function) against ARMA dependence is certainly a very important problem in time series analysis, mainly because the nonrejection of a null hypothesis of randomness implies that complicated time series analysis procedures can be safely replaced with much simpler and more traditional devices such as the usual linear model methods. As pointed out by

Dufour and Roy (1985), tests for randomness can also be used to check several important economic hypotheses, such as market efficiency, rational expectations, life cycle–permanent income hypotheses, etc.

The problem of testing a given ARMA model (with unspecified density function) against other ARMA models is, however, even more important because of its implications in the various *identification* and *validation* steps that are part of any time series model-building procedure.

Denote by $H^{(n)}(\mathbf{A}, \mathbf{B}; \cdot)$ the null hypothesis under which the observed series $\mathbf{X}^{(n)}$ is generated by the ARMA(p_1, q_1) model

$$\begin{aligned} A(L)X_t &= X_t - A_1 X_{t-1} - \cdots - A_{p_1} X_{t-p_1} \\ &= \varepsilon_t + B_1 \varepsilon_{t-1} + \cdots + B_{q_1} \varepsilon_{t-q_1} = B(L)\varepsilon_t, \qquad t \in \mathbb{Z}, \end{aligned} \tag{1.7}$$

where $\{\varepsilon_t\}$ is a white noise process with unspecified density function [L denotes the *lag operator* and $A(L) = 1 - A_1 L - \cdots - A_{p_1} L^{p_1}$, $B(L) = 1 + B_1 L + \cdots + B_{q_1} L^{q_1}$ are linear difference operators]. Consider the *filtered process* $\{Z_t = [A(L)]/[B(L)]X_t\}$ and the *filtered series* $\mathbf{Z}^{(n)} = (Z_1^{(n)}, \ldots, Z_n^{(n)})$ associated with the observed series $\mathbf{X}^{(n)}$ (at this stage, we disregard the "starting values" problem involved in the definition of $\mathbf{Z}^{(n)}$; as shown in Appendix 2, these starting values do not affect asymptotic results). Obviously, $\mathbf{X}^{(n)}$ is an ARMA(p_1, q_1) series generated by (1.7) iff $\mathbf{Z}^{(n)}$ is a white noise series. At first sight, the problem of testing $H^{(n)}(\mathbf{A}, \mathbf{B}; \cdot)$ for the original series $\mathbf{X}^{(n)}$ thus reduces to the previously solved problem of testing for white noise ($H_0^{(n)}$ for the filtered series $\mathbf{Z}^{(n)}$). This is true, actually, as far as the null hypothesis is considered; let us now turn to the alternative.

The alternative hypotheses of interest here are those under which $\mathbf{X}^{(n)}$ is an observed series from some ARMA model $\alpha(L)X_t = \beta(L)\varepsilon_t$ distinct from (1.7). In order to investigate locally optimal procedures, we shall consider sequences of alternatives that are contiguous to the null hypotheses (see Section 2.2). Let

$$\begin{aligned} \alpha^{(n)}(L)X_t &= X_t - \alpha_1^{(n)} X_{t-1} - \cdots - \alpha_{p_2}^{(n)} X_{t-p_2} \\ &= \varepsilon_t + \beta_1^{(n)} \varepsilon_{t-1} + \cdots + \beta_{q_2}^{(n)} \varepsilon_{t-q_2} = \beta^{(n)}(L)\varepsilon_t, \qquad t \in \mathbb{Z}, \end{aligned} \tag{1.8}$$

be a sequence of ARMA(p_2, q_2) models, with $p_2 \geq p_1$, $q_2 \geq q_1$ and

$$\begin{aligned} \alpha_i^{(n)} &= \begin{cases} A_i + n^{-1/2}\gamma_i, & 1 \leq i \leq p_1, \\ n^{-1/2}\gamma_i, & p_1 < i \leq p_2, \end{cases} \\ \beta_i^{(n)} &= \begin{cases} B_i + n^{-1/2}\delta_i, & 1 \leq i \leq q_1, \\ n^{-1/2}\delta_i, & q_1 < i \leq q_2. \end{cases} \end{aligned} \tag{1.9}$$

Denote by $\boldsymbol{\gamma}$ and $\boldsymbol{\delta}$ the vectors $(\gamma_1, \ldots, \gamma_{p_2})$ and $(\delta_1, \ldots, \delta_{q_2})$, respectively, and by $K^{(n)}(\mathbf{A}, \mathbf{B}; \boldsymbol{\gamma}, \boldsymbol{\delta}; f)$ the sequence of ARMA(p_2, q_2) alternatives (1.8) corresponding to specified $\boldsymbol{\gamma}$, $\boldsymbol{\delta}$ and f. If $\mathbf{X}^{(n)}$ is an observed series from (1.8), it is easy to see that the filtered series $\mathbf{Z}^{(n)}$ is generated by the model

$$\alpha^{(n)}(L)B(L)Z_t = A(L)\beta^{(n)}(L)\varepsilon_t, \qquad t \in \mathbb{Z}, \tag{1.10}$$

an ARMA(p, q) model with $p_2 - p_1 \leq p \leq p_2 + q_1$ and $q_2 - q_1 \leq q \leq p_1 + q_2$. The model (1.10) is *not* a model of the type (1.2) considered in Section 1.2, except for the *very* particular case where $\alpha^{(n)}(L)$ and $\beta^{(n)}(L)$ are of the form

$$\alpha^{(n)}(L) = A(L)\big(1 - n^{-1/2}(a_1L + \cdots + a_pL^p)\big),$$
$$(1.11)\qquad \beta^{(n)}(L) = B(L)\big(1 + n^{-1/2}(b_1L + \cdots + b_qL^q)\big),$$

yielding, for the filtered series, the ARMA$(p = p_2 - p_1, q = q_2 - q_1)$ model

$$Z_t - n^{-1/2}(a_1Z_{t-1} + \cdots + a_pZ_{t-p}) = \varepsilon_t + n^{-1/2}(b_1\varepsilon_{t-1} + \cdots + b_q\varepsilon_{t-q}).$$

Consequently, the problem of testing an ARMA model (for the original series $\mathbf{X}^{(n)}$) against another one [viz. testing $H^{(n)}(\mathbf{A}, \mathbf{B}; \cdot)$ against $K^{(n)}(\mathbf{A}, \mathbf{B}; \boldsymbol{\gamma}, \boldsymbol{\delta}; f)$] *does not*, in general, reduce to the previously studied problem of testing for white noise (for the filtered series $\mathbf{Z}^{(n)}$) against ARMA dependence [viz. testing $H_0^{(n)}$ against $K^{(n)}(\mathbf{a}, \mathbf{b}; f)$], and the results of Hallin, Ingenbleek and Puri (1985, 1987) are no longer applicable to this new problem. The optimal tests indeed will be shown to depend, in a very crucial way, on the tested model (1.7).

1.4. *Outline of the paper.* We first consider, in Section 2, the problem of testing for randomness against contiguous specified alternatives of *general linear dependence*. The asymptotic distribution of the log-likelihood ratio for this problem is derived in Proposition 2.2, and an asymptotically most powerful rank test is provided in Proposition 2.4. These results are then used in Section 3, where we show that the model (1.10) describing the behaviour of the filtered series $\mathbf{Z}^{(n)}$ under the alternative, in the problem of testing an ARMA model against some other ARMA model, is a particular case of the general linear model considered in Section 2. An asymptotically most powerful statistic for testing a given ARMA model against another one is given in Section 3.2, where we also establish the *asymptotic sufficiency* [in the sense of Le Cam (1960) and Hájek and Šidák (1967)] of a vector of $\max(p_1 + q_2, p_2 + q_1)$ rank statistics [see (1.8) and (1.9) for a definition of p_1, p_2, q_1 and q_2]. These statistics involve the f-rank autocorrelations (of all available orders) and an arbitrary *fundamental system of solutions* of the homogeneous difference equation $A(L)B(L)\Psi_t = 0$, $t \in \mathbb{Z}$, defined by the ARMA model (1.7) to be tested. The asymptotic normal distribution of this sufficient statistic is provided in Proposition 3.3.

It follows (cf. Section 3.2) from the definition of asymptotic sufficiency that, under certain conditions, asymptotically maximin most powerful tests can be based on asymptotically sufficient statistics. We therefore introduce, in Section 4.1, a *generalized quadratic serial rank statistic* extending the concept introduced in Hallin, Ingenbleek and Puri (1987). The definition of this quadratic statistic, although involving an arbitrary fundamental system of solutions of the homogeneous equation $A(L)B(L)\Psi_t = 0$, $t \in \mathbb{Z}$, is shown not to depend on the particular system adopted (Proposition 4.1); examples are provided.

The asymptotic χ^2 (under the null hypotheses) and noncentral χ^2 (under the alternative) distributions of this statistic are obtained in Proposition 4.2, and the asymptotic maximin optimality of the corresponding test is proved in Proposition 4.3. The form of the maximin-optimal statistic is discussed in some detail in Section 4.2. Section 5 provides some results on the asymptotic relative efficiencies of our rank tests with respect to some classical parametric tests (such as the classical Box–Pierce or Durbin–Watson tests).

The ranks used throughout the paper are the ranks of residuals whose derivation may require the inversion of a moving-average operator. Such residuals in this case cannot be obtained from a finite series of observations, and approximate residuals have to be substituted for the exact ones. Appendix 2 shows that this use of approximate residuals has no influence upon the results of the paper. Appendix 1 provides a brief overview of some results on linear difference equations which are used continuously throughout the paper. Appendix 3 concentrates on the proofs of the propositions.

2. General linear processes.

2.1. *The general linear alternative.* Denote by $\{\varepsilon_t;\ t \in \mathbb{Z}\}$ a discrete-time stationary white noise, i.e., a sequence of independent and identically distributed real-valued random variables with mean $E(\varepsilon_t) = 0$ and (unspecified) variance $E(\varepsilon_t^2) = \sigma^2$. Assume that these random variables have a density function $f(x)$ (with respect to the Lebesgue measure) and that the following technical conditions are satisfied:

1. ε_t has finite moments up to the sixth order.
2. $f(x)$ is (a.e.) derivable [denote by $f'(x)$ its derivative], ***absolutely continuous on finite intervals*** [see Hájek and Šidák (1967), page 15] and satisfies $E[|f'(\varepsilon_t)/f(\varepsilon_t)|^{2+\delta}] < \infty$ for some $\delta > 0$ [this implies that $f(x)$ has finite Fisher information $I(f)$: $0 < I(f) = E[(f'(\varepsilon_t)/f(\varepsilon_t))^2] < \infty$].
3. Denote by $F(x)$ the cumulative distribution function associated with $f(x)$ and let $F^{-1}(u) = \inf\{x | F(x) \geq u\}$, $0 < u < 1$. Define $\phi(x)$ as $-f'(F^{-1}(F(x)))/f(F^{-1}(F(x)))$. Assume that $\phi(x)$ is a.e. derivable and that its derivative $\phi'(x)$ is (a.e.) Lipschitzian and square integrable: $|\phi'(x) - \phi'(y)| < K|x - y|$ (K, K' and K'' will be used throughout the paper to denote constants—not necessarily the same), and $\int_0^1 \phi'^2(F^{-1}(u))\, du < \infty$. Notice that $\phi(x)$ is a.e. equal to $-f'(x)/f(x)$.
4. f is strongly unimodal [cf. Hájek and Šidák (1967), page 15].

Let $\mathbf{a} = (a_1, a_2, \ldots)$ and $\mathbf{b} = (b_1, b_2, \ldots)$ be two sequences of real coefficients such that $\sum_{i=1}^{\infty} |a_i| < \infty$ and $\sum_{i=1}^{\infty} |b_i| < \infty$. This implies that $\mathbf{a}$ and $\mathbf{b}$ are elements of the Hilbert space l^2 of the square-summable sequences, which ensures that, for n sufficiently large (we henceforth shall assume that the following holds for $n \in \mathbb{N}$), the functions $1 - n^{-1/2}\sum_{i=1}^{\infty} a_i z^i$ and $1 + n^{-1/2}\sum b_i z^i$ are both analytic inside and on the unit circle, as well as their inverses. The stochastic difference equation (of "infinite order")

$$X_t - n^{-1/2} \sum_{i=1}^{\infty} a_i X_{t-i} = \varepsilon_t + n^{-1/2} \sum_{i=1}^{\infty} b_i \varepsilon_{t-i}, \qquad t \in \mathbb{Z}, \tag{2.1}$$

therefore correctly defines a unique stationary general linear process $\{X_t^{(n)}\}$, i.e., a process of the form

$$X_t^{(n)} = \varepsilon_t + \sum_{i=1}^{\infty} c_i^{(n)} \varepsilon_{t-i}, \qquad t \in \mathbb{Z}, \tag{2.2}$$

with $\sum_{i=1}^{\infty}(c_i^{(n)})^2 < \infty$ [see, e.g., Priestley (1981), Section 3.5.7]. As can be easily seen from (2.2), $\{X_t^{(n)}\}$ has finite sixth-order moments if $\{\varepsilon_t\}$ has. Notice, however, that the coefficients $c_i^{(n)}$ in (2.2) are generally not of the form $n^{-1/2}c_i$, so that (2.2) is not a particular case of (2.1).

Let $\mathbf{x}^{(n)} = (x_1^{(n)}, \dots, x_n^{(n)})$ be the value of an observed series $\mathbf{X}^{(n)} = (X_1^{(n)}, \dots, X_n^{(n)})$ of length n. Denote by $K_f^{(n)}$ the hypothesis under which $\mathbf{X}_n$ is a finite realization of some general linear process $\{X_t^{(n)}\}$, defined by (2.1) with unspecified sequences $\mathbf{a}$ and $\mathbf{b}$ and a specified input white noise density $f(x)$ satisfying 1–3; let $K^{(n)}(\mathbf{a}, \mathbf{b}; f)$ be the subhypothesis obtained by specifying $\mathbf{a}$ and $\mathbf{b}$. Such hypotheses will be referred to as (specified or unspecified) *general linear alternatives*. Clearly, general linear alternatives constitute a generalization of the ARMA alternatives $K^{(n)}(\mathbf{a} = (a_1 \cdots a_p), \mathbf{b} = (b_1 \cdots b_q); f)$ previously considered in Section 1.2 (and this is why we adopt similar notations); they also include the hypotheses of the form $K^{(n)}(\mathbf{A}, \mathbf{B}; \boldsymbol{\gamma}, \boldsymbol{\delta}; f)$ (see Section 1.3) we shall be interested in when testing an ARMA model against other ARMA models—this latter statement will be made clearer in Section 3.1.

In Sections 2.2 and 2.3, we mainly establish for general linear alternatives some results (viz. contiguity and the form of asymptotically most powerful tests) that will be needed in the problem of testing an ARMA model against other ARMA models. First let us state a theorem due to Anderson [(1959); cf. Anderson (1971), Theorem 7.7.1] which will be used repeatedly.

THEOREM 2.1. *Let* $Y^{(n)} = Y_\kappa^{(n)} + R_\kappa^{(n)}$, $n \in \mathbb{N}$, $\kappa \in \mathbb{N}$. *Suppose that* $R_\kappa^{(n)}$ *converges to zero in probability, as* $\kappa \to \infty$, *uniformly with respect to* n [*i.e., for any* $\varepsilon' > 0$, $\varepsilon'' > 0$ *there exists a* $\kappa_0(\varepsilon', \varepsilon'')$ *not depending on* n *such that* $P[|R_\kappa^{(n)}| > \varepsilon'] < \varepsilon''$ *for any* $\kappa > \kappa_0$] *and that the distribution functions* $F_\kappa^{(n)}(y) = P[Y_\kappa^{(n)} \leq y]$ *converge, as* $n \to \infty$, *to some distribution function* $F_\kappa(y)$ *such that* $\lim_{\kappa \to \infty} F_\kappa(y) = F(y)$ *at every continuity point of* $F(y)$. *Then* $\lim_{n \to \infty} P[Y^{(n)} \leq y] = F(y)$ *at every continuity point of* $F(y)$.

2.2. *Generalized linear serial rank statistics.* In Hallin, Ingenbleek and Puri (1985), we established the contiguity of sequences of finite-order ARMA alternatives and the white noise hypothesis $H_0^{(n)}$. Proposition 2.2 extends this result to general linear hypotheses of the form $K^{(n)}(\mathbf{a}, \mathbf{b}; f)$.

Denote by $L^{(n)}(\mathbf{X}^{(n)})$ the likelihood ratio for testing $H_{0;f}^{(n)}$ (the white noise hypothesis with specified density f) against $K^{(n)}(\mathbf{a}, \mathbf{b}; f)$. Then Proposition 2.2 is proved in Appendix 3.

PROPOSITION 2.2. *The log-likelihood ratio* $\log L^{(n)}(\mathbf{X}^{(n)})$ *is asymptotically normal, with mean* $-\frac{1}{2}\sum_{i=1}^{\infty}(a_i + b_i)^2\sigma^2 I(f)$ *and variance* $\sum_{i=1}^{\infty}(a_i + b_i)^2\sigma^2 I(f)$

(under $H_{0;f}^{(n)}$). The sequences of hypotheses $K^{(n)}(\mathbf{a}, \mathbf{b}; f)$ and $H_0^{(n)}$ are therefore contiguous.

Linear serial rank statistics of order p defined in (1.1) consider successive $(p+1)$-tuples of observations and consequently cannot be expected to account for serial dependencies of orders higher than p. This is confirmed by Proposition 2.3, which motivates the introduction of *generalized linear serial rank statistics.*

Consider the linear serial rank statistics $S^{(n)}$ of order p given in (1.1). Let

$$m^{(n)} = \frac{(n-p-1)!}{n!} \sum_{1 \le i_1 \ne \cdots \ne i_{p+1} \le n} \cdots \sum a_n(i_1, \ldots, i_{p+1}) \tag{2.3}$$

be the expectation of $S^{(n)}$ under $H_0^{(n)}$. Assume that there exists a function $J(\cdots)$ defined over $(0,1)^{p+1}$ such that

$$\int_{[0,1]^{p+1}} |J(v_1, \ldots, v_{p+1})|^{2+\delta} \, dv_1 \cdots dv_{p+1} < \infty \text{ for some } \delta > 0,$$

satisfying (under $H_0^{(n)}$)

$$\lim_{n \to \infty} E\left\{\left[J\left(F\left(X_1^{(n)}\right), \ldots, F\left(X_{p+1}^{(n)}\right)\right) - a_n\left(R_1^{(n)}, \ldots, R_{p+1}^{(n)}\right)\right]^2\right\} = 0. \tag{2.4}$$

$J(\cdots)$ is called a *score-generating function associated with* $S^{(n)}$. We furthermore may assume without loss of generality [cf. Hallin, Ingenbleek and Puri 1987)] that

$$\int_{[0,1]^{p+1}} J(v_1, \ldots, v_{p+1}) \prod_{j \ne i} dv_j = 0, \qquad i = 1, \ldots, p+1.$$

The following result then generalizes Proposition 4.3 in Hallin, Ingenbleek and Puri (1985).

PROPOSITION 2.3. *$n^{1/2}(S^{(n)} - m^{(n)})$ is asymptotically normal, with mean 0 and variance V^2 under $H_0^{(n)}$, with mean $\sum_{i=1}^{p} (a_i + b_i)C_i$ and variance V^2 under $K^{(n)}(\mathbf{a}, \mathbf{b}; f)$, where*

$$V^2 = \int_{[0,1]^{2p+1}} \left\{ \left[J(v_1, \ldots, v_{p+1})\right]^2 + 2\sum_{j=1}^{p} J(v_1, \ldots, v_{p+1}) J(v_{1+j}, \ldots, v_{p+1+j}) \right\} dv_1 \cdots dv_{2p+1}$$

and

$$C_i = \int_{[0,1]^{p+1}} J(v_1, \ldots, v_{p+1}) \sum_{j=0}^{p-i} \phi\left(F^{-1}(v_{1+j})\right) F^{-1}(v_{1+j+i}) \, dv_1 \cdots dv_{p+1}.$$

For the proof, see Appendix 3.

Notice that this result implies that the asymptotic distribution of $S^{(n)}$ is exactly the same under $K^{(n)}(\mathbf{a}, \mathbf{b}; f)$ as under the pth-order *truncation* of $K^{(n)}(\mathbf{a}, \mathbf{b}; f)$ obtained by putting $0 = a_{p+1} = a_{p+2} = \cdots$ and $0 = b_{p+1} = b_{p+2} = \cdots$ in (2.1). As a consequence, a finite-order serial rank statistic generally cannot capture all the available information for testing $H_0^{(n)}$ against $K^{(n)}(\mathbf{a}, \mathbf{b}; f)$.

Accordingly unless $\mathbf{a}$ and $\mathbf{b}$ are such that $a_i = 0 = b_i$ for i larger than some $p < \infty$ [in which case the general linear alternative $K^{(n)}(\mathbf{a}, \mathbf{b}; f)$ reduces to the ARMA(p_1, q_1) alternatives considered in Hallin, Ingenbleek and Puri (1985)], the asymptotically most powerful test statistic for $H_0^{(n)}$ against $K^{(n)}(\mathbf{a}, \mathbf{b}; f)$ does not belong to the class of linear serial rank statistics anymore. Yet an asymptotically most powerful test based on ranks does exist, as asserted by Proposition 2.4, but its "order" increases with sample size n.

Define the *rank autocorrelation coefficient associated with density* f (or *f-rank autocorrelation*) of order i as

$$r_{i;f}^{(n)} = \left[(n-i)^{-1} \sum_{t=i+1}^{n} \phi\left(F^{-1}\left(\frac{R_t^{(n)}}{n+1}\right)\right) F^{-1}\left(\frac{R_{t-i}^{(n)}}{n+1}\right) - m^{(n)}\right] \bigg/ s^{(n)}, \tag{2.5}$$

with

$$m^{(n)} = (n(n-1))^{-1} \sum\sum_{1 \le i_1 \ne i_2 \le n} \phi\left(F^{-1}\left(\frac{i_1}{n+1}\right)\right) F^{-1}\left(\frac{i_2}{n+1}\right) \tag{2.6}$$

and

$$\begin{aligned}
[s^{(n)}]^2 = {} & \frac{1}{n(n-1)} \sum\sum_{1 \le i_1 \ne i_2 \le n} \left[\phi\left(F^{-1}\left(\frac{i_1}{n+1}\right)\right) F^{-1}\left(\frac{i_2}{n+1}\right)\right]^2 \\
& + 2\frac{(n-2i)}{(n-i)} \frac{1}{n(n-1)(n-2)} \sum\sum\sum_{1 \le i_1 \ne i_2 \ne i_3 \le n} \left[\phi\left(F^{-1}\left(\frac{i_1}{n+1}\right)\right)\right. \\
& \left. \times \phi\left(F^{-1}\left(\frac{i_2}{n+1}\right)\right) F^{-1}\left(\frac{i_2}{n+1}\right) F^{-1}\left(\frac{i_3}{n+1}\right)\right] \\
& + \frac{n^2 - n(2i+3) + i^2 + 5i}{n(n-1)(n-2)(n-3)(n-i)} \sum\sum\sum\sum_{1 \le i_1 \ne i_2 \ne i_3 \ne i_4 \le n} \left[\phi\left(F^{-1}\left(\frac{i_1}{n+1}\right)\right)\right. \\
& \left. \times \phi\left(F^{-1}\left(\frac{i_2}{n+1}\right)\right) F^{-1}\left(\frac{i_3}{n+1}\right) F^{-1}\left(\frac{i_4}{n+1}\right)\right] \\
& - (n-i)(m^{(n)})^2.
\end{aligned} \tag{2.7}$$

If $\mathbf{a}$ and $\mathbf{b}$ belong to l^2, we denote by $\langle \mathbf{a}, \mathbf{b} \rangle = \sum_{i=1}^{\infty} a_i b_i$ their scalar product; $\|\mathbf{a}\| = \langle a, a \rangle^{1/2}$ stands for the corresponding norm.

Proposition 2.4. *Consider the test statistic*

$$S^{(n)*} = n^{-1/2} \sum_{i=1}^{n-1} (n-i)^{1/2} (a_i + b_i) r_{i;f}^{(n)} \bigg/ \left[\sum_{i=1}^{n-1} (a_i + b_i)^2\right]^{1/2}. \tag{2.8}$$

Then:

(i) $n^{1/2}S^{(n)*}$ *is asymptotically normal, with mean* 0 *and variance* 1 *under* $H_0^{(n)}$, *and with mean* $\langle(\mathbf{a}+\mathbf{b}),(\tilde{\mathbf{a}}+\tilde{\mathbf{b}})\rangle[\sigma^2 I(f)]^{1/2}/\|\mathbf{a}+\mathbf{b}\|$ *and variance* 1 *under* $K^{(n)}(\tilde{\mathbf{a}},\tilde{\mathbf{b}}; f)$ (*with* $\tilde{\mathbf{a}},\tilde{\mathbf{b}} \in l^2$).

(ii) $S^{(n)*}$ *provides an asymptotically most powerful test for* $H_0^{(n)}$ *against the general linear alternative* $K^{(n)}(\mathbf{a},\mathbf{b}; f)$ (*among all tests of given level* α):

$$\varphi_\alpha^{(n)*}(\mathbf{a},\mathbf{b}; f) = 1$$

$$\text{(2.9)} \qquad \textit{if} \quad \sum_{i=1}^{n-1}(n-i)^{1/2}(a_i+b_i)r_{i;f}^{(n)} > \left[\sum_{i=1}^{n-1}(a_i+b_i)^2\right]^{1/2} k_{1-\alpha},$$

where $k_{1-\alpha}$ *denotes the* $(1-\alpha)$*-quantile of the standard normal distribution.*

For the proof, see Appendix 3.

Remark 2.1. Unless there exists some $p < \infty$ such that $a_i + b_i = 0$, $i > p$ [in which case (2.9) reduces to (1.3)], $S^{(n)*}$ does not belong to the class of linear serial rank statistics anymore. Its asymptotic distribution cannot be obtained by the same techniques [viz. score-generating functions) as in Hallin, Ingenbleek and Puri (1985). Call $S^{(n)*}$ a generalized linear serial rank statistic.

Remark 2.2. The asymptotic distribution of $n^{1/2}S^{(n)*}$—hence the asymptotic optimality of the corresponding test (2.9)—depends on $\mathbf{a}$ and $\mathbf{b}$ through the sum $\mathbf{a}+\mathbf{b}$ only: (2.9) is therefore asymptotically most powerful against the more general alternative $U\{K^{(n)}(\boldsymbol{\alpha},\boldsymbol{\beta}; f)|\boldsymbol{\alpha}+\boldsymbol{\beta} = \mathbf{a}+\mathbf{b}\}$.

Remark 2.3. Just as in the case of finite-order ARMA alternatives, there exist general linear alternatives $K^{(n)}(\tilde{\mathbf{a}},\tilde{\mathbf{b}}; f)$ against which the asymptotic power of (2.9) is α (viz. those for which $(\tilde{\mathbf{a}}+\tilde{\mathbf{b}}) \perp (\mathbf{a}+\mathbf{b})$, where $\perp$ denotes the orthogonality relation in l^2]. This motivates the consideration of generalized quadratic rank statistics (see Section 4.1).

3. Testing an ARMA model against another ARMA model: Asymptotic sufficiency.

3.1. *The general linear alternative for residuals.* Consider now the null hypothesis $H^{(n)}(\mathbf{A},\mathbf{B};\cdot)$ under which the observed series is generated by the ARMA(p_1, q_1) model (1.7), where

(a) $A(L)$ and $B(L)$ are of orders p_1 and q_1, respectively (i.e., $A_{p_1} \neq 0 \neq B_{q_1}$),
(b) the polynomials $A(z)$ and $B(z)$, $z \in \mathbb{C}$, have no common roots,
(c) the polynomials $A(z)$ and $B(z)$, $z \in \mathbb{C}$, have all their roots outside the unit circle,
(d) the unspecified density $f(\cdot)$ of $\{\varepsilon_t\}$ satisfies the assumptions of Section 2.1.

Denoting by Z_t the filtered observations $(A(L)/B(L))X_t$ [i.e., the observed residuals with respect to (1.7)], we have seen that, under the null hypothesis,

$\{Z_t\}$ is a white noise process. Let us show that, under the alternative $K^{(n)}(\mathbf{A}, \mathbf{B}; \boldsymbol{\gamma}, \boldsymbol{\delta}; f)$ [see (1.8) and (1.9)], $\{Z_t\}$ is generated by a general linear model of the form (2.1) we investigated in Section 2. We shall assume that $\alpha^{(n)}(L)$ and $\beta^{(n)}(L)$

(e) are such that the polynomials $\alpha^{(n)}(z)$ and $\beta^{(n)}(z)$, $z \in \mathbb{C}$, have no root which is common for any value of $n \in \mathbb{N}$.

For n sufficiently large, $\alpha^{(n)}(z)$ and $\beta^{(n)}(z)$ have all their roots outside the unit circle and (1.8) uniquely determines a sequence of stationary ARMA processes; of course, we still assume that the density function $f(\cdot)$ of $\{\varepsilon_t\}$ satisfies the assumptions of Section 2.1. Suppose $p_1 > 0$ and $q_1 > 0$, and denote by G_u and H_u, $u \in \mathbb{Z}$, the *Green's functions* of the operators $A(L)$ and $B(L)$, respectively (for definition, see Appendix 1). Under the preceding assumptions (a), (b) and (c), we have $[A(L)]^{-1} = 1 + \sum_{u=1}^{\infty} G_u L^u$ and $[B(L)]^{-1} = 1 + \sum_{u=1}^{\infty} H_u L^u$, with $\sum_{u=1}^{\infty} |G_u| < \infty$ and $\sum_{u=1}^{\infty} |H_u| < \infty$; (1.10) then can be written as

$$\left[1 - n^{-1/2} \sum_{u=1}^{\infty} \sum_{i=1}^{\min(p_2, u)} \gamma_i G_{u-i} L^u\right] Z_t = \left[1 + n^{-1/2} \sum_{u=1}^{\infty} \sum_{i=1}^{\min(q_2, u)} \delta_i H_{u-i} L^u\right] \varepsilon_t. \tag{3.1}$$

Let

$$a_i = \sum_{j=1}^{\min(p_2, i+p_1-1)} \gamma_j G_{i-j} \tag{3.2}$$

and

$$b_i = \sum_{j=1}^{\min(q_2, i+q_1-1)} \delta_j H_{i-j}, \qquad i = 1, 2, \ldots .$$

Because $G_{-1} = G_{-2} = \cdots = G_{-p_1+1} = H_{-1} = H_{-2} = \cdots = H_{-q_1+1} = 0$, (3.1) is of the form (2.1). Let us show, moreover, that $\sum_{i=1}^{\infty} |a_i| < \infty$ and $\sum_{i=1}^{\infty} |b_i| < \infty$. For $i \leq p_2 - p_1 + 1$, the a_i's are a linear combination of $G_{i-1}, G_{i-2}, \ldots, G_{i-p_2}$, which are solutions of the homogeneous equation $A(L) G_t = 0$, $t \in \mathbb{Z}$; because of assumption (c), such solutions are decaying to zero, as $i \to \infty$, at an exponential rate, which implies that $\sum_{i=1}^{\infty} |a_i|$ converges. The same holds for the b_i's and operator $B(L)$. Equation (3.1) thus constitutes a particular case of a general linear alternative, and testing $H^{(n)}(\mathbf{A}, \mathbf{B}; \cdot)$ against $K^{(n)}(\mathbf{A}, \mathbf{B}; \boldsymbol{\gamma}, \boldsymbol{\delta}; f)$ reduces to the problem of testing $H_0^{(n)}$ against $K^{(n)}(\mathbf{a}, \mathbf{b}; f)$, with $\mathbf{a}$ and $\mathbf{b}$ given in (3.2).

If $p_1 = 0$ and/or $q_1 = 0$, put $a_i = \gamma_i$, $i = 1, \ldots, p_2$, and/or $b_i = \delta_i$, $i = 1, \ldots, q_2$.

3.2. *Asymptotically most powerful test and asymptotic sufficiency of f-rank autocorrelations.* An asymptotically most powerful test for $H^{(n)}(\mathbf{A}, \mathbf{B}; \cdot)$

[ARMA model (1.7)] against $K^{(n)}(\mathbf{A}, \mathbf{B}; \boldsymbol{\gamma}, \boldsymbol{\delta}; f)$ [alternative model (1.8)] can now straightforwardly be obtained from Proposition 2.4.

PROPOSITION 3.1. *An asymptotically most powerful test for* $H^{(n)}(\mathbf{A}, \mathbf{B}; \cdot)$ *against* $K^{(n)}(\mathbf{A}, \mathbf{B}; \boldsymbol{\gamma}, \boldsymbol{\delta}; f)$ *is given by* (2.9), *where the rank autocorrelations* $r_{i;f}^{(n)}$ *given by* (2.5) *are computed from the ranks of the filtered series* $\mathbf{Z}^{(n)}$, *and the* a_i*'s and* b_i*'s are given in* (3.2). *Under the same conditions, the asymptotic distribution of the generalized linear serial rank statistic* (2.8) *remains the same as in Proposition* 2.4(i).

Unless otherwise specified, all the ranks and rank statistics we refer to in the sequel are computed from the filtered series $\mathbf{Z}^{(n)}$.

Consider the general problem of testing $H^{(n)}$ against $K^{(n)}$, where $H^{(n)} = \{l_0^{(n)}\}$ and $K^{(n)} = \{l_1^{(n)}\}$ are families of densities for some observed series $\mathbf{X}^{(n)}$ defined over a sample space $(\mathscr{X}^{(n)}, \mathscr{A}^{(n)})$. Recall that the *envelope power function* $\beta(\alpha, H^{(n)}, K^{(n)})$ for this problem is defined as

$$\beta(\alpha, H^{(n)}, K^{(n)}) = \sup_{\psi^{(n)}} \inf_{l_1^{(n)} \in K^{(n)}} E_{l_1^{(n)}}\left[\psi^{(n)}\right], \tag{3.3}$$

where the supremum is taken over all the tests $\psi^{(n)}$ satisfying, for fixed $\alpha \in (0,1)$,

$$E_{l_0^{(n)}}\left[\psi^{(n)}\right] \leq \alpha, \qquad l_0^{(n)} \in H^{(n)}. \tag{3.4}$$

Let $T^{(n)}$ be some statistic from $(\mathscr{X}^{(n)}, \mathscr{A}^{(n)})$ to some range space $(\mathscr{T}^{(n)}, \mathscr{B}^{(n)})$. Define the envelope power function $\beta_T(\alpha, H^{(n)}, K^{(n)})$ associated with $T^{(n)}$ by taking the supremum in (3.3) over the set of all $T^{(n)}$-measurable tests $\psi^{(n)}$ satisfying (3.4). If we denote by $H_T^{(n)}$ and $K_T^{(n)}$ the families of densities [on $(\mathscr{T}^{(n)}, \mathscr{B}^{(n)})$] induced from $H^{(n)}$ and $K^{(n)}$ by $T^{(n)}$, then clearly $\beta_T(\alpha, H^{(n)}, K^{(n)}) = \beta(\alpha, H_T^{(n)}, K_T^{(n)})$. Obviously, we also have

$$\beta(\alpha, H^{(n)}, K^{(n)}) \geq \beta_T(\alpha, H^{(n)}, K^{(n)}); \tag{3.5}$$

if, however, an equality in (3.5) is achieved for every $\alpha \in (0,1)$ [this happens, e.g., if $T^{(n)}$ is sufficient for the structure $(\mathscr{X}^{(n)}, \mathscr{A}^{(n)}, H^{(n)} \cup K^{(n)})$], then $T^{(n)}$ can be considered, in some sense, as *sufficient for the problem of testing* $H^{(n)}$ *against* $K^{(n)}$. Accordingly, [Hájek and Šidák (1967), Chapter 7], we say that $T^{(n)}$ *is asymptotically sufficient for the problem of testing* $H^{(n)}$ *against* $K^{(n)}$ if

$$\lim_{n \to \infty} \left[\beta(\alpha, H^{(n)}, K^{(n)}) - \beta_T(\alpha, H^{(n)}, K^{(n)})\right] = 0, \qquad \alpha \in (0,1). \tag{3.6}$$

An important consequence is that if $T^{(n)}$ is asymptotically sufficient for $H^{(n)}$ against $K^{(n)}$, then any asymptotically maximin most powerful test $\psi^{(n)*}$ for $H_T^{(n)}$ against $K_T^{(n)}$ is also asymptotically maximin most powerful for $H^{(n)}$ against $K^{(n)}$.

Let us consider now the problem of testing $H^{(n)}(\mathbf{A}, \mathbf{B}; \cdot)$ against *unspecified* ARMA alternatives. Denote by

$$K^{(n)}(\mathbf{A}, \mathbf{B}; \mathbb{R}^{p_2}, \mathbb{R}^{q_2}; f) = \bigcup \{K^{(n)}(\mathbf{A}, \mathbf{B}; \boldsymbol{\gamma}, \boldsymbol{\delta}; f) | \boldsymbol{\gamma} \in \mathbb{R}^{p_2}, \boldsymbol{\delta} \in \mathbb{R}^{q_2}\}, \tag{3.7}$$

the alternative under which the observed series $\mathbf{X}^{(n)}$ is generated by some ARMA model [(1.8) and (1.9)] where $\boldsymbol{\gamma}$ and $\boldsymbol{\delta}$ take on unspecified values, although being of specified dimensions p_2 and q_2. Let also $d \in \mathbb{R}^+$, and denote by $\mathscr{D}(d)$ the set of values of $(\boldsymbol{\gamma}, \boldsymbol{\delta}) \in \mathbb{R}^{p_2} \times \mathbb{R}^{q_2}$ such that $\|\mathbf{a} + \mathbf{b}\| \geq d$, with $\mathbf{a}$ and $\mathbf{b}$ given in (3.2). In order to obtain asymptotically maximin most powerful tests, we shall need to consider subhypotheses of the form

$$(3.8) \quad K^{(n)}(\mathbf{A},\mathbf{B};\mathbb{R}^{p_2},\mathbb{R}^{q_2}; f|d) = \bigcup\{K^{(n)}(\mathbf{A},\mathbf{B};\boldsymbol{\gamma},\boldsymbol{\delta}; f)|(\boldsymbol{\gamma},\boldsymbol{\delta}) \in \mathscr{D}(d)\}.$$

We then have the following sufficiency result.

PROPOSITION 3.2. *Let* $\{\Psi_t^{(1)},\ldots,\Psi_t^{(p_1+q_1)})$ *denote an arbitrary fundamental system of solutions of the homogeneous equation* $A(L)B(L)\Psi_t = 0$, $t \in \mathbb{Z}$. *Writing* π *for* $\max(p_2 - p_1, q_2 - q_1)$, *consider the* $\max(p_1 + q_2, p_2 + q_1)$*-tuple of rank statistics* [*assume* $n \geq \max(p_2 - p_1 + 3, q_2 - q_1 + 3, p_1 + q_1 + 2)$]

$$(3.9) \qquad \mathbf{T}_{\Psi;f}^{(n)} = \left(r_{1;f}^{(n)},\ldots,r_{\pi;f}^{(n)}, \sum_{i=\pi+1}^{n-1} \Psi_i^{(1)} r_{i;f}^{(n)},\ldots, \sum_{i=\pi+1}^{n-1} \Psi_i^{(p_1+q_1)} r_{i;f}^{(n)}\right)'.$$

Then, for any $d \in \mathbb{R}^+$, $\mathbf{T}_{\Psi;f}^{(n)}$ *is asymptotically sufficient for the problem of testing* $H^{(n)}(\mathbf{A},\mathbf{B};\cdot)$ *against* (3.8).

The proof (see Appendix 3) consists in establishing that the sufficient conditions for asymptotic sufficiency of Hájek and Šidák's (1967) Theorem 7.1.1 are satisfied.

3.3. *Asymptotic distribution of* $\mathbf{T}_{\Psi;f}^{(n)}$. We conclude this section by establishing the asymptotic multinormality of any asymptotically sufficient vector $\mathbf{T}_{\Psi;f}^{(n)}$. The proof of the following proposition essentially follows along the same lines as in Proposition 2.4 and is only briefly sketched in Appendix 3.

PROPOSITION 3.3. $n^{1/2}\mathbf{T}_{\Psi;f}^{(n)}$ *is asymptotically normal, with mean* $\mathbf{0}$ *under* $H^{(n)}(\mathbf{A},\mathbf{B};\cdot)$, *mean* $((a_1 + b_1),\ldots, (a_\pi + b_\pi), \sum_{i=\pi+1}^{\infty}\Psi_i^{(1)}(a_i + b_i), \ldots, \sum_{i=\pi+1}^{\infty}\Psi_i^{(p_1+q_1)}(a_i + b_i))'[\sigma^2 I(f)]^{1/2}$ *under* $K^{(n)}(\mathbf{A},\mathbf{B};\boldsymbol{\gamma},\boldsymbol{\delta}; f)$ *and full-rank covariance matrix*

$$\underline{\mathbf{W}}_{\Psi}^2 = \begin{pmatrix} \mathbf{I}_{\pi\times\pi} & \mathbf{0} \\ \mathbf{0} & \mathbf{W}_{\Psi}^2 \end{pmatrix}$$

under both where $\mathbf{W}_{\Psi}^2 = (W_{\Psi;kl})$ *and* $W_{\Psi;kl} = \sum_{i=\pi+1}^{\infty}\Psi_i^{(k)}\Psi_i^{(l)} < \infty$, $k, l = 1,\ldots, p_1 + q_1$.

4. Testing an ARMA model against another ARMA model: Asymptotically maximin most powerful tests.

4.1. *Generalized quadratic serial rankstatistics. Quadratic serial rank statistics*, providing rank-based versions of Box and Pierce's (1970) portmanteau statistic were introduced in Hallin, Ingenbleek and Puri (1987). Since however the asymptotically most powerful test for $H^{(n)}(\mathbf{A}, \mathbf{B}; \cdot)$ against $K^{(n)}(\mathbf{A}, \mathbf{B}; \gamma, \delta; f)$ cannot, in general, be found within the class of linear serial rank tests, quadratic serial rank statistics cannot be expected to provide optimal tests against unspecified values of γ and δ (unless, e.g., $p_1 = 0 = q_1$), and a more general type of quadratic rank statistics has to be considered: generalized quadratic serial rank statistics.

Let $\mathbf{S}^{(n)} - \mathbf{m}^{(n)} = (S_1^{(n)} - m_1^{(n)}, \ldots, S_l^{(n)} - m_l^{(n)})'$ be a column vector of generalized linear serial rank statistics centered about their means [under $H^{(n)}(\mathbf{A}, \mathbf{B}; \cdot)$]. Assuming that $n^{1/2}(\mathbf{S}^{(n)} - \mathbf{m}^{(n)})$ has an asymptotic covariance matrix $\mathbf{V}^2$ of full rank, we define a generalized quadratic serial rank statistic as a statistic of the form $Q^{(n)} = n(\mathbf{S}^{(n)} - \mathbf{m}^{(n)})'\mathbf{V}^{-2}(\mathbf{S}^{(n)} - \mathbf{m}^{(n)})$.

Since we have shown that asymptotically sufficient statistics $\mathbf{T}_{\Psi;f}^{(n)}$ exist for testing $H^{(n)}(\mathbf{A}, \mathbf{B}; \cdot)$ against $K^{(n)}(\mathbf{A}, \mathbf{B}; \gamma, \delta; f)$, it is intuitively quite natural to consider generalized quadratic serial rank statistics of the form

$$Q^{(n)*}_{A,B;f} = n\mathbf{T}_{\Psi;f}^{(n)\prime}\mathbf{W}_{\Psi}^{-2}\mathbf{T}_{\Psi;f}^{(n)}. \tag{4.1}$$

Actually, we shall establish in Section 4.2 that $Q^{(n)*}_{A,B;f}$ provides an asymptotically maximin most powerful test for our problem. We first justify the notation by establishing the important property that $Q^{(n)*}_{A,B;f}$, unlike $\mathbf{T}_{\Psi;f}^{(n)}$, does not depend on the particular fundamental system $\{\Psi_t^{(i)}\}$ adopted.

PROPOSITION 4.1. *The quadratic statistic $Q^{(n)*}_{A,B;f}$ does not depend on the fundamental system $\{\Psi_t^{(i)}\}$.*

For the proof, see Appendix 3.

Let us illustrate Proposition 4.1 by means of two examples.

EXAMPLE 4.1. Denote by $\lambda_{(1)} \cdots \lambda_{(p_1+q_1)}$ the roots of $A(z^{-1})B(z^{-1}) = 0$, $z \in \mathbb{C}$, and assume that they all are distinct (hence of multiplicity 1—also recall that they all lie inside the unit circle). Then a fundamental system of solutions is provided by $\{\Psi_t^{(j)} = \lambda_j^{t-(\pi+1)}, t \in \mathbb{Z};\ j = 1, \ldots, p_1 + q_1\}$. $Q^{(n)*}_{A,B}$ is thus a quadratic form involving the rank autocorrelations of orders 1 through π and geometrically weighted sums of rank autocorrelations of orders $\pi + 1$ through $n - 1$. $\mathbf{W}_{\Psi}^2$ is then given by $(W_{\Psi;kl}) = ((1 - \lambda_k\lambda_l)^{-1})$.

EXAMPLE 4.2. Denote by $\mathscr{G}_u = \mathscr{G}(t, t-u)$ the Green's functions associated with the difference operator $A(L)B(L)$. Then (see Appendix 1) another fundamental system of solutions is $\{\Psi_t^{(j)} = \mathscr{G}_{t-\pi-j},\ t \in \mathbb{Z};\ j = 1, \ldots, p_1 + q_1\}$. Recalling that $\mathscr{G}_{-1} = \mathscr{G}_{-2} = \cdots = \mathscr{G}_{-p_1-q_1+1} = 0$, we obtain $W_{\Psi;kl} = \sum_{i=0}^{\infty}\mathscr{G}_{-k-1+i}\mathscr{G}_{-l-1+i}$, hence

$$\mathbf{W}_{\Psi}^2 = \begin{pmatrix} \gamma^{(0)}\gamma^{(1)} & \cdots & \gamma^{(p_1+q_1-1)} \\ \gamma^{(1)} & \ddots & \vdots \\ \vdots & \ddots & \gamma^{(1)} \\ \gamma^{(p_1+q_1-1)} & \cdots & \gamma^{(1)}\gamma^{(0)} \end{pmatrix},$$

where we denote by $\gamma^{(0)}, \ldots, \gamma^{(p_1+q_1-1)}$ the first $p_1 + q_1 - 1$ autocovariances of the AR($p_1 + q_1$) process $\{\xi_t;\ t \in \mathbb{Z}\}$ characterized by the AR($p_1 + q_1$) model

$$A(L)B(L)\xi_t = \eta_t, \qquad t \in \mathbb{Z} \tag{4.2}$$

[(η_t; $t \in \mathbb{Z}$) a second-order standardized white noise]. This matrix $\mathbf{W}_\Psi^2$, which has also the nature of a *Casorati matrix* (cf. Appendix 1), is the one that appears in the usual Yule–Walker equations associated with (4.2).

We now give the asymptotic distribution of $Q_{A,B;f}^{(n)*}$.

Proposition 4.2. *Under $H^{(n)}(\mathbf{A},\mathbf{B};\cdot)$, $Q_{A,B;f}^{(n)*}$ is asymptotically χ^2 with* $\max(p_1 + q_2, p_2 + q_1)$ *degrees of freedom. Under $K^{(n)}(\mathbf{A},\mathbf{B};\boldsymbol{\gamma},\boldsymbol{\delta};f)$, it is asymptotically noncentral χ^2, with the same degrees of freedom and with noncentrality parameter*

$$\lambda_f^*(\boldsymbol{\gamma},\boldsymbol{\delta}) = \tfrac{1}{2}\|\mathbf{a}+\mathbf{b}\|^2\sigma^2 I(f), \tag{4.3}$$

where **a** *and* **b** *are given by* (3.2).

For the proof, see Appendix 3.

4.2. *Asymptotically maximin most powerful tests for testing an ARMA model against unspecified ARMA alternatives.* Let us define an asymptotically maximin most powerful test for $H^{(n)}(\mathbf{A},\mathbf{B};\cdot)$ against an unspecified ARMA(p_2, q_2) alternative as a test which is asymptotically maximin most powerful (among all tests of the same level) against any alternative of the form $K^{(n)}(\mathbf{A},\mathbf{B};\mathbb{R}^{p_2},\mathbb{R}^{q_2};f|d)$ [cf. (3.8)], $d \in \mathbb{R}^+$, f specified up to a scale parameter. We then have the following result.

Proposition 4.3. *An asymptotically maximin most powerful test for $H^{(n)}(\mathbf{A},\mathbf{B};\cdot)$ against an unspecified ARMA(p_2, q_2) alternative is provided by*

$$\psi_\alpha^{(n)*}(\mathbf{A},\mathbf{B};p_2,q_2;f) = 1 \quad \textit{if} \quad Q_{A,B;f}^{(n)*} > \chi^2_{\max(p_1+q_2,\,p_2+q_1);\,1-\alpha}. \tag{4.4}$$

Accordingly, the envelope power function

$$\beta\big(\alpha, H^{(n)}(\mathbf{A},\mathbf{B};\cdot), K^{(n)}(\mathbf{A},\mathbf{B};\mathbb{R}^{p_2},\mathbb{R}^{q_2};f|d)\big)$$

converges, as $n \to \infty$, to $1 - F_{\max(p_1+q_2,\,p_2+q_1)}(\chi^2_{\max(p_1+q_2,\,p_2+q_1);\,1-\alpha};\ \tfrac{1}{2}d^2\sigma^2 I(f))$, *where $\chi^2_{m;1-\alpha}$ denotes the* $(1-\alpha)$*-quantile of the χ^2 distribution with m degrees of freedom and $F_m(\cdot;\lambda)$ the distribution function of the noncentral χ^2 with m degrees of freedom and noncentrality parameter λ.*

Corollary 4.4. *The rank portmanteau statistic $n\sum_{i=1}^{p}(r_{i;f}^{(n)})^2$* (*cf.* *Hallin, Ingenbleek and Puri,* 1987) *is asymptotically maximin most powerful against unspecified ARMA(p_2, q_2) alternatives if and only if $p_1 = q_1 = 0$ and $p = \max(p_2, q_2)$.*

For the proof, see Appendix 3.

REMARK 4.1. Denoting by $Q_f^{(n)} = n\Sigma_{i=1}^{\pi}(r_{i;f}^{(n)})^2$ the rank portmanteau statistic of order $\pi = \max(p_2 - p_1, q_2 - q_1)$, maximin-optimal quadratic statistics $Q_{A,B;f}^{(n)*}$ can be decomposed into $Q_{A,B;f}^{(n)*} = Q_f^{(n)} + Q_{A,B;f}^{(n)}$. Because $Q_f^{(n)}$ only depends on $r_{1;f}^{(n)}, \ldots, r_{\pi;f}^{(n)}$ whereas $Q_{A,B;f}^{(n)}$ only depends on $r_{\pi+1;f}^{(n)}, r_{\pi+2;f}^{(n)}, \ldots$, $Q_f^{(n)}$ and $Q_{A,B;f}^{(n)}$ are asymptotically independent χ^2-distributed statistics [under $H^{(n)}(\mathbf{A}, \mathbf{B}; \cdot)$]. $Q_f^{(n)}$ contributes for $\max(p_2 - p_1, q_2 - q_1)$ degrees of freedom, the maximal number of possible higher-order terms (autoregressive terms of orders larger than p_1 and/or moving average terms of orders larger than q_1). $Q_{A,B;f}^{(n)}$ has the form of a *weighted rank portmanteau statistic* and contributes for $(p_1 + q_1)$ degrees of freedom, one for each tested coefficient $A_1, \ldots, A_{p_1}$, $B_1, \ldots, B_{q_1}$. The consideration of this decomposition in case $H^{(n)}(\mathbf{A}, \mathbf{B}; \cdot)$ has been rejected, may provide a useful insight into the reasons why rejection occurred, and thus suggest alternative model specifications.

REMARK 4.2. The noncentrality parameter $\lambda_f^*(\gamma, \delta)$ of the asymptotic distribution of $Q_{A,B;f}^{(n)*}$ under alternatives of the form $K^{(n)}(A, B; \gamma, \delta; f)$ also can be decomposed into $\frac{1}{2}\Sigma_{i=1}^{\pi}(a_i + b_i)^2\sigma^2 I(f)$ (contribution of the "unweighted part" $Q_f^{(n)}$) and $\frac{1}{2}\Sigma_{i=\pi+1}^{\infty}(a_i + b_i)^2\sigma^2 I(f)$ (contribution of the "weighted part" $Q_{A,B;f}^{(n)}$). A study of the relation between these two quantities and the coefficients of the tested model (1.7) provides an interesting insight into the importance of the respective contributions of the "weighted" and "unweighted" parts to the asymptotic power of (4.4). As a general rule, the contribution of the "unweighted part" does not depend on $A(L)$ and $B(L)$, whereas the contribution of the "weighted part" is an increasing function of the "closeness to the unit circle" of the roots of $A(z)B(z) = 0$, $z \in \mathbb{C}$. To show this, let us consider a simple example.

EXAMPLE 4.3. Consider the problem of testing the AR(1) model $X_t - \rho X_{t-1} = \varepsilon_t$, $t \in \mathbb{Z}$, where $|\rho| < 1$, against ARMA(2, 1) alternatives. Here $\pi = 1$ and

$$a_i = \begin{cases} \gamma_1, & i = 1, \\ \gamma_1\rho^{i-1} + \gamma_2\rho^{i-2}, & i \geq 2, \end{cases}$$

$$b_i = \begin{cases} \delta_1, & i = 1, \\ 0, & i \geq 2, \end{cases}$$

if an alternative of the form $K^{(n)}(\mathbf{A}, \mathbf{B}; \boldsymbol{\gamma} = (\gamma_1, \gamma_2), \boldsymbol{\delta} = (\delta_1, 0); f)$ is considered. The noncentrality parameter (4.3) accordingly decomposes into $(\gamma_1 + \delta_1)^2$ (contribution of the "unweighted" $Q_f^{(n)}$), which does not depend on ρ, and $\Sigma_{i=2}^{\infty}(\gamma_1\rho^{i-1} + \gamma_2\rho^{i-2})^2 = (\gamma_1\rho + \gamma_2)^2(1/(1 - \rho^2))$ (contribution of the "weighted" $Q_{A,B;f}^{(n)}$). Clearly, this latter contribution is approximately γ_2^2 if $|\rho| \simeq 0$—and the optimal test (4.4) is then approximately equivalent to the test based on an "unweighted" rank portmanteau statistic of order 2; if $|\rho| \simeq 1$, then the contribution of $Q_{A,B;f}^{(n)}$ can be arbitrarily large.

The absence of the weighted part $Q_{A,B;f}^{(n)}$ in the classical portmanteau statistic is possibly responsible for the somewhat disappointing performance of the Box–Pierce test (see also Remark 4.3).

Remark 4.3. Corollary 4.4 characterizes the only case [viz. $A(L) = I = B(L)$] where an unweighted rank portmanteau statistic $Q_f^{(n)}$ is asymptotically maximin most powerful against the whole unspecified ARMA(p_2, q_2) alternative. $Q_f^{(n)}$ can be shown however to provide maximin most powerful tests for more general model operators $A(L)$ and $B(L)$ if this alternative is restricted in an adequate manner. It can be shown, for example, that the unweighted rank portmanteau statistic of order π is asymptotically maximin most powerful for testing $H^{(n)}(\mathbf{A}, \mathbf{B}; \cdot)$ against the set of all ARMA(p_2, q_2) models of the form $A(L)a^{(n)}(L)X_t = B(L)b^{(n)}(L)\varepsilon_t$ considered in (1.11), where $a^{(n)}(L) = 1 - n^{-1/2}(a_1 L + \cdots + a_\pi L^\pi)$ and $b^{(n)}(L) = 1 + n^{-1/2}(b_1 L + \cdots + b_\pi L^\pi)$ are unspecified, i.e., against alternative models which differ from the tested one only because of the existence of π additional roots to the characteristic equations.

This latter type of alternative generalizes the ones considered for purely autoregressive processes with Gaussian generating white noise by Hosking (1978), who shows that in this particular case the parametric Box–Pierce portmanteau test asymptotically coincides with the likelihood ratio test.

Remark 4.4. $Q_{A,B;f}^{(n)*}$ depends on p_2 and q_2 only through $\Delta = \max(p_2 - p_1, q_2 - q_1)$; (4.4) is thus actually asymptotically maximin most powerful against the whole unspecified ARMA$(p_1 + \Delta, q_1 + \Delta)$ alternative.

Remark 4.5. $Q_{A,B;f}^{(n)*}$ depends on the operators $A(L)$ and $B(L)$ only through the symbolic product $A(L)B(L)$, and thus takes the same form whether the tested model is $A(L)X_t = B(L)\varepsilon_t$ or $B(L)X_t = A(L)\varepsilon_t$ or $A(L)B(L)X_t = \varepsilon_t$, etc.; but the filtered series $\mathbf{Z}^{(n)}$ from which $Q_{A,B;f}^{(n)*}$ has to be computed is not the same.

Remark 4.6. The coefficients appearing in the "weighted part" $Q_{A,B;f}^{(n)}$ of $Q_{A,B;f}^{(n)*}$ (see Remark 4.1) are computed here from the asymptotic covariance matrix $\mathbf{W}^2$ of $\mathbf{T}_{\Psi;f}^{(n)}$. For small sample sizes, an exact orthogonalizing matrix could possibly be introduced.

Example 4.4. Consider the problem of testing the null hypothesis that an observed series $\mathbf{X}^{(n)}$ was generated by the ARMA(1, 1) model $X_t - \frac{4}{5}X_{t-1} = \varepsilon_t + \frac{1}{2}\varepsilon_{t-1}$, $t \in \mathbb{Z}$, against unspecified ARMA(2, 2) alternatives [or, equivalently, against ARMA(2, 1) or ARMA(1, 2) alternatives; cf. Remark 4.4]. Let $Z_{-1}^{(n)} = Z_0^{(n)} = 0$ and $Z_t^{(n)} = X_t^{(n)} - \frac{4}{5}X_{t-1}^{(n)} - \frac{1}{2}Z_{t-1}^{(n)}$, $t = 1, \ldots, n$; denote by $r_{i;f}^{(n)}$ the corresponding f-rank autocorrelations. A fundamental system of solutions of $(1 - \frac{4}{5}L)(1 + \frac{1}{2}L)\Psi_t = 0$, $t \in \mathbb{Z}$, is $\{\Psi_t^{(1)} = (\frac{4}{5})^{t-2}; \Psi_t^{(2)} = (-\frac{1}{2})^{t-2}\}$ and $\pi = \max(p_2 - p_1, q_2 - q_1) = 1$. The covariance matrix $\underline{\mathbf{W}}^2$ (see Example 4.1) here is of the form

$$\underline{\mathbf{W}}^2 = \begin{pmatrix} \left[1 - \left(\frac{4}{5}\right)^2\right]^{-1} & \left[1 + \frac{2}{5}\right]^{-1} \\ \left[1 + \frac{2}{5}\right]^{-1} & \left[1 - \left(\frac{1}{2}\right)^2\right]^{-1} \end{pmatrix} = \begin{pmatrix} \frac{25}{9} & \frac{5}{7} \\ \frac{5}{7} & \frac{4}{3} \end{pmatrix}.$$

The statistic $Q^{(n)*}_{A,B;f}$ is thus

$$
\begin{aligned}
Q^{(n)*}_{A,B;f} &= n\left(r^{(n)}_{1;f}\right)^2 + n\left(\sum_{i=2}^{n-1}\left(\tfrac{4}{5}\right)^{i-2} r^{(n)}_{i;f},\ \sum_{i=2}^{n-1}\left(-\tfrac{1}{2}\right)^{i-2} r^{(n)}_{i;f}\right) \\
&\quad\times \begin{pmatrix} \frac{25}{9} & \frac{5}{7} \\ \frac{5}{7} & \frac{4}{3} \end{pmatrix}^{-1} \begin{pmatrix} \sum_{i=2}^{n-1}\left(\frac{4}{5}\right)^{i-2} r^{(n)}_{i;f} \\ \sum_{i=2}^{n-1}\left(-\frac{1}{2}\right)^{i-2} r^{(n)}_{i;f} \end{pmatrix} \\
&\simeq n\left(r^{(n)}_{1;f}\right)^2 + (1.436)n\left[\sum_{i=2}^{n-1}\left(\tfrac{4}{5}\right)^{i} r^{(n)}_{i;f}\right]^2 + (19.600)n\left[\sum_{i=2}^{n-1}\left(-\tfrac{1}{2}\right)^{i} r^{(n)}_{i;f}\right]^2 \\
&\quad - (3.937)n\left[\sum_{i=2}^{n-1}\left(\tfrac{4}{5}\right)^{i} r^{(n)}_{i;f}\right]\left[\sum_{i=2}^{n-1}\left(-\tfrac{1}{2}\right)^{i} r^{(n)}_{i;f}\right].
\end{aligned}
\tag{4.5}
$$

If ARMA(3, 3) [or ARMA(3, 2), ARMA(3, 1), ARMA(1, 3), ...] alternatives are to be considered, (4.5) has to be modified to

$$
\begin{aligned}
& n\left(r^{(n)}_{1;f}\right)^2 + n\left(r^{(n)}_{2;f}\right)^2 + (2.243)n\left[\sum_{i=3}^{n-1}\left(\tfrac{4}{5}\right)^{i} r^{(n)}_{i;f}\right]^2 + (78.400)n\left[\sum_{i=3}^{n-1}\left(-\tfrac{1}{2}\right)^{i} r^{(n)}_{i;f}\right]^2 \\
&\quad + (9.844)n\left[\sum_{i=3}^{n-1}\left(\tfrac{4}{5}\right)^{i} r^{(n)}_{i;f}\right]\left[\sum_{i=3}^{n-1}\left(-\tfrac{1}{2}\right)^{i} r^{(n)}_{i;f}\right].
\end{aligned}
$$

5. Comparison with existing procedures: Concluding remarks.

5.1. *A brief overview of some existing procedures.* Tests of time series models adequacy have been extensively studied, and a general survey on the literature on this subject is certainly not possible here. A common feature of the most frequently used procedures is that they do not pay much attention either to the particular model to be tested or to any specific type of alternative. The typical approach consists indeed in computing residuals from the model to be tested, then applying to those residuals some existing test for white noise. Classical examples are the *Durbin–Watson test*, its higher-order generalizations [Evans and King (1985)] or the popular *Ljung–Box–Pierce portmanteau test* [Box and Pierce (1970); Ljung and Box (1978)].

More recently, attempts have been made to bring some optimality considerations into the problem. Several authors [see, e.g., Hosking (1978)] have characterized some restricted alternatives against which existing procedures—such as Quenouille's or Box and Pierce's—achieve optimality. A more direct approach has been undertaken by some others, who derive optimal tests for some specific situations [mostly, Gaussian *Lagrange multiplier tests*—see Godfrey (1979), Hosking (1980), Poskitt and Tremayne (1980); multivariate results are also available: Hosking (1981) and Poskitt and Tremayne (1982)]. A complete result for Gaussian AR(1) series has been obtained in Dufour and King (1986), where locally best invariant tests are shown to rely on exponentially weighted sums of sample autocorrelations—a test statistic which is a parametric first-order Gaussian version of our asymptotically sufficient statistic (3.9).

Whereas existing parametric procedures thus either belong to the category of pure significance tests or address restricted classes of Gaussian alternatives, our weighted rank portmanteau tests are providing a locally asymptotically optimal solution to the problem of testing for time series models adequacy in its most general form. Moreover, the proposed solution enjoys all the additional advantages of nonparametric procedures: it is robust and distribution-free.

5.2. *Efficiency comparisons.* A quantitative comparison of the asymptotic performances of our nonparametric tests with those of existing parametric ones can be achieved through the computation of asymptotic relative efficiencies, whereas small-sample comparisons require extensive Monte Carlo experiments. We therefore limit ourselves to asymptotic considerations; a study of small-sample situations is the purpose of an ongoing research [see Hallin and Mélard (1987) for some results]. Two particular problems, for which an "optimal" parametric Gaussian procedure is available, are considered in some detail: (a) testing the coefficient of an AR(1) model and (b) testing for additional AR or MA roots. For a comparative discussion of weighted rank-based and classical portmanteau tests, we also refer to the end of Section 4.

(a) *Testing the coefficient of an AR*(1) *model.* Seven procedures are considered for testing $H_0^{(n)}$: $\rho = \rho_0$, $|\rho_0| < 1$, against $H_1^{(n)}$: $\rho \neq \rho_0$ in the AR(1) model $X_t - \rho X_{t-1} = \varepsilon_t$. All correlation coefficients are computed from the residuals $Z_t^{(n)} = X_t^{(n)} - \rho_0 X_{t-1}^{(n)}$ or from their ranks $R_t^{(n)}$. For the sake of convenience, all of the following tests have been given an asymptotic 1 degree of freedom chi-square form. The noncentrality parameters (5.2), (5.4), (5.6), (5.8) and (5.10) are those of the asymptotic noncentral chi-square distributions of the test statistics under the local sequence of alternatives associated with $X_t - (\rho_0 + n^{-1/2}d)X_{t-1} = \varepsilon_t$, where ε_t has density $f(\cdot)$. $\chi^2_{1-\alpha}$ stands for the $(1-\alpha)$-quantile of the chi-square distribution with 1 degree of freedom.

(i) Two-sided first-order sample autocorrelation test, two-sided Durbin–Watson test, von Neumann's ratio test, Ljung–Box–Pierce test of order 1 (all these tests are asymptotically equivalent):

$$\text{reject } H_0^{(n)} \text{ if } \quad (n-1)\left(r_1^{(n)}\right)^2 > \chi^2_{1-\alpha}, \tag{5.1}$$

$$\text{noncentrality parameter:} \quad d^2/2. \tag{5.2}$$

(ii) Two-sided Dufour and King test: reject $H_0^{(n)}$ if

$$(1-\rho_0^2)\left[\sum_{i=1}^{n-1}(n-i)^{1/2}\rho_0^{i-1}r_i^{(n)}\right]^2 \Big/ \left(1-\rho_0^{2(n-1)}\right) > \chi^2_{1-\alpha}, \tag{5.3}$$

$$\text{noncentrality parameter:} \quad d^2/2(1-\rho_0^2). \tag{5.4}$$

(iii)–(v) van der Waerden, Wilcoxon and Laplace (weighted) rank portmanteau tests: reject $H_0^{(n)}$ if

$$(5.5) \qquad (1-\rho_0^2)\left[\sum_{i=1}^{n-1}(n-i)^{1/2}\rho_0^{i-1}r_{i;g}^{(n)}\right]^2 \Big/ \left(1-\rho_0^{2(n-1)}\right) > \chi_{1-\alpha}^2,$$

$$(5.6) \qquad \text{noncentrality parameter:} \quad d^2\left[\int_0^1 \phi_g(G^{-1}(u))\phi_f(F^{-1}(u))\,du \times \int_0^1 G^{-1}(v)F^{-1}(v)\,dv\right]^2 \Big/ 2(1-\rho_0^2)I(g);$$

here g, G, ϕ_g and $I(g)$ denote the standard normal, logistic and double exponential densities, distribution functions, scores and Fisher information, respectively.

(vi) Spearman (weighted) rank portmanteau test: reject $H_0^{(n)}$ if

$$(5.7) \qquad (1-\rho_0^2)\left[\sum_{i=1}^{n-1}(n-i)^{1/2}\rho_0^{i-1}\tilde{r}_i^{(n)}\right]^2 \Big/ \left(1-\rho_0^{2(n-1)}\right) > \chi_{1-\alpha}^2,$$

$$(5.8) \qquad \text{noncentrality parameter:} \quad 144d^2\left[\int_0^1 u\phi(F^{-1}(u))\,du \times \int_0^1 vF^{-1}(v)\,dv\right]^2 / 2(1-\rho_0^2);$$

$\tilde{r}_i^{(n)}$ stands here [and in (5.9)] for the Spearman–Wald–Wolfowitz rank autocorrelation coefficient; see Hallin and Mélard (1987), Hallin and Puri (1987) and Dufour and Hallin (1987).

(vii) Bartels' (1982) two-sided rank version of von Neumann's ratio test:

$$(5.9) \qquad \text{reject } H_0^{(n)} \text{ if } \quad (n-1)\left(\tilde{r}_1^{(n)}\right)^2 > \chi_{1-\alpha}^2,$$

$$(5.10) \qquad \text{noncentrality parameter: } 144d^2\left[\int_0^1 u\phi(F^{-1}(u))\,du \int_0^1 vF^{-1}(v)\,dv\right]^2 \Big/ 2.$$

The *asymptotic relative efficiencies* (ARE) of all these tests with respect to each other can be obtained, under various densities f, as the ratios of their respective noncentrality parameters. Table 1 provides some numerical values.

(b) *Testing against additional AR or MA roots.* The problem considered here is a generalization of the one studied in Hosking (1978); see Remark 4.3 for description. Five procedures are compared here: the *Ljung–Box–Pierce parametric portmanteau test* (of order π if π additional roots are to be tested) which is also, asymptotically, the *Gaussian likelihood ratio test* and four *rank-based portmanteau tests* (of order π; in this particular case, they are of the *unweighted* type). According to Proposition 4.3 (and using the same notation), their asymptotic powers, against contiguous alternatives of the form considered in Remark 4.3 and such that $\Sigma_{i=1}^{\pi}(a_i+b_i)^2 = d^2$, are of the form $1 - F_\pi(\chi_{\pi;1-\alpha}^2; \lambda^2)$, where λ^2 is obtained by letting $\rho_0 = 0$ in the noncentrality parameters (5.4) (for the

TABLE 1

Asymptotic relative efficiencies of various parametric and nonparametric test procedures for problems (a) and (b) (denoted by superscripts (a) and (b), respectively) under various density types. For each density type, the ARE's are computed with respect to the locally asymptotically optimal test(s) (ARE value of 1).

	Parametric		Nonparametric				
Density type f	$r_1^{(n)}$ —	Dufour–King[a] Ljung–Box–Pierce[b]	van der Waerden[a] van der Waerden[b]	Wilcoxon[a] Wilcoxon[b]	Laplace[a] Laplace[b]	Spearman[a] Spearman[b]	Bartels[a] —
Gaussian	$(1-\rho_0^2)$	1	1	0.948	0.613	0.912	$0.912(1-\rho_0^2)$
Logistic	$0.912(1-\rho_0^2)$	0.912	0.954	1	0.741	0.912	$0.912(1-\rho_0^2)$
Double exponential	$0.500(1-\rho_0^2)$	0.500	0.613	0.741	1	0.613	$0.613(1-\rho_0^2)$

Ljung–Box–Pierce portmanteau test), (5.6) (for the van der Waerden, Wilcoxon and Laplace unweighted portmanteau tests) and (5.8) (for the Spearman unweighted portmanteau test), respectively.

Some numerical values are provided in Table 1.

An inspection of Table 1 reveals the excellent asymptotic performances of rank portmanteau tests: the van der Waerden tests perform uniformly strictly better than the optimal normal-theory tests [viz. the Dufour–King test in problem (a) and the usual Ljung–Box–Pierce portmanteau test in problem (b)] —except of course in the case of Gaussian series, where they perform equally well. The ARE of optimal normal-theory tests with respect to Laplace portmanteau tests can be as low as $\frac{1}{2}$. Note that the Wilcoxon portmanteau tests uniformly dominate the corresponding Spearman–Wald–Wolfowitz procedures. As for the Durbin–Watson and von Neumann ratio tests, their relative performances in problem (a) with respect to optimal rank-based procedures can be arbitrarily bad as $|\rho_0| \to 1$.

APPENDIX 1

Difference operators and difference equations. ARMA models characterize stochastic processes as solutions of *stochastic difference equations*. We briefly review here some basic results on difference operators and difference equations that are used throughout the present paper. These results are univariate constant-coefficient versions of more general results given in Hallin (1986) [see also Miller (1968)].

Denote by $\Theta_1, \dots, \Theta_p$ a p-tuple of real coefficients with $\Theta_p \neq 0$ and by L the *lag* or *backshift operator* ($Lz_t = z_{t-1}$). These coefficients characterize a *linear difference operator of order* p: $\Theta(L) = 1 + \Sigma_{i=1}^{p}\Theta_i L^i$. Such an operator defines *difference equations* $\Theta(L)\Psi_t = \omega_t$, $t \in \mathbb{Z}$, the solutions of which are completely determined by p successive "initial" values.

The Green's function G_t associated with the operator $\Theta(L)$ is the solution of the homogeneous equation $\Theta(L)G_t = 0$, $t \in \mathbb{Z}$, taking on initial values $G_0 = 1$, $G(-1) = \cdots = G(-p+1) = 0$.

It is easy to see that the set of all solutions of the homogeneous equation $\Theta(L)\Psi_t = 0$, $t \in \mathbb{Z}$, constitutes a vector space of dimension p. A basis of this vector space is called a *fundamental system of solutions*. Two fundamental systems are particularly convenient:

(a) Denote by $\lambda_{(1)}, \ldots, \lambda_{(p)}$ the roots of the polynomial $\Theta(z^{-1})$, i.e., the solutions of $1 + \Theta_1 z^{-1} + \cdots + \Theta_p z^{-p} = 0$, $z \in \mathbb{C}$. Assume that all these roots are of multiplicity 1. Then $\{\lambda^t_{(1)}, \ldots, \lambda^t_{(p)}\}$ constitutes a fundamental system of solutions.

(b) Another fundamental system is provided by the p-tuple of Green's functions $\{G_t, G_{t+1}, \ldots, G_{t+p-1}\}$; see Hallin [(1986), Theorem 1.1].

Using this latter fundamental system, the general solution of $\Theta(L)\Psi_t = \omega_t$ is

$$\Psi_t = \begin{cases} \sum_{u=0}^{t-1} G_u \omega_{t-u} + (G_t, G_{t+1}, \ldots, G_{t+p-1})\mathbf{C}^{-1}(\Psi_0, \Psi_{-1}, \ldots, \Psi_{-p+1})', & t \geq 1, \\ \sum_{u=t}^{-p} G_u \omega_{t-u} + (G_t, G_{t+1}, \ldots, G_{t+p-1})\mathbf{C}^{-1}(\Psi_0, \Psi_{-1}, \ldots, \Psi_{-p+1})', & t \leq -p, \end{cases}$$

where

$$\mathbf{C}^{-1} = \begin{pmatrix} G_0 & G_1 & \cdots & G_{p-1} \\ G_{-1} & G_0 & & G_{p-2} \\ \vdots & \vdots & \ddots & \vdots \\ G_{-p+1} & G_{-p+2} & \cdots & G_0 \end{pmatrix}^{-1} = \begin{pmatrix} 1 & \Theta_1 & \cdots & \Theta_{p-1} \\ 0 & \ddots & \ddots & \vdots \\ \vdots & \ddots & \ddots & \Theta_1 \\ 0 & \cdots & 0 & 1 \end{pmatrix}.$$

[$\mathbf{C}$ is the value for $t = 0$ of the *Casorati matrix* associated with the fundamental system $\{G_t, G_{t+1}, \ldots, G_{t+p-1}\}$.]

APPENDIX 2

Exact and approximate residuals. The exact residuals $\mathbf{Z}^{(n)} = (Z_1^{(n)}, \ldots, Z_n^{(n)})$, where $Z_t^{(n)} = [A(L)/B(L)]X_t^{(n)}$, $1 \leq t \leq n$, cannot be derived from the observed series $\mathbf{X}^{(n)}$ [unless $B(L) \equiv 1$]. From Appendix 1, we have indeed

$$\begin{aligned} Z_t^{(n)} = \sum_{u=0}^{t-1} H_u \left(1 - \sum_{j=1}^{p_1} A_j L^j\right) X_{t-u}^{(n)} \\ + \left(H_t, \ldots, H_{t-q_1+1}\right)\mathbf{C}^{-1}\left(Z_0^{(n)}, \ldots, Z_{-q_1+1}^{(n)}\right)', \quad t > 0, \end{aligned} \tag{A2.1}$$

where the values of $X_t^{(n)}$, $-p_1 + 1 \leq t \leq 0$ and $Z_t^{(n)}$, $-q_1 + 1 \leq t \leq 0$ are not available from the observations. Approximate residuals $\hat{Z}_t^{(n)}$ can be obtained by putting $X_0^{(n)} = X_{-1}^{(n)} = \cdots = X_{-p_1+1}^{(n)} = 0 = Z_0^{(n)} = Z_{-1}^{(n)} = \cdots = Z_{-q_1+1}^{(n)}$ in (A2.1) or, equivalently, by applying, from the previous initial values, the recursion formula $\hat{Z}_t^{(n)} = X_t^{(n)} - \sum_{j=1}^{p_1} A_j X_{t-j}^{(n)} - \sum_{i=1}^{q_1} B_i \hat{Z}_{t-i}^{(n)}$. More sophisticated devices yielding better small-sample properties such as *backforecasting* can also be used [see, e.g., Box and Jenkins (1970), pages 199–200], but the effect of the chosen initial values is asymptotically negligible. Under assumption (b) (Section 3.1), the difference $\Delta_t^{(n)} = \hat{Z}_t^{(n)} - Z_t^{(n)}$ is $O_p(\Lambda^t)$, as $t \to \infty$, where $\Lambda < 1$ is the modulus of the largest root of $A(z^{-1})B(z^{-1}) = 0$, $z \in \mathbf{C}$. Denote by $\hat{S}^{(n)*}$ the value of the optimal statistic (2.8) when the ranks $R_t^{(n)}$ of the exact residuals $Z_t^{(n)}$ are replaced by those $\hat{R}_t^{(n)}$ of the approximate residuals $\hat{Z}_t^{(n)}$. Our objective is to show that, as $n \to \infty$, $\hat{S}^{(n)*}$ and $S^{(n)*}$ are asymptotically equivalent [i.e., $n^{1/2}(\hat{S}^{(n)*} - S^{(n)*}) = o_p(1)$], so that the results of the paper are not affected if approximate residuals are used. Consider the problem of testing $H^{(n)}(\mathbf{A}, \mathbf{B}; f)$ against $K^{(n)}(\mathbf{A}, \mathbf{B}; \gamma, \boldsymbol{\delta}; f)$. Referring to the proof of Proposition 2.2 (A3.1), we have

$$\text{(A2.2)} \quad \begin{aligned} &\log L^{(n)}(\mathbf{Z}^{(n)}) - \log L^{(n)}(\hat{\mathbf{Z}}^{(n)}) \\ &\quad = \log E\left[\lambda^{(n)}(\mathbf{Z}^{(n)} + \Delta^{(n)}, \mathbf{X}_0^{(n)}) - \lambda^{(n)}(\mathbf{Z}^{(n)}, \mathbf{X}_0^{(n)}) | \mathbf{Z}^{(n)}\right]. \end{aligned}$$

Carrying on Taylor's expansion of $\lambda^{(n)}(\mathbf{Z}^{(n)} + \Delta^{(n)}, \mathbf{X}_0^{(n)})$ (with respect to the $\Delta_t^{(n)}$'s), it is easy to see that under $H^{(n)}(\mathbf{A}, \mathbf{B}; f)$, the difference (A2.2) is $o_p(1)$ as $n \to \infty$ [the arguments involved are essentially the same as in the proof of Proposition 2.2 with the additional fact that quantities of the form $a_i \Delta_{t-i}^{(n)}$ are $O_p(\Lambda^t)$]. Denote by $\hat{H}_f^{(n)}(\mathbf{A}, \mathbf{B}; f)$ the hypothesis under which $(\hat{Z}_1^{(n)}, \dots, \hat{Z}_n^{(n)})$ are independent and identically distributed with density $f(\cdot)$. $\hat{H}_f^{(n)}(\cdots)$ and $H_f^{(n)}(\cdots)$ are contiguous [because the sequences of the distributions of $\sum_{t=1}^n \log(f(\hat{Z}_t^{(n)})/f(Z_t^{(n)}))$ are relatively compact under both $H_f^{(n)}(\cdots)$ and $\hat{H}_f^{(n)}(\cdots)$; Le Cam (1960)]. Taking (A3.9) into account, it follows that

$$\text{(A2.3)} \quad \begin{aligned} n^{1/2}(\hat{S}^{(n)*} - S^{(n)*}) &= \left[\log L^{(n)}(\hat{\mathbf{Z}}^{(n)}) - \log L^{(n)}(\mathbf{Z}^{(n)})\right] \\ &\quad \times \left[\|\mathbf{a} + \mathbf{b}\|^2 \sigma^2 I(f)\right]^{-1/2} + o_p(1) \\ &= o_p(1). \end{aligned}$$

(A2.3) holds under $H^{(n)}(\mathbf{A}, \mathbf{B}; f)$ and, because of Le Cam's third lemma, it also holds under $K^{(n)}(\mathbf{A}, \mathbf{B}; \gamma, \boldsymbol{\delta}; f)$, which establishes the desired result.

APPENDIX 3

Proofs.

Proof of Proposition 2.2. The proof mainly consists in decomposing $\log L^{(n)}(\mathbf{X}^{(n)})$ into

$$\text{(A3.1)} \quad \log L^{(n)}(\mathbf{X}^{(n)}) = \mathscr{L}_\kappa^{(n)}(\mathbf{X}^{(n)}) - \tfrac{1}{2} \sum_{i=1}^{\infty} (a_i + b_i)^2 \sigma^2 I(f) + R_\kappa^{(n)} + o_p,$$

where $\mathscr{L}_\kappa^{(n)}(\mathbf{X}^{(n)}) = n^{-1/2} \sum_{t=\kappa+1}^n \phi(X_t) \sum_{i=1}^\kappa (a_i + b_i) X_{t-i}$ is asymptotically normal with mean 0 and variance $\sum_{i=1}^\kappa (a_i + b_i)^2 \sigma^2 I(f)$ [this follows from Proposition 3.1 in Hallin, Ingenbleek and Puri (1985)], and $R_\kappa^{(n)}$ converges to zero in probability [under $H_{0;f}^{(n)}$] as $\kappa \to \infty$, *uniformly* with respect to n. Since the asymptotic distribution of $\mathscr{L}_\kappa^{(n)}(\mathbf{X}^{(n)})$ clearly converges, as $\kappa \to \infty$, to the normal distribution with mean zero and variance $\Sigma_{i=1}^\infty (a_i + b_i)^2\sigma^2 I(f)$, the asymptotic normality of $\log L^{(n)}(\mathbf{X}^{(n)})$ follows from Theorem 2.1. Contiguity then immediately stems from Le Cam's first lemma [Hájek and Šidák (1967), page 204].

A decomposition of the desired (A3.1) form is obtained by considering a Taylor expansion of the log-likelihood ratio. In order to avoid purposeless intricate computations, we are treating here in some detail the case where $\mathbf{b} = (0, 0, \dots)$. (2.1) then takes the form of an AR(∞) model. The general case can be treated along the same lines by introducing the AR(∞) form of (2.1), $X_t - \Sigma_{i=1}^\infty d_i^{(n)} X_{t-i} = \varepsilon_t$, then showing that it is asymptotically equivalent with a model of the form $X_t - n^{-1/2}\Sigma_{i=1}^\infty (a_i + b_i) X_{t-i} = \varepsilon_t$ [see Hallin, Ingenbleek and Puri (1985), Appendix 2b].

Denote by $\mathbf{X}_0^{(n)} = (X_0^{(n)}, X_{-1}^{(n)}, \dots)$ the stationary solution (for $t \le 0$) of (2.1) and by $\mathbf{X}^{(n)}$ an n-tuple of i.i.d. variables with common density $f(x)$, independent from $\mathbf{X}^{(n)}$. Under $H_{0;f}^{(n)}$, the log-likelihood ratio can be expressed as

$$\log L^{(n)}(\mathbf{X}^{(n)}) = \log E\left[\lambda^{(n)}(\mathbf{X}^{(n)}, \mathbf{X}_0^{(n)}) | \mathbf{X}^{(n)}\right], \tag{A3.2}$$

with $\lambda^{(n)}(\mathbf{X}^{(n)}, \mathbf{X}_0^{(n)}) = \Pi_{t=1}^n f(X_t^{(n)} - n^{-1/2}\Sigma_{i=1}^\infty a_i X_{t-i}^{(n)})/f(X_t^{(n)})$. Expanding $\log \lambda^{(n)}(\mathbf{X}^{(n)}, \mathbf{X}_0^{(n)})$ yields

$$\begin{aligned} \log \lambda^{(n)}\big(\mathbf{X}^{(n)}, \mathbf{X}_0^{(n)}\big) &= n^{-1/2} \sum_{t=1}^n \phi\big(X_t^{(n)}\big) \sum_{i=1}^\infty a_i X_{t-i}^{(n)} \\ &\quad - (2n)^{-1} \sum_{t=1}^n \phi'\left(X_t^{(n)} - \theta n^{-1/2} \sum_{i=1}^\infty a_i X_{t-i}^{(n)}\right)\left(\sum_{i=1}^\infty a_i X_{t-i}^{(n)}\right)^2, \end{aligned} \tag{A3.3}$$

with $\theta = \theta(\mathbf{X}^{(n)}, \mathbf{X}_0^{(n)}) \in [0, 1]$. First let us consider the second-order term in (A3.3). Since ϕ' is Lipschitzian,

$$\begin{aligned} n^{-1} &\left| \sum_{t=1}^n \left[\phi'\left(X_t^{(n)} - \theta n^{-1/2} \sum_{i=1}^\infty a_i X_{t-i}^{(n)}\right) - \phi'\big(X_t^{(n)}\big)\right]\left[\sum_{i=1}^\infty a_i X_{t-i}^{(n)}\right]^2 \right| \\ &\le n^{-3/2} K \sum_{t=1}^n \left| \sum_{i=1}^\infty a_i X_{t-i}^{(n)} \right|^3. \end{aligned} \tag{A3.4}$$

But, under $H_{0;f}^{(n)}$, $X_t^{(n)}$ has finite moments of order 6, both for $t > 0$ as for $t \le 0$. Since $\Sigma_{i=1}^\infty |a_i| < \infty$, the variance of (A3.4) is $O(n^{-1})$ and the second-order term in (A3.3) reduces to

$$\begin{aligned} &-(2n)^{-1} \sum_{t=1}^n \phi'\big(X_t^{(n)}\big)\left(\sum_{i=1}^\infty a_i X_{t-i}^{(n)}\right)^2 + o_p \\ &= -(2n)^{-1} \sum_{t=1}^n \phi'\big(X_t^{(n)}\big)\left(\sum_{i=1}^\kappa a_i X_{t-i}^{(n)}\right)^2 - (2n)^{-1} \sum_{t=1}^n \phi'\big(X_t^{(n)}\big) \\ &\quad \times \left[2 \sum_{i=1}^\kappa a_i X_{t-i}^{(n)} \sum_{i=\kappa+1}^\infty a_i X_{t-i}^{(n)} + \left(\sum_{i=\kappa+1}^\infty a_i X_{t-i}^{(n)}\right)^2\right] + o_p. \end{aligned} \tag{A3.5}$$

The first sum on the right-hand side of (A3.5) is a sum of identically distributed κ-dependent terms; it therefore converges, as $n \to \infty$, to $(t > \kappa)$

$$-\tfrac{1}{2}E\left[\phi'\left(X_t^{(n)}\right)\left(\sum_{i=1}^{\kappa} a_i X_{t-i}^{(n)}\right)^2\right] = -\tfrac{1}{2}\sigma^2 I(f)\sum_{i=1}^{\kappa} a_i^2.$$

As for the remainder term, let us show that it converges to zero, as $\kappa \to \infty$, uniformly with respect to n. Taking into account the fact that $X_t^{(n)}$ has finite fourth-order moments, as well for $t > 0$ as for $t \le 0$, we obtain

$$E\left\{(2n)^{-1}\sum_{t=1}^{n}\phi'\left(X_t^{(n)}\right)\left[\sum_{i=\kappa+1}^{\infty} a_i X_{t-i}^{(n)}\right]^2\right\}^2$$

$$= \frac{1}{4n^2}\left\{\sum_{t=1}^{n} E\left[\left(\phi'\left(X_t^{(n)}\right)\right)^2\right] E\left[\left(\sum_{i=\kappa+1}^{\infty} a_i X_{t-i}^{(n)}\right)^4\right]\right.$$

$$\left. + 2\sum\sum_{1\le s<t\le n} E\left[\phi'\left(X_t^{(n)}\right)\right] E\left[\phi'\left(X_s^{(n)}\right)\left(\sum_{i=\kappa+1}^{\infty} a_i X_{t-i}^{(n)}\right)^2\left(\sum_{i=\kappa+1}^{\infty} a_i X_{s-i}^{(n)}\right)^2\right]\right\}$$

$$\le \frac{1}{4n} E\left[\left(\phi'\left(X_1^{(n)}\right)\right)^2\right] K\left(\sum_{i=0}^{\infty} |a_{\kappa+1+i}|\right)^4$$

$$+ \tfrac{1}{4}I(f) E\left[\left(\phi'\left(X_1^{(n)}\right)\right)^2\right]^{1/2} K\left(\sum_{i=0}^{\infty} |a_{\kappa+1+i}|\right)^4$$

$$\le K'\left(\sum_{i=\kappa}^{\infty} |a_i|\right)^4.$$

In a very similar way,

$$E\left\{(2n)^{-1}\sum_{t=1}^{n}\phi'\left(X_t^{(n)}\right)\left(\sum_{i=1}^{\kappa} a_i X_{t-i}^{(n)}\right)\left(\sum_{i=\kappa+1}^{\infty} a_i X_{t-i}^{(n)}\right)\right\}^2 \le K''\left(\sum_{i=\kappa}^{\infty} |a_i|\right)^4;$$

the two constants K' and K'' do not depend on n. The right-hand side of (A3.5) therefore satisfies the assumptions of Theorem 2.1 in the particular case where the limit distribution is degenerate $[F(z) = 1$ for $z \ge -\frac{1}{2}\sum_{i=1}^{\infty} a_i^2\sigma^2 I(f)$, $F(z) = 0$ elsewhere$]$ and the second-order term in (A3.3) converges in probability to $-\frac{1}{2}\sum_{i=1}^{\infty} a_i^2\sigma^2 I(f)$ as $n \to \infty$. (A3.3) consequently takes the form

$$\log \lambda^{(n)}\left(\mathbf{X}^{(n)}, \mathbf{X}_0^{(n)}\right) = n^{-1/2}\sum_{t=1}^{n}\phi\left(X_t^{(n)}\right)\sum_{i=1}^{\infty} a_i X_{t-i}^{(n)} - \tfrac{1}{2}\sum_{i=1}^{\infty} a_i^2\sigma^2 I(f) + o_p$$

$$= \mathscr{L}_\kappa^{(n)}\left(\mathbf{X}^{(n)}\right) + n^{-1/2}\sum_{t=1}^{n}\phi\left(X_t^{(n)}\right)\sum_{i=\kappa+1}^{\infty} a_i X_{t-i}^{(n)}$$

$$-\tfrac{1}{2}\sum_{i=1}^{\infty} a_i^2\sigma^2 I(f) + o_p.$$

Because $E[\phi(X_t^{(n)})\phi(X_s^{(n)})] = \delta_{ts}I(f)$, $t, s = 1, \ldots, n$, we have

$$E\left\{n^{-1/2}\sum_{t=1}^{n}\phi\left(X_t^{(n)}\right)\sum_{i=\kappa+1}^{\infty}a_iX_{t-i}^{(n)}\right\}^2$$

$$= \frac{1}{n}\sum_{t=1}^{n}E\left[\left(\phi\left(X_t^{(n)}\right)\right)^2\right]E\left[\left(\sum_{i=\kappa+1}^{\infty}a_iX_{t-i}^{(n)}\right)^2\right] \leq K'\left(\sum_{i=\kappa}^{\infty}|a_i|\right)^2;$$

the latter quantity again converges to zero as $\kappa \to \infty$, independently of n. Going back to (A3.2),

$$\log L^{(n)}(\mathbf{X}^{(n)}) = \mathscr{L}_\kappa^{(n)}(\mathbf{X}^{(n)}) - \tfrac{1}{2}\sum_{i=1}^{\infty}a_i^2\sigma^2I(f)$$

$$+\log E\left[\exp\left\{n^{-1/2}\sum_{t=1}^{n}\phi\left(X_t^{(n)}\right)\sum_{i=\kappa+1}^{\infty}a_iX_{t-i}^{(n)}\right\}\Big|\mathbf{X}^{(n)}\right] + o_p,$$

which is of the required form (A3.1). In the general case, $\Sigma_{i=1}^{\infty}a_i^2$ has to be replaced with $\Sigma_{i=1}^{\infty}(a_i + b_i)^2$. □

Proof of Proposition 2.3. Consider the log-likelihood ratio $\log L_\kappa^{(n)}(\mathbf{X}^{(n)})$ for $H_{0;f}^{(n)}$ against the κth-order "truncation" $K_\kappa^{(n)}$ of $K^{(n)}(\mathbf{a}, \mathbf{b}; f)$, $\kappa \geq p$. We know from Hallin, Ingenbleek and Puri (1985) that the joint distribution of $(\log L_\kappa^{(n)}, n^{1/2}(S^{(n)} - m^{(n)}))'$ is asymptotically normal (under $H_{0;f}^{(n)}$), with mean $(-\frac{1}{2}\Sigma_{i=1}^{\kappa}(a_i + b_i)^2\sigma^2I(f), 0)'$ and covariance matrix

$$\begin{pmatrix} \sum_{i=1}^{\kappa}(a_i+b_i)^2\sigma^2I(f) & \cdot \\ \sum_{i=1}^{p}(a_i+b_i)C_i & V^2 \end{pmatrix}.$$

Now the decomposition we obtained in (A3.1) is, up to o_p terms, of the form

$$\text{(A3.6)}\quad \log L^{(n)}(\mathbf{X}^{(n)}) = \log L_\kappa^{(n)}(\mathbf{X}^{(n)}) + R_\kappa^{(n)} - \tfrac{1}{2}\sum_{\kappa+1}^{\infty}(a_i+b_i)^2\sigma^2I(f).$$

Since V^2 is finite and does not depend on n or on κ, Theorem 2.1 can be applied again to an arbitrary linear combination of $\log L^{(n)}(\mathbf{X}^{(n)})$ and $n^{1/2}(S^{(n)} - m^{(n)})$. It follows that the joint distribution of $\log L^{(n)}$ and $n^{1/2}(S^{(n)} - m^{(n)})$ is asymptotically normal, with mean $(-\frac{1}{2}\Sigma_{i=1}^{\infty}(a_i + b_i)^2\sigma^2I(f), \mathbf{0})'$ and the same covariance matrix as before, except for the first diagonal element, which is now $\Sigma_{i=1}^{\infty}(a_i + b_i)^2\sigma^2I(f)$. Proposition 2.3 then straightforwardly follows from Le Cam's third lemma. □

Proof of Proposition 2.4. (i) The proof of part (i) of the proposition readily follows from Le Cam's third lemma by applying Theorem 2.1 and Proposition 3.2 in Hallin, Ingenbleek and Puri (1987) to an arbitrary linear combination of $\log L^{(n)}(\mathbf{X}^{(n)})$ and $n^{1/2}S_\kappa^{(n)*}$, where

$$n^{1/2}S_\kappa^{(n)*} = \frac{n^{1/2}}{\|\mathbf{a}+\mathbf{b}\|}S^{(n)*}\left[\sum_{i=1}^{n-1}(a_i+b_i)^2\right]^{1/2} - \sum_{i=\kappa+1}^{n-1}(n-i)^{1/2}(a_i+b_i)\frac{r_{i;f}^{(n)}}{\|\mathbf{a}+\mathbf{b}\|}.$$

(ii) Write $K^{(n)}$ for $K^{(n)}(\mathbf{a},\mathbf{b};f)$. It follows from (2.9) that the asymptotic power of the test based on $S^{(n)*}$ for $H_0^{(n)}$ against $K^{(n)}$ is $1-\Phi(k_{1-\alpha} - \|\mathbf{a}+\mathbf{b}\|[\sigma^2 I(f)]^{1/2})$, where Φ denotes the standard normal distribution function. Using classical notation (Section 3.2), let $\beta(\alpha, H, K)$ be the envelope power function for testing H against K at level α. Then $\beta(\alpha, H_{0;f}^{(n)}, K^{(n)})$ is the power reached by the likelihood ratio test. Using Proposition 2.2, we obtain

$$\lim_{n\to\infty}\beta\left(\alpha, H_{0;f}^{(n)}, K^{(n)}\right) = 1-\Phi\left(k_{1-\alpha} - \|\mathbf{a}+\mathbf{b}\|\left[\sigma^2 I(f)\right]^{1/2}\right).$$

Since $\beta(\alpha, H_{0;f}^{(n)}, K^{(n)}) \geq \beta(\alpha, H_0^{(n)}, K^{(n)})$ for every n, we have

$$\lim_{n\to\infty}\beta\left(\alpha, H_{0;f}^{(n)}, K^{(n)}\right) \geq \limsup_{n\to\infty}\beta\left(\alpha, H_0^{(n)}, K^{(n)}\right).$$

Therefore,

$$\begin{aligned}\limsup_{n\to\infty}\beta\left(\alpha, H_0^{(n)}, K^{(n)}\right) &\geq 1-\Phi\left(k_{1-\alpha} - \|\mathbf{a}+\mathbf{b}\|\left[\sigma^2 I(f)\right]^{1/2}\right)\\ &= \lim_{n\to\infty}\beta\left(\alpha, H_{0;f}^{(n)}, K^{(n)}\right),\end{aligned}$$

which implies that $\lim_{n\to\infty}\beta(\alpha, H_0^{(n)}, K^{(n)})$ exists and equals the asymptotic power of (2.9). □

Proof of Proposition 3.2. Let $l_{0;f}^{(n)}(\mathbf{X}^{(n)}) = \prod_{t=1}^n f(Z_t^{(n)})$. The objective is to show that there exist $\mathbf{T}_{\Psi;f}^{(n)}$-measurable variables $h_{\gamma,\delta}^{(n)}$, such that the functions

$$\bar{l}_{\gamma,\delta}^{(n)} = l_{0;f}^{(n)}h_{\gamma,\delta}^{(n)} \tag{A3.7}$$

are densities for all $(\gamma,\delta)\in\mathscr{D}(d)$ and n (n sufficiently large), and satisfy

$$\lim_{n\to\infty}\sup_{\mathscr{D}(d)}\left\|l_{\gamma,\delta}^{(n)} - \bar{l}_{\gamma,\delta}^{(n)}\right\|_{L_1} = 0, \tag{A3.8}$$

where $\|p-q\|_{L_1}$ is the L_1-distance $\int|p-q|\,d\mu$ for densities defined with respect to the σ-finite measure μ.

Consider the optimal rank statistic $S_{\mathbf{a},\mathbf{b}}^{(n)*}$ we introduced in Propositions 2.4 and 3.1. It follows from the proof of Proposition 2.4 that asymptotically the joint distribution of $\log l_{\gamma,\delta}^{(n)}/l_{0;f}^{(n)}$ and $n^{1/2}S_{\mathbf{a},\mathbf{b}}^{(n)*}$ is normal, with covariance matrix

$$\begin{pmatrix}\|\mathbf{a}+\mathbf{b}\|^2\sigma^2 I(f) & \cdot\\ \|\mathbf{a}+\mathbf{b}\|\left[\sigma^2 I(f)\right]^{1/2} & 1\end{pmatrix},$$

from which we deduce that

$$\log\left[L_{\gamma,\delta}^{(n)}/l_{0;f}^{(n)}\right] + \tfrac{1}{2}\|\mathbf{a}+\mathbf{b}\|^2\sigma^2 I(f) - n^{1/2}\|\mathbf{a}+\mathbf{b}\|\left[\sigma^2 I(f)\right]^{1/2}S_{\mathbf{a},\mathbf{b}}^{(n)*} \tag{A3.9}$$

converges to zero, in probability, under $H^{(n)}(\mathbf{A}, \mathbf{B}; f)$. Moreover, this convergence is uniform with respect to the γ_i's and δ_i's, $(\boldsymbol{\gamma}, \boldsymbol{\delta}) \in \mathscr{D}(d)$. Hence writing $s^{(n)*}_{\mathbf{a},\mathbf{b}}$ for $n^{1/2}\|\mathbf{a} + \mathbf{b}\|[\sigma^2 I(f)]^{1/2} S^{(n)*}_{\mathbf{a},\mathbf{b}}$ [with $\mathbf{a}$ and $\mathbf{b}$ given in (3.2)], we have

$$\lim_{n\to\infty} \sup_{\mathscr{D}(d)} P_{l_0^{(n)}}\left[\left|\left(l^{(n)}_{\boldsymbol{\gamma},\boldsymbol{\delta}}/l^{(n)}_{0;f}\right) - \exp\left(s^{(n)*}_{\mathbf{a},\mathbf{b}} - \tfrac{1}{2}\|\mathbf{a} + \mathbf{b}\|^2\sigma^2 I(f)\right)\right| > \varepsilon\right] = 0$$

for any $\varepsilon > 0$, which is equivalent to Hájek and Šidák's equation (10) [(1967), page 246].

Defining $h^{(n)}_{\boldsymbol{\gamma},\boldsymbol{\delta}}$ as

$$h^{(n)}_{\boldsymbol{\gamma},\boldsymbol{\delta}} = \begin{cases} B^{(n)}_{\boldsymbol{\gamma},\boldsymbol{\delta}} \exp\left[s^{(n)*}_{\mathbf{a},\mathbf{b}} - \tfrac{1}{2}\|\mathbf{a} + \mathbf{b}\|^2\sigma^2 I(f)\right], & \text{if } \left|s^{(n)*}_{\mathbf{a},\mathbf{b}}\right| \le C^{(n)}, \\ 0, & \text{if } \left|s^{(n)*}_{\mathbf{a},\mathbf{b}}\right| > C^{(n)}, \end{cases}$$

where $B^{(n)}_{\boldsymbol{\gamma},\boldsymbol{\delta}}$ is a normalizing constant such that (A3.7) is a probability density function, and reproducing Hájek and Šidák's proof of Theorem 7.1.1 [(1967), pages 247–248, including their definitions of constants $C^{(n)}$] leads to the conclusion that (A3.7) and (A3.8) are satisfied.

In order to complete the proof, it remains to show that the variables $s^{(n)*}_{\mathbf{a},\mathbf{b}}$, hence the $h^{(n)}_{\boldsymbol{\gamma},\boldsymbol{\delta}}$'s are $\mathbf{T}^{(n)}_{\Psi;f}$-measurable. Notice therefore that, for $t \ge p_2 - p_1 + 1$ and $t' \ge q_2 - q_1 + 1$, $a_t = \sum_{j=1}^{p_2}\gamma_j G_{t-j}$ and $b_{t'} = \sum_{j=1}^{q_2}\delta_j H_{t'-j}$. Now $\{G_{t-1},\dots,G_{t-p_1}\}$ and $\{H_{t-1},\dots,H_{t-q_1}\}$ constitute fundamental systems of solutions of $A(L)\Psi_t = 0$, $t \in \mathbb{Z}$, and $B(L)\Psi_t = 0$, $t \in \mathbb{Z}$, respectively (see Appendix 1), and $G_{t-p_1-1},\dots,G_{t-p_2}$ and $H_{t-q_1-1},\dots,H_{t-q_2}$ are solutions of the same equations, respectively. Hence, for $t \ge \pi + 1$, $A(L)a_t = 0 = B(L)b_t$, from which we deduce that $A(L)B(L)(a_t + b_t) = 0$, $t \ge \pi + 1$. It follows that

$$\text{(A3.10)} \quad (a_i + b_i) = \left(k_1,\dots,k_{p_1+q_1}\right)\left(\Psi_i^{(1)},\dots,\Psi_i^{(p_1+q_1)}\right)', \qquad i \ge \pi + 1,$$

where the constants $k_1,\dots,k_{p_1+q_1}$ depend on $\boldsymbol{\gamma}$ and $\boldsymbol{\delta}$ only (for given $\Psi_t^{(j)}$'s). Going back to the definition of $s^{(n)*}_{\mathbf{a}+\mathbf{b}}$, we finally obtain

$$s^{(n)*}_{\mathbf{a}+\mathbf{b}} = n^{1/2}\frac{\|\mathbf{a} + \mathbf{b}\|\left[\sigma^2 I(f)\right]^{1/2}}{\left[\sum_{i=1}^{n-1}(a_i + b_i)^2\right]^{1/2}}\left((a_1 + b_1),\dots,(a_\pi + b_\pi), k_1,\dots,k_{p_1+q_1}\right)\mathbf{T}^{(n)}_{\Psi;f},$$

which completes the proof. □

Proof of Proposition 3.3. Let $\kappa > \pi$ and consider an arbitrary linear combination of $\log L^{(n)}(\mathbf{Z}^{(n)})$, $n^{1/2}r^{(n)}_{1;f},\dots,n^{1/2}r^{(n)}_{\pi;f}$, $n^{1/2}\sum_{i=\pi+1}^{\kappa}\Psi_i^{(1)}r^{(n)}_{i;f},\dots$, and $n^{1/2}\sum_{i=\pi+1}^{\kappa}\Psi_i^{(p_1+q_1)}r^{(n)}_{i;f}$. The asymptotic normality of this linear combination can easily be established along the lines of Proposition 3.1 in Hallin, Ingenbleek and Puri (1987). Theorem 2.1 can then be applied to obtain the asymptotic normality of this linear combination in the case when $\kappa = \infty$. Proposition 3.3 finally follows from Le Cam's third lemma.

The nonsingularity of $\mathbf{W}^2_\Psi$, hence that of $\underline{\mathbf{W}}^2_\Psi$, results from the fact that the $\Psi_t^{(j)}$'s are linearly independent solutions of

$$\text{(A3.11)} \qquad A(L)B(L)\Psi_t = 0, \qquad t \in \mathbb{Z}. \qquad \square$$

Proof of Proposition 4.1. Let $\{\Theta_t^{(1)},\dots,\Theta_t^{(p_1+q_1)}\}$ and $\{\Psi_t^{(1)},\dots,\Psi_t^{(p_1+q_1)}\}$ denote two arbitrary fundamental systems for (A3.11). Then

$$\left(\Theta_t^{(1)},\dots,\Theta_t^{p_1+q_1)}\right) = \left(\Psi_t^{(1)},\dots,\Psi_t^{(p_1+q_1)}\right)\mathbf{K}',\ t \in \mathbb{Z},$$

where $\mathbf{K}$ is some nonsingular constant matrix of dimension $(p_1 + q_1) \times (p_1 + q_1)$. Hence

$$\mathbf{T}_{\theta;f}^{(n)} = \begin{pmatrix}\mathbf{I} & \mathbf{0}\\ \mathbf{0} & \mathbf{K}\end{pmatrix}\mathbf{T}_{\Psi;f}^{(n)}, \qquad \mathbf{W}_\Theta^2 = \underline{\mathbf{K}}\mathbf{W}_\Psi^2\underline{\mathbf{K}}'$$

and

$$\begin{aligned}\mathbf{T}_{\theta;f}^{(n)\prime}\underline{\mathbf{W}}_\Theta^{-2}\mathbf{T}_{\theta;f}^{(n)} &= \mathbf{T}_{\Psi;f}^{(n)\prime}\begin{pmatrix}\mathbf{I} & \mathbf{0}\\ \mathbf{0} & \mathbf{K}\end{pmatrix}\begin{pmatrix}\mathbf{I} & \mathbf{0}\\ \mathbf{0} & \mathbf{K}^{-1}\mathbf{W}_\Psi^{-2}\underline{\mathbf{K}}'^{-1}\end{pmatrix}\begin{pmatrix}\mathbf{I} & \mathbf{0}\\ \mathbf{0} & \mathbf{K}\end{pmatrix}\mathbf{T}_{\Psi;f}^{(n)}\\ &= \mathbf{T}_{\Psi;f}^{(n)\prime}\underline{\mathbf{W}}_\Psi^{-2}\mathbf{T}_{\Psi;f}^{(n)}.\end{aligned}$$

□

Proof of Proposition 4.2. The proof straightforwardly follows from Proposition 4.1. Notice indeed that, on account of (A3.10), letting $\mathbf{k}' = ((a_1 + b_1),\dots,(a_\pi + b_\pi),\ k_1,\dots,k_{p_1+q_1})$, the asymptotic mean of $n^{1/2}\mathbf{T}_{\Psi;f}^{(n)}$ under $K^{(n)}(\mathbf{A},\mathbf{B};\ \gamma,\delta;\ f)$ takes the form

$$\text{(A3.12)} \qquad \underline{\mathbf{W}}_\Psi^2\mathbf{k}\left[\sigma^2 I(f)\right]^{1/2}.$$

On the other hand, $\mathbf{k}'\underline{\mathbf{W}}_\Psi^2\mathbf{k} = \|\mathbf{a} + \mathbf{b}\|^2$, yielding the noncentrality parameter (4.3). □

Proof of Proposition 4.3. We know from Section 3.3 that $\mathbf{T}_{\Psi;f}^{(n)}$ is asymptotically sufficient for testing $H^{(n)}(\mathbf{A},\mathbf{B};\cdot)$ against (3.8). $n^{1/2}\underline{\mathbf{W}}_\Psi^{-1}\mathbf{T}_{\Psi;f}^{(n)}$ is thus also asymptotically sufficient since $\underline{\mathbf{W}}_\Psi^2$ is a full-rank covariance matrix.

Under $H^{(n)}(A, B;\ \gamma,\delta;\ f)$, the asymptotic distribution of $n^{1/2}\underline{\mathbf{W}}_\Psi^{-1}\mathbf{T}_{\Psi;f}^{(n)}$ is normal with mean $\mu(\gamma,\delta) = \underline{\mathbf{W}}_\Psi\mathbf{k}[\sigma^2 I(f)]^{1/2}$ [(A3.12)] and identity covariance matrix.

Let us show that when (γ,δ) describes $\mathbb{R}^{p_2}\times\mathbb{R}^{q_2}$, $\mathbf{k}$ and thus $\mu(\gamma,\delta)$ describe $\mathbb{R}^{\max(p_1+q_2,\,p_2+q_1)}$. Any value of $(a_1,\dots,a_{p_2-p_1}) \in \mathbb{R}^{p_2-p_1}$ can be reached by choosing appropriate values of $\gamma_1,\dots,\gamma_{p_2-p_1}$, and a_t, $t \geq p_2 - p_1 + 1$, can be made equal to any solution of $A(L)\Psi_t = 0$ by adjusting the remaining p_1 components $\gamma_{p_2-p_1+1},\dots,\gamma_{p_2}$ of γ. A similar result holds for $\mathbf{b}$ and δ. Furthermore, any solution of $A(L)B(L)\Psi_t = 0$, $t \geq \pi + 1$, can be decomposed into a solution Ψ_t^a of $A(L)\Psi_t = 0$ plus a solution Ψ_t^b of $B(L)\Psi_t = 0$. This follows indeed from the fact that $A(z)$ and $B(z)$ have no common root: A fundamental system of $A(L)B(L)\Psi_t = 0$ can thus be formed by juxtaposing a fundamental system $\{\Psi_t^{(1)},\dots,\Psi_t^{(p_1)}\}$ of $A(L)\Psi_t = 0$ with a fundamental system $\{\Psi_t^{(p_1+1)},\dots,\Psi_t^{(p_1+q_1)}\}$ of $B(L)\Psi_t = 0$.

Now, under the condition $(\gamma,\delta) \in \mathscr{D}(d)$, i.e., $\|\mathbf{a} + \mathbf{b}\|^2 \geq d^2$, the values of $\mu(\gamma,\delta)$ are restricted to those satisfying $\mu'(\gamma,\delta)\mu(\gamma,\delta) = \mathbf{k}'\underline{\mathbf{W}}_\Psi^2\mathbf{k}\sigma^2 I(f) \geq d^2\sigma^2 I(f)$. Thus the family of asymptotic distributions of $n^{1/2}\underline{\mathbf{W}}_\Psi^{-1}\mathbf{T}_{\Psi;f}^{(n)}$ under $K^{(n)}(\mathbf{A},\mathbf{B};\ \mathbb{R}^{p_2},\ \mathbb{R}^{q_2};\ f|d)$ is the set of all normal distributions with mean μ and identity covariance matrix such that $\mu'\mu \geq d^2\sigma^2 I(f)$, whereas under $H^{(n)}(\mathbf{A},\mathbf{B};\cdot)$

the asymptotic distribution of $n^{1/2}\underline{\mathbf{W}}_{\Psi}^{-1}\mathbf{T}_{\Psi;f}^{(n)}$ is normal with mean $\mathbf{0}$ and identity covariance matrix.

It follows from an invariance argument and the Hunt–Stein theorem [see, e.g., Lehmann (1959), Section 8.4] that the maximin most powerful test for this problem is based on the test statistic $n\mathbf{T}_{\Psi;f}^{(n)\prime}\underline{\mathbf{W}}_{\Psi}^{-2}\mathbf{T}_{\Psi;f}^{(n)}$—which is precisely $Q_{A,B;f}^{(n)*}$. Because of the asymptotic sufficiency of $\mathbf{T}_{\Psi;f}^{(n)}$, this test is also maximin most powerful for $H^{(n)}(\mathbf{A},\mathbf{B};\cdot)$ against any $K^{(n)}(\mathbf{A},\mathbf{B};\mathbb{R}^{p_2},\mathbb{R}^{q_2};f|d)$, $d\in\mathbb{R}^+$. □

Corollary 4.4 gives the particular form of the asymptotically maximin most powerful statistic $Q_{A,B;f}^{(n)*}$ in the case where $p_1 = q_1 = 0$ and $\pi = \max(p_2, q_2)$. This is the only case where $Q_{A,B;f}^{(n)*}$ turns out to be a quadratic serial rank statistic in the sense of Hallin, Ingenbleek and Puri (1987).

Acknowledgments. The authors acknowledge with sincere thanks the valuable comments and suggestions of the Editor, the Associate Editor and the referee.

REFERENCES

ANDERSON, T. W. (1959). On asymptotic distributions of estimates of parameters of stochastic difference equations. *Ann. Math. Statist.* **30** 676–687.

ANDERSON, T. W. (1971). *The Statistical Analysis of Time Series*. Wiley, New York.

BARTELS, R. (1982). The rank version of von Neumann's ratio test for randomness. *J. Amer. Statist. Assoc.* **77** 40–46.

BHATTACHARYYA, G. K. (1984). Tests for randomness against trend or serial correlation. In *Handbook of Statistics: Nonparametric Methods* (P. R. Krishnaiah and P. K. Sen, eds.) **4** 89–111. North-Holland, Amsterdam.

BOX, G. E. P. and JENKINS, G. M. (1970). *Time Series Analysis, forecasting and control*. Holden-Day, San Francisco.

BOX, G. E. P. and PIERCE, D. A. (1970). Distribution of residual autocorrelations in autoregressive-integrated moving average time series models. *J. Amer. Statist. Assoc.* **65** 1509–1526.

DUFOUR, J.-M. and HALLIN, M. (1987). Tests nonparamétriques optimaux pour le modèle autorégressif d'ordre un. *Annales d'Economie et de Statistique* **5** 411–434.

DUFOUR, J.-M. and KING, M. L. (1986). Optimal invariant test for the autocorrelation coefficient in linear regressions with autocorrelated errors. Technical Report, Dept. Sciences Economiques, Univ. Montréal.

DUFOUR, J.-M., LEPAGE, Y. and ZEIDAN, H. (1982). Nonparametric testing for time series: A bibliography. *Canad. J. Statist.* **10** 1–38.

DUFOUR, J.-M. and ROY, R. (1985). Some robust exact results on sample autocorrelations and tests of randomness. *J. Econometrics* **29** 257–273.

DUFOUR, J.-M. and ROY, R. (1986). Generalized portmanteau statistics and tests for randomness. *Comm. Statist. A—Theory Methods* **15** 2953–2972.

EVANS, M. A. and KING, M. L. (1985). Higher order generalizations of first order autoregressive tests. *Comm. Statist. A—Theory Methods* **14** 2907–2918.

GODFREY, L. G. (1979). Testing the adequacy of a time series model. *Biometrika* **66** 67–72.

HÁJEK, J. and ŠIDÁK, Z. (1967). *Theory of Rank Tests*. Academic, New York.

HALLIN, M. (1986). Nonstationary q-dependent processes and time-varying moving average models: Invertibility properties and the forecasting problem. *Adv. in Appl. Prob.* **18** 170–210.

HALLIN, M., INGENBLEEK, J.-FR. and PURI, M. L. (1985). Linear serial rank tests for randomness against ARMA alternatives. *Ann. Statist.* **13** 1156–1181.

HALLIN, M., INGENBLEEK, J.-FR. and PURI, M. L. (1987). Linear and quadratic rank tests for randomness against serial dependence. *J. Time Ser. Anal.* **8** 409–424.

HALLIN, M. and MÉLARD, G. (1987). Rank-based tests for randomness against first order serial dependence. Technical Report, Centre d'Economie Mathématique et d'Econométrie, Univ. Libre de Bruxelles, Brussels.

HALLIN, M. and PURI, M. L. (1987). Locally asymptotically optimal tests for randomness. In *New Perspectives in Theoretical and Applied Statistics* (M. L. Puri, J. P. Vilaplana and W. Wertz, eds.) 87–99. Wiley, New York.

HALLIN, M. and PURI, M. L. (1988). Locally asymptotically rank-based procedures for testing ARMA dependence. *Proc. Nat. Acad. Sci. U.S.A.* To appear.

HOSKING, J. R. M. (1978). A unified derivation of the asymptotic distributions of goodness-of-fit statistics for autoregressive time-series models. *J. Roy. Statist. Soc. Ser. B* **40** 341–349.

HOSKING, J. R. M. (1980). Lagrange multiplier tests of time series models. *J. Roy. Statist. Soc. Ser. B* **42** 170–181.

HOSKING, J. R. M. (1981). Lagrange multiplier tests of multivariate time series models. *J. Roy. Statist. Soc. Ser. B* **43** 219–230.

KING, M. L. and HILLIER, G. H. (1985). Locally best invariant tests of the error covariance matrix of the linear regression model. *J. Roy. Statist. Soc. Ser. B* **47** 98–102.

LE CAM, L. (1960). Locally asymptotically normal families of distributions. *Univ. California Publ. Statist.* **3** 27–98.

LEHMANN, E. L. (1959). *Testing Statistical Hypotheses*. Wiley, New York.

LJUNG, G. M. and BOX, G. E. P. (1978). On a measure of lack of fit in time series models. *Biometrika* **65** 297–303.

MILLER, K. S. (1968). *Linear Difference Equations*. Benjamin, New York.

POSKITT, D. S. and TREMAYNE, A. R. (1980). Testing the specification of a fitted autoregressive-moving average model. *Biometrika* **67** 359–363.

POSKITT, D. S. and TREMAYNE, A. R. (1982). Diagnostic tests for multiple time series models. *Ann. Statist.* **10** 114–120.

PRIESTLEY, M. B. (1981). *Spectral Analysis and Time Series*. Academic, New York.

DÉPARTEMENT DE MATHÉMATIQUE AND
INSTITUT DE STATISTIQUE
C.P. 210
UNIVERSITÉ LIBRE DE BRUXELLES
1050 BRUXELLES
BELGIUM

DEPARTMENT OF MATHEMATICS
SWAIN HALL EAST
INDIANA UNIVERSITY
BLOOMINGTON, INDIANA 47405

Reprinted by permission of the publisher,
© 1988 by The Institute of Mathematical Statistics

ON LOCALLY ASYMPTOTICALLY MAXIMIN TESTS FOR ARMA PROCESSES

MARC HALLIN and MADAN L. PURI*

The problem of testing a specified ARMA (p_1,q_1) model against ARMA (p_2,q_2) alternatives with $\Pi = \Pi(n) = \max(p_2-p_1, q_2-q_1)$ an increasing function of the series length n, is considered. It is shown that if $\lim_{n\to\infty} \Pi(n) = \infty$, then a locally asymptotically maximin test at level α is the trivial test $\phi \equiv \alpha$.

1. INTRODUCTION

Denote by $\underline{X}^{(n)} = (X_1^{(n)},\ldots,X_n^{(n)})$ an observed series of length n. As in Hallin and Puri [4], our objective is to test the null hypothesis under which $\underline{X}^{(n)}$ was generated by some specified ARMA (p_1,q_1) model--an important special case of which is the hypothesis of white noise considered in Hallin, Ingenbleek and Puri ([2],[3]), under which the $X_t^{(n)}$'s are independent and identically distributed observations. Unlike the above mentioned papers, however, where the alternative consists of ARMA models with unspecified coefficients but fixed orders p_2 and q_2, we would like to consider here sequences of ARMA alternatives in which the orders p_2 and q_2 possibly are increasing functions of the series length. Indeed, serial dependencies of (fixed but) arbitrarily high orders can be expected to be detected as $n\to\infty$. This idea of testing against alternatives of increasing orders p_2 and/or q_2 is also present in the conventional approach which consists in testing the adequacy of an ARMA model on the basis of a Box-Pierce portmanteau statistic of order m, where m is $O(n^{1/2})$.

Uniformly most powerful tests, or asymptotically uniformly most powerful ones, typically do not exist for such situations. We therefore restrict ourselves to local optimality properties. Section 2 below presents a definition of local asymptotical minimaxity inspired from Hájek and Šidák [1]. Section 3 then establishes the somewhat disappointing result that a locally asymptotically maximin procedure for the problem considered (at given probability level α) is the trivial test $\phi(\underline{X}^{(n)}) \equiv \alpha$.

2. MAXIMIN AND ASYMPTOTICALLY MAXIMIN TESTS

Denote, as usual, by $\beta(\alpha;H_o,H_1)$ the envelope power function for testing the

---

* Research supported by the Office of Naval Research Contract N00014-85-K-0648.

null hypothesis H_o against the alternative H_1:

$$\beta(\alpha;H_o,H_1) = \sup_{\phi} \inf_{Q\in H_1} E_Q(\phi) \ , \qquad \alpha\in(0,1) \ ,$$

where the supremum is taken over all tests satisfying $\sup_{P\in H_o} E_P(\phi) \leq \alpha$. A test ϕ^* is said to be maximin for H_o against H_1 at level α if simultaneously

$$E_Q(\phi^*) \geq \beta(\alpha;H_o,H_1) \ , \quad Q\in H_1 \quad \text{and} \quad E_P(\phi^*) \leq \alpha \ , \quad P\in H_o \ .$$

In order to obtain an asymptotic version of the above concept, consider sequences of null hypotheses $H_o^{(n)}$ and alternatives $H_1^{(n)}$. A sequence of tests $\phi^{(n)*}$ is said to be asymptotically maximin for $H_o^{(n)}$ against $H_1^{(n)}$ at level α if

$$\liminf_{n\to\infty} [E_{Q^{(n)}}(\phi^{(n)*}) - \beta(\alpha;H_o^{(n)},H_1^{(n)})] \geq 0 \ , \quad Q^{(n)} \in H_1^{(n)}$$

and

$$\limsup_{n\to\infty} [E_{P^{(n)}}(\phi^{(n)*}) - \alpha] \leq 0 \ , \quad P^{(n)} \in H_o^{(n)} \ .$$

If $H_1^{(n)}$ includes sequences of alternatives which in some sense can be considered as local alternatives, $\phi^{(n)*}$ will be termed a locally asymptotically maximin test. (An alternative description of these optimality concepts can be given in terms of *stringency* (cf. e.g. Wald [5])).

Clearly, in the case of a simple hypothesis $H_o = \{P\}$ against a simple alternative $H_1 = \{Q\}$, $\beta(\alpha;H_o,H_1)$ is the power $E_Q(\phi_n(\alpha;P,Q))$ of the Neyman-Pearson test for $\{P\}$ against $\{Q\}$, which is strictly larger than α unless $P\equiv Q$. More generally, the following result for composite hypotheses readily follows from the fundamental Neyman-Pearson Lemma (it is assumed that $H_o\cap H_1=\emptyset$).

Lemma 1. For any $P\in H_o$, $Q\in H_1$, $\alpha\in(0,1)$,

$$\alpha \leq \beta(\alpha;H_o,H_1) \leq E_Q(\phi_n(\alpha;P,Q)) \ ;$$

moreover, at least one of these two inequalities is strict.

An immediate consequence is that H_1 has to be bounded away from H_o (in the sense to be defined below), unless the envelope power function reduces to the trivial function $\beta(\alpha;H_o,H_1) = \alpha$, leading to the trivial maximin test $\phi \equiv \alpha$.

Lemma 2. A necessary condition for $\beta(\alpha;H_o,H_1)$ *to be strictly larger than* α *is that for any* $\alpha\in(0,1)$ *there exists some* $d>\alpha$ *such that* $E_Q[\phi_n(\alpha;P,Q)] \geq d$, *any* $P \in H_o$, $Q \in H_1$.

An asymptotic version of Lemma 2, providing a necessary (clearly, not sufficient) condition for the existence of nontrivial asymptotically maximin tests follows along the same lines.

Lemma 3. A necessary condition for $\liminf_{n\to\infty} \beta(\alpha; H_o^{(n)}, H_1^{(n)})$ *to be strictly larger than* α *is that for any* $\alpha \in (0,1)$ *there exists some* $d > \alpha$ *such that* $\liminf_{n\to\infty} E_{Q^{(n)}}[\phi_n(\alpha; P^{(n)}, Q^{(n)})] \geq d$, *for any* $P^{(n)} \in H_o^{(n)}$, $Q^{(n)} \in H_1^{(n)}$.

3. A LOCALLY ASYMPTOTICALLY MAXIMIN TEST

Going back to the testing problem described in the introduction, consider the null hypothesis $H_o^{(n)} = \{P_f^{(n)}\}$ under which $\underline{X}^{(n)}$ is a finite realization of the unique stationary solution of the stochastic difference equation (ARMA model)

$$X_t - \sum_{i=1}^{p_1} A_i X_{t-i} = \varepsilon_t + \sum_{i=1}^{q_1} B_i \varepsilon_{t-i} , \quad t \in \mathbb{Z} , \tag{3.1}$$

where $\{\varepsilon_t\}$ is an independent white noise process with mean zero, variance σ^2 and density function f ((3.1) is assumed to satisfy the usual stationarity and invertibility requirements). Denote by $P_f^{(n)}$ the corresponding likelihood function for $\underline{X}^{(n)}$. Let $\Pi(n)$ be an increasing, positive integer-valued function of n such that $\lim_{n\to\infty} \Pi(n) = \infty$. Consider the situation under which $\{X_t\}$ is generated by the ARMA (p_2, q_2) model

$$X_t - \sum_{i=1}^{p_2} \alpha_i X_{t-i} = \varepsilon_t + \sum_{i=1}^{q_2} \beta_i \varepsilon_{t-i} , \quad t \in \mathbb{Z} , \tag{3.2}$$

where $\{\varepsilon_t\}$ again is an independent white noise with density f , with unspecified orders p_2 and q_2 such that $\max(p_2, q_2) \leq \Pi(n)$ and unspecified coefficients α_i and β_i : denote by $Q_f^{(n)}$ the corresponding likelihood for $\underline{X}^{(n)}$. In order to comply with the necessary condition of Lemma 3, choose $\alpha \in (0,1)$ and $d \in (\alpha, 1)$, and define

$$H_d^{(n)} = \{Q_f^{(n)} \mid E_{Q_f^{(n)}}[\phi_n(\alpha; P_f^{(n)}, Q_f^{(n)})] \geq d\} .$$

The sequence of alternatives $H_d^{(n)}$ thus defined is a (nonempty) local sequence (with respect to $H_o^{(n)}$). Consider indeed the sequence of ARMA (p_1+1, q_1) models obtained by letting $p_2 = p_1+1$, $q_2 = q_1$,

$$\alpha_i = \begin{cases} A_i & i = 1,\ldots,p_1 \\ n^{-1/2}\delta & i = p_1+1 \end{cases} \qquad \beta_i = B_i \quad i = 1,\ldots,q_1$$

and assume that f satisfies some mild regularity conditions (see Hallin and Puri [4]). Denote by $Q^{(n)}_{f;\delta}$ the corresponding likelihood function, and by $\Lambda^{(n)}(\underline{X}^{(n)})$ the likelihood ratio $Q^{(n)}_{f;\delta}(\underline{X}^{(n)})/P^{(n)}_{f}(\underline{X}^{(n)})$. Then (Hallin and Puri [4], Proposition 2.2) $\log \Lambda^{(n)}(\underline{X}^{(n)})$ is asymptotically normal, as $n\to\infty$, with mean $-\delta^2 g^2\sigma^2 I(f)/2$ under $\{P^{(n)}_f\}$ and mean $\delta^2 g^2\sigma^2 I(f)/2$ under $\{Q^{(n)}_{f;\delta}\}$, and variance $\delta^2 g^2\sigma^2 I(f)$ under both (here σ^2 stands for the variance $\int x^2 f(x)dx$ and $I(f)$ for the Fisher's information $\int(d\log f/dx)^2 f(x)dx$, which are finite under the above mentioned regularity assumptions; $g^2 = \sum_{i=0}^{\infty} g_i^2 < \infty$, where g_i denotes the one-sided Green's function associated with the autoregressive operator $A(L)$ in (3.1)). The limit $\lim_{n\to\infty} \beta(\alpha;P^{(n)}_f, Q^{(n)}_{f;\delta})$ therefore exists, and is equal to $1-\phi(k_{1-\alpha} - |\delta|(g^2\sigma^2 I(f))^{\frac{1}{2}})$, where $\phi(\ )$ and $k_{1-\alpha}$ denote the distribution function and $(1-\alpha)$-quantile of the standard normal variable, respectively. The local alternative $Q^{(n)}_{f;\delta}$ thus belongs to $H^{(n)}_d$, for n sufficiently large, provided that $\delta^2 > [k_{1-\alpha} - \phi^{-1}(1-d)]^2[g^2\sigma^2 I(f)]^{-1}$.

Now, let us turn to the (pessimistic) conclusion of this paper: the necessary condition of Lemma 3 here is not a sufficient one, and the locally asymptotically maximin test for our problem turns out to be the trivial one, whatever the values of α and d, no matter how slowly $\Pi(n)$ increases.

Proposition. For any $\alpha\in(0,1)$, $d\in(\alpha,1)$, $\lim_{n\to\infty} \beta(\alpha;H^{(n)}_o, H^{(n)}_d) = \alpha$, *and the trivial test* $\phi(\underline{X}^{(n)}) \equiv \alpha$ *is locally asymptotically maximin.*

Proof. Consider the sequences of subhypotheses $K^{(n)}_{\Pi}$, $\Pi=1,2,\dots$, obtained from $H^{(n)}_d$ by restricting the orders p_2 and q_2 in (3.2) to $\max(p_2,q_2) \le \min(\Pi,\Pi(n))$. Since $K^{(n)}_{\Pi} \subset H^{(n)}_d$, we have

$$\beta(\alpha;H^{(n)}_o, K^{(n)}_{\Pi}) \ge \beta(\alpha;H^{(n)}_o, H^{(n)}_d) \ge \alpha \qquad (3.3)$$

for any value of Π. Now, letting $p = \max(p_1,q_1)$, it follows from Proposition 4.3 in Hallin and Puri [4] that

$$\lim_{n\to\infty} \beta(\alpha;H^{(n)}_o, K^{(n)}_{\Pi}) = 1 - F_{p+\Pi}(\chi^2_{p+\Pi;1-\alpha}; \tfrac{1}{2}[k_{1-\alpha} - \phi^{-1}(1-d)]^2) \qquad (3.4)$$

where $\chi^2_{m;1-\alpha}$ denotes the $(1-\alpha)$-quantile of the chi-square distribution with m degrees of freedom, and $F_m(\cdot;\lambda)$ the distribution function of the noncentral chi-square with m degrees of freedom and noncentrality parameter λ. When $\Pi\to\infty$, the right-hand side in (3.4) clearly tends to α, since the number of degrees of freedom increases without limit, and the noncentrality parameter remains constant. In view of (3.3), we thus have

$$\alpha = \inf_{\Pi} \lim_{n\to\infty} \beta(\alpha;H^{(n)}_o, K^{(n)}_{\Pi}) \ge \lim_{n\to\infty} \beta(\alpha;H^{(n)}_o, H^{(n)}_d) \ge \alpha ,$$

which completes the proof.

Note that the proposition remains unaffected if the innovation density f is unspecified under the null as well as under the alternative hypothesis.

ACKNOWLEDGEMENT

The authors would like to express their sincere thanks to the referee for pointing out an error in the original draft.

REFERENCES

[1] Hájek, J. and Šidák, Z. (1967). Theory of Rank Tests. Academic Press, N.Y.

[2] Hallin, M., Ingenbleek, J.-Fr. and Puri, M.L. (1985). Linear serial rank tests for randomness against ARMA alternatives. Ann. Statist. 13, 1156-1181.

[3] Hallin, M., Ingenbleek, J.-Fr. and Puri, M.L. (1987). Linear and quadratic rank tests for randomness against serial dependence. J. Time Ser. Anal. 8, 409-424.

[4] Hallin, M. and Puri, M.L. (1988). Optimal rank-based procedures for time-series analysis: testing an ARMA model against other ARMA models. Ann. Statist. (to appear, March issue).

[5] Wald, A. (1943). Tests of statistical hypotheses concerning several parameters when the number of observations is large. Trans. Amer. Math. Soc. 54, 426-482.

Université Libre de Bruxelles
Indiana University

Reprinted from
Statistical Theory and Data Analysis II,
edited by K. Matusita,
by permission of M.L. Puri's co-author

ASYMPTOTICALLY MOST POWERFUL RANK TESTS FOR MULTIVARIATE RANDOMNESS AGAINST SERIAL DEPENDENCE

MARC HALLIN, JEAN-FRANCOIS INGENBLEEK and MADAN L. PURI*

Universtité Libre de Bruxelles, Brussels, Belgium and Indiana University
Communicated by G. Kallianpur

A class of linear serial multirank statistics is introduced for the problem of testing the null hypothesis that a multivariate series of observations is white noise (with unspecified density function) against alternatives of ARMA dependence. The asymptotic distributional properties of these statistics are investigated, both under the null as well as local alternative hypotheses. These statistics are shown to provide permutationally distribution-free tests that are asymptotically most powerful against specified local alternatives of ARMA dependence. In particular, a test of the *van der Waerden* type is shown to be asymptotically as powerful as the corresponding normal theory parametric test, based on classical sample autocovariances. © 1989 Academic Press, Inc.

INTRODUCTION

The problem of testing for white noise against serial dependence is perhaps one of the most fundamental in statistical inference. The assumption of serial independence plays a crucial role indeed in most classical statistical procedures and, whenever this independence assumption is relaxed (e.g., in time series analysis), a number of hypotheses of interest (typically, all those appearing in model fitting problems) are closely related to hypotheses of white noise, to which they reduce after some adequate manipulations.

However, in spite of the importance and long history of the problem, in spite of the tremendous development of time series analysis, and in spite of the recognized need for robust and nongaussian procedures in this latter domain, the introduction of rank-based methods in the area has received little attention (for a commented bibliography, see Hallin *et al.* [5]).

Received November 4, 1987; accepted July 5, 1988.

AMS 1980 subject classifications. Primary 62M10, 62G10.

Key words and phrases. Multirank serial statistics, multivariate ARMA models, asymptotically locally most powerful tests.

* Research supported by the Office of Naval Research Contract N0014-85-K-0648.

In recent papers by the authors (Hallin *et al.* [5–7] and Hallin and Puri [9]), optimal rank-based procedures are derived for testing univariate white noise against alternatives of ARMA dependence and for the problem of testing a specified univariate ARMA model against other ARMA models. However, whereas early attempts to introduce rank methods in the problem of testing univariate white noise (with unspecified density) against serial dependence can be traced back as far as Wald and Wolfowitz [20]—the corresponding multivariate problem has never been considered so far.

This latter fact is by no means surprising, and it certainly cannot be attributed to a lack of interest in multivariate problems; indeed, it is due to the much greater complexity inherent in the nature of multivariate variables. Thus, the development of multivariate theories—especially in rank-order and time series analysis—typically has been much slower and less complete than that of their univariate counterparts.

The purpose of this paper is to provide rank-based, permutationally distribution-free (hence similar) tests which are asymptotically most powerful against local alternatives of multivariate ARMA dependence. These tests are thus multivariate analogues of the tests proposed in Hallin *et al.* [5, 7]. In order to obtain these tests, we first investigate the asymptotic distribution, under local alternatives of ARMA dependence, of a class of *linear serial multirank statistics.* The word *multirank* in this terminology emphasizes the fact that our statistics are univariate functions of *vectors* of ranks $\mathbf{R}_t^{(n)} = (R_{t,1}^{(n)} \cdots R_{t,m}^{(n)})$ (for a more precise definition, see Section 2 below), whereas the statistics considered in the existing literature are vector-valued variables, each component of which is a function of exactly *one* rank component $R_{t,i}^{(n)}$ of a rank vector $\mathbf{R}_t^{(n)}$; the word *serial* refers to the fact that the *scores* used here are functions of several consecutive vectors of ranks (viz, $\mathbf{R}_t^{(n)}, \ldots, \mathbf{R}_{t-p}^{(n)}$, $p > 0$).

The asymptotic distributional properties of linear serial multirank statistics under randomness (with unspecified density function) and under local alternatives of ARMA dependence are studied in Section 1 (where contiguity is established for the distributions of certain sequences of ARMA processes and white noise) and Section 2. Asymptotic normality is derived through permutational arguments (Sections 2.2 and 2.5), equivalence with sequences of U-statistics (Section 2.3), a central-limit theorem for weakly dependent processes and, finally, LeCam's lemmas (Section 2.4). These distributional results allow for constructing permutationally distribution-free tests for randomness, based on linear serial multirank statistics, computing their asymptotic powers against local alternatives and deriving asymptotic relative efficiencies.

Section 3 is devoted to optimality considerations. It must be noted that although the estimation theory for multiple time series has been considered by a number of authors, hypothesis testing procedures have been discussed in less detail; moreover, optimality results in the area (in a parametric approach) always refer to gaussian types of situations (even if strict gaussian assumptions need not be made—see, e.g., Hosking [11, 12] or Poskitt and Tremayne [15]). If f denotes the density of the generating white noise under the alternative of ARMA dependence, we show here that the key role is played by what we define as the *rank autocovariance matrices associated with* f, which appear to be nongaussian, rank-based, alternatives for the classical sample cross-covariance matrices, to which they actually asymptotically reduce in the case of a gaussian f and score functions of the van der Waerden type. The asymptotic joint normality of these matrices is established under a rather weak *generalized strong unimodality condition* on the density function f (Section 3.2). Finally, asymptotically locally most powerful tests are shown to be provided by appropriate linear combinations of the entries of the rank autocovariance matrices.

The optimality properties of such tests are, of course, of a local nature, and hold against the alternatives they were designed for. However, whatever the (local or nonlocal) alternatives, they have the great advantage to provide tests which are (i) of asymptotic size α for any density f, (ii) permutationally distribution-free—which, in view of classical similarity and Neyman-structure arguments, is a very important property, (iii) at least as powerful, asymptotically and locally, as their parametric counterparts, and (iv) rank-based, hence fairly robust against possible outliers.

Finally, in Section 4, we treat in some detail the case of gaussian scores. The asymptotically locally most powerful parametric test (based on usual sample autocovariances) is derived, and is shown to be equivalent to the corresponding rank-based test of the van der Waerden type.

1. Contiguous Hypotheses of Randomness and ARMA Dependence

1.1. *Notation*

In what follows, $f(\mathbf{x})$ and $F(\mathbf{x})$, $\mathbf{x} \in \mathbb{R}^m$ denote an (m-variate) probability density function, and the corresponding (m-variate) distribution function. $f_i(x)$ and $F_i(x)$, $x \in \mathbb{R}$, $i = 1, ..., m$ denote the marginal probability density functions and the corresponding marginal distribution functions. We assume that $\int \mathbf{x} f(\mathbf{x})\, d\mathbf{x} = \mathbf{0}$ and that $\int \mathbf{x}\mathbf{x}' f(\mathbf{x})\, d\mathbf{x} = \mathbf{\Sigma}$, a (finite) strictly positive definite covariance matrix with diagonal terms $(\mathbf{\Sigma})_{ii} = \sigma_i^2$, $i = 1, ..., m$. Letting $F_i^{-1}(u) = \inf\{x \mid F_i(x) \geqslant u\}$, $u \in (0, 1)$, define

$$\mathbf{F}(\mathbf{x}) = \begin{pmatrix} F_1(x_1) \\ \vdots \\ F_m(x_m) \end{pmatrix}, \qquad \mathbf{x} = (x_1 \cdots x_m)' \in \mathbb{R}^m,$$

and

$$\mathbf{F}^{-1}(\mathbf{u}) = \begin{pmatrix} F_1^{-1}(u_1) \\ \vdots \\ F_m^{-1}(u_m) \end{pmatrix}, \qquad \mathbf{u} = (u_1 \cdots u_m)' \in (0, 1)^m.$$

1.2. *ARMA Dependence*

The null hypothesis we are interested in is the hypothesis of randomness under which the observations $\mathbf{X}_t^{(n)}$ in an m-variate series $\mathbf{X}^{(n)} = (\mathbf{X}_1^{(n)}, ..., \mathbf{X}_t^{(n)}, ..., \mathbf{X}_n^{(n)})$ are independently and identically distributed, according to some (specified or unspecified) density function $f(\mathbf{x})$. Let us denote by $H_f^{(n)}$ and $H^{(n)}$ respectively this null hypothesis of randomness according as f is completely specified, or remains unspecified. The likelihood function for $\mathbf{X}^{(n)}$ under $H_f^{(n)}$ is, obviously,

$$L_f^{(n)}(\mathbf{X}^{(n)}) = \prod_{t=1}^{n} f(\mathbf{X}_t^{(n)}). \tag{1.1}$$

The serial dependence alternatives we shall consider are alternatives under which $\mathbf{X}^{(n)}$ is generated by some specified ARMA model, driven by some white noise with specified or unspecified density function. The testing procedures we are deriving in this paper are devised against all types of serial dependencies; however, as far as optimality is concerned, some parametrization of serial dependence has to be introduced, of which ARMA dependencies certainly constitute the most convenient and intuitively appealing one. Our purpose here is to derive locally asymptotically most powerful tests against ARMA dependence. Nonlocal optimality results (such as an investigation of exact Bahadur efficiency) certainly would be of great interest. However, such results seem by no means easy to derive; by the way, they are not available even for the much simpler cases of univariate linear models (with independent observations). It should be noted, also, that whereas asymptotically locally most powerful rank tests can be found in the literature for univariate linear models (see, e.g., Puri and Sen, [17]), the rank-based tests available in the multivariate case are not asymptotically locally most powerful.

In order to obtain the local optimality result, we shall use the contiguity approach introduced by LeCam [13] but popularized mostly by Hájek and Šidák [3].

Consider therefore the sequence of stochastic difference equations

$$\mathbf{Z}_t - n^{-1/2} \sum_{i=1}^{p_1} \mathbf{A}_i \mathbf{Z}_{t-i} = \boldsymbol{\varepsilon}_t + n^{-1/2} \sum_{i=1}^{p_2} \mathbf{B}_i \boldsymbol{\varepsilon}_{t-i}, \qquad t \in \mathbb{Z},\ n \in \mathbb{N}^+, \tag{1.2}$$

where $\mathbf{A}_1, ..., \mathbf{A}_{p_1}, \mathbf{B}_1, ..., \mathbf{B}_{p_2}$ are real $m \times m$ matrices, $\mathbf{A}_{p_1}$ and $\mathbf{B}_{p_2}$ are of full rank, and $\{\boldsymbol{\varepsilon}_t;\ t \in \mathbb{Z}\}$ is an m-variate white noise process, i.e., a sequence of i.i.d. random variables with common density function $f(\mathbf{x})$. For n sufficiently large, the *stationary* and *invertibility* conditions (cf., e.g., Hannan [10]) are satisfied, and (1.2) admits a uniquely defined stationary solution $\{\mathbf{Z}_t^{(n)}; t \in \mathbb{Z}\}$, say: an ARMA (p_1, p_2) *process* admitting the ARMA (p_1, p_2) representation (1.2). Moreover, denoting by $\mathscr{G}_u^{(n)}$, $u \in \mathbb{Z}$, the *Green's matrices* (see Hallin [4] or Hallin and Puri [9, Appendix 1] for a univariate version) associated with the moving average difference operator in (1.2), and by $\mathscr{C}^{(n)}(t)$ the *Casorati matrix* associated with the fundamental system $\{\mathscr{G}_t, \mathscr{G}_{t+1}, ..., \mathscr{G}_{t+p_2-1}\}$, we may write, for $t \geqslant 1$,

$$\varepsilon_t = \mathbf{Z}_t^{(n)} - n^{-1/2} \sum_{i=1}^{p_1} \mathbf{A}_i \mathbf{Z}_{t-i}^{(n)} + \sum_{u=1}^{t-1} \mathscr{G}_u^{(n)} \left(\mathbf{Z}_{t-u}^{(n)} - n^{-1/2} \sum_{i=1}^{p_1} \mathbf{A}_i \mathbf{Z}_{t-u-i}^{(n)} \right)$$
$$+ (\mathscr{G}_t^{(n)} \mathscr{G}_{t+1}^{(n)} \cdots \mathscr{G}_{t+p_2-1}^{(n)}) (\mathscr{C}^{(n)}(0))^{-1} (\varepsilon_0' \cdots \varepsilon_{-p_2+1}')' \tag{1.3}$$

with

$$[\mathscr{C}^{(n)}(0)]^{-1} = \begin{pmatrix} \mathscr{G}_0^{(n)} & \mathscr{G}_1^{(n)} & \cdots & \mathscr{G}_{p_2-1}^{(n)} \\ \mathscr{G}_{-1}^{(n)} & \mathscr{G}_0^{(n)} & \cdots & \mathscr{G}_{p_2-2}^{(n)} \\ \vdots & & & \vdots \\ \mathscr{G}_{-p_2+1}^{(n)} & & \cdots & \mathscr{G}_0^{(n)} \end{pmatrix}^{-1}$$
$$= n^{-1/2} \begin{pmatrix} \mathbf{I} & \mathbf{B}_1 & \cdots & \mathbf{B}_{p_2-1} \\ \mathbf{0} & & \ddots & \vdots \\ \vdots & \ddots & & \mathbf{B}_1 \\ \mathbf{0} & \cdots & \mathbf{0} & \mathbf{I} \end{pmatrix}.$$

Accordingly, under the alternative—denote it by $K_f^{(n)}(\mathbf{A}, \mathbf{B})$—that $\mathbf{X}^{(n)}$ constitutes a finite realization of length n of $\{\mathbf{Z}_t^{(n)}\}$, we have the likelihood

$$L_{\mathbf{A}, \mathbf{B}; f}^{(n)}(\mathbf{X}^{(n)}) = \int_{\mathbb{R}^{m(p_1+p_2)}} \Lambda_{\mathbf{A}, \mathbf{B}; f}^{(n)}(\mathbf{X}^{(n)}; \mathbf{x}_0, \mathbf{e}_0)\, dG^{(n)}(\mathbf{x}_0, \mathbf{e}_0), \tag{1.4}$$

where $G^{(n)}(\mathbf{x}_0, \mathbf{e}_0) = G^{(n)}(\mathbf{x}_0, \mathbf{x}_{-1}, ..., \mathbf{x}_{-p_1+1}, \mathbf{e}_0, \mathbf{e}_{-1}, ..., \mathbf{e}_{-p_2+1})$ stands for the joint distribution function of $\mathbf{Z}_0^{(n)}, ..., \mathbf{Z}_{-p_1+1}^{(n)}, \varepsilon_0, ..., \varepsilon_{-p_2+1}$, and

$$\Lambda_{\mathbf{A}, \mathbf{B}; f}^{(n)}(\mathbf{X}^{(n)}; \mathbf{X}_0^{(n)}, \mathbf{e}_0)$$
$$= \prod_{t=1}^{n} f \left\{ \mathbf{X}_t^{(n)} - n^{-1/2} \sum_{i=1}^{p_1} \mathbf{A}_i \mathbf{X}_{t-i}^{(n)} \right.$$
$$+ \sum_{u=1}^{t-1} \mathscr{G}_u^{(n)} \left(\mathbf{X}_{t-u}^{(n)} - n^{-1/2} \sum_{i=1}^{p_1} \mathbf{A}_i \mathbf{X}_{t-u-i}^{(n)} \right)$$
$$\left. + (\mathscr{G}_t^{(n)} \mathscr{G}_{t+1}^{(n)} \cdots \mathscr{G}_{t+p_2-1}^{(n)}) [\mathscr{C}^{(n)}(0)]^{-1} (\mathbf{e}_0' \cdots \mathbf{e}_{-p_2+1}')' \right\}. \tag{1.5}$$

1.3. *Contiguity*

Consider the log-likelihood ratio

$$\mathscr{L}^{(n)}_{\mathbf{A},\mathbf{B};f}(\mathbf{X}^{(n)}) = \begin{cases} \log L^{(n)}_{\mathbf{A},\mathbf{B};f}(\mathbf{X}^{(n)}) - \log L^{(n)}_f(\mathbf{X}^{(n)}), & \\ \qquad \text{if } L^{(n)}_f(\mathbf{X}^{(n)}) > 0 & \\ 0, \qquad \text{if } L^{(n)}_{\mathbf{A},\mathbf{B};f}(\mathbf{X}^{(n)}) = 0 = L^{(n)}_f(\mathbf{X}^{(n)}) & \\ \infty, \qquad \text{if } L^{(n)}_{\mathbf{A},\mathbf{B};f}(\mathbf{X}^{(n)}) \neq 0 = L^{(n)}_f(\mathbf{X}^{(n)}). & \end{cases} \tag{1.6}$$

In order to establish the contiguity of the sequence $K^{(n)}_f(\mathbf{A}, \mathbf{B})$ to the sequence $H^{(n)}_f$ we shall use a corollary [3, p. 204] to the so-called *Le Cam's first lemma.* To this end, we need the following regularity conditions (which are assumed to hold throughout the paper).

(i) *Finite cross-moments up to the third order*: $\int x_i^2 x_j f(\mathbf{x})\, d\mathbf{x} < \infty$, $i, j \in \{1, ..., m\}$ (a sufficient condition is $\int x^4 f_i(x)\, dx < \infty$, $i = 1, ..., m$).

(ii) *Finite Fisher's information.* $f(\mathbf{x})$ is a.e. derivable; denoting $\boldsymbol{\phi}(\mathbf{x}) = -\text{grad} \log f(\mathbf{x}) = (\phi_1(\mathbf{x}) \cdots \phi_m(\mathbf{x}))'$, we assume that $\int |\phi_i(\mathbf{x})|^{2+\delta} f(\mathbf{x})\, d\mathbf{x} < \infty$ for some $\delta > 0$. (Then $f(\mathbf{x})$ has finite Fisher's information matrix $\mathscr{I}(f) = \int \boldsymbol{\phi}(\mathbf{x})\, \boldsymbol{\phi}'(\mathbf{x}) f(\mathbf{x})\, d\mathbf{x}$).

(iii) The ϕ_i's are a.e. derivable; denoting by $\dot{\boldsymbol{\phi}}(\mathbf{x}) = (-\partial^2 \log f(\mathbf{x})/\partial x_i\, \partial x_j)$ its derivatives, assume them to satisfy (a.e.) a Lipschitz condition of the form $|\dot{\phi}_{ij}(\mathbf{x}) - \dot{\phi}_{ij}(\mathbf{y})| \leqslant k \|\mathbf{x} - \mathbf{y}\|$ (k is a constant depending on f; $\|\cdot\|$ stands for the euclidean norm).

We then have the following result.

PROPOSITION 1.1. *Under the above assumptions, letting* $p = \max(p_1, p_2)$,

$$\mathbf{D}_i = \begin{cases} \mathbf{A}_i + \mathbf{B}_i & \textit{if } 1 \leqslant i \leqslant \min(p_1, p_2) \\ \mathbf{A}_i & \textit{if } p_2 < i \leqslant p_1 \\ \mathbf{B}_i & \textit{if } p_1 < i \leqslant p_2 \\ \mathbf{0} & \textit{if } p < i \end{cases} \tag{1.7}$$

and

$$\mathscr{L}^{(n)}_0 = n^{-1/2} \sum_{t=p+1}^{n} \sum_{i=1}^{p} \text{tr}[\boldsymbol{\phi}(\mathbf{X}^{(n)}_t)\, \mathbf{X}^{(n)\prime}_{t-i} \mathbf{D}'_i], \tag{1.8}$$

the log-likelihood (1.6) *can be written as*

$$\mathscr{L}^{(n)}_{\mathbf{A},\mathbf{B};f} = \mathscr{L}^{(n)}_0 - \frac{1}{2} \sum_{i=1}^{p} \text{tr}[\mathscr{I}(f)\, \mathbf{D}_i \boldsymbol{\Sigma} \mathbf{D}'_i] + o_p(1), \tag{1.9}$$

and is asymptotically normal with mean $-\frac{1}{2}\sum_{i=1}^{p} \operatorname{tr}[\mathscr{I}(f)\,\mathbf{D}_i\boldsymbol{\Sigma}\mathbf{D}_i']$ *and variance* $\sum_{i=1}^{p} \operatorname{tr}[\mathscr{I}(f)\,\mathbf{D}_i\boldsymbol{\Sigma}\mathbf{D}_i']$, *which implies that* $K_f^{(n)}(\mathbf{A}, \mathbf{B})$ *is contiguous to* $H_f^{(n)}$.

Proof. The proof follows by considering the asymptotic behavior of a second-order Taylor expansion of $\log \Lambda^{(n)}_{\mathbf{A},\mathbf{B};f}(\mathbf{X}^{(n)}) - \sum_{t=1}^{n} \log f(\mathbf{X}_t^{(n)})$. Using the fact that

$$\sum_{u=pv+1}^{\infty} \max_{j,k} |(\mathscr{G}_u^{(n)})_{j,k}| \leqslant 2(p \max_{i,j,k} |(\mathbf{B}_i)_{j,k}|\, n^{-1/2})^{v+1} = o(n^{-v/2})$$

for any $v \in \mathbb{N}$, n sufficiently large (this property of Green's matrices can easily be adapted from the corresponding property of Green's functions, Hallin *et al.* [5, Appendix 1]), the first-order term in this expansion can be written in the form

$$\begin{aligned}
&n^{-1/2} \sum_{t=p+1}^{n} \boldsymbol{\phi}'(\mathbf{X}_t^{(n)}) \left(\sum_{i=1}^{p_1} \mathbf{A}_i - n^{1/2} \sum_{i=1}^{p_2} \mathscr{G}_i^{(n)} \right) \mathbf{X}_{t-i}^{(n)} + o_p(1) \\
&\quad = n^{-1/2} \sum_{t=p+1}^{n} \sum_{i=1}^{p} \boldsymbol{\phi}'(\mathbf{X}_t^{(n)})\, \mathbf{D}_i \mathbf{X}_{t-i}^{(n)} + o_p(1) \\
&\quad = n^{-1/2} \sum_{t=p+1}^{n} \sum_{i=1}^{p} \operatorname{tr}[\mathbf{D}_i \mathbf{X}_{t-i}^{(n)} \boldsymbol{\phi}'(\mathbf{X}_t^{(n)})] + o_p(1) \\
&\quad = \mathscr{L}_0^{(n)}(\mathbf{X}^{(n)}) + o_p(1).
\end{aligned}$$

As a sum of p-dependent identically distributed variables of the form $n^{-1/2}\zeta_t$, $\mathscr{L}_0^{(n)}$ is asymptotically normal (see, e.g., [1, p. 427]), with mean

$$E(\zeta_t) = \sum_{i=1}^{p} \operatorname{tr}[\mathbf{D}_i E\mathbf{X}_{t-i}^{(n)} E\boldsymbol{\phi}'(\mathbf{X}_t^{(n)})] = 0$$

and variance

$$\begin{aligned}
&E(\zeta_t^2) + 2 \sum_{i=1}^{p} E(\zeta_t \zeta_{t+1}) \\
&\quad = E\left\{ \boldsymbol{\phi}'(\mathbf{X}_t^{(n)}) \left(\sum_{i=1}^{p} \mathbf{D}_i \mathbf{X}_{t-i}^{(n)} \right) \left(\sum_{i=1}^{p} \mathbf{D}_i \mathbf{X}_{t-i}^{(n)} \right)' \boldsymbol{\phi}(\mathbf{X}_t^{(n)}) \right\} + 0 \\
&\quad = \sum_{i=1}^{p} E\{\boldsymbol{\phi}'(\mathbf{X}_t^{(n)})\, \mathbf{D}_i \boldsymbol{\Sigma} \mathbf{D}_i' \boldsymbol{\phi}(\mathbf{X}_t^{(n)})\} \\
&\quad = \sum_{i=1}^{p} \operatorname{tr}[\mathscr{I}(f)\, \mathbf{D}_i \boldsymbol{\Sigma} \mathbf{D}_i'].
\end{aligned}$$

The second-order term can be treated along the same lines as in the univariate case, using the Lipschitz property of $\dot{\boldsymbol{\phi}}(\mathbf{X}_t^{(n)})$ and the finiteness (under $H_f^{(n)}$) of $\mathbf{X}_t^{(n)}$'s third-order moments. This second-order term then takes the form

$$-\frac{1}{2n}\sum_{t=p+1}^{n}\left[\left(\sum_{i=1}^{p_1}\mathbf{A}_i-n^{1/2}\sum_{i=1}^{p_2}\mathscr{G}_i^{(n)}\right)\mathbf{X}_{t-i}^{(n)}\right]'\dot{\boldsymbol{\phi}}(\mathbf{X}_t^{(n)})$$
$$\times\left[\left(\sum_{i=1}^{p_1}\mathbf{A}_i-n^{1/2}\sum_{i=1}^{p_2}\mathscr{G}_i^{(n)}\right)\mathbf{X}_{t-i}^{(n)}\right]+o_p(1)$$
$$=-\frac{1}{2n}\sum_{t=p+1}^{n}\left(\sum_{i=1}^{p}\mathbf{D}_i\mathbf{X}_{t-i}^{(n)}\right)'\dot{\boldsymbol{\phi}}(\mathbf{X}_t^{(n)})\left(\sum_{i=1}^{p}\mathbf{D}_i\mathbf{X}_{t-i}^{(n)}\right)+o_p(1).$$

This second-order term thus converges in probability to

$$-\frac{1}{2}E\left\{\left(\sum_{i=1}^{p}\mathbf{D}_i\mathbf{X}_{t-i}^{(n)}\right)'\dot{\boldsymbol{\phi}}(\mathbf{X}_t^{(n)})\left(\sum_{i=1}^{p}\mathbf{D}_i\mathbf{X}_{t-i}^{(n)}\right)\right\}$$
$$=-\frac{1}{2}\sum_{i=1}^{p}E\{(\mathbf{D}_i\mathbf{X}_{t-i}^{(n)})'\mathscr{I}(f)(\mathbf{D}_i\mathbf{X}_{t-i}^{(n)})\}$$
$$=-\frac{1}{2}\sum_{i=1}^{p}\operatorname{tr}[\mathscr{I}(f)\,\mathbf{D}_i\boldsymbol{\Sigma}\mathbf{D}_i'].$$

To complete the proof, note that

$$\mathscr{L}_{\mathbf{A},\mathbf{B};f}^{(n)}(\mathbf{X}^{(n)})=\log E\left[\Lambda_{\mathbf{A},\mathbf{B};f}^{(n)}(\mathbf{X}^{(n)};\mathbf{X}_0,\boldsymbol{\epsilon}_0)\mid\mathbf{X}^{(n)}\right]-\sum_{t=1}^{n}\log f(\mathbf{X}_t^{(n)}),$$

where the expectation is taken with respect to the joint distribution $G^{(n)}(\mathbf{x}_0,\mathbf{e}_0)$ of $(\mathbf{X}_0,\boldsymbol{\epsilon}_0)$ considered independent of $\mathbf{X}^{(n)}$ itself; hence

$$\mathscr{L}_{\mathbf{A},\mathbf{B};f}^{(n)}(\mathbf{X}^{(n)})=\mathscr{L}_0^{(n)}(\mathbf{X}^{(n)})-\frac{1}{2}\sum_{i=1}^{p}\operatorname{tr}[\mathscr{I}(f)\,\mathbf{D}_i\boldsymbol{\Sigma}\mathbf{D}_i']$$
$$+\log E[\exp(\mathscr{R}^{(n)})\mid\mathbf{X}^{(n)}],$$

where the remainder term $\mathscr{R}^{(n)}$ converges to zero in quadratic mean. This in turn implies (1.9). Contiguity then follows from *Le Cam's first lemma.*

2. Distribution Theory of Linear Serial Multirank Statistics

2.1. *Linear Serial Multirank Statistics*

Let $R_{t,i}^{(n)}$ be the rank of $X_{t,i}^{(n)}$ among $\{X_{1,i}^{(n)}, ..., X_{n,i}^{(n)}\}$, $\mathbf{R}_t^{(n)} = (R_{t,1}^{(n)} \cdots R_{t,m}^{(n)})'$ and $\mathbf{R}^{(n)} = (\mathbf{R}_1^{(n)} \cdots \mathbf{R}_n^{(n)})$. $\mathbf{R}^{(n)}$ is called the *rank-collection matrix* of the series.

It is well known that (unless $m=1$) the rank vectors $\mathbf{R}_t^{(n)}$ are not distribution-free, even under the null hypothesis of randomness. If, however, we denote, as usual, by $\mathbf{R}_*^{(n)}$ the rank-collection matrix obtained from $\mathbf{R}^{(n)}$ by rearranging the columns in such a manner that the first row has elements 1, 2, ..., m (in ascending order), the $n!$ possible permutations of the columns of $\mathbf{R}_*^{(n)}$ have the same probability: $(n!)^{-1}$ times the probability of $\mathbf{R}_*^{(n)}$ itself. Consequently, the rank vectors $\mathbf{R}_t^{(n)}$ are *conditionally distribution-free*, given the "ordered" rank-collection matrix $\mathbf{R}_*^{(n)}$; we also say that they are *permutationally distribution-free.*

Define a *linear serial multirank statistic of order p* as a statistic of the form

$$S^{(n)}(\mathbf{R}^{(n)}) = (n-p)^{-1} \sum_{t=p+1}^{n} a^{(n)}(\mathbf{R}_t^{(n)}, \mathbf{R}_{t-1}^{(n)}, ..., \mathbf{R}_{t-p}^{(n)}), \tag{2.1}$$

where $\mathbf{R}^{(n)}$ is a rank-collection matrix of dimension $m \times n$, and the $a^{(n)}(\cdots)$'s are a set of $(n(n-1) \cdot \cdots \cdot (n-p))^m$ *scores*, one for each possible $(p+1)$-tuple of rank vectors. Related to these scores, assume the existence of a real-valued function $J(\mathbf{u}_1, \mathbf{u}_2, ..., \mathbf{u}_{p+1})$, with $\mathbf{u}_i \in (0, 1)^m$, satisfying

$$\int_{\mathbb{R}^{(p+1)m}} |J(\mathbf{F}(\mathbf{x}_1), ..., \mathbf{F}(\mathbf{x}_{p+1}))|^{2+\delta}\, dF(\mathbf{x}_1) \cdot \cdots \cdot dF(\mathbf{x}_{p+1}) < \infty \tag{2.2}$$

for some $\delta > 0$, and such that, under $H_f^{(n)}$,

$$\lim_{n \to \infty} E\{[J(\mathbf{F}(\mathbf{X}_t^{(n)}), ..., \mathbf{F}(\mathbf{X}_{t-p}^{(n)})) - a^{(n)}(\mathbf{R}_t^{(n)}, ..., \mathbf{R}_{t-p}^{(n)})]^2\} = 0. \tag{2.3}$$

The function $J(\mathbf{u}_1, ..., \mathbf{u}_{p+1})$ is called a *score-generating function* associated with (2.1).

Note that (2.1), as a univariate function of rank *vectors* (whence the term *multirank* statistic), is not of the type usually considered in the literature on multivariate linear models (see Puri and Sen, [17]), where multivariate linear statistics, each component of which however is a function of *one* rank at a time are usually preferred. Our statistics are thus of a more general nature, which unlike the (nonserial) existing ones allows for asymptotically locally most powerful tests.

2.2. *Some Permutational Properties of Linear Serial Multirank Statistics*

Before proceeding to the study of the asymptotic distribution of linear serial rank statistics, we first establish some of their permutational properties. Denote by $E^*(\cdot)$, $\mathrm{Var}^*(\cdot)$ and $\mathrm{Cov}^*(\cdot, \cdot)$ respectively the permutation expectations, variances, and covariances, i.e., the conditional quantities $E(\cdot \mid \mathbf{R}_*^{(n)})$, $\mathrm{Var}(\cdot \mid \mathbf{R}_*^{(n)})$, and $\mathrm{Cov}(\cdot, \cdot \mid \mathbf{R}_*^{(n)})$. The lemmas below will be used in the subsequent sections. Since under $H_f^{(n)}$, the conditional distribution of the rank vectors given $\mathbf{R}_*^{(n)}$ does not depend on the underlying f, these lemmas essentially rely on combinatorial arguments. Also, since the scores $a^{(n)}(\cdots)$ depend on several consecutive rank vectors, the enumeration techniques we have to use in the proofs are much more tedious than in the classical nonserial case, where the scores depend only on one rank vector.

LEMMA 2.1. (i)

$$E^*(S^{(n)}) = [n(n-1)\cdots(n-p)]^{-1} \sum_{1 \leqslant t_1 \neq \cdots \neq t_{p+1} \leqslant n} \sum a^{(n)}(\mathbf{R}_{t_1}^{(n)}, \ldots, \mathbf{R}_{t_{p+1}}^{(n)}).$$

(ii)

$$\begin{aligned}
\mathrm{Var}^*(S^{(n)}) &= (n-p)^{-1}\, \mathrm{Var}^*(a^{(n)}(\mathbf{R}_1^{(n)}, \ldots, \mathbf{R}_{p+1}^{(n)})) \\
&\quad + 2(n-p)^{-2} \sum_{i=1}^{p} (n-p-i) \\
&\quad \times \mathrm{Cov}^*(a^{(n)}(\mathbf{R}_1^{(n)}, \ldots, \mathbf{R}_{p+1}^{(n)}), a^{(n)}(\mathbf{R}_{1+i}^{(n)}, \ldots, \mathbf{R}_{p+1+i}^{(n)})) \\
&\quad + (n-2p)(n-2p-1)(n-p)^{-2} \\
&\quad \times \mathrm{Cov}^*(a^{(n)}(\mathbf{R}_1^{(n)}, \ldots, \mathbf{R}_{p+1}^{(n)}), a^{(n)}(\mathbf{R}_{p+2}^{(n)}, \ldots, \mathbf{R}_{2p+2}^{(n)})) \\
&\leqslant (2p+1)(n-p)^{-1}\, \mathrm{Var}^*(a^{(n)}) \\
&\quad + \frac{n^2}{(n-p)^2} \mathrm{Cov}^*(a^{(n)}(\mathbf{R}_1^{(n)}, \ldots, \mathbf{R}_{p+1}^{(n)}), (\mathbf{R}_{p+2}^{(n)}, \ldots, \mathbf{R}_{2p+2}^{(n)})).
\end{aligned} \tag{2.4}$$

Proof. (i) is obvious; (ii) follows from expanding $E^*[(\sum a^{(n)}(\cdots))^2]$.

LEMMA 2.2.

$$|\mathrm{Cov}^*(a^{(n)}(\mathbf{R}_1^{(n)}, \ldots, \mathbf{R}_{p+1}^{(n)}), a^{(n)}(\mathbf{R}_{p+2}^{(n)}, \ldots, \mathbf{R}_{2p+2}^{(n)}))| \leqslant Kn^{-1}E^*[(a^{(n)})^2].$$

Proof. The proof is similar to that of Lemmas 3 and 4 in Hallin *et al.* [5, Appendix 3].

LEMMA 2.3.

$$
\begin{aligned}
&E^*[a^{(n)}(\mathbf{R}_1^{(n)}, ..., \mathbf{R}_{p+1}^{(n)}) \mid \mathbf{R}_{p+2}^{(n)}, ..., \mathbf{R}_{p+1+k}^{(n)}] \\
&\quad = [n(n-1)\cdots(n-p)][(n-k)(n-k-1)\cdots(n-k-p)]^{-1} E^*(a^{(n)}) \\
&\quad\quad + \sum_{l=1}^{k} \sum_{p+2\leqslant\beta_1<} \cdots \sum_{<\beta_l\leqslant p+1+k} \sum_{1\leqslant\alpha_1\neq} \cdots \sum_{\neq\alpha_l\leqslant p+1} (-1)^l \\
&\quad\quad \times \frac{(n-l)(n-l-1)\cdots(n-p)}{(n-k)(n-k-1)\cdots(n-k-p)} \\
&\quad\quad \times E^*[a^{(n)}(\mathbf{R}_1^{(n)}, \mathbf{R}_2^{(n)}, ..., \mathbf{R}_{\alpha_i-1}^{(n)}, \mathbf{R}_{\beta_i}^{(n)}, \mathbf{R}_{\alpha_i+1}^{(n)}, ..., \mathbf{R}_{p+1}^{(n)}) \mid \mathbf{R}_{\beta_1}^{(n)}, \mathbf{R}_{\beta_2}^{(n)}, ..., \mathbf{R}_{\beta_l}^{(n)}]
\end{aligned}
$$

for $k = 1, ..., n-p-1$ *and with the notational convention* $(n-l)(n-l-1)\cdots(n-p) = 0$ *for* $l \geqslant p+1$.

Proof. The expectation in the left-hand side of (2.5) is actually $[(n-k)(n-k-1)\cdots(n-k-p)]^{-1} \sum a^{(n)}(\mathbf{R}_{t_1}^{(n)}, ..., \mathbf{R}_{t_{p+1}}^{(n)})$, where the summation extends over all the $(p+1)$-tuples $(\mathbf{R}_{t_1}^{(n)}, ..., \mathbf{R}_{t_{p+1}}^{(n)})$ of distinct columns of $\mathbf{R}_*^{(n)}$ not belonging to $\{\mathbf{R}_{p+2}^{(n)}, ..., \mathbf{R}_{p+1+k}^{(n)}\}$. This sum decomposes into $\sum^0 - \sum^1$, where $\sum^0$ extends over *all* possible $(p+1)$-tuples of distinct columns of $\mathbf{R}_*^{(n)}$, whereas $\sum^1$ extends over all such $(p+1)$-tuples having at least one element in the set $\{\mathbf{R}_{p+2}^{(n)}, ..., \mathbf{R}_{p+1+k}^{(n)}\}$. $\sum^1$ in turn decomposes into $\sum^{(1)} \sum^{(2)}$, where $\sum^{(1)}$ extends over all possible ordered $(p+1)$-tuples in $\sum^1$, and $\sum^{(2)}$ over the $(p+1)!$ permutations of these $(p+1)$-tuples.

Now, $\sum^{(1)}$ can be written in the form

$$
\sum^{(1)}(\cdots) = -\sum_{l=1}^{k} (-1)^l \Bigg[\sum_{p+2\leqslant\beta_1<} \cdots \sum_{<\beta_l\leqslant p+1+k} \sum^{(1)}_{(\beta_1, ..., \beta_l)} (\cdots) \Bigg],
$$

where the sum between brackets concerns all the ordered $(p+1)$-tuples of distinct columns of $\mathbf{R}_*^{(n)}$ containing at least l out of the k columns $\mathbf{R}_{p+2}^{(n)}, ..., \mathbf{R}_{p+1+k}^{(n)}$, and $\sum^{(1)}_{(\beta_1, ..., \beta_l)}$ runs over all such $(p+1)$-tuples that include $\mathbf{R}_{\beta_1}^{(n)}, ..., \mathbf{R}_{\beta_l}^{(n)}$.

A closer look at the sums $\sum^{(1)}_{(\beta_1, ..., \beta_l)} \sum^{(2)}$ shows that they extend over all possible choices of l subscripts $1 \leqslant \alpha_1 \neq \cdots \neq \alpha_l \leqslant p+1$, all the possible choices of an ordered $(p+1-l)$-tuple of $\mathbf{R}_t^{(n)}$'s in $\mathbf{R}_*^{(n)}$ not belonging to $\{\mathbf{R}_{\beta_1}^{(n)}, ..., \mathbf{R}_{\beta_l}^{(n)}\}$ and finally all the permutations of this $(p+1-l)$-tuple, i.e.,

$$\sum_{1\leqslant \alpha_1\neq}\cdots\sum_{\cdots\neq\alpha_l\leqslant p+1}\;\sum_{1\leqslant t_1\neq}\overset{\notin\{\beta_1,\ldots,\beta_l\}}{\cdots}\sum_{\cdots\neq t_{p+1-l}\leqslant p+1} a^{(n)}(\mathbf{R}^{(n)}_{t_1},\ldots,\mathbf{R}^{(n)}_{t_{\alpha_1-1}},\mathbf{R}^{(n)}_{\beta_1},\mathbf{R}^{(n)}_{t_{\alpha_1}},\ldots,\mathbf{R}^{(n)}_{t_{p+1-l}})$$
$$=\sum_{1\leqslant\alpha_1\neq}\cdots\sum_{\cdots\neq\alpha_l\leqslant p+1}(n-l)(n-l-1)\cdots(n-p)$$
$$\times E^*[a^{(n)}(\mathbf{R}^{(n)}_1,\ldots,\mathbf{R}^{(n)}_{\alpha_1-1},\mathbf{R}^{(n)}_{\beta_1},\mathbf{R}^{(n)}_{\alpha_1+1},\ldots,\mathbf{R}^{(n)}_{p+1})\mid\mathbf{R}^{(n)}_{\beta_1},\ldots,\mathbf{R}^{(n)}_{\beta_l}],$$

whereas $\sum^0$ reduces to $n(n-1)\cdots(n-p)\,E^*(a^{(n)})$. This completes the proof.

A similar type of argument can be used to establish the next lemma.

LEMMA 2.4.

$$E^*[a^{(n)}(\mathbf{R}^{(n)}_1,\ldots,\mathbf{R}^{(n)}_{t_1},\ldots,\mathbf{R}^{(n)}_{t_r},\ldots,\mathbf{R}^{(n)}_{p+1})\mid\mathbf{R}^{(n)}_{t_1},\ldots,\mathbf{R}^{(n)}_{t_r},\mathbf{R}^{(n)}_{p+2},\ldots,\mathbf{R}^{(n)}_{p+1+k}]$$
$$=\frac{(n-r)\cdots(n-p)}{(n-k)\cdots(n-k-p+r)}$$
$$\times E^*[a^{(n)}(\mathbf{R}^{(n)}_1,\ldots,\mathbf{R}^{(n)}_{t_1},\ldots,\mathbf{R}^{(n)}_{p+1})\mid\mathbf{R}^{(n)}_{t_1},\ldots,\mathbf{R}^{(n)}_{t_r}]$$
$$+\sum_{l=1}^{k}\;\sum_{p+2\leqslant\beta_1<}\cdots\sum_{<\beta_l\leqslant p+1+k}\;\sum_{1\leqslant\alpha_1\neq}\overset{\notin\{t_1,\ldots,t_r\}}{\cdots}\sum_{\neq\alpha_l\leqslant p+1}(-1)^l$$
$$\times\frac{(n-l)\cdots(n-p)}{(n-k)\cdots(n-p-k+r)}$$
$$\times E^*[a^{(n)}(\mathbf{R}^{(n)}_1,\ldots,\mathbf{R}^{(n)}_{t_i},\ldots,\mathbf{R}^{(n)}_{\alpha_{i'}-1},\mathbf{R}^{(n)}_{\beta_{i'}},\mathbf{R}^{(n)}_{\alpha_{i'}+1},\ldots,\mathbf{R}^{(n)}_{p+1})\mid\mathbf{R}^{(n)}_{t_1},\ldots,$$
$$\mathbf{R}^{(n)}_{t_r},\mathbf{R}^{(n)}_{\beta_1},\ldots,\mathbf{R}^{(n)}_{\beta_l}]$$

for $k=1,\ldots,n-p-1$ (*with the same notational convention as in Lemma* 2.3 *above*).

Define the "conditionally centered" scores $a^{(n)}_*$ as

$$a^{(n)}_*(\mathbf{R}^{(n)}_1,\ldots,\mathbf{R}^{(n)}_{p+1})$$
$$=a^{(n)}(\mathbf{R}^{(n)}_1,\ldots,\mathbf{R}^{(n)}_{p+1})$$
$$-\frac{(n-1)(n-p)}{n(n-p+1)}\sum_{t=1}^{p+1}E^*[a^{(n)}(\mathbf{R}^{(n)}_1,\ldots,\mathbf{R}^{(n)}_t,\ldots,\mathbf{R}^{(n)}_{p+1})\mid\mathbf{R}^{(n)}_t]$$
$$-\frac{(n-1)}{n(n-p-1)}\sum_{1\leqslant\alpha\neq}\sum_{\beta\leqslant p+1}E^*[a^{(n)}(\mathbf{R}^{(n)}_1,\ldots,\mathbf{R}^{(n)}_{\alpha-1},\mathbf{R}^{(n)}_{\beta},\mathbf{R}^{(n)}_{\alpha+1},\ldots,$$
$$\mathbf{R}^{(n)}_{p+1}\mid\mathbf{R}^{(n)}_{\beta}]+[np/(n-p-1)]\;E^*(a^{(n)}).$$

These $a^{(n)}_*$, however, cannot really be considered as score functions (of order p) anymore, since they depend on *all* the columns of $\mathbf{R}_*^{(n)}$. We shall use them merely as a notational convenience

LEMMA 2.5. *The conditionally centered scores* $a_*^{(n)}$ *are such that*

(i) $E^*(a_*^{(n)})=0$.

(ii) $E^*[a_*^{(n)}(\mathbf{R}_1^{(n)}, ..., \mathbf{R}_{\alpha-1}^{(n)}, \mathbf{R}_\beta^{(n)}, \mathbf{R}_{\alpha+1}^{(n)}, ..., \mathbf{R}_{p+1}^{(n)}) \mid \mathbf{R}_\beta^{(n)}]=0, \quad \alpha=1, ..., p+1, \beta=1, 2, ..., n$.

(iii) $(n-p)^{-1}\sum_{t=p+1}^{n} a_*^{(n)}(\mathbf{R}_t^{(n)}, \mathbf{R}_{t-1}^{(n)}, ..., \mathbf{R}_{t-p}^{(n)})=S^{(n)}-E^*(S^{(n)})+o_p(1)$.

Proof. (i) is a straightforward consequence of (ii). In order to prove (ii), first note that on account of Lemma 2.4, the value of

$$E^*[E^*[a^{(n)}(\mathbf{R}_1^{(n)}, ..., \mathbf{R}_{a-1}^{(n)}, \mathbf{R}_b^{(n)}, \mathbf{R}_{a+1}^{(n)}, ..., \mathbf{R}_{p+1}^{(n)}) \mid \mathbf{R}_b^{(n)}] \mid \mathbf{R}_\beta^{(n)}]$$

where the conditional expectation $E^*[\cdot \mid \mathbf{R}_\beta^{(n)}]$ is computed under $\mathbf{R}_\alpha^{(n)}=\mathbf{R}_\beta^{(n)}$ is $(n/(n-1))\ E^*(a^{(n)})-(1/(n-1))\ E^*[a^{(n)}(\mathbf{R}_1^{(n)}, ..., \mathbf{R}_{a-1}^{(n)}, \mathbf{R}_\beta^{n}, \mathbf{R}_{a+1}^{(n)}, ..., \mathbf{R}_{p+1}^{(n)}) \mid \mathbf{R}_\beta^{(n)}]$, if $b\neq\alpha$, and $E^*[a^{(n)}(\mathbf{R}_1^{(n)}, ..., \mathbf{R}_{a-1}^{(n)}, \mathbf{R}_\beta^{(n)}, \mathbf{R}_{a+1}^{(n)}, ..., \mathbf{R}_{p+1}^{(n)}) \mid \mathbf{R}_\beta^{(n)}]$ if $b=\alpha$. Therefore,

$$\begin{aligned}
&E^*[a_*^{(n)}(\mathbf{R}_1^{(n)}, ..., \mathbf{R}_{\alpha-1}^{(n)}, \mathbf{R}_\beta^{(n)}, \mathbf{R}_{\alpha+1}^{(n)}, ..., \mathbf{R}_{p+1}^{(n)}) \mid \mathbf{R}_\beta^{(n)}]\\
&\quad=E^*[a^{(n)}(\mathbf{R}_1^{(n)}, ..., \mathbf{R}_{\alpha-1}^{(n)}, \mathbf{R}_\beta^{(n)}, \mathbf{R}_{\alpha+1}^{(n)}, ..., \mathbf{R}_{p+1}^{(n)}) \mid \mathbf{R}_\beta^{(n)}]\\
&\qquad-\frac{(n-1)(n-p)}{n(n-p-1)}\left\{\frac{np}{n-1}E^*(a^{(n)})-\frac{1}{n-1}\sum_{1=t\neq\alpha}^{p+1}E^*[a^{(n)}(\mathbf{R}_1^{(n)}, ...,\right.\\
&\qquad\qquad\left.\mathbf{R}_{t-1}^{(n)}, \mathbf{R}_\beta^{(n)}, \mathbf{R}_{t+1}^{(n)}, ..., \mathbf{R}_{p+1}^{(n)}) \mid \mathbf{R}_\beta^{(n)}]\right\}\\
&\qquad-\frac{(n-1)(n-p)}{n(n-p-1)}E^*[a^{(n)}(\mathbf{R}_1^{(n)}, ..., \mathbf{R}_{\alpha-1}^{(n)}, \mathbf{R}_\beta^{(n)}, \mathbf{R}_{\alpha+1}^{(n)}, ..., \mathbf{R}_{p+1}^{(n)}) \mid \mathbf{R}_\beta^{(n)}]\\
&\qquad-\frac{(n-1)}{n(n-p-1)}\left\{\frac{np^2}{n-1}E^*(a^{(n)})-\frac{1}{n-1}\sum_{1=a\neq b}^{p+1}\sum_{1=b\neq\alpha}^{p+1}E^*[a^{(n)}(\mathbf{R}_1^{(n)}, ...,\right.\\
&\qquad\qquad\left.\mathbf{R}_{a-1}^{(n)}, \mathbf{R}_\beta^{(n)}, \mathbf{R}_{a+1}^{(n)}, ..., \mathbf{R}_{p+1}^{(n)}) \mid \mathbf{R}_\beta^{(n)}]\right\}\\
&\qquad-\frac{(n-1)}{n(n-p-1)}\sum_{1=t\neq\alpha}^{p+1}E^*[a^{(n)}(\mathbf{R}_1^{(n)}, ..., \mathbf{R}_{t-1}^{(n)}, \mathbf{R}_\beta^{(n)}, \mathbf{R}_t^{(n)}) \mid \mathbf{R}_\beta^{(n)}]\\
&\qquad+\frac{np}{n-p-1}E^*(a^{(n)})=0.
\end{aligned}$$

To establish (iii), note that $a_*^{(n)}$ is a sum of four terms. After summing over t, the first one gives $S^{(n)}$. The second one gives

$$-\frac{(n-1)}{n(n-p-1)}\sum_{t=1}^{n}\sum_{i=\max(1,\,p+2-t)}^{\min(p+1,\,n+1-t)} E^*[a^{(n)}(\mathbf{R}_1^{(n)},\ldots,\mathbf{R}_{i-1}^{(n)},\mathbf{R}_t^{(n)},\mathbf{R}_{i+1}^{(n)},\ldots,\mathbf{R}_{p+1}^{(n)})\mid \mathbf{R}_t^{(n)}]$$

$$=-\frac{(p+1)(n-1)}{(n-p-1)}E^*(S^{(n)})+o_p\left(\frac{1}{n}\right).$$

The third one gives

$$-\frac{(n-1)(p+1)p}{(n-p-1)(n-p)}E^*(S^{(n)})+O_p(n^{-2})=o_p(n^{-1/2}).$$

Regrouping with the fourth one yields

$$S^{(n)}-\left[\frac{(p+1)(n-1)(n-p)+(n-1)(p+1)p}{(n-p)(n-p-1)}-\frac{np}{n-p-1}\right]E^*(S^{(n)})$$
$$+o_p(n^{-1/2})=S^{(n)}-E^*(S^{(n)})+o_p(n^{-1/2}).$$

Note that (iii) would hold exactly (without any $o_p(1)$ term) if a circular version of $S^{(n)}$ were adopted.

LEMMA 2.6.

$$\begin{aligned}\mathrm{Var}^*(S^{(n)}) &= (n-p)^{-1}\,\mathrm{Var}^*(a_*^{(n)})\\
&\quad+2(n-p)^{-2}\sum_{i=1}^{p}(n-p-i)\\
&\quad\times\mathrm{Cov}^*(a_*^{(n)}(\mathbf{R}_1^{(n)},\ldots,\mathbf{R}_{p+1}^{(n)}),\,a_*^{(n)}(\mathbf{R}_{1+i}^{(n)},\ldots,\mathbf{R}_{p+1+i}^{(n)}))\\
&\quad+\sum_{l=1}^{p}\;\sum_{1\leqslant\beta_1<\cdots<\beta_l\leqslant p+1}\;\sum_{1\leqslant\alpha_1\neq\cdots\neq\alpha_l\leqslant p+1}(-1)^l\\
&\quad\times\frac{(n-l)\cdots(n-p+1)}{(n-p)\cdots(n-2p+1)}\\
&\quad\times E^*[a_*^{(n)}(\mathbf{R}_1^{(n)},\ldots,\mathbf{R}_{p+1}^{(n)})\\
&\quad\times E^*[a_*^{(n)}(\mathbf{R}_1^{(n)},\ldots,\mathbf{R}_{\alpha_i-1}^{(n)},\mathbf{R}_{\beta_i}^{(n)},\mathbf{R}_{\alpha_i+1}^{(n)},\ldots,\mathbf{R}_{p+1}^{(n)})\mid\\
&\qquad\mathbf{R}_{\beta_1}^{(n)},\ldots,\mathbf{R}_{\beta_l}^{(n)}]]+o_p(n^{-1/2}).\end{aligned}$$

Proof. The proof follows by using Lemma 2.5(iii), Lemma 2.1(ii), then Lemma 2.3, and finally Lemma 2.5(i).

2.3. *Linear Serial Multirank Statistics as U-Statistics*

In this section, we show that $S^{(n)} - E^*(S^{(n)})$ is asymptotically equivalent (under $H_f^{(n)}$ and up to $o_p(n^{-1/2})$ terms) to a U-statistic $\mathscr{S}^{(n)} - \mathscr{E}^{(n)}$, where

$$\mathscr{S}^{(n)} = (n-p)^{-1} \sum_{t=p+1}^{n} J(\mathbf{F}(\mathbf{X}_t^{(n)}), \ldots, \mathbf{F}(\mathbf{X}_{t-p}^{(n)})) \tag{2.6}$$

and

$$\mathscr{E}^{(n)} = [n(n-1)\cdots(n-p)]^{-1} \sum \cdots \sum_{1 \leqslant t_1 \neq \cdots \neq t_{p+1} \leqslant n} J(\mathbf{F}(\mathbf{X}_{t_1}^{(n)}), \ldots, \mathbf{F}(\mathbf{X}_{t_{p+1}}^{(n)})). \tag{2.7}$$

First, we establish that $\mathscr{S}^{(n)} - \mathscr{E}^{(n)}$ is, asymptotically, a U-statistic (still under $H_f^{(n)}$). Beginning with $\mathscr{E}^{(n)}$, define the estimable parameter

$$\theta^{\mathscr{E}} = \int_{[0,1]^{m(p+1)^2}} \gamma^{\mathscr{E}}(\mathbf{y}_1, \ldots, \mathbf{y}_{p+1})\, d\bar{F}(\mathbf{y}_1)\cdots d\bar{F}(\mathbf{y}_{p+1}), \tag{2.8}$$

where $\bar{F}(\mathbf{y})$ is the distribution function of the $m(p+1)$-dimensional random vector $\mathbf{Y}_t^{(n)} = (\mathbf{F}'(\mathbf{X}_t^{(n)})\cdots\mathbf{F}'(\mathbf{X}_{t-p}^{(n)}))'$, $t = p+1, \ldots, n$ (under $H_f^{(n)}$: $\bar{F}(\mathbf{y}) = \prod_{i=0}^{p} F[\mathbf{F}^{-1}((y_{(i-1)m+1}, \ldots, y_{im})')]$, $\mathbf{y} \in \mathbb{R}^{m(p+1)}$) and where the kernel $\gamma^{\mathscr{E}}$ (of order $p+1$) is defined as

$$\gamma^{\mathscr{E}}(\mathbf{y}_1, \ldots, \mathbf{y}_{p+1}) = [(p+1)!]^{-1} \sum_{1 \leqslant t_1 \neq \cdots \neq t_{p+1} \leqslant p+1} \sum J((y_{t_1,1}, \ldots, y_{t_1,m})', \ldots$$
$$\ldots, (y_{t_{p+1},1}, \ldots, y_{t_{p+1},m})').$$

The corresponding U-statistic is then

$$\mathscr{U}_{\mathscr{E}}^{(n)} = \binom{n-p}{p+1}^{-1} \sum_{p+1 \leqslant t_1 < \cdots < t_{p+1} \leqslant n} \cdots \sum \gamma^{\mathscr{E}}(\mathbf{Y}_{t_1}^{(n)}, \ldots, \mathbf{Y}_{t_{p+1}}^{(n)})$$
$$= \frac{n(n-1)\cdot \cdots \cdot (n-p+1)}{(n-p-1)\cdot \cdots \cdot (n-2p)}\, \mathscr{E}^{(n)} + o_p(n^{-1}).$$

It follows that $(n-p)^{1/2}(\mathscr{E}^{(n)} - \mathscr{U}_{\mathscr{E}}^{(n)}) = o_p(1)$.

As for $\mathscr{S}^{(n)}$, it will be convenient to write it as a U-statistic of the same order as $\mathscr{E}^{(n)}$. This can be achieved by putting

$$\gamma^{\mathscr{S}}(\mathbf{y}_1, \ldots, \mathbf{y}_{p+1}) = (p+1)^{-1} \sum_{i=1}^{p+1} J((y_{i,1}\cdots y_{i,m})', \ldots, (y_{i,pm+1}\cdots y_{i,(p+1)m})').$$

We now prove that $\Delta^{(n)} = (S^{(n)} - E^*(S^{(n)})) - (\mathscr{S}^{(n)} - \mathscr{E}^{(n)})$ is $o_p(n^{-1/2})$. We have

$$E[(\Delta^{(n)})^2] = E[E[(\Delta^{(n)})^2 \mid \mathbf{R}_*^{(n)}, \mathbf{F}_*^{(n)}]] = E[E^*[(\Delta^{(n)})^2 \mid \mathbf{F}_*^{(n)}]],$$

where $\mathbf{F}_*^{(n)}$ is the matrix whose columns are the vectors $\mathbf{F}(\mathbf{X}_t^{(n)})$ rearranged in such a way that the first row has elements in ascending order (i.e., the same permutation of columns $(1, ..., t, ..., n)$ as in $\mathbf{R}_*^{(n)}$). In this conditional distribution, $S^{(n)} - \mathscr{S}^{(n)}$ can be written as a (permutationally distribution-free) linear serial multirank statistic of order p,

$$\begin{aligned} S^{(n)} - \mathscr{S}^{(n)} = (n-p)^{-1} \sum_{t=p+1}^{n} & [a^{(n)}(\mathbf{R}_t^{(n)}, ..., \mathbf{R}_{t-p}^{(n)}) \\ & - J(\mathbf{F}(\mathbf{X}_t^{(n)}), ..., \mathbf{F}(\mathbf{X}_{t-p}^{(n)}))]. \end{aligned}$$

Moreover, $E^*[(\Delta^{(n)})^2 \mid \mathbf{F}_*^{(n)}]$ is precisely the conditional permutation variance of $(S^{(n)} - \mathscr{S}^{(n)})$: indeed,

$$\begin{aligned} & E^*[(S^{(n)} - \mathscr{S}^{(n)}) \mid \mathbf{F}_*^{(n)}] \\ & = E^*(S^{(n)}) - [n(n-1)\cdots(n-p)]^{-1} \\ & \quad \times \sum_{1 \leqslant i_1 \neq \cdots \neq i_{p+1} \leqslant n} \cdots \sum J((\mathbf{F}_*^{(n)})_{\cdot i_1}, ..., (\mathbf{F}_*^{(n)})_{\cdot i_{p+1}}) \\ & = E^*(S^{(n)}) - \mathscr{E}^{(n)}. \end{aligned}$$

Thus

$$E[(\Delta^{(n)})^2] = E[E^*[(\Delta^{(n)})^2 \mid \mathbf{F}_*^{(n)}]] = E[\mathrm{Var}^*[S^{(n)} - \mathscr{S}^{(n)} \mid \mathbf{F}_*^{(n)}]]. \tag{2.10}$$

Using inequality (2.4) and Lemma 2.2 we find that (2.10) is bounded by

$$\begin{aligned} & (2p+1)(n-p)^{-1} E[\mathrm{Var}^*[a^{(n)}(\mathbf{R}_1^{(n)}, ..., \mathbf{R}_{p+1}^{(n)}) - J(\mathbf{F}(\mathbf{X}_1^{(n)}), ..., \mathbf{F}(\mathbf{X}_{p+1}^{(n)})) \mid \mathbf{F}_*^{(n)}]] \\ & \qquad + n^2(n-p)^{-2} E[\mathrm{Cov}^*[(a^{(n)}(\mathbf{R}_1^{(n)}, ..., \mathbf{R}_{p+1}^{(n)}) - J(\mathbf{F}(\mathbf{X}_1^{(n)}), ..., \mathbf{F}(\mathbf{X}_{p+1}^{(n)}))), \\ & (a^{(n)}(\mathbf{R}_{p+2}^{(n)}, ..., \mathbf{R}_{2p+2}^{(n)}) - J(\mathbf{F}(\mathbf{X}_{p+2}^{(n)}), ..., \mathbf{F}(\mathbf{X}_{2p+2}^{(n)}))) \mid \mathbf{F}_*^{(n)}]] \\ & \quad \leqslant [(2p+1)(n-p) + K \cdot n](n-p)^{-2} E[E^*[(a^{(n)}(\mathbf{R}_1^{(n)}, ..., \mathbf{R}_{p+1}^{(n)}) \\ & \qquad - J(\mathbf{F}(\mathbf{X}_1^{(n)}), ..., \mathbf{F}(\mathbf{X}_{p+1}^{(n)})))^2 \mid \mathbf{F}_*^{(n)}]] \\ & \quad = [(2\mathrm{p} + K + 1) n^{-1} + o(n^{-1})] E[(a^{(n)}(\mathbf{R}_t^{(n)}, ..., \mathbf{R}_{t-p}^{(n)}) \\ & \qquad - J(\mathbf{F}(\mathbf{X}_t^{(n)}), ..., \mathbf{F}(\mathbf{X}_{t-p}^{(n)})))^2]. \end{aligned}$$

This latter quantity in view of (2.3) is $o(n^{-1})$, which establishes the desired equivalence of $S^{(n)} - E^*(S^{(n)})$ and $\mathscr{S}^{(n)} - \mathscr{E}^{(n)}$.

2.4. *Asymptotic Normality of Linear Serial Multirank Statistics*

We are now in a position to derive the asymptotic normality of linear serial multirank statistics, both under the null hypothesis and under contiguous alternatives, by using weak convergence results for U-statistics in weakly dependent processes. We therefore consider first the joint distribution under $H_f^{(n)}$ of the log-likelihood $\mathscr{L}^{(n)}_{\mathbf{A},\mathbf{B};f}$ and $(n-p)^{1/2}$ $(S^{(n)}-E^*(S^{(n)}))$.

PROPOSITION 2.1. *Under* $H_f^{(n)}$, $(\mathscr{L}^{(n)}_{\mathbf{A},\mathbf{B};f}(\mathbf{X}^{(n)})\ (n-p)^{1/2}\ \ (S^{(n)}-E^*(S^{(n)})))'$ *is asymptotically normal, with mean and covariance matrix*

$$\begin{pmatrix} -\frac{1}{2}\sum_{i=1}^{p} \operatorname{tr}[\mathscr{I}(f)\,\mathbf{D}_i\boldsymbol{\Sigma}\mathbf{D}_i'] \\ 0 \end{pmatrix}$$

and

$$\begin{pmatrix} \sum_{i=1}^{p} \operatorname{tr}[\mathscr{I}(f)\,\mathbf{D}_i\boldsymbol{\Sigma}\mathbf{D}_i'] & \sum_{i=1}^{p} \operatorname{tr}[\mathbf{C}_i\mathbf{D}_i'] \\ \sum_{i=1}^{p} \operatorname{tr}[\mathbf{C}_i\mathbf{D}_i'] & V^2 \end{pmatrix}$$

respectively, where

$$\mathbf{C}_i=\sum_{j=0}^{p-i}\int J^*(\mathbf{F}(\mathbf{x}_1),\ldots,\mathbf{F}(\mathbf{x}_{p+1}))\,\boldsymbol{\phi}(\mathbf{x}_{j+1})\,\mathbf{x}_{j+1+i}'\,dF(\mathbf{x}_1)\cdots dF(\mathbf{x}_{p+1}) \tag{2.11}$$

and

$$\begin{aligned} V^2=&\int J^{*2}(\mathbf{F}(\mathbf{x}_1),\ldots,\mathbf{F}(\mathbf{x}_{p+1}))\,dF(\mathbf{x}_1)\cdots dF(\mathbf{x}_{p+1}) \\ &+2\sum_{i=1}^{p}\int J^*(\mathbf{F}(\mathbf{x}_1),\ldots,\mathbf{F}(\mathbf{x}_{p+1})) \\ &\times J^*(\mathbf{F}(\mathbf{x}_{i+1}),\ldots,\mathbf{F}(\mathbf{x}_{i+p+1}))\,dF(\mathbf{x}_1)\cdots dF(\mathbf{x}_{p+1+i}) \end{aligned} \tag{2.12}$$

with

$$\begin{aligned} &J^*(\mathbf{u}_1,\ldots,\mathbf{u}_{p+1}) \\ &\quad=J(\mathbf{u}_1,\ldots,\mathbf{u}_{p+1}) \\ &\qquad-\sum_{i=1}^{p+1}\int J(\mathbf{F}(\mathbf{x}_1),\ldots,\mathbf{F}(\mathbf{x}_{i-1}),\mathbf{u}_1,\mathbf{F}(\mathbf{x}_i),\ldots,\mathbf{F}(\mathbf{x}_p))dF(\mathbf{x}_1)\cdots dF(\mathbf{x}_p) \\ &\qquad+p\int J(\mathbf{F}(\mathbf{x}_1),\ldots,\mathbf{F}(\mathbf{x}_{p+1}))\,dF(\mathbf{x}_1)\cdots dF(\mathbf{x}_{p+1}). \end{aligned} \tag{2.13}$$

Proof. Consider the kernel of order $p+1$ (same notation as in Section 2.3)

$$\gamma^{\mathscr{L}}(\mathbf{y}_1, ..., \mathbf{y}_{p+1}) = (p+1)^{-1} \sum_{j=1}^{p+1} \sum_{i=1}^{p} [\boldsymbol{\phi}(\mathbf{F}^{-1}((y_{j,1} \cdots y_{j,m})'))$$
$$\times (\mathbf{F}^{-1}((y_{j,im+1} \cdots y_{j,(i+1)m})'))' \mathbf{D}'_i],$$

and denote by $\mathscr{U}^{\mathscr{L}}$ the corresponding U-statistic. It is then easy to see that (under $H_f^{(n)}$) $\mathscr{L}_0^{(n)}(\mathbf{X}^{(n)}) = n^{1/2}\mathscr{U}^{\mathscr{L}} + o_p(1)$. Hence, considering an arbitrary linear combination $\alpha n^{-1/2}\mathscr{L}_0^{(n)} + \beta(S^{(n)} - E^*(S^{(n)}))$, it is asymptotically equivalent to a U-statistic with kernel $\alpha\gamma^{\mathscr{L}} + \beta(\gamma^{\mathscr{S}} - \gamma^{\mathscr{E}})$. Now, under $H_f^{(n)}$, $(\mathbf{Y}_{p+1}^{(n)} \cdots \mathbf{Y}_n^{(n)}) = (\mathbf{F}(\mathbf{X}_1^{(n)}) \cdots \mathbf{F}(\mathbf{X}_n^{(n)}))$ obviously constitutes a finite realization of a p-dependent, $m(p+1)$-variate process and therefore satisfies all the usual mixing conditions. We may thus apply one of the many available central limit theorems for U-statistics under weakly dependent observations—e.g., Yoshihara's [19] Theorem 1.

The $g_1(\mathbf{Y}_t)$ function appearing in Yoshihara [19, Theorem 1] here decomposes into $\alpha g_1^{\mathscr{L}}(\mathbf{Y}_t) + \beta g_1^{\mathscr{S}}(\mathbf{Y}_t) - \beta g_1^{\mathscr{E}}(\mathbf{Y}_t)$, say, with

$$g_1^{\mathscr{L}}(\mathbf{Y}_t) = \int \gamma^{\mathscr{L}}(\mathbf{Y}_t, \mathbf{F}(\mathbf{x}_1), ..., \mathbf{F}(\mathbf{x}_p))\, dF(\mathbf{x}_1) \cdots dF(\mathbf{x}_p)$$
$$= (p+1)^{-1} \left\{ \sum_{i=1}^{p} \operatorname{tr}[\boldsymbol{\phi}(\mathbf{F}^{-1}(Y_{t,1} \cdots Y_{t,m})') \times \mathbf{F}^{-1}(Y_{t,im+1} \cdots Y_{t,(i+1)m})' \mathbf{D}'_i] \right\},$$

$$g_1^{\mathscr{S}}(\mathbf{Y}_t) = \int \gamma^{\mathscr{S}}(\mathbf{Y}_t, \mathbf{F}(\mathbf{x}_1), ..., \mathbf{F}(\mathbf{x}_p))\, dF(\mathbf{x}_1) \cdots dF(\mathbf{x}_p)$$
$$= (p+1)^{-1} \{ J((Y_{t,1} \cdots Y_{t,m})', ..., (Y_{t,pm+1} \cdots Y_{t,(p+1)m})') + p \int J(\mathbf{F}(\mathbf{x}_1), ..., \mathbf{F}(\mathbf{x}_{p+1}))\, dF(\mathbf{x}_1) \cdots dF(\mathbf{x}_{p+1}) \},$$

and

$$g^{\mathscr{E}}(\mathbf{Y}_t) = \frac{p!}{(p+1)!} \left\{ \sum_{i=1}^{p+1} \int J(\mathbf{F}(\mathbf{x}_1), ..., \mathbf{F}(\mathbf{x}_{i-1}), \mathbf{Y}_t, \mathbf{F}(\mathbf{x}_i), ..., \mathbf{F}(\mathbf{x}_p)) \times dF(\mathbf{x}_1) \cdots dF(\mathbf{x}_p) \right\}.$$

It follows, letting $\mathbf{y}_t = (\mathbf{u}'_1, ..., \mathbf{u}'_{p+1})'$, that

$$g_1(\mathbf{y}_t) = (p+1)^{-1} \left\{ \alpha \sum_{i=1}^{p} \operatorname{tr}[\boldsymbol{\phi}(\mathbf{F}^{-1}(\mathbf{u}_1))\, \mathbf{F}^{-1}(\mathbf{u}_{i+1})\, \mathbf{D}'_i] + \beta J^*(\mathbf{u}_1, ..., \mathbf{u}_{p+1}) \right\}.$$

Since the expected value of $g_1(\mathbf{F}(\mathbf{X}_t), \ldots, \mathbf{F}(\mathbf{X}_{t-p}))$ is zero under $H_f^{(n)}$, its variance equals

$$(p+1)^{-2}\left\{\alpha^2 \sum_{i=1}^{p} \operatorname{tr}[\mathscr{I}(f)\,\mathbf{D}_i\boldsymbol{\Sigma}\mathbf{D}_i']\right.$$

$$+2\alpha\beta \sum_{i=1}^{p} \operatorname{tr}\left[\int J^*(\mathbf{F}(\mathbf{x}_1), \ldots, \mathbf{F}(\mathbf{x}_{p+1}))\right.$$

$$\left.\times \boldsymbol{\phi}(\mathbf{x}_1)\,\mathbf{x}_{i+1}'\,dF(\mathbf{x}_1)\cdots dF(\mathbf{x}_{p+1})\mathbf{D}_i'\right]$$

$$+\beta^2 \int [J^*(\mathbf{F}(\mathbf{x}_1), \ldots, \mathbf{F}(\mathbf{x}_{p+1}))]^2\,dF(\mathbf{x}_1)\cdots dF(\mathbf{x}_{p+1}),$$

whereas the covariance between $g_1(\mathbf{F}(\mathbf{X}_t), \ldots, \mathbf{F}(\mathbf{X}_{t-p}))$ and $g_1(\mathbf{F}(\mathbf{X}_{t+j}), \ldots, \mathbf{F}(\mathbf{X}_{t+j-p}))$, $j=1, \ldots, p$, is

$$(p+1)^{-2}\left\{\alpha^2\cdot 0+\alpha\beta \sum_{i=1}^{p-j} \operatorname{tr}\left[\int J^*(\mathbf{F}(\mathbf{x}_1), \ldots, \mathbf{F}(\mathbf{x}_{p+1}))\right.\right.$$

$$\left.\times \boldsymbol{\phi}(\mathbf{x}_{j+1})\,\mathbf{x}_{j+1+i}'\,dF(\mathbf{x}_1)\cdots dF(\mathbf{x}_{2p+1})\mathbf{D}_i'\right]$$

$$+\beta^2 \int J^*(\mathbf{F}(\mathbf{x}_1), \ldots, \mathbf{F}(\mathbf{x}_{p+1}))$$

$$\left.\times J^*(\mathbf{F}(\mathbf{x}_{j+1}), \ldots, \mathbf{F}(\mathbf{x}_{j+p+1}))\,dF(\mathbf{x}_1)\cdots dF(\mathbf{x}_{2p+1})\right\}.$$

On account of (2.11) and (2.12), Yoshihara's theorem ensures the asymptotic normality of $\alpha\mathscr{L}_0^{(n)}+n^{1/2}\beta(S^{(n)}-E^*(S^{(n)}))$, with mean zero and variance

$$(p+1)^2\left[\operatorname{Var}(g_1(\mathbf{F}(\mathbf{X}_1), \ldots, \mathbf{F}(\mathbf{X}_{p+1}))\right.$$

$$\left.+2\sum_{j=1}^{p} \operatorname{Cov}(g_1(\mathbf{F}(\mathbf{X}_1), \ldots, \mathbf{F}(\mathbf{X}_{p+1})), g_1(\mathbf{F}(\mathbf{X}_{j+1}), \ldots, \mathbf{F}(\mathbf{X}_{j+p+1})))\right]$$

$$=\alpha^2 \sum_{i=1}^{p} \operatorname{tr}[\mathscr{I}(f)\,\mathbf{D}_i\boldsymbol{\Sigma}\mathbf{D}_i']+2\alpha\beta \sum_{i=1}^{p} \operatorname{tr}[\mathbf{C}_i\mathbf{D}_i']+\beta^2V^2.$$

The proposition then readily follows from the usual Wold–Cramér argument.

Proposition 2.1 and *Le Cam's third lemma* now allow us to give the asymptotic joint distribution, under $K_{\mathbf{A},\mathbf{B};f}^{(n)}$, of any k-tuple of linear serial multirank statistics. We are giving here this result for $k=2$.

PROPOSITION 2.2. *Let $S_1^{(n)}$ and $S_2^{(n)}$ be two linear serial multirank statistics, or orders π_j and with score-generating functions $J_j(\mathbf{u}_1, ..., \mathbf{u}_{\pi_j+1})$, $j=1, 2$, respectively. Then*

$$((n-\pi_1)^{1/2}[S_1^{(n)}-E^*(S_1^{(n)})] \quad (n-\pi_2)^{1/2}[S_2^{(n)}-E^*(S_2^{(n)})])'$$

is asymptotically bivariate normal with mean $(0\,0)'$ under $H_f^{(n)}$ and mean $(\sum_{i=1}^{\min(p,\pi_1)} \mathrm{tr}[\mathbf{C}_{(1),i}\mathbf{D}_i'] \quad \sum_{i=1}^{\min(p,\pi_2)} \mathrm{tr}[\mathbf{C}_{(2),i}\mathbf{D}_i'])'$ under $K_{\mathbf{A},\mathbf{B};f}^{(n)}$, and with covariance matrix

$$\mathbf{V}^2 = \begin{pmatrix} V_1^2 & V_{(12)} \\ V_{(12)} & V_2^2 \end{pmatrix}$$

under both, where V_1^2, V_2^2, $\mathbf{C}_{(1),i}$, $\mathbf{C}_{(2),i}$ are defined as in Proposition 2.1, substituting

$$J_{(j)}(\mathbf{u}_1, ..., \mathbf{u}_{p+1}) = \begin{cases} J_j(\mathbf{u}_1, ..., \mathbf{u}_{\pi_j+1}) & \text{if } p \geqslant \pi_j \\ \int J_j(\mathbf{u}_1, ..., \mathbf{u}_{p+1}, \mathbf{F}(\mathbf{x}_{p+2}), ..., \mathbf{F}(\mathbf{x}_{\pi_j+1})) & \\ \quad \times dF(\mathbf{x}_{p+2}) \cdots dF(\mathbf{x}_{\pi_j+1}) & \text{if } p < \pi_j, \end{cases}$$

$j=1, 2$, for $J(\mathbf{u}_1, ..., \mathbf{u}_{p+1})$, and where the asymptotic covariance is

$$\begin{aligned} V_{(12)} = & \int J_{(1)}^*(\mathbf{F}(\mathbf{x}_1), ..., \mathbf{F}(\mathbf{x}_{p+1}))\, J_{(2)}^*(\mathbf{F}(\mathbf{x}_1), ..., \mathbf{F}(\mathbf{x}_{p+1})) \\ & \times dF(\mathbf{x}_1) \cdot \cdots \cdot dF(\mathbf{x}_{p+1}) \\ & + \sum_{i=1}^{p} \int [(J_{(1)}^*(\mathbf{F}(\mathbf{x}_1), ..., \mathbf{F}(\mathbf{x}_{p+1}))\, J_{(2)}^*(\mathbf{F}(\mathbf{x}_{i+1}), ..., \mathbf{F}(\mathbf{x}_{i+p+1})) \\ & + J_{(1)}^*(\mathbf{F}(\mathbf{x}_{i+1}), ..., \mathbf{F}(\mathbf{x}_{i+p+1}))\, J_{(2)}^*(\mathbf{F}(\mathbf{x}_1), ..., \mathbf{F}(\mathbf{x}_{p+1}))] \\ & \times dF(\mathbf{x}_1) \cdots dF(\mathbf{x}_{i+p+1})). \end{aligned} \tag{2.14}$$

Proof. The proof follows from applying Proposition 2.1 and Le Cam's third lemma to an arbitrary linear combination of $(n-p)^{-1}(n-\pi_1) S_1^{(n)}$ and $(n-p)^{-1}(n-\pi_2) S_2^{(n)}$.

2.5. *Convergence of Permutation Moments*

The asymptotic results in the preceding section were all obtained under a null hypothesis of the form $H_f^{(n)}$ (i.e., where the underlying density f is specified), and the parameters of the limiting normal distributions in Propositions 2.2 and 2.3 explicitly depend on f (cf. the asymptotic variance V^2 in (2.12), the asymptotic covariance $V_{(12)}$ in (2.14), and the "centered" score-generating functions J^* in (2.13)). These limiting distributions thus cannot, as they stand, be used for building statistical tests for the null hypothesis of randomness $H^{(n)}$ (with unspecified underlying density f).

Genuinely permutationally (conditionally) distribution-free test statistics which (unconditionally) also are asymptotically distribution-free, however, can be obtained if, in accordance with the conditional nature of rank permutation tests, the (distribution-dependent) unknown asymptotic means, variances, and covariances are replaced with their permutational analogues. Moreover, the statistics thus obtained are asymptotically equivalent (under $H^{(n)}$ as well as under contiguous alternatives) to the former ones. This follows from the fact that, as shown below, permutational first- and second-order moments converge to their unconditional counterparts.

PROPOSITION 2.4. *Under $H_f^{(n)}$, as $n \to \infty$,*

(i) $E^*(S^{(n)}) \to^P \int J(\mathbf{F}(\mathbf{x}_1), ..., \mathbf{F}(\mathbf{x}_{p+1}))\, dF(\mathbf{x}_1) \cdots dF(\mathbf{x}_{p+1})$.

(ii) *Provided that $a^{(n)}(\mathbf{R}_1^{(n)}, ..., \mathbf{R}_{p+1}^{(n)})\ a^{(n)}(\mathbf{R}_t^{(n)}, ..., \mathbf{R}_{t+p}^{(n)})$ converges in the quadratic mean to $J(\mathbf{F}(\mathbf{X}_1^{(n)}), ..., \mathbf{F}(\mathbf{X}_{p+1}^{(n)}))\, J(\mathbf{F}(\mathbf{X}_t^{(n)}), ..., \mathbf{F}(\mathbf{X}_{t+p}^{(n)}))$, $t = 1, 2, ..., p+2$, then*

$$(n-p)\operatorname{Var}^*(S^{(n)}) \xrightarrow{P} V^2.$$

(iii) *Provided that the condition in* (ii) *above is satisfied for $a_{(1)}^{(n)}(\cdots)$ and $a_{(2)}^{(n)}(\cdots)$, and provided further that $a_{(1)}^{(n)}(\mathbf{R}_1^{(n)}, ..., \mathbf{R}_{p+1}^{(n)})\ a_{(2)}^{(n)}(\mathbf{R}_t^{(n)}, ..., \mathbf{R}_{t+p}^{(n)})$ converges in the quadratic mean to $J_{(1)}(\mathbf{F}(\mathbf{X}_1^{(n)}), ..., \mathbf{F}(X_{p+1}^{(n)}))\ J_{(2)}(\mathbf{F}(\mathbf{X}_t^{(n)}), ..., \mathbf{F}(\mathbf{X}_{t+p}^{(n)}))$, then (same notation as in Proposition* 2.3)

$$((n-p_1)(n-p_2))^{1/2}\operatorname{Cov}^*(S_{(1)}^{(n)}, S_{(2)}^{(n)}) \xrightarrow{P} V_{(12)}.$$

Proof. (i). As a U-statistic, $\mathscr{E}^{(n)}$ converges a.s. to $E[J(\mathbf{F}(\mathbf{X}_1^{(n)}), ..., \mathbf{F}(\mathbf{X}_{p+1}^{(n)}))]$. In order to prove (i), it is therefore sufficient to show that $E[E^*(S^{(n)}) - \mathscr{E}^{(n)}]^2$ converges to zero. This latter expectation takes the form

$$(n(n-1)\cdot \cdots \cdot (n-p))^{-2} \sum_{1 \leqslant i_1 \neq} \cdots \sum_{\neq i_{p+1} \leqslant n} \sum_{1 \leqslant j_1 \neq} \cdots \sum_{\neq j_{p+1} \leqslant n}$$
$$\times E\{[a^{(n)}(\mathbf{R}_{i_1}^{(n)}, ..., \mathbf{R}_{i_{p+1}}^{(n)}) - J(\mathbf{F}(\mathbf{X}_{i_1}^{(n)}), ..., \mathbf{F}(\mathbf{X}_{i_{p+1}}^{(n)}))]$$
$$\times [a^{(n)}(\mathbf{R}_{j_1}^{(n)}, ..., \mathbf{R}_{j_{p+1}}^{(n)}) - J(\mathbf{F}(\mathbf{X}_{j_1}^{(n)}), ..., \mathbf{F}(\mathbf{X}_{j_{p+1}}^{(n)}))]\}$$

in which the $[n(n-1)\cdot \cdots \cdot (n-p)]^2$ summands are equal to each other. It follows from the Cauchy–Schwartz inequality that this expression is bounded by $E\{[a^{(n)}(\mathbf{R}_1^{(n)}, \ldots, \mathbf{R}_{p+1}^{(n)}) - J(\mathbf{F}(\mathbf{X}_1^{(n)}), \ldots, \mathbf{F}(\mathbf{X}_{p+1}^{(n)}))]^2\}$, whose limit, on account of assumption (2.3), is zero.

(ii). In view of (i), we may assume here, without any loss of generality, that $E[J(\mathbf{F}(\mathbf{X}_1^{(n)}), \ldots, \mathbf{F}(\mathbf{X}_{p+1}^{(n)}))] = 0$. We then have the following lemmas.

LEMMA 2.7.

$$\begin{aligned} a_*^{(n)}&(\mathbf{R}_1^{(n)}, \ldots, \mathbf{R}_{p+1}^{(n)}) \\ &= a^{(n)}(\mathbf{R}_1^{(n)}, \ldots, \mathbf{R}_{p+1}^{(n)}) \\ &\quad - \sum_{l=1}^{p+1} E^*[a^{(n)}(\mathbf{R}_1^{(n)}, \ldots, \mathbf{R}_l^{(n)}, \ldots, \mathbf{R}_{p+1}^{(n)}) \mid \mathbf{R}_l^{(n)}] + o_p(1). \end{aligned}$$

LEMMA 2.8.

$$\operatorname{Var}^*(a_*^{(n)}) \xrightarrow{P} \int J^{*2}(\mathbf{F}(\mathbf{x}_1), \ldots, \mathbf{F}(\mathbf{x}_{p+1}))\, dF(\mathbf{x}_1) \cdots dF(\mathbf{x}_{p+1}).$$

LEMMA 2.9.

$$\begin{aligned} \operatorname{Cov}^*&(a_*^{(n)}(\mathbf{R}_1^{(n)}, \ldots, \mathbf{R}_{p+1}^{(n)}), a_*^{(n)}(\mathbf{R}_{1+i}^{(n)}, \ldots, \mathbf{R}_{p+1+i}^{(n)})) \\ &\xrightarrow{P} \int J^*(\mathbf{F}(\mathbf{x}_1), \ldots, \mathbf{F}(\mathbf{x}_{p+1}))\, J^*(\mathbf{F}(\mathbf{x}_{1+i}), \ldots, \mathbf{F}(\mathbf{x}_{p+1+i})) \\ &\quad \times dF(\mathbf{x}_1) \cdots dF(\mathbf{x}_{p+1+i}). \end{aligned}$$

It follows from Lemmas 2.6, 2.8, and 2.9 that $(n-p)\operatorname{Var}^*(S^{(n)}) - (n-p) \sum_{l=1}^p \sum_\beta \sum_\alpha (\cdots)$, where $\sum_{l=1}^p \sum_\beta \sum_\alpha (\cdots)$ is the last sum in the right-hand side expression in Lemma 2.6, converges to V^2 in probability. It remains thus to show that $\sum_{l=1}^p \sum_\beta \sum_\alpha (\cdots)$ is $o_p(n^{-1})$. Now

$$\begin{aligned} E^*[a_*^{(n)}&(\mathbf{R}_1^{(n)}, \ldots, \mathbf{R}_{p+1}^{(n)})\, E^*[a_*^{(n)}(\mathbf{R}_1^{(n)}, \ldots, \mathbf{R}_{\alpha_i-1}^{(n)}, \\ &\mathbf{R}_{\beta_i}^{(n)}, \mathbf{R}_{\alpha_i+1}^{(n)}, \ldots, \mathbf{R}_{p+1}^{(n)}) \mid \mathbf{R}_{\beta_1}^{(n)}, \ldots, \mathbf{R}_{\beta_l}^{(n)}]] \\ &\leqslant [E^*(a_*^{(n)2})]^{1/2}\, [E^*[(E^*[a_*^{(n)}(\mathbf{R}_1^{(n)}, \ldots, \mathbf{R}_{\alpha_i-1}^{(n)}, \\ &\quad \mathbf{R}_{\beta_i}^{(n)}, \mathbf{R}_{\alpha_i+1}^{(n)}, \ldots, \mathbf{R}_{p+1}^{(n)} \mid \mathbf{R}_{\beta_1}^{(n)}, \ldots, \mathbf{R}_{\beta_l}^{(n)}])^2]^{1/2} \\ &\leqslant E^*(a_*^{(n)2}) \xrightarrow{P} \int J^{*2}(\mathbf{F}(\mathbf{x}_1), \ldots, \mathbf{F}(\mathbf{x}_{p+1}))\, dF(\mathbf{x}_1) \cdots dF(\mathbf{x}_{p+1}) < \infty. \end{aligned}$$

Since $\sum_{l=1}^p \sum_\beta \sum_\alpha (\cdots)$ contains a finite number of terms with coefficients $(n-l)(n-l-1) \cdots (n-p+l)(n-p)^{-1}\, (n-p-1)^{-1} \cdot \cdots \cdot (n-2p+1)^{-1}$

$=O(n^{1-p-2l})$, with $l \geqslant 1$ and $p \geqslant 1$, the proof of part (ii) of Proposition 2.4 follows.

(iii) Follows analogously.

Proof of Lemma 2.7. Without any loss of generality, we may assume here that $E(J(\cdots))=0$. Going back to the definition of $a_*^{(n)}$, it follows from Proposition 2.4(i), that $(np)(n-p-1)^{-1}$ $E^*(a^{(n)})=(np)(n-p-1)^{-1}$ $E^*(S^{(n)})$ converges to $E(J(\cdots))=0$. As for

$$(n-1)n^{-1}(n-p-1)^{-1/2}$$
$$\times E\{E^*[(a^{(n)}(\mathbf{R}_1^{(n)}, \ldots, \mathbf{R}_{\alpha-1}^{(n)}, \mathbf{R}_\beta^{(n)}, \mathbf{R}_{\alpha+1}^{(n)}, \ldots, \mathbf{R}_{p+1}^{(n)}))^2 \mid \mathbf{R}_\beta^{(n)}]\},$$

it is bounded by

$$(n-p-1)^{-2} E[E^*[a^{(n)}(\mathbf{R}_1^{(n)}, \ldots, \mathbf{R}_i^{(n)}, \ldots, \mathbf{R}_{p+1}^{(n)}) \mid \mathbf{R}_j^{(n)}]]^2$$
$$=(n-p-1)^{-2} E[(a^{(n)}(\cdots))^2],$$

which also converges to zero.

Proof of Lemma 2.8. In view of Lemma 2.7, we have

$$\begin{aligned}
\operatorname{Var}^*(a_*^{(n)}) &= E^*(a_*^{(n)2})\\
&= E^*(a^{(n)2}) + \sum_{i=1}^{p+1} E^*[(E^*[a^{(n)}(\mathbf{R}_1^{(n)}, \ldots, \mathbf{R}_i^{(n)}, \ldots, \mathbf{R}_{p+1}^{(n)}) \mid \mathbf{R}_i^{(n)}])^2]\\
&\quad + \sum_{1 \leqslant i \neq j \leqslant p+1} \sum E^*[E^*[a^{(n)}(\mathbf{R}_1^{(n)}, \ldots, \mathbf{R}_i^{(n)}, \ldots, \mathbf{R}_{p+1}^{(n)}) \mid \mathbf{R}_i^{(n)}]\\
&\quad \times E^*[a^{(n)}(\mathbf{R}_1^{(n)}, \ldots, \mathbf{R}_j^{(n)}, \ldots, \mathbf{R}_{p+1}^{(n)}) \mid \mathbf{R}_j^{(n)}]]\\
&\quad -2 \sum_{i=1}^{p+1} E^*[a^{(n)}(\mathbf{R}_1^{(n)}, \ldots, \mathbf{R}_{p+1}^{(n)})\\
&\quad \times E^*[a^{(n)}(\mathbf{R}_1^{(n)}, \ldots, \mathbf{R}_i^{(n)}, \ldots, \mathbf{R}_{p+1}^{(n)}) \mid \mathbf{R}_i^{(n)}]] + o_p(1)\\
&= E^*(a^{(n)2}) + (2) + (3) - (4) + o_p(1), \qquad \text{say.}
\end{aligned}$$

$E^*(a^{(n)2})$ converges to $\int J^2(\mathbf{F}(\mathbf{x}_1), \ldots, \mathbf{F}(\mathbf{x}_{p+1}))\, dF(\mathbf{x}_1) \cdots dF(\mathbf{x}_{p+1})$. As for the second term, Lemma 2.4 (see also the end of the proof of part (ii) of Proposition 2.4 for the $o_p(1)$ term) implies that

$$\begin{aligned}
&E^*[a^{(n)}(\mathbf{R}_1^{(n)}, \ldots, \mathbf{R}_i^{(n)}, \ldots, \mathbf{R}_{p+1}^{(n)}) \mid \mathbf{R}_i^{(n)}]\\
&\quad = E^*[a^{(n)}(\mathbf{R}_{p+2}^{(n)}, \ldots, \mathbf{R}_{p+i}^{(n)}, \mathbf{R}_i^{(n)}, \mathbf{R}_{p+i+2}^{(n)}, \ldots,\\
&\qquad \mathbf{R}_{2p+2}^{(n)}) \mid \mathbf{R}_i^{(n)}, \mathbf{R}_1^{(n)}, \ldots, \mathbf{R}_{p+1}^{(n)}] + o_p(1),
\end{aligned}$$

so that

$$
\begin{aligned}
(2) = \sum_{i=1}^{p+1} & E^*\{E^*[a^{(n)}(\mathbf{R}_1^{(n)}, \ldots, \mathbf{R}_i^{(n)}, \ldots, \mathbf{R}_{p+1}^{(n)}) \mid \mathbf{R}_i^{(n)}] \\
& \times E^*[a^{(n)}(\mathbf{R}_{p+2}^{(n)}, \ldots, \mathbf{R}_{p+i}^{(n)}, \mathbf{R}_i^{(n)}, \mathbf{R}_{p+2+i}^{(n)}, \\
& \ldots, \mathbf{R}_{2p+2}^{(n)}) \mid \mathbf{R}_i^{(n)}, \mathbf{R}_1^{(n)}, \ldots, \mathbf{R}_{p+1}^{(n)}]\} + o_p(1) \\
= \sum_{i=1}^{p+1} & E^*\{E^*[a^{(n)}(\mathbf{R}_1^{(n)}, \ldots, \mathbf{R}_{p+1}^{(n)}) \\
& \times a^{(n)}(\mathbf{R}_{p+2}^{(n)}, \ldots, \mathbf{R}_{p+i}^{(n)}, \mathbf{R}_i^{(n)}, \mathbf{R}_{p+2+i}^{(n)}, \ldots, \\
& \quad \mathbf{R}_{2p+2}^{(n)}) \mid \mathbf{R}_1^{(n)}, \ldots, \mathbf{R}_i^{(n)}, \ldots, \mathbf{R}_{p+1}^{(n)}]\} + o_p(1) \\
= \sum_{i=1}^{p+1} & E^*\{a^{(n)}(\mathbf{R}_1^{(n)}, \ldots, \mathbf{R}_{p+1}^{(n)})\, a^{(n)}(\mathbf{R}_{p+2}^{(n)}, \ldots, \mathbf{R}_i^{(n)}, \ldots, \mathbf{R}_{2p+2}^{(n)})\} + o_p(1).
\end{aligned}
$$

Since the random variable in this latter expectation converges, under the assumptions, in quadratic mean, to $J(\mathbf{F}(\mathbf{X}_1^{(n)}), \ldots, \mathbf{F}(\mathbf{X}_{p+1}^{(n)}))\ J(\mathbf{F}(\mathbf{X}_{p+2}^{(n)}), \ldots, \mathbf{F}(\mathbf{X}_{p+i}^{(n)}), \mathbf{F}(\mathbf{X}_i^{(n)}), \mathbf{F}(\mathbf{X}_{p+i+2}^{(n)}), \ldots, \mathbf{F}(\mathbf{X}_{2p+2}^{(n)}))$, we obtain that (2) converges in probability to

$$
\begin{aligned}
\sum_{i=1}^{p+1} & \int J(\mathbf{F}(\mathbf{x}_1), \ldots, \mathbf{F}(\mathbf{x}_{p+1}))\, dF(\mathbf{x}_1) \cdots dF(\mathbf{x}_{p+1}) \\
& \times \int J(\mathbf{F}(\mathbf{x}_{p+2}), \ldots, \mathbf{F}(\mathbf{x}_{p+i}), \mathbf{F}(\mathbf{x}_i), \\
& \mathbf{F}(\mathbf{x}_{p+i+2}), \ldots, \mathbf{F}(\mathbf{x}_{2p+2}))\, dF(\mathbf{x}_1) \cdots dF(\mathbf{x}_{p+i})\, dF(\mathbf{x}_{p+i+2}) \cdots dF(\mathbf{x}_{2p+2}) \\
= & \sum_{i=1}^{p+1} E\Big\{\Big[\Big(\int J(\mathbf{F}(\mathbf{x}_1), \ldots, \mathbf{F}(\mathbf{x}_{i-1}), \mathbf{F}(\mathbf{X}_i^{(n)}), \mathbf{F}(\mathbf{x}_{i+1}), \ldots, \\
& \quad \mathbf{F}(\mathbf{x}_{p+1}))\, dF(\mathbf{x}_1) \cdots dF(\mathbf{x}_{p+1})\Big)\Big]^2\Big\}.
\end{aligned}
$$

Similar reasonings lead to the results that $(3) \to^P 0$ and

$$
\begin{aligned}
(4) \xrightarrow{P} \sum_{i=1}^{p+1} & E\Big\{J(\mathbf{F}(\mathbf{X}_1^{(n)}), \ldots, \mathbf{F}(\mathbf{X}_{p+1}^{(n)})) \\
& \times \int J(\mathbf{F}(\mathbf{x}_1), \ldots, \mathbf{F}(\mathbf{X}_i^{(n)}), \ldots, \mathbf{F}(\mathbf{x}_{p+1})\, dF(\mathbf{x}_1) \cdots dF(\mathbf{x}_{p+1})\Big\}.
\end{aligned}
$$

Rearranging these terms finally yields

$$E\left\{\left[J(\mathbf{X}_1^{(n)}, ..., \mathbf{X}_{p+1}^{(n)})\right.\right.$$

$$\left.\left. - \sum_{i=1}^{p+1} \int J(\mathbf{F}(\mathbf{x}_1), ..., \mathbf{F}(\mathbf{X}_i^{(n)}), ..., \mathbf{F}(\mathbf{x}_{p+1}))\, dF(\mathbf{x}_1) \cdots dF(\mathbf{x}_{p+1})\right]^2\right\}$$

$$= E\{[J^*(\mathbf{F}(\mathbf{X}_1^{(n)}), ..., \mathbf{F}(\mathbf{X}_{p+1}^{(n)})]^2\},$$

which completes the proof.

Proof of Lemma 2.9. The proof follows proceeding as in Lemma 2.8 and is left to the reader.

2.6. *Permutationally Distribution-free Multirank Tests for Randomness*

The convergence of permutation moments in Proposition 2.4 allows us to build permutationally distribution-free, asymptotically distribution-free test statistics for testing $H^{(n)}$ (randomness with unspecified underlying density).

PROPOSITION 2.5. *Let $S_1^{(n)}$ be some linear serial multirank statistic, and let the null hypothesis $H^{(n)}$ be restricted to those white noises with densities $f(\cdot)$ such that the assumptions of Proposition 2.4 hold. Then*

(i) *the test procedure*:

$$\textit{reject } H^{(n)} \textit{if } S_1^{(n)}(\mathbf{X}^{(n)}) - E^*(S_1^{(n)}) > k_{1-\alpha}[\mathrm{Var}^*(S_1^{(n)})]^{1/2}, \quad (2.15)$$

where $k_{1-\alpha}$ is the $(1-\alpha)$-quantile of the standard normal distribution and $E^*(S_1^{(n)})$ is given in Lemma 2.1(i), *has asymptotic level α (for the sequence of null hypotheses $H^{(n)}$); its asymptotic power is*

$$1 - \Phi\left(k_{1-\alpha} - \sum_{i=1}^{p} \mathrm{tr}[\mathbf{C}_{(1),i}\mathbf{D}_i']/V_1\right)$$

against $K_{\mathbf{A},\mathbf{B};f}^{(n)}$ (where Φ stands for the standard normal distribution function and V_1 and $\mathbf{C}_{(1),i}$ are computed as in Proposition 2.2.

(ii) *denoting by $S_2^{(n)}$ some other statistic of the same type and restricting the null hypothesis $H^{(n)}$ so that the assumptions of Proposition* 2.4 *hold for both $S_1^{(n)}$ and $S_2^{(n)}$, the A.R.E. (asymptotic relative efficiency) of* (2.15) *with respect to the test constructed in a similar way from $S_2^{(n)}$ is*

$$V_2 \sum_{i=1}^{p} \mathrm{tr}[\mathbf{C}_{(1),i}\mathbf{D}_i']/V_1 \sum_{i=1}^{p} \mathrm{tr}[\mathbf{C}_{(2),i}\mathbf{D}_i']$$

(V_2 and $\mathbf{C}_{(2),i}$ defined as in Proposition 2.2).

Proof. The proof follows by applying Propositions 2.2, 2.3, and 2.4.

3. Asymptotically Most Powerful Multirank Tests against Local Alternatives

3.1. *Rank Autocovariance Matrices Associated with a Density f.*

We have shown in Hallin *et al.* [7] and Hallin and Puri [9] that in the univariate version of the problem of testing for randomness, a key role is played by serial rank statistics of the form $r_{i;f}^{(n)} = [(n-i)^{-1} \sum_{t=i+1}^{n} \phi(F^{-1}(R_t^{(n)}/(n+1)))\ F^{-1}(R_{t-i}^{(n)}/(n+1)) - m^{(n)}]/s^{(n)}$ ($m^{(n)}$ and $s^{(n)}$ being standardizing constants under $H^{(n)}$). Because of the close similarity of the asymptotic behaviour of these statistics to that of usual sample autocorrelations under Gaussian assumptions (cf. also [8]), we termed $r_{i;f}^{(n)}$ the *rank autocorrelation coefficient of order i associated with f*—although, unless f is normal, it does not take the form of an *auto*correlation coefficient. We introduce here a multivariate extension of these univariate coefficients; because, however, of notational and some technical problems inherent to their multivariate nature, we define here rank *autocovariances* instead of *rank autocorrelations.*

Definition 3.1. The *rank autocovariance matrix of order i associated with the multivariate density f* (shortly, the *f-rank autocovariance matrix of order i*) is the matrix (with elements $\gamma_{i;f;k,l}^{(n)}$)

$$\boldsymbol{\Gamma}_{i;f}^{(n)} = (n-i)^{-1} \sum_{t=i+1}^{n} \boldsymbol{\phi}\left(\mathbf{F}^{-1}\left(\frac{\mathbf{R}_t^{(n)}}{n+1}\right)\right)\left(\mathbf{F}^{-1}\left(\frac{\mathbf{R}_{t-i}^{(n)}}{n+1}\right)\right)' - \mathbf{m}_{i*}^{(n)}, \quad (3.1)$$

where

$$\mathbf{m}_{i*}^{(n)} = (n(n-1))^{-1} \sum_{1 \leqslant t_1 \neq t_2 \leqslant n} \boldsymbol{\phi}\left(\mathbf{F}^{-1}\left(\frac{\mathbf{R}_{t_1}^{(n)}}{n+1}\right)\right)\left(\mathbf{F}^{-1}\left(\frac{\mathbf{R}_{t_2}^{(n)}}{n+1}\right)\right)'. \quad (3.2)$$

$\gamma_{i;f;k,l}^{(n)}$ is a *cross*-covariance whenever $k \neq l$, an *auto*covariance if $k = l$; note that it is of the form $S^{(n)} - E^*(S^{(n)})$, with score function (for $S^{(n)}$)

$$a^{(n)}(\mathbf{R}_t^{(n)}, ..., \mathbf{R}_{t-i}^{(n)}) = \phi_k\left(\mathbf{F}^{-1}\left(\frac{\mathbf{R}_t^{(n)}}{n+1}\right)\right) F_l^{-1}\left(\frac{R_{t-i,l}^{(n)}}{n+1}\right). \quad (3.3)$$

3.2. *Asymptotic Multinormality of Rank Autocovariance Matrices*

If we are to use Propositions 2.2 and 2.5 in order to obtain the asymptotic distribution of rank autocovariance matrices, we first have to ensure that condition (2.3) and the various assumptions in Proposition 2.4

are satisfied. We therefore first establish the following convergence property.

PROPOSITION 3.1. *Let the function* $J(\mathbf{u}_1, ..., \mathbf{u}_{p+1})$, $\mathbf{u}_i \in (0, 1)^m$, $i = 1, ..., p+1$ *be expressible as a finite sum* $\sum_{k=1}^{\kappa} \prod_{i=1}^{p+1} J_{k(i)}(\mathbf{u}_i)$, *where* $J_{k(i)}$ *is monotone (with respect to all the components* $u_{i,1}, ..., u_{i,m}$ of $\mathbf{u}_i$), *and* $E[J^2_{k(i)}(\mathbf{F}(\mathbf{X}_1^{(n)}))] < \infty$, $i = 1, ..., p+1$, $k = 1, ..., \kappa$. *Then*

$$\lim_{n \to \infty} E\left\{\left[J\left(\frac{\mathbf{R}_t^{(n)}}{n+1}, \frac{\mathbf{R}_{t-1}^{(n)}}{n+1}, ..., \frac{\mathbf{R}_{t-p}^{(n)}}{n+1}\right) - J(\mathbf{F}(\mathbf{X}_t^{(n)}), ..., \mathbf{F}(\mathbf{X}_{t-p}^{(n)}))\right]^2\right\} = 0. \tag{3.5}$$

COROLLARY 3.1. *Let* $f(x)$ *be a strongly unimodal density (i.e., let* $\phi_k(\mathbf{x})$, $k = 1, ..., m$ *be non-decreasing with respect to all its argument* $x_1, ..., x_m$). *Then condition* (2.3) *and the assumptions in Propositions* 2.4 *are satisfied by the score functions (of the form* $\phi_k(\mathbf{F}^{-1}(\mathbf{R}_t^{(n)}/(n+1)))\ F_l^{-1}(R_{t-i,l}^{(n)}/(n+1)))$) *appearing in Definition* 3.1.

Proof. We may restrict ourselves, without any loss of generality, to the case $J(\mathbf{u}, \mathbf{v}) = J_{(1)}(\mathbf{u})\ J_{(2)}(\mathbf{v})$, where $J_{(1)}$ and $J_{(2)}$ are non-decreasing on $(0, 1)^m$, thus a.e. continuous. Then, $J(\mathbf{R}_t^{(n)}/(n+1), \mathbf{R}_{t-1}^{(n)}/(n+1)) - J(\mathbf{F}(\mathbf{X}_t^{(n)}), \mathbf{F}(\mathbf{X}_{t-1}^{(n)}))$ converges a.s. to zero. Now consider the (random) hypercubes (associated with elements $R_{*i,j}^{(n)}$ of $\mathbf{R}_*^{(n)}$) defined by

$$\mathscr{D}_{*(i)}^{(n)} = \left[\frac{i}{n+1}, \frac{i+1}{n+1}\right] \times \left[\frac{R_{*2,i}^{(n)}}{n+1}, \frac{R_{*2,i}^{(n)}+1}{n+1}\right] \times \cdots \times \left[\frac{R_{*m,i}^{(n)}}{n+1}, \frac{R_{*m,i}^{(n)}+1}{n+1}\right].$$

Let $\mathscr{D}_*^{(n)} = \bigcup_{i=1}^{n} \mathscr{D}_{*(i)}^{(n)}$ and $F_*^{(n)}(\mathbf{u}) = P[\mathbf{F}(\mathbf{X}_1^{(n)}) \leqslant \mathbf{u} \mid \mathbf{F}(\mathbf{X}_1^{(n)}) \in \mathscr{D}_*^{(n)}]$, $\mathbf{u} \in [0, 1]^m$. Clearly,

$$\int_{\mathscr{D}_{*(i)}^{(n)}} dF_*^{(n)}(\mathbf{u}) = 1/(n+1),\ i = 1, ..., n.$$

Monotonicity then implies

$$\frac{1}{(n+1)^2}\left[J_{(1)}\left(\frac{\mathbf{R}_t^{(n)}}{n+1}\right) J_{(2)}\left(\frac{\mathbf{R}_{t-1}^{(n)}}{n+1}\right)\right]^2$$

$$\leqslant \int_{\mathscr{D}_{*(R_{t,1}^{(n)})}^{(n)}} J_{(1)}^2(\mathbf{u})\, dF_*^{(n)}(\mathbf{u}) \cdot \int_{\mathscr{D}_{*(R_{t-1,1}^{(n)})}^{(n)}} J_{(2)}^2(\mathbf{u})\, F_*^{(n)}(\mathbf{u}).$$

Hence

$$E^*\left[J^2\left(\frac{\mathbf{R}_t^{(n)}}{n+1}, \frac{\mathbf{R}_{t-1}^{(n)}}{n+1}\right)\right]$$

$$= [n(n-1)]^{-1} \sum_{1 \leqslant t_1 \neq t_2 \leqslant n} \sum J^2_{(1)}\left(\frac{\mathbf{R}_{t_1}^{(n)}}{n+1}\right) J^2_{(2)}\left(\frac{\mathbf{R}_{t_2}^{(n)}}{n+1}\right)$$

$$\leqslant \frac{(n+1)^2}{n(n-1)} \int_{\mathscr{D}_*^{(n)}} J^2_{(1)}(\mathbf{u})\, dF_*^{(n)}(\mathbf{u}) \int_{\mathscr{D}_*^{(n)}} J^2_{(2)}(\mathbf{u})\, dF_*^{(n)}(\mathbf{u})$$

$$= \frac{(n+1)^2}{n(n-1)} E_*^{(n)}[J^2(\mathbf{F}(\mathbf{X}_t^{(n)}), \mathbf{F}(\mathbf{X}_{t-1}^{(n)}))], \qquad \text{say.}$$

Taking expectations on both sides yields

$$E\left[J^2\left(\frac{\mathbf{R}_t^{(n)}}{n+1}, \frac{\mathbf{R}_{t-1}^{(n)}}{n+1}\right)\right] \leqslant \frac{(n+1)^2}{n(n-1)} E[J^2(\mathbf{F}(\mathbf{X}_t^{(n)}), \mathbf{F}(\mathbf{X}_{t-1}^{(n)}))],$$

and thus, letting $n \to \infty$,

$$\limsup_{n \to \infty} E\left[J^2\left(\frac{\mathbf{R}_t^{(n)}}{n+1}, \frac{\mathbf{R}_{t-1}^{(n)}}{n+1}\right)\right] \leqslant E[J^2(\mathbf{F}(\mathbf{X}_t^{(n)}), \mathbf{F}(\mathbf{X}_{t-1}^{(n)}))];$$

(3.5) then follows from Theorem 1.3 on page 154 of Hájek and Šidák [3].

Corollary 3.1 is a straightforward consequence of Proposition 3.1. The score-generating functions associated with $\gamma_{i;f;j,k}^{(n)}$ are

$$J(\mathbf{u}, \mathbf{v}) = \phi_k(\mathbf{F}^{-1}(\mathbf{u}))\, F_l^{-1}(v_l), \tag{3.6}$$

and, refering to (2.13), the "centered" (under $H_g^{(n)}$, where $g(\mathbf{x})$ denotes a density with distribution function $G(\mathbf{x})$, marginal densities g_k and marginal distribution functions G_k, $k = 1, ..., m$) version of J is

$$\begin{aligned} J^*(\mathbf{u}, \mathbf{v}) &= \phi_k(\mathbf{F}^{-1}(\mathbf{u}))\, F_l^{-1}(v_l) - \int \phi_k(\mathbf{F}^{-1}(\mathbf{G}(\mathbf{x})))\, g(\mathbf{x})\, d\mathbf{x}\, F_l^{-1}(v_l) \\ &\quad - \phi_k(\mathbf{F}^{-1}(\mathbf{u})) \int F_l^{-1}(G_l(y))\, g_l(y)\, dy \\ &\quad + 2 \int \phi_k(\mathbf{F}^{-1}(\mathbf{G}(\mathbf{x})))\, F_l^{-1}(G_l(y))\, g(\mathbf{x})\, g_l(y)\, d\mathbf{x}\, dy \\ &= \left[\phi_k(\mathbf{F}^{-1}(\mathbf{u})) - \int \phi_k(\mathbf{F}^{-1}(\mathbf{G}(\mathbf{x})))\, g(\mathbf{x})\, d\mathbf{x}\right] F_l^{-1}(v_l). \end{aligned} \tag{3.7}$$

We now may derive the joint asymptotic distribution of the *cross*- and *auto*covariances $\gamma_{i;f;j,k}^{(n)}$. Let us therefore introduce some additional notation. Let $f(\mathbf{x})$ and $g(\mathbf{x})$ be two densities satisfying the technical assumptions of Section 1.3. Using the notation $\mathbf{G}(\mathbf{x})$ in an obvious way, let

$$\boldsymbol{\phi}_f(\mathbf{x}) = -\mathbf{grad} \log f(\mathbf{x}) \qquad \text{and} \qquad \boldsymbol{\phi}_g(\mathbf{x}) = -\mathbf{grad} \log g(\mathbf{x}),$$

and define

$$\mathcal{I}(f, g) = \int \boldsymbol{\phi}_f(\mathbf{F}^{-1}(\mathbf{G}(\mathbf{x})))\, \boldsymbol{\phi}'_g(\mathbf{x})\, g(\mathbf{x})\, d\mathbf{x}, \tag{3.8}$$

$$\begin{aligned}\mathcal{I}(f \mid g) = &\int \boldsymbol{\phi}_f(\mathbf{F}^{-1}(\mathbf{G}(\mathbf{x})))\, \boldsymbol{\phi}'_f(\mathbf{F}^{-1}(\mathbf{G}(\mathbf{x})))\, g(\mathbf{x})\, d\mathbf{x} \\ &- \left[\int \boldsymbol{\phi}_f(\mathbf{F}^{-1}(\mathbf{G}(\mathbf{x})))\, g(\mathbf{x})\, d\mathbf{x}\right] \\ &\times \left[\int \boldsymbol{\phi}_f(\mathbf{F}^{-1}(\mathbf{G}(\mathbf{x})))\, g(\mathbf{x})\, d\mathbf{x}\right]'\end{aligned} \tag{3.9}$$

$$\boldsymbol{\Sigma}(f, g) = \int \mathbf{F}^{-1}(\mathbf{G}(\mathbf{x}))\, \mathbf{x}' g(\mathbf{x})\, d\mathbf{x} \tag{3.10}$$

$$\boldsymbol{\Sigma}(f \mid g) = \int \mathbf{F}^{-1}(\mathbf{G}(\mathbf{x}))(\mathbf{F}^{-1}(\mathbf{G}(\mathbf{x})))'\, g(\mathbf{x})\, d\mathbf{x}; \tag{3.11}$$

clearly, $\mathcal{I}(f, f) = \mathcal{I}(f \mid f) = \mathcal{I}(f)$ and $\boldsymbol{\Sigma}(f, f) = \boldsymbol{\Sigma}(f|f) = \boldsymbol{\Sigma}$. Also, we use the classical notation $\mathbf{vec}(\mathbf{A})$, where $\mathbf{A}$ is a $q \times r$ matrix, for the $qr \times 1$ vector obtained by stacking the columns of $\mathbf{A}$ on top of each other; $\otimes$ denotes, as usual, the *Kronecker product*.

PROPOSITION 3.2. *Let $f(\mathbf{x})$ be strongly unimodal. Then*

$$\begin{pmatrix} (n-i)^{1/2}\, \mathbf{vec}(\boldsymbol{\Gamma}^{(n)}_{i;f}) \\ (n-i')^{1/2}\, \mathbf{vec}(\boldsymbol{\Gamma}^{(n)}_{i';f}) \end{pmatrix}, \quad i \neq i',$$

is asymptotically normal with mean $(\mathbf{0}, \mathbf{0})'$ under $H^{(n)}_g$ and mean

$$\begin{pmatrix} \mathbf{vec}(\mathcal{I}(f, g)\, \mathbf{D}_i \boldsymbol{\Sigma}(f, g)) \\ \mathbf{vec}(\mathcal{I}(f, g)\, \mathbf{D}_{i'} \boldsymbol{\Sigma}(f, g)) \end{pmatrix} = \begin{pmatrix} \boldsymbol{\Sigma}(f, g) \otimes \mathcal{I}(f, g)\, \mathbf{vec}\, \mathbf{D}_i \\ \boldsymbol{\Sigma}(f, g) \otimes \mathcal{I}(f, g)\, \mathbf{vec}\, \mathbf{D}_{i'} \end{pmatrix}$$

(with the convention $\mathbf{D}_j = \mathbf{0}$, $j > p$) under $K^{(n)}_{\mathbf{A}, \mathbf{B}; g}$, and with covariance matrix

$$\begin{pmatrix} \boldsymbol{\Sigma}(f|g) \otimes \mathcal{I}(f \mid g) & \mathbf{0} \\ \mathbf{0} & \boldsymbol{\Sigma}(f \mid g) \otimes \mathcal{I}(f \mid g) \end{pmatrix}$$

under both.

Note that the asymptotic covariance matrix, which is the same under the null hypothesis of randomness and under the alternative, is not distribution-free. Under $K^{(n)}_{\mathbf{A},\mathbf{B};f}$, the asymptotic mean and covariance matrices take the simpler forms

$$\begin{pmatrix} \mathbf{\Sigma} \otimes \mathscr{I}(f) \,\mathbf{vec}\, \mathbf{D}_i \\ \mathbf{\Sigma} \otimes \mathscr{I}(f) \,\mathbf{vec}\, \mathbf{D}_{i'} \end{pmatrix} \quad \text{and} \quad \begin{pmatrix} \mathbf{\Sigma} \otimes \mathscr{I}(f) & \mathbf{0} \\ \mathbf{0} & \mathbf{\Sigma} \otimes \mathscr{I}(f) \end{pmatrix}.$$

Proof of Proposition 3.2. It follows from Corollary 3.1 and the strong unimodality of $f(\mathbf{x})$ that Proposition 2.2 holds here. Taking (3.7) into account, the matrices (2.11) associated with $\gamma^{(n)}_{i;f;j,k}$ are all zero, except (provided $i \leqslant p$) for

$$\begin{aligned} \mathbf{C}^{(j,k)}_i &= \int \phi_{f;j}(\mathbf{F}^{-1}(\mathbf{G}(\mathbf{x})))\, \boldsymbol{\phi}_g(\mathbf{x})\, g(\mathbf{x})\, d\mathbf{x} \\ &\quad \times \int F_k^{-1}(G_k(\mathbf{x}))\, \mathbf{x}' g(\mathbf{x})\, d\mathbf{x} \\ &= \{[\mathbf{\Sigma}(f, g)]_{.k}\, [\mathscr{I}(f, g)]_{j.}\}'. \end{aligned}$$

Accordingly, the asymptotic mean under $K^{(n)}_{\mathbf{A},\mathbf{B};g}$ of $(n-i)^{1/2}\gamma^{(n)}_{i;f;j,k}$ is

$$\begin{aligned} \mathrm{tr}[\mathbf{C}^{(j,k)}_i \mathbf{D}'_i] &= \{\mathbf{vec}[\mathscr{I}(f, g)_{j.}]\}'\, \mathbf{vec}[\mathbf{\Sigma}(f, g)_{k.}\, \mathbf{D}'_i] \\ &= [\mathscr{I}(f, g)]_{j.}\, \mathbf{D}_i \mathbf{\Sigma}(f, g)_{.k}, \end{aligned}$$

and the asymptotic mean of $(n-i)^{1/2}\mathbf{\Gamma}^{(n)}_{i;f}$ takes the form $\mathscr{I}(f, g)\, \mathbf{D}_i \mathbf{\Sigma}(f, g)$. The identity $\mathbf{vec}[\mathscr{I}(f, g)\, \mathbf{D}_i \mathbf{\Sigma}(f, g)] = \mathbf{\Sigma}(f, g) \otimes \mathscr{I}(f, g)\, \mathbf{vec}(\mathbf{D}_i)$ is a classical property of the Kronecker product (see, e.g., Mardia *et al.* [21]). As for the asymptotic covariance matrix, it is easy to check that the off-diagonal blocks are zero, since, for $i \neq i'$, (2.14) yields terms of the form

$$\begin{aligned} &\int \left[\phi_{f,j}(\mathbf{F}^{-1}(\mathbf{G}(\mathbf{x}))) - \int \phi_{f,j}(\mathbf{F}^{-1}(\mathbf{G}(\mathbf{x})))\, g(\mathbf{x})\, d\mathbf{x}\right] \\ &\quad \times \left[\phi_{f,j'}(\mathbf{F}^{-1}(\mathbf{G}(\mathbf{x}))) - \int \phi_{f,j'}(\mathbf{F}^{-1}(\mathbf{G}(\mathbf{x})))\, g(\mathbf{x})\, d\mathbf{x}\right] g(\mathbf{x})\, d\mathbf{x} \\ &\quad \times \int F_k^{-1}(G_k(y_1))\, F_{k'}^{-1}(G_{k'}(y_2))\, g_k(y_1)\, g_{k'}(y_2)\, dy_1\, dy_2 \\ &= \int \left[\phi_{f,j}(\cdots) - \int \phi_{f,j}(\cdots)\, g(\mathbf{x})\, d\mathbf{x}\right] \\ &\quad \times \left[\phi_{f,j'}(\cdots) - \int \phi_{f,j'}(\cdots)\, g(\mathbf{x})\, d\mathbf{x}\right] g(\mathbf{x})\, d\mathbf{x} \\ &\quad \times \int F_k^{-1}(u)\, du \cdot \int F_{k'}^{-1}(u)\, du = 0, \qquad \forall j, j', k, k' \end{aligned}$$

or

$$\int \left[\phi_{f,j}(\mathbf{F}^{-1}(\mathbf{G}(\mathbf{x}))) - \int \phi_{f,j}(\cdots)\, g(\mathbf{x})\, d\mathbf{x} \right] g(\mathbf{x})\, d\mathbf{x}$$

$$\times \int \left[\phi_{f,j'}(\mathbf{F}^{-1}(\mathbf{G}(\mathbf{x}))) - \int \phi_{f,j'}(\cdots)\, g(\mathbf{x})\, d\mathbf{x} \right] g(\mathbf{x})\, d\mathbf{x}$$

$$\times \int F_k^{-1}(G_k(y_k))\, F_{k'}^{-1}(G_{k'}(y_{k'}))\, g(\mathbf{y})\, d\mathbf{y} = 0, \qquad \forall j, j', k, k'.$$

The diagonal blocks, still on account on (2.14), have entries of the form

$$\int \left[\phi_{f,j}(\mathbf{F}^{-1}(\mathbf{G}(\mathbf{x}))) - \int \phi_{f,j}(\mathbf{F}^{-1}(\mathbf{G}(\mathbf{x})))\, g(\mathbf{x})\, d\mathbf{x} \right]$$

$$\times \left[\phi_{f,j'}(\mathbf{F}^{-1}(\mathbf{G}(\mathbf{x}))) - \int \phi_{f,j'}(\mathbf{F}^{-1}(\mathbf{G}(\mathbf{x})))\, g(\mathbf{x})\, d\mathbf{x} \right] g(\mathbf{x})\, d\mathbf{x}$$

$$\times \int F_k^{-1}(G_k(x_k))\, F_{k'}^{-1}(G_{k'}(x_{k'}))\, g(\mathbf{x})\, d\mathbf{x},$$

which completes the proof.

3.3. *Locally Asymptotically Most Powerful Test for $H^{(n)}$ against Specified ARMA Alternatives*

Consider the multirank serial statistic

$$T_*^{(n)} = (n-p)^{-1/2} \sum_{i=1}^{p} (n-i)^{1/2}\, \mathrm{tr}[\mathbf{\Gamma}_{i;f}^{(n)} \mathbf{D}_i'] / \sigma_*^{(n)}, \tag{3.15}$$

where $(\sigma_*^{(n)})^2$ denotes the permutational variance $\mathrm{Var}^*(\sum_{i=1}^{p} (n-i)^{1/2}\, \mathrm{tr}[\mathbf{\Gamma}_{i;f}^{(n)} \mathbf{D}_i'])$. Then $(n-p)^{1/2} T_*^{(n)}$ is a permutationally distribution-free statistic, with (conditional and unconditional) mean zero and variance one under $H^{(n)}$; consequently, it can be used to build permutationally distribution-free tests, resulting in (unconditionally) similar, asymptotically distribution-free tests for $H^{(n)}$ (see Proposition 3.3 below).

An exact computation of the permutation variance $\sigma_*^{(n)}$, although possible, appears to be rather unpleasant, even for moderate values of the series length n. On the other hand, the corresponding asymptotic value

$$\lim_{n \to \infty} (\sigma_*^{(n)})^2 = \sum_{i=1}^{p} (\mathbf{vec}\, \mathbf{D}_i)'\, \mathbf{\Sigma}(f \mid g) \otimes \mathscr{I}(f \mid g)(\mathbf{vec}\, \mathbf{D}_i) \tag{3.16}$$

under $H_g^{(n)}$ explicitly depends on the "true" underlying density $g(\cdot)$, and thus cannot be substituted for the exact value of $(\sigma_*^{(n)})^2$ in (3.15). An approximate value, however, can be used which is easier to compute and takes advantage of the asymptotic independence (Proposition 3.2) of the autocovariances $\mathbf{\Gamma}^{(n)}_{i;f}$ and $\mathbf{\Gamma}^{(n)}_{i';f}$ for distinct lags i and i'.

LEMMA 3.1. *Define*

$$\tilde{T}^{(n)}_{*} = (n-p)^{-1/2} \sum_{i=1}^{p} (n-i)^{1/2} \operatorname{tr}[\mathbf{\Gamma}^{(n)}_{i;f} \mathbf{D}'_i]/\tilde{\sigma}^{(n)}_{*}, \tag{3.17}$$

where

$$(\tilde{\sigma}^{(n)}_{*})^2 = \sum_{i=1}^{p} (\mathbf{vec}\, \mathbf{D}_i)' \, [\mathbf{W}^{(n)}_{i*}]^2 \, (\mathbf{vec}\, \mathbf{D}_i), \tag{3.18}$$

and

$$\begin{aligned}[\mathbf{W}^{(n)}_{i*}]^2 = [n(n-1)]^{-1} \sum_{1 \leqslant t_1 \neq t_2 \leqslant n} \Bigg[\mathbf{F}^{-1}\left(\frac{\mathbf{R}^{(n)}_{t_1}}{n+1}\right)\left(\mathbf{F}^{-1}\left(\frac{\mathbf{R}^{(n)}_{t_2}}{n+1}\right)\right)' \Bigg] \\ \otimes \Bigg[\boldsymbol{\phi}\left(\mathbf{F}^{-1}\left(\frac{\mathbf{R}^{(n)}_{t_1}}{n+1}\right)\right) \boldsymbol{\phi}'\left(\mathbf{F}^{-1}\left(\frac{\mathbf{R}^{(n)}_{t_2}}{n+1}\right)\right)\Bigg] \\ - [\mathbf{vec}(\mathbf{m}^{(n)}_{i*})][\mathbf{vec}(\mathbf{m}^{(n)}_{i*})]'.\end{aligned}$$

Then $(n-p)^{1/2}(T^{(n)}_{*} - \tilde{T}^{(n)}_{*})$ *is* $o_p(1)$ *under* $H^{(n)}$ *as well as under any alternative of the form* $K^{(n)}_{\mathbf{A},\mathbf{B};g}$.

Proof. The permutation covariance matrix of $(n-i)^{1/2}\,\mathbf{vec}(\mathbf{\Gamma}^{(n)}_{i;f})$ takes (cf. part (ii) of Lemma 2.1) the form

$$\begin{aligned}&\frac{1}{n(n-1)} \sum_{1 \leqslant t_1 \neq t_2 \leqslant n} \Bigg[\mathbf{F}^{-1}\left(\frac{\mathbf{R}^{(n)}_{t_1}}{n+1}\right)\left(\mathbf{F}^{-1}\left(\frac{\mathbf{R}^{(n)}_{t_2}}{n+1}\right)\right)' \Bigg] \\ &\quad \otimes \Bigg[\boldsymbol{\phi}\left(\mathbf{F}^{-1}\left(\frac{\mathbf{R}^{(n)}_{t_1}}{n+1}\right)\right) \boldsymbol{\phi}'\left(\mathbf{F}^{-1}\left(\frac{\mathbf{R}^{(n)}_{t_2}}{n+1}\right)\right)\Bigg] \\ &\quad + 2\,\frac{n-2i}{n-i} \cdot \frac{1}{n(n-1)(n-2)} \\ &\quad \times \sum_{1 \leqslant t_1 \neq t_2 \neq t_3 \leqslant n} \Bigg[\mathbf{F}^{-1}\left(\frac{\mathbf{R}^{(n)}_{t_1}}{n+1}\right)\left(\mathbf{F}^{-1}\left(\frac{\mathbf{R}^{(n)}_{t_2}}{n+1}\right)\right)' \Bigg] \\ &\quad \otimes \Bigg[\boldsymbol{\phi}\left(\mathbf{F}^{-1}\left(\frac{\mathbf{R}^{(n)}_{t_3}}{n+1}\right)\right) \boldsymbol{\phi}'\left(\mathbf{F}^{-1}\left(\frac{\mathbf{R}^{(n)}_{t_1}}{n+1}\right)\right)\Bigg]\end{aligned}$$

$$+\frac{n^2-n(2i+3)+i^2+5i}{n-i}\cdot\frac{1}{n(n-1)(n-2)(n-3)}$$
$$\times\sum_{1\leqslant t_1\neq t_2\neq t_3\neq t_4\leqslant n}\sum\sum\sum\left[\mathbf{F}^{-1}\left(\frac{\mathbf{R}_{t_1}^{(n)}}{n+1}\right)\left(\mathbf{F}^{-1}\left(\frac{\mathbf{R}_{t_2}^{(n)}}{n+1}\right)\right)'\right]$$
$$\otimes\left[\mathbf{F}^{-1}\left(\frac{\mathbf{R}_{t_3}^{(n)}}{n+1}\right)\left(\mathbf{F}^{-1}\left(\frac{\mathbf{R}_{t_4}^{(n)}}{n+1}\right)\right)'\right]-(n-i)[\mathbf{vec}(\mathbf{m}_{i*}^{(n)})][\mathbf{vec}(\mathbf{m}_{i*}^{(n)})]'. \tag{3.19}$$

On account of Proposition 3.2, the limiting (in probability, under $H_g^{(n)}$) value of (3.19) is $\mathbf{\Sigma}(f\mid g)\otimes\mathscr{I}(f\mid g)$. Since the first term in (3.19) is the permutational expectation (under $H^{(n)}$) of

$$\left[\mathbf{F}^{-1}\left(\frac{\mathbf{R}_{t}^{(n)}}{n+1}\right)\left(\mathbf{F}^{-1}\left(\frac{\mathbf{R}_{t-1}^{(n)}}{n+1}\right)\right)'\right]$$
$$\otimes\left[\boldsymbol{\phi}\left(\mathbf{F}^{-1}\left(\frac{\mathbf{R}_{t}^{(n)}}{n+1}\right)\right)\boldsymbol{\phi}'\left(\mathbf{F}^{-1}\left(\frac{\mathbf{R}_{t-1}^{(n)}}{n+1}\right)\right)\right],$$

its limiting value (in probability, under $H_g^{(n)}$) is also (Proposition 3.2) $\mathbf{\Sigma}(f\mid g)\otimes\mathscr{I}(f\mid g)$. Also, since $\int F_k^{-1}(G_k(x))\, g_k(x)\,dx=\int F_k^{-1}(u)\,du=0$, the limiting value (still in probability, under $H_g^{(n)}$) of $\mathbf{m}_{i*}^{(n)}$ is zero. (3.19) thus reduces (up to $o_p(1)$ terms, under $H_g^{(n)}$) to its first term, which in turn is asymptotically equal to $[\mathbf{W}_*^{(n)}]^2$. Now since

$$\sum_{i=1}^{p}(n-i)^{1/2}\operatorname{tr}[\mathbf{\Gamma}_{i;f}^{(n)}\mathbf{D}_i']=\sum_{i=1}^{p}(\mathbf{vec}\,\mathbf{D}_i)'(\mathbf{vec}(n-i)^{1/2}\,\mathbf{\Gamma}_{i;f}^{(n)}),$$

and since $(n-i)^{1/2}\,\mathbf{\Gamma}_{i;f}^{(n)}$ and $(n-i')^{1/2}\,\mathbf{\Gamma}_{i';f}^{(n)}$ are asymptotically independent, $\sigma_*^{(n)}-\tilde{\sigma}_*^{(n)}$ is $o_p(1)$, both under $H^{(n)}$ and under $K_{\mathbf{A},\mathbf{B};g}^{(n)}$ $\forall\mathbf{A},\mathbf{B},g$.

We now establish that the multirank statistic $T_*^{(n)}$ (or, equivalently, $\tilde{T}_*^{(n)}$) provides a locally asymptotically most powerful test for $H^{(n)}$ against $H_{\mathbf{A},\mathbf{B};f}^{(n)}$ (not only within the class of rank-based tests, but within the whole class of tests having the same probability level).

PROPOSITION 3.3. *The permutationally distribution-free test procedure*

$$\textit{reject } H^{(n)} \qquad \textit{if } T_*^{(n)}>(n-p)^{-1/2}k_{1-\alpha}, \tag{3.20}$$

and the asymptotically distribution-free test procedure

$$\textit{reject } H^{(n)} \qquad \textit{if } \tilde{T}_*^{(n)}>(n-p)^{-1/2}k_{1-\alpha}, \tag{3.21}$$

where $k_{1-\alpha}$ denotes the $(1-\alpha)$-fractile of the standard normal distribution

(i) *have asymptotic level* α (*under* $H^{(n)}$)

(ii) *have asymptotic power*

$$1-\Phi\left(k_{1-\alpha}-\frac{\sum_{i=1}^{p}(\text{vec}\,\mathbf{D}_i)'\,\mathbf{\Sigma}(f,g)\otimes\mathcal{I}(f,g)(\text{vec}\,\mathbf{E}_i)}{\{\sum_{i=1}^{p}(\text{vec}\,\mathbf{D}_i)'\,\mathbf{\Sigma}(f\mid g)\otimes\mathcal{I}(f\mid g)(\text{vec}\,\mathbf{D}_i)\}^{1/2}}\right) \quad (3.22)$$

against $K^{(n)}_{\mathbf{E};g}=U\{K^{(n)}_{\mathbf{A},\mathbf{B};g}\mid \mathbf{A}_i+\mathbf{B}_i=\mathbf{E}_i,\ i=1,...,p\}$, *where* $\Phi(\cdot)$ *denotes the standard normal distribution function, and where* $\mathbf{E}_1, ..., \mathbf{E}_q$ *is any q-tuple of* $m\times m$ *matrices such that* $|\mathbf{E}_q|\neq 0$, *and with the convention that* $\mathbf{E}_{q+1}=\cdots=\mathbf{E}_p=\mathbf{0}$ *if* $q<p$.

(iii) *are asymptotically most powerful against*

$$K^{(n)}_{\mathbf{D};f}=U\{K^{(n)}_{\mathbf{A},\mathbf{B};f}\mid \mathbf{A}_i+\mathbf{B}_i=\mathbf{D}_i,\ i=1,...,p\}.$$

Consequently, the limit, as $n\to\infty$, *of the envelope power function for* $H^{(n)}$ *against* $K^{(n)}_{\mathbf{A},\mathbf{B};f}$ *exists, and takes the value*

$$\lim_{n\to\infty}\beta(\alpha;H^{(n)},K^{(n)}_{\mathbf{A},\mathbf{B};f})$$
$$=1-\Phi\left(k_{1-\alpha}-\left\{\sum_{i=1}^{p}(\text{vec}\,\mathbf{D}_i)'\,\mathbf{\Sigma}\otimes\mathcal{I}(f)(\text{vec}\,\mathbf{D}_i)\right\}^{1/2}\right). \quad (3.23)$$

The asymptotic power values in (3.22) and (3.23) are consistent with the corresponding univariate values (cf. Hallin *et al.* [5]), where the limit of the envelope power function was shown to be of the form $1-\Phi(k_{1-\alpha}-(\sum_{i=1}^{p}d_i^2\sigma^2 I(f))^{1/2})$.

Proof of Proposition 3.3. It is an immediate consequence of *Proposition* 3.2 that $(n-p)^{1/2}\,T^{(n)}_{*}$ is asymptotically normal with mean zero and variance one under $H^{(n)}$, with mean $\sum_{i=1}^{p}(\text{vec}\,\mathbf{D}_i)'\,\mathbf{\Sigma}(f,g)\otimes\mathcal{I}(f,g)(\text{vec}\,\mathbf{E}_i)/\lim_{n\to\infty}\sigma^{(n)}_{*}$ and variance one under $K^{(n)}_{\mathbf{E};g}$.

This establishes parts (i) and (ii) of the proposition. Now, note that against $K^{(n)}_{\mathbf{D};f}$ (3.22) reduces to

$$1-\Phi\left(k_{1-\alpha}-\left\{\sum_{i=1}^{p}(\text{vec}\,\mathbf{D}_i)'\,\mathbf{\Sigma}\otimes\mathcal{I}(f)(\text{vec}\mathbf{D}_i)\right\}^{1/2}\right)$$
$$=1-\Phi\left(k_{1-\alpha}-\left\{\sum_{i=1}^{p}\text{tr}[\mathcal{I}(f)\,\mathbf{D}_i\mathbf{\Sigma}\mathbf{D}_i']\right\}^{1/2}\right). \quad (3.24)$$

In order to prove that (3.20) and (3.21) are asymptotically most powerful against $K^{(n)}_{\mathbf{A},\mathbf{B};f}$, hence against $K^{(n)}_{\mathbf{D};f}$, it is therefore sufficient to show that (3.24) is the limit, as $n\to\infty$, of the envelope power function $\beta(\alpha;H^{(n)},K^{(n)}_{\mathbf{A},\mathbf{B};f})$, $\alpha\in(0,1)$. Consider the simple null hypothesis $H^{(n)}_f$, and the simple alternative $K^{(n)}_{\mathbf{A},\mathbf{B};f}$: the most powerful test is then the

Neyman test based on $\mathscr{L}^{(n)}_{\mathbf{A},\mathbf{B};f}$ which, on account of Proposition 1.1 and LeCam's third lemma, is asymptotically normal, with mean $-\frac{1}{2}\sum_{i=1}^{p} \operatorname{tr}[\mathscr{I}(f)\,\mathbf{D}_i\mathbf{\Sigma}\mathbf{D}'_i]$ under $H^{(n)}_f$, mean $\frac{1}{2}\sum_{i=1}^{p} \operatorname{tr}[\mathscr{I}(f)\,\mathbf{D}_i\mathbf{\Sigma}\mathbf{D}'_i]$ under $K^{(n)}_{\mathbf{A},\mathbf{B};f}$, and variance $\sum_{i=1}^{p} \operatorname{tr}[\mathscr{I}(f)\,\mathbf{D}_i\mathbf{\Sigma}\mathbf{D}'_i]$ under both. It follows that the asymptotic power of the Neyman test, hence the limit as $n\to\infty$ of the envelope power function $\beta(\alpha; H^{(n)}_f, K^{(n)}_{\mathbf{A},\mathbf{B};f})$, is (3.24). Now $\beta(\alpha; H^{(n)}_f, K^{(n)}_{\mathbf{A},\mathbf{B};f}) \geqslant \beta(\alpha, H^{(n)}, K^{(n)}_{\mathbf{A},\mathbf{B};f})$ for every n; on the other hand, it follows from (3.22) that

$$\limsup_{n\to\infty} \beta(\alpha; H^{(n)}, K^{(n)}_{\mathbf{A},\mathbf{B};f}) \geqslant \lim_{n\to\infty} \beta(\alpha; H^{(n)}_f, K^{(n)}_{\mathbf{A},\mathbf{B};f}).$$

Since this latter limit exists (and is given in (3.24)), (3.23) holds for any $\alpha\in(0, 1)$, which completes the proof.

4. Normal Theory Procedure

4.1. *Asymptotically Locally Most Powerful Parametric Test (Gaussian Case)*

Assume that the density f is normal with full-rank covariance matrix $\mathbf{\Sigma}$. Then $\boldsymbol{\phi}(\mathbf{x}) = \mathbf{\Sigma}^{-1}\mathbf{x}$, and the first-order term $\mathscr{L}^{(n)}_0$ in the expansion (1.9) of the log-likelihood $\mathscr{L}^{(n)}_{\mathbf{A},\mathbf{B};f}$ takes the form

$$\begin{aligned}\mathscr{L}^{(n)}_0 &= n^{-1/2} \sum_{t=p+1}^{n} \sum_{i=1}^{p} \operatorname{tr}[\mathbf{\Sigma}^{-1}\mathbf{X}^{(n)}_t(\mathbf{X}^{(n)}_{t-i})'\,\mathbf{D}'_i] \\ &= n^{-1/2} \sum_{i=1}^{p} (n-i) \operatorname{tr}[\mathbf{\Sigma}^{-1}\mathbf{\Gamma}^{(n)}_i\mathbf{D}'_i],\end{aligned} \tag{4.1}$$

where $\mathbf{\Gamma}^{(n)}_i$ stands for the usual sample autocovariance matrix

$$\mathbf{\Gamma}^{(n)}_i = (n-i)^{-1} \sum_{t=i+1}^{n} \mathbf{X}^{(n)}_t(\mathbf{X}^{(n)}_{t-i})'. \tag{4.2}$$

Since $(n-i)^{1|2}\,\mathbf{vec}(\mathbf{\Gamma}^{(n)}_i)$ is asymptotically normal (cf., e.g., [2]), with mean $\mathbf{0}$ (under $H^{(n)}$) and covariance matrix $\mathbf{\Sigma}\otimes\mathbf{\Sigma}$ (under $H^{(n)}$ as well as under contiguous alternatives), and since $\mathbf{\Gamma}^{(n)}_i$ and $\mathbf{\Gamma}^{(n)}_{i'}$, $i\neq i'$ are uncorrelated, the test procedure rejecting $H^{(n)}$ when

$$\sum_{i=1}^{p} (n-i)^{1/2} \operatorname{tr}[\mathbf{\Gamma}^{(n)}_i\mathbf{D}'_i\mathbf{\Sigma}^{-1}] \bigg/ \left\{\sum_{i=1}^{p} (\mathbf{vec}\,\mathbf{D}_i)'\,\mathbf{\Sigma}\otimes\mathbf{\Sigma}^{-1}(\mathbf{vec}\,\mathbf{D}_i)\right\}^{1/2} > k_{1-\alpha} \tag{4.3}$$

has asymptotical level α, $\alpha\in(0, 1)$. It is easy to gauge from (4.1) that (4.3) also provides an asymptotically most powerful test for $H^{(n)}$ against gaussian alternatives of the form $K^{(n)}_{\mathbf{D};f}$ (see Proposition 3.3(ii) for a precise definition). Actually, the (parametric) test statistic in (4.3) is asymptotically equivalent (under normality) to the optimal rank-based statistic (3.15) (here of the *van der Waerden* type to be described in Section 4.2). This easily follows from the asymptotic equivalence of (3.15) with a U-statistic (Section 2.3). The asymptotic power of (4.3) against $K^{(n)}_{\mathbf{E};\tilde{f}}$, where $\tilde{f}$ denotes a gaussian density with correlation matrix $\tilde{\mathbf{P}}$ and covariance matrix $\tilde{\mathbf{\Sigma}} = \mathbf{diag}(\tilde{\sigma}_i)\ \tilde{\mathbf{P}}\ \mathbf{diag}(\tilde{\sigma}_i)$ is

$$1-\Phi\left(k_{1-\alpha}-\frac{\sum_{i=1}^{p}(\text{vec}\,\mathbf{D}_i)'\,\mathbf{diag}(\sigma_i)\,\tilde{\mathbf{P}}\,\mathbf{diag}(\tilde{\sigma}_i)\otimes\mathbf{\Sigma}^{-1}(\text{vec}\,\mathbf{D}_i)}{\left(\begin{array}{l}\{\sum_{i=1}^{p}(\text{vec}\,\mathbf{D}_i)'\,\mathbf{diag}(\sigma_i)\,\tilde{\mathbf{P}}\,\mathbf{diag}(\sigma_i)\\ \quad\otimes\mathbf{\Sigma}^{-1}\,\mathbf{diag}(\sigma_i)\,\tilde{\mathbf{P}}\,\mathbf{diag}(\sigma_i)\,\mathbf{\Sigma}^{-1}(\text{vec}\,\mathbf{D}_i)\}^{1/2}\end{array}\right)}\right) \tag{4.4}$$

and reduces to

$$1-\Phi\left(k_{1-\alpha}-\left\{\sum_{i=1}^{p}(\text{vec}\,\mathbf{D}_i)'\,\mathbf{\Sigma}\otimes\mathbf{\Sigma}^{-1}(\text{vec}\,\mathbf{D}_i)\right\}^{1/2}\right) \tag{4.5}$$

against $K^{(n)}_{\mathbf{D};f}$.

4.2. *A van der Waerden Type Test*

Under the same assumptions as in Section 4.1, the test statistic $T^{(n)}_{\text{v.d.W.}}$ which is optimal against $K^{(n)}_{\mathbf{D};f}$ is obtained from (3.15) by considering the *van der Waerden rank autocovariance matrices*, defined as

$$\begin{aligned}\mathbf{\Gamma}^{(n)}_{i;\,\text{v.d.w.}} = (n-i)^{-1}\,\mathbf{\Sigma}^{-1}\sum_{t=i+1}^{n}\mathbf{diag}(\sigma_j)\\ \times\boldsymbol{\phi}^{-1}\left(\frac{\mathbf{R}^{(n)}_t}{n+1}\right)\left(\boldsymbol{\phi}^{-1}\left(\frac{\mathbf{R}^{(n)}_{t-i}}{n+1}\right)\right)'\mathbf{diag}(\sigma_j)-\mathbf{m}^{(n)}_{i*;\,\text{v.d.W.}}\end{aligned} \tag{4.6}$$

$(\boldsymbol{\phi}^{-1}(\mathbf{R}^{(n)}_t/(n+1)))$ denotes the vector whose components are $\phi^{-1}(R^{(n)}_{t,1}/(n+1)),\ \ldots,\ \phi^{-1}(R^{(n)}_{t,m}/(n+1))$. Since $\mathbf{\Sigma\Gamma}^{(n)}_{i;\,\text{v.d.W.}}-\mathbf{\Gamma}^{(n)}_i$ is $o_p(n^{-1/2})$ under $H^{(n)}_f$, it follows that the ARE (asymptotic relative efficiency) of the *van der Waerden test* (based on (4.6)) with respect to its parametric counterpart (4.3) is one. The *van der Waerden test*, however, has over the parametric test, the many advantages of rank-based procedures: permutational distribution-freeness, robustness, and low sensitivity to gross error—which should make it a rather attractive testing procedure, especially for relatively short series, when the underlying noise is suspected to be nongaussian, and against moving average dependencies rather than autoregressive ones (see [14]).

Now, the definition (4.6) of van der Waerden autocovariances, as well as that (4.3) of the optimal parametric test statistic explicitly involves the covariance matrix $\mathbf{\Sigma}$. Since $\mathbf{\Sigma}$ remains unspecified in most applications, it usually has to be replaced with some consistent estimator $\mathbf{\Sigma}^{(n)}$. Provided that $\mathbf{\Sigma}^{(n)}$ is a symmetric function of the observations, such a substitution does not affect the exchangeability property of the observed series (see, e.g., [18], pp. 356–357). Neither does it affect the asymptotic distribution of the test statistic. The resulting van der Waerden test accordingly remains genuinely permutationally distribution-free and asymptotically locally as powerful as its parametric counterpart.

5. Concluding Remarks

The contribution of this paper is twofold. Sections 1 and 2 mainly provide the asymptotic distribution theory of linear serial multirank statistics under randomness and under local alternatives of ARMA dependence. The results in these sections allow for constructing rank-based tests for multivariate white noise, evaluating their asymptotic power against local alternatives, and computing their AREs with respect to each other and with respect to related parametric procedures. Sections 3 and 4 show that asymptotic local optimality can be achieved by means of adequate linear multirank tests and suggest a procedure of the van der Waerden type, possessing all the distinctive features of rank-based methods (distribution-freeness, robustness against gross errors, etc.), yet asymptotically as powerful as its normal-theory counterpart. The optimality results in Section 3 also attest to the prominent role of a class of measures of dependence (closely depending on the underlying white noise density f) defined as the *rank autocovariance matrices* $\mathbf{\Gamma}^{(n)}_{i;f}$ associated with f, which provide multivariate analogues of the *rank autocorrelation* coefficients previously considered for univariate series [7, 9].

This paper constitutes a first step towards a systematic and logically coherent nonparametric approach towards multiple time series problems, an approach that has not been explored so far. Many questions remain open (in this area as well as in the corresponding parametric approach). In an ongoing research, the authors propose to derive asymptotically maximin rank-based tests against unspecified ARMA alternatives. These tests rely on quadratic forms involving the rank autocovariance matrices $\mathbf{\Gamma}^{(n)}_{i;f}$. Their asymptotic performances should be compared with those of their parametric, normal-theory competitors based on sample cross-correlations—such as Hosking's [11, 12] or Poskitt and Tremayne's [15] multivariate portmanteau tests.

On the other hand, the present paper also constitutes, to the best of our knowledge, the first attempt to obtain explicit local optimality results in a multiple time series context. In this respect, it could also serve as a basis for deriving locally asymptotically most powerful or maximin parametric (gaussian or nongaussian) procedures (e.g., substituting alternative local asymptotical maximin criteria for the gaussian Lagrange multiplier principle used in Hosking [12] in the definition of a multivariate portmanteau statistic).

REFERENCES

[1] ANDERSON, T. W. (1971). *The Statistical Analysis of Time Series.* Wiley, New York.

[2] CHITURRI, R. V. (1976). Distribution of multivariate white noise autocorrelations. *J. Amer. Statist. Assoc.* **71** 223–226.

[3] HÁJEK, J., AND ŠIDÁK, Z. (1967). *Theory of Rank Tests.* Academic Press, New York.

[4] HALLIN, M. (1986). Nonstationary q-dependent processes and time-varying moving average models: invertibility properties and the forecasting problem. *Adv. in Appl. Probab.* **18** 170–210.

[5] HALLIN, M., INGENBLEEK, J.-FR., AND PURI, M. L. (1985a). Linear serial rank tests for randomness against ARMA alternatives. *Ann. Statist.* **13** 1156–1181.

[6] HALLIN, M., INGENBLEEK, J.-FR., AND PURI, M. L. (1985b). Tests de rangs quadratiques pour une hypothèse de bruit blanc. *C.R. Acad. Sci. Paris Sér. I* **301** 935–938.

[7] HALLIN, M., INGENBLEEK, J.-FR., AND PURI, M. L. (1987). Linear and quadratic serial rank tests for randomness against serial dependence. *J. Time Ser. Anal.* **8** 409–424.

[8] HALLIN, M., AND PURI, M. L. (1987). Locally asymptotically optimal tests for randomness. In *New Perspectives in Theoretical and Applied Statistics* (M. L. Puri, J. P. Vilaplana, and W. Wertz, Eds.), pp. 87–99, Wiley, New York.

[9] HALLIN, M., AND PURI, M. L. (1988). Optimal rank-based procedures for time series analysis: Testing an ARMA model against other ARMA models. *Ann. Statist.* **16** 402–432.

[10] HANNAN, E. J. (1970). *Multiple Time Series.* Wiley, New York.

[11] HOSKING, J.R. M. (1980). The multivariate portmanteau statistic. *J. Amer. Statist. Assoc.* **75** 602–608.

[12] HOSKING, J. R. M. (1981). Lagrange multiplier tests of multivariate time series models. *J. Roy. Statist. Soc. B* **43** 219–230.

[13] LECAM, L. (1960). *Locally Asymptotically Normal Families of Distributions.* Univ. Calif. Publ. Statist. 3, pp. 37–98.

[14] LUSK, E. J., AND WRIGHT, H. (1982). Non-gaussian series and series with non-zero means: Practical Implications for time series analysis, *Statist. Probab. Lett.* **1** 2–6.

[15] POSKITT, D. S., AND TREMAYNE, A. R. (1982). Diagnostic tests for multiple time series models. *Ann. Statist.* **10** 114–120.

[16] PURI, M. L., AND SEN, P. K. (1971). *Nonparametric Methods in Multivariate Analysis.* Wiley, New York.

[17] PURI, M. L., AND SEN, P. K. (1985). *Nonparametric Methods in General Linear Models.* Wiley, New York.

[18] RANDLES, R. M., AND WOLFE, D. A. (1979). *Introduction to the Theory of Nonparametric Statistics.* Wiley, New York.

[19] YOSHIHARA, K. I. (1976). Limiting behavior of U-statistics for stationary absolutely regular processes. *Z. Wahrsch. Verw. Gebiete* **35** 237–252.

[20] WALD, A., AND WOLFOWITZ, J. (1943). An exact test for randomness in the nonparametric case based on serial correlation. *Ann. Math. Statist.* **14** 378–388.

[21] MARDIA, K. V., KENT, J. T., AND BIBBY, J. M. (1979). *Multivariate Analysis.* Academic Press, New York.

Printed by Catherine Press, Ltd., Tempelhof 41, B-8000 Brugge, Belgium
Reprinted by permission of the publisher,
© 1989 by Academic Press

WEAK CONVERGENCE OF THE *U*-STATISTIC AND WEAK INVARIANCE OF THE ONE-SAMPLE RANK ORDER STATISTIC FOR MARKOV PROCESSES AND ARMA MODELS

MICHEL HAREL and MADAN L. PURI*

Institut Universitaire de Technologie de Limoges et U.A. C.N.R.S., Rouen, France, Indiana University

Harel and Puri (1989, *J. Multivariate Anal.* **29**) studied the asymptotic behavior of the *U*-statistic and the one-sample rank order statistic for nonstationary absolutely regular processes. In this note, we present some applications of these results for Markov processes as well as ARMA processes. © 1989 Academic Press, Inc.

1. Introduction

Let X_{ni}, $1 \leqslant i \leqslant n$, $n \geqslant 1$ be real-valued r.v.'s (random variables) with continuous d.f.'s (distribution functions) $F_{ni}(x)$, $x \in \mathbb{R}$, and let $\mathbf{F}_n$ denote the d.f. of the $\mathbb{R}^n$-valued r.v. $(X_{n1}, ..., X_{nn})$.

Consider the *U*-statistic

$$U(\mathbf{F}_n) = \binom{n}{k}^{-1} \sum_{(i)}^{(n)} g(X_{ni_1}, ..., X_{ni_k}), \qquad n \geqslant k \geqslant 1, \tag{1.1}$$

where the summation $\sum_{(i)}^{(n)}$ extends over all possible $1 \leqslant i_1 < \cdots < i_k \leqslant n$ and g: $\mathbb{R}^k \to \mathbb{R}$ is a Borel measurable function which is symmetric in its $k(\geqslant 1)$ arguments.

Received December 21, 1988.

AMS 1980 subject classifications: Primary 60B10, 60F05.

Key words and phrases: *U*-statistic, rank order statistic, absolute regularity, Markov processes, ARMA processes.

* Research supported by the Office of Naval Research Contract N00014-85-K-0648.

Consider also the 1-sample rank order statistic $\mathcal{S}_{n,m}$

$$\mathcal{S}_{n,m} = \sum_{i=1}^{m} c_{ni} s(X_{ni}) \, J\left(\frac{R_{n,m,i}}{m+1}\right), \qquad n \geqslant m \geqslant 1, \tag{1.2}$$

where J is a score function, $s(x) = \operatorname{sgn}(x)$ and the c_{ni} are regression constants generated by a continuous function $h(x)$ on $[0, 1]$,

$$\begin{aligned} c_{ni} &= h\left(\frac{i}{n}\right), \; 1 \leqslant i \leqslant n, \; n \geqslant 1, \\ R_{n,m,i} &= \sum_{j=1}^{m} I_{[|X_{nj}| \leqslant |X_{ni}|]}, \; 1 \leqslant i \leqslant m \leqslant n, \end{aligned} \tag{1.3}$$

where $I_{[\;]}$ denotes the indicator function.

In Harel and Puri [3], we studied the asymptotic behavior of the statistics $U(\mathbf{F}_n)$ and $\mathcal{S}_{n,m}$ when the underlying r.v.'s are nonstationary, absolutely regular with rates

$$\beta(m) = O(m^{-(2+\delta)/\delta}) \qquad \text{for some} \quad \delta > 0 \tag{1.4}$$

in the case of the U-statistic (1.1), and

$$\beta(m) = O(m^{-4}) \tag{1.5}$$

in the case of the 1-sample rank statistic (1.2). In this paper, we provide applications of some of the results of Harel and Puri [3] for some Markov processes as well as ARMA processes.

For the ease of convenience and easy reference we state the two main results from Harel and Puri [3].

THEOREM 1.1 (Convergence of the U-statistic). *Let $F_{n,i,j}$ be the d.f. of (X_{ni}, X_{nj}), $1 \leqslant i < j \leqslant n$. Suppose that in addition to the assumption* (1.4), *the assumptions* (2.6) *and* (2.7) *of Harel and Puri* [3] *are satisfied. Furthermore, assume that for any $i, j \in \mathbb{N}^*$, with $i < j$, there exists a continuous d.f. F_{ij} on $\mathbb{R}^2$ with continuous marginals F_i^* and F_j^* such that*

$$\lim_{n \to \infty} \max_{1 \leqslant i < j \leqslant n} |F_{n,i,j}(x_1, x_2) - F_{ij}(x_1, x_2)| = 0 \quad \textit{for all} \quad (x_1, x_2) \in \mathbb{R}^2 \tag{1.6}$$

$$F_{ij} = F_{1,j-i+1} \qquad \textit{for all} \quad i < j \tag{1.7}$$

and

$$\int_{\mathbb{R}^k} |g(x_1, ..., x_k)|^r \prod_{j=1}^{k} dF(x_j) \leqslant M_0 < \infty, \tag{1.8}$$

where M_0 *is some constant* >0 *and* $F=F_i^*$ *for all* $i\in\mathbb{N}^*$ *and* g *is right continuous and has left-hand limits (r.c.l.l.) or left continuous and has right-hand limits (r.c.r.l.). Then* $n^{1/2}(U(\mathbf{F}_n)-\theta(\mathbf{F}_n))$ *converges in law to the normal distribution with mean* 0 *and variance* $k^2\sigma^2$ *if* $\sigma^2>0$, *where* $\theta(\mathbf{F}_n)=n^{-[k]}\sum_{(i_1,\dots,i_k)\in\mathbf{I}_{0,n}}\int_{\mathbb{R}^k}g(x_1,\dots,x_k)\,dF_{ni_1}\cdots dF_{ni_k}$, *where* $\mathbf{I}_{0,n}=\{i_1,\dots,i_k);\ 1\leqslant i_1\neq\cdots\neq i_k\leqslant n\}$,

$$\sigma^2=\left[\int_{\mathbb{R}^k}g^2(x_1,\dots,x_k)\prod_{l=1}^k dF(x_l)-\theta^2(F)\right]$$
$$+2\sum_{i=2}^{\infty}\left[\int_{\mathbb{R}^{2k}}g(x_1,\dots,x_k)\,g(x_{k+1},\dots,x_{2k})\right.$$
$$\times dF_{1i}(x_1,x_{k+1})\prod_{l=2}^k dF(x_l)^2$$
$$\left.\times\prod_{p=k+2}^{k}dF(x_p)-\theta^2(F)\right]<\infty, \tag{1.9}$$

where $\theta(F)=\int_{\mathbb{R}^k}g(x_1,\dots,x_k)\prod_{l=1}^k dF(x_l)$.

We now assume that $F_{ni}=F_n$ for any i $(1\leqslant i\leqslant n)$. For any real number x, define $H_n(|x|)$ as $H_n(|x|)=F_n(|x|)-F_n(-|x|)$. Let F be a d.f. on $\mathbb{R}$ and define the d.f. H on $\mathbb{R}^+$ by $H(|x|)=F(|x|)-F(-|x|)$.

For a score function $J(u)$ which is square integrable put

$$\mu_n=\mu_J(F_n)=\int_{\mathbb{R}}s(x)\,J(H_n(|x|)\,dF_n(x). \tag{1.10}$$

For h defined in (1.3) put

$$\mu_{n,m}=\mu_n\sum_{i=1}^m c_{ni}=\mu_n\sum_{i=1}^m h\left(\frac{i}{n}\right),\qquad m\leqslant n. \tag{1.11}$$

For any sequence of d.f.'s $\{F_{1i};\ i\geqslant 2\}$ on $\mathbb{R}^2$ with marginals F we denote

$$\sigma^2(F_{1i}\})=\lim_{n\to\infty}\left\{\int_{\mathbb{R}}g^2(x)\,dF(x)\right.$$
$$\left.+2\sum_{l=2}^n\int_{\mathbb{R}^2}g(x)\,g(y)\,dF_{1l}(x,y)\right\} \tag{1.12}$$

if the limit exists, where

$$g(x)=\int_{\mathbb{R}}s(y)\,\{I_{[|x|\leqslant|y|]}-H(|y|)\}\,J'(H(|y|)\,dF(y)$$
$$+s(x)\,J(H(|x|))-\int_{\mathbb{R}}s(y)\,J(H(|y|)\,dF(y).$$

For every $n \geqslant 1$, let

$$V_n(t) = \begin{cases} 0, & \text{for} \quad t = 0 \\ (\mathscr{S}_{n,k} - \mu_{n,k})/\sigma^{1/2}, & \text{for} \quad t = k/n,\ k = 1, \ldots, n \\ \text{linearly interpolated}, & \text{for} \quad t \in [k - 1/n, k/n],\ k = 1, \ldots, n, \end{cases} \tag{1.13}$$

where σ is the positive constant defined in (1.12).

The proces $V_n(t) = \{V_n(t),\ 0 \leqslant t \leqslant 1\}$ belongs to the space C_1 of all continuous functions on $[0, 1]$ with which we associate the usual uniform metric. Then, we have

THEOREM 1.2 (Convergence of the rank order statistic). *Suppose the sequence $\{X_{ni}\}$ is absolutely regular with rate* (1.5) *and the sequence $\{F_n\}$ satisfies the conditions* (1.6) *and* (1.7) *of Theorem* 1.1. *Let J be a score function having a bounded second derivative. If $\sigma^2(\{F_{1i}\})$ defined by* (1.12) *is strictly positive, then V_n defined in* (1.13) *converges weakly in the uniform topology on C_1 to the process $V_0 = \{V_0(t), 0 \leqslant t \leqslant 1\}$, where*

$$V_0(t) = \int_0^t h(u)\, dW(u), \qquad 0 \leqslant t \leqslant 1, \tag{1.14}$$

and $W = \{W(t),\ 0 \leqslant t \leqslant 1\}$ is a standard Brownian motion process, and $\sigma^2(\{F_{1i}\} < \infty$.

2. APPLICATIONS TO MARKOV PROCESSES AND ARMA PROCESSES

2.1. *Markov Processes*

Consider a sequence $\{X_{ni};\ i \in \mathbb{Z}\}$ *of* $\mathbb{R}$-valued processes such that for all $n \in \mathbb{N}^*$, $\{X_{ni}\}$ is a Markov process with stationary transition probabilities $P_n(x; A)$, where $A \in \mathscr{B}$, $\mathscr{B}$ is the Borel σ-field of $\mathbb{R}$, and $x \in \mathbb{R}$.

Recall that the Markov process is *geometrically ergodic* if it is ergodic and if there exists $0 < \rho_n < 1$ such that

$$\|P_n^m(x; \cdot) - \mu_n(\cdot)\| = O(\rho_n^m) \qquad \text{for all} \quad \text{a.s. } x \in \mathbb{R},$$

where $\|\cdot\|$ denotes the norm of total variation. (ρ_n is called the rate) and P_n^m is the m-step transition probability.

THEOREM 2.1. *Let $\{X_{ni};\ i \in \mathbb{Z}\}$ be a Markov process such that for every $n \in \mathbb{N}^*$, $\{X_{ni}\}$ is either* (a) *aperiodic, Harris recurrent, and geometrically* *ergodic with rates* $0<\rho_n<\rho_0$, $\rho_0 \in (0,1)$ *or* (b) *aperiodic and Doeblin recurrent.*

Suppose there exists a probability measure μ_0 *on* $\mathbb{R}$ *and a transition probability* P_0 *such that*

$$\mu_n((-\infty, x]) \to \mu_0((-\infty, x]) \quad \text{as} \quad n \to \infty \tag{2.1}$$

for all $x \in \mathbb{R}$;

$$P_n(x; (-\infty, y]) \to P_0(x; (-\infty, y]) \quad \text{as} \quad n \to \infty \tag{2.2}$$

for all $(x, y) \in \mathbb{R}^2$. *Then for a function g satisfying the conditions of Theorem* 1.1 *and for a U-statistic defined in* (1.1), *the conclusion of Theorem* 1.1 *holds. Also for a score function J having a second bounded derivative and the process* V_n *defined in* (1.13), *the conclusion of Theorem* 1.2 *holds.*

Proof. (i) Suppose (a) holds. From Theorems 1.1 and 1.2, we have to show that the conditions (1.6) and (1.7) are verified and the sequence $\{X_{ni}\}$ *is absolutely regular with the rate* (1.4) *or* (1.5).

From Nummelin and Tuominen [5], a Markov process which is aperiodic, Harris recurrent, and gemetrically ergodic with rates ρ_n, satisfies

$$\int_{\mathbb{R}} \|P_n^m(x;\cdot) - \mu_n(\cdot)\| \, \mu_n(dx) = O(\rho_n^m).$$

From Proposition 1 of Davydov [2] we have

$$\beta(m) = \sup_{n \in \mathbb{N}^*} \int_{\mathbb{R}} \|P_n^m(x;\cdot) - \mu_n(\cdot)\| \, \mu_n(dx)$$

and we deduce that the sequence $\{X_{ni}\}$ is absolutely regular with a geometrical rate ρ_0 which implies (1.4) or (1.5).

We now show that (1.6) is satisfied. For any $n \geqslant 1$, it can easily be seen that the d.f. $F_{n,i,j}$ has the same marginals F_n, say. For any $m \geqslant 1$, let P_0^m be the m-step transition probability of P_0.

Let G_m be the d.f. associated with the probability measure Q_m, where

$$Q_m(A_1 \times A_2) = \int_{A_1} \int_{A_2} P_0^m(x; dy) \, \mu_0(dx) \qquad \forall (A_1, A_2) \in \mathscr{B}^2.$$

For any $(i, j) \in (\mathbb{N}^*)^2$ with $i<j$, we consider the d.f. F_{ij} defined by $F_{ij} = G_{j-i}$. From the definition of F_{ij}, it is clear that the marginal d.f.'s of F_{ij} are identical. We denote them by F.

Let $(x, y) \in \mathbb{R}^2$ be fixed, and note the inequality

$$\begin{aligned}|F_{n,i,j}(x, y) - F_{ij}(x, y)| \leqslant & |F_{n,i,j}(x, y) - F_n(x)\, F_n(y)| \\ & + |F_n(x) - F(x)|\, F_n(y) \\ & + F(x)\, |F_n(y) - F(y)| \\ & + |F_{ij}(x, y) - F(x)\, F(y)|. \end{aligned} \tag{2.3}$$

As the sequence $\{X_{ni}\}$ is absolutely regular with the geometrical rate ρ_0, it is also strong mixing with the same rate and, from the definition of strong mixing, we deduce that $\forall \varepsilon > 0\; \exists m_0 \in \mathbb{N}^*$ such that $\forall (i, j) \in (\mathbb{N}^*)^2$ with $i < j$ and $j - i \geqslant m_0$,

$$|F_{n,i,j}(x, y) - F_n(x)\, F_n(y)| < A\rho_0^m \leqslant \varepsilon/4, \tag{2.4}$$

where A is some positive constant, and from (2.1) and (2.2), we also obtain

$$|F_{ij}(x, y) - F(x)\, F(y)| < \varepsilon/4. \tag{2.5}$$

Now from conditions (2.1) and (2.2), we also deduce that $\forall \varepsilon > 0\; \exists n_0$ such that $\forall n \geqslant n_0$ and $\forall (i, j) \in (\mathbb{N}^*)^2$ with $i < j$ and $j - i < m_0$,

$$|F_{n,i,j}((x, y) - F_{ij}(x, y)| < \varepsilon \tag{2.6}$$

$$|F_n(x) - F(x)| < \varepsilon/4 \tag{2.7}$$

$$|F_n(y) - F(y)| < \varepsilon/4. \tag{2.8}$$

From (2.3)–(2.5) and (2.7), (2.8) we deduce that $\forall n \geqslant n_0$, $\forall (i, j) \in (\mathbb{N}^*)^2$ with $i < j$ and $j - i \geqslant m_0$,

$$|F_{n,i,j}(x, y) - F_{ij}(x, y)| < \varepsilon/4 + \varepsilon/4 + \varepsilon/4 + \varepsilon/4 = \varepsilon. \tag{2.9}$$

(2.6) and (2.9) yield (1.6). As the d.f. F_{ij} has the same marginals F, we obtain (1.7).

(ii) Suppose (b) holds. From Davydov [2], a Markov process is Doeblin recurrent and aperiodic is geometrically φ-mixing. This implies that $\{X_{ni}\}$ is also absolutely regular with a geometrical rate.

EXAMPLE 2.1. Consider the process $\{X_{ni};\ i \in \mathbb{Z}\}$, where $X_{n,i+1} = a_1^{(n)} X_{ni} + a_2^{(n)} X_{ni} \varepsilon_{i+1} + a_3^{(n)} \varepsilon_{i+1} + a_4^{(n)} \varepsilon_{i+1}^2 + a_5^{(n)}$ where the a's are real numbers and $\{\varepsilon_i; i \in \mathbb{Z}\}$ is a white noise with strictly positive density. Then Mokkadem [4] has shown that if $(a_1^{(n)})^2 + (a_2^{(n)})^2\, E(\varepsilon_1^2) < 1$ and $E(\varepsilon_1^4) < \infty$, then the process $\{X_{ni}; i \in \mathbb{Z}\}$ is geometrically ergodic.

If we have

$$\forall j \in \{1, ..., 5\} \qquad \exists a_j \in \mathbb{R}$$

such that

$$\lim_{n \to \infty} a_j^{(n)} = a_j$$

and

$$a_1^2 + a_2^2 E(\varepsilon_1^2) < 1$$

then the conditions (2.1) and (2.2) are satisfied and we can apply Theorem 2.1.

EXAMPLE 2.2. Consider the process $\{X_{ni}; i \in \mathbb{Z}\}$, where $X_{n,i+1} = f_n(X_{ni}) + \varepsilon_{i+1}$ where the ε's are independent and identically distributed random variables with strictly positive density and f_n: $\mathbb{R} \to \mathbb{R}$ is bounded and continuous. (This model was studied by Collomb and Doukhan [1].) It is easy to check that this model is Doeblin recurrent and aperiodic and we deduce that $\{X_{ni}; i \in \mathbb{Z}\}$ is geometrically φ-mixing and if f_n converges simply to a bounded and continuous function f_0, we can apply Theorem 2.1.

2.2. *ARMA Processes*

Consider a sequence of ARMA processes $\{X_{ni}; i \in \mathbb{Z}\}$, $n \in \mathbb{N}^*$,

$$X_{n,i+1} = \tilde{a}_1^{(n)} X_{ni} + \tilde{a}_2^{(n)} \varepsilon_{i+1} + \varepsilon_i \tag{2.10}$$

where $\{\varepsilon_i;\ i \in \mathbb{Z}\}$ is a sequence of independent random variables such that $E(\varepsilon_i) = 0$.

THEOREM 2.2. *Let $\{X_{ni}; i \in \mathbb{Z}\}$ be a sequence of ARMA processes given by* (2.10). *Suppose $\{X_{ni}\}$ satisfies the conditions*:

> *$\{\varepsilon_i;\ i \in \mathbb{Z}\}$ is a sequence of independent and identically distributed random variables with strictly positive density*; (2.11)

$$\exists(\tilde{a}_1, \tilde{a}_2) \in (-1, 1) \times \mathbb{R} \tag{2.12}$$

such that

$$\lim_{n \to \infty} \tilde{a}_1^{(n)} = \tilde{a}_1 \qquad \text{and} \qquad \lim_{n \to \infty} \tilde{a}_2^{(n)} = \tilde{a}_2.$$

Then for a function g satisfying the conditions of Theorem 1.1 *and for a U-statistic defined in* (1.1) *the conclusions of Theorem* 1.1 *hold.*

Also for a score function J having a second bounded derivative and for the process V_n defined in (1.13) *the conclusions of Theorem* 1.2 *hold.*

Proof. Particular case of Theorem 2.1.

REFERENCES

[1] COLLOMB, G., AND DOUKHAN, P. (1983). Estimation de la fonction d'autorégression d'un processus stationnaire et φ mélangeant: risque quadratique pour la méthode du noyau. *C.R. Acad. Sci. Paris* **296** 859–862.

[2] DAVYDOV, YU. A. (1973). Mixing conditions for Markov chains. *Theory Probab. Appl.* **2** 312–328.

[3] HAREL, M., AND PURI, M. L. (1989). Limiting behavior *U*-statistics, *V*-statistics and one sample rank order statistics for nonstationary absolutely regular processes. *J. Multivariate Anal.* **29**.

[4] MOKKADEM, A. (1978). Sur un modèle autorégressif non linéaire. Ergodicité et ergodicité géométrique. *J. Time Ser. Anal.* **8** 195–204.

[5] NUMMELIN, E., AND TUOMINEN, P. (1982). Geometric ergodicity of Harris recurrent Markov chains with applications to renewal theory. *Stochastic Process. Appl.* **12** 187–202.

Reprinted by permission of the publisher,
© 1989 by Academic Press

WEAK CONVERGENCE OF SERIAL RANK STATISTICS UNDER DEPENDENCE WITH APPLICATIONS IN TIME SERIES AND MARKOV PROCESSES

MICHEL HAREL and MADAN L. PURI[1]

Insitut Universitaire de Technologie de Limoges and U.A. C.N.R.S., Rouen, and Indiana University

The asymptotic normality of linear serial rank statistics introduced by Hallin, Ingenbleek and Puri (1985) for the problem of testing white noise against ARMA alternatives is established for φ-mixing as well as strong mixing sequences of random variables. Applications in Markov processes and ARMA processes in time series are provided.

1. Introduction. Let $\{X_{n,i},\ 1 \le i \le n,\ n \ge 1\}$ be real-valued random variables with continuous distribution functions $F_n(x) = P(X_{n,i} \le x)$, $1 \le i \le n$, $n \ge 1$. Consider the statistics

$$\mathscr{S}_n = (n-k)^{-1} \sum_{i=k+1}^{n} c_{n,i} a_n(R_{n,i-k}, \dots, R_{n,i}), \tag{1.1}$$

where the $c_{n,i}$ are known constants, $a_n(\cdot, \dots, \cdot)$ are the scores, $R_{n,i}$ denotes the rank of $X_{n,i}$ among $(X_{n,1}, \dots, X_{n,n})$ and $k \ge 1$ is a fixed integer $(< n)$. Our aim is to study the asymptotic behavior of $\mathscr{S}_n$ when the sequence $\{X_{n,i}\}$ is φ-mixing with rates

$$\varphi(m) = O(m^{-1-\varepsilon}) \quad \text{for some } \varepsilon > 0,\ m \ge 1 \tag{1.2}$$

or

$$\sum_{m=1}^{\infty} m^{-1} \varphi^{1/2(3+k)}(m) < \infty, \tag{1.3}$$

or is strong-mixing with rates

$$\sum_{m=1}^{\infty} m^{2(k+2)} \alpha^{\varepsilon}(m) < \infty \quad \text{for some } \varepsilon \in (0, 1/2(3+k)). \tag{1.4}$$

Recall that the array $\{X_{n,i},\ 1 \le i \le n,\ n \ge 1\}$ is φ-mixing if

$$\sup_{m \le n} \sup_{1 \le j \le n-m} \{|P(A|B) - P(A)| : B \in \sigma(X_{n,i}, 1 \le i \le j),$$
$$A \in \sigma(X_{n,i}, i \ge j+m)\} = \varphi(m) \downarrow 0$$

Received February 1988; revised July 1989.

[1]Research supported by Office of Naval Research Contract N00014-85-K-0648.

AMS 1980 *subject classifications*. 60F05, 60J05, 62M10.

Key words and phrases. Serial rank statistics, φ-mixing, strong mixing, graduate empirical process, graduate rank process, Skorohod topology, weak convergence, Markov process, ARMA process.

as $m \uparrow \infty$ for positive integers j and m, and it is strong-mixing if

$$\sup_{m \le n} \sup_{1 \le j \le n-m} \{|P(A \cap B) - P(A)P(B)|, A \in \sigma(X_{n,i}, 1 \le i \le j),$$
$$B \in \sigma(X_{n,i}, i \ge j+m)\}$$
$$= \alpha(m) \downarrow 0 \quad \text{as } m \uparrow \infty,$$

for positive integers j and m. Here $\sigma(X_{n,i}, i \le j)$ and $\sigma(X_{n,i}, i \ge j + m)$ are the σ-fields generated by $(X_{n,1}, \dots, X_{n,j})$ and $(X_{n,j+m}, X_{n,j+m+1}, \dots)$, respectively. The asymptotic behavior of the statistic $\mathscr{S}_n$ under strong-mixing conditions leads to interesting applications in ARMA processes in time series as well as in Markov processes (Section 6). In passing we may mention that Hallin, Ingenbleek and Puri (1985) established the asymptotic normality of linear *serial* rank statistics $\mathscr{S}_n$ defined in (1.1) for an ARMA process contiguous to white noise. We show (in Section 2) that contiguity is not necessary for the derivation of the asymptotic distribution theory derived in Hallin, Ingenbleek and Puri (1985) and our results also lead to applications in some Markov processes which are either geometrically ergodic or Doeblin recurrent, and to some ARMA processes. For a related problem dealing with the applications of U-statistics [see Harel and Puri (1989a, 1990)] to some Markov processes and ARMA models, the reader is referred to Harel and Puri (1989b).

2. Asymptotic normality. We start with a few preliminaries.

Denote by $\hat{F}_n(x)$, the right continuous empirical distribution function of $X_{n,i}$, $i = 1, \dots, n$; i.e., let $\hat{F}_n(x) = n^{-1}\sum_{i=1}^n I_{\{X_{n,i} \le x\}}$ where $I_{\{\cdot\}}$ denotes the indicator function. Denote by G_n the distribution function of the $k + 1$ of the successive random variables $X_{n,1}, \dots, X_{n,n}$. Let H_n (for each $n \ge k + 1$) be a sequence of continuous distribution functions on $(0, 1)^{k+1}$, defined by

$$H_n(\mathbf{t}) = G_n\big(F_n^{-1}(t_1), \dots, F_n^{-1}(t_{k+1})\big) \quad \text{for all } \mathbf{t} = (t_1, \dots, t_{k+1}) \in (0,1)^{k+1}, \tag{2.1}$$

where $F_n^{-1}(u) = \inf\{t\colon F_n(t) \ge u\}$, $0 < u < 1$. Since H_n is continuous, it is actually well defined on $[0, 1]^{k+1}$. Though G_n, H_n and $\mathbf{t}$ depend on k, we have suppressed this fact for notational convenience.

Denote by C_{k+2}, the space of all continuous maps $f\colon [0, 1]^{k+2} \to \mathbb{R}$, and by $C_{k+1}(j)$, $1 \le j \le k + 1$, the space of all continuous and bounded maps $f\colon A(j) \to \mathbb{R}$, where $A(j) = [0, 1]^{j-1} \times (0, 1) \times [0, 1]^{k+1-j}$.

Definition. We say that the sequence $\{H_n\}$ satisfies the differentiability condition if (a) $\partial H_n/\partial t_j$ exists on $A(j)$ and belongs to $C_{k+1}(j)$, $1 \le j \le k + 1$, and (b) $\partial H_n/\partial t_j \to l_j$ in the uniform topology on any compact subset of $A(j)$ as $n \to \infty$, and l_j belongs to $C_{k+1}(j)$.

We define the *graduate empirical process* [also called the *copula process*, see, e.g., Gaenssler and Stute (1987), Chapter V] W_n as

$$(2.2)\qquad W_n(t) = (n-k)^{-1/2} \sum_{i=k+1}^{[nt_0]} \left\{ \prod_{j=1}^{k+1} I_{\{F_n(X_{n,i+j-k-1}) \le t_j\}} - H_n(\mathbf{t}) \right\}$$

for all $t = (t_0, \mathbf{t}) = (t_0, t_1, \ldots, t_{k+1}) \in (0,1)^{k+2}$, where $[nt_0]$ denotes the integral part of the real number nt_0.

We also consider the rank process L_n (called the *graduate rank process*) defined as

$$(2.3)\qquad L_n(t) = (n-k)^{-1/2} \sum_{i=k+1}^{[nt_0]} \left\{ \prod_{j=1}^{k+1} I_{\{\hat{F}_n(X_{n,i+j-k-1}) \le t_j\}} - H_n(\mathbf{t}) \right\}.$$

For any $n \ge 1$, we define a signed measure λ_n concentrated on $\{1/n, \ldots, (n-1)/n, 1\}^{k+2}$ by setting

$$\lambda_n\left(\prod_{j=0}^{k+1} \left[\frac{i_j}{n}, 1 \right] \right) = c_{n,i_0} a_n(i_1, \ldots, i_{k+1}),$$

for all $(i_0, \ldots, i_{k+1}) \in \{1, \ldots, n\}^{k+2}$. (By convention, $c_{n,i_0} = 0$ if $i_0 < k+1$.)

We also define a centering coefficient b_n by

$$(2.4)\qquad b_n = \int_{[0,1]^{k+2}} \hat{H}_n(t) \lambda_n(dt),$$

where $\hat{H}_n$ is the function $[0,1]^{k+2} \to \mathbb{R}^+$ such that $\hat{H}_n(t) = ([nt_0] - k) H_n(\mathbf{t})$.

We now state the following theorem, the proof of which is given in Section 5.

THEOREM 2.1. *Assume that there exists a Radon measure λ_0 on $[0,1]^{k+2}$ such that*

$$(2.5)\qquad \lim_{n\to\infty} \int f\,d\lambda_n = \int f\,d\lambda_0 \quad \text{for all } f \in C_{k+2}$$

and

$$(2.6)\qquad \sup_{n\in\mathbb{N}} \int f\,d|\lambda_n| < \infty; \qquad \mathbb{N} = \{0,1,2,\ldots\},$$

where $|\lambda_n|$ denotes the measure of total variation.

Assume that the sequence $\{X_{n,i}\}$ is (a) *φ-mixing with rates* (1.2) *or* (b) *strong-mixing with rates* (1.4). *Furthermore, assume that* (c) *the covariance functions $\{K_n,\ n \ge 1\}$ of the empirical processes $\{W_n,\ n \ge 1\}$ defined in* (2.2) *converge to a function $K(\cdot,\cdot)$ in pointwise topology as $n \to \infty$ and* (d) *$\{H_n\}$ satisfies the differentiability conditions. Then L_n converges weakly in uniform topology to a Gaussian process L_∞ with trajectories a.s. in C_{k+2}, and*

$(n-k)^{1/2}(\mathscr{S}_n - b_n)$ converges in law to the normal distribution with mean 0 and variance σ^2, where

$$(2.7)\quad \sigma^2 = \int_{[0,1]^{k+2}} \cdots \int_{[0,1]^{k+2}} E[L_\infty(t), L_\infty(t')]\, d\lambda_0(t)\, d\lambda_0(t') < \infty.$$

REMARK 2.1. Theorem 2.1 is proved under the assumption that the sequence $\{X_{n,i}\}$ is nonstationary and either φ-mixing with rates (1.2) or strong-mixing with rates (1.4). The theorem does not hold with the φ-mixing rates (1.3) unless one assumes stationarity (which implies that the distribution functions F_n, G_n and H_n are equal to unique distribution functions F, G and H, respectively) and the special case when $c_{n,i} = 1$ for all i.

Let $\tilde{\mathscr{S}}_n$ denote the statistics $\mathscr{S}_n$ when $c_{n,i} = 1$ for all i, i.e., let

$$(2.8)\qquad \tilde{\mathscr{S}}_n = \sum_{i=k+1}^{n} a_n(R_{n,i-k}, \ldots, R_{n,i}),$$

and let $\tilde{b}_n$ denote the corresponding centering constant, i.e.,

$$(2.9)\qquad \tilde{b}_n = \int_{[0,1]^{k+1}} H_n(\mathbf{t})\tilde{\lambda}_n(d\mathbf{t}),$$

where $\tilde{\lambda}_n$ is a measure concentrated on $\{1/n, \ldots, (n-1)/n, 1\}^{k+1}$ and

$$\tilde{\lambda}_n\left(\prod_{j=1}^{k+1}\left[\frac{i_j}{n}, 1\right]\right) = a_n(i_1, \ldots, i_{k+1}).$$

Then, we have the following theorem.

THEOREM 2.2. *Assume there exists a Radon measure $\tilde{\lambda}_0$ on $[0,1]^{k+1}$ such that*

$$(2.10)\qquad \lim_{n\to\infty} \int \tilde{f}\, d\tilde{\lambda}_n = \int \tilde{f}\, d\tilde{\lambda}_0$$

and

$$(2.11)\qquad \sup_{n\in\mathbb{N}} \int \tilde{f}\, d|\tilde{\lambda}_n| < \infty,$$

where $|\tilde{\lambda}_n|$ denotes the measure of total variation.

Assume that the sequence $\{X_{n,i}\}$ is (a′) *φ-mixing with rates* (1.3) *and* (b′) *H satisfies the differentiability condition. Then $L_n(1,\mathbf{t})$ converges weakly in uniform topology to a Gaussian process $\tilde{L}_\infty$ with trajectories a.s. in C_{k+1}, and $(n-k)^{1/2}(\tilde{\mathscr{S}}_n - \tilde{b}_n)$ converges in law to the normal distribution with mean 0 and variance $\tilde{\sigma}^2$, where*

$$(2.12)\quad \tilde{\sigma}^2 = \int_{[0,1]^{k+1}} \cdots \int_{[0,1]^{k+1}} E[\tilde{L}_\infty(\mathbf{t}), \tilde{L}_\infty(\mathbf{t}')]\, d\tilde{\lambda}_0(\mathbf{t})\, d\tilde{\lambda}_0(\mathbf{t}') \qquad (<\infty).$$

The proof follows from Theorem 2.1 by putting $t_0 = 1$ for the processes W_n and L_n, and showing that the finite projections of W_n converge to a normal law (the proof of which is given in Proposition 3.4).

The following corollary gives sufficient conditions under which the conditions (2.5) and (2.6) are satisfied.

COROLLARY 2.2. *Let J be a function on $[0,1]^{k+2}$ such that*

$$J(i_0/n,\dots,i_{k+1}/n) = c_{n,i_0} a_n(i_1,\dots,i_{k+1})$$

for all $(i_0,\dots,i_{k+1}) \in \{1,\dots,n\}^{k+2}$, $J = J_d + J_c$, where J_d is a step function taking only a finite number of jumps, and where for any $I \subset \{0,\dots,k+1\}$, J_c has a continuous derivative $\partial^I J_c/(\partial t_j)_{j\in I}$, then the conditions (2.5) and (2.6) are satisfied.

PROOF. It suffices to prove the above corollary in the case when J_d has only one jump, say at $a = (a_0,\dots,a_{k+1}) \in [0,1]^{k+2}$. Let λ'_n and λ''_n be measures on $[0,1]^{k+2}$ defined by

$$\lambda'_n\left(\prod_{j=0}^{k+1}\left[\frac{i_j}{n},1\right]\right) = J_c\left(\frac{i_0}{n},\dots,\frac{i_{k+1}}{n}\right)$$

and

$$\lambda''_n\left(\prod_{j=0}^{k+1}\left[\frac{i_j}{n},1\right]\right) = J_d\left(\frac{i_0}{n},\dots,\frac{i_{k+1}}{n}\right),$$

for all $(i_0,\dots,i_{k+1}) \in \{1/n,\dots,(n-1)/n,1\}^{k+2}$.

It is easy to check that

$$\lim_{n\to\infty}\int_{[0,1]^{k+2}} f\,d\lambda'_n = \sum_{I\subset\{0,\dots,k+1\}}\int_{[0,1]^{k+2}} f\frac{\partial^I J_c}{(\partial t_j)_{j\in I}}\Big((t_j)_{j\in I},(1)^{k+2-i}\Big),$$

for all $f \in C_{k+2}$ whose $i = \operatorname{card} I$.

Thus, we obtain a measure λ'_0 satisfying

$$\lim_{n\to\infty}\int_{[0,1]^{k+2}} f\,d\lambda'_n(t) = \int_{[0,1]^{k+2}} f\,d\lambda'_0(t).$$

Analogously, we obtain

$$\lim_{n\to\infty}\int_{[0,1]^{k+2}} f\,d\lambda''_n(t) = f(a)\sum_{I\subset\{0,\dots,k+1\}}(-1)^i J_d\Big(\underset{i\in I}{(a_{i^-})},\underset{i\notin I}{(a_{i^+})}\Big),$$

for all $f \in C_{k+2}$, where $i = \operatorname{card} I$. □

3. Weak convergence of the graduate empirical process and the graduate rank process. We start with preliminaries.

3.1A. *The spaces D_{k+2} and C_{k+2}.* Let f: $[0,1]^{k+2} \to \mathbb{R}$. For $\rho \in \{0,1\}^{k+2}$, define

$$f_\rho(t) = \lim_{\substack{s_i \uparrow t_i,\, \rho(i)=1 \\ s_i \downarrow t_i,\, \rho(i)=0}} f(s), (s,t) \in \left([0,1]^{k+2}\right)^2, \qquad i = 0, 1, \ldots, k+1,$$

if it exists, in which case, call $f_\rho(t)$ the ρ-limit of f at t. Denote by D_{k+2}, the space of all maps f: $[0,1]^{k+2} \to \mathbb{R}$ such that for all $\rho \in \{0,1\}^{k+2}$, f_ρ exists and $f_\rho = f$ for $\rho = (0, \ldots, 0)$.

We say that we have *special Skorohod topology* on D_{k+2} if we have the uniform topology for the first coordinate and the J_1-Skorohod topology for the other coordinates. [For definition of Skorohod topology, cf. Skorohod (1956) and Billingsley (1968).]

We define a modulus of continuity for any bounded function f: $[0,1]^{k+2} \to \mathbb{R}^+$ to be denoted by $\omega(f, \delta)$, $(\delta > 0)$, by setting

$$\omega(f, \delta) = \sup_{(t,t') \in ([0,1]^{k+2})^2} |f(t) - f(t')|, \qquad \|t - t'\| < \delta, \tag{3.1}$$

where $\|t\| = \sup\{|t_j|, 0 \le j \le k+1\}$. Note that f belongs to C_{k+2} if and only if $\lim_{\delta \to 0} \omega(f, \delta) = 0$.

The following proposition, which is a variant of Theorem 1.2 of Dudley (1978), will be used to prove Proposition 3.4.

PROPOSITION 3.1. *Let Y_n be a process with values in D_{k+2} and measurable with respect to $\mathscr{U}_{k+2}$, the σ-field generated by the uniform topology (on D_{k+2}). Let P_n denote the law of Y_n. Then, there exists a probability measure P, with $P(C_{k+2}) = 1$, for which P_n converges weakly with respect to the uniform topology if and only if*

(a) *for all finite subsets U of $[0,1]^{k+2}$, $\phi_U(P_n)$ converges weakly to $\phi_U(P)$ (ϕ_U is the projection of D_{k+2} on $\mathbb{R}^U$),*

(b) $\forall\, \varepsilon > 0$, $\lim_{\delta \to 0} \limsup_{n \to \infty} P_n[\{f;\, \omega(f, \delta) \ge \varepsilon\}] = 0$.

The proof is given in the Appendix.

3.1B. *Grid accompanying a sequence of probability measures.* We call a *grid T* of $[0,1]^{k+2}$ a subset of $[0,1]^{k+2}$ such that $T = \prod_{j=0}^{k+1} T^{(j)}$, where $T^{(j)}$ is a finite subset of $[0,1]$ which includes 0 and 1.

We call a *pace τ* of a grid $T = \prod_{j=0}^{k+1} T^{(j)}$ the number $\tau = \max_{0 \le j \le k+1} \tau_j$, where $\tau_j = \max\{|t_j' - t_j|$, t_j' and t_j are successive elements in $T^{(j)}\}$.

We denote the *lower boundary* of T by $\underset{\leftarrow}{T}$ where

$$\underset{\leftarrow}{T} = \bigcup_{j=0}^{k+1} \left(\prod_{l=0}^{j-1} T^{(l)} \times \{0\} \times \prod_{l=j+1}^{k+1} T^{(l)} \right).$$

We call *block* B of T any part of T in the form

$$B = \prod_{j=0}^{k+1} \{(t_j, t'_j] \cap T^{(j)}, \text{ where } t_j \text{ and } t'_j \text{ belong to } T^{(j)} \text{ and } t_j < t'_j\}.$$

We call *evaluation* $e_T^{(B)}$ of B into T, the operator $e_T^{(B)}$: $D_{k+2} \to \mathbb{R}^+$ such that

$$e_T^{(B)}(f) = \sum_{(\varepsilon_0, \dots, \varepsilon_{k+1}) \in \{0,1\}^{k+2}} (-1)^{\sum_{i=0}^{k+1} \varepsilon_i} f[(1-\varepsilon_0)t_0 + \varepsilon_0 t'_0, \dots, (1-\varepsilon_{k+1})t_{k+1} + \varepsilon_{k+1} t'_{k+1}]. \tag{3.2}$$

Let ν be a finite measure on $[0,1]^{k+2}$ and let T be a grid of $[0,1]^{k+2}$. We call *reduction* $\tilde{\nu}$ of ν on T the measure on T defined by

$$\tilde{\nu}(\{t\}) = \begin{cases} 0 & \text{if } t \in \underset{\leftarrow}{T}, \\ \nu\left(\prod_{j=0}^{k+1} (t'_j, t_j]\right) & \text{if } t \notin \underset{\leftarrow}{T}, \end{cases}$$

where

$$t'_j = \max\{x;\, x \in T^{(j)};\, x < t_j,\, t_j \in T^{(j)}\}.$$

For any $\delta > 0$, we set

$$\omega_T(f, \delta) = \sup\{|f(t) - f(t')|;\, (t, t') \in T^2,\, \|t - t'\| \le \delta\}.$$

We say that a sequence $\{T_n\}_{n \in \mathbb{N}^*}$ of grids is asymptotically dense in $[0,1]^{k+2}$ if the pace τ_n of T_n satisfies $\lim_{n\to\infty} \tau_n = 0$, $\mathbb{N}^* = \mathbb{N} - \{0\}$, $\mathbb{N} = 0, 1, 2, \dots$.

Let P_n, $n \in \mathbb{N}^*$, be a sequence of probability measures on $(D_{k+2}, \mathscr{D}_{k+2})$, where $\mathscr{D}_{k+2}$ is the σ-field generated by the Skorohod topology (on D_{k+2}). We say that the sequence $\{T_n\}$ of grids accompanies the sequence $\{P_n\}$ if and only if $\forall\, \varepsilon > 0$, $\exists\, \varepsilon' > 0$ and $\forall\, \delta \in [0, \frac{1}{2})$, $\exists\, N_0 \in N^*$, we have

$$P_n[\{f \in D_{k+2},\, \omega(f, \delta) \ge \varepsilon \text{ and } \omega_{T_n}(f, 2\delta) < \varepsilon'\}] = 0 \qquad \forall\, n \ge N_0.$$

The following propositions [Propositions (3.2) and (3.3)] are variants of a result of Neuhaus (1971) [see, e.g., Theorems 2 and 4 in Balacheff and Dupont (1980)] and will be used in Section 4.

PROPOSITION 3.2. *Let P_n, $n \in \mathbb{N}$, be probability measures on $(D_{k+2}, \mathscr{D}_{k+2})$ such that the following conditions are satisfied*:

(3.3) $\phi_U(P_n)$ *converges weakly to some probability measure P_U on $\mathbb{R}^U$ for any finite subset U of $[0,1]^{k+2}$*

and

$$\forall\, \varepsilon > 0, \qquad \lim_{\delta\to 0} \limsup_{n\to\infty} P_n[f \in D_{k+2};\, \omega(f, \delta) \ge \varepsilon] = 0. \tag{3.4}$$

Then P_n converges weakly with respect to the Skorohod topology to some probability measure P and $P(C_{k+2}) = 1$.

PROPOSITION 3.3. *Let ν be a positive finite measure on $[0,1]^{k+2}$ with continuous marginals. Let P_n be a sequence of probability measures on $(D_{k+2}, \mathscr{D}_{k+2})$ such that $\forall\, n \in \mathbb{N}$, $P_n[f \in D_{k+2};\, f|[0,1]^{k+2} = 0] = 1$. Let T_n be a sequence of grids asymptotically dense in $[0,1]^{k+2}$ and accompanying P_n. Furthermore, suppose that for any block B_n of T_n,*

$$(3.5)\qquad P_n\left[f \in D_{2+k};\, |e_{T_n}^{(B_n)}(f)| \geq \lambda\right] \leq \lambda^{-\gamma}(\tilde{\nu}_n(B_n))^{\beta},$$

where $\tilde{\nu}_n$ is the reduction of ν on T_n, and $\beta > 1$ and $\gamma > 0$. Then, we have $\forall\, \varepsilon > 0$, $\exists\, \delta \in (0,1)$ and $N_0 \in \mathbb{N}$, such that

$$(3.6)\qquad P_n\left[f \in D_{k+2};\, \omega(f,\delta) \geq \varepsilon\right] \leq \varepsilon \qquad \forall\, n \geq N_0.$$

3.2A. *Convergence of the graduate empirical processes.*

PROPOSITION 3.4. *Under the conditions* (a) *and* (c) *or* (b) *and* (c) *of Theorem* 2.1, *W_n converges weakly in the special Skorohod topology to a Gaussian process W_∞ with trajectories a.s. in C_{k+2}. Under the conditions* (a′) *of Theorem* 2.2, *$\tilde{W}_n = W_n(1,\mathbf{t})$ converges in the Skorohod topology to a Gaussian process $\tilde{W}_\infty$ with trajectories a.s. in C_{k+1}.*

3.2B. *Convergence of the graduate rank process.*

PROPOSITION 3.5. *Under the conditions* (a), (c) *and* (d) *or* (b), (c) *and* (d) *of Theorem* 2.1, *L_n converges weakly in uniform topology to a Gaussian process L_∞ with trajectories a.s. in C_{k+2}. Under the conditions* (a′) *and* (b′) *of Theorem* 2.2, *$\tilde{L}_n = L_n(1,\mathbf{t})$ converges weakly in uniform topology to a Gaussian process $\tilde{L}_\infty$ with trajectories a.s. in C_{k+1}.*

4. Proofs of Propositions 3.4 and 3.5. Our proofs of the Propositions 3.4 and 3.5 are based on the ideas of Balacheff and Dupont (1980), who considered the asymptotic normality of the truncated empirical processes under φ-mixing with rates $\Sigma_{m=1}^{\infty} m\varphi^{1/2}(m) < \infty$. Here in this paper, we consider the rates (1.2) and (1.3), which are slower than the one considered by them. In addition, we also derive results under strong mixing (1.4) which have not been considered in the literature. To establish their result, Balacheff and Dupont (1980) used a slight modification of an inequality due to Rüschendorf (1974) which is not applicable in our situation. Our proofs are based on the following two lemmas.

LEMMA 4.1. *Let the sequence $\{X_{n,i}\}$ of real-valued random variables (centered at its expectation) be φ-mixing with rates $\Sigma_{m=1}^{\infty} m^{-1}\varphi^{1/2q}(m) < \infty$, where q is an integer. Denote by N_n the number of indices i, $1 \leq i \leq n$, for which $X_{n,i}$ is not identical to zero. Set $S_n = \Sigma_{i=1}^{n} X_{n,i}$ and $\|X_{n,i}\|_l =$*

$(\int|X_{n,i}|^{2l}\,dP_n)^{1/2l}$. *Then, for any* $q \ge 1$, *there exists a constant* $C_q(\varphi)$ *depending only on* q *and* φ *such that*

$$E(S_n^{2q}) \le C_q(\varphi)\sum_{l=1}^{q} N_n^{q/l}\Big(\sup_{1\le j\le n}\|X_{n,j}\|_l\Big)^{2q}. \tag{4.1}$$

The proof is a slight modification of Theorem 2.1 of Neumann (1982) and is sketched briefly in the Appendix.

LEMMA 4.2. *Let the sequence* $\{X_{n,i}\}$ *of real-valued random variables (centered at its expectation) be strong-mixing with rates* $\sum_{m=1}^{\infty} m^{2q-2}\alpha^{\varepsilon}(m) < \infty$, $\varepsilon \in (0, 1/2q)$ *and* $|X_{n,i}| \le 1$, $1 \le i \le n$, $n \ge 1$, *where* q *is an integer. Let* N_n *be the number of indexes* i, $1 \le i \le n$, *for which* $X_{n,i}$ *is not identical to zero. Set* $S_n = \sum_{i=1}^{n} X_{n,i}$ *and* $\|X_{n,i}\|_{\varepsilon} = (\int|X_{n,i}|^{2/(1-\varepsilon)}\,dP_n)^{1-\varepsilon}$. *Then, for any* $q \ge 1$, *there exists a constant* $C_q(\alpha)$ *depending only on* q *and* α *such that*

$$E(S_n^{2q}) \le C_q(\alpha)\sum_{l=1}^{q} N_n^{l}\Big(\sup_{1\le i\le n}\|X_{n,i}\|_{\varepsilon}\Big)^{l}. \tag{4.2}$$

The proof is similar to that of Theorem II.10 of Doukhan and Portal (1987) and is therefore omitted.

LEMMA 4.3 (Neumann, 1982). *Let* $\{Y_i,\ i \ge 1\}$ *be a stationary sequence of real-valued random variables centered at its expectation and with finite second moment. Assume that the sequence is* φ*-mixing with rates* $\sum_{m=1}^{\infty} m^{-1}\varphi^{1/2}(m) < \infty$. *Then, there exists a positive constant* K *such that* $n^{-1}E(\sum_{i=1}^{n} Y_i)^2 \to K^2$ *as* $n \to \infty$.

Since the reference Neumann (1982) is not readily available, we have (at the suggestion of one of the referees) given the proof in the Appendix.

PROOF OF PROPOSITION 3.4. Consider a sequence $Z_{m,i}$, $1 \le i \le m$, $m \ge 1$ of $\mathbb{R}^{k+1}$-valued random variables defined by $Z_{m,i} = (X_{m+k,i}, \dots, X_{m+k,i+k}) = (Z_{m,i}^{(1)}, \dots, Z_{m,i}^{(k+1)})$, $1 \le i \le m$, $m \ge 1$. Then the $(k+1)$-variate truncated empirical process $\tilde W_m$ associated with this sequence is given by

$$\tilde W_m(t_0, \mathbf{t}) = m^{-1/2}\sum_{i=1}^{[(m+k)t_0]-k}\left[\prod_{j=1}^{k+1} I_{\{F_{m+k}(Z_{m,i}^{(j)}) \le t_j\}} - H_{m+k}(\mathbf{t})\right] \tag{4.3}$$

and this is the same as the graduate process W_n defined in (2.2). Now the process W_n defines a probability measure Q_n on $(D_{k+2}, \mathscr{D}_{k+2})$.

To prove this proposition we have to verify (3.3) and (3.4). Following Withers [(1975), Corollary 1], it can be shown that $\phi_U(Q_n)$ converges weakly to a Gaussian measure Q_U if (i) $K_n \to$ some function K, (ii) $\sum_{m\ge 1}\alpha(m) < \infty$ and (iii) $m^{1-a}\alpha([m^b]) \to 0$ as $m \to \infty$, where $0 < 2b < a < 1 - b$. Now, in our situation (i) holds by assumption (c), (ii) follows from (1.2) or (1.4) and (iii) follows from (1.2) or (1.4) by taking $a = 3/4 - \varepsilon/8$, $b = 1/4$ and ε sufficiently small. (Since taking $\alpha(m) = m^{-1-\varepsilon}$, $m^{1-a}\alpha([m^b]) \le Am^{-\varepsilon/8} \to 0$ as $m \to \infty$.) Thus, (3.3) holds whenever conditions (a) and (c) or (b) and (c) of Theorem 2.1 are satisfied.

Now suppose that the condition (a′) of Theorem 2.2 holds with $X_{n,i} \equiv X_i$. Then, for any $p \in \mathbb{N}^*$, any $\mathbf{t}^{(l)} \in [0,1]^{k+1}$ and any $\lambda_l \in \mathbb{R}$, $1 \le l \le p$, let $g_i^{(l)}(X_i)$ and $g_i(X_i)$ be the random variables defined by

$$g_i^{(l)}(X_i) = \prod_{j=1}^{k+1} \left[I_{\{F(X_{i+j-k-1}) \le t_j\}} - H(\mathbf{t}^{(l)}) \right] \quad \text{and} \quad g_i(X_i) = \sum_{l=1}^{p} \lambda_l g_i^{(l)}(X_i),$$

where $X_i = (X_{i-k}, X_{i-k+1}, \ldots, X_i)$. Then, we have $\sum_{l=1}^{p} \lambda_l \tilde{W}_n(\boldsymbol{t}^{(l)}) = (n-k)^{-1/2} \sum_{i=k+1}^{n} g_i(X_i)$, and so (3.3) also holds by Lemma 4.3 and the central limit theorem for the stationary and φ-mixing case [cf. Ibragimov and Linnik (1971), Theorem 18.5.1 and Lemma 4.3]. Now, to prove (3.4), we shall use Proposition 3.3 and verify (3.5) [which will imply (3.4)].

Let $T_n = \{i/m;\ 0 \le i \le m\}^{k+2}$ be a sequence of grids with $n = m + k$. T_n is asymptotically dense in $[0,1]^{k+2}$ and we prove that T_n accompanies Q_n. Now for every $\mathbf{t} \in (0,1)^{k+1}$, let $(\underline{\mathbf{t}}, \bar{\mathbf{t}})$ be the points of $\pi(T_n)$, where π is the projection defined by $\pi(t) = \mathbf{t}$ such that $\underline{\mathbf{t}} \le \mathbf{t} \le \bar{\mathbf{t}}$ and $\|\bar{\mathbf{t}} - \underline{\mathbf{t}}\| \le 1/m$. Let us write $\underline{t}_0 = [nt_0]/n$ for every $t_0 \in [0,1]$. As the marginals of H_n are uniform, we obtain (after some computations) that

$$|W_n(t_0, \mathbf{t}) - W_n(t_0', \mathbf{t}')| \le \frac{2K}{\sqrt{m}} + |W_n(\underline{t}_0, \bar{\mathbf{t}}) - W_n(t_0', \underline{\mathbf{t}}')|,$$

$\forall\ (t_0, \mathbf{t}) \in [0,1]^{k+2}$ and $\forall\ (t_0', \mathbf{t}') \in [0,1]^{k+2}$. Consequently, $\forall\ \delta \in (0, \frac{1}{2}]$, we have $\omega(W_n, \delta) \le 2k/\sqrt{m} + \omega_{T_n}(W_n, 2\delta)$. It follows that T_n accompanies Q_n. It remains to show that Q_n satisfies (3.5).

Let $\sum_{m=1}^{\infty} m^{-1} \varphi^{1/2(3+k)}(m) < \infty$ [see (1.3)], and let B_n be a block of T_n defined in Section 3.1.b. Using Lemma 4.1 with $q = k + 3$, we obtain [see (3.2)]

$$\begin{aligned} E\left[e_{T_n}^{(B_n)}(W_n)\right]^{2(k+3)} &\le C_{k+3}(\varphi) \sum_{l=1}^{k+3} m^{-(k+3)} [(m+k)(t_0 - t_0')]^{(k+3)/l} \\ &\quad \times \left[\prod_{j=1}^{k+1} (t_j - t_j') \right]^{(k+3)/(k+1)l} \\ &\le C_{k+3}(\varphi) \sum_{l=1}^{k+3} m^{-(k+3)} \times (m+k)^{(k+3)/l} \\ &\quad \times \left[\prod_{j=0}^{k+1} (t_j - t_j') \right]^{(k+3)/(k+2)l} \\ &\le C_{k+3}(\varphi)(k+3) \left[\prod_{j=0}^{k+1} (t_j - t_j') \right]^{(k+3)/(k+2)}. \end{aligned}$$

Now let $\nu = (C_{k+3}(\varphi)(k+3))^{1/\beta}U$, where U is a uniform measure on $[0,1]^{k+2}$ and $\beta = (k+3)/(k+2)$. Then, by the Markov inequality, we obtain [see (3.5)]

$$Q_n\Big[f \in D_{k+2}; \big|e_{T_n}^{(B_n)}(f)\big| \geq \lambda\Big] \leq \lambda^{-2(k+3)}(\tilde{\nu}_n(B_n))^{\beta},$$

which implies (3.6) for the φ-mixing rates (1.3) [and so also for (1.2)]. For the strong-mixing case with rates (1.4), we use Lemma 4.2 with $\varepsilon < (2(k+3))^{-1}$ and obtain from (4.2),

$$E\big[e_{T_n}^{(B_n)}(W_n)\big]^{2(k+3)}$$

$$\leq C_{k+3}(\alpha)\sum_{l=1}^{k+3} m^{-(k+3-l)}(t_0 - t_0')^l\left(\prod_{j=1}^{k+1}(t_j - t_j')\right)^{l(1-\varepsilon)/(k+1)},$$

which [with $\beta = ((k+2)(1-\varepsilon)+1)/k+2$] implies (3.5) and hence (3.6). We derive the convergence with respect to the special Skorohod topology because W_n is measurable with respect to this topology and we use Proposition 3.1 to the first coordinate (of W_n). □

Proof of Proposition 3.5. The main line of proof is as follows: We consider a map G_n: $\mathscr{V}_n \to D_{k+2}$, where $\mathscr{V}_n$ is a subset of D_{k+2} and is such that $L_n = G_n \circ W_n$, $n \geq 1$. We show that G_n: $(\mathscr{V}_n, d) \to (D_{k+2}, \rho)$ is a continuous map, where d is the special Skorohod metric and ρ is the uniform metric.

Let $\mathscr{V}$ be a subset of D_{k+2} such that for any $v \in \mathscr{V}$, v equals zero on the lower boundary of $[0,1]^{k+2}$ and also for $\mathbf{t} = (1,\ldots,1)$. It will be noted that $\mathscr{V}_n \subset \mathscr{V}$ for $\forall\, n \geq 1$.

Let G: $\mathscr{V} \to D_{k+2}$ be a map defined by

$$\text{(4.4)}\quad G(v)(t) = v(t) - t_0\sum_{j=1}^{k+1}\big[v(1,\ldots,t_j,\ldots,1) \times l_j(t_1,\ldots,t_{k+1})\big],$$

where l_j is the limit of $\partial H_n/\partial t_j$ as $n \to \infty$. We will show that $\forall\, (v_n)_{n\in\mathbb{N}^*} \in (\Pi_{n\in\mathbb{N}^*}\mathscr{V}_n)$ and $\forall\, v \in \mathscr{V} \cap C_{k+2}$, $v_n \to_d v \Rightarrow G_n(v_n) \to_\rho G(v)$ as $n \to \infty$. Now, using Lemma 3 of Balacheff and Dupont (1980), we get the desired convergence.

Let $\mathscr{Y}_n = \{y \in [0,1]^n$: $(y^{(1)},\ldots,y^{(n)})$ are distinct points of $(0,1)\}$. We define Y_n: $[0,1]^n \to D_{k+2}$ by setting

$$Y_n(y)(t) = (n-k)^{-1/2}\sum_{i=k+1}^{[nt_0]}\left[\prod_{j=1}^{k+1} I_{\{y^{(i+j-k-1)} \leq t_j\}} - H_n(\mathbf{t})\right],$$

for all $y = (y^{(1)},\ldots,y^{(n)}) \in \mathscr{Y}_n$ and $t = (t_0, \mathbf{t}) \in [0,1]^{k+2}$.

We define the space $\mathscr{V}_n$ by $\mathscr{V}_n = Y_n(\mathscr{Y}_n)$. For any $j \in \{1,\ldots,k+1\}$ we define an operator τ_j: $\mathscr{V}_n \to D_1$ as follows.

Let $y_{(1)} < \cdots < y_{(n)}$ be the order values of $(y^{(1)}, \ldots, y^{(n)})$ (by convention, $y_{(0)} = 0$, $y_{(n+1)} = 1$), and let $v_n = Y_n(y)$. Then,

$$\tau_j(v_n)(t_j) = \begin{cases} y^{(l)}, & \text{where } y^{(l)} = \max\{y^{(m)}; m \in \{j, \ldots, j+n-k-1\}\} \\ & \text{if } t_j = 1, \\ y^{(q)}, & \text{where } y^{(q)} = \max\{y^{(m)} \le y_{(i)}; \\ & \qquad m \in \{0, j, \ldots, j+n-k-1\}\} \\ & \qquad \text{if } t_j \in \left[\frac{i}{n}, \frac{i+1}{n}\right), \end{cases} \tag{4.5}$$

where $i = \{0, 1, \ldots, n-1\}$.

Now the map $G_n\colon \mathscr{V}_n \to D_{k+2}$ is given by

$$\begin{aligned} G_n(v_n)(t) &= v_n(t_0, \tau_1(v_n)(t_1), \ldots, \tau_{k+1}(v_n)(t_{k+1})) \\ &\quad + (n-k)^{-1/2} \sum_{i=k+1}^{[nt_0]} \left[H_n(\tau_1(v_n)(t_1), \ldots, \tau_{k+1}(v_n)(t_{k+1}))\right. \\ &\qquad \left. - H_n(t_1, \ldots, t_{k+1})\right]. \end{aligned} \tag{4.6}$$

We now give the formal proof.

The first thing we have to show is that G_n is continuous for every n.

Let $\{v_{n,l}\}n \ge 1$, $l \ge 1$, be a sequence of functions in $\mathscr{V}_n$ and let $v_{n,l} \to v_n (\in \mathscr{V}_n)$ with respect to special Skorohod topology. We show that $G_n(v_{n,l}) \to G_n(v_n)$ in uniform topology. From the definition of the special Skorohod topology, we have a sequence $\{\lambda_{j,l}\}_{1 \le j \le k+1, l \ge 1} \in \Lambda^{k+1}$ such that $\forall\, \varepsilon > 0$, $\exists\, l_\varepsilon \in \mathbb{N}$ such that $\max_{1 \le j \le k+1} |\lambda_{j,l}(t_j) - t_j| \le \varepsilon$ and

$$\begin{aligned} &|v_{n,l}(t) - v_n(t_0, \lambda_{1,l}(t_1), \ldots, \lambda_{k+1,l}(t_{k+1}))| < \varepsilon \\ &\qquad \forall\, l \le l_\varepsilon \text{ and } \forall\, t \in [0,1]^{k+2}, \end{aligned} \tag{4.7}$$

where Λ denotes the space of maps $h\colon [0,1] \to [0,1]$ which are nondecreasing, continuous and bijective, and Λ^{k+1} denotes the space of maps $\lambda\colon [0,1]^{k+1} \to [0,1]^{k+1}$, where $\lambda(t_1, \ldots, t_{k+1}) = (\lambda_1(t_1), \ldots, \lambda_{k+1}(t_{k+1}))$, $\lambda_j \in \Lambda$, $1 \le j \le k+1$. Then, we have the following lemma.

LEMMA 4.4. *$\exists\, l_0 > 0$ such that $\forall\, l \ge l_0$, $\forall\, j \in \{1, \ldots, k+1\}$ and $\forall\, t_j \in [0,1]$,*

$$\lambda_{j,l}(\tau_j(v_{n,l})(t_j)) = \tau_j(v_n)(t_j).$$

PROOF. For fixed j, let $(y^{l,1}, \ldots, y^{l,n-k})$ be a nondecreasing sequence of discontinuity points of $\tau_j(v_{n,l})$, and let $(y^{0,1}, \ldots, y^{0,n-k})$ be a (nondecreasing) sequence of discontinuity points of $\tau_j(v_n)$. (By convention, $y^{l,0} = y^{0,0} = 0$, $y^{l,n-k+1} = y^{0,n-k+1} = 1$.) For $i \in \{0, 1, \dots, n-k+1\}$, let $t_j \in [y^{l,i}, y^{l,i+1})$. Then,

$$(n-k)^{-1/2}\left[v_{n,l}(1,\dots,t_j,\dots,1) + H_n(1,\dots,t_j,\dots,1)\right] = i(n-k)^{-1}. \tag{4.8}$$

Let $h \in \{0, 1, \dots, n-k+1\} \to \lambda_{j,l}(t_j) \in [y^{0,h}, y^{0,h+1})$. Then, we have

$$\begin{aligned}(n-k)^{-1/2}&\left[v_n(1,\dots,\lambda_{j,l}(t_j),\dots,1) + H_n(1,\dots,\lambda_{j,l}(t_j),\dots,1)\right]\\ &= h(n-k)^{-1}.\end{aligned} \tag{4.9}$$

From (4.8) and (4.9), we deduce

$$\begin{aligned}\left|\frac{h}{n-k} - \frac{i}{n-k}\right| &\le |t_j - \lambda_{j,l}(t_j)| + (n-k)^{-1/2}\\ &\quad \times |v_{n,l}(1,\dots,t_j,\dots,1) - v_n(1,\dots,\lambda_{j,l}(t_j),\dots,1)|\\ &< \frac{1}{n-k} \qquad \forall\, l \ge \text{some } l_j.\end{aligned} \tag{4.10}$$

Thus, $|h/(n-k) - i/(n-k)| < 1/(n-k)$ and this implies that $h = i$. Now let $l_0 = \max_{1 \le j \le k} l_j$. Then, $\forall\ l \ge l_0$ and $\forall\ t_j \in [y^{l,i}, y^{l,i+1})$, we have $\lambda_{j,l}(t_j) \in [y^{0,i}, y^{0,i+1})$. Since the functions $\lambda_{j,l}$ are continuous and strictly nondecreasing, the proof follows. □

We now decompose G_n defined in (4.7) as $G_n = \gamma_n + \delta_n$, where $\gamma_n(v_n)(t) = v_n(t_0, \tau_1(v_n)(t_1), \dots, \tau_{k+1}(v_n)(t_{k+1}))$ and $\delta_n = G_n - \gamma_n$.

Lemma 4.5. (a) $\gamma_n\colon (\mathscr{V}_n, d) \to (D_{k+2}, \rho)$ *is continuous.*
(b) $\delta_n\colon (\mathscr{V}_n, d) \to (D_{k+2}, \rho)$ *is continuous.*

Proof. For $t \in [0,1]^{k+2}$, for $\forall\ \varepsilon > 0$, $\exists\ l_\varepsilon \ni \forall\ l \ge l_\varepsilon$, we have (using Lemma 4.4)

$$\begin{aligned}&|v_{n,l}(t_0, \tau_1(v_{n,l})(t_1), \dots, \tau_{k+1}(v_{n,l})(t_{k+1}))\\ &\quad - v_n(t_0, \tau_1(v_n)(t_1), \dots, \tau_{k+1}(v_n)(t_{k+1}))| < \varepsilon.\end{aligned}$$

The proof follows. Part (b) follows analogously, noting that H_n has uniform marginals. □

We now prove the convergence of the sequence $\{G_n\}$.

Let $v_n \in \mathscr{V}_n$, $n \in \mathbb{N}^*$, and suppose that $v_n \to_d v \in C_{k+2}$ and $v = 0$ on the lower boundary of $[0,1]^{k+2}$ and also when $\mathbf{t} = (1, \dots, 1)$. We have to prove that $G_n(v_n) \to_\rho G(v)$. The proof is based on the following lemmas.

Lemma 4.6. $\forall\, j \in \{1, \dots, k+1\}$:

(a) $\tau_j(v_n) \to \mathrm{id}_{[0,1]}$ *in uniform topology.*
(b) $(n-k)^{1/2}(\tau_j(v_n) - \mathrm{id}_{[0,1]}) \to -v(1, \dots, \mathrm{id}_{[0,1]}, \dots, 1)$ *in uniform topology, where* $\mathrm{id}_{[0,1]}$ *is the identity function on* $[0,1]$.

PROOF. Note that $\forall\, v_n$, $\exists\, y_n = (y_n^{(1)}, \ldots, y_n^{(n)})$ such that $v_n = Y_n(y_n)$. Now, for fixed j and for each $n \geq k+1$, define $v_n^{(j)}(t_j)$ as $v_n^{(j)}(t_j) = n^{-1/2}\sum_{i=1}^n \{I_{\{y_n^{(i)} \leq t_j\}} - t_j\}$ and note that $v_n^{(j)}(t_j)$ can also be written as

$$v_n^{(j)}(t_j) = ((n-k)/n)^{1/2} v_n(1, \ldots, t_j, \ldots, 1) + n^{-1/2} \sum_{i=1}^{k-j} \left[I_{\{y_n^{(i)} \leq t_j\}} - t_j \right]$$

$$+ n^{-1/2} \sum_{i=n-j}^{n} \left[I_{\{y_n^{(i)} \leq t_j\}} - t_j \right].$$

Since $v_n \to_d v$ (which implies $v_n \to_\rho v$), it follows that $v_n^{(j)}(t_j) \to_\rho v(1, \ldots, t_j, \ldots, 1)$. Thus, we can write

$$|\tau_j(v_n)(t_j) - t_j| = \left| n^{-1} \sum_{i=1}^{n} \{ I_{\{y_n^{(i)} \leq \tau_j(v_n)t_j)\}} - t_j \} - n^{-1/2} v_n^{(j)}\big(\tau_j(v_n)(t_j)\big) \right|$$

$$\leq \frac{k}{n} + n^{-1/2}\big[\rho\big(v_n^{(j)}, v(1, \ldots, \cdot\,, \ldots, 1)\big)$$

$$+\rho\big(v(1, \ldots, \cdot\,, \ldots, 1), g\big)\big] \to 0 \quad \text{as} \quad n \to \infty, \text{ where } g \equiv 0.$$

This proves part (a). The proof of part (b) is similar. □

LEMMA 4.7. *$\gamma_n(v_n) \to v$ in uniform topology.*

PROOF. Follows by definition and Lemma 4.6(a). □

LEMMA 4.8. *$\delta_n(v_n) \to \delta(v) = G(v) - v$ in uniform topology.*

PROOF. For $t \in [0,1]^{k+2}$, we have

$$\delta_n(v_n)(t) = \frac{[nt_0] - k}{(n-k)^{1/2}}$$

$$\times \{ H_n(\tau_1(v_n)(t_1), \ldots, \tau_{k+1}(v_n)(t_{k+1})) - H_n(t_1, \ldots, t_{k+1}) \}.$$

If there exists a $j \in \{1, \ldots, k+1\}$, $t_j < n^{-1}$, then

$$\delta_n(v_n)(t) = \frac{k - [nt_0]}{(n-k)^{1/2}} H_n(t_1, \ldots, t_{k+1}) \leq \frac{k - [nt_0]}{(n-k)^{1/2}} \frac{1}{n},$$

and so $\delta_n(v_n) \to 0$ as $n \to \infty$.

If $\forall\, j \in \{1, \ldots, k+1\}$, $t_j \geq n^{-1}$, then by the Taylor expansion,

$$\delta_n(v_n)(t) = \frac{[nt_0] - k}{(n-k)^{1/2}} \sum_{j=1}^{k+1} \{\tau_j(v_n)(t_j) - t_j\}$$

$$\times \frac{\partial}{\partial t_j} H_n(\theta_{n,1}(t_1), \ldots, \theta_{n,k+1}(t_{k+1})),$$

where $\theta_{n,j}(t_j) \in [t_j \wedge \tau_j(v_n)(t_j), t_j \vee \tau_j(v_n)(t_j)]$. Since $\{H_n\}$ satisfies the differentiability condition, we deduce from Lemma 4.6, the desired result.

Now, since $G_n(v_n) = \gamma_n(v_n) + \delta_n(v_n)$, we obtain (using Lemmas 4.7 and 4.8) that $G_n(v_n) \to G(v) = v + \delta(v)$. The proof of Proposition 3.5 follows. □

5. Proof of Theorem 2.1. First, we show $\mathscr{S}_n$ can be written as

$$\mathscr{S}_n = (n-k)^{-1/2}\left[\int_{[0,1]^{k+2}} L_n(t)\lambda_n(dt)\right] + b_n, \tag{5.1}$$

where λ_n is a signed measure on $[0,1]^{k+2}$ and b_n is the centering constant defined in Section 2.

$$\begin{aligned}
&(n-k)^{1/2}(\mathscr{S}_n - b_n)\\
&= (n-k)^{-1/2}\left[\sum_{i=k+1}^{n} c_{n,i} a_n(R_{n,i-k},\ldots,R_{n,i}) - b_n\right]\\
&= (n-k)^{-1/2}\left[\left(\sum_{i=k+1}^{n} c_{n,i}\left(\sum_A a_n(i_1,\ldots,i_{k+1})\prod_{j=1}^{k+1} I_{[R_{n,i+j-k-1}=i_j]}\right)\right)\right.\\
&\qquad\left. - \sum_B \lambda_n\left(\frac{i_0}{n},\ldots,\frac{i_{k+1}}{n}\right)\hat{H}_n\left(\frac{i_0}{n},\ldots,\frac{i_{k+1}}{n}\right)\right]\\
&= (n-k)^{-1/2}\left[\sum_B \lambda_n\left(\frac{i_0}{n},\ldots,\frac{i_{k+1}}{n}\right)\right.\\
&\qquad\left.\times\left(\sum_{i=k+1}^{i_0}\left(\prod_{j=1}^{k+1} I_{[R_{n,i+j-k-1}\le i_j]} - \hat{H}_n\left(\frac{i_0}{n},\ldots,\frac{i_{k+1}}{n}\right)\right)\right)\right]\\
&= \sum_B \lambda_n\left(\frac{i_0}{n},\ldots,\frac{i_{k+1}}{n}\right)L_n\left(\frac{i_0}{n},\ldots,\frac{i_{k+1}}{n}\right)\\
&= \int_{[0,1]^{k+2}} L_n(t)\lambda_n(dt),
\end{aligned}$$

where $\sum_A$ is the sum over all $(i_1,\ldots,i_{k+1})$ in $\{1,\ldots,n\}^{k+1}$ and $\sum_B$ is the sum over all $(i_0,\ldots,i_{k+1})$ in $\{1,\ldots,n\}^{k+2}$, where λ_n is defined in Section 2 and L_n is given by (2.3). We now prove that

$$\int_{[0,1]^{k+2}} L_n(t)\lambda_n(dt) \to \int_{[0,1]^{k+2}} L_\infty(t)\lambda_0(dt) \quad \text{as } n \to \infty. \tag{5.2}$$

Let $h_n\colon D_{k+2} \to \mathbb{R}$ be defined as

$$h_n(f) = \int_{[0,1]^{k+2}} f\lambda_n(dt), \qquad n \ge 0. \tag{5.3}$$

Let $\{f_n,\ n \ge 1\}$ be a sequence of functions in D_{k+2}, and suppose that $f_n \to f_0$ in uniform topology, where $f_0 \in C_{k+2}$. We show that

$$h_n(f_n) \to h_0(f_0). \tag{5.4}$$

We have

$$\left|\int_{[0,1]^{k+2}} f_n \lambda_n(dt) - \int_{[0,1]^{k+2}} f_0 \lambda_0(dt)\right|$$
$$\le \left|\int_{[0,1]^{k+2}} |f_n - f_0| \lambda_n(dt)\right| + \left|\int_{[0,1]^{k+2}} f_0(\lambda_n - \lambda_0)(dt)\right|$$
$$\le \sup_{t\in[0,1]^{k+2}} |f_n(t) - f(t)| \left|\int_{[0,1]^{k+2}} \lambda_n(dt)\right| + \left|\int_{[0,1]^{k+2}} f_0(\lambda_n - \lambda_0)(dt)\right|.$$

(5.4) follows using (2.5), (2.6) and (5.3), and (5.2) follows using Billingsley [(1968), Theorem 5.5)] and Proposition 3.5.

Now we prove that condition (2.6) of Theorem 2.1 is satisfied. By using (5.1) and (5.2) we deduce that

$$\sigma^2 = \int_{[0,1]^{k+2}} \int_{[0,1]^{k+2}} E[L_\infty(t) L_\infty(t')] \lambda_0(dt) \lambda_0(dt'). \tag{5.5}$$

We have [see (4.4)]

$$L_\infty(t) = W_\infty(t) - \sum_{j=1}^{k+1} t_0 W_\infty(1, \dots, t_j, \dots, 1) l_j(\mathbf{t}). \tag{5.6}$$

From (5.5) and (5.6) the equality in (2.6) holds.

It remains to show that $\sigma^2 < \infty$. By assumption (d) of Proposition 3.4,

$$\begin{aligned}
\lim_{n\to\infty} \Bigg| E\Bigg[& \left\{ W_n(t) - \sum_{j=1}^{k+1} t_0 W_n(1,\dots,t_j,\dots,1) l_j(\mathbf{t}) \right\} \\
& \times \left\{ W_n(t') - \sum_{j=1}^{k+1} t_0' W_n(1,\dots,t_j',\dots,1) l_j(\mathbf{t}) \right\} \Bigg] \Bigg| \\
&= |E[L_\infty(t) L_\infty(t')]| \\
&\le \lim_{n\to\infty} \left[E\left\{ W_n(t) - \sum_{j=1}^{k+1} t_0 W_n(1,\dots,t_j,\dots,1) l_j(\mathbf{t}) \right\}^2 \right]^{1/2} \\
&\quad \times \left[E\left\{ W_n(t') - \sum_{j=1}^{k+1} t_0' W_n(1,\dots,t_j',\dots,1) l_j(\mathbf{t}') \right\}^2 \right]^{1/2} \\
&= \lim_{n\to\infty} [A_n B_n],
\end{aligned} \tag{5.7}$$

by the Schwarz inequality.

Let now $\{X_{n,i}\}$ be φ-mixing with rates (1.2) or (1.3). Then from Lemma 4.1 with $q = 1$, we obtain

$$
\begin{aligned}
A_n \le \Bigg[E\Bigg\{ W_n^2(t) + 2|W_n(t)| t_0 \sum_{j=1}^{k+1} |W_n(1,\dots,t_j,\dots,1)| l_j(\mathbf{t}) \\
+ t_0^2 \sum_{s=1}^{k+1} \sum_{j=1}^{k+1} l_j(\mathbf{t}) l_s(\mathbf{t}) |W_n(1,\dots,t_j,\dots,1) W_n(1,\dots,t_s,\dots,1)| \Bigg\} \Bigg]^{1/2} \\
\le C_1 \Bigg[\left(\prod_{m=1}^{k+1} t_m \right)^{1/(k+2)} + 2t_0 \left(\prod_{m=1}^{k+1} t_m \right)^{1/2(k+1)} l_j(\mathbf{t}) \\
+ t_0^2 \sum_{s=1}^{k+1} \sum_{j=1}^{k+1} l_j(\mathbf{t}) l_s(\mathbf{t}) t_j^{1/2} t_s^{1/2} \Bigg]^{1/2},
\end{aligned}
\tag{5.8}
$$

where $C_1 > 0$ is some constant.

Similarly, B_n is less than or equal to some inequality with t's changed to t's. Thus $|E[L_0(t)L_0(t')]|$ is bounded by a function which is $\lambda_0 \times \lambda_0$ integrable, and so $|E[L_0(t)L_0(t')]|$ is also $\lambda_0 \times \lambda_0$ integrable.

Let now $\{X_{n,i}\}$ be strong-mixing with rates (1.4). Then, using Lemma 4.2 with $q = 2$, we obtain

$$
\begin{aligned}
A_n \le C_2 \Bigg[\left(\prod_{m=0}^{k+1} t_m \right)^{(1-\varepsilon)/(k+2)} + 2t_0 \left(\prod_{m=1}^{k+1} t_m \right)^{(1-\varepsilon)/2(k+1)} l_j(\mathbf{t}) \\
+ t_0^2 \sum_{s=1}^{k+1} \sum_{j=1}^{k+1} l_j(\mathbf{t}) l_s(\mathbf{t}) t_j^{(1-\varepsilon)/2} t_s^{(1-\varepsilon)/2} \Bigg]^{1/2}
\end{aligned}
$$

and a similar inequality for B_n, and the result follows as in the case of φ-mixing.

The proof of Theorem 2.2 follows analogously.

6. Applications to Markov processes and ARMA processes.

6.1. *Markov processes.* Consider a sequence $\{X_{t,n};\ n \in \mathbb{Z}\}$ of $\mathbb{R}$-valued processes such that for all $t \in \mathbb{N}^*$, $\{X_{t,n}\}$ is a k-Markov process with stationary transition probabilities $P_t(x_1,\dots,x_k; A)$, where $A \in \mathscr{B}$, $\mathscr{B}$ is the Borel σ-field of $\mathbb{R}$ and $(x_1,\dots,x_k) \in \mathbb{R}^k$.

We say that the Markov process is *ergodic* if there exists a unique probability measure μ_t on $\mathbb{R}^k$ with marginals Π_t on $\mathbb{R}$ such that

$$
\Pi_t(A) = \int_{\mathbb{R}^k} P_t(x_1,\dots,x_k; A) \mu_t(dx_1,\dots,dx_k) \quad \text{for all } A \in \mathscr{B}.
$$

We denote by P_t^m the m-step transition probability defined by

$$P_t^{m+l}(x_1,\ldots,x_k;A)=\int_{\mathbb{R}}P_t^l(x_2,\ldots,x_k,y;A)\,P_t^m(x_1,\ldots,x_k;dy),$$

for all $A\in\mathscr{B}$ and $(x_1,\ldots,x_k)\in\mathbb{R}^k$.

For a transition probability $P_t(\cdots,\cdot)$ and invariant measure μ_t and marginal Π_t, we denote by $P_t^*(\cdot,\cdot)$ the transition probability defined by

$$\int_{\prod_{i=1}^k(-\infty,y_i]}P_t(u_1,\ldots,u_k;(-\infty,y_{k+1}])\mu_t(du_1,\ldots,du_k)$$
$$=\int_{-\infty}^{y_{k+1}}P_t^*\left(u_{k+1};\prod_{i=1}^k(-\infty,y_i]\right)\Pi_t(du_{k+1}).$$

We say that the Markov process is *geometrically ergodic* if it is ergodic and if there exists $0<\rho_t<1$ such that

$$\|P_t^m(x_1,\ldots,x_k;\cdot)-\Pi_t(\cdot)\|=O(\rho_t^m)\quad\text{for all a.s. }(x_1,\ldots,x_k)\in\mathbb{R}^k,$$

where $\|\cdot\|$ denotes the norm of total variation and ρ_t is called the rate.

The Markov process is *Harris recurrent* if there exists a σ-finite measure ν_t on $\mathbb{R}$ with $\nu_t(\mathbb{R})>0$ such that $\nu_t(A)>0$ implies $P_t(x_1,\ldots,x_k;X_{t,n}\in A\text{ i.o.})=1$ for all $(x_1,\ldots,x_k)\in\mathbb{R}^k$.

Finally, the Markov process is *Doeblin recurrent* if it is ergodic and there exists a finite measure ν_t on $\mathbb{R}$ with $\nu_t(\mathbb{R})>0$, an $m\ge 1$ and $\varepsilon>0$ such that $P_t^m(x_1,\ldots,x_k;A)\le 1-\varepsilon$ if $\nu_t(A)\le\varepsilon$ for all $(x_1,\ldots,x_k)\in\mathbb{R}^k$ and $A\in\mathscr{B}$.

Let us denote $\forall\, j\in\{1,\ldots,k+1\}$ and $\forall\, M>0$,

$$R_j(M)=(-\infty,+\infty)^{j-1}\times[-M,M]\times(-\infty,+\infty)^{k-j+1}.$$

Then we have the following theorem.

THEOREM 6.1. *Let $\{X_{t,n},\ n\in\mathbb{Z}\}$ be a Markov process such that for every $t\in N^*$, $\{X_{t,n}\}$ is either* (a) *aperiodic, Harris recurrent and geometrically ergodic with rates $0<\rho_t<\rho_0$, $\rho_0\in(0,1)$ or* (b) *aperiodic and Doeblin recurrent. Suppose there exists a probability μ_0 on $\mathbb{R}^k$ and a transition probability $P_0(\cdots;\cdot)$ such that*

$$\sup_{A\in\mathscr{B}^k}|\mu_t(A)-\mu_0(A)|=O(t^{-\alpha}),\qquad\alpha>0,\tag{6.1}$$

$$\sup|P_t(x_1,\ldots,x_k;A)-P_0(x_1,\ldots,x_k;A)|\to 0\quad\text{as } t\to\infty,\tag{6.2}$$

where sup *is over $A\in\mathscr{B}$ and $(x_1,\ldots,x_k)\in R_j(M)$ for every $j\in\{1,\ldots,k+1\}$, $\forall\, M>0$, and*

$$\sup|P_t^*(x_{k+1};A_k)-P_0^*(x_{k+1};A_k)|\to 0\quad\text{as } t\to\infty,\tag{6.3}$$

where sup *is over $|x_{k+1}|\le M$ and $A_k\in\mathscr{B}^k$.*

Then, under the assumptions (3.2) *and* (3.3), $(n-k)^{1/2}(\mathscr{S}_n-b_n)$ *converges in law to a normal distribution with mean* 0 *and variance σ^2, where b_n*

and σ^2 *are given by* (2.4) *and* (2.7), *respectively*. (It is assumed that P_t, P_0, P_t^*, P_0^* have densities *continuous* in x's and μ_0 have densities).

PROOF. (i) Suppose (a) holds. First, we show that the process is geometrically strong mixing. It is well known [see Nummelin and Tuominen (1982)] that if a Markov chain is aperiodic, Harris recurrent and geometrically ergodic with rate ρ_t, then

$$\int \|P_t^m(x_1,\dots,x_k;\cdot) - \Pi_t(\cdot)\|\mu_t(dx_1,\dots,dx_k) = O(\rho_t^m)$$

and this property is equivalent to strong mixing with rate ρ_t^m [see Rosenblatt (1971), page 199]. Next, we show that the covariance functions of the associated graduate empirical process (2.2), converge to a function K, but this is a consequence of Lemma 6 of Rüschendorf (1974) which remains true for strong-mixing conditions with a geometric rate.

Let G_t be the distribution function of the $k+1$ successive random variables of $\{X_{tn}\}$ and let H_t be the measure on $[0,1]^{k+1}$ defined by $H_t(y_1,\dots,y_{k+1}) = G_t(\Pi_t^{-1}(y_1),\dots,\Pi_t^{-1}(y_{k+1}))$, where Π_t is the marginal of μ_t for all $(y_1,\dots,y_{k+1}) \in [0,1]^{k+1}$ and $t \geq 0$ (note that we also denote by Π_t the distribution function associated with the measure Π_t).

We have to show that $\{H_t\}_{t>0}$ satisfies the differentiability condition (given in Section 2).

Set $l_t^{(j)} = \partial H_t/\partial t_j$, and let $F_t^{(j)}$ be the conditional distribution function defined as

$$F_t^{(j)}(u_j; y_1,\dots,y_{j-1},y_{j+1},\dots,y_{k+1})$$
$$= \int_R P_t(u_1,\dots,u_k;(-\infty,y_{k+1}])\mu_t^j(du_1,\dots,du_{j-1},du_{j+1},du_k),$$

where $R = \prod_{l=1}^{j-1}(-\infty,y_l]\prod_{l=j+1}^{k}(-\infty,y_l]$, μ_t^j is the measure associated with the distribution function $H_t(u_1,\dots,u_{j-1},1,u_{j+1},\dots,u_k,1)$ if $j \leq k$, and $F_t^{(k+1)}$ is the conditional distribution function associated with P_t^*. We have

$$l_t^{(j)}(y_1,\dots,y_{k+1})$$
$$= F_t^{(j)}\big(\Pi_t^{-1}(y_j);\Pi_t^{-1}(y_1),\dots,\Pi_t^{-1}(y_{j-1}),\Pi_t^{-1}(y_{j+1}),\dots,\Pi_t^{-1}(y_{k+1})\big).$$

Also,

$$l_t^{(j)}(y_1,\dots,y_k) - l_0^{(j)}(y_1,\dots,y_k)$$
$$= \big[F_t^{(j)}(\Pi_t^{-1}(y_j);\Pi_t^{-1}(y_1),\dots,\Pi_t^{-1}(y_{j-1}),\Pi_t^{-1}(y_{j+1}),\dots,\Pi_t^{-1}(y_{k+1}))$$
$$-F_0^{(j)}(\Pi_t^{-1}(y_j);\Pi_t^{-1}(y_1),\dots,\Pi_t^{-1}(y_{j-1}),\Pi_t^{-1}(y_{j+1}),\dots,\Pi_t^{-1}(y_{k+1}))\big]$$
$$+\big[F_0^{(j)}(\Pi_t^{-1}(y_j);\Pi_t^{-1}(y_1),\dots,\Pi_t^{-1}(y_{j-1}),\Pi_t^{-1}(y_{j+1}),\dots,\Pi_t^{-1}(y_{k+1}))$$
$$-F_0^{(j)}(\Pi_0^{-1}(y_j);\Pi_0^{-1}(y_1),\dots,\Pi_0^{-1}(y_{j-1}),\Pi_0^{-1}(y_{j+1}),\dots,\Pi_0^{-1}(y_{k+1}))\big]$$
$$= A + B.$$

To simplify the notation, take $j = 1$ (for $j \neq 1$, the method is exactly the same). Then, we have

$$\begin{aligned} A &= F_t^{(1)}\big(\Pi_t^{-1}(y_1); \Pi_t^{-1}(y_2), \dots, \Pi_t^{-1}(y_{k+1})\big) \\ &\quad - F_0^{(1)}\big(\Pi_t^{-1}(y_1); \Pi_t^{-1}(y_2), \dots, \Pi_t^{-1}(y_{k+1})\big) \\ &= \int_{\Pi_{l=2}^k(-\infty, x_l]} P_t(x_1; u_2, \dots, u_k; (-\infty, x_{k+1}])\mu_t^1(du_2, \dots, du_k) \\ &\quad - \int_{\Pi_{l=2}^k(-\infty, x_l]} P_0(x_1, u_2, \dots, u_k; (-\infty, x_{k-1}])\mu_0^1(du_2, \dots, du_k) \\ &\qquad\qquad\qquad \big[\text{for } \Pi_t^{-1}(y_l) = x_l,\, l \in \{1, \dots, k+1\}\big] \\ &\leq \varepsilon + \left| \int_{\Pi_{l=2}^k(-\infty, x_l]} \mu_t^1(du_2, \dots, u_k) - \mu_0^1(du_2, \dots, du_k) \right| \leq 2\varepsilon \end{aligned}$$

for all $(x_1, \dots, x_{k+1}) \in R_1(M)$.

We also have

$$\begin{aligned} B &= F_0^{(1)}\big(\Pi_t^{-1}(y_1); \Pi_t^{-1}(y_2), \dots, \Pi_t^{-1}(y_{k+1})\big) \\ &\quad - F_0^{(1)}\big(\Pi_0^{-1}(y_1); \Pi_0^{-1}(y_2), \dots, \Pi_0^{-1}(y_{k+1})\big) \\ &= F_0^{(1)}\big(\Pi_0^{-1} \circ \Pi_0 \circ \Pi_t^{-1}(y_1); \Pi_0^{-1} \circ \Pi_0 \circ \Pi_t^{-1}(y_2), \dots, \Pi_0^{-1} \circ \Pi_0 \circ \Pi_t^{-1}(y_{k+1})\big) \\ &\quad - F_0^{(1)}\big(\Pi_0^{-1} \circ \Pi_0 \circ \Pi_0^{-1}(y_1); \Pi_0^{-1} \circ \Pi_0 \circ \Pi_0^{-1}(y_2), \dots, \Pi_0^{-1} \circ \Pi_0 \circ \Pi_0^{-1}(y_{k+1}). \end{aligned}$$

Noting that

$$\begin{aligned} \sup_{y_1 \in [0,1]} |\Pi_0 \circ \Pi_t^{-1}(y_1) - y_1| &= \sup_{y_1 \in [0,1]} |y_1 - \Pi_t \circ \Pi_0^{-1}(y_1)| \\ &= \sup_{y_1 \in [0,1]} |\Pi_0 \circ \Pi_0^{-1}(y_1) - \Pi_t \circ \Pi_0^{-1}(y_1)|, \end{aligned}$$

we find [using (6.1)] that $B < \varepsilon$ for sufficiently large t. Thus, $l_t^{(j)}(y_1, \dots, y_{k+1}) \to l_0^{(j)}(y_1, \dots, y_{k+1})$ as $t \to \infty$ uniformly in $(y_1, \dots, y_k) \in R_j(M)$ for any $M > 0$, and so $\{H_t\}_{t>0}$ satisfies the differentiability condition.

(ii) Suppose (b) is satisfied. Then the proof follows from Davydov (1973), who proved that a Markov process which is Doeblin recurrent and aperiodic is geometrically φ-mixing. □

EXAMPLE 6.1. Consider the process $\{X_n, n \in \mathbb{Z}\}$, where $X_{n+1} = a_1 X_n + a_2 X_n \varepsilon_{n+1} + a_3 \varepsilon_{n+1} + a_4 \varepsilon_{n+1}^2 + a_5$, where the a's are real numbers and $\{\varepsilon_n, n \in \mathbb{Z}\}$ is a white noise with strictly positive density. Then Mokkadem (1985) has shown that if $a_1^2 + a_2^2 E(\varepsilon_1^2) < 1$ and $E(\varepsilon_1^4) < \infty$, then the process $\{X_n, n \in \mathbb{Z}\}$ is geometrically ergodic and geometrically strong mixing. Thus, the asymptotic normality of the statistic $\mathscr{S}_n$ based on the ranks of $\{X_n\}$ follows.

EXAMPLE 6.2. Consider the process $\{X_n, n \in \mathbb{Z}\}$, where $X_{n+1} = f(X_n) + \varepsilon_{n+1}$, where the ε's are independent and identically distributed random variables with strictly positive density, and $f\colon \mathbb{R} \to \mathbb{R}$ is bounded, nondecreasing and continuous. [This model was studied by Collomb and Doukhan (1983).] It is easy to check that this model is Doeblin recurrent and aperiodic, and we deduce that $\{X_n\}$ is geometrically φ-mixing and we can apply Theorem 6.1.

6.2. *ARMA processes.* Consider a sequence of ARMA (k_1, k_2) processes,

$$\prod_{j=1}^{k_1}\left(1 - a_j^{(n)}U\right)X_{n,i} = Q_{k_2}^{(n)}(U)\varepsilon_i, \qquad i \in \mathbb{Z}, n \in \mathbb{N}^*, \tag{6.4}$$

where $U\lambda_i = \lambda_{i-1}$, $Q_{k_2}^{(n)}(U) = \sum_{l=0}^{k_2} b_l^{(n)}U^l$, $b_0^{(n)} = 1$ and $\{\varepsilon_i,\ i \in \mathbb{Z}\}$ is a sequence of independent random variables such that $E(\varepsilon_i) = 0$ and ε_i has a density $g_i(x)$, $i \in \mathbb{Z}$.

Then we have the following lemma.

LEMMA 6.1 [Gorodetskii (1977), Withers (1981)]. *Let the sequence $\{X_{n,i}$, $i \in \mathbb{Z}\}$ satisfy the following conditions*:

$$\sup_{i\in\mathbb{Z}} \int_{-\infty}^{\infty} |g_i(x+\beta) - g_i(x)|\, dx \le c_1|\beta|, \quad \forall\, \beta \text{ and some } c_1 > 0; \tag{6.5}$$

$$\sup_{i\in\mathbb{Z}} E|\varepsilon_i| < c_2 < \infty \quad \text{and} \quad \sup_{n\in\mathbb{N}}\ \sup_{1\le j\le k_1} |a_j^{(n)}| < \rho < 1, \tag{6.6}$$

where c_2 and ρ are some constants. Then for any $n \in \mathbb{N}^*$, the process $\{X_{n,i}$; $i \in \mathbb{Z}\}$ is strong-mixing with rate $\alpha(m) = O(\rho_0^{m/2})$ for each $\rho_0 > \rho$.

THEOREM 6.2. *Let the sequence $\{X_{n,i},\ i \in \mathbb{Z}\}$ of ARMA (k_1, k_2) process given by* (6.4) *satisfy the following conditions*:

(6.7) *$\{\varepsilon_i,\ i \in \mathbb{Z}\}$ is a sequence of independent and identically distributed random variables, each having $\mathcal{N}(0, \sigma^{*2})$ distribution.*

(6.8) *$\forall\, j \in \{1, \dots, k_1\}$, $\exists\ \alpha > 0$ and $a_j \in (-1, 1)$, $a_j \neq 0$, such that $|a_j^{(n)} - a_j| = O(n^{-\alpha})$, and $\forall\ l \in \{1, \dots, k_2\}$, $\exists\ \beta > 0$, and $b_l \in \mathbb{R}$ such that $|b_l^{(n)} - b_l| = O(n^{-\beta})$.*

Then for the rank statistic $\mathcal{S}_n$ associated with the sequence $\{X_{n,1}, \dots, X_{n,n}\}$ and the score functions satisfying the assumptions of Theorem 2.1, *$(n-k)^{1/2}(\mathcal{S}_n - b_n)$ converges in law to the $\mathcal{N}(0, \sigma^2)$ distribution, where b_n and σ^2 are given by* (2.4) *and* (2.7), *respectively.*

PROOF. To prove this theorem, we first note, using Lemma 6.1, that the sequence $\{X_{n,i}\}$ is geometrically strong-mixing. Now, let F_n be the distribution function of $X_{n,i}$ and F_0 the distribution function of a stationary random variable X_{oi} defined by an ARMA (k_1, k_2) process with coefficients a_j, $1 \le j \le k_1$ and b_l, $1 \le l \le k_2$. Now we prove the differentiability condition for $H_n(\mathbf{t})$ defined in (2.1) be verifying (6.1), (6.2) and (6.3).

Let P_n^j be the transition distribution function of $X_{n,1}, \dots, X_{n,j-1}$, $X_{n,j+1}, \dots, X_{n,k_1}$, and G_n^j the distribution function of $(X_{n,1}, \dots, X_{n,k_1-1})$,

$n \geq 0$. Then (6.1), (6.2) and (6.3) are satisfied in view of the following well-known result.

LEMMA 6.2. *Let $\{G_n, n \geq 0\}$ be a sequence of k-dimensional normal distribution functions each with mean vector* $\mathbf{0}$. *Let the covariance matrices of G_n and G_0 be $\mathbf{\Sigma}_n = ((\sigma_{il}^{(n)}))$ and $\mathbf{\Sigma}_0 = ((\sigma_{il}^{*}))$ and assume that $|\sigma_{il}^{(n)} - \sigma_{il}^{*}| = O(n^{-\alpha})$ for each $i, l = 1, \ldots, k$. Then G_n converges uniformly to G_0.*

7. Appendix.

7A. *Proof of Proposition* 3.1. (a) and (b) are sufficient conditions. They follow immediately from Proposition 3.2 by using a result from Billingsley [(1968), page 151, line 15].

We have only to prove that (a) and (b) are necessary conditions.

Let $\mathscr{U}_{k+2}^{*}$ be the σ-field generated by the uniform topology on C_{k+2}.

As P is concentrated on a separable space $(C_{k+2}, \mathscr{U}_{k+2}^{*})$, it follows from Wichura [(1970), Theorem 1] that there exists a probability space $(\Omega, \mathscr{A}, \mu)$ and a sequence of random variables $\{Y_n^*\}$, $n \in \mathbb{N}^*$, and a random variable Y^* such that $\mu(Y_n^*) = P_n$, $\mu(Y^*) = P$ and $Y_n^* \to Y^*$ a.s. μ. For any δ (> 0), we consider the map T_δ: $D_{k+2} \to \mathbb{R}$ defined by

$$T_\delta(f) = \sup\{|f(t) - f(t'); \|t - t'\| \leq \delta\}.$$

Then, T_δ is a continuous map for the uniform topology on $\mathscr{U}_{k+2}$.

Now, consider a sequence of random variables $\{Z_{n,\delta}\}$, $n \in \mathbb{N}^*$, and a random variable Z_δ defined as

$$Z_{n,\delta} = T_\delta \circ Y_n^*, \qquad Z_\delta = T_\delta \circ Y^*.$$

As Y_n^* converges a.s. to Y^*, it follows that $\forall\, \varepsilon > 0$, $\exists\, N_0 \in \mathbb{N}$ such that

$$\mu\{|Z_{n,\delta} - Z| > \varepsilon/2\} < \varepsilon/2, \qquad \forall\, n \geq N_0. \tag{7.1}$$

As Y^* is concentrated on C_{k+2}, we have also $\forall\, \varepsilon > 0$, $\exists\, \delta > 0$ such that

$$\mu\{|Z_\delta| > \varepsilon/2\} < \varepsilon/2. \tag{7.2}$$

(7.1) and (7.2) imply

$$\mu\{|Z_{n,\delta}| > \varepsilon\} < \varepsilon$$

or

$$P_n[\,f; \omega(\,f, \delta) \geq \varepsilon] = \mu\{|Z_{n,\delta}| > \varepsilon\} < \varepsilon, \tag{7.3}$$

and from (7.3) we obtain condition (b). Condition (a) is immediate. Proposition 3.1. is proved.

7B. *Proof of Lemma* 4.1. Without loss of generality, we can take $N_n = n$. First, we prove that for any p, $1 \leq p \leq n$,

$$E\left(\sum_{i=1}^{p} X_{n,i}\right)^{2q} \leq C_q \sum_{l=1}^{q} p^{q/l}\left(\sup_{1 \leq j \leq n} \|X_{n,j}\|_l\right)^{2q} (h(p,l))^{2q}, \tag{7.4}$$

where C_q is a constant depending only on q and φ and

$$h(p,l) = \exp\left\{\sum_{l=1}^{s} \varphi^{1/2q}([2^{1/2(l+1)}])\right\} \quad \text{for } 2^s \le p < 2^{s+1}.$$

For any (l, p), $1 \le l \le p \le n$, we define $S(l, p)$ by

$$S(l,p) = \sum_{i=l\wedge(n+1)}^{(l+p-1)\wedge(n+1)} X_{n,i}, \quad \text{where, by convention, } X_{n,n+1} \equiv 0.$$

Denote $S = S(l, p)$, $S' = S(l + p + r, p)$, $R = S(l + p, r) - S(2p + l, r)$ for $r \ge 1$,

$$a(p,q) = \sup_{l\ge 1}\left[E\left(\sum_{i=l\wedge(n+1)}^{(l+p-1)\wedge(n+1)} X_{n,i}\right)^{2q}\right]^{1/2q}, \quad \text{and} \quad m_l = \sup_{1\le j\le n} \|X_{n,j}\|_l.$$

Then, after some computations, we obtain the inequality,

$$\text{(7.5)} \qquad E(S + S')^{2q} \le 2(a(p,q))^{2q} \exp\{(2q\varphi(r))^{1/2q}\} + (2a(p,q-1))^{2q}.$$

From the Minkowski inequality, it follows that

$$\left\|\sum_{l=1}^{(l+2p-1)\wedge(n+1)} X_{n,i}\right\|_{2q} = \|S + S' + R\|_{2q} \le \|S + S'\|_{2q} + 2rm_q$$

$$\le 2^{1/2q}a(p,q)\exp\{\varphi(r)^{1/2q}\} + 2a(p,q-1) + 2rm_q.$$

Now take $p = 2^s$, $s \ge 1$, and put $r = r(s,q) = [2^{s/2(q+1)}]$, $\varphi(s,q) = (\varphi(r))^{1/2q}$.
Then, from (7.5), we can write

$$a(2^s,q) \le 2^{1/2q}a(2^{s-1},q)\exp\{\psi(s,q)\} + 2a(2^{s-1},q-1) + 2r(s,q)m_q,$$

$$\text{(7.6)} \qquad \begin{aligned} a(2^s,q) &\le 2^{s/2q}\left(1 + 2\sum_{i=1}^{s} 2^{-i/2q}r(i,q)\right)m_q \exp\left\{\sum_{j=1}^{s} \psi(j,q)\right\} \\ &\quad + 2\sum_{i=1}^{s} 2^{(s-i)/2q}a(2^{i-1},q-1)\exp\left\{\sum_{j=i+1}^{s} \psi(j,q)\right\}, \end{aligned}$$

where, by convention, $\sum_{j=s+1}^{s}\psi(j,q) = 0$.
For $q = 1$, we have

$$a(2^s,1) \le K_1 2^{s/2}h(2^s,1)m_1,$$

where K_1 is a positive constant.
We give a proof by recurrence on q. Suppose that, for all $q \ge 2$ and $p = 2^s$, we have

$$a(2^s,q-1) \le K_{q-1}\sum_{l=1}^{q-1} 2^{s/2l}h(2^s,l)m_l,$$

where K_{q-1} is a positive constant. From (7.6), we deduce

$$a(2^s, q) \leq 2^{s/2q}\left(\exp\left\{\sum_{j=1}^{s} \psi(j,q)\right\}\right)\left(1 + 2\sum_{j=1}^{s} 2^{-1/2(j/q - j/(q+1))}\right) m_q$$

$$+ 2A_q K_{q-1} \sum_{l=1}^{q-1} 2^{s/2l} h(2^s, l) m_l,$$

where A_q is a constant depending only on q and φ. That is,

$$a(2^s, q) \leq K_q \sum_{l=1}^{q} 2^{s/2l} h(2^s, l) m_l. \tag{7.7}$$

Finally, for each $p \leq n$, we can write the binary decomposition as

$$p = \sum_{i=0}^{s} v_i 2^i, \qquad v_i \in \{0,1\}.$$

From the equality $h(p, l) = h(2^s, l)$ for $2^s \leq p < 2^{s+1}$ and (7.7), it follows that

$$a(p,q) \leq \sum_{i=0}^{s} v_i a(2^i, q) \leq \sum_{i=0}^{s} a(2^i, q),$$

$$a(p,q) \leq K_q \sum_{i=0}^{s} \sum_{l=1}^{q} 2^{i/2l} h(2^i, l) m_l, \tag{7.8}$$

$$(a(p,q))^{2q} \leq C_q \sum_{l=1}^{q} p^{q/l} (h(p,l) m_l)^{2q},$$

and (7.4) is proved.

(4.1) now follows by putting $p = n = N_n$ and by using the relation

$$\sum_{l=0}^{\infty} (\varphi(2^l))^{1/2q} < +\infty \quad \Leftrightarrow \quad \sum_{m=1}^{\infty} m^{-1} (\varphi(m))^{1/2q} < +\infty. \tag{7.9}$$

Lemma 4.1 is proved.

7C. *Proof of Lemma* 4.3. For every $p, N, r \in \mathbb{N}$, we define $S_N = \sum_{i=1}^{N} Y_i$, $T_{N,j} = \sum_{i=1}^{N} Y_{j(N+r)+i}$, $R_{N,j} = \sum_{i=1}^{r} (Y_{j(N+r)+N+i} - Y_{pN+jr+i})$ for $j = 0, \ldots, p-1$. For every $l \in \mathbb{N}$, we denote $K_l^2 = E(\sum_{i=1}^{l} Y_i)^2$.

From the property of stationarity, we have

$$K_N^2 = E(S_N^2) = E(T_{N,j})^2 \quad \text{for } j = 0, \ldots, p-1$$

and

$$\left| E(S_N)^2 - \frac{1}{p} E\left(\sum_{j=0}^{p-1} T_{N,j}\right)^2 \right| \leq p(\varphi(r))^{1/2} K_N^2. \tag{7.10}$$

We have

$$S_{pN} = \sum_{i=1}^{pN} Y_i = \sum_{j=0}^{p-1} T_{N,j} + \sum_{j=0}^{p-1} R_{N,j}.$$

We deduce from (7.10)

$$(7.11) \qquad \left| E(S_N)^2 - \frac{1}{p} E(S_{pN})^2 \right| \le p(\varphi(r))^{1/2} K_N^2 + 4pK_r K_N.$$

From Lemma 4.1, there exists a constant C depending only on φ such that

$$K_r^2 \le CrK_1^2 \quad \text{and} \quad K_N^2 \le CNK_1^2.$$

Now, taking $r = r(N) = [N^{1/2}]$ and using (7.10) for $p = 2$, we obtain

$$\begin{aligned}\left| \frac{1}{2^s} E(S_{2^s})^2 - \frac{1}{2^{s+m}} E(S_{2^{s+m}})^2 \right| &\le \frac{1}{2^s} \sum_{k=0}^{m-1} \frac{1}{2^k} \left| E(S_{2^{s+k}})^2 - \frac{1}{2} E(S_{2^{s+k+1}})^2 \right| \\ &\le 8CK_1^2 \sum_{k=s}^{s+m-1} \left((\varphi([2^{k/2}]) + 2^{-k/4}\right).\end{aligned}$$

It follows that $(1/2^s)E(S_{2^s})^2$ is a Cauchy sequence. Hence there exists a constant K such that

$$\frac{1}{2^s} E(S_{2^s})^2 \to K^2 \quad \text{as } s \to \infty.$$

We deduce that for every $p_0 \in \mathbb{N}$,

$$\sup_{p \ge p_0} \left| K^2 - \frac{1}{p2^s} E(S_{p2^s})^2 \right| \to 0 \quad \text{as } s \to \infty.$$

Let l be such that $pq \le l \le (p+1)q$. Then

$$\left| \frac{1}{pq} E(S_{pq})^2 - \frac{1}{l} E(S_l)^2 \right| \le 2CK_1^2 \left(\frac{1}{p} + \frac{1}{p^{1/2}} \right) \quad \text{for all } q \ge 1.$$

Consequently, $\forall\, \varepsilon > 0$, $\exists\, p_0 \in \mathbb{N}$, such that

$$\sup_{l \ge p} \left| \frac{1}{p[1/p]} E(S_{p[1/p]})^2 - \frac{1}{l} E(S_l)^2 \right| < \frac{\varepsilon}{2}, \qquad \forall\, p \ge p_0.$$

Now if we choose s_0 such that $2^{s_0} \ge p_0$ and

$$\sup_{p \le 2p_0} \left| K^2 - \frac{1}{p2^s} E(S_{p2^s})^2 \right| < \frac{\varepsilon}{2} \quad \text{for all } s \ge s_0,$$

then there exists for every $n \ge n_0 = p_0 2^{s_0}$, an $s \ge s_0$ and a p for which $p_0 \le p < 2p_0$ and $p2^s \le n < (p+1)2^s$.

We deduce that $\forall\ n \geq n_0$,

$$\left|K^2 - \frac{1}{n}E(S_n)^2\right| \leq \left|K^2 - \frac{1}{p2^s}E(S_{p2^s})^2\right| + \left|\frac{1}{p2^s}E(S_{p2^s})^2 - \frac{1}{n}E(S_n)^2\right|$$

$$< \frac{\varepsilon}{2} + \frac{\varepsilon}{2} = \varepsilon,$$

which implies

$$\frac{1}{n}E(S_n)^2 \to K^2 \quad \text{as } n \to \infty$$

and Lemma 4.3 is proved.

Acknowledgments. The authors would like to express their sincere gratitude to Professor Peter Ney, the Associate Editor, and the referees for their critical examination of the first draft. Their constructive criticisms and suggestions for improvements are gratefully acknowledged.

REFERENCES

BALACHEFF, S. and DUPONT, G. (1980). Normalité asymptotique des processus empiriques tronqués et des processus de rang. *Statistique non Paramétrique Asymptotique. Lecture Notes in Math.* **821** 19–45. Springer, Berlin.

BILLINGSLEY, P. (1968). *Convergence of Probability Measures*. Wiley, New York.

COLLOMB, G. and DOUKHAN, P. (1983). Estimation de la fonction d'autoregression d'un processus stationnaire et φ mélangeant: Risque quadratique pour la méthode du noyau. *C. R. Acad. Sci. Paris Sér. I Math.* **296** 859–862.

DAVYDOV, YU. A. (1973). Mixing conditions for Markov chains. *Theory Probab. Appl.* **18** 312–328.

DOUKHAN, P. and PORTAL, F. (1987). Principe d'invariance faible pour la fonction de répartition empirique dans un cadre multidimensionnel et mélangeant. *Probab. Math. Statist.* **8** 117–132.

DUDLEY, R. M. (1978). Central limit theorems for empirical measures. *Ann. Probab.* **6** 899–929.

GAENSSLER, P. and STUTE, W. (1987). *Seminar on Empirical Processes*. Birkäuser, Boston.

GORODETSKII, V. V. (1977). On the strong mixing property for linear sequences. *Theory Probab. Appl.* **22** 411–413.

HALLIN, M., INGENBLEEK, J.-F. and PURI, M. L. (1985). Linear serial rank tests for randomness against ARMA alternatives. *Ann. Statist.* **13** 1156–1181.

HAREL, M. (1980). Convergence en loi pour la topologie de Skorohod du processus empirique multidimensionnel normalisé tronqué et semi-corrigé. *Statistique non Paramétrique Asymptotique. Lecture Notes in Math.* **821** 46–85. Springer, Berlin.

HAREL, M. and PURI, M. L. (1989a). Limiting behavior of *U*-statistics, *V*-statistics and one-sample rank order statistics for non-stationary absolutely regular processes. *J. Multivariate Anal.* **31** 181–204.

HAREL, M. and PURI, M. L. (1989b). Weak convergence of the *U*-statistic and weak invariance of the one-sample rank order statistic for Markov processes and ARMA models. *J. Multivariate Anal.* **31** 259–265.

HAREL, M. and PURI, M. L. (1990). Weak invariance of generalized *U*-statistics for nonstationary absolutely regular processes. *Stochastic Process. Appl.* **34** 341–360.

IBRAGIMOV, J. A. and LINNIK, YU. V. (1971). *Independent and Stationary Sequences of Random Variables*. Wolters-Nordhoff, Groningen.

MOKKADEM, A. (1985). Le modèle non linéaire AR(1) général. Ergodicité et ergodicité géométrique. *C. R. Acad. Sci. Paris Sér. I Math.* **301** 889–892.

Neuhaus, G. (1971). On weak convergence of stochastic processes with multidimensional time parameter. *Ann. Math. Statist.* **42** 1285–1295.

Neumann, N. (1982). Ein schwaches Invarianzprinzip für den gewichten empirischen Prozess von gleichmässig mischenden Zufallsvariablen. Ph.D. dissertation, Göttingen.

Nummelin, E. and Tuominen, P. (1982). Geometric ergodicity of Harris recurrent Markov chains with applications to renewal theory. *Stochastic Process. Appl.* **12** 187–202.

Rosenblatt, M. (1971). *Markov Processes: Structure and Asymptotic Behavior*. Springer, Berlin.

Rüschendorf, L. (1974). On the empirical process of multivariate, dependent random variables. *J. Multivariate Anal.* **4** 469–478.

Rüschendorf, L. (1976). Asymptotic distributions of multivariate rank order statistics. *Ann. Statist.* **4** 912–923.

Shorack, G. R. and Wellner, J. A. (1986). *Empirical Processes with Applications to Statistics*. Wiley, New York.

Skorohod, A. V. (1956). Limit theorems for stochastic processes. *Theory Probab. Appl.* **1** 261–290.

Wichura, M. J. (1970). On the construction of almost uniformly convergent random variables with given weakly convergent image laws. *Ann. Math. Statist.* **41** 1101–1115.

Withers, C. S. (1975). Convergence of empirical processes of mixing rv's on [0, 1]. *Ann. Statist.* **3** 1101–1108.

Withers, C. S. (1981). Conditions for linear processes to be strong mixing. *Z. Wahrsch. Verw. Gebiete* **57** 477–480.

Département Technique de Commercialisation
Institut Universitaire de Technologie de Limoges
Allée André Maurois
87065 Limoges Cedex
France

Department of Mathematics
Indiana University
Bloomington, Indiana 47405

Reprinted by permission of the publisher,
© 1990 by The Institute of Mathematical Statistics

ASYMPTOTIC NORMALITY OF L-STATISTICS BASED ON $m(n)$-DECOMPOSABLE TIME SERIES

K. C. CHANDA, M. L. PURI* and F. H. RUYMGAART

Department of Mathematics, Texas Tech University, Lubbock, Texas 79409
Department of Mathematics, Indiana University, Bloomington, Indiana 47405
Mathematics Institute, Katholieke Universiteit Nijmegen, 6525ED Nijmegen, The Netherlands

In this paper we introduce the concept of $m(n)$-decomposability as an alternative to classical mixing concepts. We illustrate how to handle the ensuing technicalities by proving asymptotic normality of L-statistics, based on such a decomposable time series, as a typical example. © 1990 Academic Press, Inc.

1. Introduction: The Model and the Problem

A. $m(n)$-*Decomposable Processes*

For each $n \in \mathbf{N}$ let $X_1, ..., X_n$ be a stochastic process of real valued random variables defined on the same probability space such that we have a decomposition $X_i = X_{i,m} + \bar{X}_{i,m}$, for some $m = m(n) \in \mathbf{N}_0$, where

$$\text{the } X_{i,m(n)} \text{ are } m(n)\text{-dependent and } \bar{X}_{i,m(n)} \rightarrow_p 0, \quad \text{as} \quad n \rightarrow \infty. \tag{1.1}$$

Received July 20, 1989; revised April 16, 1990.

AMS 1980 subject classifications: primary 62M10; secondary 62G30.

Key words and phrases: $m(n)$-decomposable processes; mixing; empirical processes for $m(n)$-decomposable samples; L-statistics.

* Research supported by the Office of Naval Research Contract N00014–85-K-0648.

For brevity such a process will be referred to as $m(n)$-decomposable. Working with $m(n)$-decomposable processes for large n typically requires a specification of the orders of magnitude of $m(n)$, $\delta(n)$, and $\varepsilon(n)$ in

$$P(|\bar{X}_{i,m(n)}| \geqslant \varepsilon(n)) \leqslant \delta(n). \tag{1.2}$$

The idea of writing an arbitrary process as the sum of a "nice" one and a negligible error has been inspired by the manner in which asymptotics are dealt with for linear processes (see, e.g., Ibragimov and Linnik [16, Theorem 18.6.5]). However, we believe that it is expedient to recognize that many important processes (including nonlinear ones; see also the examples below) satisfy our general $m(n)$-decomposability condition for reasonable choices of $\varepsilon(n)$ and $\delta(n)$. Moreover, it might be possible to extend existing results for the m-dependent case (m fixed) to $m(n)$-decomposable processes that are extremely difficult to extend to the mixing case. One might, in particular, think of extending results on rates of convergence, large deviations, quantile processes, and applications etc. for m-dependent processes (Heinrich [14], Aly [1], Chaganty [5]) to $m(n)$-decomposable processes. The concept seems natural as well as tractable and might provide a useful alternative to the classical mixing concepts.

It has also been recognized that mixing conditions are not always suitable for research (Pham and Tran [24]). Decomposability reduces to $m(n)$-dependence in a convenient and straightforward way. Although in interesting cases we have $m(n)$ to allow to tend to infinity with the sample size n, tools for such $m(n)$-dependent sequences exist like, e.g., a central limit theorem (Berk [4]; see also M. M. Rao [2, Chap. 8]), or may be developed fairly easily. The technicalities may be considerable but not prohibitive. We do not known the position of decomposability in the spectrum of mixing properties. We do know, however, that there exist linear processes that are decomposable but not strongly mixing [2]. The concept seems tailormade for processes with Volterra expansions of a given finite order [28]. Just for curiosity we show that the ranks of an i.i.d. sample are also decomposable. It is useful to note that in linear models with decomposable errors, the residuals are decomposable as well.

EXAMPLE 1.1 (Linear processes). Let $\mathbf{Z}$ be the set of all integers and $\{Z_k, k \in \mathbf{Z}\}$ an i.i.d. sequence of real valued innovations. For each $n \in \mathbf{N}$ consider the so-called linear process

$$X_i = \sum_{k \in \mathbf{Z}} a_k Z_{i-k}, \qquad i \in \{1, ..., n\}. \tag{1.3}$$

As is well known such processes contain ARMA-models as special cases. In order to show that the process is decomposable let us write

$$X_{i,m(n)} = \sum_{|k| \leqslant \theta m(n)} a_k Z_{i-k}, \qquad \bar{X}_{i,m(n)} = \sum_{|k| > \theta m(n)} a_k Z_{i-k}, \tag{1.4}$$

for some suitable $0 < \theta < 1$. Let us assume that $E|Z_i|^\gamma < \infty$ for some $0 < \gamma \leqslant 1$ and that, for given parameters $\tau > \frac{1}{2}$ and $0 < \rho < \frac{1}{2}$, we have

$$|a_k| = \mathcal{O}(|k|^{-\lambda}), \qquad \text{as } |k| \to \infty, \text{ for some } \lambda > (2 + \tau\gamma + \rho)/(\rho\gamma). \tag{1.5}$$

According to Chanda and Ruymgaart [8, Sect. 4] we have, letting $m(n) = \mathcal{O}(n^\rho)$ and $\varepsilon(n) = \mathcal{O}(n^{-\tau})$,

$$P(|\bar{X}_{i,m(n)}| \geqslant \varepsilon(n)) = \delta(n) = \mathcal{O}(n^{\rho(1-\lambda\gamma)+\tau\gamma}), \qquad \text{as} \quad n \to \infty. \tag{1.6}$$

It follows that for $m(n)$ and $\varepsilon(n)$ of prescribed order, $\delta(n)$ can be made arbitrarily small, provided λ may be assumed sufficiently large. Similar conditions on λ may be found for $1 < \gamma < 2$ and $\gamma \geqslant 2$.

Example 1.2 (A non-linear process). In addition to the notation of the previous example let $a_{l,r} \in \mathbf{R}$ be given numbers. The process

$$X_i = \sum_{k \in \mathbf{Z}} a_k Z_{i-k} + \sum_{l,r \in \mathbf{Z}}\sum a_{lr} Z_{i-l} Z_{i-r}, \tag{1.7}$$

with quadratic Volterra expansion is obviously non-linear but still decomposable in a way similar to (1.4), viz.,

$$X_{i,m(n)} = \sum_{|k| \leqslant \theta m(n)} a_k Z_{i-k} + \sum_{|l|,|r| \leqslant \theta m(n)}\sum a_{lr} Z_{i-l} Z_{i-r}, \tag{1.8}$$

$\bar{X}_{i,m(n)} = X_i - X_{i,m(n)}$. Of course, higher order Volterra expansions might be also considered.

Example 1.3 (Linear regression). Let $Y_i = x_i'\theta + E_i$, for $i = 1, ..., n$, where the x_i are given vectors in $\mathbf{R}^p$, $\theta \in \mathbf{R}^p$ is an unknown parameter, and the errors $E_1, ..., E_n$ are an $m(n)$-decomposable process that in this context will usually satisfy certain stationarity conditions. Suppose that $\hat{\theta}_n$ is a consistent estimator and $t \in \mathbf{R}^p$ is the true parameter. Then the residuals may be written as

$$\begin{aligned} X_i &= Y_i - x_i'\hat{\theta}_n = E_i - x_i'(\hat{\theta}_n - t) \\ &= E_{i,m(n)} + \bar{E}_{i,m(n)} - x_i'(\hat{\theta}_n - t) = X_{i,m(n)} + \bar{X}_{i,m(n)}, \end{aligned} \tag{1.9}$$

where the $X_{i,m(n)} = E_{i,m(n)}$ are $m(n)$-dependent and the $\bar{X}_{i,m(n)} = \bar{E}_{i,m(n)} - x_i'(\hat{\theta}_n - t)$ are asymptotically negligible. Hence the empirical process of the residuals is a special case of the empirical process based on an $m(n)$-decomposable series of observables. The same remark holds true, of course, for certain useful functionals of the empirical residual process, like linear combinations of order statistics, to be used for diagnostics.

EXAMPLE 1.4 (Ranks). Let $\xi_1, ..., \xi_n$ be i.i.d. uniform $(0, 1)$ random variables and $R_1, ..., R_n$ be the corresponding ranks. Let us simply decompose the standardized ranks $R_i/(n+1)$ into

$$\frac{R_i}{n+1} = \xi_i + \frac{R_i}{n+1} - \xi_i = X_{i,0} + \bar{X}_{i,0}. \tag{1.10}$$

It is clear that the $X_{i,0}$ are independent (i.e., 0-dependent). According to Hájek and Šidák [11], we have

$$P(|\bar{X}_{i,0}| \geqslant \varepsilon(n)) \leqslant E\left(\frac{R_i}{n+1} - \xi_i\right)^2 \Big/ \varepsilon(n)^2 \leqslant C/n\{\varepsilon(n)\}^2 = \delta(n). \tag{1.11}$$

This entails that, even with $m(n) = 0$ for all $n \in \mathbf{N}$, we can achieve $\delta(n) \to 0$, provided that $\varepsilon(n) = o(n^{-1/2})$, as $n \to \infty$.

B. *Asymptotics for L-Statistics*

When dealing with $m(n)$-decomposable processes, the technicalities will be different from those under classical mixing conditions. The purpose of this paper is to introduce the decomposable model and to give a typical illustration of how to handle the technicalities by proving asymptotic normality of L-statistics. This kind of statistics is important in view of the growing interest in robustness for time series [21]. Linear combinations of (a function of) order statistics play an important role in robust estimation of location and dispersion, and certain quadratic forms of the order statistics might be useful for robust estimation of serial correlation; here we restrict ourselves to the linear case. The cornerstone for asymptotic efficiency and robustness considerations is the variance of the limiting normal distribution. Here we give a full proof of the asymptotic normality, assuming a certain stationarity condition in addition to specific orders for $m(n)$, $\delta(n)$, and $\varepsilon(n)$, and restricting to smooth score functions that are zero in neighborhoods of 0 and 1. A proof for score functions that are allowed to tend to infinity near 0 and 1 is available [7], but the technicalities are rather formidable and for robustness the former type is much more adequate. We will use the classical Chernoff–Savage approach which was first applied by Moore [22] and subsequently employed by Ruymgaart and van Zuijlen [31] in the study of L-statistics. This method has also been used in

Nieuwenhuis and Ruymgaart [22] for serial rank statistics, but in the more restrictive model (1.4), and our approach will be patterned on that paper. For serial rank statistics under mixing conditions see Harel and Puri [13] and see Harel and Puri [12] for other robust statistics in nonstationary absolutely regular models. For the special subclass of linear processes in (1.4), density estimation has been considered in Chanda and Ruymgaart [8], and for general decomposable processes, curve estimation will be considered in a forthcoming paper. For this approach, some results on the empirical process will be needed that might be of independent interest. Aly [1] considers quantile and rank processes for m-dependent random variables (m fixed) and gives an application to the asymptotics of two-sample rank statistics. We hope to extend some of these results to $m(n)$-decomposable sample elements by exploiting a well-known simple relation between the empirical and quantile process.

Gastwirth and Rubin [10] prove asymptotic normality of L-statistics when a fixed finite number of order statistics is involved, assuming a strongly mixing strictly stationary model. Despite considerable overlap, this model differs from ours. Moreover, unboundedness of the score function—which is one of the major theoretical difficulties—cannot occur in such a situation. However, Gastwirth and Rubin [10] only heuristically obtain an expression for the asymptotic variance of general L-statistics and focus on some interesting efficiency and robustness results. Since asymptotic normality of L-statistics is a classical problem in statistics [9, 34, 32], it seems worthwhile to present a full proof and thus provide the theoretical background for the considerations in Gastwirth and Rubin [10]. Portnoy [25] shows that certain M-estimators are asymptotically minimax for location in an m-dependent model for fixed $m \in \mathbf{N}$; these results are extended in Portnoy [26] assuming asymptotic normality. Martin [20] proves robustness of certain generalized M-estimators for location in ARMA-models, and Koul [19] derives asymptotic normality for certain robust statistics, including a special class of L-statistics, in a regression model with strongly mixing errors. Chanda [6] introduces linear combinations of the unordered sample elements as location estimators and derives the—not necessarily normal—limiting distribution assuming certain stable distributions for the innovations. An interesting line of further research is the estimation, using L-statistics, of the parameter θ in the general linear regression model of Example 1.3 with $m(n)$-decomposable errors, following the asymptotic linearity results of Jurečková [17, 18]; see also Ruppert and Carroll [30], Welsh [35], and Portnoy and Koenker [27].

C. *Organization*

In Section 2 we formulate the assumptions and main results. In passing we give an outline of the proof. The necessary tools from empirical process theory are summarized in Section 3. We deal with the asymptotic normality of the first-order terms in Section 4. Section 5 is devoted to the asymptotic negligibility of the remainder term. Finally, in Section 6 we simplify the expression for the variance under an additional weak-stationarity condition and compare this expression with the one in Gastwirth and Rubin [10].

2. Assumptions and Main Result

Assumption 2.1. The underlying process $X_1, ..., X_n$ is $m(n)$-decomposable. Moreover, the X_i have common d.f. F (independent of n) and the $X_{i,m(n)}$ have common d.f. $F_{m(n)}$. The d.f. F is continuously differentiable on $\mathbf{R}$ with derivative f satisfying $\|f\|_\infty < \infty$, and $\{x \in \mathbf{R}: f(x) > 0\}$ is convex.

Let $\hat{F}_n$ be the empirical d.f. of the X_i at stage n and $\hat{\Gamma}_n$ be the empirical d.f. of the transformed random variables $\xi_i = F(X_i)$. Assumption 2.1 entails the $m(n)$-decomposability of the ξ_i, since by the mean value theorem we have

$$\xi_i = F(X_{i,m}) + \bar{X}_{i,m} f(X_{i,m} + \Theta \bar{X}_{i,m}) = \xi_{i,m} + \bar{\xi}_{i,m}, \tag{2.1}$$

for some random $\Theta \in (0, 1)$. The common d.f. of the ξ_i is uniform (0, 1). As usual we define the corresponding reduced empirical process by

$$U_n = \{U_n(t) = n^{1/2}(\hat{\Gamma}_n(t) - t), t \in [0, 1]\}. \tag{2.2}$$

Throughout this paper the numbers

$$A, B, C \in (0, \infty) \tag{2.3}$$

will be used as generic constants that are independent of all the relevant parameters, as, in particular, the sample size n, and that may differ from line to line.

To describe the class of L-statistics that we are going to consider let J: $[0, 1] \to \mathbf{R}$ and Ψ: $(0, 1) \to \mathbf{R}$ be given functions. Assumption 2.1 implies the continuity of F^{-1} on $\{x \in \mathbf{R} : 0 < F(x) < 1\}$; we write $\Psi_F = \Psi(F^{-1})$: $(0, 1) \to \mathbf{R}$. We will focus on statistics of the type

$$T_n = \int_{-\infty}^{\infty} \Psi(x)\, J(\hat{F}_n(x))\, d\hat{F}_n(x) = \frac{1}{n} \sum_{i=1}^{n} c_{ni} \Psi(X_{i:n}), \tag{2.4}$$

where $c_{ni} = J(i/n)$. It is obvious that

$$T_n =_d \int_0^1 \Psi_F(t)\, J(\hat{\Gamma}_n(t))\, d\hat{\Gamma}_n(t) = \frac{1}{n} \sum_{i=1}^{n} c_{ni} \Psi_F(\xi_{i:n}). \tag{2.5}$$

The latter representation will be used throughout.

Assumption 2.2. The function $J\colon [0,1]\to \mathbf{R}$ has two continuous derivatives J', J'' on $[0,1]$ and is of the form

$$J=0 \text{ on } [0,\Delta]\cup[1-\Delta,1], \text{ for some } 0<\Delta<\tfrac{1}{2}. \tag{2.6}$$

The function $\Psi_F\colon (0,1)\to\mathbf{R}$ has one continuous derivative Ψ_F' on $(0,1)$.

Assumption 2.2 clearly guarantees the finiteness of

$$\mu=\int_0^1 J(t)\,\Psi_F(t)\,dt. \tag{2.7}$$

It follows from the mean value theorem that, for $t\in[0,1]$ and $\Theta_n(t)\in(0,1)$ random, we have

$$J(\hat{\Gamma}_n(t))=J(t)+n^{-1/2}U_n(t)\,J'(t+\Theta_n(t)\,n^{-1/2}U_n(t)). \tag{2.8}$$

By substitution in (2.5) we find the decomposition

$$n^{1/2}(T_n-\mu)=A_{0n}+A_{1n}+B_n, \tag{2.9}$$

where the A- and B-terms are first- and second-order terms, respectively, given by

$$A_{0n}=\int_0^1 \Psi_F(t)\,J(t)\,dU_n(t), \tag{2.10}$$

$$A_{1n}=\int_0^1 U_n(t)\,\Psi_F(t)\,J'(t)\,dt, \tag{2.11}$$

$$B_n=\int_0^1 U_n(t)\,\Psi_F(t)\,J'(t+\Theta_n(t)\,n^{-1/2}U_n(t))\,d\hat{\Gamma}_n(t)-A_{1n}. \tag{2.12}$$

In order to pass from the ξ_i to the $m(n)$-dependent $\xi_{i,m(n)}$ we need to specify the asymptotic negligibility of the $\bar{\xi}_{i,m(n)}$. Let us consider the sequences defined in (1.3) and write

$$\Omega_{\varepsilon(n)}=\{\max_{1\leqslant i\leqslant n}|X_{i,m(n)}|<\varepsilon(n)\}. \tag{2.13}$$

The next assumption prescribes suitable orders of magnitude for these sequences.

Assumption 2.3. There exist $m(n)=\mathcal{O}(n^{\rho})$, $\varepsilon(n)=\mathcal{O}(n^{-\tau})$, as $n\to\infty$, with

$$0\leqslant\rho<\tfrac{1}{4},\qquad \tau>2(1-\rho), \tag{2.14}$$

such that $\delta(n) = \mathcal{O}(n^{-\sigma})$, as $n \to \infty$, with

$$\sigma > 2(1-\rho). \tag{2.15}$$

Let us write, for brevity,

$$K(t) = \Psi_F(t)\, J(t) + \int_0^1 \{1_{[0,s]}(t) - s\}\, \Psi_F(s)\, J'(s)\, ds,\ t \in [0, 1], \tag{2.16}$$

and introduce the $m(n)$-dependent centered random variables

$$Z_{ni} = K(\xi_{i,m(n)}) - EK(\xi_{i,m(n)}) = K(\xi_{i,m(n)}) - \mu_n. \tag{2.17}$$

Under the assumptions made above we will prove (Sections 3 and 4)

$$A_{0n} + A_{1n} - n^{-1/2} \sum_{i=1}^{n} Z_{ni}$$

$$= n^{-1/2} \sum_{i=1}^{n} \{(K(\xi_i) - \mu) - (K(\xi_{i,m(n)}) - \mu_n)\} \to_p 0, \quad \text{as } n \to \infty, \tag{2.18}$$

$$B_n \to_p 0, \qquad \text{as} \quad n \to \infty. \tag{2.19}$$

To prove the asymptotic normality of $n^{-1/2} \sum_{i=1}^{n} Z_{ni}$, we may apply Berk [4], provided that the next assumption is satisfied.

ASSUMPTION 2.4. The following conditions are fulfilled:

$$\operatorname{Var}\left(\sum_{i=j+1}^{k} Z_{ni}\right) \leqslant C(k-j), \qquad \text{for all} \quad j = 0, 1, ..., n; \tag{2.20}$$

$$\lim_{n \to \infty} \frac{1}{n} \operatorname{Var}\left(\sum_{i=1}^{n} Z_{ni}\right) = v^2 \in (0, \infty). \tag{2.21}$$

THEOREM 2.1. *Let Assumptions* 2.1–2.4 *be satisfied. Then we have*

$$n^{1/2}(T_n - \mu) \to_d \mathcal{N}(0, v^2), \qquad \textit{as} \quad n \to \infty, \tag{2.22}$$

with μ *as in* (2.7) *and* v^2 *given by* (2.21).

3. TOOLS FROM EMPIRICAL PROCESS THEORY

Some properties of empirical processes based on decomposable models will be reviewed here. Proofs will not be given since they are very similar to those in Nieuwenhuis and Ruymgaart [23]; see also Chanda and

Ruymgaart [8]. Since these properties hold true for arbitrary n, there is no need to specify the dependence on n of the parameters; in particular, we will write Ω_ε for the set in (2.13). In applications, however, these parameters will usually depend on n in a suitable way. We will then, e.g., choose $\varepsilon=\varepsilon(n)$ in such a way that $P(\Omega^c_{\varepsilon(n)})\leqslant n\delta(n)=\mathcal{O}(n^{1-\sigma})\to 0$, as $n\to\infty$. To formulate the results we need the function

$$\psi(\lambda)=2\lambda^{-2}\int_0^\lambda \log(1+x)\,dx, \qquad \lambda>0; \quad \psi(0)=1. \tag{3.1}$$

It should be noted that $\psi(\lambda)\downarrow 0$ as $\lambda\uparrow\infty$. For further properties and the role this function plays in precise asymptotic considerations, the reader is referred to, e.g., Shorack and Wellner [33]. We start with a basic fluctuation inequality that might be of independent interest.

THEOREM 3.1. *Suppose that Assumption* 2.1 *is fulfilled. For all* $n\in\mathbf{N}$ *we have*

$$\mathrm{P}\left(\Omega_\varepsilon\cap\left\{\sup_{a\leqslant s<t\leqslant b}|U_n(t)-U_n(s)|\geqslant\lambda\right\}\right)$$
$$\leqslant Cm\exp\left(\frac{-A\lambda^2}{m(b-a)}\psi\left(\frac{B\lambda}{n^{1/2}(b-a)}\right)\right), \tag{3.2}$$

provided that

$$\lambda\geqslant Cn^{1/2}(\varepsilon+\delta), \qquad b-a\geqslant C(\varepsilon+\delta). \tag{3.3}$$

Because Nieuwenhuis and Ruymgaart [23] only exploit the decomposability property of the linear process mentioned in Example 1.1, the proof for the present Theorem 2.1 can be just copied. As a corollary to Theorem 3.1 let us note that for $m(n)$, $\varepsilon(n)$, $\delta(n)$ satisfying Assumption 2.3 and, by choosing $a=0$, $b=1$, we obtain

$$P(\Omega_{\varepsilon(n)}\cap\{\sup_{0\leqslant t\leqslant 1}|U_n(t)|\geqslant n^\nu\})$$
$$\leqslant Cn^\rho\exp\left(\frac{-An^{2\nu}}{n^\rho}\psi\left(\frac{Bn^\nu}{n^{1/2}}\right)\right)\to 0, \qquad \text{as} \quad n\to\infty, \tag{3.4}$$

provided that

$$\tfrac{1}{2}\rho<\nu<\tfrac{1}{2}. \tag{3.5}$$

Note that (3.3) is clearly fulfilled for the present choice of λ, a, and b.

4. The Leading Terms

In this section we will be concerned with proving (2.18) and the asymptotic normality of the standardized sum of the Z_{ni}. It follows from Assumptions 2.1–2.3, (2.1), and (2.13) that

$$\{\max_{1 \leqslant i \leqslant n} \bar{\xi}_{i,m(n)} \leqslant \|f\|_\infty \varepsilon(n)\} \supset \Omega_{\varepsilon(n)}, \qquad E\bar{\xi}_{i,m(n)} \leqslant Cn^{-\sigma \wedge \tau}. \tag{4.1}$$

The definition of K in (2.16) entails that K has a continuous derivative K' and that both K and K' are bounded on $(0, 1)$ and zero on $(0, \Delta) \cup (1-\Delta, 1)$. Employing (4.1) and the mean value theorem we see that

$$n^{-1/2} \sum_{i=1}^{n} E|K(\xi_i) - K(\xi_{i,m(n)})|$$

$$\leqslant Cn^{-1/2} \sum_{i=1}^{n} E\bar{\xi}_{i,m(n)} \leqslant Cn^{1/2-\sigma \wedge \tau} \to 0, \qquad \text{as} \quad n \to \infty, \tag{4.2}$$

according to Assumption 2.3, so that (2.18) follows.

Moreover, because K is bounded we have

$$\sup_{i,n} E|Z_{ni}|^{2+\gamma} \leqslant C \sup_{i,n} E|K(\xi_{i,m(n)})|^{2+\gamma} \leqslant M_\gamma, \tag{4.3}$$

where M_γ is a finite number for each $\gamma > 0$. Furthermore, let us note that there exists $\gamma > 0$ such that

$$\{m(n)\}^{2+2/\gamma}/n \leqslant Cn^{2\rho(1+1/\gamma)-1} \to 0, \qquad \text{as} \quad n \to \infty, \tag{4.4}$$

since by (2.14) we have $2\rho(1+1/\gamma) < 1$ for γ sufficiently large. Assumption 2.4, jointly with (4.3) and (4.4), suffices for application of Berk [4], which yields $n^{-1/2} \sum_{i=1}^{n} Z_{ni} \to_d \mathcal{N}(0, v^2)$, as $n \to \infty$. Since we have also proved (2.18), it follows that

$$A_{0n} + A_{1n} \to_d \mathcal{N}(0, v^2), \qquad \text{as} \quad n \to \infty, \tag{4.5}$$

which settles the asymptotic normality of the leading terms.

5. The Remainder Term

This section is devoted to a proof of (2.19). This may be done with the help of the properties in Section 3 in a way similar to that in the i.i.d. case (see, e.g., Ruymgaart and van Zuijlen [31]), the case of spacings (see, e.g.,

Beirlant *et al.* [3]), of U-statistics (see, e.g., Helmers and Ruymgaart [15]). As in Nieuwenhuis and Ruymgaart [23], however, the present model of a decomposable process is a little more complicated but, on the other hand, the score function is of a much simpler type.

Let us write, for brevity,

$$t + \Theta_n(t)\, n^{-1/2} U_n(t) = t + \Theta_n(t)(\hat{\Gamma}_n(t) - t) = \tilde{\Gamma}_n(t), \tag{5.1}$$

see (2.8), and note that $\tilde{\Gamma}_n(t)$ is a random point between t and $\hat{\Gamma}_n(t)$ so that $\sup_{0\leqslant t\leqslant 1} |\tilde{\Gamma}_n(t) - t| \leqslant \sup_{0\leqslant t\leqslant 1} |\hat{\Gamma}_n(t) - t|$. Now the remainder term will be decomposed into $B_n = B_{n1} + B_{n2}$, where

$$B_{n1} = \int_0^1 U_n(t)\{J'(\hat{\Gamma}_n(t)) - J'(t)\}\, \Psi_F(t)\, d\hat{\Gamma}_n(t), \tag{5.2}$$

$$B_{n2} = \int_0^1 U_n(t)\, J'(t)\, \Psi_F(t)\, d\{\hat{\Gamma}_n(t) - t\}. \tag{5.3}$$

Let us single out the subsets, for $v \geqslant 0$,

$$\Omega_{n,v} = \{ \sup_{0\leqslant t\leqslant 1} |U_n(t)| \leqslant n^v \}, \qquad \Omega_n = \Omega_{\varepsilon(n)} \cap \Omega_{n,v}, \tag{5.4}$$

and note that (3.4) entails

$$P(\Omega_n) \to 1, \qquad \text{as } n \to \infty, \text{ for each } \tfrac{1}{2}\rho < v < \tfrac{1}{2}. \tag{5.5}$$

For (5.5) we only need to observe that for such values of v condition (3.3) is amply satisfied due to Assumption 2.3. Consequently it suffices to prove that

$$C_{nj} = \mathbf{1}_{\Omega_n} B_{nj} \to_p 0, \qquad \text{as} \quad n \to \infty, \qquad \text{for} \quad j = 1, 2. \tag{5.6}$$

Let us first consider C_{n1} and note that by the mean value theorem,

$$J'(\tilde{\Gamma}_n(t)) - J(t) = (\tilde{\Gamma}_n(t) - t)\, J''(\tilde{\tilde{\Gamma}}_n(t)), \tag{5.7}$$

where $\tilde{\tilde{\Gamma}}_n(t)$ is a random point between t and $\tilde{\Gamma}_n(t)$. It follows that

$$\begin{aligned} |C_{n1}| &= \left| \int_0^1 U_n(t)(\tilde{\Gamma}_n(t) - t)\, J''(\tilde{\tilde{\Gamma}}_n(t))\, \Psi_F(t)\, d\hat{\Gamma}_n(t) \right| \\ &\leqslant C n^{-1/2} \mathbf{1}_{\Omega_n} \{ \sup_{0\leqslant t\leqslant 1} |U_n(t)| \}^2 \\ &= \mathcal{O}_p(n^{-1/2+2v}) = o_p(1), \quad \text{as} \quad n \to \infty, \end{aligned} \tag{5.8}$$

provided that we choose

$$\nu = \tfrac{1}{8}. \tag{5.9}$$

Let us now consider C_{n2} and introduce the parameters

$$\phi = \tfrac{3}{8} + \tfrac{1}{2}\rho, \qquad \zeta = \tfrac{1}{8}. \tag{5.10}$$

Partitioning the unit interval into subintervals of equal length $[n^\phi]^{-1}$, we define the left continuous step functions

$$S_n(t) = \begin{cases} \dfrac{1}{[n^\phi]}, & \text{for } t = 0 = t_0; \\[2ex] \dfrac{k}{[n^\phi]}, & \text{for } t_{k-1} = \dfrac{k-1}{[n^\phi]} < t \leqslant \dfrac{k}{[n^\phi]},\ 1 \leqslant k \leqslant \tfrac{1}{2}[n^\phi]; \\[2ex] \dfrac{k-1}{[n^\phi]}, & \text{for } \dfrac{k-1}{[n^\phi]} < t \leqslant \dfrac{k}{[n^\phi]},\ \tfrac{1}{2}[n^\phi] < k \leqslant [n^\phi]. \end{cases} \tag{5.11}$$

We need to decompose C_{n2} into

$$D_{n1} = \mathbf{1}_{\Omega_n} \int_0^1 U_n(S_n(t))\, \Psi_F(S_n(t))\, J'(S_n(t))\, d\{\hat{\Gamma}_n(t) - t\}, \tag{5.12}$$

$$D_{n2} = \mathbf{1}_{\Omega_n} \int_0^1 \{U_n(t)\, \Psi_F(t)\, J'(t) - U_n(S_n(t))\, \Psi_F(S_n(t))\, J'(S_n(t))\}\, d\hat{\Gamma}_n(t), \tag{5.13}$$

$$D_{n3} = \mathbf{1}_{\Omega_n} \int_0^1 \{U_n(S_n(t))\, \Psi_F(S_n(t))\, J'(S_n(t)) - U_n(t)\, \Psi_F(t)\, J'(t)\}\, dt. \tag{5.14}$$

Let us first derive the useful relations

$$\begin{aligned} &P(\mathbf{1}_{\Omega_n} \cdot \max_k |U_n(t_k) - U_n(t_{k-1})| \geqslant n^{-\zeta}) \\ &\leqslant P(\max_k \sup_{t_{k-1} \leqslant t \leqslant t_k} |U_n(S_n(t)) - U_n(t)| \geqslant n^{-\zeta}) \\ &\leqslant \sum_{k=1}^{[n^\phi]} P(\sup_{t_{k-1} \leqslant s < t \leqslant t_k} |U_n(s) - U_n(t)| \geqslant n^{-\zeta}) \\ &\leqslant C n^{\phi+\rho} \exp\left(\frac{-An^{-2\zeta}}{n^{\rho-\phi}}\, \psi\left(\frac{Bn^{-\zeta}}{n^{1/2-\phi}}\right)\right) \to 0, \qquad \text{as } n \to \infty, \end{aligned} \tag{5.15}$$

by applying (3.2) with $a=t_{k-1}$ and $b=t_k$. Condition (3.3) is fulfilled because $\lambda=n^{-\zeta}\geqslant Cn^{1/2}n^{-\sigma\wedge\tau}(\zeta=\frac{1}{8}<\sigma\wedge\tau-\frac{1}{2})$ and $b-a=n^{-\phi}\geqslant Cn^{-\sigma\wedge\tau}$ $(\phi<\frac{1}{2}<\sigma\wedge\tau)$. Moreover, $\phi-\rho-2\zeta=\frac{3}{8}+\frac{1}{2}\rho-\rho-\frac{2}{8}=\frac{1}{8}-\frac{1}{2}\rho>0$ and $\phi-\frac{1}{2}-\zeta=\frac{3}{8}+\frac{1}{2}\rho-\frac{1}{2}-\frac{1}{8}=\frac{1}{2}\rho-\frac{1}{4}<0$, which entails the convergence to zero in (5.15).

For D_{n1} relation (5.15) yields

$$\begin{aligned}|D_{n1}| &\leqslant Cn^{-1/2}n^{\nu}n^{\phi}\mathbf{1}_{\Omega_n}\max_{k}|U_n(t_k)-U_n(t_{k-1})|\\ &=\mathcal{O}_p(n^{-1/2+\nu+\phi}n^{-\zeta})=o_p(1),\qquad \text{as}\quad n\to\infty,\end{aligned}\tag{5.16}$$

since $\nu+\phi-\frac{1}{2}-\zeta=\frac{1}{8}+\frac{3}{8}+\frac{1}{2}\rho-\frac{1}{2}-\frac{1}{8}<0$.

Writing $L(t)=J'(t)\,\Psi_F(t)$, $t\in(0,1)$, it is clear that

$$\begin{aligned}|D_{n2}|\vee|D_{n3}| &\leqslant \mathbf{1}_{\Omega_n}\cdot\sup_{0<t<1}|U_n(S_n(t))-U_n(t)|\,|L(S_n(t))|\\ &\quad+\mathbf{1}_{\Omega_n}\cdot\sup_{0<t<1}|U_n(t)|\,|L(S_n(t))-L(t)|\\ &=\mathcal{O}_p(n^{-\zeta})+\mathcal{O}_p(n^{\nu-\phi})=o_p(1),\qquad \text{as}\quad n\to\infty,\end{aligned}\tag{5.17}$$

because $\nu-\phi=\frac{1}{8}-\frac{8}{8}-\frac{1}{2}\rho<0$. For (5.17) we use (3.4), (5.15), the fact that L has a bounded and continuous derivative, the definition of S_n in (5.11), and the mean value theorem to deal with $L(S_n(t))-L(t)$.

6. Simplification of Variance

It is clear that in the special models of Examples 1.1 and 1.2 the process $X_{1,m(n)},\dots,X_{n,m(n)}$ and hence the process $\xi_{1,m(n)},\dots,\xi_{n,m(n)}$ are strictly stationary. For a simplification of the expression for the asymptotic variance in a general $m(n)$-decomposable model the weaker assumption that

$$\xi_{1,m(n)},\dots,\xi_{n,m(n)}\ \text{is stationary up to order 2,}\tag{6.1}$$

suffices. Note that this condition is not implied by (1.1) and (1.2).

Theorem 6.1. *Let* (6.1) *be satisfied. Under the conditions of Theorem* 2.1 *we have*

$$v^2=\operatorname{Var}(K(\xi_1))+2\sum_{j=2}^{\infty}\operatorname{Cov}(K(\xi_1),K(\xi_j)).\tag{6.2}$$

Proof. Relation (4.3) entails that $E|Z_{n1}Z_{nj}|\leqslant C$, for some $C\in(0,\infty)$ independent of n, where the Z_{nj} are defined in (2.17). Due to the

$m(n)$-dependence, $EZ_{nk}Z_{nj}=0$ for $|k-j|>m(n)$. Writing $v_n^2=EZ_{n1}^2+2\sum_{j=2}^{m(n)} EZ_{n1}Z_{nj}$, we see that

$$\left|\frac{1}{n}\operatorname{Var}\left(\sum_{i=1}^{n} Z_{ni}\right)-v_n^2\right|=\mathcal{O}(n^{\rho-1})\rightarrow 0, \qquad \text{as} \quad n\rightarrow\infty, \tag{6.3}$$

because $1-\rho>\frac{3}{4}$.

The next step is to show that we may replace $Z_{ni}=K_n(\xi_{i,m(n)})-\mu_n$ by $K_n(\xi_i)-\mu_n$. For arbitrary $1\leqslant j\leqslant n$ we have

$$\begin{aligned}
&|E\{(K(\xi_1)-\mu)(K(\xi_j)-\mu)-(K(\xi_{1,m(n)})-\mu_n)(K(\xi_{j,m(n)})-\mu_n)\}|\\
&\quad\leqslant |E\{K(\xi_1)\,K(\xi_j)-K(\xi_{1,m(n)})\,K(\xi_{j,m(n)})\}|+|\mu-\mu_n|\,|\mu+\mu_n|\\
&\quad\leqslant E|K(\xi_j)|\,|K(\xi_1)-K(\xi_{1,m(n)})|\\
&\qquad+E|K(\xi_{1,m(n)})|\,|K(\xi_j)-K(\xi_{j,m(n)})|+|\mu-\mu_n|\,|\mu+\mu_n|.
\end{aligned} \tag{6.4}$$

By application of the mean value theorem as in (4.2) we find that both $|\mu-\mu_n|\,|\mu+\mu_n|$ and the expectations on the right in (6.4) are bounded by $n^{-\sigma\wedge\tau}$. It is clear that we may carry out the replacement mentioned above, because $\rho-\sigma\wedge\tau<0$. Q.E.D.

The asymptotic variance in Gastwirth and Rubin [10] is of the same type as our v^2. These authors use an alternative expression for the covariance and we may proceed in a similar way. Let us write F_j for the joint cumulative d.f. of (ξ_1, ξ_j) and note that both marginals are uniform (0, 1). Partial integration yields

$$\mu=\int_0^1 K(t)\,dt=-\int_0^1 t\,dK(t), \tag{6.5}$$

because K is zero in neighborhoods of 0 and 1. A similar partial integration can be obtained for double integrals applying e.g. Abel's summation technique to the approximating Riemann–Stieltjes double sums. We thus find

$$\begin{aligned}
&\operatorname{Cov}(K(\xi_1), K(\xi_j))\\
&\quad=EK(\xi_1)\,K(\xi_j)-\mu^2=\int_0^1\int_0^1 K(s)\,K(t)\,dF_j(s,t)-\mu^2\\
&\quad=\int_0^1\int_0^1 F_j(s,t)\,dK(s)\,dK(t)-\int_0^1\int_0^1 st\,dK(s)\,dK(t)\\
&\quad=\int_0^1\int_0^1 \{F_j(s,t)-st\}\,dK(s)\,dK(t),
\end{aligned} \tag{6.6}$$

which is the expression occurring in Gastwirth and Rubin [10, formula (2.10)].

Acknowledgments

The authors thank the editor and a referee for some comments, in particular, regarding the use of the $m(n)$-decomposability concept and its potential application in the linear model.

References

[1] Aly, E.-E.A.A. (1987). On quantile processes for m-dependent rv's. *Math. Oper. Forsch. Stat., Statistics* **18** 423–435.

[2] Andrews, D. W. K. (1984). Non-strong mixing autoregressive processes. *J. Appl. Probab.* **21** 930–934.

[3] Beirlant, J., van der Meulen, E. C., Ruymgaart, F. H., and van Zuijlen, M. C. A. (1982). On functions bouding the empirical distribution of uniform spacings. *Z. Wahrsch. Verw. Gebiete* **61** 417–430.

[4] Berk, K. N. (1973). A central limit theorem for m-dependent random variables with unbounded m. *Ann. Probab.* **2** 352–354.

[5] Chaganty, N. R. (1989). *Strong Moderate Deviation Theorems for m-Dependent Random Variables.* Tech. Report, Old Dominion University, Norfolk, Virginia.

[6] Chanda, K. C. (1985). Sampling distribution for a class of estimators for nonregular linear processes. *Statist. Probab. Lett.* **3** 261–268.

[7] Chanda, K. C., Puri, M. L., and Ruymgaart, F. H. (1989). *Asymptotic Normality of L-Statistics Based on Decomposable Time Series.* Report 8922, Dept. Mathematics, Katholieke Universiteit Nijmegen.

[8] Chanda, K. C., and Ruymgaart, F. H. (1987). General linear processes: A property of the empirical process applied to density and mode estimation. *J. Time Ser. Anal.* **11** 185–199.

[9] Chernoff, H., Gastwirth, J. L., and Johns, M. V. (1967). Asymptotic distribution of linear combinations of functions of order statistics with applications to estimation. *Ann. Math. Statist.* **38** 52–72.

[10] Gastwirth, J. L., and Rubin, H. (1975). The behavior of robust estimators on dependent data. *Ann. Statist.* **3** 1070–1100.

[11] Hájek, J., and Šidák, Z. (1967). *Theory of Rank Tests.* Academic Press, New York.

[12] Harel, M., and Puri, M. L. (1989). Limiting behavior of U-statistics, V-statistics, and one sample rank order statistics for nonstationary absolutely regular processes. *J. Multivariate Anal.* **30** 181–204.

[13] Harel, M., and Puri, M. L. (1988). Convergence faible de la statistique sérielle linéaire de rang avec des fonctions de scores et des constantes de régression non bornées en condition de mélange. *C. R. Acad. Sci. Paris Sér. I,* **307** 617–620.

[14] Heinrich, L. (1985). Non-uniform estimates, moderate and large deviations in the central limit theorem for m-dependent random variables. *Math. Nachr.* **121** 107–121.

[15] Helmers, R., and Ruymgaart, F. H. (1988). Asymptotic normality of generalized L-statistics with unbounded scores. *J. Statist. Plann. Infer.* **19**, 43–53.

[16] Ibragimov, I. A., and Linnik, Yu. V. (1971). *Independent and Stationary Sequences of Random Variables.* Wolters-Noordhoff, Groningen.

[17] JUREČKOVÁ, J. (1971). Nonparametric estimate of regression coefficients. *Ann. Math. Statist.* **42** 1328–1338.

[18] JUREČKOVÁ, J. (1980). Asymptotic representation of M-estimators of location. *Math. Oper. Forsch. Stat., Statistics* **11** 61–73.

[19] KOUL, H. L. (1977). Behavior of robust estimators in the regression model with dependent errors. *Ann. Statist.* **5** 681–699.

[20] MARTIN, R. D. (1978). Robust estimation of autoregressive models. In *Directions in Time Series* (D. R. Brillinger and G. C. Tiao, Eds.), pp. 228–254, IMS, Hayward.

[21] MARTIN, R. D., AND YOHAI, V. J. (1986). Influence functionals for time series. *Ann. Statist.* **14** 781–818.

[22] MOORE, D. S. (1968). An elementary proof of asymptotic normality of linear functions of order statistics. *Ann. Math. Statist.* **39** 263–265.

[23] NIEUWENHUIS, G., AND RUYMGAART, F. H. (1988). Some stochastic inequalities and asymptotic normality of serial rank statistics in general linear processes. *J. Statist. Plann. Infer.* **25** 53–79.

[24] PHAM, T. D., AND TRAN, L. T. (1985). Some mixing properties of time series models. *Stochastic Process. Appl.* **19** 297–303.

[25] PORTNOY, S. L. (1977). Robust estimation in dependent situations. *Ann. Statist.* **5** 22–43.

[26] PORTNOY, S. L. (1979). Further remarks on robust estimation in dependent situations. *Ann. Statist.* **7** 224–231.

[27] PORTNOY, S., AND KOENKER, R. (1989). Adaptive L-estimation for linear models. *Ann. Statist.* **17** 362–381.

[28] PRIESTLEY, M. B. (1981). *Spectral Analysis and Time Series.* Vol. 1, Academic Press, New York.,

[29] RAO, M. M. (1984). *Probability Theory with Applications.* Academic Press, New York.

[30] RUPPERT, D., AND CARROLL, R. J. (1980). Trimmed least squares estimation in the linear model. *J. Amer. Statist. Assoc.* **75** 828–838.

[31] RUYMGAART, F. H., AND VAN ZUYLEN, M. C. A. (1977). Asymptotic normality of linear combinations of functions of order statistics in the non-i.d.d. case. *Proc. Kon. Ned. Akad. Wetensch.* **80** 432–447.

[32] SHORACK, G. R. (1972). Functions of order statistics. *Ann. Math. Statist.* **43** 412–427.

[33] SHORACK, G. R., AND WELLNER, J. A. (1986). *Empirical Processes with Applications in Statistics.* Wiley, New York.

[34] STIGLER, S. M. (1969). Linear functions of order statistics. *Ann. Math. Statist.* **40** 770–788.

[35] WELSH, A. H. (1987). The trimmed mean in the linear model. *Ann. Statist.* **15** 20–36.

Reprinted by permission of the publisher,
© 1990 by Academic Press

TIME SERIES ANALYSIS VIA RANK ORDER THEORY: SIGNED-RANK TESTS FOR ARMA MODELS

MARC HALLIN* and MADAN L. PURI†

Université Libre de Bruxelles, Brussels, Belgium, Indiana University
Communicated by C. R. Rao

An asymptotic distribution theory is developed for a general class of signed-rank serial statistics, and is then used to derive asymptotically locally optimal tests (in the maximin sense) for testing an ARMA model against other ARMA models. Special cases yield Fisher–Yates, van der Waerden, and Wilcoxon type tests. The asymptotic relative efficiencies of the proposed procedures with respect to each other, and with respect to their normal theory counterparts, are provided. © 1991 Academic Press, Inc.

1. INTRODUCTION

1.1. *Invariance, Ranks and Signed-Ranks*

Invariance arguments constitute the theoretical cornerstone of rank-based methods: whenever weaker unbiasedness or similarity properties can be considered satisfying or adequate, *permutation tests*, which generally follow from the usual *Neyman structure* arguments, are theoretically preferable to rank tests, since they are less restrictive and thus allow for more powerful results (though, of course, their practical implementation may be more problematic).

Which type of ranks (*signed* or *unsigned*) should be adopted—if

Received May 3, 1989; revised October 23, 1990.

AMS 1980 subject classifications: 62G10, 62M10.

Key words and phrases: signed ranks, serial rank statistics, ARMA models, asymptotic relative efficiency, locally asymptotically optimal tests.

* Research supported by the the Fonds National de la Recherche Scientifique and the Ministère de la Communauté Française de Belgique.

† Research supported by the Office of Naval Research Contracts N00014–85–K–0648 and N00014–91–J–1020.

rank-based techniques, hence invariance properties are to be considered—depends on the invariance features of the testing problem and maximal invariants at hand. If, for instance, the observations (or some function thereof) under the null hypothesis to be tested are i.i.d. with unspecified density, it is well known that the vector of (ordinary) ranks constitutes a maximal invariant (for the group of continuous, order-preserving transformations). If the common unspecified density further can be assumed symmetric with respect to the origin or any other specified median (call this the homogeneous symmetry hypothesis), signed ranks are maximal invariant (for the subgroup of continuous, order-preserving, *odd* transformations). If the assumption of identical distributions is dropped (call this the non-homogeneous symmetry hypothesis) then a maximal invariant is the vector of signs, etc. Clearly, ordinary ranks and the vector of signs both remain invariant under the symmetric i.i.d. case, but they no longer are *maximal* invariant, and using a rank test or a sign test instead of a signed-rank one generally results in a loss of relevant information, hence a loss of power. This loss can be dramatic (e.g., when using ranks in testing against a shift alternative) or it may be nil (e.g., when using signs in testing against double exponential shifts). One nevertheless always should conclude in favor of a maximal invariant, whenever a choice is possible.

Now, the effectiveness of the choice between signed and unsigned ranks, in the symmetric i.i.d. case, is somewhat obscured, in practice, by the fact that (unsigned) ranks are totally insensitive to a variety of alternatives. Testing the slope of a regression line, e.g., can be achieved (in a strictly unbiased manner) through either ordinary ranks or signed ones, whereas testing the intercept using ordinary ranks is impossible. A rather perverse consequence of this fact is that the type of alternative at hand, in most cases, apparently dictates which type of ranks (ordinary ranks for the slope, signed ranks for the intercept) should be adopted, and thus which invariance argument should be invoked—see, e.g., Puri and Sen [38, Sections 5.2 and 5.3] for a typical example of such a questionable attitude.

Serial dependence problems, such as that of testing an ARMA model with unspecified innovation density provide an interesting example where the choice between signed and unsigned ranks cannot be eluded and where power considerations do not supersede the much more fundamental invariance principles underlying this choice.

For a review of rank-based techniques in the time series analysis context, we refer to Dufour *et al.* [10], Bhattacharyya [2], and Hallin and Puri [23].

1.2. *Outline of the Paper*

Optimal rank-based tests for serial dependence problems were derived in Hallin *et al.* [17, 18] and Hallin and Puri [22]. Our objective here is to obtain signed-rank analogues of these optimal procedures. Indeed, it is often the case, in practice, that the innovation density of a series under study can be assumed to be symmetrical with respect to some known median—zero in most cases. This is the situation considered in several classical quick tests for randomness [41, 32, 14, 15], as well as in more recent papers (Dufour [6],—where the ranks are those of the products X_tX_{t-1} of successive observations—or Tran [39], where *nonserial* signed-rank statistics are considered under various mixing conditions). As emphasized in Section 1.1, invariance considerations then lead to signed-rank techniques instead of unsigned-rank ones.

Accordingly, we develop here the asymptotic distribution theory required for constructing signed-rank tests based on a general class of *linear serial signed-rank statistics* (Section 2). We therefore establish the asymptotic equivalence of such statistics with adequate parametric ones, the asymptotic distribution of which we then obtain under the null hypothesis of homogeneous symmetry, as well as under local alternatives of second-order linear dependence. A special case of serial signed-rank statistic—the so-called signed-rank autocorrelation coefficients—is introduced in Section 3. In Section 4, we show that asymptotically optimal signed-rank tests for the problem of testing an ARMA model with unspecified, symmetric innovation density can be obtained on substituting these signed autocorrelations for the unsigned ones in the optimal test statistics described in Hallin and Puri [22]. Special cases yield statistics of the *Fisher–Yates–Fraser*, *van der Waerden*, *Wilcoxon*, or *Laplace* types, which are asymptotically equivalent (up to $o_P(n^{-1/2})$ terms) under the hypothesis of homogeneous symmetry as well as under local alternatives, to their unsigned counterparts: the power comparisons and asymptotic relative efficiencies discussed in Hallin and Puri [22] and Hallin and Mélard [20] thus remain valid here.

The fact that their mutual AREs are one, however, does not imply that the advantage of using signed-rank autocorrelations instead of unsigned ones is nil, or negligible, either for short series lengths, or even asymptotically. Numerical investigations of the performances of signed-rank tests [21] indicate that, for fixed values of n, the power of signed-rank procedures can be substantially larger than that of unsigned ones, an empirical finding that should be confirmed by a theoretical investigation of the corresponding deficiencies.

Because of their ease of application, excellent overall performance or broader invariance and distribution-freeness properties, one may also be interested in non-optimal testing procedures, such as the signed version of the Spearman–Wald–Wolfowitz autocorrelation coefficient (see [40 or 1]), or the generalized runs tests considered in Dufour and Hallin [8]. The AREs of such tests can be obtained from Proposition 2.2. Explicit ARE values (with respect to normal-theory procedures) are provided, which for these generalized runs tests, are unexpectedly high. A numerical application is discussed in Section 5.

2. Asymptotic Distribution Theory of Linear Serial Signed-Rank Statistics

2.1. *Linear Serial Signed-Rank Statistics*

Consider a series $\mathbf{Z}^{(n)} = (Z_1^{(n)}, ..., Z_t^{(n)}, ..., Z_n^{(n)})$ of length n; depending on the particular problem at hand, $\mathbf{Z}^{(n)}$ will be the observed series $\mathbf{X}^{(n)} = (X_1^{(n)}, ..., X_t^{(n)}, ..., X_n^{(n)})$ itself, or it will result from $\mathbf{X}^{(n)}$ by means of some adequate transformation. Denote by $R_{+,t}^{(n)}$ the rank of $|Z_t^{(n)}|$ among the absolute values $|Z_1^{(n)}|, ..., |Z_n^{(n)}|$, by $\mathbf{R}_+^{(n)}$ the vector of ranks $(R_{+,1}^{(n)}, ..., R_{+,n}^{(n)})$, by $\text{sgn}(Z_t^{(n)})$ the sign of $Z_t^{(n)}$ and by $\mathbf{sgn}(\mathbf{Z}^{(n)})$ the vector of signs $(\text{sgn}(Z_1^{(n)}), ..., \text{sgn}(Z_n^{(n)}))$. Under the null hypothesis $H_0^{(n)}$ that $Z_1^{(n)}, ..., Z_n^{(n)}$ are independently and identically distributed, according to some absolutely continuous distribution function with unspecified density $f(x)$ satisfying $f(x) = f(-x)$, $\mathbf{R}_+^{(n)}$ constitutes (with probability one) a random permutation of $\{1, ..., n\}$, $\mathbf{sgn}(\mathbf{Z}^{(n)})$ is uniformly distributed on $\{-1, 1\}^n$ (still with probability one), and the two vectors $\mathbf{R}_+^{(n)}$ and $\mathbf{sgn}(\mathbf{Z}^{(n)})$ are mutually independent. $H_0^{(n)}$ is usually known as the hypothesis of (homogeneous) symmetry although in the present context, randomness will be emphasized rather than symmetry. This same hypothesis will be denoted by $H_f^{(n)}$ whenever the density f is specified (up to a scale parameter). Both $\mathbf{R}_+^{(n)}$ and $\mathbf{sgn}(\mathbf{Z}^{(n)})$ are indeed invariant with respect to scale transformations.

Letting $a_+^{(n)}(i_1, ..., i_{k+1})$, where $\{i_1, ..., i_{k+1}\}$ is a $(k+1)$-tuple of elements of $\{\pm 1, \pm 2, ..., \pm n\}$ such that all absolute values $|i_1|, ..., |i_{k+1}|$ are distinct, denote a set of $2^{k+1}n!/(n-k-1)!$ *scores*, define a *linear serial signed-rank statistic* of order k as a statistic of the form

$$S_+^{(n)} = (n-k)^{-1} \sum_{t=k+1}^{n} a_+^{(n)}(\text{sgn}(Z_t^{(n)})\, R_{+,t}^{(n)}, \text{sgn}(Z_{t-1}^{(n)})\, R_{+,t-1}^{(n)}, ..., \text{sgn}(Z_{t-k}^{(n)})\, R_{+,t-k}^{(n)}). \tag{2.1}$$

Classical examples are

(i) the *classical runs test statistic* (with runs defined with respect to zero), of order one, characterized by

$$a_+^{(n)}(i_1, i_2) = \begin{cases} 1 & \text{if } i_1 i_2 < 0 \\ 0 & \text{if } i_1 i_2 \geqslant 0. \end{cases} \tag{2.2}$$

The resulting test is Goodman's [14] *simplified runs test* (with constant $k=0$—see also [15]), which coincides with Dufour's [6] runs test (although the latter is described in terms of the ranks of the products $|Z_t^{(n)} Z_{t-1}^{(n)}|$).

(ii) the *turning point* or *runs up and down test statistic*, of order two, with scores

$$a_+^{(n)}(i_1, i_2, i_3) = \begin{cases} 1 & \text{if} \quad i_1 > i_2 < i_3 \quad \text{or} \quad i_1 < i_2 > i_3 \\ 0 & \text{elsewhere.} \end{cases} \tag{2.3}$$

This test statistic is also measurable with respect to the usual (un-signed) ranks, and was studied by Wallis and Moore [41], Moore and Wallis [32], and Levene [28].

Both the runs test and the turning point test have very desirable properties of quickness, ease of use, and robustness. However, being *quick* tests, they cannot be hoped to be very powerful (see [15] for a study of the power function under Markovian alternatives and Sections 4.2 and 5 below for exact ARE values and numerical examples). Better results can be expected from, e.g.,

(iii) the signed version of the *Spearman–Wald–Wolfowitz rank serial correlation coefficient*, of order $k \geqslant 1$—call it the *signed Spearman autocorrelation of order k*—with scores (up to multiplicative and additive constants)

$$a_+^{(n)}(i_1, i_2, ..., i_{k+1}) = i_1 i_{k+1}. \tag{2.4}$$

This signed version, to the best of our knowledge, has not been considered so far. Its superiority over its unsigned counterpart [40, 1] is attested in Section 5 below.

Locally asymptotically optimal procedures for time-series analysis testing problems, however, typically involve signed versions of the *f-rank autocorrelation coefficients* introduced in Hallin *et al.* [18] and Hallin and Puri [22]. The *signed f-rank autocorrelation of order k* is characterized (up to multiplicative and additive constants—see (3.1) and (3.3) for a more precise definition) by scores $a_+^{(n)}(i_1, i_2, ..., i_{k+1})$ which are proportional to

$$\phi[F^{-1}[(n+1+i_1)/2(n+1)]]\, F^{-1}[(n+1+i_{k+1})/2(n+1)], \tag{2.5}$$

where $f(z)$, $F(z)$, and $\phi(z)$ denote a symmetric probability density function (satisfying a few regularity assumptions, such as being derivable—see Section 2.2), the corresponding distribution function, and the logarithmic derivative $-(d/dz)\log f(z)$, respectively. An alternative, asymptotically equivalent, version is obtained by replacing (2.5) with

$$E\{\phi[F^{-1}[(1+\text{sgn}(i_1)\,U_n^{(|i_1|)})/2]]\}\,E\{F^{-1}[(1+\text{sgn}(i_{k+1})\,U_n^{(|i_{k+1}|)})/2]]\}, \tag{2.6}$$

where $U_n^{(1)} \leqslant U_n^{(2)} \leqslant \cdots \leqslant U_n^{(n)}$ are the ordered random variables of a sample of size n from the rectangular $[0, 1]$ distribution function.

Let also $V_1, \ldots, V_n$ denote an n-tuple of i.i.d. rectangular $[-1, +1]$ variables. Considering a signed-rank statistic $S_+^{(n)}$ of order k, with scores $a_+^{(n)}(\cdots)$, assume the existence of a function J_+, defined on the $(k+1)$-dimensional open square $(-1, 1)^{k+1}$, such that for some strictly positive δ

$$E[|J_+(V_1, V_2, \ldots, V_{k+1})|^{2+\delta}] < \infty. \tag{2.7}$$

Assume, furthermore, that

$$\lim_{n\to\infty} E\{[a_+^{(n)}(\text{sgn}(V_1)\,R_{+,1}^{(n)}, \ldots, \text{sgn}(V_{k+1})\,R_{+,k+1}^{(n)}) - J_+(V_1, \ldots, V_{k+1})]^2\} = 0, \tag{2.8}$$

where $R_{+,i}^{(n)}$ denotes the rank of $|V_i|$ among $|V_1|, \ldots, |V_n|$. Such a function $J_+(\cdots)$ is called a *score-generating function* for $S_+^{(n)}$.

Associated with $J_+(\cdots)$, define the *normalized* score-generating function $J_+^*(\cdots)$ as

$$\begin{aligned}
J_+^*(v_1, v_2, \ldots, v_{k+1})
&= J_+(v_1, \ldots, v_{k+1}) - \sum_{l=1}^{k+1} E[J_+(V_1, \ldots, V_{k+1}) \mid |V_l| = |v_1|] \\
&\quad + kE[J_+(V_1, \ldots, V_{k+1})] \\
&= J_+(v_1, v_2, \ldots, v_{k+1}) \\
&\quad - 2^{-(k+1)} \sum_{s\in\{-1,1\}} \sum_{l=0}^{k} \int_{[-1,1]^k} J_+(w_1, \ldots, w_l, sv_1, w_{l+1}, \ldots, w_k)\, d\mathbf{w} \\
&\quad + 2^{-(k+1)}k \int_{[-1,1]^{k+1}} J_+(w_1, w_2, \ldots, w_{k+1})\, d\mathbf{w},
\end{aligned} \tag{2.9}$$

where $\int_{[-1,1]^k}(\cdots)\,d\mathbf{v}$ stands for the k-fold integral $\int_{-1}^{1}\cdots\int_{-1}^{1}(\cdots)\,\text{sgn}(v_1)\cdots\text{sgn}(v_k)\,dv_1\cdots dv_k$.

Finally, denote by F the distribution function associated with the symmetric density function f, and put $F_+ = 2F - 1$. Clearly, under $H_f^{(n)}$, $F_+(Z_1^{(n)}), \ldots, F_+(Z_n^{(n)})$ are i.i.d. rectangular $[-1, +1]$ random variables. Also note that $\text{sgn}(x)\,F_+(|x|) = F_+(x)$, and $F_+^{-1}(v) =$

$F^{-1}((v+1)/2)$, $v \in (-1, 1)$. Here, as in the sequel, the inverse H^{-1} of a nondecreasing function H is taken as $H^{-1}(v) = \inf\{x \mid H(x) \geqslant v\}$.

2.2. *Asymptotic Distribution of Linear Serial Signed-Rank Statistics*

The asymptotic distribution of a linear serial signed-rank statistic $S_+^{(n)}$ with score-generating function J_+ is provided here under the null hypothesis $H_0^{(n)}$ as well as under contiguous alternatives of serial dependence. Although the results of Hallin *et al.* [17] do not apply (the statistics $\mathscr{S}_+^{(n)}$ and $\mathscr{E}_+^{(n)}$ below are not of the same type as $\mathscr{S}^{(n)}$ and $\mathscr{E}^{(n)}$ in this latter reference), the technical ideas used in the proofs are essentially similar, and, except for Proposition 2.1, the details are left to the reader.

Define

$$\mathscr{S}_+^{(n)} = (n-k)^{-1} \sum_{t=k+1}^{n} J_+(F_+(Z_t^{(n)}), \ldots, F_+(Z_{t-k}^{(n)})) \tag{2.10}$$

and

$$\mathscr{E}_+^{(n)} = [2^{k+1} n(n-1) \cdot \cdots \cdot (n-k)]^{-1} \sum_{\mathbf{s} \in \{-1, 1\}^{k+1}} \quad \sum_{1 \leqslant t_1 \neq \cdots \neq t_{k+1} \leqslant n} \cdots \sum A(\mathbf{s}, F_+)$$

where

$$A(\mathbf{s}, F_+) = J_+(s_1 F_+(Z_{t_1}^{(n)}), \ldots, s_{k+1} F_+(Z_{t_{k+1}}^{(n)})). \tag{2.11}$$

Note that $\mathscr{E}_+^{(n)}$ actually constitutes the mean of $\mathscr{S}_+^{(n)}$, conditional upon the order statistic of the series of absolute values $|Z_1^{(n)}|, \ldots, |Z_n^{(n)}|$. Finally, denote by $m_+^{(n)}$ the mean of $S_+^{(n)}$ under $H_0^{(n)}$:

$$m_+^{(n)} = [2^{k+1} n(n-1) \cdot \cdots \cdot (n-k)]^{-1} \times \sum_{\mathbf{s} \in \{-1, 1\}^{k+1}} \quad \sum_{1 \leqslant i_k \neq \cdots \neq i_{k+1} \leqslant n} \cdots \sum a_+^{(n)}(s_1 i_1, \ldots, s_{k+1} i_{k+1}). \tag{2.12}$$

Proposition 2.1. *Under $H_f^{(n)}$, $(S_+^{(n)} - m_+^{(n)}) - (\mathscr{S}_+^{(n)} - \mathscr{E}_+^{(n)}) = o_p(n^{-1/2})$, as $n \to \infty$.*

Proof. Let $\Delta^{(n)} = (n-k)^{1/2} (S_+^{(n)} - m_+^{(n)} - \mathscr{S}_+^{(n)} + \mathscr{E}_+^{(n)})$, and denote by $|\mathbf{Z}_{(\cdot)}^{(n)}| = (|Z^{(n)}|_{(1)}, \ldots, |Z^{(n)}|_{(n)})$ the order statistic for the series of absolute values $|Z_1^{(n)}|, \ldots, |Z_n^{(n)}|$. Under $H_f^{(n)}$,

$$\begin{aligned} E[(\Delta^{(n)})^2] &= E\{E\{[(n-k)^{1/2} (S_+^{(n)} - \mathscr{S}_+^{(n)}) \\ &\quad - (n-k)^{1/2} (m_+^{(n)} - \mathscr{E}_+^{(n)})]^2 \mid |\mathbf{Z}_{(\cdot)}^{(n)}|\}\} \\ &= (n-k)\, E\{D^2(S_+^{(n)} - \mathscr{S}_+^{(n)} \mid |\mathbf{Z}_{(\cdot)}^{(n)}|)\}, \end{aligned}$$

where $D^2(S_+^{(n)} - \mathscr{S}_+^{(n)} \mid |\mathbf{Z}_{(\cdot)}^{(n)}|)$ stands for the variance of $S_+^{(n)} - \mathscr{S}_+^{(n)}$ conditional upon $|\mathbf{Z}_{(\cdot)}^{(n)}|$. Denote by $(t(1), \ldots, t(n))$ the vector of *antiranks*—namely, the vector whose ith component $t(i)$ is such that $R^{(n)}_{+,t(i)}=i$, and by $\mathbf{s}^{(n)}_{(\cdot)}$ the vector with ith component $s^{(n)}_{(i)}=\operatorname{sgn}(Z^{(n)}_{t(i)})$. Then

$$\begin{aligned}D^2(S^{(n)}_+ - \mathscr{S}^{(n)}_+ \mid |\mathbf{Z}^{(n)}_{(\cdot)}|) &= E\{D^2[S^{(n)}_+ - \mathscr{S}^{(n)}_+ \mid \mathbf{s}^{(n)}_{(\cdot)}, |\mathbf{Z}^{(n)}_{(\cdot)}|] \mid |\mathbf{Z}^{(n)}_{(\cdot)}|\}\\ &\quad + D^2[E[S^{(n)}_+ - \mathscr{S}^{(n)}_+ \mid \mathbf{s}^{(n)}_{(\cdot)}, |\mathbf{Z}^{(n)}_{(\cdot)}|] \mid |\mathbf{Z}^{(n)}_{(\cdot)}|]\\ &= D^2_{(1)} + D^2_{(2)}, \qquad \text{say.}\end{aligned}$$

Conditional upon $|\mathbf{Z}^{(n)}_{(\cdot)}|$ and $\mathbf{s}^{(n)}_{(\cdot)}$, $S^{(n)}_+ - \mathscr{S}^{(n)}_+$ is a linear serial rank statistic of order k (with respect to the ranks $R^{(n)}_{+,t}$), with scores

$$\begin{aligned}A^{(n)}_{\mathbf{s}(\cdot),|\mathbf{Z}|}(i_1, i_2, \ldots, i_{k+1}) &= a^{(n)}_+(s_{(i_1)}i_1, \ldots, s_{(i_{k+1})}i_{k+1})\\ &\quad - J_+(s_{(i_1)}(2F(|Z^{(n)}|_{(i_1)})-1), \ldots, s_{(i_{k+1})}(2F(|Z^{(n)}|_{(i_{k+1})})-1)).\end{aligned}$$

Using Lemmas 2 and 4 of Hallin *et al.* [17], $D^2_{(1)}$ is bounded by

$$\begin{aligned}&(n-k)^{-1}\, KE\{E[[A^{(n)}_{\mathbf{s}(\cdot),|\mathbf{Z}|}(R^{(n)}_{+,1}, \ldots, R^{(n)}_{+,k+1})]^2 \mid \mathbf{s}^{(n)}_{(\cdot)}, |\mathbf{Z}^{(n)}_{(\cdot)}|] \mid |\mathbf{Z}^{(n)}_{(\cdot)}|\}\\ &\quad = (n-k)^{-1}\, KE\{[A^{(n)}_{\mathbf{s}(\cdot),|\mathbf{Z}|}(R^{(n)}_{+,1}, \ldots, R^{(n)}_{+,k+1})]^2 \mid |\mathbf{Z}^{(n)}_{(\cdot)}|\}\end{aligned}$$

(still under $H^{(n)}_f$), where K denotes a constant. Turning to $D^2_{(2)}$,

$$\begin{aligned}&E[S^{(n)}_+ - \mathscr{S}^{(n)}_+ \mid \mathbf{s}^{(n)}_{(\cdot)}, |\mathbf{Z}^{(n)}_{(\cdot)}|]\\ &\quad = [n(n-1)\cdot\cdots\cdot(n-k)]^{-1} \sum_{1\leqslant t_1\neq\cdots\neq t_{k+1}\leqslant n}\cdots\sum A^{(n)}_{\mathbf{s}(\cdot),|\mathbf{Z}|}(t_1, \ldots, t_{k+1}).\end{aligned}$$

Hence,

$$\begin{aligned}D^2_{(2)} &= [n(n-1)\cdot\cdots\cdot(n-k)]^{-2}\\ &\quad \times \sum_{1\leqslant t_1\neq\cdots\neq t_{k+1}\leqslant n}\cdots\sum \Bigg[D^2(A^{(n)}_{\mathbf{s}(\cdot),|\mathbf{Z}|}(t_1, \ldots, t_{k+1}) \mid |\mathbf{Z}^{(n)}_{(\cdot)}|)\\ &\quad + \sum_{1\leqslant t'_1\neq\cdots\neq t'_{k+1}\leqslant n}\cdots\sum \operatorname{Cov}(A^{(n)}_{\mathbf{s}(\cdot),|\mathbf{Z}|}(t_1,\ldots,t_{k+1}), A^{(n)}_{\mathbf{s}(\cdot),|\mathbf{Z}|}(t'_1,\ldots,t'_{k+1}) \mid |\mathbf{Z}^{(n)}_{(\cdot)}|)\Bigg],\end{aligned}$$

where the summation over $t'_1 \cdots t'_{k+1}$ runs over all *ordered* $(k+1)$-tuples of *distinct* integers $(t'_1, \ldots, t'_{k+1})$ such that $\{t'_1, \ldots, t'_{k+1}\} \neq \{t_1, \ldots, t_{k+1}\}$. Using the fact that for each ordered $(k+1)$-tuple $(t_1, \ldots, t_{k+1})$, there exist $(n-k-1)(n-k-2)\cdots(n-2k-1)$ $(k+1)$-tuples $(t'_1, \ldots, t'_{k+1})$ such that $\{t_1, \ldots, t_{k+1}\} \cap \{t'_1, \ldots, t'_{k+1}\} = \varnothing$, yielding a covariance value of zero in the above sum, and bounding $\mathrm{Cov}(\xi, \eta)$ with the arithmetic mean $\frac{1}{2}(D^2(\xi)+D^2(\eta))$ of the corresponding variances, we obtain

$$D^2_{(2)} \leqslant \frac{n(n-1)\cdot \cdots \cdot (n-k) - (n-k-1)\cdot \cdots \cdot (n-2k-1)}{[n(n-1)\cdot \cdots \cdot (n-k)]^2}$$

$$\times \sum_{1\leqslant t_1 \neq \cdots \neq t_{k+1} \leqslant n} \cdots \sum D^2(A^{(n)}_{\mathbf{s}(\cdot), |\mathbf{Z}|}(t_1, ..., t_{k+1}) \,||\, \mathbf{Z}^{(n)}_{(\cdot)}|)$$

$$= \frac{n(n-1)\cdot \cdots \cdot (n-k) - (n-k-1)\cdot \cdots \cdot (n-2k-1)}{n(n-1)\cdot \cdots \cdot (n-k)}$$

$$\times E[D^2(A^{(n)}_{\mathbf{s}(\cdot), |\mathbf{Z}|}(R^{(n)}_{+,1}, ..., R^{(n)}_{+,k+1}) \,|\, |\mathbf{Z}^{(n)}_{(\cdot)}|)].$$

Finally,

$$E[(\Delta^{(n)})^2] = (n-k)\, E[D^2_{(1)} + D^2_{(2)}]$$

$$\leqslant \left[K + (n-k) \frac{n(n-1)\cdot \cdots \cdot (n-k) - (n-k-1)\cdot \cdots \cdot (n-2k-1)}{n(n-1)\cdot \cdots \cdot (n-k)} \right]$$

$$\times E\{[A^{(n)}_{\mathbf{s}(\cdot), |\mathbf{Z}|}(R^{(n)}_{+,1}, ..., R^{(n)}_{+,k+1})]^2\}.$$

The coefficient in this latter expression is $O(1)$, whereas, because of (2.8), the expectation converges to zero, as $n \to \infty$, under $H^{(n)}_f$. This completes the proof.

Note that $\mathscr{E}^{(n)}_+$ in general does not reduce to the same form as the statistic $\mathscr{E}^{(n)}$ appearing in Hallin *et al.* [17, formula (4.4)]. Accordingly, the asymptotic equivalence between signed-rank statistics and an unsigned one we are establishing in Section 3 below for the particular case of autocorrelation coefficients does not hold for general signed-rank statistics.

The alternatives $K^{(n)}(\mathbf{a}, \mathbf{b}; g)$ we are considering in part (ii) of Proposition 2.2 below are general linear serial dependence alternatives, under which the observed series $\mathbf{Z}^{(n)}$ is a finite realization of some stochastic process satisfying

$$X_t - n^{-1/2} \sum_{i=1}^{\infty} a_i Z_{t-i} = \varepsilon_t + n^{-1/2} \sum_{i=1}^{\infty} b_i \varepsilon_{t-i}, \qquad t \in \mathbb{Z},$$

where $\mathbf{a} = (a_1, a_2, ...)$ and $\mathbf{b} = (b_1, b_2, ...)$ are such that $\sum_{i=1}^{\infty} |a_i| < \infty$ and $\sum_{i=1}^{\infty} |b_i| < \infty$ (see, e.g., [37, Section 3.5.7]), and where $\{\varepsilon_t\}$ is an independent white noise, with density g satisfying the following assumptions:

(A1) $g(x) = g(-x)$, and $\int x^6 g(x)\, dx < \infty$.

(A2) $g(x)$ is *absolutely continuous on finite intervals* (see [16]). Hence there exists a function $\dot{g}$ such that $g(b) - g(a) = \int_a^b \dot{g}(x)\, dx$, $-\infty < a < b < \infty$, and $\dot{g}(x) = dg(x)/dx$ a.e. Defining $\varphi_g = -\dot{g}/g$, assume that $\int |\varphi_g(x)|^{2+\delta} g(x)\,dx < \infty$ for some $\delta > 0$; this implies finite Fisher information $I(g) = \int \varphi_g^2(x)\, g(x)\, dx$.

(A3) A finite derivative $\dot{\varphi}_g(x)) = d(\varphi_g(x))/dx$ exists at all but a finite number of points, and satisfies the Lipschitz condition $|\dot{\varphi}_g(x) - \dot{\varphi}(y)| < K_g |x-y|$ for all x, y such that $\dot{\varphi}_g(z)$ exists $\forall z \in (x, y)$.

PROPOSITION 2.2. (i) *Assume that* (2.7) *and* (2.8) *hold. Then* $(n-k)^{1/2}(S_+^{(n)} - m_+^{(n)})$ *is asymptotically normal under* $H_0^{(n)}$, *with mean zero and variance*

$$V_+^2 = E[(J_+^*(V_1, ..., V_{k+1}))^2] + 2\sum_{i=1}^{k} E[J_+^*(V_1, ..., V_{k+1})\, J_+^*(V_{1+i}, ..., V_{k+1+i})]. \tag{2.13}$$

(ii) *Provided, moreover, that the density g satisfies the technical assumptions* (A1)–(A3), $(n-k)^{1/2}(S_+^{(n)} - m_+^{(n)})$ *is asymptotically normal under* $K^{(n)}(\mathbf{a}, \mathbf{b}; g)$, *still with variance* V_+^2, *but with mean* $\sum_{i=1}^{k}(a_i + b_i)\, C_i^+$, *where*

$$C_i^+ = E[J_+(V_1, ..., V_{k+1}) \sum_{j=0}^{k-i} \varphi_g(G_+^{-1}(V_{1+j}))\, G_+^{-1}(V_{1+j+i})] \tag{2.14}$$

(*notation G and* G_+ *are used here in an obvious fashion*); *whether* J_+ *or* J_+^* *is used in* (2.14) *does not affect the value of* C_i^+.

Proof. Let $\mathbf{Y}_t^{(n)} = (Z_t^{(n)}, Z_{t-1}^{(n)}, ..., Z_{t-k}^{(n)})'$, $k+1 \leqslant t \leqslant n$, and consider the kernels (of order k)

$$\begin{aligned}\Psi_{\mathscr{S}}(\mathbf{y}_1, ..., \mathbf{y}_{k+1}) &= (k+1)^{-1} \sum_{j=1}^{k+1} J_+(\mathrm{sgn}(y_{j,1})(2F(|y_{j,1}|) - 1),\\ &\quad ..., \mathrm{sgn}(y_{j,k+1})(2F(|y_{j,k+1}|) - 1))\end{aligned}$$

and

$$\begin{aligned}\Psi_{\mathscr{E}}(\mathbf{y}_1, ..., \mathbf{y}_{k+1}) = [2^k(k+1)!]^{-1} \sum_{\mathscr{P}} \sum_{\mathbf{s}} J_+(s_1(2F(|y_{j,1}|) - 1),\\ ..., s_{k+1}(2F(|y_{j,k+1}|) - 1)),\end{aligned}$$

where $\mathbf{y}_j = (y_{j,1}, ..., y_{j,k+1})' \in \mathbb{R}^{k+1}$, and the summations $\sum_{\mathscr{P}}$ and $\sum_{\mathbf{s}}$ run over all $(k+1)!$ permutations $(j_1, ..., j_{k+1})$ of $(1, ..., k+1)$ and all elements $\mathbf{s} = (s_1, ..., s_{k+1})$ of $\{-1, 1\}^{k+1}$, respectively. Then, denoting by $\mathscr{U}_{\mathscr{S}}^{(n)}$ and $\mathscr{U}_{\mathscr{E}}^{(n)}$ the U-statistics induced from the $(k+1)$-dimensional series

$\mathbf{Y}_{k+1}^{(n)}, ..., \mathbf{Y}_n^{(n)}$, of length $n-k$, by the kernels $\Psi_{\mathscr{S}}$ and $\Psi_{\mathscr{E}}$, respectively, we have under $H_f^{(n)}$

$$(\mathscr{S}_+^{(n)} - \mathscr{E}_+^{(n)}) - (\mathscr{U}_{\mathscr{S}}^{(n)} - \mathscr{U}_{\mathscr{E}}^{(n)})] = o_p((n-k)^{-1/2}). \tag{2.15}$$

The proof of (2.15) follows roughly along the same lines as in Hallin *et al.* [17, Section 1.2], and is left to the reader. Being $(k+1)$-dependent, the process $\{\mathbf{Y}_t^{(n)}\}$ is, still under $H_f^{(n)}$, absolutely regular and, in view of (2.7), the conditions of Yoshihara's [42] central limit theorem for U-statistics hold for $\mathscr{U}_{\mathscr{S}}^{(n)} - \mathscr{U}_{\mathscr{E}}^{(n)}$. The g_1 function in Yoshihara's notation is here

$$\begin{aligned}
g_1(\mathbf{Y}_t^{(n)}) &= E\{\Psi_{\mathscr{S}}(\mathbf{Y}_t^{(n)}, \mathbf{Y}_{k+1}^{(n)}, ..., \mathbf{Y}_{2k}^{(n)}) - \Psi_{\mathscr{E}}(\mathbf{Y}_t^{(n)}, \mathbf{Y}_{k+1}^{(n)}, ..., \mathbf{Y}_{2k}^{(n)}) \mid Y_t^{(n)}\} \\
&= (k-1)^{-1}\Bigg[J_+(\operatorname{sgn}(Z_t^{(n)})(2F(|Z_t^{(n)}|)-1), \\
&\quad ..., \operatorname{sgn}(Z_{t-k}^{(n)})(2F(|Z_{t-k}^{(n)}|)-1)) \\
&\quad + 2^{-(k+1)}k \sum_{\mathbf{s}} \int_0^1 \cdots \int_0^1 J_+(s_1 u_1, ..., s_{k+1}u_{k+1})\, du_1 \cdots du_{k+1} \\
&\quad - 2^{-(k+1)} \sum_{\mathbf{s}} \sum_{l=0}^{k} \int_0^1 \cdots \int_0^1 J_+(s_1 u_1, ..., s_l u_l, s_{l+1}(2F(|Z_t^{(n)}|)-1), \\
&\quad s_{l+2}u_{l+1}, ..., s_{k+1}u_k)\, du_1 \cdots du_k \Bigg] \\
&= (k+1)^{-1} J_+^*(\operatorname{sgn}(Z_t^{(n)})(2F(|Z_t^{(n)}|)-1), \\
&\quad ..., \operatorname{sgn}(Z_{t-k}^{(n)})(2F(|Z_{t-k}^{(n)}|)-1)) \\
&= (k+1)^{-1} J_+^*(F_+(Z_t^{(n)}), ..., F_+(Z_{t-k}^{(n)})).
\end{aligned}$$

The asymptotic normality under $H_f^{(n)}$, with mean zero and variance (2.13), of $(n-k)^{1/2}(\mathscr{U}_{\mathscr{S}}^{(n)} - \mathscr{U}_{\mathscr{E}}^{(n)})$—hence, on account of Proposition 2.1, of $(n-k)^{1/2}(S_+^{(n)} - m_+^{(n)})$ follows. Since the limit distribution does not depend on the density f, on which no assumption whatever has been made (except for symmetry), the result holds under $H_0^{(n)}$, which completes the proof of part (i) of the proposition. As for part (ii), it essentially follows along the same lines as in Hallin and Puri [22, Section 2.1], by establishing the asymptotic joint normality of $(n-k)^{1/2}(S_+^{(n)} - m_+^{(n)})$ and the log-likelihood function, then applying LeCam's third lemma (in Hájek and Šidák's [16] terminology). The covariance terms in this joint asymptotic distribution yield, under $K^{(n)}(\mathbf{a}, \mathbf{b}; g)$ (note that assumptions (A1)–(A3) on g are necessary to ensure contiguity), the desired form for the mean, although with constants

$$\begin{aligned}
C_i^+ = \sum_{j=0}^{k-i} \int_{[0,1]^{k+1}} & J_+^*(2u_1 - 1, ..., 2u_{k+1} - 1) \\
& \times \varphi_g(G^{-1}(u_{j+1}))\, G^{-1}(u_{j+i+1})\, du_1 \cdots du_{k+1}.
\end{aligned}$$

However, it follows from (2.9) and the fact that $\int_0^1 \varphi_g(G^{-1}(u))\, du = 0 = \int_0^1 G^{-1}(u)\, du$ that

$$\int_0^1 [J_+(2u_1 - 1, ..., 2u_{k+1} - 1) - J_+^*(2u_1 - 1, ..., 2u_{k+1} - 1)]$$
$$\times \varphi_g(G^{-1}(u_l))\, G^{-1}(u_{l'})\, du_1 \cdots du_{k+1} = 0,$$

for all $1 \leqslant l \neq l' \leqslant k+1$. The details are left to the reader.

3. Signed *f*-Rank Autocorrelations

3.1. *Definitions*

A particular type of serial rank statistic has been shown [22] to play an essential role in rank-based inference for linear time-series models: the so-called *f-rank autocorrelation coefficients*, whose role is comparable to that of ordinary parametric autocorrelations in classical, Gaussian time-series analysis. We introduce here two asymptotically equivalent definitions of a concept of *signed f*-rank autocorrelation coefficients. Denote by g a symmetric density function satisfying assumptions (A1)–(A3), by G and $\varphi_g = -\dot{g}/g$ the corresponding distribution and score functions. Let $G_+ = 2G - 1$. Assume furthermore that

(A4) g is *strongly unimodal*; i.e., φ_g is a nondecreasing function.

The most convenient definition of a *signed-rank autocorrelation of order k associated with density g* is

$$r_{k;g}^{(n)+} = [(n-k)\,\sigma_{g+}^{(n)}]^{-1} \sum_{t=k+1}^{n} \operatorname{sgn}(Z_t^{(n)} Z_{t-k}^{(n)})$$
$$\times \varphi_g\left(G_+^{-1}\left(\frac{R_{+,t}^{(n)}}{n+1}\right)\right) G_+^{-1}\left(\frac{R_{+,t-k}^{(n)}}{n+1}\right), \tag{3.1}$$

where

$$(\sigma_{g+}^{(n)})^2 = [n(n-1)]^{-1} \sum\sum_{1 \leqslant i_1 \neq i_2 \leqslant n} \left[\varphi_g\left(G_+^{-1}\left(\frac{i_1}{n+1}\right)\right) G_+^{-1}\left(\frac{i_2}{n+1}\right)\right]^2$$
$$= [n(n-1)]^{-1} \left\{ \sum_{i=1}^{n} \left[\varphi_g\left(G_+^{-1}\left(\frac{i}{n+1}\right)\right)\right]^2 \sum_{i=1}^{n} \left[G_+^{-1}\left(\frac{i}{n+1}\right)\right]^2 \right.$$
$$\left. - \sum_{i=1}^{n} \left[\varphi_g\left(G_+^{-1}\left(\frac{i}{n+1}\right)\right) G_+^{-1}\left(\frac{i}{n+1}\right)\right]^2 \right\} \tag{3.2}$$

denotes the exact variance (under $H_o^{(n)}$) of $(n-k)^{-1/2}$ times the sum $\sum_{t=k+1}^{n}\cdots$ in (3.1): $(n-k)^{1/2}\, r_{k;g}^{(n)+}$ is therefore exactly standardized under $H_0^{(n)}$. In order to see that (3.1) constitutes a serial signed-rank statistic of the form (2.1), note that, because of the symmetry of g,

$$\operatorname{sgn}(Z_t^{(n)} Z_{t-k}^{(n)})\, \varphi_g\left(G_+^{-1}\left(\frac{R_{+,t}^{(n)}}{n+1}\right)\right) G_+^{-1}\left(\frac{R_{+,t-k}^{(n)}}{n+1}\right)$$
$$= \varphi_g(G_+^{-1}(\operatorname{sgn}(Z_t^{(n)})\, R_{+,t}^{(n)}/(n+1)))\, G_+^{-1}(\operatorname{sgn}(Z_{t-k}^{(n)})\, R_{+,t-k}^{(n)}/(n+1)); \tag{3.3}$$

(3.1) is thus characterized by the scores

$$a_+^{(n)}(i_1, \ldots, i_{k+1}) = (\sigma_{g+}^{(n)})^{-1}\, \varphi_g(G_+^{-1}(i_1/(n+1)))\, G_+^{-1}(i_{k+1}/(n+1)). \tag{3.4}$$

The form of $r_{k,g}^{(n)+}$ is very similar to that of its unsigned counterpart

$$r_{k;g}^{(n)} = [(n-k)\,\sigma_{k;g}^{(n)}]^{-1}\left\{\sum_{t=k+1}^{n} \varphi_g\left(G^{-1}\left(\frac{R_t^{(n)}}{n+1}\right)\right) G^{-1}\left(\frac{R_{t-k}^{(n)}}{n+1}\right) - m_g^{(n)}\right\},$$

where $R_t^{(n)}$ denotes the (unsigned) rank of $Z_t^{(n)}$ among $Z_1^{(n)}, \ldots, Z_n^{(n)}$,

$$m_g^{(n)} = [n(n-1)]^{-1} \sum\sum_{1 \leqslant i_1 \neq i_2 \leqslant n} \varphi_g\left(G^{-1}\left(\frac{i_1}{n+1}\right)\right) G^{-1}\left(\frac{i_2}{n+1}\right),$$

and

$$\begin{aligned}
(\sigma_{k;g}^{(n)})^2 = {} & [n(n-1)]^{-1} \sum\sum_{1 \leqslant i_1 \neq i_2 \leqslant n} \left[\varphi_g\left(G^{-1}\left(\frac{i_1}{n+1}\right)\right) G^{-1}\left(\frac{i_2}{n+1}\right)\right]^2 \\
& + 2(n-2k)[n(n-1)(n-2)(n-k)]^{-1} \\
& \times \sum\sum\sum_{1 \leqslant i_1 \neq i_2 \neq i_3 \leqslant n} \left[\varphi_g\left(G^{-1}\left(\frac{i_1}{n+1}\right)\right) \varphi_g\left(G^{-1}\left(\frac{i_2}{n+1}\right)\right)\right. \\
& \left.\times G^{-1}\left(\frac{i_2}{n+1}\right) G^{-1}\left(\frac{i_3}{n+1}\right)\right] \\
& + (n^2 - n(2k+3) + k^2 + 5k)[n(n-1)(n-2)(n-3)(n-k)]^{-1} \\
& \times \sum\sum\sum\sum_{1 \leqslant i_1 \neq i_2 \neq i_3 \neq i_4 \leqslant n} \left[\varphi_g\left(G^{-1}\left(\frac{i_1}{n+1}\right)\right) \varphi_g\left(G^{-1}\left(\frac{i_2}{n+1}\right)\right)\right. \\
& \left.\times G^{-1}\left(\frac{i_3}{n+1}\right) G^{-1}\left(\frac{i_4}{n+1}\right)\right] - (n-k)(m_g^{(n)})^2;
\end{aligned}$$

$(\sigma_{k;g}^{(n)})^2$ again denotes the variance (under $H_o^{(n)}$) of $(n-k)^{-1/2}$ times the sum

$\sum_{t=k+1} \cdots$ (see [22, (2.5)]). Note that $\sigma_{g+}^{(n)}$ is considerably simpler to compute than its unsigned counterpart $\sigma_{k;g}^{(n)}$, due to the fact that all the covariance terms in (3.2) cancel out; moreover, $\sigma_{g+}^{(n)}$ does not depend on k.

Particularizing the density type, we obtain the *van der Waerden signed rank autocorrelations* (associated with normal densities)

$$r_{k;\,\mathrm{vdW}}^{(n)+} = [\sigma_{\mathrm{vdW}+}^{(n)}(n-k)]^{-1} \sum_{t=k+1}^{n} \mathrm{sgn}(Z_t^{(n)} Z_{t-k}^{(n)}) \times \Phi^{-1}\left(\frac{1}{2} + \frac{R_{+,t}^{(n)}}{2(n+1)}\right) \Phi^{-1}\left(\frac{1}{2} + \frac{R_{+,t-k}^{(n)}}{2(n+1)}\right), \tag{3.5}$$

where $\Phi(x) = (2\pi)^{-1/2} \int_{-\infty}^{x} \exp(-u^2/2)\, du$, the *Wilcoxon signed autocorrelations* (associated with logistic densities)

$$r_{k;\mathrm{W}}^{(n)+} = [\sigma_{\mathrm{W}+}^{(n)}(n-k)(n+1)]^{-1} \times \sum_{t=k+1}^{n} \mathrm{sgn}(Z_t^{(n)} Z_{t-k}^{(n)})\, R_{+,t}^{(n)} \log\left(\frac{n+1+R_{+,t-k}^{(n)}}{n+1-R_{+,t-k}^{(n)}}\right), \tag{3.6}$$

and, the *Laplace signed rank autocorrelations* (associated with double exponential densities)

$$r_{k;\mathrm{L}}^{(n)+} = -[\sigma_{\mathrm{L}+}^{(n)}(n-k)]^{-1} \sum_{t=k+1}^{n} \mathrm{sgn}(Z_t^{(n)} Z_{t-k}^{(n)}) \log\left(1 - \frac{R_{+,t-k}^{(n)}}{n+1}\right). \tag{3.7}$$

An inspection of formulas (3.5) through (3.7) reveals that the van der Waerden autocorrelation (3.5) is the usual autocorrelation coefficient computed from the series of the "signed standard normal quantiles" of the original series $\mathbf{Z}^{(n)}$: $\mathrm{sgn}(Z_t^{(n)})\, \Phi^{-1}(\frac{1}{2} + R_{+,t}^{(n)}/2(n+1))$ is indeed equal to $\mathrm{sgn}(Z_t^{(n)})\, \zeta_{(n+1-R_{+,t}^{(n)})/2(n+1)}$, where ζ_λ denotes the $(1-\lambda)$-quantile of the standard normal distribution. The Wilcoxon autocorrelation (3.6) is a weighted version of the usual *Wilcoxon signed rank statistic*, and the Laplace autocorrelation (3.7) is a weighted version of order k of the traditional *runs test statistic*.

An alternative definition for the signed-rank autocorrelation of order k associated with g is

$$\tilde{r}_{k;g}^{(n)+} = [(n-k)\,\tilde{\sigma}_{g+}^{(n)}]^{-1} \sum_{t=k+1}^{n} \mathrm{sgn}(Z_t^{(n)} Z_{t-k}^{(n)}) \times E\{\varphi_g(G_+^{-1}(U_n^{(R_{+,t}^{(n)})}))\}\, E\{G_+^{-1}(U_n^{(R_{+,t-k}^{(n)})})\}, \tag{3.8}$$

where $U_n^{(1)} \leqslant U_n^{(2)} \leqslant \cdots \leqslant U_n^{(n)}$ denote an ordered sample of i.i.d. rectangular [0, 1] random variables, and

$$(\tilde{\sigma}_{g+}^{(n)})^2 = [n(n-1)]^{-1} \left\{ \sum_{i=1}^{n} \{E[\varphi_g(G_+^{-1}(U_n^{(i)}))]\}^2 \sum_{i=1}^{n} \{E[G_+^{-1}(U_n^{(i)})]\}^2 - \sum_{i=1}^{n} \{E[\varphi_g(G_+^{-1}(U_n^{(i)}))]\, E[G_+^{-1}(U_n^{(i)})]\}^2 \right\}.$$

The corresponding scores are

$$\tilde{a}_+^{(n)}(i_1, ..., i_{k+1}) = (\tilde{\sigma}_{g+}^{(n)})^{-1} E[\varphi_g(G_+^{-1}(\mathrm{sgn}(i_1)\, U_n^{(|i_1|)})] \times E[G_+^{-1}(\mathrm{sgn}(i_{k+1})\, U^{(|i_{k+1}|)})]. \tag{3.9}$$

Particularizing (3.8) to the Gaussian case yields a *Fraser–Fisher–Yates autocorrelation coefficient*,

$$r_{k,\,\mathrm{FY}}^{(n)+} = [\sigma_{\mathrm{FY}}^{(n)}(n-k)]^{-1} \sum_{t=k+1}^{n} \mathrm{sgn}(Z_t^{(n)} Z_{t-k}^{(n)}) \times E[(\chi_n^{(R_{+,t}^{(n)})})^{1/2}]\, E[(\chi_n^{(R_{+,t-k}^{(n)})})^{1/2}],$$

where $\chi_n^{(1)} \leqslant \chi_n^{(2)} \leqslant \cdots \leqslant \chi_n^{(n)}$ denote an ordered sample from the chi-square distribution with one degree of freedom.

3.2. *Asymptotic Distribution of Signed-Rank Autocorrelations*

In order to justify the above definitions of signed-rank autocorrelations (as well as the claimed asymptotic equivalence of $r_{k;f}^{(n)+}$ and $\tilde{r}_{k;f}^{(n)+}$), it is sufficient to establish the asymptotic equivalence of $r_{k;f}^{(n)+}$ and $\tilde{r}_{k;f}^{(n)+}$ with their unsigned counterparts $r_{k;f}^{(n)}$.

PROPOSITION 3.1. *Assume that assumptions* (A1)–(A4) *hold. Then both* $(n-k)^{1/2}(r_{k;g}^{(n)+} - r_{k;g}^{(n)})$ *and* $(n-k)^{1/2}(\tilde{r}_{k;g}^{(n)+} - r_{k;g}^{(n)})$ *are* $o_p(1)$, *as* $n \to \infty$, *under* $H_0^{(n)}$, *and thus under any sequence of alternatives contiguous to* $H_0^{(n)}$.

Before proceeding to the proof of Proposition 3.1, we first state the following lemma, the proof of which easily follows by using a multivariate version of Hájek and Šidák's [16] proof of Lemma V.1.6a.

LEMMA 3.1. *If the function* $J_+(v_1, ..., v_{k+1})$ *is a.e.* (*on* $(-1, 1)^{k+1}$) *continuous, non-decreasing with respect to all its arguments, and if either*

$$a^{(n)}(i_1, ..., i_{k+1}) = J_+\left(\frac{i_1}{n+1}, ..., \frac{i_{k+1}}{n+1}\right)$$

or

$$a^{(n)}(i_1, ..., i_{k+1}) = E\{J_+(\mathrm{sgn}(i_1)\, U_n^{(|i_1|)}, ..., \mathrm{sgn}(i_{k+1})\, U_n^{(|i_{k+1}|)})\},$$

where $U_n^{(1)}, \ldots, U_n^{(n)}$ *denote an ordered sample of i.i.d. rectangular* $[0, 1]$ *variables, then condition* (2.8) *is satisfied.*

Proof of Proposition 3.1. Clearly,

$$\begin{aligned}\operatorname{sgn}(zz')\,\varphi_g\left(G_+^{-1}\left(\frac{i}{n+1}\right)\right)G_+^{-1}\left(\frac{i'}{n+1}\right)\\ = \varphi_g\left(G_+^{-1}\left(\operatorname{sgn}(z)\frac{i}{n+1}\right)\right)G_+^{-1}\left(\operatorname{sgn}(z')\frac{i'}{n+1}\right).\end{aligned}$$

Also the strong unimodality property of g implies that φ_g is a monotonically increasing function. In view of Lemma 3.1, $\sigma_{g+}^{(n)}r_{k;g}^{(n)+}$ and $\tilde{\sigma}_{g+}^{(n)}\tilde{r}_{k;g}^{(n)+}$ thus both admit

$$\varphi_g(G_+^{-1}(v_1))\,G_+^{-1}(v_{k+1}) = \varphi_g\left(G^{-1}\left(\frac{1+v_1}{2}\right)\right)G^{-1}\left(\frac{1+v_{k+1}}{2}\right)$$

as their score-generating function. It then follows from (2.8) that

$$\begin{aligned}\lim_{n\to\infty}(\sigma_{g+}^{(n)})^2 &= \lim_{n\to\infty}(\tilde{\sigma}_{g+}^{(n)})^2\\ &= \int_0^1\int_0^1[\varphi_g(G^{-1}(u_1))\,G^{-1}(u_2)]^2\,du_1\,du_2 = \sigma_g^2 I(g),\end{aligned}$$

where σ_g^2 stands for $\int x^2 g\,dx$. Accordingly, $r_{k;g}^{(n)+}$ and $\tilde{r}_{k;g}^{(n)+}$ admit

$$J_+(v_1, \ldots, v_{k+1}) = [\sigma_g^2 I(g)]^{-1/2}\,\varphi_g\left(G^{-1}\left(\frac{1+v_1}{2}\right)\right)G^{-1}\left(\frac{1+v_{k+1}}{2}\right)$$

as their score-generating function. Definitions (2.10) and (2.11) then yield (under $H_f^{(n)}$)

$$\mathcal{S}_+^{(n)} = [(n-k)\,\sigma_g^2 I(f)]^{-1}\sum_{t=k+1}^{n}\varphi_g(G^{-1}(F(Z_t^{(n)})))\,G^{-1}(F(Z_{t-k}^{(n)})), \qquad (3.10)$$

and $\mathcal{E}_+^{(n)} = 0$. Now (3.10) exactly coincides with the $\mathcal{S}^{(n)}$ function associated with $r_{k;g}^{(n)}$ (cf. [18, Proposition 3.1]), whereas the corresponding $\mathcal{E}^{(n)}$ is [17, formula (4.4)])

$$\begin{aligned}\mathcal{E}^{(n)} &= [n(n-1)]^{-1}\sum_{t_1\neq t_2}\sum\varphi_g(G^{-1}(F(Z_{t_1}^{(n)})))\,G^{-1}(F(Z_{t_2}^{(n)}))\\ &= n^{-2}\sum_{t=1}^{n}\varphi_g(G^{-1}(F(Z_t^{(n)})))\sum_{t=1}^{n}G^{-1}(F(Z_t^{(n)})) + o_p(n^{-1/2})\\ &= o_p(n^{-1/2}),\end{aligned}$$

since

$$n^{-1}\sum \varphi_g(G^{-1}(F(Z_t^{(n)}))) \xrightarrow{P} \int \varphi_g(G^{-1}(u))\,du = 0$$

and

$$n^{-1/2}\sum G^{-1}(F(Z_t^{(n)})) \xrightarrow{\mathscr{L}} N(0, \sigma_g),$$

as $n \to \infty$. The proposition then follows from the fact that $r_{k;g}^{(n)} - (\mathscr{S}^{(n)} - \mathscr{E}^{(n)})$ is $o_p(n^{-1/2})$.

4. Locally Asymptotically Optimal Signed-Rank Tests for ARMA Models

4.1. *Locally Asymptotically Maximin Tests*

Testing the adequacy of an ARMA model is certainly one of the most fundamental testing problems in time-series analysis, both in its own right and because of its implications in the identification (see e.g. [35]) and validation steps of the "Box–Jenkins methodology." We show here how signed-rank statistics provide locally asymptotically optimal tests (in the maximin sense) for this important problem, when the innovation density under the null hypothesis remains unspecified (but symmetric).

Let us denote by $H^{(n)}(\mathbf{A}, \mathbf{B}; \cdot)$ (resp. $H^{(n)}(\mathbf{A}, \mathbf{B}; f)$) the hypothesis under which an observed series $\mathbf{X}^{(n)} = (X_{-p_1+1}^{(n)}, ..., X_0^{(n)}, X_1^{(n)}, ..., X_n^{(n)})$ is generated by the ARMA (p_1, q_1) model

$$X_t - \sum_{i=1}^{p_1} A_i X_{t-i} = \varepsilon_t + \sum_{i=1}^{q_1} B_i \varepsilon_{t-i}, \qquad t \in \mathbb{Z}, \tag{4.1}$$

where $\{\varepsilon_t\}$ is an i.i.d. innovation process with unspecified symmetric density (resp. with symmetric density f satisfying assumptions (A1)–(A4)). The polynomials $A(z) = 1 - \sum_{i=1}^{p_1} A_i z^i$ and $B(z) = 1 + \sum_{i=1}^{q_1} B_i z^i$, $z \in \mathbb{C}$, are required to have distinct roots, all lying outside the unit-circle, so that the usual stationarity and invertibility conditions be satisfied.

In order to obtain local alternatives, consider the sequence of ARMA (p_2, q_2) models $(p_2 \geqslant p_1, q_2 \geqslant q_1)$

$$X_t - \sum_{i=1}^{p_1} (A_i + n^{-1/2}\gamma_i)\, X_{t-i} - n^{-1/2} \sum_{i=p_1+1}^{p_2} \gamma_i X_{t-i}$$

$$= \varepsilon_t + \sum_{i=1}^{q_1} (B_i + n^{-1/2}\delta_i)\, \varepsilon_{t-i} + n^{-1/2} \sum_{i=q_1+1}^{q_2} \delta_i \varepsilon_{t-i}, \qquad t \in \mathbb{Z}, \tag{4.2}$$

where $\{\varepsilon_t\}$ again denotes an i.i.d. innovation process with density f

satisfying assumptions (A1)–(A4). Denote by $K^{(n)}(\mathbf{A}+\boldsymbol{\gamma}, \mathbf{B}+\boldsymbol{\delta}; f)$ the hypothesis under which $\mathbf{X}^{(n)}$ is generated by (4.2).

Let $\mathbf{a}=(a_1, a_2, ..., a_i, ...)$ and $\mathbf{b}=(b_1, b_2, ..., b_i, ...)$ be defined by the polynomial ratios

$$\left(\sum_{i=1}^{p_2} \gamma_i z^i\right) \Big/ \left(1-\sum_{i=1}^{p_1} A_i z^i\right) = -\sum_{i=1}^{\infty} a_i z^i$$

and

$$\left(\sum_{i=1}^{q_2} \delta_i z^i\right) \Big/ \left(1+\sum_{i=1}^{q_1} B_i z^i\right) = \sum_{i=1}^{\infty} b_i z^i$$

($\mathbf{a}$ and $\mathbf{b}$ also can be expressed in terms of $\boldsymbol{\gamma}$, $\boldsymbol{\delta}$, and the Green's functions associated with the difference operators $A(L)$ and $B(L)$, where L denotes the lag operator—see [22, Section 3.1]). It follows from the stationarity and invertibility properties of (4.1) that the sequences $\mathbf{a}$ and $\mathbf{b}$ are absolutely summable: denote by $\|\mathbf{a}+\mathbf{b}\| = [\sum_{i=1}^{\infty} (a_i+b_i)^2]^{1/2}$, the l^2 norm of $\mathbf{a}+\mathbf{b}$.

The power of the most powerful test (viz. the *Neyman test*) for $H^{(n)}(\mathbf{A}, \mathbf{B}; f)$ against $K^{(n)}(\mathbf{A}+\boldsymbol{\gamma}, \mathbf{B}+\boldsymbol{\delta}; f)$ at level α can be shown [22] to converge to the asymptotic value

$$1-\Phi(k_{1-\alpha}-\|\mathbf{a}+\mathbf{b}\| [\sigma^2 I(f)]^{1/2}), \tag{4.3}$$

where $\Phi(\cdot)$ and $k_{1-\alpha}$ denote the standard normal distribution function and $(1-\alpha)$-quantile, respectively. The norm $\|\mathbf{a}+\mathbf{b}\|$ thus appears as a natural distance between the sequences $H^{(n)}(\mathbf{A}, \mathbf{B}; \cdot)$ and $K^{(n)}(\mathbf{A}+\boldsymbol{\gamma}, \mathbf{B}+\boldsymbol{\delta}; f)$, ($H^{(n)}(\mathbf{A}, \mathbf{B}; f)$ actually is a least favorable distribution in the problem of testing $H^{(n)}(\mathbf{A}, \mathbf{B}; \cdot)$ against $K^{(n)}(\mathbf{A}+\boldsymbol{\gamma}, \mathbf{B}+\boldsymbol{\delta}; f)$). Accordingly, if optimal tests are to be derived against alternatives under which $\boldsymbol{\gamma} \in \mathbb{R}^{p_2}$ and $\boldsymbol{\delta} \in \mathbb{R}^{q_2}$ remain unspecified,. the most natural idea is to look for an *asymptotical maximin property* against local alternatives that remain "bounded away from the null hypothesis" in the sense of the above distance. Consider therefore the *unspecified* ARMA (p_2, q_2) alternative

$$\begin{aligned} &K^{(n)}(\mathbf{A}+\mathbb{R}^{p_2}, \mathbf{B}+\mathbb{R}^{q_2}; f \mid d) \\ &\quad = \bigcup \{K^{(n)}(\mathbf{A}+\boldsymbol{\gamma}, \mathbf{B}+\boldsymbol{\delta}; f) \mid \boldsymbol{\gamma} \in \mathbb{R}^{p_2}, \boldsymbol{\delta} \in \mathbb{R}^{q_2}, \|\mathbf{a}+\mathbf{b}\| \geqslant d\}, \end{aligned} \tag{4.4}$$

where $d>0$ is some arbitrary positive constant.

Finally, denote by $\mathbf{Z}^{(n)}$ the filtered series obtained from $\mathbf{X}^{(n)}$ by applying the linear recursion

$$Z_t^{(n)} = X_t^{(n)} - \sum_{i=1}^{p_1} A_i X_{t-i}^{(n)} - \sum_{i=1}^{q_1} B_i Z_{t-i}^{(n)}, \qquad t=1, ..., n,$$

with initial values $Z_0^{(n)} = Z_{-1}^{(n)} = \cdots = Z_{-q_1+1}^{(n)} = 0 : \mathbf{Z}^{(n)}$ is the series from which the ranks $R_{+,t}^{(n)}$, and the signs $\operatorname{sgn}(Z_t^{(n)})$ in the following proposition are to be computed.

PROPOSITION 4.1. *Let $\{\psi_t^{(1)}, \psi_t^{(2)}, ..., \psi_t^{(p_1+q_1)}\}$ be an arbitrary fundamental system of solutions of the homogeneous difference equation*

$$\psi_t - \sum_{j=1}^{p_1+q_1} \sum_{i=\max(0,j-q_1)}^{\min(j,p_1)} A_i B_{j-i} \psi_{t-j} = 0, \qquad t \in \mathbb{Z}, \tag{4.5}$$

with the convention $A_0 = -1$, $B_0 = 1$. Putting $\pi = \max(p_2 - p_1, q_2 - q_1)$, denote by $\mathbf{W}_\psi^{(n)}$ the (non-singular) $(p_1+q_1) \times (p_1+q_1)$ matrix with elements

$$W_{\psi;k,l}^{(n)} = \sum_{i=\pi+1}^{n-1} \psi_i^{(k)} \psi_i^{(l)}, \qquad k, l = 1, ..., p_1 + q_1, \tag{4.6}$$

and consider the (p_1+q_1)-dimensional vector of signed-rank statistics

$$n^{1/2} \mathbf{T}_{\psi;f}^{(n)+} = \begin{pmatrix} \sum_{i=\pi+1}^{n-1} (n-i)^{1/2} \psi_i^{(1)} r_{i;f}^{(n)+} \\ \vdots \\ \sum_{i=\pi+1}^{n-1} (n-i)^{1/2} \psi_i^{(p_1+q_1)} r_{i;f}^{(n)+} \end{pmatrix}. \tag{4.7}$$

Then, the quadratic test statistic

$$Q_{A,B;f}^{(n)+} = \sum_{i=1}^{\pi} (n-i)(r_{i;f}^{(n)+})^2 + n(\mathbf{T}_{\psi;f}^{(n)+})' (\mathbf{W}_\psi^{(n)})^{-1} (\mathbf{T}_{\psi;f}^{(n)+}) \tag{4.8}$$

(i) *does not depend on the particular fundamental system $\{\psi_i^{(1)}, ..., \psi_i^{(p_1+q_1)}\}$ adopted but only on the coefficients A_i and B_i of the tested model* (4.1) (*hence the notation $Q_{A,B;f}^{(n)+}$*);

(ii) *is asymptotically chi-square, with* $\max(p_1+q_2, p_2+q_1)$ *degrees of freedom under $H^{(n)}(\mathbf{A}, \mathbf{B}; \cdot)$*;

(iii) *is asymptotically noncentral chi-square, with* $\max(p_1+q_2, p_2+q_1)$ *degrees of freedom and noncentrality parameter*

$$\tfrac{1}{2} \|\mathbf{a}+\mathbf{b}\|^2 \left[\int \phi(F^{-1}(u))\, \phi_g(G^{-1}(u))\, du \int F^{-1}(v)\, G^{-1}(v)\, dv \right]^2 [\sigma^2 I(f)]^{-1}$$

under $K^{(n)}(\mathbf{A}+\boldsymbol{\gamma}, \mathbf{B}+\boldsymbol{\delta}; g)$ (g here denotes an arbitrary symmetric density, satisfying assumptions (A1)–(A4); *the notation G and ϕ_g are used in an obvious fashion*);

(iv) *provides an asymptotically maximin most powerful test for $H^{(n)}(\mathbf{A}, \mathbf{B}; \cdot)$ against the unspecified ARMA (p_2, q_2) alternative* (4.4), *for* *any $d>0$; the corresponding envelope power function accordingly converges to $1-F_{\max(p_1+q_2,\,p_2+q_1)}(\chi^2_{1-\alpha};\, d^2\sigma^2 I(f)/2)$, where $\chi^2_{1-\alpha}$ is the $(1-\alpha)$-quantile of the chi-square distribution with* $\max(p_1+q_2, p_2+q_1)$ *degrees of freedom, and $F_v(\cdot;\lambda)$ denotes the noncentral chi-square distribution function with v degrees of freedom and noncentrality parameter λ.*

Proof. Since $\{\psi_t^{(1)}, \ldots, \psi_t^{(p_1, q_1)}\}$ is a fundamental system of solutions of (4.2), its p_1+q_1 elements are linearly independent, which ensures the nonsingularity of $\mathbf{W}_\psi^{(n)}$. Now, because of the vector structure of the space of solutions of (4.2), any fundamental system of solutions $\{\tau_t^{(1)}, \ldots, \tau_t^{(p_1+q_1)}\}$ is of the form

$$(\tau_t^{(1)}, \ldots, \tau_t^{(p_1+q_1)})' = \mathbf{K}(\psi_t^{(1)}, \ldots, \psi_t^{(p_1+q_1)})',$$

with $\mathbf{K}$ a full-rank $(p_1+q_1)\times(p_1+q_1)$ matrix of constants. Using obvious notation, we have

$$\mathbf{W}_\tau^{(n)} = \begin{pmatrix} \tau_{\pi+1}^{(1)} & \cdots & \tau_{n-1}^{(1)} \\ \vdots & & \vdots \\ \tau_{\pi+1}^{(k)} & \cdots & \tau_{n-1}^{(k)} \\ \vdots & & \vdots \\ \tau_{\pi+1}^{(p_1+q_1)} & \cdots & \tau_{n-1}^{(p_1+q_1)} \end{pmatrix} \begin{pmatrix} \tau_{\pi+1}^{(1)} & \cdots & \tau_{n-1}^{(1)} \\ \vdots & & \vdots \\ \tau_{\pi+1}^{(k)} & \cdots & \tau_{n-1}^{(k)} \\ \vdots & & \vdots \\ \tau_{\pi+1}^{(p_1+q_1)} & \cdots & \tau_{n-1}^{(p_1+q_1)} \end{pmatrix}'$$

$$= \mathbf{K} \begin{pmatrix} \psi_{\pi+1}^{(1)} & \cdots & \psi_{n-1}^{(1)} \\ \vdots & & \vdots \\ \psi_{\pi+1}^{(k)} & \cdots & \psi_{n-1}^{(k)} \\ \vdots & & \vdots \\ \psi_{\pi+1}^{(p_1+q_1)} & \cdots & \psi_{n-1}^{(p_1+q_1)} \end{pmatrix} \begin{pmatrix} \psi_{\pi+1}^{(1)} & \cdots & \psi_{n-1}^{(1)} \\ \vdots & & \vdots \\ \psi_{\pi+1}^{(k)} & \cdots & \psi_{n-1}^{(k)} \\ \vdots & & \vdots \\ \psi_{\pi+1}^{(p_1+q_1)} & \cdots & \psi_{n-1}^{(p_1+q_1)} \end{pmatrix}' \mathbf{K}'$$

$$= \mathbf{K}\mathbf{W}_\psi \mathbf{K}'.$$

Hence

$$n(\mathbf{T}_{\tau;f}^{(n)+})' (\mathbf{W}_\tau^{(n)})^{-1} (\mathbf{T}_{\tau;f}^{(n)+}) = n(\mathbf{T}_{\psi;f}^{(n)+})' \mathbf{K}'(\mathbf{K}')^{-1} (\mathbf{W}_\psi^{(n)})^{-1} \mathbf{K}^{-1}\mathbf{K}(\mathbf{T}_{\psi;f}^{(n)+})$$
$$= n(\mathbf{T}_{\psi;f}^{(n)+})' (\mathbf{W}_\psi^{(n)})^{-1} (\mathbf{T}_{\psi;f}^{(n)+}),$$

which completes part (i) of the proof. The other parts of the Proposition follow from the fact that, because of Proposition 3.1 (ii),

$$\Big[(n-1)^{1/2} r_{1;f}^{(n)+}, \ldots, (n-\pi)^{1/2} r_{\pi;f}^{(n)+}, \sum_{i=\pi+1}^{n-1} (n-i)^{1/2} \psi_i^{(1)} r_{i;f}^{(n)+}, \ldots,$$
$$\sum_{i=\pi+1}^{n-1} (n-i)^{1/2} \psi_i^{(p_1+q_1)} r_{i;f}^{(n)+}\Big]' - n^{1/2}\mathbf{T}_{\psi;f}^{(n)} = o_p(1),$$

under $H^{(n)}(\mathbf{A}, \mathbf{B}; \cdot)$ as well as under (4.4), where $\mathbf{T}^{(n)}_{\psi;f}$ is the rank-based statistic considered in Hallin and Puri [22, Proposition 3.2], whereas, due to the stationarity and invertibility properties of (4.1), $\mathbf{W}^{(n)}_{\psi}$ converges componentwise to the matrix $\mathbf{W}$ with elements $\sum_{i=\pi+1}^{\infty} \psi_i^{(k)} \psi_i^{(l)}$. Parts (ii), (iii), and (iv) of the proposition then follow from Proposition 4.2 and 4.3 in Hallin and Puri [22].

Note that an asymptotical form of the optimal test of the above Proposition consists in rejecting $H^{(n)}(\mathbf{A}, \mathbf{B}; \cdot)$ whenever $Q^{(n)+}_{A,B;f}$ is larger than the $(1-\alpha)$-quantile of the chi-square distribution with $\max(p_1+q_2, p_2+q_1)$ degrees of freedom. Note also that the test statistic $Q^{(n)+}_{A,B;f}$ decomposes into two asymptotically independent parts. The first one, as the sum of squared signed-rank autocorrelation coefficients of orders 1 through π, is a signed-rank version of the well-known Ljung–Box–Pierce *portmanteau* statistic [3, 29], and does not depend on the tested model (4,1). The second part, on the other hand, crucially depends on the coefficients A_i and B_i of (4.1) (through the fundamental system of solutions), and constitutes a *weighted*, signed-rank version of the portmanteau statistic.

All of the above results still hold, of course, without any modification, if the coefficients $\tilde{r}^{(n)+}_{i;f}$ are substituted for the $r^{(n)+}_{i;f}$ ones.

4.2. *Asymptotic Relative Efficiencies*

The parametric version of the problem treated in this paper (viz. that of obtaining locally optimal tests for general, possibly non-Gaussian, time-series models) has not been explicitly considered so far in the literature. The existing results deal with special cases, such as testing a specified Gaussian $AR(1)$ model [9], testing for randomness [22], testing against simple pth order autoregressive disturbances [11], testing general one-parameter Gaussian models [27], testing against additional AR or MA roots in Gaussian models [24], testing the adequacy of fitted Gaussian ARMA (p, q) models ([13, 25, 26, 33, 34]—see also [35, 36]).

It is, however, easy to see that the optimal (in the sense of Proposition 4.1) normal-theory procedure here can be obtained by substituting classical autocorrelation coefficients

$$r_k^{(n)} = n \sum_{t=k+1}^{n} Z_t^{(n)} Z_{t-k}^{(n)} / (n-k) \sum_{t=1}^{n} (Z_t^{(n)})^2 \tag{4.9}$$

for the signed-rank ones in definitions (4.7) and (4.8). The resulting quadratic test statistic $Q^{(n)}_{A,B}$ (coinciding with the appropriate Gaussian Lagrange multiplier test statistic) is then asymptotically equal (up to $o_P(1)$ terms) to the corresponding van der Waerden or Fisher–Yates–Fraser statistic, for Gaussian innovation densities, both under the null hypothesis as well as under contiguous Gaussian alternatives. Proposition 4.1 (iii) applies to this parametric test, but the noncentrality parameter under the alternative $K^{(n)}(A+\gamma, B+\delta; g)$ now is simply $\|a+b\|^2/2$ (this is a consequence of [17, Proposition 5.2]).

The mutual asymptotic relative efficiencies (AREs) of the various tests just described can be obtained as the ratios of their respective noncentrality parameters. Some numerical values, under Gaussian, logistic, and double exponential densities, respectively, are provided in Table I.

The results of Sections 3 and 4 also can be used to derive asymptotic normality and asymptotic relative efficiency results for the *signed Spearman autocorrelations*

$$r_{k;S}^{(n)+} = (n-k)^{-1} \sum_{t=k+1}^{n} \operatorname{sgn}(Z_t^{(n)} Z_{t-k}^{(n)})\, R_{+,t}^{(n)} R_{+,t-k}^{(n)} / \sigma_{+;S}^{(n)},$$

where

$$\begin{aligned}(\sigma_{+;S}^{(n)})^2 &= [n(n-1)]^{-1} \sum\sum_{1 \leqslant i_1 \neq i_2 \leqslant n} (i_1 i_2)^2 \\ &= [n(n-1)]^{-1} \left[\left(\sum_{i=1}^{n} i^2 \right)^2 - \sum_{i=1}^{n} i^4 \right] \\ &= (n+1)(20n^3 + 24n^2 - 5n - 6)/180\end{aligned}$$

TABLE I

Mutual AREs for the Various Quadratic Parametric and Rank-Based Tests Described Above—Under Gaussian, Logistic, and Double Exponential Densities, Respectively

	Classical parametric	van der Waerden–Fisher–Yates	Wilcoxon	Laplace	
Classical parametric	1.000	1.000	1.005	1.634	Normal
	1.000	0.954	0.911	1.232	Logistic
	1.000	0.816	0.675	0.500	Double exp.
van der Waerden–Fisher–Yates	1.000	1.000	1.055	1.634	Normal
	1.048	1.000	0.954	1.291	Logistic
	1.226	1.000	0.827	0.613	Double exp.
Wilcoxon	0.948	0.948	1.000	1.550	Normal
	1.098	1.048	1.000	1.352	Logistic
	1.482	1.209	1.000	0.741	Double exp.
Laplace	0.612	0.612	0.646	1.000	Normal
	0.812	0.775	0.740	1.000	Logistic
	2.000	1.631	1.350	1.000	Double exp.

TABLE II

AREs of Tests Based on Signed Spearman Autocorrelations or Runs with Respect to Their Parametric Counterparts Based on Classical Autocorrelation Coefficients, under Gaussian, Logistic, and Double Exponential Densities, Respectively

	Normal	Logistic	Double exponential
Signed Spearman	0.912	1.000	1.266
runs	0.405	0.438	0.500

is such that $(n-k)^{1/2}\, r_{k;S}^{(n)+}$ is exactly standardized. Just as signed f-rank autocorrelation coefficients, the signed Spearman ones can be shown to be asymptotically standard normal (viz. $(n-k)^{1/2}\, r_{k;S}^{(n)+} \xrightarrow{\mathscr{L}} \mathscr{N}(0, 1)$, under $H_0^{(n)}$). Asymptotic relative efficiencies of tests based on signed Spearman autocorrelations, with respect, e.g., to the corresponding parametric ones, based on ordinary sample autocorrelations, again follow from applying Proposition 2.2. Numerical values are given in Table II.

As another application of the results obtained in Sections 3 and 4, consider the kth-order generalization of the classical runs test statistic

$$S_k^{(n)} = (n-k)^{-1} \sum_{t=k+1}^{n} I[Z_t^{(n)} Z_{t-k}^{(n)} < 0], \qquad k = 1, ..., n-1, \quad (4.10)$$

where $I[\cdot]$ denotes the indicator of the set $[\cdot]$. This statistic has been introduced in Dufour and Hallin [8], where portmanteau and optimal quadratic runs tests of the form (4.8) are studied in some detail. The main advantage of such generalized runs statistics is that they remain invariant under the extremely general null hypothesis that $\mathbf{Z}^{(n)}$ is a vector of non-homogeneous (i.e., possibly nonidentically distributed), independent variables with zero median (even discrete distributions are allowed). This allows for heteroskedasticity, discrete distributions, ties, etc., making (4.10) very attractive in such fields as econometrics, for example. Though the resulting tests belong to the category of "easy and quick" tests, their asymptotic efficiencies with respect to the parametric procedures based on $r_k^{(n)}$ are not bad at all. The ARE values shown in Table II follow as an application of Proposition 2.2.

5. Examples

Testing the hypothesis that an observed series $\mathbf{X}^{(n)}$ was generated by an AR(1) model $X_t - \rho X_{t-1} = \varepsilon_t$ with autoregressive coefficient $\rho = \rho_o \in (-1, 1)$ and unspecified, symmetric innovation density against an alternative with $\rho > \rho_o$ and innovation density f can be achieved (in a locally most powerful manner, provided that f satisfies the adequate technical assumptions A1–A3) by rejecting $\rho = \rho_o$ whenever the signed-rank statistic

$$(1-\rho_0^2)^{1/2} \sum_{i=1}^{n-1} (n-i)^{1/2} \rho_o^{i-1} r_{i;f}^{(n)+} \tag{5.1}$$

exceeds the $(1-\alpha)$-standard normal quantile. The ranks to be considered in (5.1) are those of the residuals $Z_t^{(n)} = X_t^{(n)} - \rho_o X_{t-1}^{(n)}$, $t = 1, \ldots, n$. Spearman autocorrelations $r_{i;S}^{(n)+}$ also can be used in (5.1), either if no particular innovation density is to be emphasized under the alternative, or if particular attention is devoted to alternatives under which the technical assumptions A1–A3 are not satisfied: e.g., Cauchy or stable innovation densities.

Much attention has been devoted to processes with heavy-tailed innovation densities: see Fama [12], Mandelbrot [30], McCullough [31], among many others, for economic and financial applications. It is well known that classical results on the asymptotic distributions of usual autocorrelation coefficients do not hold anymore if the innovation variance is infinite (see, e.g., [5, 4].) Classical identification and diagnostic checking techniques, based on correlograms, consequently cannot be used safely in this context and can be totally misleading if this possibility of an infinite variance is overlooked: numerical investigation actually suggests that they behave extremely poorly. This is in very sharp contrast with our rank-based procedures, which remain perfectly valid and quite powerful—though, due to the lack of contiguity results, their local powers cannot be obtained explicitly.

In order to illustrate this point, an artificial series of length 16 has been generated from the AR(1) model $X_t - 0.65X_{t-1} = \varepsilon_t$, where the ε_t's are i.i.d. with Cauchy density. The resulting series is shown in Table III, along with the residuals $Z_t = X_t - 0.5X_{t-1}$, and the signed ranks $\operatorname{sgn}(Z_t)\, R_{+,t}^{(15)}$ required for testing $H_0 : \rho = 0.5$. Note the significant trough starting with X_{13}—a behavior very typical in Cauchy series.

The locally optimal parametric statistic

$$(1-(0.5)^2)^{1/2} \sum_i (15-i)^{1/2} (0.5)^{i-1} r_i^{(15)},$$

where $r_i^{(15)}$ denotes the usual autocorrelation coefficient, its unsigned and signed rank-based counterparts (of the form

$$(1-(0.5)^2)^{1/2} \sum_i (15-i)^{1/2} (0.5)^{i-1} r_{i;s}^{(15)+},$$

TABLE III

Sixteen Observations of the AR(1) Process $X_t - 0.65X_{t-1} = \varepsilon_t$, ε_t i.i.d. with Cauchy Density, along with the Residuals $Z_t = X_t - 0.5X_{t-1}$ and the Signed Ranks $\text{sgn}(Z_t)\,R^{(15)}_{+,t}$

t	X_t	Z_t	$\text{sgn}(Z_t)\,R_{+,t}$
0	0.949123		
1	0.280853	−0.193708	−2
2	0.814711	0.674284	7
3	0.837240	0.429885	4
4	−0.826629	−1.245249	−10
5	−0.847275	−0.433961	−5
6	−1.952889	−1.529251	−11
7	−1.933655	−0.957211	−8
8	−1.986299	−1.019471	−9
9	−1.495813	−0.502664	−6
10	−0.949718	−0.201811	−3
11	−2.440600	−1.965651	−12
12	−1.132867	0.087388	1
13	−68.528602	−67.962169	−15
14	−46.126992	−11.862691	−14
15	−33.056494	−9.992998	−13

e.g., in the signed Spearman case) have been computed for this residual series Z_t, $t = 1, \ldots, 15$, yielding the figures shown in Table IV.

An inspection of Table IV shows how unreliable classical techniques are in the presence of Cauchy innovations: if considered in its finite-variance-theory distribution, the parametric statistic is totally inefficient. Quite on the contrary, in spite of the short series length, all rank-based statistics significantly reject the null hypothesis (except for the unsigned Laplace one). Finally, signed-rank statistics appear to be substantially more powerful here than their unsigned counterparts (they all reject at the $\alpha = 1\%$ level).

Rank-based tests thus seem to bring the best alternative to normal theory parametric techniques in the presence of infinite innovation variance.

The same hypothesis ($H_0 : \rho = \rho_o$ in the AR(1) model $X_t - \rho X_{t-1} = \varepsilon_t$) could be tested against a more general alternative of ARMA (p, q) dependence

$$X_t - A_1 X_{t-1} - A_2 X_{t-2} = \varepsilon_t + B\varepsilon_{t-1}$$

with orders $1 \leqslant p \leqslant 2$ and $q \leqslant 1$ (viz. $p = 2$ if $A_2 \neq 0$, $p = 1$ if $A_2 = 0$ and

TABLE IV

Observed Values of the Signed and Unsigned Locally Optimal Spearman, van der Waerden, Wilcoxon, and Laplace Test Statistics, along with the Corresponding P-values, Computed from the Residual Series $Z_1 \cdots Z_{15}$ in Table III

Test statistic	unsigned		signed	
	Observed value	P-value	Observed value	P-value
Spearman	1.388062	0.0826	3.018533	0.0013
van der Waerden	1.674870	0.0470	2.432446	0.0075
Wilcoxon	1.690271	0.0453	2.839421	0.0023
Laplace	0.976343	0.1648	3.789278	0.0003
Parametric	0.294180	0.3844		

Note. The observed value and P-value of the parametric normal theory test statistic are also provided—though the normal approximation is not valid: this latter P-value thus is not correct. All P-values follow from normal approximation.

$A_1 \neq 0$; $q=1$ if $B \neq 0$, $q=0$ if $B=0$) and unspecified coefficients A_1, A_2, and B. For $\rho_o = 0.5$, the test statistic to be used would be of the form

$$(n-1)(r_{1;f}^{(n)+})^2 + \left[\sum_{i=2}^{n-1} (n-i)^{1/2} \rho_o^{i-2} r_{i;f}^{(n)+}\right]^2 (1-\rho_o^2), \tag{5.2}$$

where $\rho_o = 0.5$ and the ranks still are those shown in Table III. The observed value of (5.2) should be compared with the $(1-\alpha)$-quantile of a chi-square variable with two degrees of freedom. If a Spearman version of (5.2) is to be considered, the resulting ARE, with respect to the corresponding Gaussian Lagrange multiplier test (see [26]) is 0.912 under Gaussian innovations, but 1.000 under logistic and 1.266 under double-exponential ones (see Table II).

The numerical values of the signed Spearman statistic

$$14(r_{1;S}^{(15)+})^2 + \left[\sum_{i \geqslant 2} (15-i)^{1/2} (0.5)^{i-2} r_{i;S}^{(15)+}\right]^2 (1-(0.5)^2),$$

its unsigned and Gaussian parametric counterparts are shown in Table V, along with their P-values.

All the previous remarks remain valid here. The figures in Table V show that the tests based on (5.2) are not as powerful (against $A_1 = 0.65$, $A_2 = B_1 = 0$) as those based on (5.1); this is not very surprising, since (5.1) is locally optimal against this type of alternative, whereas (5.2) is not.

TABLE V

Observed Values, and P-values, of the Signed Spearman, Unsigned Spearman, and Parametric Statistics (5.2), for the Series Given in Table III

	Observed value	P-value
Signed Spearman	4.2491	0.1195
Unsigned Spearman	3.2779	0.1942
Parametric	0.1250	0.9394

Note. P-values are derived from asymptotic chi-square approximation (2 of freedom).

A more systematic Monte Carlo study of the advantage of signed-rank techniques over unsigned and parametric ones can be found in Hallin *et al.* [21].

Acknowledgments

The authors gratefully acknowledge several suggestions of the referees, which greatly improved the final presentation of the paper. The numerical example was prepared with A. Laforet and G. Mélard, to whom we express our sincere thanks. Computer facilities were provided by the Computing Center, Université Libre de Bruxelles and Vrije Universiteit Brussel.

References

[1] Bartels, R. (1982). The rank version of von Neumann's ratio test for randomness. *J. Amer. Statist. Assoc.* **77** 40–46.

[2] Bhattacharyya, G. K. (1984). Tests for randomness against trend or serial correlations. *Handbook of Statistics, Vol. 4*, pp. 89–111. North-Holland, Amsterdam/New York.

[3] Box, G. E. P., and Pierce, D. A. (1970). Distribution of residual autocorrelations in autoregressive-integrated moving average time-series models. *J. Amer. Statist. Assoc.* **65** 1509–1526.

[4] Chan, N. H., and Tran, L. T. (1989). On the first-order autoregressive process with infinite variance. *Econom. Theory* **5** 354–362.

[5] Davis, R. D., and Resnick, S. (1986). Limit theory for the sample covariance and correlation functions of moving averages. *Ann. Statist.* **14** 533–558.

[6] Dufour, J.-M. (1981). Rank tests for serial dependence. *J. Time Series Anal.* **2** 117–128.

[7] Dufour, J.-M., and Hallin, M. (1987). Tests non paramétriques optimaux pour le modèle autorégressif d'ordre un. *Ann. Econom. Statist.* **6–7** 411–434.

[8] Dufour, J.-M., and Hallin, M. (1990). *Runs tests for ARMA dependence.* Technical Report, Département de Sciences Economiques, Université de Montréal, Montréal.

[9] Dufour, J.-M., and King, M. L. (1990). Optimal invariant test for the autocorrelation coefficient in linear regressions with autocorrelated erors. *J. Econometrics*, to appear.

[10] Dufour, J.-M., Lepage, Y., and Zeidan, H. (1982). Nonparametric testing for time series: a bibliography. *Canad. J. Statist.* **10** 1–38.

[11] EVANS, M. A., AND KING, M. L. (1985). Higher order generalizations of first order autoregressive tests. *Comm. Statist. Theory Methods* **14** 2907–2918.

[12] FAMA, E. (1965). The behavior of stock market prices. *J. Business* **38** 34–105.

[13] GODFREY, L. G. (1979). Testing the adequacy of a time series model. *Biometrika* **66** 67–72.

[14] GOODMAN, L. A. (1958). Simplified runs tests and likelihood ratio tests for Markov chains. *Biometrika* **45** 181–197.

[15] GRANGER, C. W. J. (1963). A quick test for serial correlation suitable for use with non-stationary time series. *J. Amer. Statist. Assoc.* **58** 728–736.

[16] HÁJEK, J., AND ŠIDÁK Z. (1967). *Theory of Rank Tests.* Academic Press, New York.

[17] HALLIN, M., INGENBLEEK, J.-FR., AND PURI, M. L. (1985). Linear serial rank tests for randomness against ARMA alternatives. *Ann. Statist.* **13** 1156–1181.

[18] HALLIN, M., INGENBLEEK, J.-FR., AND PURI, M. L. (1987). Linear and quadratic serial rank tests for randomness against serial dependence. *J. Time Series Anal.* **8** 409–424.

[19] HALLIN, M., INGENBLEEK, J.-FR., AND PURI, M. L. (1989). Asymptotically most powerful rank tests for multivariate randomness against serial dependence. *J. Multivariate Anal.* **30** 34–71.

[20] HALLIN, M., AND MÉLARD, G. (1988). Rank-based tests for randomness against first-order serial dependence. *J. Amer. Statist. Assoc.* **83** 1117–1128.

[21] HALLIN, M., LAFORET, A., AND MELARD, G. (1990). Distribution-free tests against serial dependence: signed or unsigned ranks? *J. Statist. Plann. Inference* **24** 151–165.

[22] HALLIN, M., AND PURI, M. L. (1988). Optimal rank-based procedures for time-series analysis: Testing an ARMA model against other ARMA models. *Ann. Statist.* **16** 402–432.

[23] HALLIN, M., AND PURI, M. L. (1990). Rank tests in time series analysis: A review, unpublished.

[24] HOSKING, J. R. M. (1978). A unified derivation of the asymptotic distributions of goodness-of-fit statistics for autoregressive time series models. *J. Roy. Statist. Soc. Ser. B* **40** 341–349.

[25] HOSKING, J. R. M. (1980). Lagrange multiplier tests of time series models. *J. Roy. Statist. Soc. Ser. B* **42** 170–181.

[26] HOSKING, J. R. M. (1981). Lagrange multiplier tests of multivariate time series models. *J. Roy. Statist. Soc. Ser. B* **43** 219–230.

[27] KING, M. L., AND HILLIER, G. H. (1985). Locally best invariant tests of the error covariance matrix of the linear regression model. *J. Roy. Statist. Soc. Ser. B* **47** 98–102.

[28] LEVENE, H. (1952). On the power function of tests of randomness based on runs up and down. *Ann. Math. Statist.* **23** 34–56.

[29] LJUNG, G. M., AND BOX, G. E. P. (1978). On a measure of lack of fit in time-series models. *Biometrika* **65** 197–303.

[30] MANDELBROT, B. (1969). Long-run linearity, locally Gaussian process, *H*-spectra, and infinite variances. *Internat. Econom. Rev.* **10** 82–111.

[31] MCCULLOUGH, H. (1978). Continuous time processes with stable increments. *J. Business* **51** 601–619.

[32] MOORE, G. H., AND WALLIS, W. A. (1943). Time-series significance tests based on signs of differences. *J. Amer. Statist. Assoc.* **38** 153–164.

[33] POSKITT, D. S., AND TREMAYNE, A. R. (1980). Testing the specification of a fitted autoregressive-moving average model. *Biometrika* **67** 359–363.

[34] POSKITT, D. S., AND TREMAYNE, A. R. (1982). Diagnostic tests for multiple time series models. *Ann. Statist.* **10** 114–120.

[35] PÖTSCHER, B. M. (1983). Order estimation in ARMA models by Lagrangian multiplier tests. *Ann. Statist.* **11** 872–885.

[36] PÖTSCHER, B. M. (1985). The behaviour of the Lagrangian multiplier test in testing the orders of an ARMA model. *Metrika* **32** 129–150.

[37] PRIESTLEY, M. B. (1981). *Spectral Analysis and Time Series.* Academic Press, New York.

[38] PURI, M. L., AND SEN, P. K. (1985). *Nonparametric Methods in General Linear Models.* Wiley, New York.

[39] TRAN, L. T. (1988). Rank order statistics for time series models. *Ann. Inst. Statist. Math.* **40** 247–260.

[40] WALD, A., AND WOLFOWITZ, J. (1943). An exact test for randomness in the non-parametric case based on serial correlation, *Ann. Math. Statist.* **14** 378–388.

[41] WALLIS, W. A., AND MOORE, G. H. (1941). A significance test for time series. *J. Amer. Statist. Assoc.* **36** 401–409.

[42] YOSHIHARA, K. I. (1976). Limiting behavior of *U*-statistics for stationary, absolutely regular processes. *Z. Wahrsch. Verw. Gebiete* **35** 237–252.

Reprinted by permission of the publisher,
© 1991 by Academic Press

RANK TESTS FOR TIME SERIES ANALYSIS. A SURVEY

MARC HALLIN* and MADAN L. PURI†

Abstract. Rank-based testing procedures have proven quite efficient in classical linear models with independent observations; they long ago have entered daily practice in such fields as biostatistics or experimental planning. Still, despite the fact that many "historical" rank statistics (runs, signs of differences, turning point, Wald and Wolfowitz's rank autocorrelation coefficient, ...) actually were devised for time series situations, and despite the recognized need for nonparametric, robust or non Gaussian methods in the area, most time series analysts completely ignore rank-based procedures. Our objective with the present survey is to show that rank-based techniques very successfully can handle most of the testing problems occurring in time series context, such as testing for white noise, testing ARMA (p, q) dependence against ARMA $(p+d, q+d)$ dependence, or testing linear hypotheses about the coefficients of an ARMA model—always, under unspecified innovation density. The resulting tests of course are distribution-free. But they also are as powerful as (often, strictly more powerful than) their classical, correlogram-based counterparts. In addition, they are considerably more robust: whereas classical parametric methods can yield extremely misleading diagnostic information when the data have outliers, atypical startup behavior or heavy-tailed distributions, rank-based tests exhibit much better resistance to aberrations of this type. All these properties should make them extremely attractive, e.g. in the identification and diagnostic checking process, where the conclusions drawn from rank-based techniques are likely to be much more reliable than those resulting from an inspection of traditional correlograms.

We start (Section 1) with a general introduction, where we show how invariance arguments naturally lead to (signed or unsigned) rank-based methods. Section 2 provides a bibliographical survey of rank-based methods in serial dependence problems. The so-called linear serial rank statistics and their asymptotic distributions are introduced in Section 3. Section 4 deals with the basic theoretical result from which the optimal testing procedures described in the subsequent sections follow: local asymptotic normality of ARMA likelihood families and the local sufficiency of ranks. Locally most powerful and locally maximin rank-based tests are investigated in Section 5. Their asymptotic performance is discussed; explicit ARE (asymptotic relative efficiency) values (with respect to Gaussian parametric methods) are provided. Finally, Section 6 deals with the important problem of aligned rank tests: asymptotic invariant aligned rank tests, asymptotically most stringent within the class of all asymptotically similar tests at given probability level are derived for linear hypotheses about the coefficients of an ARMA model (with unspecified innovation density). The ARE of these tests with respect to the corresponding Gaussian Lagrange multiplier method is still the same as in Section 5, and can be as high as 2, e.g. when using median-test (or Laplace) scores under double-exponential innovation density.

AMS(MOS) subject classifications. Primary 62G10, 62M10

Key words and phrases. Rank tests, signed rank tests, invariance, aligned rank tests, permutation tests, nonhomogeneous white noise, linear trend, serial and nonserial linear rank statistics, local asymptotic normality, local asymptotic sufficiency, rank autocorrelations, locally optimal rank tests, asymptotic relative efficiencies, asymptotic invariance, asymptotically similar tests.

1. INVARIANCE, THE HYPOTHESIS OF WHITE NOISE AND RANKS

1.1 Testing for white noise: invariance arguments. Consider the null hypothesis H under which a series $\mathbf{X} = (X_1, \ldots, X_n)$ of observations is white

*Département de Mathématique and Institut de Statistique, C.P. 210, Université Libre de Bruxelles, 1050 Bruxelles, Belgium

†Department of Mathematics, Indiana University, Swain Hall East, Bloomington, Indiana 47405 U.S.A. Research supported by the Office of Naval Research Contract N00014-91-J-1020

noise. More precisely, let H denote the hypothesis under which observations $X_1, \ldots, X_t, \ldots, X_n$ are independent and identically distributed, with unspecified density (all densities here are with respect to the Lebesgue measure on $\mathbb{R}$). As an alternative, take the extremely vast and general collection K of all n-dimensional distributions under which the observed X_t's are not white noise anymore—either because of distributional heterogeneity, serial dependence, or both.

The whole theory of rank tests originates in this almost trivial observation that, letting $g : \mathbb{R} \to \mathbb{R}$ denote an arbitrary, continuous, one-to-one (hence monotonic: without any loss of generality, it can be assumed order-preserving) transformation, $\mathbf{X} = (X_1, \ldots, X_n)$ is white noise if and only if $g(\mathbf{X}) = (g(X_1), \ldots, g(X_n))$ is. More rigorously, it is easy to see that the problem of testing H against K is invariant with respect to the group $\mathcal{G}$ of continuous, order-preserving transformations $g : \mathbf{x} \mapsto g(\mathbf{x})$.

The classical attitude in such a situation consists in restricting oneself to the consideration of invariant tests, i.e. tests which are measurable with respect to some maximal invariant statistic. And a maximal invariant statistic for H consists of the vector of ranks $\mathbf{R} = (R_1, \ldots, R_n)$, where R_t denotes the rank of X_t among $X_1, \ldots, X_n$. This is how ranks come into the picture: invariance is the theoretical cornerstone of rank-based inference. Distribution-freeness (hence similarity and unbiasedness), increased power, robustness, ..., however attractive they are, are just by-products, and generally can be reached through other, more specific techniques.

As a consequence, there is no point in using ranks if, under the null hypothesis to be tested, the quantities from which ranks are computed do not constitute a white noise series (or, at least, an exchangeable series).

The above invariance argument is well accepted in such statistical areas as life testing or experimental planning. It probably sounds highly unfamiliar, theoretical and abstract to most time series analysts. Time series analysis indeed is very deeply marked with hidden Gaussian assumptions — more deeply, perhaps, than any other statistical area: the $\mathcal{L}^2$ approach, pervasive use of autocovariances and autocorrelations, Gaussian likelihood or least squares estimates, Gaussian Lagrange multipliers, ... all arise from (explicit or implicit) Gaussian assumptions. Though most of such Gaussian methods generally remain *asymptotically* valid under fairly general assumptions, it is usually agreed that the data preferably should be at least *approximately* Gaussian. Whenever they really do not look Gaussian , various preliminary transformations are generally performed before the analysis is started: logarithms, Box-Cox transformations, etc... Now, if the objective is "to normalize" (under H) a series $X_1, \ldots, X_n$, the exact transformation to be applied is $\Phi^{-1} \circ F$, where F denotes the cumulated distribution function of the X_t's and Φ stands for the standard normal distribution function. Of course, the trouble is that F in practice is unknown. However, if the empirical distribution function $\widehat{F}$ is substituted for the unknown one, $\Phi^{-1} \circ \widehat{F}$ computed at X_t is nothing else than $\Phi^{-1}(R_t/(n+1))$, where R_t is the rank of X_t among $X_1, \ldots, X_n$; $\widehat{F}(X_t)$ here is taken as $R_t/(n+1)$ in order to avoid trivial complications in the definition of $\Phi^{-1} \circ \widehat{F}$. The transformed series then is measurable with respect to the vector of ranks, and any subsequent inference procedure (of the van der Waerden type, in rank–order theory terminology) will be rank–based, hence invariant. The idea of invariance thus is not that remote from time series practice where the parameters e.g. of a Box–Cox transformation are also estimated from the data.

1.2 Testing for white noise: unbiasedness arguments. Still, one might be reluctant in enforcing the invariance principle underlying the use of rank–based tests. A more commonly accepted hypothesis testing principle then is the principle of unbiasedness. It is well known that (under fairly general continuity assumptions) a necessary condition for unbiasedness is similarity. Whenever a sufficient, boundedly complete statistic exists for the submodel consisting of the common border $\overline{H} \cap \overline{K}$ between the null hypothesis and the alternative (in any topology for which expectation is continuous with respect to the distributions in H and K), α-similarity in turn is equivalent to Neyman's α–structure property, with respect to the latter statistic. And test statistics possessing this latter property are obtained by conditioning upon the sufficient statistic at hand, then considering conditional tests.

Clearly, in the present problem of testing H against K , $\overline{H} \cap \overline{K} = H$, and a sufficient, complete statistic is the order statistic $\mathbf{X}_{(\cdot)} = (X_{(1)} \leq X_{(2)} \leq \cdots \leq X_{(n)})$. Conditioning upon $\mathbf{X}_{(\cdot)}$ yields the class of *permutation tests*.

Whereas invariance arguments lead to rank–based tests, unbiasedness arguments thus lead to permutation tests. Since rank tests are a particular case of permutation tests, unbiasedness conditions are uniformly weaker than invariance ones.

As we shall see, the class of rank–based tests and hence, a fortiori, that of permutation tests are locally asymptotically essentially complete for problems dealing e.g. with ARMA models, in the sense that they both contain locally asymptotically optimal tests for all usual testing problems in the area.

1.3 Hypotheses reducing to white noise. Aligned ranks. The hypothesis of white noise of course plays a very fundamental role, in its own right, in a variety of classical problems, such as testing against location shifts in two–or k-sample situations, testing against trend or serial dependence, etc. This role is even more fundamental in view of the host of problems which after adequate transformation reduce to that of testing for white noise.

In most statistical models, an observed series $\mathbf{X} = (X_1, \ldots, X_n)$ indeed is such that, denoting by $\boldsymbol{\theta}$ some parameter, the transformed series $\mathbf{Z}(\boldsymbol{\theta}) = (Z_1, \ldots, Z_m) = \mathcal{F}_{\boldsymbol{\theta}}(\mathbf{X})$—call it a residual series—is white noise. In a "parametric" setting, the distribution of this residual series is assumed to be known (in most cases, it is assumed to be Gaussian); very often however, a "nonparametric" approach, where the residual density remains unspecified, would be much more realistic. The problem of testing $H_0 : \boldsymbol{\theta} = \boldsymbol{\theta}_0$ then reduces to that of testing for white noise (with unspecified density) in terms of the residual series $\mathbf{Z}(\boldsymbol{\theta}_0)$.

Accordingly, if invariance arguments are to be considered, rank tests should be used, where the ranks R_t are computed from the residual series $\mathbf{Z}(\boldsymbol{\theta}_0)$. Similarly, permutation tests, satisfying weaker unbiasedness requirements, should be based on the $m!$ permutations of $\mathbf{Z}(\boldsymbol{\theta}_0)$.

In the specific area of time series analysis, ARMA models, or linear models with ARMA error terms, or bilinear models, fall within the above class of statistical models whenever the innovation density remains unspecified. Consider for example the case of an autoregressive process with linear trend.

Example 1.1. AR(1) model with linear trend. The observations X_t, $t = 0, 1, \ldots, n$, satisfy

$$X_t = \alpha + \beta c_t + e_t, \qquad t = 0, 1, \ldots, n$$

where α and β are regression parameters, $c_0, c_1, \ldots, c_n$ are known regression constants, and e_t is some solution of

$$e_t - \rho e_{t-1} = \varepsilon_t \ , \qquad t = 1, \ldots, n \ ,$$

with $\{\varepsilon_t\}$ an iid process with unspecified density, mean zero and variance one, and $|\rho| < 1$. Clearly, the null hypothesis

$$H_0 : (\alpha, \beta, \rho) = (\alpha_0, \beta_0, \rho_0)$$

is equivalent to

$$H : \mathbf{Z}(\alpha_0, \beta_0, \rho_0) \quad \text{is white noise (unspecified density)}$$

where

$$\mathbf{Z}(\alpha_0, \beta_0, \rho_0) = (Z_1, \ldots, Z_n) \ ,$$

with

$$Z_t = X_t - \rho_0 X_{t-1} - \alpha_0(1 - \rho_0) - \beta_0(c_t - \rho_0 e_{t-1}) \ , \quad t = 1, \ldots, n \ .$$

Still, hypotheses to be tested in practice seldom are of the simple form $H_0 : \boldsymbol{\theta} = \boldsymbol{\theta}_0$, under which the parameter $\boldsymbol{\theta}$ has a completely specified value $\boldsymbol{\theta}_0$. In most problems, null hypotheses take the more general form $H_0 : \boldsymbol{\theta} \in \Theta_0$, where Θ_0 is some subset of the parametric space Θ. Common examples in time series analysis are: testing an ARMA (p, q) model (with unspecified innovation density and unspecified autoregressive and moving average coefficients) against alternatives of ARMA $(p + \pi, q + \pi)$ dependence (π a given positive integer), testing for the absence of trend in an ARMA (p, q) process with unspecified covariance structure and unspecified innovation density, testing for ARMA (p, q) dependence (specified orders p and q, unspecified coefficients, unspecified innovation density) against bilinear dependence, etc.

Though a transformation $\mathcal{F}_{\boldsymbol{\theta}}$ still exists which transforms the observed series $\mathbf{X}$ into a residual white noise $\mathbf{Z}(\boldsymbol{\theta})$, the value of $\boldsymbol{\theta}$ remains unspecified also under H_0.

A tempting, intuitively natural idea then consists in substituting some "good" estimate $\widehat{\boldsymbol{\theta}}$ for the unknown paramether $\boldsymbol{\theta}$, and an "estimated" residual series $\mathbf{Z}(\widehat{\boldsymbol{\theta}})$

for the "true" one $\mathbf{Z}(\boldsymbol{\theta})$. By a "good" estimate $\widehat{\boldsymbol{\theta}}$, it is meant that $\widehat{\boldsymbol{\theta}}$ has values in Θ_0 and enjoys, under H_0, such standard properties as being root n-consistent. Estimated residuals can be expected to be close to the exact ones. Treating them as the latter yields ***aligned ranks*** $\widehat{R}_t$ and ***aligned rank tests***, or ***aligned permutation tests***.

The big trouble with aligned ranks is that they generally do not enjoy (even "approximately") the fundamental invariance properties of exact ranks: the estimation of $\boldsymbol{\theta}$ indeed most of time destroys the exchangeability structure of $\mathbf{Z}(\boldsymbol{\theta})$ which constitutes the justification of rank–based methods, so that from a decision–theoretical point of view, there is no point anymore in using (aligned) ranks. By the way, the distribution–freeness property of rank–based statistics under the null hypothesis, which was a consequence of invariance, also disappears, hence the similarity and unbiasedness features of (genuinely distribution-free) rank tests. As a consequence, the computation of exact critical values for aligned rank tests is generally impossible. Even worse: the asymptotic results available for rank statistics (based on exact ranks R_t) generally cannot be used for the corresponding aligned ones (based on the aligned ranks $\widehat{R}_t$) since in general exact and aligned rank statistics are not even asymptotically equivalent.

From a theoretical point of view as from the point of view of practice, aligned rank methods (as well as aligned permutation tests) thus are settling quite a number of nontrivial problems. Section 6 below is devoted to providing some methodology in solving these problems in time series context.

1.4 Testing for symmetric white noise. The density of the observations X_t (of the residuals Z_t in section 1.3) under the null hypothesis H so far has been assumed to be completely unspecified (except perhaps for moment conditions such as, e.g. the existence of a finite variance or finite Fisher information). In many practical problems, this density further can be assumed to be symmetric (with respect to some known median, which can be set equal to zero without any loss of generality). Denote by H' this hypothesis of symmetric white noise.

The vector of ranks $\mathbf{R}$ then loses its maximal invariant status for the benefit of the couple $(\mathbf{s}, \mathbf{R}_+)$, where $\mathbf{s} = (s_1, \ldots, s_n)$ is the vector of signs $s_t = \text{sign}\,(X_t)$ and $\mathbf{R}_+ = (R_{+,1}, \ldots, R_{+,n})$ is the vector of the ranks $R_{+,t}$ of the absolute values $|X_t|$ among $|X_1| \ldots |X_n|$ (invariance here is with respect to the group of continuous, even, order-preserving transformations).

Statistics which are measurable with respect to this maximal invariant are known as ***signed–rank statistics***.

The unbiasedness approach to the same problem leads to conditioning upon the order statistic of absolute values $|\mathbf{X}|_{(\cdot)} = (|X|_{(1)} \leq \cdots \leq |X|_{(n)})$, which is sufficient complete here, yielding a class of sign-assignment – absolute value permutation tests, based on the $2^n n!$ possible combination of a permutation of $|\mathbf{X}|_{(\cdot)}$ and a vector of n signs.

1.5 Testing for nonhomogeneous white noise. A further case frequently occurs in applications, in which the observations, or adequate residual values, under the null hypotheses to be tested, are still independent and symmetrically distributed, though their distributions possibly might not be identical anymore—denote this by H''. This hypothesis H'' of nonhomogeneous symmetric white noise allows for heteroskedasticity, contamination, etc. Symmetric discrete distributions are also allowed.

A maximal invariant (with respect to the group of componentwise continuous, even and order-preserving transformations of $\mathbf{X}$) is the vector $\mathbf{s}$ of signs, leading to sign tests.

The unbiasedness approach—conditioning upon the vector of absolute values $(|X_1|, \ldots, |X_n|)$—yields the broader class of tests which are *conditionally* measurable with respect to the vector signs.

2. NONPARAMETRIC TESTS FOR SERIAL DEPENDENCE: A BIBLIOGRAPHICAL SURVEY

The nonparametric, rank-based approach to time series analysis problems actually has a pretty long history, since rank tests against serial dependence and runs tests, which are a particular case of rank tests, can be traced back to the very beginnings of rank-based inference. Wald and Wolfowitz (1943) already suggest to substitute the ranks, or some function thereof, for the observations in the problem of testing randomness against serial dependence. Tests based on runs, runs up and down, signs of first differences or turning points already had been considered, for the same problem, in Fisher (1926), Kermack and McKendrick (1937), Mood (1940) and Wallis and Moore (1941); they subsequently have been developed in Moore and Wallis (1943), Levene and Wolfowitz (1944), Wolfowitz (1944), David (1947), Goodman (1958), Edgington (1961) and Granger (1963).

Jogdeo (1968) derives asymptotical results for a very general class of rank statistics, which includes the rank autocorrelation coefficients to be used repeatedly in subsequent sections. The conditions he puts on the score functions however are too restrictive for most purposes (excluding, e.g. the so-called van der Waerden autocorrelations).

In none of these early papers is any particular alternative considered, nor any optimality question addressed. The first attempts to investigate the power of serial rank procedures against specific alternatives are due to Beran (1972), who introduces rank analogues of integrated periodogram spectral processes, and, in a more applied context, to Knoke (1977), who conducts a Monte Carlo study of the asymptotic relative efficiencies of several tests based on serial rank statistics (namely, the Wald-Wolfowitz rank autocorrelation coefficient, the turning point statistic and a Kolmogorov-Smirnov one) with respect to the classical first-order sample autocorrelation coefficient, for first-order autoregressive alternatives. This, in some sense, was a first step towards the introduction of rank methods in time series practice—but still the statistics studied there are not new, are not specifically devised against any particular alternative, and cannot handle the more general problem of testing, e.g. an ARMA model against other ARMA models (as described in section 1.3).

Locally most powerful rank tests for randomness against Gaussian autoregressive or moving average alternatives are derived in Gupta and Govindarajulu (1980). The test statistics depend on ranks via the expected values of products of order statistics, and are asymptotically equivalent to our van der Waerden autocorrelation coefficients. Aiyar (1981) also investigates first–order statistics of the Wald–Wolfowitz and van der Waerden types, and derives their asymptotic relative efficiencies with respect to the corresponding Gaussian procedure, still under first–order autoregressive alternatives.

In a similar situation, Dufour (1981) suggests (traditional nonserial) signed–rank tests for H' based on the signs and ranks of products of the form $X_t X_{t-k}$. This way of reducing a serial problem to a nonserial one however may result in a loss of relevant information and the ranks used there are not those following from invariance arguments (except in the case of nonhomogeneous white noise—see Section 1.5). In a more applied context, Bartels (1982) introduces a rank–based version of von Neumann's ratio statistic which is asymptotically equivalent to the Wald–Wolfowitz autocorrelation coefficient of order one, and investigates the power of the resulting test through Monte Carlo techniques.

Runs also have been used for testing the absence of cross–correlation between two series: see Goodman and Grunfeld (1961) and Yang and Schreckengost (1981). In the more complex domain of dynamic econometric models, Campbell and Dufour (1991) present a promising rank-based approach to the Mankiw–Shapiro rational expectation model. Switzer (1984) proposes a Wald–Wolfowitz version of the spatial variogram—a rather isolated attempt to introduce ranks in the area of spatial processes.

In a somewhat different, nonserial context, rank tests against trend alternatives have been intensively investigated (see Mann (1945) or Savage (1957) for early papers on the subject, Aiyar, Guillier and Albers (1979) for a more recent one), as well as the asymptotic behavior of nonserial rank statistics, including multivariate ones, under various mixing conditions (Serfling (1968), Albers (1978), Tran (1988), Harel (1988) and Harel and Puri (1989a,b; 1990a; 1991; 1992)).

For more details, we refer to the review papers by Dufour, Lepage and Zeidan (1982) and Bhattacharyya (1984); see also Govindarajulu (1983).

Up to this point, and in spite of a long history, the subject of rank–based inference for time series problems thus remained largely unexplored. Quite a number of partial results were scattered around, but no coherent and structured theory was available. The few existing central limit theorems for serial rank statistics were too restrictive for most purposes; except for a few exceptions, the only problem that had been considered was that of testing for white noise; optimality problems remained essentially untouched... the rank–order section of the time series analyst's tool kit was pretty limited and nearly empty.

A systematic and coherent treatment of time series problems, based on a LeCam–Hájek approach, since then has been undertaken by the authors in a series of papers, starting with Hallin, Ingenbleek and Puri (1985, 1987), where the problem of testing for white noise against ARMA alternatives is considered in its full generality (yielding locally maximin, rank–based portmanteau tests). The long term objective is to obtain a logically consistent methodology for classical time series problems (identification, diagnostic checking,...), relying on adequate, rank based substitutes for time series analysis familiar tools such as correlograms, partial correlograms, Lagrange multipliers, etc.

Locally asymptotically maximin rank–based tests for testing an ARMA model (with unspecified innovation density) against other ARMA models are derived in Hallin and Puri (1988). The particular case of first–order autoregressive models is treated in detail in Dufour and Hallin (1987). The small sample performance of rank–based tests is investigated in Hallin and Mélard (1989), and appears to be surprisingly good, even for pretty short series. Signed rank techniques, for ARMA models with symmetric, otherwise unspecified innovation densities, are considered in Hallin and Puri (1991a) and Hallin, Laforet and Mélard (1989). Optimal runs tests, allowing for nonhomogeneous innovation processes, are treated in Dufour and Hallin (1990a), the problem of testing multivariate white noise against alternatives of (multivariate) ARMA dependence in Hallin, Ingenbleek and Puri (1989) and Hallin and Puri (1991b), and that of testing white noise against first–order, superdiagonal bilinear dependence in Benghabrit and Hallin (1992). Finally, the very general case of ARMA models with a regression trend (also known as the dynamic regression model) is studied in Hallin and Puri (1991c), where locally optimal, aligned, rank and signed–rank, tests are derived for a variety of problems, including that of testing an ARMA (p, q) model with unspecified trend component, unspecified ARMA coefficients and unspecified innovation density, against ARMA $(p + \pi, q + \pi)$ alternatives.

A review of the main results in the above papers is the subject of the present survey.

Related results on the asymptotic distribution of serial rank statistics under dependence can be found in Harel and Puri (1990b,c), Tran (1990), Nieuwenhuis and Ruymgaart (1990). Permutation tests against serial dependence are considered in David and Fix (1966), Ghosh (1954), Dufour and Roy (1985, 1986), Dufour and Hallin (1990b) and Hallin, Mélard and Milhaud (1992). A somewhat hybrid test, with mixed parametric and nonparametric (rank-based) features has been considered by Kreiss (1990a) for testing AR models.

Rank–based techniques have proven very efficient in classical linear model problems (regression, trend, analysis of variance,...), and have entered daily practice in biostatistics and experimental planning. When the required technology is available, there is no reason for time series analysis to escape this rule—even more so in view of the recognized need for robust and non–Gaussian methods in the area.

3. LINEAR RANK STATISTICS

3.1 Serial and nonserial linear rank statistics. Denote by $\mathbf{Z}^{(n)} = Z_1^{(n)}, \ldots, Z_t^{(n)}, \ldots, Z_n^{(n)})$ a series of length n. Depending on the problem, $\mathbf{Z}^{(n)}$ either may be the original observation, or some residual series. Let $R_t^{(n)}$ be the rank of $Z_t^{(n)}$ among $Z_1^{(n)}, \ldots, Z_n^{(n)}$. If $\mathbf{Z}^{(n)}$ is white noise, then, with probability one, $\mathbf{R}^{(n)} = (R_1^{(n)}, \ldots, R_n^{(n)})$ is uniformly distributed over the $n!$ permutations of $(1, 2, \ldots, n)$, whatever the underlying density of the $Z_t^{(n)}$'s.

It is well known (Hájek and Šidák 1967; Puri and Sen 1971 and 1985) that locally asymptotically best tests for linear models (two or k samples, regression, analysis of variance,...) can be based upon the class of (*simple nonserial) linear rank statistics*, of the form

$$S^{(n)} = n^{-1} \sum_{t=1}^{n} c_t a^{(n)}(R_t^{(n)}) , \tag{3.1}$$

where $a^{(n)}$ denotes some *score function*, and the c_t's are known (*regression*) constants. In a time series context, they also can be used in testing against trend (Mann 1945; Savage 1957). Such nonserial statistics however cannot capture serial dependence features: actually, it can be shown that their asymptotic distribution is the same under white noise as under local alternatives of serial dependence. Hallin et al. (1985) therefore suggest the consideration of *serial linear rank statistics*, of the form

$$S^{(n)} = (n-k)^{-1} \sum_{t=k+1}^{n} a^{(n)}(R_t^{(n)}, R_{t-1}^{(n)}, \ldots, R_{t-k}^{(n)}) , \tag{3.2}$$

where the score function $a^{(n)}$ depends on the values of $(k+1)$ successive ranks: (3.2) is a serial linear rank statistic *of order* k, and can be expected to be sensitive to serial dependencies of orders 1 through k. Nonserial statistics can be considered as a particular case ($k = 0$) of the serial ones (or linear combinations thereof).

Classical examples are the *runs statistic* (where runs are taken with respect to the median), or order one, with scores

$$a^{(n)}(i_1, i_2) = I[(2i_1 - n - 1)(2i_2 - n - 1) < 0] ,$$

the *turning point* statistic, or order three, with scores

$$a^{(n)}(i_1, i_2, i_3) = I[i_1 > i_2 < i_3 \quad \text{or} \quad i_1 < i_2 > i_3]$$

or Wald and Wolfowitz' rank autocorrelation coefficient of order k (a serial version of Spearman's statistic, here, unlike in Wald and Wolfowitz (1943)'s paper, in a noncircular version) with scores

$$a^{(n)}(i_1, \ldots, i_{k+1}) = i_1 i_{k+1} . \tag{3.3}$$

The mean $m^{(n)}$ of (3.2) under the assumption that $\mathbf{Z}^{(n)}$ is white noise is easy to compute:

$$m^{(n)} = [n(n-1)\ldots(n-k)]^{-1} \sum \sum_{i_1 \neq \cdots \neq i_{k+1}} a^{(n)}(i_1, \ldots, i_{k+1}) , \tag{3.4}$$

where $\sum \cdots \sum_{i_1 \neq \cdots \neq i_{k+1}}$ stands for a summation running over all $(k+1)$-tuples of *distinct* integers between 1 and n. The variance of (3.2) also is easy to obtain from combinatorial arguments—though a general closed form is rather tedious.

Most serial statistics of practical interest can be decomposed into a linear combination of simpler statistics of the form

$$S^{(n)} = (n-k)^{-1} \sum_{t=k+1}^{n} a^{(n)}(R_t^{(n)}) b^{(n)}(R_{t-k}^{(n)}) \,. \tag{3.5}$$

The runs statistic for example, is of the form $S_1^{(n)} + S_2^{(n)}$, where $S_1^{(n)}$ and $S_2^{(n)}$ rely on the scores

$$a_1^{(n)}(i_1, i_2) = I[2i_1 < n+1].I[2i_2 > n+1]$$

and

$$a_2^{(n)}(i_1, i_2) = I[2i_2 > n+1].I[2i_1 < n+1] \,.$$

A statistic of the form (3.5) will be called a *simple* serial linear rank statistic of order k. The Wald–Wolfowitz autocorrelation coefficient, as well as the f–rank autocorrelation coefficients to be introduced later, are *simple* rank statistics.

Letting

$$S_{\ell m}^{(n)} = \sum_{i=1}^{n} [a^{(n)}(i)]^{\ell} [b^{(n)}(i)]^{m} \,,$$

we have, for the mean $m^{(n)}$ of the simple statistic $S^{(n)}$ and the variance $(\sigma^{(n)})^2$ of $(n-k)^{\frac{1}{2}} S^{(n)}$ under white noise,

$$m^{(n)} = [n(n-1)]^{-1} [S_{10}^{(n)} S_{01}^{(n)} - S_{11}^{(n)}] \tag{3.6}$$

and

$$\begin{aligned}(\sigma^{(n)})^2 &= [n(n-1)]^{-1} [S_{20}^{(n)} S_{02}^{(n)} - S_{22}^{(n)}] \\ &+ 2\frac{(n-2k)^+}{n-k} [n(n-1)(n-2)]^{-1} [S_{10}^{(n)} S_{01}^{(n)} S_{11}^{(n)} - S_{21}^{(n)} S_{01}^{(n)} - S_{12}^{(n)} S_{10}^{(n)} - (S_{11}^{(n)})^2 + 2S_{22}^{(n)}] \\ &+ \frac{(n-k)(n-k-1) - 2(n-2k)^+}{n-k} [n(n-1)(n-2)(n-3)]^{-1} \\ &\times [(S_{10}^{(n)} S_{01}^{(n)})^2 + 2(S_{11}^{(n)})^2 + S_{20}^{(n)} S_{02}^{(n)} - 6S_{22}^{(n)} \\ &\quad - 4S_{11}^{(n)} S_{10}^{(n)} S_{01}^{(n)} - S_{20}^{(n)} (S_{01}^{(n)})^2 - S_{02}^{(n)} (S_{10}^{(n)})^2 + 4S_{21}^{(n)} S_{01}^{(n)} + 4S_{12}^{(n)} S_{10}^{(n)}] \\ &\quad - (n-k)(m^{(n)})^2, \qquad \text{where } (\cdot)^+ = \max(.,0).\end{aligned} \tag{3.7}$$

Accordingly, $(n-k)^{\frac{1}{2}}(S^{(n)})/\sigma^{(n)}$ is exactly standardized (still, under the hypothesis that $\mathbf{Z}^{(n)}$ is white noise).

3.2 Serial and nonserial linear signed–rank statistics. As mentioned in Section 1.4, whenever symmetry assumptions can be made, the vector of ranks loses its maximal invariance properties for the benefit of $(\mathbf{s}^{(n)}, \mathbf{R}_+^{(n)})$, where $\mathbf{s}^{(n)} = (s_1, \ldots, s_n)$ is the vector of signs $s_t = \text{sign}(Z_t^{(n)}) = Z_t^{(n)}/|Z_t^{(n)}|$ (with the convention $0/0 = 1$), and $\mathbf{R}_+^{(n)}$ denotes the vector of the ranks $R_{+,t}^{(n)}$ of absolute values $|Z_t^{(n)}|$ among $|Z_1^{(n)}|, \ldots, |Z_n^{(n)}|$. If $\mathbf{Z}^{(n)}$ is absolutely continuous symmetric white noise, then $\mathbf{s}^{(n)}$ and $\mathbf{R}_+^{(n)}$ are independently distributed, $\mathbf{s}^{(n)}$ uniformly over the 2^n elements of $\{-1, 1\}^n$, and $\mathbf{R}_+^{(n)}$ uniformly over the $n!$ permutations of $\{1, \ldots, n\}$.

Here again, locally best tests for linear models with independent and identically distributed symmetric error (unspecified density) terms can be based on (nonserial) linear, signed–rank statistics, of the form

$$S_+^{(n)} = n^{-1} \sum_{t=1}^{n} c_t a_+^{(n)}(s_t R_{+,t}^{(n)}) , \tag{3.8}$$

where the c_t's are known constants, and $a_+^{(n)}$ denotes a score function, defined over $\{\pm 1, \pm 2, \ldots, \pm n\}$. Time series problems require the more general class of serial linear signed–rank statistics (Hallin and Puri 1991a), of the form

$$S_+^{(n)} = (n-k)^{-1} \sum_{t=k+1}^{n} a_+^{(n)}(s_t R_{+,t}^{(n)}, s_{t-1} R_{+,t-1}^{(n)}, \ldots, s_{t-k} R_{+,t-k}^{(n)}). \tag{3.9}$$

Simple examples are Goodman (1958)'s simplified runs test (with $a_+^{(n)}(i_1, i_2) = I[i_1 i_2 < 0]$), which coincides (up to additive and multiplicative constants) with Dufour (1981)'s runs test, or the signed version of the Spearman–Wald–Wolfowitz serial correlation tests, with $a_+^{(n)}(i_1, \ldots, i_{k+1}) = i_1 i_{k+1}$.

A linear signed–rank statistic will be called *simple* if it can be written as

$$S_+^{(n)} = (n-k)^{-1} \sum_{t=k+1}^{n} s_t s_{t-k} a_+^{(n)}(R_{+,t}^{(n)}) b_+^{(n)}(R_{+,t-k}^{(n)}) . \tag{3.10}$$

The signed Spearman–Wald–Wolfowitz autocorrelations, as well as the signed f–rank autocorrelations to be described later, belong to this family of simple serial statistics. The mean of (3.10) under the hypothesis of symmetric white noise obviously is zero, whereas, due to the fact that all cross–products have expectation zero, the variance of $(n-k)^{\frac{1}{2}} S_+^{(n)}$ takes the very simple form

$$\begin{aligned} (\sigma_+^{(n)})^2 &= [n(n-1)]^{-1} \sum_{i_1 \neq i_2} \sum (a_+^{(n)}(i_1) b_+^{(n)}(i_2))^2 \\ &= [n(n-1)]^{-1} \left\{ \sum_i [a_+^{(n)}(i)]^2 \sum_i [b_+^{(n)}(i)]^2 - \sum_i [a_+^{(n)}(i) b_+^{(n)}(i)]^2 \right\} . \end{aligned} \tag{3.11}$$

Here again, $(n-k)^{\frac{1}{2}} S_+^{(n)}/\sigma_+^{(n)}$ is exactly standardized (provided that $\mathbf{Z}^{(n)}$ is symmetric white noise).

3.3 Asymptotic normality. Assuming that the series $\mathbf{Z}^{(n)}$ from which the ranks are taken is white noise, consider the serial statistic $S^{(n)}$ in (3.2). A real-valued function J, defined over the $(k+1)$-dimensional unit square $(0,1)^{k+1}$, will be called a ***score-generating function*** for $S^{(n)}$ if

$$\int_{(0,1)^{k+1}} J^{2+\delta}(u_1,\ldots,u_{k+1})du_1\ldots du_{k+1} < \infty \tag{3.12}$$

for some $\delta > 0$ and

$$\lim_{n\to\infty} E\left[(a^{(n)}(R_{U,1}^{(n)},\ldots,R_{U,k+1}^{(n)}) - J(U_1,\ldots,U_{k+1}))^2\right] = 0\ , \tag{3.13}$$

where $R_{U,1}^{(n)}\ldots,R_{U,k+1}^{(n)}$ denote the ranks of $U_1,\ldots U_{k+1}$ in a random sample $U_1,\ldots,$ U_n of independent and identically distributed, uniform (over [0,1]) random variables.

Define

$$\begin{aligned} J^*(u_1,\ldots,u_{k+1}) = J(u_1,\ldots,u_{k+1}) &- \sum_{i=1}^{k+1} E[J(U_1,\ldots,U_{i-1},u_1,U_i,\ldots,U_k)] \\ &+k\ E[J(U_1,\ldots,U_{k+1})]\ . \end{aligned} \tag{3.14}$$

Then, relying on a central limit theorem (e.g., Yoshihara, 1976) for U-statistics under absolutely regular processes, the following asymptotic normality result has been proved (cf. Hallin et al., 1985).

PROPOSITION 3.1. *Let J denote a score-generating function for the serial rank statistic $S^{(n)}$ in (3.2). Then $(n-k)^{\frac{1}{2}}(S^{(n)}-m^{(n)})$ is asymptotically normal (under white noise), with mean zero and variance V^2, where*

$$\begin{aligned} V^2 = &\int_{[0,1]^{k+1}} [J^*(u_1,\ldots,u_{k+1})]^2 du_1\ldots du_{k+1} \\ &+ 2\sum_{i=1}^{k} \int_{[0,1]^{k+i+1}} J^*(u_1,\ldots,u_{k+1})J^*(u_{1+i},\ldots,u_{k+1+i})du_1\ldots du_{k+1+i}\ . \end{aligned} \tag{3.15}$$

This result allows for explicit normal approximations, under the null hypotheses of white noise, for P-values and critical points of tests based on linear serial rank statistics.

Example 3.1. The Spearman autocorrelation coefficient of order one is often defined (cf. e.g. Kendall and Stuart, 1968) as

$$r_S^{(n)} = \frac{(n-1)^{-1}\Sigma_{t=2}^{n} R_t^{(n)} R_{t-1}^{(n)} - (n+1)^2/4}{(n^2-1)/12}\ ,$$

though an exactly standardized version (see Hallin and Mélard 1988) is more convenient for practical purposes. Letting $S^{(n)} = (n-1)^{-1}\sum_{t=2}^{n} R_t^{(n)} R_{t-1}^{(n)}/(n+1)^2$, with $m^{(n)} = (3n+2)/12(n+1)$, it is easy to check that

$$(3.16) \qquad r_S^{(n)} - 12(S^{(n)} - m^{(n)}) = 24(n-1)^{-1}(S^{(n)} - m^{(n)}) - (n-1)^{-1} .$$

$S^{(n)}$ is a linear serial rank statistic of order one. A score–generating function for $S^{(n)}$ is $J(u,v) = uv$, yielding

$$J^*(u,v) = uv - u + \frac{1}{4} .$$

It follows from Proposition 3.1 that $(n-1)^{\frac{1}{2}}(S^{(n)} - m^{(n)})$ is asymptotically normal, with mean zero and variance

$$V^2 = \int_{[0,1]^2} \left(uv - u + \frac{1}{4}\right)^2 du dv = 1/144 .$$

This, with (3.16), entails that $r_S^{(n)} - 12(S^{(n)} - m^{(n)})$ is $o_P(n^{-\frac{1}{2}})$, as $n \to \infty$, and confirms the classical result that $(n-1)^{\frac{1}{2}} r_S^{(n)}$ (as well as $12(n-1)^{\frac{1}{2}}(S^{(n)} - m^{(n)})$) is asymptotically standard normal.

As in the nonserial case, (3.13) can be shown to hold for J satisfying (3.12) and scores defined by

$$a^{(n)}(i_1,\ldots,i_{k+1}) = E\left[J(U_1,\ldots,U_{k+1}) | R_1^{(n)} = i_1,\ldots,R_{k+1}^{(n)} = i_{k+1}\right]$$

(*exact scores*) or, for J monotone with respect to each argument, satisfying (3.12), and scores

$$a^{(n)}(i_1,\ldots,i_{k+1}) = J\left(\frac{i_1}{n+1},\ldots,\frac{i_{k+1}}{n+1}\right)$$

(*approximate scores*).

The case of signed–rank statistics is roughly similar. A *score–generating function* J^+ for $S_+^{(n)} = (n-k)^{-1}\Sigma a_+^{(n)}(s_t R_t^{(n)},\ldots,s_{t-k}R_{t-k}^{(n)})$ is a real-valued function, defined over the $(k+1)$–dimensional open square $(-1,1)^{k+1}$, such that, denoting by $V_1,\ldots,V_n$ an n–tuple of independent and identically distributed rectangular $[-1,1]$ variables,

$$(3.17) \qquad E[|J_+(V_1,V_2,\ldots,V_{k+1})|^{2+\delta}] < \infty$$

for some $\delta > 0$, and

$$(3.18) \quad \lim_{n\to\infty} E\{[a_+^{(n)}(\mathrm{sgn}(V_1)R_{+,1}^{(n)},\ldots,\mathrm{sgn}(V_{k+1})R_{+,k+1}^{(n)}) - J_+(V_1,\ldots,V_{k+1})]^2\} = 0 ,$$

as $n \to \infty$. $R^{(n)}_{+,i}$ here denotes the rank of $|V_i|$ among $|V_1|, \dots, |V_n|$; the notation $\mathrm{sgn}(V_i)$ is used in an obvious fashion. Associated with J_+, define J^*_+ as

$$\begin{aligned}(3.19)\qquad J^*_+(v_1, \dots, v_{k+1}) &= J_+(v_1, \dots, v_{k+1}) \\ &- 2^{-(k+1)} \sum_{s\in\{-1,1\}} \sum_{\ell=0}^{k} \int_{[-1,1]^k} J_+(w_1, \dots, w_\ell, sv_1, w_{\ell+1}, \dots, w_k) dw \\ &+ 2^{-(k+1)} k \int_{[-1,1]^k} J_+(w_1, w_2, \dots, w_{k+1}) dw \ ,\end{aligned}$$

where $\int_{[-1,1]^k} (\dots) dv$ stands for $\int_{-1}^{1} \dots \int_{-1}^{1} (\dots)\, \mathrm{sgn}(v_1) \dots \mathrm{sgn}(v_k) dv_1 \dots dv_k$.

Proposition 3.2. *Let J_+ denote a score-generating function for the signed-rank statistic $S^{(n)}_+$ in (3.9). Then $(n-k)^{\frac{1}{2}} S^{(n)}_+$ is asymptotically normal (under symmetric white noise), with mean zero and variance V^2_+, where (same notation as in (3.18))*

$$(3.20)\ V^2_+ = E[(J^*_+(V_1, \dots, V_{k+1}))^2] + 2 \sum_{i=1}^{k} E[J^*_+(V_1, \dots, V_{k+1}) J^*_+(V_{1+i}, \dots, V_{k+1+i})].$$

Example 3.2. Parallel to the unsigned statistic considered in Example 3.1, a signed version of the Spearman autocorrelation coefficient of order one can be defined as

$$r^{(n)}_{+,S} = \frac{(n-1)^{-1} \Sigma^n_{t=2} s_t s_{t-1} R^{(n)}_{+,t} R^{(n)}_{+,t-1}}{(n+1)(2n+1)/6}$$

(though an exactly standardized version might be preferable for practical purposes—see Hallin et al. 1990). Letting

$$S^{(n)}_+ = (n-1)^{-1} \Sigma^n_{t=2} s_t s_{t-1} R^{(n)}_{+,t} R^{(n)}_{+,t-1} / (n+1)^2,$$

we have $r^{(n)}_{+,S} = 6 S^{(n)}_+ (n+1)^2/(n+1)(2n+1)$. A score-generating function for $S^{(n)}_+$ is $J_+(u,v) = uv$. Since $E[V_1 V_2 \| V_\ell| = |v\|] = 0, \ell = 1,2, v \in [0,1]$ and $E[V_1 V_2] = 0$, $J^*_+(u,v) = J_+(u,v)$ (under symmetric white noise), and

$$V^2_+ = \frac{1}{4} \int_{[-1,1]^2} (uv)^2 du\, dv = 1/9 \ .$$

It follows that $(n-1)^{\frac{1}{2}} S^{(n)}$ is asymptotically normal, with mean zero and variance 1/9, and that

$$r^{(n)}_{+,S} - 3S^{(n)}_+ = S^{(n)}_+ \left[6 \frac{(n+1)^2}{(n+1)(2n+1)} - 3\right] = o_P(n^{-\frac{1}{2}}) \ .$$

Consequently, $(n-1)^{\frac{1}{2}} r_{+,S}^{(n)}$ is asymptotically standard normal (under symmetric white noise). This statistic, to the best of our knowledge, never had been considered in the literature. Its asymptotic performance is investigated in Hallin et al. (1990).

As in the unsigned case, (3.18) holds, for $(2+\delta)$-integrable score-generating functions J_+ and *exact scores*

$$a_+^{(n)}(i_1, \ldots, i_{k+1}) = E\left[J_+(V_1, \ldots, V_{k+1}) \middle| \operatorname{sgn}(V_1) R_{+,1}^{(n)} = i_1, \ldots, \operatorname{sgn}(V_{k+1}) R_{+,k+1}^{(n)}\right],$$

or, provided in addition that J_+ is monotone with respect to each argument, with *approximate scores*

$$a_+^{(n)}(i_1, \ldots, i_{k+1}) = J_+\left(\frac{i_1}{n+1}, \ldots, \frac{i_{k+1}}{n+1}\right).$$

4. LOCAL ASYMPTOTIC NORMALITY OF ARMA PROCESSES AND THE LOCAL SUFFICIENCY OF RANKS

4.1 Local asymptotic normality results for ARMA models. The asymptotic results of Section 3 are valid under white noise assumptions, and thus allow for constructing invariant, distribution–free tests based on signed or unsigned serial linear rank statistics, for a variety of time series problems. If, however, power or optimality issues are to be addressed, more information is needed on the structure of the specific problem at hand. Since uniformly most powerful tests cannot be expected to exist—such strong optimality results in general are not available, in time series analysis, even in a more restricted parametric, Gaussian context—weaker optimality properties have to be considered. These weaker properties, of a local and asymptotic nature, rely on the local structure of the families of likelihood functions involved.

This local structure has been studied by several authors, who under various technical assumptions have established the locally asymptotically normal (LAN) structure (LeCam, 1960) of families of ARMA likelihood functions. Davies (1973) and Dzhaparidze (1986) investigate the LAN property for Gaussian ARMA processes. Akritas and Johnson (1982) deal with AR processes only. Hallin et al. (1985) and Hallin and Puri (1987) established a LAN result for general ARMA processes, using classical, Cramér-type, technical assumptions. Swensen (1985) considers the case of AR processes with a regression trend, and Garel (1989) that of MA ones, still with trend. Kreiss (1987) deals with ARMA processes without trend, and Kreiss (1990b) with AR(∞) process. The results of Swensen and Kreiss rely on martingale central limit theorems, and those by Akritas and Johnson on quadratic mean differentiability conditions which are less restrictive than the Cramér–type assumptions made by Hallin et al. (1985), Hallin and Puri (1987) and Garel (1989).

We shall not attempt here to describe in detail the technical conditions under which the LAN property holds. The assumptions we are giving here are not the weakest; but they are quite simple, and they are satisfied by most densities used in practice (not all of them: e.g. Cauchy).

4.2 Notation and main assumptions. All densities below—denoted by f, g, with distribution functions F, G—are of the form

$$g(x) = g_\sigma(x) = \sigma^{-1} g_1(x/\sigma) > 0\ , \quad x \in \mathbf{R}$$

with

$$\int x\ g_1(x)dx = 0 \quad \text{and} \quad \int x^2 g_1(x)dx = 1\ .$$

The variance σ^2 will always remain unspecified; since however it plays no role in the sequel, it will be dropped in the notation. Still for simplicity, we shall assume that g is absolutely continuous, so that the derivative $\dot{g}(x) = dg(x)/dx$ exists for almost all x. Defining $\varphi_g(x) = -\dot{g}(x)/g(x)$, we also assume that the Fisher information

$$\int \varphi_g^2(x) g(x) dx = \sigma^2 I(g) = I(g_1) = \int \varphi_{g_1}^2(x) g_1(x) dx$$

is finite. Since $g(x)$ is always strictly positive, G is strictly increasing, and the inverse G^{-1} is well defined.

Consider the stochastic difference equation (ARMA (p_1, q_1) model)

$$X_t - \sum_{i=1}^{p_1} A_i X_{t-i} = \varepsilon_t + \sum_{i=1}^{q_1} B_i \varepsilon_{t-i}\ , \tag{4.1}$$

in short $A(L)X_t = B(L)\varepsilon_t$, where L stands for the lag operator and $A(z) = 1 - \Sigma A_i z^i$, $B(z) = 1 + \Sigma B_i z^i$. In all ARMA models below, it is always assumed that $A(z)$ and $B(z)$, $z \in \mathbf{C}$ have no common roots, and satisfy the usual invertibility and causality conditions. We denote by $H_g^{(n)}(A, B)$ the hypothesis under which an observed series $\mathbf{X}^{(n)} = (X_1^{(n)}, \ldots, X_t^{(n)}, \ldots, X_n^{(n)})$ is a finite realization of some solution of (4.1), where $\{\varepsilon_t\}$ is a white noise process (i.e., an iid process) with density g. The notation $H^{(n)}(A, B) = \bigcup_g H_g^{(n)}(A, B)$ is used whenever g remains completely unspecified (except for the few general technical conditions above); the notation $H_+^{(n)}(A, B) = \bigcup_g \{H_g^{(n)}(A, B) |\ g$ symmetric with respect to zero$\}$ is used whenever g is symmetric but otherwise unspecified (still, apart from the few general technical conditions).

4.3 Signed and unsigned f–rank autocorrelations. Just as parametric autocorrelation coefficients play an essential role in the classical parametric analysis of time series models, a special case of serial rank statistics will play an essential role in our nonparametric approach. These nonparametric counterparts of usual autocorrelation coefficients have been introduced in Hallin et al. (1987) as ***f–rank autocorrelations*** and in Hallin and Puri (1991a) as ***signed f-rank autocorrelations***.

Denote by $f, \varphi = -\dot{f}/f$ and F a probability density function, the corresponding score function and distribution function, respectively. The f–rank and signed f–rank autocorrelations (see below for a definition) being invariant with respect to scale transformations, f can be chosen as f_1, so that $\int x^2 f(x)dx = 1$. The (unsigned) f–rank autocorrelation of lag i is then defined as

$$(4.2)\quad r_{i;f}^{(n)} = (n-i)^{-1}\left[\sum_{t=i+1}^{n} \varphi\left(F^{-1}\left(\frac{R_t^{(n)}}{n+1}\right)\right) F^{-1}\left(\frac{R_{t-i}^{(n)}}{n+1}\right) - m^{(n)}\right] / s^{(n)} ,$$

where $m^{(n)}$ and $s^{(n)}$ are given by (3.6) and (3.7) with

$$S_{\ell m}^{(n)} = \sum_{i=1}^{n} \varphi\left(F^{-1}\left(\frac{i}{n+1}\right)\right)^{\ell}\left(F^{-1}\left(\frac{i}{n+1}\right)\right)^{m} .$$

Particular cases are

(a) the ***van der Waerden autocorrelation coefficients***, associated with Gaussian densities,

$$r_{i;vdW}^{(n)} = \left[(n-i)^{-1} \sum_{t=i+1}^{n} \Phi^{-1}\left(\frac{R_t^{(n)}}{n+1}\right) \Phi^{-1}\left(\frac{R_{t-i}^{(n)}}{n+1}\right) - m_{vdW}^{(n)}\right] / s_{vdW}^{(n)}$$

(as usual $\Phi(x) = (2\pi)^{-\frac{1}{2}} \int_{-\infty}^{x} e^{-y^2/2}dy$ denotes the standard normal distribution function; note that, due to the symmetry of the Gaussian distribution, quite a number of terms in (3.6) and (3.7) cancel out),

(b) the ***Wilcoxon autocorrelation coefficients***, associated with logistic densities,

$$r_{i;W}^{(n)} = \left[(n-i)^{-1} \sum_{t=i+1}^{n} \left(\frac{R_t^{(n)}}{n+1} - \frac{1}{2}\right) \log \frac{R_{t-i}^{(n)}}{n+1-R_{t-i}^{(n)}} - m_W^{(n)}\right] / s_W^{(n)} ,$$

(c) the ***Laplace autocorrelation coefficients***, associated with double exponential densities,

$$r_{i;L}^{(n)} = \left[(n-i)^{-1} \sum_{t=i+1}^{n} \mathrm{sgn}\left(\frac{R_t^{(n)}}{n+1} - \frac{1}{2}\right) \left[\log(2\frac{R_{t-i}^{(n)}}{n+1}\right) I\left[R_t^{(n)} \geq \frac{n+1}{2}\right] - \log\left(2-2\frac{R_{t-i}^{(n)}}{n+1}\right) I\left[R_t^{(n)} < \frac{n+1}{2}\right]\right] - m_L^{(n)}\right] / s_L^{(n)}.$$

Note that only the van der Waerden statistic can be considered, *stricto sensu*, as an autocorrelation coefficient. Actually, up to normalizing constants, $r_{i;vdW}^{(n)}$ is the classical autocorrelation coefficient resulting from substituting the transformed observations $(\Phi^{-1}(R_t^{(n)}/n+1))$ for the original one ($X_t^{(n)}$ or $Z_t^{(n)}$). The symmetric structure of this measure of serial dependence, where the past and future play exchangeable roles, is a consequence of the fact that (except for a few, very particular MA cases) Gaussian ARMA processes are the only time–reversible ones (Weiss, 1975; Hallin et al., 1988). The past and future, in non Gaussian processes, do not play symmetric roles; accordingly, they do not play symmetric roles in the definition of f-rank autocorrelations either.

The importance of f-rank autocorrelations will follow from the asymptotic decomposition of log likelihood ratios (section 4.4 below).

Under symmetric densities, a signed version of f-rank autocorrelations, involving signed ranks, can be used with the same notation as above. Letting $F_+ = 2F-1$, define the signed f-rank autocorrelation of lag i as

$$(4.3)\ r_{i;f}^{(n)+} = (n-i)^{-1} \sum_{t=i+1}^{n} \mathrm{sgn}(Z_t^{(n)} Z_{t-i}^{(n)}) \varphi\left(F_+^{-1}\left(\frac{R_{+,t}^{(n)}}{n+1}\right)\right) F_+^{-1}\left(\frac{R_{+,t-i}^{(n)}}{n+1}\right) / s_{+,f}^{(n)}$$

where $\{Z_t^{(n)}\}$ is the series from which the ranks are computed, and

$$(s_{+,f}^{(n)})^2 = [n(n-1)]^{-1}\left\{\sum_{i=1}^{n}\varphi\left(F_+^{-1}\left(\frac{i}{n+1}\right)\right)^2 \sum_{i=1}^{n}\left(F_+^{-1}\left(\frac{i}{n+1}\right)\right)^2 - \sum_{i=1}^{n}\left[\varphi\left(F_+^{-1}\left(\frac{i}{n+1}\right)\right) F_+^{-1}\left(\frac{i}{n+1}\right)\right]^2\right\}.$$

Here again, provided the $Z_t^{(n)}$'s are symmetric white noise, $(n-i)^{\frac{1}{2}} r_{i;f}^{(n)+}$ is exactly standardized. Particular cases are

(d) the ***signed van der Waerden autocorrelation coefficients***

$$r_{i;vdW}^{(n)+} = (n-i)^{-1} \sum_{t=i+1}^{n} \mathrm{sgn}(Z_t^{(n)} Z_{t-i}^{(n)}) \Phi^{-1}\left(\frac{n+1+R_{+,t}^{(n)}}{2(n+1)}\right) \Phi^{-1}\left(\frac{n+1+R_{+,t-i}^{(n)}}{2(n+1)}\right)$$
$$\times [n(n-1)]^{\frac{1}{2}} \left\{\left[\sum_{i=1}^{n}(\Phi^{-1}(i/(n+1)))^2\right]^2 - \sum_{i=1}^{n}(\Phi^{-1}(i/(n+1)))^4\right\}^{-\frac{1}{2}},$$

(e) the ***signed Wilcoxon autocorrelation coefficients***

$$r_{i;W}^{(n)+} = (n-i)^{-1} \sum_{t=i+1}^{n} \mathrm{sgn}(Z_t^{(n)} Z_{t-i}^{(n)}) R_{+,t}^{(n)} \log\left(\frac{n+1+R_{+,t-i}^{(n)}}{n+1-R_{+,t-i}^{(n)}}\right) / s_{+,W}^{(n)},$$

(f) the ***signed Laplace autocorrelation coefficients***

$$r_{i;L}^{(n)+} = -(n-i)^{-1} \sum_{t=i+1}^{n} \mathrm{sgn}(Z_t^{(n)} Z_{t-i}^{(n)}) \log\left(1 - \frac{R_{+,t-i}^{(n)}}{n+1}\right) \Big/ \left\{n^{-1}\sum_{i=1}^{n}\left[\log\left(\frac{i}{n+1}\right)\right]^2\right\}^{\frac{1}{2}}.$$

A signed van der Waerden autocorrelation is thus the usual autocorrelation computed from the series of "signed standard normal quantiles" $\mathrm{sgn}(Z_t^{(n)})\Phi^{-1}\left(\frac{1}{2} + R_t^{(n)}/2(n+1)\right)$ associated with the residual series $\mathbf{Z}^{(n)}$. Wilcoxon and Laplace autocorrelations constitute weighted versions of the traditional Wilcoxon signed rank and runs test statistics respectively.

The following asymptotic results can be established for all f-rank autocorrelations. If $H^{(n)}(A,B)$ (resp. $H_+^{(n)}(A,B)$) is the hypothesis of interest, the ranks (resp. signed ranks) to be used are those of the residuals $Z_t^{(n)} = [A(L)/B(L)]X_t^{(n)}$ (the "starting values" used in the inversion of $B(L)$ can be chosen arbitrarily.)

Proposition 4.1. (Hallin et al., 1987; Hallin and Puri, 1991).

(i) Under $H^{(n)}(A,B)$, $(n-i)^{\frac{1}{2}}r^{(n)}_{i;f}$ *and* $(n-j)^{\frac{1}{2}}r^{(n)}_{j;f}$, $i \neq j$, *are asymptotically jointly normal, with mean zero and unit covariance matrix.*

(ii) Under $H^{(n)}_{+}(A,B)$, $(r^{(n)+}_{i;f} - r^{(n)}_{i;f})$ *is* $o_P(n^{-\frac{1}{2}})$; *accordingly,* $(n-i)^{\frac{1}{2}}r^{(n)+}_{i;f}$ *and* $(n-j)^{\frac{1}{2}}r^{(n)+}_{j;f}$, $i \neq j$ *also are asymptotically jointly normal, with mean zero and unit covariance matrix.*

4.4 Local asymptotic normality. Let $p_2 \geq p_1$, $q_2 \geq q_1$, $\mathbf{A} = (A_1 \ldots A_{p_1} 0 \ldots 0)' \in \mathbb{R}^{p_2}$, $\mathbf{B} = (B_1 \ldots B_{q_1} 0 \ldots 0)' \in \mathbb{R}^{q_2}$, $\boldsymbol{\theta} = (\mathbf{A}', \mathbf{B}')' \in \mathbb{R}^{p_2+q_2}$. Denote by $\boldsymbol{\Theta}$ the (open) subset of $\mathbb{R}^{p_2+q_2}$ for which (4.1) constitutes a valid, causal and invertible, ARMA model of orders p_1 and q_1 (i.e. $A_{p_1} \neq 0 \neq B_{q_1}$, no common roots, causality and invertibility). Similarly, let $\boldsymbol{\gamma} \in \mathbb{R}^{p_2}$, $\boldsymbol{\delta} = \mathbb{R}^{q_2}$, $\boldsymbol{\tau} = (\boldsymbol{\gamma}', \boldsymbol{\delta}')' \in \mathbb{R}^{p_2+q_2}$. $H^{(n)}_g(\boldsymbol{\theta} + n^{-\frac{1}{2}}\boldsymbol{\tau})$ accordingly constitutes a sequence of hypotheses, approaching, in some sense to be made clearer in the sequel, to $H^{(n)}_g(\boldsymbol{\theta})$. Under this sequence $\mathbf{X}^{(n)}$ is generated by some ARMA (p,q) model, $p_1 \leq p \leq p_2$, $q_1 \leq q \leq q_2$, with coefficients of the form $\mathbf{A} + n^{-\frac{1}{2}}\boldsymbol{\gamma}, \mathbf{B} + n^{-\frac{1}{2}}\boldsymbol{\delta}$ and innovation density g.

Denote by $L^{(n)}_{\boldsymbol{\theta};g}$ the likelihood function of $\mathbf{X}^{(n)}$ under $H^{(n)}_g(\boldsymbol{\theta})$, hence by $L^{(n)}_{\boldsymbol{\theta}+n^{-\frac{1}{2}}\boldsymbol{\tau};g}$ the likelihood of $\mathbf{X}^{(n)}$ under $H^{(n)}_g(\boldsymbol{\theta}+n^{-\frac{1}{2}}\boldsymbol{\tau})$. Define the random variable

$$\Lambda^{(n)}_{\boldsymbol{\theta};\boldsymbol{\tau};g}(\mathbf{X}^{(n)}) = \log\left[L^{(n)}_{\boldsymbol{\theta}+n^{-\frac{1}{2}}\boldsymbol{\tau};g}(\mathbf{X}^{(n)})/L^{(n)}_{\boldsymbol{\theta};g}(\mathbf{X}^{(n)})\right]$$

(whenever $L^{(n)}_{\boldsymbol{\theta}+n^{-\frac{1}{2}}\boldsymbol{\tau};g}$ and $L^{(n)}_{\boldsymbol{\theta};g}$ are zero, Λ can be left arbitrary).

Denote by g_u and h_u the Green's functions associated with the difference operators $A(L)$ and $B(L)$ respectively (i.e. characterized by

$$\sum_{u=0}^{\infty} g_u L^u = [A(L)]^{-1} \quad \text{and} \quad \sum_{u=0}^{\infty} h_u L^u = [B(L)]^{-1}) .$$

Let

$$a_i = \sum_{j=1}^{\min(p_2,i+p_1-1)} \gamma_j g_{i-j} \tag{4.4a}$$

and

$$b_i = \sum_{j=1}^{\min(q_2,i+q_1-1)} \delta_j h_{i-j}, \tag{4.4b}$$

so that

$$a(L) = \sum_{i=1}^{\infty} a_i L^i = \left[\sum_{i=1}^{p_2} \gamma_i L^i\right] / A(L)$$

and

$$b(L) = \sum_{i=1}^{\infty} b_i L^i = \left(\sum_{i=1}^{q_2} \delta_i L^i \right) / B(L).$$

It follows from the causality and invertibility properties of (4.1) that the sequences $\mathbf{a} = (a_i)$ and $\mathbf{b} = (b_i)$ are absolutely summable: denote by $\|\mathbf{a} + \mathbf{b}\|$ their ℓ^2 norm $\left[\sum_{i=1}^{\infty} (a_i + b_i)^2 \right]^{\frac{1}{2}}$. Also, letting $\pi = \max(p_2 - p_1, q_2 - q_1)$, introduce the $(\pi + p_1 + q_1) \times (p_2 + q_2)$ matrix

$$\text{(4.5)}\quad \mathbf{M}(\boldsymbol{\theta}) = \begin{pmatrix} 1 & 0 & \dots & 0 & 1 & 0 & \dots & 0 \\ g_1 & 1 & & & h_1 & 1 & & \\ \vdots & & \ddots & \vdots & \vdots & & \ddots & \vdots \\ g_{p_2-1} & & \dots & 1 & h_{q_2-1} & & \dots & 1 \\ g_{p_2} & & \dots & g_1 & h_{q_2} & & \dots & h_1 \\ \vdots & & & \vdots & \vdots & & & \vdots \\ g_{\pi+p_1+q_1-1} & & \dots & g_{\pi+p_1+q_1-p_2} & h_{\pi+p_1+q_1-1} & & \dots & h_{\pi+p_1+q_1-q_2} \end{pmatrix}.$$

Finally, denote by $\{\psi_t^{(1)} \dots \psi_t^{(p_1+q_1)}\}$ an arbitrary fundamental system of solutions of the homogeneous equation (or order $p_1 + q_1$) $A(L)B(L)\psi_t = 0$, $t \in \mathbf{Z}$; by $\mathbf{C}_\psi$ the Casorati matrix

$$\text{(4.6)}\qquad \mathbf{C}_\psi = \begin{pmatrix} \psi_{\pi+1}^{(1)} & \dots & \psi_{\pi+1}^{(p_1+q_1)} \\ \psi_{\pi+2}^{(1)} & \dots & \psi_{\pi+2}^{(p_1+q_1)} \\ \vdots & & \vdots \\ \psi_{\pi+p_1+q_1}^{(1)} & \dots & \psi_{\pi+p_1+q_1}^{(p_1+q_1)} \end{pmatrix}$$

and by $n^{\frac{1}{2}} \mathbf{T}_{\psi;g}^{(n)}$ the vector of rank statistics

$$\text{(4.7)}\qquad n^{\frac{1}{2}} \mathbf{T}_{\psi;g}^{(n)} = \begin{pmatrix} (n-1)^{\frac{1}{2}}\ r_{1;g}^{(n)} \\ \vdots \\ (n-\pi)^{\frac{1}{2}}\ r_{\pi;g}^{(n)} \\ \sum_{i=\pi+1}^{n-1} (n-i)^{\frac{1}{2}} \psi_i^{(1)}\ r_{i;g}^{(n)} \\ \vdots \\ \sum_{i=\pi+1}^{n-1} (n-i)^{\frac{1}{2}}\ \psi_i^{(p_1+q_1)}\ r_{i;g}^{(n)} \end{pmatrix},$$

where the ranks are those of the residuals $Z_t^{(n)} = [A(L)/B(L)]X_t^{(n)}$. The covariance matrix of $n^{\frac{1}{2}}\mathbf{T}_{\psi;g}^{(n)}$ under $H^{(n)}(A,B)$ is

$$\mathbf{W}_{\psi}^{(n)} = \left(\begin{array}{c|c} \mathbf{I} & \mathbf{0} \\ \hline \mathbf{0} & \mathbf{w}_{\psi}^{(n)} \end{array}\right), \tag{4.8}$$

with
$$w_{\psi;k\ell}^{(n)} = \sum_{i=\pi+1}^{n-1} \psi_i^{(k)}\psi_i^{(\ell)} \ , \ k,\ell = 1,\ldots,p_1+q_1 \ ,$$

which converges, as $n \to \infty$, to

$$\mathbf{W}_{\psi} = \left(\begin{array}{c|c} \mathbf{I} & \mathbf{0} \\ \hline \mathbf{0} & \mathbf{w}_{\psi} \end{array}\right), \tag{4.9}$$

with $w_{\psi,k\ell} = \sum_{i=\pi+1}^{\infty} \psi_i^{(k)}\psi_i^{(\ell)}$ (convergence again follows from the causality and invertibility of $A(L)$ and $B(L)$).

We may now state the main LAN result.

PROPOSITION 4.2. *For each $\boldsymbol{\theta} \in \boldsymbol{\Theta}$ (and under mild technical assumptions on g)*

(i)
$$\Lambda_{\boldsymbol{\theta},\boldsymbol{\tau}^{(n)};g}^{(n)}(\mathbf{X}^{(n)}) = n^{\frac{1}{2}}(\boldsymbol{\tau}^{(n)})'\mathbf{M}'(\boldsymbol{\theta}) \left(\begin{array}{c|c} \mathbf{I}_{\pi\times\pi} & \mathbf{0} \\ \hline \mathbf{0} & \mathbf{C}_{\psi}'^{-1} \end{array}\right) \mathbf{T}_{\psi;g}^{(n)}(I(g_1))^{\frac{1}{2}}$$
$$-\frac{1}{2}(\boldsymbol{\tau}^{(n)})'\mathbf{M}'(\boldsymbol{\theta}) \left(\begin{array}{c|c} \mathbf{I}_{\pi\times\pi} & \mathbf{0} \\ \hline \mathbf{0} & \mathbf{C}_{\psi}'^{-1}\mathbf{w}_{\psi}\mathbf{C}_{\psi}^{-1} \end{array}\right) \mathbf{M}(\boldsymbol{\theta})\boldsymbol{\tau}^{(n)}\ I(g_1) + o_P(1)\ ,$$

under $H_g^{(n)}(\boldsymbol{\theta})$, as $n \to \infty$, for $\boldsymbol{\tau}^{(n)}$ such that $\sup_n(\boldsymbol{\tau}^{(n)})'\boldsymbol{\tau}^{(n)} < \infty$.

(ii) *for all $\boldsymbol{\tau} \in \mathrm{R}^{p_2+q_2}$, $n^{\frac{1}{2}}\boldsymbol{\tau}'\mathbf{M}'(\boldsymbol{\theta}) \left(\begin{array}{c|c} \mathbf{I}_{\pi\times\pi} & \mathbf{0} \\ \hline \mathbf{0} & \mathbf{C}_{\psi}'^{-1} \end{array}\right) \mathbf{T}_{\psi;g}^{(n)}$ is asymptotically normal, under $H^{(n)}(\boldsymbol{\theta})$, as $n \to \infty$, with mean zero and variance*

$$\boldsymbol{\tau}'\mathbf{M}'(\boldsymbol{\theta}) \left(\begin{array}{c|c} \mathbf{I}_{\pi\times\pi} & \mathbf{0} \\ \hline \mathbf{0} & \mathbf{C}_{\psi}'^{-1}\mathbf{w}_{\psi}\mathbf{C}_{\psi}^{-1} \end{array}\right) \mathbf{M}(\boldsymbol{\theta})\boldsymbol{\tau}\ I(g_1) = \|\mathbf{a}+\mathbf{b}\|^2 I(g_1).$$

(iii) *Denote by $r_{i;g}^{(n)}(\boldsymbol{\tau})$ the f-rank autocorrelations computed from the residuals*

$$Z_t^{(n)}(\boldsymbol{\tau}) = \left[\sum_{i=1}^{p_2}(A_i + n^{-\frac{1}{2}}\gamma_i)L^i \Big/ \sum_{i=1}^{q_2}(B_i + n^{-\frac{1}{2}}\delta_i)L^i\right] X_t^{(n)}\ .$$

Then, under $H_f^{(n)}(\boldsymbol{\theta})$, as $n \to \infty$,

$$n^{\frac{1}{2}}(r_{i;g}^{(n)}(\tau) - r_{i,g}^{(n)}) = \tag{4.10}$$

$$-(a_i+b_i)\left\{\int_0^1 \varphi_f(F^{-1}(u))\varphi_g(G^{-1}(u))du \int_0^1 F^{-1}(u)G^{-1}(u)du\right\} I(g_1)^{-\frac{1}{2}} + o_P(1)$$

$$= -(a_i + b_i)I(g|f) + o_P(1)$$

$$\text{with } I(g|f) = \left\{\int_0^1 \varphi_f(F^{-1}(u))\varphi_g(G^{-1}(u))du \int F^{-1}(u)G^{-1}(u)du\right\} I(g_1)^{-\frac{1}{2}}.$$

It follows from (i) and (ii) that $\Lambda_{\boldsymbol{\theta};\tau;g}^{(n)}(\mathbf{X}^{(n)})$ is asymptotically normal, under $H_g^{(n)}(\boldsymbol{\theta})$, with mean $-\frac{1}{2}\|\mathbf{a}+\mathbf{b}\|^2\mathbf{I}(g_1)$ and variance $\|\mathbf{a}+\mathbf{b}\|^2 I(g_1)$.

The family of likelihoods $\{L_{\boldsymbol{\theta};g},\ \boldsymbol{\theta} \in \boldsymbol{\Theta}\}$ is thus locally asymptotically normal (more precisely, the family $\{L_{\boldsymbol{\theta};g}, \boldsymbol{\theta} \in \mathbb{R}^{p_2+q_2}\}$ is *restricted locally asymptotically normal* for $\boldsymbol{\theta} \in \boldsymbol{\Theta}$) in the sense of LeCam (1960)'s conditions (DN1 to DN6), and $\mathbf{T}_{\psi;f}^{(n)}$—hence the corresponding f–rank autocorrelations, or the ranks themselves are locally sufficient. Note that the dimension of the locally sufficient statistic $\mathbf{T}_{\psi;f}^{(n)}$ is $\pi + p_1 + q_1 = \max(p_1 + q_2, p_2 + q_1)$, whereas the dimension of the parameter space is $p_2 + q_2$. This corresponds to the well-known fact that the information matrix of an ARMA model is singular.

In view of Proposition 4.1, Proposition 4.2 remains valid, under symmetric densities, if signed ranks and signed rank autocorrelations are substituted for the unsigned ones. This latter fact is of particular interest.

The type of ranks (signed or unsigned) to be adopted—*if* rank–based techniques are to be considered—indeed depends, as explained in Section 1, on the invariance features of the testing problem at hand. Now, the effectiveness of the choice between signed unsigned ranks, in the classical symmetric i.i.d. case, is obscured, in practice, by the fact that (unsigned) ranks are totally insensitive to a variety of alternatives. Testing the slope of a regression line e.g. can be achieved (in a strictly unbiased manner) by means of either unsigned or signed rank techniques, whereas testing the intercept using unsigned ranks is impossible. A highly perverse consequence of this fact is that the type of alternative at hand in most cases apparently dictates which type of ranks (unsigned ones for the slope, signed ones for the intercept) should be adopted, and thus which invariance argument is relevant—see e.g. Puri and Sen (1985, Sections 5.2 and 5.3) for a typical example of this questionable attitude. Serial dependence problems and, more particularly, that of testing an ARMA model with unspecified innovation density, provide an interesting instance where an effective choice between signed and unsigned ranks cannot be eluded, and where power considerations do not supersede the much more fundamental invariance principles underlying this choice.

The following consequences of Proposition 4.2 will be particularly useful in the sequel.

PROPOSITION 4.3. *(i) Let J denote a score–generating function for the serial rank statistic $S^{(n)}$ in (3.2). Then, under $H_g^{(n)}(\boldsymbol{\theta}+n^{-\frac{1}{2}}\boldsymbol{\tau})$, $(n-k)^{\frac{1}{2}}(S^{(n)}-m^{(n)})$ is asymptotically normal, as $n\to\infty$, with mean $-\frac{1}{2}\sum_{i=1}^{k}(a_i+b_i)C_i$, where*

$$C_i=\sum_{j=0}^{k-i}\int_{[0,1]^{k+1}} J^*(u_1,\ldots,u_{k+1})\varphi_g(G^{-1}(u_{1+j}))G^{-1}(u_{1+j+i})du_1\ldots du_{k+1}\ ,$$

and variance V^2 given in (3.15).

(ii) Let J_+ denote a score–generating function for the serial signed–rank statistic $S_+^{(n)}$ in (3.9). Then, under $H_g^{(n)}(\boldsymbol{\theta}+n^{-\frac{1}{2}}\boldsymbol{\tau})$ (g, a symmetric density function), $(n-k)^{\frac{1}{2}}S_+^{(n)}$ is asymptotically normal, as $n\to\infty$, with mean $\sum_{i=1}^{k}(a_i+b_i)C_i^+$, where

$$C_i^+=\sum_{j=0}^{k-i}\int_{[0,1]^{k+1}} J_+(2u_1-1,\ldots,2u_{k+1}-1)\varphi_g(G^{-1}(u_{1+j}))G^{-1}(u_{1+j+i})du_1\ldots du_{k+1}\ ,$$

and variance V_+^2 given in (3.18) (due to the fact that

$$\int_0^1[J_+(2u-1,\ldots,2u_{k+1}-1)-J_+^*(2u_1-1,\ldots,2u_{k+1}-1)]\varphi_g(G^{-1}(u_\ell))G^{-1}(u_{\ell'})du_1\ldots du_{k+1}=0$$

for all $1\le\ell\ne\ell'\le k+1$, whether J_+ or J_+^* is used does not affect the value of C_i^+).

(iii) Both $(n-i)^{\frac{1}{2}}r_{i;f}^{(n)}$ and $(n-i)^{\frac{1}{2}}r_{i,f}^{(n)+}$ (which are asymptotically standard normal under $H^{(n)}(\boldsymbol{\theta})$ and $H_+^{(n)}(\boldsymbol{\theta})$, respectively), are asymptotically normal under $H_g^{(n)}(\boldsymbol{\theta}+n^{-\frac{1}{2}}\boldsymbol{\tau})$, as $n\to\infty$, with mean $(a_i+b_i)I(f|g)$, where

$$I(f|g)=\left\{\int_0^1\varphi_f(F^{-1}(u))\varphi_g(G^{-1}(u))du\int_0^1F^{-1}(u)G^{-1}(u)du\right\}[I(f_1)]^{-\frac{1}{2}}$$

and variance one (for $r_{i;f}^{(n)+}$, of course, f and g are to be symmetric).

PROPOSITION 4.4. *(i) $n^{\frac{1}{2}}\mathbf{T}_{\psi;f}^{(n)}$ is asymptotically normal, as $n\to\infty$, with mean $\mathbf{0}$ under $H^{(n)}(\boldsymbol{\theta})$, mean*

$$\begin{pmatrix} a_1+b_1\\ \vdots\\ a_\pi+b_\pi\\ \sum_{i=\pi+1}^{\infty}(a_i+b_i)\psi_i^{(1)}\\ \vdots\\ \sum_{i=\pi+1}^{\infty}(a_i+b_i)\psi_i^{(p_1+q_1)}\end{pmatrix}I(f|g)=\boldsymbol{\mu}_\psi(\boldsymbol{\tau})I(f|g) \tag{4.11}$$

under $H_g^{(n)}(\boldsymbol{\theta}+n^{-\frac{1}{2}}\boldsymbol{\tau})$ *and full-rank covariance matrix* $\mathbf{W}_\psi$ *(see (4.9)) under both.*

(ii) The quadratic rank statistic $n(\mathbf{T}_{\psi;f}^{(n)})'\begin{pmatrix}\mathbf{I} & \mathbf{0}\\ \mathbf{0} & \mathbf{W}_\psi^{-1}\end{pmatrix}\mathbf{T}_{\psi;f}^{(n)}$ *does not depend on the particular fundamental system* $\{\psi_t^{(1)},\dots,\psi_t^{(p_1+q_1)}\}$ *adopted; it is asymptotically chi-square, with* $\pi+p_1+q_1=\max(p_1+q_2,p_2+q_1)$ *degrees of freedom under* $H^{(n)}(\boldsymbol{\theta})$*, and asymptotically noncentral chi-square, still with* $\max(p_1+q_2,p_2+q_1)$ *degrees of freedom, but with noncentrality parameter* $\frac{1}{2}[\|\mathbf{a}+\mathbf{b}\|I(f|g)]^2$ *under* $H_g^{(n)}(\boldsymbol{\theta}+n^{-\frac{1}{2}}\boldsymbol{\tau})$.

The consequences of Proposition 4.2 are of primary importance for all inference problems in the area: hypothesis testing, estimation, model selection and identification, among others. In the hypothesis testing context, which we develop in some detail in Sections 5 and 6, Proposition 4.2, in a very intuitive interpretation, implies that, asymptotically and locally, testing problems about $\boldsymbol{\theta}$ can be treated as testing problems about the mean $\boldsymbol{\mu}_\psi$ of the central sequence of statistics $\mathbf{T}_{\psi;f}^{(n)}$. A complex problem about ARMA parameters $\boldsymbol{\theta}$ thus turns out to be asymptotically equivalent to a hopefully much simpler one about the mean of a multinormal statistic with locally constant covariance structure.

5. LOCALLY ASYMPTOTICALLY OPTIMAL RANK TESTS

5.1 Locally asymptotically most powerful tests. The theoretical results of Section 4.4 allow for the construction of locally asymptotically optimal rank tests for a variety of problems. The problems treated in the present section are those dealing with ARMA models which, under the null hypothesis and except for the innovation density, are completely specified. Depending upon the alternative (which may have a one-dimensional or multidimensional structure, locally (one-sided) most powerful or locally maximin rank-based tests can be derived.

Denote by $H^{(n)}$ and $K^{(n)}$ two sequences of hypotheses, i.e. two sequences of non-overlapping subsets of some parameter space. A sequence $\phi_*^{(n)}$ of tests is called *asymptotically most powerful for* $H^{(n)}$ against $K^{(n)}$ at probability level α if

$$\limsup_{n\to\infty}\left[E_{\boldsymbol{\theta}^{(n)}}(\phi_*^{(n)})-\alpha\right]\le 0\,,\quad \boldsymbol{\theta}^{(n)}\in H^{(n)} \tag{5.1}$$

and, for any sequence $\phi^{(n)}$ satisfying (5.1),

$$\liminf_{n\to\infty}\left[E_{\boldsymbol{\theta}^{(n)}}(\phi_*^{(n)}-\phi^{(n)})\right]\ge 0\,,\quad \boldsymbol{\theta}^{(n)}\in K^{(n)}\,. \tag{5.2}$$

Whenever $K^{(n)}$ can be considered as a *local alternative* with respect to $H^{(n)}$ (e.g. $K^{(n)}$ contiguous to $H^{(n)}$), $\phi_*^{(n)}$ will also be termed *locally asymptotically most powerful.*

PROPOSITION 5.1. *The (sequence of) rank-based test(s)*

$$\phi^{(n)}=1\quad\text{iff}\sum_{i=1}^{n-1}(n-i)^{\frac{1}{2}}(a_i+b_i)r_{i;f}^{(n)}>k_{1-\alpha}\left[\sum_{i=1}^{n-1}(a_i+b_i)^2\right]^{\frac{1}{2}}\,, \tag{5.3}$$

where $k_{1-\alpha} = \Phi^{-1}(1-\alpha)$ *denotes the* $(1-\alpha)$*-standard normal quantile,*

(i) is locally asymptotically most powerful (at probability level α*) for testing* $H^{(n)}(\boldsymbol{\theta})$ *against* $H_f^{(n)}(\boldsymbol{\theta} + n^{-\frac{1}{2}} k\boldsymbol{\tau})$, $k > 0$ *arbitrary.*

(ii) has asymptotic power

$$1 - \Phi(k_{1-\alpha} - \|\mathbf{a}+\mathbf{b}\| \; I(f \mid g)) \tag{5.4}$$

against $H_g^{(n)}(\boldsymbol{\theta} + n^{-\frac{1}{2}}\boldsymbol{\tau})$.

Note that (as expected) the asymptotic power of the rank-based test (5.3) against $H_f^{(n)}(\boldsymbol{\theta}+n^{-\frac{1}{2}}\boldsymbol{\tau})$ —namely, $1-\Phi(k_{1-\alpha}-\|\mathbf{a}+\mathbf{b}\|[I(f_1)]^{\frac{1}{2}})$ —equals that of the Neyman test $\phi_{NP}^{(n)}$ (for $H_f^{(n)}(\boldsymbol{\theta})$ against $H_f^{(n)}(\boldsymbol{\theta}+n^{-\frac{1}{2}}\boldsymbol{\tau})$). The latter indeed consists in rejecting $H_f^{(n)}(\boldsymbol{\theta})$ whenever $\Lambda_{\boldsymbol{\theta};\boldsymbol{\tau};f}^{(n)}$ is "too large", i.e., in view of Proposition 4.2, asymptotically reduces to

$$\phi_{NP}^{(n)} = 1 \text{ iff } \Lambda_{\boldsymbol{\theta};\boldsymbol{\tau};f}^{(n)} + \frac{1}{2}\|\mathbf{a}+\mathbf{b}\|^2 I(f_1) > k_{1-\alpha}\|\mathbf{a}+\mathbf{b}\|[I(f_1)]^{\frac{1}{2}} . \tag{5.5}$$

The asymptotic distribution of $\Lambda_{\boldsymbol{\theta};\boldsymbol{\tau};f}^{(n)}$ under $H_f^{(n)}(\boldsymbol{\theta}+n^{-\frac{1}{2}}\boldsymbol{\tau})$ can be shown to be normal, with mean $\frac{1}{2}\ \|\mathbf{a}+\mathbf{b}\|^2 I(f_1)$ and the same variance as under $H_f^{(n)}(\boldsymbol{\theta})$. The asymptotic power of (5.5) therefore is $1-\Phi(k_{1-\alpha} - \|\mathbf{a}+\mathbf{b}\|[I(f_1)]^{\frac{1}{2}})$. The norm $\|\mathbf{a}+\mathbf{b}\|$ accordingly can be interpreted as a "natural" distance between the sequences $H^{(n)}(\boldsymbol{\theta})$ and $H_f^{(n)}(\boldsymbol{\theta}+n^{\frac{1}{2}}\boldsymbol{\tau})$, equivalent to the L^1–distance $2\sup_\alpha\{1-\alpha-\Phi(k_{1-\alpha} - \|\mathbf{a}+\mathbf{b}\|[I(f_1)]^{-\frac{1}{2}})\}$.

Here again, in case the innovation densities are restricted to symmetric ones, signed ranks and signed autocorrelation coefficients can be substituted for the unsigned ones: Proposition 5.1 can be formulated without modification with $r_{i;f}^{(n)+}$ instead of $r_{i;f}^{(n)}$ (and, of course, symmetric densities f and g). If a parametric, Gaussian approach, is adopted (all densities then, under the null hypothesis as well as under the alternative, are assumed to be Gaussian), Proposition 5.1 takes the form

PROPOSITION 5.2. *Denote by* $r_i^{(n)}$ *the classical autocorrelation coefficient of order* i*. The parametric test*

$$\phi^{(n)} = 1 \text{ iff } \sum_{i=1}^{n-1}(n-i)^{\frac{1}{2}}(a_i+b_i)\, r_i^{(n)} > k_{1-\alpha}\left[\sum_{i=1}^{n-1}(a_i+b_i)^2\right]^{\frac{1}{2}}$$

(i) is locally most powerful (at probability level α*) for testing* $H^{(n)}(\boldsymbol{\theta})$ *(or* $H_+^{(n)}(\boldsymbol{\theta})$*) against* $H_{\mathcal{N}}^{(n)}(\boldsymbol{\theta}+n^{-\frac{1}{2}}k\boldsymbol{\tau})$ *(where* $\mathcal{N}$ *stands for a normal density with mean zero and arbitrary variance),* $k>0$ *arbitrary.*

(ii) has asymptotic power

$$1-\Phi(k_{1-\alpha} - \|\mathbf{a}+\mathbf{b}\|) \tag{5.6}$$

against $H_g^{(n)}(\boldsymbol{\theta}+n^{-\frac{1}{2}}\boldsymbol{\tau})$*, whatever* g *may be.*

This confirms the analogy between f–rank autocorrelations (under innovation density f) and usual autocorrelation coefficients (under Gaussian innovation densities). The eventual superiority of f–rank autocorrelations comes from the $\mathbf{I}(\mathbf{f}|\mathbf{g})$ factor appearing in (5.4), but not in (5.6).

Example 5.1. Consider the problem of testing the null hypotheses $H^{(n)}(0)$ that $\mathbf{X}^{(n)}$ is white noise (with unspecified density) against an alternative of possible ARMA (p,q) dependence, $\max(p,q) = 1$, i.e. $X_t - AX_{t-1} = \varepsilon_t + B\varepsilon_{t-1}$, $A+B > 0$. Here $Z_t^{(n)} = X_t^{(n)}$ and the ranks are those of the observed series $\mathbf{X}^{(n)}$ itself. The rank–based test rejecting $H^{(n)}(0)$ whenever $(n-1)^{\frac{1}{2}} r_{1;f}^{(n)} > k_{1-\alpha}$ is locally most powerful (against $X_t - n^{-\frac{1}{2}}\alpha X_{t-1} = \varepsilon_t + n^{-\frac{1}{2}}\beta\varepsilon_{t-1}$, $\alpha + \beta > 0$ arbitrary) if ε_t, under the alternative, has density f (with arbitrary variance). If Gaussian (logistic, double exponential) alternatives are to be privileged or are to be feared most, then a van der Waerden (Wilcoxon, Laplace) autocorrelation $r_{1;vdW}^{(n)}$ $(r_{1;W}^{(n)}, r_{1;L}^{(n)})$ should be adopted. If the density under $H^{(n)}(0)$ can be assumed symmetric, then signed ranks and signed autocorrelations can be substituted for the unsigned ones.

Example 5.2. Consider the problem of testing $\rho = \rho_0$ against $\rho > \rho_0$ in the $AR(1)$ model $X_t - \rho X_t = \varepsilon_t$ with unspecified but symmetric innovation density. A signed–rank, genuinely distribution–free test can be performed, which is locally most powerful against Gaussian alternatives, among all tests at (asymptotic) probability level α, by rejecting $\rho = \rho_0$ whenever

$$(1-\rho_0^2)^{\frac{1}{2}} \sum_{i=1}^{n-1} (n-i)^{\frac{1}{2}} \rho_0^{i-1} r_{i;f}^{(n)+} > k_{1\alpha} \, .$$

The signs and ranks here are those of the residuals $Z_t^{(n)} = X_t^{(n)} - \rho_0 X_{t-1}^{(n)}$ and their absolute values $|Z_t^{(n)}|$ (assume $\mathbf{X}^{(n)} = (X_0^{(n)}, X_1^{(n)}, \ldots, X_n^{(n)})$). If the density (under $\rho = \rho_0$) cannot be assumed symmetric, unsigned ranks and unsigned autocorrelations should be used instead of the signed ones.

The asymptotic relative efficiencies of the signed and unsigned rank procedures described in this section with respect to each other, and with respect to their parametric counterparts, are computed in Section 5.3.

5.2 Locally maximin tests. The local alternatives considered in Section 5.2 are basically one–dimensional, one–sided alternatives: the tests provided by Propositions 5.1 and 5.2 are optimal against a specified $(\boldsymbol{\gamma}, \boldsymbol{\delta})$ direction in the parameter space, along with specified innovation density. In many situations of practical interest, the alternative is inherently multidimensional, and no particular $(\boldsymbol{\gamma}, \boldsymbol{\delta})$ direction is to be privileged. A natural idea then consists in considering tests which are ***locally asymptotically maximin***.

Informally speaking, a maximin test, at given probability level α, is a test whose worst performance is best within the class of all tests at level α. More precisely, let $H^{(n)}$ and $K^{(n)}$ denote two sequences of (nonoverlapping) hypotheses. The corresponding sequence of ***envelope power functions*** is

$$\beta(\alpha; H^{(n)}, K^{(n)}) = \sup_{\phi} \inf_{\theta} E_{\theta}(\phi) , \quad \alpha \in (0,1) ,$$

where the $\sup_{\phi}$ and $\inf_{\theta}$ are taken over all tests ϕ satisfying $E_{\theta}\phi \leq \alpha$, $\theta \in H^{(n)}$, and all values of θ in $K^{(n)}$, respectively. A sequence $\phi_*^{(n)}$ of tests is called ***asymptotically*** (***locally***, if $K^{(n)}$ is a local alternative to $H^{(n)}$) maximin for $H^{(n)}$ against $K^{(n)}$, at (asymptotic) level α, if

$$\limsup_{n\to\infty} \left[E_{\theta^{(n)}}(\phi_*^{(n)}) - \alpha \right] \leq 0 , \quad \theta^{(n)} \in H^{(n)} \tag{5.7}$$

and

$$\liminf_{n\to\infty} \left[E_{\theta^{(n)}}(\phi_*^{(n)}) - \beta(\alpha; H^{(n)}, K^{(n)}) \right] \geq 0 , \quad \theta^{(n)} \in K^{(n)} . \tag{5.8}$$

If however nontrivial maximin tests are to be obtained, $K^{(n)}$ has to be bounded away from $H^{(n)}$. If Indeed the L^1–distance between $H^{(n)}$ and $K^{(n)}$ (for fixed n) would be zero, the envelope power function trivially would reduce to $\beta(\alpha; H^{(n)}, K^{(n)}) = \alpha$. The simplest idea, if $H^{(n)}$ and a local alternative $K^{(n)}$ are to be considered, consists in defining a local alternative whose L^1–distance to $H^{(n)}$ remains (asymptotically) bounded from below by some fixed, strictly positive, constant d. In the problem of testing $H^{(n)}(\theta)$ (or $H_+^{(n)}(\theta)$) —i.e. a specified ARMA (p_1, q_1) model with unspecified (or symmetric unspecified) innovation density —against unspecified ARMA (p, q) alternatives, with $p_1 \leq p \leq p_2$ and $q_1 \leq q \leq q_2$, such local alternatives with bounded L_1–distance from $H^{(n)}(\theta)$ (or $H_+^{(n)}(\theta)$) are (for fixed f) of the form

$$K_f^{(n)}(d) = \bigcup \{ H_f^{(n)}(\theta + n^{-\frac{1}{2}}\tau) \mid \|\mathbf{a} + \mathbf{b}\| \geq d \} , \tag{5.9}$$

where the union is taken over all $\tau \in \mathbb{R}^{p_2+q_2}$ such that $\|\mathbf{a} + \mathbf{b}\|$ (with $\mathbf{a}$ and $\mathbf{b}$ as defined in (4.4) is larger than or equal to d. The alternative (5.9) accordingly can be interpreted as the outside of a L^1–hypersphere. Due to the particular L^1 topology of ARMA likelihoods, $K_f^{(n)}(d)$ however resembles a cylinder rather than a sphere in the $\mathbb{R}^{p_2+q_2}$ space. The choice of d, as we shall see, does not affect the final result and the form of locally maximin tests.

PROPOSITION 5.3. *(i) The (sequence of) rank–based test(s)*

$$\phi^{(n)} = 1 \text{ iff } n(\mathbf{T}_{\psi;f}^{(n)})' \begin{pmatrix} \mathbf{I} & \mathbf{0} \\ \mathbf{I} & \mathbf{w}_{\psi}^{-1} \end{pmatrix} \mathbf{T}_{\psi;f}^{(n)} > \chi^2_{1-\alpha} \tag{5.10}$$

(which, from Proposition 4.4, do not depend on the particular fundamental system adopted), where $\chi^2_{1-\alpha}$ *denotes the* $(1-\alpha)$*–quantile of a chi–square variable with*

$\pi + p_1 + q_1 = \max(p_1 + q_2, p_2 + q_1)$ degrees of freedom, is locally asymptotically maximin, at probability level α, against any alternative of the form $K_f^{(n)}(d)$, $d > 0$.

(ii) The asymptotic power of (5.10) against $H_g^{(n)}(\boldsymbol{\theta} + n^{-\frac{1}{2}}\boldsymbol{\tau})$ is

$$1 - F(\chi^2_{1-\alpha}; \tfrac{1}{2}\|\mathbf{a} + \mathbf{b}\|^2(I(f|g))^2) ,$$

where $F(\ ; \lambda)$ denotes the distribution function of a noncentral chi–square variable with $\pi + p_1 + q_1$ degrees of freedom and noncentrality parameter λ.

(iii) The envelope power function $\beta(\alpha; H^{(n)}(\boldsymbol{\theta}), K_f^{(n)}(d))$ converges to (same notation as above) $1 - F(\chi^2_{1-\alpha}; \frac{1}{2}\|\mathbf{a} + \mathbf{b}\|^2 I(f_1))$.

The same result of course still holds if signed ranks are substituted for unsigned ones. Its parametric counterpart involves a parametric version $\mathbf{T}^{(n)}_{\psi;\mathcal{N}}$, say, of $\mathbf{T}^{(n)}_{\psi;f}$, where classical parametric autocorrelations $r_i^{(n)}$ are substituted for the rank-band ones $r^{(n)}_{i,f}$.

PROPOSITION 5.4. *(i) Substituting $\mathbf{T}^{(n)}_{\psi;\mathcal{N}}$ for $\mathbf{T}^{(n)}_{\psi;f}$ in (5.10) yields a parametric test which is locally maximin, at probability level α, for $H^{(n)}(\boldsymbol{\theta})$ (for $H_+^{(n)}(\boldsymbol{\theta})$) against any $K_{\mathcal{N}}^{(n)}(d)$, $d > 0$ (where $\mathcal{N}$ stands for a normal density with mean zero and arbitrary variance).*

(ii) The asymptotic power of this test against $H_g^{(n)}(\boldsymbol{\theta} + n^{-\frac{1}{2}}\boldsymbol{\tau})$ is (same notation as above) $1 - F(\chi^2_{1-\alpha}; \frac{1}{2}\|\mathbf{a} + \mathbf{b}\|^2)$.

Example 5.3. As in Example 5.1, consider the problem of testing the null hypothesis $H^{(n)}(0)$ that $\mathbf{X}^{(n)}$ is white noise, with unspecified (or unspecified symmetric) density. The alternative now is the whole class of possible ARMA (p, q) dependencies, with $0 < \max(p, q) \leq \pi$ and unspecified coefficients, but the subset of it at which optimality is desired is that with innovation density f. The *f–rank* (signed or unsigned) ***portmanteau test*** of order π

$$\phi^{(n)} = 1 \quad \text{if and only if} \quad \sum_{i=1}^{\pi}(n - i)(r^{(n)}_{i;f})^2 > \chi^2_{1-\alpha} ,$$

where $\chi^2_{1-\alpha}$ is the $(1-\alpha)$–quantile of a chi–square variable with π degrees of freedom, is then locally asymptotically maximin. The same property holds for the parametric portmanteau test (where the parametric autocorrelations $r_i^{(n)}$ are substituted for the rank–based $r^{(n)}_{i;f}$) against normal alternatives.

Example 5.4. Consider the problem of testing the null hypothesis that an observed series $\mathbf{X}^{(n)}$ was generated by the ARMA (1,1) model $X_t - \dfrac{4}{5} X_{t-1} = \varepsilon_t + \dfrac{1}{2} \varepsilon_{t-1}$, $t \in \mathbb{Z}$, against unspecified ARMA (2,2) alternatives (or, equivalently, against ARMA (2,1) or ARMA (1,2) alternatives). Let $Z^{(n)}_{-1} = Z^{(n)}_0 = 0$ and

$Z_t^{(n)} = X_t^{(n)} - \frac{4}{5} X_{t-1}^{(n)} - \frac{1}{2} Z_{t-1}^{(n)}$, $t = 1, \ldots, n$; denote by $r_{i,f}^{(n)}$ the corresponding f-rank autocorrelations. A fundamental system of solutions of

$$\left(1 - \frac{4}{5} L\right)\left(1 + \frac{1}{2}\right)\psi_t = 0 \ , \ t \in \mathbb{Z}$$

is

$$\left\{\psi_t^{(1)} = \left(\frac{4}{5}\right)^{t-2} ; \psi_t^{(2)} = \left(-\frac{1}{2}\right)^{t-2}\right\};$$

here $\pi = \max(p_2 - p_1, q_2 - q_1) = 1$. The covariance matrix $\mathbf{W}^2$ is of the form

$$\mathbf{W}^2 = \begin{pmatrix} \left[1 - \left(\frac{4}{5}\right)^2\right]^{-1} & \left[1 + \frac{2}{5}\right]^{-1} \\ \left[1 + \frac{2}{5}\right]^{-1} & \left[1 - \left(\frac{1}{2}\right)^2\right]^{-1} \end{pmatrix} = \begin{pmatrix} \frac{25}{9} & \frac{5}{7} \\ \frac{5}{7} & \frac{4}{3} \end{pmatrix} .$$

The test statistic of Proposition 5.3(i) is thus

$$\begin{aligned} Q_{\psi;f}^{(n)} &= n(r_{1;f}^{(n)})^2 + \\ & n\left(\sum_{i=2}^{n-1}\left(\frac{4}{5}\right)^{i-2} r_{i;f}^{(n)}, \sum_{i=2}^{n-1}\left(-\frac{1}{2}\right)^{i-2} r_{i;f}^{(n)}\right) \begin{pmatrix} \frac{25}{9} & \frac{5}{7} \\ \frac{5}{7} & \frac{4}{3} \end{pmatrix} \begin{pmatrix} \sum_{i=2}^{n-1}\left(\frac{4}{5}\right)^{i-2} r_{i;f}^{(n)} \\ \sum_{i=2}^{n-1}\left(-\frac{1}{2}\right)^{i-2} r_{i;f}^{(n)} \end{pmatrix} \\ &\simeq n(r_{1;f}^{(n)})^2 + (1.436)n\left[\sum_{i=2}^{n-1}\left(\frac{4}{5}\right)^{i} r_{i;f}^{(n)}\right]^2 + (19.600)n\left[\sum_{i=2}^{n-1}\left(-\frac{1}{2}\right)^{i} r_{i;f}^{(n)}\right]^2 \\ & \quad - (3.937)n\left[\sum_{i=2}^{n-1}\left(\frac{4}{5}\right)^{i} r_{i;f}^{(n)}\right]\left[\sum_{i=2}^{n-1}\left(-\frac{1}{2}\right)^{i} r_{i;f}^{(n)}\right] . \end{aligned}$$

If ARMA (3,3) (or ARMA (3,2), ARMA (3,1), ARMA (1,3), ...) alternatives are to be considered, $Q_{\psi;f}^{(n)}$ has to be modified to

$$\begin{aligned} & n(r_{1;f}^{(n)})^2 + n(r_{2;f}^{(n)})^2 + (2.243)n\left[\sum_{i=3}^{n-1}\left(\frac{4}{5}\right)^{i} r_{i;f}^{(n)}\right]^2 + (78.400)n\left[\sum_{i=3}^{n-1}\left(-\frac{1}{2}\right)^{i} r_{i;f}^{(n)}\right]^2 \\ & \quad + (9.844)n\left[\sum_{i=3}^{n-1}\left(\frac{4}{5}\right)^{i} r_{i;f}^{(n)}\right]\left[\sum_{i=3}^{n-1}\left(-\frac{1}{2}\right)^{i} r_{i;f}^{(n)}\right] . \end{aligned}$$

5.3 Asymptotic Relative Efficiencies. Asymptotic, local power comparisons between tests are usually made on the basis of asymptotic relative efficiencies (ARE's). Recall that the ARE of a test ψ_1 with respect to another one ψ_2, both at probability level α, can be defined as the limiting value of the ratio N_2/N_1, where N_1 is the number of observations required for the power of ψ_1 to equal the power of

ψ_2 based on N_2 observations, as $\min(N_1, N_2) \to \infty$ and the alternative converges to the null hypothesis (see e.g. Puri and Sen 1985, section 3.8, for a more rigorous definition).

ARE computations are greatly simplified in situations where ψ_1 and ψ_2 are based on test statistics which are asymptotically normal or chi-square under the null hypothesis and contiguous consequences of alternatives. If ψ_1 and ψ_2 consist in rejecting the null hypothesis $H^{(n)}$ for "large values of" test statistics, S_1 and S_2 respectively, which are asymptotically normal, $\mathcal{N}(0, \sigma_1)$ and $\mathcal{N}(0, \sigma_2)$ respectively, under $H^{(n)}$, and $\mathcal{N}(\mu_1, \sigma_1)$ and $\mathcal{N}(\mu_2, \sigma_2)$ respectively, under the alternative $K^{(n)}$, then the ARE of ψ_1 with respect to ψ_2 is

$$\text{ARE } (\psi_1/\psi_2) = \left(\frac{\mu_1\sigma_2}{\sigma_1\mu_2}\right)^2 .$$

If S_1 and S_2 are both asymptotically chi-square with d degrees of freedom under $H^{(n)}$, and asymptotically noncentral chi-square with d degrees of freedom and noncentrality parameters λ_1 and λ_2, respectively, then

$$\text{ARE } (\psi_1/\psi_2) = (\lambda_1/\lambda_2)^2 .$$

The results of Sections 5.1 and 5.2 thus allow for an explicit computation of the mutual ARE's of the various parametric and nonparametric tests proposed.

PROPOSITION 5.5. *The asymptotic efficiency of the optimal procedures (as described in Propositions 5.1 and (5.3) based on (signed or unsigned) f–rank autocorrelations with respect to their counterparts based on (signed or unsigned) g–rank autocorrelations, against alternatives of the form $H_h^{(n)}(\boldsymbol{\theta} + n^{-\frac{1}{2}}\boldsymbol{\tau})$, is*

$$[I(f|h)/I(g|h)]^2 = \left[\frac{\int_0^1 \phi_f(F^{-1}(u))\phi_h(H^{-1}(u))du \int_0^1 F^{-1}(u)H^{-1}(u)du}{\int_0^1 \phi_g(G^{-1}(u))\phi_h(H^{-1}(u))du \int_0^1 G^{-1}(u)H^{-1}(u)du}\right]^2 ,$$

where notations $\phi_f, \phi_g, \phi_h, F, G, H$ are used in an obvious fashion. Under the same conditions, the ARE of procedures based on f–rank autocorrelations with respect to the optimal Gaussian parametric procedures, based on classical parametric autocorrelation coefficients (as described in Propositions 5.2 and 5.4) is

$$\left[\int_0^1 \phi_f(F^{-1}(u))\phi_h(H^{-1}(u))du \int_0^1 F^{-1}(u)H^{-1}(u)du\right]^2 /I(f_1) ,$$

which reduces to $I(f_1) \geq 1$ for $h = f$ and $I(f_1) = 1$ if and only if f and h are Gaussian.

Other asymptotic efficiencies can be derived from Proposition 4.3. Explicit numerical values are provided in Table 5.1 below.

Table 5.1. *Mutual ARE's for the various parametric and nonparametric tests described in Sections 5.1 and 5.2, under normal, logistic and double exponential densities, respectively.*

		(1)	(2)	(3)	(4)	(5)	
	Classical	1.000	1.000	1.005	1.634	1.096	normal
(1)	parametric	1.000	0.954	0.911	1.232	1.000	logistic
		1.000	0.816	0.675	0.500	0.790	double exp.
	Signed or	1.000	1.000	1.055	1.634	1.096	normal
(2)	unsigned	1.048	1.000	0.954	1.291	1.048	logistic
	van der Waerden	1.226	1.000	0.827	0.613	0.968	double exp.
	Signed or	0.948	0.948	1.000	1.550	1.091	normal
(3)	unsigned	1.098	1.048	1.000	1.352	1.098	logistic
	Wilcoxon	1.482	1.209	1.000	0.741	1.170	double exp.
	Signed or	0.612	0.612	0.646	1.000	0.671	normal
(4)	unsigned	0.812	0.775	0.740	1.000	0.812	logistic
	Laplace	2.000	1.631	1.350	1.000	1.580	double exp.
	Signed or	0.912	0.912	0.917	1.490	1.000	normal
(5)	unsigned	1.000	0.954	0.911	1.232	1.000	logistic
	Spearman	1.266	1.033	0.855	0.633	1.000	double exp.

An inspection of Table 5.1 reveals the excellent asymptotic performances of rank tests: the van der Waerden tests perform uniformly and strictly better than the corresponding normal-theory tests—except of course under normal densities, where they perform equally well. The ARE of optimal normal theory tests with respect to the corresponding Laplace procedure can be as low as 0.500. Spearman tests (which are never optimal, since the Spearman autocorrelation coefficients do not belong to the class of f-rank autocorrelations) are uniformly (though not very significantly) dominated by Wilcoxon tests.

As expected from Proposition 4.3, signed and unsigned optimal tests are asymptotically equivalent. The fact that their mutual ARE's are one however does not imply that the advantage of using signed-rank tests instead of unsigned ones (whenever innovation densities are symmetric) is nil, or negligible, neither for short series lengths, nor even asymptotically. Numerical investigations of the performance of signed-rank tests (Hallin et al. 1990) indicate that, for moderate, fixed n, the power of signed-rank procedures can be substantially larger than that of unsigned ones, an empirical finding that should be confirmed by a theoretical investigation of the corresponding deficiencies.

A systematic investigation of the finite sample behavior of (unsigned) rank based tests for randomness, both under the null hypothesis (tables of exact critical values) and under alternatives of $AR(1)$ serial dependence (Monte Carlo study of the power function) has been conducted in Hallin and Mélard (1988). This study reveals that rank tests often are substantially more powerful than traditional parametric procedures, even for pretty short series ($n = 20$, e.g.); of course, they are much more reliable under the null hypothesis, and theoretically provide exact tests (whereas the probability level of parametric tests is only approximative) even when traditional procedures are not valid — e.g. under Cauchy innovation densities (see Hallin and Puri (1991a) for a numerical illustration). In addition, they also are considerably more robust, and less sensitive to the presence of outliers, atypical startup behavior, etc.

All these properties show how useful a general rank based methodology would be in the identification and validation steps of time series analysis. Such a methodology however requires additional results allowing for the treatment of nuisance parameters. This is the subject of the next section.

6. ALIGNED RANK TESTS

6.1 Ranks and aligned ranks. The test described in Section 5 are valid if and only if the residuals $Z_t^{(n)}$, the ranks of which are used, under the null hypothesis to be tested, are *exact* residuals. These tests thus allow for testing ARMA models with unspecified innovation densities but completely specified coefficients.

Unfortunately, in most practical problems, one has to test ARMA models with *unspecified* coefficients. This is the case, typically, in diagnostic checking and model validation situations. It is also the case in identification problems, where the main tools (partial correlograms, corner method tables, ...) actually are test statistics— even though the identification process does not exactly reduce to any hypothesis testing problem.

More generally, one might like to test the null hypothesis $H^{(n)}(\mathcal{E}_r)$ (or $H_+^{(n)}(\mathcal{E}_r)$, in the case of unspecified, symmetric innovations) that the parameters $A_1, \ldots, A_{p_1}$, $B_1, \ldots, B_{q_1}$ of the ARMA (p_1, q_1) model underlying the observed series $\mathbf{X}^{(n)}$ satisfy some given linear constraints (to be precise, $p_1 + q_1 - r$ linearly independent ones) and thus belong to some r-dimensional linear subspace $\mathcal{E}_r$ of $\mathbb{R}^{p_1+q_1}$. Innovation densities in $H^{(n)}(\mathcal{E}_r)$ and $H_+^{(n)}(\mathcal{E}_r)$ remain unspecified; whenever they need to be specified, the notation $H_f^{(n)}(\mathcal{E}_r)$ will be used. The alternative consists of unrestricted ARMA (p, q) models with $p_1 \leq p \leq p_2$, $q_1 \leq q \leq q_2$. More specific alternatives however might be considered when optimality properties are to be described.

In what follows, we assume the existence of a root n consistent sequence $\widehat{\boldsymbol{\theta}}^{(n)} = (\widehat{A}_1^{(n)}, \ldots, \widehat{A}_{p_1}^{(n)}, 0, \ldots, 0, \widehat{B}_1^{(n)}, \ldots, \widehat{B}_{q_1}^{(n)}, 0, \ldots, 0)'$ of constrained estimates of $\boldsymbol{\theta} = (A_1, \ldots, A_{p_1}, 0, \ldots, 0, B_1, \ldots, B_{q_1}, 0, \ldots, 0)'$, i.e. a sequence of statistics such that $(\widehat{A}_1^{(n)}, \ldots, \widehat{A}_{p_1}^{(n)}, \widehat{B}_1^{(n)}, \ldots, \widehat{B}_{q_1}^{(n)})' \in \mathcal{E}_r$ (in the sequel, we also write $\widehat{\boldsymbol{\theta}}^{(n)} \in \mathcal{E}_r$ or $\boldsymbol{\theta} \in \mathcal{E}_r$) and the distributions of $n^{\frac{1}{2}}(\widehat{\boldsymbol{\theta}}^{(n)} - \boldsymbol{\theta})$ form, under $H^{(n)}(\mathcal{E}_r)$ (or under $H_+^{(n)}(\mathcal{E}_r)$), a relatively compact sequence.

Root n consistency however is not sufficient for establishing the asymptotic results stated below: additional uniformity properties are required, which are satisfied if "smoother" versions of $\widehat{\boldsymbol{\theta}}^{(n)}$ are used (see LeCam 1960, Appendix 1). One of the "smoothed" versions of $\widehat{\boldsymbol{\theta}}^{(n)}$ is as follows (same reference). Denote by $\mathcal{A}^{(n)}$ the σ-field of Borel sets of $\mathbb{R}^{(n)}$, by V an open convex symmetric neighborhood of the origin in $\mathcal{E}_r$, by $\mathcal{A}_*^{(n)}$ the product σ-field of $\mathcal{A}^{(n)}$ by the σ-field of Borel subsets of $\mathcal{E}_r$. On V, letting ν denote a probability measure having a bounded continuous density with respect to the Lebesgue measure, define $\mathbf{v}^{(n)}$ as a random variable having distribution ν on V, and put

$$\boldsymbol{\theta}_*^{(n)} = \widehat{\boldsymbol{\theta}}^{(n)} + n^{-\frac{1}{2}}\mathbf{v}^{(n)} .$$

The sequence of estimates $\boldsymbol{\theta}_*^{(n)}$ is still root n consistent, and meets the uniformity requirements that might not hold for $\widehat{\boldsymbol{\theta}}^{(n)}$ and are technically necessary for establishing Propositions 6.1 and 6.2 below.

Other smoothing methods are also described in LeCam (1960).

We do insist however that these smoothing procedures are of little practical relevance, if any at all. First, because they typically have the nature of analytical procedures guaranteeing ad hoc probabilistic limit results, whereas in statistical practice, one is interested in approximation results only: for given V and ν, and for fixed, "reasonably large" n, it makes extremely little difference whether $\boldsymbol{\theta}_*^{(n)}$ or $\widehat{\boldsymbol{\theta}}^{(n)}$ is used. A second reason is that the actual computation of $\widehat{\boldsymbol{\theta}}^{(n)}$, with a finite number of decimal values, *automatically* provides a "smooth", approximate value $\boldsymbol{\theta}_{**}^{(n)}$, say, of $\widehat{\boldsymbol{\theta}}^{(n)}$ (hence of $\boldsymbol{\theta}_*^{(n)}$).

For all these reasons, this smoothing problem should not be given too much attention and, in the sequel, we shall make no notational difference between $\widehat{\boldsymbol{\theta}}^{(n)}$, $\boldsymbol{\theta}_*^{(n)}$ and $\boldsymbol{\theta}_{**}^{(n)}$: the simple notation $\widehat{\boldsymbol{\theta}}^{(n)}$ will be used throughout, and $\widehat{\boldsymbol{\theta}}^{(n)}$ will be assumed "smooth enough".

Denote by $\widehat{Z}_t^{(n)} = Z_t^{(n)}(\widehat{\boldsymbol{\theta}}^{(n)}) = (\Sigma_i \widehat{A}_i^{(n)} L^i / \Sigma_i \widehat{B}_i^{(n)} L^i) X_t^{(n)}$ the "estimated" residuals associated with $\widehat{\boldsymbol{\theta}}^{(n)}$, by $\widehat{R}_t^{(n)} = R_t^{(n)}(\widehat{\boldsymbol{\theta}}^{(n)})$, $\widehat{R}_{+,t}^{(n)} = R_{+,t}^{(n)}(\widehat{\boldsymbol{\theta}}^{(n)})$ and $\widehat{s}_t = s_t(\widehat{\boldsymbol{\theta}}^{(n)})$ the corresponding ranks and signs (namely, the *aligned* ranks, *aligned* signed ranks and *aligned* signs for the problem considered). The notation $\widehat{r}_{i;f}^{(n)} = r_{i;f}^{(n)}(\widehat{\boldsymbol{\theta}}^{(n)})$, $\widehat{r}_{i;f}^{(n)+} = r_{i;f}^{(n)+}(\widehat{\boldsymbol{\theta}}^{(n)})$, $\widehat{a}_i, \widehat{b}_i, \widehat{\psi}_t^{(1)}, \ldots, \widehat{\psi}_t^{(p_1+q_1)}, \widehat{\mathbf{w}}_{\widehat{\psi}}^{(n)}, \widehat{\mathbf{W}}_{\widehat{\psi}}^{(n)}, \widehat{\mathbf{T}}_{\widehat{\psi};f}^{(n)}, \widehat{\mathbf{T}}_{\widehat{\psi};f}^{(n)+}, \ldots$ will be used in a similar fashion.

Since the f-rank autocorrelations $r_{i;f}^{(n)}$ and the rank-based statistics $\mathbf{T}_{\psi;f}^{(n)}$ (or the signed ones $\mathbf{T}_{\psi;f}^{(n)+}$) have been shown locally asymptotically sufficient for $H^{(n)}(\boldsymbol{\theta})$ (or $H_+^{(n)}(\boldsymbol{\theta})$) against $H_f^{(n)}(\boldsymbol{\theta} + n^{-\frac{1}{2}}\boldsymbol{\tau})$, $\boldsymbol{\tau} \in \mathbb{R}^{p_2+q_2}$, a natural idea would consist in using the aligned autocorrelations $\widehat{r}_{i;f}^{(n)}$ (or $\widehat{r}_{i;f}^{(n)+}$) and aligned rank statistics $\widehat{\mathbf{T}}_{\widehat{\psi};f}^{(n)}$ (or $\widehat{\mathbf{T}}_{\widehat{\psi};f}^{(n)+}$) for $H^{(n)}(\mathcal{E}_r)$ (or $H_+^{(n)}(\mathcal{E}_r)$) against $\bigcup\{H_f^{(n)}(\boldsymbol{\theta} + n^{-\frac{1}{2}}\boldsymbol{\tau}), \boldsymbol{\theta} \in \mathcal{E}_r, \boldsymbol{\theta} + n^{-\frac{1}{2}}\boldsymbol{\tau} \notin \mathcal{E}_r\}$. A closer look at this apparently simple idea however reveals a number of apparently unredeemable theoretical drawbacks.

First, invariance is lost. Substituting estimated parameters $\widehat{\boldsymbol{\theta}}^{(n)}$ for exact ones $\boldsymbol{\theta}$ induces among the residuals $\widehat{Z}_t^{(n)}$ complex interrelations that destroy their exchangeability features, hence the invariance property of ranks. This is not just a slight defect in view of the fact, stressed in Section 1, that invariance is the cornerstone of rank-based inference: *if invariance is lost, there is not point in using ranks anymore.*

Second, distribution-freeness, and even asymptotic distribution-freeness, is lost as well: it follows indeed from Proposition 6.1 below that the asymptotic mean of aligned rank autocorrelation coefficients under $H_g^{(n)}(\boldsymbol{\theta})$ depends on g. This again is not just a detail without importance, since distribution-freeness (or, at least, asymptotic distribution-freeness) is crucially necessary if a testing procedure valid under unspecified densities g is to be carried out.

Last and not the least, the local optimality properties of Section 5 obviously cannot be expected to hold anymore in the case of aligned rank tests. Even in the parametric, normal-theory context, additional requirements of unbiasedness or similarity have to be invoked, when nuisance parameters are present, if optimality results are to be obtained. Some form of similarity is likely to be necessary here, too.

6.2 Asymptotic invariance. Since strict invariance apparently is too restrictive a requirement here, some weaker, asymptotic form of it might be more appropriate. Let us define an asymptotically invariant statistic as a statistic asymptotically equivalent (under the null hypothesis considered, hence under contiguous alternatives) to an invariant (hence, in the present context, rank-based) one. More precisely, let $S^{(n)}$ denote a sequence of rank-based statistics such that, for some appropriate centering sequence $m^{(n)}, n^{-\frac{1}{2}}(S^{(n)} - m^{(n)})$ is relatively compact under a sequence of null hypotheses $H^{(n)}$. The sequence $\widetilde{S}^{(n)}$ obtained on substituting aligned ranks (aligned signs, aligned signed ranks) for the exact ones in $S^{(n)}$ is said to be ***asymptotically invariant*** under $H^{(n)}$ if $\widetilde{S}^{(n)} - S^{(n)} = o_P(n^{-\frac{1}{2}})$, under $H^{(n)}$, as $n \to \infty$.

This concept of asymptotic invariance here would be helpful if, e.g. $\widehat{r}_{i;f}^{(n)} - r_{i;f}^{(n)}$ or $\widehat{r}_{i;f}^{(n)+} - r_{i;f}^{(n)+}$ would be $o_P(n^{-\frac{1}{2}})$ under $H^{(n)}(\boldsymbol{\theta})$, i.e. if aligned rank autocorrelation coefficients would be asymptotically equivalent to the genuine ones. Unfortunately, this does not hold, since, from Proposition 4.2, it can be shown that

PROPOSITION 6.1. *Under $H_g^{(n)}(\boldsymbol{\theta})$,*

$$n^{\frac{1}{2}}(\widehat{r}_{i;f}^{(n)} - r_{i;f}^{(n)}) = -(\widehat{a}_i^{(n)} + \widehat{b}_i^{(n)})I(f|g) + o_P(1)\,, \quad n \to \infty\,, \tag{6.1}$$

where $\widehat{a}_i^{(n)}$ and $\widehat{b}_i^{(n)}$ result from substituting $n^{1/2}(\widehat{\boldsymbol{\theta}}^{(n)} - \boldsymbol{\theta})$ for $\boldsymbol{\tau} = (\boldsymbol{\gamma}', \boldsymbol{\delta}')'$ in (4.4). The same result also holds for $\widehat{r}_{i;f}^{(n)+}$.

It follows that $\widehat{r}_{i;f}^{(n)}$ cannot be asymptotically invariant: $I(f|g)$ indeed is not a distribution-free quantity. Nor can $\widehat{r}_{i;f}^{(n)+}$ be.

Though however no aligned-rank autocorrelation individually is asymptotically invariant, some specific linear combinations of them are. More precisely, denote by $\mathbf{T}_{\psi;f}^{(n)}$ the locally (at $\boldsymbol{\theta}$) sufficient statistic associated with the fundamental system $\{\psi_t^{(1)}, \ldots, \psi_t^{(p_1+q_1)}\}$, by $\widehat{\psi}_t^{(j)}$ the value at t of the solution of $(\Sigma_i \widehat{A}_i L^i)(\Sigma_i \widehat{B}_i L^i)\psi_t =$

0 characterized by the same starting values $\widehat{\psi}_s^{(j)} = \psi_s^{(j)}$, $s = 1, 2, \ldots, p_1 + q_1$ as $\psi_t^{(j)}$, and by $\widehat{\mathbf{T}}^{(n)}_{\widehat{\psi};f}$ the locally (at $\boldsymbol{\theta} = \widehat{\boldsymbol{\theta}}^{(n)}$) sufficient statistic associated with the fundamental system $\{\widehat{\psi}_t^{(n)}, \ldots, \widehat{\psi}_t^{(p_1+q_1)}\}$ but computed from the aligned ranks $\widehat{R}_t^{(n)}$ · $\widehat{\mathbf{T}}^{(n)+}_{\widehat{\psi};f}$ can be defined similarly in the signed rank case.

The asymptotic mean under $H_g^{(n)}(\boldsymbol{\theta}+n^{-\frac{1}{2}}\boldsymbol{\tau})$ of $n^{1/2}\mathbf{T}^{(n)}_{\psi;f}$ is $\boldsymbol{\mu}_{\psi;f}(\boldsymbol{\tau})I(f|g)$ (Proposition 4.4) where $\boldsymbol{\mu}_{\psi;f}$ is a linear transform (of rank $\pi + p_1 + q_1$) of $\boldsymbol{\tau}$, depending on $\boldsymbol{\theta}$ and the choice of the fundamental system $\{\psi_t^{(j)}\}$, but not on g. When $\boldsymbol{\tau}$ unrestrictedly takes its values in $\mathbf{R}^{p_2+q_2}$, $\boldsymbol{\mu}_{\psi;f}$ can take any value in $\mathbf{R}^{\pi+p_1+q_1}$. If $\boldsymbol{\tau}$ is of the form $(\gamma_1, \ldots, \gamma_{p_1}, 0, \ldots, 0, \delta_1, \ldots, \delta_{q_1}, 0, \ldots, 0)'$, where $(\gamma_1, \ldots, \gamma_{p_1}, \delta_1, \ldots, \delta_{q_1})'$ belongs to $\mathcal{E}_r$, thus satisfying $p_1 + q_1 - r$ independent linear restrictions, it can be shown that $\boldsymbol{\mu}_{\psi;f}(\boldsymbol{\tau})$ also satisfies $\pi + p_1 + q_1 - r$ independent linear restrictions, of the form

$$\boldsymbol{\Omega}(\boldsymbol{\theta})\boldsymbol{\mu}_{\psi;f} = \mathbf{0}\,, \tag{6.2}$$

where $\boldsymbol{\Omega}(\boldsymbol{\theta})$, a $(\pi + p_1 + q_1 - r) \times (\pi + p_1 + q_1)$ matrix of rank $\pi + p_1 + q_1 - r = \max(p_2 + q_1, p_1 + q_2) - r$, is a continuous function of $\boldsymbol{\theta}$. The vector $\boldsymbol{\mu}_{\psi;f}$ accordingly lies in a r-dimensional linear subspace of $\mathbf{R}^{\pi+p_1+q_1}$.

The following result can then be established.

PROPOSITION 6.2. *For any $\boldsymbol{\theta} \in \mathcal{E}_r$, any fundamental system $\{\psi_t^{(i)}\}$ and any f ,*

$$\boldsymbol{\Omega}(\widehat{\boldsymbol{\theta}}^{(n)}) \left[\widehat{\mathbf{T}}^{(n)}_{\widehat{\psi};f} - \mathbf{T}^{(n)}_{\psi;f}\right] = o_P(n^{-\frac{1}{2}}) \tag{6.3}$$

under $H^{(n)}(\boldsymbol{\theta})$, $\boldsymbol{\theta} \in \mathcal{E}_r$, as $n \to \infty$. Accordingly, $\boldsymbol{\Omega}(\widehat{\boldsymbol{\theta}}^{(n)})\widehat{\mathbf{T}}^{(n)}_{\widehat{\psi};f}$ is a vector of $\pi + p_1 + q_1 - r$ asymptotically invariant statistics (under $H^{(n)}(\boldsymbol{\theta})$).

A similar result can be proved, under $H_+^{(n)}(\boldsymbol{\theta})$, for $\boldsymbol{\Omega}(\widehat{\boldsymbol{\theta}}^{(n)})\widehat{\mathbf{T}}^{(n)+}_{\widehat{\psi};f}$.

6.3 Locally asymptotically similar tests. The following definition of local asymptotic similarity has been proposed by LeCam (1960) (under the terminology *differential asymptotic similarity*). A sequence $\phi^{(n)}$ of tests is said to be locally asymptotically similar under $H_f^{(n)}(\mathcal{E}_r)$, at asymptotic probability level α, if for every $\boldsymbol{\theta} \in \mathcal{E}_r$ and every bounded $B \subset \mathcal{E}_r$

$$\lim_{n\to\infty} \sup_{\boldsymbol{\tau}\in B} \left|E_{H_f^{(n)}(\boldsymbol{\theta}+n^{-\frac{1}{2}}\boldsymbol{\tau})}(\phi^{(n)}) - \alpha\right| = 0\,. \tag{6.4}$$

LeCam (loc. cit.) then shows how locally optimal tests, asymptotically enjoying, within the class of locally asymptotically similar tests, the same optimal properties (stringency or minimaxity) as the usual normal theory tests for Gaussian linear hypotheses, follow on applying the Gaussian likelihood ratio principle at the local asymptotical level.

For the particular problem of testing $H_f^{(n)}(\mathcal{E}_r)$, this leads to a test statistic of the form

$$
\begin{aligned}
Q_f^{(n)} &= \inf_{\tilde{\mu}:\Omega(\hat{\theta}^{(n)})\tilde{\mu}=0} [(n^{\frac{1}{2}}\hat{\mathbf{T}}^{(n)}_{\hat{\psi};f} - \tilde{\mu})'\widehat{\mathbf{W}}_{\hat{\psi}}^{-1}(n^{\frac{1}{2}}\hat{\mathbf{T}}^{(n)}_{\hat{\psi};f} - \tilde{\mu})] \\
&= n(\hat{\mathbf{T}}^{(n)}_{\hat{\psi};f})'\left[\widehat{\mathbf{W}}_{\hat{\psi}}^{-1} - \widehat{\mathbf{W}}_{\hat{\psi}}^{-1}\mathbf{K}(\mathbf{K}'\widehat{\mathbf{W}}_{\hat{\psi}}^{-1}\mathbf{K})^{-1}\mathbf{K}'\widehat{\mathbf{W}}_{\hat{\psi}}^{-1}\right]\hat{\mathbf{T}}^{(n)}_{\hat{\psi};f},
\end{aligned}
\tag{6.5}
$$

where $\mathbf{K} = \mathbf{K}(\hat{\theta}^{(n)})$ is any $(\pi + p_1 + q_1) \times r$ full rank matrix such that $\Omega(\hat{\theta}^{(n)})\mathbf{K} = \mathbf{0}$ (so that $\Omega(\hat{\theta}^{(n)})\tilde{\mu} = \mathbf{0}$ if and only if $\tilde{\mu} = \mathbf{K}\lambda$ for some $\lambda \in \mathbf{R}^r$, and $\mathbf{N}\ \mathbf{K} = \mathbf{0}$ if and only if $\mathbf{N} = \mathbf{M}\Omega(\hat{\theta}^{(n)})$ for some matrix $\mathbf{M}$ of appropriate dimension).

Denoting by $\widehat{\mathbf{W}}_{\hat{\psi}}^{\frac{1}{2}}$ an arbitrary symmetric square root of $\widehat{\mathbf{W}}_{\hat{\psi}}$, (6.5) also takes the form

$$
\begin{aligned}
&n(\hat{\mathbf{T}}^{(n)}_{\hat{\psi};f})'\widehat{\mathbf{W}}_{\hat{\psi}}^{-\frac{1}{2}}\left[\mathbf{I} - \widehat{\mathbf{W}}_{\hat{\psi}}^{-\frac{1}{2}}\mathbf{K}(\mathbf{K}'\widehat{\mathbf{W}}_{\hat{\psi}}^{-1}\mathbf{K})^{-1}\mathbf{K}'\widehat{\mathbf{W}}_{\hat{\psi}}^{-\frac{1}{2}}\right]\widehat{\mathbf{W}}_{\hat{\psi}}^{-\frac{1}{2}}\hat{\mathbf{T}}^{(n)}_{\hat{\psi};f} \\
&= n(\hat{\mathbf{T}}^{(n)}_{\hat{\psi};f})'\widehat{\mathbf{W}}_{\hat{\psi}}^{-\frac{1}{2}}\mathbf{P}\widehat{\mathbf{W}}_{\hat{\psi}}^{-\frac{1}{2}}\hat{\mathbf{T}}^{(n)}_{\hat{\psi};f},
\end{aligned}
\tag{6.6}
$$

where $\mathbf{P}$ is a symmetric, idempotent matrix of rank $\pi + p_1 + q_1 - r$. $Q_f^{(n)}$ therefore is asymptotically chi–square under $H_f^{(n)}(\theta)$, $\theta \in \mathcal{E}_r$. But, since $\mathbf{P}\ \widehat{\mathbf{W}}_{\hat{\psi}}^{-\frac{1}{2}}\mathbf{K} = \mathbf{0}$, $Q_f^{(n)}$ is also of the form

$$
n(\hat{\mathbf{T}}^{(n)}_{\hat{\psi};f})'\Omega'(\hat{\theta}^{(n)})\mathbf{M}'\ \mathbf{M}\Omega(\hat{\theta}^{(n)})\hat{\mathbf{T}}^{(n)}_{\hat{\psi};f}
$$

(for some matrix $\mathbf{M}$, the explicit form of which is not needed here); hence, in view of Proposition 6.2, $Q_f^{(n)}$ is asymptotically invariant under $H^{(n)}(\theta)$, $\theta \in \mathcal{E}_r$. Accordingly, $Q_f^{(n)}$ is also asymptotically chi–square, with $\pi + p_1 + q_1 - r$ degrees of freedom, under any hypothesis of the form $H_g^{(n)}(\theta)$, hence under $H^{(n)}(\theta)$.

A similar result holds for the signed version $Q_f^{(n)+}$ of $Q_f^{(n)}$.

6.4 Optimal aligned rank tests. Summing up the findings of Sections 6.2 and 6.3, we may state the following result.

PROPOSITION 6.3. *The sequence of aligned–rank tests rejecting $H^{(n)}(\mathcal{E}_r)$ whenever*

$$
Q_f^{(n)} > \chi^2_{1-\alpha},
$$

where $\chi^2_{1-\alpha}$ denotes the $(1-\alpha)$–quantile of a chi–square variable with $\pi + p_1 + q_1 - r$ degrees of freedom and $Q_f^{(n)}$ is given in (6.5)

(i) is asymptotically invariant

(ii) is locally asymptotically similar under $H^{(n)}(\mathcal{E}_r)$, at probability level α

(iii) is asymptotically most stringent against $\bigcup\{H_f^{(n)}(\theta + n^{-\frac{1}{2}}\tau)|\theta \in \mathcal{E}_r,\ \tau \notin \mathcal{E}_r\}$, within the class of all asymptotically similar (under $H_f^{(n)}(\mathcal{E}_r)$, at probability level α) tests.

A similar proposition holds in the signed–rank case.

The test thus proposed possesses all the asymptotic optimality properties one can expect when testing $H_f^{(n)}(\mathcal{E}_r)$ (with specified innovation density f). In addition, it is asymptotically invariant, similar and distribution–free under the much broader null hypothesis $H^{(n)}(\mathcal{E}_r)$ (with unspecified innovation density); it has much better robustness features than its usual Gaussian parametric competitors, with ARE values (with respect to the latter) as shown in Table 5.1. The amount of computation involved is not heavier than in the case e.g. of traditional Lagrange multiplier tests (see also Kreiss (1990 a) for optimal parametric AR procedures).

For all these reasons, they should be very attractive in a variety of practical problems when innovation densities are suspected to be severely non–Gaussian, or contaminated by possible outliers. For instance, it advantageously could be substituted for parametric Lagrange multiplier statistics in Pötscher (1983, 1985)'s recursive identification procedure.

6.5 Example: testing AR (1) versus ARMA (2,1) dependence. As an illustration, consider the problem of testing a null hypothesis of first–order autoregressive dependence

$$X_t - \rho X_{t-1} = \varepsilon_t \ , \quad 0 < |\rho| < 1 \ ,$$

where the parameter ρ and density of ε_t remain unspecified. Denote by $\widehat{\rho}^{(n)}$ a root n consistent estimate of ρ (the least square estimate, for instance), and let $\widehat{Z}_t^{(n)} = X_t^{(n)} - \widehat{\rho}^{(n)} X_{t-1}^{(n)}$, $t = 1, \ldots, n$ (for simplicity, assume that $X_0^{(n)}$ is available). Assume that an ARMA (2,1) alternative is considered. Suppose however that, for some reason, one is willing to be particularly powerful against ARMA processes with innovation density f (f specified up to a scale transformation). The (aligned) ranks $\widehat{R}_t^{(n)}$ here are those of the residuals $\widehat{Z}_t^{(n)}$, and the autocorrelation coefficients to be used are the (aligned) f–rank autocorrelations $\widehat{r}_{i;f}^{(n)}$.

A simple fundamental system (a fundamental system here consists of any non-identically zero solution, since the dimension of the solution space is one) of solutions of $(1 - \widehat{\rho}^{(n)} L)\psi_t = 0$ is $\widehat{\psi}_t = (\widehat{\rho}^{(n)})^{t-2}$, and the corresponding (aligned) locally sufficient statistic is ($p_1 = 1$, $q_1 = 0$; hence $\pi = 1$ and $r = 1$)

$$n^{\frac{1}{2}} \widehat{\mathbf{T}}_{\widehat{\rho};f}^{(n)} = \begin{pmatrix} (n-1)^{\frac{1}{2}} \ \widehat{r}_{1;f}^{(n)} \\ \sum_{j=2}^{n-1} (n-j)^{\frac{1}{2}} (\widehat{\rho}^{(n)})^{j-2} \widehat{r}_{j;f}^{(n)} \end{pmatrix} .$$

The corresponding, *exact* $\mathbf{T}_{\psi;f}^{(n)}$ (not a statistic since ρ is unknown) is

$$n^{\frac{1}{2}} \mathbf{T}_{\rho;f}^{(n)} = \begin{pmatrix} (n-1)^{\frac{1}{2}} \ r_{1;f}^{(n)} \\ \sum_{j=2}^{n-1} (n-j)^{\frac{1}{2}} \ \rho^{j-2} r_{j;f}^{(n)} \end{pmatrix}$$

which is asymptotically bivariate normal, with mean zero and covariance matrix

$$\mathbf{W}_\psi = \begin{pmatrix} 1 & 0 \\ 0 & (1-\rho^2)^{-1} \end{pmatrix}$$

under $H^{(n)}(\rho)$.

Local ARMA (2,1) alternatives $H_g^{(n)}(\rho,\boldsymbol{\gamma},\delta)$ are of the form

$$X_t - (\rho + n^{-\frac{1}{2}}\gamma_1)X_{t-1} - n^{-\frac{1}{2}}\gamma_2 X_{t-2} = \varepsilon_t + n^{-\frac{1}{2}}\delta\ \varepsilon_{t-1}\ ,$$

where $\boldsymbol{\gamma} = (\gamma_1, \gamma_2)'$ and the innovation density is g ; $n^{\frac{1}{2}}\mathbf{T}_{\psi;f}^{(n)}$ under such an alternative is asymptotically bivariate normal, still with covariance matrix $\mathbf{W}$, and with mean

$$\boldsymbol{\mu}_{\rho;f}(\boldsymbol{\gamma},\delta)I(f|g) = \begin{pmatrix} \gamma_1 + \delta \\ \\ [\rho(\gamma_1 + \delta) + \gamma_2]/(1-\rho^2) \end{pmatrix} I(f|g)$$

since $$a_i = \begin{cases} \gamma_1 & \text{if} \quad i = 1 \\ \\ \rho^{i-2}(\rho\gamma_1 + \gamma_2) & \text{if} \quad i > 1 \end{cases} \quad \text{and} \quad b_i = \delta\rho^{i-1}\ ,\ i = 1,2,\ldots\ .$$

Clearly, if $\gamma_2 = \delta = 0$, $\boldsymbol{\mu}_{\rho;f}(\boldsymbol{\gamma},\delta)$ is of the form $\gamma_1(1 \quad \rho/(1-\rho^2))'$, and accordingly satisfies the (unique, since $\pi + p_1 + q_1 - r = 1 + 1 + 0 - 1 = 1$) linear constraint

$$(-\rho \qquad 1-\rho^2)\boldsymbol{\mu}_{\rho;f}(\boldsymbol{\gamma},\delta) = 0\ .$$

Here $\boldsymbol{\Omega}(\boldsymbol{\theta}) = \boldsymbol{\Omega}(\rho)$ is the 1×2 row matrix $(-\rho \quad 1-\rho^2)$. It follows that

$$\begin{aligned} &(-\widehat{\rho}^{(n)} \qquad 1-(\widehat{\rho}^{(n)})^2)\ \widehat{\mathbf{T}}_{\rho;f}^{(n)} \\ (6.7) \qquad &= -(n-1)^{\frac{1}{2}}\widehat{\rho}^{(n)}\widehat{r}_{1;f}^{(n)} + [1-(\widehat{\rho}^{(n)})^2]\sum_{j=2}^{n-1}(n-j)^{\frac{1}{2}}(\widehat{\rho}^{(n)})^{j-2}\widehat{r}_{j;f}^{(n)} \end{aligned}$$

is asymptotically invariant, and asymptotically equivalent to its exact, non-aligned counterpart. The optimal, asymptotically similar aligned rank test of Proposition 6.3 consists in rejecting the null hypothesis whenever the quadratic statistic

$$(6.8) \qquad Q_f^{(n)} = \left\{ -(n-1)^{\frac{1}{2}}\widehat{\rho}^{(n)}\widehat{r}_{1;f}^{(n)} + [1-(\widehat{\rho}^{(n)})^2]\sum_{j=2}^{n-1}(n-j)^{\frac{1}{2}}(\widehat{\rho}^{(n)})^{j-2}\ \widehat{r}_{j;f}^{(n)} \right\}^2$$

exceeds the $(1-\alpha)$ quantile of a chi–square variable with one degree of freedom (since the asymptotic variance of (6.7) is one).

The form of the test statistic (6.8) is not quite familiar, and its relation with classical, parametric time series procedures does not straightforwardly appear from (6.8). Denote by $Q_{\mathcal{N}}^{(n)}$ the Gaussian counterpart of $Q_f^{(n)}$

$$(6.9) \qquad Q_{\mathcal{N}}^{(n)} = \left\{ -(n-1)^{\frac{1}{2}}\widehat{\rho}^{(n)}\widehat{r}_{1}^{(n)} + [1-(\widehat{\rho}^{(n)})^2]\sum_{j=2}^{n-1}(n-j)^{\frac{1}{2}}(\widehat{\rho}^{(n)})^{j-2}\widehat{r}_{j}^{(n)} \right\}^2 ,$$

where $\widehat{\rho}^{(n)}$ denotes the Gaussian maximum likelihood estimator of the AR (1) parameter ρ, and $\widehat{r}_j^{(n)}$ stands for the corresponding classical residual autocorrelations.

The Gaussian Lagrange multiplier approach to the problem (Godfrey 1979; Hosking 1980) leads to the test statistic (Hosking 1980, Theorem 1)

$$Q_{\mathcal{L}}^{(n)} = n \left\{ \sum_{j=2}^{n-1} (\widehat{\rho}^{(n)})^{j-2} \widehat{r}_j^{(n)} \right\}^2 , \tag{6.10}$$

to be compared also with the quantiles of a chi-square variable with one degree of freedom.

$Q_{\mathcal{N}}^{(n)}$ and $Q_{\mathcal{L}}^{(n)}$ apparently are distinct test statistics. However, it follows from asymptotic linear relationships among (parametric) estimated residual autocorrelations (McLeod 1978, Theorem 1) that

$$\sum_{j=1}^{n-1} (\widehat{\rho}^{(n)})^{j-1} \widehat{r}_j^{(n)} = o_P(n^{-\frac{1}{2}}) , \qquad n \to \infty . \tag{6.11}$$

Accordingly, our locally optimal Gaussian statistic $Q_{\mathcal{N}}^{(n)}$ satisfies

$$Q_{\mathcal{N}}^{(n)} = n(\widehat{r}_1^{(n)})^2 + o_P(1) , \qquad n \to \infty . \tag{6.12}$$

Similarly, for the Lagrange multiplier statistic, (6.11) implies

$$Q_{\mathcal{L}}^{(n)} = n(\widehat{r}_1^{(n)})^2 + o_P(1) , \qquad n \to \infty , \tag{6.13}$$

which in turn entails

$$Q_{\mathcal{N}}^{(n)} - Q_{\mathcal{L}}^{(n)} = o_P(1) , \qquad n \to \infty .$$

The parametric, Gaussian counterpart of our rank-based statistic (6.8) is thus asymptotically equivalent to the Gaussian Lagrange multiplier test statistic. It follows that the ARE's of rank tests based on (6.8) with respect either to their Gaussian parametric counterpart (based on (6.9)) or the more familiar Lagrange multiplier test (based on (6.10)) are still those provided in Table 5.1.

Whether an asymptotic equivalence of the form (6.11) or (6.12) also holds for aligned rank-based residual autocorrelations is not known. The asymptotic invariance of (6.7) implies that (6.8) is asymptotically equivalent to its exact unaligned version $\overline{Q}_f^{(n)}$, say. The van der Waerden version of the latter, under Gaussian densities, in turn is asymptotically equivalent to the parametric $Q_{\mathcal{N}}^{(n)}$, hence to $n(\widehat{r}_1^{(n)})^2$. Note however that, whereas $Q_{vdW}^{(n)}$, $Q_{\mathcal{N}}^{(n)}$ and $Q_{\mathcal{L}}^{(n)}$ are asymptotically equivalent to the corresponding quantities computed from exact residuals, these equivalences do not hold for $n(\widehat{r}_1^{(n)})^2$ which is asymptotically distinct from $n(r_1^{(n)})^2$, where $r_1^{(n)}$ denotes the exact first-order residual autocorrelation. Nor is $\widehat{r}_{1;vdW}^{(n)}$ asymptotically invariant. The linear relation (6.11) indeed does not hold for exact residual autocorrelations.

Acknowledgement. This work has greatly benefited from the pleasant environment and peaceful landscapes of La Martinie-Sergeac. Marc Hallin would like to thank Mr. and Mrs. Baudoux for their warm hospitality during the summer of 1990, as well as Dominique, Marie, Catherine and Françoise for their enduring patience.

REFERENCES

AIYAR, R.J. (1981), *Asymptotic efficiency of rank tests of randomness against autocorrelation*, Ann. Inst. Statist. Math **33 A**, 255–262.

AIYAR, R.J., GUILLIER, C.L., and ALBERS, W. (1979), *Asymptotic relative efficiencies of rank tests for trend alternatives*, Jour. Amer. Statist. Assoc. **74**, 226–231.

AKRITAS, M.G., and JOHNSON, R.A. (1982), *Efficiencies of tests and estimators for p-order autoregressive processes when the error is non normal*, Annals Inst. Statist. Math. **34**, 579–589.

ALBERS, W. (1978), *One–sample rank tests under autoregressive dependence*, Ann. Statist. **6**, 836–845.

BARTELS, R. (1982), *The rank version of von Neumann's ratio test for randomness*, Jour. Amer. Statist. Assoc. **77**, 40–46.

BENGHABRIT, Y. and HALLIN, M. (1992), *Optimal rank–based tests against first–order superdiagonal bilinear dependence*, Jour. Statist. Plan. and Inf., to appear.

BERAN, R.J. (1972), *Rank spectral processes and tests for serial dependence*, Ann. Math. Statist. **43**, 1749–1766.

BHATTACHARYYA, G.K. (1984), *Tests for randomness against trend or serial correlations*, In Handbook of Statistics, **4**, (P.R. Krishnaiah and P.K. Sen, eds.), 89–111. North–Holland, Amsterdam and New York.

BROCKWELL, P.J., and DAVIS, R.A. (1987), *Time Series: Theory and Applications*, Springer Verlag, New York.

CAMPBELL, P.J., and DUFOUR, J.–M. (1991), *Over-rejections in rational expectations models: a nonparametric approach to the Mankiw–Shapiro problem*, Economics Letters, (35), 285–290.

DAVID, F.N. (1947), *A power function for tests of randomness in a sequence of alternatives*, Biometrika **34**, 335–339.

DAVID, F.N. and FIX, E. (1966), *Randomization and the serial correlation coefficient*, In Research Papers in Statistics, Festschrift for J. Neyman, (F.N. David, ed.), 461–468. J. Wiley, New York.

DAVIES, R.B. (1973), *Asymptotic inference in stationary Gaussian time series*, Adv. Appl. Prob. **5**, 469–497.

DUFOUR, J.–M. (1981), *Rank tests for serial dependence*, Jour. Time Series Anal. **2**, 117–128.

DUFOUR, J.–M. and HALLIN, M. (1987), *Tests non paramétriques optimaux pour le modèle autorégressif d'ordre un*, Ann. Econ. Stat. **5**, 411–434.

DUFOUR, J.–M. and HALLIN, M. (1990a), *Runs tests for ARMA dependence*, Technical Report, Département de Sciences Economiques, Université de Montréal, Montréal.

DUFOUR, J.–M. and HALLIN, M. (1990b), *An exponential bound for the permutational distribution of a first–order autocorrelation coefficient*, Statistique et Analyse des Données **15**, 45–56.

DUFOUR, J.–M., and HALLIN, M., (1991), *Improved Eaton bounds for linear combinations of bounded random variables, with statistical applications*, Technical Report, C.R.D.E., Université de Montréal, Montréal.

DUFOUR, J.–M., LEPAGE, Y. and ZEIDAN, H. (1982), *Nonparametric testing for time series: a bibliography*, Canad. Jour. Statist., **10**, 1–38.

DUFOUR, J.–M., and ROY, R. (1985), *Some robust exact results on sample autocorrelations and tests of randomness*, Journal of Econometrics **29** , 257–273 and **41**, 279–281.

DUFOUR, J.–M., and ROY, R.(1986), *Generalized portmanteau statistics and tests for randomness*, Communications in Statistics A, Theory and Methods **15**, 2953–2972.

DZHAPARIDZE, K.O. (1986), *Parameter Estimation and Hypothesis Testing in Spectral Analysis of Stationary Time Series*, Springer Verlag, New York.

EDGINGTON, E.S. (1961), *Probability table for number of runs of signs of first differences in ordered series*, J. Amer. Statist. Assoc. **56**, 156–159.

FISHER, R.A. (1926), *On the random sequence*, Quart. J. Roy. Meteorol. Soc. **52**, 250.

GAREL, B. (1989), *The asymptotic distribution of the likelihood ratio for MA processes with a regression trend*, Statist. Decisions **7**, 167–184.

GHOSH, M.N. (1954), *Asymptotic distribution of serial statistics and applications to problems of nonparametric tests of hypotheses*, Ann. Math. Statist. **25**, 218–251.

GODFREY, L.G. (1979), *Testing the adequacy of a time series model*, Biometrika **66**, 67–72.

GOODMAN, L.A. (1958), *Simplified runs tests and likelihood ratio tests for Markov chains*, Biometrika **45**, 181–197.

GOODMAN, L.A. and GRUNFELD, Y. (1961), *Some nonparametric tests for comovements between time series*, J. Amer. Statist. Assoc. **56**, 11–26.

GOVINDARAJULU, Z. (1983), *Rank tests for randomness against autocorrelated alternatives*, In Time Series Analysis: Theory and Practice 4. (O.D. Anderson, ed.) 65–73. North–Holland, Amsterdam and New York.

GRANGER, C.W.J. (1963), *A quick test for serial correlation suitable for use with non–stationary time series*, Jour. Amer. Statist. Assoc. **58**, 728–736.

GUPTA, G.D. and GOVINDARAJULU, Z. (1980), *Nonparametric tests of randomness against autocorrelated normal alternatives*, Biometrika **67**, 375–379.

HÁJEK, J. and ŠIDÁK, Z. (1967), *Theory of Rank Tests*, Academic Press, New York.

HALLIN, M. (1986), *Nonstationary q-dependent processes and time-varying moving-average models: invertibility properties and the forecasting problem*, Advances in Applied Probability **18**, 170–210.

HALLIN, M., INGENBLEEK, J.–FR. and PURI, M.L. (1985), *Linear serial rank tests for randomness against ARMA alternatives*, Ann. Statist. **13**, 1156–1181.

HALLIN, M., INGENBLEEK, J.–FR. and PURI, M.L. (1987), *Linear and quadratic serial rank tests for randomness against serial dependence*, Jour. Time Series Anal. **8**, 409–424.

HALLIN, M., INGENBLEEK, J.–FR. and PURI, M.L. (1989), *Asymptotically most powerful rank tests for multivariate randomness against serial dependence*, Jour. Multivariate Analysis **30**, 34–71.

HALLIN, M., LAFORET, A. and MÉLARD, G. (1990), *Distribution–free tests against serial dependence: signed or unsigned ranks?*, Jour. Statist. Plan. Inf. **24**, 151–165.

HALLIN, M., LEFEVRE, C. and PURI, M.L. (1988), *On time–reversibility and the uniqueness of moving average representations for non–Gaussian stationary time series*, Biometrika **75**, 170–171.

HALLIN, M. and MÉLARD, G. (1988), *Rank–based tests for randomness against first order serial dependence*, Jour. Amer. Statist. Assoc. **83**, 1117–1129.

HALLIN, M., MÉLARD, G. and MILHAUD, X. (1992), *Permutational extreme values of autocorrelation coefficients and a Pitman test against serial dependence*, Ann. Statist. **20**, to appear.

HALLIN, M. and PURI, M.L. (1988), *On locally asymptotically maximin tests for ARMA processes*, in Statistical Theory and Data Analysis II, K. Matusita, Ed., North–Holland, Amsterdam and New-York, 495–500.

HALLIN, M. and PURI, M.L. (1988), *Optimal rank–based procedures for time–series analysis: testing an ARMA model against other ARMA models*. *Ann. Statist.*, **16**, 402–432.

HALLIN, M. and PURI, M.L. (1991a), *Time series analysis via rank–order theory: signed–rank tests for ARMA models*, Jour. Multivariate Analysis **39**, 1–29.

HALLIN, M. and PURI, M.L. (1991b), *A multivariate Wald–Wolfowitz rank test against serial dependence*, Technical Report.

HALLIN, M. and PURI, M.L. (1991c), *Aligned rank tests for linear models with autocorrelated errors and time series with a linear trend*, Technical Report.

HAREL, M. (1988), *Weak convergence of multidimensional rank statistics under φ–mixing conditions*, Jour. Statist. Plan. Inf. **20**, 41–63.

HAREL, M. and PURI, M.L. (1989a), *Limiting behavior of U–statistics, V–statistics and one sample rank order statistics for nonstationary absolutely regular processes*, Jour. Multivariate Analysis **30**, 181–204.

HAREL, M. and PURI, M.L. (1989b), *Weak convergence of the U–statistic and weak invariance of the one sample rank order statistic for Markov processes and ARMA models*, Jour. Multivariate Analysis **31**, 258–265.

HAREL, M. and PURI, M.L. (1990a), *Limiting behavior of one sample rank–order statistics with unbounded scores for nonstationary absolutely regular processes*, Jour. Statist. Plan. Inf. **27**, 1–23.

HAREL, M. and PURI, M.L. (1990b), *Weak convergence of serial rank statistics under dependence with applications in time series and Markov processes*, Ann. Prob. **18**, 1361–1387.

HAREL, M. and PURI, M.L.(1990c), *Weak convergence of the serial linear rank statistic with unbounded scores and regression constants under mixing conditions*, Jour. Statist. Plan. Inf. **25**, 163–186.

HAREL, M. and PURI, M.L. (1991), *Weak invariance of the multidimensional rank statistic for nonstationary absolutely regular processes*, Jour. Multivariate Analysis **36**, 204–221.

HAREL, M. and PURI, M.L. (1992), *Weak convergence of the simple linear rank statistic under mixing conditions in the nonstationary case*, Teor. Veroyatnost. i Premenen., to appear.

HOSKING, J.R.M. (1980), *Lagrange multiplier tests of time series models*, Jour. Roy. Statist. Soc. B. **42**, 170–181.

JOGDEO, K. (1968), *Asymptotic normality in nonparametric methods*, Ann. Math. Statist. **39**, 905–922.

KENDALL, M., and STUART, A. (1968), *The advanced Theory of Statistics*, Vol. 3, Griffin, London.

KERMACK, W.O. and MCKENDRICK, A.G. (1937), *Tests for randomness in a series of numerical observations*, Proc. Roy. Soc. Edinburgh **57**, 228–240.

KNOKE, D.J. (1977), *Testing for randomness against autocorrelation: alternative tests*, Biometrika **64**, 523–529.

KREISS, J.–P. (1987), *On adaptive estimation in stationary ARMA processes*, Ann. Statist. **15**, 112–133.

KREISS, J.–P. (1990a), *Testing linear hypotheses in autoregressions*, Annals of Statistics **18**, 1470–1482.

KREISS, J.–P. (1990b), *Local asymptotic normality for autoregression with infinite order*, Jour. of Statist. Plan. and Inf., **26**, 185–219.

LECAM, L. (1960), *Locally asymptotically normal families of distributions*, Univ. Calif. Publ. Statist. **3**, 37–98.

LECAM, L.(1986), *Asymptotic Methods in Statistical Decision Theory*, Springer–Verlag, New York.

LEVENE, H. and WOLFOWITZ, J. (1944), *The covariance matrix of runs up and down*, Ann. Math. Statist. **15**, 58–69.

MANN, H.B. (1945), *Nonparametric test against trend*, Econometrica **13**, 245–259.

MC.LEOD, A.I. (1978), *On the distribution of residual autocorrelations in Box–Jenkins models*, J.R.S.S. ser. B, **40**, 296–302.

MOOD, A.M. (1940), *The distribution theory of runs*, Ann. Math. Statist. **11**, 367–392.

MOORE, G.H. and WALLIS, W.A. (1943), *Time–series significance tests based on signs of differences*, Jour. Amer. Statist. Assoc. **38**, 153–164.

NIEUWENHUIS, G. and RUYMGAART, F. (1990), *Some stochastic inequalities and asymptotic normality of serial rank statistics in general linear processes*, Jour. Statist. Plan. Inf. **25**, 53–80.

PHAM, D.T. (1987), *Exact maximum likelihood estimate and Lagrange multiplier test statistic for ARMA models*, Jour. Time Series Anal. **8**, 61–78.

PÖTSCHER, B.M. (1983), *Order estimation in ARMA models by Lagrangian multiplier tests*, Annals of Statistics, **11**, 872–885.

PÖTSCHER, B.M. (1985), *The behaviour of the Lagrange multiplier test in testing the orders of an ARMA model*, Metrika, **32**, 129–150.

PURI, M.L. and SEN, P.K. (1971), *Nonparametric Methods in Multivariate Analysis*, Wiley, New York.

PURI, M.L. and SEN, P.K. (1985), *Nonparametric Methods in General Linear Models*, Wiley, New York.

SAVAGE, I.R. (1957), *Contributions to the theory of rank order statistics—the "trend" case*, Ann. Math. Statist. 28, 968–977.

SERFLING, R. (1968), *The Wilcoxon two-sample statistic on strong mixing process*, Ann. Math. Statist. 39, 1202–1209.

SWENSEN, A.R. (1985), *The asymptotic distribution of the likelihood ratio for autoregressive time series with a regression trend*, Jour. Multivariate Anal. 16, 54–70.

SWITZER, P. (1984), *Inference for spatial autocorrelation functions*, In Geostatistics for Natural Resources Characterization, (G. Verly et al., eds.). D. Reidel, Dordrecht,127–140.

TRAN, L.T. (1988), *Rank order statistics for time series models*, Annals Inst. Stat. Math. 40, 247–260.

TRAN, L.T. (1990), *Rank statistics for testing randomness against serial dependence*, Jour. Statist. Plan. and Inf. 24, 215–232.

WALD, A. (1943), *Tests of statistical hypotheses concerning several parameters when the number of observations is large*, Trans. Amer. Math. Soc. 54, 426–482.

WALD, A. and WOLFOWITZ, J. (1943), *An exact test for randomness in the nonparametric case based on serial correlation*, Ann. Math. Statist. 14, 378–388.

WALLIS, W.A. and MOORE, G.H. (1941), *A significance test for time series*, Jour. Amer. Statist. Assoc. 36, 401–409.

WEISS, G. (1975), *Time-series reversibility of linear stochastic processes*, Jour. of Appl. Probability, 12, 831–836.

WOLFOWITZ, J. (1944), *Asymptotic distribution of runs up and down*, Ann. Math. Statist. 15, 163–172.

YANG, M.C.K. and SCHRECKENGOST, J.F. (1981), *Difference sign test for comovements between two time series*, Comm. Statist. Theor. Meth. A10 4, 355–369.

Reprinted from
New Directions in Time Series Analysis (the IMA
Volume in Mathematics and its Applications),
edited by David Brillinger, Emanuel Parzen, Murray
Rosenblat et al., Volume 45 (1992), pages 111-153,
by permission of the publisher,
© 1992 by Springer Verlag GmbH & Co. KG

NONPARAMETRIC PREDICTION FOR RANDOM FIELDS

MADAN L. PURI* and FRITS H. RUYMGAART

Indiana University, Bloomington, IN, USA
Texas Tech University, Lubbock, TX, USA

We study prediction for vector valued random fields in a nonparametric setting. The prediction problem is formulated as the problem of estimating certain conditional expectations and a speed of uniform a.s. convergence is obtained, modifying results for conditional empirical processes derived from series with one-dimensional time. As an alternative to the usual mixing conditions we model the dependence by asymptotic decomposability. This includes linear (which generalizes ARMA) fields and random fields with a finite order Volterra expansion. As an example of a linear field we briefly discuss the finite-difference simulation of the heat equation blurred by additive random noise.

AMS 1980 Subject Classifications: Primary 62G07; Secondary 60G60.

random fields * prediction * conditional expectations * asymptotic decomposability

1. Introduction and basic notation

In this paper we consider some nonparametric prediction problems for random fields with a dependence structure that differs from the usual mixing conditions. Throughout the random field will be defined on all of $\mathbb{Z}^k$, where $\mathbb{Z}$ denotes the set of all integers and $k \in \mathbb{N}$ is fixed, and it will be observed on a finite part $\mathbb{T} \subset \mathbb{Z}^k$. We will refer to the indices $i \in \mathbb{Z}^k$ as (multidimensional) 'time points' although we realize that in practice they might be points on the line, in the plane or in space, possibly combined with a 'real' (one-dimensional) time coordinate. If such a real time coordinate is not included, the field describes a steady state as, e.g., in geophysics when the spatial distribution of a certain raw material is considered. On the other hand, the option of including such a real time coordinate allows for modeling dynamical systems that vary both in location and time as, e.g., the heat flow equation.

Correspondence to: Dr. Frits H. Ruymgaart, Department of Mathematics, Texas Tech. University, P.O. Box 41042, Lubbock, TX 79409-1042, USA.

* Research supported by the Office of Naval Research Contract N00014-91-J-1020.

The random field will consist of $(d+1)$-dimensional random vectors $\{Z_i = (X_i, Y_i), i \in \mathbb{Z}^k\}$ with the same distribution and generic element $Z = (X, Y)$, all defined on the same probability space $(\Omega, \mathscr{W}, \mathbb{P})$ and with $X_i : \Omega \to \mathbb{R}^d$, $Y_i : \Omega \to \mathbb{R}$ $(d \in \mathbb{N})$. In many practical situations the field $\{Z_i\}$ is obtained by regrouping the elements of an underlying random field $\{\xi_i, i \in \mathbb{Z}^k\}$, $k \in \mathbb{N}$, of actual interest so as to conveniently describe the desired prediction problem of the latter. This problem then boils down to estimation of

$$\mathbb{E}(Y \mid X = x) = \int y \, \mathrm{d}\mathbb{P}\{Y \leqslant y \mid X = x\}, \quad x \in I \subset \mathbb{R}^d, \tag{1.1}$$

where I is an interval. This problem and in particular the problem of estimating the conditional probabilities on the right in (1.1) will be considered in Section 4 where results in Ruymgaart (1991) for one-dimensional time will be modified for multidimensional time to yield a speed of uniform a.s. convergence.

The condition that the Z_i have the same distribution is almost the weakest stationarity assumption that still allows for statistical inference, and we need only a little more: see Section 3 for the details. However, in order for this condition to be satisfied typically a stronger stationarity condition for the underlying random field $\{\xi_i\}$ is required as will be clear from some examples in Section 2.

Besides stationarity we will need the condition that the dependence between any two elements in the field becomes small if a distance or pseudodistance between the corresponding time points gets large. This kind of condition is usually described in terms of a mixing assumption on the field (Yadrenko, 1983), that has been verified in particular for linear random fields to be discussed below (Nakhapetyan, 1980; Guyon and Richardson, 1984; see also Rosenblatt, 1985, and Tran, 1990, for additional relevant references). However, we will avoid the classical mixing concepts which are intuitively hard to precisely understand since, even for one-dimensional time, they may fail to hold in very decent cases (Andrews, 1984). To remedy such pathologies one may add a technical condition like absolute regularity (Gorodetskii, 1977; Pham and Tran, 1985; and Doukhan and Guyon, 1991, for an extension to linear random fields). We prefer to work with a concept, introduced for one-dimensional time in Chanda, Puri and Ruymgaart (1990), that might be called $m(n)$- or asymptotic decomposability. This concept seems wide enough to cover a great variety of processes ranging from mixing to long range dependent and even non-mixing processes. A precise definition for multidimensional time will be given in Section 3.

Although in this note we adopt the point of view that a random field is a well-established concept for which no motivation is required, we would like to briefly discuss in Section 2 some models of practical interest that give rise to random fields where, moreover, the question of prediction makes sense. An important class of random fields is the class of (real valued) linear fields (Rosenblatt, 1985) that can be extended to fields with a finite order Volterra expansion (see Priestley, 1981, for a discussion in one-dimensional time). In one-dimensional time such linear processes arise as stationary solutions to linear difference equations, subject to additive noise, and processes with a Volterra expansion might serve as an approximation to a solution of certain non-linear difference equations, like, e.g.,

$$\xi_{i+p} = \psi(\xi_i, \xi_{i+1}, \ldots, \xi_{i+p-1}) + \zeta_{i+p}, \tag{1.2}$$

for fixed $p \in \mathbb{N}$, innovations ζ_{i+p}, and some function ψ. Without being rigorous we sketch how a linear field may arise as a stationary solution of the discrete time simulation of the partial differential equation, subject to additive noise, that physically describes the heat flow but that is also useful in modeling traffic flows (see Weits, 1990, for recent results and references) and image processing. As a typical prediction problem, given a stationary field, we consider 'Kriging'. Finally we very sketchily touch upon Dirichlet's problem assuming that a stationary process on the boundary, taken to be random noise for convenience, can be observed. If only part of that process is observed we are faced with a prediction problem in a natural way. A difficulty with the latter problem and with steady-state boundary problems in general is the stationarity and, if a relatively small region is to be considered, the decay of the dependence. Dynamical systems are better structured for asymptotics since, at least in principle, one may let the process indefinitely continue in time. For the numerical aspects of these examples we refer to Hildebrand (1968).

2. Examples of random fields and prediction

2.1. General classes

Throughout this subsection let $\{\zeta_i, i \in \mathbb{Z}^k\}$ be a random field of i.i.d. real valued random variables called the innovations field, and let $\{a_i, i \in \mathbb{Z}^k\}$ be real numbers. The random field

$$\xi_i = \sum_{j \in \mathbb{Z}^k} a_j \zeta_{i-j}, \quad i \in \mathbb{Z}^k, \tag{2.1}$$

is called a linear random field. The random field

$$\xi_i = \sum_{j \in \mathbb{Z}^k} a_j \zeta_{i-j} + \sum_{l,m \in \mathbb{Z}^k}\sum b_{l,m} \zeta_{i-l} \zeta_{i-m}, \quad i \in \mathbb{Z}^k, \tag{2.2}$$

where the $b_{l,m}$ are real numbers, generalizes (2.1) and is a random field with second order Volterra expansion. Further generalization to arbitrary finite order is straightforward.

The i.i.d. assumption on the innovations entails that the processes in (2.1) and (2.2) satisfy the strongest possible stationarity condition. In fact, for any finite set $S \subset \mathbb{Z}^k$ let ξ_s denote the vector of all $\{\xi_i, i \in S\}$ numbered in some serial order. Then we have

$$\mathscr{L}(\xi_S) = \mathscr{L}(\xi_{j+S}), \quad j \in \mathbb{Z}^k, \tag{2.3}$$

where $j+S=\{j+i\colon i\in S\}$ and the same type of serial order is used on the right in (2.3). We will see in Section 3 that the concept of asymptotic decomposability is tailormade for dealing with the dependence structure in random fields of type (2.1) or (2.2).

To formulate the prediction problem let V, $W\subset\mathbb{Z}^k$ be arbitrary finite subsets, and let $|\cdot|$ denote the cardinality of finite sets. For a measurable mapping $\psi:\mathbb{R}^{|W|}\to\mathbb{R}$ the nonparametric prediction problem will be interpreted as the problem of estimating

$$\mathbb{E}(\psi(\xi_W)\,|\,\xi_V=x),\quad x\in\mathbb{R}^{|V|}, \tag{2.4}$$

where we will assume ψ to be bounded for convenience. The strong stationarity in (2.3) entails that the random vectors

$$Z_i=(X_i,\,Y_i)\quad\text{with } X_i=\xi_{i+V},\ Y_i=\psi(\xi_{i+W}), \tag{2.5}$$

have the same distribution for all $i\in\mathbb{Z}^k$. (Even the stronger stationarity property (2.3) holds true for the Z_i-field but we don't need that.) Hence estimating (2.4) reduces to estimating the common conditional expectation

$$\mathbb{E}(Y_i\,|\,X_i=x),\quad x\in\mathbb{R}^{|V|}, \tag{2.6}$$

given a sample of the vectors Z_i in (2.5).

2.2. *Dynamical systems*

By way of an example let us focus here on the one-dimensional heat flow equation where the real variable x represents location and the real variable t represents time. A function $u(x,t)$, $(x,t)\in\mathcal{R}=\mathbb{R}\times(0,\infty)$, has to be determined that satisfies

$$\frac{\partial u(x,t)}{\partial t}=c^2\frac{\partial^2u(x,t)}{\partial x^2},\quad (x,t)\in\mathcal{R}, \tag{2.7}$$

subject to the initial value condition

$$u(x,0)=f(x),\quad x\in\mathbb{R}. \tag{2.8}$$

We will consider a finite difference simulation of this problem following Hildebrand (1968, Section 3.2) and meanwhile randomize the equation and initial value condition.

For this purpose let $\{\zeta_{i,j},(i,j)\in\mathbb{Z}^2\}$ denote an i.i.d. innovations field and let us identify the starting time with $N\in\mathbb{Z}$. The corresponding partial difference equation is of the form $u_{i,j}=\rho u_{i-1,j-1}+(1-2\rho)u_{i,j-1}+\rho u_{i+1,j-1}$, where $\rho=c^2k/h^2$ with h the spacing in the x-direction and k the spacing in the t-direction. Let us for simplicity assume that the initial values are given by

$$f(x_i)=\zeta_{i,N},\quad i\in\mathbb{Z}. \tag{2.9}$$

Adding the noise $\zeta_{i,j}$ at (i,j) and choosing $\rho=\frac{1}{2}$ (sometimes $\rho=\frac{1}{6}$ is preferred) we arrive at the equation

$$\xi_{i,j}=\tfrac{1}{2}(\xi_{i-1,j-1}+\xi_{i+1,j-1})+\zeta_{i,j}. \tag{2.10}$$

To solve (2.10) we start at any time $j > N$ and arbitrary $i \in \mathbb{Z}$ and make successive substitutions until the time N is reached. We thus find the finite sum

$$\begin{aligned}\xi_{i,j} = \zeta_{i,j} &+ \tfrac{1}{2}(\zeta_{i-1,j-1} + \zeta_{i+1,j-1}) \\ &+ \tfrac{1}{4}(\zeta_{i-2,j-2} + 2\zeta_{i,j-2} + \zeta_{i+2,j-2}) \\ &+ \tfrac{1}{8}(\zeta_{i-3,j-3} + 3\zeta_{i-1,j-3} + 3\zeta_{i+1,j-3} + \zeta_{i+3,j-3}) \\ &+ \cdots, \end{aligned} \tag{2.11}$$

where the last term contains a linear combination of the $\zeta_{i,N}$. If the system is at work for a long time the irregularities of the beginning will disappear in the end. In (2.11) we see that letting $N \to -\infty$ yields an infinite sum in the limit for each (i, j). This is a linear process and it obviously satisfies the strong stationarity property (2.3) since changing the point (i, j) does not effect the pattern of coefficients.

Various prediction patterns are conceivable, in particular those for which future values are to be estimated from earlier observations. We might consider

$$V = \{(m, j) : j = k - p, \ldots, k - 1\}, \qquad W = \{(m, k)\}, \tag{2.12}$$

with (bounded) $\psi : \mathbb{R} \to \mathbb{R}$, as an example.

2.3. *A particular prediction problem: Kriging*

Another type of prediction pattern that is of practical importance is the prediction of yield of a certain part of a deposit; such problems are referred to as 'Kriging'. In geophysics observations at just a single point of the deposit cannot in general be obtained and therefore a local averaging procedure is applied (Näther, 1985). In this way the actual underlying field is regularized, and we will assume this field to be stationary. Here we will not engage in the discussion whether such a model assumption really makes sense in this context.

Based on observations in a region $V \subset \mathbb{Z}^k$ the average $(\sum_{j \in W} \xi_j)/|W|$ is to be predicted, where often $V \subset W$. Here we will again interpret prediction as the problem of estimating

$$\mathbb{E}\left(\frac{1}{|W|} \sum_{j \in W} \xi_j \,\middle|\, \xi_V = x\right), \quad x \in \mathbb{R}^{|V|}. \tag{2.13}$$

We assume that the sampling region $\mathbb{T}$ is such that for a sufficiently large number of time points $i \in \mathbb{T}$ we have $i + V \in \mathbb{T}$ and $i + W \in \mathbb{T}$. For such time points we can observe

$$Z_i = (X_i, Y_i) \quad \text{with } X_i = \xi_{i+V}, \ \ Y_i = \frac{1}{|W|} \sum_{j \in W} \xi_{i+j}, \tag{2.14}$$

and the problem is once more reduced to one of estimating (2.6) given a sample of such vectors.

2.4. Steady-state systems

As an example of a steady-state problem let us consider the two-dimensional Dirichlet problem in which a function $u(x, y)$, $(x, y) \in \mathcal{R} \subset \mathbb{R}^2$ is to be determined in such a way that it satisfies Laplace's equation

$$(\nabla^2 u)(x, y) = \frac{\partial^2 u(x, y)}{\partial x^2} + \frac{\partial^2 u(x, y)}{\partial y^2} = 0, \quad (x, y) \in \mathcal{R}^0, \tag{2.15}$$

where $\mathcal{R}^0$ is the interior of $\mathcal{R}$ (the region $\mathcal{R}$ should be suitably defined), and assumes prescribed values

$$u(x, y) = f(x, y), \quad (x, y) \in \partial\mathcal{R}, \tag{2.16}$$

on the boundary. Again we will consider a finite difference simulation following Hildebrand (1968, Section 3.16).

Let us assume that the region is a square which, after discretization, reduces to $\{0, \dots, n\}^2 \backslash \{(0, 0), (n, 0), (n, n), (0, n)\}$ and let $\{\zeta_{i,j}, (i, j) \in \mathbb{Z}^2\}$ be an i.i.d. innovations field. The corresponding partial difference equation can be obtained, e.g., from the difference scheme that relates the value at (i,j) to four neighboring points according to $u_{i,j} = \frac{1}{4}\{u_{i+1,j} + u_{i,j+1} + u_{i-1,j} + u_{i,j-1}\}$. A well-known difficulty of this type of problem is that the values in the interior cannot be successively found from the boundary values but have to be found simultaneously as the solution of a set of linear equations. For simplicity let us assume that the values on the boundary are given by the values of the innovations at those points, i.e.,

$$f(i, j) = \zeta_{i,j} \quad \text{for } (i, j) \text{ on the boundary.} \tag{2.17}$$

Let us take $n = 3$, e.g., and number the points on the boundary as $Q_1 = (1, 0)$, $Q_2 = (2, 0)$, $Q_3 = (3, 1)$, $Q_4 = (4, 1)$, $Q_5 = (2, 3)$, $Q_6 = (1, 3)$, $Q_7 = (0, 2)$, $Q_8 = (0, 1)$, and the interior points according to the diagonal ordering as $P_1 = (1, 1)$, $P_2 = (1, 2)$, $P_3 = (2, 1)$, $P_4 = (2, 2)$. In this notation the boundary values can be written

$$f(P_i) = \zeta_i, \quad i \in \{1, \dots, 8\}. \tag{2.18}$$

Writing, similarly, the values in the interior as $u(Q_j) = \xi_j$ the system of linear equations by which these values are determined are

$$\begin{aligned}
4\xi_1 - \xi_2 - \xi_3 &= \zeta_1 + \zeta_8 = \eta_1, \\
4\xi_2 - \xi_1 - \xi_4 &= \zeta_6 + \zeta_7 = \eta_2, \\
4\xi_3 - \xi_4 - \xi_1 &= \zeta_2 + \zeta_3 = \eta_3, \\
4\xi_4 = \xi_2 - \xi_3 &= \zeta_4 + \zeta_5 = \eta_4,
\end{aligned} \tag{2.19}$$

with the solution

$$\begin{aligned}
\xi_1 &= \tfrac{1}{192}(56\eta_1 + 16\eta_2 + 16\eta_3 + 8\eta_4), \\
\xi_2 &= \tfrac{1}{192}(16\eta_1 + 56\eta_2 + 8\eta_3 + 16\eta_4), \\
\xi_3 &= \tfrac{1}{192}(16\eta_1 + 8\eta_2 + 56\eta_3 + 16\eta_4), \\
\xi_4 &= \tfrac{1}{192}(8\eta_1 + 16\eta_2 + 16\eta_3 + 56\eta_4),
\end{aligned} \tag{2.20}$$

According to Hildebrand (1968, footnote p. 283) the diagonal ordering always leads to a system (as in (2.19)) with a matrix which is 'block tridiagonal'; see also Forsythe and Wasow (1960).

We see that the four ξ_i have the same distribution and that also all the pairs and all the triples have the same distribution. For prediction purposes it might be useful to observe that all the pairs (ζ_1, ξ_1), (ζ_2, ξ_3), (ζ_3, ξ_3), (ζ_4, ξ_4), (ζ_5, ξ_4), (ζ_6, ξ_2), (ζ_7, ξ_2), (ζ_8, ξ_1) have the same distribution as well. This kind of stationarity might also be inferred from the available symmetries.

However, for larger squares all the variables in the field defined on the interior will not in general have the same distribution. Symmetry considerations again suggest that on the boundary of a specific subsquare (symmetrically located with respect to the boundary and with the exception of the four vertices) the variables have the same distribution, but that this common distribution may very well be different for another such subsquare. In the case of pure random noise on the boundary, it seems possible that refinement of the grid entails a decay of the dependence for time points far away, but this needs further specification.

A natural prediction problem arises in the case where all the boundary values have not been observed. Since all the boundary values need to be known for computation of the values in the interior, first the unobserved boundary values have to be predicted. If the boundary process is supposed to be pure random noise, as in our example, prediction is of course not very promising. If, on the other hand, we assume a stationary process on the boundary with a nontrivial cyclical dependence structure, the decay of the dependence may not be strong enough. Embedding the region under investigation in a sequence of expanding regions might produce the proper context for asymptotics with the desired decay of dependence, but this might not be a realistic approach.

3. Stationarity and dependence assumptions

Let us now return to the d-dimensional random field $\{Z_i,\ i\in\mathbb{Z}^k\}$ described in the introduction. Suppose that for $n\in\mathbb{N}$ we observe the field on the time set $\mathbb{T}_n\subset\mathbb{Z}^k$, with $\mathbb{T}_1\subset\mathbb{T}_2\subset\cdots$ and $|\mathbb{T}_n|\to\infty$, as $n\to\infty$. The distance in any Euclidean space $\mathbb{R}^m$ will be derived from the norm

$$\|x-y\| = \max_{1\leq\nu\leq m} |x_\nu - y_\nu|, \tag{3.1}$$

$x=(x_1,\ldots,x_m)$, $y=(y_1,\ldots,y_m)\in\mathbb{R}^m$, in the usual way. The field will be called *asymptotically decomposable* if for each sampling stage $n\in\mathbb{N}$ there exists $r=r(n)\geq 1$ and a decomposition $Z_i = Z_{i,n}+\bar{Z}_{i,n}$ such that

$$\mathscr{L}(Z_i),\ \mathscr{L}(Z_{i,n}) \text{ independent of } i\in\mathbb{T}_n, \tag{3.2}$$

$$Z_{i,n} \perp\!\!\!\perp Z_{j,n} \quad \text{for all } i,j\in\mathbb{T}_n \text{ with } \|i-j\|\geq r(n), \tag{3.3}$$

$$\max_{i\in\mathbb{T}_n} \mathbb{P}\{\|\bar{Z}_{i,n}\|\geq\varepsilon(n)\}\leq\delta(n) \quad \text{with } \varepsilon(n),\ \delta(n)\to 0, \text{ as } n\to\infty. \tag{3.4}$$

To achieve (3.4) we will usually have to let $r(n) \to \infty$, as $n \to \infty$, unless the dependence structure is trivial. For asymptotics the rates at which $r(n) \to \infty$ and $\varepsilon(n)$, $\delta(n) \to 0$ as $n \to \infty$, need to be specified. In many interesting cases suitable rates for $\varepsilon(n)$ and $\delta(n)$ can be obtained for $r(n) \to \infty$, where still $r(n) \ll n$.

By way of an important example let us check these conditions for the linear random field (2.1) based on i.i.d. innovations. Let us assume that the following conditions are satisfied:

$$\mathbb{E}|\zeta_i|^{\mu} < \infty \quad \text{for some } 0 < \mu \leq 1, \tag{3.5}$$

$$|a_j| = \mathrm{O}(\|j\|^{-\lambda}) \text{ as } \|j\| \to \infty \quad \text{for some } \lambda > 0. \tag{3.6}$$

Here we focus on small values for μ which is an interesting case from the standpoint of robustness. If the ζ_i have a Cauchy distribution they still satisfy (3.5) for any $0 < \mu < 1$ and on the other hand such distributions for the innovations may generate outliers in the random field of the ξ_i. The classical assumption is $\mu = 2$ and then quite other techniques apply (Rosenblatt, 1985, Chapter VIII). For values $\mu > 1$ a result very similar to the one below can be obtained by a slight modification of the proof.

Also a little more notation is needed. For any two sequences $\{\alpha_n\}$, $\{\beta_n\}$ of real numbers it will be convenient to write

$$\alpha_n \approx \beta_n \text{ as } n \to \infty, \quad \text{if } 0 < \liminf_{n\to\infty} \alpha_n/\beta_n \leq \limsup_{n\to\infty} \alpha_n/\beta_n < \infty. \tag{3.7}$$

We will consider a C'-mapping $\Psi : \mathbb{R}^{|S|} \to \mathbb{R}^m$, for some $m \in \mathbb{N}$, and at any $x \in \mathbb{R}^m$ write the differential as $[\mathrm{d}\Psi]_x$. For this matrix we will use the matrix norm $\|[\mathrm{d}\Psi]_x\|$ which is defined as the maximum of the absolute values of its elements. The condition

$$\sup_{x \in \mathbb{R}^{|S|}} \|[\mathrm{d}\Psi]_x\| \leq M < \infty, \tag{3.8}$$

will be useful. We are now prepared for a verification of asymptotic decomposability for a class of vector fields, derived from the linear field in a way suitable for prediction problems of the latter field (see (2.5)).

Theorem 3.1. *For a finite subset $S \subset \mathbb{Z}^k$ and the mapping Ψ mentioned above, satisfying in particular* (3.8), *let us introduce the vector field*

$$\{Z_i = \Psi(\xi_{i+S}), i \in \mathbb{Z}^k\}. \tag{3.9}$$

If the innovations $\{\zeta_i\}$ defining the linear field $\{\xi_i\}$ according to (2.1) *satisfy* (3.5) *and if the coefficients satisfy* (3.6), *then $\{Z_i\}$ satisfies the asymptotic decomposability conditions* (3.2)–(3.4) *with $r(n) \approx n^{\rho}$, $\varepsilon(n) \approx n^{-\sigma}$, $\delta(n) \approx n^{-\tau}$ for any $\rho > 0$, $\sigma > 0$, $\tau > 0$, provided that*

$$\lambda \geq (\tau + \sigma\mu + k\rho)/(\rho\mu). \tag{3.10}$$

Proof. Let us first prove the claim for the special case $S=\{0\}$ and $\Psi(x)=x, x\in\mathbb{R}$, i.e., for the linear field $\{\xi_i\}$ itself. For $q(n)>1$ let us define

$$\xi_{i,n}=\sum_{\|j\|\leq q(n)} a_j\zeta_{i-j}, \qquad \bar{\xi}_{i,n}=\sum_{\|j\|>q(n)} a_j\zeta_{i-j}. \tag{3.11}$$

Then it is clear that (3.3) is satisfied with $r(n)=2q(n)+1$. In a way very similar to Chanda and Ruymgaart (1990) for one-dimensional time it follows that, for numbers $0<A<B<\infty$,

$$\begin{aligned}\mathbb{P}\{|\bar{\xi}_{i,n}|\geq\varepsilon(n)\}&\leq\{\varepsilon(n)\}^{-\mu}\left\{\sum_{\|j\|>q(n)}|a_j|^{\mu}\mathbb{E}(|\zeta_{i-j}|^{\mu})\right\}\\&\leq\frac{\mathbb{E}(|\zeta_i|^{\mu})}{\{\varepsilon(n)\}^{\mu}}\sum_{q=q(n)}^{\infty}\sum_{\|j\|=q}\|j\|^{-\lambda\mu}\\&\leq\frac{A}{\{\varepsilon(n)\}^{\mu}}\sum_{q=q(n)}^{\infty}q^{k-1}\cdot q^{-\lambda\mu}\leq\frac{B}{\{\varepsilon(n)\}^{\mu}}\int_{q(n)}^{\infty}x^{k-1-\lambda\mu}\,\mathrm{d}x\\&=B\{\varepsilon(n)\}^{-\mu}\{q(n)\}^{k-\lambda\mu}.\end{aligned} \tag{3.12}$$

It follows that choosing λ according to (3.10) yields the desired order for the $\delta(n)$.

Let us next prove the result for arbitrary finite S but for the special function $\Psi(x)=x, x\in\mathbb{R}^{|S|}$, and let us define the random vectors $\xi_{i+S,n}=\{\xi_{i+j,n}, j\in S\}$, $\bar{\xi}_{i+S,n}=\{\bar{\xi}_{i+j,n}, j\in S\}$ with the $\xi_{i,n}$ and $\bar{\xi}_{i,n}$ as in (3.11). Let $Q(S)$ be the smallest 'square' in $\mathbb{Z}^k$ which contains S and let $l(S)$ denote the common length of the sides of this square. Then it is immediate that

$$\xi_{i+S,n} \perp\!\!\!\perp \xi_{j+S,n} \quad \text{for all } i,j \text{ with } \|i-j\|\geq r(n)+l(S). \tag{3.13}$$

Moreover, using the norm in (3.1) we have

$$\mathbb{P}\{\|\bar{\xi}_{i+S,n}\|\geq\varepsilon(n)\}\leq\sum_{j\in S}\mathbb{P}\{|\bar{\xi}_{i+j,n}|\geq\varepsilon(n)\}\leq|S|\cdot\delta(n). \tag{3.14}$$

Since $l(S)$ and $|S|$ are fixed finite numbers all the parameters remain of the same order as in the first case.

Finally let us consider the general case of the theorem. Then let us define $Z_{i,n}=\Psi(\xi_{i+S,n})$, $\bar{Z}_{i,n}=Z_i-Z_{i,n}$. Since the $Z_{i,n}$ involve only the variables $\xi_{i+S,n}$ it is obvious that (3.13) remains true with $\xi_{i+S,n}$, $\xi_{j+S,n}$ replaced by $Z_{i,n}$ and $Z_{j,n}$ respectively. The mean value theorem entails that in terms of the matrix-norm defined above (3.8) and the norm in (3.1) we have

$$\begin{aligned}\|\bar{Z}_{i,n}\|&=\|\Psi(\xi_{i+S})-\Psi(\xi_{i+S,n})\|\\&\leq\left\{\sup_{x\in\mathbb{R}^{|S|}}\|[\mathrm{d}\Psi]_x\|\right\}\|\bar{\xi}_{i+S,n}\|\leq M\|\bar{\xi}_{i+S,n}\|.\end{aligned} \tag{3.15}$$

Hence $\mathbb{P}\{\|\bar{Z}_{i,n}\|\geq\varepsilon(n)\}$ is of the same asymptotic order as the probabilities in (3.14). □

It is clear from the obtained expressions that an extension to the case where $S=S_n$ with $|S_n|\to\infty$, as $n\to\infty$, at a suitable rate is possible. In the examples below we briefly collect some other remarks that illustrate the asymptotic decomposability.

Example 3.1. The approach of Theorem 3.1 could be generalized for fields of the type (2.2). For the field itself the decomposition

$$\xi_{i,n} = \sum_{\|j\| \leq q(n)} a_j \zeta_{i-j} + \sum_{\|l\|,\|m\| \leq q(n)} b_{l,m} \zeta_{i-l} \zeta_{i-m}, \tag{3.16}$$

and $\bar{\xi}_{i,n} = \xi_i - \xi_{i,n}$ should work. Omitting the technical details that are very similar to those above and that are not very interesting we see that again (3.3) is satisfied with $\mu(n) = 2q(n) + 1$. The asymptotic negligibility of $\bar{\xi}_{i,n}$ follows by proving the slightly stronger claim that each of the random variables

$$\bar{\xi}^{(1)}_{i,n} = \sum_{\|j\| > q(n)} a_j \zeta_{i-j}, \tag{3.17}$$

$$\bar{\xi}^{(2)}_{i,n} = \sum_{l \in \mathbb{Z}^k, \|m\| > q(n)} b_{l,m} \zeta_{i-l} \zeta_{i-m}, \tag{3.18}$$

$$\bar{\xi}^{(3)}_{i,n} = \sum_{\|l\| > q(n), m \in \mathbb{Z}^k} b_{l,m} \zeta_{i-l} \zeta_{i-m}, \tag{3.19}$$

is asymptotically negligible. To prove this a suitable extension of condition (3.6) will be needed including a statement on the order of magnitude of the coefficient $b_{l,m}$.

Example 3.2. We have treated the k components in the same way in (3.3) where we use the norm (3.1). When modeling certain practical situations we might find the condition that the decay of the dependence is somewhat the same in each direction, unrealistic. In the example of the heat equation it might be argued that for fixed 'real' time the innovations in the space direction should not be considered as independent; in such a direction they might rather form a stationary process with dependence. If in addition the space coordinate remains bounded to a finite interval, the dependence cannot in fact become arbitrarily small for points with the same 'real' time coordinate, even though the grid could be made finer. In the direction of 'real' time, however, this is different because we may let this coordinate tend to infinity (at least in principle) and make the usual assumption of small dependence between points far away in 'real' time. Formally this boils down to using a pseudonorm like

$$\|i - j\| = \max_{1 \leq \nu \leq l} |i_\nu - j_\nu| \quad \text{for some } 1 \leq l < k. \tag{3.20}$$

We will not consider the case of pseudonorms here since it requires only minor technical modifications.

4. Convergence rate of estimators

In order to introduce the estimators some notation will be needed. By a left-half open interval in finite dimensional Euclidean space we mean the Cartesian product of left-half open intervals in $\mathbb{R}$, including $\mathbb{R}$ itself. We write

$$\mathscr{I}, \mathscr{J} \text{ for the set of all left-half open intervals in } \mathbb{R}^d, \mathbb{R}, \tag{4.1}$$

respectively. Let us also write, for $A \in \mathscr{B}^{d+1}$,

$$P(A) = \mathbb{P}\{Z \in A\}, \qquad \hat{P}_n(A) = \frac{1}{|\mathbb{T}_n|} \sum_{i \in \mathbb{T}_n} \mathbb{1}_A(Z_i), \tag{4.2}$$

and, for given $J \in \mathscr{J}$ and $B \in \mathscr{B}^d$, let

$$P_J(B) = \mathbb{P}\{X \in B, Y \in J\}, \qquad \hat{P}_{J,n}(B) = \frac{1}{|\mathbb{T}_n|} \sum_{i \in \mathbb{T}_n} \mathbb{1}_B(X_i)\mathbb{1}_J(Y_i). \tag{4.3}$$

Although $P_J(B) = P(B \times J)$ we prefer the first notation because X and Y play an asymmetric role. In general P_J is a defective probability measure on $(\mathbb{R}^d, \mathscr{B}^d)$ for each $J \in \mathscr{J}$. Of course, for $J = \mathbb{R}$, the measures

$$P_{\mathbb{R}} \text{ and } \hat{P}_{\mathbb{R},n}, \tag{4.4}$$

are precisely the marginal distribution of X and its empirical analogue.

It is clear that $P_J \ll P_{\mathbb{R}}$ (even $P_J \leqslant P_{\mathbb{R}}$) for each $J \in \mathscr{J}$ and it is well-known that

$$P(J \mid x) = \mathbb{P}\{Y \in J \mid X = x\} = \frac{\mathrm{d}P_J}{\mathrm{d}P_{\mathbb{R}}}(x), \quad x \in \mathbb{R}^d, \ J \in \mathscr{J}. \tag{4.5}$$

As in Stute (1986) we will use naive kernels for estimating the conditional probability in (4.5). Therefore we introduce half-open squares $Q_{n,x}$ in $\mathbb{R}^d$, with center at x and sides of length l_n, i.e.

$$Q_{n,x} \in \mathscr{I}, \quad \Lambda(Q_{n,x}) = l_n^d, \tag{4.6}$$

where Λ denotes Lebesgue measure in Euclidean space. A non-random approximation of (4.6) is, for sufficiently small l_n, provided by

$$P_n(J \mid x) = \frac{P_J(Q_{n,x})}{P_{\mathbb{R}}(Q_{n,x})}, \quad x \in \mathbb{R}^d, \ J \in \mathscr{J}. \tag{4.7}$$

As a natural estimator we will consider

$$\hat{P}_n(J \mid x) = \frac{\hat{P}_{J,n}(Q_{n,x})}{\hat{P}_{\mathbb{R},n}(Q_{n,x})}, \quad x \in \mathbb{R}^d, \ J \in \mathscr{J}. \tag{4.8}$$

Comparison of $\hat{P}_n(J \mid x)$ and $P(J \mid x)$ will be through comparison of $\hat{P}_n(J \mid x)$ with $P_n(J \mid x)$ and of $P_n(J \mid x)$ with $P(J \mid x)$. The latter concerns the non-random part which has nothing to do with the particular statistical model and which entirely follows the usual pattern, well-known from the literature. Therefore we will focus on the random part. The first assumption that we need is just a regularity condition on P.

Remark 4.1. Naive kernels are easy to work with in practice. In first order asymptotics they usually yield the same result as smooth non-negative kernels. For theoretical considerations it is convenient that the curve estimators are directly related to the empirical process indexed by squares, when naive kernels are used.

Assumption 4.1. The probability distribution P of $Z=(X, Y)$ has density f with respect to Lebesgue measure on $\mathbb{R}^{d+1}$. This density and all its marginals are continuous and bounded on their respective supports. We focus on a bounded interval $I_0 \subset \mathbb{R}^d$ such that the density of the marginal $P_{\mathbb{R}}$ of X in (4.4) is strictly positive on the closure $\bar{I}_0$.

Remark 4.2. It should be stressed that the density f in Assumption 4.1 is the density of (X, Y), the common density of the observable elements (X_i, Y_i) in the random field. It is *not* the common density of the unobservable innovations ζ_i, if such an innovation process generates the random field as in Section 2.2. It is well-known from Andrews (1984), e.g., that the innovations may have discrete densities and yet produce a process for which the density f is smooth in the way we require it. Hence our situation covers nonmixing fields. It is most likely that in a similar way examples of unbounded innovation densities can be constructed that yield smooth f, in other words that asymptotically decomposable random fields contain certain long range dependent random fields as special cases. Many mixing fields are certainly covered. The precise relation between asymptotic decomposability and traditional mixing concepts, however, is not obvious and requires further study.

Since our properties will be of an asymptotic nature we will refer to a sequence of sampling schemes $\mathbb{T}_n$ with $|\mathbb{T}_n| \to \infty$, as $n \to \infty$. Before presenting some sufficient conditions on the orders of magnitude of the parameters in the asymptotic decomposability conditions (3.2)-(3.4), let us consider an example displaying the pattern that, apart from minor modifications, can also be followed for general sampling schemes. In cases where the sample elements are derived from a sample $\{\xi_i, i \in \mathbb{U}_n\}$ from a process of actual interest, according to $Z_i = \{\xi_{i+S}, i \in \mathbb{U}_n\}$ for some finite subset $S \subset \mathbb{Z}^k$ (see (2.5) and (3.9)), we have the relation

$$\mathbb{T}_n = \{i \in \mathbb{U}_n \,;\, i + S \subset \mathbb{U}_n\}. \tag{4.9}$$

Example 4.1. Let us take $\mathbb{T}_n = \{-n, \ldots, -1, 0, 1, \ldots, n\}^k$, $n \in \mathbb{N}$, and for simplicity assume that (see (3.3))

$$(2n+1)/r(n) = \nu(n) \in \mathbb{N}. \tag{4.10}$$

Let us consider the subset

$$S_n = \{-n, -n+1, \ldots, -n+r(n)-1\}^k, \tag{4.11}$$

and the subgrids

$$\mathbb{T}_{n,j} = j + \{-n, -n+r(n), \ldots, n-r(n)\}^k, \quad j \in S_n. \tag{4.12}$$

Note that we have

$$\mathbb{T}_n = \bigcup_{j \in S_n} \mathbb{T}_{n,j}. \tag{4.13}$$

Then, for each $j \in S_n$, the collection of random elements

$$\{Z_{i,n}, i \in \mathbb{T}_{n,j}\}, \tag{4.14}$$

is an i.i.d. family of size $|\mathbb{T}_{n,j}| = \{\nu(n)\}^k$, and the total number of such i.i.d. families is $|S_n| = \{r(n)\}^k$. For more generally shaped sampling schemes we will just assume that (4.13) holds true with specified orders for the $|\mathbb{T}_{n,j}|$ and $|S_n|$.

Assumption 4.2. If $|\mathbb{T}_n| \to \infty$, as $n \to \infty$, the random field $\{Z_i, i \in \mathbb{T}_n\}$ satisfies the asymptotic decomposability conditions in (3.2)-(3.4) with parameters satisfying the following conditions. There exist $0 \leq \rho < \frac{1}{2}$, $\sigma > 1$, $\tau > 1$ such that (3.4) and (4.13) are satisfied with

$$|S_n| = O(|\mathbb{T}_n|^{\rho}), \quad \min_{j \in S_n} |\mathbb{T}_{n,j}| \approx \max_{j \in S_n} |\mathbb{T}_{n,j}|, \quad \text{as } n \to \infty; \tag{4.15}$$

$$\varepsilon(n) = O(|\mathbb{T}_n|^{-\sigma}), \ \delta(n) = O(|\mathbb{T}_n|^{-1-\tau}), \quad \text{as } n \to \infty. \tag{4.16}$$

Theorem 4.1. *Let Assumptions* 4.1, 4.2 *be satisfied. Choosing* $l_n^d = |\mathbb{T}_n|^{\alpha}$ *in* (4.6) *for some* $0 < \alpha < \frac{1}{2}(1-\rho)$, *we have*

$$|\mathbb{T}_n|^{\mu} \sup_{x \in I_0, J \subset \mathcal{J}} |\hat{P}_n(J\,|\,x) - P_n(J\,|\,x)| \xrightarrow{\text{a.s.}} 0 \quad \text{as } n \to \infty, \tag{4.17}$$

for any $0 < \mu < \frac{1}{2} - \frac{1}{2}(\alpha + \rho)$.

Proof. Let us introduce the function

$$\psi(x) = 2\lambda^{-2} \int_0^x \log(1+x)\, dx, \quad \lambda > 0, \ \psi(0) = 1, \tag{4.18}$$

and the sets

$$\Omega_n = \bigcap_{i \in \mathbb{T}_n} \{\|\bar{Z}_{i,n}\| < \varepsilon(n)\} \subset \Omega, \quad n \in \mathbb{N}. \tag{4.19}$$

As a special case of Chanda and Ruymgaart (1991, Theorem 2.1) we have, for any $J_0 \in \mathcal{J}$ and sufficiently large $n \in \mathbb{N}$,

$$\mathbb{P}\left(\left\{\sup_{I \subset I_0, J \subset J_0} |\mathbb{T}_n|^{1/2} |\hat{P}_{J,n}(I) - P_J(I)| \geq \lambda\right\} \cap \Omega_n\right)$$
$$\leq C|\mathbb{T}_n|^{\rho} \exp\left(\frac{-A\lambda^2|\mathbb{T}_n|^{-\rho}}{P_{J_0}(I_0)}\, \psi\left(\frac{B\lambda|\mathbb{T}_n|^{-1/2}}{P_{J_0}(I_0)}\right)\right), \tag{4.20}$$

for $\lambda \geq 0$ satisfying

$$\lambda \geq C|\mathbb{T}_n|^{1/2}\{\varepsilon(n) \text{ diameter } (I_0 \times J_0) + \varepsilon^2(n) + |\mathbb{T}_n|^{-\tau}\}, \tag{4.21}$$

$$P_{J_0}(I_0) \geq C\{\varepsilon(n) \text{ diameter } (I_0 \times J_0) + \varepsilon^2(n) + |\mathbb{T}_n^{-\tau}|\}, \tag{4.22}$$

and with A, B, $C \in (0, \infty)$ generic constants.

Let us write $\hat{P}_{J,n}(Q_{n,x})/\Lambda(Q_{n,x}) = |\mathbb{T}_n|^{\alpha} \hat{P}_{J,n}(Q_{n,x}) = \hat{f}_{J,n}(x)$ and $P_J(Q_{n,x})/\Lambda(Q_{n,x}) = |\mathbb{T}_n|^{\alpha} P_J(Q_{n,x}) = f_{J,n}(x)$, $x \in I_0$, $J \in \mathcal{J}$. In this notation we have

$$\hat{P}_n(J\,|\,x) = \frac{\hat{f}_{J,n}(x)}{\hat{f}_{\mathbb{R},n}(x)}, \quad P_n(J\,|\,x) = \frac{f_{J,n}(x)}{f_{\mathbb{R},n}(x)}, \qquad x \in \mathbb{R}^d, \ J \in \mathcal{J}. \tag{4.23}$$

Next we will first show that

$$|\mathbb{T}_n|^{\mu} \sup_{x\in I_0, J\in\mathscr{J}} |\hat{f}_{J,n}(x)-f_{J,n}(x)| \xrightarrow{\text{a.s.}} 0 \quad \text{as } n\to\infty. \tag{4.24}$$

For this purpose, at stage n let $I_n=\bigcup_{x\in I_0} Q_{n,x}$ and let $\mathscr{I}_n$ be a disjoint covering of I_n by squares $I_{n,k}$ having $\Lambda(I_{n,k})=|\mathbb{T}_n|^{-\alpha}$ so that $|\mathscr{I}_n|=O(|\mathbb{T}_n|^{\alpha})$, as $n\to\infty$. Because each $Q_{n,x}$ is a square with the same Lebesgue measure as the squares in $\mathscr{I}_n$, it follows that each $Q_{n,x}$ intersects at most 2^d elements of this covering. Therefore we have, for $0<\mu<\frac{1}{2}-\frac{1}{2}(\alpha+\rho)$,

$$\begin{aligned}|\mathbb{T}_n|^{\mu} &\sup_{x\in I_0, J\in\mathscr{J}} |\hat{f}_{J,n}(x)-f_{J,n}(x)| \\ &\leqslant 2^d|\mathbb{T}_n|^{\mu+\alpha-1/2} \max_k \sup_{I\subset I_{n,k}, J\in\mathscr{J}} |\mathbb{T}_n|^{1/2}|\hat{P}_{J,n}(I)-P_J(I)|,\end{aligned} \tag{4.25}$$

where $I\in\mathscr{I}$. This entails that, for any $\lambda>0$,

$$\begin{aligned}p_n &= \mathbb{P}\left\{|\mathbb{T}_n|^{\mu} \sup_{x\in I_0, J\in\mathscr{J}} |\hat{f}_{J,n}(x)-f_{J,n}(x)|\geqslant\lambda\right\} \\ &\leqslant \sum_k \mathbb{P}\left(\left\{\sup_{I\subset I_{n,k}, J\in\mathscr{J}} |\mathbb{T}_n|^{1/2}|\hat{P}_{J,n}(I)-P_J(I)|\geqslant|\mathbb{T}_n|^{1/2-\mu-\alpha}\lambda\right\}\cap\Omega_n\right) \\ &\quad +\mathbb{P}(\Omega_n^c).\end{aligned} \tag{4.26}$$

Conditions (4.21) and (4.22) for application of (4.20) are easily checked, and application yields

$$p_n\leqslant C|\mathscr{I}_n||\mathbb{T}_n|^{\rho}\exp(-A|\mathbb{T}_n|^{1-2\mu-\alpha-\rho})+C|\mathbb{T}_n|^{-\tau}, \tag{4.27}$$

where we use that $P_{\mathbb{R}}(I_{n,k})\leqslant C|\mathbb{T}_n|^{-\alpha}$, because of $f_{\mathbb{R}}\geqslant\varepsilon>0$ on $I_n\supset I$ for n sufficiently large. Since $\tau>1$ we have $\sum_n p_n<\infty$ so that (4.24) follows.

Next it is clear that (4.24) entails as a special case that

$$|\mathbb{T}_n|^{\mu} \sup_{x\in I_0} |\hat{f}_{\mathbb{R},n}(x)-f_{\mathbb{R},n}(x)| \xrightarrow{\text{a.s.}} 0 \quad \text{as } n\to\infty, \tag{4.28}$$

because $\mathbb{R}\in\mathscr{J}$. To finish the proof let us note that for sufficiently large n we must have

$$\inf_{x\in I_0} f_{\mathbb{R},n}(x)=\inf_{x\in I_0}\frac{1}{\Lambda(Q_{n,x})}\int_{Q_{n,x}} f_{\mathbb{R}}(y)\,\mathrm{d}y\geqslant\varepsilon>0. \tag{4.29}$$

It follows that

$$\begin{aligned}&\sup_{x\in I_0, J\in\mathscr{J}} |\hat{P}_n(J\,|\,x)-P_n(J\,|\,x)| \\ &\quad\leqslant \frac{1}{\inf_x f_{\mathbb{R},n}(x)} \sup_{x,J} |\hat{f}_{J,n}(x)-f_{J,n}(x)| \\ &\quad\leqslant \frac{\sup_{x,J}\hat{f}_{J,n}(x)}{\{\inf_x f_{\mathbb{R},n}(x)\}\{\inf_x \hat{f}_{\mathbb{R},n}(x)\}} \sup_{x,J}|\hat{f}_{J,n}(x)-f_{J,n}(x)|.\end{aligned} \tag{4.30}$$

The proof is now easily completed by using (4.24) and (4.28). □

Let us now turn to the actual prediction problem of estimating (1.1). Since the random variable Y will not in general be bounded, some further assumptions will be needed.

Assumption 4.3. There exists a measurable function $g:\mathbb{R}^d \to [0, \infty)$ such that

$$\sup_{x \in I_*} f(x, y) \leqslant g(y), \quad y \in \mathbb{R}, \tag{4.31}$$

$$\int_{\mathbb{R}} |y|^q g(y)\, \mathrm{d}y = C < \infty \quad \text{for some } q > 1, \tag{4.32}$$

where I_* is any interval the interior of which contains $\bar{I}_0$. (Hence, for sufficiently large n, we have $Q_{n,x} \subset I_*$ for each $x \in \bar{I}_0$.)

Let us write (cf. (4.7) and (4.8))

$$P_n((-\infty, y]|x) = F_n(y \mid x), \qquad \hat{P}_n((-\infty, y]|x) = \hat{F}_n(y \mid x). \tag{4.33}$$

Assumptions 4.1 and 4.3 entail that for any measurable $\varphi:\mathbb{R} \to \mathbb{R}$ with $|\varphi(y)| \leqslant |y|^q$, the conditional expectation

$$E_n(\varphi(Y) \mid x) \int_{\mathbb{R}} \varphi(y)\, \mathrm{d}F_n(y \mid x), \tag{4.34}$$

exists for $x \in I_0$. For each $n \in \mathbb{N}$, let us now introduce a strictly increasing continuous and bounded function $\varphi_n:\mathbb{R} \to \mathbb{R}$ such that $|\varphi_n(y)| \leqslant y$, $y \in \mathbb{R}$, $\varphi_n(y) = y$ for $-n^a \leqslant y \leqslant n^a$, and $|\varphi_n(y)| \leqslant 2n^a$ for $y \in \mathbb{R}$, where

$$a = k\{\tfrac{1}{2} - \tfrac{1}{2}(\alpha + \rho)\}/q. \tag{4.35}$$

Theorem 4.2. *Let Assumptions* 4.1–4.3 *be fulfilled and let us choose again* $l_n^d = |\mathbb{T}_n|^\alpha$ *in* (4.6) *for some* $0 < \alpha < \frac{1}{2}(1 - \rho)$. *Then we have*

$$|\mathbb{T}_n|^\mu \sup_{x \in I_0} |\hat{E}_n(\varphi_n(Y)|x) - E_n(Y|x)| \xrightarrow{\text{a.s.}} 0 \quad \textit{as } n \to \infty \tag{4.36}$$

for any $\mu < ((q-1)/q)\{\frac{1}{2} - \frac{1}{2}(\alpha + \rho)\}$.

Proof. Since $\varphi_n(Y)$ is bounded it follows easily from Theorem 4.1 that

$$\sup_{x \in I_0} |\hat{E}_n(\varphi_n(Y)|x) - E_n(\varphi_n(Y)|x)|$$
$$= \mathrm{o}(|\mathbb{T}_n|^{-\gamma} \cdot n^a) = \mathrm{o}(|\mathbb{T}_n|^{-\gamma + a/k}) \quad \text{as } n \to \infty \tag{4.37}$$

for any $0 < \gamma < \frac{1}{2} - \frac{1}{2}(\alpha + \rho)$; cf. Ruymgaart (1991).

Next let c_0 denote the minimum of the marginal density of X on $\bar{I}_0$. With $B_n=(-\infty,-n^a]\cup[n^a,\infty)$ it follows from Assumptions 4.1 and 4.3 that

$$\begin{aligned}
&\sup_{x\in I_0}|E_n(\varphi_n(Y)|x)-E_n(Y|x)|\\
&\quad\leqslant \sup_{x\in I_0} E_n(|Y-\varphi_n(Y)|\,|x)\\
&\quad\leqslant \sup_{x\in I_0}\int_{\mathbb{R}}|y-\varphi_n(y)|\,\mathrm{d}F_n(y|x)\\
&\quad\leqslant \sup_{x\in I_0}\int_{B_n}|y|\left\{\frac{\int_{Q_{n,x}} f(u,y)\,\mathrm{d}u}{\int_{\mathbb{R}}\int_{Q_{n,x}} f(u,y)\,\mathrm{d}u\,\mathrm{d}y}\right\}\mathrm{d}y\\
&\quad\leqslant 2\int_{n^a}^{\infty}|y|\frac{g(y)\Lambda(Q_{n,x})}{c_0\Lambda(Q_{n,x})}\,\mathrm{d}y\\
&\quad\leqslant (2C/c_0)n^{-a(q-1)}\int_{\mathbb{R}}|y|^q g(y)\,\mathrm{d}y\leqslant(2C/c_0)n^{-a(q-1)}.
\end{aligned}\tag{4.38}$$

The triangle inequality jointly with (4.37) and (4.38) yields for the left-hand side in (4.36) the asymptotic order of magnitude $\mathrm{o}(|\mathbb{T}|^{-\gamma+a/k})+\mathrm{o}(|\mathbb{T}|^{-(q-1)a/k})=\mathrm{o}(|\mathbb{T}|^{-\mu})$, as $n\to\infty$, for any $\mu<qa/k-a/k=(q-1)\{\frac{1}{2}-\frac{1}{2}(\alpha+\rho)\}/q$. □

Remark 4.3. Let us take $k=d=1$ and $\rho=0$ in Theorem 4.2. Then the problem reduces to estimating $E(Y|\cdot)$ given an i.i.d. sequence $(X_1,Y_1),\ldots,(X_n,Y_n)$ of bivariate random vectors. Assuming $E(Y|\cdot)$ to have a continuous second derivative it is known from Stone (1982) that the optimal rate of a.s. uniform convergence over compact sets for estimators of $E(Y|\cdot)$ is $\mathrm{O}(n^{-2/5}(\log n)^{2/5})$ as $n\to\infty$, provided some further regularity conditions are fulfilled. One of these is the existence of a bounded conditional variance. Choosing $\alpha=\frac{1}{5}$ our estimators satisfy

$$\sup_{x\in I_0}|\hat{E}_n(\varphi_n(Y)|x)-E_n(\varphi_n(Y)|x)|=\mathrm{o}(n^{-\mu})\quad\text{as } n\to\infty \tag{4.39}$$

for any $0<\mu<\frac{2}{5}(1-1/q)$. Since for this α the speed of convergence of the nonrandom biased part $\sup_{x\in I_0}|E_n(Y|x)-E(Y|x)|$ is at least as good for this α it follows that

$$\sup_{x\in I_0}|\hat{E}_n(\varphi_n(Y)|x)-E(Y|x)|=\mathrm{o}(n^{-\mu})\quad\text{as } n\to\infty \tag{4.40}$$

for any $0<\mu<\frac{2}{5}(1-1/q)$. For large q the upper bound on μ is close to $\frac{2}{5}$ (which yields the optimal rate apart from the logarithmic factor). See also Collomb and Härdle (1986), and Györfi et al. (1989) for related results in a time series setting.

Example 4.1. Let us return to the Kriging problem of Section 2.3 and assume that $V\subset W$. According to (2.14) in this example $\Psi:\mathbb{R}^{|W|}\to\mathbb{R}^{|V|+1}$ (i.e. $S=W$, $m=|V|+1$ in Theorem 3.1), where

$$\Psi(x)=\left(x_1,\ldots,x_{|V|},\frac{1}{|W|}\sum_{j=1}^{|W|}x_j\right),\quad x=(x_1,\ldots,x_{|W|})\in\mathbb{R}^{|W|}, \tag{4.41}$$

or in the notation of (2.4),

$$\psi(x)=\frac{1}{|W|}\sum_{j=1}^{|W|} x_j, \quad x=(x_1,\ldots,x_{|W|})\in\mathbb{R}^{|W|}, \tag{4.42}$$

It follows that $[\mathrm{d}\Psi]_x$ is a $(|V|+1)\times|W|$-matrix given by

$$[\mathrm{d}\Psi]_x=\begin{pmatrix} 1 & 0 & \cdots & 0 & 1/|W| \\ 0 & \ddots & \ddots & \vdots & \vdots \\ \vdots & \ddots & \ddots & 0 & \vdots \\ 0 & \cdots & 0 & 1 & 1/|W| \\ 0 & \cdots & \cdots & 0 & 1/|W| \\ \vdots & & & \vdots & \vdots \\ \vdots & & & \vdots & \vdots \\ 0 & \cdots & \cdots & 0 & 1/|W| \end{pmatrix}, \quad \text{so that } \|[\mathrm{d}\Psi]\|_x=1. \tag{4.43}$$

Since this norm does not depend on x, it follows that (3.8) is satisfied (with $M=1$). However, the function ψ in (4.42) is not bounded, so that a moment condition like Assumption 4.3 will be required.

Acknowledgments

The authors are indebted to Professor Clyde Martin for pointing out the importance of the heat flow equation in image processing. They also wish to thank the Editor for his careful handling of the manuscript and the referees for their useful and constructive criticism.

References

D.W.K. Andrews, Non-strong mixing autoregressive processes, J. Appl. Probab. 21 (1984) 930–934.

K.C. Chanda, M.L. Puri and F.H. Ruymgaart, Asymptotic normality of L-statistics based on $m(n)$-decomposable time series, J. Multivariate Anal. 35 (1990) 260–275.

K.C. Chanda and F.H. Ruymgaart, General linear processes: a property of the empirical process applied to density and mode estimation, J. Time Ser. Anal. 11 (1990) 185–199.

K.C. Chanda and F.H. Ruymgaart, Curve estimation for $m(n)$-decomposable time series including bilinear processes, J. Multivariate Anal. 38 (1991) 149–166.

G. Collomb and W. Härdle, Strong mixing convergence rates in robust nonparametric time series analysis, Stochastic Process. Appl. 23 (1986) 77–89.

P. Doukhan and X. Guyon, Mixing for linear random fields, C.R. Acad. Sci. Paris Sér. A 313 (1991) 465–470.

G.E. Forsythe and W.R. Wasow, Finite-Difference Methods for Partial Differential Equations (Wiley, New York, 1960).

V.V. Gorodetskil, On the strong mixing property for linear sequences, Theor. Probab. Appl. 22 (1977) 411–413.

X. Guyon and S. Richardson, Vitesse de convergence du théorème de la limite centrale pour des champs faiblement dépendants, Z. Wahrsch. Verw. Gebiete 66 (1984) 297–314.

L. Györfi, W. Härdle, P. Sarda and P. Vieu, Nonparametric Curve estimation from Time Series. Lecture Notes in Statist. No. 60 (Springer, New York, 1989).

F.B. Hildebrand, Finite-Difference Equations and Simulations (Prentice-Hall, Englewood Cliffs, NY, 1968).

B.S. Nakhapetyan, The central limit theorem for random fields with mixing conditions, in: R.L. Dobrushin and Ya.G. Sinai, eds., Multicomponent Systems. Adv. in Probab. No. 6 (Dekker, New York, 1980) pp. 531–548.

W. Näther, Effective Observation of Random Fields (Teubner, Leipzig, 1985).

D.T. Pham and L.T. Tran, Some strong mixing properties of time series models, Stochastic Process. Appl. 19 (1985) 297–303.

M.B. Priestley, Spectral Analysis and Time Series, Vol. 1 (Academic Press, New York, 1981).

M. Rosenblatt, Stationary Sequences and Random Fields (Birkhäuser, Boston, MA, 1985).

F.H. Ruymgaart, Conditional empirical, quantile and difference processes for a large class of time series with applications, Rep., Dept. of Math., Texas Tech. Univ. (Lubbock, TX, 1991).

C. Stone, Optimal global rates of convergence of nonparametric regression, Ann. Statist. 10 (1982) 1040–1053.

W. Stute, On almost sure convergence of conditional empirical distribution functions, Ann. Probab. 14 (1986) 891–901.

L.T. Tran, Kernel density estimation on random fields, J. Multivariate Anal. 34 (1990) 37–53.

E.A.G. Weits, A stochastic heat equation for freeway traffic flow, Thesis, Tech. Univ. of Delft (Delft, The Netherlands, 1990).

M.I. Yadrenko, Spectral Theory of Random fields (Optimization Software, New York, 1983).

Reprinted by permission of the publisher,
© 1993 by Elsevier Science

ASYMPTOTIC BEHAVIOR OF *L*-STATISTICS FOR A LARGE CLASS OF TIME SERIES

MADAN L. PURI[1]* and FRITS H. RUYMGAART[2]**

[1]*Department of Mathematics, Indiana University, Bloomington, IN 47405, U.S.A.*

[2]*Department of Mathematics, Texas Tech University, Lubbock, TX 79409-1042, U.S.A.*

Abstract. In this paper we derive the asymptotic normality of L-statistics with unbounded scores for a large class of time series. To handle the dependence structure, we use the concept of $m(n)$-decomposability as an alternative to classical mixing concepts.

Key words and phrases: Decomposable processes, mixing, empirical processes for decomposable samples, L-statistics.

1. Introduction

Linear combinations of functions of order statistics (L-statistics) are a well-established class of statistics in robust and nonparametric theory. Their usefulness does not remain restricted to the i.i.d. case but extends to other models like time series; see e.g. Chernoff *et al.* (1967), Stigler (1969), Shorack (1972), Shorack and Wellner (1986), respectively Gastwirth and Rubin (1975), Koul (1977), Portnoy (1977, 1979), Martin (1978) and Martin and Yohai (1986).

In this paper we consider asymptotic normality of L-statistics for a wide class of time series where we use a new way ($m(n)$- or *asymptotic decomposability*) to specify the dependence structure. On the one hand, this new concept allows us to avoid the usual mixing conditions that are hard to understand intuitively since they may fail to hold in very decent cases (Andrews (1984)), and that are occasionally hard to work with (Pham and Tran (1985)). On the other hand, the concept is tailor-made for dealing with linear processes or, more generally, processes with a finite order Volterra expansion (Priestley (1981)), and bilinear processes (Rao and Gabr (1984), Chanda and Ruymgaart (1990)).

We will focus on L-statistics with score functions that are allowed to grow indefinitely near the endpoints of the unit interval. This case is technically more

* Research supported by the Office of Naval Research Contract N00014-91-J-1020.

** Part of this work was done while the author was at the Department of Mathematics, KUN, Nijmegen, The Netherlands.

interesting—and more difficult—than the case of scores that are zero in neighborhoods of the endpoints of the unit interval. For the latter simpler case and a more elaborate discussion and illustration of $m(n)$-decomposability we refer to Chanda *et al.* (1990).

For ease of reference, however, let us recall the definition of $m(n)$-decomposability for a real valued time series. Let $X_1, X_2, \ldots$ be a stochastic process where the X_i are real valued with a common d.f. F. Suppose that for each $n \in \mathbb{N}$ we have a decomposition $X_i = X_{i,m(n)} + \bar{X}_{i,m(n)}$, for some $m(n) \in \{0, 1, \ldots, n\}$, where the $X_{i,m(n)}$ have a common d.f. $F_{m(n)}$, and where

(1.1) $$\text{the } X_{i,m(n)} \text{ are } m(n)\text{-dependent},$$

(1.2) $$\max_{1 \le i \le n} P(|\bar{X}_{i,m(n)}| \ge \epsilon(n)) \le \delta(n),$$

$$\text{for some } \epsilon(n), \delta(n) \to 0 \text{ as } n \to \infty.$$

Such a process is defined as $m(n)$-decomposable. It should be noted that the $m(n)$-dependence mentioned in (1.1) is the ordinary concept for fixed n. For different values of the sample size n, however, we allow the order $m(n)$ of the dependence to be different. In many cases $m(n)$ will increase as a power of n. Working with decomposable processes for large n typically requires a specification of the orders of magnitude of $m(n)$, $\epsilon(n)$ and $\delta(n)$. In interesting examples of time series it is inevitable that $m(n) \to \infty$, as $n \to \infty$; yet in many cases suitable rates of $\epsilon(n)$ and $\delta(n)$ can be obtained for $m(n) \ll n$.

To prove the asymptotic normality we use the classical Chernoff-Savage approach which was first applied by Moore (1968) to L-statistics. In Section 2 we formulate the assumptions and give the main result. In passing we give a first outline of the proof. The necessary tools from empirical process theory that might be of independent interest are summarized in Section 3, and the asymptotic normality of the first order terms is dealt with in Section 4. In Section 5 we derive the asymptotic negligibility of the remainder term and, finally, in Section 6 we simplify the expression for the variance under an additional weak-stationarity condition and give some examples.

2. Assumptions and main result

ASSUMPTION 2.1. The underlying process $X_1, \ldots, X_n$ is decomposable and hence by definition satisfies (1.1) and (1.2). The common d.f. F of the X_i is continuously differentiable on $\mathbb{R}$ with derivative f satisfying $\|f\|_\infty < \infty$, and $\{x \in \mathbb{R} : f(x) > 0\}$ convex.

Let $\hat{F}_n$ be the empirical d.f. of the X_i at stage n and $\hat{\Gamma}_n$ the empirical d.f. of the transformed random variables $\xi_i = F(X_i)$. Assumption 2.1 entails the decomposability of the ξ_i, since by the mean value theorem we have

(2.1) $$\xi_i = F(X_{i,m}) + \bar{X}_{i,m} f(X_{i,m} + \theta \bar{X}_{i,m}) = \xi_{i,m} + \bar{\xi}_{i,m},$$

for random $\theta \in (0, 1)$. The common d.f. of the ξ_i is uniform $(0, 1)$. As usual we define the corresponding reduced empirical process by

(2.2) $$U_n = \{U_n(t) = n^{1/2}(\hat{\Gamma}_n(t) - t), t \in [0, 1]\}.$$

Throughout this paper the numbers

$$A, B, C \in (0, \infty), \tag{2.3}$$

will be used as generic constants that are independent of all the relevant parameters, as in particular the sample size n, and that may differ from line to line.

For each $q > 0$ and $n \in \mathbb{N}$ there exists a continuously differentiable increasing function $\ell_q : [0,1] \to [0,1]$ with

$$\begin{aligned} &\ell_q([0,1]) \subset \left[\frac{1}{2}n^{-q}, 1 - \frac{1}{2}n^{-q}\right]; \\ &\ell_q(t) = t, \ \ t \in [n^{-q}, 1 - n^{-q}]; \qquad 0 \leq \ell_q^{(1)} \leq 1. \end{aligned} \tag{2.4}$$

The dependence on n is suppressed in the notation. Given a function $K : (0,1) \to \mathbb{R}$ we define

$$K_q(t) = K(\ell_q(t)), \qquad t \in [0,1]. \tag{2.5}$$

When K is differentiable we have according to the chain rule that $K_q^{(1)} = (d/dt)K(\ell_q(t)) = K^{(1)}(\ell_q(t))\ell_q^{(1)}(t)$, $t \in (0,1)$, and hence

$$|K_q^{(1)}(t)| \leq |K^{(1)}(\ell_q(t))|, \qquad t \in (0,1). \tag{2.6}$$

A particularly useful function is

$$R(t) = \{t(1-t)\}^{-1}, \qquad t \in (0,1). \tag{2.7}$$

Let us observe that

$$0 < \max_{0 \leq t \leq 1} R_q^{\gamma}(t) \leq C n^{\gamma q}. \tag{2.8}$$

To describe the class of L-statistics that we are going to consider, let $J : (0,1) \to \mathbb{R}$ and $\Psi : (0,1) \to \mathbb{R}$ be given functions. Assumption 2.1 implies the continuity of F^{-1} on $\{x \in \mathbb{R} : 0 < F(x) < 1\}$; we write $\Psi_F = \Psi(F^{-1}) : (0,1) \to \mathbb{R}$. We will focus on statistics of the type

$$T_n = \int_{-\infty}^{\infty} \Psi(x) J_\zeta(\hat{F}_n(x)) d\hat{F}_n(x) = \frac{1}{n}\sum_{i=1}^{n} c_{ni}\Psi(X_{i:n}), \tag{2.9}$$

where J_ζ is obtained from J via (2.5) with $q = \zeta$ and $c_{ni} = J_\zeta(i/n)$. It is obvious that

$$T_n \stackrel{d}{=} \int_0^1 \Psi_F(t) J_\zeta(\hat{\Gamma}_n(t)) d\hat{\Gamma}_n(t) = \frac{1}{n}\sum_{i=1}^{n} c_{ni}\Psi_F(\xi_{i:n}). \tag{2.10}$$

The latter representation will be used throughout.

ASSUMPTION 2.2. The function $J : (0,1) \to \mathbb{R}$ is twice and the function $\Psi_F : (0,1) \to \mathbb{R}$ is once continuously differentiable. For numbers

$$\alpha, \beta > 0 \text{ such that } \alpha + \beta < \Delta, \qquad \text{for some } 0 < \Delta < \frac{1}{2}, \tag{2.11}$$

they satisfy ($J^{(0)} = J$, $\Psi_F^{(0)} = \Psi_F$)

$$\begin{aligned} |J^{(j)}| &\leq CR^{\alpha+j} \text{ on } (0,1), \quad j \in \{0,1,2\}; \\ |\Psi_F^{(j)}| &\leq CR^{\beta+j} \text{ on } (0,1), \quad j \in \{0,1\}. \end{aligned} \tag{2.12}$$

The parameter ζ in (2.10) satisfies

$$0 < \zeta < 1/\{5(1-\Delta)\}. \tag{2.13}$$

Because we use J_ζ rather than J as a score function, it should be noted that the scores stay bounded for each n, but are allowed to tend to ∞ (near 0 and 1) at a certain rate when $n \to \infty$. Since we don't want to impose any further condition on the distribution of the $\xi_{i,m(n)}$, this additional control of the scores is technically very convenient. As it will turn out, under Assumption 2.1 we may replace Ψ_F by $\Psi_{F,r}$ (obtained via (2.5) for a suitable $q = r$) without affecting the weak limiting behavior of the statistics in (2.10). This replacement has similar technical advantages. Let us choose

$$r = 1/\{2(1-\Delta)\}, \tag{2.14}$$

and consider $\Psi_{F,r}$ obtained from Ψ_F via (2.5) with $q = r$. Assumption 2.1 guarantees the finiteness of the numbers

$$\mu_n = \int_0^1 \Psi_F(t) J_\zeta(t)dt, \qquad \tilde{\mu}_n = \int_0^1 \Psi_{F,r}(t) J_\zeta(t)dt. \tag{2.15}$$

Let us introduce

$$\tilde{T}_n = \int_0^1 \Psi_{F,r}(t) J_\zeta(\hat{\Gamma}_n(t)) d\hat{\Gamma}_n(t). \tag{2.16}$$

Jointly with (2.14) it can easily be shown that

$$n^{1/2}(T_n - \mu_n) - n^{1/2}(\tilde{T}_n - \tilde{\mu}_n) \xrightarrow{p} 0 \quad \text{as} \quad n \to \infty. \tag{2.17}$$

Because $\Psi_{F,r}$ is bounded, the statistics $\tilde{T}_n$ are easier to deal with.

It follows from the mean value theorem that for $t \in [0,1]$ and $\theta_n(t) \in (0,1)$ random, we have

$$J_\zeta(\hat{\Gamma}_n(t)) = J_\zeta(t) + n^{-1/2} U_n(t) J_\zeta^{(1)}(t + \theta_n(t) n^{-1/2} U_n(t)). \tag{2.18}$$

By substitution in (2.16) we find

$$n^{1/2}(\tilde{T}_n - \tilde{\mu}_n) = \tilde{A}_{0n} + \tilde{A}_{1,n} + \tilde{B}_n, \tag{2.19}$$

where the $\tilde{A}$- and $\tilde{B}$-terms are first and second order terms respectively given by

$$\tilde{A}_{0n} = \int_0^1 \Psi_{F,r}(t)J_\varsigma(t)dU_n(t), \tag{2.20}$$

$$\tilde{A}_{1n} = \int_0^1 U_n(t)\Psi_{F,r}(t)J_\varsigma^{(1)}(t)dt, \tag{2.21}$$

$$\tilde{B}_n = \int_0^1 U_n(t)\Psi_{F,r}(t)J_\varsigma^{(1)}(t+\theta_n(t)n^{-1/2}U_n(t))d\hat{\Gamma}_n(t) - \tilde{A}_{1n}. \tag{2.22}$$

The next assumption prescribes suitable orders of magnitude for the sequences of parameters in (1.2), depending on Δ in (2.11).

ASSUMPTION 2.3. Let ς satisfy (2.13). There exist $m(n) = O(n^\rho)$, $\epsilon(n) = O(n^{-\tau})$, as $n \to \infty$, with

$$0 \le \rho < \varsigma(1-2\Delta), \qquad \tau > 1 + 1/\{2(1-\Delta)\}, \tag{2.23}$$

such that $\delta(n) = n^{-\sigma}$, as $n \to \infty$, with

$$\sigma > 1 + 1/\{2(1-\Delta)\}. \tag{2.24}$$

Let us write, for brevity,

$$\tilde{K}_n(t) = \Psi_{F,r}(t)J_\varsigma(t) + \int_0^1 \left\{1_{[0,s]}(t) - s\right\}\Psi_{F,r}(s)J_\varsigma^{(1)}(s)ds, \qquad t \in [0,1], \tag{2.25}$$

so that $\tilde{A}_{0n} + \tilde{A}_{1n} = \sum_{i=1}^n \{\tilde{K}_n(\xi_i) - \tilde{\mu}_n\}$, and consider the triangular array of $m(n)$-dependent centered random variables

$$\tilde{Z}_{ni} = \tilde{K}_n(\xi_{i,m(n)}) - E\tilde{K}_n(\xi_{i,m(n)}) = \tilde{K}_n(\xi_{i,m(n)}) - \tilde{\tilde{\mu}}_n. \tag{2.26}$$

Under the assumptions made above, we will prove (Sections 3 and 4)

$$\begin{aligned} &\tilde{A}_{0n} + \tilde{A}_{1n} - n^{-1/2}\sum_{i=1}^n \tilde{Z}_{ni} \\ &= n^{-1/2}\sum_{i=1}^n \{(\tilde{K}_n(\xi_i) - \tilde{\mu}_n) - (\tilde{K}_n(\xi_{i,m(n)}) - \tilde{\tilde{\mu}}_n)\} \xrightarrow{p} 0 \\ &\qquad \text{as} \quad n \to \infty, \end{aligned} \tag{2.27}$$

$$\tilde{B}_n \xrightarrow{p} 0 \quad \text{as} \quad n \to \infty. \tag{2.28}$$

To prove the asymptotic normality of $n^{-1/2}\sum_{i=1}^n \tilde{Z}_{ni}$ we may apply Berk (1973) (see also Rao (1984)), provided that the next assumption is satisfied.

ASSUMPTION 2.4. The following conditions are fulfilled:

$$\text{Var}\left(\sum_{i=j+1}^{k} \tilde{Z}_{ni}\right) \leq C(k-j) \quad \text{for all} \quad j,k \in \{0,1,\ldots,n\}. \tag{2.29}$$

$$\lim_{n\to\infty} \frac{1}{n}\text{Var}\left(\sum_{i=1}^{n} \tilde{Z}_{ni}\right) = \tilde{v}^2 \in (0,\infty). \tag{2.30}$$

THEOREM 2.1. *When Assumptions* 2.1–2.4 *are satisfied we have*

$$n^{1/2}(T_n - \tilde{\mu}_n) \xrightarrow{d} \mathcal{N}(0, \tilde{v}^2) \quad \text{as} \quad n \to \infty, \tag{2.31}$$

where $\tilde{\mu}_n$ *may be replaced by* μ_n *(see* (2.15)) *and* $\tilde{v}^2$ *is defined in* (2.31).

3. Tools from empirical process theory

Some properties of empirical processes based on decomposable models will be reviewed here. Proofs will not be given since they are very similar to those in Nieuwenhuis and Ruymgaart (1990); see also Chanda and Ruymgaart (1990). Since these properties hold true for arbitrary n there is no need to specify the dependence on n of the parameters; in particular we will write m, ϵ, δ for the parameters in (1.2). In applications, however, these parameters will usually depend on n in a suitable way. We will then e.g. choose $\epsilon = \epsilon(n)$ in such a way that $P(\Omega^c_{\epsilon(n)}) \leq n\delta(n) = O(n^{1-\sigma\wedge\tau}) \to 0$, as $n \to \infty$, where $\sigma \wedge \tau = \min\{\sigma, \tau\}$. To formulate the results we need the function

$$\psi(\lambda) = 2\lambda^{-2}\int_0^\lambda \log(1+x)dx, \quad \lambda > 0; \qquad \psi(0) = 1. \tag{3.1}$$

It should be noted that $\psi(\lambda) \downarrow 0$ as $\lambda \uparrow \infty$. For further properties and the role this function plays in precise asymptotic considerations the reader is referred to e.g. Shorack and Wellner (1986). We start with a basic fluctuation inequality that might be of independent interest.

THEOREM 3.1. *Suppose that Assumption* 2.1 *is fulfilled. For all* $n \in \mathbb{N}$ *we have*

$$P\left(\Omega_\epsilon \cap \left\{\sup_{a\leq s<t\leq b} |U_n(t) - U_n(s)| \geq \lambda\right\}\right) \leq Cm\exp\left(\frac{-A\lambda^2}{m(b-a)}\psi\left(\frac{B\lambda}{n^{1/2}(b-a)}\right)\right), \tag{3.2}$$

where $\Omega_\epsilon = \{\max_{1\leq i\leq n} |\bar{X}_{i,m}| < \epsilon\}$, *provided that*

$$\lambda \geq Cn^{1/2}(\epsilon + \delta), \qquad b - a \geq C(\epsilon + \delta). \tag{3.3}$$

Because Nieuwenhuis and Ruymgaart (1990) only exploit the decomposability property of the linear process the proof for the present Theorem 3.1 can be just copied. The proof of the next theorem requires only a minor modification of the slightly different weight functions that will be used. For arbitrary $0 \le \gamma \le 1/2$ and $q > 0$ consider the function R_q^γ (see (2.4)–(2.8)). For the proof of the next theorem the pattern of the proof of Nieuwenhuis and Ruymgaart ((1990), Theorem 3.1) may be followed by using (2.8) in particular and by partitioning $[0,1]$ into intervals of length $[n^q]^{-1}$.

THEOREM 3.2. *Suppose that Assumption* 2.1 *is satisfied. For any* $0 \le \gamma \le 1/2$ *and* $q > 0$ *we have*

$$P\left(\Omega_\epsilon \cap \left\{\sup_{0\le t\le 1} R_q^\gamma(t)|U_n(t)| \ge \lambda\right\}\right) \le Cmn^q \exp\left(\frac{-A\lambda^2}{m}\psi\left(\frac{B\lambda n^{q(1-\gamma)}}{n^{1/2}}\right)\right), \tag{3.4}$$

where Ω_ϵ *is as above, provided that*

$$\lambda \ge Cn^{1/2}n^{\gamma q}(\epsilon+\delta), \qquad n^{-q} \ge C(\epsilon+\delta). \tag{3.5}$$

As a corollary to Theorem 3.2 let us prove the useful property that, for any $0 < \beta < 1$,

$$P\left(\Omega_\epsilon \cap \left\{\hat{\Gamma}_n(t) \ge \beta t, \text{ for } n^{-q} \le t \le 1\right\}\right) \ge 1 - Cmn^q \exp\left(\frac{-An^{1-q}}{m}\right), \tag{3.6}$$

provided that (3.5) is satisfied with $\lambda = An^{(1-q)/2}$. As a matter of fact we have $P(\hat{\Gamma}_n(t) \ge \beta t, \text{ for } n^{-q} \le t \le 1) = P(\inf_{n^{-q}\le t\le 1} \hat{\Gamma}_n(t)/t \ge \beta) = P(\sup_{n^{-q}\le t\le 1}(t - \hat{\Gamma}_n(t))/t \le 1-\beta) \ge 1 - P(\sup_{n^{-q}\le t\le 1}|U_n(t)|/t \ge n^{1/2}(1-\beta)) \ge 1 - P(\sup_{n^{-q}\le t\le 1}|U_n(t)|/t^{1/2} \ge n^{(1-q)/2}(1-\beta)) \ge 1 - P(\sup_{0\le t\le 1} R_q^{1/2}|U_n(t)| \ge An^{(1-q)/2})$. The lower bound in (3.6) follows by applying (3.4) with $\gamma = 1/2$ and $\lambda = An^{(1-q)/2}$.

4. The leading terms

In this section we will be concerned with proving (2.27) and the asymptotic normality of the standarized sum of the $\tilde{Z}_{ni}$. It follows from Assumptions 2.1–2.3, (2.1) and (2.23) that

$$\left\{\max_{1\le i\le n} \bar{\xi}_{i,m(n)} \le \|f\|_\infty \epsilon(n)\right\} \supset \Omega_{\epsilon(n)}, \qquad E\bar{\xi}_{i,m(n)} \le Cn^{-\sigma\wedge\tau}. \tag{4.1}$$

The definition of $\tilde{K}_n$ in (2.25) implies that, for $i = 0, 1$,

$$|\tilde{K}_n^{(i)}| \le CR_\zeta^{-\alpha}R_r^{-\beta-i} \quad \text{on } (0,1), \tag{4.2}$$

$$\max_{0\le t\le 1} |K_n^{(i)}(t)| \le Cn^{\zeta\alpha+r(\beta+i)}, \tag{4.3}$$

where we use the fact that $0 < \zeta < r$.

Employing (4.1)–(4.3) and the mean value theorem we see that

$$n^{-1/2}\sum_{i=1}^{n} E|\tilde{K}_n(\xi_i) - \tilde{K}_n(\xi_{i,m(n)})| \le n^{-1/2} n^{\zeta\alpha + r(\beta+1)} \sum_{i=1}^{n} E(\bar{\xi}_{i,m(n)}) \le n^{1/2+\zeta\alpha+r(\beta+1)-\sigma\wedge\tau}. \tag{4.4}$$

According to the asssumptions and (2.14) we have $\sigma \wedge \tau - 1/2 - \zeta\alpha - r(\beta+1) > 1/2 + 1/2(1-\Delta) - r\Delta - 1/2(1-\Delta) = 1/2 - r\Delta > 0$, so that the upper bound in (4.4) converges to 0, as $n \to \infty$. Because $|E\tilde{K}_n(\xi_i) - E\tilde{K}_n(\xi_{i,m(n)})| \le E|\tilde{K}_n(\xi_i) - \tilde{K}_n(\xi_{i,m(n)})|$ we have proved (2.27).

Let us now choose an incidental parameter

$$\gamma = 1/\Delta - 2. \tag{4.5}$$

For such a γ we have, again using $0 < \zeta < r$,

$$\begin{aligned}\sup_{i,n} E|\tilde{Z}_{ni}|^{2+\gamma} &\le C \sup_{i,n} E|\tilde{K}_n(\xi_{i,m(n)})|^{2+\gamma} \\ &\le C \sup_{i,n} E(R_\zeta^{-\alpha(2+\gamma)} \cdot R_r^{-\beta(2+\gamma)})(\xi_i - \bar{\xi}_{i,m(n)}) \\ &\le C \int_0^1 R^{-(\alpha+\beta)(2+\gamma)}(t)dt \\ &\quad + Cn^{-\sigma\wedge\tau} n^{\zeta\alpha(2+\gamma)+r\beta(2+\gamma)+r}.\end{aligned} \tag{4.6}$$

Since $\sigma\wedge\tau - r - (\zeta\alpha + r\beta)(2+\gamma) > \sigma\wedge\tau - r - r\Delta(2 + 1/\Delta - 2) = \sigma\wedge\tau - 2r > 1 + 1/2(1-\Delta) - 1/(1-\Delta) = 1 - 1/2(1-\Delta) > 0$, it follows that the upper bound in (4.6) is bounded by a finite number because, moreover, $(\alpha+\beta)(2+\gamma) < \Delta(2+\gamma) = 1$.

Finally, let us note that the choice of ρ in (2.23) implies

$$n^{-1}\{m(n)\}^{2+2/\gamma} \to 0 \quad \text{as} \quad n \to \infty, \tag{4.7}$$

because $\rho(2+2/\gamma) - 1 < \{(1-2\Delta)/5(1-\Delta)\}\{2(1-\Delta)/(1-2\Delta)\} - 1 = 2/5 - 1 < 0$.

Assumption 2.4 jointly with (4.6) and (4.7) suffice for application of Berk (1973), which yields $n^{-1/2}\sum_{i=1}^{n}\tilde{Z}_{ni} \xrightarrow{d} \mathcal{N}(0,\tilde{v}^2)$ as $n \to \infty$. Since we have also proved (2.27), it follows that $\tilde{A}_{0n} + \tilde{A}_{1n} \xrightarrow{d} \mathcal{N}(0,\tilde{v}^2)$ as $n \to \infty$, which settles the asymptotic normality of the leading terms.

5. The remainder term

This section is devoted to a proof of (2.28). This may be done with the help of the properties in Section 3 in a way similar to that in the i.i.d. case. As in Nieuwenhuis and Ruymgaart (1990), however, the present situation of a decomposable process is more complicated and requires control over a larger set of parameters.

Let us introduce the intervals

$$(5.1)\qquad I(1) = [0, 2n^{-\varsigma}], \qquad I(2) = [2n^{-\varsigma}, 1 - 2n^{-\varsigma}], \qquad I(3) = [1 - 2n^{-\varsigma}, 1],$$

and write, for brevity,

$$(5.2)\qquad t + \theta_n(t)n^{-1/2}U_n(t) = t + \theta_n(t)(\hat{\Gamma}_n(t) - t) = \tilde{\Gamma}_n(t),$$

and note that $\tilde{\Gamma}_n(t)$ is a random point between t and $\hat{\Gamma}_n(t)$. Now the remainder term will be decomposed into $\tilde{B}_n = \sum_{j=1}^{4} \tilde{B}_{nj}$, where

$$(5.3)\ \tilde{B}_{nj} = \int_{I(j)} U_n(t)\Psi_{F,r}(t)\left\{J_\varsigma^{(1)}(\tilde{\Gamma}_n(t)) - J_\varsigma^{(1)}(t)\right\} d\hat{\Gamma}_n(t), \qquad j \in \{1, 2, 3\};$$

$$(5.4)\ \tilde{B}_{n4} = \int_0^1 U_n(t)\Psi_{F,r}(t)J_\varsigma^{(1)}(t)d\{\hat{\Gamma}_n(t) - t\}.$$

Let us single out the subsets

$$(5.5)\qquad \Omega_{n1} = \left\{\sup_{0\le t\le 1} R_r^{1/2}(t)|U_n(t)| \le n^{\nu/2}\right\},$$

$$(5.6)\qquad \Omega_{n2} = \left\{\hat{\Gamma}_n(t) \ge \frac{1}{2}t\ \forall t \in [n^{-\varsigma}, 1]\right\},$$

where the incidental parameter ν satisfies

$$(5.7)\qquad \nu = \varsigma(1 - 2\Delta).$$

Jointly with (2.14) and (5.7) conditions (2.23) and (2.24) imply that the conditions for application of (3.4) and (3.6) are satisfied and that

$$(5.8)\qquad P(\Omega_n) \to 1 \text{ as } n \to \infty, \qquad \text{with } \Omega_n = \Omega_{\epsilon(n)} \cap \Omega_{n1} \cap \Omega_{n2}.$$

Consequently it suffices to prove that

$$(5.9)\qquad C_{nj} = 1_{\Omega_n}\tilde{B}_{nj} \xrightarrow{p} 0 \quad \text{as} \quad n \to \infty, \qquad \text{for} \quad j = 1, 2, 3, 4.$$

Our assumptions entail that (note that $\varsigma < r$)

$$(5.10)\ E|C_{n1}| \le Cn^{\nu/2}\int_{I(1)} \frac{R_r^\beta(t)R_\varsigma^{\alpha+1}(t)}{R_r^{1/2}(t)}dt \le Cn^{\nu/2}\int_{I(1)} R_\varsigma^{\alpha+\beta+1/2}(t)dt$$
$$\le Cn^{\nu/2+\varsigma(\alpha+\beta)-\varsigma/2} \to 0 \quad \text{as} \quad n \to \infty,$$

because $\nu/2 + \varsigma(\alpha + \beta) - \varsigma/2 = \varsigma(\alpha + \beta) - \Delta < 0$. Since $J_\varsigma = J$ on $I(2)$, we may apply the mean value theorem once more, which yields

$$(5.11)\qquad J_\varsigma^{(1)}(\tilde{\Gamma}_n(t)) - J_\varsigma^{(1)}(t) = (\tilde{\Gamma}_n(t) - t)J_\varsigma^{(2)}(\tilde{\tilde{\Gamma}}_n(t)),$$

where $\tilde{\tilde{\Gamma}}_n(t)$ is also a random point between t and $\hat{\Gamma}_n(t)$. It follows that

$$(5.12)\qquad \begin{aligned} 1_{\Omega_n}|J_\zeta^{(1)}(\tilde{\Gamma}_n(t)) - J_\zeta^{(1)}(t)| &\leq C 1_{\Omega_n}|\tilde{\Gamma}_n(t) - t| R_\zeta^{\alpha+2}(\tilde{\tilde{\Gamma}}_n(t)) \\ &\leq C n^{-1/2}|U_n(t)| R_\zeta^{\alpha+2}(t), \qquad t \in I(2). \end{aligned}$$

Hence we find by substitution (note that $\zeta < r$)

$$(5.13)\qquad \begin{aligned} E|C_{n2}| &\leq C n^{-1/2} n^{\nu} \int_{I(2)} \frac{R_r^{\beta}(t) R_\zeta^{\alpha+2}(t)}{R_r(t)} dt \\ &\leq C n^{-1/2+\nu} \int_{I(2)} R_\zeta^{\alpha+\beta+1}(t) dt \\ &\leq C n^{-1/2+\nu+\zeta(\alpha+\beta)} \to 0 \quad \text{as} \quad n \to \infty, \end{aligned}$$

because $-1/2 + \nu + \zeta(\alpha + \beta) < -1/2 + \zeta(1 - 2\Delta) + \zeta\Delta = -1/2 + \zeta(1 - \Delta) < 0$ by (2.13). The random variable C_{n3} may be dealt with in a way similar to that of C_{n1}.

Let us now consider C_{n4} and introduce the parameter

$$(5.14)\qquad \phi = 1 - \frac{5}{2}\zeta(1 - 2\Delta) + \frac{1}{2}\rho > 1 - \frac{1 - 2\Delta}{2(1 - \Delta)} + \frac{1}{2}\rho > 0.$$

Partitioning the unit interval into subintervals of equal length $[n^\phi]^{-1}$, we define the left continuous step functions

$$(5.15)\qquad S_n(t) = \begin{cases} \dfrac{1}{[n^\phi]}, & \text{for } t = 0 = t_0; \\ \dfrac{k}{[n^\phi]}, & \text{for } t_{k-1} = \dfrac{k-1}{[n^\phi]} < t \leq \dfrac{k}{[n^\phi]},\ 1 \leq k \leq \dfrac{1}{2}[n^\phi]; \\ \dfrac{k-1}{[n^\phi]}, & \text{for } \dfrac{k-1}{[n^\phi]} < t \leq \dfrac{k}{[n^\phi]},\ \dfrac{1}{2}[n^\phi] < k \leq [n^\phi]. \end{cases}$$

We need to decompose C_{n4} into

$$(5.16)\qquad D_{n1} = 1_{\Omega_n} \int_0^1 U_n(S_n(t)) \Psi_{F,r}(S_n(t)) J_\zeta^{(1)}(S_n(t)) d\{\hat{\Gamma}_n(t) - t\},$$

$$(5.17)\qquad \begin{aligned} D_{n2} = 1_{\Omega_n} \int_0^1 &\{U_n(t)\Psi_{F,r}(t) J_\zeta^{(1)}(t) \\ &- U_n(S_n(t))\Psi_{F,r}(S_n(t)) J_\zeta^{(1)}(S_n(t))\} d\hat{\Gamma}_n(t), \end{aligned}$$

$$(5.18)\qquad \begin{aligned} D_{n3} = 1_{\Omega_n} \int_0^1 &\{U_n(S_n(t))\Psi_{F,r}(S_n(t)) J_\zeta^{(1)}(S_n(t)) \\ &- U_n(t)\Psi_{F,r}(t) J_\zeta^{(1)}(t)\} dt. \end{aligned}$$

Let us note that

$$
\begin{aligned}
(5.19) \qquad |D_{n1}| &= 1_{\Omega_n} n^{-1/2} \left| \sum_{k=1}^{[n^\phi]} \{U_n(S_n(t_k))\Psi_{F,r}(S_n(t_k)) \right. \\
&\qquad \left. \cdot J_\zeta^{(1)}(S_n(t_k))\}\{U_n(t_k) - U_n(t_{k-1})\} \right| \\
&\le Cn^{-1/2} n^{\nu/2} n^\phi \left\{ 1_{\Omega_n} \max_k |U_n(t_k) - U_n(t_{k-1})| \right\} \\
&\quad \cdot \left\{ \frac{1}{[n^\phi]} \sum_{k=1}^{[n^\phi]} \frac{R_r^\beta(S_n(t_k)) R_\zeta^{\alpha+1}(S_n(t_k))}{R_r^{1/2}(S_n(t_k))} \right\} \\
&\le Cn^{-1/2+\nu/2+\phi} \left\{ 1_{\Omega_n} \max_k |U_n(t_k) - U_n(t_{k-1})| \right\} \\
&\quad \cdot \left\{ \int_0^1 R^{\alpha+\beta+1/2}(t)dt \right\}.
\end{aligned}
$$

Choosing another parameter

$$
(5.20) \qquad \lambda = \frac{1}{2} - \frac{3}{2}\zeta(1-2\Delta) > \frac{1}{2} - \frac{3(1-2\Delta)}{5(1-\Delta)} > 0,
$$

it follows from (3.2) that

$$
\begin{aligned}
(5.21) \qquad & P\left(\Omega_n \cap \left\{ \max_k |U_n(t_k) - U_n(t_{k-1})| > n^{-\lambda} \right\} \right) \\
&\le P\left(\max_k \sup_{t_{k-1} \le t \le t_k} |U_n(S_n(t)) - U_n(t)| \ge n^{-\lambda} \right) \\
&\le \sum_{k=1}^{[n^\phi]} P\left(\sup_{t_{k-1} \le s < t \le t_k} |U_n(t) - U_n(s)| \ge n^{-\lambda} \right) \\
&\le Cn^{\phi+\rho} \exp\left(\frac{-An^{-2\lambda}}{n^{\rho-\phi}} \phi\left(\frac{Bn^{-\lambda}}{n^{1/2-\phi}} \right) \right) + n^\phi n n^{-\sigma}.
\end{aligned}
$$

Condition (3.3) is indeed fulfilled, since $-\lambda - 1/2 + \sigma \wedge \tau > 0$ and $-\phi + \sigma \wedge \tau > 0$ as follows easily from the choices we made. In order to show that the upper bound in (5.21) converges to 0 as $n \to \infty$ note that $\phi - \rho - 2\lambda = 1 - (5/2)\zeta(1-2\Delta) - \rho/2 - 1 + 3\zeta(1-2\Delta) > 0$ by (2.23), and that

$$
\phi - \frac{1}{2} - \lambda = 1 - \frac{5}{2}\zeta(1-2\Delta) - \frac{1}{2} - \frac{1}{2} + \frac{3}{2}\zeta(1-2\Delta) < 0.
$$

Combining these results we may finally conclude that

$$
(5.22) \qquad |D_{n1}| = O_p(n^{-1/2+\nu/2+\phi-\lambda}) = o_p(1) \quad \text{as} \quad n \to \infty.
$$

because

$$\frac{1}{2}\nu+\phi-\frac{1}{2}-\lambda=\frac{1}{2}\zeta(1-2\Delta)+1-\frac{5}{2}\zeta(1-2\Delta)+\frac{1}{2}\rho-\frac{1}{2}-\frac{1}{2}+\frac{3}{2}\zeta(1-2\Delta)$$
$$<\frac{1}{2}\zeta(1-2\Delta)-\frac{5}{2}\zeta(1-2\Delta)+\frac{1}{2}\zeta(1-2\Delta)+\frac{3}{2}\zeta(1-2\Delta)=0.$$

By taking expectations of a suitable upper bound it follows that D_{n2} may be dealt with in a similar way as D_{n3}, so that we restrict attention to the latter random variable. With the intervals $I(j)$ as in (5.1) it can be shown as in the proof of (5.10) that the parts of D_{n3} corresonding to a restriction of the integral to either $I(1)$ or $I(3)$ converge to 0 in probability as $n\to\infty$. Hence it remains to consider the part of D_{n3} corresonding to a restriction of the integral to $I(2)$; this part will be decomposed into the sum of the random variables

$$H_{n1}=1_{\Omega_n}\int_{I(2)}\{U_n(S_n(t))-U_n(t)\}\,L_n(S_n(t))dt, \tag{5.23}$$

$$H_{n2}=1_{\Omega_n}\int_{I(2)}U_n(t)\,\{L_n(S_n(t))-L_n(t)\}\,dt, \tag{5.24}$$

where $L_n(t)=\Psi_{F,r}(t)J_\zeta^{(1)}(t)$, $t\in[0,1]$.

Due to the way in which we have defined S_n in (5.15) it follows that $|L_n(S_n(t))|\le CR_r^\beta(S_n(t))R_\zeta^{\alpha+1}(S_n(t))\le CR^{\alpha+\beta+1}(S_n(t))\le CR^{\alpha+\beta+1}(t)$, for all $t\in I(2)$. Hence it is clear that

$$|H_{n1}|\le C1_{\Omega_n}\left\{\sup_{0\le t\le 1}|U_n(S_n(t))-U_n(t)|\right\}\left\{\int_{I(2)}R^{\alpha+\beta+1}(t)dt\right\} \tag{5.25}$$
$$\le C1_{\Omega_n}\left\{\sup_{0\le t\le 1}|U_n(S_n(t))-U_n(t)|\right\}n^{\zeta(\alpha+\beta)}.$$

Reasoning as in (5.21) we see that

$$|H_{n1}|=O_p(n^{\zeta(\alpha+\beta)-\lambda})=o_p(1)\qquad\text{as}\quad n\to\infty, \tag{5.26}$$

because $\lambda-\zeta(\alpha+\beta)>1/2-(3/2)\zeta(1-2\Delta)-\zeta\Delta>(2-\Delta)/\{10(1-\Delta)\}>0$ by substituting the upper bound for ζ in (2.13).

For H_{n2} let us apply the mean value theorem and employ the special construction of S_n to find

$$|L_n(S_n(t))-L_n(t)|\le Cn^{-\phi}n^{\zeta(\alpha+\beta+2)}. \tag{5.27}$$

This entails

$$|H_{n2}|=O_p(n^{\nu/2-\phi+\zeta(\alpha+\beta+2)})=o_p(1)\qquad\text{as}\quad n\to\infty, \tag{5.28}$$

since $\phi-\zeta(\alpha+\beta+2)-\nu/2\ge 1-(5/2)\zeta(1-2\Delta)-\zeta(\Delta+2)-(1/2)\zeta(1-2\Delta)>0$ by substituting the upper bound for ζ in (2.13).

6. Concluding remarks

A. *Sinplification variance*

In many special cases (see, e.g., Chanda *et al.* (1990)) the process $X_{1,m(n)}, \ldots, X_{n,m(n)}$ and hence the process

$$\xi_{1,m(n)}, \ldots, \xi_{n,m(n)} \text{ is strictly stationary.} \tag{6.1}$$

Note that this condition which is not implied by (1.1) and (1.2) will be needed below.

THEOREM 6.1. *Let* (6.1) *be satisfied. Under the conditions of Theorem* 2.1 *we have*

$$\tilde{v}^2 = \mathrm{Var}(\tilde{K}_n(\xi_1)) + 2 \lim_{n\to\infty} \sum_{j=2}^{m(n)} \mathrm{Cov}(\tilde{K}_n(\xi_1), \tilde{K}_n(\xi_j)). \tag{6.2}$$

PROOF. Relation (4.6) entails that $E|\tilde{Z}_{n1}\tilde{Z}_{nj}| \leq C$, for some $C \in (0,\infty)$ independent of n, where the $\tilde{Z}_{nj}$ are defined in (2.26). Due to the $m(n)$-dependence $E\tilde{Z}_{nk}\tilde{Z}_{nj} = 0$ for $|k-j| > m(n)$. Writing $\tilde{v}_n^2 = E\tilde{Z}_{n1}^2 + 2\sum_{j=2}^{m(n)} E\tilde{Z}_{n1}\tilde{Z}_{nj}$ we see that

$$\left| \frac{1}{n} \mathrm{Var}\left(\sum_{i=1}^{n} \tilde{Z}_{ni} \right) - \tilde{v}_n^2 \right| = O(n^{\rho-1}) \to 0 \quad \text{as} \quad n \to \infty, \tag{6.3}$$

because $1-\rho > 1-(1-2\Delta)/\{5(1-\Delta)\} > 0$.

The next step is to show that we may replace $\tilde{Z}_{ni} = \tilde{K}_n(\xi_{i,m(n)}) - \tilde{\tilde{\mu}}_n$ by $\tilde{K}_n(\xi_i) - \tilde{\mu}_n$. For arbitrary $1 \leq j \leq n$, we have

$$\begin{aligned}
&|E\{(\tilde{K}_n(\xi_1) - \tilde{\mu}_n)(\tilde{K}_n(\xi_j) - \tilde{\mu}_n) \\
&\quad - (\tilde{K}_n(\xi_{1,m(n)}) - \tilde{\tilde{\mu}}_n)(\tilde{K}_n(\xi_{j,m(n)}) - \tilde{\tilde{\mu}}_n)\}| \\
&\quad \leq |E\{\tilde{K}_n(\xi_1)\tilde{K}_n(\xi_j) - \tilde{K}_n(\xi_{1,m(n)})\tilde{K}_n(\xi_{j,m(n)})\}| \\
&\qquad + |\tilde{\mu}_n - \tilde{\tilde{\mu}}_n||\tilde{\mu}_n + \tilde{\tilde{\mu}}_n| \\
&\quad \leq E|\tilde{K}_n(\xi_j)||\tilde{K}_n(\xi_1) - \tilde{K}_n(\xi_{1,m(n)})| \\
&\qquad + E|\tilde{K}_n(\xi_{1,m(n)})||\tilde{K}_n(\xi_j) - \tilde{K}_n(\xi_{j,m(n)})| \\
&\qquad + |\tilde{\mu}_n - \tilde{\tilde{\mu}}_n||\tilde{\mu}_n + \tilde{\tilde{\mu}}_n|.
\end{aligned} \tag{6.4}$$

By application of the mean value theorem we find that $|\tilde{\mu}_n - \tilde{\tilde{\mu}}_n||\tilde{\mu}_n + \tilde{\tilde{\mu}}_n| \leq Cn^{-\sigma\wedge\tau + r(\beta+1)+\zeta\alpha}$, and that both expectations on the right in (6.4) are bounded by $Cn^{-\sigma\wedge\tau+2(r\beta+\zeta\alpha)+\tau}$. It is clear that we may carry out the replacement mentioned above, provided that

$$n^{-\sigma\wedge\tau+2(r\beta+\zeta\alpha)+r+\rho} \to 0 \quad \text{as} \quad n \to \infty. \tag{6.5}$$

The assumed restrictions on the parameters entail that $\sigma \wedge \tau - 2(r\beta + \zeta\alpha) - r - \rho > 1 + 1/\{2(1-\Delta)\} - 2\beta/\{2(1-\Delta)\} - 2\alpha/\{5(1-\Delta)\} - 1/\{2(1-\Delta)\} - (1-2\Delta)/\{5(1-\Delta)\} > (4-8\Delta)/\{5(1-\Delta)\} > 0$ indeed. □

B. *Relations among parameters*

The parameter ρ measures how strong the dependence is among the first order parts of the sample elements, the i.i.d. case being covered by $\rho = 0$. Even in the i.i.d. case the value $\alpha + \beta$ or Δ should remain strictly below $1/2$ in order to make sure that the limiting variance exists. It is clear that this asymptotic variance will increase with ρ for constant $\alpha + \beta$. Therefore we expect that a large ρ requires a small Δ, a relationship which is indeed expressed by (2.23). The parameters τ, σ control the amount of noise around the tractable $m(n)$-dependent first order components of the sample elements. These noise components have a bearing on the values of random functions like $J_\zeta(\hat{F}_n(x))$ in (2.9). The larger τ and σ, the smaller the influence of the noisy parts and the larger the value of Δ that we can afford.

C. *Examples*

In the examples below we consider L-estimation of the symmetry point $\vartheta \in \mathbb{R}$, assuming that the X_i have the d.f. $F(\cdot - \vartheta)$ for some F symmetric around 0. We take for simplicity

$$\Psi(x) = x, \quad x \in \mathbb{R}, \qquad \text{so that} \quad \Psi_F(t) = F^{-1}(t), \quad t \in (0,1). \tag{6.6}$$

Due to symmetry we may focus on the left-hand tail of the distributions and on values of t near 0. Let α and β be the parameters in Assumption 2.2 that control the growth of $|J|$ and $|F^{-1}|$ near 0.

First let us take $f(x) \sim a|x|^{-1-1/\beta}$, as $x \to -\infty$, for some $0 < a < \infty$, which yields $|F^{-1}(t)| \sim bt^{-\beta}$, as $t \downarrow 0$, for some $0 < b < \infty$. For densities of this type the efficient score remains bounded, so that we may take any $0 < \zeta < 2/5$ and α arbitrarily close to 0 and hence β in the entire range $(0, \Delta)$.

Now let us take $f(x) \sim a\exp(-|x|^\gamma)$, as $x \to -\infty$, for some $0 < a < \infty$ and $\gamma > 1$. In this case both F^{-1} and the efficient scores are of logarithmic order, as $t \downarrow 0$, and hence any small but strictly positive value of α and any $0 < \zeta < 2/5$ are suitable where also a small strictly positive value of β suffices. We can exploit this freedom by allowing a strong dependence i.e. a large value for ρ in the range $[0, 2/5)$, see (2.23).

Finally let f be concentrated on a finite interval $[-A, A]$, so that the density has zero tails. More specifically let $f(x) \sim a(x+A)^{-1+1/\alpha}$. Then F^{-1} is bounded, of course, with $F^{-1}(t) \sim -A + (t/a)^\alpha$, as $t \downarrow 0$. The efficient score function is of order $t^{-\alpha}$, as $t \downarrow 0$. Since we can take β arbitrarily small positive, for α the range $(0, \Delta)$ is available.

Acknowledgments

The authors would like to thank the referees for their valuable comments that have led to an improvement of the presentation.

References

Andrews, D. W. K. (1984). Non-strong mixing autoregressive processes, *J. Appl. Probab.*, **21**, 930–934.

Beirlant, J., van der Meulen, E. C., Ruymgaart, F. H. and van Zuijlen, M. C. A. (1982). On functions bounding the empirical distribution of uniform spacings, *Z. Wahrsch. Th. Verw. Gebiete*, **61**, 417–430.

Berk, K. N. (1973). A central limit theorem for m-dependent random variables with unbounded m, *Ann. Probab.*, **2**, 352–354.

Chanda, K. C. and Ruymgaart, F. H. (1990). General linear processes: a property of the empirical process applied to density and mode estimation, *J. Time Ser. Anal.*, **11**, 185–199.

Chanda, K. C., Puri, M. L. and Ruymgaart, F. H. (1990). Asymptotic normality of L-statistics based on $m(n)$-decomposable time series, *J. Multivariate Anal.*, **35**, 260–275.

Chernoff, H., Gastwirth, J. L. and Johns, M. V. (1967). Asymptotic distribution of linear combinations of functions of order statistics with applications to estimation, *Ann. Math. Statist.*, **38**, 52–72.

Gastwirth, J. L. and Rubin, H. (1975). The behavior of robust estimators on dependent data, *Ann. Statist.*, **3**, 1070–1100.

Koul, H. L. (1977). Behavior of robust estimators in the regression model with dependent errors, *Ann. Statist.*, **5**, 681–699.

Martin, R. D. (1978). Robust estimation of autoregressive models, *Directions in Time Series* (eds. D. R. Brillinger and G. C. Tiao), 228–254, IMS, Hayward.

Martin, R. D. and Yohai, V. J. (1986). Influence functionals for time series, *Ann. Statist.*, **14**, 781–818.

Moore, D. S. (1968). An elementary proof of asymptotic normality of linear functions of order statistics, *Ann. Math. Statist.*, **39**, 263–265.

Nieuwenhuis, G. and Ruymgaart, F. H. (1990). Some stochastic inequalities and asymptotic normality of serial rank statistics in general linear processes, *J. Statist. Plann. Inference*, **25**, 53–79.

Pham, T. D. and Tran, L. T. (1985). Some mixing properties of time series models, *Stochastic Process. Appl.*, **19**, 297–303.

Portnoy, S. L. (1977). Robust estimation in dependent situations, *Ann. Statist.*, **5**, 22–43.

Portnoy, S. L. (1979). Further remarks on robust estimation in dependent situations, *Ann. Statist.*, **7**, 224–231.

Priestley, M. B. (1981). *Spectral Analysis and Time Series*, Vol. 1, Academic Press, New York.

Rao, M. M. (1984). *Probability Theory with Applications*, Academic Press, New York.

Rao, T. S. and Gabr, M. M. (1984). An introduction to bispectral analysis and bilinear time series models, *Lecture Notes in Statist.*, **24**.

Shorack, G. R. (1972). Functions of order statistics, *Ann. Math. Statist.*, **43**, 412–427.

Shorack, G. R. and Wellner, J. A. (1986). *Empirical Processes with Applications in Statistics*, Wiley, New York.

Stigler, S. M. (1969). Linear functions of order statistics, *Ann. Math. Statist.*, **40**, 770–788.

Reprinted from
Annals of Statistical Mathematics,
Volume 45 (1993), pages 687-701,
By permission of M.L. Puri's co-author

ALIGNED RANK TESTS FOR LINEAR MODELS WITH AUTOCORRELATED ERROR TERMS

MARC HALLIN and MADAN L. PURI*

Université Libre de Bruxelles, Bruxelles, Belgium, Indiana University

Linear models in which the unobserved error constitutes a realization of some stationary ARMA process or, equivalently, ARMA processes with a linear regression trend, are considered under unspecified innovation densities. Due to serial dependence among the observations, the classical rank-based techniques, which have been developed for linear models with independent observations and unspecified error densities, do not apply; nor do the existing rank-based procedures for serial dependence problems, where the observations are assumed to be stationary in the mean (or the median). Moreover, all problems of practical interest (testing the significance of a subset of regression coefficients, identifying the orders p and q of the ARMA (p, q) dependence, overall diagnostic checking of the model, ...) involve nuisance parameters. Typically, one is interested either in the regression trend, and the serial dependence parameters are nuisance parameters; or the serial dependence structure is the main concern, and trend somehow has to be removed. A rank-based approach to such problems thus not only requires extending the classical Hájek-type theory of rank tests to serially dependent situations, it also requires a generalized theory of aligned rank tests. This is the purpose of the present paper. The key result is a local asymptotic normality (LAN) result involving a rank-measurable central sequence; depending on the model considered (with symmetric or totally unspecified innovation densities), the ranks to be used are either signed or unsigned. This LAN result, along with a particular local asymptotic linearity property, implies the local asymptotic sufficiency of (aligned) ranks for a broad class of testing problems—mainly, testing linear restrictions on the parameters of the model. Asymptotically invariant aligned rank tests which are locally asymptotically most stringent similarly are

Received January 28, 1992; revised November 11, 1993

AMS 1980 subject classifications: 62M10, 62G10.

Key words and phrases: ARMA models with linear trend, time series regression, linear models, autocorrelated errors, rank tests, signed rank tests, aligned rank tests, local asymptotic normality, asymptotic linearity, most stringent test, invariant test.

* Research supported by the Office of Naval Research Contract N00014-91-J-1020.

derived. Unlike former results on aligned rank tests for linear models with independent observations, the present ones are entirely obtained from the LAN structure of the underlying problem, and provide a remarkable illustration of the power and appropriateness of LeCam's approach in nonparametric inference. They generalize and reinforce most previous results on rank-based tests for general linear models, as well as the few available ones on rank-based inference for time series analysis. The resulting testing procedures are invariant or asymptotically invariant, hence (asymptotically) distribution-free. They also are as powerful as (often, strictly more powerful than) their classical, correlogram-based counterparts. In addition, they are considerably more robust, and remain fully reliable in the presence of outliers, heavy-tailed distributions, atypical startup behavior, etc. © 1994 Academic Press, Inc.

1. Introduction

1.1. *Rank-Based Inference for Autocorrelated Data.* ARMA models with a regression trend—alternative terminologies are *time series regression models*, or *linear models with autocorrelated errors*—constitute a very large and very rich class of statistical models, the most fundamental one, perhaps, in some applications such as econometrics, business or environmental studies. Their importance has long been recognized; see, e.g., Cochrane and Orcutt (1949), Anderson (1954), Durbin (1960), Pierce (1971a, 1971b), Tsay (1984), Judge *et al.* (1985), and Swensen (1985), to quote only a few references, since even a brief bibliographical survey is not possible here.

In spite of the importance of such models and their applications in several fields, and in spite of the recognized need for robust and nonparametric inference methods in this area, little has been done so far to go beyond the classical Gaussian, least-squares, or Lagrange multiplier approaches, except perhaps for functional estimation methods, which are of an entirely different nature and scope, and will not be considered here. More particularly, no attempt has been made to introduce nonparametric methods based on (signed or unsigned) ranks in the area. This is all the more surprising in view of the well-developed theory of rank-based tests and *R*-estimation procedures for classical linear models—a theory which is essentially complete; see Hájek and Šidák (1967) and Puri and Sen (1971, 1985) for a systematic account and state-of-the-art. Rank methods have been shown to be quite flexible and efficient in this context, more powerful, often, than their normal-theory counterparts. Unfortunately, they also have been shown rather sensitive to the presence of serial dependence among the data; see Serfling (1968) and Gastwirth and Rubin (1971) for the effects of serial correlation on the usual Wilcoxon and sign test statistics. Interesting but isolated attempts to extend classical, nonserial rank tests to the context of serially dependent observations are due to Serfling (1968), and Albers (1978) who shows how one-sample signed rank tests should be modified in the presence of autoregressive observations. Tran (1988), Harel (1988), and Harel and Puri (1989a, 1989b, 1990a, 1990b, 1991) consider the asymptotic behavior of a variety of rank-based statistics under mixing conditions, but their results do not allow for any specific hypothesis testing procedure for the problems at hand. The presence of serial correlation among linear model observations thus far essentially rules out any recourse to rank-based methods.

The statistical theory of rank tests for stationary time series analysis, i.e., for problems dealing with the serial dependence structure of observed data without trend, on the other hand, has a pretty long history (it can be traced back as far as Fisher, 1926, or Wald and Wolfowitz, 1943): see Dufour *et al.* (1982) or Hallin and Puri (1992) for a review. The first step towards a systematic study of rank tests in this area was taken in Hallin *et al.* (1985, 1987) and Hallin and Puri (1988, 1991a), where the problems of testing for white noise or testing one ARMA model against other ARMA models using either rank or signed rank tests are considered. In all of these papers, however, as well as in some earlier ones, such as Wald and Wolfowitz (1943), Beran (1972), Knoke (1977), Bartels (1982), Dufour (1981), or Nieuwenhuis and Ruymgaart (1990), to quote only a few, the mean of the observed series is assumed to be a constant, precluding the presence of any trend or seasonal component, and thus severely restricting the applicability of the proposed tests in most practical situations.

Our objective here is to amalgamate the nonserial features of classical linear models and the serial ones of time series models, and to propose a unified framework for a nonparametric approach to hypothesis testing problems in the analysis of linear models with ARMA error terms. The resulting methodology would be of little practical interest if it were restricted to testing problems in which the null model is (except, of course, for the innovation density) completely specified. The main practical problems in the area indeed consist in testing whether some given subset of explanatory variables can be removed from the trend part of the model, or in identifying the orders of the ARMA part of the model, or in checking the validity of some estimated model. Such problems typically involve nuisance parameters which somehow are to be estimated. The ranks or signed ranks to be used then are *aligned ranks*, i.e., they are computed from estimated residuals.

The trouble with aligned ranks is that the invariance property, which constitutes the theoretical foundation for rank-based inference (see Subsection 2.2), is lost in the alignment process. Without this fundamental invariance property, there is little theoretical motivation, if any at all, for restricting oneself to rank-measurable test statistics. Moreover, distribution-freeness and, in most cases, even asymptotic distribution-freeness is lost as well, which precludes the definition of asymptotically valid critical values. Fortunately, it can be shown that optimal test statistics (in a sense to be defined below) can be found within the class of *asymptotically invariant* aligned rank statistics, i.e., statistics that are asymptotically equivalent to a genuinely invariant one. Invariance and distribution-freeness for such statistics are thus recovered under asymptotic form.

The asymptotically invariant aligned rank tests we are deriving address a very broad class of testing problems, including all the usual problems of practical relevance. They are either asymptotically locally maximin (among *all* tests of given asymptotic probability level), or asymptotically locally most stringent (still, of course, at given asymptotic probability level). This type of local asymptotic optimality is the best that can be expected in the context, either for parametric or for rank-based inference; see Wald (1943), Wolfowitz (1949), and Lehmann (1986, Chap. 8). The corresponding locally asymptotically optimal normal-theory tests are also obtained here, as a by-product—a result which, to the best of our knowledge, so far has been derived only for purely autoregressive processes without trend (Kreiss, 1990a). The asymptotic relative efficiencies of our rank-based methods with respect to these optimal normal theory tests are investigated, and appear to be very close to the values classically obtained in the case of independent observations.

The key technical result is a local asymptotic normality (LAN) property, with a (signed or unsigned) rank-based central sequence, which constitutes a dynamic extension of Hájek's pioneering "rank-based LAN result" for a simple independent regression model; see Strasser (1985, Sects. 79–80). Starting with Hájek (1962), this LAN approach (or, in most cases, its contiguity implications) has been very successfully applied in a variety of problems connected with rank tests for independent observations: see Adichie (1967), Beran (1970), Puri and Sen (1985), and the recent monograph by Janssen and Mason (1990) where more general situations (still for independent observations, though) are considered. In all these papers, however, the many implications of LAN are not fully exploited; typically, contiguity is used to derive local asymptotic powers and asymptotic relative efficiencies. LAN so far has never been used for a systematic treatment of aligned rank tests (see, e.g., Adichie, 1978; McKean and Hettmansperger, 1976; Chiang and Puri, 1984), as it is here.

Our "rank-based LAN result" also extends various existing "parametric LAN results" for time series situations: see Davies (1973) and Dzhaparidze (1986) for Gaussian ARMA processes; Akritas and Johnson (1982), Garel (1989), Kreiss (1987, 1990a) for AR, MA, ARMA, and AR(∞) processes with arbitrary innovation densities, respectively; Hallin *et al.* (1985) and Hallin and Puri (1988) for a rank-based approach to the ARMA case; and Swensen (1985) for AR processes with a regression trend. More general

LAN for more general models are reviewed in Jeganathan (1991). The main difference between all these results and ours lies in the form of the central sequence appearing in the LAQ (Local Asymptotic Quadratic approximation of log-likelihood ratios) part of LAN. This central sequence in our case has to be measurable with respect to residual ranks; as a matter of fact, this central sequence appears to be measurable with respect to the rank-based residual autocorrelations, (e.g., of the van der Waerden, Wilcoxon, or Laplace types) introduced in Hallin *et al.* (1987) and Hallin and Puri (1991a). This property of being correlogram-based certainly constitutes an important advantage over the central sequences considered previously, since so much of time series practice and the insight of time series analysts is connected with correlogram inspection. From the technical point of view, we mainly combine a Hájek-type projection result (generalized from Hallin *et al.*, 1985) with (an adaptation of) a method of Swensen (1985), itself relying on a martingale central limit result by LeCam (1986, Chap. 10).

1.2. *Outline of the Paper.* After describing (Section 2) the model to be considered throughout and stating the main technical assumptions to be made, we introduce (Section 3) the class of serial and nonserial, signed and unsigned rank statistics that will play the main role in the sequel. These statistics mainly are lagged versions of the simple linear rank statistics used in linear models with independent observations, and the rank-based autocorrelation coefficients introduced in Hallin *et al.* (1987) and Hallin and Puri (1991a). The fundamental LAN property, with a rank-based central sequence, is stated and proved in Section 4.

The results of Section 4 are used in Section 5 for constructing locally asymptotically most stringent similar aligned rank tests for a variety of hypotheses. Moreover, we show that, unlike the aligned versions of central sequences themselves, the locally optimal aligned test statistics thus obtained are asymptotically invariant, i.e., asymptotically equivalent to their exact, non-aligned, and genuinely distribution-free counterparts. The null hypotheses considered consist of linear constraints on the parameters of the model (either from the trend part, the ARMA part, or from both). All problems of practical relevance belong to this type.

Finally, some important examples are treated in some detail (Section 6). Asymptotic relative efficiencies are also provided, which show that in addition to being asymptotically invariant, asymptotically distribution-free, and naturally robust, the testing procedures we are proposing here are also often more powerful than their parametric Gaussian theory competitors (based on traditional correlograms). All results are provided both for the signed and unsigned rank cases. As in the problem of testing an ARMA model against another ARMA model, the choice between signed and unsigned ranks here has to be made beforehand, from invariance considerations (symmetric or totally arbitrary innovation densities)—not according to the hypotheses to be tested, as most people do; see Hallin and Puri (1991a) for a discussion of this questionable practice.

2. Definitions and Main Technical Assumptions

2.1. *Linear Models with ARMA Error Terms.* Consider the linear model

$$\mathbf{Y}^{(n)} = \mathbf{X}^{(n)}\boldsymbol{\beta} + \mathbf{e}^{(n)}, \tag{2.1}$$

where $\mathbf{Y}^{(n)} = (Y_1^{(n)}, ..., Y_n^{(n)})'$ is an observed series of length n,

$$\mathbf{X}^{(n)} = \begin{pmatrix} x_{1,1}^{(n)} & \cdots & x_{1,k}^{(n)} \\ \vdots & & \vdots \\ x_{n,1}^{(n)} & \cdots & x_{n,k}^{(n)} \end{pmatrix} = \begin{pmatrix} (\mathbf{x}_1^{(n)})' \\ \vdots \\ (\mathbf{x}_n^{(n)})' \end{pmatrix} \tag{2.2}$$

is a sequence of full-rank $n \times k$ matrices of known (regression) constants, $\boldsymbol{\beta} = (\beta_1, ..., \beta_k)'$ is a vector of regression parameters, and $\mathbf{e}^{(n)} = (e_1^{(n)}, ..., e_n^{(n)})'$ is the vector of (unobserved) error terms. For $t \leqslant 0$, let $\mathbf{x}_t^{(n)} = \mathbf{0}$.

Instead of the classical assumption of independence, we assume that $\mathbf{e}^{(n)}$ is a realization of length n of some solution of a stochastic difference equation (an ARMA (p, q) model) of the form $(A_p \neq 0 \neq B_q)$

$$e_t - \sum_{i=1}^{p} A_i e_{t-i} = \varepsilon_t + \sum_{i=1}^{q} B_i \varepsilon_{t-i}, \qquad t \in \mathbb{Z}, \tag{2.3}$$

where $\{\varepsilon_t;\ t \in \mathbb{Z}\}$ is an independent white noise process. The orders p and q in (2.3) for the moment are not specified, but we assume that $p \leqslant p_1$ and $q \leqslant q_1$. Letting

$$\mathbf{A}' = (A_1 \cdots A_p 0 \cdots 0) \in \mathbb{R}^{p_1}, \qquad \mathbf{B}' = (B_1 \cdots B_q 0 \cdots 0) \in \mathbb{R}^{q_1},$$

and $\boldsymbol{\theta}' = (\boldsymbol{\beta}'\mathbf{A}'\mathbf{B}') \in \mathbb{R}^{k_1}$, $k_1 = k + p_1 + q_1$, denote by $H_f^{(n)}(\boldsymbol{\theta})$ the hypothesis under which $\mathbf{Y}^{(n)}$ is generated by (2.1) and (2.3), with parameter value $\boldsymbol{\theta}$ and a probability density f for ε_t (below referred to as the *innovation density*). The notation $H^{(n)}(\boldsymbol{\theta})$ will be used when the density f in $H_f^{(n)}(\boldsymbol{\theta})$ remains *totally* unspecified, and the notation $H_+^{(n)}(\boldsymbol{\theta})$ whenever f is unspecified but symmetric with respect to the origin.

Though all hypotheses of the form $H_f^{(n)}(\boldsymbol{\theta})$ and all the likelihood functions to be considered in the sequel are concerned with the observed series $\mathbf{Y}^{(n)}$ of length n, we provisionally also assume that the values of

$e_0^{(n)}, e_{-1}^{(n)}, ..., e_{-p+1}^{(n)}$ and $\varepsilon_0^{(n)}, \varepsilon_{-1}^{(n)}, ..., \varepsilon_{-q+1}^{(n)}$ are available; moreover, for the sake of notational simplicity, $e_i^{(n)}$, $i \leq 0$, will also be expressed as $Y_i^{(n)} - (x_{i,1}^{(n)} \cdots x_{i,k}^{(n)})\boldsymbol{\beta}$. This assumption will be relaxed later on (Lemma 4.2). Denote by L the lag operator, and let $A(L) = 1 - \sum_i A_i L^i$, $B(L) = 1 + \sum_i B_i L^i$. Provided **A** and **B** satisfy the traditional causality and invertibility conditions (see assumptions (**A2**), we have, under $H^{(n)}(\boldsymbol{\theta})$,

$$\varepsilon_t^{(n)} = \frac{A(L)}{B(L)} (Y_t^{(n)} - (x_{t,1}^{(n)} \cdots x_{t,k}^{(n)})\boldsymbol{\beta}), \qquad t = 1, ..., n. \tag{2.4}$$

2.2. *Invariance, Ranks, and Signed Ranks.* The whole theory of rank tests originates in the fact that the problem of testing for white noise with unspecified density is invariant under the group of continuous, order-preserving transformations. A maximal invariant statistic for this problem is the vector of ranks, so that the classical invariance argument leads to the consideration of rank-based tests. Distribution-freeness (hence similarity and unbiasedness), increased power, robustness, ..., however attractive they are, are just by-products, and generally can be reached through other, more specific techniques.

Invariance is thus the theoretical cornerstone of rank-based inference, and there is no point in using ranks if, under the null hypothesis to be tested, the quantities from which ranks are computed do not constitute at least an exchangeable series.

This however does not imply that the use of rank tests should be restricted to testing for white noise. In most statistical models, an observed series $\mathbf{Y} = (Y_1, ..., Y_n)$ indeed is such that, denoting by $\boldsymbol{\theta}$ some parameter, some transformed series $\mathbf{Z}(\boldsymbol{\theta}) = (Z_1(\boldsymbol{\theta}), ..., Z_n(\boldsymbol{\theta}))' = \mathscr{F}_{\boldsymbol{\theta}}(\mathbf{Y})$ (call it a *residual* series) is white noise. In a parametric setting, the distribution of this residual series is assumed to be known (in most cases, it is explicitly or implicitly assumed to be Gaussian), though a "nonparametric" approach, where this residual density remains unspecified, would be much more realistic. Provided that $\boldsymbol{\theta} = \boldsymbol{\theta}_0$ if and only if $\mathbf{Z}(\boldsymbol{\theta}_0)$ is white noise, invariance arguments lead to considering rank tests for H: $\boldsymbol{\theta} = \boldsymbol{\theta}_0$ (with unspecified residual density) based on the ranks of the residuals $\mathbf{Z}(\boldsymbol{\theta})$.

The linear model considered here ((2.1) and (2.2)) falls within the above class of statistical models, with $Z_t(\boldsymbol{\theta}) = (A(L)/B(L))(Y_t - \mathbf{x}_t'\boldsymbol{\beta})$. In order for $\boldsymbol{\theta} = \boldsymbol{\theta}_0$ to be equivalent with $\mathbf{Z}(\boldsymbol{\theta}_0)$ being white noise, some additional care is needed. A variation of $\boldsymbol{\theta}$ indeed might well have the same effect on the residuals Z_t than some continuous, order-preserving transformation; namely, it might not affect the ranks of the Z_t's. Consider for example the case where $\boldsymbol{\beta}$, thus $\boldsymbol{\theta}$, comprises an intercept, and assume that $\boldsymbol{\theta}_1$ and $\boldsymbol{\theta}_0$ are equal, except for this intercept: then $Z_t(\boldsymbol{\theta}_1) - Z_t(\boldsymbol{\theta}_0)$ is a constant (not depending on t), and the corresponding rank vectors are identical.

Invariant inference about the intercept, if any in (2.1), is then impossible. In order to avoid any problem in the sequel, whenever (unsigned) ranks are considered, we assume (assumption (**A3**)(iv)) that $\bar{\mathbf{x}}^{(n)} = (1/n) \sum_{t=1}^{n} \mathbf{x}_t^{(n)} = \mathbf{0}\ \forall n$, so that $\boldsymbol{\beta}$ does not contain any intercept; this of course can be done without any loss of generality by recentering the **x**'s and Y's.

More generally, the residual ranks would not be affected if the difference between $\boldsymbol{\theta}_1$ and $\boldsymbol{\theta}_0$ just would have the effect of adding an arbitrary solution ψ_t of the nonhomogeneous difference equation

$$A(L)\psi_t = \mathscr{C}_0, \qquad t \in \mathbb{Z} \tag{2.5}$$

to the trend $(\mathbf{x}_t^{(n)})' \boldsymbol{\beta}$ in (2.1). Taking into account the assumption just made that $\bar{\mathbf{x}}^{(n)} = \mathbf{0}\ \forall n$, we may restrict ourselves to considering arbitrary solutions ψ_t of the homogeneous equation $A(L)\psi_t = 0$, $t \in \mathbb{Z}$. Assume that for some ψ_t there exists a $\boldsymbol{\eta} \in \mathbb{R}^k$ such that $\boldsymbol{\eta} \neq \mathbf{0}$ and $(\mathbf{x}_t^{(n)})' \boldsymbol{\beta} + \psi_t = (\mathbf{x}_t^{(n)})' (\boldsymbol{\beta} + \boldsymbol{\eta})$, $t = 1, \ldots, n$. Then $A(L)(\mathbf{x}_t^{(n)})' \boldsymbol{\eta} = \mathbf{0}$, $t = 1, \ldots, n$, and the matrix

$$\begin{pmatrix} A(L)(\mathbf{x}_1^{(n)})' \\ \vdots \\ A(L)(\mathbf{x}_n^{(n)})' \end{pmatrix} = \begin{pmatrix} -A_{p_0} & \cdots & -A_1 & 1 & \cdots & 0 \\ \vdots & \ddots & & \ddots & \ddots & \vdots \\ 0 & \cdots & -A_{p_0} & -A_1 & & 1 \end{pmatrix} \begin{pmatrix} (\mathbf{x}_{-p_0+1}^{(n)})' \\ \vdots \\ (\mathbf{x}_n^{(n)})' \end{pmatrix}$$

would not be of full rank. This however cannot happen if $\mathbf{X}^{(n)}$ itself is assumed to be of full rank; this latter assumption will be satisfied under (**A3**)(ii) below.

Still, in practice, hypotheses to be tested seldom are of the simple form under which the parameter has a completely specified value $\boldsymbol{\theta} = \boldsymbol{\theta}_0$. In most problems, null hypotheses take the more general form $\boldsymbol{\theta} \in \Theta_0$, where Θ_0 is some subset of the parameter space Θ: though a transformation $\mathscr{T}_{\boldsymbol{\theta}}$ still exists which transforms the observed series **Y** into a residual one $\mathbf{Z}(\boldsymbol{\theta})$, the value of $\boldsymbol{\theta}$ remains unspecified within Θ_0. An intuitively natural idea then consists in substituting a "good" estimate $\hat{\boldsymbol{\theta}}$ for $\boldsymbol{\theta}$, and an estimated residual series $\mathbf{Z}(\hat{\boldsymbol{\theta}})$ for the genuine one $\mathbf{Z}(\boldsymbol{\theta})$. Ranking these estimated residuals yields *aligned ranks*. The trouble with the latter is that they generally do not enjoy, even approximately, the fundamental invariance properties of exact ranks: estimating $\boldsymbol{\theta}$ in general destroys the exchangeability structure of $\mathbf{Z}(\boldsymbol{\theta})$ and, subsequently, the distribution-freeness of rank statistics, hence the invariance, unbiasedness, and similarity characteristics of rank tests. From a theoretical point of view as well as from the point of view of applications, aligned rank methods thus are settling quite a number of nontrivial problems. These are the specific subject of Section 5 below.

The density f of the ε_t's in (2.2) so far has been assumed to be completely unspecified (except perhaps for technical requirements: see assumptions (**A1**)). In many practical problems, this density (the innovation density, under assumptions (**A2**)) further can be assumed to be symmetric with respect to zero. The vector of ranks then loses its maximal invariant status for the benefit of the vector of signs $\mathbf{s} = (\text{sgn}(Z_1(\boldsymbol{\theta})), ..., \text{sgn}(Z_n(\boldsymbol{\theta})))$ and the vector $\mathbf{R}_+ = (R_{+,1}, ..., R_{+,n})$, where $R_{+,t}$ is the rank of $|Z_t(\boldsymbol{\theta})|$ among $|Z_1(\boldsymbol{\theta})|, ..., |Z_n(\boldsymbol{\theta})|$ (invariance here is with respect to the group $\mathscr{G}^+$ of continuous, even, order-preserving transformations). Invariant tests then are the signed rank tests, based on signed rank statistics.

2.3. *Main Technical Assumptions.* The LAN and asymptotic linearity results to be established in Sections 4 and 5 below, as well as the asymptotic normality results for rank-based statistics derived in Section 3 of course require some regularity assumptions, which we concentrate in the present section. These assumptions slightly differ, depending on the type of invariance assumed, hence on the type of ranks to be used: we adopt the notation (**A1**), (**A2**), (**A3**) for the unsigned case, (**A1**$^+$), (**A2**), (**A3**$^+$) for the signed case. These assumptions either concern: (**A1**) and (**A1**$^+$) the innovation density (the use of this terminology will be justified later, from assumption (**A2**)); (**A2**) the ARMA model (2.3); and (**A3**) and (**A3**$^+$) the regression constants in (2.2). They are standard in the context of either time series analysis, the asymptotic theory of linear models, or the theory of rank tests.

(**A1**)(i) $\int xf(x)\,dx = 0$, and $0 < \int x^2 f(x)\,dx = \sigma^2 < \infty$.

(ii) f is absolutely continuous, i.e., there exists a function $\dot{f}$ such that $f(b) - f(a) = \int_a^b \dot{f}(x)\,dx$ for all $-\infty < a < b < \infty$; $\dot{f}$ then a.e. coincides with df/dx.

(iii) Letting $\phi_f = -\dot{f}/f$, $\int [\phi_f(x)]^4 f(x)\,dx < \infty$; this implies finite Fisher information $\mathscr{I}(f) = \int \phi_f^2(x) f(x)\,dx$.

(iv) f is *strongly unimodal*, i.e., ϕ_f is nondecreasing.

(v) $f(x) > 0$ $\forall x \in \mathbb{R}$, and $(\varepsilon_{-q_1+1} \cdots \varepsilon_0 e_{-p_0+1} \cdots e_0)$ possesses a nowhere vanishing joint density $g^0(\cdot\,; \boldsymbol{\theta})$ satisfying $g^0(\cdot\,; \boldsymbol{\theta}^{(n)}) - g^0(\cdot\,; \boldsymbol{\theta}) = o_P(1)$, as $\boldsymbol{\theta}^{(n)} \to \boldsymbol{\theta}$, under $H_f^{(n)}(\boldsymbol{\theta})$.

(vi) ϕ_f is piecewise Lipschitz, i.e., there exists a finite partition of $\mathbb{R}$ into J nonoverlapping intervals I_j such that $|\phi_f(x) - \phi_f(y)| < A_f\,|x - y|$ $\forall j = 1, ..., J$, $\forall x, y \in I_j$.

Assumptions (**A1**$^+$) are exactly the same as (**A1**), except that under (**A1**$^+$)(i), f in addition is supposed to be symmetric with respect to zero ($f(x) = f(-x)$).

Assumptions (i) through (iv) are used in proving the asymptotic normality of the serial and nonserial rank statistics considered in Section 3, as well as for the LAN result in Section 4; assumptions (v) and (vi) mainly take care of the initial value problem. As pointed out by Swensen (1985),

(iii) could be weakened into $\int (\phi_f(x))^2 f(x)\,dx < \infty$ in the absence of a trend component. A scale family $\{f_\sigma(\cdot) = (1/\sigma) f_1(\cdot/\sigma) \mid \sigma \in \mathbb{R}^+\}$ with $\int x^2 f_1(x)\,dx = 1$ is called a *density type*. Obviously, assumptions (**A1**) and (**A1**$^+$) deal with a whole density type rather than with an isolated density f.

The second set of assumptions deals with the ARMA part of the model. Unlike (**A1**) and (**A3**), it puts restrictions on the set of possible values of $\boldsymbol{\theta}$ under the null hypotheses to be considered below, and does not depend on the type of invariance argument used: hence (**A2**$^+$) and (**A2**) are identical. As usual, the orders p and q in (2.2) play a special role in the parametrization. The information matrices (more precisely, the covariance matrices $\boldsymbol{\Gamma}(\boldsymbol{\theta})$ appearing in the LAQ property 4.1(i)) indeed are not continuous functions of p and q; also, the way we are expressing locally asymptotically sufficient statistics in the sequel requires p and q to be fixed under the null hypotheses considered. Therefore, under all null hypotheses to be considered below, p and q are assumed to be specified constants, p_0 and q_0, say; under the alternatives, p and q remain unspecified, with $p_0 \leqslant p \leqslant p_1$ and $q_0 \leqslant q \leqslant q_1$. This allows for testing against higher order ARMA alternatives. In case the orders p and q are not relevant to the testing problem under study, simply let $p_0 = p_1$, $q_0 = q$.

(**A2**) Let L denote the lag operator, $A(L) = 1 - \sum_{i=1}^{p_0} A_i L^i$, $B(L) = 1 + \sum_{i=1}^{q_0} B_i L^i$. Define Θ as the subset of all $\boldsymbol{\theta}$ in $\mathbb{R}^{k_1}$ such that

(i) $A(L)$ and $B(L)$ have degrees p_0 and q_0, respectively, i.e., $A_{p_0} \neq 0 \neq B_{q_0}$, $A_i = 0$, $i > p_0$, $B_i = 0$, $i > q_0$.

(ii) All the roots of the characteristic polynomials $A(z)$ and $B(z)$, $z \in \mathbb{C}$, lie outside the unit disc.

(iii) $A(z)$ and $B(z)$ have no common roots.

Assumptions (**A2**) then can be expressed as $\boldsymbol{\theta} \in \Theta$.

Let g_u and h_u, $u = 0, 1, \ldots$, denote the Green's functions associated with the difference operators $A(L)$ and $B(L)$, respectively. These functions can be characterized as a particular basis of the vector space of solutions of the corresponding homogeneous difference equations (see Hallin, 1986, or Hallin and Puri, 1988), or by means of the polynomial ratios

$$
\begin{aligned}
(A(z))^{-1} &= \left(1 - \sum_{i=1}^{p_0} A_i z^i\right)^{-1} = \sum_{u=0}^{\infty} g_u z^u \\
(B(z))^{-1} &= \left(1 + \sum_{i=1}^{q_0} B_i z^i\right)^{-1} = \sum_{u=0}^{\infty} h_u z^u, \qquad z \in \mathbb{C}, \quad |z| < 1;
\end{aligned} \tag{2.6}
$$

for $p_0 = 0$ and/or $q_0 = 0$, define g_u and/or h_u as $\delta(u, 0)$ (the Kronecker delta). Assumptions (**A2**) ensure that g_t and h_t converge to zero, as $t \to \infty$, at exponential rate. For the same reason, letting $c(L)=\sum_{i=0}^{\infty} c_i L^i = A(L)(B(L))^{-1}$, i.e.,

$$c_i = I[i=0] + \left(h_i - \sum_{j=1}^{\min(p_0,\, i+q_0-1)} A_j h_{i-j}\right) I[i>0], \qquad (2.7)$$

$c_t \to 0$, as $t \to \infty$, exponentially fast, since $B(L)c_t = 0$. Whenever dependence on $\mathbf{A}$ and $\mathbf{B}$ is to be emphasized, the notation $g_i(\boldsymbol{\theta})$, $h_i(\boldsymbol{\theta})$, $c_i(\boldsymbol{\theta})$ will be used.

(A3) Letting $\mathbf{x}_t^{(n)} = (x_{t,1}^{(n)} \cdots x_{t,k_0}^{(n)})'$, and $\mathbf{C}_i^{(n)} = (n-i)^{-1} \sum_{t=i+1}^{n} \mathbf{x}_t^{(n)}(\mathbf{x}_{t-i}^{(n)})'$, $i = 0, 1, ..., n-1$, denote by $\mathbf{D}^{(n)}$ the diagonal matrix with elements $(\mathbf{C}_0^{(n)})_{11} \cdots (\mathbf{C}_0^{(n)})_{kk}$.

(i) $(\mathbf{C}_0^{(n)})_{jj} > 0$, $j = 1, ..., k$; put $\mathbf{R}_i^{(n)} = (\mathbf{D}^{(n)})^{-1/2}\, \mathbf{C}_i^{(n)} (\mathbf{D}^{(n)})^{-1/2}$.

(ii) The limits $\lim_{n\to\infty} \mathbf{R}_i^{(n)} = \mathbf{R}_i$ exist for all i; $\mathbf{R}_0$ is positive definite and therefore can be factorized into $\mathbf{R}_0 = (\mathbf{K}\mathbf{K}')^{-1}$ with $\mathbf{K}$ a full-rank $k \times k$ matrix. Letting $\mathbf{K}^{(n)} = (\mathbf{D}^{(n)})^{-1/2}\, \mathbf{K}$, $\mathbf{K}^{(n)}$ is also of full rank.

(iii) The classical Noether conditions hold letting $\bar{x}_j^{(n)} = n^{-1} \sum_{t=1}^{n} x_{t,j}^{(n)}$, $j = 1, ..., k$:

$$\lim_{n\to\infty} \left[\max_{1 \leqslant t \leqslant n} (x_{t,j}^{(n)} - \bar{x}_j^{(n)})^2 \Big/ \sum_{t=1}^{n} (x_{t,j}^{(n)} - \bar{x}_j^{(n)})^2 \right] = 0.$$

Note that this also implies $\lim_{n\to\infty} [\max_{1\leqslant t\leqslant n} (x_{t,j}^{(n)})^2 / \sum_{t=1}^{n} (x_{t,j}^{(n)})^2] = 0$, $j = 1, ..., k$ (the converse obviously is not true).

(iv) Let $\bar{\mathbf{x}}^{(n)} = (\bar{x}_1^{(n)}, ..., \bar{x}_k^{(n)})'$: $\bar{\mathbf{x}}^{(n)} = \mathbf{0}\ \forall n$.

Assumptions $(\mathbf{A3}^+)$ coincide with **(A3)** except for **(A3)**(iv), which is not required anymore.

3. Serial and Nonserial Rank Statistics

3.1. *Unsigned Rank Statistics.* The whole of Section 3 is more or less self-contained, so that it can be read independently of the other sections of the paper. Denote by $\mathbf{R}^{(n)} = (R_1^{(n)}, ..., R_n^{(n)})$ a vector of ranks, i.e., some random permutation of $(1, ..., n)$, and by $H^{(n)}$ the hypothesis under which $\mathbf{R}^{(n)}$ is uniformly distributed over its $n!$ possible values. Denote by f_1 a density function satisfying assumptions **(A1)**(i) through (iv) with however $\int x^2 f_1(x)\, dx = 1$, by F_1 the corresponding distribution function, and let $\phi_1 = -\dot{f}_1/f_1$. The inverse F_1^{-1} of a nondecreasing function F_1 is defined as $F_1^{-1}(u) = \inf\{x \mid F_1(x) \geqslant u\}$.

The rank statistics we are using in the sequel are of two types: serial and nonserial. The nonserial ones are k-dimensional vectors of the form

$$\underset{\sim}{\mathbf{t}}_{i;f}^{(n)} = (n-i)^{-1}\,[\mathscr{I}(f_1)]^{-1/2} \sum_{t=i+1}^{n} \left[\phi_1\left(F_1^{-1}\left(\frac{R_t^{(n)}}{n+1}\right)\right) - \bar{\phi}_1^{(n)}\right] (\mathbf{K}^{(n)})'\,\mathbf{x}_{t-i}^{(n)}, \tag{3.1}$$

$i = 0, 1, \ldots, n-1$, with $\bar{\phi}_1^{(n)} = n^{-1}\sum_{t=1}^{n}\phi_1(F_1^{-1}(t/(n+1)))$. Each $\underset{\sim}{\mathbf{t}}_{i;f}^{(n)}$ is a vector of simple linear rank statistics, centered about its mean. The components of $\underset{\sim}{\mathbf{t}}_{o;f}^{(n)}$ are of the form considered in classical regression models, and play the same role as the numerators of parametric Student statistics; the statistics $\underset{\sim}{\mathbf{t}}_{i;f}^{(n)}$ for $i > 0$ are lagged versions of $\underset{\sim}{\mathbf{t}}_{o;f}^{(n)}$, and will be needed (in testing hypotheses about the regression coefficients $\boldsymbol{\beta}$ in (2.1)) as soon as serial dependence is present among the data. Obviously, the $\underset{\sim}{\mathbf{t}}_{i;f}^{(n)}$'s are not mutually independent. Exact cross-covariances under $H^{(n)}$ can be computed from combinatorial arguments similar to those in Hájek and Šidák (1967, II.3.1),

$$\begin{aligned}
&\mathrm{cov}((n-i)^{1/2}\,\underset{\sim}{\mathbf{t}}_{i;f}^{(n)}, (n-j)^{1/2}\,\underset{\sim}{\mathbf{t}}_{j;f}^{(n)}) \\
&\quad = [(n-1)(n-i)^{1/2}(n-j)^{1/2}\,\mathscr{I}(f_1)]^{-1}\,(n - \max(i,j)) \\
&\quad \times \left[\sum_{t=1}^{n}\left(\phi_1\left(F_1^{-1}\left(\frac{t}{n+1}\right)\right) - \bar{\phi}_1^{(n)}\right)^2\right] (\mathbf{K}^{(n)})'\,\mathbf{C}_{|i-j|}^{(n-\max(i,j))}\mathbf{K}^{(n)},
\end{aligned} \tag{3.2}$$

converging, as $n \to \infty$, and under assumptions (**A1**)(i)–(iv), (**A3**) to $\mathbf{K}'\mathbf{R}_{|i-j|}\mathbf{K}$.

The serial statistics we are considering are the *rank autocorrelation coefficients* introduced in Hallin *et al.* (1987) and Hallin and Puri (1988),

$$\underset{\sim}{r}_{i;f}^{(n)} = \left[(n-i)^{-1}\sum_{t=i+1}^{n}\phi\left(F^{-1}\left(\frac{R_t^{(n)}}{n+1}\right)\right)F^{-1}\left(\frac{R_{t-i}^{(n)}}{n+1}\right) - m_f^{(n)}\right]\Big/ s_{i;f}^{(n)}, \tag{3.3}$$

$i = 1, 2, \ldots, n-1$, where

$$m_f^{(n)} = [n(n-1)]^{-1}\,[S_{10}^{(n)}S_{01}^{(n)} - S_{11}^{(n)}], \tag{3.4}$$

and

$$\begin{aligned}
(s_{i;f}^{(n)})^2 &= [n(n-1)]^{-1}\,[S_{20}^{(n)}S_{02}^{(n)} - S_{22}^{(n)}] \\
&\quad + 2(n-2i)^{+}\,[n(n-1)(n-2)(n-i)]^{-1} \\
&\quad \times [S_{10}^{(n)}S_{01}^{(n)}S_{11}^{(n)} - S_{21}^{(n)}S_{01}^{(n)} - S_{12}^{(n)}S_{10}^{(n)} \\
&\quad - (S_{11}^{(n)})^2 + 2S_{22}^{(n)}]
\end{aligned}$$

$$+[(n-i)(n-i-1)-2(n-2i)^{+}]$$
$$\times [n(n-1)(n-2)(n-3)(n-i)]^{-1}$$
$$\times [(S_{10}^{(n)}S_{01}^{(n)})^2+2(S_{11}^{(n)})^2+S_{20}^{(n)}S_{02}^{(n)}-6S_{22}^{(n)}$$
$$-4S_{11}^{(n)}S_{10}^{(n)}S_{01}^{(n)}-S_{20}^{(n)}(S_{01}^{(n)})^2$$
$$-S_{02}^{(n)}(S_{10}^{(n)})^2+4S_{21}^{(n)}S_{01}^{(n)}+4S_{12}^{(n)}S_{10}^{(n)}]-(n-i)(m_f^{(n)})^2, \quad (3.5)$$

with $(\cdot)^{+}=\max(\cdot, 0)$ and $S_{lm}^{(n)}=\sum_{i=1}^{n} [\phi(F^{-1}(i/(n+1)))]^l\,[F^{-1}(i/(n+1))]^m$; see Dufour and Roy (1985, 1986) for related results on permutational variances. The rank-based autocorrelation coefficients have been shown (above references) to enjoy most of the intuitive and inferential properties of classical autocorrelation coefficients. In order to facilitate the interpretation of the corresponding correlogram, they have been exactly standardized; viz. $(n-i)^{1/2}\,\underset{\sim}{r}_{i;f}^{(n)}$ under $H^{(n)}$ has *exactly* mean zero and variance one. Also observe that rank autocorrelations are associated with a density type rather than with a density function, i.e., they are invariant under scale transformations of f. This invariance is not simply a consequence of standardization, but results from the intrinsic form of the score function itself, since (with obvious notation)

$$\phi_\sigma\left(F_\sigma^{-1}\left(\frac{i}{n+1}\right)\right)F_\sigma^{-1}\left(\frac{j}{n+1}\right)=\phi_1\left(F_1^{-1}\left(\frac{i}{n+1}\right)\right)F_1^{-1}\left(\frac{j}{n+1}\right).$$

The F, f, and ϕ functions in definitions (3.3) through (3.6) thus do not have to be taken as the standard ones F_1, f_1, and $\phi_1=-\dot{f}_1/f_1$, respectively. The nonserial statistics $\underset{\sim}{t}_{i;f}^{(n)}$ do not enjoy such a property, since

$$\phi_1\left(F_1^{-1}\left(\frac{j}{n+1}\right)\right)=\sigma\phi_\sigma\left(\sigma F_1^{-1}\left(\frac{j}{n+1}\right)\right)=\sigma\phi_\sigma\left(F_\sigma^{-1}\left(\frac{j}{n+1}\right)\right). \quad (3.6)$$

Particularizing the analytical form of f in (3.1) and (3.3) yields statistics of the van der Waerden (Gaussian f), Wilcoxon (logistic f), or Laplace (double exponential f) forms (see Hallin *et al.*, 1987). For the sake of simplicity, we have adopted *approximate score* definitions. *Exact scores* also could be used, just substituting $E[\phi_1(F_1^{-1}(U_1)) \mid R_1^{(n)}=i]$ and $E[\phi(F^{-1}(U_1))\,F^{-1}(U_2) \mid R_1^{(n)}=i,\ R_2^{(n)}=j]$ for $\phi(F^{-1}(i/(n+1)))$ and $\phi(F^{-1}(i/(n+1)))\,F^{-1}(j/(n+1))$, respectively, where $R_t^{(n)}$ denotes the rank of U_t, $t=1, ..., n$, in an n-tuple of iid variables $U_1, U_2, ..., U_n$ uniformly distributed over $[0, 1]$.

PROPOSITION 3.1. *Let f satisfy assumptions* (**A1**)(i) *through* (iv), *and assume that assumptions* (**A3**)(i) *through* (iv) *also hold. Then, for all integers l', l'', the vector*

$$\begin{aligned}&(n^{1/2}(\underset{\sim}{\mathbf{t}}_{o;f}^{(n)})', (n-1)^{1/2} (\underset{\sim}{\mathbf{t}}_{1;f}^{(n)})', \ldots, (n-l'+1)^{1/2} (\underset{\sim}{\mathbf{t}}_{l'-1;f}^{(n)})',\\ &(n-1)^{1/2} \underset{\sim}{r}_{1;f}^{(n)}, \ldots, (n-l'')^{1/2} \underset{\sim}{r}_{l'';f}^{(n)})'\end{aligned} \tag{3.7}$$

is asymptotically multinormal under $H^{(n)}$, as $n \to \infty$, with mean **0** *and covariance matrix*

$$\begin{pmatrix} \begin{pmatrix} \mathbf{I}_{k\times k} & \mathbf{K}'\mathbf{R}_1\mathbf{K} & \cdots & \mathbf{K}'\mathbf{R}_{l'-1}\mathbf{K} \\ \mathbf{K}'\mathbf{R}_1\mathbf{K} & \mathbf{I}_{k\times k} & \ddots & \\ \vdots & \ddots & \ddots & \vdots \\ & & & \mathbf{K}'\mathbf{R}_1\mathbf{K} \\ \mathbf{K}'\mathbf{R}_{l'-1}\mathbf{K} & \cdots & \mathbf{K}'\mathbf{R}_1\mathbf{K} & \mathbf{I}_{k\times k} \end{pmatrix} & \mathbf{O}_{kl'\times l''} \\ \mathbf{O}_{l''\times kl'} & \mathbf{I}_{l''\times l''} \end{pmatrix}. \tag{3.8}$$

(Note that the assumptions here are on the density f characterizing the scores in (3.1) and (3.3); no restriction is made on the density of the observations from which ranks are computed.)

Proof. The proof follows by piecing together the projection methods of Hájek (Hájek and Šidák, 1967, V.1) and Hallin *et al.* (1985, Sect. 4), and a central limit theorem for sums of l''-dependent, heteroskedastic variables; due to the special form of rank-autocorrelation coefficients, the more general central limit theorem of Yoshihara (1976) for U-statistics under absolute regularity, used in Hallin *et al.* (1985), is not necessary here. Denote by $U_1^{(n)}, \ldots, U_n^{(n)}$ an n-tuple of iid random variables, uniformly distributed over [0, 1]. Without any loss of generality, we may assume that $R_t^{(n)}$ under $H^{(n)}$ is the rank of $U_t^{(n)}$ among $U_1^{(n)}, \ldots, U_n^{(n)}$. Since (**A1**)(i)–(iv) hold, $\phi_1(F_1^{-1}(\cdot))$ is nondecreasing and square integrable over [0, 1], and since the Noether conditions (**A3**)(iii) are satisfied by the regression constants, it follows from Hájek and Šidák (1967, Lemma V.1.6.a and Proof of V.1.5.5.a) that $\underset{\sim}{\mathbf{t}}_{i;f}^{(n)} - \mathbf{t}_{i;f}^{(n)}$, $i = 0, \ldots, l'-1$, is $o_P(n^{-1/2})$ under $H^{(n)}$, as $n \to \infty$, where

$$\mathbf{t}_{i;f}^{(n)} = (n-i)^{-1} [\mathscr{I}(f_1)]^{-1/2} \sum_{t=i+1}^{n} [\phi_1(F_1^{-1}(U_t^{(n)}))(\mathbf{K}^{(n)})' (\mathbf{x}_{t-i}^{(n)} - \bar{\mathbf{x}}^{(n)})]. \tag{3.9}$$

Similarly, it follows from Hallin *et al.* (1985, Sect. 4), and the fact that $\phi(F^{-1}(u))\,F^{-1}(v)$ is nondecreasing with respect to both u and v, that $\underset{\sim}{r}{}^{(n)}_{i;f} - r^{(n)}_{i;f} = o_P(n^{-1/2})$, with

$$r^{(n)}_{i;f} = (n-i)^{-1}\,[\mathscr{I}(f_1)]^{-1/2} \sum_{t=i+1}^{n} \phi(F^{-1}(U^{(n)}_t))\,F^{-1}(U^{(n)}_{t-i}); \quad (3.10)$$

we also use here the fact that $m^{(n)}_f = o(n^{-1/2})$ and $\lim_{n\to\infty}(s^{(n)}_{i;f})^2 = \sigma^2\mathscr{I}(f) = \mathscr{I}(f_1)$. Letting $X^{(n)}_t = F^{-1}(U^{(n)}_t)$, consider the arbitrary linear combination of the $kl' + l''$ variables in (3.7) ($\boldsymbol{\alpha}_i$, b_i are arbitrary constants; $l = \max(l'-1, l'')$)

$$S^{(n)} = n^{-1/2}\left\{\sum_{i=0}^{l'-1} (n-i)^{1/2}\,\boldsymbol{\alpha}'_i \mathbf{t}^{(n)}_{i;f} + \sum_{i=1}^{l''} (n-i)^{1/2}\,\beta_i r^{(n)}_{i;f}\right\}$$

$$= (n-l)^{-1}\,\mathscr{I}(f_1)^{-1/2} \sum_{t=l+1}^{n} \phi_1(X^{(n)}_t)$$

$$\times\left[\sum_{i=0}^{l'-1} \boldsymbol{\alpha}'_i(\mathbf{K}^{(n)})'\,(\mathbf{x}^{(n)}_{t-i} - \bar{\mathbf{x}}^{(n)}) + \sum_{i=1}^{l''} \beta_i X^{(n)}_{t-i}\right] + o_P(n^{-1/2})$$

$$= (n-l)^{-1} \sum_{t=l+1}^{n} \xi_t + o_P(n^{-1/2}), \qquad \text{say.}$$

Using the classical Hoeffding–Robbins treatment of l-dependent sequences (though the present one is not stationary, so that classical central limit results for l-dependent summands do not apply), let $j \geqslant 2l+1$, $r = [(n-l)/j]$,

$$S^{(n)}_{*j} = (n-l)^{-1} \sum_{s=0}^{r-1} (\xi_{sj+1} + \cdots + \xi_{(s+1)j-l}) = (n-l)^{-1} \sum_{s=0}^{r-1} \Xi^j_s, \qquad \text{say,}$$

and $S^{(n)}_{0j} = S^{(n)} - S^{(n)}_{*j}$; $[n/j]$ stands for the integer part of n/j. Clearly, $S^{(n)}_{*j}$ under $H^{(n)}$ is a sum of independent variables Ξ^j_s, with variances

$$\mathrm{Var}(\Xi^j_s) = \sum_{i_1,\,i_2=0}^{l'-1} \boldsymbol{\alpha}'_{i_1}(\mathbf{K}^{(n)})' \sum_{t=sl+1}^{(s+1)l-j} (\mathbf{x}^{(n)}_{t-i_1} - \bar{\mathbf{x}}^{(n)})$$

$$\times (\mathbf{x}^{(n)}_{t-i_2} - \bar{\mathbf{x}}^{(n)})'\,\mathbf{K}^{(n)}\boldsymbol{\alpha}_{i_2} + (l-j) \sum_{i=1}^{l''} \beta_i^2.$$

It follows that $S^{(n)}_{*j}$ is of the form

$$S^{(n)}_{*j} = (n-l)^{-1} \sum_{s=0}^{r-1} [\mathrm{Var}(\Xi^j_s)]^{1/2}\,\Xi^0_j, \quad (3.11)$$

a linear combination of iid variables Ξ_s^0 with mean zero and variance one. Because of assumptions (**A3**), the constants $(n-l)^{-1}[\mathrm{Var}(\Xi_s^j)]^{1/2}$ satisfy the Noether condition required for the central limit theorem V.1.2 of Hájek and Šidák (1967) (at least for a subsequence of values of j), so that $n^{1/2}S_{*j}^{(n)}$ is asymptotically normal, as $n\to\infty$, with mean zero and variance $n^{-1}\sum_{s=0}^{\infty}\mathrm{Var}(\Xi_s^j)$ (still under $H^{(n)}$). Now,

$$\begin{aligned}\lim_{j\to\infty} n^{-1}\sum_{s=0}^{\infty}\mathrm{Var}(\Xi_s^j) &= \sum_{i_1,i_2=0}^{l'-1}\boldsymbol{\alpha}_{i_1}'\lim_{n\to\infty}(\mathbf{K}^{(n)})'n^{-1}\sum_{t=1}^{n}(\mathbf{x}_{t-i_1}^{(n)}-\bar{\mathbf{x}}^{(n)})\\ &\quad\times(\mathbf{x}_{t-i_2}^{(n)}-\bar{\mathbf{x}}^{(n)})'\mathbf{K}^{(n)}\boldsymbol{\alpha}_{i_2}+\sum\beta_i^2\\ &= \sum_{i_1,i_2=0}^{l'-1}\boldsymbol{\alpha}_{i_1}'\mathbf{K}'\mathbf{R}_{|i_1-i_2|}\mathbf{K}\boldsymbol{\alpha}_{i_2}+\sum\beta_i^2,\end{aligned}\tag{3.12}$$

since $\bar{\mathbf{x}}^{(n)}=\mathbf{0}$. On the other hand, the remainder term $S_{0j}^{(n)}$ is $o_P(n^{-1/2})$ as $j\to\infty$, uniformly in n, i.e., for all $\varepsilon>0$, $\lim_{j\to\infty}\limsup_{n\to\infty}P(n^{1/2}|S_{0j}^{(n)}|>\varepsilon)=0$. It follows that $n^{1/2}S^{(n)}$ is asymptotically normal, as $n\to\infty$ (see, e.g., Anderson, 1971, 7.7.1) with mean zero and variance (3.12). This, along with the classical Cramér–Wold argument, completes the proof.

3.2. *Signed Rank Statistics.* Denote by $\mathbf{s}^{(n)}=(s_1^{(n)},\dots,s_n^{(n)})$ a vector of independent symmetric Bernoulli variables: $P(s_t^{(n)}=-1)=0.5=P(s_t^{(n)}=1)$, $t=1,\dots,n$. Let $\mathbf{R}_+^{(n)}=(R_{+,1}^{(n)},\dots,R_{+,n}^{(n)})$ be a vector of ranks, and denote by $H_+^{(n)}$ the hypothesis under which $\mathbf{R}_+^{(n)}$ is uniformly distributed over the $n!$ permutations of $(1,\dots,n)$, and $\mathbf{s}^{(n)}$ and $\mathbf{R}_+^{(n)}$ are independent. This of course occurs whenever the signs are those of an exchangeable n-tuple of variables whose distribution is symmetric with respect to zero, and the ranks $R_{+,t}^{(n)}$ are computed from the absolute values of these variables. The signed counterparts of the nonserial rank-based statistics (3.1) are the nonserial signed rank statistics

$$\underset{\sim}{\mathbf{t}}_{i;f}^{(n)+}=(n-i)^{-1}[\mathscr{I}(f_1)]^{-1/2}\sum_{t=i+1}^{n}s_t^{(n)}\phi_1\left(F_{1;+}^{-1}\left(\frac{R_{+,t}^{(n)}}{n+1}\right)\right)(\mathbf{K}^{(n)})'\mathbf{x}_{t-i}^{(n)},\tag{3.13}$$

$i=0,1,\dots,n-1$, with $F_+=2F-1$. Thus $F_{1;+}^{-1}(v)=F_1^{-1}((v+1)/2)$, $v\in(-1,1)$. The serial statistics to be used are the signed rank autocorrelation coefficients introduced in Hallin and Puri (1991a),

$$\underset{\sim}{r}_{i;f}^{(n)+}=[(n-i)s_f^{(n)+}]^{-1}\sum_{t=i+1}^{n}s_t^{(n)}s_{t-i}^{(n)}\phi\left(F_+^{-1}\left(\frac{R_{+,t}^{(n)}}{n+1}\right)\right)F_+^{-1}\left(\frac{R_{+,t-i}^{(n)}}{n+1}\right),\tag{3.14}$$

where

$$(s_f^{(n)+})^2=[n(n-i)]^{-1}\left\{\sum_{j=1}^{n}\left[\phi\left(F_+^{-1}\left(\frac{j}{n+1}\right)\right)\right]^2\sum_{j=1}^{n}\left[F_+^{-1}\left(\frac{j}{n+1}\right)\right]^2 - \sum_{j=1}^{n}\left[\phi\left(F_+^{-1}\left(\frac{j}{n+1}\right)\right)F_+^{-1}\left(\frac{j}{n+1}\right)\right]^2\right\}. \tag{3.15}$$

Similar comments can be made about $\underset{\sim}{\mathbf{t}}_{i;f}^{(n)+}$ and $\underset{\sim}{r}_{i;f}^{(n)+}$ as about their unsigned counterparts $\underset{\sim}{\mathbf{t}}_{i;f}^{(n)}$ and $\underset{\sim}{r}_{i;f}^{(n)}$. Here again $(n-i)^{1/2}\,\underset{\sim}{r}_{i;f}^{(n)+}$ is exactly standardized under $H_+^{(n)}$, and $\underset{\sim}{\mathbf{t}}_{i;f}^{(n)+}$ is centered about its expectation; exact score versions could be used as well. The exact cross-covariance matrix (under $H_+^{(n)}$) of $(n-i)^{1/2}\,\underset{\sim}{\mathbf{t}}_{i;f}^{(n)+}$ and $(n-j)^{1/2}\,\underset{\sim}{\mathbf{t}}_{j;f}^{(n)+}$ is

$$[n(n-i)^{1/2}(n-j)^{1/2}\mathscr{I}(f_1)]^{-1}(n-\max(i,j)) \times\left[\sum_{t=1}^{n}\phi_1^2\left(F_{1;+}^{-1}\left(\frac{t}{n+1}\right)\right)\right](\mathbf{K}^{(n)})'\,\mathbf{C}_{|i-j|}^{(n-\max(i,j))}\mathbf{K}^{(n)} \tag{3.16}$$

converging, as $n\to\infty$, and under assumptions $(\mathbf{A1}^+)$(i)–(iii), $(\mathbf{A3})$, to the same limit as (3.2).

The asymptotic distribution, under $H_+^{(n)}$, of any $(l'+l'')$-tuple of signed statistics similar to (3.7) is exactly the same as in Proposition 3.1. Note however that the assumption that $\bar{\mathbf{x}}^{(n)}=\mathbf{0}$ is not made here.

PROPOSITION 3.2. *Let f satisfy assumptions* $(\mathbf{A1}^+)$(i) *through* (iv), *and assume that assumptions* $(\mathbf{A3}^+)$ *also hold. Then the signed rank version of the* $(l'+l'')$*-tuple* (3.7) *is asymptotically multinormal under* $H_+^{(n)}$, *as* $n\to\infty$, *with mean* $\mathbf{0}$ *and covariance matrix* (3.8).

Proof. The proof is very similar to that of Proposition 3.1, and is only briefly sketched here. Denoting by $V_1, ..., V_n$ an n-tuple of iid variables uniformly distributed over $[-1;1]$, assume that the signs $s_t^{(n)}$ and signed ranks $R_{+,t}^{(n)}$ are those of V_t among $V_1, ..., V_n$. Hájek's projection method for signed rank statistics yields

$$\underset{\sim}{\mathbf{t}}_{i;f}^{(n)+}=(n-i)^{-1}[\mathscr{I}(f_1)]^{-1/2} \times\sum_{t=i+1}^{n}\phi_1\left(F_1^{-1}\left(\frac{V_t+1}{2}\right)\right)(\mathbf{K}^{(n)})'\,\mathbf{x}_{t-i}^{(n)}+o_P(n^{-1/2}). \tag{3.17}$$

Now, $U_t=(V_t+1)/2$, $t=1, ..., n$, are iid, uniformly distributed over $[0, 1]$, and (Hallin and Puri, 1991a, Proposition 3.1) $n^{1/2}(\underset{\sim}{r}_{i;f}^{(n)+}-\underset{\sim}{r}_{i;f}^{(n)})$ is o_P as $n\to\infty$. The end of the proof of Proposition 3.1 thus also applies here.

Whereas signed and unsigned autocorrelations under $H_+^{(n)}$ are asymptotically equivalent up to $o_P(n^{-1/2})$ terms, as $n \to \infty$, the same is not true, in general, with nonserial statistics. Denoting again by $V_1^{(n)}, ..., V_n^{(n)}$ a n-tuple of iid variables uniformly distributed over $[-1; 1]$, assume that the ranks $R_t^{(n)}$, signs $s_t^{(n)}$, and signed ranks $R_{+,t}^{(n)}$ are those of $V_t^{(n)}$ among $V_1^{(n)}, ..., V_n^{(n)}$.

Proposition 3.3. *Under $H_+^{(n)}$, assumptions* (**A1**$^+$)(i)–(iv) and (**A3**$^+$), *as* $n \to \infty$,

(i) $$n^{1/2}(\underset{\sim}{\mathbf{t}}_{i;f}^{(n)+} - \underset{\sim}{\mathbf{t}}_{i;f}^{(n)}) = (n-i)^{-1/2}[\mathscr{I}(f_1)]^{-1/2} \times \sum_{t=i+1}^{n} \phi_1(F_1^{-1}((V_t^{(n)}+1)/2))(\mathbf{K}^{(n)})'\,\bar{\mathbf{x}}^{(n)} + o_P(1);$$

accordingly, the difference $\underset{\sim}{\mathbf{t}}_{i;f}^{(n)+} - \underset{\sim}{\mathbf{t}}_{i;f}^{(n)}$ *is* $o_P(n^{-1/2})$ *if and only if* $\lim_{n\to\infty} \bar{\mathbf{x}}^{(n)} = \mathbf{0}$.

(ii) *The difference between* $\underset{\sim}{r}_{i;f}^{(n)+}$ *and* $\underset{\sim}{r}_{i;f}^{(n)}$ *is (always)* $o_P(n^{-1/2})$.

Proof. Part (i) is an immediate consequence of (3.9) and (3.16); part (ii) is proved in Hallin and Puri (1991a, Proposition 3.1).

The difference in the asymptotic behaviors of $\underset{\sim}{\mathbf{t}}_{i;f}^{(n)}$ and $\underset{\sim}{\mathbf{t}}_{i;f}^{(n)+}$ is not surprising. Clearly, if the same constant is added to a n-tuple of observations, their ranks $R_t^{(n)}$ are not affected, whereas their signs and signed ranks, $s_t^{(n)}$ and $R_{+,t}^{(n)}$, are. The asymptotic behaviors of $\underset{\sim}{\mathbf{t}}_{i;f}^{(n)}$ and $\underset{\sim}{\mathbf{t}}_{i;f}^{(n)+}$ are consistent with this fact, which has been discussed in Subsection 2.2.

4. Local Asymptotic Normality

4.1. *Preparation.* Going back to the model described in Section 2, denote by $\mathbf{Z}^{(n)}(\boldsymbol{\theta})$, or simply $\mathbf{Z}^{(n)}$ the filtered series $(Z_1^{(n)}, ..., Z_n^{(n)})$ obtained from the observations $\mathbf{Y}^{(n)}$ by means of the linear recursion

$$Z_t^{(n)} = Y_t^{(n)} - (\mathbf{x}^{(n)})'\,\boldsymbol{\beta} - \sum_{i=1}^{p_0} A_i(Y_{t-i}^{(n)} - (\mathbf{x}_{t-i}^{(n)})'\,\boldsymbol{\beta}) - \sum_{i=1}^{q_0} B_i Z_{t-i}^{(n)}, \quad t = 1, ..., n, \tag{4.1}$$

with initial values $Z_t^{(n)} = \varepsilon_t$, $t = -q_0+1, ..., 0$, and $Y_t^{(n)} - \mathbf{x}'^{(n)}\boldsymbol{\beta} = e_t^{(n)}$, $t = -p_0+1, ..., 0$. An alternative expression for $Z_t^{(n)}(\boldsymbol{\theta})$ is in terms of the Green's functions $h_u(\boldsymbol{\theta})$ and the constants $c_u(\boldsymbol{\theta})$ defined in (2.7):

$$Z_t^{(n)} = \sum_{u=0}^{t-1} c_u(\mathbf{Y}_{t-u}^{(n)} - (\mathbf{x}_{t-u}^{(n)})'\,\boldsymbol{\beta}) - \sum_{j=0}^{p_0-1} e_{-j}\left[\sum_{i=j+1}^{p} A_i h_{t-p_0+i+j}\right] + \sum_{j=0}^{q_0-1} \varepsilon_{-j}\left[1 + \sum_{i=1}^{j} B_i h_{t+j-i}\right]. \tag{4.2}$$

Clearly, under $H^{(n)}(\boldsymbol{\theta})$ (or $H_{+}^{(n)}(\boldsymbol{\theta})$), $Z_t^{(n)}(\boldsymbol{\theta})$ coincides with $\varepsilon_t^{(n)}$. In practice, the initial values $e_0, \ldots, e_{-p_0+1}, \varepsilon_0, \ldots, \varepsilon_{-q_0+1}$ of course are not available, and approximate residuals will be used instead of exact ones (letting, e.g., $e_t = \varepsilon_t = 0$, $t \leqslant 0$ in (4.2)). The differences $|Z_t^{(n)} - \varepsilon_t^{(n)}|$ then are $O_P(|\lambda(\boldsymbol{\theta})|^{-t})$, where $\lambda(\boldsymbol{\theta})$ denotes the root of $B(z) = 0$, $z \in \mathbb{C}$, which is closest to the unit circle. As far as asymptotic results are concerned, this difference under assumptions (**A1**) (or (**A1**$^+$)) can safely be neglected: see Lemma 4.2 for a formal proof. More sophisticated devices, such as backforecasting procedures, also can be used to choose initial values in (4.1) and (4.2); they may yield better small sample performance, but have no influence upon asymptotic results.

All the ranks and signs to be used in the sequel are computed from residuals of the form $Z_t^{(n)}(\boldsymbol{\theta})$. When the dependence on $\boldsymbol{\theta}$ is to be emphasized, we write $R_t^{(n)}(\boldsymbol{\theta})$, $R_{+t}^{(n)}(\boldsymbol{\theta})$, $s_t^{(n)}(\boldsymbol{\theta})$ for the rank of $Z_t^{(n)}(\boldsymbol{\theta})$ among $Z_1^{(n)}(\boldsymbol{\theta}), \ldots, Z_n^{(n)}(\boldsymbol{\theta})$, the rank of $|Z_t^{(n)}(\boldsymbol{\theta})|$ among $|Z_1^{(n)}(\boldsymbol{\theta})|, \ldots, |Z_n^{(n)}(\boldsymbol{\theta})|$, and the sign of $Z_t^{(n)}(\boldsymbol{\theta})$; $\boldsymbol{\theta}$ will be omitted whenever no confusion is possible.

Let $\boldsymbol{\eta}^{(n)}$, $\boldsymbol{\gamma}^{(n)}$, and $\boldsymbol{\delta}^{(n)}$ denote sequences of real vectors, of dimensions k, p_1, and q_1, respectively, and put

$$\boldsymbol{\tau}^{(n)} = ((\boldsymbol{\eta}^{(n)})', (\boldsymbol{\gamma}^{(n)})', (\boldsymbol{\delta}^{(n)})')',$$
$$\mathbf{v}(n) = n^{-1/2} \begin{pmatrix} \sigma_f \mathbf{K}^{(n)} & \mathbf{0}_{k \times (p_1+q_1)} \\ \mathbf{0}_{(p_1+q_1)\times k} & \mathbf{I}_{(p_1+q_1)\times(p_1+q_1)} \end{pmatrix}. \tag{4.3}$$

Let $\|\cdot\|$ stand for the Euclidean norm. If $\boldsymbol{\theta} \in \boldsymbol{\Theta}$ and $\sup_n \|\boldsymbol{\tau}^{(n)}\| < \infty$, $\boldsymbol{\theta} + \mathbf{v}(n)\boldsymbol{\tau}^{(n)}$ for n sufficiently large also satisfies (**A2**)(ii) and (iii) (see Kreiss, 1987, Lemma 6.1 for a more precise statement), and $H_f^{(n)}(\boldsymbol{\theta} + \mathbf{v}(n)\boldsymbol{\tau}^{(n)})$ constitutes a sequence of hypotheses approaching $H_f^{(n)}(\boldsymbol{\theta})$, in a sense (contiguity) to be made clearer below, as $n \to \infty$.

Associated with $\boldsymbol{\gamma}^{(n)}$ and $\boldsymbol{\delta}^{(n)}$, the following sequences will play a role in the sequel,

$$a_i^{(n)} = \sum_{j=1}^{\min(p_1,\, i+p_0-1)} \gamma_j^{(n)} g_{i-j} \qquad b_i^{(n)} = \sum_{j=1}^{\min(q_1,\, i+q_0-1)} \delta_j^{(n)} h_{i-j}; \tag{4.4}$$

for $p_0 = 0$ and/or $q_0 = 0$, this reduces to $a_i^{(n)} = \gamma_i^{(n)}$ and/or $b_i^{(n)} = \delta_i^{(n)}$. When the presence of $\boldsymbol{\theta}$ has to be emphasized, the notation $a_i^{(n)}(\boldsymbol{\theta})$ and $b_i^{(n)}(\boldsymbol{\theta})$ will be used. Assumptions (**A2**) ensure that the Green's functions g_t and h_t, as well as any solution ψ_t of the homogeneous equation $A(L)\,B(L)\psi_t = 0$, $t \in \mathbb{Z}$, converge to zero, as $t \to \infty$, at exponential rate. As a linear combination of $g_{i-1}, g_{i-2}, \ldots, g_{i-p_1}$, $a_i^{(n)}$ for $i \geqslant p_1 - p_0 + 1$ and fixed n is a solution of $A(L)\psi_t = 0$. Similarly, for $i \geqslant q_1 - q_0 + 1$, $b_i^{(n)}$ is a solution of $B(L)\psi_t = 0$. Therefore, $\sum_{i=1}^{\infty} |a_i^{(n)}|$, $\sum_{i=1}^{\infty} |b_i^{(n)}|$, $\sum_{i=1}^{\infty} |a_i^{(n)} + b_i^{(n)}|$, as well as $\|\mathbf{a}^{(n)}\|^2 = \sum_{i=1}^{\infty} (a_i^{(n)})^2$, $\|\mathbf{b}^{(n)}\|^2 = \sum_{i=1}^{\infty} (b_i^{(n)})^2$, and $\|\mathbf{a}^{(n)} + \mathbf{b}^{(n)}\|^2 = \sum_{i=1}^{n} (a_i + b_i)^2$ are finite quantities. Letting $\gamma^{(n)}(L)=\sum_i \gamma_i^{(n)} L^i$, $\delta^{(n)}(L)=\sum_i \delta_i^{(n)} L^i$, $a^{(n)}(L)=\sum_i a_i^{(n)} L^i$, $b^{(n)}(L)=\sum_i b_i^{(n)} L^i$, we also have $a^{(n)}(L)=\gamma^{(n)}(L)/A(L)$ and $b^{(n)}(L)=\delta^{(n)}(L)/B(L)$. Note that each $a_i^{(n)}(\cdot)$ and $b_i^{(n)}(\cdot)$ (for fixed sequences $\boldsymbol{\tau}^{(n)}$) is continuous with respect to $\boldsymbol{\theta}$.

Also, denote by $\{\psi_t^{(1)}, \ldots, \psi_t^{(p_0+q_0)}\}$ a fundamental system of solutions of the homogeneous equation

$$A(L)\,B(L)\psi_t=0, \qquad t\in\mathbb{Z}. \tag{4.5}$$

Letting $\pi=\max(p_1-p_0, q_1-q_0)$ and $\pi_0=\pi+p_0+q_0=\max(p_1+q_0, q_1+p_0)$, define

$$\mathbf{C}_\psi=\begin{pmatrix} \psi_{\pi+1}^{(1)} & \cdots & \psi_{\pi+1}^{(p_0+q_0)} \\ \psi_{\pi+2}^{(1)} & \cdots & \psi_{\pi+2}^{(p_0+q_0)} \\ \vdots & & \vdots \\ \psi_{\pi_0}^{(1)} & \cdots & \psi_{\pi_0}^{(p_0+q_0)} \end{pmatrix}. \tag{4.6}$$

Up to a rearrangement of its rows, $\mathbf{C}_\psi$ is a *Casorati matrix*; hence it has full rank p_0+q_0. Also associated with the same fundamental system, define the $(p_0+q_0)\times(p_0+q_0)$ matrix $\mathbf{w}_\psi^2$ with elements

$$(\mathbf{w}_\psi^2)_{i,j}=\sum_{t=\pi+1}^{\infty} \psi_t^{(i)}\psi_t^{(j)}, \qquad i, j=1, \ldots, p_0+q_0 \tag{4.7}$$

(convergence again follows from the above remark on the asymptotic behavior of solutions of $A(L)\,B(L)\psi_t=0$). It results from the mutual linear independence of the elements of a fundamental system of solutions that $\mathbf{w}_\psi^2$ is strictly positive definite.

All the above quantities of course depend on $\boldsymbol{\theta}$. We assume that the fundamental system is characterized by a given (p_0+q_0)-tuple of initial values $(\psi_{\pi+1}^{(i)}, \ldots, \psi_{\pi_0}^{(i)}, i=1, \ldots, p_0+q_0$, say) which are *given* continuous functions of $\boldsymbol{\theta}$. Then, for all $i=1, \ldots, p_0+q_0$ and all $t\in\mathbb{Z}$, $\psi_t^{(i)}$ is a continuous function of $\boldsymbol{\theta}$. Similarly, $\mathbf{C}_\psi$ and $\mathbf{w}_\psi^2$ are continuous functions of $\boldsymbol{\theta}$; the notation $\psi_t^{(i)}(\boldsymbol{\theta})$, $\mathbf{C}_\psi(\boldsymbol{\theta})$, $\mathbf{w}_\psi^2(\boldsymbol{\theta})$ however will be avoided as much as possible.

Convenient choices of a fundamental system are as follows.

(a) (*Green's functions*) Denote by $\mathscr{G}_u$, $u\in\mathbb{Z}$, the Green's function associated with the difference operator $A(L)\,B(L)$: a fundamental system of solutions of (4.5) is $\{\psi_t^{(i)}=\mathscr{G}_{t-\pi-i}, i=1, \ldots, p_0+q_0\}$, with

$$\mathscr{C}_\psi=\begin{pmatrix} 1 & 0 & \cdots & 0 \\ \mathscr{G}_1 & 1 & & 0 \\ \vdots & \vdots & \ddots & \vdots \\ \mathscr{G}_{p_0+q_0-1} & \mathscr{G}_{p_0+q_0-2} & \cdots & 1 \end{pmatrix}=\begin{pmatrix} 1 & 0 & \cdots & 0 \\ \alpha_1 & 1 & & 0 \\ \vdots & \vdots & \ddots & \vdots \\ \alpha_{p_0+q_0-1} & \alpha_{p_0+q_0-2} & \cdots & 1 \end{pmatrix}^{-1} \tag{4.8}$$

and

$$\mathbf{w}_\psi^2 = \begin{pmatrix} \gamma(0) & \gamma(1) & \cdots & \gamma(p_0+q_0-1) \\ \gamma(1) & & & \vdots \\ & \ddots & \ddots & \\ \vdots & \ddots & & \gamma(1) \\ \gamma(p_0+q_0-1) & \cdots & \gamma(1) & \gamma(0) \end{pmatrix}, \tag{4.9}$$

where the α_i's are defined by $A(L)\,B(L) = 1 + \sum_{i=1}^{p_0+q_0} \alpha_i L^i$, and (4.9) clearly is the covariance matrix of (p_0+q_0) successive observations of the stationary $AR(p_0+q_0)$ process $\{\xi_t\}$ characterized by $A(L)\,B(L)\xi_t = \varepsilon_t$, $t \in \mathbb{Z}$, $E(\varepsilon_t \varepsilon_s) = \delta_{ts}$.

(b) (*Characteristic roots*) Denote by $\lambda^{(1)}, ..., \lambda^{(p_0+q_0)}$ the roots of $A(z^{-1})\,B(z^{-1}) = 0$, $z \in \mathbb{C}$, and assume for simplicity that they are all distinct: a fundamental system of solutions of (4.5) is $\{\psi_t^{(i)} = (\lambda^{(i)})^{t-(\pi+1)}, i = 1, ..., p_0+q_0\}$, with

$$\mathscr{C}_\psi = \begin{pmatrix} 1 & \cdots & 1 \\ \lambda^{(1)} & & \lambda^{(p_0+q_0)} \\ \vdots & & \vdots \\ (\lambda^{(1)})^{p_0+q_0-1} & \cdots & (\lambda^{(p_0+q_0)})^{p_0+q_0-1} \end{pmatrix}$$

(a Vandermonde matrix) and $(\mathbf{w}_\psi^2)_{i,j} = (1 - \lambda^{(i)}\lambda^{(j)})^{-1}$.

In both cases, the continuity requirements (with respect to the coefficients A_i and B_i) are clearly satisfied.

Finally, associated with the Green's functions $g_u(\boldsymbol{\theta})$ and $h_u(\boldsymbol{\theta})$, consider the following $\pi_0 \times (p_1+q_1)$ matrices $\mathbf{M}(\boldsymbol{\theta})$, of rank π_0,

$$\mathbf{M} = \begin{pmatrix} 1 & 0 & \cdots & 0 & 1 & 0 & \cdots & 0 \\ g_1 & 1 & & \vdots & h_1 & 1 & & \vdots \\ \vdots & & \ddots & \vdots & \vdots & & \ddots & \vdots \\ g_{p_1-1} & & \cdots & 1 & h_{q_1-1} & & \cdots & 1 \\ g_{p_1} & & \cdots & g_1 & h_{q_1} & & \cdots & h_1 \\ \vdots & & & \vdots & \vdots & & & \vdots \\ g_{\pi_0-1} & & & g_{\pi_0-p_1} & h_{\pi_0-1} & \cdots & & h_{\pi_0-q_1} \end{pmatrix}; \tag{4.10}$$

for $p_0 = 0$ $(q_0 = 0)$, the first p_1 (the last q_1) columns of $\mathbf{M}$ reduce to $\binom{\mathbf{I}}{\mathbf{0}}$, $\mathbf{I}$ of dimension $p_1 \times p_1$ $(q_1 \times q_1)$ and $\mathbf{0}$ of dimension $(q_0+\pi-p_1) \times p_1$ $((p_0+\pi-q_1) \times q_1)$. Note that $\mathbf{M}(\boldsymbol{\theta})$ is continuous with respect to $\boldsymbol{\theta} \in \Theta$.

4.2. *A Rank-Based LAN Result.* Denote by $\kappa(n)$ an arbitrary non-decreasing integer valued function of n such that $\kappa(n) \leqslant n$ and

$\lim_{n \to \infty} \kappa(n) = \infty$. Recalling that $\pi = \max(p_1 - p_0, q_1 - q_0)$ and $\pi_0 = \pi + p_0 + q_0$, define the $(\pi_0 + k)$-dimensional rank statistic

$$n^{1/2} \underset{\sim}{\mathbf{T}}^{(n)}_{\psi;f} = \begin{pmatrix} \sum_{i=0}^{\kappa(n)} (n-i)^{1/2} c_i \underset{\sim}{\mathbf{t}}^{(n)}_{i;f} \\ (n-1)^{1/2} \underset{\sim}{r}^{(n)}_{1;f} \\ \vdots \\ (n-\pi)^{1/2} \underset{\sim}{r}^{(n)}_{\pi;f} \\ \sum_{i=\pi+1}^{\kappa(n)} (n-i)^{1/2} \psi^{(1)}_i \underset{\sim}{r}^{(n)}_{i;f} \\ \vdots \\ \sum_{i=\pi+1}^{\kappa(n)} (n-i)^{1/2} \psi^{(p_0+q_0)}_i \underset{\sim}{r}^{(n)}_{i;f} \end{pmatrix} = n^{1/2} \begin{pmatrix} \underset{\sim}{\mathbf{T}}^{(n)}_{I;f} \\ \underset{\sim}{\mathbf{T}}^{(n)}_{II;f} \end{pmatrix}, \tag{4.11}$$

where all rank-based statistics $\underset{\sim}{\mathbf{t}}^{(n)}_{i;f} = \underset{\sim}{\mathbf{t}}^{(n)}_{i;f}(\boldsymbol{\theta})$ and $r^{(n)}_{i;f} = r^{(n)}_{i;f}(\boldsymbol{\theta})$ are computed from the ranks $R^{(n)}_t(\boldsymbol{\theta})$ of the residuals $Z^{(n)}_1(\boldsymbol{\theta}), \ldots, Z^{(n)}_n(\boldsymbol{\theta})$; when the dependence on $\boldsymbol{\theta}$ is to be emphasized, we write $\underset{\sim}{\mathbf{T}}^{(n)}_{I;f}(\boldsymbol{\theta})$, $\underset{\sim}{\mathbf{T}}^{(n)}_{II;f}(\boldsymbol{\theta})$, $\underset{\sim}{\mathbf{T}}^{(n)}_{\psi;f}(\boldsymbol{\theta})$. The same statistic, where however the unsigned rank statistics $\underset{\sim}{\mathbf{t}}^{(n)}_{i;f}(\boldsymbol{\theta})$ and $\underset{\sim}{r}^{(n)}_{i;f}(\boldsymbol{\theta})$ are replaced by their signed rank versions $\underset{\sim}{\mathbf{t}}^{(n)+}_{i;f}(\boldsymbol{\theta})$ and $\underset{\sim}{r}^{(n)+}_{i;f}(\boldsymbol{\theta})$, computed from the signed ranks $R^{(n)}_{+;t}(\boldsymbol{\theta})$ and signs of the residuals $Z^{(n)}_t(\boldsymbol{\theta})$, will be denoted by $\underset{\sim}{\mathbf{T}}^{(n)+}_{\psi;f}(\boldsymbol{\theta})$.

We now may state the main LAN result.

PROPOSITION 4.1. *Let* $\Lambda^{(n)}_{f;\boldsymbol{\theta}+\mathbf{v}(n)\boldsymbol{\tau}^{(n)}/\boldsymbol{\theta}}$ *denote the logarithm of the likelihood ratio for* $H^{(n)}_f(\boldsymbol{\theta} + \mathbf{v}(n)\boldsymbol{\tau}^{(n)})$ *with respect to* $H^{(n)}_f(\boldsymbol{\theta})$. *For all* $\boldsymbol{\theta} \in \Theta$, *all f satisfying assumptions* (**A1**)(i)–(v), *all* $\mathbf{X}^{(n)}$ *satisfying assumptions* (**A3**), *and all sequences* $\boldsymbol{\tau}^{(n)}$ *such that* $\sup_n \|\boldsymbol{\tau}^{(n)}\| < \infty$,

(i) (LAQ)

$$\Lambda^{(n)}_{f;\boldsymbol{\theta}+\mathbf{v}(n)\boldsymbol{\tau}^{(n)}/\boldsymbol{\theta}}(\mathbf{Y}^{(n)}) = (\boldsymbol{\tau}^{(n)})' \underset{\sim}{\boldsymbol{\Delta}}^{(n)}_f(\boldsymbol{\theta}) - \tfrac{1}{2}(\boldsymbol{\tau}^{(n)})' \boldsymbol{\Gamma}_f(\boldsymbol{\theta}) \boldsymbol{\tau}^{(n)} + o_P(1) \tag{4.12}$$

under $H^{(n)}_f(\boldsymbol{\theta})$, *as* $n \to \infty$, *where*

$$\underset{\sim}{\boldsymbol{\Delta}}^{(n)}_f(\boldsymbol{\theta}) = \begin{pmatrix} \underset{\sim}{\boldsymbol{\Delta}}^{(n)}_{I;f}(\boldsymbol{\theta}) \\ \underset{\sim}{\boldsymbol{\Delta}}^{(n)}_{II;f}(\boldsymbol{\theta}) \end{pmatrix} = [n\mathscr{I}(f_1)]^{1/2} \begin{pmatrix} \underset{\sim}{\mathbf{T}}^{(n)}_{I;f}(\boldsymbol{\theta}) \\ \mathbf{M}'(\boldsymbol{\theta}) \begin{pmatrix} \mathbf{I}_{\pi\times\pi} & \mathbf{0} \\ \mathbf{0} & \mathbf{C}'_\psi \end{pmatrix}^{-1} \underset{\sim}{\mathbf{T}}^{(n)}_{II;\psi,f}(\boldsymbol{\theta}) \end{pmatrix} \tag{4.13}$$

and

$$\boldsymbol{\Gamma}_f(\boldsymbol{\theta}) = \begin{pmatrix} \boldsymbol{\Gamma}_{I;f}(\boldsymbol{\theta}) & \mathbf{0}_{k\times\pi_0} \\ \mathbf{0}_{\pi_0\times k} & \boldsymbol{\Gamma}_{II;f}(\boldsymbol{\theta}) \end{pmatrix}, \tag{4.14}$$

with

$$\mathbf{\Gamma}_{I;f}(\boldsymbol{\theta}) = \mathscr{I}(f_1)\,\mathbf{K}'\left(\sum_{u=0}^{\infty}\sum_{v=0}^{\infty} c_u c_v \mathbf{R}_{|u-v|}\right)\mathbf{K} \tag{4.15}$$

and

$$\mathbf{\Gamma}_{II;f}(\boldsymbol{\theta}) = \mathscr{I}(f_1)\,\mathbf{M}'(\boldsymbol{\theta})\begin{pmatrix}\mathbf{I}_{\pi\times\pi} & \mathbf{0}\\ \mathbf{0} & (\mathbf{C}_\psi^{-1})'\,\mathbf{w}_\psi^2\,\mathbf{C}_\psi^{-1}\end{pmatrix}\mathbf{M}(\boldsymbol{\theta}). \tag{4.16}$$

(ii) (*Asymptotic normality of central sequences*) $\underset{\sim}{\mathbf{\Delta}}_f^{(n)}(\boldsymbol{\theta})$ *under* $H_f^{(n)}(\boldsymbol{\theta})$ *is asymptotically normal as* $n \to \infty$, *with mean* $\mathbf{0}$ *and asymptotic covariance matrix* $\mathbf{\Gamma}_f(\boldsymbol{\theta})$.

(iii) *If f is symmetric* (*thus satisfying assumptions* ($\mathbf{A1}^+$) *instead of* ($\mathbf{A1}$)), *and with* ($\mathbf{A3}^+$) *instead of* ($\mathbf{A3}$) (*hence without the assumption that* $\bar{\mathbf{x}}^{(n)} = \mathbf{0}$), (i) *and* (ii) *still hold if the signed rank statistics* $\underset{\sim}{\mathbf{T}}_{\psi;f}^{(n)+}(\boldsymbol{\theta})$ *are substituted for the unsigned ones. The resulting central sequence will be denoted by* $\underset{\sim}{\mathbf{\Delta}}_f^{(n)+}(\boldsymbol{\theta})$; *the covariance matrix* $\mathbf{\Gamma}$ *is the same as in* (i) *and* (ii).

Proof of Proposition 4.1. The main point of the proof consists in transforming the proposed rank-based central sequences into more familiar parametric ones, similar to those obtained by Swensen (1985) and Kreiss (1987). For (i), up to $o_P(1)$ terms, $\Lambda^{(n)}$ under (4.12) decomposes into $\Lambda^{(n)} = \Lambda_I^{(n)} + \Lambda_{II}^{(n)}$, with

$$\Lambda_I^{(n)} = (\boldsymbol{\eta}^{(n)})'\,\underset{\sim}{\mathbf{\Delta}}_{I;f}^{(n)}(\boldsymbol{\theta}) - \tfrac{1}{2}(\boldsymbol{\eta}^{(n)})'\,\mathbf{\Gamma}_{I;f}(\boldsymbol{\theta})\boldsymbol{\eta}^{(n)} \tag{4.17}$$

and

$$\Lambda_{II}^{(n)} = ((\boldsymbol{\gamma}^{(n)})'\,(\boldsymbol{\delta}^{(n)})')\,\underset{\sim}{\mathbf{\Delta}}_{II;f}^{(n)}(\boldsymbol{\theta}) - \tfrac{1}{2}((\boldsymbol{\gamma}^{(n)})'\,(\boldsymbol{\delta}^{(n)})')\,\mathbf{\Gamma}_{II;f}(\boldsymbol{\theta})((\boldsymbol{\gamma}^{(n)})'\,(\boldsymbol{\delta}^{(n)})')'. \tag{4.18}$$

Clearly, $\Lambda_I^{(n)}$ accounts for the trend part of the model, $\Lambda_{II}^{(n)}$ for the ARMA part. First consider the trend part:

$$\begin{aligned}\Lambda_I^{(n)} = {} & [\mathscr{I}(f_1)]^{1/2}\,(\boldsymbol{\eta}^{(n)})'\sum_{i=0}^{\kappa(n)}(n-i)^{1/2}\,c_i\underset{\sim}{\mathbf{t}}_{i;f}^{(n)}\\ & -\tfrac{1}{2}\mathscr{I}(f_1)(\boldsymbol{\eta}^{(n)})'\,\mathbf{K}'\left(\sum_{u,v} c_u c_v \mathbf{R}_{|u-v|}\right)\mathbf{K}\boldsymbol{\eta}^{(n)}.\end{aligned} \tag{4.19}$$

It follows from (3.9) and the fact that $\sigma^{-1}\phi_1(F_1^{-1}(\cdot)) = \phi_\sigma(F_\sigma^{-1}(\cdot))$ that the linear term in the right hand side of (4.19) is of the form

$$\begin{aligned}&\sigma_f(\boldsymbol{\eta}^{(n)})'\,(\mathbf{K}^{(n)})'\sum_{i=0}^{\kappa(n)} c_i(n-i)^{-1/2}\sum_{t=i+1}^{n}\phi_\sigma(\varepsilon_t^{(n)})\,\mathbf{x}_{t-i}^{(n)} + o_P(1)\\ &= n^{-1/2}\sigma_f(\mathbf{K}^{(n)}\boldsymbol{\eta}^{(n)})'\sum_{t=1}^{n}\sum_{i=0}^{n-1}\phi_\sigma(\varepsilon_t^{(n)})\,c_i\mathbf{x}_{t-i}^{(n)} + o_P(1)\end{aligned} \tag{4.20}$$

(under $H_f^{(n)}(\boldsymbol{\theta})$, as $n \to \infty$, with the convention $\mathbf{x}_t^{(n)} = \mathbf{0}$, $t \leqslant 0$). Except for the fact that $c_i^{(n)}$, $i = 1, \ldots, (n-1)$, replaces the autoregressive coefficients $-A_i$, $i = 1, \ldots, p_0$ (to which they reduce in the purely AR case ($q_0 = 0$)), $\Lambda_I^{(n)}$ thus coincides with the corresponding trend part in the purely AR result of Swensen (1985, Theorem 2); note that $n^{-1/2}\sigma_f \mathbf{K}^{(n)}\boldsymbol{\eta}^{(n)}$ in our notation is $\mathbf{K}^{(n)}\mathbf{b}$ in Swensen's. Turning to the ARMA part,

$$
\begin{aligned}
&((\boldsymbol{\gamma}^{(n)})' \, (\boldsymbol{\delta}^{(n)})') \, \underset{\sim}{\boldsymbol{\Delta}}{}_{II;f}^{(n)}(\boldsymbol{\theta})[\mathscr{I}(f_1)]^{-1/2} \\
&\quad = ((\boldsymbol{\gamma}^{(n)})' \, (\boldsymbol{\delta}^{(n)})') \, \mathbf{M}'(\boldsymbol{\theta}) \begin{pmatrix} \mathbf{I} & \mathbf{0} \\ \mathbf{0} & \mathbf{C}'_\psi \end{pmatrix}^{-1} (n^{1/2} \underset{\sim}{\mathbf{T}}{}_{II;\psi,f}^{(n)}(\boldsymbol{\theta})) \\
&\quad = \begin{pmatrix} \boldsymbol{\gamma}^{(n)} \\ \boldsymbol{\delta}^{(n)} \end{pmatrix}' \mathbf{M}'(\boldsymbol{\theta}) \\
&\qquad \times \begin{pmatrix} (n-1)^{1/2} \, \underset{\sim}{r}{}_{1;f}^{(n)} \\ \vdots \\ (n-\pi)^{1/2} \, \underset{\sim}{r}{}_{\pi;f}^{(n)} \\ (\mathbf{C}'_\psi)^{-1} \begin{pmatrix} \psi_{\pi+1}^{(1)} & \cdots & \psi_{\kappa(n)}^{(1)} \\ \vdots & & \vdots \\ \psi_{\pi+1}^{(p_0+q_0)} & \cdots & \psi_{\kappa(n)}^{(p_0+q_0)} \end{pmatrix} \begin{pmatrix} (n-\pi-1)^{1/2} \, \underset{\sim}{r}{}_{\pi+1;f}^{(n)} \\ \vdots \\ (n-\kappa(n))^{1/2} \, \underset{\sim}{r}{}_{\kappa(n);f}^{(n)} \end{pmatrix} \end{pmatrix}
\end{aligned}
\tag{4.21}
$$

$$
\begin{aligned}
&= \begin{pmatrix} \boldsymbol{\gamma}^{(n)} \\ \boldsymbol{\delta}^{(n)} \end{pmatrix}' \mathbf{M}'(\boldsymbol{\theta}) \begin{pmatrix} \mathbf{I}_{\pi \times \pi} & \mathbf{O}_{\pi \times (\kappa(n)-\pi)} \\ \mathbf{0}_{(p_0+q_0) \times \pi} & (\mathbf{C}'_\psi)^{-1} \begin{pmatrix} \psi_{\pi+1}^{(1)} & \cdots & \psi_{\kappa(n)}^{(1)} \\ \vdots & & \vdots \\ \psi_{\pi+1}^{(p_0+q_0)} & \cdots & \psi_{\kappa(n)}^{(p_0+q_0)} \end{pmatrix} \end{pmatrix} \\
&\quad \times \begin{pmatrix} (n-1)^{1/2} \, \underset{\sim}{r}{}_{1;f}^{(n)} \\ \vdots \\ (n-\kappa(n))^{1/2} \, \underset{\sim}{r}{}_{\kappa(n);f}^{(n)} \end{pmatrix}.
\end{aligned}
\tag{4.22}
$$

Now, from the definition (4.4),

$$
\begin{pmatrix} a_1^{(n)} + b_1^{(n)} \\ \vdots \\ a_{\pi_0}^{(n)} + b_{\pi_0}^{(n)} \end{pmatrix} = \mathbf{M}(\boldsymbol{\theta}) \begin{pmatrix} \boldsymbol{\gamma}^{(n)} \\ \boldsymbol{\delta}^{(n)} \end{pmatrix}, \tag{4.23}
$$

whereas for $t \geqslant \pi + 1$, $a_t^{(n)} + b_t^{(n)}$ is a solution of $A(L)\,B(L)\psi_t = 0$ and hence can be expressed as a linear combination of the fundamental system $(\psi_t^{(1)} \cdots \psi_t^{(p_0+q_0)})$; more precisely,

$$
a_i^{(n)} + b_i^{(n)} = (\psi_i^{(1)} \cdots \psi_i^{(p_0+q_0)}) \, \mathbf{C}_\psi^{-1} (a_{\pi+1}^{(n)} + b_{\pi+1}^{(n)} \cdots a_{\pi_0}^{(n)} + b_{\pi_0}^{(n)})', \qquad i \geqslant \pi + 1,
$$

which entails

$$\begin{pmatrix} a^{(n)}_{\pi+1} + b^{(n)}_{\pi+1} \\ \vdots \\ a^{(n)}_{\kappa(n)} + b^{(n)}_{\kappa(n)} \end{pmatrix}$$

$$= \left(\mathbf{0}_{(\kappa(n)-\pi)\times\pi} \quad \begin{pmatrix} \psi^{(1)}_{\pi+1} & \cdots & \psi^{(p_0+q_0)}_{\pi+1} \\ \vdots & & \vdots \\ \psi^{(1)}_{\kappa(n)} & \cdots & \psi^{(p_0+q_0)}_{\kappa(n)} \end{pmatrix} \mathbf{C}^{-1}_{\psi} \right) \begin{pmatrix} a^{(n)}_{1} + b^{(n)}_{1} \\ \vdots \\ a^{(n)}_{\pi_0} + b^{(n)}_{\pi_0} \end{pmatrix}. \tag{4.24}$$

It thus follows from (4.23) and (4.24) that

$$\begin{pmatrix} a^{(n)}_{1} + b^{(n)}_{1} \\ \vdots \\ a^{(n)}_{\kappa(n)} + b^{(n)}_{\kappa(n)} \end{pmatrix}$$

$$= \begin{pmatrix} \mathbf{I}_{\pi\times\pi} & \mathbf{O}_{\pi\times(p_0+q_0)} \\ \mathbf{0}_{(\kappa(n)-\pi)\times\pi} & \begin{pmatrix} \psi^{(1)}_{\pi+1} & \cdots & \psi^{(p_0+q_0)}_{\pi+1} \\ \vdots & & \vdots \\ \psi^{(1)}_{\kappa(n)} & \cdots & \psi^{(p_0+q_0)}_{\kappa(n)} \end{pmatrix} \mathbf{C}^{-1}_{\psi} \end{pmatrix} \mathbf{M}(\boldsymbol{\theta}) \begin{pmatrix} \boldsymbol{\gamma}^{(n)} \\ \boldsymbol{\delta}^{(n)} \end{pmatrix} \tag{4.25}$$

which along with (4.22) in turn implies

$$((\boldsymbol{\gamma}^{(n)})' \; (\boldsymbol{\delta}^{(n)})') \, \underset{\sim}{\boldsymbol{\Delta}}^{(n)}_{II;f}(\boldsymbol{\theta}) = \sum_{i=1}^{\kappa(n)} (a^{(n)}_i + b^{(n)}_i)(n-i)^{1/2} \, \underset{\sim}{r}^{(n)}_{i;f} [\mathscr{I}(f_1)]^{1/2}. \tag{4.26}$$

From (4.26) and (3.9), we then obtain

$$((\boldsymbol{\gamma}^{(n)})' \; (\boldsymbol{\delta}^{(n)})') \, \underset{\sim}{\boldsymbol{\Delta}}^{(n)}_{II;f}(\boldsymbol{\theta})$$

$$= n^{-1/2} \sum_{i=1}^{\kappa(n)} (a^{(n)}_i + b^{(n)}_i) \sum_{t=i+1}^{n} \phi_\sigma(\varepsilon^{(n)}_t)\varepsilon^{(n)}_{t-i} + o_P(1) \tag{4.27}$$

$$= n^{-1/2} \sum_{t=2}^{n} \phi_\sigma(\varepsilon^{(n)}_t) \sum_{i=1}^{\infty} (a^{(n)}_i + b^{(n)}_i)\varepsilon^{(n)}_{t-i}$$

$$- n^{-1/2} \sum_{t=2}^{n} \phi_\sigma(\varepsilon^{(n)}_t) \sum_{i=\min(t,\,\kappa(n)+1)}^{\infty} (a^{(n)}_i + b^{(n)}_i)\varepsilon^{(n)}_{t-i} + o_P(1). \tag{4.28}$$

The first term in this latter expression also can be written as

$$n^{-1/2} \sum_{t=1}^{n} \phi_\sigma(\varepsilon^{(n)}_t) \left(\frac{\gamma^{(n)}(L)}{A(L)} + \frac{\delta^{(n)}(L)}{B(L)} \right) \varepsilon^{(n)}_t$$

$$= n^{-1/2} \sum_{t=1}^{n} \phi_\sigma(\varepsilon_t^{(n)})((\boldsymbol{\gamma}^{(n)})' \, (\boldsymbol{\delta}^{(n)})') \begin{pmatrix} \dfrac{1}{A(L)} \begin{pmatrix} \varepsilon_{t-1}^{(n)} \\ \vdots \\ \varepsilon_{t-p_1}^{(n)} \end{pmatrix} \\ \dfrac{1}{B(L)} \begin{pmatrix} \varepsilon_{t-1}^{(n)} \\ \vdots \\ \varepsilon_{t-q_1}^{(n)} \end{pmatrix} \end{pmatrix}, \tag{4.29}$$

which, using the notation $\mathbf{Z}(t; \boldsymbol{\theta}, \boldsymbol{\theta})$ from Kreiss (1987, Lemma 2.2), is $n^{-1/2} \sum_{t=1}^{n} \phi_\sigma(\varepsilon_t^{(n)})((\boldsymbol{\gamma}^{(n)})' \, (\boldsymbol{\delta}^{(n)})') \, \mathbf{Z}(t-1; \boldsymbol{\theta}, \boldsymbol{\theta})$. As for the second term in (4.28), for all N and sufficiently large n, it decomposes into

$$n^{-1/2} \sum_{t=2}^{N} \left[\phi_\sigma(\varepsilon_t^{(n)}) \sum_{i=t}^{\infty} (a_i^{(n)} + b_i^{(n)}) \varepsilon_{t-i}^{(n)} \right]$$

$$+ n^{-1/2} \sum_{t=N+1}^{\infty} \phi_\sigma(\varepsilon_t^{(n)}) \sum_{i=\kappa(n)+1}^{\infty} (a_i^{(n)} + b_i^{(n)}) \varepsilon_{t-i}^{(n)}$$

$$= R_{N,1}^{(n)} + R_{N,2}^{(n)}, \qquad \text{say.}$$

As a sum of N terms with finite variance divided by $n^{1/2}$, $R_{N,1}^{(n)}$ is $o_P(1)$, as $n \to \infty$, for all N. The variance of $R_{N,2}^{(n)}$ is bounded by $\sigma^2 \mathscr{I}(f) \sum_{i=N}^{\infty} (a_i^{(n)} + b_i^{(n)})^2 \leqslant K \, |\lambda|^{-2N}$, uniformly in n, where λ denotes the root of $A(z) \, B(z) = 0$ closest to the unit circle and K is some constant. Hence $R_{N,2}^{(n)}$ is $o_P(1)$, as $N \to \infty$, uniformly in n. It follows (Brockwell and Davis, 1987, Proposition 6.3.9) that the rank-based linear part $((\boldsymbol{\gamma}^{(n)})' \, (\boldsymbol{\delta}^{(n)})') \, \underset{\sim}{\boldsymbol{\Delta}}_{II;f}^{(n)}$ of $\Lambda_{II}^{(n)}$ coincides, up to $o_P(1)$ terms, with the corresponding parametric quantity appearing in Kreiss (1987, Theorem 3.1); in the purely autoregressive case, of course, we retrieve the central sequence obtained by Swensen (1985).

A similar reasoning yields, for the quadratic part of $\Lambda_{II}^{(n)}$,

$$((\boldsymbol{\gamma}^{(n)})' \, (\boldsymbol{\delta}^{(n)})') \, \boldsymbol{\Gamma}_{II;f}(\boldsymbol{\theta}) \begin{pmatrix} \boldsymbol{\gamma}^{(n)} \\ \boldsymbol{\delta}^{(n)} \end{pmatrix} = \sum_{i=1}^{\infty} (a_i^{(n)} + b_i^{(n)})^2 \, \mathscr{I}(f_1), \tag{4.30}$$

which does not depend on the particular system $(\psi_t^{(1)}, \ldots, \psi_t^{(p_0+q_0)})$ adopted; convergence in (4.30) again follows from assumptions (**A2**). Since for constant sequences of the form $\boldsymbol{\gamma}^{(n)} = \boldsymbol{\gamma}$, $\boldsymbol{\delta}^{(n)} = \boldsymbol{\delta}$, we have, from (4.26) and (Proposition 3.1) the asymptotic independence of $(n-i)^{1/2} \, \underset{\sim}{r}_{i;f}^{(n)}$ and $(n-j)^{1/2} \, \underset{\sim}{r}_{j;f}^{(n)}$,

$$\lim_{n \to \infty} E\{[(\boldsymbol{\gamma}'\boldsymbol{\delta}') \, \underset{\sim}{\boldsymbol{\Delta}}_{II;f}^{(n)}(\boldsymbol{\theta})]^2\} = \lim_{n \to \infty} \sum_{i=1}^{\kappa(n)} (a_i + b_i)^2 \, \mathscr{I}(f_1) = \sum_{i=1}^{\infty} (a_i + b_i)^2 \, \mathscr{I}(f_1),$$

the quadratic terms in $\Lambda_{II}^{(n)}$ and Kreiss (1987, Theorem 3.1) also coincide.

Summing up, our rank-based LAQ result, as far as the ARMA part of the model is concerned, is equivalent with the Kreiss (1987) parametric LAQ result for ARMA processes without trend, whereas our trend part is of the same form as in the Swensen (1985) result for AR processes with a linear trend. The end of the proof of part (i) of the proposition consists in adapting the Swensen (1985) proof to the present situation, and checking that conditions (1.2) through (1.7) (Lemma 1, same reference) are satisfied. In order not to overload the present paper, we leave this to the reader. The Kreiss (1987) method, which relies on Roussas (1979), presumably also could be adapted to the presence of an additional trend term. For a multivariate version of the proof, see Garel and Hallin (1991).

(ii) Part (ii), which deals with the asymptotic normality of the central sequence (4.13) follows from (4.19), (4.26), and the asymptotic distribution derived in Proposition 3.1.

(iii) The signed-rank case follows along the same lines as the unsigned one, though under slightly different assumptions (mainly, $\bar{\mathbf{x}}^{(n)} = \mathbf{0}$ is not required). Actually, (3.16) and part (ii) of Proposition 3.3 formally lead to the same asymptotic equivalences (4.20) and (4.27) between rank-based and parametric statistics as above.

Note that the covariance matrix $\mathbf{\Gamma}_f(\boldsymbol{\theta})$ of the central sequence $\underset{\sim}{\mathbf{\Delta}}_f^{(n)}(\boldsymbol{\theta})$ depends on f_1 only, i.e., on the density type of f; $\mathbf{\Gamma}_f(\boldsymbol{\theta})$ actually is of the form

$$\mathbf{\Gamma}_f(\boldsymbol{\theta}) = \mathscr{I}(f_1)\,\mathbf{\Gamma}(\boldsymbol{\theta}) = \sigma_f^2 \mathscr{I}(f)\,\mathbf{\Gamma}(\boldsymbol{\theta}), \tag{4.31}$$

where $\mathbf{\Gamma}(\boldsymbol{\theta})$ is the covariance matrix associated with normal densities.

The residuals $Z_t^{(n)}(\boldsymbol{\theta})$ so far have been assumed to be "exact residuals," computed from the nonavailable "exact initial values" $\varepsilon_0, \dots, \varepsilon_{-q+1}, e_0, \dots, e_{-p+1}$. The following lemma shows that asymptotic results are not affected if approximate residuals $\tilde{Z}_t^{(n)}(\boldsymbol{\theta})$ computed from arbitrary starting values $\tilde{\varepsilon}_0, \dots, \tilde{\varepsilon}_{-q+1}, \tilde{e}_0, \dots, \tilde{e}_{-p+1}$ are substituted for the exact ones—which of course is essential for practical applications. The main assumption here is (**A1**)(vi), which is weaker than the Kreiss (1987) global Lipschitz condition (p. 118); unlike Kreiss' one, it is satisfied by double exponential densities.

LEMMA 4.2. *Let* $\mathbf{t}_{i;f}^{(n)}$, $r_{i;f}^{(n)}$, $\underset{\sim}{\mathbf{t}}_{i;f}^{(n)}$, *and* $\tilde{r}_{i;f}^{(n)}$ *be defined as in* (3.9) *and* (3.10), *with* $U_t^{(n)} = F(Z_t^{(n)}(\boldsymbol{\theta}))$ *and* $\tilde{U}_t^{(n)} = F(\tilde{Z}_t^{(n)}(\boldsymbol{\theta}))$, *respectively. Also, let* $\tilde{\underset{\sim}{\mathbf{t}}}_{i;f}^{(n)}$, $\tilde{\underset{\sim}{\mathbf{t}}}_{i;f}^{(n)+}$, $\underset{\sim}{\tilde{r}}_{i;f}^{(n)}$, $\underset{\sim}{\tilde{r}}_{i;f}^{(n)+}$ *denote the rank and signed rank statistics computed from the approximate residuals* $\tilde{Z}_t^{(n)}(\boldsymbol{\theta})$. *Assume that* (**A1**) *or* (**A1**$^+$)(i)–(vi), (**A2**), *and* (**A3**) *or* (**A3**$^+$) *hold. Then, under* $H_f^{(n)}(\boldsymbol{\theta})$ *and hence also under contiguous alternatives, as* $n \to \infty$,

(i) $\tilde{\mathbf{t}}^{(n)}_{i;f} - \mathbf{t}^{(n)}_{i;f}$ *and* $\tilde{r}^{(n)}_{i;f} - r^{(n)}_{i;f}$ *are* $o_P(n^{-1/2})$,

(ii) $\tilde{\underset{\sim}{\mathbf{t}}}^{(n)}_{i;f} - \underset{\sim}{\mathbf{t}}^{(n)}_{i;f}$ *and* $\tilde{\underset{\sim}{r}}^{(n)}_{i;f} - \underset{\sim}{r}^{(n)}_{i;f}$ *are* $o_P(n^{-1/2})$, *and*

(iii) $\tilde{\underset{\sim}{\mathbf{t}}}^{(n)+}_{i;f} - \underset{\sim}{\mathbf{t}}^{(n)+}_{i;f}$ *and* $\tilde{\underset{\sim}{r}}^{(n)+}_{i;f} - \underset{\sim}{r}^{(n)+}_{i;f}$ *are* $o_P(n^{-1/2})$,

so that asymptotic results remain valid if approximate residuals are substituted for the exact ones.

Proof. The proof can be found in Hallin and Puri (1991b).

4.3. *Serial and Nonserial Rank Statistics under Local Alternatives.* The local power of tests based on the rank statistics considered in Section 3 can be obtained from the following result.

PROPOSITION 4.3. (i) *Let f and g satisfy assumptions* (**A1**)(i) *through* (v), *and assume that* (**A2**) *and* (**A3**) *also hold. Then, for all integers l', l'', and all $\boldsymbol{\tau}^{(n)} = ((\boldsymbol{\eta}^{(n)})' (\boldsymbol{\gamma}^{(n)})' (\boldsymbol{\delta}^{(n)})')' \in \mathbb{R}^{k_1}$ such that $\sup_n \|\boldsymbol{\tau}^{(n)}\| < \infty$, the vector of rank statistics* (3.7), *where the ranks are those of the residuals $Z^{(n)}_t(\boldsymbol{\theta})$, is asymptotically multinormal, under $H^{(n)}_g(\boldsymbol{\theta} + \mathbf{v}(n)\boldsymbol{\tau}^{(n)})$, as $n \to \infty$, with mean*

$$\frac{\mathscr{I}(f_1, g_1)}{[\mathscr{I}(f_1)]^{1/2}} \begin{pmatrix} \begin{pmatrix} \mathbf{K}' \sum_{j=0}^{\infty} c_j(\boldsymbol{\theta}) \mathbf{R}_j \mathbf{K} \\ \vdots \\ \mathbf{K}' \sum_{j=0}^{\infty} c_j(\boldsymbol{\theta}) \mathbf{R}_{|i-j|} \mathbf{K} \\ \vdots \\ \mathbf{K}' \sum_{j=0}^{\infty} c_j(\boldsymbol{\theta}) \mathbf{R}_{|l'-j|} \mathbf{K} \end{pmatrix} \boldsymbol{\eta}^{(n)} \\ \sigma(f_1, g_1) \begin{pmatrix} a^{(n)}_1(\boldsymbol{\theta}) + b^{(n)}_1(\boldsymbol{\theta}) \\ \vdots \\ a^{(n)}_{l''}(\boldsymbol{\theta}) + b^{(n)}_{l''}(\boldsymbol{\theta}) \end{pmatrix} \end{pmatrix} \tag{4.32}$$

and covariance matrix (3.8), *where*

$$\mathscr{I}(f, g) = \int_0^1 \phi_f(F^{-1}(u))\, \phi_g(G^{-1}(u))\, du, \tag{4.33}$$

and

$$\sigma(f, g) = \int_0^1 F^{-1}(u)\, G^{-1}(u)\, du, \tag{4.34}$$

so that $\mathscr{I}(f, f) = \mathscr{I}(f)$, $\sigma(f, f) = \sigma_f^2$, *and* $\sigma(f, g)\, \mathscr{I}(f, g) = \sigma(f_1, g_1)\, \mathscr{I}(f_1, g_1)$.

(ii) *Similar results hold if f and g satisfy assumptions* (**A1**$^+$)(i) *through* (v), *and* (**A2**) *and* (**A3**$^+$) *hold, for the signed rank version of* (3.7).

(iii) *Let g satisfy assumptions* (**A1**)(i) *through* (v), *and assume that* (**A2**) *and* (**A3**) *hold, or let g satisfy assumptions* (**A1**$^+$)(i) *through* (v), *and assume that* (**A2**) *and* (**A3**$^+$) *hold. Consider the vector of parametric statistics*

$$\begin{aligned}&(n^{1/2}(\mathbf{t}_0^{(n)})', (n-1)^{1/2}(\mathbf{t}_1^{(n)})', ..., (n-l'+1)^{1/2}(\mathbf{t}_{l'-1}^{(n)})',\\ &(n-1)^{1/2} r_1^{(n)}, ..., (n-l'')^{1/2} r_{l''}^{(n)})',\end{aligned} \tag{4.35}$$

where

$$\mathbf{t}_i^{(n)} = \mathbf{t}_i^{(n)}(\boldsymbol{\theta}) = (n-i)^{-1}\hat{\sigma}^{-1} \sum_{t=i+1}^{n} Z_t^{(n)}(\boldsymbol{\theta})(\mathbf{K}^{(n)})' \mathbf{x}_{t-i}^{(n)}, \qquad i=0, 1, ..., n-1, \tag{4.36}$$

$\hat{\sigma}^2 = n^{-1}\sum_{t=1}^{n}[Z_t^{(n)}(\boldsymbol{\theta})]^2$, *and* $r_i^{(n)}$ *is the usual autocorrelation coefficient of order i*

$$r_i^{(n)} = r_i^{(n)}(\boldsymbol{\theta}) = \sum_{t=i+1}^{n} Z_t^{(n)}(\boldsymbol{\theta})\, Z_{t-i}^{(n)}(\boldsymbol{\theta}) \Big/ \sum_{t=1}^{n} [Z_t^{(n)}(\boldsymbol{\theta})]^2. \tag{4.37}$$

Then (4.35) *is asymptotically normal as* $n \to \infty$, *with mean* **0** *under* $H_g^{(n)}(\boldsymbol{\theta})$ *and mean*

$$\left(\begin{array}{c}\left(\begin{array}{c}\mathbf{K}' \sum_{j=0}^{\infty} c_j(\boldsymbol{\theta})\mathbf{R}_j\mathbf{K}\\ \vdots \\ \mathbf{K}' \sum_{j=0}^{\infty} c_j(\boldsymbol{\theta})\mathbf{R}_{|l'-j|}\mathbf{K}\end{array}\right)\boldsymbol{\eta}^{(n)}\\ a_1^{(n)}(\boldsymbol{\theta}) + b_1^{(n)}(\boldsymbol{\theta})\\ \vdots \\ a_{l''}^{(n)}(\boldsymbol{\theta}) + b_{l''}^{(n)}(\boldsymbol{\theta})\end{array}\right) \tag{4.38}$$

under $H_g^{(n)}(\boldsymbol{\theta} + \mathbf{v}(n)\boldsymbol{\tau}^{(n)})$; *under both* $H_g^{(n)}(\boldsymbol{\theta})$ *and* $H_g^{(n)}(\boldsymbol{\theta} + \mathbf{v}(n)\boldsymbol{\tau}^{(n)})$, *the asymptotic covariance matrix is the same as in* (3.8).

Proof. The proof of (i) and (ii) follows by substituting $\mathbf{t}_{i;f}^{(n)}$ and $r_{i;f}^{(n)}$, as defined in (3.9) and (3.10), with $U_t = G(Z_t^{(n)}(\boldsymbol{\theta}))$, for $\underset{\sim}{\mathbf{t}}_{i;f}^{(n)}$ and $\underset{\sim}{r}_{i;f}^{(n)}$ in (3.7). The joint distribution, under $H_g^{(n)}(\boldsymbol{\theta})$, of (3.7) and the log-likelihood ratio $\Lambda_{g;\boldsymbol{\theta}+\mathbf{v}(n)\boldsymbol{\tau}^{(n)}/\boldsymbol{\theta}}^{(n)}$, where the latter is taken under the asymptotic form resulting from (4.20) and (4.27), then essentially follows from the same arguments as in the proof of Proposition 3.1. Application of LeCam's so-called *third Lemma* (Hájek and Šidák, 1967) completes the proof. Part (iii) of the proposition constitutes the parametric, normal-theory version of Propositions 3.1, 3.2, and 4.3(i); it can be derived by means of the same techniques as above—though the first step of substituting a parametric statistic for a rank-based one of course is not necessary.

Note that the k components of $\mathbf{t}_0^{(n)}$ are proportional to classical Student statistics, whence the notation. The major difference between the parametric result (iii) and the rank-based ones (i) and (ii) is that the asymptotic mean (4.35) of (4.33) under local alternatives does not depend on the underlying density g. Also remark that assumptions (**A1**) (or (**A1**$^+$)) on g are necessary for the asymptotic normality of (4.33) under $H_g^{(n)}(\boldsymbol{\theta})$, whereas the same result (Propositions 3.1 and 3.2) holds in the rank-based case without any assumption whatever on the underlying density g.

5. Locally Asymptotically Optimal Aligned Rank Tests

5.1. *Local Asymptotic Linearity of Serial and Nonserial Rank Statistics.* The local asymptotic linearity property established here plays a fundamental role in the treatment of aligned rank statistics. It generalizes several earlier results for nonserial statistics (Jurečková 1969, 1971a, 1971b; Koul, 1969, 1977; van Eeden 1972), and extends them to the serial case, for which they are new. The notation $\underset{\sim}{\mathbf{t}}_{i;f}^{(n)}(\boldsymbol{\theta}+\mathbf{v}(n)\boldsymbol{\tau}^{(n)})$, $\underset{\sim}{\mathbf{t}}_{i;f}^{(n)+}(\boldsymbol{\theta}+\mathbf{v}(n)\boldsymbol{\tau}^{(n)})$, ... will be used in an obvious fashion when the ranks and signs are those of $Z_t^{(n)}(\boldsymbol{\theta}+\mathbf{v}(n)\boldsymbol{\tau}^{(n)})$. Note that (5.3) and (5.4) extend the result (4.5) of Kreiss (1987).

Proposition 5.1. (i) *Assume that* (**A1**) ((i)–(vi) *for* f; (i)–(v) *for* g), (**A2**), *and* (**A3**) *hold. Then, under* $H_g^{(n)}(\boldsymbol{\theta})$, *for all* i *and all* $\boldsymbol{\tau}^{(n)}$ *such that* $\sup_n \|\boldsymbol{\tau}^{(n)}\| < \infty$,

$$n^{1/2}[\underset{\sim}{\mathbf{t}}_{i;f}^{(n)}(\boldsymbol{\theta})-\underset{\sim}{\mathbf{t}}_{i;f}^{(n)}(\boldsymbol{\theta}+\mathbf{v}(n)\boldsymbol{\tau}^{(n)})]-\frac{\mathscr{I}(f_1,g_1)}{[\mathscr{I}(f_1)]^{1/2}}\mathbf{K}'\sum_{j=0}^{\infty}c_j(\boldsymbol{\theta})\mathbf{R}_{|i-j|}\mathbf{K}\boldsymbol{\eta}^{(n)} \tag{5.1}$$

and

$$n^{1/2}[\underset{\sim}{r}_{i;f}^{(n)}(\boldsymbol{\theta})-\underset{\sim}{r}_{i;f}^{(n)}(\boldsymbol{\theta}+\mathbf{v}(n)\boldsymbol{\tau}^{(n)})] \\ -\sigma(f_1,g_1)\mathscr{I}(f_1,g_1)[\mathscr{I}(f_1)]^{-1/2}[a_i^{(n)}(\boldsymbol{\theta})+b_i^{(n)}(\boldsymbol{\theta})] \tag{5.2}$$

are $o_P(1)$ *as* $n\to\infty$; $\mathscr{I}(f,g)$ *and* $\sigma(f,g)$ *are defined in* (4.33) *and* (4.34), *respectively.*

(ii) *Similar results hold for* $\underset{\sim}{\mathbf{t}}^{(n)+}_{i;f}$ *and* $\underset{\sim}{r}^{(n)+}_{i;f}$ *under* (**A1**$^+$) ((i)–(vi) *for* f; (i)–(v) *for* g), (**A2**), *and* (**A3**$^+$).

(iii) *Under assumptions* (**A1**)(i)–(v) *for* g, (**A2**), *and* (**A3**$^+$), *for the parametric statistics defined in* (4.36) *and* (4.37), *we have* (*still under* $H_g^{(n)}(\boldsymbol{\theta})$, *as* $n \to \infty$)

$$n^{1/2}[\mathbf{t}_i^{(n)}(\boldsymbol{\theta}) - \mathbf{t}_i^{(n)}(\boldsymbol{\theta} + \mathbf{v}(n)\boldsymbol{\tau}^{(n)})] - \mathbf{K}' \sum_{j=0}^{\infty} c_j(\boldsymbol{\theta})\mathbf{R}_{|i-j|}\mathbf{K}\boldsymbol{\eta}^{(n)} = o_P(1) \quad (5.3)$$

and

$$n^{1/2}[r_i^{(n)}(\boldsymbol{\theta}) - r_i^{(n)}(\boldsymbol{\theta} + \mathbf{v}(n)\boldsymbol{\tau}^{(n)})] - [a_i^{(n)}(\boldsymbol{\theta}) + b_i^{(n)}(\boldsymbol{\theta})] = o_P(1). \quad (5.4)$$

Proof. The parameter $\boldsymbol{\theta}$ and the sequence $\boldsymbol{\tau}^{(n)}$ being fixed throughout, the following simplified notation will be considered here and in Lemmas 5.2 through 5.8: let $\boldsymbol{\theta}^{(n)} = \boldsymbol{\theta} + \mathbf{v}(n)\boldsymbol{\tau}^{(n)}$, $Z_t^0 = Z^{(n)}(\boldsymbol{\theta})$, and $Z_t^{(n)} = Z_t^{(n)}(\boldsymbol{\theta}^{(n)})$. Also, we write $\phi \circ F^{-1} \circ G(\cdot)$ instead of $\phi(F^{-1}(G(\cdot)))$. The proofs for the nonserial case (5.1) and for the parametric cases (5.3) and (5.4) are much easier than that of (5.2), and rely on the same truncating technique: accordingly, we leave them to the reader, and concentrate on (5.2). We first establish the following three lemmas for the process $Z_t^{(n)}$.

LEMMA 5.2. *Under* $H_g^{(n)}(\boldsymbol{\theta})$ *and for sufficiently large* n, $\{Z_t^{(n)} - E(Z_t^{(n)}); t \in \mathbb{Z}\}$ *is a stationary causal and invertible ARMA process, with uniformly* (*in* n) *bounded variance.*

LEMMA 5.3. *Under* $H_g^{(n)}(\boldsymbol{\theta})$ *and for sufficiently large* n, $\{Z_t^{(n)}; t \in \mathbb{Z}\}$ *is an absolutely regular process with mixing rates* $\beta^{(n)}(k)$, $k \in \mathbb{N}$, *satisfying* $\beta^{(n)}(k) \leqslant \beta(k)$, *where* $\beta(k)$ *is exponentially decreasing in* k.

LEMMA 5.4. *Under* $H_g^{(n)}(\boldsymbol{\theta})$, *as* $n \to \infty$, $\max_{1 \leqslant t \leqslant n} |Z_t^{(n)} - Z_t^0| = o_P(1)$.

Proof of Lemma 5.2. By the definition of $Z_t^{(n)}$, for fixed n we have

$$\begin{aligned} Z_t^{(n)} &= \frac{A(L) - n^{-1/2}\gamma^{(n)}(L)}{B(L) + n^{-1/2}\delta^{(n)}(L)} [Y_t^{(n)} - (\mathbf{x}_t^{(n)})' (\boldsymbol{\beta} + n^{-1/2}\mathbf{K}^{(n)}\boldsymbol{\eta}^{(n)})] \\ &= \frac{A(L) - n^{-1/2}\gamma^{(n)}(L)}{B(L) + n^{-1/2}\delta^{(n)}(L)} [e_t^{(n)} + n^{-1/2}\mathbf{K}^{(n)}\boldsymbol{\eta}^{(n)}] \\ &= \frac{B(L)}{A(L)} \frac{A(L) - n^{-1/2}\gamma^{(n)}(L)}{B(L) + n^{-1/2}\delta^{(n)}(L)} \varepsilon_t^{(n)} \\ &\quad + n^{-1/2} \frac{A(L) - n^{-1/2}\gamma^{(n)}(L)}{B(L) + n^{-1/2}\delta^{(n)}(L)} (\mathbf{x}_t^{(n)})' \mathbf{K}^{(n)}\boldsymbol{\eta}^{(n)}. \end{aligned} \quad (5.5)$$

The first term in the latter expression is a zero-mean ARMA $(p^{(n)}, q^{(n)})$ process, with $p^{(n)} \leqslant p_1 + q_0$, $q^{(n)} \leqslant q_1 + p_0$, whereas the second term is non-random; causality, invertibility, and a uniformly bounded variance follow from Lemma 6.1 of Kreiss (1987).

Proof of Lemma 5.3. This lemma is an immediate consequence of Theorem 3.1 of Pham and Tran (1985). Their condition on the eigenvalues of the Markovian representation matrix $\mathbf{F}^{(n)}$ follows from the causality of the AR operator in (5.5), and (still from Kreiss, 1987, Lemma 6.1) holds uniformly with respect to n (i.e., there exists N and $\eta > 0$ such that $\|\mathbf{F}^{(n)}\| < 1 - \eta$ for all $n > N$; $\|\cdot\|$ here denotes the spectral norm). Also, due to the absolute continuity of the density function g of ε_t, for all $\Delta > 0$

$$\begin{aligned}\int_{-\infty}^{\infty} |g(x+\Delta) - g(x)|\, dx &\leqslant \int_{-\infty}^{\infty} \int_{x}^{x+\Delta} |\dot{g}(z)|\, dz\, dx \\ &= \int_{-\infty}^{\infty} \int_{z-\Delta}^{z} |\phi_g(z)|\, g(z)\, dx\, dz \\ &= \Delta \int_{-\infty}^{\infty} |\phi_g(z)|\, g(z)\, dz\end{aligned}$$

which is $O(\Delta)$ as $\Delta \downarrow O$, so that Pham and Tran's condition on g is satisfied with $\delta = \gamma = 1$. The uniformity in n of the exponential decrease of $\beta^{(n)}(\cdot)$ follows from the uniformity of the bound on spectral norms.

Proof of Lemma 5.4. From (5.5), we have

$$\begin{aligned}Z_t^{(0)} - Z_t^{(n)} &= \frac{A(L)}{B(L)}\left[1 - \frac{B(L)}{A(L)}\, \frac{A(L) - n^{-1/2}\gamma^{(n)}(L)}{B(L) + n^{-1/2}\delta^{(n)}(L)}\right] e_t^{(n)} \\ &\quad + n^{-1/2}\, \frac{A(L) - n^{-1/2}\gamma^{(n)}(L)}{B(L) + n^{-1/2}\delta^{(n)}(L)}\, (\mathbf{x}_t^{(n)})'\, \mathbf{K}^{(n)} \boldsymbol{\eta}^{(n)}. \qquad (5.6)\end{aligned}$$

Still from Lemma 6.1 of Kreiss (1987), the sums of the absolute values of the coefficients of $(A(L) - n^{-1/2}\gamma^{(n)}(L))/(B(L) + n^{-1/2}\delta^{(n)}(L))$ are uniformly bounded (for n sufficiently large). Accordingly, for some $M > 0$,

$$\begin{aligned}&\max_{1 \leqslant t \leqslant n} n^{-1/2}\, \frac{A(L) - n^{-1/2}\gamma^{(n)}(L)}{B(L) + n^{-1/2}\delta^{(n)}(L)}\, (\mathbf{x}_t^{(n)})'\, \mathbf{K}^{(n)} \boldsymbol{\eta}^{(n)} \\ &\quad \leqslant n^{-1/2} M \max_{1 \leqslant t \leqslant n} |(\mathbf{x}_t^{(n)})'\, \mathbf{K}^{(n)} \boldsymbol{\eta}^{(n)}|,\end{aligned}$$

which is $o(1)$ as $n\to\infty$ since, due to the Noether conditions **(A3)**(iii),

$$n^{-1}\max_{1\leqslant t\leqslant n}|(\mathbf{x}_t^{(n)})'\,\mathbf{K}^{(n)}\boldsymbol{\eta}^{(n)}(\boldsymbol{\eta}^{(n)})'\,(\mathbf{K}^{(n)})'\,\mathbf{x}_t^{(n)}|$$
$$\leqslant \sup_n \|\boldsymbol{\tau}^{(n)}\|\,n^{-1}\max_{1\leqslant t\leqslant n}(\mathbf{x}_t^{(n)})'\,(\mathbf{D}^{(n)})^{-1/2}\,\mathbf{R}_0^{-1}(\mathbf{D}^{(n)})^{-1/2}\,\mathbf{x}_t^{(n)}$$
$$\leqslant \sup_n \|\boldsymbol{\tau}^{(n)}\|\,\|\mathbf{R}_0^{-1}\|\sum_{j=1}^{k}\left[\max_{1\leqslant t\leqslant n}(x_{t,j}^{(n)})^2\Big/\sum_{t=1}^{n}(x_{t,j}^{(n)})^2\right],$$

itself $o(1)$ as $n\to\infty$. Turning to the random term in the right-hand side of (5.6), a little symbolic algebra yields

$$\frac{A(L)}{B(L)}\left[1-\frac{B(L)}{A(L)}\frac{A(L)-n^{-1/2}\gamma^{(n)}(L)}{B(L)+n^{-1/2}\delta^{(n)}(L)}\right]=n^{-1/2}A(L)\left[\frac{a^{(n)}(L)+b^{(n)}(L)}{B(L)+n^{-1/2}\delta^{(n)}(L)}\right]. \tag{5.7}$$

As already mentioned, the sums $\sum_i |a_i^{(n)}+b_i^{(n)}|$ are bounded, uniformly in n, and so is the sum of the absolute values of the coefficients of $(B(L)+n^{-1/2}\delta^{(n)}(L))^{-1}$ (still, for n sufficiently large). Hence, $\max_{1\leqslant t\leqslant n}|A(L)[(a^{(n)}(L)+b^{(n)}(L))/(B(L)+n^{-1/2}\delta^{(n)}(L))]\,e_t^{(n)}|$, as the maximum of a stationary process with finite variance, is $o_P(n^{1/2})$ as $n\to\infty$, so that (5.6) is $o_P(1)$ as was to be proved.

Now, for every $m\in\mathbb{N}$, consider the truncation

$$F_{(m)}^{-1}(u)=\begin{cases}F^{-1}(u_m^-), & u<u_m^-\\ F^{-1}(u), & u_m^-\leqslant u<1-u_m^+\\ F^{-1}(1-u_m^+), & 1-u_m^+\leqslant u\end{cases} \tag{5.8}$$

of F^{-1}, where $u_m^{\pm}$ are chosen so that $\int_0^{u_m^-}F^{-1}(u)\,du+\int_{1-u_m^+}^1 F^{-1}(u)\,du=0$, and $u_m^-+u_m^+=2/(m+1)$, which implies

$$\int_0^1 F_{(m)}^{-1}(u)\,du=\int_0^1 F^{-1}(u)\,du=0. \tag{5.9}$$

We then prove the following three lemmas.

Lemma 5.5. *For any fixed* m,

$$\begin{aligned}&n^{-1/2}\sum_{t=i+1}^{n}[\phi\circ F_{(m)}^{-1}\circ G(Z_t^0)][F_{(m)}^{-1}\circ G(Z_{t-i}^0)]\\ &\quad-n^{-1/2}\sum_{t=i+1}^{n}[\phi\circ F_{(m)}^{-1}\circ G(Z_t^{(n)})][F_{(m)}^{-1}\circ G(Z_{t-i}^{(n)})]\\ &\quad+E_{\boldsymbol{\theta}}\left[n^{-1/2}\sum_{t=i+1}^{n}[\phi\circ F_{(m)}^{-1}\circ G(Z_t^{(n)})][F_{(m)}^{-1}\circ G(Z_{t-i}^{(n)})]\right]=o_P(1),\end{aligned} \tag{5.10}$$

under $H_g^{(n)}(\boldsymbol{\theta})$, *as* $n \to \infty$ (ϕ *here stands for* ϕ_f; *the expectation* $E_{\boldsymbol{\theta}}$ *is taken under* $H_g^{(n)}(\boldsymbol{\theta})$).

LEMMA 5.6. *For all* $\varepsilon > 0$ *and* $\eta > 0$, *there exists* $M_1 = M_1(\varepsilon, \eta)$ *such that for all* $m > M_1$ *and* $n > N_1(m)$, *under* $H_g^{(n)}(\boldsymbol{\theta})$,

$$P\left[\left| \sum_{t=i+1}^{n} [[\phi \circ F^{-1} \circ G(Z_t^0)][F^{-1} \circ G(Z_{t-i}^0)] - [\phi \circ F_{(m)}^{-1} \circ G(Z_t^0)][F_{(m)}^{-1} \circ G(Z_{t-i}^0)]]\right| > n^{1/2}\varepsilon\right] < \eta. \tag{5.11}$$

LEMMA 5.8. *For all* $\varepsilon > 0$ *and* $\eta > 0$, *there exists* $M_2 = M_2(\varepsilon, \eta)$ *such that for all* $m > M_2$ *and* $n > N_2(m)$, *under* $H_g^{(n)}(\boldsymbol{\theta})$,

$$P\left[\left| \sum_{t=i+1}^{n} [[\phi \circ F^{-1} \circ G(Z_t^{(n)})][F^{-1} \circ G(Z_{t-i}^{(n)})] - [\phi \circ F_{(m)}^{-1} \circ G(Z_t^{(n)})][F_{(m)}^{-1} \circ G(Z_{t-i}^{(n)})]]\right| > n^{1/2}\varepsilon\right] < \eta. \tag{5.12}$$

LEMMA 5.7. *For all* $\varepsilon > 0$, *there exists* $M_3 = M_3(\varepsilon)$ *such that for all* $m > M_3$ *and* n

$$\left| E_{\boldsymbol{\theta}}\left[\sum_{t=i+1}^{n} [[\phi \circ F^{-1} \circ G(Z_t^{(n)})][F^{-1} \circ G(Z_{t-i}^{(n)})] - [\phi \circ F_{(m)}^{-1} \circ G(Z_t^{(n)})][F_{(m)}^{-1} \circ G(Z_{t-i}^{(n)})]]\right]\right| < \varepsilon n^{1/2}.$$

Proof of Lemma 5.5. Expression (5.10) also can be written as

$$\begin{aligned} n^{-1/2}&\left[\sum_{t=i+1}^{n} [\phi \circ F_{(m)}^{-1} \circ G(Z_t^0)[(F_{(m)}^{-1} \circ G(Z_{t-i}^0) - F_{(m)}^{-1} \circ G(Z_{t-i}^{(n)})]]\right] \\ &+ n^{-1/2}\left[\sum_{t=i+1}^{n} [\phi \circ F_{(m)}^{-1} \circ G(Z_t^0) - \phi \circ F_{(m)}^{-1} \circ G(Z_t^{(n)})]\, F_{(m)}^{-1} \circ G(Z_{t-i}^{(n)})\right] \\ &+ E_{\boldsymbol{\theta}}\left[n^{-1/2} \sum_{t=i+1}^{n} [\phi \circ F_{(m)}^{-1} \circ G(Z_t^{(n)})][F_{(m)}^{-1} \circ G(Z_{t-i}^{(n)})]\right] \\ &= D_1^{(n)} + D_2^{(n)} - E_{\boldsymbol{\theta}}(D_2^{(n)}), \qquad \text{say.} \end{aligned} \tag{5.13}$$

Due to the independence, under $H_g^{(n)}(\boldsymbol{\theta})$, between $Z_t^0 = \varepsilon_t^{(n)}$ and $(Z_{t-i}^0, Z_{t-i}^{(n)})$, the expectation of $D_1^{(n)}$ is indeed zero, and that of $D_2^{(n)}$ coincides with the third term in (5.13). First, let us consider the variance of $D_1^{(n)}$ under $H_g^{(n)}(\boldsymbol{\theta})$:

$$\mathrm{Var}_{\boldsymbol{\theta}}(D_1^{(n)}) = E_{\boldsymbol{\theta}}[(D_1^{(n)})^2]$$
$$= n^{-1}\mathscr{I}(f)(n-i)\, E_{\boldsymbol{\theta}}[(F_{(m)}^{-1}\circ G(Z_t^0) - F_{(m)}^{-1}\circ G(Z_t^{(n)}))^2]. \qquad (5.14)$$

Continuity of $F_{(m)}^{-1}G$ and Lemma 5.4 imply that $F_{(m)}^{-1}\circ G(Z_t^0) - F_{(m)}^{-1}\circ G(Z_t^{(n)})$ is $o_P(1)$ as $n\to\infty$; since it is bounded, for all n, by $F^{-1}(1-u_m^+) - F^{-1}(u_m^-)$, it also converges to 0 in the quadratic mean, so that $\lim_{n\to\infty} E_{\boldsymbol{\theta}}[(D_1^{(n)})^2] = 0$ and $D_1^{(n)}$ is $o_P(1)$. Turning to $D_2^{(n)}$, we show that its variance under $H_g^{(n)}(\boldsymbol{\theta} + \mathbf{v}(n)\boldsymbol{\tau}^{(n)})$ tends to zero, which implies that $D_2^{(n)} - E_{\boldsymbol{\theta}^{(n)}}(D_2^{(n)})$ tends to zero in $P_{\boldsymbol{\theta}^{(n)}}$-probability, hence also in $P_{\boldsymbol{\theta}}$-probability; however, due to contiguity,

$$E_{\boldsymbol{\theta}^{(n)}}(D_2^{(n)}) = E_{\boldsymbol{\theta}^{(n)}}\Bigg\{n^{-1/2}\sum_{t=i+1}^{n}[\phi\circ F_{(m)}^{-1}\circ G(Z_t^0)$$
$$- \phi\circ F_{(m)}^{-1}\circ G(Z_t^{(n)})]\, F_{(m)}^{-1}\circ G(Z_{t-i}^0)\Bigg\}$$
$$= E_{\boldsymbol{\theta}^{(n)}}\Bigg\{n^{-1/2}\sum_{t=i+1}^{n}[\phi\circ F_{(m)}^{-1}\circ G(Z_t^0)][F_{(m)}^{-1}\circ G(Z_{t-i}^0)]\Bigg\}$$
$$= -E_{\boldsymbol{\theta}}\Bigg\{n^{-1/2}\sum_{t=i+1}^{n}[\phi\circ F_{(m)}^{-1}\circ G(Z_t^{(n)})][F_{(m)}^{-1}\circ G(Z_{t-i}^{(n)})]\Bigg\}$$
$$= E_{\boldsymbol{\theta}}(D_2^{(n)}),$$

so that $\lim_{n\to\infty}\mathrm{Var}_{\boldsymbol{\theta}^{(n)}}(D_2^{(n)}) = 0$ implies $D_2^{(n)} - E_{\boldsymbol{\theta}}(D_2^{(n)}) = o_{P_{\boldsymbol{\theta}}}(1)$. Defining $\bar{D}_2^{(n)} = n^{-1/2}[\sum_{t=i+1}^{n}[\phi\circ F_{(m)}^{-1}\circ G(Z_t^{(n)}) - \phi\circ F_{(m)}^{-1}\circ G(Z_t^0)]\, F_{(m)}^{-1}\circ G(Z_{t-i}^0)]$, $\lim_{n\to\infty}\mathrm{Var}_{\boldsymbol{\theta}^{(n)}}(D_2^{(n)}) = 0$ iff $\lim_{n\to\infty}\mathrm{Var}_{\boldsymbol{\theta}}(\bar{D}_2^{(n)}) = 0$, which we now proceed to show. Clearly,

$$\mathrm{Var}_{\boldsymbol{\theta}}(\bar{D}_2^{(n)}) = \frac{n-i}{n}\mathrm{Var}_{\boldsymbol{\theta}}[[\phi\circ F_{(m)}^{-1}\circ G(Z_t^{(n)})$$
$$- \phi\circ F_{(m)}^{-1}\circ G(Z_t^0)]\, F_{(m)}^{-1}\circ G(Z_{t-i}^0)]$$
$$+ 2\sum_{k=1}^{n-i-1}\frac{n-i-k}{n}\mathrm{Cov}_{\boldsymbol{\theta}}(k), \qquad (5.15)$$

where $\mathrm{Cov}_{\boldsymbol{\theta}}(k)$ stands for the autocovariance

$$\mathrm{Cov}_{\boldsymbol{\theta}}[(\phi\circ F_{(m)}^{-1}G(Z_t^{(n)}) - \phi\circ F_{(m)}^{-1}\circ G(Z_t^0))\, F_{(m)}^{-1}\circ G(Z_{t-i}^0),$$
$$(\phi\circ F_{(m)}^{-1}\circ G(Z_{t-k}^{(n)}) - \phi\circ F_{(m)}^{-1}\circ G(Z_{t-k}^0))\, F_{(m)}^{-1}\circ G(Z_{t-k-i}^0)].$$

The variance term in (5.15) (denote it by $\mathrm{Cov}_{\boldsymbol{\theta}}(0)$) is bounded by

$$E_{\boldsymbol{\theta}}\{[(\phi\circ F_{(m)}^{-1}G(Z_t^{(n)})-\phi\circ F_{(m)}^{-1}\circ G(Z_t^0))\,F_{(m)}^{-1}\circ G(Z_{t-i}^0)]^2\}$$
$$\leqslant A_f^2 E_{\boldsymbol{\theta}}\{[(F_{(m)}^{-1}\circ G(Z_t^{(n)})-F_{(m)}^{-1}\circ G(Z_t^0))\,F_{(m)}^{-1}\circ G(Z_{t-i}^0)]^2\}$$
$$\leqslant [A_f F^{-1}(1-u_m^+)]^2\, E_{\boldsymbol{\theta}}\{[(F_{(m)}^{-1}\circ G(Z_t^{(n)})-F_{(m)}^{-1}\circ G(Z_t^0)]^2\}. \tag{5.16}$$

Here A_f denotes the Lipschitz constant in assumption (**A1**)(vi), where for the sake of simplicity we have assumed a uniform Lipschitz condition instead of a piecewise one—the piecewise case can be taken care of along the same lines as in the proof of Lemma 4.2(i). The expectation in the last bound in (5.16) being the same as in (5.14), it similarly tends to zero, and so does $\mathrm{Cov}_{\boldsymbol{\theta}}(0)$. As for the covariance terms, note that Z_t^0 under $H_g^{(n)}(\boldsymbol{\theta})$ coincides with the innovation process of $Z_t^{(n)}$. Accordingly, $(Z_t^{(n)}, Z_t^0)$, $(Z_{t-1}^{(n)}, Z_{t-1}^0)$, ... generate the same σ-algebra as $Z_t^{(n)}, Z_{t-1}^{(n)}, \ldots$, whereas Z_{t+k}^0 is independent of $(Z_t^{(n)}, Z_t^0)$, $(Z_{t-1}^{(n)}, Z_{t-1}^0)$, ..., $k>0$. It follows that the bivariate process $(Z_t^{(n)}, Z_t^0)$ is absolutely regular, with the same mixing rates $\beta^{(n)}(k)$ as $Z_t^{(n)}$ in Lemma 5.3. Lemma 1 of Yoshihara (1976) (with $\delta=1$) then yields $|\mathrm{Cov}_{\boldsymbol{\theta}}(k)|\leqslant 4\,\mathrm{Cov}_{\boldsymbol{\theta}}(0)[\beta^{(n)}(k)]^{1/2}$ which in turn, from Lemma 5.3, is bounded by $4\,\mathrm{Cov}_{\boldsymbol{\theta}}(0)[\beta(k)]^{1/2}$. It follows that

$$\left|\sum_{k=1}^{n-i-1}\frac{n-i-k}{n}\mathrm{Cov}_{\boldsymbol{\theta}}(k)\right|\leqslant\sum_{k=1}^{\infty}|\mathrm{Cov}_{\boldsymbol{\theta}}(k)|\leqslant 4\,\mathrm{Cov}_{\boldsymbol{\theta}}(0)\sum_{k=1}^{\infty}[\beta(k)]^{1/2};$$

since $\beta(k)$ is decreasing exponentially fast, and since we have shown that $\mathrm{Cov}_{\boldsymbol{\theta}}(0)$ is $o(1)$ as $n\to\infty$, the covariance term in (5.15) also tends to zero, which completes the proof of the lemma.

Proof of Lemma 5.6. For fixed m and n,

$$\mathrm{Var}_{\boldsymbol{\theta}}\Bigg[n^{-1/2}\sum_{t=i+1}^{n}[[\phi\circ F^{-1}\circ G(Z_t^0)][F^{-1}\circ G(Z_{t-i}^0)]$$
$$-[\phi\circ F_{(m)}^{-1}\circ G(Z_t^0)][F_{(m)}^{-1}\circ G(Z_{t-i}^0)]]\Bigg]$$
$$=\frac{n-i}{n}\mathrm{Var}_{\boldsymbol{\theta}}([\phi\circ F^{-1}\circ G(Z_t^0)][F^{-1}\circ G(Z_{t-i}^0)])$$
$$+\frac{n-i}{n}\mathrm{Var}_{\boldsymbol{\theta}}([\phi\circ F_{(m)}^{-1}\circ G(Z_t^0)][F_{(m)}^{-1}\circ G(Z_{t-i}^0)])$$
$$-2\frac{n-i}{n}\mathrm{Cov}_{\boldsymbol{\theta}}([\phi\circ F^{-1}\circ G(Z_t^0)][F^{-1}\circ G(Z_{t-i}^0)],$$
$$[\phi\circ F_{(m)}^{-1}\circ G(Z_t^0)][F_{(m)}^{-1}\circ G(Z_{t-i}^0)])$$

$$= \frac{n-i}{n}\left[\sigma_f^2 \mathscr{I}(f) + \int_0^1 [\phi \circ F_{(m)}^{-1}(u)]^2\, du \int_0^1 [F_{(m)}^{-1}(u)]^2\, du\right]$$

$$-2\frac{n-i}{n}\int_0^1 [\phi \circ F^{-1}(u)][\phi \circ F_{(m)}^{-1}(u)]\, du \int_0^1 [F^{-1}(u)\, F_{(m)}^{-1}(u)]\, du. \tag{5.17}$$

As $F_{(m)}^{-1}(u)$ converges to $F^{-1}(u)$ from below, limits for $m \to \infty$ can be taken under the integral signs, and this variance clearly converges to zero, as $m \to \infty$, uniformly in n. Since the quantity inside the variance in (5.17) has expectation zero under $H_g^{(n)}(\boldsymbol{\theta})$, Chebyshev's inequality leads to the desired result.

Proof of Lemma 5.7. The claim in Lemma 5.7 is exactly the same as in Lemma 5.6, except that $Z_t^{(n)}$ replaces Z_t^0; accordingly, it holds under $H_g^{(n)}(\boldsymbol{\theta}^{(n)})$. That it also holds under $H_g^{(n)}(\boldsymbol{\theta})$ follows from an application of Lemma 3.5 of Jurečková (1969).

Proof of Lemma 5.8. Lemma 5.8 immediately follows from Lebesgue's dominated convergence theorem.

We now may proceed to the proof of (5.2). Due to the mutual contiguity of $H_g^{(n)}(\boldsymbol{\theta})$ and $H_g^{(n)}(\boldsymbol{\theta}^{(n)})$, the differences $\underset{\sim}{r}_{i;f}^{(n)}(\boldsymbol{\theta}^{(n)}) - r_{i;f}^{(n)}(\boldsymbol{\theta}^{(n)})$, which (see the proof of Proposition 3.1) are $o_P(n^{-1/2})$ under $H_g^{(n)}(\boldsymbol{\theta}^{(n)})$, still are $o_P(n^{-1/2})$ under $H_g^{(n)}(\boldsymbol{\theta})$; the definition (3.10) of $r_{i;f}^{(n)}$ here of course takes the form

$$r_{i;f}^{(n)}(\boldsymbol{\theta}) = (n-i)^{-1}\, [\mathscr{I}(f_1)]^{-1/2} \sum_{t=i+1}^{n} [\phi \circ F^{-1} \circ G(Z_t^0)][F^{-1} \circ G(Z_{t-i}^0)].$$

Also, from Proposition 4.3(i), $\sigma(f, g)\, \mathscr{I}(f, g)[\mathscr{I}(f_1)]^{-1/2}\, (a_i^{(n)}(\boldsymbol{\theta}) + b_i^{(n)}(\boldsymbol{\theta}))$ is the asymptotic mean, under $H_g^{(n)}(\boldsymbol{\theta})$, of $-n^{1/2} r_{i;f}^{(n)}(\boldsymbol{\theta}^{(n)})$. Accordingly, the left-hand side in (5.2) under $H_g^{(n)}(\boldsymbol{\theta})$ decomposes into

$$n^{-1/2} \sum_{t=i+1}^{n} ([\phi \circ F_{(m)}^{-1} \circ G(Z_t^0)][F_{(m)}^{-1} \circ G(Z_{t-i}^0)]$$

$$- [\phi \circ F_{(m)}^{-1} \circ G(Z_t^{(n)})][F_{(m)}^{-1} \circ G(Z_{t-i}^{(n)})])$$

$$+ E_{\boldsymbol{\theta}}\left[n^{-1/2} \sum_{t=i+1}^{n} [\phi \circ F_{(m)}^{-1} \circ G(Z_t^{(n)})][F_{(m)}^{-1} \circ G(Z_{t-i}^{(n)})]\right]$$

$$+ R_{(m)}^{(n)} + o_P(1), \tag{5.18}$$

which, from Lemma 5.5, reduces to $R_{(m)}^{(n)} + o_P(1)$ (o_P, here are as $n \to \infty$, for fixed m), where

$$\begin{aligned} R_{(m)}^{(n)} = n^{-1/2} \sum_{t=i+1}^{n} & [[\phi \circ F^{-1} \circ G(Z_t^0)][F^{-1} \circ G(Z_{t-i}^0)] \\ & - [\phi \circ F^{-1} \circ G(Z_t^{(n)})][F^{-1} \circ G(Z_{t-i}^{(n)})] \\ & - [\phi \circ F_{(m)}^{-1} \circ G(Z_t^0)][F_{(m)}^{-1} \circ G(Z_{t-i}^0)] \\ & + [\phi \circ F_{(m)}^{-1} \circ G(Z_t^{(n)})][F_{(m)}^{-1} \circ G(Z_{t-i}^{(n)})]] \\ & + E_{\boldsymbol{\theta}} \left\{ n^{-1/2} \sum_{t=i+1}^{n} [[\phi \circ F^{-1} \circ G(Z_t^{(n)})][F^{-1} \circ G(Z_{t-i}^{(n)})] \right. \\ & \left. - [\phi \circ F_{(m)}^{-1} \circ G(Z_t^{(n)})][F_{(m)}^{-1} \circ G(Z_{t-i}^{(n)})]] \right\}. \end{aligned}$$

It follows from Lemmas 5.6, 5.7, and 5.8 that $R_{(m)}^{(n)}$ converges to zero, in probability, as $m \to \infty$, uniformly in n. Application of a classical result (Anderson, 1971, Theorem 7.7.1), here in a degenerate form, yields the desired result that (5.18) is $o_P(1)$ as $n \to \infty$. This completes the proof of assertion (5.2).

As already mentioned, the proofs of (5.1), (5.3), and (5.4) are of the same type as above, though slightly easier; they are left to the reader. Due to their asymptotic equivalence (though, for $\underset{\sim}{\mathbf{t}}_{i;f}^{(n)+}$, under different assumptions) with $r_{i;f}^{(n)}$ and $\mathbf{t}_{i;f}^{(n)}$, the signed-rank case can be treated exactly the same way.

5.2. *Linear Hypotheses.* Denote by $\mathcal{M}(\mathbf{E})$ the linear subspace of $\mathbb{R}^n$ spanned by the columns of a $n \times m$ matrix $\mathbf{E}$, by $\mathcal{M}(\mathbf{E})_\perp$ the orthogonal complement of $\mathcal{M}(\mathbf{E})$ (with respect to $\mathbb{R}^n$ and the ordinary scalar product), and by $\mathbf{E}_\perp$ an arbitrary basis of $\mathcal{M}(\mathbf{E})_\perp$. When $\mathbf{E}(\cdot)$ is a continuous function of $\boldsymbol{\theta}$, $\mathbf{E}_\perp(\boldsymbol{\theta})$ can and will be chosen so that $\mathbf{E}_\perp(\cdot)$ is continuous, too. Let $\boldsymbol{\Omega}$ be a real matrix of dimension $k_1 \times (k + \pi_0 - r)$, $\pi \leqslant r \leqslant k + \pi_0$, with zeros in rows $k + p_0 + 1, ..., k + p_1$ and $k + p_1 + q_0 + 1, ..., k + p_1 + q_1 = k_1$. Assume that $\boldsymbol{\Omega}$ has rank $k + \pi_0 - r$, and denote by $H_f^{(n)}(\boldsymbol{\Omega}; \boldsymbol{\theta}_0)$, where $\boldsymbol{\theta}_0$ is a particular element of Θ, the null hypothesis under which the observation $\mathbf{Y}^{(n)}$ is generated by (2.1) and (2.3), with innovation density f and parameter value $\boldsymbol{\theta} = (\boldsymbol{\beta}', \mathbf{A}', \mathbf{B}')' \in \Theta$ satisfying

$$\boldsymbol{\theta} = \boldsymbol{\Omega}\boldsymbol{\omega} + \boldsymbol{\theta}_0 \tag{5.19a}$$

for some unspecified $\boldsymbol{\omega} \in \mathbb{R}^{k+\pi_0-r}$ or, equivalently, the $r + \min(p_1 - p_0, q_1 - q_0)$ linear restrictions

$$\boldsymbol{\Omega}_\perp'(\boldsymbol{\theta} - \boldsymbol{\theta}_0) = \mathbf{0}. \tag{5.19b}$$

The reasons behind our notation for the dimensions of $\mathbf{\Omega}$ and $\mathbf{\Omega}_\perp$ will become clear later on. Whenever f remains unspecified (but satisfying assumptions (**A1**)(i)–(v)), or unspecified but symmetric with respect to zero (satisfying (**A1**$^+$)(i)–(v)), we use the notation $H^{(n)}(\mathbf{\Omega}; \boldsymbol{\theta}_0)$ and $H_+^{(n)}(\mathbf{\Omega}; \boldsymbol{\theta}_0)$, respectively.

If however the aligned rank tests we plan to construct for $H^{(n)}(\mathbf{\Omega}; \boldsymbol{\theta}_0)$ and $H_+^{(n)}(\mathbf{\Omega}; \boldsymbol{\theta}_0)$ are to be asymptotically invariant (see Subsection 5.3), the linear restrictions (5.19) cannot "mix" the serial parameters **A** and **B** with the nonserial ones $\boldsymbol{\beta}$. This is due to the differential reaction of serial and nonserial statistics to a variation of the innovation density underlying the data (see Proposition 4.3 and the proof of Lemma 5.11 below). We therefore make another assumption that $\mathbf{\Omega}$ and $\mathbf{\Omega}'_\perp$ are of a special block-diagonal form.

(**A4**) The matrices $\mathbf{\Omega}$ and $\mathbf{\Omega}'_\perp$ in (5.19) are of the form

$$\mathbf{\Omega} = \begin{pmatrix} \mathbf{\Omega}_I & \mathbf{0} \\ \mathbf{0} & \mathbf{\Omega}_{II} \end{pmatrix} \quad \text{and} \quad \mathbf{\Omega}'_\perp = \begin{pmatrix} \mathbf{\Omega}'_{I;\perp} & \mathbf{0} \\ \mathbf{0} & \mathbf{\Omega}'_{II;\perp} \end{pmatrix},$$

where $\mathbf{\Omega}_I$, $\mathbf{\Omega}_{II}$, $\mathbf{\Omega}'_{I;\perp}$, and $\mathbf{\Omega}'_{II;\perp}$ are of dimensions $k \times (k - r_1)$, $(p_1 + q_1) \times (\pi_0 - r_2)$, $r_1 \times k$ and $(r_2 + \min(p_1 - p_0, q_1 - q_0)) \times (p_1 + q_1)$, respectively ($\pi \leqslant r_2 \leqslant \pi_0$; $r_1 + r_2 = r$).

Such an assumption is not really restrictive, since linear combinations involving both regression and ARMA parameters have little practical meaning, and the class of hypotheses just described covers all cases of null hypotheses generally considered in linear models and time series analysis (such as testing the significance of a subset of explanatory variables in the trend part, without specifying the ARMA parameters, or testing the orders of the ARMA model without specifying the regression trend), plus their combinations.

EXAMPLE 5.1. Let the orders of the ARMA model for the error terms be fixed ($p_0 = p_1$, $q_0 = q_1$, $\pi = 0$, $k_1 = k + p_0 + q_0$), and consider the null hypothesis under which $\beta_1 = \beta_2 = \cdots = \beta_l = 0$, $l \leqslant k$: this corresponds to $r_1 = l$, $r_2 = 0$, $\boldsymbol{\theta}_0 = \mathbf{0}$, and

$$\mathbf{\Omega}_I = \begin{pmatrix} \mathbf{0}_{l \times (k-l)} \\ \mathbf{I}_{(k-l) \times (k-l)} \end{pmatrix}, \qquad \mathbf{\Omega}_{II} = \mathbf{I}_{(p_1+q_1) \times (p_1+q_1)}, \qquad \mathbf{\Omega}'_{I;\perp} = (\mathbf{I}_{l \times l} \quad \mathbf{0}_{l \times (k-l)}).$$

EXAMPLE 5.2. Assume that ARMA (p_0, q_0) dependence is to be tested against ARMA $(p_0 + \pi, q_0 + \pi)$ dependence, with unspecified trend. Then $r_1 = 0$, $r_2 = \pi$, $\boldsymbol{\theta}_0 = \mathbf{0}$, and

$$\mathbf{\Omega}_I=\mathbf{I}_{k\times k},\qquad \mathbf{\Omega}_{II}=\begin{pmatrix}\mathbf{I}_{p_0\times p_0} & & \mathbf{O}_{p_0\times q_0}\\ & \mathbf{O}_{\pi\times(p_0+q_0)} & \\ \mathbf{O}_{q_0\times p_0} & & \mathbf{I}_{q_0\times q_0}\\ & \mathbf{O}_{\pi\times(p_0+q_0)} & \end{pmatrix},$$

$$\mathbf{\Omega}'_{II;\perp}=\begin{pmatrix} & \mathbf{I}_{\pi\times\pi} & & \mathbf{O}_{\pi\times\pi}\\ \mathbf{O}_{2\pi\times p_0} & & \mathbf{O}_{2\pi\times q_0} & \\ & \mathbf{O}_{\pi\times\pi} & & \mathbf{I}_{\pi\times\pi}\end{pmatrix}.$$

The special case where $\pi=1$ plays an important role in several model identification procedures (see, e.g., Pötscher 1983, 1985). Our rank-based test statistics in this context are providing an interesting, robust, invariant, and powerful alternative to classical Lagrange multiplier statistics (see Subsection 6.3.2).

EXAMPLE 5.3. The two examples above can be combined, yielding $(r_1=l;\ r_2=\pi;\ \mathbf{\theta}_0=\mathbf{0})$

$$\mathbf{\Omega}'_{\perp}=\begin{pmatrix}\mathbf{I}_{l\times l} & \mathbf{O}_{l\times(k-l)} & & \mathbf{O}_{l\times\pi} & & \mathbf{O}_{l\times\pi}\\ & & \mathbf{O}_{(l+2\pi)\times p_0} & \mathbf{I}_{\pi\times\pi} & \mathbf{O}_{(l+2\pi)\times q_0} & \mathbf{O}_{\pi\times\pi}\\ \mathbf{O}_{2\pi\times l} & \mathbf{O}_{2\pi\times(k-l)} & & \mathbf{O}_{\pi\times\pi} & & \mathbf{I}_{\pi\times\pi}\end{pmatrix}.$$

5.3. *Aligned Ranks, Aligned Rank Statistics, and Asymptotic Invariance.* Aligned ranks are the ranks computed from some estimated residuals $\hat{Z}_t^{(n)}=Z_t^{(n)}(\hat{\mathbf{\theta}}^{(n)})$, where $\hat{\mathbf{\theta}}^{(n)}=((\hat{\beta}^{(n)})',(\hat{\mathbf{A}}^{(n)})',(\hat{\mathbf{B}}^{(n)})')'$ is some estimate of $\mathbf{\theta}$. This estimate should be "reasonably good": in the problem of testing $H^{(n)}(\mathbf{\Omega};\mathbf{\theta}_0)$ or $H_+^{(n)}(\mathbf{\Omega};\mathbf{\theta}_0)$, the following assumptions will be made on the sequence $\hat{\mathbf{\theta}}^{(n)}$.

(A5) (i) $\hat{\mathbf{\theta}}^{(n)}$ is a constrained estimate, i.e., it satisfies (5.19) for all n.

(ii) $\hat{\mathbf{\theta}}^{(n)}$ is $\mathbf{v}(n)$-consistent: for all f satisfying (**A1**)(i) and (ii), all $\mathbf{\theta}$ satisfying (5.19), and all $\varepsilon>0$, there exist $b=b(f,\mathbf{\theta},\varepsilon)$ and $N=N(f,\mathbf{\theta},\varepsilon)$ such that under $H_f^{(n)}(\mathbf{\theta})$, $P[\|[\mathbf{v}(n)]^{-1}(\hat{\mathbf{\theta}}^{(n)}-\mathbf{\theta})\|>b]<\varepsilon$ for all $n\geqslant N$ ($\|\cdot\|$, here and in (**A5**)(iii), stands for the Euclidean norm in $\mathbb{R}^{k_1}$).

(iii) $\hat{\mathbf{\theta}}^{(n)}$ is *locally discrete*, i.e., the number of possible values of $\hat{\mathbf{\theta}}^{(n)}$ in balls of the form $\{\mathbf{t}\in\Theta \mid \|[\mathbf{v}(n)]^{-1}(\mathbf{t}-\mathbf{t}_0)\leqslant c\}$, $c>0$, remains bounded, as $n\to\infty$, for all $\mathbf{t}_0$ satisfying (5.19) and all fixed c.

Assumption (**A5**)(ii), which requires $\hat{\mathbf{\theta}}^{(n)}$ to converge at the appropriate rate, is pretty mild, and is satisfied by all estimates usually considered in this context—least square estimates, (pseudo-)maximum likelihood estimates and more general M-estimates, Yule–Walker estimates, etc. The local discreteness assumption (**A5**)(iii) is less classical, but typical in the LAN context (see LeCam, 1960, and also Kreiss, 1987), and ensures the required uniformity properties for Proposition 5.9 below; other possible ways of enforcing such uniformity properties consist in smoothing $\hat{\boldsymbol{\theta}}^{(n)}$ in an appropriate way (see LeCam, 1960, where various discretization and smoothing techniques are described). It should be insisted however that such assumptions, as well as those (**A3**) concerning the asymptotic behavior of regression constants, are of little practical relevance, if any. They typically have the nature of analytical conditions ensuring ad hoc probabilistic limit results, whereas statistical practice is essentially interested in approximation results. As a matter of fact, any estimate, for fixed n, could be considered as part of a sequence satisfying (**A5**)(iii), whether it is discretized or not. Moreover, the actual numerical computation of $\hat{\boldsymbol{\theta}}^{(n)}$, with a finite number of decimal values resulting (especially in the time series context) from some numerical optimization technique, automatically provides a smooth, discrete version $\tilde{\boldsymbol{\theta}}^{(n)}$ of $\hat{\boldsymbol{\theta}}^{(n)}$.

As a rule, hats will be used for aligned quantities, computed from aligned residuals $\hat{Z}_t^{(n)} = Z_t^{(n)}(\hat{\boldsymbol{\theta}}^{(n)})$: aligned ranks $\hat{R}_t^{(n)} = R_t^{(n)}(\hat{\boldsymbol{\theta}}^{(n)})$, aligned signs $\hat{s}_t^{(n)} = \operatorname{sgn}(\hat{Z}_t^{(n)})$, aligned signed-ranks $\hat{R}_{+,t}^{(n)}, \ldots$ The notation $\hat{\underset{\sim}{r}}_{i;f}^{(n)} = \underset{\sim}{r}_{i;f}^{(n)}(\hat{\boldsymbol{\theta}}^{(n)})$, $\hat{\underset{\sim}{\mathbf{t}}}_{i;f}^{(n)} = \underset{\sim}{\mathbf{t}}_{i;f}^{(n)}(\hat{\boldsymbol{\theta}}^{(n)})$, $\hat{\psi}_t^{(i)} = \psi_t^{(i)}(\hat{\boldsymbol{\theta}}^{(n)})$, $\hat{c}_i^{(n)} = c_i^{(n)}(\hat{\boldsymbol{\theta}}^{(n)})$, $\hat{T}_{\hat{\psi};f}^{(n)} = T_{\hat{\psi};f}^{(n)}(\hat{\boldsymbol{\theta}}^{(n)})$, $\mathbf{C}_{\hat{\psi};f}$, $\hat{\mathbf{w}}_{\hat{\psi};f}$, $\hat{\mathbf{M}}$, ..., will be used in a similar way.

Since the rank-based statistics $\underset{\sim}{\mathbf{t}}_{i;f}^{(n)}(\boldsymbol{\theta})$ and $\underset{\sim}{r}_{i;f}^{(n)}(\boldsymbol{\theta})$ (or their signed counterparts) have been shown (this follows from the LAN property) asymptotically sufficient for testing $H^{(n)}(\boldsymbol{\theta})$ $(H_+^{(n)}(\boldsymbol{\theta}))$ against local alternatives of the form $H_f^{(n)}(\boldsymbol{\theta} + \mathbf{v}(n)\boldsymbol{\tau})$, $\boldsymbol{\tau} \in \mathbb{R}^{k_1}$, a very natural idea would consist in considering the aligned rank statistics $\hat{\underset{\sim}{\mathbf{t}}}_{i;f}^{(n)}$ and $\hat{\underset{\sim}{r}}_{i;f}^{(n)}$ (the aligned signed rank statistics $\hat{\underset{\sim}{\mathbf{t}}}_{i;f}^{(n)+}$ and $\hat{\underset{\sim}{r}}_{i;f}^{(n)+}$) in testing $H^{(n)}(\boldsymbol{\Omega}; \boldsymbol{\theta}_0)$ (or $H_+^{(n)}(\boldsymbol{\Omega}; \boldsymbol{\theta}_0)$) against

$$\bigcup \{H_f^{(n)}(\boldsymbol{\theta} + \mathbf{v}(n)\boldsymbol{\tau}) \mid \boldsymbol{\Omega}'_\perp(\boldsymbol{\theta} - \boldsymbol{\theta}_0) = \mathbf{0},\ \boldsymbol{\Omega}'_\perp \mathbf{v}(n)\boldsymbol{\tau} \neq \mathbf{0}\}. \tag{5.20}$$

A closer look at this apparently simple idea however raises a number of theoretical objections. First, invariance is lost. Substituting estimated parameters $\hat{\boldsymbol{\theta}}^{(n)}$ for the exact ones $\boldsymbol{\theta}$ induces among the residuals $\hat{Z}_t^{(n)}$ complex interrelations which destroy their exchangeability features, and hence the invariance property of their ranks. This is not just a minor shortcoming in view of the basic role of invariance in the justification of rank-based inference: *if invariance is lost, there is no point in using ranks anymore*! Second, distribution-freeness, and even asymptotic distribution-freeness, is lost as well; it follows indeed from Proposition 5.9 below that the asymptotic distributions under $H_g^{(n)}(\boldsymbol{\theta})$ of $\hat{\underset{\sim}{\mathbf{t}}}_{i;f}^{(n)}$, $\hat{\underset{\sim}{\mathbf{t}}}_{i;f}^{(n)+}$, $\hat{\underset{\sim}{r}}_{i;f}^{(n)}$, and $\hat{\underset{\sim}{r}}_{i;f}^{(n)+}$ depend on g.

PROPOSITION 5.9. (i) *Assume that* (**A1**) ((i)–(vi) *for* f; (i)–(v) *for* g), (**A2**) (*for* $\boldsymbol{\theta} = (\boldsymbol{\beta}', \mathbf{A}', \mathbf{B}')'$), (**A3**), *and* (**A5**)(ii) *and* (iii) (*for* $\hat{\boldsymbol{\theta}}^{(n)} = ((\hat{\boldsymbol{\beta}}^{(n)})', (\hat{\mathbf{A}}^{(n)})', (\hat{\mathbf{B}}^{(n)})')'$) *hold. Then, under* $H_g^{(n)}(\boldsymbol{\theta})$, *as* $n \to \infty$,

$$n^{1/2}\hat{\underset{\sim}{\mathbf{t}}}{}^{(n)}_{i;f} = n^{1/2}\underset{\sim}{\mathbf{t}}{}^{(n)}_{i;f} - \frac{\mathcal{I}(f_1, g_1)}{\sigma_g[\mathcal{I}(f_1)]^{1/2}}\mathbf{K}'\left[\sum_{j=0}^{\infty} c_j(\boldsymbol{\theta})\mathbf{R}_{|i-j|}\right] \times (\mathbf{D}^{(n)})^{1/2}\,[n^{1/2}(\hat{\boldsymbol{\beta}}^{(n)} - \boldsymbol{\beta})] + o_P(1) \tag{5.21}$$

and

$$n^{1/2}\hat{\underset{\sim}{r}}{}^{(n)}_{i;f} = n^{1/2}\underset{\sim}{r}{}^{(n)}_{i;f} - \sigma(f_1, g_1)\,\mathcal{I}(f_1, g_1)[\mathcal{I}(f_1)]^{-1/2} \times n^{1/2}\left[\sum_{j=1}^{p_0} g_{i-j}(\boldsymbol{\theta})(\hat{A}_j^{(n)} - A_j) + n^{1/2}\sum_{j=1}^{q_0} h_{i-j}(\boldsymbol{\theta})(\hat{B}_j^{(n)} - B_j)\right] + o_P(1). \tag{5.22}$$

(ii) *Similar results hold for* $\hat{\underset{\sim}{\mathbf{t}}}{}^{(n)+}_{i;f}$ *and* $\hat{\underset{\sim}{r}}{}^{(n)+}_{i;f}$, *under* (**A1**$^+$) ((i)–(vi) *for* f; (i)–(v) *for* g), (**A2**), (**A3**$^+$), *and* (**A5**) ((ii) *and* (iii)).

(iii) *Under* (**A1**) ((i)–(v)) *for* g, (**A2**), (**A3**$^+$), (**A5**) ((ii) *and* (iii)), *and* $H_g^{(n)}(\boldsymbol{\theta})$, *as* $n \to \infty$, *for the parametric statistics* (4.40) *and* (4.41),

$$n^{1/2}\hat{\mathbf{t}}_i^{(n)} = n^{1/2}\mathbf{t}_i^{(n)} - \sigma_g^{-1}\mathbf{K}'\left[\sum_{j=0}^{\infty} c_j(\boldsymbol{\theta})\mathbf{R}_{|i-j|}\right] \times (\mathbf{D}^{(n)})^{1/2}\,[n^{1/2}(\hat{\boldsymbol{\beta}}^{(n)} - \boldsymbol{\beta})] + o_P(1) \tag{5.23}$$

and

$$n^{1/2}\hat{r}_i^{(n)} = n^{1/2}r_i^{(n)} - n^{1/2}\left[\sum_{j=1}^{p_0} g_{i-j}(\boldsymbol{\theta})(\hat{A}_j^{(n)} - A_j) + \sum_{j=1}^{q_0} h_{i-j}(\boldsymbol{\theta})(\hat{B}_j^{(n)} - B_j)\right] + o_P(1). \tag{5.24}$$

Proof. The proof readily follows from the asymptotic linearity of the statistics considered (Proposition 5.1), the local discreteness of the estimator $\hat{\boldsymbol{\theta}}^{(n)}$, and an argument which goes back to LeCam (see Kreiss, 1987, Lemma 4.4, for a proof). Note that the random variables in the right-hand side of (5.22) are obtained from (4.4) upon replacing $\gamma_j^{(n)}$ and $\delta_j^{(n)}$ by $n^{1/2}(\hat{A}_j^{(n)} - A_j)$ and $n^{1/2}(\hat{B}_j^{(n)} - B_j)$, respectively.

Since invariance clearly is too restrictive a requirement here, some weaker, asymptotic form of invariance might be more appropriate.

DEFINITION 5.1. Denote by $(\mathcal{X}^{(n)}, \mathcal{A}^{(n)}, \mathcal{P}^{(n)})$, $n \in \mathbb{N}$, a sequence of statistical models, and by $\mathcal{G}^{(n)}$ a sequence of groups of $\mathcal{A}^{(n)}$-measurable transformations acting on $\mathscr{X}^{(n)}$. Assume that $\mathscr{P}^{(n)}$ is invariant under $\mathscr{G}^{(n)}$. A sequence of statistics $S^{(n)}$ defined on $(\mathscr{X}^{(n)}, \mathscr{A}^{(n)})$ is said to be *asymptotically invariant* for $\mathscr{G}^{(n)}$ under $\mathscr{P}^{(n)}$, as $n \to \infty$, if there exists a sequence $S_0^{(n)}$ of random variables (defined on the same $(\mathscr{X}^{(n)}, \mathscr{A}^{(n)})$) such that for some $d \in \mathbb{R}$

(i) $n^d S_0^{(n)}$ is a nondegenerate relatively compact sequence, under $\mathscr{P}^{(n)}$, as $n \to \infty$, i.e., $\forall \varepsilon > 0$ there exists $b(\varepsilon)$ such that $P[|S_0^{(n)}| > n^{-d} b(\varepsilon)] < \varepsilon$ for all n, but there exists no $c \in \mathbb{R}$ such that $n^d(S_0^{(n)} - c) = o_P(1)$;

(ii) $S_0^{(n)}$ is (exactly) invariant under $\mathscr{G}^{(n)}$; and

(iii) $S^{(n)} - S_0^{(n)} = o_P(n^{-d})$ under $\mathscr{P}^{(n)}$, as $n \to \infty$.

It is a direct consequence of Proposition 5.9 that, except maybe in some particular cases, neither $\hat{\underset{\sim}{\mathbf{t}}}_{i;f}^{(n)}$ ($\hat{\underset{\sim}{\mathbf{t}}}_{i;f}^{(n)+}$) nor $\hat{\underset{\sim}{r}}_{i;f}^{(n)}$ ($\hat{\underset{\sim}{r}}_{i;f}^{(n)+}$) are asymptotically invariant—since $[\mathbf{v}(n)]^{-1}(\hat{\boldsymbol{\theta}}^{(n)} - \boldsymbol{\theta})$ under **(A5)** in general is a nondegenerate $O_P(1)$ sequence as $n \to \infty$; invariance here and in the sequel of course is with respect to the groups of order-preserving transformations described in Subsection 2.2. However, asymptotically invariant aligned rank statistics do exist, as shown in Lemma 5.11; moreover, they fortunately will happen to be exactly the ones that are needed in order to carry on asymptotically locally optimal tests (Proposition 5.10 and Lemma 5.11).

5.4. *Asymptotically Invariant, Locally Asymptotically Most Stringent Aligned Rank Tests.* The main consequence of the LAN result in Proposition 4.1 is that for each $\boldsymbol{\theta} \in \Theta$, and for given f, the local experiments $\mathscr{E}_f^{(n)}(\boldsymbol{\theta}) = \{P_{\boldsymbol{\theta} + \mathbf{v}(n)\boldsymbol{\tau}};\ \boldsymbol{\tau} \in \mathbb{R}^{k_1}\}$ converge weakly to the k_1-dimensional Gaussian shift experiment $\mathscr{G}_f(\boldsymbol{\theta}) = \{\mathscr{N}(\boldsymbol{\Gamma}_f(\boldsymbol{\theta})\boldsymbol{\tau}, \boldsymbol{\Gamma}_f(\boldsymbol{\theta}));\ \boldsymbol{\tau} \in \mathbb{R}^{k_1}\}$. The local version of the sequence of null hypotheses $H_f^{(n)}(\boldsymbol{\Omega}; \boldsymbol{\theta}_0)$ is the sequence of linear hypotheses $\mathbf{v}(n)\boldsymbol{\tau} \in \mathscr{M}(\boldsymbol{\Omega})$, i.e., denoting by $\boldsymbol{\mu}_\Delta$ the mean of a random variable $\boldsymbol{\Delta}$ with distribution in $\mathscr{G}_f(\boldsymbol{\theta})$, and dropping the subscript n in $\mathbf{v}(n)$,

$$\boldsymbol{\mu}_\Delta \in \mathscr{M}(\boldsymbol{\Gamma}_f(\boldsymbol{\theta})\mathbf{v}^{-1}\boldsymbol{\Omega}) = \mathscr{M}\left(\boldsymbol{\Gamma}(\boldsymbol{\theta})\begin{pmatrix}\mathbf{K}^{(n)} & \mathbf{0}\\ \mathbf{0} & \mathbf{I}\end{pmatrix}^{-1}\boldsymbol{\Omega}\right) \tag{5.25a}$$

or, equivalently,

$$[\boldsymbol{\Gamma}_f(\boldsymbol{\theta})\mathbf{v}^{-1}\boldsymbol{\Omega}]'_\perp \boldsymbol{\mu}_\Delta = \mathbf{0}, \qquad \text{or} \qquad \left[\boldsymbol{\Gamma}(\boldsymbol{\theta})\begin{pmatrix}\mathbf{K}^{(n)} & \mathbf{0}\\ \mathbf{0} & \mathbf{I}\end{pmatrix}^{-1}\boldsymbol{\Omega}\right]'_\perp \boldsymbol{\mu}_\Delta = \mathbf{0}. \tag{5.25b}$$

Since loss functions in hypothesis testing are finite, weak convergence implies that the convergence of risk functions associated with $\mathscr{E}_f^{(n)}(\boldsymbol{\theta})$ to the corresponding risk functions associated with $\mathscr{G}_f(\boldsymbol{\theta})$, for given $\boldsymbol{\theta}$, f, and $\boldsymbol{\tau}$, is uniform among all possible testing procedures (LeCam, 1990, Chap. 2; LeCam, 1986, 7.4). Though the tests derived below vary with the possible values $\mathbf{v}$ of $\mathbf{v}(n)$ (actually, with the values of $\mathbf{K}^{(n)}$), their optimality properties thus are valid uniformly with respect to $\mathbf{v}$ which accordingly can be considered fixed in the derivation of optimal procedures.

Optimal tests in the Gaussian shift model for linear hypotheses of the form (5.25) are well known to follow from the application of likelihood ratio procedures (Wald, 1943; Wolfowitz, 1949; LeCam, 1960, Sect. 7; LeCam, 1986, 11.9). Optimality here means *stringency*, and also similarity and unbiasedness; equivalently, it can be rephrased in terms of best constant or average power over adequate hypersurfaces (in $\mathbb{R}^{k_1}$), or minimaxity against adequate alternatives. The optimality properties (in the Gaussian shift experiment) of such tests then can be translated into local asymptotic optimality properties (see LeCam, 1960, Sect. 7; 1986, 11.9) for the original problem. Most interestingly, the resulting test statistic happens to be an asymptotically invariant aligned-rank statistic; as already emphasized, without this asymptotic invariance property, there would be extremely little theoretical justification in considering aligned rank procedures. Proposition 5.10 summarizes all these conclusions.

PROPOSITION 5.10. *Assume that* **(A1)** ((**A1**$^+$)) (i)–(v), **(A2)**, *and* **(A3)** ((**A3**$^+$)) *hold under* $H^{(n)}(\mathbf{\Omega};\boldsymbol{\theta}_0)$ (*under* $H_+^{(n)}(\mathbf{\Omega};\boldsymbol{\theta}_0)$), *and that* f, $\mathbf{\Omega}$, *and* $\hat{\boldsymbol{\theta}}^{(n)}$ *satisfy* **(A1)** ((**A1**$^+$)) (i)–(vi), **(A4)**, *and* **(A5)**, *respectively. Denote by* $\mathring{\mathbf{M}}(\boldsymbol{\theta})$ *the full-rank* $(\pi+p_0+q_0)\times(p_0+q_0)$ *matrix resulting from* $\mathbf{M}(\boldsymbol{\theta})$ *by deleting columns* $p_0+1, \ldots, p_1$ *and* $p_1+q_0+1, \ldots, p_1+q_1$; *similarly, denote by* $\mathring{\mathbf{\Omega}}_{II}$ *the full-rank* $(p_0+q_0)\times(p_1+q_1-r_2)$ *matrix resulting from* $\mathbf{\Omega}_{II}$ *by deleting rows* $p_0+1, \ldots, p_1$ *and* $p_1+q_0+1, \ldots, p_1+q_1$. *Let*

$$\mathbf{Q}^{(n)}(\boldsymbol{\theta})=\begin{pmatrix}\mathbf{Q}_I^{(n)}(\boldsymbol{\theta}) & \mathbf{0}\\ \mathbf{0} & \mathbf{Q}_{II}(\boldsymbol{\theta})\end{pmatrix},$$

with

$$\mathbf{Q}_I^{(n)}(\boldsymbol{\theta})=\mathbf{\Gamma}_I^{-1}(\boldsymbol{\theta})-(\mathbf{K}^{(n)})^{-1}\mathbf{\Omega}_I\{\mathbf{\Omega}_I'[(\mathbf{K}^{(n)})^{-1}]'\,\mathbf{\Gamma}_I(\boldsymbol{\theta})(\mathbf{K}^{(n)})^{-1}\mathbf{\Omega}_I\}^{-1}\times\mathbf{\Omega}_I'[(\mathbf{K}^{(n)})^{-1}]'$$

(*where* $\mathbf{\Gamma}_I(\boldsymbol{\theta})=[\mathscr{I}(f_1)]^{-1}\mathbf{\Gamma}_{I;f}(\boldsymbol{\theta})$, $\mathbf{\Gamma}_{II}(\boldsymbol{\theta})=[\mathscr{I}(f_1)]^{-1}\mathbf{\Gamma}_{II;f}(\boldsymbol{\theta})$), *and*

$$\mathbf{Q}_{II}(\boldsymbol{\theta})=\begin{pmatrix}\mathbf{I} & \mathbf{0}\\ \mathbf{0} & \mathbf{w}_\psi^2\end{pmatrix}^{-1}-\begin{pmatrix}\mathbf{I} & \mathbf{0}\\ \mathbf{0} & \mathbf{C}_\psi\end{pmatrix}^{-1}\mathring{\mathbf{M}}(\boldsymbol{\theta})\mathring{\mathbf{\Omega}}_{II}\times\left[\mathring{\mathbf{\Omega}}'\mathring{\mathbf{M}}'(\boldsymbol{\theta})\begin{pmatrix}\mathbf{I} & \mathbf{0}\\ \mathbf{0} & \mathbf{C}_\psi\mathbf{w}_\psi^{-2}\mathbf{C}_\psi'\end{pmatrix}^{-1}\mathring{\mathbf{M}}(\boldsymbol{\theta})\mathring{\mathbf{\Omega}}_{II}\right]\times\mathring{\mathbf{\Omega}}_{II}'\mathring{\mathbf{M}}'(\boldsymbol{\theta})\begin{pmatrix}\mathbf{I} & \mathbf{0}\\ \mathbf{0} & \mathbf{C}_\psi'\end{pmatrix}^{-1}.$$

Consider the sequence of aligned rank tests $\underset{\sim}{\phi}_f^{(n)}$ *rejecting* $H^{(n)}(\mathbf{\Omega}; \mathbf{\theta}_0)$ *whenever*

$$n(\underset{\sim}{\mathbf{T}}_{\psi;f}^{(n)}(\hat{\mathbf{\theta}}^{(n)}))' \, \mathbf{Q}^{(n)}(\hat{\mathbf{\theta}}^{(n)}) \, \underset{\sim}{\mathbf{T}}_{\psi;f}^{(n)}(\hat{\mathbf{\theta}}^{(n)}) > \chi^2_{r;1-\alpha} \tag{5.26}$$

and the sequence of aligned signed rank tests $\underset{\sim}{\phi}_f^{(n)+}$ *rejecting* $H_+^{(n)}(\mathbf{\Omega}; \mathbf{\theta}_0)$ *whenever*

$$n(\underset{\sim}{\mathbf{T}}_{\psi;f}^{(n)+}(\hat{\mathbf{\theta}}^{(n)}))' \, \mathbf{Q}^{(n)}(\hat{\mathbf{\theta}}^{(n)}) \, \underset{\sim}{\mathbf{T}}_{\psi;f}^{(n)+}(\hat{\mathbf{\theta}}^{(n)}) > \chi^2_{r;1-\alpha}, \tag{5.27}$$

where $\chi^2_{r;1-\alpha}$ *denotes (any quantity converging to) the* $(1-\alpha)$*-quantile of the chi-square distribution with* r *degrees of freedom,* $r = r_1 + r_2$*. Then*

(i) $\underset{\sim}{\phi}_f^{(n)}$ $(\underset{\sim}{\phi}_f^{(n)+})$ *is locally asymptotically most stringent, for* $H^{(n)}(\mathbf{\Omega}; \mathbf{\theta}_0)$ *and for* $H_+^{(n)}(\mathbf{\Omega}; \mathbf{\theta}_0)$*, respectively, against* $\bigcup \{H_f^{(n)}(\mathbf{\theta}); \mathbf{\theta} - \mathbf{\theta}_0 \notin \mathcal{M}(\mathbf{\Omega}); f(\cdot) = \sigma^{-1} f(\cdot/\sigma), \sigma \in \mathbb{R}^+\}$*, at probability level* α*;*

(ii) $\underset{\sim}{\phi}_f^{(n)}$ *and* $\underset{\sim}{\phi}_f^{(n)+}$ *are locally asymptotically* α*-similar, under* $H^{(n)}(\mathbf{\Omega}; \mathbf{\theta}_0)$ *and* $H_+^{(n)}(\mathbf{\Omega}; \mathbf{\theta}_0)$*, respectively;*

(iii) *the test statistics in (5.26) and (5.27) are asymptotically invariant, under* $H^{(n)}(\mathbf{\Omega}; \mathbf{\theta}_0)$ *and* $H_+^{(n)}(\mathbf{\Omega}; \mathbf{\theta}_0)$*, respectively;*

(iv) *for given values of* p_0 *and* q_0*, the statistics in (5.26) and (5.27) depend on* p_1 *and* q_1 *only through* $\pi = \max(p_1 - p_0, q_1 - q_0)$*.*

In order to prove Proposition 5.10, we need the following three lemmas.

LEMMA 5.11. *Assume that* f *satisfies* **(A1)** *or* **(A1$^+$)**(i)–(v)*, and that* **(A2)**, **(A3)** *or* **(A3$^+$)**, **(A4)**, *and* **(A5)** *hold. Then the* $(k + \pi_0 - r)$*-dimensional aligned rank statistics*

$$\left[\mathbf{\Gamma}(\hat{\mathbf{\theta}}^{(n)}) \begin{pmatrix} \mathbf{K}^{(n)} & \mathbf{0} \\ \mathbf{0} & \mathbf{I} \end{pmatrix}^{-1} \mathbf{\Omega}\right]_{\perp}' \underset{\sim}{\mathbf{\Delta}}_f^{(n)}(\hat{\mathbf{\theta}}^{(n)}) \tag{5.28}$$

and

$$\left[\mathbf{\Gamma}(\hat{\mathbf{\theta}}^{(n)}) \begin{pmatrix} \mathbf{K}^{(n)} & \mathbf{0} \\ \mathbf{0} & \mathbf{I} \end{pmatrix}^{-1} \mathbf{\Omega}\right]_{\perp}' \underset{\sim}{\mathbf{\Delta}}_f^{(n)+}(\hat{\mathbf{\theta}}^{(n)}) \tag{5.29}$$

are asymptotically invariant, under $H^{(n)}(\mathbf{\Omega}; \mathbf{\theta}_0)$ *and* $H_+^{(n)}(\mathbf{\Omega}; \mathbf{\theta}_0)$*, respectively, as* $n \to \infty$*.*

Proof of Lemma 5.11. First consider $\hat{\underset{\sim}{\mathbf{\Delta}}}_{I;f}^{(n)} = \underset{\sim}{\mathbf{\Delta}}_{I;f}^{(n)}(\hat{\mathbf{\theta}}^{(n)})$ under $H_g^{(n)}(\mathbf{\theta})$. For fixed κ (an n sufficiently large, so that $\kappa(n) > \kappa$),

$$\begin{aligned}
\hat{\underset{\sim}{\mathbf{\Delta}}}_{I;f}^{(n)} - \underset{\sim}{\mathbf{\Delta}}_{I;f}^{(n)} &= n^{1/2}[\mathcal{I}(f_1)]^{1/2} (\hat{\underset{\sim}{\mathbf{T}}}_{I;f}^{(n)} - \underset{\sim}{\mathbf{T}}_{I;f}^{(n)}) \\
&= [\mathcal{I}(f_1)]^{1/2} \left\{ \sum_{i=0}^{\kappa} (n-i)^{1/2} [\hat{c}_i^{(n)} \hat{\underset{\sim}{\mathbf{t}}}_{i;f}^{(n)} - c_i \underset{\sim}{\mathbf{t}}_{i;f}^{(n)}] \right. \\
&\quad \left. + \sum_{i=\kappa+1}^{\kappa(n)} (n-i)^{1/2} [\hat{c}_i^{(n)} \hat{\underset{\sim}{\mathbf{t}}}_{i;f}^{(n)} - c_i \underset{\sim}{\mathbf{t}}_{i;f}^{(n)}] \right\}
\end{aligned}$$

$$= -\sigma_g^{-1}\mathscr{I}(f_1, g_1)\,\mathbf{K}' \sum_{i=0}^{\infty}\sum_{j=0}^{\infty} c_i c_j \mathbf{R}_{|i-j|}(\mathbf{D}^{(n)})^{1/2}\, n^{1/2}(\hat{\boldsymbol{\beta}}^{(n)} - \boldsymbol{\beta})$$

$$+ [\mathscr{I}(f_1)]^{1/2} \sum_{i=\kappa+1}^{\kappa(n)} (n-i)^{1/2}\,[\hat{c}_i^{(n)}\hat{\underset{\sim}{\mathbf{t}}}{}_{i;f}^{(n)} - c_i \underset{\sim}{\mathbf{t}}{}_{i;f}^{(n)}]$$

$$+ \sigma_g^{-1}\mathscr{I}(f_1, g_1)\,\mathbf{K}' \sum_{i=\kappa+1}^{\infty}\sum_{j=0}^{\infty} c_i c_j \mathbf{R}_{|i-j|}$$

$$\times (\mathbf{D}^{(n)})^{1/2}\, n^{1/2}(\hat{\boldsymbol{\beta}}^{(n)} - \boldsymbol{\beta}) + \mathbf{R}(n; \kappa),$$

where $\mathbf{R}(n; \kappa)$ for fixed κ is $o_P(1)$ in the norm as $n \to \infty$ and the sums $\sum_{\kappa+1}^{\kappa(n)}$ and $\sum_{\kappa+1}^{\infty}$ are $o_P(1)$ as $\kappa \to \infty$, uniformly in n: this results from the continuity of the $c_i(\cdot)$'s, their exponential decrease in i, (5.21) applied to the finite sum $\sum_{i=0}^{\kappa}$, and (**A5**). It follows that

$$\hat{\underset{\sim}{\boldsymbol{\Delta}}}{}_{I;f}^{(n)} - \underset{\sim}{\boldsymbol{\Delta}}{}_{I;f}^{(n)} = -\sigma_g^{-1}\mathscr{I}(f_1, g_1)\,\mathbf{K}' \sum_{i=0}^{\infty}\sum_{j=0}^{\infty} c_i c_j \mathbf{R}_{|i-j|}\mathbf{K}\mathbf{K}^{-1}$$

$$\times (\mathbf{D}^{(n)})^{1/2}\, n^{1/2}(\hat{\boldsymbol{\beta}}^{(n)} - \boldsymbol{\beta}) + o_P(1)$$

$$= -\sigma_g^{-1}\mathscr{I}(f_1, g_1)[\mathscr{I}(f_1)]^{-1}\,\boldsymbol{\Gamma}_{I;f}(\boldsymbol{\theta})(n^{-1/2}\mathbf{K}^{(n)})^{-1}$$

$$\times (\hat{\boldsymbol{\beta}}^{(n)} - \boldsymbol{\beta}) + o_P(1), \qquad \text{as} \quad n \to \infty. \tag{5.30}$$

As for $\hat{\underset{\sim}{\boldsymbol{\Delta}}}{}_{II;f}^{(n)} = \underset{\sim}{\boldsymbol{\Delta}}{}_{II;f}^{(n)}(\hat{\boldsymbol{\theta}}^{(n)})$, omitting the details related to the non-uniformity in i of (5.22) (which can be treated exactly as above, by decomposing $\sum_{i=1}^{\kappa(n)}$ into $\sum_{i=1}^{\kappa} + \sum_{i=\kappa+1}^{\kappa(n)}$), we have, from the continuity in $\boldsymbol{\theta}$ of $\mathbf{M}$, $\mathbf{C}_\psi$, the Green's functions, etc., and from (4.13) and (4.22),

$$\hat{\underset{\sim}{\boldsymbol{\Delta}}}{}_{II;f}^{(n)} - \underset{\sim}{\boldsymbol{\Delta}}{}_{II;f}^{(n)}$$

$$= [\mathscr{I}(f_1)]^{1/2}\left\{\mathbf{M}'(\hat{\boldsymbol{\theta}}^{(n)})\begin{pmatrix}\mathbf{I} & \mathbf{O}\\ \mathbf{O} & \mathbf{C}'_{\hat{\psi}}\end{pmatrix}^{-1} n^{1/2}\hat{\underset{\sim}{\mathbf{T}}}{}_{II;f}^{(n)}\right.$$

$$\left. - \mathbf{M}'(\boldsymbol{\theta})\begin{pmatrix}\mathbf{I} & \mathbf{O}\\ \mathbf{O} & \mathbf{C}'_{\psi}\end{pmatrix}^{-1} n^{1/2}\underset{\sim}{\mathbf{T}}{}_{II;f}^{(n)}\right\}$$

$$= -\sigma(f_1, g_1)\,\mathscr{I}(f_1, g_1)\,\mathbf{M}'(\boldsymbol{\theta})$$

$$\times \begin{pmatrix} \mathbf{I}_{\pi\times\pi} & \mathbf{0}_{\pi\times(\kappa(n)-\pi)} \\ \mathbf{0}_{(p_0+q_0)\times\pi} & (\mathbf{C}'_\psi)^{-1}\begin{pmatrix}\psi_{\pi+1}^{(1)} & \cdots & \psi_{\kappa(n)}^{(1)} \\ \vdots & & \vdots \\ \psi_{\pi+1}^{(p_0+q_0)} & \cdots & \psi_{\kappa(n)}^{(p_0+q_0)}\end{pmatrix}\end{pmatrix}$$

$$\times n^{1/2}\begin{pmatrix} \sum_{j=1}^{p_0} g_{1-j}(\boldsymbol{\theta})(\hat{A}_j^{(n)}-A_j)+\sum_{j=1}^{q_0} h_{1-j}(\boldsymbol{\theta})(\hat{B}_j^{(n)}-B_j) \\ \vdots \\ \sum_{j=1}^{p_0} g_{\kappa(n)-j}(\boldsymbol{\theta})(\hat{A}_j^{(n)}-A_j)+\sum_{j=1}^{q_0} h_{\kappa(n)-j}(\boldsymbol{\theta})(\hat{B}_j^{(n)}-B_j) \end{pmatrix}$$
$$+ o_P(1).$$

The elements of the random vector in this latter expression result from substituting $n^{1/2}(\hat{A}_j^{(n)}-A_j)$ and $n^{1/2}(\hat{B}_j^{(n)}-B_j)$ for $\gamma_j^{(n)}$ and $\delta_j^{(n)}$ in (4.4). From (4.25), we thus obtain

$$\hat{\underset{\sim}{\boldsymbol{\Delta}}}{}^{(n)}_{II;f}-\underset{\sim}{\boldsymbol{\Delta}}{}^{(n)}_{II;f}$$
$$= -\sigma(f_1, g_1)\,\mathscr{I}(f_1, g_1)\,\mathbf{M}'(\boldsymbol{\theta})\begin{pmatrix}\mathbf{I} & \mathbf{O}\\ \mathbf{O} & (\mathbf{C}_\psi^{-1})'\,\mathbf{w}_\psi^2\mathbf{C}_\psi^{-1}\end{pmatrix}$$
$$\times\mathbf{M}(\boldsymbol{\theta})\left[n^{1/2}\begin{pmatrix}\hat{\mathbf{A}}^{(n)}-\mathbf{A}\\ \hat{\mathbf{B}}^{(n)}-\mathbf{B}\end{pmatrix}\right]+o_P(1),$$

which, on behalf of (4.16), in turn takes the form

$$-\frac{\sigma(f_1, g_1)\,\mathscr{I}(f_1, g_1)}{\mathscr{I}(f_1)}\boldsymbol{\Gamma}_{II;f}(\boldsymbol{\theta})\left[n^{1/2}\begin{pmatrix}\hat{\mathbf{A}}^{(n)}-\mathbf{A}\\ \hat{\mathbf{B}}^{(n)}-\mathbf{B}\end{pmatrix}\right]+o_P(1). \qquad (5.31)$$

Assembling (5.30) and (5.31), we thus have shown that under $H_g^{(n)}(\boldsymbol{\theta})$, where $\boldsymbol{\theta}$ and $\hat{\boldsymbol{\theta}}^{(n)}$ satisfy (5.19),

$$\hat{\underset{\sim}{\boldsymbol{\Delta}}}{}^{(n)}_{f}-\underset{\sim}{\boldsymbol{\Delta}}{}^{(n)}_{f}$$
$$= -\begin{pmatrix}\dfrac{\mathscr{I}(f_1, g_1)}{\mathscr{I}(f_1)}\mathbf{I} & \mathbf{O}\\ \mathbf{O} & \dfrac{\sigma(f_1, g_1)\,\mathscr{I}(f_1, g_1)}{\mathscr{I}(f_1)}\mathbf{I}\end{pmatrix}\boldsymbol{\Gamma}_f(\boldsymbol{\theta})[\mathbf{v}(n)]^{-1}$$
$$\times(\hat{\boldsymbol{\theta}}^{(n)}-\boldsymbol{\theta})+o_P(1)$$
$$=\begin{pmatrix}\mathscr{I}(f_1, g_1)\mathbf{I} & \mathbf{O}\\ \mathbf{O} & \sigma(f_1, g_1)\,\mathscr{I}(f_1, g_1)\mathbf{I}\end{pmatrix}\boldsymbol{\Gamma}(\hat{\boldsymbol{\theta}}^{(n)})[\mathbf{v}(n)]^{-1}\,\boldsymbol{\Omega}\boldsymbol{\omega}^{(n)}+o_P(1)$$

for some (random) sequence $\boldsymbol{\omega}^{(n)}$ of $\mathbb{R}^{k+\pi_0-r}$. Now, due to the block-diagonal structure of $\boldsymbol{\Gamma}$, $\mathbf{v}(n)$, and $\boldsymbol{\Omega}$,

$$\mathscr{M}\left[\begin{pmatrix}\mathscr{I}(f_1, g_1)\mathbf{I} & \mathbf{O}\\ \mathbf{O} & \sigma(f_1, g_1)\,\mathscr{I}(f_1, g_1)\mathbf{I}\end{pmatrix}\boldsymbol{\Gamma}(\hat{\boldsymbol{\theta}}^{(n)})[\mathbf{v}(n)]^{-1}\,\boldsymbol{\Omega}\right]$$
$$=\mathscr{M}\left[\boldsymbol{\Gamma}(\hat{\boldsymbol{\theta}}^{(n)})\begin{pmatrix}\mathbf{K}^{(n)} & \mathbf{0}\\ \mathbf{0} & \mathbf{I}\end{pmatrix}^{-1}\boldsymbol{\Omega}\right],$$

which does not depend on g anymore. The desired asymptotic invariance property of (5.28) then follows from (5.25), and the (exact) invariance and asymptotic normality of $\underset{\sim}{\boldsymbol{\Delta}}_f^{(n)}$. The signed-rank case (5.29) is entirely similar.

LEMMA 5.12. *Let* $\boldsymbol{\Delta} \sim \mathcal{N}(\boldsymbol{\Gamma}\boldsymbol{\tau}; \boldsymbol{\Gamma})$. *A most stringent* α*-level test of* (5.25) *is* $\phi(\boldsymbol{\Delta}) = 1$ *iff*

$$\boldsymbol{\Delta}'(\mathbf{I} - \boldsymbol{\Gamma}\tilde{\boldsymbol{\Omega}}(\tilde{\boldsymbol{\Omega}}'\boldsymbol{\Gamma}\tilde{\boldsymbol{\Omega}})^{-}\,\tilde{\boldsymbol{\Omega}}'\boldsymbol{\Gamma}\boldsymbol{\Gamma}^{-})'\,\boldsymbol{\Gamma}^{-}(\mathbf{I} - \boldsymbol{\Gamma}\tilde{\boldsymbol{\Omega}}(\tilde{\boldsymbol{\Omega}}'\boldsymbol{\Gamma}\tilde{\boldsymbol{\Omega}})^{-}\,\tilde{\boldsymbol{\Omega}}'\boldsymbol{\Gamma}\boldsymbol{\Gamma}^{-})\boldsymbol{\Delta} > \chi^2_{r^0;\,1-\alpha}, \tag{5.32}$$

where $\tilde{\boldsymbol{\Omega}} = \mathbf{v}^{-1}\boldsymbol{\Omega}$, $\boldsymbol{\Gamma}^{-}$ *and* $(\tilde{\boldsymbol{\Omega}}'\boldsymbol{\Gamma}\tilde{\boldsymbol{\Omega}})^{-}$ *are arbitrary generalized inverses,* $r^0 = \text{rank}(\boldsymbol{\Gamma}) - \text{rank}(\tilde{\boldsymbol{\Omega}}'\boldsymbol{\Gamma}\tilde{\boldsymbol{\Omega}})$, *and* $\chi^2_{r;\,1-\alpha}$ *is the* $(1-\alpha)$*-quantile of the chi-square distribution with r degrees of freedom.*

The fact that $\boldsymbol{\Gamma}$ may be singular of course complicates the form of (5.32) which, in case of a full-rank $\boldsymbol{\Gamma}$ (viz. when $\min(p_1 - p_0, q_1 - q_0) = 0$) reduces to the more familiar form

$$\boldsymbol{\Delta}'(\boldsymbol{\Gamma}^{-1} - \tilde{\boldsymbol{\Omega}}(\tilde{\boldsymbol{\Omega}}'\boldsymbol{\Gamma}\tilde{\boldsymbol{\Omega}})^{-1}\,\tilde{\boldsymbol{\Omega}}')\boldsymbol{\Delta}, \tag{5.33}$$

with r degrees of freedom.

Proof. The likelihood ratio test here consists in rejecting (5.25) for large values of

$$\min_{\boldsymbol{\mu} \in \mathcal{M}(\boldsymbol{\Gamma}\tilde{\boldsymbol{\Omega}})} (\boldsymbol{\Delta} - \boldsymbol{\mu})'\,\boldsymbol{\Gamma}^{-}(\boldsymbol{\Delta} - \boldsymbol{\mu}) - \min_{\boldsymbol{\mu} \in \mathcal{M}(\boldsymbol{\Gamma})} (\boldsymbol{\Delta} - \boldsymbol{\mu})'\,\boldsymbol{\Gamma}^{-}(\boldsymbol{\Delta} - \boldsymbol{\mu}) \tag{5.34}$$

(for the likelihood of a singular distribution, see Rao and Mitra, 1971, 10.5). Since however $\boldsymbol{\Delta}$ itself with probability one belongs to $\mathcal{M}(\boldsymbol{\Gamma})$, the second term in (5.34) is zero. As for the first term, the minimum is provided by a $\boldsymbol{\Gamma}^{-}$-semi norm-least squares generalized inverse $[\boldsymbol{\Gamma}\tilde{\boldsymbol{\Omega}}]_{l(\Gamma^{-})}$ of $\boldsymbol{\Gamma}\tilde{\boldsymbol{\Omega}}$ (Rao and Mitra, 1971, 3.2), yielding (5.32). The chi-square distribution of (5.32) under (5.25) follows on checking (Rao and Mitra, 1971, Theorem 9.2.1) that, letting $\tilde{\mathbf{Q}} = \mathbf{I} - \boldsymbol{\Gamma}\tilde{\boldsymbol{\Omega}}(\tilde{\boldsymbol{\Omega}}'\boldsymbol{\Gamma}\tilde{\boldsymbol{\Omega}})^{-}\,\tilde{\boldsymbol{\Omega}}'\boldsymbol{\Gamma}\boldsymbol{\Gamma}^{-}$, under (5.25),

(a) $\boldsymbol{\Gamma}\tilde{\mathbf{Q}}'\boldsymbol{\Gamma}^{-}\tilde{\mathbf{Q}}\boldsymbol{\Gamma}\tilde{\mathbf{Q}}'\boldsymbol{\Gamma}^{-}\tilde{\mathbf{Q}}\boldsymbol{\Gamma} = \boldsymbol{\Gamma}\tilde{\mathbf{Q}}'\boldsymbol{\Gamma}^{-}\tilde{\mathbf{Q}}\boldsymbol{\Gamma}$,

(b) $\mathcal{M}(\boldsymbol{\Gamma}\tilde{\mathbf{Q}}\boldsymbol{\Gamma}\boldsymbol{\tau}) \subseteq \mathcal{M}(\boldsymbol{\Gamma}\tilde{\mathbf{Q}}\boldsymbol{\Gamma})$,

(c) $(\tilde{\mathbf{Q}}'\boldsymbol{\Gamma}^{-}\tilde{\mathbf{Q}}\boldsymbol{\Gamma}\boldsymbol{\tau})'\,\boldsymbol{\Gamma}(\tilde{\mathbf{Q}}\boldsymbol{\Gamma}^{-}\tilde{\mathbf{Q}}\boldsymbol{\Gamma}\boldsymbol{\tau}) = \boldsymbol{\tau}'\boldsymbol{\Gamma}\tilde{\mathbf{Q}}'\boldsymbol{\Gamma}^{-}\tilde{\mathbf{Q}}\boldsymbol{\Gamma}\boldsymbol{\tau}$,

(d) $\text{tr}(\tilde{\mathbf{Q}}'\boldsymbol{\Gamma}^{-}\tilde{\mathbf{Q}}\boldsymbol{\Gamma}) = r^0$,

(e) $\boldsymbol{\tau}'\boldsymbol{\Gamma}(\tilde{\mathbf{Q}}'\boldsymbol{\Gamma}^{-}\tilde{\mathbf{Q}})'\,\boldsymbol{\Gamma}(\tilde{\mathbf{Q}}'\boldsymbol{\Gamma}^{-}\tilde{\mathbf{Q}})\,\boldsymbol{\Gamma}(\tilde{\mathbf{Q}}'\boldsymbol{\Gamma}^{-}\tilde{\mathbf{Q}})\boldsymbol{\Gamma}\boldsymbol{\tau} = \boldsymbol{\tau}'\boldsymbol{\Gamma}\tilde{\mathbf{Q}}'\boldsymbol{\Gamma}^{-}\tilde{\mathbf{Q}}\boldsymbol{\Gamma}\boldsymbol{\tau} = \mathbf{0}$.

All these five conditions follow from standard computation and application of the usual properties of generalized inverses (Rao and Mitra, 1971, 2.2); note that, since (5.32) is invariant for any choice of the generalized inverses

Γ^- and $(\tilde{\mathbf{\Omega}}'\mathbf{\Gamma}\tilde{\mathbf{\Omega}})^-$, they (in the above proof of (a)–(e)) can be assumed to be reflexive generalized inverses.

LEMMA 5.13. *Assume that the assumptions of Proposition* 5.10 *hold. Letting*

$$\bar{\mathbf{Q}}^{(n)}(\boldsymbol{\theta}) = \begin{pmatrix} \mathbf{Q}_I^{(n)}(\boldsymbol{\theta}) & \mathbf{0} \\ \mathbf{0} & \bar{\mathbf{Q}}_{II}(\boldsymbol{\theta}) \end{pmatrix},$$

with $\mathbf{Q}_I^{(n)}(\boldsymbol{\theta})$ *as defined in Proposition* 5.10, *and*

$$\begin{aligned}\bar{\mathbf{Q}}_{II}^{(n)}(\boldsymbol{\theta}) = {} & [\mathbf{I} - \mathbf{\Gamma}_{II}(\boldsymbol{\theta})\,\mathbf{\Omega}_{II}(\mathbf{\Omega}'_{II}\mathbf{\Gamma}_{II}(\boldsymbol{\theta})\,\mathbf{\Omega}_{II})^-\,\mathbf{\Omega}'_{II}\mathbf{\Gamma}_{II}(\boldsymbol{\theta})\,\mathbf{\Gamma}_{II}^-(\boldsymbol{\theta})]' \\ & \times \mathbf{\Gamma}_{II}^-[\mathbf{I} - \mathbf{\Gamma}_{II}(\boldsymbol{\theta})\,\mathbf{\Omega}_{II}(\mathbf{\Omega}'_{II}\mathbf{\Gamma}_{II}(\boldsymbol{\theta})\,\mathbf{\Omega}_{II})^-\,\mathbf{\Omega}'_{II}\mathbf{\Gamma}_{II}(\boldsymbol{\theta})\,\mathbf{\Gamma}_{II}^-(\boldsymbol{\theta})]',\end{aligned} \tag{5.35}$$

denote by $\bar{\phi}_f^{(n)}$ *the sequence of aligned rank tests rejecting* $H^{(n)}(\mathbf{\Omega}; \boldsymbol{\theta}_0)$ *whenever*

$$[\mathscr{I}(f_1)]^{-1}\,(\mathbf{\Delta}_f^{(n)}(\hat{\boldsymbol{\theta}}^{(n)}))'\,\bar{\mathbf{Q}}^{(n)}(\hat{\boldsymbol{\theta}}^{(n)})\,\mathbf{\Delta}_f^{(n)}(\hat{\boldsymbol{\theta}}^{(n)}) > \chi^2_{r^0;\,1-\alpha}, \tag{5.36}$$

where $r^0 = \text{rank}(\mathbf{\Gamma}(\boldsymbol{\theta})) - \text{rank}(\mathbf{\Omega}'(\mathbf{v}'(n))^{-1}\,\mathbf{\Gamma}(\boldsymbol{\theta})(\mathbf{v}(n))^{-1}\,\mathbf{\Omega})$. *Then Proposition* 5.10(i)–(iii) *hold for* $\bar{\phi}_f^{(n)}$ (r^0 *will be shown equal to r in the proof of Proposition* 5.10).

Proof of Lemma 5.13. (i) It follows from the classical application to hypothesis testing problems of LAN and the weak convergence of $\mathscr{E}_f^{(n)}(\boldsymbol{\theta})$ to $\mathscr{G}_f(\boldsymbol{\theta})$ (LeCam, 1960, Sect. 7; 1986, 11.9), that the stringency of the test $\phi(\mathbf{\Delta})$ described in Lemma 5.12 extends, in local, asymptotic form, to the tests $\phi(\mathbf{\Delta}_f^{(n)}(\hat{\boldsymbol{\theta}}^{(n)}))$ obtained from (5.32) on substituting $\mathbf{\Delta}_f^{(n)}(\hat{\boldsymbol{\theta}}^{(n)})$ for $\mathbf{\Delta}$ and $\mathbf{\Gamma}_f(\hat{\boldsymbol{\theta}}^{(n)})$ for $\mathbf{\Gamma}$. Due to the special block-diagonal structure of the matrices in (5.32), the resulting quadratic form is invariant if $\mathbf{v}(n)$ is replaced with $\left(\begin{smallmatrix}\mathbf{K}^{(n)} & \mathbf{0} \\ \mathbf{0} & \mathbf{I}\end{smallmatrix}\right)$, and breaks into two parts. The first one is a full-rank quadratic form in the nonserial part $\mathbf{\Delta}_{I;f}^{(n)}$ of $\mathbf{\Delta}_f^{(n)}$; it takes (see (5.33)) the simpler form

$$[\mathscr{I}(f_1)]^{-1}\,(\mathbf{\Delta}_{I;f}^{(n)}(\hat{\boldsymbol{\theta}}^{(n)}))'\,\mathbf{Q}_I^{(n)}(\hat{\boldsymbol{\theta}}^{(n)})(\mathbf{\Delta}_{I;f}^{(n)}(\hat{\boldsymbol{\theta}}^{(n)})). \tag{5.37}$$

The second part involves the serial components $\mathbf{\Delta}_{II;f}^{(n)}$ of $\mathbf{\Delta}_f^{(n)}$; whenever $\mathbf{\Gamma}_{II;f}$ is singular, the more general form

$$[\mathscr{I}(f_1)]^{-1}\,(\mathbf{\Delta}_{II;f}^{(n)}(\hat{\boldsymbol{\theta}}^{(n)}))'\,\bar{\mathbf{Q}}_{II}(\hat{\boldsymbol{\theta}}^{(n)})(\mathbf{\Delta}_{II;f}^{(n)}(\hat{\boldsymbol{\theta}}^{(n)})) \tag{5.38}$$

has to be maintained. Adding (5.37) and (5.38) yields the quadratic form in (5.36): $\bar{\phi}_f^{(n)}$ thus is locally asymptotically most stringent for $H_f^{(n)}(\mathbf{\Omega}; \boldsymbol{\theta}_0)$ against $\bigcup \{H_f^{(n)}(\boldsymbol{\theta}); \boldsymbol{\theta} - \boldsymbol{\theta}_0 \notin \mathscr{M}(\mathbf{\Omega})\}$. That this property can be extended to $H^{(n)}(\mathbf{\Omega}; \boldsymbol{\theta}_0)$ or $H_+^{(n)}(\mathbf{\Omega}; \boldsymbol{\theta}_0)$ against $\bigcup \{H_f^{(n)}(\boldsymbol{\theta}); \boldsymbol{\theta} - \boldsymbol{\theta}_0 \notin \mathscr{M}(\mathbf{\Omega});\ f(\cdot) = \sigma^{-1} f(\cdot/\sigma),\ \sigma \in \mathbb{R}^+\}$, is a consequence of asymptotic invariance, to be proved in (iii) below, and of the fact that $\bar{\phi}_f^{(n)}$ depends on the density type of f only, i.e., on f_1.

(ii) This is established, in a general setting, by LeCam (1960, Sect. 7), under the terminology *differential asymptotic similarity*.

(iii) Referring to Lemma 5.11, a sufficient condition for the asymptotic invariance of the test statistic in (5.36) is that it be expressible as a quadratic form in (5.28). This in turn holds provided the rows of $\bar{\mathbf{Q}}^{(n)}(\hat{\boldsymbol{\theta}}^{(n)})$ belong to $\mathscr{M}[\boldsymbol{\Gamma}(\hat{\boldsymbol{\theta}}^{(n)})\begin{pmatrix}\mathbf{K}^{(n)} & \mathbf{0}\\ \mathbf{0} & \mathbf{I}\end{pmatrix}^{-1}\boldsymbol{\Omega}]_\perp$; i.e., provided that $\bar{\mathbf{Q}}^{(n)}(\hat{\boldsymbol{\theta}}^{(n)})\,\boldsymbol{\Gamma}(\hat{\boldsymbol{\theta}}^{(n)})\begin{pmatrix}\mathbf{K}^{(n)} & \mathbf{0}\\ \mathbf{0} & \mathbf{I}\end{pmatrix}^{-1}\boldsymbol{\Omega}=\mathbf{0}$. Dropping all subscripts, this latter quantity takes the form

$$\boldsymbol{\Gamma}\begin{pmatrix}\mathbf{K}^{-1} & \mathbf{0}\\ \mathbf{0} & \mathbf{I}\end{pmatrix}\boldsymbol{\Omega}-\boldsymbol{\Gamma}\begin{pmatrix}\mathbf{K}^{-1} & \mathbf{0}\\ \mathbf{0} & \mathbf{I}\end{pmatrix}\boldsymbol{\Omega}\left(\boldsymbol{\Omega}\begin{pmatrix}\mathbf{K}^{-1} & \mathbf{0}\\ \mathbf{0} & \mathbf{I}\end{pmatrix}'\boldsymbol{\Gamma}\begin{pmatrix}\mathbf{K}^{-1} & \mathbf{0}\\ \mathbf{0} & \mathbf{I}\end{pmatrix}\boldsymbol{\Omega}\right)^{-}$$
$$\times\boldsymbol{\Omega}'\begin{pmatrix}\mathbf{K}^{-1} & \mathbf{0}\\ \mathbf{0} & \mathbf{I}\end{pmatrix}'\boldsymbol{\Gamma}\begin{pmatrix}\mathbf{K}^{-1} & \mathbf{0}\\ \mathbf{0} & \mathbf{I}\end{pmatrix}\boldsymbol{\Omega}$$

which, applying Lemma 2.2.6 of Rao and Mitra (1971), finally reduces to

$$\boldsymbol{\Gamma}\begin{pmatrix}\mathbf{K}^{-1} & \mathbf{0}\\ \mathbf{0} & \mathbf{I}\end{pmatrix}\boldsymbol{\Omega}-\boldsymbol{\Gamma}\begin{pmatrix}\mathbf{K}^{-1} & \mathbf{0}\\ \mathbf{0} & \mathbf{I}\end{pmatrix}\boldsymbol{\Omega}=\mathbf{0},$$

as was to be shown.

We now turn to the proof of the main proposition.

Proof of Proposition 5.10. In order to prove (i)–(iii) (in the unsigned case), it is sufficient, in view of Lemma 5.13, to show that $\phi_f^{(n)}=\bar{\phi}_f^{(n)}$, i.e., (5.26) and (5.36) actually are the same test. The test statistics in both cases clearly are a sum of two quadratic forms, the first of which (dealing with the nonserial parts of $\mathbf{T}_{\psi;f}^{(n)}$ and $\boldsymbol{\Delta}_f^{(n)}$) obviously coincide. For the second of them (dealing with the serial parts $\mathbf{T}_{II;f}^{(n)}$ and $\boldsymbol{\Delta}_{II;f}^{(n)}$), noting that, due to the fact that $\mathbf{M}[\mathbf{M}'(\cdots)\mathbf{M}]^{-}[\mathbf{M}'(\cdots)\mathbf{M}]=\mathbf{M}$ (Rao and Mitra, 1971, Lemma 2.26), $\mathbf{M}\boldsymbol{\Gamma}_{II}^{-}\boldsymbol{\Gamma}_{II}=\mathbf{M}$ and $\boldsymbol{\Gamma}_{II}\boldsymbol{\Gamma}_{II}^{-}\mathbf{M}'=\mathbf{M}'$, we obtain, after some routine algebra,

$$\begin{aligned}
&[\mathscr{I}(f_1)]^{-1}(\hat{\boldsymbol{\Delta}}_{II;f}^{(n)})'\,\bar{\mathbf{Q}}_{II}(\hat{\boldsymbol{\theta}}^{(n)})\,\hat{\boldsymbol{\Delta}}_{II;f}^{(n)}\\
&=n(\hat{\mathbf{T}}_{II;f}^{(n)})'\begin{pmatrix}\mathbf{I} & \mathbf{0}\\ \mathbf{0} & \mathbf{C}_{\hat{\psi}}\end{pmatrix}^{-1}\hat{\mathbf{M}}\bar{\mathbf{Q}}_{II}(\hat{\boldsymbol{\theta}}^{(n)})\,\hat{\mathbf{M}}'\begin{pmatrix}\mathbf{I} & \mathbf{0}\\ \mathbf{0} & \mathbf{C}'_{\hat{\psi}}\end{pmatrix}^{-1}\mathbf{T}_{II;f}^{(n)}\\
&=n(\hat{\mathbf{T}}_{II;f}^{(n)})'\begin{pmatrix}\mathbf{I} & \mathbf{0}\\ \mathbf{0} & \mathbf{C}_{\hat{\psi}}\end{pmatrix}^{-1}\\
&\quad\times\hat{\mathbf{M}}\left\{\hat{\boldsymbol{\Gamma}}_{II}^{-}-\boldsymbol{\Omega}_{II}\left[\boldsymbol{\Omega}'_{II}\hat{\mathbf{M}}'\begin{pmatrix}\mathbf{I} & \mathbf{0}\\ \mathbf{0} & \mathbf{C}_{\hat{\psi}}\hat{\mathbf{w}}^{-2}\mathbf{C}'_{\hat{\psi}}\end{pmatrix}^{-1}\hat{\mathbf{M}}\boldsymbol{\Omega}_{II}\right]^{-}\boldsymbol{\Omega}'_{II}\right\}\\
&\quad\times\hat{\mathbf{M}}'\begin{pmatrix}\mathbf{I} & \mathbf{0}\\ \mathbf{0} & \mathbf{C}'_{\hat{\psi}}\end{pmatrix}^{-1}\hat{\mathbf{T}}_{II;f}^{(n)}.
\end{aligned}\tag{5.39}$$

Still from Rao and Mitra (1971, Lemma 2.2.5),

$$\begin{pmatrix}\mathbf{I} & \mathbf{0}\\ \mathbf{0} & \mathbf{C}_{\hat{\psi}}\end{pmatrix}^{-1}\hat{\mathbf{M}}\hat{\mathbf{\Gamma}}_{II}^{-}\hat{\mathbf{M}}'\begin{pmatrix}\mathbf{I} & \mathbf{0}\\ \mathbf{0} & \mathbf{C}'_{\hat{\psi}}\end{pmatrix}^{-1}$$
$$=\begin{pmatrix}\mathbf{I} & \mathbf{0}\\ \mathbf{0} & \mathbf{C}_{\hat{\psi}}\end{pmatrix}^{-1}\hat{\mathbf{M}}\left[\hat{\mathbf{M}}'\begin{pmatrix}\mathbf{I} & \mathbf{0}\\ \mathbf{0} & \mathbf{C}_{\hat{\psi}}\hat{\mathbf{w}}^{-2}\mathbf{C}'_{\hat{\psi}}\end{pmatrix}^{-1}\hat{\mathbf{M}}\right]^{-}\hat{\mathbf{M}}'\begin{pmatrix}\mathbf{I} & \mathbf{0}\\ \mathbf{0} & \mathbf{C}'_{\hat{\psi}}\end{pmatrix}^{-1}$$
$$=\begin{pmatrix}\mathbf{I} & \mathbf{0}\\ \mathbf{0} & \hat{\mathbf{w}}^{2}\end{pmatrix}^{-1}. \tag{5.40}$$

On the other hand, since $\mathbf{\Omega}_{II}$ has zeros in rows $p_0+1, \ldots, p_1$ and $p_1+q_0+1, \ldots, p_1+q_1$,

$$\hat{\mathbf{M}}\mathbf{\Omega}_{II}\left[\mathbf{\Omega}'_{II}\hat{\mathbf{M}}'\begin{pmatrix}\mathbf{I} & \mathbf{0}\\ \mathbf{0} & \mathbf{C}_{\hat{\psi}}\hat{\mathbf{w}}^{-2}\mathbf{C}'_{\hat{\psi}}\end{pmatrix}^{-1}\hat{\mathbf{M}}\mathbf{\Omega}_{II}\right]^{-1}\mathbf{\Omega}'_{II}\hat{\mathbf{M}}'$$
$$=\mathring{\mathbf{M}}(\hat{\boldsymbol{\theta}}^{(n)})\,\mathring{\mathbf{\Omega}}_{II}\left[\mathring{\mathbf{\Omega}}'_{II}\mathring{\mathbf{M}}'(\hat{\boldsymbol{\theta}}^{(n)})\begin{pmatrix}\mathbf{I} & \mathbf{0}\\ \mathbf{0} & \mathbf{C}_{\hat{\psi}}\hat{\mathbf{w}}^{-2}\mathbf{C}'_{\hat{\psi}}\end{pmatrix}^{-1}\right.$$
$$\left.\times\mathring{\mathbf{M}}(\hat{\boldsymbol{\theta}}^{(n)})\mathring{\mathbf{\Omega}}_{II}\right]^{-}\mathring{\mathbf{\Omega}}'_{II}\mathring{\mathbf{M}}'(\hat{\boldsymbol{\theta}}^{(n)}). \tag{5.41}$$

Substituting (5.40) and (5.41) into (5.39) yields the desired result that

$$[\mathscr{I}(f_1)]^{-1}(\hat{\underset{\sim}{\mathbf{\Delta}}}{}^{(n)}_{II;f})'\,\bar{\mathbf{Q}}_{II}(\hat{\boldsymbol{\theta}}^{(n)})\,\hat{\underset{\sim}{\mathbf{\Delta}}}{}^{(n)}_{II;f}=n(\hat{\underset{\sim}{\mathbf{T}}}{}^{(n)}_{II;f})'\,\mathbf{Q}_{II}(\hat{\boldsymbol{\theta}}^{(n)})\hat{\underset{\sim}{\mathbf{T}}}{}^{(n)}_{II;f}. \tag{5.42}$$

The asymptotic invariance (Lemma 5.13) of $[\mathscr{I}(f_1)]^{-1}(\hat{\underset{\sim}{\mathbf{\Delta}}}{}^{(n)}_{f})'\,\bar{\mathbf{Q}}(\hat{\boldsymbol{\theta}}^{(n)})\hat{\underset{\sim}{\mathbf{\Delta}}}{}^{(n)}_{f}$ of course implies that of $n(\hat{\underset{\sim}{\mathbf{T}}}{}^{(n)}_{\hat{\psi};f})'\,\mathbf{Q}^{(n)}(\hat{\boldsymbol{\theta}}^{(n)})\hat{\underset{\sim}{\mathbf{T}}}{}^{(n)}_{\hat{\psi};f}$, so that their common asymptotic distribution under $H^{(n)}(\mathbf{\Omega};\boldsymbol{\theta}_0)$, as well as under contiguous alternatives, is the same as that of $n(\underset{\sim}{\mathbf{T}}{}^{(n)}_{\psi;f})'\,\mathbf{Q}^{(n)}(\boldsymbol{\theta})\underset{\sim}{\mathbf{T}}{}^{(n)}_{\psi;f}$. The nonserial part of this latter quadratic form contributes r_1 degrees of freedom, since $n^{1/2}\underset{\sim}{\mathbf{T}}{}^{(n)}_{I;f}$ under $H^{(n)}(\boldsymbol{\theta})$ is asymptotically normal, with mean zero and positive definite covariance matrix $\mathscr{I}(f_1)\,\mathbf{\Gamma}_I(\boldsymbol{\theta})$, and $[\mathscr{I}(f_1)]^{-1}\,\mathbf{\Gamma}_I^{1/2}(\boldsymbol{\theta})\,\mathbf{Q}_I^{(n)}(\boldsymbol{\theta})\,\mathbf{\Gamma}_I^{1/2}(\boldsymbol{\theta})$ is easily checked to be idempotent with rank r_1. Similarly, $n^{1/2}\underset{\sim}{\mathbf{T}}{}^{(n)}_{II;f}$ has positive definite asymptotic covariance matrix $\left(\begin{smallmatrix}\mathbf{I} & \mathbf{0}\\ \mathbf{0} & \mathbf{w}_\psi^2\end{smallmatrix}\right)$, and the rank of the idempotent $\left(\begin{smallmatrix}\mathbf{I} & \mathbf{0}\\ \mathbf{0} & \mathbf{w}_\psi\end{smallmatrix}\right)\mathbf{Q}_{II}(\boldsymbol{\theta})\left(\begin{smallmatrix}\mathbf{I} & \mathbf{0}\\ \mathbf{0} & \mathbf{w}_\psi\end{smallmatrix}\right)$ is

$$\pi+p_0+q_0-\operatorname{tr}\left\{\left[\mathring{\mathbf{\Omega}}'_{II}\mathring{\mathbf{M}}'\begin{pmatrix}\mathbf{I} & \mathbf{0}\\ \mathbf{0} & \mathbf{C}_{\psi}\mathbf{w}_{\psi}^{-2}\mathbf{C}_{\psi}\end{pmatrix}^{-1}\mathring{\mathbf{M}}\mathring{\mathbf{\Omega}}_{II}\right]\right.$$
$$\left.\times\left[\mathring{\mathbf{\Omega}}'_{II}\mathring{\mathbf{M}}'\begin{pmatrix}\mathbf{I} & \mathbf{0}\\ \mathbf{0} & \mathbf{C}_{\psi}\mathbf{w}_{\psi}^{-2}\mathbf{C}_{\psi}\end{pmatrix}^{-1}\mathring{\mathbf{M}}\mathring{\mathbf{\Omega}}_{II}\right]^{-}\right\}$$
$$=\pi_0-\operatorname{rank}\left[\mathring{\mathbf{\Omega}}'_{II}\mathring{\mathbf{M}}'\begin{pmatrix}\mathbf{I} & \mathbf{0}\\ \mathbf{0} & \mathbf{C}_{\psi}\mathbf{w}_{\psi}^{-2}\mathbf{C}_{\psi}\end{pmatrix}^{-1}\mathring{\mathbf{M}}\mathring{\mathbf{\Omega}}_{II}\right]$$
$$=\pi_0-\operatorname{rank}(\mathring{\mathbf{M}}\mathring{\mathbf{\Omega}}_{II})=\pi_0-\min(\pi_0,\pi_0-r_2)=r_2,$$

since $\mathring{\mathbf{M}}$ and $\mathring{\mathbf{\Omega}}_{II}$ are of maximal rank. The serial part of the test statistic thus accounts for r_2 additional degrees of freedom, yielding a total of $r = r_1 + r_2$ degrees of freedom in (5.26) and (5.36). Now part (iv) of the proposition follows on noting that $\mathring{\mathbf{T}}^{(n)}_{\psi;f}$, as well as $\mathring{\mathbf{M}}$ and $\mathring{\mathbf{\Omega}}_{II}$ (but unlike $\mathbf{M}$ and $\mathbf{\Omega}_{II}$ themselves) depend on p_1 and q_1 only through π.

The proof for the signed rank case is entirely similar.

5.5. *Local Asymptotic Power.* We conclude this section with a result on the local asymptotic performance of the optimal tests described in Proposition 5.10.

PROPOSITION 5.14. *The power* $1 - \beta^{(n)}_{f:g}(\boldsymbol{\theta} + \mathbf{v}(n)\boldsymbol{\tau})$ *of* $\phi^{(n)}_{f}$ *at* $H^{(n)}_{g}(\boldsymbol{\theta} + \mathbf{v}(n)\boldsymbol{\tau})$, *where* $\boldsymbol{\tau} = (\boldsymbol{\eta}', \boldsymbol{\gamma}', \boldsymbol{\delta}')'$ *and* $\boldsymbol{\theta} - \boldsymbol{\theta}_0 \in \mathcal{M}(\mathbf{\Omega})$, *satisfies*

$$\lim_{n \to \infty} \{[1 - \beta^{(n)}_{f:g}(\boldsymbol{\theta} + \mathbf{v}(n)\boldsymbol{\tau})] - [1 - F^{r}_{\chi}(\chi^2_{r;1-\alpha}; \rho^{(n)}_{f:g}(\boldsymbol{\theta}; \boldsymbol{\tau}))]\} = 0, \tag{5.43}$$

where $F^{r}_{\chi}(\cdot; \rho)$ *denotes the noncentral chi-square distribution function with* r *degrees of freedom and noncentrality parameter* ρ, *and*

$$\rho^{(n)}_{f:g}(\boldsymbol{\theta}; \boldsymbol{\tau}) = [\mathcal{I}(f_1)]^{-1}\boldsymbol{\tau}' \begin{pmatrix} \mathcal{I}(f_1, g_1)\mathbf{I} & \mathbf{O} \\ \mathbf{O} & \sigma(f_1, g_1)\mathcal{I}(f_1, g_1)\mathbf{I} \end{pmatrix}$$

$$\times \mathbf{\Gamma}(\boldsymbol{\theta})\,\bar{\mathbf{Q}}^{(n)}(\boldsymbol{\theta})\,\mathbf{\Gamma}(\boldsymbol{\theta}) \begin{pmatrix} \mathcal{I}(f_1, g_1)\mathbf{I} & \mathbf{O} \\ \mathbf{O} & \sigma(f_1, g_1)\mathcal{I}(f_1, g_1)\mathbf{I} \end{pmatrix} \boldsymbol{\tau} \tag{5.44}$$

$$= \frac{[\sigma(f_1, g_1)\mathcal{I}(f_1, g_1)]^2}{\mathcal{I}(f_1)} \begin{pmatrix} [\sigma(f_1, g_1)]^{-1}\mathbf{\Gamma}_I(\boldsymbol{\theta})\boldsymbol{\eta}' \\ a_1 + b_1 \\ \vdots \\ a_\pi + b_\pi \\ \sum_{i=\pi+1}^{\infty} (a_i + b_i)\psi_i^{(1)} \\ \vdots \\ \sum_{i=\pi+1}^{\infty} (a_i + b_i)\psi_i^{(p_0+q_0)} \end{pmatrix}'$$

$$\times \mathbf{Q}^{(n)}(\boldsymbol{\theta}) \begin{pmatrix} [\sigma(f_1, g_1)]^{-1} \boldsymbol{\Gamma}_I(\boldsymbol{\theta}) \boldsymbol{\eta} \\ a_1 + b_1 \\ \vdots \\ a_\pi + b_\pi \\ \sum_{i=\pi+1}^{\infty} (a_i + b_i) \psi_i^{(1)} \\ \vdots \\ \sum_{i=\pi+1}^{\infty} (a_i + b_i) \psi_i^{(p_0+q_0)} \end{pmatrix}. \tag{5.45}$$

A similar result holds for the power $1 - \beta_{f;g}^{(n)+}(\boldsymbol{\theta} + \mathbf{v}(n)\boldsymbol{\tau})$ *of* $\underset{\sim}{\phi}_f^{(n)+}$.

Proof. Due to asymptotic invariance and contiguity, the asymptotic distribution under the alternative considered of the test statistic in (5.26) (or in (5.36)) is the same as that of its non-aligned counterpart $n(\underset{\sim}{\mathbf{T}}_{\psi;f}^{(n)})' \mathbf{Q}^{(n)}(\boldsymbol{\theta}) \underset{\sim}{\mathbf{T}}_{\psi;f}^{(n)}$. It follows from Proposition 4.3 that $n^{1/2}\underset{\sim}{\mathbf{T}}_{\psi;f}^{(n)}$ is still asymptotically normal under $H_g^{(n)}(\boldsymbol{\theta} + \mathbf{v}(n)\boldsymbol{\tau})$, with the same asymptotic covariance matrix as under $H^{(n)}(\boldsymbol{\theta})$, but with mean

$$\frac{\mathscr{I}(f_1, g_1)}{[\mathscr{I}(f_1)]^{1/2}} \left(\boldsymbol{\eta}' \boldsymbol{\Gamma}_I(\boldsymbol{\theta}) \quad \sigma(f_1, g_1) \begin{pmatrix} a_1 + b_1 \\ \vdots \\ a_\pi + b_\pi \\ \sum_{i=\pi+1}^{\infty} (a_i + b_i) \psi_i^{(1)} \\ \vdots \\ \sum_{i=\pi+1}^{\infty} (a_i + b_i) \psi_i^{(p_0+q_0)} \end{pmatrix}' \right)',$$

yielding the noncentral chi-square distribution with noncentrality parameter (5.45). The equivalent form (5.44) is easily obtained from the asymptotic distribution, under $H_g^{(n)}(\boldsymbol{\theta} + \mathbf{v}(n)\boldsymbol{\tau})$, of $\underset{\sim}{\boldsymbol{\Delta}}_f^{(n)}(\boldsymbol{\theta})$ which, still from Proposition 4.3, is normal with covariance matrix $\boldsymbol{\Gamma}_f(\boldsymbol{\theta})$, but with mean

$$\begin{pmatrix} \frac{\mathscr{I}(f_1, g_1)}{\mathscr{I}(f_1)} \mathbf{I} & \mathbf{O} \\ \mathbf{O} & \frac{\sigma(f_1, g_1)\, \mathscr{I}(f_1, g_1)}{\mathscr{I}(f_1)} \mathbf{I} \end{pmatrix} \boldsymbol{\Gamma}_f(\boldsymbol{\theta}) \boldsymbol{\tau}$$

$$= \begin{pmatrix} \mathscr{I}(f_1, g_1) \mathbf{I} & \mathbf{O} \\ \mathbf{O} & \sigma(f_1, g_1)\, \mathscr{I}(f_1, g_1) \mathbf{I} \end{pmatrix} \boldsymbol{\Gamma}(\boldsymbol{\theta}) \boldsymbol{\tau}. \tag{5.46}$$

The conditions (a) through (d) in the proof of Lemma 5.12 hold without modification. Plugging (5.46) into condition (e) yields the noncentrality parameter (5.44) and the noncentral chi-square distribution functions $F^r_\chi(\cdot;\rho^{(n)}_{f;g}(\boldsymbol{\theta};\boldsymbol{\tau}))$. Note that the convergence in (5.43) and (5.44), though nonuniform in $\boldsymbol{\theta}$ and $\boldsymbol{\tau}$, is uniform in the possible values of the sequences $\mathbf{v}(n)$ or $\mathbf{K}^{(n)}$.

6. Comparison with Parametric Procedures and Examples

6.1. *Asymptotic Relative Efficiencies (AREs)*

Parametric versions of Propositions 5.10 and 5.14 can be established, along the same lines, though of course no asymptotic invariance can be expected to hold here (note, however, that asymptotic distribution-freeness holds). They generalize and reinforce earlier results, e.g., by Kreiss (1990a) or Dufour and King (1991). Define $\mathbf{T}^{(n)}_\psi(\boldsymbol{\theta})$ as the Gaussian parametric statistics obtained by substituting in (4.11) the parametric statistics (4.40) and (4.41) for their respective rank-based counterparts. The notation is that of Proposition 5.10.

Proposition 6.1. *Assume that* **(A1)**(i)–(v), **(A2)**, *and* **(A3$^+$)** *hold under* $H^{(n)}(\boldsymbol{\Omega};\boldsymbol{\theta}_0)$, *and that* $\boldsymbol{\Omega}$ *and* $\hat{\boldsymbol{\theta}}^{(n)}$ *satisfy* **(A4)** *and* **(A5)**, *respectively. The sequence of parametric tests* $\phi^{(n)}$ *rejecting* $H^{(n)}(\boldsymbol{\Omega};\boldsymbol{\theta}_0)$ *whenever*

$$(\mathbf{T}^{(n)}_\psi(\hat{\boldsymbol{\theta}}^{(n)}))'\,\mathbf{Q}^{(n)}(\hat{\boldsymbol{\theta}}^{(n)})\,\mathbf{T}^{(n)}_\psi(\hat{\boldsymbol{\theta}}^{(n)}) > \chi^2_{r;1-\alpha} \tag{6.1}$$

(i) *has asymptotic level* α

(ii) *is locally asymptotically most stringent for* $H^{(n)}(\boldsymbol{\Omega};\boldsymbol{\theta}_0)$ *against the Gaussian alternative* $\bigcup\{H^{(n)}_f(\boldsymbol{\theta});\ (\boldsymbol{\theta}-\boldsymbol{\theta}_0)\notin\mathcal{M}(\boldsymbol{\Omega});\ f(\cdot)=(2\pi\sigma^2)^{-1/2}\exp(-(\cdot/\sigma)^2/2),\ \sigma\in\mathbb{R}^+\}$, *at probability level* α,

(iii) *at* $H^{(n)}_g(\boldsymbol{\theta}+\mathbf{v}(n)\boldsymbol{\tau})$ *has power* $1-\beta^{(n)}(\boldsymbol{\theta}+\mathbf{v}(n)\boldsymbol{\tau})$ *satisfying*

$$\lim_{n\to\infty}\{[1-\beta^{(n)}(\boldsymbol{\theta}+\mathbf{v}(n)\boldsymbol{\tau})]-[1-F^r_\chi(\chi^2_{r;1-\alpha};\rho^{(n)}(\boldsymbol{\theta};\boldsymbol{\tau}))]\}=0,$$

with

$$\begin{aligned}\rho^{(n)}(\boldsymbol{\theta};\boldsymbol{\tau}) &= \boldsymbol{\tau}'\boldsymbol{\Gamma}(\boldsymbol{\theta})\,\bar{\mathbf{Q}}^{(n)}(\boldsymbol{\theta})\,\boldsymbol{\Gamma}(\boldsymbol{\theta})\boldsymbol{\tau}\\ &= \Bigg(\boldsymbol{\eta}'\boldsymbol{\Gamma}_I(\boldsymbol{\theta}),\quad a_1+b_1,\dots,a_\pi+b_\pi,\ \sum_{i=\pi+1}^{\infty}(a_i+b_i)\psi_i^{(1)},\dots,\\ &\qquad \sum_{i=\pi+1}^{\infty}(a_i+b_i)\psi_i^{(p_0+q_0)}\Bigg)\,\mathbf{Q}^{(n)}(\boldsymbol{\theta})\\ &\qquad \times\Bigg(\boldsymbol{\eta}'\boldsymbol{\Gamma}_I(\boldsymbol{\theta}),\quad a_1+b_1,\dots,a_\pi+b_\pi,\ \sum_{i=\pi+1}^{\infty}(a_i+b_i)\psi_i^{(1)},\dots,\\ &\qquad \sum_{i=\pi+1}^{\infty}(a_i+b_i)\psi_i^{(p_0+q_0)}\Bigg)'.\end{aligned} \tag{6.2}$$

The asymptotic relative efficiencies (AREs) of the aligned-rank tests of Proposition 5.10 with respect to the corresponding parametric normal-theory tests of Proposition 6.1 follow as a direct consequence. Indeed, the asymptotic relative efficiency under $H_g^{(n)}(\boldsymbol{\theta}+\nu(n)\boldsymbol{\tau})$ of the aligned rank tests (5.26) and (5.27) with respect to their parametric competitor (6.1) is the limit, when it exists, of the corresponding ratio of noncentrality parameters $ARE_g(\boldsymbol{\theta}, \boldsymbol{\tau}; \underset{\sim}{\phi}_f, \phi) = \lim_{n\to\infty}[\rho_{f;g}^{(n)}(\boldsymbol{\theta}; \boldsymbol{\tau})/\rho^{(n)}(\boldsymbol{\theta}; \boldsymbol{\tau})]$. Depending on the hypothesis to be tested (i.e., depending on $\boldsymbol{\Omega}$), the existence of such a limit may require an additional assumption ((**A3**)(v) or (**A3**$^+$)(iv)) on the asymptotic behavior of the regression constants.

(**A3**)(v) $\lim_{n\to\infty}[\mathbf{D}^{(n)}/\mathrm{tr}\, \mathbf{D}^{(n)}] = \mathbf{D}^2$, where $\mathbf{D}^2$ is a finite, positive definite diagonal matrix.

Note that (**A3**)(v) implies the existence of

$$\begin{aligned}\lim_{n\to\infty} \mathbf{Q}_I^{(n)}(\boldsymbol{\theta}) &= \boldsymbol{\Gamma}_I^{-1}(\boldsymbol{\theta}) - \mathbf{K}^{-1}\mathbf{D}\boldsymbol{\Omega}_I\{\boldsymbol{\Omega}_I'(\mathbf{K}^{-1}\mathbf{D})' \\ &\quad\times \boldsymbol{\Gamma}_I(\boldsymbol{\theta})\mathbf{K}^{-1}\mathbf{D}\boldsymbol{\Omega}_I\}^{-1}\boldsymbol{\Omega}_I'(\mathbf{K}^{-1}\mathbf{D})' \\ &= \mathbf{Q}_I(\boldsymbol{\theta}), \qquad \text{say.}\end{aligned} \tag{6.3}$$

PROPOSITION 6.2. *Assume that the assumptions of Proposition* 5.10 *hold, and let* $\boldsymbol{\tau} = (\boldsymbol{\eta}', \boldsymbol{\gamma}', \boldsymbol{\delta}')'$. *Then*

(i) *either* $\lim_{n\to\infty}[\boldsymbol{\Omega}_I]_\perp' \mathbf{K}^{(n)}\boldsymbol{\eta} = \mathbf{0}$, *and* $ARE_g(\boldsymbol{\theta}, \boldsymbol{\tau}; \underset{\sim}{\phi}_f, \phi) = [\sigma(f_1, g_1)\mathscr{I}(f_1, g_1)]^2/\mathscr{I}(f_1)$,

(ii) *or* $[\boldsymbol{\Omega}_{II}]_\perp' \binom{\boldsymbol{\gamma}}{\boldsymbol{\delta}} = \mathbf{0}$, *and* $ARE_g(\boldsymbol{\theta}, \boldsymbol{\tau}; \underset{\sim}{\phi}_f, \phi) = [\mathscr{I}(f_1, g_1)]^2/\mathscr{I}(f_1)$,

(iii) *or* (**A3**)(v) *holds, and*

$$ARE_g(\boldsymbol{\theta}, \boldsymbol{\tau}; \underset{\sim}{\phi}_f, \phi) = \frac{[\mathscr{I}(f_1, g_1)]^2\, \boldsymbol{\eta}'\boldsymbol{\Gamma}_I(\boldsymbol{\theta})\,\mathbf{Q}_I(\boldsymbol{\theta})\,\boldsymbol{\Gamma}_I(\boldsymbol{\theta})\boldsymbol{\eta} + [\sigma(f_1, g_1)]^2 \times (\boldsymbol{\gamma}', \boldsymbol{\delta}')\,\boldsymbol{\Gamma}_{II}(\boldsymbol{\theta})\,\bar{\mathbf{Q}}_{II}(\boldsymbol{\theta})\,\boldsymbol{\Gamma}_{II}(\boldsymbol{\theta})(\boldsymbol{\gamma}', \boldsymbol{\delta}')'}{\mathscr{I}(f_1)[\boldsymbol{\eta}'\boldsymbol{\Gamma}_I(\boldsymbol{\theta})\,\mathbf{Q}_I(\boldsymbol{\theta})\,\boldsymbol{\Gamma}_I(\boldsymbol{\theta})\boldsymbol{\eta} + (\boldsymbol{\gamma}', \boldsymbol{\delta}')\,\boldsymbol{\Gamma}_{II}(\boldsymbol{\theta})\,\bar{\mathbf{Q}}_{II}(\boldsymbol{\theta})\,\boldsymbol{\Gamma}_{II}(\boldsymbol{\theta})(\boldsymbol{\gamma}', \boldsymbol{\delta}')']} \tag{6.4}$$

(iv) *Moreover, in either of the three cases above*

$$[\sigma(f_1, g_1)\mathscr{I}(f_1, g_1)]^2/\mathscr{I}(f_1) \leqslant ARE_g(\boldsymbol{\theta}, \boldsymbol{\tau}; \underset{\sim}{\phi}_f, \phi) \leqslant [\mathscr{I}(f_1, g_1)]^2/\mathscr{I}(f_1) \tag{6.5}$$

yielding, for $f = g$, $ARE_f(\boldsymbol{\theta}, \boldsymbol{\tau}; \underset{\sim}{\phi}_f, \phi) = \mathscr{I}(f_1)$.

Proof. Parts (i), (ii), and (iii) readily follow from inspecting the ratio of noncentrality parameters (5.44) and (6.2); (6.4) implies that the ARE

TABLE 6.1
Numerical Implications of Proposition 6.2

Scores (signed or unsigned)	Density types: Normal	Logistic	Double exp.	
van der Waerden	1.000	1.048	1.226	Lower bound
$\phi_1(F_1^{-1}(\cdot))=\Phi^{-1}(\cdot)$	1.000	1.049	1.280	Upper bound
Wilcoxon	0.948	1.098	1.483	Lower bound
$\phi_1(F_1^{-1}(\cdot))=(\cdot)$	0.955	1.098	1.500	Upper bound
Laplace	0.612	0.812	2.000	Lower bound
$\phi_1(F_1^{-1}(\cdot))=\text{sgn}(\cdot)$	0.637	0.823	2.000	Upper bound

Note. Lower and upper bounds (cf. (6.5)) for the AREs of the (signed and unsigned) aligned rank tests described in Proposition 5.10 with respect to their normal theory counterparts (Proposition 6.1), under normal, logistic, and double exponential densities. The van der Waerden, Wilcoxon, and Laplace scores are those associated with normal, logistic, and double-exponential densities, respectively.

under study always lies between the minimal and the maximum of $[\mathscr{I}(f_1, g_1)]^2/\mathscr{I}(f_1)$ and $[\sigma(f_1, g_1)\,\mathscr{I}(f_1, g_1)]^2/\mathscr{I}(f_1)$. However, from Chebyshev's inequality, we have $\sigma(f_1, g_1) \leqslant [\sigma(f_1, f_1)\,\sigma(g_1, g_1)]^{1/2} = \sigma_{f_1}\sigma_{g_1} = 1$, with equality at $f_1 = g_1$ only, which entails (6.5).

Table 6.1 summarizes the numerical implications of Proposition 6.2 for some usual densities and score functions. As asserted by (6.5), the asymptotic relative efficiencies related to the nonserial part of the problem are larger than those related with serial dependence aspects. For the densities and score functions in the table, this difference however is extremely small, and the AREs we can expect from the tests described in Proposition 5.10 are thus very close to the AREs of classical van der Waerden, Terry–Fisher–Yates–Fraser, Wilcoxon, sign or median tests for, e.g., location problems.

6.2. *Example* 5.1 (*Continued*)

Denote by $\mathring{\mathbf{R}}_i$ the $(k-l)\times(k-l)$ matrix obtained by deleting rows and columns 1 through l in $\mathbf{R}_i$, $i=0, 1, \ldots$, and by $\mathring{\mathbf{K}}^{-1}$ the $k\times(k-l)$ matrix similarly resulting from $\mathbf{K}^{-1}$ by deleting the first l columns. Assume that $\hat{\boldsymbol{\theta}}^{(n)}$ is a constrained estimate satisfying (**A5**) (see, e.g., Durbin, 1960, or Pierce, 1971a). Standard algebra yields

$$\mathbf{Q}_I^{(n)}(\hat{\boldsymbol{\theta}}^{(n)}) = \mathbf{Q}_I(\hat{\boldsymbol{\theta}}^{(n)})$$
$$= \mathbf{K}^{-1}\left(\sum_{u=0}^{\infty}\sum_{v=0}^{\infty} \hat{c}_u\hat{c}_v\mathbf{R}_{|u-v|}\right)^{-1}(\mathbf{K}^{-1})'$$
$$-\mathring{\mathbf{K}}^{-1}\left(\sum_{u=0}^{\infty}\sum_{v=0}^{\infty} \hat{c}_u\hat{c}_v\mathring{\mathbf{R}}_{|u-v|}\right)^{-1}(\mathring{\mathbf{K}}^{-1})'$$

and $\mathbf{Q}_{II}(\hat{\boldsymbol{\theta}}^{(n)}) = \mathbf{0}$. The test (5.26) thus takes the form

$$\underset{\sim}{\phi}_f^{(n)} = 1$$

iff

$$\left(\sum_{i=0}^{\kappa(n)} (n-i)^{1/2}\, \hat{c}_i \underset{\sim}{\mathbf{t}}_{i;f}^{(n)}\right)' \mathbf{Q}_I(\hat{\boldsymbol{\theta}}^{(n)}) \left(\sum_{i=0}^{\kappa(n)} (n-i)^{1/2}\, \hat{c}_i \underset{\sim}{\mathbf{t}}_{i;f}^{(n)}\right) > \chi^2_{l;\,1-\alpha}.$$

This test is obviously very different from the one that would result from overlooking the serial dependence structure of the error process (viz., letting $c_0 = 1$, and $\hat{c}_i = 0$, $i \geqslant 1$ in the above expression). The statistical implications of treating correlated observations as if they were independent have been discussed at length, particularly in the econometric literature (see, e.g., Anderson, 1954, or Judge *et al.*, 1985).

6.3. *Example* 5.2 (*Continued*)

6.3.1. *Testing for Independent Error Terms.* The particular case when $p_0 = q_0 = 0$—viz. the problem of testing that the error term, in the general linear model (2.1) with unspecified parameter $\boldsymbol{\beta}$, is independent white noise with unspecified density, is of primary practical importance and so far could not be solved in a rank-based context. Here $\pi = \max(p_1, q_1)$—though, from Proposition 5.10(iv) we know that we may as well consider $\pi = p_1 = q_1$; $\hat{\boldsymbol{\theta}}^{(n)} = ((\hat{\boldsymbol{\beta}}^{(n)})', \mathbf{0}_{1\times p_1}, \mathbf{0}_{1\times q_1})'$ can be chosen so that $\hat{\boldsymbol{\beta}}^{(n)}$ is the usual least squares estimate $(\mathbf{X}'\mathbf{X})^{-1}\mathbf{X}'\mathbf{Y}$: the residuals $Z_t^{(n)}(\hat{\boldsymbol{\theta}}^{(n)})$ from which the aligned ranks $\hat{R}_t^{(n)}$ are computed then are the familiar least squares residuals. Other estimates, among which R-estimates, however, are equally valid. The asymptotically locally sufficient statistic $n^{1/2}\underset{\sim}{\mathbf{T}}_f^{(n)}$ reduces to the vector of rank-based residual autocorrelations $((n-1)^{1/2}\,\underset{\sim}{\hat{r}}_{1;f}^{(n)} \cdots (n-\pi)^{1/2}\,\underset{\sim}{\hat{r}}_{\pi;f}^{(n)})$, $\mathbf{Q}_I^{(n)} = \mathbf{0}_{k\times k}$, $\mathbf{Q}_{II}^{(n)} = \mathbf{I}_{\pi\times\pi}$, $r = \pi$ and the optimal test of Proposition 5.10 is thus the aligned-rank version of the Box and Pierce (1970) portmanteau test of order π

$$\underset{\sim}{\phi}_f^{(n)} = 1 \qquad \text{iff} \quad \sum_{i=1}^{\pi} (n-i)(\underset{\sim}{\hat{r}}_{i;f}^{(n)})^2 > \chi^2_{\pi;\,1-\alpha}. \tag{6.6}$$

The (aligned) signed-rank version of (6.6) is entirely similar (except, of course, that $\underset{\sim}{\hat{r}}_{i;f}^{(n)+}$ is used instead of $\underset{\sim}{\hat{r}}_{i;f}^{(n)}$).

Contrary to Example 5.1, where the presence of autocorrelation deeply modifies the treatment of inference problems dealing with the nonserial features of the problem, the presence and preliminary estimation of the trend here does not alter the form of the optimal statistic, neither in the unsigned nor in the signed case.

6.3.2. *Testing ARMA* (p_0, q_0) *against ARMA* $(p_0+\pi, q_0+\pi)$ *Dependence.* Here $\mathbf{\Omega}_I=\mathbf{I}_{k\times k}$ and $\mathring{\mathbf{\Omega}}_{II}=\mathbf{I}_{(p_0+q_0)\times(p_0+q_0)}$, so that $\mathbf{Q}_I^{(n)}=\mathbf{0}_{k\times k}$, *and*

$$\mathbf{Q}_{II}(\boldsymbol{\theta})=\begin{pmatrix}\mathbf{I} & \mathbf{0}\\ \mathbf{0} & \mathbf{w}_\psi^{-2}\end{pmatrix}-\begin{pmatrix}\mathbf{I} & \mathbf{0}\\ \mathbf{0} & \mathbf{C}_\psi^{-1}\end{pmatrix}\mathring{\mathbf{M}}(\boldsymbol{\theta})\left(\mathring{\mathbf{M}}'(\boldsymbol{\theta})\begin{pmatrix}\mathbf{I} & \mathbf{0}\\ \mathbf{0} & \mathbf{C}_\psi\mathbf{w}_\psi^{-2}\mathbf{C}_\psi'\end{pmatrix}\mathring{\mathbf{M}}(\boldsymbol{\theta})\right)^{-1}$$
$$\times\mathring{\mathbf{M}}'(\boldsymbol{\theta})\begin{pmatrix}\mathbf{I} & \mathbf{0}\\ \mathbf{0} & \mathbf{C}_\psi^{-1}\end{pmatrix}'. \tag{6.7}$$

The resulting test statistics have π degrees of freedom.

The particular case where $\pi=1$ is of particular interest because of its possible use in the model identification context (see Pötscher, 1983, 1985). Particularizing the fundamental system, let

$$\psi_t^{(i)}=\begin{cases} g_{t-i}, & i=1,\ldots,p_0\\ h_{t-i}, & i=p_0+1,\ldots,p_0+q_0.\end{cases}$$

Then $\mathring{\mathbf{M}}(\boldsymbol{\theta})=\begin{pmatrix}1\,0\cdots0\,1\,0\cdots0\\ \mathbf{C}_\psi\end{pmatrix}$ and

$$\mathbf{Q}_{II}(\boldsymbol{\theta})=\begin{pmatrix}1 & \mathbf{0}\\ \mathbf{0} & \mathbf{w}_\psi^{2}\end{pmatrix}^{-1}-\begin{pmatrix}1\,0\cdots0\,1\,0\cdots0\\ \mathbf{I}_{(p_0+q_0)\times(p_0+q_0)}\end{pmatrix}\mathbf{W}_\psi^{-2}\begin{pmatrix}1\,0\cdots0\,1\,0\cdots0\\ \mathbf{I}_{(p_0+q_0)\times(p_0+q_0)}\end{pmatrix}' \tag{6.8}$$

with $(\mathbf{W}_\psi^2)_{i,j}=\sum_{t=1}^\infty\psi_t^{(i)}\psi_t^{(j)}$ (whereas $(\mathbf{w}_\psi^2)_{i,j}=\sum_{t=2}^\infty\psi_t^{(i)}\psi_t^{(j)}$). The resulting test statistic $\hat{Q}_f^{(n)}=n(\underset{\sim}{\hat{\mathbf{T}}}{}_{II;f}^{(n)})'\,\mathbf{Q}_{II}(\hat{\boldsymbol{\theta}}^{(n)})\underset{\sim}{\hat{\mathbf{T}}}{}_{II;f}^{(n)}$, with one degree of freedom, is not quite familiar, and its relation to classical parametric time series procedures is not obvious at first sight. Denoting by $\hat{\mathbf{T}}_{II}^{(n)}$ the parametric version of $\underset{\sim}{\hat{\mathbf{T}}}{}_{II;f}^{(n)}$, it follows from McLeod (1978, Theorem 1) that

$$n^{1/2}\hat{\mathbf{T}}_{II}^{(n)}=\begin{pmatrix}1 & 0 & \cdots & 0\\ 0 & g_1 & \cdots & g_{\kappa(n)-1}\\ \vdots & & & \vdots\\ 0 & 0 & \cdots & g_{\kappa(n)-p_0}\\ 0 & h_1 & \cdots & h_{\kappa(n)-1}\\ \vdots & \vdots & & \vdots\\ 0 & 0 & \cdots & h_{\kappa(n)-q_0}\end{pmatrix}\begin{pmatrix}(n-1)^{1/2}\,\hat{r}_1^{(n)}\\ (n-2)^{1/2}\,\hat{r}_2^{(n)}\\ \vdots\\ (n-\kappa(n))^{1/2}\,\hat{r}_{\kappa(n)}^{(n)}\end{pmatrix} \tag{6.9}$$
$$=(n-1)^{1/2}\,(\hat{r}_1^{(n)},-\hat{r}_2^{(n)},\quad 0,\ldots,0)'+o_P(1),$$

as $n\to\infty$. Hence the parametric Gaussian counterpart $\hat{Q}_{\mathcal{N}}^{(n)}$ of the aligned-rank test statistic $\hat{Q}_f^{(n)}$ is asymptotically equivalent to the squared, first-order estimated residual autocorrelation:

$$\hat{Q}_{\mathcal{N}}^{(n)}=(n-1)(\hat{r}_1^{(n)})^2+o_P(1),\qquad \text{as}\quad n\to\infty.$$

Now, turning to the Gaussian Lagrange multiplier test statistic $\hat{Q}_{\mathcal{L}}^{(n)}$ for ARMA (p_0, q_0) against ARMA (p_0+1, q_0) dependence (Godfrey, 1979;

Hosking, 1980; Pham, 1987), the Lagrange multiplier is (Hosking, 1980, Theorem 1)

$$\begin{aligned}\hat{d}^{(n)} &= -n(\hat{r}^{(n)}_{p_0+1} + \hat{g}_1 \hat{r}^{(n)}_{p_0+2} + \cdots + \hat{g}_{n-p_0} \hat{r}^{(n)}_{n-1}) \\ &= -n(\hat{g}^{(n)}_{-p_0+1} \hat{r}^{(n)}_2 + \cdots + \hat{g}_{-1} \hat{r}^{(n)}_{p_0} + \hat{g}_0 \hat{r}^{(n)}_{p_0+1} + \cdots + \hat{g}_{n-p_0} \hat{r}^{(n)}_{n-1})\end{aligned}$$

which, using Theorem 1 of McLeod (1978) again, implies $n^{-1/2}\hat{d}^{(n)} = (n-1)^{1/2}(\hat{g}_{-p}\hat{r}^{(n)}_1) + o_P(1)$, as $n \to \infty$. The Lagrange multiplier test thus asymptotically consists in rejecting ARMA (p_0, q_0) dependence whenever $\hat{Q}^{(n)}_{\mathscr{L}} = (n-1)(\hat{r}^{(n)}_1)^2 = \hat{Q}^{(n)}_{\mathscr{N}}$ is larger than the appropriate quantile of the chi-square distribution with one degree of freedom. Our aligned-rank test based on $\hat{\underset{\sim}{Q}}^{(n)}_f$ thus generalizes the corresponding Gaussian Lagrange multiplier test based on $\hat{Q}^{(n)}_{\mathscr{L}}$.

As an illustration, consider the problem of testing a null hypothesis of first-order autoregressive dependence, with unspecified autoregressive parameter $\rho \in (-1, 1)$, unspecified linear trend, and unspecified innovation density, against AR(2) or, equivalently in view of Proposition 5.10(iv), ARMA(2, 1) dependence. The estimated residuals from which the aligned ranks are to be computed are $\hat{Z}^{(n)}_t = (Y^{(n)}_t - (\mathbf{x}^{(n)}_t)' \hat{\boldsymbol{\beta}}^{(n)}) - \hat{\rho}(Y^{(n)}_{t-1} - (\mathbf{x}^{(n)}_{t-1})' \hat{\boldsymbol{\beta}}^{(n)})$ (for simplicity, assume that $Y^{(n)}_0 - (\mathbf{x}^{(n)}_0)' \hat{\boldsymbol{\beta}}^{(n)} = 0$). A simple fundamental system (here consisting of one solution) of $\psi_t - \rho\psi_{t-1} = 0$ is $\{\psi_t\} = \{\rho^{t-1}\}$, with $\mathbf{w}^2_\psi = \rho^2/(1-\rho^2)$, $\mathbf{W}^2_\psi = 1/(1-\rho^2)$, $\mathbf{C}_\psi = \rho$, and $\underset{0}{\mathbf{M}}(\boldsymbol{\theta}) = 1$. Hence $n^{1/2}\underset{\sim}{\mathbf{T}}^{(n)}_{II;f} = ((n-1)^{1/2}\underset{\sim}{r}^{(n)}_{1;f}, \sum_{i=2}^{\kappa(n)} (n-i)^{1/2} \rho^{i-1} \underset{\sim}{r}^{(n)}_{1;f})'$, so that, letting $\kappa(n) = n-1$, the test statistic $n(\hat{\underset{\sim}{\mathbf{T}}}^{(n)}_{\psi;f})' \mathbf{Q}^{(n)}(\hat{\boldsymbol{\theta}}^{(n)}) \hat{\underset{\sim}{\mathbf{T}}}^{(n)}_{\psi;f} = \hat{\underset{\sim}{Q}}^{(n)}_f$ takes the simple form

$$\hat{\underset{\sim}{Q}}^{(n)}_f = \left((n-1)^{1/2} \hat{\rho} \hat{\underset{\sim}{r}}^{(n)}_{1;f} - \frac{[1-\hat{\rho}^2]}{\hat{\rho}} \sum_{i=2}^{n-1} (n-i)^{1/2} \hat{\rho}^{i-1} \hat{\underset{\sim}{r}}^{(n)}_{i;f} \right)^2. \tag{6.10}$$

As for the Gaussian counterpart of $\hat{\underset{\sim}{Q}}^{(n)}_f$, we have

$$Q^{(n)}_{\mathscr{N}} = \left\{ -(n-1)^{1/2} \hat{\rho} \hat{r}^{(n)}_1 + \frac{[1-\hat{\rho}^2]}{\hat{\rho}} \sum_{i=2}^{n-1} (n-i)^{1/2} \hat{\rho}^{i-1} \hat{r}^{(n)}_i \right\}^2, \tag{6.11}$$

where $\hat{\rho} = \hat{\rho}^n$ denotes the Gaussian maximum likelihood estimator of the AR(1) parameter ρ, and $\hat{r}^{(n)}_i$ stands for the corresponding classical residual autocorrelations. On the other hand, the Gaussian Lagrange multiplier approach leads to

$$Q^{(n)}_{\mathscr{L}} = n \left\{ \sum_{i=2}^{n-1} \hat{\rho}^{i-1} \hat{r}^{(n)}_i \right\}^2 \Big/ \hat{\rho}^2, \tag{6.12}$$

still to be compared with the quantiles of a chi-square variable with one degree of freedom.

$Q^{(n)}_{\mathscr{N}}$ and $Q^{(n)}_{\mathscr{L}}$ apparently are distinct test statistics. However, from Theorem 1 of McLeod (1978),

$$\sum_{i=1}^{n-1} \hat{\rho}^{i-1} \hat{r}_i^{(n)} = o_P(n^{-1/2}), \qquad n \to \infty, \tag{6.13}$$

so that

$$Q_{\mathcal{N}}^{(n)} = n(\hat{r}_1^{(n)})^2 + o_P(1), \qquad n \to \infty, \tag{6.14}$$

$Q_{\mathcal{L}}^{(n)} = n(\hat{r}_1^{(n)})^2 + o_P(1)$, $n \to \infty$, and $Q_{\mathcal{N}}^{(n)} - Q_{\mathcal{L}}^{(n)} = o_P(1)$, $n \to \infty$. The parametric, Gaussian counterpart of our rank-based statistic (6.10) is thus asymptotically equivalent to the Gaussian Lagrange multiplier test statistic. Whether an asymptotic equivalence of the form (6.13) or (6.14) also holds for aligned rank-based residual autocorrelations is not known. Asymptotic invariance implies that (6.10) is asymptotically equivalent to its exact unaligned version $Q_f^{(n)}$, say. The van der Waerden version $Q_{vdW}^{(n)}$ of the latter, under Gaussian densities, in turn is asymptotically equivalent to the parametric $Q_{\mathcal{N}}^{(n)}$, hence to $n(\hat{r}_1^{(n)})^2$. Note however that, whereas $Q_{vdW}^{(n)}$, $Q_{\mathcal{N}}^{(n)}$, and $Q_{\mathcal{L}}^{(n)}$ are asymptotically equivalent to the corresponding quantities computed from exact residuals, these equivalences do not hold for $n(\hat{r}_1^{(n)})^2$ which is asymptotically distinct from $n(r_1^{(n)})^2$, where $r_1^{(n)}$ denotes the exact first-order residual autocorrelation. Nor is $\hat{r}_{1;vdW}^{(n)}$ asymptotically invariant. The linear relation (6.13) indeed does not hold for exact residual autocorrelations. Still, this provides for (6.10) an interpretation as an extended, rank-based, non-Gaussian version of the traditional Lagrange multiplier test statistic.

Acknowledgments

Marc Hallin would like to express his deep gratitude to Professor Henry H. H. Remak, the Director of the Indiana University Institute for Advanced Study, for providing a most congenial and stimulating environment for research during his fellowship at the Institute in the summer of 1991.

References

[1] Adichie, J. N. (1967). Estimation of regression parameters based on rank tests. *Ann. Math. Statist.* **38** 894–904.

[2] Adichie, J. N. (1978). Rank tests of sub-hypotheses in the general linear regression. *Ann. Statist.* **6** 1012–1026.

[3] Akritas, M. G., and Johnson, R. A. (1982). Efficiencies of tests and estimators for p-order autoregressive processes when the error is non normal. *Ann. Inst. Statist. Math.* **34** 579–589.

[4] Albers, W. (1978). One-sample rank tests under autoregressive dependence. *Ann. Statist.* **6** 836–845.

[5] Anderson, R. L. (1954). The problem of autocorrelation in regression analysis. *J. Amer. Statist. Assoc.* **49** 113–129.

[6] Anderson, T. W. (1971). *The Statistical Analysis of Time Series.* Wiley, New York.

[7] Bartels, R. (1982). The rank version of von Neumann's ratio test for randomness. *J. Amer. Statist. Assoc.* **77** 40–46.

[8] Beran, R. J. (1970). Linear rank statistics under alternatives indexed by a vector parameter. *Ann. Math. Statist.* **41** 1896–1905.

[9] BERAN, R. J. (1972). Rank spectral processes and tests for serial dependence. *Ann. Math. Statist.* **43** 1749–1766.

[10] BOX, G. E. P., AND PIERCE, D. A. (1970). Distribution of residual autocorrelations in autoregressive-integrated moving-average time series models. *J. Amer. Statist. Assoc.* **65** 1509–1526.

[11] BROCKWELL, P. J., AND DAVIS, R. A. (1987). *Time Series: Theory and Methods.* Springer-Verlag, New York.

[12] CHIANG, C. Y., AND PURI, M. L. (1984). Rank procedures for testing subhypotheses in linear regression. *Ann. Inst. Statist. Math.* **36** 35–50.

[13] COCHRANE, D., AND ORCUTT, G. H. (1949). Application of least squares regression to relationships containing autocorrelated error terms. *J. Amer. Statist. Assoc.* **44** 32–61.

[14] DAVIES, R. B. (1973). Asymptotic inference in stationary Gaussian time series. *Adv. Appl. Prob.* **5** 469–497.

[15] DUFOUR, J.-M. (1981). Rank tests for serial dependence. *J. Time Ser. Anal.* **2** 117–128.

[16] DUFOUR, J.-M., AND KING, M. (1991). Optimal invariant tests for the autocorrelation coefficient in linear regressions with stationary or nonstationary AR(1) errors. *J. Econometrics* **47** 115–143.

[17] DUFOUR, J.-M., LEPAGE, Y., AND ZEIDAN, H. (1982). Nonparametric testing for time series: A bibliography. *Canad. J. Statist.* **10** 1–38.

[18] DUFOUR, J.-M., AND ROY, R. (1985). Some robust exact results on sample autocorrelations and tests of randomness. *J. Econometrics* **29** 257–273; and **41** 279–281.

[19] DUFOUR, J.-M., AND ROY, R. (1986). Generalized portmanteau statistics and tests for randomness. *Comm. Statist. A Theory Methods* **15** 2953–2972.

[20] DURBIN, J. (1960). Estimation of parameters in time-series regression models. *J. Roy. Statist. Soc. Ser. B* **22** 139–153.

[21] DZHAPARIDZE, K. O. (1986). *Parameter Estimation and Hypothesis Testing in Spectral Analysis of Stationary Time Series.* Springer-Verlag, New York.

[22] FISHER, R. A. (1926). On the random sequence. *Quart. J. Roy. Meteorol. Soc.* **52** 250.

[23] GAREL, B. (1989). The asymptotic distribution of the likelihood ratio for MA processes with a regression trend. *Statist. Decisions* **7** 167–184.

[24] GAREL, B., AND HALLIN, M. (1991). Local asymptotic linearity of multivariate ARMA processes with a linear trend. Unpublished manuscript.

[25] GASTWIRTH, J. L., AND RUBIN, H. (1971). Effect of dependence on the level of some one-sample tests. *J. Amer. Statist. Assoc.* **66** 816–820.

[26] GODFREY, L. G. (1979). Testing the adequacy of a time series model. *Biometrika* **66** 67–72.

[27] HÁJEK, J. (1962). Asymptotically most powerful rank order tests. *Ann. Math. Statist.* **33** 1124–1147.

[28] HÁJEK, J., AND ŠIDÁK, Z. (1967). *Theory of Rank Tests.* Academic Press, New York.

[29] HALLIN, M. (1986). Nonstationary q-dependent processes and time-varying moving-average models: Invertibility properties and the forecasting problem. *Adv. Appl. Probab.* **18** 170–210.

[30] HALLIN, M., INGENBLEEK, J.-FR., AND PURI, M. L. (1985). Linear serial rank tests for randomness against ARMA alternatives. *Ann. Statist.* **13** 1156–1181.

[31] HALLIN, M., INGENBLEEK, J.-FR., AND PURI, M. L. (1987). Linear and quadratic serial rank tests for randomness against serial dependence. *J. Time Ser. Anal.* **8** 409–424.

[32] HALLIN, M., AND PURI, M. L. (1988). Optimal rank-based procedures for time-series analysis: Testing an ARMA model against other ARMA models. *Ann. Statist.* **16** 402–432.

[33] HALLIN, M., AND PURI, M. L. (1991). Time series analysis via rank-order theory: Signed-rank tests for ARMA models. *J. Multivariate Anal.* **39** 1–29.

[34] Hallin, M., and Puri, M. L. (1991b). *Aligned Rank Tests for Linear Models with Autocorrelated Error Terms.* Technical Report. Indiana University, Bloomington.

[35] Hallin, M., and Puri, M. L. (1992). Rank tests for time series analysis: A survey. In *New Directions in Time Series Analysis*, Vol. 45 (D. Brillinger *et al.*, Eds.), pp. 111–153, Springer-Verlag, New York.

[36] Harel, M. (1988). Weak convergence of multidimensional rank statistics under ϕ-mixing conditions. *J. Statist. Plann. Inference* **20** 41–63.

[37] Harel, M., and Puri, M. L. (1989a). Limiting behavior of U-statistics, V-statistics and one sample rank order statistics for nonstationary absolutely regular processes. *J. Multivariate Anal.* **30** 181–204.

[38] Harel, M., and Puri, M. L. (1989b). Weak convergence of the U-statistic and weak invariance of the one-sample rank-order statistic for Markov processes and ARMA models. *J. Multivariate Anal.* **31** 258–265.

[39] Harel, M., and Puri, M. L. (1990a). Limiting behavior of one-sample rank-order statistics with unbounded scores for nonstationary absolutely regular processes. *J. Statist. Plann. Inference* **27** 1–23.

[40] Harel, M., and Puri, M. L. (1990b). Weak convergence of serial rank statistics under dependence with applications in time series and Markov processes. *Ann. Probab.* **18** 1361–1387.

[41] Harel, M., and Puri, M. L. (1991). Weak invariance of the multidimensional rank statistic for nonstationary absolutely regular processes. *J. Multivariate Anal.* **36** 204–221.

[42] Hosking, J. R. M. (1980). Lagrange multiplier tests of time series models. *J. Roy. Statist. Soc. Ser. B* **42** 170–181.

[43] Janssen, A., and Mason, D. M. (1990). *Non-standard Rank Tests.* Springer-Verlag, New York.

[44] Jeganathan, P. (1991). *Some Aspects of Asymptotic Theory with Applications to Time Series Models.* Technical Report.

[45] Judge, C. G., Griffiths, W. E., Hill, R. C., Lütkepohl, H., and Lee, T. C. (1985). *The Theory and Practice of Econometrics.* Wiley, New York.

[46] Jurečková, J. (1969). Asymptotic linearity of a rank statistic in regression parameter. *Ann. Math. Statist.* **40** 1889–1900.

[47] Jurečková, J. (1971a). Nonparametric estimate of regression coefficients. *Ann. Math. Statist.* **42** 1328–1338.

[48] Jurečková, J. (1971b). Asymptotic independence of rank test statistic for testing symmetry on regression. *Sankhyā Ser. A* **33** 1–18.

[49] Knoke, D. J. (1977). Testing for randomness against autocorrelation: Alternative tests. *Biometrika* **64** 523–529.

[50] Koul, H. L. (1969). Asymptotic behavior of Wilcoxon type confidence regions in multiple linear regression. *Ann. Math. Statist.* **40** 1950–1979.

[51] Koul, H. L. (1977). Behavior of robust estimators in the regression model with dependent errors. *Ann. Statist.* **5** 681–699.

[52] Kreiss, J.-P. (1987). On adaptive estimation in stationary ARMA processes. *Ann. Statist.* **15** 112–133.

[53] Kreiss, J.-P. (1990a). Testing linear hypotheses in autoregressions. *Ann. Statist.* **18** 1470–1482.

[54] Kreiss, J.-P. (1990b). Local asymptotic normality for autoregression with infinite order. *J. Statist. Plann. Inference* **26** 185–219.

[55] LeCam, L. (1960). Locally asymptotically normal families of distributions. *Univ. Calif. Publ. Statist.* **3** 37–98.

[56] LeCam, L. (1986). *Asymptotic Methods in Statistical Decision Theory.* Springer-Verlag, New York.

[57] LeCam, L., and Yang, Gr.L. (1990). *Asymptotics in Statistics.* Springer-Verlag, New York.

[58] Lehmann, E. L. (1986). *Testing Statistical Hypotheses*, 2nd ed. Wiley, New York.

[59] McKean, J. W., and Hettmansperger, T. P. (1976). Tests of hypotheses based on ranks in the general linear model. *Comm. Statist. A* **5** 693–709.

[60] McLeod, A. I. (1978). On the distribution of residual autocorrelations in Box-Jenkins models. *J. Roy. Statist. Soc. Ser. B* **40** 296–302.

[61] Nieuwenhuis, G., and Ruymgaart, F. (1990). Some stochastic inequalities and asymptotic normality of serial rank statistics in general linear processes. *J. Statist. Plann. Inference* **25** 53–80.

[62] Pham, D. T. (1987). Exact maximum likelihood estimate and Lagrange multiplier test statistic for ARMA models. *J. Time Ser. Anal.* **8** 61–78.

[63] Pham, D. T., and Tran, L. T. (1985). Some mixing properties of time series models. *Stochastic Process. Appl.* **19** 297–303.

[64] Pierce, D. A. (1971a). Least squares estimation in the regression model with autoregressive-moving average errors. *Biometrika* **58** 299–312.

[65] Pierce, D. A. (1971b). Distribution of residual autocorrelation in the regression model with autoregressive-moving average errors. *J. Roy. Statist. Soc. Ser. B* **33** 140–146.

[66] Pötscher, B. M. (1983). Order estimation in ARMA models by Lagrangian multiplier tests. *Ann. Statist.* **11** 872–885.

[67] Pötscher, B. M. (1985). The behaviour of the Lagrange multiplier test in testing the orders of an ARMA model. *Metrika* **32** 129–150.

[68] Puri, M. L., and Sen, P. K. (1971). *Nonparametric Methods in Multivariate Analysis.* Wiley, New York.

[69] Puri, M. L., and Sen, P. K. (1985). *Nonparametric Methods in General Linear Models.* Wiley, New York.

[70] Rao, C. R., and Mitra, S. K. (1971). *Generalized Inverses of Matrices and Its Applications.* Wiley, New York.

[71] Roussas, G. G. (1979). Asymptotic distribution of the log-likelihood function for stochastic processes. *Z. Wahrsch. Verw. Gebiete* **47** 31–46.

[72] Serfling, R. (1968). The Wilcoxon two-sample statistic on strong mixing process. *Ann. Math. Statist.* **39** 1202–1209.

[73] Strasser, H. (1985). *Mathematical Theory of Statistics.* de Gruyter, Berlin.

[74] Swensen, A. R. (1985). The asymptotic distribution of the likelihood ratio for autoregressive time series with a regression trend. *J. Multivariate Anal.* **16** 54–70.

[75] Tran, L. T. (1988). Rank order statistics for time series models. *Ann. Inst. Statist. Math.* **40** 247–260.

[76] Tran, L. T. (1990). Rank statistics for testing randomness against serial dependence. *J. Statist. Plann. Inference* **24** 215–232.

[77] Tsay, R. (1984). Regression models with time series errors. *J. Amer. Statist. Assoc.* **79** 118–124.

[78] van Eeden, C. (1972). An analogue, for signed rank statistics, for Jurečková's asymptotic linearity theorem for rank statistics. *Ann. Math. Statist.* **43** 791–802.

[79] Wald, A. (1943). Tests of statistical hypotheses concerning several parameters when the number of observations is large. *Trans. Amer. Math. Soc.* **54** 426–482.

[80] Wald, A., and Wolfowitz, J. (1943). An exact test for randomness in the non-parametric case based on serial correlation. *Ann. Math. Statist.* **14** 378–388.

[81] Wolfowitz, J. (1949). The power of the classical tests associated with the normal distribution. *Ann. Math. Statist.* **20** 540–551.

[82] Yoshihara, K. I. (1976). Limiting behavior of U-statistics for stationary absolutely regular processes. *Z. Wahrsch. Verw. Gebiete* **35** 237–252.

Reprinted by permission of the publisher,
© 1994 by Academic Press

CHANGE CURVES IN THE PRESENCE OF DEPENDENT NOISE

MADAN L. PURI* and FRITS H. RUYMGAART

Indiana University and Texas Tech University

In image processing the assumption of independent errors is not always realistic. Therefore we consider the problem of change curve estimation when the error process is in a class of stationary random fields. This class contains ARMA fields as special cases. A speed of a.s. uniform convergence is obtained.

1. Introduction and Outline. In this paper a nonparametric method will be discussed to detect the curve along which a random field in the plane is supposed to change its average level. Let us assume that we observe random variables

$$X_{i,j} = \mu_{i,j} + E_{i,j}, \quad (i,j) \in \{1, ..., n\}^2, \tag{1.1}$$

defined for all i, j, and n on one and the same probability space $(\Omega, \mathcal{W}, \mathbb{P})$. The indices (i, j) divide into two groups: within each group the $\mu_{i,j}$ are constant but between groups they are different. The error terms $E_{i,j}$ form a stationary asymptotically decomposable random field; such random fields include in particular linear random fields and more specifically ARMA fields as special cases. Further specifications will be given below. The assumption that the indices i and j have the same range is purely a matter of convenience.

For a further specification of the numbers $\mu_{i,j}$ as well as for asymptotic considerations it turns out to be useful to rescale the two-dimensional "time" and to define the observable process ($n \in \mathbb{N}$, $t = (t_1, t_2)$)

$$X_n(t) = X_{i,j}, \text{ for } \frac{i-1}{n} < t_1 \leq \frac{i}{n}, \ \frac{j-1}{n} < t_2 \leq \frac{j}{n}, \ (i,j) \in \{1, ..., n\}^2, \tag{1.2}$$

on $(0, 1]^2$. It may occasionally be convenient to define $X_n = 0$ elsewhere. Similarly we may construct $\mu_n(t)$ and $E_n(t)$ for $t = (0, 1]^2$ from the $\mu_{i,j}$ and the $E_{i,j}$, setting them 0 elsewhere.

---

* Research supported by the Office of Naval Research Grant N0014-91-J-1020

AMS 1991 Subject Classification: Primary 62G05; Secondary 62M40

Key words and phrases: Change curves, random fields, nearly black objects.

The curve Γ is assumed to be the graph of a sufficiently smooth monotonically nonincreasing function from a to b ($0 \le a < b \le 1$). We also assume that $(0,1]^2 \setminus \Gamma$ is a disconnected set with two components C_1, C_2 having interiors C_1°, C_2° and closures $\overline{C}_1, \overline{C}_2$. We index these sets in such a way that $(0,0) \in C_1$, $(1,1) \in C_2$; note that $\overline{C}_1 \cup \overline{C}_2 = [0,1]^2$. These assumptions entail in particular that, for any $t \in \Gamma$ and $\varepsilon > 0$,

$$\begin{cases} (0,1]^2 \cap [t_1 - \varepsilon, t_1] \times [t_2 - \varepsilon, t_2] \subset \overline{C}_1, \\ (0,1]^2 \cap [t_1, t_1 + \varepsilon] \times [t_2, t_2 + \varepsilon] \subset \overline{C}_2. \end{cases} \tag{1.3}$$

For μ_n, we assume that

$$\mu_n(t) \longrightarrow \mu(t) = c_1 I_{C_1}(t) + c_2 I_{C_2}(t), \quad \text{as } n \to \infty,\ t \in (0,1]^2 \setminus \Gamma, \tag{1.4}$$

where $I(\cdot)$ denotes the indicator function.

The assumption that the $E_{i,j}$ form a stationary asymptotically decomposable random field means that there exist parameters $\rho, \sigma, \tau \in (0, \infty)$ such that for every $n \in \mathbb{N}$ the following holds: there exist $r(n) \ge 1$, $\gamma(n) \in (0,\infty)$, $\delta(n) \in (0,\infty)$ and decompositions satisfying

$$\begin{cases} E_{i,j} = E_{i,j}^{(n)} + \overline{E}_{i,j}^{(n)}, \\ \text{with } \mathcal{L}(E_{i,j}),\ \mathcal{L}(E_{i,j}^{(n)}) \text{ independent of } (i,j) \in \{1, \ldots, n\}^2; \end{cases} \tag{1.5}$$

$$E_{i,j}^{(n)} \perp E_{k,\ell}^{(n)} \text{ for all } (i,j), (k,\ell) \in \{1, \ldots, n\}^2 \quad \text{with } |i-k| \vee |j-\ell| \ge r(n); \tag{1.6}$$

$$\mathbb{P}\left\{ |\overline{E}_{i,j}^{(n)}| \ge \gamma(n) \right\} \le \delta(n); \tag{1.7}$$

$$r(n) = O(n^{2\rho}), \quad \gamma(n) = O(n^{-2\sigma}), \quad \delta(n) = O(n^{-2-\tau}), \quad \text{as } n \to \infty. \tag{1.8}$$

If the $E_{i,j}$ happen to be independent, the conditions are trivially satisfied with all the $\overline{E}_{i,j}^{(n)} = 0$. For applications it is convenient to further specify the orders in (1.8). We will assume that

$$0 \le \rho < \frac{1}{2}, \quad \sigma > 1, \quad \tau > 1. \tag{1.9}$$

Although at first sight the definition of asymptotic decomposability might appear rather technical, the conditions are tailor-made and typically fulfilled for random fields with a Volterra expansion of finite order (Priestley (1981)) and thus it covers a wide range of interesting random fields. By way of an example, let us give simple sufficient conditions for a linear random field to satisfy the conditions; a proof can be found in Puri and Ruymgaart (1991). Let $\xi_{i,j}$, $(i,j) \in \mathbb{Z}^2$, be i.i.d. random variables with

$$\mathbb{E}|\xi_{i,j}|^\nu < \infty, \quad \text{for some } 0 < \nu \leq 1, \tag{1.10}$$

and let us consider the linear field

$$E_{i,j} = \sum_{k\in\mathbb{Z}} \sum_{\ell\in\mathbb{Z}} a_{k,\ell}\xi_{i-k,j-\ell}, \quad (i,j) \in \{1,...,n\}^2, \tag{1.11}$$

where the real numbers $a_{k,\ell}$ satisfy

$$|a_{k,\ell}| = \mathcal{O}((k \vee \ell)^{-\lambda}), \quad \text{as } k \vee \ell \to \infty, \quad \text{for some } \lambda \geq \frac{\tau + \sigma\nu + 2\rho}{\nu\rho}. \tag{1.12}$$

Then the field satisfies the asymptotic decomposability conditions (1.5)–(1.9). Of course the conditions will also be satisfied for $\nu > 1$ in (1.10) but we focussed on the interesting case where not even a first moment might exist. See also Chanda et al. (1990), Chanda and Ruymgaart (1991), and Ruymgaart (1991) for further information about asymptotic decomposability.

Having specified the assumptions regarding the random field, let us now turn to the problem that we want to consider, viz. the construction of estimators for the change curve Γ and investigation of the convergence of the estimators. Rather than estimating the curve itself our procedure boils down to constructing a sequence of stochastic processes $\{\hat{\Gamma}_n,\ n \in \mathbb{N}\}$ on $(0,1]^2$. For almost every $\omega \in \Omega$ there appears to exist $n(\omega) \in \mathbb{N}$ such that $\hat{\Gamma}_n$ is 0 everywhere except on a strip around the curve Γ, provided that $n \geq n(\omega)$. As the sample size tends to ∞ the width of these strips tends to zero. On Γ itself the processes $\hat{\Gamma}_n$ are almost everywhere uniformly close to the difference in level $c_1 - c_2$ in the case where the error distribution is symmetric about 0. Various methods are proposed in the literature to detect change but usually for one–dimensional time and i.i.d. errors; see, e.g., Pettitt (1979), Wolfe and Schechtman (1984), and Csörgö and Horvath (1989). Van de Geer (1988) considers, more generally, change in regression including some results for multidimensional regression but with independent errors. For independent variables and one dimensional time, Carlstein (1988) proposed a very general method not restricted to differences in location. Recently Tsybakov (1991) studied the problem of image estimation, which subsumes multidimensional time, but also this author assumes the errors to be i.i.d.. A closely related reference is also Carlstein and Krishnamoorthy (1992). See Parzen (1991) for general remarks regarding the analysis of change.

The method proposed here consists in reducing the model to a "nearly black object" as studied in Donoho et al. (1991). The procedure consists of three steps where the first is a robust smoothing of the data like in Tsybakov (1991). See also the survey paper by Pitas and Venetsanopoulos (1992) on nonlinear smoothing and the use of linear combinations of order statistics. The smoothed field is supposed to be pretty much stabilized around a certain fixed value on C_1° and around a different fixed value on C_2°. In the second step the smoothed process is convoluted with a special type of kernel, having an effect similar to differentiation but applicable to any function. Consequently the result of this convolution is supposed to be a process which is close to 0 everywhere except for a strip around the curve Γ. In fact a more sophisticated form of this second step is known as a wavelet transform. In the recent literature such wavelet transforms are extensively studied. For the relation with change detection in a deterministic setting we refer in particular to Mallat and Zhong (1992). The purpose of the third step is to further enhance the features of the last process by replacing each value by 0 except when it exceeds a certain suitably chosen threshold. The use of such a threshold is intuitively clear and can be mathematically justified as arising from a nonlinear (L_1−) penalty function in a least squares setting (Donoho et al. (1991)).

Section 2 is devoted to a precise description of the procedure sketched above, and in Section 3 we present results on the speed of uniform almost sure convergence and some other asymptotic results. In Section 4 we briefly comment on the results and the assumptions.

2. Description of the Estimation Procedure. At stage n, let us choose $\varepsilon = \varepsilon_n \in (0,1)$ such that

$$\lim_{n\to\infty} n^{\delta/2}\varepsilon = 1, \quad \text{for some } 0<\delta<2 \qquad \text{and, for convenience,} \quad n\,\varepsilon,\ \varepsilon^{-1} \in \mathbb{N}. \tag{2.1}$$

See below for a choice of δ. We divide $(0,1]^2$ into intervals

$$T_n(i,j) = (\varepsilon(i-1), \varepsilon\, i] \times (\varepsilon(j-1), \varepsilon\, j], \quad (i,j) \in \{1,\ldots,1/\varepsilon\}^2. \tag{2.2}$$

Let us introduce the index sets

$$\mathcal{I}_{n,\Gamma} = \{(i,j) : T_n(i,j) \cap \Gamma \neq \phi\}; \text{ note that } \#\mathcal{I}_{n,\Gamma} \le 2\, n^{\delta/2}; \tag{2.3}$$

$$\mathcal{I}_{n,\alpha} = \{(i,j) : T_n(i,j) \cap C_\alpha \neq \phi\} \setminus \mathcal{I}_{n,\Gamma}, \quad \text{for } \alpha = 1,2. \tag{2.4}$$

Of course we have $\mathcal{I}_{n,\Gamma} \cup \mathcal{I}_{n,1} \cup \mathcal{I}_{n,2} = \{1,\ldots,n\}^2$. It will also be convenient to define the subsets

$$\Gamma_\varepsilon = \bigcup_{(i,j)\in\mathcal{I}_{n,\Gamma}} T_n(i,j); \tag{2.5}$$

$$C_{\alpha,\varepsilon} = \bigcup_{(i,j)\in\mathcal{I}_{n,\alpha}} T_n(i,j), \quad \text{for } \alpha = 1,2. \tag{2.6}$$

It should be noted that

$$\Gamma_\varepsilon \downarrow \Gamma, \quad C_{\alpha,\varepsilon} \uparrow C_\alpha^\circ, \quad \text{for } \alpha = 1,2, \text{ as } n \uparrow \infty. \tag{2.7}$$

Step 1. Let $J : [0,1] \to [0,\infty)$ be a score function which is 0 in neighborhoods of 0 and 1 (but not identically equal to 0) and which has a bounded continuous first derivative on $(0,1)$. We define

$$\hat{F}_{\varepsilon,t}(x) = \left(\frac{1}{\varepsilon n}\right)^2 \sum\sum_{(\frac{k}{n},\frac{\ell}{n})\in T_n(i,j)} I_{(-\infty,x]}(X_{k,\ell}), \quad x \in I\!R, \tag{2.8}$$

for $t \in T_n(i,j)$ and $(i,j) \in \{1,\ldots,1/\varepsilon\}^2$. This is the empirical d.f. for the block of observables with $(k,\ell) \in \{(i-1)\varepsilon\, n+1,\ldots,i\,\varepsilon\, n\} \times \{(j-1)\varepsilon\, n+1,\ldots,j\,\varepsilon\, n\}$, containing $(\varepsilon\, n)^2$ elements. The original process X_n will now be replaced with

$$Y_n(t) = \int_0^1 \hat{F}_{\varepsilon,t}^{-1}(s)J(s)ds, \quad t \in T_n(i,j), \ (i,j) \in \{1,\ldots,1/\varepsilon\}^2. \tag{2.9}$$

This process is a step function with constant values on the $T_n(i,j)$.

For $t \in C_{\alpha,\varepsilon}$, the $X_n(t)$ have a common d.f. F_α, say, and the $Y_n(t)$ are robust estimators of $\int_0^1 F_\alpha^{-1}(s)J(s)ds$, $\alpha = 1,2$. The precise values of these location functionals is not important: the only thing that matters is that they are different. Under the present assumptions F_1 and F_2 are obviously translates of each other and hence we have indeed

$$\int_0^1 F_1^{-1}(s)J(s)ds \neq \int_0^1 F_2^{-1}(s)J(s)ds. \tag{2.10}$$

More specifically if F_1 is continuous and symmetric about some point (so that F_2 enjoys these properties as well) we even have

$$\int_0^1 F_\alpha^{-1}(s)J(s)ds = c_\alpha, \quad \alpha = 1,2, \tag{2.11}$$

provided that J is symmetric about $1/2$, where the c_α are defined in (1.4).

Step 2. For the same ε as in (2.1), let us define the kernel $K_\varepsilon : I\!R^2 \to I\!R$ by

$$K_\varepsilon(t) = \frac{1}{\varepsilon^{2q}}\left\{I_{[0,\varepsilon^q]^2}(t) - I_{[-\varepsilon^q,0]^2}(t)\right\} = K_\varepsilon^+(t) - K_\varepsilon^-(t), \quad t \in I\!R^2, \tag{2.12}$$

for some $0 < q < 1$, a suitable value of which will be given below. This function is particularly well suited to detect discontinuities along curves running "northwest", like the present curve Γ, when used as a convolution kernel. It satisfies $K_\varepsilon * c = 0$ for any constant function c on $\mathbb{R}^2$ and it has the property that $K_\varepsilon * \mu$, with μ defined in (1.4), will be 0 outside a strip with a width of order $\varepsilon^q = \varepsilon_n^q$ (as $n \to \infty$). The smooth process Y_n that stabilizes around the two values in (2.10) is now replaced by the process

$$Z_n(t) = K_\varepsilon * Y_n(t), \quad t \in (0,1]^2, \tag{2.13}$$

$Z_n = 0$ elsewhere. The present choice of kernel is easy to work with but a smooth version with essentially the same properties might occasionally be more desirable since convolution with a smooth kernel yields a process Z_n that is also smooth.

Step 3. The process Z_n is likely to reveal the position of the change curve Γ by a ridge in its surface. It is possible to further improve on the signal to noise ratio by applying a nonlinear L_1-smoothing technique employed by Donoho et al. (1991), leading to

$$\hat{\Gamma}_n(t) = Z_n(t) I_{[\lambda_n, \infty)}(|Z_n(t)|), \quad t \in (0,1]^2, \tag{2.14}$$

for a suitably chosen threshold $\lambda_n \in (0, \infty)$. It is intuitively clear that $\hat{\Gamma}_n$ should usually better display the change curve than Z_n.

3. Some Asymptotics. The following facts about linear combinations of order statistics that turn out to be expedient in the analysis of the process Y_n can be found in Boos (1979); see also Ruymgaart (1981). Let $\mathcal{F}$ denote the class of all d.f.'s on $\mathbb{R}$ and consider the functional $\theta : \mathcal{F} \to \mathbb{R}$, defined by

$$\theta(F) = \int_0^1 F^{-1}(s) J(s) ds, \quad F \in \mathcal{F}. \tag{3.1}$$

Note that $Y_n(t) = \theta(\hat{F}_{\varepsilon,t})$. For any $F \in \mathcal{F}$ we have

$$Y_n(t) - \theta(F) = \int_{-\infty}^{\infty} \left\{ \hat{F}_{\varepsilon,t}(x) - F(x) \right\} J(F(x)) dx + R_{\varepsilon,t}, \tag{3.2}$$

where the remainder $R_{\varepsilon,t}$ equals

$$R_{\varepsilon,t} = \frac{1}{2} \int_{-\infty}^{\infty} \left\{ \hat{F}_{\varepsilon,t}(x) - F(x) \right\}^2 J'(\tilde{F}_{\varepsilon,t}(x)) dx, \tag{3.3}$$

with $\tilde{F}_{\varepsilon,t}(x)$ between $\hat{F}_{\varepsilon,t}(x)$ and $F(x)$ for every $x \in \mathbb{R}$.

Assumption 3.1. It will be assumed, often without explicit reference that (1.4)–(1.9), as well as (2.1) and the assumption that F_1 and F_2 are continuous, are fulfilled.

Lemma 3.1. *Provided that* $0 < \zeta < (1-\rho)/2$ *we have, for* $\alpha = 1, 2$,

$$n^{(2-\delta)\zeta} \sup_{t \in C_{\alpha,\epsilon}} \sup_{x \in \mathbb{R}} \left| \hat{F}_{\epsilon,t}(x) - F_\alpha(x) \right| \xrightarrow{a.s.} 0, \quad \text{as } n \to \infty. \tag{3.4}$$

Proof. Recall that $t \mapsto \hat{F}_{\epsilon,t}(x)$ is constant for $t \in T_n(i,j)$. Let us take $T_n(i,j) \subset C_{\alpha,\epsilon}$ and observe that to prove the present asymptotic properties we may as well assume that $c_\alpha = 0$ so that we are dealing with the $E_{i,j}$. Each $\hat{F}_{\epsilon,t}$ being based on $(\varepsilon\, n)^2$ observations, it follows from a modification for two-dimensional indices of Chanda and Ruymgaart (1991, Theorem 2.1) that, for any $c \in (0,\infty)$,

$$\begin{aligned} &\mathbb{P}\left(\left\{\sup_{x \in \mathbb{R}} n^{(2-\delta)\zeta} |\hat{F}_{\epsilon,t}(x) - F_\alpha(x)| \geq c\right\} \cap \Omega_n\right) \\ &\leq C n^{2-\delta} \exp\left(-Ac^2 n^{(2-\delta)(1-\rho-2\zeta)}\right), \end{aligned} \tag{3.5}$$

where $\Omega_n = \{\max_{i,j} |\bar{E}_{i,j}^{(n)}| < \gamma(n)\}$ (see (1.7)) and $A, C \in (0,\infty)$ are fixed numbers (independent of n).

Writing $(i\varepsilon, j\varepsilon) = t_{i,j}$, it follows that

$$\begin{aligned} &\mathbb{P}\left(\left\{\sup_{t \in C_{\alpha,\epsilon}} \sup_{x \in \mathbb{R}} n^{(2-\delta)\zeta} |\hat{F}_{\epsilon,t}(x) - F_\alpha(x)| \geq c\right\} \cap \Omega_n\right) \\ \leq\ & n^\delta \mathbb{P}\left(\left\{\sup_{x \in \mathbb{R}} n^{(2-\delta)\zeta} |\hat{F}_{\epsilon,t_{i,j}}(x) - F_\alpha(x)| \geq c\right\} \cap \Omega_n\right) \\ \leq\ & C\, n^\delta n^{2-\delta} \exp\left(-Ac^2\, n^{(2-\delta)(1-\rho-2\zeta)}\right). \end{aligned} \tag{3.6}$$

Because $\mathbb{P}(\Omega_n^c) = O(n^{-2\tau})$, as $n \to \infty$, and by assumption $(2-\delta)(1-\rho-2\zeta) > 0$ the claimed almost sure convergence follows. ∎

Lemma 3.2. *For* $0 < \zeta < (1-\rho)/2$, *the processes* Y_n *satisfy*

$$n^{(2-\delta)\zeta} \sup_{t \in C_{\alpha,\epsilon}} |Y_n(t) - \theta(F_\alpha)| \xrightarrow{a.s.} 0, \quad \text{as } n \to \infty. \tag{3.7}$$

Proof. For each $t \in C_{\alpha,\varepsilon}$, the absolute value of the first term on the right in (3.2), with $\theta(F)$ replaced by $\theta(F_\alpha)$, is obviously bounded by

$$\begin{aligned} &\sup_{t \in C_{\alpha,\epsilon}} \sup_{x \in \mathbb{R}} |\hat{F}_{\epsilon,t}(x) - F_\alpha(x)| \int_{-\infty}^{\infty} J(F(x))dx \\ &= o\left(n^{-(2-\delta)\zeta}\right), \quad \text{as } n \to \infty, \text{ a.s.}, \end{aligned} \tag{3.8}$$

according to (3.4). For $R_{n,t}$ in (3.3), let us note that the assumptions on J entail that $J' = 0$ on $[0,a] \cup [1-a,1]$ for some $0 < a < \frac{1}{2}$. Since $\tilde{F}_{\varepsilon,t}$ is between $\hat{F}_{\varepsilon,t}$ and F_α, it follows once more from (3.4) that the integration in (3.3) is effectively restricted to an interval contained in $[F_\alpha^{-1}(a/2),\ F_\alpha^{-1}(1-a/2)]$ for n sufficiently large, a.s. It follows that

$$\begin{aligned}\sup_{t \in C_{\alpha,\varepsilon}} |R_{\varepsilon,t}| &\leq \left\{ \frac{1}{2} \sup_{t \in C_{\alpha,\varepsilon}} \sup_{x \in \mathbb{R}} |\hat{F}_{\varepsilon,t}(x) - F_\alpha(x)|^2 \right\} \\ &\quad \times \left\{ F_\alpha^{-1}(1-a/2) - F_\alpha^{-1}(a/2) \right\} \sup_{a/2 \leq s \leq 1-a/2} |J'(s)| \\ &= o(n^{-2(2-\delta)\zeta}), \quad \text{as } n \to \infty, \text{ a.s.}\end{aligned} \tag{3.9}$$

Hence the order in (3.8) prevails. ∎

Let us define strips

$$\Gamma_{\varepsilon,k} = \bigcup_{(t_1,t_2) \in \Gamma} [t_1 - k\varepsilon^q,\ t_1 + k\varepsilon^q] \times [t_2 - k\varepsilon^q,\ t_2 + k\varepsilon^q], \quad k \in \mathbb{N}, \tag{3.10}$$

around the curve and briefly write

$$\bar{\Gamma}_{\varepsilon,k} = (0,1] \times (0,1] \setminus \Gamma_{\varepsilon,k}. \tag{3.11}$$

Lemma 3.3. *For $0 < \zeta < (1-\rho)/2$, the processes Z_n satisfy*

$$n^{(2-\delta)\zeta} \sup_{t \in \bar{\Gamma}_{\varepsilon,2}} |Z_n(t)| \xrightarrow{a.s.} 0, \quad \text{as } n \to \infty. \tag{3.12}$$

Proof. According to (3.7), there exists a measurable $N \subset \Omega$ with $\mathbb{P}(N) = 0$, and $c_n = c_n(\omega) \downarrow 0$ (as $n \to \infty$) for each $\omega \in \Omega \setminus N$ such that we have

$$\begin{aligned}\theta(F_\alpha) - c_n n^{-(2-\delta)\zeta} &\leq K_\varepsilon^\pm * (\theta(F_\alpha) - c_n n^{-(2-\delta)\zeta}) \\ &\leq K_\varepsilon^\pm * Y_n \leq K_\varepsilon^\pm * (\theta(F_\alpha) + c_n n^{-(2-\delta)\zeta}) \leq \theta(F_\alpha) + c_n n^{-(2-\delta)\zeta},\end{aligned} \tag{3.13}$$

for $t \in C_{\alpha,\varepsilon} \cap \bar{\Gamma}_{\varepsilon,2}$. This entails

$$|Z_n(t)| \leq 2\, c_n n^{-(2-\delta)\zeta}, \quad t \in \bar{\Gamma}_{\varepsilon,2}, \tag{3.14}$$

and the lemma follows. ∎

In order to investigate the behavior of the estimators at points t on the curve Γ we need some further preparations. First of all it should be noted that all we have been doing so far for the observations $X_{i,j}$ could also be done – mutatis mutandis – for the absolute values $|X_{i,j}|$. Let us denote the successive processes based on the absolute values by $|X|_n(t)$, $|Y|_n(t)$, and $|Z|_n(t)$, $t \in (0,1]^2$, and let us write $|F|_\alpha$ for the d.f. of $|X|_n(t)$ for $t \in C_{\alpha,\varepsilon}$. In particular we have

$$n^{(2-\delta)\zeta} \sup_{t \in C_{\alpha,\varepsilon}} \left| |Y|_n(t) - \theta(|F|_\alpha) \right| \xrightarrow{a.s.} 0, \quad \text{as } n \to \infty, \tag{3.15}$$

and of course $|Y_n(t)| \le |Y|_n(t)$ for all t.

For an arbitrary $t \in \Gamma$ let us write

$$\begin{cases} S_{1,\varepsilon}(t) = [t_1 - \varepsilon^q, t_1] \times [t_2 - \varepsilon^q, t_2], \\ S_{2,\varepsilon}(t) = [t_1, t_1 + \varepsilon^q] \times [t_2, t_2 + \varepsilon^q]; \end{cases} \tag{3.16}$$

$$\begin{cases} Q_{1,\varepsilon}(t) = [t_1 - \varepsilon^q, t_1 - \varepsilon] \times [t_2 - \varepsilon^q, t_2 - \varepsilon], \\ Q_{2,\varepsilon}(t) = [t_1 + \varepsilon, t_1 + \varepsilon^q] \times [t_2 + \varepsilon, t_2 + \varepsilon^q]. \end{cases} \tag{3.17}$$

Note the relationship between the $S_{1,\varepsilon}(t)$, $S_{2,\varepsilon}(t)$ and $\Gamma_{\varepsilon,1}$. Finally, let us write

$$\Gamma^{(\varepsilon)} = \Gamma \cap [\varepsilon^q, 1 - \varepsilon^q]^2, \tag{3.18}$$

for the part of the curve that doesn't come too close to the boundary of the unit square and let us note that

$$C_{\alpha,\varepsilon} \supset S_{\alpha,\varepsilon}(t) \cap \Gamma_\varepsilon^c \supset Q_{\alpha,\varepsilon}(t), \quad t \in \Gamma^{(\varepsilon)}, \ \alpha = 1, 2. \tag{3.19}$$

Lemma 3.4. *For $0 < \xi < \delta(1-q)/2 \wedge (2-\delta)\zeta$, the processes Z_n satisfy*

$$n^{\xi} \sup_{t \in \Gamma^{(\varepsilon)}} |Z_n(t) - \{\theta(F_1) - \theta(F_2)\}| \xrightarrow{a.s.} 0, \quad \text{as } n \to \infty. \tag{3.20}$$

Proof. First observe that, for $t \in \Gamma^{(\varepsilon)}$,

$$\begin{aligned} &|Z_n(t) - \{\theta(F_1) - \theta(F_2)\}| \\ &\le \left| \varepsilon^{-2q} \iint_{S_{1,\varepsilon}(t)} Y_n(s) ds - \theta(F_1) \right| + \left| \varepsilon^{-2q} \iint_{S_{2,\varepsilon}(t)} Y_n(s) ds - \theta(F_2) \right|. \end{aligned} \tag{3.21}$$

Since both terms on the right in (3.21) can be dealt with similarly, let us focus on the first one. Application of (3.7), (3.15), and (3.19) yields that, almost surely,

$$\left| \varepsilon^{-2q} \iint_{S_{1,\varepsilon}(t)} Y_n(s) ds - \theta(F_1) \right| \tag{3.22}$$

$$\leq \left|\varepsilon^{-2q}\iint_{Q_{1,\varepsilon}(t)} Y_n(s)ds - \theta(F_1)\right| + \varepsilon^{-2q}\iint_{S_{1,\varepsilon}(t)\setminus Q_{1,\varepsilon}(t)} |Y|_n(s)ds$$

$$\leq \left\{1-\varepsilon^{-2q}(\varepsilon^q-\varepsilon)^2\right\}\theta(F_1) + \varepsilon^{-2q}(\varepsilon^q-\varepsilon)^2 n^{-(2-\delta)\zeta} + \varepsilon^{-2q}\iint_{S_{1,\varepsilon}(t)\setminus Q_{1,\varepsilon}(t)} |Y|_n(s)ds$$

$$\leq 2\varepsilon^{1-q}\theta(F_1) + n^{-(2-\delta)\zeta} + \varepsilon^{-2q}\iint_{S_{1,\varepsilon}(t)\setminus Q_{1,\varepsilon}(t)} |Y|_n(s)ds.$$

It remains to obtain an a.s. upper bound for the process $|Y|_n(s)$ for $s \in \Gamma_\varepsilon$. For any such s, this process is based on $(\varepsilon n)^2$ sample elements $|X_{i,j}|$ a fraction of which has d.f. $|F|_1$ and the remainder fraction of which has d.f. $|F|_2$. Since all observations are nonnegative, so is the linear combination of order statistics. By completing the fractions to two samples each of size $(\varepsilon n)^2$, and one with sample elements having d.f. $|F|_1$ and the other with sample elements having d.f. $|F|_2$, we may add the corresponding linear combinations of order statistics and arrive at a process $|Y^*|_n(s)$ which clearly satisfies, for almost every $\omega \in \Omega$,

$$0 \leq |Y|_n(s) \leq |Y^*|_n(s) \leq \theta(|F|_1) + \theta(|F|_2) + c_n^*(\omega)n^{-(2-\delta)\zeta}, \tag{3.23}$$

where $c_n^*(\omega) \downarrow 0$, as $n \to \infty$, and where $s \in \Gamma_\varepsilon$. It follows that, almost surely,

$$\begin{aligned}\varepsilon^{-2q}\iint_{S_{1,\varepsilon}(t)\setminus Q_{1,\varepsilon}(t)} &|Y|_n(s)ds \\ &\leq 2\varepsilon^{1-q}\left\{\theta(|F|_1) + \theta(|F|_2) + n^{-(2-\delta)\zeta}\right\}.\end{aligned} \tag{3.24}$$

The range for ξ is easily obtained by combining the orders of magnitude appearing in (3.22) and (3.24). ∎

Theorem 3.1. *Let Assumption 3.1 be satisfied. Then for almost every $\omega \in \Omega$,*

$$\begin{cases} \text{there exists } n(\omega) \text{ such that } \sup_{t\in\bar{\Gamma}_{\varepsilon,2}} |\hat{\Gamma}_n(t)| = 0, & \text{for } n \geq n(\omega), \\ \sup_{t\in\Gamma^{(\varepsilon)}} n^{\xi}|\hat{\Gamma}_n(t) - \{\theta(F_1)-\theta(F_2)\}| \longrightarrow 0, & \text{as } n \to \infty, \end{cases} \tag{3.25}$$

for any $0 < \zeta < (1-\rho)/2$, and $0 < \xi < \delta(1-q)/2 \wedge (2-\delta)\zeta$, provided that we choose the threshold $\lambda_n = \lambda n^{-\eta}$, for some $0 < \eta < (2-\delta)\zeta$ and $0 < \lambda < \infty$.

Proof. This is immediate from the lemmas. For $t \in \bar{\Gamma}_{\varepsilon,2}$ we have $|\hat{\Gamma}_n(t)| \leq |Z_n(t)|$ so that the first part of the equality in (3.25) is immediate from (3.12). We see from (3.20) that for n sufficiently large $|Z_n(t)| \geq |\theta(F_1)-\theta(F_2)| - n^{-\xi}$, uniformly for $t \in \Gamma^{(\varepsilon)}$, a.s. Hence eventually we have

$|Z_n(t)| > \lambda_n$ for all $t \in \Gamma^{(\varepsilon)}$ simultaneously, meaning that $\hat{\Gamma}_n(t) = Z_n(t)$, a.s., for all $t \in \Gamma^{(\varepsilon)}$ simultaneously provided that n is sufficiently large. ∎

The theorem enables us to eventually identify the strip $\Gamma_{\varepsilon,3}$ in which $\Gamma^{(\varepsilon)}$ must lie. As an estimator of $\Gamma^{(\varepsilon)}$ we might take any curve in this strip that satisfies the prior assumptions. Since the width of the strip is of order $n^{-\delta q/2}$, as $n \to \infty$, this settles a speed of almost sure convergence. For practical purposes data-driven parameter selection is one of the problems that remains to be considered. Although a slow convergence rate of the process Z_n due to strong dependence (i.e. large ρ) doesn't seem to affect the order of ε asymptotically, it is very likely to have an impact on the constants determining the actual strip width.

4. Some Comments.

1. The curves Γ that we consider here run "northwest". When we know this we should choose kernels that run "northeast". Conversely, curves that run "northeast" can be better detected by kernels running "northwest". When we don't know what kind of curve we are dealing with, or when the regions of constant values are separated by a simple closed curve we might screen the entire domain twice: once with a northeast and once with a northwest kernel.

2. As we observed already in the introduction, the asymptotic decomposability condition is trivially fulfilled in the case of i.i.d. errors. Under mild assumptions, linear random fields satisfy the conditions, as we have seen. Since bilinear processes for one-dimensional time are shown to be asymptotically decomposable (Chanda and Ruymgaart (1991)) we may conjecture that bilinear fields also satisfy that condition, in which case an important class of nonlinear processes would be included. Of course, asymptotic decomposability in its present generality is hard to verify. Our main purpose was, however, to show that our search procedure has rather good asymptotic properties for a broad class of error fields that goes far beyond the i.i.d. case.

Acknowledgments. We are grateful to Manny Parzen for sending a reprint of his paper and to David Donoho for pointing out the usefulness of kernels like those in (2.12) to detect boundaries and for sending a copy of his paper. We would also like to thank Ed Carlstein for some useful remarks.

REFERENCES

Boos, D. D. (1979). A differential for L-statistics. *Ann. Statist.* **7**, 955–959.

Carlstein, E. (1988). Nonparametric change point estimation. *Ann. Statist.* **16**, 188–197.

CARLSTEIN, E. and KRISHNAMOORTHY, C. (1992). Boundary estimation. *J. Amer. Statist. Assoc.* **87**, 430–438.

CHANDA, K. C., PURI, M. L. and RUYMGAART, F. H. (1990). Asymptotic normality of L-statistics based on $m(n)$-decomposable time series. *J. Multivariate Anal.* **35**, 260–275.

CHANDA, K. C. and RUYMGAART, F. H. (1991). Curve estimation for $m(n)$-decomposable time series including bilinear processes. *J. Multivariate Anal.* **38**, 149–166.

CSÖRGÖ, M. and HORVÁTH, L. (1988). Invariance principles for changepoint problems. *J. Multivariate Anal.* **27**, 151–168.

DONOHO, D. L., JOHNSTONE, I. M., HOCH, J. C. and STERN, A. S. (1991). Maximum entropy and the nearly black object. *J. Roy. Statist. Soc. Ser. B* **53**, 41–81.

MALLAT, S. and ZHONG, S. (1992). Wavelet transform maxima and multiscale edges. In: *Wavelets and Their Applications,* pp. 67–104 (M.B. Ruskai, Ed.). Jones and Bartlett, Boston.

PARZEN, E. (1991). Comparison change analysis. To appear in: *Proc. Internat. Symp. Nonparametric Statist.* (A. K. Saleh, Ed.). Elsevier, Amsterdam.

PETTITT, A. N. (1979). A non-parametric approach to the change point problem. *Appl. Statist.* **28**, 126–135.

PITAS, I. and VENETSANOPOULOS, A. N. (1992). Order statistics in digital image processing. *Proc. IEEE* **80**, 1892–1921.

PRIESTLEY, M. B. (1981). *Spectral Analysis and Time Series,* Vol. 1. Academic Press, New York.

PURI, M. L. and RUYMGAART, F. H. (1991). Nonparametric prediction for random fields. *Stochastic Process. Appl.* **48**, 139–156.

RUYMGAART, F. H. (1981). A robust principal component analysis. *J. Multivariate Anal.* **11**, 485–497.

RUYMGAART, F. H. (1991). Conditional empirical, quantile and difference processes for a large class of time series with applications. *J. Statist. Pl. Inf.*, to appear.

TSYBAKOV, A. B. (1991). Nonparametric techniques in image estimation. In: *Nonparametric Functional Estimation and Related Topics,* pp. 669–677 (G. Roussas, Ed.). Kluwer, Dordrecht.

VAN DE GEER, S. (1988). *Regression Analysis and Empirical Processes.* CWI-tract 45, Centre for Mathematics and Computer Science, Amsterdam.

WOLFE, D. A. and SCHECHTMAN, E. (1984). Nonparametric statistical procedures for the changepoint problem. *J. Statist. Plann. Inference* **9**, 389–396.

DEPARTMENT OF MATHEMATICS
INDIANA UNIVERSITY
BLOOMINGTON, IN 47405

DEPARTMENT OF MATHEMATICS
TEXAS TECH UNIVERSITY
LUBBOCK, TX 79409

Reprinted by permission of the publisher,
© 1994 by The Institute of Mathematical Statistics

A MULTIVARIATE WALD-WOLFOWITZ RANK TEST AGAINST SERIAL DEPENDENCE

MARC HALLIN and MADAN L. PURI*

Université Libre de Bruxelles and Indiana University

Key words and phrases: Wald-Wolfowitz rank test, rank cross-covariance matrix, multivariate ARMA models, multivariate portmanteau test, multivariate model identification.
AMS 1985 subject classifications: Primary 62M10, 62G10.

ABSTRACT

Rank-based cross-covariance matrices, extending to the case of multivariate observed series the (univariate) rank autocorrelation coefficients introduced by Wald and Wolfowitz (1943), are considered. A permutational central limit theorem is established for the joint distribution of such matrices, under the null hypothesis of (multivariate) randomness as well as under contiguous alternatives of (multivariate) ARMA dependence. A rank-based, permutationally distribution-free test of the portmanteau type is derived, and its asymptotic local power is investigated. Finally, a modified rank-based version of Tiao and Box's model specification procedure is proposed, which is likely to be more reliable under non-Gaussian conditions, and more robust against gross errors.

RÉSUMÉ

Des matrices de covariances croisées fondées sur les rangs, généralisant au cas des séries multivariées les coefficients d'autocorrélation de rangs introduits par Wald et Wolfowitz (1943) sont considérées. Un théorème central-limite permutationnel est établi pour ces matrices, sous l'hypothèse que la série sous-jacente constitue la réalisation d'un bruit blanc multivarié, ainsi que sous des contre-hypothèses contiguës de dépendance ARMA. Un test de rangs du type portemanteau est également construit, et sa puissance asymptotique locale est explicitement calculée. Enfin, une version fondée sur les rangs de la procédure d'identification de Tiao et Box est proposée. Celle-ci est plus fiable que la procédure usuelle sous des conditions non gaussiennes, et plus robuste par rapport à la présence de valeurs aberrantes.

1. RANK-BASED CROSS-COVARIANCE MATRICES OF THE SPEARMAN-WALD-WOLFOWITZ TYPE

Denote by $\mathbf{X}^{(n)} = (\mathbf{X}_1^{(n)}, \ldots, \mathbf{X}_t^{(n)}, \ldots, \mathbf{X}_n^{(n)})$ a series of m-variate observations $\mathbf{X}_t^{(n)} = (X_{t,1}^{(n)}, \ldots, X_{t,m}^{(n)})^\top$, $t = 1, \ldots, n$. We are interested in the null hypothesis $\mathcal{H}^{(n)}$ of randomness under which the observations $\mathbf{X}_t^{(n)}$ are independently and identically distributed, according to some unspecified density $f(\mathbf{x})$, $\mathbf{x} = (x_1, \ldots, x_m)^\top \in \mathbb{R}^m$. Whenever f needs to be specified, the resulting subhypothesis will be denoted by $\mathcal{H}_f^{(n)}$. The alternatives we are considering are serial-dependence alternatives, with emphasis on multivariate ARMA dependence.

Much attention has been devoted recently to the one-dimensional version of this nonparametric problem; see, e.g., Dufour (1981), Dufour and Roy (1985, 1986), Chan and Tran (1992) or Shaug and Tjøstheim (1993). In this univariate version [$m = 1$; observed

*Research supported by the Office of Naval Research Contract N00014-91-J-1020.

series $\mathbf{X}^{(n)} = (X_1^{(n)}, \ldots, X_n^{(n)})$], usual invariance arguments lead to the consideration of rank-based testing procedures; see Hallin and Puri (1992). In this latter context, Wald and Wolfowitz (1943) proposed test statistics of the form (up to additive and multiplicative constants) $S_i^{(n)} = n^{-1} \sum_{t=1}^{n} R_t^{(n)} R_{t-i}^{(n)}$, where $R_t^{(n)}$ denotes, as usual, the rank of $X_t^{(n)}$ among $X_1^{(n)}, \ldots, X_n^{(n)}$, and with the (convenient but rather arbitrary) convention that $R_s^{(n)}$, $s = 0, -1, \ldots, -i+1$, is set equal to $R_{n+s}^{(n)}$. Clearly, $S_i^{(n)}$ is a (circular) serial version, of lag i, of Spearman's ordinary rank correlation coefficient (still, up to additive and multiplicative constants). The performance of such test statistics in a time series context have been investigated by Knoke (1977) and Bartels (1982), among others—see also Hallin and Mélard (1988). A systematic theoretical study of serial, rank-based statistics is provided in Hallin *et al.* (1985, 1987) and Hallin and Puri (1988, 1991, 1994).

Our objective here is to extend Wald and Wolfowitz's idea to the problem of testing for *multivariate* randomness. Due to the greater complexity inherent in the nature of multivariate variables, the development of multivariate theories—especially in the area of rank-based procedures—typically has been much slower than that of their univariate counterparts. Multivariate rank-based statistics for serial-dependence problems have been introduced in Hallin *et al.* (1989), from which several asymptotic results are used here.

The multivariate generalization of Wald and Wolfowitz's rank-based autocorrelation coefficients quite naturally takes the form of rank-based cross-covariance matrices. Letting $R_{t,i}^{(n)}$ denote the rank of $X_{t,i}^{(n)}$ among $X_{1,i}^{(n)}, \ldots, X_{n,i}^{(n)}$, put $\mathbf{R}_t^{(n)} = (R_{t,1}^{(n)}, \ldots, R_{t,m}^{(n)})^{\mathsf{T}}$, and $\mathbf{R}^{(n)} = (\mathbf{R}_1^{(n)}, \ldots, \mathbf{R}_n^{(n)})$: $\mathbf{R}^{(n)}$ is called the *rank-collection matrix* associated with the observed series $\mathbf{X}^{(n)}$.

It is well known that, unless $m = 1$, the rank vectors $\mathbf{R}_t^{(n)}$ are not distribution-free, even under the null hypothesis of randomness. If however we denote by $\mathbf{R}_*^{(n)}$ the rank-collection matrix resulting from $\mathbf{R}^{(n)}$ on rearranging the columns in such a manner that the first row yields elements $1, 2, \ldots, n$ in ascending order, then the $n!$ possible permutations of the columns of $\mathbf{R}_*^{(n)}$ are conditionally equiprobable: the rank vectors $\mathbf{R}_t^{(n)}$ are thus conditionally distribution-free, given the "ordered" rank-collection matrix $\mathbf{R}_*^{(n)}$.

The consideration of this conditional distribution actually also follows from unbiasedness and Neyman-structure arguments. The matrix $\mathbf{X}_*^{(n)}$ resulting from $\mathbf{X}^{(n)}$ on rearranging the columns as in $\mathbf{R}_*^{(n)}$ indeed can be shown to be sufficient complete under $\mathcal{H}^{(n)}$. Conditioning, as usual, upon this sufficient complete statistic yields the permutational distribution of $(\mathbf{X}_1^{(n)}, \ldots, \mathbf{X}_n^{(n)})$. Additional componentwise invariance arguments then lead to the permutational distribution of $\mathbf{R}^{(n)}$, which is precisely the conditional distribution of the rank vectors given in $\mathbf{R}_*^{(n)}$.

Define the *rank cross-covariance matrix of lag i* as

$$\boldsymbol{\Gamma}_i^{(n)} = (n-i)^{-1} \sum_{t=i+1}^{n} (\mathbf{R}_t^{(n)})(\mathbf{R}_{t-i}^{(n)})^{\mathsf{T}} - \mathbf{m}_{i*}^{(n)}, \qquad i \geq 1, \tag{1.1}$$

where $m_{i*}^{(n)}$ denotes the permutational mean of $(n-i)^{-1} \sum (\mathbf{R}_t^{(n)})(\mathbf{R}_{t-i}^{(n)})^{\mathsf{T}}$. Denote by $\gamma_{i;k,l}^{(n)}$ the elements of $\boldsymbol{\Gamma}_i^{(n)}$: for $k \neq l$, $\gamma_{i;k,l}^{(n)}$ is a rank *cross*-covariance coefficient of lag i, and for $k = l$ it is a (univariate) rank *auto*covariance coefficient, of the type studied in Hallin *et al.* (1985, 1987). Note that a similar notation is adopted in Hallin *et al.* (1989) for a more general type of rank cross-covariance matrix, which we do not consider here.

Section 2 below is devoted to the asymptotic distribution theory of rank cross-covariance matrices: $(n-i)^{\frac{1}{2}}(n+1)^{-2}$ vec $\boldsymbol{\Gamma}_i^{(n)}$ will be shown to be asymptotically multinormal, under the null hypothesis of randomness as well as under contiguous alternatives of ARMA dependence. Section 3 discusses statistical applications in time-series analysis. Before proceeding to asymptotic results, however, we start with an explicit expression for the permutational mean matrices $\mathbf{m}_{i*}^{(n)}$. Here $\mathbf{1}$ stands for a matrix of ones of appropriate dimensions; O_P and o_P are unconditional O_P's and o_P's.

PROPOSITION 1.1. *The permutational mean matrices $\mathbf{m}_{i*}^{(n)}$ in (1.1) are of the form*

$$\mathbf{m}_{i*}^{(n)} = \frac{n(n+1)^2}{4(n-1)}\mathbf{1} - \frac{\mathbf{m}_{0*}^{(n)}}{n-1}, \tag{1.2}$$

with $\mathbf{m}_{0*}^{(n)} = n^{-1}\sum_{t=1}^{n}(\mathbf{R}_t^{(n)})(\mathbf{R}_t^{(n)})^{\mathsf{T}}$. Their elements uniformly satisfy

$$\frac{n^2}{4} < \frac{(n+1)(3n+2)}{12} \leq (\mathbf{m}_{i*}^{(n)})_{kl} < \frac{n(n+1)^2}{4(n-1)}. \tag{1.3}$$

Moreover, as $n \longrightarrow \infty$,

$$n^{-2}\mathbf{m}_{i*}^{(n)} = \{\tfrac{1}{4} + O_P(n^{-1})\}\mathbf{1}, \qquad i \geq 1. \tag{1.4}$$

Proof. For $i = 1, 2, \ldots, n-1$, we have

$$\begin{aligned}
\mathbf{m}_{i*}^{(n)} &= \{n(n-1)\}^{-1}\sum_{1<t_1\neq t_2\leq n}\sum(\mathbf{R}_{t_1}^{(n)})(\mathbf{R}_{t_2}^{(n)})^{\mathsf{T}} \\
&= \{n(n-1)\}^{-1}\left(\sum_{t=1}^{n}(\mathbf{R}_t^{(n)})\sum_{t=1}^{n}(\mathbf{R}_t^{(n)})^{\mathsf{T}} - n\mathbf{m}_{0*}^{(n)}\right) \\
&= \{n(n-1)\}^{-1}\left\{\left(\sum_{r=1}^{n} r\right)^2\mathbf{1} - n\mathbf{m}_{0*}^{(n)}\right\} \\
&= \frac{n(n-1)^2}{4(n-1)}\mathbf{1} - \frac{\mathbf{m}_{0*}^{(n)}}{n-1}.
\end{aligned}$$

Now, $(n\mathbf{m}_{0*}^{(n)})_{k,l} = \sum_{t=1}^{n} R_{t,k}^{(n)}R_{t,l}^{(n)} \leq \sum_{t=1}^{n}(t^2) = n(n+1)(2n+1)/6$; (1.3) and (1.4) straightforwardly follow. □

Note that $\mathbf{m}_{i*}^{(n)}$, $i \geq 1$, does not depend on i.

2. ASYMPTOTIC DISTRIBUTION THEORY

Rank cross-covariance matrices can be used for testing $\mathcal{H}^{(n)}$ against arbitrary alternatives of serial dependence. If however local asymptotic power results are desired, more specific alternatives are to be considered. Denote by $\mathcal{K}_f^{(n)}(\mathbf{A}, \mathbf{B})$ the local alternative under which $\mathbf{X}^{(n)}$ constitutes a finite realization of some solution of the stochastic difference equation (ARMA model)

$$\mathbf{X}_t - n^{-\frac{1}{2}}\sum_{i=1}^{p}\mathbf{A}_i\mathbf{X}_{t-i} = \boldsymbol{\varepsilon}_t + n^{-\frac{1}{2}}\sum_{i=1}^{q}\mathbf{B}_i\boldsymbol{\varepsilon}_{t-i}, \qquad t \in \mathbb{Z}, \tag{2.1}$$

where $\boldsymbol{\varepsilon}_t$ is an independent m-variate white noise with density function f. Assume that f satisfies some mild technical conditions, such as being twice differentiable [put $\boldsymbol{\phi}(\mathbf{x}) = -\text{grad} \log f(\mathbf{x})$] and having finite Fisher information [$\int \boldsymbol{\phi}(\mathbf{x})\boldsymbol{\phi}^{\mathsf{T}}(\mathbf{x})f(\mathbf{x})\,d\mathbf{x} = \mathbf{I}(f) < \infty$].

Then the sequences $\mathcal{K}_f^{(n)}(\mathbf{A},\mathbf{B})$ and $\mathcal{H}_f^{(n)}$ are contiguous [see Hallin *et al.* (1989) for details].

Denote by F, f_i and F_i ($i = 1,\dots,m$), respectively, the m-dimensional distribution function, marginal densities and marginal distribution functions associated with F. Assume that $\int \mathbf{x}f(\mathbf{x})\,d\mathbf{x} = 0$, and that $\int \mathbf{x}\mathbf{x}^{\mathsf{T}}f(\mathbf{x})\,d\mathbf{x} = \mathbf{\Sigma}$, a finite, strictly positive definite covariance matrix. Finally, define $\mathbf{F}(\mathbf{x}) = \left(F_1(x_1),\dots,F_m(x_m)\right)^{\mathsf{T}}$, $\mathbf{x} = (x_1,\dots,x_m)^{\mathsf{T}} \in \mathbb{R}^m$.

PROPOSITION 2.1. *Let $i' \neq i''$. As $n \longrightarrow \infty$, the joint distribution of $(n+1)^{-2}\left((n-i')^{\frac{1}{2}}\right.$ $\operatorname{vec}^{\mathsf{T}}\mathbf{\Gamma}_{i'}^{(n)}, \left.(n-i'')^{\frac{1}{2}}\operatorname{vec}^{\mathsf{T}}\mathbf{\Gamma}_{i''}^{(n)}\right)^{\mathsf{T}}$ is asymptotically multinormal, with mean $\mathbf{0}$ under $\mathcal{H}^{(n)}$ and mean*

$$\begin{pmatrix} \{\int \mathbf{F}(\mathbf{x})\mathbf{x}^{\mathsf{T}}\,dF(\mathbf{x})\} \otimes \{\int \mathbf{F}(\mathbf{x})\boldsymbol{\phi}^{\mathsf{T}}(\mathbf{x})\,dF(\mathbf{x})\}\operatorname{vec}(\mathbf{A}_{i'} + \mathbf{B}_{i'}) \\ \{\int \mathbf{F}(\mathbf{x})\mathbf{x}^{\mathsf{T}}\,dF(\mathbf{x})\} \otimes \{\int \mathbf{F}(\mathbf{x})\boldsymbol{\phi}^{\mathsf{T}}(\mathbf{x})\,dF(\mathbf{x})\}\operatorname{vec}(\mathbf{A}_{i''} + \mathbf{B}_{i''}) \end{pmatrix} \tag{2.2}$$

under $\mathcal{K}_f^{(n)}(\mathbf{A},\mathbf{B})$, and with covariance matrix

$$\begin{pmatrix} [\int \mathbf{F}(\mathbf{x})\{\mathbf{F}(\mathbf{x})\}^{\mathsf{T}}\,dF(\mathbf{x}) - \frac{1}{4}]^{\otimes 2} & \mathbf{0} \\ \mathbf{0} & [\int \mathbf{F}(\mathbf{x})\{\mathbf{F}(\mathbf{x})\}^{\mathsf{T}}\,dF(\mathbf{x}) - \frac{1}{4}]^{\otimes 2} \end{pmatrix} \tag{2.3}$$

under both (all diagonal elements in this latter matrix are equal to $\frac{1}{144}$).

Proof. Letting $I = \max(i', i'')$, we see that $(n+1)^{-2}(\gamma_{j';k,l}^{(n)} + m_{j^*;k,l}^{(n)})$ is of the form

$$(n-I)^{-1}\sum_{t=I+1}^{n} J_k\left(\frac{\mathbf{R}_t^{(n)}}{n+1}\right) J_l\left(\frac{\mathbf{R}_{t-j}^{(n)}}{n+1}\right) + o_P(n^{-\frac{1}{2}}),$$

where $J_h(\mathbf{u}) = u_h$ ($h = 1,\dots,m$)—a nondecreasing function with respect to all the components of $\mathbf{u} \in (0,1)^m$—satisfies $\mathcal{E}J_h^2\left(\mathbf{F}(\mathbf{X}_1^{(n)})\right) = \int_0^1 u^2\,du < \infty$. The assumptions of Proposition 3.1 in Hallin *et al.* (1989) are thus fulfilled, and $J(\mathbf{u}_1,\dots,\mathbf{u}_{I+1}) = u_{1,k}u_{j+1,l}$ is a score-generating function for $(n+1)^{-2}(\gamma_{j;k,l}^{(n)} + m_{j^*;k,l}^{(n)})$. The corresponding "centered scores" (Proposition 2.1, same reference) are

$$J_{jkl}^*(\mathbf{u}_1,\,\dots,\,\mathbf{u}_{I+1}) = u_{1,k}u_{j+1,l} - \tfrac{1}{2}(u_{1,k} + u_{j+1,l}) - \tfrac{1}{4}.$$

Proposition 2.2 (same reference as above) then entails the asymptotic joint binormality of $(n+1)^{-2}\left((n-I)^{\frac{1}{2}}\gamma_{j';k',l'}^{(n)}, (n-I)^{\frac{1}{2}}\gamma_{j'';k'',l''}^{(n)}\right)^{\mathsf{T}}$, with mean $(\mathbf{0},\mathbf{0})^{\mathsf{T}}$ under $\mathcal{H}_f^{(n)}$, mean

$$\left(\sum_{i=1}^{\min(j',\kappa)} \operatorname{tr}\{\mathbf{C}_{j'k'l'}^{i}(\mathbf{A}_i + \mathbf{B}_i)\} \quad \sum_{i=1}^{\min(j'',\kappa)} \operatorname{tr}\{\mathbf{C}_{j''k''l''}^{i}(\mathbf{A}_i + \mathbf{B}_i)\}\right)^{\mathsf{T}}$$

under $\mathcal{K}_f^{(n)}(\mathbf{A},\mathbf{B})$, and covariance matrix

$$\begin{pmatrix} V_{j'k'l'}^2 & V_{12} \\ V_{12} & V_{j''k''l''}^2 \end{pmatrix}$$

under both, where $\kappa = \max(p, q)$, $\mathbf{A}_i = \mathbf{0}$ for $i = p+1, \ldots, \kappa$, $\mathbf{B}_i = \mathbf{0}$ for $i = q+1, \ldots, \kappa$, and with (all expectations below are taken under $\mathcal{H}_f^{(n)}$)

$$\mathbf{C}^i_{jkl} = \sum_{h=0}^{\kappa-i} \mathcal{E}\{ J^*_{jkl}(\mathbf{F}(\mathbf{X}^{(n)}_1), \ldots, \mathbf{F}(\mathbf{X}^{(n)}_{l-1}))\boldsymbol{\phi}(\mathbf{X}^{(n)}_{h+1})(\mathbf{X}^{(n)}_{h+i+1})^\top \}$$

$$= \begin{cases} \mathcal{E}\{ F_k(X^{(n)}_{1,k}) F_l(X^{(n)}_{j+1,l}) \boldsymbol{\phi}(\mathbf{X}^{(n)}_1)(\mathbf{X}^{(n)}_{j+1})^\top \} & \text{if } i = j, \\ \mathbf{0} & \text{if } i \neq j \end{cases}$$

$$= \begin{cases} \int F_k(x_k)\boldsymbol{\phi}(\mathbf{x})\, dF(\mathbf{x}) \int F_l(y_l)\mathbf{y}^\top\, dF(\mathbf{y}) & \text{if } i = j, \\ \mathbf{0} & \text{if } i \neq j, \end{cases}$$

$$V^2_{jkl} = \mathcal{E}\Big(J^{*2}_{jkl}(\mathbf{F}(\mathbf{X}^{(n)}_1), \ldots, \mathbf{F}(\mathbf{X}^{(n)}_{l+1}))$$

$$+ 2\sum_{h=1}^{l} J^*_{jkl}(\mathbf{F}(\mathbf{X}^{(n)}_1), \ldots, \mathbf{F}(\mathbf{X}^{(n)}_{l+1})) J^*_{jkl}(\mathbf{F}(\mathbf{X}^{(n)}_{1+h}), \ldots, \mathbf{F}(\mathbf{X}^{(n)}_{l+1+h}))\Big)$$

$$= \mathcal{E}\left(\left[F_k(X^{(n)}_{1,k})\, F_l(X^{(n)}_{1+j,l}) - \tfrac{1}{2}\{ F_k(X^{(n)}_{1,k}) + F_l(X^{(n)}_{1+j,l})\} + \tfrac{1}{4}\right]^2\right)$$

$$= \int\int \{uv - \tfrac{1}{2}(u+v) + \tfrac{1}{4}\}^2\, du\, dv = \tfrac{1}{144},$$

and

$$V_{12} = \mathcal{E}\Big(J^*_{j'k'l'}(\mathbf{F}(\mathbf{X}^{(n)}_1), \ldots, \mathbf{F}(\mathbf{X}^{(n)}_{l+1})) J^*_{j''k''l''}(\mathbf{F}(\mathbf{X}^{(n)}_1), \ldots, \mathbf{F}(\mathbf{X}^{(n)}_{l+1}))$$

$$+ \sum_{h=1}^{l} \{ J^*_{j'k'l'}(\mathbf{F}(\mathbf{X}^{(n)}_1), \ldots, \mathbf{F}(\mathbf{X}^{(n)}_{l+1})) J^*_{j''k''l''}(\mathbf{F}(\mathbf{X}^{(n)}_{1+h}), \ldots, \mathbf{F}(\mathbf{X}^{(n)}_{l+1+h}))$$

$$+ J^*_{j'k'l'}(\mathbf{F}(\mathbf{X}^{(n)}_{1+h}), \ldots, \mathbf{F}(\mathbf{X}^{(n)}_{l+1+h})) J^*_{j''k''l''}(\mathbf{F}(\mathbf{X}^{(n)}_1), \ldots, \mathbf{F}(\mathbf{X}^{(n)}_{l+1}))\}\Big)$$

$$= \begin{cases} \mathcal{E}\left(\left[F_{k'}(X^{(n)}_{1,k'}) F_{l'}(X^{(n)}_{1+j,l'}) - \tfrac{1}{2}\{ F_{k'}(X^{(n)}_{1,k'}) + F_{l'}(X^{(n)}_{1+j,l'})\} + \tfrac{1}{4}\right] \right. & \\ \quad \left. \times \left[F_{k''}(X^{(n)}_{1,k''}) F_{l''}(X^{(n)}_{1+j,l''}) - \tfrac{1}{2}\{ F_{k''}(X^{(n)}_{1,k''}) + F_{l''}(X^{(n)}_{1+j,l''})\} + \tfrac{1}{4}\right]\right) & \text{if } j' = j'' = j, \\ 0 & \text{if } j' \neq j'' \end{cases}$$

$$= \begin{cases} \left\{ \int F_{k'}(x_{k'}) F_{k''}(x_{k''})\, dF(\mathbf{x}) - \tfrac{1}{4}\right\} \left\{ \int F_{l'}(x_{l'}) F_{l''}(x_{l''})\, dF(\mathbf{x}) - \tfrac{1}{4}\right\} & \text{if } j' = j'' = j, \\ 0 & \text{if } j' \neq j''. \end{cases}$$

Proposition 2.1 follows. □

The asymptotic distribution of $\boldsymbol{\Gamma}^n_i$, as provided by Proposition 2.1, unfortunately cannot be used for inferential purposes, since the asymptotic covariance matrix (2.3) is not distribution-free, and clearly depends on the underlying joint density $f(\mathbf{x})$. Genuinely permutationally distribution-free rank statistics with asymptotic distribution-free unconditional distributions, however, can be obtained if conditional (permutational) covariances are substituted for the unconditional ones. The exact permutational covariance of vec $\mathbf{\Gamma}_i^{(n)}$ is (for $n \geq 2i$)

$$\begin{aligned}\mathbf{W}_i^* &= \{n(n-1)(n-i)\}^{-1}\sum\sum_{t_1\neq t_2}\{\mathbf{R}_{t_1}^{(n)}(\mathbf{R}_{t_2}^{(n)})^\top\}^{\otimes 2}\\ &\quad + 2(n-2i)\{n(n-1)(n-2)(n-i)^2\}^{-1}\sum\sum\sum_{t_1\neq t_2\neq t_3}\{\mathbf{R}_{t_1}^{(n)}(\mathbf{R}_{t_2}^{(n)})^\top\}\otimes\{\mathbf{R}_{t_3}^{(n)}(\mathbf{R}_{t_1}^{(n)})^\top\}\\ &\quad + \{n^2-n(2i+3)+i^2+5i\}\{n(n-1)(n-2)(n-3)(n-i)^2\}^{-1}\\ &\quad \times\sum\sum\sum\sum_{t_1\neq t_2\neq t_3\neq t_4}\{\mathbf{R}_{t_1}^{(n)}(\mathbf{R}_{t_2}^{(n)})^\top\}\otimes\{\mathbf{R}_{t_3}^{(n)}(\mathbf{R}_{t_4}^{(n)})^\top\} - (\text{vec}\,\mathbf{m}_{i*}^{(n)})(\text{vec}\,\mathbf{m}_{i*}^{(n)})^\top. \qquad (2.4)\end{aligned}$$

This exact, somewhat cumbersome form results from combinatorial arguments [see Lemma 2.1 in Hallin *et al.* (1989). In practice, it can be replaced by the following much simpler one, without affecting asymptotic results.

PROPOSITION 2.2. *Let*

$$\tilde{\mathbf{W}}_i^* = \{n(n-1)(n-i)\}^{-1} \times\sum\sum_{t_1\neq t_2}\left[\mathbf{R}_{t_1}^{(n)}(\mathbf{R}_{t_1}^{(n)})^\top - \frac{(n+1)^2}{4}\mathbf{1}\right]\otimes\left[\mathbf{R}_{t_2}^{(n)}(\mathbf{R}_{t_2}^{(n)})^\top - \frac{(n+1)^2}{4}\mathbf{1}\right].$$

Then $\tilde{\mathbf{W}}_i^* - \mathbf{W}_i^*$ *is* $o_P(n^3)$, *and* $(\text{vec}\,\mathbf{\Gamma}_i^{(n)})^\top(\tilde{\mathbf{W}}_i^*)^{-1}(\text{vec}\,\mathbf{\Gamma}^{(n)})$ *is asymptotically chi-square (under* $\mathcal{H}^{(n)}$*), with* m^2 *degrees of freedom, provided that*

$$\mathbf{V} = \int \mathbf{F}(\mathbf{x})\{\mathbf{F}(\mathbf{x})\}^\top\, dF(\mathbf{x}) - \frac{\mathbf{1}}{4}$$

is a full-rank matrix.

Proof. It follows from Proposition 2.4(ii) in Hallin *et al.* (1989) that the permutational covariance matrix of $(n-i)^{\frac{1}{2}}(n+1)^{-2}$ vec $\mathbf{\Gamma}_i^{(n)}$, viz. $(n-i)(n+1)^{-4}\mathbf{W}_i^*$, converges in probability to its asymptotical counterpart; $(n-i)(n+1)^{-4}\mathbf{W}_i^* - \mathbf{V}^{\otimes 2} = o_P(1)$ in (2.3). On the other hand, $(n-i)(n+1)^{-4}\tilde{\mathbf{W}}_i^*$ is the permutational mean of the (matrix of) linear serial multirank statistic(s) of order one

$$(n-1)^{-1}\sum_{t=2}^{n}\left\{\frac{\mathbf{R}_t^{(n)}}{n+1}\left(\frac{\mathbf{R}_t^{(n)}}{n+1}\right)^\top - \frac{\mathbf{1}}{4}\right\}\otimes\left\{\frac{\mathbf{R}_{t-1}^{(n)}}{n+1}\left(\frac{\mathbf{R}_{t-1}^{(n)}}{n+1}\right)^\top - \frac{\mathbf{1}}{4}\right\},$$

with a (matrix of) score-generating function(s) $\mathbf{J}(\mathbf{u},\mathbf{v}) = (\mathbf{u}\mathbf{u}^\top - \frac{\mathbf{1}}{4})\otimes(\mathbf{v}\mathbf{v}^\top - \frac{\mathbf{1}}{4})$ [this is a consequence of Proposition 3.1 in Hallin *et al.* (1989)]. Applying again Proposition 2.4 (same reference), this permutational mean converges in probability to its unconditional asymptotical counterpart:

$$(n-i)(n+1)^{-4}\tilde{\mathbf{W}}_i^* - \mathbf{V}^{\otimes 2} = o_P(1).$$

It follows that $\tilde{\mathbf{W}}_i^* - \mathbf{W}_i^*$ is $o_P(n^3)$. Now, from Proposition 2.1 above, and provided that $\mathbf{V}$ (hence $\mathbf{V}^{\otimes 2}$) is of full rank, $Q_i^{(n)} = (n-i)(n+1)^{-4}(\text{vec}\,\mathbf{\Gamma}_i^{(n)})^\top(\mathbf{V}^{\otimes 2})_i^{-1}$ vec $\mathbf{\Gamma}_i^{(n)}$ is asymptotically chi-square, under $\mathcal{H}^{(n)}$, with m^2 degrees of freedom. Since

$(\text{vec}\,\boldsymbol{\Gamma}_i^{(n)})^{\mathsf{T}}(\mathbf{W}_i^*)^{-1}(\text{vec}\,\boldsymbol{\Gamma}_i^{(n)})$ and $(\text{vec}\,\boldsymbol{\Gamma}_i^{(n)})^{\mathsf{T}}(\tilde{\mathbf{W}}_i^*)^{-1}(\text{vec}\,\boldsymbol{\Gamma}_i^{(n)})$ are asymptotically equivalent to $Q_i^{(n)}$, their asymptotic distribution is also chi-square—with however the additional property that they do not involve unknown quantities such as $\mathbf{V}$. This completes the proof. □

3. STATISTICAL APPLICATIONS

3.1. Testing Randomness against Specified ARMA Dependence Alternatives.

The distributional results of Section 2 of course can be exploited in the problem of testing the null hypothesis $\mathcal{H}^{(n)}$ of randomness. Two testing procedures are investigated here. The first one deals with the case of a specified alternative of ARMA dependence; the second one is a general, rank-based, portmanteau-type test, to be used against unspecified alternatives of serial dependence of order less than or equal to κ. Finally, a rank-based version of Tiao and Box's specification method is presented in Section 3.3.

If a particular local alternative of ARMA dependence, of the form (2.1), is considered, a "directional" test can be based on a linear combination of rank cross-covariances of the form

$$T_{\mathrm{WW}}^{(n)} = \frac{\sum_{i=1}^{\kappa} \operatorname{tr}\{\mathbf{W}_0^{*-1}\boldsymbol{\Gamma}_i^{(n)}(\mathbf{A}_i+\mathbf{B}_i)^{\mathsf{T}}\}}{[\sum_{i=1}^{\kappa}\{\text{vec}\,\mathbf{W}_0^{*-1}(\mathbf{A}_i+\mathbf{B}_i)\}^{\mathsf{T}}\mathbf{W}_i^*\{\text{vec}\,\mathbf{W}_0^{*-1}(\mathbf{A}_i+\mathbf{B}_i)\}]^{\frac{1}{2}}}, \tag{3.1}$$

where

$$\mathbf{W}_0^* = n^{-1}\sum_{t=1}^{n}\mathbf{R}_t^{(n)}(\mathbf{R}_t^{(n)})^{\mathsf{T}} - \frac{(n+1)^2}{4}\mathbf{1}. \tag{3.2}$$

If necessary, $\tilde{\mathbf{W}}_i^*$ can be substituted for $\mathbf{W}_i^*$; as for $\mathbf{W}_0^*$, it is easy to see that $(n+1)^{-2}\mathbf{W}_0^*$ is a consistent, permutationally invariant estimator of $\mathbf{V}$; it is supposed here to be of full rank.

It follows from Proposition 2.1 that $T_{\mathrm{WW}}^{(n)}$ is exactly (i.e., permutationally) standardized, and hence asymptotically standard normal (under the assumptions of Proposition 2.1). Indeed, letting $\mathcal{E}_*$ denote permutational expectations,

$$\begin{aligned}
&\mathcal{E}_*\{\operatorname{tr}\mathbf{W}_0^{*-1}\boldsymbol{\Gamma}_i^{(n)}(\mathbf{A}_i+\mathbf{B}_i)^{\mathsf{T}}\}^2\\
&\quad= \mathcal{E}_*\{\text{vec}^{\mathsf{T}}(\mathbf{A}_i+\mathbf{B}_i)\ \text{vec}(\mathbf{W}_0^{*-1}\boldsymbol{\Gamma}_i^{(n)})\ \text{vec}^{\mathsf{T}}(\mathbf{W}_0^{*-1}\boldsymbol{\Gamma}_i^{(n)})\ \text{vec}(\mathbf{A}_i+\mathbf{B}_i)\}\\
&\quad= \{\text{vec}^{\mathsf{T}}(\mathbf{A}_i+\mathbf{B}_i)\}(\mathbf{I}\otimes\mathbf{W}_0^{*-1})\mathbf{W}_i^*(\mathbf{I}\otimes\mathbf{W}_0^{*-1})\{\text{vec}(\mathbf{A}_i+\mathbf{B}_i)\}\\
&\quad= \text{vec}^{\mathsf{T}}\{\mathbf{W}_0^{*-1}(\mathbf{A}_i+\mathbf{B}_i)\}\ \mathbf{W}_i^*\ \text{vec}\{\mathbf{W}_0^{*-1}(\mathbf{A}_i+\mathbf{B}_i)\}.
\end{aligned}$$

Still on account of Proposition 2.1, the asymptotic power, under $\mathcal{K}_f^{(n)}(\mathbf{A},\mathbf{B})$, of the test rejecting $\mathcal{H}^{(n)}$ whenever $T_{\mathrm{WW}}^{(n)}$ exceeds the standard normal $(1-\alpha)$-quantile $k_{1-\alpha}$ is

$$\beta_{\mathrm{WW}} = 1 - \Phi\left(k_{1-\alpha} - \frac{\sum_{i=1}^{\kappa}\operatorname{tr}[\mathcal{E}\{\mathbf{F}(\mathbf{X})\boldsymbol{\phi}^{\mathsf{T}}(\mathbf{X})\}(\mathbf{A}_i+\mathbf{B}_i)\mathcal{E}\{\mathbf{F}(\mathbf{X})\mathbf{X}^{\mathsf{T}}\}(\mathbf{A}_i+\mathbf{B}_i)^{\mathsf{T}}\mathbf{V}^{-1}]}{(\sum_{i=1}^{\kappa}\operatorname{tr}\{(\mathbf{A}_i+\mathbf{B}_i)\mathbf{V}(\mathbf{A}_i+\mathbf{B}_i)^{\mathsf{T}}\mathbf{V}^{-1}\})^{\frac{1}{2}}}\right), \tag{3.3}$$

where the expectations are computed under $\mathcal{H}_f^{(n)}$, and Φ denotes, as usual, the standard normal distribution function.

The corresponding normal-theory parametric procedure relies (Hallin *et al.* 1989, Section 4.1) on the usual sample cross-covariance matrices

$$\hat{\mathbf{\Gamma}}_i^{(n)} = (n-i)^{-1} \sum_{t=i+1}^{n} \mathbf{X}_t^{(n)} (\mathbf{X}_{t-i}^{(n)})^{\mathsf{T}} \tag{3.4}$$

[if the assumption that $\mathcal{E}\mathbf{X}_t = \mathbf{0}$ is not to be made, the observations in (3.4) and in the subsequent formulas have to be centered about their mean] and rejects the null hypothesis of (supposedly full-rank Gaussian) randomness whenever

$$T_{\mathrm{G}}^{(n)} = \frac{\sum_{i=1}^{\kappa}(n-i)^{\frac{1}{2}} \operatorname{tr}\{(\hat{\mathbf{\Gamma}}_0^{(n)})^{-1}\hat{\mathbf{\Gamma}}_i^{(n)}(\mathbf{A}_i+\mathbf{B}_i)^{\mathsf{T}}\}}{\{\sum_{i=1}^{\kappa}\{\operatorname{vec}(\hat{\mathbf{\Gamma}}_0^{(n)})^{-1}(\mathbf{A}_i+\mathbf{B}_i)\}^{\mathsf{T}}(\hat{\mathbf{\Gamma}}_0^{(n)})^{\otimes 2}\{\operatorname{vec}(\hat{\mathbf{\Gamma}}_0^{(n)})^{-1}(\mathbf{A}_i+\mathbf{B}_i)\}\}^{\frac{1}{2}}} > k_{1-\alpha}. \tag{3.5}$$

The resulting asymptotic power, still under $\mathcal{K}_f^{\mathsf{T}}(\mathbf{A},\mathbf{B})$, is

$$\beta_{\mathrm{G}} = 1 - \Phi\left(k_{1-\alpha} - \left\{\sum_{i=1}^{\kappa} \operatorname{tr}\{(\mathbf{A}_i+\mathbf{B}_i)\mathbf{\Sigma}(\mathbf{A}_i+\mathbf{B}_i)^{\mathsf{T}}\mathbf{\Sigma}^{-1}\}\right\}^{\frac{1}{2}}\right). \tag{3.6}$$

The asymptotic relative efficiency of the Wald-Wolfowitz test based on $T_{\mathrm{WW}}^{(n)}$ with respect to the Gaussian one based on $T_{\mathrm{G}}^{(n)}$ accordingly is

$$\operatorname{ARE}(T_{\mathrm{WW}}^{(n)};\, T_{\mathrm{G}}^{(n)}) = \frac{\left(\sum_{i=1}^{\kappa}\operatorname{tr}[\mathcal{E}\{\mathbf{F}(\mathbf{X})\boldsymbol{\phi}^{\mathsf{T}}(\mathbf{X})\}(\mathbf{A}_i+\mathbf{B}_i)\mathcal{E}\{\mathbf{F}(\mathbf{X})\mathbf{X}^{\mathsf{T}}\}(\mathbf{A}_i+\mathbf{B}_i)^{\mathsf{T}}\mathbf{V}^{-1}]\right)^2}{\left(\sum_{i=1}^{\kappa}\operatorname{tr}\{(\mathbf{A}_i+\mathbf{B}_i)\mathbf{V}(\mathbf{A}_i+\mathbf{B}_i)^{\mathsf{T}}\mathbf{V}^{-1}\}\right)\left(\sum_{i=1}^{\kappa}\operatorname{tr}\{(\mathbf{A}_i+\mathbf{B}_i)\mathbf{\Sigma}(\mathbf{A}_i+\mathbf{B}_i)^{\mathsf{T}}\mathbf{\Sigma}^{-1}\}\right)}. \tag{3.7}$$

The matrices $\mathcal{E}\{\mathbf{F}(\mathbf{X})\mathbf{X}^{\mathsf{T}}\}$, $\mathcal{E}\{\mathbf{F}(\mathbf{X})\boldsymbol{\phi}^{\mathsf{T}}(\mathbf{X})\}$ and $\mathbf{V} = \mathcal{E}\{\mathbf{F}(\mathbf{X})\mathbf{F}^{\mathsf{T}}(\mathbf{X})\} - \frac{1}{4}$ in general are not easy to compute. If however f itself is Gaussian, with covariance matrix $\mathbf{\Sigma} = (\sigma_{ij})$, then $\boldsymbol{\phi}(\mathbf{X}) = \mathbf{\Sigma}^{-1}\mathbf{X}$,

$$\mathcal{E}\{F_i(X_i)X_j\} = \frac{\sigma_{ij}}{\sigma_{ii}}\mathcal{E}\{F_i(X_i)X_i\} = \frac{\sigma_{ij}}{\sigma_{ii}^{\frac{1}{2}}}\int_0^1 u\Phi^{-1}(u)\,du = \frac{\sigma_{ij}}{\sigma_{ii}^{\frac{1}{2}}}(4\pi)^{-\frac{1}{2}},$$

which implies

$$\mathcal{E}\{\mathbf{F}(\mathbf{X})\mathbf{X}^{\mathsf{T}}\} = \operatorname{diag}(2\sqrt{\pi\sigma})^{-1}\,\mathbf{\Sigma},$$

and

$$\mathcal{E}\{\mathbf{F}(\mathbf{X})\boldsymbol{\phi}^{\mathsf{T}}(\mathbf{X})\} = \mathcal{E}\{\mathbf{F}(\mathbf{X})\mathbf{X}^{\mathsf{T}}\}\mathbf{\Sigma}^{-1} = \operatorname{diag}(2\sqrt{\pi\sigma})^{-1},$$

where $\operatorname{diag}(2\sqrt{\pi\sigma})^{-1}$ stands for the diagonal matrix with ith diagonal element $(2\sqrt{\pi\sigma_{ii}^{\frac{1}{2}}})^{-1}$. If moreover $\mathbf{\Sigma} = \sigma^2\mathbf{I}$, then

$$\mathbf{V} = \mathcal{E}\{\mathbf{F}(\mathbf{X})\,\mathbf{F}^{\mathsf{T}}(\mathbf{X})\} - \frac{1}{4} = \left[\int u^2\,du - \frac{1}{4}\right]\mathbf{I} = \frac{\mathbf{I}}{12},$$

and (3.7) reduces to $(3/\pi)^2$, which is exactly the ARE, under Gaussian assumptions, of the univariate Wald-Wolfowitz test against serial dependence with respect to its parametric counterpart, based on usual sample autocorrelations (see, e.g., Hallin *et al.* 1985).

3.2. A Wald-Wolfowitz Multivariate Portmanteau Test.

Specified alternatives such as (2.1) seldom are considered in practice. In the case of an omnibus alternative of ARMA(p, q) dependence, with unspecified coefficients $\mathbf{A}_i$ and $\mathbf{B}_i$ but with orders p and q smaller than κ, one might like to consider the following rank-based *portmanteau* procedure, which consists in rejecting $\mathcal{H}^{(n)}$ whenever

$$Q_{\mathrm{WW}}^{(n)} = \sum_{i=1}^{\kappa} (\mathrm{vec}\, \mathbf{\Gamma}_i^{(n)})^{\mathsf{T}} \{\tilde{\mathbf{W}}_i^*\}^{-1} (\mathrm{vec}\, \mathbf{\Gamma}_i^{(n)}) > \chi^2_{\kappa m^2;1-\alpha}, \tag{3.8}$$

where $\chi^2_{\kappa m^2;1-\alpha}$ denotes the $(1-\alpha)$-quantile of a chi-square variable with κm^2 degrees of freedom. The corresponding Gaussian portmanteau test relies on

$$Q_{\mathrm{G}}^{(n)} = \sum_{i=1}^{\kappa} (n-i)(\mathrm{vec}\, \hat{\mathbf{\Gamma}}_i^{(n)})^{\mathsf{T}} \{(\hat{\mathbf{\Gamma}}_0^{(n)})^{\otimes 2}\}^{-1} (\mathrm{vec}\, \hat{\mathbf{\Gamma}}_i^{(n)}), \tag{3.9}$$

which is also (Chiturri 1976) asymptotically chi-square, under the hypothesis of Gaussian, full-rank white noise, still with κm^2 degrees of freedom.

The asymptotic powers and relative efficiencies of these two portmanteau procedures under $\mathcal{K}_f^{(n)}(\mathbf{A}, \mathbf{B})$ depend on the noncentrality parameters

$$\begin{aligned}
&\lambda^2_{\mathrm{WW}}(\mathbf{A}, \mathbf{B}; f) \\
&= \sum_{i=1}^{\kappa} \mathrm{vec}^{\mathsf{T}}(\mathbf{A}_i + \mathbf{B}_i) \\
&\quad \times [\mathcal{E}\{\mathbf{X}\mathbf{F}^{\mathsf{T}}(\mathbf{X})\} \otimes \mathcal{E}\{\boldsymbol{\phi}(\mathbf{X})\mathbf{F}^{\mathsf{T}}(\mathbf{X})\}](\mathbf{V}^{-1})^{\otimes 2}[\mathcal{E}\{\mathbf{F}(\mathbf{X})\mathbf{X}^{\mathsf{T}}\} \otimes \mathcal{E}\{\mathbf{F}(\mathbf{X})\boldsymbol{\phi}^{\mathsf{T}}(\mathbf{X})\}] \\
&\quad \times \mathrm{vec}(\mathbf{A}_i + \mathbf{B}_i) \\
&= \sum_{i=1}^{\kappa} \mathrm{tr}[\mathbf{V}^{-1}\mathcal{E}\{\mathbf{F}(\mathbf{X})\boldsymbol{\phi}^{\mathsf{T}}(\mathbf{X})\}(\mathbf{A}_i + \mathbf{B}_i)\mathcal{E}\{\mathbf{F}(\mathbf{X})\mathbf{X}^{\mathsf{T}}\}\mathbf{V}^{-1} \\
&\quad \times \mathcal{E}\{\mathbf{X}\mathbf{F}^{\mathsf{T}}(\mathbf{X})\}(\mathbf{A}_i + \mathbf{B}_i)^{\mathsf{T}}\mathcal{E}\{\boldsymbol{\phi}(\mathbf{X})\mathbf{F}^{\mathsf{T}}(\mathbf{X})\}] \\
&= \sum_{i=1}^{\kappa} \mathrm{vec}^{\mathsf{T}}(\mathbf{A}_i + \mathbf{B}_i)\, [\mathcal{E}\{\mathbf{F}(\mathbf{X})\mathbf{X}^{\mathsf{T}}\}\mathbf{V}^{-1}\mathcal{E}\{\mathbf{X}\mathbf{F}^{\mathsf{T}}(\mathbf{X})\}] \\
&\qquad \otimes [\mathcal{E}\{\boldsymbol{\phi}(\mathbf{X})\mathbf{F}^{\mathsf{T}}(\mathbf{X})\}\mathbf{V}^{-1}\mathcal{E}\{\mathbf{F}(\mathbf{X})\boldsymbol{\phi}^{\mathsf{T}}(\mathbf{X})\}]\, \mathrm{vec}(\mathbf{A}_i + \mathbf{B}_i)
\end{aligned}$$

and

$$\begin{aligned}
\lambda^2_{\mathrm{G}}(\mathbf{A}, \mathbf{B}; f) &= \sum_{i=1}^{\kappa} \mathrm{vec}^{\mathsf{T}}(\mathbf{A}_i + \mathbf{B}_i)\, (\mathbf{\Sigma} \otimes \mathbf{I})(\mathbf{\Sigma}^{-1})^{\otimes 2}(\mathbf{\Sigma} \otimes \mathbf{I})\, \mathrm{vec}(\mathbf{A}_i + \mathbf{B}_i) \\
&= \sum_{i=1}^{\kappa} \mathrm{tr}\{\mathbf{\Sigma}^{-1}(\mathbf{A}_i + \mathbf{B}_i)\mathbf{\Sigma}(\mathbf{A}_i + \mathbf{B}_i)^{\mathsf{T}}\} \\
&= \sum_{i=1}^{\kappa} \mathrm{vec}^{\mathsf{T}}(\mathbf{A}_i + \mathbf{B}_i)\, (\mathbf{\Sigma} \otimes \mathbf{\Sigma}^{-1})\, \mathrm{vec}(\mathbf{A}_i + \mathbf{B}_i).
\end{aligned}$$

The asymptotic relative efficiency of the Wald-Wolfowitz procedure with respect to the Gaussian one, under $\mathcal{K}_f^{(n)}(\mathbf{A}, \mathbf{B})$, accordingly is

$$\mathrm{ARE}(Q_{\mathrm{WW}}^{(n)};\ Q_{\mathrm{G}}^{(n)}) = \frac{\lambda_{\mathrm{WW}}^2(\mathbf{A},\ \mathbf{B}; f)}{\lambda_{\mathrm{G}}^2(\mathbf{A},\ \mathbf{B}; f)}. \tag{3.10}$$

Even under Gaussian assumptions, there is no guarantee that this ARE will be smaller than one; if, however, F is Gaussian with diagonal covariance matrix $\sigma^2\mathbf{I}$, (3.10) again reduces to $9/\pi^2$.

3.3. A Rank-Based Version of Tiao and Box's Tentative Specification Method.

Tiao and Box (1981) suggest basing a tentative specification of multivariate ARMA models on the following device. Denoting by $\hat{\rho}_{i;k,l}^{(n)}$ the sample cross-correlation

$$\sum_{t=i+1}^{n} X_{t,k}^{(n)} X_{t-i,l}^{(n)} \left(\sum_t (X_{t,k}^{(n)})^2 \sum_t (X_{t,l}^{(n)})^2 \right)^{-\frac{1}{2}},$$

they propose considering, for each lag $i = 1, 2, \ldots$, a matrix where element (k, l) is replaced with

a *plus* sign if $\hat{\rho}_{i;k,l}^{(n)}$ is greater than $2n^{-\frac{1}{2}}$,
a *minus* sign if $\hat{\rho}_{i;k,l}^{(n)}$ is less than $-2n^{-\frac{1}{2}}$,
a *dot* if $\hat{\rho}_{i;k,l}^{(n)}$ lies in $[\pm 2n^{-\frac{1}{2}}]$.

These indicator symbols of course cannot be considered as resulting from formal (5%) significance tests, but (quoting the authors) "as a rather crude signal-to-noise ratio guide (...) providing useful and assimilable indicators of the general correlation pattern."

From Proposition 2.1, it follows that the rank-based cross-covariance coefficients $\gamma_{i;k,l}^{(n)}$ are asymptotically normal under $\mathcal{H}^{(n)}$, with mean zero and variance $(n+1)^4/144(n-i)$. Accordingly, this widely used device of Tiao and Box can be modified in an obvious way, substituting rank-based cross-covariances for sample autocorrelations, and replacing $\gamma_{i;k,l}^{(n)}$ in $\mathbf{\Gamma}_i^{(n)}$ with

a *plus* sign if $\gamma_{i;k,l}^{(n)}$ is greater than $(n+1)^2/6(n-i)^{\frac{1}{2}}$,
a *minus* sign if $\gamma_{i;k,l}^{(n)}$ is less than $-(n+1)^2/6(n-i)^{\frac{1}{2}}$,
a *dot* if $\gamma_{i;k,l}^{(n)}$ lies in $[\pm(n+1)^2/6(n-i)^{\frac{1}{2}}]$.

This Wald-Wolfowitz rank-based version of Tiao and Box's procedure is likely to be more reliable under non-Gaussian conditions, and more robust against gross error. Contrary to the methods described in Sections 3.1 and 3.2 above, its practical implementation is quite straightforward.

ACKNOWLEDGEMENT

The authors gratefully acknowledge the very careful reading of the manuscript by an Associate Editor and a referee, which significantly improved the final presentation of the paper.

REFERENCES

Bartels, R. (1982). The rank version of von Neumann's ratio test for randomness. *J. Amer. Statist. Assoc.*, 77, 40–46.

Chan, N.H., and Tran, L.T. (1992). Nonparametric tests for serial dependence. *J. Time Ser. Anal.*, 13, 19–28.

Chiturri, R.V. (1976). Distribution of multivariate white noise autocorrelations. *J. Amer. Statist. Assoc.*, 71, 223–226.

Dufour, J.-M. (1981). Rank tests for serial dependence. *J. Time Ser. Anal.*, 2, 117–128.

Dufour, J.-M., and Roy, R. (1985). Some robust exact results on sample autocorrelations and tests of randomness. *J. Econometrics*, 29, 257–273.

Dufour, J.-M., and Roy, R. (1986). Generalized portmanteau statistics and tests of randomness. *Comm. Statist. Theory Methods*, 15, 2953–2972.

Hallin, M., Ingenbleek, J.-Fr., and Puri, M.L. (1985). Linear serial rank tests for randomness against ARMA alternatives. *Ann. Statist.*, 13, 1156–1181.

Hallin, M., Ingenbleek, J.-Fr., and Puri, M.L. (1987). Linear and quadratic rank tests for randomness against serial dependence. *J. Time Ser. Anal.*, 8, 409–424.

Hallin, M., Ingenbleek, J.-Fr., and Puri, M.L. (1989). Asymptotically most powerful rank tests for multivariate randomness against serial dependence. *J. Multivariate Anal.*, 30, 34–71.

Hallin, M., and Mélard, G. (1988). Rank-based tests for randomness against first-order serial dependence. *J. Amer. Statist. Assoc.*, 83, 1117–1128.

Hallin, M., and Puri, M.L. (1988). Optimal rank-based procedures for time series analysis: Testing an ARMA model against other ARMA models. *Ann. Statist.*, 16, 402–432.

Hallin, M., and Puri, M.L. (1991). Times series analysis via rank-order theory: Signed rank tests for ARMA models. *J. Multivariate Anal.*, 39, 1–29.

Hallin, M., and Puri, M.L. (1992). Rank tests for time series analysis: A survey. *New Directions in Time Series Analysis* (D. Brillinger *et al., eds.*), Springer-Verlag, 111–153.

Hallin, M., and Puri, M.L. (1994). Aligned rank tests for linear models with autocorrelated error terms. *J. Multivariate Anal.*, 50, 175–237.

Knoke, D.J. (1977). Testing for randomness against autocorrelation: Alternative tests. *Biometrika*, 64, 523–529.

Skaug, H.J., and Tjøstheim, D. (1993). Nonparametric tests of serial independence. *Developments in Time Series Analysis*, Volume in Honour of M.B. Priestley (T. Subba Rao, *ed*), Chapman and Hall, 207–229.

Tiao, G.C., and Box, G.E.P. (1981). Modeling multiple time series with applications. *J. Amer. Statist. Assoc.*, 76, 802–816.

Wald, A., and Wolfowitz, J. (1943). An exact test for randomness in the nonparametric case based on serial correlation. *Ann. Math. Statist.*, 14, 378–388.

Received 15 November 1993
Revised 26 May 1994
Accepted 13 June 1994

Institut de Statistique et Département de Mathématique
Université Libre de Bruxelles
Campus de la Plaine C.P. 210
B 1050 Bruxelles
Belgium

Department of Mathematics
Indiana University
Bloomington, Indiana
U.S.A. 47405

Reprinted by permission of
The Statistical Society of Canada,
© 1995 by The Statistical Society of Canada

NONPARAMETRIC APPROACH FOR NON-GAUSSIAN VECTOR STATIONARY PROCESSES

MASANOBU TANIGUCHI, MADAN L. PURI* and MASAO KONDO

Osaka University, Toyonaka, Japan
Indiana University
Kagoshima University, Korimoto, Japan

Suppose that $\{\mathbf{z}(t)\}$ is a non-Gaussian vector stationary process with spectral density matrix $f(\lambda)$. In this paper we consider the testing problem H: $\int_{-\pi}^{\pi} K\{f(\lambda)\}\, d\lambda = c$ against A: $\int_{-\pi}^{\pi} K\{f(\lambda)\}\, d\lambda \neq c$, where $K\{\cdot\}$ is an appropriate function and c is a given constant. For this problem we propose a test T_n based on $\int_{-\pi}^{\pi} K\{\hat{f}_n(\lambda)\}\, d\lambda$, where $\hat{f}_n(\lambda)$ is a nonparametric spectral estimator of $f(\lambda)$, and we define an efficacy of T_n under a sequence of nonparametric contiguous alternatives. The efficacy usually depnds on the fourth-order cumulant spectra f_4^Z of $\mathbf{z}(t)$. If it does not depend on f_4^Z, we say that T_n is non-Gaussian robust. We will give sufficient conditions for T_n to be non-Gaussian robust. Since our test setting is very wide we can apply the result to many problems in time series. We discuss interrelation analysis of the components of $\{\mathbf{z}(t)\}$ and eigenvalue analysis of $f(\lambda)$. The essential point of our approach is that we do not assume the parametric form of $f(\lambda)$. Also some numerical studies are given and they confirm the theoretical results. © 1996 Academic Press, Inc.

Received September 8, 1994.

AMS 1980 subject classifications: primary 62M15, 62G10.

Key words and phrases: non-Gaussian vector stationary process, nonparametric hypothesis testing, spectral density matrix, fourth-order cumulant spectral density, non-Gaussian robustness, efficacy, measure of linear dependence, principal components, analysis of time series, nonparametric spectral estimator, asymptotic theory.

* Research supported by the Office of Naval Research Contract N00014-91-J-1020.

1. Introduction

The ordinary nonparametric approach for independently and identically distributed random variables has developed in various directions. For example Hallin, Ingenbleek, and Puri [9] and Hallin and Puri [10] introduced a class of linear serial rank statistics for the problem of testing a given ARMA model against other ARMA models. They derived the asymptotic distributions of the proposed test statistics under the null as well as alternative hypotheses and gave an explicit formulation of the asymptotically most powerful test under a sequence of contiguous ARMA alternatives. Dzhaparidze [6] considered a class $\mathscr{F}$ of goodness-of-fit tests for testing a simple hypothesis about the form of the spectral density. He investigated the asymptotic properties of test $T \in \mathscr{F}$ under a sequence of nonparametric contiguous alternatives.

Suppose that $\{z(t);\ t=0, \pm 1, ...\}$ is a p-dimensional non-Gaussian vector stationary process with spectral density matrix $f(\lambda)$. In this paper we consider the testing problem

$$H: \int_{-\pi}^{\pi} K\{f(\lambda)\}\, d\lambda = c,$$

against

$$A: \int_{-\pi}^{\pi} K\{f(\lambda)\}\, d\lambda \neq c, \tag{1.1}$$

where $K\{\cdot\}$ is a holomorphic function defined on $\mathbf{C}^{p^2}$ and c is a given constant. For this problem we propose a test statistic T_n based on $\int_{-\pi}^{\pi} K\{\hat{f}_n(\lambda)\}\, d\lambda$, where $\hat{f}_n(\lambda)$ is a nonparametric spectral estimator of $f(\lambda)$. In Section 2 we investiage the asymptotic properties of T_n, and introduce an efficacy of T_n, eff(T_n), which measures a goodness of T_n. Usually eff(T_n) depends on the fourth-order cumulant spectra of $\{\mathbf{z}(t)\}$. If it does not depend on the fourth-order cumulant spectra, then we say that T_n is non-Gaussian robust. For a Gaussian scalar process, Taniguchi and Kondo [20] and Kondo and Taniguchi [14] proved some superiority of T_n to the existing methods. In this paper we develop the discussion beyond their scope. In Section 3 we give sufficient conditions for T_n to be non-Gaussian robust in typical examples of $K\{\cdot\}$. Our test setting is unexpectedly wide and can be applied to many problems in time series. In Section 4 we discuss interrelation analysis of the components of $\{\mathbf{z}(t)\}$. The measure of linear dependence $F_{X,Y}$ is known to be an important quantity in econometrics and is related to the causality (e.g., Geweke [8]). We can see that our setting (1.1) includes the testing problem

$$H: F_{X,Y} = c \qquad \text{against} \qquad A: F_{X,Y} \neq c.$$

In Section 5 we deal with a testing problem for the integral of certain function of the eigenvalues of $f(\lambda)$. It is shown that our setting (1.1) also includes this problem. We apply this to the principal components analysis of $\{\mathbf{z}(t)\}$.

The functional of the spectral density matrix $\int_{-\pi}^{\pi} K\{f(\lambda)\}\, d\lambda$ represents so many important indices in time series analysis. Therefore we will be able to find many other unexpected applications of (1.1). Here it may be noted that we will establish the $\sqrt{n}$-consistent asymptotic theory for $\int_{-\pi}^{\pi} K\{\hat{f}_n(\lambda)\}\, d\lambda$.

Our approach is designed for essentially nonparametric hypotheses and so is different from that of Hallin *et al.* [9, 10]. Also, since Dzhaparidze's class $\mathscr{F}$ of tests is based on $\int_{-x}^{x} \{I_n(\lambda) f(\lambda)^{-1}\}\, d\lambda$, where $I_n(\lambda)$ is the periodogram, his problem does not include our testing problem (1.1).

As for the notations used in this paper, we denote the set of all integers by J, and denote Kronecker's delta by $\delta(t, s)$.

2. Basic Theory

In this section we formulate some basic theorems concerning the nonparametric testing problem. Let $\{\mathbf{z}(t);\ t \in J\}$ be a vector-valued linear process generated as

$$\mathbf{z}(t) = \sum_{s=0}^{\infty} G(s)\, \mathbf{e}(t-s), \qquad t \in J,$$

where the $\mathbf{z}(t)$'s have p components and the $\mathbf{e}(t)$'s are p-vectors such that $E\{\mathbf{e}(t)\} = 0$ and $E\{\mathbf{e}(t)\, \mathbf{e}(s)'\} = \delta(t, s)\, \Omega$, with Ω a $p \times p$ positive definite matrix; the $G(s)$'s are $p \times p$ matrices. We denote the (a, b) component of Ω and $G(s)$ by Ω_{ab} and $G_{ab}(s)$, respectively, and denote the ath component of $\mathbf{z}(t)$ and $\mathbf{e}(t)$ by $z_a(t)$ and $e_a(t)$, respectively. Initially, we make the following assumption.

Assumption 1. (i) For $a, b = 1, ..., p$,

$$\sum_{s=0}^{\infty} (1 + s^2) |G_{ab}(s)| < \infty.$$

(ii) In the unit disc $|z| \leqslant 1$,

$$\det \left\{ \sum_{s=0}^{\infty} G(x)\, z^s \right\} \neq 0.$$

Under this assumption the process $\{\mathbf{z}(t)\}$ is a second-order stationary process with the spectral density matrix

$$f(\lambda)=\frac{1}{2\pi}A(\lambda)\,\Omega A(\lambda)^*, \qquad -\pi<\lambda\leqslant\pi,$$

where $A(\lambda)=\sum_{s=0}^{\infty}G(s)\,e^{i\lambda s}$. The (a,b) component of $f(\lambda)$ and $A(\lambda)$ are denoted by $f_{ab}(\lambda)$ and $A_{ab}(\lambda)$, respectively. For a partial realization $\{\mathbf{z}(1),\ldots,\mathbf{z}(n)\}$, the periodogram matrix is defined as

$$I_n(\lambda)=\frac{1}{2\pi n}\left\{\sum_{i=1}^{n}\mathbf{z}(t)\,e^{i\lambda t}\right\}\left\{\sum_{t=1}^{n}\mathbf{z}(t)\,e^{i\lambda t}\right\}^*.$$

Since Gaussianity is not assumed for the process we need the following assumption.

Assumption 2. (i) $\{\mathbf{e}(t)\}$ is fourth-order stationary. (ii) The joint fourth-order cumulants $c^e_{abcd}(t_1,t_2,t_3)$ of $e_a(t)$, $e_b(t+t_1)$, $e_c(t+t_2)$, $e_d(t+t_3)$ satisfy

$$\sum_{t_1,t_2,t_3=-\infty}^{\infty}|c^e_{abcd}(t_1,t_2,t_3)|<\infty, \qquad a,b,c,d=1,\ldots,p.$$

Then $\{e(t)\}$ has the fourth-order cumulant spectral density

$$f^e_{abcd}(\lambda_1,\lambda_2,\lambda_3)=\left(\frac{1}{2\pi}\right)^3\sum_{t_1,t_2,t_3=-\infty}^{\infty}c^e_{abcd}(t_1,t_2,t_3)\,e^{-i(\lambda_1t_1+\lambda_2t_2+\lambda_3t_3)}.$$

Similarly we can define $c^{z}_{abcd}(t_1,t_2,t_3)$ and $f^{z}_{abcd}(\lambda_1,\lambda_2,\lambda_3)$, respectively, the fourth-order cumulant and spectral density of the process $\{\mathbf{z}(t)\}$.

Assumption 3. Let D be an open subset of $\mathbf{C}^{p^2}$. $K\colon D\to\mathbb{R}$ is holomorphic.

Consider the testing problem

$$H\colon\int_{-\pi}^{\pi}K\{f(\lambda)\}\,d\lambda=c,$$

against

$$A\colon\int_{-\pi}^{\pi}K\{f(\lambda)\}\,d\lambda\neq c, \tag{2.1}$$

where c is a given constant. This test setting is unexpectedly wide and can be applied to many problems in time series. Several important applications will be given in Sections 3–5. For the problem (2.1) we propose a test based on $\int_{-\pi}^{\pi} K\{\hat{f}_n(\lambda)\}\,d\lambda$, where $\hat{f}_n(\lambda)$ is a nonparametric spectral estimator of $f(\lambda)$. In this paper we use

$$\hat{f}_n(\lambda)=\int_{-\pi}^{\pi} W_n(\lambda-\mu)\, I_n(\mu)\, d\mu.$$

Here $W_n(\cdot)$ satisfies the following.

Assumption 4. (i) $W(x)$ is bounded, even, nonnegative and such that

$$\int_{-\infty}^{\infty} W(x)\, dx=1.$$

(ii) For $M=0(n^{\alpha})(\frac{1}{4}<\alpha<\frac{1}{2})$, the function $W_n(\lambda)=MW(M\lambda)$ can be expanded as

$$W_n(\lambda)=\frac{1}{2\pi}\sum_{l} w\left(\frac{l}{M}\right)\exp(-il\lambda),$$

where $w(x)$ is a continuous, even function with $w(0)=1$, $|w(x)|\leqslant 1$ and $\int_{-\infty}^{\infty} w(x)^2\, dx<\infty$, and satisfies

$$\lim_{|x|\to 0}\frac{1-w(x)}{|x|^2}=\kappa_2<\infty.$$

Under our assumptions it is not difficult to check that the assumptions of Theorems 9 and 10 in Hannan [11, Section V] are satisfied, whence

$$\hat{f}_n(\lambda)-f(\lambda)=O_p\{\sqrt{M/n}\}, \tag{2.2}$$

uniformly in $\lambda\in[-\pi,\pi]$.

We proceed to discuss the asymptotic theory for $\int_{-\pi}^{\pi} K\{\hat{f}_n(\lambda)\}\,d\lambda$. For this we impose the following conditions on the process $\{\mathbf{z}(t)\}$, as in Hosoya and Taniguchi [12]. We denote the σ-field generated by $\{e(s)\colon s\leqslant t\}$ by $B(t)$.

Assumptions 5. (i) For $a, b=1, ..., p$ and $m\in J$,

$$\operatorname{Var}[E\{e_a(t)\, e_b(t+m)\mid B(t-\tau)\}-\delta(m,0)\,\Omega_{ab}]=O(\tau^{-2-\varepsilon}),\qquad \varepsilon>0,$$

uniformly in t.

(ii) $E\,|E\{e_a(t_1)\,e_b(t_2)\,e_c(t_3)\,e_d(t_4)|\,B(t_1-\tau)\}-E\{e_a(t_1)\;e_b(t_2)\;e_c(t_3)\;e_d(t_4)\}|=O(\tau^{-1-\eta})$, uniformly in t_1, where $t_1\leqslant t_2\leqslant t_3\leqslant t_4$ and $\eta>0$.

First, we state the following basic result, whose proof we have put in Section 6.

THEOREM 1. *Suppose that Assumptions* 1–5 *hold. Under the null hypothesis H,*

$$S_n = \sqrt{n}\left[\int_{-\pi}^{\pi} K\{\hat{f}_n(\lambda)\}\, d\lambda - c\right]$$

has asymptotically a normal distribution with mean zero and variance $v_1(f) + v_2(f_4^z)$, *where*

$$v_1(f) = 4\pi \int_{-\pi}^{\pi} \mathrm{tr}[f(\lambda)\, K^{(1)}\{f(\lambda)\}]^2\, d\lambda$$

and

$$v_2(f_4^z) = 2\pi \sum_{r,t,u,s=1}^{p} \iint_{-\pi}^{\pi} K_{rt}^{(1)}(\lambda_1)\, K_{us}^{(1)}(\lambda_2) f_{rtus}^z(-\lambda_1, \lambda_2, -\lambda_2)\, d\lambda_1\, d\lambda_2.$$

Here $K^{(1)}\{f(\lambda)\}$ *is the first derivative of* $K\{f(\lambda)\}$ *at* $f(\lambda)$ *(see* [15]*), and* $K_{rt}^{(1)}(\lambda)$ *is the* (r, t) *component of* $K^{(1)}\{f(\lambda)\}$.

Here it should be noted that $\sqrt{n}$ consistency holds in Theorem 1 despite (2.2). This is due to the fact that integration of $\hat{f}_n$ recovers $\sqrt{n}$ consistency. For the testing problem (2.1) we are led to estimate the asymptotic variance $v_1(f) + v_2(f_4^z)$ of S_n. In view of Theorem 1 we can propose $v_1(\hat{f}_n)$ as a consistent estimator of $v_1(f)$. Regarding estimation of $v_2(f_4^z)$, Taniguchi [18] and Keenan [13] gave consistent estimators of $v_2(f_4^z)$ when the process concerned is scalar-valued. In what follows we extend Taniguchi's estimator to the case when the process is vector-valued. This extension is not straightforward and requires a modification of the scalar case.

All moments of $\{\mathbf{z}(t)\}$ up to eighth-order are assumed to exist and we set

$$c_{a_1 \cdots a_k}^z(t_1, ..., t_{k-1}) = \mathrm{cum}\{z_{a_1}(0), z_{a_2}(t_1), ..., z_{a_k}(t_{k-1})\},$$

$a_1, ..., a_k = 1, ..., p;\ k = 1, 2, ..., 8.$

Assumption 6. For each $j = 1, 2, ..., k-1$ and any k-tuple $a_1, a_2, ..., a_k$ we have

$$\sum_{t_1, ..., t_{k-1} = -\infty}^{\infty} |t_j c_{a_1 \cdots a_k}^z(t_1, ..., t_{k-1})| < \infty,$$

$k = 2, 3, ..., 8.$

Under Assumption 6 we may define the kth order ($k = 2, ..., 8$) cumulant spectral density by

$$f^{z}_{a_1 \cdots a_k}(\lambda_1, ..., \lambda_{k-1})$$
$$= (2\pi)^{-k+1} \sum_{t_1, ..., t_{k-1} = -\infty}^{\infty} c^{z}_{a_1 \cdots a_k}(t_1, ..., t_{k-1}) \exp\left(-i \sum_{j=1}^{k-1} \lambda_j t_j\right).$$

Define

$$d_{a_j}(\lambda) = \sum_{t=1}^{n} z_{a_j}(t)\, e^{it\lambda}.$$

We impose a further assumption.

Assumption 7. (i) $H(x)$ is a real-valued function, even, of bounded variation, and such that

$$\int_{-\pi}^{\pi} H(x)\, dx = 1,$$

and $H(x) = 0$ for $x \notin [-\pi, \pi]$.

(ii) $\{B_n\}$ is a sequence which satisfies $B_n \to 0$, $B_n^2 n \to \infty$ as $n \to \infty$.

Henceforth we set $H_n(x) = B_n^{-1} H(B_n^{-1} x)$. The following proposition gives a consistent estimator of

$$\iint_{-\pi}^{\pi} K(\lambda_1, \lambda_2) f^{z}_{a_1 a_2 a_3 a_4}(-\lambda_1, \lambda_2, -\lambda_2)\, d\lambda_1\, d\lambda_2, \tag{2.3}$$

where $K(\lambda_1, \lambda_2)$ is continuous on $[-\pi, \pi] \times [-\pi, \pi]$.

Proposition 1. *Under Assumptions* 1–7, *the following is a consistent estimator of* (2.3):

$$U_n = \left(\frac{2\pi}{n}\right)^3 \sum_{j_1} \sum_{j_2} \sum_{j_3} H_n\left\{\frac{2\pi(j_2 + j_3)}{n}\right\} K\left(\frac{2\pi j_1}{n}, \frac{2\pi j_2}{n}\right)$$
$$\times \frac{1}{8\pi^3 n} d_{a_1}\left\{\frac{2\pi(-j_1 + j_2 + j_3)}{n}\right\} d_{a_2}\left(\frac{2\pi j_1}{n}\right) d_{a_3}\left(\frac{-2\pi j_2}{n}\right) d_{a_4}\left(\frac{-2\pi j_3}{n}\right) \tag{2.4.1}$$

$$-\int_{-\pi}^{\pi} \left[K(\lambda, -\lambda) \hat{f}_{a_1 a_3}(-\lambda) \hat{f}_{a_2 a_4}(\lambda) + K(\lambda, \lambda) \hat{f}_{a_1 a_4}(-\lambda) \hat{f}_{a_3 a_3}(\lambda)\right] d\lambda \tag{2.4.2}$$

$$-\frac{H(0)}{B_n}\left(\frac{2\pi}{n}\right)^2 \sum_{j_1}\sum_{j_2} K\left(\frac{2\pi j_1}{n}, \frac{2\pi j_2}{n}\right)$$

$$\times\left(\frac{1}{2\pi n}\right)^2 d_{a_1}\left(\frac{-2\pi j_1}{n}\right) d_{a_2}\left(\frac{2\pi j_1}{n}\right) d_{a_3}\left(\frac{-2\pi j_2}{n}\right) d_{a_4}\left(\frac{2\pi j_2}{n}\right), \quad (2.4.3)$$

where, here and subsequently, sums with respect to $j_1, \ldots$ *are from* $[-n/2]+1$ *to* $[n/2]$, *and* $\hat{f}_{a_k a_j}(\lambda)$ *is the nonparametric spectral estimator of* $f_{a_k a_j}(\lambda)$ *defined previously.*

The proof is given in Section 6.

Let $\hat{v}_2$ be the consistent estimator of $v_2(f_4^z)$ given in the manner of Proposition 1. Then, Theorem 1, Proposition 1, and Slutsky's theorem together yield the following.

THEOREM 2. *Suppose that Assumptions* 1–7 *hold. Then, under the null hypothesis H,*

$$T_n = S_n/\sqrt{v_1(\hat{f}_n) + \hat{v}_2}$$

has asymptotically the standard normal distribution $\mathcal{N}(0, 1)$. *In particular, if the process is Gaussian,*

$$T_n^G = S_n/\sqrt{v_1(\hat{f}_n)}$$

converges in distribution to $\mathcal{N}(0, 1)$.

From Theorem 2 we can propose the test of H given by the critical region

$$[|T_n| > t_\alpha], \quad (2.5)$$

where t_α is defined by

$$\int_{t_\alpha}^{\infty} (2\pi)^{-1/2} \exp\left(-\frac{x^2}{2}\right) dx = \frac{\alpha}{2}.$$

Next we introduce a measure of goodness of our test. Let $a(\lambda)$ be a $p \times p$ matrix whose entries $a_{kl}(\lambda)$ $(k, l = 1, \ldots, p)$ are square integrable functions on $[-\pi, \pi]$. We assume that $a(\lambda)$ is positive definite for each $\lambda \in [-\pi, \pi]$. Consider a sequence of alternative spectral density matrices

$$g_n(\lambda) = f(\lambda) + \frac{1}{\sqrt{n}} a(\lambda). \quad (2.6)$$

Let $E_{g_n}(\cdot)$ and $V_f(\cdot)$ denote the expectation under $g_n(\lambda)$ and the variance under $f(\lambda)$, respectively. It is natural to define an efficacy of T_n by

$$\text{eff}(T_n) = \lim_{n\to\infty} \frac{E_{g_n}(S_n)}{\sqrt{V_f(S_n)}}, \tag{2.7}$$

in line with the usual definition for a sequence of "parametric alternatives" (e.g., Randles and Wolfe [16, pp. 147–149]). Then we see that

$$\begin{aligned}\text{eff}(T_n) &= \lim_{n\to\infty} \frac{\sqrt{n}\int_{-\pi}^{\pi} [K\{g_n(\lambda)\} - K\{f(\lambda)\}]\, d\lambda}{\sqrt{v_1(f) + v_2(f_4^z)}} \\ &= \frac{\int_{-\pi}^{\pi} \text{tr}[K^{(1)}\{f(\lambda)\}\, a(\lambda)]\, d\lambda}{\sqrt{v_1(f) + v_2(f_4^z)}}.\end{aligned} \tag{2.8}$$

For another test T_n^* we can define an asymptotic relative efficiency (ARE) of T_n relative to T_n^* by

$$\text{ARE}(T_n, T_n^*) = \left\{\frac{\text{eff}(T_n)}{\text{eff}(T_n^*)}\right\}^2. \tag{2.9}$$

For a Gaussian scalar process, Taniguchi and Kondo [20] and Kondo and Taniguchi [14] proved some superiority of T_n to the existing methods. In this paper we develop the discussion beyond their scope. In later sections we will discuss non-Gaussian robustness of T_n, interrelation analysis for the components of $\{\mathbf{z}(t)\}$ and eigenvalue analysis of the spectral density matrix.

3. Non-Gaussian Robustness

In the previous section we evaluated the efficacy of T_n. If the process is not Gaussian, it depends on a non-Gaussian quantity $v_2(f_4^z)$. Henceforth we say that T_n is non-Gaussian robust if $v_2(f_4^z) = 0$. This definition means that if T_n is non-Gaussian robust then its efficacy attains the "Gaussian efficacy"

$$\int_{-\pi}^{\pi} \text{tr}[K^{(1)}\{f(\lambda)\}\, a(\lambda)]\, dy/\sqrt{v_1(f)} \tag{3.1}$$

although the process is not Gaussian. We deal with the same process $\{\mathbf{z}(t)\}$ as in Section 2. We impose a further assumption on the innovation process $\{\mathbf{e}(t)\}$.

Assumption 8. (i) $\{\mathbf{e}(t); t \in J\}$ is a family of i.i.d. random vectors with $E\{\mathbf{e}(t)\} \equiv 0$ and the variance matrix Ω.

(ii) The fourth-order cumulants

$$\kappa_{abcd} = \mathrm{cum}\{e_a(t), e_b(t), e_c(t), e_d(t)\},$$

exist for $a, b, c, d = 1, \ldots, p$, and $t \in J$.

Then we get the following proposition, whose proof we have put in Section 6.

PROPOSITION 2. *Suppose that Assumptions* 1 *and* 8 *hold. If*

$$\int_{-\pi}^{\pi} A^*(\lambda)\, K^{(1)}\{f(\lambda)\}\, A(\lambda)\, d\lambda = 0 \qquad (p \times p\text{-null matrix}), \tag{3.2}$$

then $v_2(f_4^-) = 0$.

In this section we consider the testing problem related to the following three measures of nearness between $p \times p$-spectral density matrices $f(\lambda)$ and $g(\lambda)$.

(I) Likelihood ratio measure. Define

$$D_{LR}(f, g) = \int_{-\pi}^{\pi} \left[\log \frac{\det\{g(\lambda)\}}{\det\{f(\lambda)\}} + \mathrm{tr}\{f(\lambda)\, g(\lambda)^{-1}\} - p \right] d\lambda. \tag{3.3}$$

Altough the process concerned is not Gaussian, we can formally make the Gaussian likelihood ratio GLR. The above measure is an approximation of $n^{-1}E(GLR)$. We refer to it as a likelihood ratio measure between $f(\lambda)$ and $g(\lambda)$. If we are interested in the testing problem

$$H_{\mathrm{LR}} : D_{\mathrm{LR}}\left(f, \frac{1}{2\pi} I_p\right) = c,$$

against

$$A_{\mathrm{LR}} : D_{\mathrm{LR}}\left(f, \frac{1}{2\pi} I_p\right) \neq c, \tag{3.4}$$

where c is a given constant and I_p is the $p \times p$-identity matrix, we can set down

$$D_{\mathrm{LR}}\left(f, \frac{1}{2\pi} I_p\right) = \int_{-\pi}^{\pi} K_{\mathrm{LR}}\{f(\lambda)\}\, d\lambda,$$

where $K_{\mathrm{LR}}\{f(\lambda)\} = -\log\det\{f(\lambda)\} + 2\pi \operatorname{tr}\{f(\lambda)\} - p - p\log 2\pi$. Let $T_n^{(\mathrm{LR})}$ be the test statistic for (3.4) which is given in the manner of Theorem 2. By $T_n^{(\mathrm{LR})}$, the nearness of $f(\lambda)$ to the white noise will be examined.

(II) α-entropy measure. For $\alpha \in (0, 1)$, Albrecht [1] introduced the α-entropy measure between $f(\lambda)$ and $g(\lambda)$,

$$D_\alpha(f, g) = \frac{1}{4\pi}\int_{-\pi}^{\pi} [\log\det\{(1-\alpha) I_p + \alpha f(\lambda)\, g(\lambda)^{-1}\}$$

$$-\alpha \log\det\{f(\lambda)\, g(\lambda)^{-1}\}]\, d\lambda,$$

which measures a nearness of $f(\lambda)$ to $g(\lambda)$. We set down

$$D_\alpha\left(f, \frac{1}{2\pi} I_p\right) = \int_{-\pi}^{\pi} K_\alpha\{f(\lambda)\}\, d\lambda,$$

where

$$K_\alpha\{f(\lambda)\} = (1/4\pi)[\log\det\{(1-\alpha) I_p + 2\pi\alpha f(\lambda)\} - \alpha\log\det\{2\pi f(\lambda)\}].$$

Consider the testing problem

$$H_\alpha\colon \int_{-\pi}^{\pi} K_\alpha\{f(\lambda)\}\, d\lambda = c$$

against

$$A_\alpha\colon \int_{-\pi}^{\pi} K_\alpha\{f(\lambda)\}\, d\lambda \neq c. \tag{3.5}$$

Similarly we can make the test statistic $T_n^{(\alpha)}$ for (3.5) in the manner of Theorem 2.

(III) Log-square distance. Define

$$D_{\mathrm{LOG}} = \int_{-\pi}^{\pi} [\log\det\{f(\lambda)\} - \log\det\{g(\lambda)\}]^2\, d\lambda.$$

We set down

$$D_{\mathrm{LOG}}(f, I_p) = \int_{-\pi}^{\pi} K_{\mathrm{LOG}}\{f(\lambda)\}\, d\lambda,$$

where $K_{\mathrm{LOG}}\{f(\lambda)\} = [\log\det\{f(\lambda)\}]^2$. For the testing problem

$$H_{\mathrm{LOG}}\colon \int_{-\pi}^{\pi} K_{\mathrm{LOG}}\{f(\lambda)\}\, d\lambda = c,$$

against

$$A_{\mathrm{LOG}}\colon \int_{-\pi}^{\pi} K_{\mathrm{LOG}}\{f(\lambda)\}\, d\lambda \neq c, \tag{3.6}$$

we can construct the test statistic $T_n^{(\mathrm{LOG})}$ in the manner of Theorem 2.

The above three measures $D_{\mathrm{LR}}(\cdot)$, $D_\alpha(\cdot)$, and $D_{\mathrm{LOG}}(\cdot)$ enable us to test how $f(\lambda)$ is distant from the white noise process nonparametrically. The following theorem is concerned with the non-Gaussian robustness of $T_n^{(\mathrm{LR})}$, $T_n^{(\alpha)}$, and $T_n^{(\mathrm{LOG})}$.

THEOREM 3. *Suppose that Assumptions* 1–8 *hold.*

(i) *For the testing problem* (3.4), $T_n^{(\mathrm{LR})}$ *is non-Gaussian robust if*

$$\sum_{j=0}^{\infty} G(j)'\, G(j) = \Omega^{-1}. \tag{3.7}$$

(ii) *For the testing problem* (3.5), $T_n^{(\alpha)}$ *is non-Gaussian robust if*

$$\int_{-\pi}^{\pi} \{(1-\alpha)\, A(\lambda)^{-1}\, A^*(\lambda)^{-1}\, \Omega^{-1} + \alpha I_p\}^{-1}\, d\lambda = 2\pi I_p. \tag{3.8}$$

(iii) *For the testing problem* (3.6), $T_n^{(\mathrm{LOG})}$ *is non-Gaussian robust if the spectral density matrix is expressed as*

$$f(\lambda) = \exp\left\{\sum_{j\neq 0} A_j \cos(j\lambda)\right\}, \tag{3.9}$$

where the A_j*'s are* $p\times p$*-matrices and* $\exp\{\cdot\}$ *is the matrix exponential (for the definition, see Bellman* [2, p. 169]).

Now we give a numerical example related to Theorem 3.

EXAMPLE 1. Let $\{\mathbf{z}(t);\, t\in J\}$ be a scalar process with spectral density $f(\lambda)=\exp\{a\cos\lambda\}$ and $Ez(t)=0$. For this process we consider the testing problem (3.6), i.e.,

$$\begin{aligned} H_{\mathrm{LOG}}&\colon \int_{-\pi}^{\pi} \{\log f(\lambda)\}^2\, d\lambda = a^2\pi \\ A_{\mathrm{LOG}}&\colon \int_{-\pi}^{\pi} \{\log f(\lambda)\}^2\, d\lambda \neq a^2\pi. \end{aligned} \tag{3.10}$$

Then the test statistic is

$$T_n^{(\mathrm{LOG})} = \frac{\sqrt{n}\,[\int_{-\pi}^{\pi} \{\log \hat{f}_n(\lambda)\}^2\, d\lambda - a^2\pi]}{4\sqrt{\pi}\,[\int_{-\pi}^{\pi} \{\log \hat{f}_n(\lambda)\}^2\, d\lambda]^{1/2}}.$$

TABLE I

a		Case (1)	Case (2)
0.5	Mean of $T_n^{(\mathrm{LOG})}$	−0.028	0.040
	Variance of $T_n^{(\mathrm{LOG})}$	0.972	1.078
	Frequency of $T_n^{(\mathrm{LOG})} < -1.64$	0.050	0.058
	Frequency of $T_n^{(\mathrm{LOG})} > 1.64$	0.040	0.056
0.6	Mean of $T_n^{(\mathrm{LOG})}$	−0.036	0.016
	Variance of $T_n^{(\mathrm{LOG})}$	1.066	1.015
	Frequency of $T_n^{(\mathrm{LOG})} < -1.64$	0.056	0.062
	Frequency of $T_n^{(\mathrm{LOG})} > 1.64$	0.060	0.042

In view of Theorem 3, $T_n^{(\mathrm{LOG})}$ is evidently non-Gaussian robust. For this fact we deal with the following two cases. We can express $\mathbf{z}(t)$ as

$$\mathbf{z}(t) = \sum_{s=0}^{\infty} g(s)\, e(t-s), \qquad t \in J.$$

Case (1). $e(t)$; $t \in J$, are i.i.d. as $\mathcal{N}(0, 1)$ (Gaussian case).
Case (2). $e(t)$; $t \in J$, are i.i.d. with probability density

$$p(x) = \exp(-x+1)$$

(non-Gaussian case).

For Cases (1) and (2) we generated $z(1), ..., z(1024)$, respectively. Then we calculated $T_n^{(\mathrm{LOG})}$ with $M = 100$ for Cases (1) and (2) and iterated this procedure 1000 times. We get Table I. The table agrees with our theoretical results approximately.

4. Interrelation Analysis of the Components of the Process

Our test can be applied to testing for interrelation of the components of $\{\mathbf{z}(t)\}$. Suppose that the process $\{\mathbf{z}(t); t \in J\}$ has the form

$$\mathbf{z}(t) = \begin{pmatrix} \mathbf{x}(t) \\ \mathbf{y}(t) \end{pmatrix}$$

with $\mathbf{x}(t)$, q vector-valued, and $\mathbf{y}(t)$, r vector-valued; $q + r = p$ and has the spectral density matrix

$$f(\lambda) = \begin{bmatrix} f_{xx}(\lambda) & f_{xy}(\lambda) \\ f_{yx}(\lambda) & f_{yy}(\lambda) \end{bmatrix}. \tag{4.1}$$

Denote by $H\{\cdot\}$ the linear closed manifold generated by $\{\cdot\}$ and denote by $\text{proj}[\mathbf{x}(t) \mid H\{\cdot\}]$ the projection of $\mathbf{x}(t)$ on $H\{\cdot\}$. We consider the residual process

$$\mathbf{u}_1(t) = \mathbf{x}(t) - \text{proj}[\mathbf{x}(t) \mid H\{\mathbf{x}(t-1), \mathbf{x}(t-2), \ldots\}],$$
$$\mathbf{v}_1(t) = \mathbf{y}(t) - \text{proj}[\mathbf{y}(t) \mid H\{\mathbf{y}(t-1), \mathbf{y}(t-2), \ldots\}],$$
$$\mathbf{u}_2(t) = \mathbf{x}(t) - \text{proj}[\mathbf{x}(t) \mid H\{\mathbf{x}(t-1), \mathbf{x}(t-2), \ldots; \mathbf{y}(t-1), \mathbf{y}(t-2), \ldots\}],$$
$$\mathbf{v}_2(t) = \mathbf{y}(t) - \text{proj}[\mathbf{y}(t) \mid H\{\mathbf{x}(t-1), \mathbf{x}(t-2), \ldots; \mathbf{y}(t-1), \mathbf{y}(t-2), \ldots\}],$$

and

$$\mathbf{u}_3(t) = \mathbf{x}(t) - \text{proj}[\mathbf{x}(t) \mid H\{\mathbf{x}(t-1), \mathbf{x}(t-2), \ldots; \mathbf{y}(t), \mathbf{y}(t-1), \ldots\}].$$

The measure of linear feedback from $Y = \{\mathbf{y}(t)\}$ to $X = \{\mathbf{x}(t)\}$ is defined by

$$F_{Y \to X} = \log[\det\{\text{Var}(\mathbf{u}_1(t))\}/\det\{\text{Var}(\mathbf{u}_2(t))\}].$$

Symmetrically,

$$F_{X \to Y} = \log[\det\{\text{Var}(\mathbf{v}_1(t))\}/\det\{\text{Var}(\mathbf{v}_2(t))\}].$$

The measure of instanteous linear feedback

$$F_{X \cdot Y} = \log[\det\{\text{Var}(\mathbf{u}_2(t))\}/\det\{\text{Var}(\mathbf{u}_3(t))\}]$$

has motivation similar to that of the above two measures. The following

$$F_{X, Y} = \log[\det\{\text{Var}(\mathbf{u}_1(t))\} \det\{\text{Var}(\mathbf{v}_1(t))\}/\det \Omega]$$

is called the measure of linear dependence. Then it is known that

$$F_{X, Y} = F_{Y \to X} + F_{X \to Y} + F_{X \cdot Y} \tag{4.2}$$

and

$$F_{X, Y} = \int_{-\pi}^{\pi} K\{f(\lambda)\}\, d\lambda, \tag{4.3}$$

where

$$K\{f(\lambda)\} = -\frac{1}{2\pi} \log[\det\{I_q - f_{xy}(\lambda) f_{yy}^{-1}(\lambda) f_{yx}(\lambda) f_{xx}^{-1}(\lambda)\}]$$

(see [8]). Since $F_{Y \to X}$, $F_{X \to Y}$, and $F_{X \cdot Y}$ are important econometric measures which represent "strength of causality," the testing problem

$$H: F_{X, Y} = c,$$

against

$$A: F_{X,Y} \neq c \tag{4.4}$$

is important. This is exactly an example of our testing problem. Hence we can test (4.4) by using T_n in Theorem 2.

EXAMPLE 2. Let $\mathbf{z}(t) = \{x(t), y(t)\}'$ be a two-dimensional linear process generated by

$$\mathbf{z}(t) = \mathbf{e}(t) + \begin{pmatrix} 1 & \varepsilon \\ \varepsilon & 1 \end{pmatrix} \mathbf{e}(t-1), \qquad t \in J, \tag{4.5}$$

where $\mathbf{e}(t)$'s are i.i.d. $\mathcal{N}(0, I_2)$. We denote the spectral density matrix of $\mathbf{z}(t)$ by

$$f(\lambda) = \begin{pmatrix} f_{xx}(\lambda) & f_{xy}(\lambda) \\ f_{yx}(\lambda) & f_{yy}(\lambda) \end{pmatrix}.$$

Consider the testing problem

$$H: F_{X,Y} = \int_{-\pi}^{\pi} K\{f(\lambda)\}\, d\lambda = c,$$

against,

$$A: F_{X,Y} \neq c, \tag{4.6}$$

where $K\{f(\lambda)\} = -(1/2\pi) \log\{1 - f_{xy}(\lambda) f_{yy}^{-1}(\lambda) f_{yx}(\lambda) f_{xx}^{-1}(\lambda)\}$. For this problem we generated $\mathbf{z}(1), ..., \mathbf{z}(1024)$ given by (4.5) for $\varepsilon = 1.15$, 1.20, 1.25. Then we calculated T_n in Theorem 2 and iterated this procedure 500 times. The results are given by Table II. We can see that they agree with the theoretical results.

Next, we turn to an investigation of another interrelation analysis. Let $\mathbf{z}(t) = \{\mathbf{u}(t)', \mathbf{v}(t)', \mathbf{w}(t)'\}'$ be a p-dimensional process satisfying all the

TABLE II

ε	Mean of T_n	Variance of T_n	Frequency of $T_n < -1.64$	Frequency of $T_n > 1.64$
1.15	0.077	1.073	0.058	0.058
1.20	0.067	1.006	0.042	0.056
1.25	0.076	1.000	0.050	0.052

assumptions in Section 2, where $\mathbf{u}(t)$, $\mathbf{v}(t)$, and $\mathbf{w}(t)$ are q, r, and s component processes, respectively. Correspondingly we have the partition

$$f(\lambda)=\begin{bmatrix} f_{uu}(\lambda) & f_{uv}(\lambda) & f_{uw}(\lambda) \\ f_{vu}(\lambda) & f_{vv}(\lambda) & f_{vw}(\lambda) \\ f_{wu}(\lambda) & f_{wv}(\lambda) & f_{ww}(\lambda) \end{bmatrix}$$

and the spectral representation

$$\mathbf{z}(t)=\int_{-\pi}^{\pi} e^{-it\lambda}\, d\xi(\lambda),$$

with $d\xi(\lambda)=(d\xi_u(\lambda)', d\xi_v(\lambda)', d\xi_w(\lambda)')'$. Hannan [11] considered a test for association for $\mathbf{u}=\{\mathbf{u}(t)\}$ with $\mathbf{v}=\{\mathbf{v}(t)\}$ (at frequency λ) after allowing for any effects of $\mathbf{w}=\{\mathbf{w}(t)\}$. The hypothesis is given by

$$H_\lambda: f_{uv}(\lambda)-f_{uw}(\lambda)f_{ww}(\lambda)^{-1}f_{wv}(\lambda)=0, \tag{4.7}$$

which means that $d\xi_u(\lambda)-f_{uw}(\lambda)f_{ww}(\lambda)^{-1}\,d\xi_w(\lambda)$ is incoherent with $d\xi_v(\lambda)-f_{vw}(\lambda)f_{ww}(\lambda)^{-1}\,d\xi_w(\lambda)$ and all of the apparent association between $\mathbf{u}$ and $\mathbf{v}$ is truly due only to their common association with $\mathbf{w}$. For a given λ, Hannan [11] developed the testing theory for (4.7) based on the asymptotic normality of the finite Fourier transformations of $\{\mathbf{z}(t)\}$ in a neighborhood of λ. In view of our testing problem we can consider a test for association for $\mathbf{u}$ with $\mathbf{v}$ at "all the frequency $\lambda\in[-\pi,\pi]$" after allowing for any effects of $\mathbf{w}$. The hypothesis is written as

$$H: f_{uv}(\lambda)-f_{uw}(\lambda)f_{ww}(\lambda)^{-1}f_{wv}(\lambda)=0 \qquad \text{for all} \quad \lambda\in[-\pi,\pi],$$

which is equivalent to

$$H: \int_{-\pi}^{\pi} K\{f(\lambda)\}\, d\lambda=0, \tag{4.8}$$

where

$$K\{f(\lambda)\}=\mathrm{tr}[\{f_{uv}(\lambda)-f_{uw}(\lambda)f_{ww}(\lambda)^{-1}f_{wv}(\lambda)\} \times\{f_{vu}(\lambda)-f_{vw}(\lambda)f_{ww}(\lambda)^{-1}f_{wu}(\lambda)\}].$$

Therefore we can apply all the results in Section 2 to this testing problem.

5. Eigenvalue Analysis of the Spectral Density Matrix

It is well known that the eigenvalues play a fundamental role in multivariate problems. In this section we investigate the principal components analysis of vector time series, which is related to the eigenvalues of the spectral density matrix. First, we state the following lemma which is due to Magnus and Neudecker [15].

LEMMA 1. *Let $\mu_1, \ldots, \mu_p$ be the eigenvalues of a matrix $M_0 \in \mathbf{C}^{p^2}$ and assume that μ_i is simple (i.e., $\mu_1 > \mu_2 > \cdots > \mu_p$). Then a scalar function $\mu_{(i)}$ exists, defined in a neighborhood $\mathcal{N}(M_0) \subset \mathbf{C}^{p^2}$ of M_0, such that $\mu_{(i)}(M_0) = \mu_i$ and $\mu_{(i)}(M)$ is a simple eigenvalue of M for every $M \in \mathcal{N}(M_0)$. Moreover, $\mu_{(i)}$ is ∞ time differentiable on $\mathcal{N}(M_0)$, and the differential is*

$$d\mu_{(i)} = \operatorname{tr}\left[\left\{\prod_{\substack{j=1\\ j\neq 1}}^{p}\left(\frac{\mu_j I_p - M_0}{\mu_j - \mu_i}\right)\right\} dM\right] \tag{5.1}$$

Next we turn to discuss the principal components analysis of the vector linear process $\mathbf{z}(t)$ defined in Section 2. Suppose that the spectral density matrix $f(\lambda)$ has the simple eigenvalues $\mu_1(\lambda) > \mu_2(\lambda) > \cdots > \mu_p(\lambda)$. Then the variance of the error series by the q $(q < p)$ principal components is given by

$$\int_{-\pi}^{\pi}\left\{\sum_{j=q+1}^{p} \mu_j(\lambda)\right\} d\lambda. \tag{5.2}$$

If we are interested in the degree of the above measure we can set down the testing problem

$$H\colon \int_{-\pi}^{\pi} K\{f(\lambda)\}\, d\lambda = c,$$

against

$$A\colon \int_{-\pi}^{\pi} K\{f(\lambda)\}\, d\lambda \neq c, \tag{5.3}$$

where $K\{f(\lambda)\} = \sum_{j=q+1}^{p} \mu_j(\lambda)$. In view of Lemma 1 the derivative of $K\{\cdot\}$ is

$$K^{(1)}\{f(\lambda)\} = \sum_{i=q+1}^{p}\left[\prod_{\substack{j=1\\ j\neq 1}}^{p}\left\{\frac{\mu_j(\lambda)\, I_p - f(\lambda)}{\mu_j(\lambda) - \mu_i(\lambda)}\right\}\right]$$

Thus we may construct the test T_n of Theorem 2 for (5.3), whence we can test (5.3) by using T_n.

In Sections 3, 4, and 5 we gave several applications of our testing problem. Since the functional of the spectral density matrix $\int_{-\pi}^{\pi} K\{f(\lambda)\}\, d\lambda$ represents so many important indices in time series analysis, we will be able to find many other unexpected applications of our testing problem.

At this end of this section we summarize the merits of our approach:

(1) Our approach is designed for essentially nonparametric hypotheses. So we do not need any parametric assumptions on the spectral density matrix.

(2) Our test statistic is based on the nonparametric spectral estimator. So we do not need iterative methods to calculate it.

(3) Since our approach is based on the integral of the spectral density matrix $f(\lambda)$ we can develop the $\sqrt{n}$-consistent asymptotic theory (although the nonparametric spectral estimator $\hat{f}_n(\lambda)$ is not a $\sqrt{n}$-consistent estimator of $f(\lambda)$).

(4) We do not assume the Gaussianity of the process $\{\mathbf{z}(t)\}$.

6. Some Proofs

Proof of Theorem 1. Let $\|a\|$ be the norm of $a=(a_1, ..., a_{p^2})' \in \mathbf{C}^{p^2}$ defined by

$$\|a\| = \left\{ \sum_{j=1}^{p^2} |a_j|^2 \right\}^{1/2}.$$

Since the function K is holomorphic in D it has the Taylor series expansion in an open neighborhood $U \subset D$ (e.g., Brillinger [5]; Stewart and Tall [17]). Let

$$\hat{H}(\lambda) = K\{\hat{f}_n(\lambda)\} - K\{f(\lambda)\} - \mathrm{tr}[K^{(1)}\{f(\lambda)\}\{\hat{f}_n(\lambda) - f(\lambda)\}]. \quad (6.1)$$

Using Cauchy's estimate for the derivatives of K (see Bhattacharya and Rao [3, p. 68]) we can see that there exists $\delta > 0$ such that

$$\hat{H}(\lambda) = O\{\|\hat{f}_n(\lambda) - f(\lambda)\|^2\}, \quad (6.2)$$

in the open ball $B(\delta) = \{\|\hat{f}_n(\lambda) - f(\lambda)\| \leqslant \delta\}$.

For every $\varepsilon > 0$ there exists $c > 0$ such that

$$P\left\{|\hat{H}(\lambda)| \geqslant c\frac{M}{n}\right\} \leqslant P\left\{|\hat{H}(\lambda)| \geqslant c\frac{M}{n} \cap B(\delta)\right\} + P\{\|\hat{f}_n(\lambda) - f(\lambda)\| > \delta\} < \varepsilon,$$

because of (2.2). Hence,

$$\hat{H}(\lambda) = O_p(M/n), \quad (6.3)$$

uniformly in λ. Under the null hypothesis, S_n is written as

$$S_n = \sqrt{n} \int_{-\pi}^{\pi} [K\{\hat{f}_n(\lambda)\} - K\{f(\lambda)\}] \, d\lambda.$$

It follows from (6.3) that

$$\begin{aligned} S_n &= \sqrt{n} \int_{-\pi}^{\pi} \mathrm{tr}[K^{(1)}\{f(\lambda)\}\{\hat{f}_n(\lambda) - f(\lambda)\}] \, d\lambda + o_p(1) \\ &= L_n + o_p(1) \qquad \text{(say)}. \end{aligned} \tag{6.4}$$

Putting

$$J_n = \sqrt{n} \int_{-\pi}^{\pi} \mathrm{tr}[K^{(1)}\{f(\lambda)\}\{I_n(\lambda) - f(\lambda)\}] \, d\lambda,$$

we next show that

$$|L_n - J_n| = o_p(1). \tag{6.5}$$

From the definition of $\hat{f}_n$, L_n is written as

$$\begin{aligned} L_n &= \sqrt{n} \int_{-\pi}^{\pi} \mathrm{tr}\left[K^{(1)}\{f(\lambda)\} \int_{-\pi}^{\pi} \{I_n(\mu) - f(\mu)\} \, W_n(\lambda - \mu) \, d\mu \right] d\lambda \\ &\quad + \sqrt{n} \int_{-\pi}^{\pi} \mathrm{tr}\left[K^{(1)}\{f(\lambda)\} \left\{ \int_{-\pi}^{\pi} f(\mu) \, W_n(\lambda - \mu) \, d\mu - f(\lambda) \right\} \right] d\lambda \\ &= L_n^{(1)} + L_n^{(2)} \qquad \text{(say)}. \end{aligned}$$

By changing variable $M(\lambda - \mu) \to \eta$ we obtain

$$L_n^{(1)} = \sqrt{n} \int_{-\pi}^{\pi} \int_{M(-\pi-\mu)}^{M(\pi-\mu)} \mathrm{tr}\left[K^{(1)} \left\{ f\left(\mu + \frac{\eta}{M}\right) \right\} W(\eta) \, d\eta \{I_n(\mu) - f(\mu)\} \right] d\mu.$$

From Assumption 4(i), we can write

$$|J_n - L_n^{(1)}| = \left| \sqrt{n} \int_{-\pi}^{\pi} \mathrm{tr}[A_M(\mu)\{I_n(\mu) - f(\mu)\}] \, d\mu \right|,$$

where

$$A_M(\mu) = \int_{M(-\pi-\mu)}^{M(\pi-\mu)} K^{(1)} \left\{ f\left(\mu + \frac{\eta}{M}\right) \right\} W(\eta) \, d\eta - K^{(1)}\{f(\mu)\} \int_{-\infty}^{\infty} W(\eta) \, d\eta.$$

In view of Lemma A.3.3 in Hosoya and Taniguchi [12],

$$E\,|J_n - L_n^{(1)}|^2 = 4\pi \int_{-\pi}^{\pi} \operatorname{tr}[f(\mu)\, A_M(\mu) f(\mu)\, A_M(\mu)]\, d\mu$$
$$+2\pi \sum_{r,t,u,v=1}^{p} \iint_{-\pi}^{\pi} A_M^{rt}(\mu_1)\, A_M^{uv}(\mu_2)$$
$$\times f_{rtuv}^{z}(-\mu_1, \mu_2, -\mu_2)\, d\mu_1\, d\mu_2 + o(1), \qquad (6.6)$$

where $A_M^{rt}(\mu)$ is the (r, t) component of $A_M(\mu)$. By the dominated convergence theorem it is shown that, for every $\varepsilon > 0$,

$$\lim_{M\to\infty} \|A_M(\mu)\| = 0 \qquad \text{for} \quad \mu \in B_\varepsilon,$$

where $B_\varepsilon = [-\pi+\varepsilon, \pi-\varepsilon]$. Hence, for every $\varepsilon > 0$,

$$I(B_\varepsilon) = \int_{B_\varepsilon} \operatorname{tr}[f(\mu)\, A_M(\mu) f(\mu)\, A_M(\mu)]\, d\mu \to 0,$$
$$I^{rtuv}(B_\varepsilon \times B_\varepsilon) = \iint_{B_\varepsilon \times B_\varepsilon} A_M^{rt}(\mu_1)\, A_M^{uv}(\mu_2) f_{rtuv}^{z}(-\mu_1, \mu_2, -\mu_2)\, d\mu_1\, d\mu_2 \to 0, \qquad (6.7)$$

$r, t, u, v = 1, ..., p$, as $M \to \infty$. While all the components of $A_M(\mu)$, $f(\mu)$, and $f_{rtuv}^{z}(\ ,\ ,\ )$ are bounded on $B = [-\pi, \pi]$, so there exists $d > 0$ such that

$$|I(B - B_\varepsilon)| \leqslant d\varepsilon, \qquad |I^{rtuv}(B\times B - B_\varepsilon \times B_\varepsilon)| \leqslant d\varepsilon \qquad \text{for} \quad r, t, u, v = 1, ..., p. \qquad (6.8)$$

Since ε is chosen arbitrarily, (6.6), (6.7), and (6.8) imply $|J_n - L_n^{(1)}| = o_P(1)$. Thus the proof of (6.5) is complete if we show $L_n^{(2)} = o(1)$. The bias evaluation method (e.g., Hannan [11, p. 283]) yields

$$\int_{-\pi}^{\pi} f(\mu)\, W_n(\lambda - \mu)\, d\mu - f(\lambda) = O(M^{-2}),$$

uniformly in $\lambda \in B$, which implies

$$L_n^{(2)} = O(\sqrt{n}/M^2) \to 0 \qquad \text{as} \quad n \to \infty.$$

Thus we have proved that the asymptotic distribution of S_n is equivalent to that of J_n. Applying Lemma A.3.3 of Hosoya and Taniguchi [12] to J_n, the proof is completed.

Proof of Proposition 1. First, we evaluate the expectation of the term (2.4.1):

$$E\{(2.4.1)\}=\left(\frac{2\pi}{n}\right)^3\sum_{j_1}\sum_{j_2}\sum_{j_3}H_n\left\{\frac{2\pi(j_2+j_3)}{n}\right\}K\left(\frac{2\pi j_1}{n},\frac{2\pi j_2}{n}\right)\frac{1}{8\pi^3 n}$$

$$\times\left[\operatorname{cum}\left[d_{a_1}\left\{\frac{2\pi(-j_1+j_2+j_3)}{n}\right\},d_{a_2}\left(\frac{2\pi j_1}{n}\right),\right.\right.$$

$$\left.\times d_{a_3}\left(\frac{-2\pi j_2}{n}\right),d_{a_4}\left(\frac{-2\pi j_3}{n}\right)\right] \tag{6.9.1}$$

$$+\operatorname{cum}\left[d_{a_1}\left\{\frac{2\pi(-j_1+j_2+j_3)}{n}\right\},d_{a_2}\left(\frac{2\pi j_1}{n}\right)\right]$$

$$\times\operatorname{cum}\left[d_{a_3}\left(\frac{-2\pi j_2}{n}\right),d_{a_4}\left(\frac{-2\pi j_3}{n}\right)\right] \tag{6.9.2}$$

$$+\operatorname{cum}\left[d_{a_1}\left\{\frac{2\pi(-j_1+j_2+j_3)}{n}\right\},d_{a_3}\left(\frac{-2\pi j_2}{n}\right)\right]$$

$$\times\operatorname{cum}\left[d_{a_2}\left(\frac{2\pi j_1}{n}\right),d_{a_4}\left(\frac{-2\pi j_3}{n}\right)\right] \tag{6.9.3}$$

$$+\operatorname{cum}\left[d_{a_1}\left\{\frac{2\pi(-j_1+j_2+j_3)}{n}\right\},d_{a_4}\left(\frac{-2\pi j_3}{n}\right)\right]$$

$$\left.\times\operatorname{cum}\left[d_{a_2}\left(\frac{2\pi j_1}{n}\right),d_{a_3}\left(\frac{-2\pi j_2}{n}\right)\right]\right]. \tag{6.9.4}$$

It is known that

$$\begin{aligned}&\operatorname{cum}\{d_{a_1}(\lambda_1),\ldots,d_{a_k}(\lambda_k)\}\\&\quad=(2\pi)^{k-1}\Delta_n(\lambda_1+\cdots+\lambda_k)f^z_{a_1\cdots a_k}(-\lambda_2,\ldots,-\lambda_k)+O(1),\end{aligned} \tag{6.10}$$

where $\Delta_n=\sum_{t=1}^n e^{i\lambda t}$ and $O(1)$ is uniform in $\lambda_1,\ldots,\lambda_k$ (e.g., [5]). Using (6.10), (6.9.1) is written as

$$\left(\frac{2\pi}{n}\right)^3\sum_{j_1}\sum_{j_2}\sum_{j_3}\left[H_n\left\{\frac{2\pi(j_2+j_3)}{n}\right\}K\left(\frac{2\pi j_1}{n},\frac{2\pi j_2}{n}\right)\right.$$

$$\left.\times f^z_{a_1a_2a_3a_4}\left(-\frac{2\pi j_1}{n},\frac{2\pi j_2}{n},\frac{2\pi j_3}{n}\right)+O(n^{-1})\right]$$

$$=\left(\frac{2\pi}{n}\right)^2\sum_{j_1}\sum_{j_2}{}^*\left(\frac{2\pi}{B_n n}\right)\sum_j{}^{**}\left[H\left(\frac{2\pi j}{B_n n}\right)K\left(\frac{2\pi j_1}{n},\frac{2\pi j_2}{n}\right)\right.$$

$$\left.\times f^z_{a_1a_2a_3a_4}\left(-\frac{2\pi j_1}{n},\frac{2\pi j_2}{n},\frac{2\pi(j-j_2)}{n}\right)+O(n^{-1})\right], \tag{6.11}$$

where $\sum_{j_2}^{*}$ is the sum for j_2, satisfying

$$-\left[\frac{n}{2}\right]+1\leqslant j_2\leqslant\left[\frac{n}{2}\right] \quad \text{and} \quad -\left[\frac{n}{2}\right]\leqslant j-j_2\leqslant\left[\frac{n}{2}\right],$$

and $\sum_{j}^{**}$ is the sum for j satisfying

$$-\left[\frac{B_n n}{2}\right]+1\leqslant j\leqslant\left[\frac{B_n n}{2}\right].$$

We can see that (6.11) is equal to

$$\iint_{-\pi}^{\pi} K(\lambda_1, \lambda_2) f^{z}_{a_1a_2a_3a_4}(-\lambda_1, \lambda_2, -\lambda_2)\, d\lambda_1\, d\lambda_2 + o(1).$$

Similarly,

$$(6.9.2) = \frac{H(0)}{B_n} \iint_{-\pi}^{\pi} K(\lambda_1, \lambda_2) f_{a_1a_2}(-\lambda_1) f_{a_3a_4}(-\lambda_2)\, d\lambda_1\, d\lambda_2 + o(1),$$

$$\{(6.9.3)+(6.9.4)\} = \int_{-\pi}^{\pi} [K(\lambda, -\lambda) f_{a_1a_3}(-\lambda) f_{a_2a_4}(\lambda) + K(\lambda, \lambda) \times f_{a_1a_4}(-\lambda) f_{a_2a_3}(\lambda)]\, d\lambda + o(1). \tag{6.12}$$

In view of Theorem 1, (2.4.2) is a consistent estimator of (6.12). Also it is easily shown that

$$\lim_{n\to\infty} [E\{(2.4.3)\} + (6.9.2)] = 0,$$

which implies that U_n is an asymptotically unbiased estimator of (2.3). From (6.10), it is not difficult to show that

$$\lim_{n\to\infty} \mathrm{Var}\{(2.4.1)\} = 0, \qquad \lim_{n\to\infty} \mathrm{Var}\{(2.4.3)\} = 0,$$

which completes the proof. In our vector-valued case, the term (2.4.2) differs from the corresponding term of the scalar-valued case.

Proof of Proposition 2. Since the process $\{\mathbf{z}(t)\}$ is a linear process satisfying Assumptions 1 and 8, by using fundamental properties of the cumulant (e.g., [5]), we have

$$f^{z}_{rtus}(-\lambda_1, \lambda_2, -\lambda_2)$$
$$= \left(\frac{1}{2\pi}\right)^3 \sum_{a,b,c,d=1}^{p} \kappa_{abcd} A_{ra}(-\lambda_1)\, A_{tb}(\lambda_1)\, A_{uc}(-\lambda_2)\, A_{sd}(\lambda_2).$$

From the definition of $v_2(f_4^z)$ we obtain

$$v_2(f_4^z) = \left(\frac{1}{2\pi}\right)^2 \sum_{a,b,c,d=1}^{p} \kappa_{abcd} \left\{ \sum_{r,t=1}^{p} \int_{-\pi}^{\pi} K_{rt}^{(1)}(\lambda_1) A_{ra}(-\lambda_1) A_{tb}(\lambda_1) \, d\lambda_1 \right\}$$

$$\times \left\{ \sum_{u,s=1}^{p} \int_{-\pi}^{\pi} K_{us}^{(1)}(\lambda_2) A_{uc}(-\lambda_2) A_{sd}(\lambda_2) \, d\lambda_2 \right\}. \tag{6.13}$$

From (3.2) it follows that

$$\sum_{r,t=1}^{p} \int_{-\pi}^{\pi} A_{ra}(-\lambda_1) K_{rt}^{(1)}(\lambda_1) A_{tb}(\lambda_1) \, d\lambda_1 = 0,$$

for all $a, b = 1, ..., p$, which implies that $(6.13) = 0$.

Proof of Theorem 3. (i) From Magnus and Neudecker [15] it is easy to see that

$$K_{\mathrm{LR}}^{(1)}\{f(\lambda)\} = 2\pi I_p - f(\lambda)^{-1}. \tag{6.14}$$

Since $f(\lambda) = (1/2\pi) A(\lambda) \Omega A(\lambda)^*$ with $A(\lambda) = \sum_{j=0}^{\infty} G(j) e^{i\lambda j}$, we have

$$\int_{-\pi}^{\pi} A^*(\lambda) K_{\mathrm{LR}}^{(1)}\{f(\lambda)\} A(\lambda) \, d\lambda = \int_{-\pi}^{\pi} A^*(\lambda)[2\pi I_p - f(\lambda)^{-1}] A(\lambda) \, d\lambda$$

$$= \int_{-\pi}^{\pi} 2\pi [A^*(\lambda) A(\lambda) - \Omega^{-1}] \, d\lambda. \tag{6.15}$$

By (3.7), we can show that $(6.15) = 0$. Hence, the assertion follows from Proposition 2.

(ii) Similarly, we obtain

$$K_{\alpha}^{(1)}\{f(\lambda)\} = \frac{\alpha}{4\pi} \left[\left\{ (1-\alpha) \frac{I_p}{2\pi} + \alpha f(\lambda) \right\}^{-1} - f(\lambda)^{-1} \right], \tag{6.16}$$

which shows that

$$\int_{-\pi}^{\pi} A^*(\lambda) K_{\alpha}^{(1)}\{f(\lambda)\} A(\lambda) \, d\lambda$$

$$= \frac{\alpha}{4\pi} \int_{-\pi}^{\pi} \left[\left\{ (1-\alpha) \frac{1}{2\pi} A(\lambda)^{-1} A^*(\lambda)^{-1} + \alpha \frac{\Omega}{2\pi} \right\}^{-1} - 2\pi \Omega^{-1} \right] d\lambda. \tag{6.17}$$

Evidently $(6.17) = 0$ if (3.8) is satisfied, whence the assertion follows from Proposition 2.

(iii) It is shown that

$$K_{\mathrm{LOG}}^{(1)}\{f(\lambda)\} = 2\log\det\{f(\lambda)\}\, f(\lambda)^{-1}. \tag{6.18}$$

Then

$$\int_{-\pi}^{\pi} A^*(\lambda)\, K_{\mathrm{LOG}}^{(1)}\{f(\lambda)\}\, A(\lambda)\, d\lambda = 4\pi \int_{-\pi}^{\pi} \log\det\{f(\lambda)\}\, \Omega^{-1}\, d\lambda. \tag{6.19}$$

For a $p \times p$-matrix A, it is known that

$$\det\{\exp A\} = e^{\operatorname{tr} A}, \tag{6.20}$$

which implies that (6.19) is equal to

$$4\pi \int_{-\pi}^{\pi} \operatorname{tr}\left\{\sum_{j \neq 0} A_j \cos(j\lambda)\right\} \Omega^{-1}\, d\lambda = 0.$$

Thus $T_n^{(\mathrm{LOG})}$ is non-Gaussian robust.

References

[1] Albrecht, V. (1983). On the convergence rate of probability of error in Bayesian discrimination between two Gaussian processes. In *Asymptotic Statistics* 2, *Proceedings of the Third Prague Symposium on Asymptotic Statistics*, pp. 165–175.

[2] Bellman, R. (1974). *Introduction to Matrix Analysis*. McGraw–Hill, New York.

[3] Bhattacharya, R. N., and Rao, R. R. (1976). *Normal Approximation and Asymptotic Expansions*. Wiley, New York.

[4] Bloomfield, P. (1973). An exponential model for the spectrum of a scalar time series. *Biometrika* **60** 217–226.

[5] Brillinger, D. R. (1975). *Time Series Data Analysis and Theory*. Holt, Rinehart, & Winston, New York.

[6] Dzhaparidze, K. (1986). *Parameter Estimation and Hypothesis Testing in Spectral Analysis of Stationary Time Series*. Springer-Verlag, New York.

[7] Fuller, W. A. (1976). *Introduction to Statistical Time Series*. Wiley, New York.

[8] Geweke, J. (1982). Measurement of linear dependence and feedback between multiple time series. *J. Amer. Statist. Assoc.* **77** 304–324.

[9] Hallin, M., Ingenbleek, J. F., and Puri, M. L. (1985). Linear serial rank tests for randomness against ARMA alternatives. *Ann. Statist.* **13** 1156–1181.

[10] Hallin, M., and Puri, M. L. (1988). Optimal rank-based procedures for time series analysis: Testing an ARMA model against other ARMA models. *Ann. Statist.* **16** 402–432.

[11] Hannan, E. J. (1970). *Multiple Time Series*. Wiley, New York.

[12] Hosoya, Y., and Taniguchi, M. (1982). A central limit theorem for stationary processes and the parameter estimation of linear processes. *Ann. Statist.* **10** 132–153.

[13] Keenan, D. M. (1987). Limiting behavior of functionals of higher-order sample cumulant spectra. *Ann. Statist.* **15** 134–151.

[14] Kondo, M., and Taniguchi, M. (1993). Two sample problem in time series analysis. In *Proceedings of the Third Pacific Area Statistical Conference*, pp. 165–174.

[15] MAGNUS, J. R., AND NEUDECKER, H. (1988). *Matrix Differential Calculus with Applications in Statistics and Econometrics.* Wiley, New York.

[16] RANDLES, R. H. AND WOLFE, D. A. (1979). *Introduction to the Theory of Nonparametric Statistics.* Wiley, New York.

[17] STEWART, I. AND TALL, D. (1983). *Complex Analysis.* Cambridge Univ. Press, Cambridge, UK.

[18] TANIGUCHI, M. (1982). On estimation of the integrals of the fourth order cumulant spectral density. *Biometrika* **69** 117–122.

[19] TANIGUCHI, M. (1987). Minimum contrast estimation for spectral densities of stationary processes. *J. Roy. Statist. Soc. Ser. B* **49** 315–325.

[20] TANIGUCHI, M., AND KONDO, M. (1993). Non-parametric approach in time series analysis. *J. Time Ser. Anal.* **14** 397–408.

Reprinted by permission of the publisher,
© 1996 by Academic Press

THE LIMITING DENSITY OF UNIT ROOT TEST STATISTICS: A UNIFYING TECHNIQUE

MITHAT GÖNEN, MADAN L. PURI, FRITS H. RUYMGAART and MARTIEN C. A. VAN ZUIJLEN

Logic Associates, Indiana University, Texas Tech University and Katholieke Universiteit Nijmegen

Abstract. In this note we introduce a simple principle to derive a constructive expression for the density of the limiting distribution, under the null hypothesis, of unit root statistics for an AR(1)-process in a variety of situations. We consider the case of unknown mean and reconsider the well-known situation where the mean is zero. For long-range dependent errors we indicate how the principle might apply again. We also show that in principle the method also works for a near unit root case. Weak convergence and subsequent Karhunen–Loève expansion of the weak limit of the partial sum process of the errors plays an important role, along with the evaluation of a certain normal type integral with complex mean and variance. For independent and long range dependent errors this weak limit is ordinary and fractional Brownian motion respectively.

Keywords and phrases. unit root tests; weak convergence; partial sum process; ordinary and fractional Brownian motion; Karhunen–Loève expansions; wavelets; normal type integrals with complex parameters.

AMS 1991 subject classification. Primary 62M10; secondary 62E20.

1. INTRODUCTION

There is a renewed interest in the asymptotics of unit or, more generally, near unit root statistics. Such statistics are used to test the nonstationarity hypothesis in certain autoregressive (AR) models. For a summary and recent developments with a focus on theoretical aspects see, for example, Fuller (1996) and Tanaka (1996). For a special view towards econometric applications we refer, for instance, to Maddala and Kim (1996), and Charemsa (1996). In this note we are primarily interested in presenting a simple, unified approach to obtain a more or less explicit form of the density of the limiting distribution in a variety of cases. To the best of our knowledge this principle has not yet been recognized in the literature, so that even in cases where an explicit expression is already known, our method still provides a new approach. We also include a case where such an expression doesn't yet seem to exist.

In its simplest form the problem is to find the density of the asymptotic distribution of the least squares estimator $\hat{\rho}$, say, of the parameter ρ in the AR(1)-model

$$X_0 = 0, \quad X_k = \rho X_{k-1} + \varepsilon_k, \quad k \in \mathbb{N}, \tag{1.1}$$

with IID $(0, \sigma^2)$ errors $\varepsilon_0, \varepsilon_1, \varepsilon_2, \ldots$, under the null hypothesis $\rho = 1$. This hypothesis corresponds to the random walk model. White (1958) gave a heuristic proof of the weak convergence of the statistics to a functional of the Wiener process, exploiting the weak convergence of the partial sum process of the errors to Brownian motion. Dickey and Fuller (1979) obtained a limiting distribution in terms of a quotient of quadratic forms of independent normal variables. This expression was used for simulations to tabulate the limiting distribution. Later Chan and Wei (1985, 1987) formalized White's (1958) heuristic approach. These authors, moreover, observed that subsequent Karhunen–Loève expansion of the Wiener process yields the same expression as the one obtained by Dickey and Fuller (1979). Even in the simple model (1.1) finding an explicit expression for the limiting density is a difficult problem. Under the assumption of symmetry of the error distribution, Rao (1978, 1980) obtained such an expression. A different expression in closed form, without this restriction, was obtained by Abadir (1993). An exact explicit formula for the limiting density of a studentized version of this statistic can be found in Abadir (1995).

An extension of this problem is to investigate the limiting density under local alternatives where ρ is near 1. Chan and Wei (1987) proved that under such local alternatives the weak limit equals in law a functional involving the Ornstein–Uhlenbeck process; see also Perron (1991). An explicit form of the limiting density under local alternatives was derived by Larsson (1995), both without and with studentization.

More realistically, one may want to allow the mean of the observable to be unknown and extend (1.1) to the model

$$X_0 = \mu + \varepsilon_0, \qquad X_k - \mu = \rho(X_{k-1} - \mu) + \varepsilon_k, \qquad k \in \mathbb{N}, \tag{1.2}$$

where all three parameters μ, ρ, and σ^2 are unknown. The estimator of ρ and its weak limit are more complicated now. Its limiting distribution under the null hypothesis is given in Fuller (1996, Theorem 10.1.3), where it is implicitly given as a rather complicated functional of the Wiener process. We are not aware of a result regarding a constructive expression for the density of this distribution.

It has been shown by Sowell (1990) that under certain long-range dependence (Beran (1994)) in model (1.1) the limiting distribution is a – slightly different – functional of fractional Brownian motion. An expression for the limiting density, however, was not included.

The general principle that we want to exploit here applies whenever the limiting distribution is the law of a function of a finite number of suitable functionals of a Gaussian process with known Karhunen–Loève expansion. These functionals should be a linear combination of either the independent normal variables in this expansion or of their squares. This is the case in all the examples mentioned above where the errors are independent. If the Karhunen–

Loève expansion for the fractional Brownian motion involved were known, the example with long range dependent errors would also be included. This example will be used, however, to illustrate how a wavelet expansion might lead to a good approximation (Tewfik and Kim (1992)). In Section 2 we will present the main tool. For model (1.1) the limiting density of the statistic and its studentized version will be considered in Section 3 under the null hypothesis and local alternatives. In Section 4 we will discuss the limiting density of the statistic under the null hypothesis in model (1.2). Finally, in Section 5 we briefly comment on the long range dependent error case.

2. A SIMPLE GENERAL PRINCIPLE

Let $Z_1, Z_2, \ldots$ denote i.i.d. standard normal random variables and let $a_{i,1}, a_{i,2}, \ldots, (i = 1, \ldots, I)$ and $b_{j,1}, b_{j,2}, \ldots (j = 1, \ldots, J)$ be given numbers such that

$$\sum_{m=1}^{\infty} a_{i,m}^2 < \infty, \qquad \sum_{m=1}^{\infty} b_{j,m}^2 < \infty, \qquad \text{for all } i \text{ and } j. \tag{2.1}$$

Furthermore let us introduce the random variables

$$U_i := \sum_{m=1}^{\infty} a_{i,m} Z_m, \qquad V_j := \sum_{m=1}^{\infty} b_{j,m} Z_m^2. \tag{2.2}$$

Consider a real valued function ξ, defined and continuous on the actual range Δ, say, of $(U, V) := (U_1, \ldots, U_I, V_1, \ldots, V_J)$. We will be interested in the density of the random variable

$$R := \xi(U, V). \tag{2.3}$$

For brevity let us also write $a_m := (a_{1,m}, \ldots, a_{I,m})$, $b_m := (b_{1,m}, \ldots, b_{J,m})$, $s := (s_1, \ldots, s_I)$, $t := (t_1, \ldots, t_J)$. Under the usual regularity conditions the density of R can be obtained from the joint density of (U, V) by a simple change of variables. Therefore in this section we focus on this joint density. The independence of the Z_m entails that the joint characteristic function of the U_i and V_j satisfies

$$\mathscr{F}_{U,V}(s, t) = \prod_{m=1}^{\infty} \mathbf{E}\, e^{i(\langle s, a_m\rangle Z_m + \langle t, b_m\rangle Z_m^2)}, \; (s, t) \in \mathbb{R}^{I+J}. \tag{2.4}$$

For further evaluation we need the following well-known result which can be easily verified with the help of the residue theorem.

LEMMA 2.1. *For any complex numbers a, b with* $\operatorname{Re} b > 0$ *we have*

$$\int_{-\infty}^{\infty} \mathrm{e}^{-b(x-a)^2}\, dx = \sqrt{\frac{\pi}{|b|}}\, e^{-\frac{1}{2} i \arg(b)}. \tag{2.5}$$

Applying (2.5) to the m-th factor on the right in (2.4) we obtain

$$\mathbf{E}\exp(i\langle s, a_m\rangle Z_m + \langle t, b_m\rangle Z_m^2)$$

$$= (2\pi)^{-\frac{1}{2}}\int_{-\infty}^{\infty} \exp\left(i(\langle s, a_m\rangle x + \langle t, b_m\rangle x^2) - \frac{1}{2}x^2\right) dx$$

$$= (1 + 4\langle t, b_m\rangle^2)^{-\frac{1}{4}} \exp\left(-\frac{\langle s, a_m\rangle^2}{2 - 4i\langle t, b_m\rangle} + \frac{1}{2} i \arctan 2\langle t, b_m\rangle \right).$$

Hence for the characteristic function we have the following result.

THEOREM 2.1. *The joint characteristic function of* (U, V) *equals*

$$\mathscr{F}_{U,V}(s, t) = \left\{ \prod_{m=1}^{\infty} (1 + 4\langle t, b_m\rangle^2) \right\}^{-\frac{1}{4}}$$

$$\times \exp\left(-\frac{1}{2}\sum_{m=1}^{\infty} \left\{ \frac{\langle s, a_m\rangle^2 (1 + 2i\langle t, b_m\rangle)}{1 + 4\langle t, b_m\rangle^2} - i \arctan 2\langle t, b_m\rangle \right\} \right),$$

for $(s, t) \in \mathbb{R}^{I+J}$. (2.6)

Consequently, the joint density is given by

$$f_{U,V}(u, v) = \left(\frac{1}{2\pi}\right)^{(I+J)/2} \int_{\mathbb{R}^I}\int_{\mathbb{R}^J} e^{-i(\langle u,s\rangle + \langle v,t\rangle)} \mathscr{F}_{U,V}(s, t)\, ds\, dt,$$

for $(u, v) \in \mathbb{R}^J \times (0, \infty)^J$. (2.7)

REMARK 2.1. Expansions like those in (2.2) may arise from certain functionals over a Gaussian process $\mathscr{G}$ in $\mathscr{C}([0, 1])$, like Brownian or fractional Brownian motion in this paper. If $\mathscr{G}$ has Karhunen–Loève expansion $\mathscr{G}(t) = \sum_{m=1}^{\infty} \sqrt{\lambda_m} Z_m \phi_m(t)$, where $\{\varphi_m\}$ is an orthonormal basis of $L^2(0, 1)$ we have, for example, $\|\mathscr{G}\|^2 = \sum_{m=1}^{\infty} \lambda_m Z_m^2$.

REMARK 2.2. It is, of course, much harder to include general quadratic forms like $\sum_k \sum_m c_{k,m} Z_k Z_m$. In the examples of the present paper, however, such forms don't show up.

REMARK 2.3. Forms of the type $\sum_m a_m Z_m^k$ for powers $k = 3, 4, \ldots$ are hard because evaluation of the integral in (2.5) is only possible for a quadratic exponent. In none of the present examples, however, such powers occur.

3. KNOWN MEAN

3.1. *Null hypothesis*

In the situation of model (1.1) we have the estimators

$$\hat{\rho} := \frac{\sum_{k=1}^{n} X_k X_{k-1}}{\sum_{k=1}^{n} X_{k-1}^2}, \qquad \hat{\sigma}^2 := \frac{1}{n}\sum_{k=1}^{n}(X_k - \hat{\rho}X_{k-1})^2. \tag{3.1}$$

Let $W(t)$, $0 \leqslant t \leqslant 1$, denote the standard Wiener process on the unit interval. It is well-known that the nonstudentized and studentized statistics converge weakly according to

$$n(\hat{\rho} - 1) \to \mathbf{D}\frac{W^2(1) - 1}{2\int_0^1 W^2(t)\,dt}, \qquad \text{as } n \to \infty, \tag{3.2}$$

$$(\hat{\rho} - 1)\sqrt{\frac{\sum_{k=1}^{n} X_{k-1}^2}{\hat{\sigma}^2}} \to \mathbf{D}\frac{W^2(1) - 1}{2\sqrt{\int_0^1 W^2(t)\,dt}}, \qquad \text{as } n \to \infty, \tag{3.3}$$

respectively.

The Wiener process has the Karhunen–Loève expansion

$$W(t) = \sum_{m=1}^{\infty} \sqrt{\lambda_m} Z_m \varphi_m(t), \qquad 0 \leqslant t \leqslant 1, \tag{3.4}$$

where the Z_m are IID standard normal and

$$\varphi_m(t) := \sqrt{2}\sin\left(m - \frac{1}{2}\right)\pi t, \qquad 0 \leqslant t \leqslant 1,$$

$$\lambda_m := \left\{\frac{1}{(m - \frac{1}{2})\pi}\right\}^2, \qquad m \in \mathbb{N}. \tag{3.5}$$

Let us now introduce

$$U := \sqrt{2}\sum_{m=1}^{\infty} \sqrt{\lambda_m}(-1)^{m+1} Z_m = W(1). \tag{3.6}$$

$$V := \sum_{m=1}^{\infty} \lambda_m Z_m^2 = \int_0^1 W^2(t)\,dt. \tag{3.7}$$

Consequently the weak limits in (3.2) and (3.3) can be written in the form (cf. (2.3))

$$R_1 := \frac{U^2 - 1}{2V}, \qquad R_2 := \frac{U^2 - 1}{2\sqrt{V}}. \tag{3.8}$$

The joint density of U and V can be obtained from (2.6) and (2.7) with $I = J = 1$, λ_m as in (3.5), and

$$\langle s, a_m \rangle = (-1)^{m+1}\sqrt{2\lambda_m s}, \qquad s \in \mathbb{R}, \tag{3.9}$$

$$\langle t, b_m \rangle = \lambda_m t, \qquad t \in \mathbb{R}. \tag{3.10}$$

The actual domain of this density is

$$\Delta := \{(u, v) \in \mathbb{R}^2 : \ v > 0\}. \tag{3.11}$$

Now R_1 in (3.8) is obtained from (2.3) by choosing

$$\xi = (u^2 - 1)/2v, \qquad (u, v) \in \Delta. \tag{3.12}$$

By adding

$$\eta = v, \qquad (u, v) \in \Delta, \tag{3.13}$$

we obtain the vector function $(\xi, \eta) : \Delta \to \Gamma_1 := \{(\xi, \eta) \in \mathbb{R}^2 : \xi \geqslant -1/2\eta, \eta > 0\}$. This function is one-to-one on the two parts $\Delta^- := \{(u, v) : u \leqslant 0, v > 0\}$ and $\Delta^+ = \{(u, v) : u \geqslant 0, v > 0\}$, and the inverses are

$$u = \pm\sqrt{1 + 2\xi\eta}, \qquad v = \eta, (\xi, \eta) \in \Gamma. \tag{3.14}$$

For each inverse the absolute value of the Jacobian equals

$$|\mathscr{J}(\xi, \eta)| = \frac{\eta}{\sqrt{1 + 2\xi\eta}}, \tag{3.15}$$

so that the joint density of R_1 and $S := V$ equals

$$f_{R_1,S}(\xi, \eta) = \frac{\eta}{\sqrt{1 + 2\xi\eta}} \sum_{j=1}^{2} f_{U,V}((-1)^j\sqrt{1 + 2\xi\eta}, \eta), \qquad (\xi, \eta) \in \Gamma_1. \tag{3.16}$$

This entails at once the expression for the density of R_1 below, which is different from the expressions in Rao (1978) or Abadir (1993).

THEOREM 3.1. *The density of the limiting distribution of the statistic in (3.2) under the null hypothesis $\rho = 1$ is given by*

$$f_{R_1}(\xi) = \begin{cases} \int_{\eta=0}^{-1/2\xi} f_{R_1,S}(\xi, \eta)\, d\eta, & \xi < 0, \\ \int_{\eta=0}^{\infty} f_{R_1,S}(\xi, \eta)\, d\eta, & \xi \geqslant 0, \end{cases} \tag{3.17}$$

where $f_{R_1,S}$ is given by (3.16) with $f_{U,V}$ specified by (3.9) and (3.10).

Similarly, in order to deal with the studentized version, for the function in (2.3) we choose $\xi_2 = (u^2 - 1)/2\sqrt{v}$, $(u, v) \in \Delta$. By adding again $\eta = v$, $(u, v) \in \Delta$, we obtain the vector function $(\xi, \eta) : \Delta \to \Gamma_2 := \{(\xi, \eta \in \mathbb{R}^2 : \xi \geqslant -1/2\sqrt{\eta}, \ \eta > 0\}$. This function is one-to-one on the two parts $\Delta^- := \{(u, v) : u \leqslant 0, \ v > 0\}$ and $\Delta^+ = \{(u, v) : u \geqslant 0, v > 0\}$, and the inverses are $u = \pm\sqrt{1+2\xi\sqrt{\eta}}$, $v = \eta$, $(\xi, \eta) \in \Gamma_2$. For each inverse the absolute value of the Jacobian equals $|\mathscr{J}(\xi, \eta)| = \sqrt{\eta}/\sqrt{1+2\xi\sqrt{\eta}}$. The joint density of R_2 and $S := V$ equals

$$f_{R_2,S}(\xi, \eta) = \frac{\sqrt{\eta}}{\sqrt{1+2\xi\sqrt{\eta}}} \sum_{j=1}^{2} f_{U,V}\left((-1)^j\sqrt{1+2\xi\sqrt{\eta}}, \eta\right), \qquad (\xi, \eta) \in \Gamma_2. \tag{3.18}$$

This yields the following expression for the density of R_2, which is different from the one in Abadir (1995).

THEOREM 3.2. *The density of the limiting distribution of the statistic in (3.3) under the null hypothesis $\rho = 1$ is given by*

$$f_{R_2}(\xi) = \begin{cases} \int_{\eta=0}^{1/4\xi^2} f_{R_2,S}(\xi, \eta)\, d\eta, & \xi < 0, \\ \int_0^\infty f_{R_2,S}(\xi, \eta)\, d\eta, & \xi \geqslant 0. \end{cases} \tag{3.19}$$

3.2. *Local alternatives*

More generally, the asymptotics can be studied under the model $\rho_n = 1 - \theta/n$, $\theta \geqslant 0$, including the null hypothesis for $\theta = 0$ and local alternatives for $\theta > 0$. It is known from Chan and Wei (1987) and Larsson (1995) that $n(\hat{\rho} - \rho_n) \rightarrow_{\mathbf{D}} \int_0^1 Y(t)\, dW(t)/\int_0^1 Y^2(t)\, dt$, as $n \to \infty$, where $Y(t)$ is the Ornstein–Uhlenbeck process: the solution of the equation $dY(t) = -\theta Y(t)\, dt + dW(t)$, $\theta \geqslant 0$, with $Y(0) = 0$. It is well-known from the theory of Itô-integration (Arnold (1974)) that $\int_0^1 Y(t)\, dW(t) = \frac{1}{2}Y^2(1) - \frac{1}{2} + \theta \int_0^1 Y^2(t)\, dt$, so that the weak limit of the statistics in (32.) under local alternatives can be written as

$$\theta + \frac{Y^2(1) - 1}{2\int_0^1 Y^2(t)\, dt}. \tag{3.20}$$

The studentized statistics in (3.3) have weak limit

$$\frac{Y^2(1) - 1 + 2\theta \int_0^1 Y^2(t)\, dt}{\sqrt{\int_0^1 Y^2(t)\, dt}}, \tag{3.21}$$

under local alternatives. Larsson (1995) obtained the results (3.20) and (3.21) and used them to derive the limiting density in an explicit form.

For us the important point is that the expressions in (3.20) and (3.21) can be dealt with in a way very similar to the method used for the null hypothesis. The Ornstein–Uhlenbeck process is given by $Y(t) = \int_0^1 e^{-\theta(t-s)}\, dW(s)$, $0 \leqslant t \leqslant 1$, and has zero mean and covariance function

$$\mathbf{E}Y(s)Y(t) = \frac{e^{2\theta(s\wedge t)} - 1}{2\theta} e^{-\theta(s+t)}, \qquad (s, t) \in [0, 1] \times [0, 1]. \tag{3.22}$$

If $\theta = 0$ we have $Y(t) = W(t)$. For $\theta > 0$ this process has a Karhunen–Loève expansion

$$Y(t) = \sum_{m=1}^{\infty} \sqrt{\lambda_{\theta,m}} Z_m \varphi_{\theta,m}(t), \qquad 0 \leqslant t \leqslant 1, \tag{3.23}$$

for certain $\varphi_{\theta,m}$ and $\lambda_{\theta,m}$ that will not be specified here (see, e.g., Grenander (1981)). The weak limits in (3.20) and (3.21) are functions of

$$U_\theta := \sum_{m=1}^{\infty} \sqrt{\lambda_{\theta,m}} \varphi_{\theta,m}(1) Z_m = Y(1), \tag{3.24}$$

$$V_\theta := \sum_{m=1}^{\infty} \lambda_{\theta,m} Z_m^2 = \int_0^1 Y^2(t)\, dt, \tag{3.25}$$

so that indeed the results of Section 2 apply again.

4. UNKNOWN MEAN

Let us now consider model (1.2). We will be concerned with the limiting density of the test statistic

$$\hat{\rho} := \frac{\sum_{k=1}^{n}(X_k - \overline{X})(X_{k-1} - \overline{X})}{\sum_{k=1}^{n}(X_{k-1} - \overline{X})^2}, \qquad \overline{X} := \frac{1}{n}\sum_{k=1}^{n} X_k, \tag{4.1}$$

under the null hypothesis $\rho = 1$. Let us write $S_0 := \varepsilon_0$, $S_k := \sum_{j=0}^{k} \varepsilon_j$, so that $Y_k := X_k - \mu = S_k$, $k = 0, \ldots, n$. We see from the quotient structure of $\hat{\rho}$ that without real loss of generality we may assume that $\sigma^2 = 1$, as will be done henceforth. By simple algebra we see, moreover, that

$$\hat{\rho} - 1 = \frac{\left(\frac{1}{n}\right)^2 \sum_{k=1}^{n} \varepsilon_k S_{k-1} - \left(\frac{1}{n}\right)^2 S_n \overline{Y}}{\left(\frac{1}{n}\right)^2 \sum_{k=1}^{n} S_{k-1}^2 - \frac{1}{n}\overline{Y}^2 + 2\left(\frac{1}{n}\right)^2 S_n \overline{Y}}, \qquad \overline{Y} := \frac{1}{n}\sum_{k=1}^{n} Y_k. \tag{4.2}$$

Next let us write (cf. (3.6) and (3.7))

$$U_1 := U := \sqrt{2} \sum_{m=1}^{\infty} \sqrt{\lambda_m} (-1)^{m+1} Z_m = W(1), \tag{4.3}$$

$$U_2 := \sqrt{2} \sum_{m=1}^{\infty} \lambda_m Z_m = \int_0^1 W(t)\, dt, \tag{4.4}$$

$$V := \sum_{m=1}^{\infty} \lambda_m Z_m^2 = \int_0^1 W^2(t)\, dt. \tag{4.5}$$

It has been shown in Fuller (1996, Theorem 10.1.3) that representation (4.2) coupled with the weak convergence of the partial sum process of the S_k yields

$$n(\hat{\rho} - 1) \to_{\mathbf{D}} \frac{U_1^2 - 1 - 2U_1U_2}{2(V - U_2^2)}, \qquad \text{as } n \to \infty. \tag{4.6}$$

Again the principle of Section 2 applies.

REMARK 4.1. One might consider other versions of $\hat{\rho}$ and for instance replace $X_{k-1} - \overline{X}$ with $X_{k-1} - \overline{X}_{-1}$, where $\overline{X}_{-1} = 1/n\sum_{k=1}^{n} X_{k-1}$. Such modifications might change the limiting distribution.

The joint density of (U_1, U_2, V) is obtained from (2.6) and (2.7) with $I = 2$, $J = 1$, λ_m as in (3.5), and

$$\langle s, a_m \rangle = \sqrt{2\lambda_m}\{(-1)^{m+1}s_1 + s_2\}, \qquad (s_1, s_2) \in \mathbb{R} \times \mathbb{R}, \tag{4.7}$$

$$\langle t, b_m \rangle = \lambda_m t, \qquad t \in \mathbb{R}. \tag{4.8}$$

Because $U_2^2 = \{\int_0^1 W(t)\, dt\}^2 \leqslant \int_0^1 W^2(t)\, dt = V$, the actual range of the random vector (U_1, U_2, V) and hence the actual domain of the density $f_{U_1,U_2,V}$ remains restricted to

$$\Delta := \{(u_1, u_2, v) \in \mathbb{R}^3 : v > u_2^2\}. \tag{4.9}$$

In this case the function ξ in (2.3) assumes the form

$$\xi = \frac{u_1^2 - 1 - 2u_1u_2}{2(v - u_2^2)}, \qquad (u_1, u_2, v) \in \Delta, \tag{4.10}$$

and is clearly continuous on Δ. By adding the functions

$$\eta = u_2, \qquad \zeta = v, \qquad (u_1, u_2, v) \in \Delta, \tag{4.11}$$

we obtain the vector function

$$(\xi, \eta, \zeta) : \Delta \to \Gamma := \{(\xi, \eta, \zeta) \in \mathbb{R}^3 : \xi \geqslant (1 + \eta^2)/2(\eta^2 - \zeta), \zeta \geqslant \eta^2\}.$$

This function is one-to-one on the two parts $\Delta^- := \{(u_1, u_2, v) : u_1 \leqslant u_2, v > u_2^2\}$ and $\Delta^+ := \{(u_1, u_2, v) : u_1 \geqslant u_2, v > u_2^2\}$ with inverses given by

$$u_1 = \eta \pm \sqrt{1 + \eta^2 + 2\xi(\zeta - \eta^2)}, \qquad u_2 = \eta, \qquad v = \zeta, \qquad (\xi, \eta, \zeta) \in \Gamma. \tag{4.12}$$

For each inverse the absolute value of the Jacobian equals

$$|\mathcal{J}(\xi, \eta, \zeta)| = (\zeta - \eta^2)/\sqrt{1 + \eta^2 + 2\xi(\zeta - \eta^2)}. \tag{4.13}$$

Hence we see that the joint density of $R := (U_1^2 - 1 - 2U_1U_2)/2(V - U_2^2)$, $S := U_2$, and $T := V$ equals

$$f_{R,S,T}(\xi, \eta, \zeta) = \frac{\zeta - \eta^2}{\sqrt{1 + \eta^2 + 2\xi(\zeta - \eta^2)}} \sum_{j=1}^{2} f_{U_1,U_2,V}(\eta + (-1)^j \sqrt{1 + \eta^2 + 2\xi(\zeta - \eta^2)}, \eta, \zeta),$$

$$(\xi, \eta, \zeta) \in \Gamma. \quad (4.14)$$

Our final result of this section is summarized in the next theorem.

THEOREM 4.1. *The density of the limiting distribution of the statistic $n(\hat{\rho} - 1)$ under the null hypothesis $\rho = 1$ is given by*

$$f_R(\xi) = \begin{cases} \int_{\eta=-\infty}^{\infty} \int_{\zeta=\eta^2}^{\eta^2-(1+\eta^2)/2\xi} f_{R,S,T}(\xi, \eta, \zeta)\, d\zeta\, d\eta, & \xi < 0, \\ \int_{\eta=-\infty}^{\infty} \int_{\zeta=\eta^2}^{\infty} f_{R,S,T}(\xi, \eta, \zeta)\, d\zeta\, d\eta, & \xi \geqslant 0, \end{cases} \quad (4.15)$$

where $f_{R,S,T}$ is given by (4.14) with $f_{U_1,U_2,V}$ specified by (4.7) and (4.8).

Note that, formally, f_R in (3.17) could be obtained from f_R in (4.15) by substituting $\eta = 0$ and omitting the integration over η in the latter, and replacing ζ with η in the former.

5. DEPENDENT ERRORS

In this section we will be concerned with the extension of model (1.1) to the case of dependent errors, assuming again that the null hypothesis $\rho = 1$ holds true. The errors ε_k are supposed to form a stationary Gaussian process with $\mathbf{E}\varepsilon_k = 0$ and $\mathbf{Var}\,\varepsilon_k = 1$, and

$$\mathbf{Var}\left(\sum_{k=1}^{n} \varepsilon_k\right) \sim cn^{2H}, \qquad \text{as } n \to \infty, \qquad \text{for some } 3/4 < H < 1, \quad (5.1)$$

and $0 < c < \infty$. Let $W_H(t)$, $0 \leqslant t \leqslant 1$, denote fractional Brownian motion with self-similarity parameter H as in (5.1); see Beran (1994). Weak convergence of the partial sum process of the errors coupled with a suitable representation of the statistic entails the weak convergence

$$n(\hat{\rho} - 1) \to \mathbf{D} \frac{W_H^2(1)}{2\int_0^1 W_H^2(t)\, dt} \qquad \textit{as } n \to \infty, \quad (5.2)$$

a result that is contained in Sowell (1990) with a correction in Marinucci and Robinson (1998). Note, however, that the latter authors write B_H rather than W_H for the fractional Brownian motion that occurs in (5.2).

If the Karhunen–Loève expansion of W_H were known we could proceed in much the same way as in Section 3 to derive an expression for the density of the distribution on the right in (5.2). An alternative might be to employ a Karhunen–Loève-type of expansion by means of a suitable orthonormal wavelet basis of $L^2([0, 1])$. Let φ be a compactly supported smooth scaling function and ψ the corresponding wavelet, and write

$$\varphi_{l,m}(t) := 2^{l/2}\varphi(2^l t - m), \qquad \psi_{l,m}(t) := 2^{l/2}\psi(2^l t - m).$$

Fixing a resolution level L we have

$$W_H(t) = \textstyle\sum_m \langle W_H, \psi_{L,m}\rangle \varphi_{L,m}(t) + \sum_{l>L}\sum_k \langle W_H, \psi_{l,m}\rangle \psi_{l,m}(t),$$

and for L sufficiently large we have

$$W_H(t) \approx \sum_m \sqrt{\lambda_{L,m}(H)} Z_m \varphi_{L,m}(t), \qquad 0 \leqslant t \leqslant 1, \tag{5.3}$$

where the Z_m are standard normal variables and the $\lambda_{L,m}(H)$ are chosen in such a way that $\sqrt{\lambda_{L,m}(H)}Z_m$ has the same variance as $\langle W_H, \varphi_{L,m}\rangle$. It has been shown by Tewfik and Kim (1992) that the random coefficients in the actual wavelet expansion are almost all independent, provided that ψ has a sufficiently large number of vanishing moments (see also Flandrin (1992)). Although the approximation in (5.3) is based on the scaling function rather than the wavelet, one might use these results or Cohen and Ryan (1995, Theorem 5.1) to try to establish low correlations among the Z_m. In that case one might proceed with the right-hand side of (5.3) as with (3.4) in the preceding sections. All these considerations, however, are based on the assumption that H is known, an assumption which might be quite unrealistic. Here we will not pursue these questions any further.

ACKNOWLEDGMENTS

The authors are grateful to Professor W. A. Fuller for his suggestion to consider the unit root test statistics for $AR(1)$-processes with unknown mean. They are also indebted to an associate editor and two referees for very useful comments.

REFERENCES

ABADIR, K. M. (1993). The limiting distribution of the autocorrelation coefficient under a unit root. *Ann. Statist.* 21, 1058–70.

ABADIR, K. M. (1995). The limiting distribution of the t ratio under a unit root. *Econometric Theory* 11, 775–93.

ARNOLD, L. (1974). *Stochastic Differential Equations: Theory and Applications*. Wiley, New York.

BERAN, J. (1994). *Statistics for Long-Memory Processes*. Chapman and Hall, New York.

CHAN, H. N. and WEI, C. Z. (1985). Limiting distributions of least squares estimate of unstable autoregressive processes. Tech. Report, Univ. of Maryland.

CHAN, H. N. and WEI, C. Z. (1987). Asymptotic inference for nearly nonstationary $AR(1)$-processes. *Ann. Statist.* 15, 1050–63.

CHAREMSA, W. W. (1986). Unit root econometrics and econometric nonlinearites. *J. Statist. Pl. Inf.* 49, 105–26.
COHEN, A. and RYAN, R. D. (1995). *Wavelets and Multiscale Signal Processing.* Chapman and Hall, New York.
DICKEY, D. A. and FULLER, W. A. (1979). Distribution of the estimators for autoregressive time series with a unit root. *J. Amer. Statist. Assoc.* 74, 427–31.
FLANDRIN, P. (1992). Wavelet analysis and synthesis of fractional Brownian motion. *IEEE Trans. Inf. Th.* 35, 197–9.
FULLER, W. A. (1996). *Introduction to Statistical Time Series*, Second Edition. Wiley, New York.
GRENANDER, U. (1981). *Abstract Inference.* Wiley, New York.
LARSSON, R. (1995). The asymptotic distributions of some test statistics in near-integrated AR processes. *Econometric Theory* 11, 306–30.
MADDALA, G. S. and KIM, I.-M. (1996). Structural change and unit roots. *J. Statist. Pl. Inf.* 49, 93–103.
MARINUCCI, D. and ROBINSON, P. M. (1998). Alternative forms of fractional Brownian motion. Discussion Paper EM/98/354, London School of Economics.
PERRON, P. (1991). A continuous time approximation to the unstable first-order autoregressive process: the case without an intercept. *Econometrica* 59, 211–36.
RAO, M. M. (1978). Asymptotic distribution of an estimator of the boundary parameter of an unstable process. *Ann. Statist.* 6, 185–90.
RAO, M. M. (1980). Correction to "Asymptotic distribution of an estimator of the boundary parameter of an unstable process". *Ann. Statist.* 8, 1403.
SOWELL, F. (1990). The fractional unit root distribution. *Econometrica* 58, 495–505.
TANAKA, K. (1996). *Time Series Analysis: Nonstationary and Noninvertible Distribution Theory.* Wiley, New York.
TEWFIK, A. H. and KIM, M. (1992). Correlation structure of discrete wavelet coefficients of fractional Brownian motion. *IEEE Trans. Inf. Th.* 38, 904–9.
WHITE, J. S. (1958). The limiting distribution of the serial correlation coefficient in the explosive case. *Ann. Math. Statist.* 29, 1188–97.

Reprinted by permission of the publisher,
© 2000 by Blackwell Publishing, LTD

Part II

Fuzzy Set Theory and Related Topics

INTEGRATION ON FUZZY SETS*

MADAN L. PURI and DAN A. RALESCU

Department of Mathematics, Indiana University, Bloomington, Indiana 47405
Department of Mathematics, University of Cincinnati, Cincinnati, Ohio 45221

In this paper we define the integral of a function on a fuzzy set. Such an integral generalizes the integral over ordinary (nonfuzzy) sets. Some of the basic properties of this integral are stated. The main result shows that under suitable hypotheses, the integral on a fuzzy set equals the integral over some level set. Some applications are indicated.

1. Introduction

The concept of fuzzy set as introduced by Zadeh [7] was proved to be useful in modelling inexact systems [2], where subjective judgments and the nonstatistical nature of inexactness make classical methods of probability theory unsuitable. The concept of fuzzy integral as developed in [5] and [3] has been used in the problems of optimization with inexact constraints [6, 4].

In this paper we define the integral of a (measurable) function over a fuzzy set. The measure with respect to which this integral is defined is a classical measure.

Throughout this paper, $(X, \mathcal{A}, \mu)$ will denote a measure space consisting of a set X, a σ-algebra $\mathcal{A}$ of subsets of X, and a positive measure μ on $\mathcal{A}$. If $f\colon X \to [0, \infty]$ is a positive extended real-valued measurable function, we denote the integral of f w.r.t. μ by $\int_X f\, d\mu$. Also $\int_A f\, d\mu = \int_X f \chi_A d\mu$, where χ_A is the indicator (characteristic) function of $A \in \mathcal{A}$.

A fuzzy subset of X (or fuzzy set) is a function $u\colon X \to [0, 1]$. Let $F(X)$ denote the set of all fuzzy subsets of X. It is well known [2] that $F(X)$ is a

*Research supported by the National Science Foundation Grant IST-7918468.

completely distributive lattice with operations

$$(u \vee v)(x) = \max[u(x), v(x)], \qquad u, v \in F(X).$$
$$(u \wedge v)(x) = \min[u(x), v(x)],$$

These operations generalize the union and intersection of classical subsets of X (i.e., characteristic function χ: $X \to \{0, 1\}$).

A fuzzy set $u \in F(X)$ is called measurable if, for any $\alpha \in [0, 1]$, the α-level set $L_\alpha(u) = \{x \in X: u(x) \geqslant \alpha\} \in \mathcal{Q}$.

Let f: $X \to [0, \infty]$ be a measurable function, and let $u \in F_m(X)$, where $F_m(X) \subset F(X)$ is the set of all measurable fuzzy subsets of X.

DEFINITION. The integral of f over u (with respect to μ) denoted by $\int_u f\, d\mu$ is $\int_X f \cdot u\, d\mu$.

There are two reasons for defining the integral on a fuzzy set in this manner: (i) If $u = \chi_A$ is a characteristic function of a (nonfuzzy) measurable set A, then $\int_{\chi_A} f\, d\mu = \int_A f\, d\mu$. Thus the integral on a fuzzy set generalizes the integral on a classical set. (ii) Let us denote $\nu(A) = \int_A f\, d\mu$, $A \in \mathcal{Q}$. Then, it is well known that ν is a positive measure. Furthermore, if g: $X \to [0, \infty]$ is measurable, the integral of g with respect to ν is given by $\int_X g\, d\nu = \int_X g \cdot f\, d\mu$. Now if we take $g = u \in F_m(X)$, then $\int_u f\, d\mu = \int_X u\, d\nu$ and $\int_X u\, d\nu$ "generalizes" $\nu(A)$ (i.e., if $u = \chi_A$, then $\int_X \chi_A\, d\nu = \nu(A)$). We may also think of $\int_X u\, d\nu$ as the extension of ν to $F_m(X)$. This gives rise to the possibility of "measuring" fuzzy sets.

The following proposition sums up some of the basic properties of the integral defined above.

PROPOSITION. *Let* $u, v \in F_m(X)$, $A \in \mathcal{Q}$, *and let* $\alpha \geqslant 0$, $\beta \geqslant 0$ *be real numbers. Then*

(a) $$u \leqslant v \Rightarrow \int_u f\, d\mu \leqslant \int_v f\, d\mu,$$

(b) $$u \wedge v = 0 \Rightarrow \int_{u \vee v} f\, d\mu = \int_u f\, d\mu + \int_v f\, d\mu,$$

(c) $$\int_u (\alpha f + \beta g)\, d\mu = \alpha \int_u f\, d\mu + \beta \int_u g\, d\mu,$$

(d) $$\int_u \chi_A\, d\mu = \int_A u\, d\mu.$$

Proof. Trivial. (Note that $u \wedge v = 0$ means that u and v are disjoint fuzzy sets).

The integral over a fuzzy set makes sense even if the function f is not positive. Indeed, we may take a function $f \in L^1(\mu)$ and define $\int_u f\, d\mu = \int_X f \cdot u\, d\mu$. The last integral makes sense since $0 \leqslant u \leqslant 1$. Obviously $f \to \int_u f\, d\mu$ is a linear functional for fixed u.

2. The Main Result

For $u \in F_m(X)$, let us denote $L_\alpha(u) = \{x \in X\colon u(x) \geqslant \alpha\}$, the α-level sets of u for $\alpha \in [0, 1]$. The following problem arises in a natural way: Is there any relationship between $\int_u f\, d\mu$ and $\int_{L_\alpha(u)} f\, d\mu$ for some α?

We shall prove that under suitable assumptions on f and u, the answer is in the affirmative. Precisely, we show that $\int_u f\, d\mu = \int_{L_\alpha(u)} f\, d\mu$ for some $\alpha \in (0, 1)$. (Intuitively speaking, even though inexactness is involved in $\int_u f\, d\mu$ through u, we can restrict our attention to a suitable "threshold" α.)

Let $f\colon X \to [0, \infty)$ be measurable. Let $u \in F_m(X)$ and suppose measure μ is σ-finite.

Theorem 2.1. *If $f = 0$ a.e. μ on $\{x \in X\colon u(x) = \alpha\}$ for any $\alpha \in (0, 1)$, then there exists an $\bar{\alpha} \in (0, 1)$ such that $\int_u f\, d\mu = \int_{L_{\bar{\alpha}}(u)} f\, d\mu$.*

Proof. Set $\nu(A) = \int_A f\, d\mu$, $A \in \mathcal{Q}$. Obviously ν is a positive measure on $\mathcal{Q}$. If $g\colon X \to [0, \infty)$ is measurable, then $\int_X g\, d\nu = \int_X gf\, d\mu$.

Thus

$$\int_u f\, d\mu = \int_X fu\, d\mu = \int_X u\, d\nu. \tag{2.1}$$

Since $u \geqslant 0$, we may write

$$\int_X u\, d\nu = \int_0^1 \nu(L_\alpha(u))\, d\alpha, \tag{2.2}$$

where $d\alpha$ is the Lebesgue measure on $[0, 1]$. This follows from the Fubini theorem (since f is finite and μ is σ-finite, and so ν is also σ-finite).

Consider now the function $M(\alpha) = \nu(L_\alpha(u))$ and assume for the moment that M is continuous on $(0, 1)$. Then using the mean-value theorem [1], we find that there exists an $\bar{\alpha} \in (0, 1)$ such that

$$\int_0^1 \nu(L_\alpha(u))\, d\alpha = \int_0^1 M(\alpha)\, d\alpha = M(\bar{\alpha}) = \nu(L_{\bar{\alpha}}(u)) = \int_{L_{\bar{\alpha}}(u)} f\, d\mu. \tag{2.3}$$

Equations (2.1), (2.2) and (2.3) yield the desired result. The only point to show is that $M(\alpha)$ is indeed continuous on $(0, 1)$.

By definition, M is nondecreasing, and so the left-hand and the right-hand limits viz. $\lim_{\alpha \to \alpha_0^-} M(\alpha)$ and $\lim_{\alpha \to \alpha_0^+} M(\alpha)$ exist and are finite. In fact, M is always left-continuous. Indeed, let us take $\alpha_0 \in (0,1)$ and $\{\alpha_n, n \geqslant 1\}$ a sequence increasing to α_0. Then, $L_{\alpha_1}(u) \supset L_{\alpha_2}(u) \supset \cdots$ and $L_{\alpha_0}(u) = \cap_{n=1}^{\infty} L_{\alpha_n}(u)$. Consequently,

$$M(\alpha_0) = \nu\big(L_{\alpha_0}(u)\big) = \lim_{n\to\infty} \nu\big(L_{\alpha_n}(u)\big) = \lim_{n\to\infty} M(\alpha_n).$$

We now show that if $f = 0$ a.e. μ on $\{x \in X: u(x) = \alpha\}$, then M is also right continuous.

Take $\alpha_0 \in (0,1)$ and $\{\alpha_n, n \geqslant 1\}$ a sequence decreasing to α_0. Then $L_{\alpha_1}(u) \subset L_{\alpha_2}(u) \subset \ldots$ and $\cup_{n=1}^{\infty} L_{\alpha_n}(u) = \{x \in X: u(x) > \alpha_0\}$.
It follows that

$$\lim_{n\to\infty} \nu\big(L_{\alpha_n}(u)\big) = \nu\{x \in X: u(x) > \alpha_0\} = \int_{\{x: u(x) > \alpha_0\}} f\,d\mu$$

$$= \int_{L_{\alpha_0}(u)} f\,d\mu - \int_{\{x: u(x) = \alpha_0\}} f\,d\mu = \int_{L_{\alpha_0}(u)} f\,d\mu, \quad (2.4)$$

since $f = 0$ a.e. μ on $u^{-1}\{\alpha_0\}$. The proof follows.

We now find a condition for the fuzzy set u, such that for any measurable and positive f, we have $\int_u f\,d\mu = \int_{L_{\bar\alpha}(u)} f\,d\mu$ for some $\bar\alpha \in (0,1)$. Note that $\bar\alpha$ in general depends on f. It turns out that such a condition in fact characterizes the class of fuzzy sets for which the above result is true. We have the following theorem.

THEOREM 2.2. *Let $u \in F_m(X)$ and μ be σ-finite. Then the following are equivalent*:

(i) $\mu\{x \in X: u(x) = \alpha\} = 0$ *for any* $\alpha \in (0,1)$.

(ii) *For any measurable and positive f, there exists an $\bar\alpha \in (0,1)$ such that* $\int_u f\,d\mu = \int_{L_{\bar\alpha}(u)} f\,d\mu$.

Proof. Using Theorem 2.1, it follows that (i) $\Rightarrow$ (ii). We now prove that (ii) $\Rightarrow$ (i). Take $f = \chi_{\{u=\alpha\}}$ for a fixed $\alpha \in (0,1)$. Then

$$\int_u \chi_{\{u=\alpha\}}\,d\mu = \int_{L_{\bar\alpha}(u)} \chi_{\{u=\alpha\}}\,d\mu \qquad \text{for some } \bar\alpha = \bar\alpha(\alpha) \in (0,1).$$

Consequently,

$$\mu\big(L_{\bar\alpha}(u) \cap \{u = \alpha\}\big) = \int_X u \cdot \chi_{\{u=\alpha\}}\,d\mu$$

$$= \int_{\{u=\alpha\}} u\chi_{\{u=\alpha\}}\,d\mu + \int_{\{u\neq\alpha\}} u\chi_{\{u=\alpha\}}\,d\mu$$

$$= \alpha \cdot \mu\{u = \alpha\}.$$

If we assume $\alpha \geqslant \bar{\alpha}$, then $\alpha\mu\{u = \alpha\} = \mu\{u = \alpha)$, and since $\alpha \neq 1$, this implies $\mu\{u = \alpha\} = 0$.

If $\alpha < \bar{\alpha}$, then $\alpha \cdot \mu\{u = \alpha) = 0$, and so $\mu\{u = \alpha\} = 0$, since $\alpha \neq 0$. Thus $\mu(u^{-1}\{\alpha\}) = 0$ for any $\alpha \in (0, 1)$. The proof follows.

3. CONCLUDING REMARKS

(i) The integral over a fuzzy set can be used to define the area of a fuzzy set. Let u: $X \to [0, 1]$ be a fuzzy set. Then its measure is given by $\mu(u) = \int_u d\mu = \int_X u d\mu$. (This definition is related to an extension of a probability measure to fuzzy sets as in Zadeh [8]). By Theorem 2.2, if $\mu\{x \in X: u(x) = \alpha\} = 0\ \forall\alpha \in (0, 1)$, then $\mu(u) = \mu(L_{\bar{\alpha}}(u))$ for some $\bar{\alpha} \in (0, 1)$.

(ii) The integral over a fuzzy set can also be used in some other problems of practical interest. Consider for example a control problem where often we are faced with a mixture of fuzziness and randomness. A system modelling a physical process may be deterministic while in choosing the initial time t_0 and/or the final time t_f some subjectivity is involved.

Suppose, also, that a performance function $\Phi(x(t), u(t), t)$ is given, where $x(t)$ is the state of the system at time t and $u(t)$ is the input at time t. Due to the inexactness involved in this process, the time interval is no more $[t_0, t_f]$, but a fuzzy subset $\mathfrak{T}$ of $[0, \infty)$. The optimization problem consists in finding $u(t)$ which maximizes $\int_{\mathfrak{T}} \Phi(x(t), u(t), t)\, dt$, where dt is the Lebesgue measure in $[0, \infty)$. The integral over a fuzzy set as defined above, and Theorems 2.1 and 2.2 are useful in solving such a fuzzy control problem.

REFERENCES

1. E. HEWITT AND K. STROMBERG, "Real and Abstract Analysis," Springer-Verlag, New York, 1965.
2. C. V. NEGOITA AND D. A. RALESCU, "Applications of Fuzzy Sets to Systems Analysis," Wiley, New York, 1975.
3. D. RALESCU AND G. ADAMS, The fuzzy integral, *J. Math. Anal. Appl.* **75** (1980), 562–570.
4. D. RALESCU, Toward a general theory of fuzzy variables, *J. Math. Anal. Appl.*, **86** (1982), 176–193.
5. M. SUGENO, "Theory of Fuzzy Integral and Its Applications," Ph.D. Dissertation, Tokyo Institute of Technology.
6. H. TANAKA, T. OKUDA, AND K. ASAI, On fuzzy mathematical programming, *J. Cybernet.* **3** (1974), 37–46.
7. L. A. ZADEH, Fuzzy sets, *Inform. Contr.* **8** (1965), 338–353.
8. L. A. ZADEH, Probability measures of fuzzy events, *J. Math. Anal. Appl.* **23** (1968), 421–427.

Reprinted by permission of the publisher,
© 1982 by Academic Press

A POSSIBILITY MEASURE IS NOT A FUZZY MEASURE*

MADAN L. PURI and DAN RALESCU

Department of Mathematics, Indiana University, Bloomington, IN 47405, U.S.A.
Department of Mathematics, University of Cincinnati, Cincinnati, OH 45221, U.S.A.

Received December 1980
Revised January 1981

In this note we show that a possibility measure is not a particular type of fuzzy measure, except in trivial cases.

Keywords: Possibility measure, Fuzzy measure, Capacity.

1. Introduction

Let X be a set and let $\mathscr{A}$ be a σ-algebra of subsets of X. A *fuzzy measure* (Sugeno [6]) is a set function $\mu:\mathscr{A}\to[0,1]$ with the properties:

(FM1) $\mu(\emptyset)=0$,

(FM2) $A\subset B\Rightarrow\mu(A)\leqslant\mu(B)$,

(FM3) $A_1\subset A_2\subset\cdots\Rightarrow\mu\left(\bigcup_{n=1}^{\infty}A_n\right)=\lim_{n\to\infty}\mu(A_n)$,

(FM4) $A_1\supset A_2\supset\cdots\Rightarrow\mu\left(\bigcap_{n=1}^{\infty}A_n\right)=\lim_{n\to\infty}\mu(A_n)$.

Let $\mathscr{P}(X)$ denote the set of all subsets of X. A *possibility measure* (Zadeh [7]) is a set function $\pi:\mathscr{P}(X)\to[0,1]$ with the properties:

(P1) $\pi(\emptyset)=0$,

(P2) $A\subset B\Rightarrow\pi(A)\leqslant\pi(B)$,

(P3) $\pi\left(\bigcup_{i\in I}A_i\right)=\sup_{i\in I}\pi(A_i)$.

Clearly, any possibility measure is uniquely determined by a function $f:X\to[0,1]$, via the formula

$$\pi(A)=\sup_{x\in A}f(x),\quad A\subset X. \qquad (*)$$

* This work was supported by the National Science Foundation grant number IST–7918468.

0165-0114/82/0000–0000/$02.75 © 1982 North-Holland

Indeed, it suffices to take $f(x)=\pi(\{x\})$, $x\in X$.

The next question arises, then, naturally: is any possibility measure a fuzzy measure? It is easy to see that any possibility measure satisfies (FM3). It was pointed out in [1, p. 1629; 3, p. 665; 6, p. 12] that the set function π defined by $(*)$ is a fuzzy measure. We prove in the next section that this happens only in trivial cases.

2. Main result

We first give two counterexamples which show that, even in 'nice' cases, a possibility measure is *not* a fuzzy measure.

Example 1. Take $X=\mathbb{R}$, $f(x)=1$, and π as defined by $(*)$. Consider $A_n=(n,\infty)$; observe that $\pi(\bigcap_{n=1}^{\infty} A_n)=0$, while $\lim_{n\to\infty}\pi(A_n)=1$, thus (FM4) is violated.

Example 2. Take $X=[0,1]$, $f(x)=1$ for $x\in[0,1)$, and $f(1)=0$. Consider $A_n=[1-1/n,1]$; thus $\pi(\bigcap_{n=1}^{\infty} A_n)=0$, while $\lim_{n\to\infty}\pi(A_n)=1$.

In the rest of this section, we suppose that $X=\mathbb{R}^k$ (the k-dimensional euclidean space). The reader who feels that this assumption is too restrictive may take X to be any metric space without isolated points.

Theorem. *Let* $f:\mathbb{R}^k\to[0,1]$ *and* $\pi(A)=\sup_{x\in A} f(x)$, $A\subset\mathbb{R}^k$ *be the associated possibility measure. If* π *is a fuzzy measure, then* $f(x)=0$ *at every point of continuity of* f.

Proof. Take a point $x_0\in\mathbb{R}^k$ such that f is continuous at x_0. Define $A_n=\{x\in\mathbb{R}^k \mid \|x-x_0\|<1/n,\ x\neq x_0\}$ (here $\|\ \|$ denotes the euclidean norm in $\mathbb{R}^k$). Obviously $A_n\neq\emptyset$; also $A_1\supset A_2\supset\cdots$ and $\bigcap_{n=1}^{\infty} A_n=\emptyset$. Since π satisfies (FM4), it follows that $\lim_{n\to\infty}\pi(A_n)=0$.

Let $(x_j)_j$ be a sequence such that $x_0=\lim_{j\to\infty} x_j$, $x_j\neq x_0$. If $n\geqslant 1$ is fixed, then $x_j\in A_n$ for all $j\geqslant j_n$. Thus $f(x_j)\leqslant\sup_{x\in A_n} f(x)$ for $j\geqslant j_n$. It follows that $0\leqslant \limsup_{j\to\infty} f(x_j)\leqslant\sup_{x\in A_n} f(x)=\pi(A_n)$. Since this is true for any $n\geqslant 1$, we conclude that $\limsup_{j\to\infty} f(x_j)=0$, thus $\lim_{j\to\infty} f(x_j)=0$. Finally, since f is continuous at x_0, it follows that $f(x_0)=\lim_{y\to x_0} f(y)=0$, which ends the proof.

Corollary. *Let* π *be a possibility measure with a continuous 'density'* f. *If* π *is a fuzzy measure, then* $\pi=0$.

Proof. Obvious.

The next question is: when is a possibility measure a fuzzy measure? The implication is always true if the set X is finite. In general, as it can be seen from the above, π fails to satisfy (FM4). However, if f is upper semicontinuous, any possibility measure π (given by $(*)$) is a capacity (i.e. satisfies (FM4), where the A_n's are *compact* subsets of the topological space X; for more details, see [4, 5]).

References

[1] A. Kandel, Fuzzy sets, fuzzy algebra, and fuzzy statistics, Proc. of the IEEE 66 (1978) 1619–1639.

[2] C. V. Negoita and D. A. Ralescu, Applications of Fuzzy Sets to Systems Analysis (Wiley, New York, 1975).

[3] H. T. Nguyen, On fuzziness and linguistic probabilities, J. Math. Anal. Appl. 61 (1977) 658–671.

[4] H. T. Nguyen, Some mathematical tools for linguistic probabilities, Fuzzy Sets and Systems 2 (1979) 53–65.

[5] D. Ralescu, Measures, capacities, and optimization with inexact constraints, to appear.

[6] M. Sugeno, Theory of fuzzy integrals and its applications, Ph.D. Thesis, Tokyo Inst. of Technology (1974).

[7] L. A. Zadeh, Fuzzy sets as a basis for a theory of possibility, Fuzzy Sets and Systems 1 (1978) 3–28.

Reprinted by permission of the publisher,
© 1982 by Elsevier Science

STRONG LAW OF LARGE NUMBERS FOR BANACH SPACE VALUED RANDOM SETS[1]

MADAN L. PURI and DAN A. RALESCU

Indiana University and University of Cincinnati

In this paper we prove a strong law of large numbers for random sets whose values are compact convex subsets of a Banach space.

1. Introduction and preliminaries. The study of the theory of random sets initiated by Robbins (1944, 1945) and defined as multivalued measurable mappings on a probability space has received considerable attention in recent years (see e.g. Kendall (1974), Matheron (1975) and, Fortet and Kambouzia (1975), among others). Motivation for studying random sets is both theoretical (since they generalize random variables and random vectors) and practical (since they represent geometrical objects in certain models of growth).

Artstein and Vitale (1975) proved a strong law of large numbers (SLLN) for $\mathbb{R}^p$-valued random sets (i.e. random sets whose values are compact subsets of $\mathbb{R}^p$). Later, using different methods, Cressie (1978) proved a SLLN for some particular class of $\mathbb{R}^p$-valued random sets.

The methods of Cressie (1978) do not seem suitable for extension to more general random sets. Artstein and Vitale (1975) use an embedding of all compact convex subsets of $\mathbb{R}^p$ into $C(S^{p-1})$, the Banach space of continuous functions on the unit sphere in $\mathbb{R}^p$. However, their result is not valid if $\mathbb{R}^p$ is replaced by a Banach space.

In this paper we consider random sets as (measurable) mappings from a probability space into the set of compact convex subsets of a Banach space. In this setting, we prove a SLLN for sequences of independent and identically distributed random sets. Our main tool in proving this limit theorem is a result due to Rådström (1952) which states that the collection of compact convex subsets of a Banach space can be embedded as a convex cone in a normed space.

Our result generalizes all previous SLLN for random sets. Note, however, that if the random sets take values in an infinitely dimensional space, the hypothesis concerning the convexity of values cannot be dropped.

Let F be a real Banach space. Denote by $\mathscr{K}(F)$, the collection of all nonempty compact convex subsets of F. The following operations are defined in $\mathscr{K}(F)$:

$$A + B = \{a + b \mid a \in A, b \in B\}, \qquad \alpha A = \{\alpha a \mid a \in A\}$$

where $A, B \in \mathscr{K}(F)$, $\alpha \in \mathbb{R}$. Note that $\mathscr{K}(F)$ is *not* a vector space but it becomes a complete metric space when endowed with the Hausdorff distance

$$d_H(A, B) = \max\{\sup_{a\in A}\inf_{b\in B}\|a - b\|, \qquad \sup_{b\in B}\inf_{a\in A}\|a - b\|\}$$

where $\|\cdot\|$ denotes the norm in F and $A, B \in \mathscr{K}(F)$.

We use the notation $\|A\| = d_H(\{0\}, A) = \sup\{\|a\| \mid a \in A\}$ where $A \in \mathscr{K}(F)$.

Let $(\Omega, \mathscr{A}, P)$ be a probability space. A *random set* (F-valued random set) is a Borel measurable function $X: \Omega \to \mathscr{K}(F)$. Different concepts of measurability can be introduced, but they are equivalent in this setting (see, e.g. Debreu (1966)).

The expected value of a random set is defined by using the integral due to Debreu (1966). (For another definition, see Aumann (1965), and for the equivalence of two

[1] Research supported in part by the National Science Foundation Grant IST-7918468.

Received January 1982; revised February 1982.

AMS 1970 *subject classification.* Primary 60D05; secondary 60F15.

Key words and phrases. Strong law of large numbers, random sets.

definitions, see Byrne (1978)). To recall the definition due to Debreu (1966), suppose first that X is a *simple* random set, i.e. $X = \sum_{i=1}^{k} K_i\chi_{A_i}$ where $K_i \in \mathscr{K}(F)$, $A_i \in \mathscr{A}$ and χ_{A_i} is the characteristic function of A_i. Then, the expected value of X is defined as $E(X) = \int_\Omega X\,dP = \sum_{i=1}^{k} P(A_i)K_i$. To extend this integral to more general random sets, one needs to use the Rådström (1952) embedding theorem: There exists a normed space $\mathscr{F}$ and an isometry $j:\mathscr{K}(F) \to \mathscr{F}$ such that addition in $\mathscr{F}$ induces addition in $\mathscr{K}(F)$ and multiplication by nonnegative scalars in $\mathscr{F}$ induces the corresponding operation in $\mathscr{K}(F)$. It is clear that if $X:\Omega \to \mathscr{K}(F)$ is a random set, then $E(X) = \int_\Omega X\,dP$ can be defined as a Bochner integral. It can be shown (see Debreu (1966), pages 354 and 363)) that $E(X) \in \mathscr{K}(F)$ if X is integrable, i.e., if $E(\|X\|) < \infty$.

2. Strong law of large numbers. To prove the SLLN for the Banach space valued random sets, we will use the Rådström embedding theorem together with the SLLN for the Banach space valued random elements. To adopt this approach, we need the following lemma:

LEMMA 2.1. *Let $X:\Omega \to \mathscr{K}(F)$ be a random set such that $E(\|X\|) < \infty$. If $j:\mathscr{K}(F) \to \mathscr{F}$ is the isometry given by the Rådström embedding theorem, then $E(j\circ X) = j(E(X))$.*

PROOF. Suppose first that X is a simple random set, i.e., $X = \sum_{i=1}^{k} K_i\chi_{A_i}$ where $K_i \in \mathscr{K}(F)$, $A_i \in \mathscr{A}$ and χ_{A_i} is the characteristic function of A_i. Then, since $E(X) = \sum_{i=1}^{k} P(A_i)K_i$ and $j\circ X = \sum_{i=1}^{k} j(K_i)\chi_{A_i}$, the result follows.

Now suppose that X is a random set such that $E(\|X\|) < \infty$. Then there exists a sequence of simple random sets X_n such that $X_n \to X$ a.e. and

$$\lim_{m,n\to\infty} \int_\Omega |j\circ X_m - j\circ X_n|\,dP = 0$$

(Debreu (1966, page 365)). The result follows by noting that $E(X) = \lim_{n\to\infty} E(X_n)$, $j\circ X_n \to j\circ X$ a.e., and using the definition of the Bochner integral.

We now prove the SLLN for random sets.

THEOREM 2.1. *If X_n, $n \geq 1$ is a sequence of independent and identically distributed random sets such that $E(\|X_1\|) < \infty$, then $\sum_{i=1}^{n} X_i/n \to E(X_1)$ a.e., the convergence being in the Hausdorff metric.*

PROOF. From the Rådström embedding theorem, there exists a normed space $\mathscr{F}$ and an isometry $j:\mathscr{K}(F) \to \mathscr{F}$.

Thus $j\circ X_1, j\circ X_2, \cdots, j\circ X_n$ are independent and identically distributed, and $E(\|j\circ X_1\|) = E(\|X_1\|) < \infty$. From the standard SLLN in Banach space (cf. Mourier (1955)), it follows that $(\sum_{i=1}^{n} j\circ X_i)/n \to E(j\circ X_1)$ a.e. which implies that $j\circ[(\sum_{i=1}^{n} X_i)/n] \to E(j\circ X_1)$ a.e.

By Lemma 2.1, $j\circ[(\sum_{i=1}^{n} X_i)/n] \to j(E(X_1))$ a.e. Finally, since j is an isometry, it follows that $(\sum_{i=1}^{n} X_i)/n \to E(X_1)$ a.e., the convergence being in the Hausdorff metric.

REFERENCES

[1] ARTSTEIN, Z. and VITALE, R. (1975). A strong law of large numbers for random compact sets. *Ann. Probability* **3** 879–882.

[2] AUMANN, R. J. (1965). Integrals of set-valued functions. *J. Math. Analysis and Appl.* **12** 1–12.

[3] BILLINGSLEY, P. (1968). *Convergence of Probability Measures.* Wiley, New York.

[4] BYRNE, C. L. (1978). Remarks on the set-valued integrals of Debreu and Aumann. *J. Math. Anal. and Appl.* **62** 243–246.

[5] CRESSIE, N. (1978). A strong limit theorem for random sets. *Suppl. Adv. Appl. Probability.* **10** 36–46.

[6] DEBREU, G. (1966). Integration of correspondences. *Proc. Fifth Berkeley Symp. Math. Statist. and Probability* **2** 351–372. Univ. of California.

[7] FORTET, R. and KAMBOUZIA, M. (1975). Ensembles aléatoires induits par une répartition ponctuelle aléatoire. *C. R. Acad. Sc. Paris,* Série A. **280** 1447.

[8] Kendall, D. G. (1974). Foundations of a theory of random sets. In *Stochastic Geometry*, ed. E. F. Harding and D. G. Kendall. Wiley, New York.
[9] Matheron, G. (1975). *Random Sets and Integral Geometry*. Wiley, New York.
[10] Mourier, E. (1955). L-random elements and L^*-random elements in Banach spaces. *Proc. Third Berkeley Symp. Math. Statist. and Probability* **2** 231–242. University of California.
[11] Rådström, H. (1952). An embedding theorem for spaces of convex sets. *Proc. Amer. Math. Soc.* **3** 165–169.
[12] Robbins, H. E. (1944). On the measure of a random set. *Ann. Math. Statist.* **15** 70–74.
[13] Robbins, H. E. (1945). On the measure of a random set, II. *Ann. Math. Statist.* **16** 342–347.

Department of Mathematics
Indiana University
Bloomington, Indiana 47405

Department of Mathematics
University of Cincinnati
Cincinnati, Ohio 45221

Reprinted by permission of the publisher,
© 1983 by The Institute of Mathematical Statistics

STRONG LAW OF LARGE NUMBERS WITH RESPECT TO A SET-VALUED PROBABILITY MEASURE[1]

MADAN L. PURI and DAN A. RALESCU

Indiana University and University of Cincinnati

In this paper we define the expected value of a random vector with respect to a set-valued probability measure. The concepts of independent and identically distributed random vectors are appropriately defined, and a strong law of large numbers is derived in this setting. Finally, an example of a set-valued probability useful in Bayesian inference is provided.

1. Introduction. This research is motivated by the following consideration: there are instances in Bayesian estimation when the prior probability is not known precisely. In such situations DeRobertis and Hartigan (1981) suggest using an interval of measures rather than a single prior, and extend the Bayes theorem in this setting. This idea is reminiscent of upper and lower probabilities (see Koopman, 1940, and Dempster, 1967). The risk $R(\theta, \delta)$ associated with a decision function δ is a random variable in the Bayes setting (since the unknown parameter θ is assumed to be a random variable). The main question then is: how one can evaluate the average risk when the prior measure is not known precisely. The concept which seems to be useful in such situations is that of a set-valued measure (see Debreu and Schmeidler, 1970, and Artstein, 1972) with respect to which the expectation of a random variable is evaluated.

In Section 2 we give some preliminaries on set-valued measures, and we define the expected value. In Section 3 we prove a strong law of large numbers with respect to a set-valued probability. In Section 4 we give an example of a set-valued probability measure.

2. Expectation with respect to a set-valued probability measure. The concept of a set-valued measure was defined in connection with the integral of a set-valued function (see Debreu and Schmeidler, 1970).

Let Ω be a set, $\mathscr{A}$ a σ-algebra of subsets of Ω, and $\mathscr{P}(\mathbb{R}^n)$ the collection of all subsets of $\mathbb{R}^n$. A set-valued measure is a function $\Pi : \mathscr{A} \to \mathscr{P}(\mathbb{R}^n)$, such that (i) $\Pi(A) \neq \phi$ for every $A \in \mathscr{A}$, (ii) $\Pi(\cup_{j=1}^{\infty} A_j) = \sum_{j=1}^{\infty} \Pi(A_j)$ for every disjoint family $\{A_j\}_j$, $A_j \in \mathscr{A}$.

Here the sum $\sum_{j=1}^{\infty} B_j$ of subsets of $\mathbb{R}^n$ is defined as the collection of all vectors $b = \sum_{j=1}^{\infty} b_j$ where $b_j \in B_j$ and $\sum_{j=1}^{\infty} \| b_j \| < \infty$.

In what follows we consider only bounded set-valued measures (such that $\Pi(\Omega)$ is bounded). It follows that for such measures, $\Pi(\phi) = \{0\}$.

A *selection* μ of Π is a vector-valued measure $\mu : \mathscr{A} \to \mathbb{R}^n$, such that $\mu(A) \in \Pi(A)$ for every $A \in \mathscr{A}$.

An atom of the set-valued measure Π is an event $A \in \mathscr{A}$ with $\Pi(A) \neq \{0\}$ and such that $A_1 \subset A$ implies $\Pi(A_1) = \{0\}$ or $\Pi(A \backslash A_1) = \{0\}$. A set-valued measure with no atoms is called nonatomic.

The following theorem due to Artstein (1972) will be used in the sequel:

THEOREM 2.1. (a) *If Π is bounded, nonatomic set-valued measure, then $\Pi(A)$ is convex for every $A \in \mathscr{A}$.*

(b) *If Π is bounded set-valued measure, then, for every $A \in \mathscr{A}$ and $x \in \Pi(A)$, there exists a selection μ of Π such that $\mu(A) = x$.*

Received October 1982.

[1] Research supported by the National Science Foundation Grant IST-7918468.

AMS 1980 *subject classification.* Primary, 60B12; secondary, 60F15.

Key words and phrases. Set-valued measure, strong law of large numbers, interval of measures.

A *set-valued probability* on Ω is a set-valued measure $\Pi:\mathscr{A} \to \mathscr{P}([0, 1])$ such that $1 \in \Pi(\Omega)$.

A *set-valued probability space* is a triple $(\Omega, \mathscr{A}, \Pi)$ where Π is a set-valued probability. Without loss of generality, one can assume that Π is absolutely continuous with respect to a probability measure P on Ω; $\Pi \ll P$, that is, for a set $A \in \mathscr{A}$ for which $P(A) = 0$, we have $\Pi(A) = \{0\}$. (see Artstein, 1972).

Let $X:\Omega \to \mathbb{R}^n$ be a random vector such that $E_P(\|X\|) = \int_\Omega \|X\| \, dP < \infty$. The *expected value* of X with respect to Π is defined as $\int_\Omega X \, d\Pi = \{\int_\Omega X \, d\mu : \mu \text{ is a selection of } \Pi\}$. According to Theorem 2.1 (b) it is clear that $\int_\Omega X d\Pi \neq \phi$ if $E_P(\|X\|) < \infty$.

3. Strong law of large numbers. Let $(\Omega, \mathscr{A}, \Pi)$ be a set-valued probability space, and let $X:\Omega \to \mathbb{R}^n$ be a random vector. Then X induces a set-valued probability on the Borel sets in $\mathbb{R}^n$ (denoted by $\mathscr{B}_n$) in the following way:

$$B \in \mathscr{B}_n, \quad \Pi_X(B) = \Pi(X \in B).$$

The random vectors X_i, $i \geq 1$ defined on $(\Omega, \mathscr{A}, \Pi)$ are *independent* if $\Pi(X_1 \in B_1, X_2 \in B_2, \cdots, X_i \in B_i) = \Pi(X_1 \in B_1) \cdots \Pi(X_i \in B_i)$ where the product of subsets M and N of $[0, 1]$ is defined by $MN = \{mn : m \in M, n \in N\}$. They are identically distributed if $\Pi_{X_1} = \cdots = \Pi_{X_i} = \cdots$. Clearly these concepts generalize the classical concepts of independent and identically distributed random vectors (with respect to an ordinary probability measure).

Finally, we need another notation: if $x \in \mathbb{R}^n$, and $A \subset \mathbb{R}^n$, then

$$d(x, A) = \inf_{a \in A} \| x - a \|.$$

We now prove our main theorem.

THEOREM 3.1. *Let X_i, $i \geq 1$ be independent and identically distributed random vectors defined on a set-valued probability space $(\Omega, \mathscr{A}, \Pi)$ such that $\Pi \ll P$ where P is a probability measure. If $E_P(\|X_1\|) < \infty$, then $d((1/n) \sum_{j=1}^n X_j, \int_\Omega X_1 \, d\Pi) \to 0$ almost everywhere with respect to Π.*

PROOF. Clearly, if μ is a selection of Π (which exists according to Theorem 2.1 (b)), it is not true in general that X_i, $i \geq 1$ are independent and identically distributed with respect to μ.

To prove the theorem, we will show the following:

There exists a probability measure Q on Ω which is a selection of Π and such that X_i, $i \geq 1$ are independent and identically distributed with respect to Q, and $E_Q(\|X_1\|) < \infty$.

Let $Q(A) = \sup \Pi(A)$ for every $A \in \mathscr{A}$. The fact that Q is a probability measure follows from Proposition 3.1 of Artstein (1972). Also it is clear that $Q(X_i \in B) = Q(X_1 \in B)$ for every $B \in \mathscr{B}$, and $i \geq 1$ i.e., X_i, $i \geq 1$ are identically distributed.

To prove that X_i, $i \geq 1$ are independent, it suffices to show that $Q(X_1 \in B_1, X_2 \in B_2) = Q(X_1 \in B_1)Q(X_2 \in B_2)$ for every $B_1, B_2 \in \mathscr{B}_n$. By the definition of independence, it suffices to show that $\sup(MN) = \sup M \sup N$ where $M, N \subset [0, 1]$. This being easy to establish, the desired independence follows.

Since $\Pi \ll P$, it follows from classical results that $E_P(\|X_1\|) < \infty$ implies $\int_\Omega \|X_1\| \, dQ < \infty$.

We now prove that Q is a selection of Π. From Theorem 2.1(b) there exists a probability measure Q_1 which is a selection of Π. Clearly $Q_1(A) \leq Q(A)$ for every $A \in \mathscr{A}$, but this implies that $Q_1 = Q$.

Now from the classical law of large numbers, it follows that $(1/n) \sum_{j=1}^n X_j \to E_Q(X_1)$ almost everywhere with respect to Q. Thus

$$d\left(\frac{1}{n} \sum_{j=1}^n X_j, \int_\Omega X_1 \, d\Pi\right) \leq \left\| \frac{1}{n} \sum_{j=1}^n X_j - E_Q(X_1) \right\| \to 0,$$

and so $d((1/n) \sum_{j=1}^n X_j, \int_\Omega X_1 \, d\Pi) \to 0$ almost everywhere (Q).

The definition of Q implies that the above convergence actually holds almost everywhere with respect to Π, that is, $\Pi(\sum_{j=1}^{n} X_j/n \not\to \int_\Omega X_1 \, d\Pi) = \{0\}$. This completes the proof.

4. An example. The strong law of large numbers proved in Section 3 is a generalization of the classical law of large numbers.

The simplest example of a set-valued probability measure is provided by an interval of measures (as studied by DeRobertis and Hartigan, 1981). More precisely let P_1 and P_2 be two finite measures on $(\Omega, \mathscr{A})$ such that $P_1(A) \le P_2(A)$ for every $A \in \mathscr{A}$, and let P_2 be a probability measure. Let $\Pi : \mathscr{A} \to \mathscr{P}([0, 1])$ be defined as

$$\Pi(A) = [P_1(A), P_2(A)], \quad A \in \mathscr{A}. \tag{4.1}$$

Clearly $\Pi(\phi) = \{0\}$ and $1 \in \Pi(\Omega)$.

Let $\{A_j\}_j$ be a disjoint family, $A_j \in \mathscr{A}$. We must show that $\Pi(\cup_{j=1}^{\infty} A_j) = \sum_{j=1}^{\infty} \Pi(A_j)$. This is equivalent to

$$\textstyle\sum_{j=1}^{\infty} [P_1(A_j), P_2(A_j)] = [\sum_{j=1}^{\infty} P_1(A_j), \sum_{j=1}^{\infty} P_2(A_j)].$$

The above equality follows from the formulas

$$\inf \textstyle\sum_{j=1}^{\infty} [P_1(A_j), P_2(A_j)] = \sum_{j=1}^{\infty} P_1(A_j),$$

$$\sup \textstyle\sum_{j=1}^{\infty} [P_1(A_j), P_2(A_j)] = \sum_{j=1}^{\infty} P_2(A_j),$$

and from the convexity of $\sum_{j=1}^{\infty} [P_1(A_j), P_2(A_j)]$.

Thus Π defined by (4.1) is a set-valued probability. Also Π is absolutely continuous with respect to P_2.

If $X : \Omega \to \mathbb{R}^n$ is a random vector such that $E_{P_2}(\|X\|) < \infty$, then the expected value of X (as defined in Section 2) is given by $\int_\Omega X \, d\Pi = \{E_P(X) : P_1 \le P \le P_2\}$ where P is a finite measure.

If X_i, $i \ge 1$ are independent and identically distributed with respect to Π (given by (4.1)) and note that the latter condition is equivalent to the fact that X_i, $i \ge 1$ are identically distributed with respect to P_1 and P_2, then the law of large numbers given by Theorem 3.1 implies

$$\inf_{P_1 \le P \le P_2} \left\| \frac{1}{n} \textstyle\sum_{j=1}^{n} X_j - E_P(X_1) \right\| \to 0 \quad \text{almost everywhere with respect to } P_2.$$

It is interesting to note that, under certain hypotheses, every set-valued probability is of the form (4.1).

THEOREM 4.1. *Let $\Pi : \mathscr{A} \to \mathscr{P}([0, 1])$ be a nonatomic set-valued probability measure such that $\Pi(\Omega)$ is closed. Then $\Pi(A) = [P_1(A), P_2(A)]$ for every $A \in \mathscr{A}$, where P_1 is a measure and P_2 is a probability measure such that $P_1(A) \le P_2(A)$, $A \in \mathscr{A}$.*

PROOF. Denote $P_1(A) = \inf \Pi(A)$ and $P_2(A) = \sup \Pi(A)$, $A \in \mathscr{A}$. We show that P_1 and P_2 are measures. Let $\{A_j\}_j$ be a disjoint family of sets in $\mathscr{A}$. Then, clearly

$$P_1(\cup_{j=1}^{\infty} A_j) = \inf(\textstyle\sum_{j=1}^{\infty} \Pi(A_j)) \ge \sum_{j=1}^{\infty} \inf \Pi(A_j) = \sum_{j=1}^{\infty} P_1(A_j). \tag{4.2}$$

Let $\varepsilon > 0$. Then there exists $x_j \in \Pi(A_j)$ such that $x_j < \inf \Pi(A_j) + \varepsilon/2^j$, $j \ge 1$. Thus $\sum_{j=1}^{\infty} x_j \le \sum_{j=1}^{\infty} \inf \Pi(A_j) + \varepsilon$. From (4.2), the series $\sum_{j=1}^{\infty} \inf \Pi(A_j)$ is convergent. So $\sum_{j=1}^{\infty} x_j \in \sum_{j=1}^{\infty} \Pi(A_j)$. Consequently

$$\inf(\textstyle\sum_{j=1}^{\infty} \Pi(A_j)) \le \sum_{j=1}^{\infty} \inf \Pi(A_j) + \varepsilon. \tag{4.3}$$

Since $\varepsilon > 0$ is arbitrary, it follows from (4.2) and (4.3) that

$$P_1(\cup_{j=1}^{\infty} A_j) = \textstyle\sum_{j=1}^{\infty} P_1(A_j).$$

Now $P_1(\Omega) = \inf \Pi(\Omega) \in \Pi(\Omega)$ since $\Pi(\Omega)$ is closed. From Theorem 2.1(b), there exists a selection Q_1 of Π such that $Q_1(\Omega) = P_1(\Omega)$. Since $P_1 \le Q_1$, we have $P_1 = Q_1$, so P_1 is a selection of Π. Similarly P_2 is a selection of Π, and obviously $P_2(\Omega) = 1$. Finally, since (from Theorem 2.1 (a)) $\Pi(A)$ is convex for every $A \in \mathscr{A}$, it follows that $\Pi(A) = [P_1(A), P_2(A)]$, $A \in \mathscr{A}$, which was to be proved.

REMARK. It may be noted that the intervals of probability measures (DeRobertis and Hartigan, op. cit.) are restandardized when computing expectations and posterior probabilities, so that the ranges of expectations for set-valued probabilities and for intervals of probabilities do not coincide.

REFERENCES

[1] ARTSTEIN, Z. (1972). Set-valued measures. *Trans. Amer. Math. Soc.* **165** 103–125.

[2] DEBREU, G. and SCHMEIDLER, D. (1970). The Radon-Nikodym derivative of a correspondence. *Proc. Sixth Berkeley Symp. Math. Statist. Probab.*, 41–56. Univ. of California Press.

[3] DEMPSTER, A. P. (1967). Upper and lower probabilities induced by a multivalued mapping. *Ann. Math. Statist.* **36** 325–339.

[4] DEROBERTIS, L. and HARTIGAN, J. A. (1981). Bayesian inference using intervals of measures. *Ann. Statist.* **9** 235–244.

[5] KOOPMAN, B. O. (1940). The axioms and algebra of intuitive probability. *Ann. Math.* **41** 269–278.

DEPARTMENT OF MATHEMATICS
INDIANA UNIVERSITY
BLOOMINGTON, INDIANA 47405

DEPARTMENT OF MATHEMATICS
UNIVERSITY OF CINCINNATI
CINCINNATI, OHIO 45221

**Reprinted by permission of the publisher,
© 1983 by The Institute of Mathematical Statistics**

DIFFERENTIALS OF FUZZY FUNCTIONS*

MADAN L. PURI and DAN A. RALESCU

Department of Mathematics, Indiana University, Bloomington, Indiana 47405
Department of Mathematics, University of Cincinnati, Cincinnati, Ohio 45221

IN MEMORY OF JOSEPH KAMPÉ DE FÉRIET

In this paper the Rådström embedding theorem (*Proc. Amer. Math. Soc.* **3** (1952), 165) is generalized and is used to define the concept of the differential of a fuzzy function.

1. Introduction

The purpose of this paper is to define the concept of the differential of a fuzzy function which extends the differential of a set-valued function. The latter concept was first defined by Hukuhara [8]. Later, Banks and Jacobs [2] defined a more general concept of differential and investigated its properties.

Set-valued functions and their calculus were found useful in some of the problems of economics [1] and control theory [7]. From a probabilistic viewpoint, random sets (as a particular case of set-valued functions) have a rather well-developed theory [11].

Zadeh [16] introduced the notion of fuzzy set and later its relationships with random sets were investigated by Fortet and Kambouzia [5], Féron [4], and Goodman [6]. Fuzzy random variables [13] as a generalization of random sets can be used to represent inexactness due to both randomness and fuzziness.

On the other hand, the analysis of evidence [15] was related to the fuzzy

* Research supported by National Science Foundation Grant IST-7918468.

analysis and to the theory of possibility by Zadeh [17] and, more recently, by Kampé de Fériet [9].

In defining the differential of a set-valued function [2] and the integral of such a function [3], a key result is an embedding theorem due to Rådström [14]. This theorem, which states that the collection of nonempty closed bounded and convex subsets of a Banach space can be embedded in a normed space, makes it possible to define the differential of a set-valued function as a differential of a function between normed spaces. To extend the definition of differential to fuzzy functions, it is natural to extend first the Rådström embedding theorem by considering an appropriate space of fuzzy subsets of a Banach space. This extension is given in Section 2 by using a suitable generalization of the Hausdorff metric. As an application of this embedding theorem, we introduce the concept of the differential of a fuzzy function in Section 3 and study some of its properties.

2. The Embedding Theorem

Let X be a reflexive Banach space, and let A, B be two nonempty bounded subsets of X. The *Hausdorff distance* between A and B is

$$d_{\mathrm{H}}(A, B) = \max\left[\sup_{a\in A} \inf_{b\in B} \|a-b\|, \sup_{b\in B} \inf_{a\in A} \|a-b\|\right], \tag{2.1}$$

where $\| \ \|$ denotes the norm in X.

If $Q(X)$ denotes the collection of all nonempty compact and convex subsets of X, it is well known [10] that $(Q(X), d_{\mathrm{H}})$ is a complete metric space.

If M is a set, a *fuzzy subset* of M is a function $u: M \to [0, 1]$. The set of all fuzzy subsets of M, $\mathscr{F}(M)$ is a completely distributive lattice which includes the ordinary subsets of M (viewed as characteristic functions $M \to \{0, 1\}$).

For any fuzzy subset $u: M \to [0, 1]$, denote by $L_\alpha(u) = \{m \in M \mid u(m) \geqslant \alpha\}$, $\alpha \in [0, 1]$, the α-level set of u.

If M is a vector space, a fuzzy subset $u \in \mathscr{F}(M)$ is called a *fuzzy convex subset* (see [16]), if

$$u(\lambda m_1 + (1-\lambda) m_2) \geqslant \min[u(m_1), u(m_2)] \tag{2.2}$$

for every $m_1, m_2 \in M$, $\lambda \in [0, 1]$.

If X is a reflexive Banach space, in order to extend the Hausdorff distance, we shall consider the subset $\mathscr{F}_0(X)$ of $\mathscr{F}(X)$, containing all fuzzy sets $u: X \to [0, 1]$ with properties:

(i) u is upper semicontinuous,

(ii) u is fuzzy convex,

(iii) $L_\alpha(u)$ is compact, for every $\alpha \neq 0$.

If $u, v \in \mathscr{F}_0(X)$, define the *distance* between u and v by

$$d(u, v) = \sup_{\alpha > 0} d_{\mathrm{H}}(L_\alpha(u), L_\alpha(v)), \tag{2.3}$$

where d_{H} denotes the Hausdorff distance.

We shall use the following result: The proof, given in [13] for the case $X = \mathbb{R}^n$, works in the general case, and will be omitted here.

THEOREM 2.1. *$(\mathscr{F}_0(X), d)$ is a complete metric space.*

The embedding theorem of Rådström will be extended to $\mathscr{F}_0(X)$. To do this, a linear structure is defined in $\mathscr{F}_0(X)$ by

$$(u + v)(x) = \sup\{\alpha \in [0, 1] \mid x \in L_\alpha(u) + L_\alpha(v)\} \tag{2.4}$$

$$\begin{aligned} (\lambda u)(x) &= u(\lambda^{-1}x), && \text{if } \lambda \neq 0, \\ &= 0, && \text{if } \lambda = 0,\ x \neq 0, \\ &= \sup_{y \in X} u(y), && \text{if } \lambda = 0,\ x = 0, \end{aligned} \tag{2.5}$$

for $u, v \in \mathscr{F}_0(X)$, $\lambda \in \mathbb{R}$.

It is clear that these definitions extend the corresponding operations in $Q(X)$ (addition of sets and multiplication of a set by a scalar). In the process of generalizing the embedding theorem, Lemma 2.1 will be useful.

LEMMA 2.1. *Let M be a set and let $\{M_\alpha \mid \alpha \in [0, 1]\}$ be a family of subsets of M such that:*

(a) $M_0 = M$,

(b) $\alpha \leqslant \beta \Rightarrow M_\beta \subseteq M_\alpha$,

(c) $\alpha_1 \leqslant \alpha_2 \leqslant \cdots$, $\lim_{n \to \infty} \alpha_n = \alpha \Rightarrow M_\alpha = \bigcap_{n=1}^{\infty} M_{\alpha_n}$.

Then the fuzzy subset $\phi: M \to [0, 1]$ defined by $\phi(m) = \sup\{\alpha \in [0, 1] \mid x \in M_\alpha\}$ has the property that $L_\alpha(\phi) = M_\alpha$ for every $\alpha \in [0, 1]$.

The proof is given in [12].

Propositions 2.1–2.3 will lead to Theorem 2.2, which is the desired embedding result.

PROPOSITON 2.1. *If $u, v \in \mathscr{F}_0(X)$, then $L_\alpha(u + v) = L_\alpha(u) + L_\alpha(v)$ for every $\alpha \in [0, 1]$.*

Proof. Denote $X_\alpha = L_\alpha(u) + L_\alpha(v)$. Obviously, $X_0 = X$, and $\alpha \leqslant \beta \Rightarrow X_\beta \subseteq X_\alpha$. To apply Lemma 2.1, we check that $\alpha_1 \leqslant \alpha_2 \leqslant \cdots$, $\lim_{n \to \infty} \alpha_n = \alpha_0 \Rightarrow X_{\alpha_0} = \bigcap_{n=1}^{\infty} X_{\alpha_n}$.

Take $x \in \bigcap_{n=1}^{\infty} X_{\alpha_n}$; then $x = x_n + y_n$, $x_n \in L_{\alpha_n}(u)$, $y_n \in L_{\alpha_n}(v)$. Since $L_{\alpha_1}(u) \supset L_{\alpha_2}(u) \supset \cdots$, it follows that $\{x_n\}_n \subset L_{\alpha_1}(u)$; analogously, $\{y_n\}_n \subset L_{\alpha_1}(v)$. Since $L_{\alpha_1}(u)$, $L_{\alpha_1}(v)$ are compact, there are subsequences such that $x_{n_k} \to x_0$, $y_{n_k} \to y_0$. By using the upper semicontinuity of u and v, and $u(x_n) \geqslant \alpha_n$, $v(y_n) \geqslant \alpha_n$, it follows easily that $u(x_0) \geqslant \alpha_0$, $v(y_0) \geqslant \alpha_0$. From this and $x = x_n + y_n$, by taking subsequences we conclude that $x = x_0 + y_0 \in X_{\alpha_0}$.

From Lemma 2.1 it follows that $u+v$ as defined by (2.4) satisfies $L_\alpha(u+v) = L_\alpha(u) + L_\alpha(v)$ for every $\alpha \in [0, 1]$.

Proposition 2.2 shows that the cancellation law holds in $(\mathscr{F}_0(X), +)$.

PROPOSITION 2.2. *If $u, v, w \in \mathscr{F}_0(X)$ and $u + w = v + w$, it follows that $u = v$.*

Proof. This follows from Proposition 2.1 and [14, Lemma 2]. When scalar multiplication (2.5) is concerned, it is easy to see that $L_\alpha(\lambda u) = \lambda L_\alpha(u)$ for $u \in \mathscr{F}_0(X)$, $\lambda \in \mathbb{R}$, and every $\alpha \in [0, 1]$. Note, however, that $(\mathscr{F}_0(X), +, \cdot)$ is *not* a vector space.

PROPOSITION 2.3. *If $u, v, w \in \mathscr{F}_0(X)$, then $d(u+w, v+w) = d(u, v)$.*

Proof. This follows easily from the definition of d (2.3) and from [14, Lemma 3].

Theorem 2.2 gives the desired embedding result.

THEOREM 2.2. *There exists a normed space $\mathscr{X}$ such that $\mathscr{F}_0(X)$ can be embedded isometrically into $\mathscr{X}$.*

Proof. This follows from Propositions 2.2 and 2.3 and from [14, Theorem 1].

Remark. The knowledge of the structure of the normed space $\mathscr{X}$ will be necessary in the next section. It can be described as follows: Define in $\mathscr{F}_0(X) \times \mathscr{F}_0(X)$ the equivalence relation

$$(u, v) \sim (u', v') \Leftrightarrow u + v' = v + u'. \tag{2.6}$$

The equivalence class of (u, v) will be denoted by $\langle u, v\rangle$. The space $\mathscr{X}$ is the set of equivalence classes. A vector space structure is defined in $\mathscr{X}$ by

$$\langle u, v\rangle + \langle u', v'\rangle = \langle u + u', v + v'\rangle, \tag{2.7}$$

$$\begin{aligned} \lambda\langle u, v\rangle &= \langle \lambda u, \lambda v\rangle, && \text{if } \lambda \geqslant 0, \\ &= \langle (-\lambda) v, (-\lambda) u\rangle, && \text{if } \lambda < 0. \end{aligned} \tag{2.8}$$

The embedding $j: \mathscr{F}_0(X) \to \mathscr{X}$ is defined by

$$j(u) = \langle u, 0 \rangle, \tag{2.9}$$

where 0 is the fuzzy subset $0(x) = 0$.

Finally, the norm in $\mathscr{X}$ is defined by

$$\|\langle u, v \rangle\| = d(u, v). \tag{2.10}$$

3. The Differential of a Fuzzy Function

Let X be a normed space and U be an open subset of X. Let Y be a reflexive Banach space.

By *fuzzy function* we mean a function $F: U \to \mathscr{F}_0(Y)$; such a function associates to each point $x \in U$ a fuzzy subset $F(x)$ of Y (with properties (i)–(iii) described in Section 2). Clearly, such fuzzy functions generalize set-valued functions $U \to Q(Y)$.

To define the differential of a fuzzy function, we shall use the embedding Theorem 2.2 and the classical concept of a differential in normed spaces.

By Theorem 2.2, $\mathscr{F}_0(Y)$ can be embedded isometrically in a normed space $\mathscr{Y}$; let $j: \mathscr{F}_0(Y) \to \mathscr{Y}$ denote this embedding.

Definition 3.1. The fuzzy function $F: U \to \mathscr{F}_0(Y)$ is called *differentiable* at $x_0 \in U$ if the map $\hat{F} = j \circ F$ is differentiable at x_0.

More precisely, F is differentiable at $x_0 \in U$ if there exists a linear bounded operator $\hat{F}'(x_0): X \to \mathscr{Y}$, such that

$$\lim_{x \to x_0} [\|\hat{F}(x) - \hat{F}(x_0) - \hat{F}'(x_0)(x - x_0)\| / \|x - x_0\|] = 0. \tag{3.1}$$

This concept of differential generalizes the differential of a set-valued function $U \to Q(Y)$ as studied by Banks and Jacobs [2].

If X is a finite-dimensional vector space with basis $e_1, e_2, ..., e_n$, and if $\hat{F}'(x_0)(e_k) \in j(\mathscr{F}_0(Y)) \subseteq \mathscr{Y}$ for $k = 1, 2, ..., n$, we say that the fuzzy function F is *conically differentiable* at $x_0 \in U$. This concept will be related to the differential defined by Hukuhara [8].

To define the Hukuhara differential of a fuzzy function, we shall consider a more particular context: $X = \mathbb{R}$, $Y = \mathbb{R}^n$, and U is an open interval of the real line. A fuzzy function in this context is a function $F: U \to \mathscr{F}_0(\mathbb{R}^n)$.

If $u, v \in \mathscr{F}_0(\mathbb{R}^n)$, and if there exists a fuzzy subset $\xi \in \mathscr{F}_0(\mathbb{R}^n)$ such that $\xi + u = v$, then ξ is *unique* by Proposition 2.2. In this case, ξ is called the *Hukuhara difference* of v and u and is denoted by $v - u$.

DEFINITION 3.2. A fuzzy function $F: U \to \mathscr{F}_0(\mathbb{R}^n)$ is called *H-differentiable* at $x_0 \in U$ if there exists $DF(x_0) \in \mathscr{F}_0(\mathbb{R}^n)$, such that the limits $\lim_{h\to 0+} [(F(x_0+h) - F(x_0))/h]$ and $\lim_{h\to 0+} [(F(x_0) - F(x_0-h))/h]$ both exist and are equal to $DF(x_0)$.

The relationship between conical differentiability and H-differentiability of fuzzy functions is given in the following proposition which generalizes a corresponding result of Banks and Jacobs [2]:

PROPOSITION 3.1. *If the fuzzy function $F: U \to \mathscr{F}_0(\mathbb{R}^n)$ is H-differentiable at $x_0 \in U$, then F is conically differentiable at x_0 and*

$$\hat{F}'(x_0)(h) = h\langle DF(x_0), 0\rangle. \tag{3.2}$$

Proof. Observe that if $v - u$ exists for $u, v \in \mathscr{F}_0(\mathbb{R}^n)$, then $\langle v, 0\rangle - \langle u, 0\rangle = \langle v - u, 0\rangle$. We have

$$\left\| \frac{\hat{F}(x_0 + \Delta x) - \hat{F}(x_0)}{\Delta x} - \langle DF(x_0), 0\rangle \right\| = d\left(\frac{F(x_0+\Delta x) - F(x_0)}{\Delta x}, DF(x_0)\right) \tag{3.3}$$

and the result follows as in [2].

The above theorem shows that differentiability as given in Definition 3.1 is a more general concept than H-differentiability.

REFERENCES

1. R. J. AUMANN AND M. PERLES, A variational problem arising in economics, *J. Math. Anal. Appl.* **11** (1965), 488–503.
2. H. T. BANKS AND M. Q. JACOBS, A differential calculus for multifunctions, *J. Math. Anal. Appl.* **29** (1970), 246–272.
3. G. DEBREU, Integration of correspondences, *in* "Proc. Fifth Berkeley Symp. Math. Statist. Probability," pp. 351–372, Univ. of California Press, Berkeley, 1966.
4. R. FÉRON, Ensembles aléatoires flous, *C. R. Acad. Sci. Paris Sér. A* **282** (1976), 903–906.
5. R. FORTET AND M. KAMBOUZIA, Ensembles aléatoires et ensembles flous, *Publ. Économétriques* (1976).
6. I. R. GOODMAN, Fuzzy sets as equivalence classes of random sets, *in* "Recent Developments in Fuzzy Set and Possibility Theory" (R. Yager, Ed.), Pergamon, Elmsford, N.Y., 1982.
7. H. HERMES, Calculus of set-valued functions and control, *J. Math. and Mech.* **18** (1968), 47–59.
8. M. HUKUHARA, Intégration des applications mesurables dont la valuer est un compact convexe, *Funkcial. Ekvac.* **10** (1967), 205–223.
9. J. KAMPÉ DE FÉRIET, Une interpretation des mesures de plausibilité et de credibilité au sens de G. Shafer et de la fonction d'appartenance définissant un ensemble flou de L. Zadeh, *in* "Publ. IRMA 2," Université de Lille, France, 1980.

10. C. Kuratowski, "Topologie I," Monografie Matematyczne, Warsaw, 1948.
11. G. Matheron, "Random Sets and Integral Geometry," Wiley, New York, 1975.
12. C. V. Negoita and D. A. Ralescu, "Applications of Fuzzy Sets to Systems Analysis," Wiley, New York, 1975.
13. M. L. Puri and D. A. Ralescu, "Fuzzy Random Variables," submitted.
14. H. Rådström, An embedding theorem for spaces of convex sets, *Proc. Amer. Math. Soc.* **3** (1952), 165–169.
15. G. Shafer, "A Mathematical Theory of Evidence," Princeton Univ. Press, Princeton, N.J., 1976.
16. L. A. Zadeh, Fuzzy sets, *Inform. and Control* **8** (1965), 338–353.
17. L. A. Zadeh, Fuzzy sets and information granularity, *in* "Advances in Fuzzy Set Theory and Applications" (M. Gupta, R. Ragade, and R. Yager, Eds.), p. 3–18, North-Holland, Amsterdam, 1979.

Reprinted by permission of the publisher,
© 1983 by Academic Press

THE CONCEPT OF NORMALITY FOR FUZZY RANDOM VARIABLES[1]

MADAN L. PURI and DAN A. RALESCU
Indiana University and University of Cincinnati

In this paper we define the concept of a normal fuzzy random variable and we prove the following representation theorem: Every normal fuzzy random variable equals the sum of its expected value and a mean zero random vector.

1. Introduction. The concept of a fuzzy random variable [cf., Puri and Ralescu (1984)] was defined as a tool for representing relationships between the outcomes of a random experiment and inexact data. By inexactness here we mean nonstatistical inexactness that is due to subjectivity and imprecision of human knowledge rather than to the occurrence of random events. Fuzzy random variables generalize the concept of a random variable as well as that of a random set [Matheron (1975)].

To make this theory applicable to statistical analysis of inexact data, we consider in this paper the concept of normality for fuzzy random variables. [For the concept of normality of random sets, the reader is referred to Lyashenko (1980). For a characterization of p-stable random sets, the reader is referred to Giné and Hahn (1985).]

After some preliminaries on fuzzy random variables given in Section 2, we describe in Section 3 an embedding theorem for the space of fuzzy sets with compact convex levels that satisfy a certain Lipschitz condition. In Section 4, we define the concept of a normal fuzzy random variable in $\mathbb{R}^p$ and represent such a variable as the sum of a constant fuzzy set and a random vector in $\mathbb{R}^p$. This result extends the corresponding result for random sets [Lyashenko (1980)] and, analogous to his result, it shows that a normal fuzzy random variable is "degenerate" in the sense that it is a random translate of a fixed fuzzy set. Our result is also analogous to the one given in Giné and Hahn (1985) which deals with random sets in Banach space.

2. Fuzzy random variables. Let $\mathscr{K}(\mathbb{R}^p)$ denote the nonempty compact subsets of $\mathbb{R}^p$, and let $\mathscr{K}_c(\mathbb{R}^p)$ denote the nonempty compact convex subsets of $\mathbb{R}^p$. The Hausdorff distance

$$d(A,B) = \max\left\{\sup_{a\in A}\inf_{b\in B}\|a-b\|,\ \sup_{b\in B}\inf_{a\in A}\|a-b\|\right\}$$

gives $\mathscr{K}(\mathbb{R}^p)$ a structure of a metric space. Moreover, this space is complete and separable, and $\mathscr{K}_c(\mathbb{R}^p)$ is a closed subspace [Debreu, (1967)].

Received May 1984; revised February 1985.

[1] Research supported by a National Science Foundation Grant.

AMS 1980 *subject classifications*. Primary 60D05; secondary 03E72.

Key words and phrases. Fuzzy set, fuzzy random variable, embedding, Gaussian element.

Denote $\|A\| = d(A, \{0\})$ for $A \in \mathscr{K}(\mathbb{R}^p)$. A linear structure is defined in $\mathscr{K}(\mathbb{R}^p)$ by the operations

$$A + B = \{a + b | a \in A,\ b \in B\}, \qquad \lambda A = \{\lambda a | a \in A\},$$

where $A, B \in \mathscr{K}(\mathbb{R}^p)$ and $\lambda \in \mathbb{R}$. Note that $\mathscr{K}(\mathbb{R}^p)$ is not a vector space.

Let $(\Omega, \mathscr{A}, P)$ be a probability space. A *random set* is a Borel measurable function f: $\Omega \to \mathscr{K}(\mathbb{R}^p)$. The *expected value* Ef [Aumann (1965)] is defined by

$$Ef = \{E\phi | \phi \in L^1(\Omega, \mathscr{A}, P),\ \phi(\omega) \in f(\omega) \text{ a.e.}\},$$

where ϕ: $\Omega \to \mathbb{R}^p$ is a selection of f.

Note that $E\|f\| < \infty$ implies that $Ef \in \mathscr{K}(\mathbb{R}^p)$.

A *fuzzy set* in $\mathbb{R}^p$ is a function u: $\mathbb{R}^p \to [0,1]$ [Zadeh (1965)]. Denote by $L_\alpha(u) = \{x \in \mathbb{R}^p | u(x) \geq \alpha\}$ for $0 \leq \alpha \leq 1$, the level sets of u. Further, denote by $\operatorname{supp} u$, the support of u, i.e., the closure of the set $\{x \in \mathbb{R}^p | u(x) \neq 0\}$.

As an extension of $\mathscr{K}(\mathbb{R}^p)$ we define the space $\mathscr{F}(\mathbb{R}^p)$ of fuzzy sets u: $\mathbb{R}^p \to [0,1]$ with the properties:

(a) u is upper semicontinuous;

(b) $\operatorname{supp} u$ is compact;

(c) $\{x \in \mathbb{R}^p | u(x) = 1\} \neq \phi$.

The linear structure in $\mathscr{F}(\mathbb{R}^p)$ is defined by the operations $(u + v)(x) = \sup_{y+z=x} \min[u(y), v(z)]$,

$$(\lambda u)(x) = \begin{cases} u(\lambda^{-1}x), & \text{if } \lambda \neq 0, \\ \chi_{\{0\}}(x), & \text{if } \lambda = 0, \end{cases}$$

where $u, v \in \mathscr{F}(\mathbb{R}^p)$, $\lambda \in \mathbb{R}$, and χ_A denotes the characteristic function of $A \subset \mathbb{R}^p$.

Note that $L_\alpha(u + v) = L_\alpha(u) + L_\alpha(v)$ and $L_\alpha(\lambda u) = \lambda L_\alpha(u)$ for every $0 \leq \alpha \leq 1$.

We extend the Hausdorff metric by defining $d_\infty(u, v) = \sup_{\alpha > 0} d(L_\alpha u, L_\alpha v)$; $u, v \in \mathscr{F}(\mathbb{R}^p)$ (see Puri and Ralescu, 1981 and 1983).

It is possible to show that $(\mathscr{F}(\mathbb{R}^p), d_\infty)$ is complete [Puri and Ralescu (1984)].

Denote by $\mathscr{F}_c(\mathbb{R}^p)$ the space of fuzzy sets $u \in \mathscr{F}(\mathbb{R}^p)$ such that $L_\alpha(u)$ is convex for each $\alpha \geq 0$.

A *fuzzy random variable* is a Borel measurable function X: $\Omega \to (\mathscr{F}(\mathbb{R}^p), d_\infty)$. If $E\|\operatorname{supp} X\| < \infty$, then the *expected value* EX is the unique fuzzy set satisfying the property

$$L_\alpha(EX) = E(L_\alpha X), \qquad 0 < \alpha \leq 1$$

[Puri and Ralescu (1984), Theorem 3.1]. It also follows that $E\|\operatorname{supp} X\| < \infty$ implies $EX \in \mathscr{F}(\mathbb{R}^p)$.

A strong law of large numbers and a central limit theorem for fuzzy random variables were derived in Klement, Puri, and Ralescu (1985a, b). A law of large numbers was also given by Kruse (1982) using a different concept of expected value.

3. The embedding theorem. Let $S^{p-1} = \{x \in \mathbb{R}^p \mid \|x\| = 1\}$ denote the unit sphere in $\mathbb{R}^p$. It is well known that $\mathscr{K}_c(\mathbb{R}^p)$ can be embedded isometrically into $C(S^{p-1})$ (the Banach space of continuous functions on S^{p-1}). This result goes back to Minkowski [see, e.g., Artstein and Vitale (1975)].

Such an embedding is realized via the support function of a compact convex set. More precisely, if $K \in \mathscr{K}_c(\mathbb{R}^p)$, its support function is defined by

$$s_K(x) = \sup_{a \in K} \langle x, a \rangle, \qquad x \in S^{p-1},$$

where $\langle \cdot, \cdot \rangle$ is the inner product in $\mathbb{R}^p$.

In the case of the space $\mathscr{F}_c(\mathbb{R}^p)$ of fuzzy sets, we seek a similar embedding. However, more restrictions need to be imposed to do this.

Consider the space $\mathscr{F}_{CL}(\mathbb{R}^p)$ of fuzzy sets $u \in \mathscr{F}_c(\mathbb{R}^p)$ such that the map $\alpha \to L_\alpha(u)$ is Lipschitz with respect to the Hausdorff distance. More precisely, $u \in \mathscr{F}_{CL}(\mathbb{R}^p)$ if there exists a constant $M > 0$, such that $d(L_\alpha u, L_\beta u) \le M|\alpha - \beta|$ for every $\alpha, \beta \in (0,1]$.

The following theorem gives the desired embedding, which will be used in the next section to define and characterize fuzzy random variables.

THEOREM 3.1. *There exists a function*

$$j\colon \mathscr{F}_{CL}(\mathbb{R}^p) \to C([0,1] \times S^{p-1})$$

such that:

(i) *j is an isometry* (*i.e.*, $\|j(u) - j(v)\|_\infty = d_\infty(u, v)$);
(ii) $j(u + v) = j(u) + j(v)$;
(iii) $j(\lambda u) = \lambda j(u)$, $\lambda \ge 0$.

PROOF. Define $j(u) = s_u$, where

$$s_u(\alpha, x) = \begin{cases} s_{L_\alpha u}(x), & \text{if } \alpha > 0, \\ s_{\operatorname{supp} u}(x), & \text{if } \alpha = 0. \end{cases}$$

The function s_u is Lipschitz on $[0,1] \times S^{p-1}$:

$$\begin{aligned} |s_u(\alpha, x) - s_u(\beta, y)| &\le \left|s_{L_\alpha u}(x) - s_{L_\beta u}(x)\right| + \left|s_{L_\beta u}(x) - s_{L_\beta u}(y)\right| \\ &\le \|s_{L_\alpha u} - s_{L_\beta u}\|_\infty + \|L_\beta u\| \, \|x - y\| \\ &\le \|u\|_L(|\alpha - \beta| + \|x - y\|), \end{aligned} \tag{3.1}$$

where

$$\|u\|_L = \sup_{\alpha \ne \beta} \frac{d(L_\alpha u, L_\beta u)}{|\alpha - \beta|} + \sup_{\alpha > 0} \|L_\alpha u\|$$

and we have used the fact that $K \in \mathscr{K}_c(\mathbb{R}^p) \to s_K$ is an isometry, as well as the fact that s_K satisfies a Lipschitz condition with constant $\|K\|$.

Since any two norms in $\mathbb{R}^{p+1}$ are equivalent, inequality (3.1) becomes

$$|s_u(\alpha, x) - s_u(\beta, y)| \leq q\|u\|_L(|\alpha - \beta|^2 + \|x - y\|^2)^{1/2},$$

where q is a constant.

Thus $s_u = j(u) \in C([0,1] \times S^{p-1})$.

It is now easy to show that j is an isometry, i.e.,

$$\|j(u) - j(v)\|_\infty = \sup_{\alpha, x} |s_u(\alpha, x) - s_v(\alpha, x)| = d_\infty(u, v),$$

as well as properties (ii) and (iii), and the proof is completed. □

NOTE. s_u can also be defined on $[0,1] \times \mathbb{R}^p$ by the same formula as above, and it satisfies the following properties

(a) $s_u(\alpha, \cdot)$ is subadditive, i.e., $s_u(\alpha, x + y) \leq s_u(\alpha, x) + s_u(\alpha, y)$ for $x, y \in \mathbb{R}^p$.

(b) $s_u(\alpha, \cdot)$ is positively homogeneous, i.e., $s_u(\alpha, \lambda x) = \lambda s_u(\alpha, x)$ for $x \in \mathbb{R}^p$, $\lambda \geq 0$.

(c) $s_u(\cdot, x)$ is decreasing.

(d) $s_u(\cdot, x)$ is left-continuous.

Properties (a), (b), and (c) will be used in the proof of our representation theorem in Section 4.

From Theorem 3.1, it follows that every fuzzy random variable $X: \Omega \to \mathscr{F}_{CL}(\mathbb{R}^p)$ can be thought of as a random element of $C([0,1] \times S^{p-1})$ by considering $j(X) = s_X: \Omega \to C([0,1] \times S^{p-1})$ where $s_X(\omega) = s_{X(\omega)}$.

The following result will also be useful.

LEMMA 3.1. *If* $X: \Omega \to \mathscr{F}_{CL}(\mathbb{R}^p)$ *satisfies* $E\|\operatorname{supp} X\| < \infty$, *then* $E[s_X(\alpha, x)] = s_{EX}(\alpha, x)$, *for* $\alpha \in [0,1]$, $x \in \mathbb{R}^p$.

PROOF. Let $\alpha_0 \in [0,1]$ be fixed and consider the random set $f = L_{\alpha_0}(X)$. We first show that $E[s_f(x)] = s_{Ef}(x)$ for each $x \in S^{p-1}$. Indeed, it is easy to see that this result holds if $f = \sum_{i=1}^n \chi_{A_i} K_i$ where $A_i \in \mathscr{A}$, $K_i \in \mathscr{K}_c(\mathbb{R}^p)$, and χ_{A_i} denotes the characteristic function of A_i (i.e., if f is a simple function). The desired formula follows by using the expected value of Debreu (1967), the fact that $K \in \mathscr{K}_c(\mathbb{R}^p) \to s_K$ is an isometry, and the Lebesgue dominated convergence theorem. Thus

$$E\left[s_{L_{\alpha_0}X}(x)\right] = s_{E(L_{\alpha_0}X)}(x) = s_{L_{\alpha_0}(EX)}(x) = s_{EX}(\alpha_0, x),$$

implying $E[s_X(\alpha_0, x)] = s_{EX}(\alpha_0, x)$, the desired result. We have used the property $E(L_\alpha X) = L_\alpha(EX)$ of the expected value EX. □

4. Normal fuzzy random variables. In this section we define the concept of a normal fuzzy random variable (by using the embedding of Section 3), and we prove that such a variable can be uniquely represented as the sum of a constant fuzzy set and a normal random vector in $\mathbb{R}^p$.

Before doing this, we shall need the following preliminaries: Let $C(M)$ denote the space of continuous functions on the compact set M.

A random element $f: \Omega \to C(M)$ is Gaussian if the vector $(f(t_1), \dots, f(t_n))$ is normal for any $t_1, \dots, t_n \in M$.

DEFINITION 4.1. A fuzzy random variable $X: \Omega \to \mathscr{F}_{CL}(\mathbb{R}^p)$ is *normal* if s_X is a Gaussian random element of $C([0,1] \times S^{p-1})$.

This definition and the comment above imply that $(s_X(\alpha_1, x_1), \dots, s_X(\alpha_n, x_n))$ is normal for every $\alpha_1, \dots, \alpha_n \in [0,1]$ and $x_1, \dots, x_n \in S^{p-1}$. Since $s_X(\alpha, x)$ is positively homogeneous in x, it follows that $(s_X(\alpha_1, x_1), \dots, s_X(\alpha_n, x_n))$ is normal for every $\alpha_1, \dots, \alpha_n \in [0,1]$ and $x_1, \dots, x_n \in \mathbb{R}^p$. It also follows from Definition 4.1 that if X and Y are independent normal fuzzy random variables, then $X + Y$ is normal. Also λX is normal whenever X is normal and $\lambda \in \mathbb{R}$.

REMARK. This definition of normality for fuzzy random variables is an extension of the concept of normality of random vectors in $\mathbb{R}^p$ as well as the concept of a normal random set [Lyashenko (1980)]. Definition 4.1 is a natural one, in view of the embedding Theorem 3.1, since in this way, normality is defined via the random element s_X that takes values in the familiar Banach space $C([0,1] \times S^{p-1})$.

We now prove the characterization theorem for normal fuzzy random variables.

THEOREM 4.1. *Let $X: \Omega \to \mathscr{F}_{CL}(\mathbb{R}^p)$ be a fuzzy random variable with $E\|\text{supp}\, X\| < \infty$. The following statements are equivalent.*

(1) *X is normal.*

(2) *$X = EX + \{\xi\}$ where ξ is a normal random vector with mean zero.*

PROOF. (2) $\Rightarrow$ (1). It is easily seen that if $\alpha_1, \dots, \alpha_n \in [0,1]$ and $x_1, \dots, x_n \in S^{p-1}$, then $(s_X(\alpha_1, x_1), \dots, s_X(\alpha_n, x_n)) = M\xi + b$ where M is a constant matrix and b is a constant vector. Since $M\xi + b$ is normal, it follows that s_X is Gaussian in $C([0,1] \times S^{p-1})$.

(1) $\Rightarrow$ (2). Let s_X be Gaussian. Fix $\omega \in \Omega$ and $\alpha \in [0,1]$. Define $\psi(\omega, \alpha) = s_{X(\omega)}(\alpha, \cdot) - s_{EX}(\alpha, \cdot)$. Thus $\psi(\omega, \alpha): \mathbb{R}^p \to \mathbb{R}$.

CLAIM 1. $\psi(\omega, \alpha)$ is linear.

Let $\lambda, \mu \geq 0$, and $x_1, x_2 \in \mathbb{R}^p$ be fixed. Then

$$\lambda s_X(\alpha, x_1) + \mu s_X(\alpha, x_2) - s_X(\alpha, \lambda x_1 + \mu x_2) \geq 0, \tag{4.1}$$

since $s_X(\alpha, x)$ is positively homogeneous in x as well as subadditive [properties (a) and (b) of Section 3].

By hypothesis, the vector $(s_X(\alpha, x_1), s_X(\alpha, x_2), s_X(\alpha, \lambda x_1 + \mu x_2))$ is normal, which implies that $\lambda s_X(\alpha, x_1) + \mu s_X(\alpha, x_2) - s_X(\alpha, \lambda x_1 + \mu x_2)$ is a normal random variable. From (4.1), it follows that this random variable is degenerate, i.e.,

$\lambda s_X(\alpha, x_1) + \mu s_X(\alpha, x_2) - s_X(\alpha, \lambda x_1 + \mu x_2) = E[\lambda s_X(\alpha, x_1) + \mu s_X(\alpha, x_2) - s_X(\alpha, \lambda x_1 + \mu x_2)] = \lambda s_{EX}(\alpha, x_1) + \mu s_{EX}(\alpha, x_2) - s_{EX}(\alpha, \lambda x_1 + \mu x_2)$. The last equality follows from Lemma 3.1. In terms of our function $\psi(\omega, \alpha)$, this implies that $\psi(\omega, \alpha)(\lambda x_1 + \mu x_2) = \lambda\psi(\omega, \alpha)(x_1) + \mu\psi(\omega, \alpha)(x_2)$. Since $\psi(\omega, \alpha)(0) = 0$, this proves the linearity of $\psi(\omega, \alpha)$.

It is well known that such a linear map from $\mathbb{R}^p$ into $\mathbb{R}$ can be represented as $\psi(\omega, \alpha)(x) = \langle x, \xi_\alpha(\omega)\rangle$, $x \in \mathbb{R}^p$ for some $\xi_\alpha(\omega) \in \mathbb{R}^p$.

CLAIM 2. $\xi_\alpha(\omega)$ does not depend on $\alpha \in [0,1]$.

Let $\alpha, \beta \in [0,1]$, $\alpha \le \beta$ fixed. From property (c) of Section 3, it follows that $s_X(\alpha, x) \ge s_X(\beta, x)$ for every $x \in \mathbb{R}^p$. But (using the same argument as before) $s_X(\alpha, x) - s_X(\beta, x)$ is a normal random variable, which implies that $s_X(\alpha, x) - s_X(\beta, x) = E[s_X(\alpha, x) - s_X(\beta, x)] = s_{EX}(\alpha, x) - s_{EX}(\beta, x)$. Thus $\psi(\omega, \alpha)(x) = \psi(\omega, \beta)(x)$ for every $x \in \mathbb{R}^p$. This shows that $\xi_\alpha(\omega) = \xi(\omega)$ is independent of α.

Thus we have the representation $\psi(\omega, \alpha)(x) = \langle x, \xi(\omega)\rangle$. Since $\langle x, \xi\rangle$ is measurable for each $x \in \mathbb{R}^p$, it follows that $\xi\colon \Omega \to \mathbb{R}^p$ is measurable. Now, from the representation $\langle x, \xi\rangle = s_X(\alpha, x) - s_{EX}(\alpha, x)$, it follows that $\langle x, \xi\rangle$ is a normal mean zero random variable for each $x \in \mathbb{R}^p$. Thus ξ is a normal mean zero random vector.

To conclude the proof of the theorem, we write $s_X(\alpha, x) = s_{EX}(\alpha, x) + \langle x, \xi\rangle = s_{EX}(\alpha, x) + s_{\{\xi\}}(\alpha, x) = s_{EX+\{\xi\}}(\alpha, x)$, implying that $X = EX + \{\xi\}$. □

REFERENCES

[1] ARTSTEIN, Z. and VITALE, R. A. (1975). A strong law of large numbers for random compact sets. *Ann. Probab.* **3** 879–882.

[2] AUMANN, R. J. (1965). Integrals of set-valued functions. *J. Math. Anal. Appl.* **12** 1–12.

[3] DEBREU, G. (1967). Integration of correspondences. *Proc. Fifth Berkeley Symp. Math. Statist. Probability* **II** 351–372. Univ. Calif. Press.

[4] GINÉ E. and HAHN, M. G. (1985). Characterization and domains of attraction of p-stable random compact sets. *Ann. Probab.* **13** 447–468 .

[5] KLEMENT, E. P. PURI, M. L. and RALESCU, D. A. (1985a). Law of large numbers for fuzzy random variables. Technical report, Department of Mathematics, Indiana University.

[6] KLEMENT, E. P., PURI, M. L. and RALESCU, D.A. (1985b). Central limit theorems for fuzzy random variables. Technical report, Department of Mathematics, Indiana University.

[7] KRUSE, R. (1982). The strong law of large numbers for fuzzy random variables. *Inform. Sci.* **28** 233–241.

[8] LYASHENKO, N. N. (1980). Statistics of random compacts in euclidean space. *Zap. Naucn. Sem. Leningrad. Otdel. Mat. Inst. Steklov.* **98** 115–139. (English translation in *J. Soviet Math.* **21** (1983) 76–92.

[9] MATHERON, G. (1975). *Random Sets and Integral Geometry*. Wiley, New York.

[10] PURI, M. L. and RALESCU, D. A. (1981). Différentielle d'une fonction floue. *C. R. Acad. Sc. Paris*, t. 293, Série I, 237–239.

[11] PURI, M. L. and RALESCU, D. A. (1982). Integration on fuzzy sets. *Adv. in Appl. Math.* **3** 430–434.

[12] PURI, M. L. and RALESCU, D. A. (1983). Differentials of fuzzy functions. *J. Math. Anal. Appl.* **91** 552–558.

[13] PURI, M. L. and RALESCU, D. A. (1984). Fuzzy random variables. To appear in *J. Math. Anal. Appl.* (January 1986 issue).

[14] WEIL, W. (1982). An application of the central limit theorem for Banach space-valued random variables to the theory of random sets. *Z. Wahrsch. verw. Gebiete* **60** 203–208.

[15] ZADEH, L. A. (1965). Fuzzy sets. *Inform. and Control* **8** 338–353.

DEPARTMENT OF MATHEMATICS
INDIANA UNIVERSITY
BLOOMINGTON, INDIANA 47405

DEPARTMENT OF MATHEMATICS
UNIVERSITY OF CINCINNATI
CINCINNATI, OHIO 45221

Reprinted by permission of the publisher,
© 1985 by The Institute of Mathematical Statistics

LIMIT THEOREMS FOR RANDOM COMPACT SETS IN BANACH SPACE

MADAN L. PURI and DAN A. RALESCU

Department of Mathematics, Indiana University, Bloomington, Indiana 47405, U.S.A.
Department of Mathematics, University of Cincinnati, Cincinnati, Ohio 45221, U.S.A.

1. *Introduction.* The concept of random set, though vaguely known for a long time (possibly since Buffon's needle problem), did not develop until Robbins [**25**, **26**] provided for the first time a solid mathematical formulation of this concept and investigated relationships between random sets and geometric probabilities. Later on (in a different context) this concept gave rise to a more general concept of set-valued function in topology, and applications were also found in several areas such as economics (see, for example, Aumann [**3**] and Debreu [**10**]) and control theory (see, for example, Hermes [**13**]), among others. Recently in two independent formulations, D. G. Kendall [**17**] and Matheron [**21**] provided a comprehensive mathematical theory of this concept influenced by the geometric probability point of view. Actually, much of the research in this area falls under the heading of stochastic geometry (see, for example, M. G. Kendall and P. A. P. Moran [**18**]). The D. G. Kendall and Matheron theories have been compared and 'reconciled' by Ripley [**24**]. In the past few years random sets have been investigated as extensions of random variables and random vectors, and in this framework the problems of deriving limit theorems have received a great deal of attention. This approach is benefiting greatly from probability results in Banach spaces.

The first limit theorem was a strong law of large numbers for random compact sets in $\mathbb{R}^p$, derived by Artstein and Vitale [**2**]. The central limit theorem for random sets was derived for special cases by Cressie [**9**] and Lyashenko [**19**], and in full generality by Weil [**29**], again for random compact sets in $\mathbb{R}^p$. Puri and Ralescu [**22**] proved the strong law of large numbers for random compact convex sets in Banach space. Recently, Giné, Hahn and Zinn [**12**] initiated the study of the central limit theorem and the law of iterated logarithm for random compact convex sets in a Banach space. The purpose of this paper is twofold: we show that the hypothesis of convexity can be dropped in Banach spaces of type $p > 1$, and we prove a central limit theorem for random sets which generalizes a result of Giné, Hahn and Zinn [**12**].

The paper is organized as follows. In Section 2 we state some preliminary results about random sets which are needed later. In Section 3 we discuss different embedding results for compact convex sets which are essential for limit theorems. In Section 4 we generalize the Shapley–Folkman theorem (see Arrow and Hahn [**1**]) to Banach space of type p. This generalization is then used to study non-convex random sets. In

Section 5 we prove the strong law of large numbers for random compact sets in a Banach space. Section 6 deals with a central limit theorem for random sets which generalizes a result of Giné, Hahn and Zinn [**12**].

2. *Random sets: Preliminaries.* Let $\mathscr{X}$ be a separable Banach space with norm $\|\ \|$, and let $\mathscr{X}^*$ be the dual of $\mathscr{X}$. Let $\mathscr{K}(\mathscr{X})$ denote the collection of non-empty compact subsets of $\mathscr{X}$, and let $\operatorname{co}\mathscr{K}(\mathscr{X})$ denote the non-empty compact convex subsets of $\mathscr{X}$. If $A \in \mathscr{K}(\mathscr{X})$, $\overline{\operatorname{co}}\,A$ will denote the closed convex hull of A. Note that $\overline{\operatorname{co}}\,A \in \operatorname{co}\mathscr{K}(\mathscr{X})$ by Mazur's theorem ([**11**], p. 416). The space $\mathscr{K}(\mathscr{X})$ has a linear structure induced by the (Minkowski) addition and scalar multiplication:

$$A+B = \{a+b \,|\, a\in A,\ b\in B\},$$
$$\lambda A = \{\lambda a \,|\, a \in A\},$$

for $A, B \in \mathscr{K}(\mathscr{X})$ and $\lambda \in \mathbb{R}$. However (and this is crucial), $\mathscr{K}(\mathscr{X})$ is *not* a vector space (it is not a group with respect to addition).

If $A, B \in \mathscr{K}(\mathscr{X})$, the *Hausdorff distance* is defined as

$$d(A,B) = \max\{\sup_{a\in A}\inf_{b\in B}\|a-b\|,\ \sup_{b\in B}\inf_{a\in A}\|a-b\|\}, \tag{2·1}$$

and it is well known that $(\mathscr{K}(\mathscr{X}), d)$ is a complete, separable metric space.

If $A \in \mathscr{K}(\mathscr{X})$, then we shall write $\|A\| = d(A,\{0\}) = \sup_{a\in A}\|a\|$.

Let $(\Omega, \mathscr{A}, P)$ be a probability space. A *random set* (random compact set) is a Borel measurable function $X\colon \Omega \to \mathscr{K}(\mathscr{X})$. Different concepts of measurability can be defined, but they are all equivalent in this setting (see Debreu [**10**]).

The expected value of the random set was defined by Aumann [**3**] and later by Debreu [**10**]. These definitions were shown to be equivalent by Byrne [**5**].

If X is a random compact *convex* set (i.e., $\operatorname{co}\mathscr{K}(\mathscr{X})$-valued), then EX is defined as

$$EX = \{Ef \,|\, f \in L^1(\Omega,\mathscr{A},P), f(\omega)\in X(\omega) \text{ a.s.}\}. \tag{2·2}$$

Here $f\colon \Omega \to \mathscr{X}$ is a *selection* of X and Ef denotes the classical expectation (via the Bochner integral). In general EX may be empty, but if $E\|X\| < \infty$, then $EX \in \operatorname{co}\mathscr{K}(\mathscr{X})$ (Debreu [**10**], Aumann [**3**]).

If X is a random compact set, then, by *definition*

$$EX = E(\overline{\operatorname{co}}\,X) \tag{2·3}$$

and so $EX \in \operatorname{co}\mathscr{K}(\mathscr{X})$. Actually, if X is finite-dimensional and if P is a *non-atomic* probability measure, then EX is always convex and (2·3) follows [**3**]. These facts, however, depend strongly on the Lyapounov theorem on the range of a vector-valued measure (see, for example, [**27**], p. 114, theorem 5·5) and so, in an infinite-dimensional space $\mathscr{X}$, definition (2·3) is needed.

The general method of approach in proving limit theorems for random sets initiated by Artstein and Vitale [**2**] is as follows:

(*a*) Consider first random compact convex sets, and embed $\operatorname{co}\mathscr{K}(\mathscr{X})$ into a Banach space.

(*b*) Use a limit theorem for Banach space valued random elements.

(*c*) Drop the convexity assumption.

Step (*b*) is usually the easiest to perform due to the vast literature on probability in Banach space; step (*a*) can be dealt with in several ways (according to the cases $\dim \mathscr{X} < \infty$ and $\dim \mathscr{X} = \infty$), while step (*c*), to the best of our knowledge, has not been dealt with in infinite-dimensional Banach spaces.

In the next section, we deal with step (*a*) in detail. (Since embedding results are crucial in proving limit theorems, detailed discussion is necessary.)

3. *Embedding of compact convex sets.* Here we focus on the possibility of embedding $\text{co}\,\mathscr{K}(\mathscr{X})$ into a Banach space. Two embedding theorems exist: one due to Rådström [**23**] (see Puri and Ralescu [**22**]), and one due to Hörmander [**15**] (see Giné, Hahn and Zinn [**12**]). The latter goes back to Minkowski if $\dim \mathscr{X} < \infty$.

To point out the difficulties which arise when $\mathscr{X}$ is infinite-dimensional, we shall state here both the embedding theorems.

THEOREM 3·1 (Rådström [**23**]). *There exist a normed space* $\mathscr{U}$ *and an isometry* $\alpha: \text{co}\,\mathscr{K}(\mathscr{X}) \to \mathscr{U}$ *such that*

(i) $\alpha(A+B) = \alpha(A) + \alpha(B)$, $\alpha(\lambda A) = \lambda\alpha(A)$ $(\lambda \geq 0)$,

(ii) $\mathscr{U} = \text{Sp}\,\text{Im}\,\alpha$.

Here $\text{Sp}\,\text{Im}\,\alpha$ denotes the linear subspace spanned by the image of α in $\mathscr{U}$. Note that $\mathscr{U}$ is not complete in general, but one can always take the completion of $\mathscr{U}$ and thus embed $\text{co}\,\mathscr{K}(\mathscr{X})$ into a Banach space.

To state the second theorem, we need some additional notation.

We shall denote the unit ball in $\mathscr{X}^*$ by $\text{ball}\,\mathscr{X}^*$; $\text{ball}\,\mathscr{X}^* = \{x^* \in \mathscr{X}^* \mid \|x^*\| \leq 1\}$. If T is a topological space, we denote by $C_b(T)$ the Banach space of all bounded and continuous functions $\phi: T \to \mathbb{R}$, the norm being $\|\phi\| = \sup_{t \in T} |\phi(t)|$.

In particular, $C_b(\text{ball}\,\mathscr{X}^*)$ denotes all bounded and (norm) continuous functions $\phi: \text{ball}\,\mathscr{X}^* \to \mathbb{R}$.

We now state the Hörmander embedding theorem.

THEOREM 3·2 (Hörmander [**15**]). *There exists an isometry* $j: \text{co}\,\mathscr{K}(\mathscr{X}) \to C_b(\text{ball}\,\mathscr{X}^*)$ *such that*

$$j(A+B) = j(A) + j(B), \quad j(\lambda M) = \lambda j(M) \quad (\lambda \geq 0).$$

Note. The embedding j is defined by the *support function* of a compact convex set:

$$A \in \text{co}\,\mathscr{K}(\mathscr{X}), \quad s_A(x^*) = \sup_{a \in A} \langle x^*, a\rangle,$$

where $\langle x^*, a\rangle = x^*(a)$. More precisely, $j(A) = s_A$.

There are some advantages and some disadvantages in using either Theorem 3·1 or 3·2. Therem 3·1 provides an abstract space $\mathscr{U}$ (via an algebraic construction) and this is inconvenient for a central limit theorem. However, the space $\mathscr{U}$ is separable (see Debreu [**10**]). The space $C_b(\text{ball}\,\mathscr{X}^*)$ is more familiar, but it is *not* separable (unless $\dim \mathscr{X} < \infty$). Most of the probability results in Banach space are true for separable spaces only. Actually, separability is a key issue and so we give some details about this embedding.

As Giné, Hahn and Zinn [**12**] have observed, $\text{co}\,\mathscr{K}(\mathscr{X})$ can be embedded into $C(\text{ball}\,\mathscr{X}^*, w^*)$ – the Banach space of functions defined on $\text{ball}\,\mathscr{X}^*$ and continuous in the weak* topology. It is well known that $(\text{ball}\,\mathscr{X}^*, w^*)$ is compact and metrizable, so $C(\text{ball}\,\mathscr{X}^*, w^*)$ is separable. The embedding $j: \text{co}\,\mathscr{K}(\mathscr{X}) \to C(\text{ball}\,\mathscr{X}^*, w^*)$ is performed as follows:

Let $\{x_n: n \geqslant 1\}$ be a countable dense subset of $\mathscr{X}$. Then

$$\{F|F \text{ finite subset of } \{x_n: n \geqslant 1\}\}$$

is dense in $\mathscr{K}(\mathscr{X})$ (Debreu [**10**]). Thus $\{\operatorname{co} F|F$ finite subset of $\{x_n: n \geqslant 1\}\}$ is dense in $\operatorname{co}\mathscr{K}(\mathscr{X})$. It is easy to see that $\{s_{\operatorname{co} F}|F$ finite subset of $\{x_n: n \geqslant 1\}\} \subseteq C(\operatorname{ball}\mathscr{X}^*, w^*)$; since $j: \operatorname{co}\mathscr{K}(\mathscr{X}) \to C_b(\operatorname{ball}\mathscr{X}^*)$ is an isometry, it follows that $j(A) = s_A \in C(\operatorname{ball}\mathscr{X}^*, w^*)$ for every $A \in \operatorname{co}\mathscr{K}(\mathscr{X})$.

Note. Let j: $\operatorname{co}\mathscr{K}(\mathscr{X}) \to C(\operatorname{ball}\mathscr{X}^*, w^*)$ denote this isometry; clearly, the image of j, i.e. $\operatorname{Im} j = j(\operatorname{co}\mathscr{K}(\mathscr{X}))$ is the closure of $\{s_{\operatorname{co} F}|F$ finite subset of $\{x_n: n \geqslant 1\}\}$ in $C(\operatorname{ball}\mathscr{X}^*, w^*)$.

Actually $\operatorname{Im} j$ is a closed convex cone in $C(\operatorname{ball}\mathscr{X}^*, w^*)$; let $\mathscr{Y} = \operatorname{Sp}(\operatorname{Im} j)$ denote the linear subspace generated by $\operatorname{Im} j$. Clearly $\mathscr{Y} = \operatorname{Im} j - \operatorname{Im} j = \{s_A - s_B|A, B \in \operatorname{co}\mathscr{K}(\mathscr{X})\}$.

In general, $\mathscr{Y}$ is not closed; from the Stone–Weierstrass theorem, it follows that $\mathscr{Y} + \mathbb{R}1 = \{s_A - s_B + c|A,\ B \in \operatorname{co}\mathscr{K}(\mathscr{X}), c \in \mathbb{R}\}$ is dense in $C(\operatorname{ball}\mathscr{X}^*, w^*)$, where $\mathbb{R}1$ denotes all constant functions.

It is not difficult to show that $\mathscr{Y}$ is isometrically isomorphic to the space $\mathscr{U}$ provided by the Rådström embedding theorem 3.1.

4. *Generalization of the Shapley–Folkman–Starr theorem.* Here we are concerned with step (*c*) as described in Section 2, so we consider the space $\mathscr{K}(\mathscr{X})$. How 'non-convex' are the sets in $\mathscr{K}(\mathscr{X})$? One possible measure of nonconvexity is $d(A, \overline{\operatorname{co}} A)$, $A \in \mathscr{K}(\mathscr{X})$. Another such measure is the *inner radius*, as defined by Starr [**28**]: if $A \in \mathscr{K}(\mathscr{X})$, then the inner radius $r(A)$ is given by

$$r(A) = \sup_{a \in \overline{\operatorname{co}} A} \inf\{R \geqslant 0|\exists a_1, \ldots, a_s \in A, a \in \operatorname{co}\{a_1, \ldots, a_s\} \text{ and } \|a - a_i\| \leqslant R\}.$$

Obviously, if A is already convex, then $d(A, \overline{\operatorname{co}} A) = r(A) = 0$.

If $\dim \mathscr{X} < \infty$, the Shapley–Folkman theorem (see Arrow and Hahn [**1**]) can be used to show that addition of sets is, in some sense, 'convexifying': for any

$$A_1, A_2, \ldots, A_n \in \mathscr{K}(\mathscr{X})$$

we have $$d(A_1 + \ldots + A_n, \overline{\operatorname{co}}(A_1 + \ldots + A_n)) \leqslant (\dim \mathscr{X})^{\frac{1}{2}} \max_{1 \leqslant i \leqslant n} \|A_i\|.$$

Obviously, this result cannot be generalized as such to infinite-dimensional spaces. However, an inequality involving the inner radius was obtained by Cassels [**6**] in the case when $\mathscr{X}$ is a Hilbert space. We extend here this result to Banach spaces of type p, $p > 1$.

A Banach space $\mathscr{X}$ is of *type p* if there exists a constant $K > 0$ such that

$$E\left\|\sum_{i=1}^{n} f_i\right\|^p \leqslant K \sum_{i=1}^{n} E\|f_i\|^p$$

for every $f_1, f_2, \ldots, f_n$ independent $\mathscr{X}$-valued random elements with mean 0.

Every Hilbert space is of type 2, the spaces L^p with $1 < p < \infty$ are of type $\min(p, 2)$, while $C[0, 1]$ and L^∞ are *not* of type p for any $p > 1$ (they are of type 1).

Our generalization of the Shapley–Folkman–Starr theorem is as follows:

THEOREM 4·1. *Let $\mathscr{X}$ be a Banach space of type p, $p > 1$. There exists a constant $K > 0$ such that*

$$d(A_1+A_2+\ldots+A_n, \overline{\mathrm{co}}\,(A_1+A_2+\ldots+A_n)) \leqslant K^{1/p}[r^p(A_1)+\ldots+r^p(A_n)]^{1/p}$$

for every $A_1, A_2, \ldots, A_n \in \mathscr{K}(\mathscr{X})$.

Proof. Adapting the ideas of Cassels [**6**], we consider simple functions $f = \Sigma_{i=1}^{s} x_i \chi_{A_i}$, where $x_i \in \mathscr{X}$ and $A_i \in \mathscr{A}$. Denote $V(f) = E\|f-Ef\|^p$ where $Ef = \Sigma_{i=1}^{s} x_i P(A_i)$.

Since $\mathscr{X}$ is of type p, there exists $K > 0$, such that

(*a*) $V(f_1+\ldots+f_n) \leqslant K[V(f_1)+\ldots+V(f_n)]$

whenever f_i $(1 \leqslant i \leqslant n)$ are independent. Now define $V(A)$, for $A \in \mathscr{K}(\mathscr{X})$, by

$$V(A) = \sup_{a \in \overline{\mathrm{co}}\, A} \inf_{\substack{f \in A \\ Ef = a}} V(f),$$

where $f \in A$ means that $f = \Sigma_i x_i \chi_{A_i}$ $(x_i \in A)$. Using (*a*), one can show that

(*b*) $V(A_1+\ldots+A_n) \leqslant K[V(A_1)+\ \ldots\ +(A_n)]$

for any A_i $(1 \leqslant i \leqslant n)$ in $\mathscr{K}(\mathscr{X})$. The proof follows, using (*b*) and the inequalities

$$d(A, \overline{\mathrm{co}}\, A) \leqslant [V(A)]^{1/p} \leqslant r(A), \quad A \in \mathscr{K}(\mathscr{X}).$$

5. *Strong law of large numbers* (*SLLN*). We will consider a sequence $\{X_n\}_n$ of random compact sets, which are independent and identically distributed (i.i.d.). These concepts make sense in view of the metric structure of $\mathscr{K}(\mathscr{X})$ (see Billingsley [**4**]).

Puri and Ralescu [**22**] proved the SLLN for random compact convex sets by using the Rådström embedding (Theorem 3·1). Giné, Hahn and Zinn [**12**] showed that the SLLN can be proved by using Theorem 3·2 instead. For related work, see Hess [**14**]. (We would like to thank Richard Vitale for bringing this reference to our attention.)

Next is our SLLN for random compact sets.

THEOREM 5·1. *Let $\mathscr{X}$ be a Banach space of type $p(>1)$, and let $\{X_n\}_n$ be i.i.d. random compact sets in $\mathscr{X}$ with $E\|X_1\| < \infty$. Then*

$$(X_1+X_2+\ldots+X_n)/n \to EX_1 \text{ a.s.}$$

the convergence being in the Hausdorff metric.

Remark. That this theorem holds for any Banach space has recently been established independently by Artstein and Hanson in unpublished reports. Their methods are different from ours.

Proof. We will perform steps (*a*), (*b*), (*c*) as described in Section 2.

(1) Assume first the $\{X_n\}_n$ are i.i.d. random compact convex sets, i.e.

$$X_n \colon \Omega \to \mathrm{co}\,\mathscr{K}(\mathscr{X}).$$

From Puri and Ralescu [**22**] it follows that $(X_1+\ldots+X_n)/n \to EX_1$ a.s.

(2) Consider now $X_n \colon \Omega \to \mathscr{K}(\mathscr{X})$; obviously

$$d\left(\frac{1}{n}\sum_{i=1}^{n} X_i, E(\overline{\mathrm{co}}\, X_1)\right) \leqslant d\left(\frac{1}{n}\sum_{i=1}^{n} X_i, \frac{1}{n}\sum_{i=1}^{n} \overline{\mathrm{co}}\, X_i\right) + d\left(\frac{1}{n}\sum_{i=1}^{n} \overline{\mathrm{co}}\, X_i, E(\overline{\mathrm{co}}\, X_1)\right).$$

The second term on the right-hand side tends to 0 by (1); for the first term we apply Theorem 4·1:

$$d\left(\frac{1}{n}\sum_{i=1}^{n} X_i, \frac{1}{n}\sum_{i=1}^{n} \overline{\mathrm{co}}\, X_i\right) \leqslant \frac{K^{1/p}}{n}\left[\sum_{n=1}^{n} r^p(X_i)\right]^{1/p}.$$

Now $r(X_1) \leqslant 2\|X_1\|$ so $Er(X_1) < \infty$; this implies $(1/n^p)\sum_{i=1}^n r^p(X_i) \to 0$ a.s. by the Marcinkiewicz–Zygmund SLLN for random variables (see, for example, Chung [**7**]). Finally, $d((1/n)\sum_{i=1}^n X_i, E(\operatorname{co} X_1)) \to 0$ a.s. and, since $E(\operatorname{co} X_1) = EX_1$, this concludes the proof.

Remark. Theorem 5·1 can be applied whenever $\mathscr{X}$ is a Hilbert space, or $\mathscr{X} = L^p$ with $1 < p < \infty$.

6. *Central Limit Theorem* (*CLT*). The CLT was given by Cressie [**9**], in the particular case of random sets assuming only finitely many values. Independently, Lyashenko [**19**] gave the CLT in the general case of $\mathscr{K}(\mathbb{R}^p)$, even for nonidentically distributed random sets. However, his hint of proof only seems to provide a CLT in a metric different from the Hausdorff metric.

By using the methods described in Section 2, Weil [**29**] proved the CLT for random compact sets in $\mathbb{R}^p$.

The infinite-dimensional case is more difficult. The first to prove such a CLT were Giné, Hahn and Zinn [**12**], who showed in their interesting paper that, under some conditions, the CLT holds for compact convex i.i.d. random sets $\{x_i; i \geqslant 1\}$ with $X_i \subseteq \bigcup_{n=1}^{\infty} nK$ a.s. (where K is a *fixed* compact, convex, symmetric set).

We shall generalize this CLT, in the following sense: the values of our random sets are not necessarily included in some multiple of a fixed compact set K.

Let $\mathbb{R}^+ = [0, \infty)$ and let us consider a function $\phi\colon \mathbb{R}^+ \to \mathbb{R}^+$ which is non-decreasing, continuous, sublinear (i.e. $\phi(r+s) \leqslant \phi(r) + \phi(s)$), and satisfies $\phi(0) = 0$. Let also $K \subseteq \mathscr{X}$ be a fixed compact, convex, symmetric set. For any set $M \subseteq \mathbb{R}$, we denote by $|M| = \{|m| \mid m \in M\} \subseteq \mathbb{R}^+$.

For $A \in \operatorname{co}\mathscr{K}(\mathscr{X})$, we denote by

$$\|A\|_{\phi, K} = \inf\{\alpha \geqslant 0 \mid |z^*(A)| \subseteq \alpha\phi(|z^*(K)|), \forall \|z^*\| \leqslant 1, Z^* \in X^*\}$$

and, obviously, we are interested in sets A such that $\|A\|_{\phi, K} < \infty$.

Note that a semimetric can be defined on ball $\mathscr{X}^*$ by the formula:

$$d_{\phi, K}(x^*, y^*) = \phi(\sup_{u \in K} |\langle x^* - y^*, u\rangle|).$$

If $\bar{d}$ is any semimetric on ball $\mathscr{X}^*$ such that (ball $\mathscr{X}^*, \bar{d}$) is compact, let $N(\text{ball } \mathscr{X}^*, \bar{d}, \epsilon)$ denote the minimal number of balls of radius $\leqslant \epsilon$, with centres in ball $\mathscr{X}^*$, which cover ball $\mathscr{X}^*$.

Let $H(\text{ball } \mathscr{X}^*, \bar{d}, \epsilon) = \log N(\text{ball } \mathscr{X}^*, \bar{d}, \epsilon)$ denote the *metric entropy* of ball $\mathscr{X}^*$ with respect to $\bar{d}$.

We can now state our CLT for random sets.

Theorem 6·1. *Let $\{X_n\}_n$ be i.i.d. random compact convex subsets of $\mathscr{X}$ such that the following hypotheses are satisfied:*

(1) $E\|X_1\|^2_{\phi, K} < \infty$,

(2) $\int_0^1 H^{\frac{1}{2}}(\text{ball } \mathscr{X}^*, d_{\phi, K}, \epsilon)\, d\epsilon < \infty$.

Then $\sqrt{n}\, d((1/n)\sum_{i=1}^n X_i, EX_1)$ converges in distribution to $\|Z\|$, where Z is a Gaussian random element of $C(\text{ball } \mathscr{X}^*, w^*)$.

Proof. First embed co $\mathscr{K}(\mathscr{X})$ into C (ball $\mathscr{X}^*, w^*$) as discussed in Section 3. We need to apply the Jain and Marcus CLT [**16**]. To this end we must first check a Lipschitz condition for s_{X_1} (the support function, viewed as a random element of C(ball $\mathscr{X}^*, w^*$)). If $A \in \text{co}\,\mathscr{K}(\mathscr{X})$, it is easy to check that

$$|s_A(x^*) - s_A(y^*)| \leqslant \sup_{a \in A} |\langle x^* - y^*, a\rangle|.$$

Assume $\|A\|_{\phi, K} < \infty$, so we obtain

$$|s_A(x^*) - s_A(y^*)| \leqslant \alpha\phi(\sup_{u \in K} |\langle x^* - y^*, u\rangle|)$$

for some $\alpha \geqslant 0$. Therefore

$$|s_A(x^*) - s_A(y^*)| \leqslant \|A\|_{\phi, K}\, d_{\phi, K}(x^*, y^*).$$

Obviously hypothesis (1) implies that $\|X_1\|_{\phi, K} < \infty$ a.s., so we get

$$|s_{X_1}(x^*) - s_{X_1}(y^*)| \leqslant \|X_1\|_{\phi, K}\, d_{\phi, K}(x^*, y^*)$$

and, again, (1) implies that the Lipschitz condition is satisfied.

The metric entropy condition required here is actually (2) since, as can easily be seen, $d_{\phi, K}$ is continuous in the weak*-topology. Therefore

$$\sqrt{n}\left[\frac{1}{n}\sum_{i=1}^{n} s_{X_i} = E(s_{X_1})\right] \to Z \quad \text{in distribution,}$$

where Z is a Gaussian element of C(ball $\mathscr{X}^*, w^*$).

Now the fact that our embedding is an isometry, together with the continuous mapping theorem, enables us to conclude that

$$\sqrt{n}\, d\left(\frac{1}{n}\sum_{i=1}^{n} X_i, EX_1\right) \to \|Z\|$$

in distribution. The proof follows.

Remark. If, for a fixed compact set K, it happens that $A \subseteq \alpha K$, for some $\alpha \geqslant 0$, then $|z^*(A)| \subseteq \alpha|z^*(K)|$ for $\|z^*\| \leqslant 1$. If we take $\phi(r) = r$ (the identity), then we obtain the CLT of Giné, Hahn and Zinn [**12**].

Example. Let $A \in \text{co}\,\mathscr{K}(\mathscr{X})$ and assume that there is a constant $0 < \lambda \leqslant 1$, such that $\|A\|_{\lambda, K} = \inf\{\alpha \geqslant 0 \,|\, |z^*(A)| \subseteq \alpha|z^*(K)|^\lambda$ for every $\|z^*\| \leqslant 1\} < \infty$. Here we denoted by $M^\lambda = \{m^\lambda | m \in M\}$ for every $M \subseteq \mathbb{R}^+$.

It follows from Theorem 6.1 that if $\{X_n\}_n$ are i.i.d. random sets with $E\|X_1\|^2_{\lambda, K} < \infty$ and $\int_0^1 H^{\frac{1}{2}}(\text{ball}\,\mathscr{X}^*, d_{\lambda, K}, \epsilon)\, d\epsilon < \infty$, then

$$\sqrt{n}\, d\left(\frac{1}{n}\sum_{i=1}^{n} X_i, EX_1\right) \to \|Z\| \quad \text{in distribution.}$$

This research was supported in part by the National Science Foundation Grants IST-7918468 and MCS-8301409.

REFERENCES

[1] K. J. Arrow and F. H. Hahn. *General Competitive Analysis*. Holden-Day (1971).

[2] Z. Artstein and R. A. Vitale. A strong law of large numbers for random compact sets. *Ann. Prob.* **3** (1975), 879–882.

[3] R. J. Aumann. Integrals of set-valued functions. *J. Math. Analysis Applic.* **12** (1965), 1–12.

[4] P. Billingsley. *Convergence of Probability Measures*. Wiley (1968).

[5] C. L. Byrne. Remarks on the set-valued integrals of Debreu and Aumann. *J. Math. Anal. Appl.* **62** (1978), 243–246.

[6] J. W. S. Cassels. Measures of the non-convexity of sets and the Shapley–Folkman–Starr theorem. *Math. Proc. Cambridge Philos. Soc.* **78** (1975), 433–436.

[7] K. L. Chung. *A Course in Probability Theory*. Academic Press (1974).

[8] N. Cressie. A strong limit theorem for random sets. *Suppl. Adv. Appl. Prob.* **10** (1978), 36–46.

[9] N. Cressie. A central limit theorem for random sets. *Z. Wahrsch. Verw. Gebiete* **49** (1979), 37–47.

[10] G. Debreu. Integration of correspondences. *Proc. Fifth Berkeley Symp. Math. Statist. Probability* **2**, 351–372. University of California Press (1966).

[11] N. Dunford and J. T. Schwartz. *Linear Operators*, part I. Interscience (1958).

[12] E. Giné, M. G. Hahn and J. Zinn. Limit theorems for random sets: an application of probability in Banach space results. *Proc. Fourth Int. Conf. on Prob. in Banach Space, Oberwolfach*, Springer Lecture Notes in Math. **990** (1983), 112–135.

[13] H. Hermes. Calculus of set-valued functions and control. *J. Math. Mech.* **18** (1968), 47–59.

[14] C. Hess.Théorème ergodique et loi forte des grands nombres pour des ensembles aléatoires. *C. R. Acad. Sci. Paris, Sér. A* **288** (1979), 519–522.

[15] L. Hörmander. Sur la fonction d'appui des ensembles convexes dans un espace localement convexe. *Ark. Mat.* **3** (1954), 181–186.

[16] N. C. Jain and M. B. Marcus. Central limit theorem for $C(S)$-valued random variables. *J. Funct. Anal.* **19** (1975), 216–231.

[17] D. G. Kendall. Foundations of a theory of random sets. *In Stochastic Geometry* (ed. E. F. Harding and D. G. Kendall). Wiley (1974).

[18] M. G. Kendall and P. A. P. Moran. *Geometrical Probability*. Griffin (1963).

[19] N. N. Lyashenko. On limit theorems for sums of independent compact random subsets in the Euclidean space. *Zap. Nauchn. Sem. Leningrad. Otdel. Mat. Inst. Steklov* **85** (1979), 113–128 (in Russian).

[20] S. Mase. Random convex sets which are infinitely divisible with respect to Minkowski addition. *Adv. Appl. Prob.* **11** (1979), 834–850.

[21] G. Matheron. *Random Sets and Integral Geometry*. Wiley (1975).

[22] M. L. Puri and D. A. Ralescu. Strong law of large numbers for Banach space valued random sets. *Ann. Probab.* **11** (1983), 222–224.

[23] H. Rådström. An embedding theorem for spaces of convex sets. *Proc. Amer. Math. Soc.* **3** (1952), 165–169.

[24] B. D. Ripley. The foundations of stochastic geometry. *Ann. Probab.* **4** (1976), 995–998.

[25] H. E. Robbins. On the measure of a random set. *Ann. Math. Statist.* **14** (1944), 70–74.

[26] H. E. Robbins. On the measure of a random set, II. *Ann. Math. Statist.* **15** (1945), 342–347.

[27] W. Rudin. *Functional Analysis*. McGraw-Hill (1973).

[28] R. Starr. Quasi-equilibria in markets with nonconvex preferences. *Econometrica* **37** (1969), 25–38.

[29] W. Weil. An application of the central limit theorem for Banach space-valued random variables to the theory of random sets. *Z. Wahrsch. Verw. Gebiete* **60** (1982), 203–208.

**Reprinted by permission of the publisher,
© 1985 by Cambridge University Press**

FUZZY RANDOM VARIABLES*

MADAN L. PURI and DAN A. RALESCU

Department of Mathematics, Indiana University, Bloomington, Indiana 47405
Department of Mathematics, University of Cincinnati, Cincinnati, Ohio 45221

In this paper we define the concepts of fuzzy random variable and the expectation of a fuzzy random variable. The new definition of expectation generalizes the integral of a set-valued function. We derive some properties of these new concepts. By considering a suitable generalization of the Hausdorff metric, we derive the Lebesgue-dominated convergence type theorem. © 1986 Academic Press, Inc.

1. Introduction

In practice we are often faced with random experiments whose outcomes are not numbers (or vectors in $\mathbb{R}^n$) but are expressed in inexact linguistic terms. As an example, consider a group of individuals chosen at random who are questioned about the weather on a particular city on a particular winter day. Some possible answers would be "cold," "more or less cold," "very cold," "extremely cold," and so on. A natural question which arises with reference to this example is: What is the average opinion about the weather in that particular city?

A possible way of handling situations like this is by using the concepts of fuzzy sets and fuzzy functions [24] found useful in many applications, notably in pattern recognition, clustering, information retrieval, and systems analysis (cf. [16]).

Motivated by examples of the type given above and related problems (especially concerning group opinions), we introduce fuzzy random variables and their expectations, and we investigate some of their properties.

* This research was supported by the National Science Foundation Grant IST-7918468 and MCS-8301409.

Fuzzy random variables (or fuzzy variables) generalize random variables and random vectors; they also generalize random sets [15]. The expected value of a fuzzy variable is a natural generalization of the integral of a set-valued function [2].

Kwakernaak [14] introduced the notion of a fuzzy random variable as a function $F: \Omega \to \bar{\mathscr{F}}(\mathbb{R})$ (subject to certain measurability conditions), where $(\Omega, \mathscr{A}, P)$ is a probability space, and $\bar{\mathscr{F}}(\mathbb{R})$ denotes all piecewise continuous functions $u: \mathbb{R} \to [0, 1]$.

Féron [8] defined a fuzzy random set as a measurable function $F: \Omega \to \mathscr{F}(\mathscr{X})$, where $\mathscr{X}$ is a topological space, $\mathscr{F}(\mathscr{X}) = \{u: \mathscr{X} \to [0, 1]\}$, and $\{x \in \mathscr{X}: F(\omega)(x) \geqslant \alpha\}$ are closed subsets of $\mathscr{X}$ for each $0 \leqslant \alpha \leqslant 1$, $\omega \in \Omega$. Relationships between fuzzy sets and random sets were studied by Fortet and Kambouzia [9] and by Goodman [10].

However, in the work of the authors mentioned above, no attempt is made to define the expected value of a fuzzy variable and to study its properties. Our definition of the expected value is new and it provides a natural generalization of the set valued function (that is, random set) setting.

In Section 2, we briefly state some results related to the integral calculus for set-valued functions. These results will be frequently referred to in the subsequent sections.

In Section 3, we introduce the notion of a fuzzy random variable slightly different than that of Kwakernaak [14]. We define it as a function (subject to certain measurability requirements) $X: \Omega \to \mathscr{F}_0(\mathbb{R}^n)$, where $(\Omega, \mathscr{A}, P)$ is a probability space, and $\mathscr{F}_0(\mathbb{R}^n)$ denotes all functions (fuzzy subsets of $\mathbb{R}^n$) $u: \mathbb{R}^n \to [0, 1]$ such that $\{x \in \mathbb{R}^n: u(x) \geqslant \alpha\}$ is nonempty and compact for each $0 < \alpha \leqslant 1$. In this setting we define the expected value $E(X)$ of a fuzzy variable X.

In Section 4 we study some properties of this expected value. To this end, we first define a metric in $\mathscr{F}_0(\mathbb{R}^n)$ which generalizes the Hausdorff metric in the space of compact subsets of $\mathbb{R}^n$. We show that, under certain conditions $E(X)$ is a fuzzy convex set. The main result of this section is a Lebesgue-dominated convergence type theorem. This generalizes the corresponding results of Aumann [2] and Debreu [6]. Some of the results of this section can possibly be extended to Banach space valued fuzzy variables, that is, functions $X: \Omega \to \mathscr{F}_1(\mathscr{Y})$, where $(\Omega, \mathscr{A}, P)$ is a probability space, $\mathscr{Y}$ is a Banach space, and $\mathscr{F}_1(\mathscr{Y})$ denotes all functions (fuzzy sets) $u: \mathscr{Y} \to [0, 1]$ whose levels $\{y \in \mathscr{Y}: u(y) \geqslant \alpha\}$, $\alpha \neq 0$ are closed, bounded, and convex subsets of $\mathscr{Y}$. However, this generalization will not be carried out in this paper.

Finally, in Section 5 we state some problems of further research interest.

The concept of an integral of a set-valued function was first introduced by Kudō [12] in connection with the theory of experiments in statistics. Later, using Kudō [12] and Richter [23], this concept of an integral was extended by Aumann [2] who proved some important convergence properties. Debreu [6] defined another concept of an integral of a set-valued function in a more general context, studied its properties, and showed that under suitable hypotheses, it coincided with the integral of Aumann [2].

Applications of these integrals are found in economics [3], control theory [11], and probability theory [1, 5, 20].

We define two different types of convergence for sequences of sets. Let A and B be two nonempty bounded subsets of $\mathbb{R}^n$. The distance between A and B is defined by the *Hausdorff metric*,

$$d_H(A, B) = \max[\sup_{a \in A} \inf_{b \in B} \|a-b\|, \sup_{b \in B} \inf_{a \in A} \|a-b\|], \tag{2.1}$$

where $\|\cdot\|$ denotes the usual euclidean norm in $\mathbb{R}^n$.

We denote the *Hausdorff semimetric* by $\rho(A, B) = \sup_{a \in A} \inf_{b \in B} \|a-b\|$. It is clear that

$$\rho(A, B) = 0 \Leftrightarrow A \subset \bar{B} \tag{2.2}$$

and

$$\rho(A, C) \leqslant \rho(A, B) + \rho(B, C), \tag{2.3}$$

where A, B, C are nonempty bounded subsets of $\mathbb{R}^n$, and $\bar{B}$ denotes the closure of B.

Also $d_H(A, B) = \max[\rho(A, B), \rho(B, A)]$ and

$$d_H(A, B) = 0 \Leftrightarrow \bar{A} = \bar{B}. \tag{2.4}$$

If $Q(\mathbb{R}^n)$ denotes the set of all nonempty, compact subsets of $\mathbb{R}^n$, it is clear that $(Q(\mathbb{R}^n), d_H)$ becomes a metric space. The following theorem gives a more precise result.

THEOREM 2.1. *The metric space $(Q(\mathbb{R}^n), d_H)$ is complete and separable.*

Another type of convergence for a sequence of sets was defined by Kuratowski [13].

We say that a sequence of sets $\{C_k\}_k$, $C_k \subset \mathbb{R}^n$, *converges* to a set $C \subset \mathbb{R}^n$, denoted by $C = \lim_k C_k$, if

$$C = \liminf C_k = \limsup C_k, \tag{2.5}$$

where

$$\liminf C_k = \{x \in \mathbb{R}^n \colon x = \lim_{k \to \infty} x_k,\ x_k \in C_k\}, \tag{2.6}$$

$$\limsup C_k = \bigcap_{k=1}^{\infty} \left(\overline{\bigcup_{m=k}^{\infty} C_m} \right). \tag{2.7}$$

We mention that for sequences of closed sets, convergence in the Hausdorff metric implies convergence in the sense of Kuratowski. On $Q(\mathbb{R}^n)$, both types of convergence are equivalent provided the limit set is nonempty, that is, the sequence is bounded (cf. [15]).

Now, let $(\Omega, \mathscr{A}, P)$ be a probability space where the probability measure P is assumed to be *nonatomic*.

A *set-valued function* is a function $F\colon \Omega \to \mathscr{P}(\mathbb{R}^n)$ such that $F(\omega) \neq \varnothing$ for every $\omega \in \Omega$. By $L^1(P)$ (or by $L^1(P, \mathbb{R}^n)$) we denote the space of P-integrable functions $f\colon \Omega \to \mathbb{R}^n$. We denote by $S(F)$ the set of all $L^1(P)$ selections of F, that is,

$$S(F) = \{ f \in L^1(P) \colon f(\omega) \in F(\omega) \text{ a.e.} \} \tag{2.8}$$

The *Aumann integral* of F is defined by

$$(A) \int F = \left\{ \int_{\Omega} f\, dP \colon f \in S(F) \right\} \tag{2.9}$$

For easy reference, we state the following results due to Richter [23], Aumann [2], and Debreu [6].

THEOREM 2.2. *If $F\colon \Omega \to \mathscr{P}(\mathbb{R}^n)$ is a set-valued function, then $(A)\int F$ is a convex subset of $\mathbb{R}^n$.*

Observe, however, that $(A)\int F$ may be empty in general.

A function $F\colon \Omega \to \mathscr{P}(\mathbb{R}^n)$ is called *measurable* if its graph $\{(\omega, x)\colon x \in F(\omega)\}$ belongs to $\mathscr{A} \times \mathscr{B}$ (where $\mathscr{B}$ denotes the Borel subsets of $\mathbb{R}^n$). A function $F\colon \Omega \to \mathscr{P}(\mathbb{R}^n)$ is called *integrably bounded* if there exists a function $h\colon \Omega \to \mathbb{R}$, $h \in L^1(P, \mathbb{R})$ such that $\|x\| \leqslant h(\omega)$ for all x, ω with $x \in F(\omega)$.

THEOREM 2.3 [2]. *If $F\colon \Omega \to \mathscr{P}(\mathbb{R}^n)$ is measurable and integrably bounded, then $(A)\int F \neq \varnothing$.*

Another result about the structure of $(A)\int F$ is given by

THEOREM 2.4 [2]. *If $F\colon \Omega \to \mathscr{P}(\mathbb{R}^n)$, $F(\omega)$ is closed for every $\omega \in \Omega$, and if F is integrably bounded, then $(A)\int F$ is a compact subset of $\mathbb{R}^n$.*

The following theorem is a generalization of the Lebesgue dominated convergence theorem.

THEOREM 2.5 [2]. *If* $F_k\colon \Omega \to \mathscr{P}(\mathbb{R}^n)$ *are measurable and if there exists* $h \in L^1(P, \mathbb{R})$ *such that* $\sup_{k \geqslant 1} \|f_k(\omega)\| \leqslant h(\omega)$ *for every* $f_k \in S(F_k)$, *and if* $F_k(\omega) \to F(\omega)$ (*in the sense of Kuratowski*), *then* $(A)\int F_k \to (A)\int F$.

In Debreu [6], a concept of an integral is defined for more general set-valued functions $F\colon \Omega \to K(\mathscr{X})$, where $\mathscr{X}$ is a Banach space and $K(\mathscr{X})$ denotes all nonempty compact convex subsets of $\mathscr{X}$. The key result used in defining the Debreu integral is an embedding theorem for $K(\mathscr{X})$ due to Rådström (1952).

Without going through the details of this construction here, we may mention that the Aumann integral can also be defined in this more general setting and it is possible to prove the equivalence of the Aumann and Debreu integrals without assuming that $\mathscr{X}$ is reflexive (see Byrne [4]).

Remark. It is important to observe that Theorem 2.5 can be stated in a different form by replacing convergence in the sense of Kuratowski by convergence in the Hausdorff metric. The statement of the theorem remains unchanged provided we assume that all functions take values in $Q(\mathbb{R}^n)$. We shall use this version of Theorem 2.5 in Section 4, where we shall generalize the Lebesgue dominated convergence theorem to fuzzy variables.

3. FUZZY VARIABLES AND THEIR EXPECTATIONS

Let $(\Omega, \mathscr{A}, P)$ be a probability space where P is a probability measure. Let $\mathscr{F}_0(\mathbb{R}^n)$ denote the set of fuzzy subsets $u\colon \mathbb{R}^n \to [0, 1]$ with the following properties:

(a) $\{x \in \mathbb{R}^n\colon u(x) \geqslant \alpha\}$ is compact for each $\alpha > 0$, (3.1)

(b) $\{x \in \mathbb{R}^n\colon u(x) = 1\} \neq \varnothing$. (3.2)

DEFINITION 3.1. A *fuzzy random variable* (or *fuzzy variable*) is a function $X\colon \Omega \to \mathscr{F}_0(\mathbb{R}^n)$ such that

$$\{(\omega, x)\colon x \in X_\alpha(\omega)\} \in \mathscr{A} \times \mathscr{B} \tag{3.3}$$

for every $\alpha \in [0, 1]$, where $X_\alpha\colon \Omega \to \mathscr{P}(\mathbb{R}^n)$ is defined by

$$X_\alpha(\omega) = \{x \in \mathbb{R}^n\colon X(\omega)(x) \geqslant \alpha\}. \tag{3.4}$$

A fuzzy variable X is called *integrably bounded* if X_α is integrably bounded for all $\alpha \in (0, 1]$, i.e., if for any $\alpha \in (0, 1]$, there exists $h_\alpha \in L^1(\Omega)$ such that $\|x\| \leqslant h_\alpha(\omega)$ for each x, ω with $x \in X_\alpha(\omega)$.

Here $L^1(\Omega)$ denotes all functions $h: \Omega \to \mathbb{R}$ which are integrable with respect to the probability measure P.

Motivated by examples similar to the one given in the introduction, we define the expected value $E(X)$ of a fuzzy variable $X: \Omega \to \mathscr{F}_0(\mathbb{R}^n)$ in such a way that the following conditions are satisfied:

$$E(X) \in \mathscr{F}_0(\mathbb{R}^n), \tag{3.5}$$

$$\{x \in \mathbb{R}^n: (E(X))(x) \geqslant \alpha\} = \int X_\alpha \text{ for each } \alpha \in [0, 1]. \tag{3.6}$$

The next theorem shows that under certain assumptions, there is a unique fuzzy set satisfying these requirements. The proof is based on the following lemma.

LEMMA 3.1. *Let M be a set and let $\{M_\alpha: \alpha \in [0, 1]\}$ be a family of subsets of M such that*

(i) $M_0 = M$,

(ii) $\alpha \leqslant \beta$ *implies* $M_\beta \subseteq M_\alpha$,

(iii) $\alpha_1 \leqslant \alpha_2 \leqslant \cdots$, $\lim_{n \to \infty} \alpha_n = \alpha$ *implies* $M_\alpha = \bigcap_{n=1}^{\infty} M_{\alpha_n}$.

Then, the function $\phi: M \to [0, 1]$ defined by $\phi(x) = \sup\{\alpha \in [0, 1]: x \in M_\alpha\}$ has the property that $\{x \in M: \phi(x) \geqslant \alpha\} = M_\alpha$ for every $\alpha \in [0, 1]$.

Proof. See [16].

THEOREM 3.1. *If $X: \Omega \to \mathscr{F}_0(\mathbb{R}^n)$ is an integrably bounded fuzzy variable, there exists a unique fuzzy set $v \in \mathscr{F}_0(\mathbb{R}^n)$ such that*

$$\{x \in \mathbb{R}^n: v(x) \geqslant \alpha\} = \int X_\alpha \textit{ for every } \alpha \in [0, 1] \tag{3.7}$$

Proof. Let $M_\alpha = \int X_\alpha$, $\alpha \in [0, 1]$. Since X_α is measurable and integrably bounded, it follows from Theorem 2.3 that $M_\alpha \neq \varnothing$. Since $X_\alpha(\omega) = \{x: X(\omega)(x) \geqslant \alpha\}$ are closed subsets of $\mathbb{R}^n$ for all $\omega \in \Omega$, it follows from Theorem 2.4 that $M_\alpha = \int X_\alpha$ is compact.

Consider now the family $\{M_\alpha: \alpha \in [0, 1]\}$ of subsets of $\mathbb{R}^n$. Note that $X_0(\omega) = \{x: X(\omega)(x) \geqslant 0\} = \mathbb{R}^n$, for all $\omega \in \Omega$. Thus $\int X_0 = \mathbb{R}^n$. If $\alpha \leqslant \beta$, then clearly $X_\alpha(\omega) \supseteq X_\beta(\omega)$ for $\omega \in \Omega$. Thus $M_\alpha \supseteq M_\beta$.

We now apply Lemma 3.1. To do so we have to check that $\alpha_1 \leqslant \alpha_2 \leqslant \cdots$, $\lim_{n \to \infty} \alpha_n = \alpha$, $\alpha \neq 0$ implies $M_\alpha = \bigcap_{n=1}^{\infty} M_{\alpha_n}$. Observe that $\bigcap_{n=1}^{\infty} X_{\alpha_n}(\omega) =$

$X_{\alpha_0}(\omega)$ for all $\omega \in \Omega$. Since $X_{\alpha_n}(\omega)$ are compact, $n \in \mathbb{N}$, it follows by a simple argument that

$$X_{\alpha_n}(\omega) \to X_{\alpha_0}(\omega), \tag{3.8}$$

where converge is in the Hausdorff metric.

Now $X_{\alpha_1}(\omega) \supseteq X_{\alpha_2}(\omega) \supseteq \cdots$, and since X_{α_1} is integrably bounded there exists $h \in L^1(\Omega)$ such that $\|f(\omega)\| \leqslant h(\omega)$ for every $f \in S(X_{\alpha_1})$. It follows that $\|g(\omega)\| \leqslant h(\omega)$ for every $g \in S(X_{\alpha_n})$, $n \in \mathbb{N}$. Thus $\{X_{\alpha_n}, n \geqslant 1\}$ is bounded by the same integrable function, and since the X_{α_n} are also measurable, it follows from Theorem 2.5 that $\int X_{\alpha_n} \to \int X_{\alpha_0}$ (in the Hausdorff metric). Observe that $\{\int X_{\alpha_n}, n \geqslant 1\}$ is a decreasing sequence of compact sets and so it must converge to its intersection. Thus we obtain

$$\bigcap_{n=1}^{\infty} \int X_{\alpha_n} = \int X_{\alpha_0}. \tag{3.9}$$

By Lemma 3.1, the fuzzy set defined by $v(x) = \sup\{\alpha \in [0, 1]: x \in M_\alpha\}$ satisfies

$$\{x \in \mathbb{R}^n: v(x) \geqslant \alpha\} = M_\alpha = \int X_\alpha, \qquad \alpha \in [0, 1]. \tag{3.10}$$

The uniqueness of v is obvious, since if two fuzzy sets v and w satisfy (3.12), then $\{x: v(x) \geqslant \alpha\} = \{x: w(x) \geqslant \alpha\}$ for every α and this implies $v = w$.

Finally, $v \in \mathscr{F}_0(\mathbb{R}^n)$ since $\{x: v(x) \geqslant \alpha\} = M_\alpha$ is a compact set and $\{x: v(x) = 1\} = M_1 = \int X_1 \neq \varnothing$. This completes the proof.

We use the above theorem to define the expected value of a fuzzy random variable $X: \Omega \to \mathscr{F}_0(\mathbb{R}^n)$ which is integrably bounded.

DEFINITION 3.2. The *expected value* of X, denoted by $E(X)$, is the fuzzy set $v \in \mathscr{F}_0(\mathbb{R}^n)$ such that $\{x \in \mathbb{R}^n: v(x) \geqslant \alpha\} = \int X_\alpha$ for every $\alpha \in [0, 1]$.

Note that the existence and uniqueness of v are established in Theorem 3.1.

Thus

$$(E(X))(x) = \sup\left\{\alpha \in [0, 1]: x \in \int X_\alpha\right\} \tag{3.11}$$

and its level sets are given by

$$\{x: (E(X))(x) \geqslant \alpha\} = \int X_\alpha, \qquad \alpha \in [0, 1] \tag{3.12}$$

4. Properties of the Expectation

Our aim is to extend the Lebesgue dominated convergence theorem to fuzzy random variables. To this end, we first define a metric in $\mathscr{F}_0(\mathbb{R}^n)$ which generalizes the Hausdorff metric.

Let $u, v \in \mathscr{F}_0(\mathbb{R}^n)$, and set

$$d(u, v) = \sup_{\alpha > 0} d_H(L_\alpha(u), L_\alpha(v)), \tag{4.1}$$

where d_H is the Hausdorff metric, and we denote by $L_\alpha(u) = \{x: u(x) \geqslant \alpha\}$, $L_\alpha(v) = \{x: v(x) \geqslant \alpha\}$.

Proposition 4.1. *$(\mathscr{F}_0(\mathbb{R}^n), d)$ is a metric space.*

Proof. (i) $d(u, v) = 0$ implies $d_H(L_\alpha(u), L_\alpha(v)) = 0$ for each $\alpha > 0$, and this implies $L_\alpha(u) = L_\alpha(v)$, $\alpha > 0$, which implies $u = v$. (Recall that $L_\alpha(u)$ and $L_\alpha(v)$ are compact).

(ii) Obviously $d(u, v) = d(v, u)$.

(iii) The triangle inequality $d(u, v) \leqslant d(u, w) + d(w, v)$ follows from the corresponding inequality for d_H.

Note that if A and B are compact subsets of $\mathbb{R}^n$, and χ_A and χ_B are their respective characteristic functions, then $d(\chi_A, \chi_B) = d_H(A, B)$.

The following result generalizes Theorem 2.1.

Theorem 4.1. *The metric space $(\mathscr{F}_0(\mathbb{R}^n), d)$ is complete.*

The proof of this theorem is given in the Appendix.

Theorem 2.2 can easily be extended to fuzzy variables by using the concept of fuzzy convexity [24].

A fuzzy set $u: \mathbb{R}^n \to [0, 1]$ is called a *fuzzy convex set*, if

$$u(\lambda x + (1-\lambda)\, y) \geqslant \min[u(x), u(y)] \quad \text{for ever} \quad x, y \in \mathbb{R}^n, \lambda \in [0, 1]. \tag{4.2}$$

Theorem 4.2. *If the probability measure P is nonatomic, and if $X: \Omega \to \mathscr{F}_0(\mathbb{R}^n)$ is integrably bounded fuzzy variable, then $E(X)$ is a fuzzy convex set.*

Proof. Let $v = E(X)$. Since $\{v \geqslant \alpha\} = \int X_\alpha$, it follows from Theorem 2.2 that $\{v \geqslant \alpha\}$ is a convex set, $\alpha \in [0, 1]$. The proof now follows by noting that (4.2) is equivalent to the convexity of $\{u \geqslant \alpha\}$ for every $\alpha \in [0, 1]$.

The following theorem extends the Lebesgue dominated convergence Theorem 2.5 (see the remark at the end of Section 2). We assume that the probability measure P is nonatomic.

THEOREM 4.3. *Let* $\{X_k, k \geqslant 1\}$ *and* X *be integrably bounded fuzzy variables such that* $X_k(\omega) \to^d X(\omega)$ *for almost every* $\omega \in \Omega$. *Suppose there exists an* $h \in L^1(\Omega)$ *such that* $\sup_{x \in X_{k,\alpha}(\omega)} \|x\| \leqslant h(\omega)$ *for all* $k \geqslant 1$ *and* $\alpha > 0$, *where* $X_{k,\alpha}(\omega) = L_\alpha(X_k(\omega)) = \{X_k(\omega) \geqslant \alpha\}$. *Then* $E(X_k) \to^d E(X)$.

Proof. To prove this theorem, we shall use a technique similar to the one in Debreu [7, pp. 366; 367]). Assume first that X_k, X are such that $L_\alpha(X_k(\omega))$ and $L_\alpha(X(\omega))$ are convex subsets of $\mathbb{R}^n$ for every $\alpha \geqslant 0$.

Then for every $\alpha > 0$,

$$
\begin{aligned}
d_H(L_\alpha(E(X_k)), L_\alpha(E(X))) &= d_H\left(\int L_\alpha(X_k), \int L_\alpha(X)\right) \\
&\leqslant \int d_H(L_\alpha(X_k), L_\alpha(X)) \leqslant \int d(X_k(\omega), X(\omega))\, dP(\omega)
\end{aligned}
\tag{4.3}
$$

implying $d(E(X_k), E(X)) \leqslant \int d(X_k(\omega), X(\omega))\, dP(\omega)$. Now $d(X_k(\omega), X(\omega)) \to 0$, a.e. and also, $d(X_k(\omega), X(\omega)) \leqslant d(X_k(\omega), 0) + d(0, X(\omega))$, where 0 denotes the set $\{0\}$. It follows that

$$
\begin{aligned}
d(X_k(\omega), 0) &= \sup_{\alpha > 0} d_H(L_\alpha(X_k(\omega)), 0) \\
&= \sup_{\alpha > 0} \sup_{x \in X_{k,\alpha}(\omega)} \|x\| \leqslant h(\omega), \ h \in L^1(\Omega).
\end{aligned}
$$

Using now the classical Lebesgue dominated convergence theorem, we obtain $d(E(X_k), E(X)) \to 0$.

Assume now that X_k, X are as in the statement of the theorem. For any subset A of $\mathbb{R}^n$, denote by $\operatorname{Co} A$ its convex hull. For a random set $F: \Omega \to \mathcal{P}(\mathbb{R}^n)$, denote by $(\operatorname{Co} F)(\omega) = \operatorname{Co}(F(\omega))$, $\omega \in \Omega$. Note that, since P is nonatomic, $\int F = \int \operatorname{Co} F$ (provided the integral exists).

Also, if F and G are random sets (not necessarily convex valued), we have $d_H(\int F, \int G) = d_H(\int \operatorname{Co} F, \int \operatorname{Co} G) \leqslant \int d_H(\operatorname{Co} F, \operatorname{Co} G) \leqslant \int d_H(F, G)$, the last inequality following from Price [17]. Now a simple argument similar to the one above (see (4.3)) concludes the proof.

5. COMPUTATION OF $E(X)$

In this section, we provide some examples to compute the expected value of a fuzzy random variable.

EXAMPLE 1. Toss a fair coin. Denote the outcomes "Tail" by T and "Head" by H. Suppose a player loses *approximately* \$10 if the outcome is T, and wins an amount *much larger* than \$100 but *not much larger* than \$1000 if the outcome is H. The fuzzy r.v. then is $X\colon \{T, H\} \to \mathscr{F}_0(\mathbb{R})$, where $X(T)=$ "approximately -10," and $X(H)=$ "much larger than 100 but not much larger than 1000." For (intuitively plausible) technical reasons, let us write $X(T)=u$ and $X(H)=v$, where $u, v\colon \mathbb{R} \to [0, 1]$ are given by, say, $u(x)=[1-(x+10)^2/4]^+$ and $v(x)=[1-(x-630)^2/380^2]^+$, where $f^+ = \max(f, 0)$. Since u and v are continuous with compact support, it is easy to show by using (3.13) that $E(X)(x)=\sup_{y+z=2x}\min\{[1-(y+10)^2/4]^+, [1-(z-630)^2/380^2]^+\}$.

In particular, the support of $E(X)$ is included in the interval $[119, 501]$.

EXAMPLE 2 (Extension of Example 1). Let $X\colon \Omega \to \mathscr{F}_0(\mathbb{R})$ be a fuzzy variable such that $P(X=u_i)=p_i$, $i=1,\dots, r$, where $u_i\colon \mathbb{R} \to [0, 1]$ are continuous with compact support. Then

$$E(X)=\sum_{i=1}^{r} p_i u_i \tag{5.1}$$

The sum of two fuzzy sets is defined as $(u+v)(x)= \sup_{y+z=x}\min[u(y), v(z)]$, $x \in \mathbb{R}$, and the product of a scalar and a fuzzy set is defined as

$$(\lambda u)(x)=\begin{cases} u(\lambda^{-1}x) & \text{if } \lambda \neq 0 \\ 0 & \text{if } \lambda=0,\ x\neq 0 \\ \sup\limits_{y\in\mathbb{R}} u(y) & \text{if } \lambda=0,\ x=0. \end{cases}$$

(These definitions generalize the corresponding operations with sets).

EXAMPLE 3. Consider now a fuzzy r.v. of *discrete* type, i.e., $X\colon \Omega \to \mathscr{F}_0(\mathbb{R})$ such that $P(X=u_i)=p_i$, $i=1, 2,\dots$, where $u_i\colon \mathbb{R} \to [0, 1]$ are continuous and have compact support. The infinite sum of fuzzy sets is defined by $(\sum_{j=1}^{\infty} v_j)(x)=\sup \inf_{j\geqslant 1}[v_j(y_j)]$, where the supremum is taken over all sequence $\{y_1, y_2,\dots\}$ such that $x=\sum_{j=1}^{\infty} y_j$. Then by using Theorem 3.1 and the fact that $v(A)=\int_A F\,dP$ is a set-valued measure for every random set F, (see [7]), it is easy to check (as a generalization of (5.1)) that

$$E(X)=\sum_{i=1}^{\infty} p_i u_i. \tag{5.2}$$

EXAMPLE 4 (Nondiscrete case). Although the theory developed in Sections 3 and 4 applies to fuzzy r.v.'s of general type, the computation of the expected value becomes somewhat complicated in the case of nondiscrete r.v.

Let X denote the diameter of a hole made by a drillpress and, because of the errors of measurement, let the values of X be fuzzy sets rather than numbers. Let $X\colon \Omega \to \mathscr{F}_0(\mathbb{R})$ such that $X(\omega)$ is continuous and has compact support, $\omega \in \Omega$. Then, to compute $E(X)$ we use Example 2 along with Theorem 4.3.

If $s = \sum_{i=1}^{r} u_i \chi_{A_i}$ is a simple function, $u_i \in \mathscr{F}_0(\mathbb{R})$, $A_i \in \mathscr{A}$, then $E(s) = \sum_{i=1}^{r} u_i P(A_i)$. If X is an integrably bounded fuzzy r.v., then $E(X) = \lim_{k \to \infty} E(s_k)$, where s_k is a simple function and $s_k \to X$, (all these limits are in the metric d defined by (4.1)).

Remark. In the above examples, we have made a somewhat restrictive assumption that the fuzzy r.v.'s take on values which have compact support. If this assumption is dropped, then the addition of fuzzy sets defined in Example 2 (and the corresponding formula (5.1)) should be replaced by a different operation defined as follows: $(u+v)(x) = \sup\{\alpha \in [0,1]\colon x \in L_\alpha(u) + L_\alpha(v)\}$, where $u, v \in \mathscr{F}_0(\mathbb{R})$, $L_\alpha(u) = \{x \in \mathbb{R}\colon u(x) \geqslant \alpha\}$, $L_\alpha(v) = \{x \in \mathbb{R}\colon v(x) \geqslant \alpha\}$. For details see Puri and Ralescu [18, 21].

6. Concluding Remarks

Motivated by the study of asymptotic or limiting opinion (see the example in the Introduction), it is desirable to explore different limit theorems for sequences of independent fuzzy random variables. Of particularly great interests would be the theorems which generalize the classical law of large numbers and the central limit theorem. (See Artstein and Vitale [1], Cressie [5], and Puri and Ralescu [20] for generalizations of some of these theorems to random sets.)

7. Appendix

Proof of Theorem 4.1. Let $\{u_n, n \geqslant 1\}$ be a Cauchy sequence in $\mathscr{F}_0(\mathbb{R}^n)$. Consider a fixed $\alpha > 0$. Then $\{L_\alpha(u_n), n \geqslant 1\}$ is a Cauchy sequence in $(Q(\mathbb{R}^n), d_H)$ where $Q(\mathbb{R}^n)$ denotes all nonempty compact subsets of $\mathbb{R}^n$.

Since $(Q(\mathbb{R}^n), d_H)$ is complete (Theorem 2.1), it follows that

$$L_\alpha(u_n) \xrightarrow{d_H} M_\alpha \in Q(\mathbb{R}^n).$$

Actually, from the Definition (4.1) of d and from the continuity of d_H, it is easy to see that $L_\alpha(u_n) \to^{d_H} M_\alpha$, uniformly in $\alpha \in [0, 1]$.

Consider now the family $\{M_\alpha: \alpha \in [0, 1]\}$, where $M_0 = \mathbb{R}^n$. Take $\alpha \leqslant \beta$ and denote by ρ the Hausdorff semimetric in $Q(\mathbb{R}^n)$ (see Sect. 2). We then have

$$\rho(M_\beta, M_\alpha) \leqslant \rho(M_\beta, L_\beta(u_n)) + \rho(L_\beta(u_n), L_\alpha(u_n)) + \rho(L_\alpha(u_n), M_\alpha).$$

Since $L_\beta(u_n) \subseteq L_\alpha(u_n)$, it follows that $\rho(L_\beta(u_n), L_\alpha(u_n)) = 0$. Thus $\rho(M_\beta, M_\alpha) \leqslant \rho(M_\beta, L_\beta(u_n)) + \rho(L_\alpha(u_n), M_\alpha) < \varepsilon$ if n is large enough. Hence $\rho(M_\beta, M_\alpha) = 0$, and since M_β, M_α are closed, we have $M_\beta \subseteq M_\alpha$.

Now take $\alpha > 0$, $\alpha_n \uparrow$, and $\lim_{n\to\infty} \alpha_n = \alpha$. We have to show that $M_\alpha = \bigcap_{n=1}^\infty M_{\alpha_n}$.

It is clear that

$$M_\alpha \subseteq \bigcup_{n=1}^\infty M_{\alpha_n}. \tag{7.1}$$

Using again the Hausdorff semimetric, we get

$$\rho\left(\bigcap_{n=1}^\infty M_{\alpha_n}, M_\alpha\right) \leqslant \rho\left(\bigcap_{n=1}^\infty M_{\alpha_n}, \bigcap_{n=1}^\infty (L_{\alpha_n}(u_j)\right) + \rho\left(\bigcap_{n=1}^\infty L_{\alpha_n}(u_j), L_\alpha(u_j)\right)$$
$$+ \rho(L_\alpha(u_j), M_\alpha) \qquad \text{for fixed } j.$$

But $\rho(\bigcap_{n=1}^\infty L_{\alpha_n}(u_j), L_\alpha(u_j)) = 0$. Consequently, for every $\varepsilon > 0$, there exists j_ε such that

$$\rho\left(\bigcap_{n=1}^\infty M_{\alpha_n}, M_\alpha\right) \leqslant \varepsilon + \rho\left(\bigcap_{n=1}^\infty M_{\alpha_n}, \bigcap_{n=1}^\infty L_{\alpha_n}(u_j)\right)$$

for $j \geqslant j_\varepsilon$, since $L_\alpha(u_j) \to M_\alpha$.

Now

$$\rho\left(\bigcap_{n=1}^\infty M_{\alpha_n}, \bigcap_{n=1}^\infty L_{\alpha_n}(u_j)\right) \leqslant \rho\left(\bigcap_{n=1}^\infty M_{\alpha_n}, M_{\alpha_p}\right) + \rho(M_{\alpha_p}, L_{\alpha_p}(u_j))$$
$$+ \rho\left(L_{\alpha_p}(u_j), \bigcap_{n=1}^\infty L_{\alpha_n}(u_j)\right) \qquad \text{for any } p \geqslant 1.$$

Since $\bigcap_{n=1}^\infty M_{\alpha_n} \subseteq M_{\alpha_p}$, we obtain

$$\rho\left(\bigcap_{n=1}^\infty M_{\alpha_n}, \bigcap_{n=1}^\infty L_{\alpha_n}(u_j)\right) \leqslant \rho(M_{\alpha_p}, L_{\alpha_p}(u_j)) + \rho\left(L_{\alpha_p}(u_j), \bigcap_{n=1}^\infty L_{\alpha_n}(u_j)\right).$$

Now $\rho(M_{\alpha_p}, L_{\alpha_p}(u_j)) < \varepsilon$ for $j \geqslant j_0$. Note that (since the convergence $L_\alpha(u_j) \to M_\alpha$ is uniform in α) j_0 does not depend on p. Since $\{L_{\alpha_p}(u_j), p \geqslant 1\}$ decreases to $\bigcap_{n=1}^\infty L_{\alpha_n}(u_j)$, if follows that $\rho(L_{\alpha_{p_0}}(u_j), \bigcap_{n=1}^\infty L_{\alpha_n}(u_j)) < \varepsilon$ for some p_0 (depending on j). Thus $\rho(\bigcap_{n=1}^\infty M_{\alpha_n}, \bigcap_{n=1}^\infty L_{\alpha_n}(u_j)) < 2\varepsilon$, if j is large.

Finally, by taking j large enough, we obtain

$$\rho\left(\bigcap_{n=1}^\infty M_{\alpha_n}, M_\alpha\right) \leqslant 3\varepsilon, \qquad \text{i.e., } \bigcap_{n=1}^\infty M_{\alpha_n} \subseteq M_\alpha. \tag{7.2}$$

Equations (7.1) and (7.2) yield $\bigcap_{n=1}^\infty M_{\alpha_n} = M_\alpha$. Thus Lemma 3.1 is applicable and there exists $u \in \mathscr{F}_0(\mathbb{R}^n)$ with $L_\alpha(u) = M_\alpha$ for every $\alpha \in [0, 1]$. It follows that $L_\alpha(u_n) \to^{d_H} L_\alpha(u)$. It remains to show that $u_n \to u$ in $(\mathscr{F}_0(\mathbb{R}^n), d)$.

Let $\varepsilon > 0$. Then, since $\{u_n\}$ is Cauchy, there exists n_ε such that $n, m > n_\varepsilon$ implies $d(u_n, u_m) < \varepsilon$.

Let $n (> n_\varepsilon)$ be fixed. Then

$$\begin{aligned} d_H(L_\alpha(u_n), L_\alpha(u)) &= \lim_{m \to \infty} d_H(L_\alpha(u_n), L_\alpha(u_m)) \\ &\leqslant \overline{\lim_{m \to \infty}} \sup_{\alpha > 0} d_H(L_\alpha(u_n), L_\alpha(u_m)) \\ &= \overline{\lim_{m \to \infty}} d(u_n, u_m) < \varepsilon. \end{aligned}$$

Thus $\sup_{\alpha > 0} d_H(L_\alpha(u_n), L_\alpha(u)) \leqslant \varepsilon$, i.e., $d(u_n, u) \leqslant \varepsilon$ for $n > n_\varepsilon$, implying that $u_n \to u$ in the metric d. The proof terminates.

References

1. Z. Artstein and R. A. Vitale, A strong law of large numbers for random compact sets, *Ann. Probab.* (1975), 879–882.
2. R. J. Aumann, Integrals of set-valued functions, *J. Math. Anal. Appl.* **12** (1965), 1–12.
3. R. J. Aumann and M. Perles, A variational problem arising in economic. *J. Math. Anal. Appl.* **11** (1965), 488–503.
4. C. L. Byrne, Remarks on the set-valued integrals of Debreu and Aumann, *J. Math. Anal. Appl.* **62** (1978), 243–246.
5. N. Cressie, A central limit theorem for random sets, *Z. Wahrsch. Verw. Gebiete* **49** (1979), 37–47.
6. G. Debreu, Integration of correspondences, *in* "Proc. Fifth Berkeley Symp. Math. Statist. Probability," pp. 351–372, Univ. of California Press, Berkeley, 1966.
7. G. Debreu and D. Schmeidler, The Radon–Nikodým derivative of a correspondence, *in* "Proc. Sixth Berkeley Symp. Math. Statist. Probability," pp. 41–56, Univ. of California Press, Berkeley, 1970.
8. R. Féron, Ensembles aléatoires flous. *C. R. Acad. Sci. Paris Ser. A* **282** (1976), 903–906.

9. R. FORTET AND M. KAMBOUZIA, Ensembles aléatoires et ensembles flous, *Publ. Économétriques* **9** (1976), 1–23.
10. I. R. GOODMAN, Fuzzy sets as equivalence classes of random sets, *in* "Recent Developments in Fuzzy Set and Possibility Theory" (R. Yager, Ed.), Pergamon, Elmsford, N.Y., 1980.
11. H. HERMES, Calculus of set-valued functions and control, *J. Math. Mech.* **18** (1968), 47–59.
12. H. KUDŌ, Dependent experiments and sufficient statistics, *Natur. Sci. Rep. Ochanomizu Univ.* **4** (1953), 151–163.
13. C. KURATOWSKI, "Topologie I," Monografie Matematyczne, Warsaw, 1948.
14. H. KWAKERNAAK, Fuzzy random variables: Definition and theorems, *Inform. Sci.* **15** (1978), 1–29.
15. G. MATHERON, "Random Sets and Integral Geometry," Wiley, New York, 1975.
16. C. V. NEGOITA AND D. RALESCU, "Applications of Fuzzy Sets to Systems Analysis," Wiley, New York, 1975.
17. G. B. PRICE, The theory of integration, *Trans. Amer. Math. Soc.* **47** (1940), 1–50.
18. M. L. PURI AND D. RALESCU, Différentielle d'une fonction floue. *C. R. Acad. Sci. Paris Sér. I.* **293** (1981), 237–239.
19. M. L. PURI AND D. RALESCU, Integration on fuzzy sets, *Advan. Appl. Math.* **3** (1982), 430–434.
20. M. L. PURI AND D. RALESCU, Strong law of large numbers for Banach space valued random sets, *Ann. Probab.* **11** (1983), 222–224.
21. M. L. PURI AND D. RALESCU, Differentials of fuzzy functions, *J. Math. Anal. Appl.* **91** (1983), 552–558.
22. H. RÅDSTRÖM, An embedding theorem for spaces of convex sets, *Proc. Amer. Math. Soc.* **3** (1952), 165–169.
23. H. RICHTER, Verallgemeinerung eines in der Statistik benötigten Satzes der Masstheorie, *Math. Ann.* **150** (1963), 85–90.
24. L. A. ZADEH, Fuzzy sets, *Inform. Control* **8** (1965), 338–353.

Reprinted by permission of the publisher,
© 1986 by Academic Press

GAUSSIAN RANDOM SETS IN BANACH SPACE*

M. L. PURI,[†] D. A. RALESCU[†] and S. S. RALESCU

Abstract. We define a Gaussian random set in a Banach space, and we prove the following characterization theorem: *Every Gaussian random set can be represented as the sum of its expected value and a Gaussian mean zero random element.*

1. Introduction. The concept of *random set* (Robbins [16], [17]), Matheron [11]) has received considerable attention in recent years, partly motivated by geometric probability considerations and partly as an application of probability techniques in Banach space. The strong law of large numbers for random sets in $\mathbf{R}^p$ was proved by Artstein and Vitale [1] and for random sets in Banach space by Puri and Ralescu [12], [13]. The central limit theorem (CLT)

* Received by the editors November 8, 1985.

† Research supported by a National Science Foundation Grant.

for random sets in $\mathbf{R}^p$ was proved in a particular case by Cressie [5] and in a general set-up by Weil [18]. In a Banach space setting, the CLT was given by Giné, Hahn and Zinn [8] and a more general result by Puri and Ralescu [13].

In view of statistical applications, the concept of normality for random sets should play an important role. Such a concept was defined by Lyashenko [10] in $\mathbf{R}^p$, who also gave a characterization of normal random sets.

In this paper, we introduce the concept of *a Gaussian random set* in a separable Banach space $\mathscr{X}$ and characterize such a set as the sum of a constant set and a Gaussian random element (of $\mathscr{X}$).

In § 2 we give some preliminaries on random sets and their expectations. Section 3 deals with an embedding of random compact sets into random elements of the space of continuous functions on some compact set. Finally, in § 4 we define Gaussian random sets in $\mathscr{X}$ and prove the characterization theorem.

2. Random sets. Let $\mathscr{X}$ denote a separable Banach space with norm $\|\cdot\|$, $\mathscr{X}^*$ the dual of $\mathscr{X}$, and $\langle\cdots\rangle: \mathscr{X}^*\times\mathscr{X}\to\mathbf{R}$ the duality map: $\langle x^*, x\rangle = x^*(x)$ for $x^*\in\mathscr{X}^*$, $x\in\mathscr{X}$.

If $(\Omega, \mathscr{A}, \mathbf{P})$ is a probability space, a *random element* in $\mathscr{X}$ is a Borel measurable function $f: \Omega\to\mathscr{X}$. The *expected value* $\mathbf{E}f$ is defined in terms of the Bochner integral; it exists if $\mathbf{E}\|f\|<\infty$.

A random element f is *Gaussian* if $\langle x^*, f\rangle$ is a normal random variable for any $x^*\in\mathscr{X}^*$.

We will be interested in random elements in $C(M)$, the Banach space of continuous functions on the compact, Hausdorff space M. It follows easily from the Riesz representation theorem that if $f: \Omega\to C(M)$ is Gaussian, then the vector $(f(t_1), \cdots, f(t_n))$ is (multivariate) normal, for any $t_1, \cdots, t_n\in M$. Let $\mathscr{K}(\mathscr{X})$ denote all the nonempty compact subsets of $\mathscr{X}$, and let $\mathscr{K}_c(\mathscr{X})$ denote the nonempty compact convex subsets of $\mathscr{X}$. The space $\mathscr{K}(\mathscr{X})$ is a metric space with respect to the Hausdorff metric:

$$d(A, B) = \max\{\sup_{a\in A}\inf_{b\in B}\|a-b\|, \sup_{b\in B}\inf_{a\in A}\|a-b\|\} \text{ for } A, B\in\mathscr{K}(\mathscr{X}).$$

More precisely, $(\mathscr{K}(\mathscr{X}), d)$ is a complete, separable metric space, and $\mathscr{K}_c(\mathscr{X})$ is a closed subspace (see Debreu [6]).

The space $\mathscr{K}(\mathscr{X})$ has a linear structure induced by the operations:

$$A+B=\{a+b \mid a\in A, b\in B\}, \lambda A=\{\lambda a \mid a\in A\} \text{ for } A, B\in\mathscr{K}(\mathscr{X}), \lambda\in\mathbf{R}.$$

Note however that $\mathscr{K}(\mathscr{X})$ is *not* a vector space (it is not a group with respect to addition).

Denote $\|A\| = d(A, \{0\})$ for $A\in\mathscr{K}(\mathscr{X})$.

As before, let $(\Omega, A, \mathbf{P})$ be a probability space. A *random set* (in $\mathscr{X}$) is a Borel measurable function $X: \Omega\to\mathscr{K}(\mathscr{X})$. The *expected value* $\mathbf{E}X$ is defined as $\mathbf{E}X=\{\mathbf{E}f \mid f\in L^1(\Omega, A, \mathbf{P}), f(\omega)\in X(\omega) \text{ a.e.}\}$ where $f: \Omega\to\mathscr{X}$ is a selection of X (Aumann [2]). If $\mathbf{E}\|X\|<\infty$, then $\mathbf{E}X\in\mathscr{K}(\mathscr{X})$. Another definition of the expected value is due to Debreu [6]: first, the space $\mathscr{K}_c(\mathscr{X})$ is embedded into a Banach space (via the Rådström embedding theorem [14]) and then one proceeds as in the case of the Bochner integral. If $\mathbf{E}\|X\|<\infty$, the two definitions of $\mathbf{E}X$ are equivalent (Byrne [3]).

In what follows we consider random sets with values in $\mathscr{K}_c(\mathscr{X})$, i.e., random convex sets.

3. The embedding theorem. An isometric embedding of $\mathscr{K}_c(\mathscr{X})$ into a separable Banach space of functions can be obtained from a theorem of Hörmander [9] (see also Giné, Hahn and Zinn [8]).

Let ball $\mathscr{X}^*=\{x^*\in\mathscr{X}^* \mid \|x^*\|\leq 1\}$ denote the unit ball of $\mathscr{X}^*$, let w^* denote the weak* topology in $\mathscr{X}^*$, and C (ball $\mathscr{X}^*$, w^*), the Banach space of functions defined on ball $\mathscr{X}^*$ and continuous in weak* topology.

THEOREM 3.1. *There exists a function* $j: \mathscr{K}_c(\mathscr{X})\to C$ *(ball* $\mathscr{X}^*$, w^*) *such that*

(a) *j is an isometry* (i.e., $\|j(K)-j(L)\|_\infty = d(K, L)$).

(b) $j(K+L)=j(K)+j(L)$.

(c) $j(\lambda K)=\lambda j(K)$, $\lambda\geqq 0$.

Proof. From Hörmander [9, Theorem 8], it follows that there is an isometry j satisfying (b) and (c), from $\mathscr{K}_c(\mathscr{X})$ into the space of bounded continuous functions on ball $\mathscr{X}^*$. Since each $K\in\mathscr{K}_c(\mathscr{X})$ is totally bounded, it follows that $j(K)\in C$ (ball $\mathscr{X}^*$, w^*).

Note. Since (ball $\mathscr{X}^*$, w^*) is compact and metrizable, the space C (ball $\mathscr{X}^*$, w^*) is separable. The isometry j is defined via the *support function* of a compact set: $j(K)=s_K$ where $s_K(x^*) = \sup_{a\in K}\langle x^*, a\rangle$, $x^*\in$ ball $\mathscr{X}^*$.

Actually, s_K can be defined on $\mathscr{X}^*$ as above, and has the following simple properties:

(i) s_K is subadditive, i.e., $s_K(x^*+y^*) \leqq s_K(x^*)+s_K(y^*)$ for $x^*, y^* \in X^*$.

(ii) s_K is positively homogeneous, i.e., $s_K(\lambda x^*)=\lambda s_K(x^*)$ for $x^* \in X^*$, $\lambda \geqq 0$.

In view of this embedding theorem, any random set $X: \Omega \to \mathscr{K}_c(\mathscr{X})$ can be thought of as a random element of C (ball X^*, w^*) by considering $j(X)=s_X: \Omega \to C$ (ball $\mathscr{X}^*$, w^*), where $s_X(\omega)=s_{X(\omega)}$.

The following result will be useful.

LEMMA 3.1. *If* $X: \Omega \to \mathscr{K}_c(\mathscr{X})$ *satisfies* $\mathbf{E}\|X\|<\infty$, *then* $\mathbf{E}[s_X(x^*)]=s_{\mathbf{E}X}(x^*)$ *for* $x^* \in$ ball X^*.

Proof. It is easily seen that the result holds if $X=\sum_{i=1}^n \chi_{A_i} K_i$ where $A_i \in \mathscr{A}$, $K_i \in \mathscr{K}_c(\mathscr{X})$ and χ_{A_i} denotes the characteristic function of A_i (i.e., if X is a simple function). The proof follows by using the definition of the Debreu expected value, the fact that j is an isometry, and the Lebesgue dominated convergence theorem.

4. Gaussian random sets. The concept of a normal random set in $\mathbf{R}^p$ was defined by Lyashenko [10] who also characterized such random sets. His proof of characterization, however, is strongly dependent on the finite dimensionality assumption. We define here a Gaussian random set in an arbitrary separable Banach space, and we represent such a random set as the sum of a constant set and a Gaussian random element.

Definition 4.1. A random set $X: \Omega \to \mathscr{K}_c(\mathscr{X})$ is *Gaussian* if s_X is a Gaussian random element of C (ball $\mathscr{X}^*$, w^*).

This definition implies (see § 2) that $(s_X(x_1^*), \cdots, s_X(x_n^*)$ is normal for every $x_1^*, \cdots, x_n^* \in$ ball $\mathscr{X}^*$. Consequently $(s_X(x_1^*), \cdots, s_X(x_n^*))$ is normal for every $x_1^*, \cdots, x_n^* \in \mathscr{X}^*$ (since s_X is positively homogeneous; see property (ii) in § 3).

It is obvious that if X and Y are independent Gaussian random sets, then $X+Y$ is Gaussian. Also λX is Gaussian ($\lambda \geqq 0$) whenever X is Gaussian.

The following theorem characterizes the Gaussian random sets in $\mathscr{X}$.

THEOREM 4.1. *Let* $X: \Omega \to \mathscr{K}_c(\mathscr{X})$ *be a random set with* $\mathbf{E}\|X\|<\infty$. *The following statements are equivalent.*

(i) X *is Gaussian.*

(ii) $X=\mathbf{E}X+\{\xi\}$ *where* ξ *is a Gaussian random element in* X *with mean zero.*

Proof. (ii)$\Rightarrow$(i). From $X=\mathbf{E}X+\{\xi\}$, it follows that $s_X=s_{\mathbf{E}X}+\langle\cdot, \xi\rangle$. Thus, if $\varphi^* \in (C(\text{ball } \mathscr{X}^*, w^*))^*$, then $\varphi^*(s_X)=\varphi^*(s_{\mathbf{E}X})+\varphi^*(\langle\cdot, \xi\rangle)$.

Denote $\Phi(x)=\varphi^*(\langle\cdot, x\rangle)$. Clearly Φ is linear, and since $|\Phi(x)| \leqq \|\varphi^*\| \|x\|$, it follows that $\Phi \in \mathscr{X}^*$. Thus $\langle\Phi, \xi\rangle=\varphi^*(\langle\cdot, \xi\rangle)$ is a normal random variable which implies that $\varphi^*(s_X)$ is normal. Consequently, X is Gaussian.

We now prove that (i)$\Rightarrow$(ii). Let s_X be Gaussian. Define $\psi: \Omega \to C$ (ball $\mathscr{X}^*$, w^*) by $\psi=s_X-s_{\mathbf{E}X}$. Fix an $\omega \in \Omega$. Then $\psi(w)=s_{X(\omega)}-s_{\mathbf{E}X}$.

CLAIM 1. $\psi(\omega)$ *is linear and weak*-continuous on* $\mathscr{X}^*$.

Let $\lambda_1, \cdots, \lambda_n \geqq 0$ and $x_1^*, \cdots, x_n^* \in \mathscr{X}^*$ be fixed. Then $\sum_{i=1}^n \lambda_i s_X(x_i^*)-s_X(\sum_{i=1}^n \lambda_i x_i^*) \geqq 0$ (from properties (i) and (ii) in § 3). By hypothesis, $(s_X(x_1^*), \cdots, s_X(x_n^*), s_X(\sum_{i=1}^n \lambda_i x_i^*))$ is normal which implies that $\sum_{i=1}^n \lambda_i s_X(x_i^*)-s_X(\sum_{i=1}^n \lambda_i x_i^*)$ is a normal random variable. From the above inequality, it follows that this random variable must be degenerate, i.e.,

$$\sum_{i=1}^n \lambda_i s_X(x_i^*)-s_X\left(\sum_{i=1}^n \lambda_i x_i^*\right)=\mathbf{E}\left[\sum_{i=1}^n \lambda_i s_X(x_i^*)-s_X\left(\sum_{i=1}^n \lambda_i x_i^*\right)\right]$$
$$=\sum_{i=1}^n \lambda_i s_{\mathbf{E}X}(x_i^*)-s_{\mathbf{E}X}\left(\sum_{i=1}^n \lambda_i x_i^*\right),$$

the last equality following from Lemma 3.1.

Thus $\sum_{i=1}^n \lambda_i \psi(\omega)(x_i^*)=\psi(\omega) \sum_{i=1}^n (\lambda_i x_i^*)$. Take now $\lambda_i=1$, $i=1, \cdots, n$. We get $\psi(\omega)(\sum_{i=1}^n x_i^*)=\sum_{i=1}^n \psi(\omega)(x_i^*)$, implying that $\psi(\omega)$ is additive. Take $x_1^*=x^*$, $x_2^*=-x^*$, $\lambda_1=\lambda_2=1$. We get $\psi(\omega)(-x^*)=-\psi(\omega)(x^*)$, implying that $\psi(\omega)$ is linear.

Since $\psi(\omega)$ is weak*-continuous on ball $\mathscr{X}^*$, it follows from Lemma 5.1 (in the Appendix) that $\psi(\omega)$ is weak*-continuous on $\mathscr{X}^*$. It is well known that such a $\psi(\omega)$ is representable as $\psi(\omega)(x^*)=\langle x^*, \xi(\omega)\rangle$, $x^* \in \mathscr{X}^*$ for some $\xi(\omega) \in \mathscr{X}$ (see, for example, Yoshida [19, p. 112]).

CLAIM 2. $\xi: \Omega \to \mathscr{X}$ *is measurable.*

Indeed, since $\langle x^*, \xi\rangle=\psi(\cdot)(x^*)=s_X(x^*)-s_{\mathbf{E}X}(\tau^*)$, it follows that $\langle x^*, \xi\rangle$ is measurable for each $x^* \in \mathscr{X}^*$. Using the Pettis theorem (Yoshida [19, p. 131]), it follows that ξ is strongly measurable (since $\mathscr{X}$ is separable).

CLAIM 3. *ξ is mean zero Gaussian.*

Since $\langle x^*, \xi\rangle = s_X(x^*) - s_{\mathbf{E}X}(x^*)$, it follows that $\langle x^*, \xi\rangle$ is a normal mean zero random variable, for each $x^* \in \mathscr{X}^*$.

To conclude the proof of the theorem, we write $s_X(x^*) = s_{\mathbf{E}X}(x^*) + \langle x^*, \xi\rangle = s_{\mathbf{E}X}(x^*) + s_{\{\xi\}}(x^*) = s_{\mathbf{E}X+\{\xi\}}(x^*)$, implying that $X = \mathbf{E}X + \{\xi\}$, giving the desired representation.

5. Appendix. The following lemma is used in the proof of Theorem 4.1.

LEMMA 5.1. *Let φ: $\mathscr{X}^* \to \mathbf{R}$ be linear and weak*-continuous on ball $\mathscr{X}^*$. Then φ is weak*-continuous on $\mathscr{X}^*$.*

Proof. It suffices to show that $\{\varphi \leqq c\}$ is weak*-closed for any $c \in \mathbf{R}$. Obviously $\{\varphi \leqq c\}$ is convex, and so, by the Krein-Smulian theorem (see Dunford and Schwartz [7, p. 429]), it is enough to show that $\{\varphi \leqq c\} \cap r$ (ball $\mathscr{X}^*$) is weak*-closed for all $r > 0$, where r (ball $\mathscr{X}^*$) is the ball with radius r.

Now $\{\varphi \leqq c\} \cap r$ (ball $\mathscr{X}^*$) $= r[\{\varphi \leqq c/r\} \cap \text{ball } \mathscr{X}^*]$ and $\{\varphi \leqq c/r\} \cap \text{ball } \mathscr{X}^*$ is weak*-closed since φ is weak*-continuous on ball $\mathscr{X}^*$. The proof follows.

REFERENCES

[1] Z. ARTSTEIN AND R. A. VITALE, *A strong law of large numbers for random compact sets*, Ann. Probab., 3 (1975), pp. 879–882.

[2] R. J. AUMANN, *Integrals of set-valued functions*, J. Math. Anal. Appl., 12 (1965), pp. 1–12.

[3] C. L. BYRNE, *Remarks on the set-valued integrals of Debreu and Aumann*, J. Math. Anal. Appl., 62 (1978), pp. 243–246.

[4] N. CRESSIE, *A strong limit theorem for random sets*, Suppl. Adv. Appl. Probab., 10 (1978), pp. 36–46.

[5] ———, *A central limit theorem for random sets*, Z. Wahrsch. verw. Geb., 49 (1979), pp. 37–47.

[6] G. DEBREU, *Integration of correspondences*, Proc. Fifth Berkeley Symposium on Mathematical Statistics and Probability. Vol. II. Univ. California Press, Berkeley, Los Angeles, 1966, pp. 351–372.

[7] N. DUNFORD AND J. SCHWARTZ, *Linear Operators, Part 1: General Theory*, Wiley Interscience, New York, 1958.

[8] E. GINÉ, M. C. HAHN AND J. ZINN, *Limit Theorems for Random Sets: An Application of Probability in Banach Space Results*, Lecture Notes in Mathematics 990, 1983, pp. 112–135.

[9] L. HÖRMANDER, *Sur la fonction d'appui des ensembles convexes dans un espace localement convexe*, Ark. Math., 3 (1954), pp. 181–186.

[10] N. N. LYASHENKO, *The statistics of random compact sets of Euclidean space*, Zap. Nauchn. Sem. Leningrad Otdel. Mat. Inst. Steklov, 98 (1980), pp. 115–139. (In Russian.)

[11] G. MATHERON, *Random Sets and Integral Geometry*, John Wiley, New York, 1975.

[12] M. L. PURI AND D. A. RALESCU, *Strong law of large number for Banach space-valued random sets*, Ann. Probab., 11 (1983), pp. 222–224.

[13] ———, *Limit theorems for random compact sets in Banach space*, Math. Proc. Cambridge Phil. Soc., 97 (1985), pp. 151–158.

[14] H. RÅDSTRÖM, *An embedding theorem for spaces of convex sets*, Proc. Amer. Math. Soc., 3 (1952) pp. 165–169.

[15] B. D. RIPLEY, *Foundations of random set theory*, Ann. Probab., 4 (1976), pp. 995–998.

[16] H. E. ROBBINS, *On the measure of a random set*, Ann. Math. Statist., 15 (1944), pp. 70–74.

[17] ———, *On the measure of a random set*. II, Ann. Math. Statist., 16 (1945), pp. 342–347.

[18] W. WEIL, *An application of the central limit theorem for Banach space-valued random variables to the theory of random sets*, Z. Wahrsch. verw. Geb., 60 (1982), pp. 203–208.

[19] K. YOSHIDA, *Functional Analysis*, Springer-Verlag, New York–Berlin–Heidelberg, 1965.

Reprinted from
Theory of Probability and Its Applications, Volume 31
(1986), pages 526-529, by permission of the publisher,
© 1986 by SIAM

LIMIT THEOREMS FOR FUZZY RANDOM VARIABLES

E. P. KLEMENT[1†], M. L. PURI[2] and D. A. RALESCU[1]

[1]*Department of Mathematics, University of Cincinnati, Cincinnati, Ohio 45221, U.S.A.*
[2]*Department of Mathematics, Indiana University, Bloomington, Indiana, 47405, U.S.A.*

(*Communicated by I. N. Sneddon, F.R.S. – Received* 7 *November* 1985)

A strong law of large numbers and a central limit theorem are proved for independent and identically distributed fuzzy random variables, whose values are fuzzy sets with compact levels. The proofs are based on embedding theorems as well as on probability techniques in Banach space.

1. Introduction

The concept of fuzzy set was introduced by Zadeh (1965). Subsequent developments focused on applications of this concept to pattern recognition and system analysis, among other areas (see Negoita & Ralescu 1975). Puri & Ralescu (1986) studied fuzzy random variables as a generalization of random sets. The purpose of this generalization was the introduction of statistical techniques (such as estimation) for pattern recognition. Consider, for example, the problem of recognition of a handwritten character. A random sample is taken from that character. How can the expected character (or a prototype) be defined, and how can it be estimated? The data here are represented by fuzzy sets rather than numbers. The strong law of large numbers (SLLN) and the central limit theorem (CLT) for fuzzy random variables are first steps in the direction of the estimation of vague parameters. After some preliminaries on random sets introduced in §2, we introduce various spaces of fuzzy sets and distances on these spaces in §3. We prove, with respect to one of these distances, that the space of fuzzy sets with compact levels is separable. We also extend the Rådström (1952) embedding theorem to the space of fuzzy sets with compact convex levels. In §4 we recall some facts about fuzzy random variables and their expected value. These concepts were defined by Puri & Ralescu (1986). We prove the Lebesgue dominated convergence type theorem. The result concerning SLLN for fuzzy random variables is given in §5. This result generalizes the corresponding result for random sets due to Artstein & Vitale (1975). In §6, we prove an embedding theorem for the space of fuzzy sets with compact convex levels satisfying a Lipschitz condition. This embedding is a key tool in proving our CLT for fuzzy random variables in §7. Our results extends the

† Permanent address: Institut für Mathematik, Johannes Kepler Universität, A-4040 Linz, Austria.

CLT for random sets (cf. Weil 1982; Giné *et al.* 1982). Finally, in §8, we comment on some spaces of fuzzy sets where the CLT can be applied.

Our approach relies heavily on probability techniques in Banach space. As with random sets, our SLLN is not an immediate application of those techniques, since the spaces of fuzzy sets are not Banach spaces. In fact, they are not even vector spaces, so we rely on embedding theorems as mentioned in §§3 and 6. Our framework is more complex than that of random sets: on the one hand, separability and therefore measurability problems arise; on the other hand, since an integral similar to that of Debreu (1967) is not available for fuzzy random variables, our proof of the SLLN uses a truncation argument and a Lebesgue dominated convergence type theorem. Also, due to the complexity of spaces of fuzzy sets (as compared to spaces of sets), our CLT has a more complicated form than the CLT for random sets.

2. Preliminaries on random sets

Let $\mathcal{K}(\mathbb{R}^p)$ denote the collection of non-empty compact subsets of the Euclidean space $\mathbb{R}^p$, and let $\mathcal{K}_c(\mathbb{R}^p)$ denote the non-empty compact convex subsets of $\mathbb{R}^p$. The space $\mathcal{K}(\mathbb{R}^p)$ has a linear structure induced by the (Minkowski) addition and scalar multiplication:

$$A+B = \{a+b \mid a \in A, b \in B\}, \quad \lambda A = \{\lambda a \mid a \in A\}$$

for $A, B \in \mathcal{K}(\mathbb{R}^p), \lambda \in \mathbb{R}$. However, $\mathcal{K}(\mathbb{R}^p)$ is not a vector space (it is not a group with respect to addition).

The space $\mathcal{K}(\mathbb{R}^p)$ is metrizable by the Hausdorff distance

$$d(A,B) = \max\{\sup_{a \in A} \inf_{b \in B} \|a-b\|, \quad \sup_{b \in B} \inf_{a \in A} \|a-b\|\},$$

where $\| \ \|$ denotes the euclidean norm and $A, B \in \mathcal{K}(\mathbb{R}^p)$. It is well known that $(\mathcal{K}(\mathbb{R}^p), d)$ is a complete separable metric space (Debreu (1967)). If $A \in \mathcal{K}(\mathbb{R}^p)$, we write $\|A\| = d(A, \{0\})$.

Let $(\Omega, \mathcal{A}, P)$ be a probability space. A *random set* is a Borel measurable function $f\colon \Omega \to \mathcal{K}(\mathbb{R}^p)$.

The *expected value Ef* of a random set was defined by Aumann (1965) as:

$$Ef = \{E\phi \mid \phi \in L^1(\Omega, \mathcal{A}, P), \quad \phi(\omega) \in f(\omega) \text{ a.e.}\}$$

where $\phi\colon \Omega \to \mathbb{R}^p$ is a *selection* of f and $E\phi$ denotes the expectation of the random vector ϕ. If f is $\mathcal{K}_c(\mathbb{R}^p)$-valued and if $E\|f\| < \infty$, then $Ef \in \mathcal{K}_c(\mathbb{R}^p)$.

Various properties related to the calculus of set-valued functions (including the Lebesgue dominated convergence type theorems) are discussed in Aumann (1965) and Debreu (1967). A comprehensive theory of random sets is given in Matheron (1975).

3. Spaces of fuzzy sets

A *fuzzy subset* of $\mathbb{R}^p$ (*fuzzy set*) is a function of $u: \mathbb{R}^p \to [0, 1]$. For each such fuzzy set u, we denote by $L_\alpha u = \{x \in \mathbb{R}^p \mid u(x) \geqslant \alpha\}, 0 \leqslant \alpha \leqslant 1$, its α-level set. By $\operatorname{supp} u$ we denote the support of u, i.e. the closure of the set $\{x \in \mathbb{R}^p \mid u(x) > 0\}$.

We consider the collection $\mathscr{F}(\mathbb{R}^p)$ of those fuzzy sets $u: \mathbb{R}^p \to [0, 1]$ with the following properties.

(1) u is upper semicontinuous; (2) $\operatorname{supp} u$ is compact; (3) $\{x \in \mathbb{R}^p \mid u(x) = 1\} \neq \varnothing$.

The space $\mathscr{F}(\mathbb{R}^p)$ extends $\mathscr{K}(\mathbb{R}^p)$ in the sense that for each $A \in \mathscr{K}(\mathbb{R}^p)$, its characteristic function $\chi_A \in \mathscr{F}(\mathbb{R}^p)$.

A linear structure in $\mathscr{F}(\mathbb{R}^p)$ is defined via the following operations:

$$(u+v)(x) = \sup_{y+z=x} \min[u(y), v(z)],$$

$$(\lambda u)(x) = \begin{cases} u(\lambda^{-1}x) & \text{if } \lambda \neq 0 \\ \chi_{\{0\}}(x) & \text{if } \lambda = 0 \end{cases}$$

for $u, v \in \mathscr{F}(\mathbb{R}^p), \lambda \in \mathbb{R}$.

By using simple topological arguments and properties of upper semicontinuous functions, it is easy to see that $u+v, \lambda u \in \mathscr{F}(\mathbb{R}^p)$. The following properties will be needed later:

$$L_\alpha(u+v) = L_\alpha u + L_\alpha v \quad \text{and} \quad L_\alpha(\lambda u) = \lambda L_\alpha u$$

for every $0 \leqslant \alpha \leqslant 1$.

There is no unique metric in $\mathscr{F}(\mathbb{R}^p)$ which extends the Hausdorff distance.

In this paper we will mainly be concerned with the metric defined as

$$d_1(u, v) = \int_0^1 d(L_\alpha u, L_\alpha v)\,\mathrm{d}\alpha.$$

We will also use the metric

$$d_\infty(u, v) = \sup_{\alpha > 0} d(L_\alpha u, L_\alpha v)$$

which was studied in Puri & Ralescu (1983*b*)

Clearly, if $A, B \in \mathscr{K}(\mathbb{R}^p)$, then

$$d_1(\chi_A, \chi_B) = d_\infty(\chi_A, \chi_B) = d(A, B).$$

Proposition 3.1. *$(\mathscr{F}(\mathbb{R}^p), d_1)$ is a metric space.*

Proof: Obviously $d_1(u, v) < \infty$ since $\operatorname{supp} u$ and $\operatorname{supp} v$ are bounded.

The function $\phi(\alpha) = d(L_\alpha u, L_\alpha v)$ is measurable: if $\alpha_1 \leqslant \alpha_2 \leqslant \ldots$ with $\lim \alpha_n = \alpha$, then $L_\alpha u = \bigcap_{n=1}^\infty L_{\alpha_n} u$, $L_\alpha v = \bigcap_{n=1}^\infty L_{\alpha_n} v$. Thus $d(L_{\alpha_n} u, L_\alpha u) \to 0$, $d(L_{\alpha_n} v, L_\alpha v) \to 0$ and, from the continuity of the Hausdorff distance, we conclude that ϕ is left-continuous and therefore measurable.

The triangle inequality and symmetry property of d_1 are clear. It remains to show that $d_1(u, v) = 0$ implies $u = v$.

If $d_1(u, v) = 0$, it follows that $L_\alpha u = L_\alpha v$ a.e., therefore u and v have equal levels for $\alpha \in [0, 1] \backslash A$ where $\lambda(A) = 0$ (λ denotes the Lebesgue measure).

It is well known that $[0, 1] \backslash A$ is dense in $[0, 1]$. If $\alpha_0 \in A, \alpha_0 > 0$, there exists an increasing sequence $\alpha_n \in [0,1]\backslash A, \alpha_n \to \alpha_0$. Then

$$L_{\alpha_0} u = \bigcap_{n=1}^{\infty} L_{\alpha n} u = \bigcap_{n=1}^{\infty} L_{\alpha n} v = L_{\alpha_0} v.$$

So $L_\alpha u = L_\alpha v$ for each $\alpha \in [0,1]$, which implies $u = v$.

A crucial property in connection with the strong law of large numbers is that of separability (see, e.g. Taylor 1978). It turns out that the space $(\mathscr{F}(\mathbb{R}^p), d_1)$ is separable. The space $(\mathscr{F}(\mathbb{R}^p), d_\infty)$, on the other hand, is not separable, which means that the distance d_1 is preferable to d_∞. These statements are proved in the following.

PROPOSITION 3.2. *The metric space $(\mathscr{F}(\mathbb{R}^p), d_1)$ is separable.*

Proof. Suppose that $u \in \mathscr{F}(\mathbb{R}^p)$ and $\epsilon > 0$ are given.

1. Since $\operatorname{supp} u$ is compact, there exist p-dimensional cubes $S_i = \prod_{j=1}^p [a_{ij}, b_{ij})$, $i = 1, \ldots, r$ such that $a_{ij}, b_{ij} \in \mathbb{Q}, 0 < b_{ij} - a_{ij} \leqslant (2\epsilon)/\sqrt{p}$, and $\operatorname{supp} u \subseteq \bigcup_{i=1}^r S_i$. Consider the corner point $T_i = (a_{i1}, \ldots, a_{ip})$ of S_i and define the fuzzy set ϕ: $\mathbb{R}^p \to [0,1]$ by

$$\phi(x) = \begin{cases} \operatorname{supp}_{x \in \bar{S}_i} u(x) & \text{if} \quad x = T_i, i = 1, \ldots, r, \\ 0 & \text{otherwise.} \end{cases}$$

Obviously we have $\phi \in \mathscr{F}(\mathbb{R}^p)$. Putting $\alpha_i = \phi(T_i)$, we relabel $S_1, \ldots, S_r$ and $T_1, \ldots, T_r$ (if necessary) such that $0 = \alpha_0 \leqslant \alpha_1 \leqslant \alpha_2 \leqslant \ldots \leqslant \alpha_r = 1$.

2. We claim that $d_\infty(u, \phi) \leqslant \frac{1}{2}\epsilon$. To this end we choose an $\alpha > 0$. Since $\alpha_{i_0 - 1} < \alpha \leqslant \alpha_{i_0}$ for some $1 \leqslant i_0 \leqslant r$ we get $L_\alpha u \supseteq L_{\alpha i_0} u$ and $L_\alpha \phi = L_{\alpha i_0} \phi = \{T_{i_0}, T_{i_0+1}, \ldots, T_r\}$. As $x \in L_\alpha u$ implies $x \in S_{i_1}$ for some $i_1 \geqslant i_0$, it follows that $\min_{i_0 \leqslant i \leqslant r} \|x - T_i\| \leqslant \|x - T_{i_1}\| \leqslant \frac{1}{2}\epsilon$.

On the other hand, for any $i \geqslant i_0$, we have $\phi(T_i) = \operatorname{supp}_{x \in S_i} u(x) \geqslant \alpha$. Since u is upper semicontinuous, it attains its supremum at some point $x_0 \in \bar{S}_i \cap L_\alpha u$. This gives

$$\inf_{x \in L_\alpha u} \|T_i - x\| \leqslant \|T_i - x_0\| \geqslant \tfrac{1}{2}\epsilon.$$

Therefore

$$d_\infty(u, \phi) = \sup_{\alpha > 0} d(L_\alpha u, L_\alpha \phi) = \sup_{\alpha > 0} d(L_\alpha u, \{T_{i_0}, \ldots, T_r\}) \leqslant \tfrac{1}{2}\epsilon.$$

3. If necessary, we relabel $\alpha_1, \ldots, \alpha_r$ to obtain $0 \leqslant \alpha_1 < \alpha_2 < \ldots < \alpha_s = 1$ with $s \leqslant r$. If $\alpha_k \notin \mathbb{Q}$, we choose $\beta_k \in \mathbb{Q}$ such that $\max(\alpha_{k-1}, \alpha_k - \epsilon/M) < \beta_k < \alpha_k$ with $M > 2(r-1)\operatorname{diam}(\operatorname{supp} u)$, while if $\alpha_k \in \mathbb{Q}$, we set $\beta_k = \alpha_k (k = 1, \ldots, s)$. Here $\operatorname{diam}(\operatorname{supp} u)$ stands for the diameter of $\operatorname{supp} u$.

Defining $\psi \in \mathscr{F}(\mathbb{R}^p)$ by

$$\psi(x) = \begin{cases} \beta_k & \text{if} \quad \phi(x) = \alpha_k, \\ 0 & \text{otherwise} \end{cases}$$

yields

$$\begin{aligned} d_1(\phi, \psi) &= \int_0^1 d(L_\alpha \phi, L_\alpha \psi)\,\mathrm{d}\alpha = \sum_{i=1}^{s-1} \int_{\beta_i}^{\alpha_i} d(L_\alpha \phi, L_\alpha \psi)\,\mathrm{d}\alpha \\ &\leqslant \operatorname{diam}(\operatorname{supp} u) \cdot \sum_{i=1}^{s-1} (\alpha_i - \beta_i) < \tfrac{1}{2}\epsilon. \end{aligned}$$

4. By the triangle inequality, we get

$$d_1(u,\psi) \leqslant d_1(u,\phi)+d_1(\phi,\psi) \leqslant d_\infty(u,\phi)+d_1(\phi,\psi) < \epsilon.$$

Since the set of functions ψ is countable, separability of $(\mathscr{F}(\mathbb{R}^p), d_1)$ follows.

PROPOSITION 3.3 *$(\mathscr{F}(\mathbb{R}^p), d_\infty)$ is not separable.*

Proof. Put $a = (0,0,\ldots,0)\in\mathbb{R}^p$ and $b = (1,0,0,\ldots,0)\in\mathbb{R}^p$. For each $\alpha\in(0,1)$, define $\phi_\alpha\in\mathscr{F}(\mathbb{R}^p)$ by

$$\phi_\alpha(x) = \begin{cases} 1 & \text{if} \quad x = b \\ \alpha & \text{if} \quad x\in \operatorname{co}\{a,b\}\backslash\{b\} \\ 0 & \text{otherwise,} \end{cases}$$

where $\operatorname{co}\{a,b\}$ is the convex hull of $\{a,b\}$.

It is easy to check that $d_\infty(\phi_\alpha,\phi_\beta) = 1$ for $\alpha \neq \beta$.

Denote by $\mathscr{F}_c(\mathbb{R}^p)$ the space of fuzzy sets $u\in\mathscr{F}(\mathbb{R}^p)$ such that $L_\alpha u$ is convex for each $\alpha \geqslant 0$. Alternatively, $u\in\mathscr{F}_c(\mathbb{R}^p)$ if $u\in\mathscr{F}(\mathbb{R}^p)$ and u is *fuzzy convex* (i.e. $u(\lambda x+(1-\lambda)y) \geqslant \min[u(x),u(y)]$ for all $x,y\in\mathbb{R}^p$, $\lambda\in[0,1]$; see Zadeh (1965)).

The space $\mathscr{F}_c(\mathbb{R}^p)$ plays an important role since it can be embedded isometrically into a Banach space. Actually this embedding generalizes the Rådström embedding theorem (Rådström 1952) of $\mathscr{K}_c(\mathbb{R}^p)$ into a Banach space.

Such an embedding for fuzzy sets is given in Puri & Ralescu (1983*b*), in a more general framework. For completeness we state this result here and sketch its proof.

PROPOSITION 3.4. *There exists a normed space χ and a function $j\colon \mathscr{F}_c(\mathbb{R}^p)\to\chi$ with properties*:

(*a*) *j is an isometry (i.e. $\|j(u)-j(v)\| = d_1(u,v)$)*,

(*b*) *$j(u+v) = j(u)+j(v)$*,

(*c*) *$j(\lambda u) = \lambda j(u), \lambda \geqslant 0$.*

Proof. Define an equivalence relation in $\mathscr{F}_c(\mathbb{R}^p)\times\mathscr{F}_c(\mathbb{R}^p)$ by

$$(u,v) \sim (u',v') \Leftrightarrow u+v' = u'+v.$$

A general theorem of Rådström (1952) can be used; the necessary hypotheses are verified as in Puri & Ralescu (1983*b*). Ths space of equivalence classes $[u,v]$ of pairs (u,v) is denoted by χ. The norm in χ is defined by $\|[u,v]\| = d_1(u,v)$. It is easy to check that $j\colon\mathscr{F}_c(\mathbb{R}^p)\to\chi$ defined by $j(u) = [u,\chi_{\{0\}}]$ is an isometry, i.e. $\|j(u)-j(v)\| = d_1(u,v)$, and properties (*b*), (*c*) follow from the definitions.

4. FUZZY RANDOM VARIABLES

The concept of a random set was generalized by Puri & Ralescu (1986). A *fuzzy random variable* (FRV) is a Borel measurable function $X\colon \Omega\to(\mathscr{F}(\mathbb{R}^p), d_\infty)$. Note that we use d_∞ here and not d_1; the reason for this will soon become clear.

If X is a fuzzy random variable such that $E\|\operatorname{supp} X\| < \infty$, then the *expected value EX* is the (unique) fuzzy set satisfying the property

$$L_\alpha(EX) = E(L_\alpha X), 0 < \alpha \leqslant 1$$

(Puri & Ralescu 1986, Theorem 3.1).

It follows that $EX\colon \mathbb{R}^p \to [0,1]$ is upper semicontinuous and $L_1(EX) \neq \varnothing$.

Also $\operatorname{supp} EX$ is compact; to see this, note that it suffices to show $L_\alpha(EX) \subseteq K$ for every $\alpha > 0$ and for some compact set K. But $L_\alpha(EX) = E(L_\alpha X) \subseteq E(\operatorname{supp} X)$. Since $E\|\operatorname{supp} X\| < \infty$, it follows that $E(\operatorname{supp} X)$ is a non-empty compact set, so we can take $K = E(\operatorname{supp} X)$.

In the discussion above, it is implicit that $\operatorname{supp} X$ is measurable, i.e. it is a random set. This fact follows from the following.

PROPOSITION 4.1. *If X is a fuzzy random variable, then* $\operatorname{supp} X$ *is a random set.*

Proof. It is enough to show that the function $\phi\colon (\mathscr{F}(\mathbb{R}^p), d_\infty) \to (\mathscr{K}(\mathbb{R}^p), d), \phi(u) = \operatorname{supp} u$ is Borel measurable. Note that $d(L_{1/n} u, \operatorname{supp} u) \to 0$ as $n \to \infty$. So, if we define $\phi_n(u) = L_{1/n} u$, then $\phi = \lim_{n\to\infty} \phi_n$ pointwise. Now ϕ_n are obviously continuous functions, so ϕ is measurable.

Note. A similar argument will not hold if d_∞ is replaced by d_1.

From the above result, it follows that if X is a fuzzy random variable, then $EX \in \mathscr{F}(\mathbb{R}^p)$.

An important property of the expected value, which will be needed later, is the Lebesgue dominated convergence type theorem. Such a result is contained in Puri & Ralescu (1986), but for the metric d_∞.

We state and prove the new version here, in the form that we need in the next section. By $\{0\}$ we denote the one point set containing 0 (identified with its characteristic function $\chi_{\{0\}}$).

THEOREM 4.1. *Let $\{X_k \mid k \geqslant 1\}$, X be fuzzy random variables with values in $\mathscr{F}_c(\mathbb{R}^p)$ and such that $E\|\operatorname{supp} X_k\| < \infty, E\|\operatorname{supp} X\| < \infty$. Suppose that $X_k \to X$* a.e. *in the metric d_1, and that $d_1(X_k(\omega), \{0\}) \leqslant h(\omega)$ for all $k \geqslant 1$, where $h\colon \Omega \to \mathbb{R}$ is integrable. Then $EX_k \to EX$ in the metric d_1.*

Proof.

$$d_1(EX_k, EX) = \int_0^1 d[L_\alpha(EX_k), L_\alpha(EX)]\,\mathrm{d}\alpha$$

$$= \int_0^1 d[E(L_\alpha X_k), E(L_\alpha X)]\,\mathrm{d}\alpha \leqslant \int_0^1 E[d(L_\alpha X_k, L_\alpha X)]\,\mathrm{d}\alpha$$

by using an inequality of Debreu (1967, p. 366–367). From the Fubini theorem we conclude that

$$d_1(EX_k, EX) \leqslant E[d_1(X_k, X)].$$

From the hypothesis, $d_1(X_k, X) \to 0$ a.e. Also,

$$d_1(X_k, X) \leqslant d_1(X_k, \{0\}) + d_1(\{0\}, X) \leqslant h + d_1(\{0\}, X).$$

From the classical Lebesgue dominated convergence theorem, it follows that $d_1(EX_k, EX) \to 0$, concluding the proof.

5. LAW OF LARGE NUMBERS

The strong law of large numbers (SLLN) for random sets was derived by Artstein & Vitale (1975). This was extended to random convex sets in a Banach space by Puri & Ralescu (1983*a*, 1985).

Here we generalize the Artstein–Vitale result to fuzzy random variables. Note that such a SLLN was stated by Féron (1979) by using a different metric, but no proof was provided.

We will consider fuzzy random variables with values in $\mathscr{F}(\mathbb{R}^p)$. Note that it makes sense to talk about independent and identically distributed FRVS, since $(\mathscr{F}(\mathbb{R}^p), d_\infty)$ is a metric space (see Billingsley 1968).

We will also need the concept of a convex hull of a fuzzy set (see Lowen 1980).

If $u \in \mathscr{F}(\mathbb{R}^p)$, then $\operatorname{co} u \in \mathscr{F}_c(\mathbb{R}^p)$, the *convex hull* of u is defined by

$$\operatorname{co} u = \inf\{v \in \mathscr{F}_c(\mathbb{R}^p) \mid v \geqslant u\}.$$

It is possible to show that

$$L_\alpha(\operatorname{co} u) = \operatorname{co}(L_\alpha u), 0 \leqslant \alpha \leqslant 1.$$

If $X: \Omega \to \mathscr{F}(\mathbb{R}^p)$ is a fuzzy random variable, then $\operatorname{co} X: \Omega \to \mathscr{F}_c(\mathbb{R}^p)$ is defined by $(\operatorname{co} X)(\omega) = \operatorname{co} X(\omega)$.

Our main result is as follows.

THEOREM 5.1. *Let $\{X_k \mid k \geqslant 1\}$ be independent and identically distributed fuzzy random variables such that $E\|\operatorname{supp} X_1\| < \infty$. Then*

$$(X_1 + X_2 + \ldots + X_n)/n \to E(\operatorname{co} X_1) \text{ a.e.},$$

the convergence being in the metric d_1.

Proof. (1) Consider first $X_k: \Omega \to \mathscr{F}_c(\mathbb{R}^p)$ and let $j: \mathscr{F}_c(\mathbb{R}^p) \to \chi$ be the isometry provided by Proposition 3.4.

Since $(\mathscr{F}_c(\mathbb{R}^p), d_1)$ is separable, it is easy to show that χ is separable. Then $\{j \circ X_k \mid k \geqslant 1\}$ are IID χ-valued random elements. By a standard SLLN in Banach space, it follows that $1/n \sum_{k=1}^n (j \circ X_k) \to E(j \circ X_1)$ a.e.

The main point now is to show that $E(j \circ X_1) = j(EX_1)$ if $E\|\operatorname{supp} X_1\| < \infty$.

Assume first that X_1 is a simple function, i.e. $X_1 = \sum_{i=1}^l u_i \chi_{A_i}$, $u_i \in \mathscr{F}_c(\mathbb{R}^p)$, $A_i \in \mathscr{A}$. It is easy to check that $E(j \circ X_1) = j(EX_1)$ in this case.

Since X_1 is measurable, there exists a sequence of simple functions (FRVS) s_m with $s_m \to X_1$ a.e. in the metric d_1 (note that we can not assert the existence of such a sequence which converges in d_∞, unless X_1 is separably valued).

Also $d_1(s_m, \{0\}) \to d_1(X_1, \{0\})$ a.e. from the continuity of d_1. Consider the truncated FRVS t_m as follows:

$$t_m(\omega) = \begin{cases} s_m(\omega), & \text{if } \; d_1(s_m(\omega), \{0\}) \leqslant 2d_1(X_1(\omega), \{0\}) \\ \{0\}, & \text{otherwise.} \end{cases}$$

Note that t_m are simple functions. It is easy to see that $d_1(t_m, X_1) \to 0$ a.e. and that $d_1(t_m(\omega), \{0\}) \leqslant 2d_1(X_1(\omega), \{0\})$.

The hypotheses of Theorem 4.1 are satisfied, so $d_1(Et_m, EX_1) \to 0$. Therefore $j(Et_m) \to j(EX_1)$ in χ. It is easy to see that $j \circ t_m \to j \circ X_1$ and, from properties of the Bochner integral, that $E(j \circ t_m) \to E(j \circ X_1)$. Since $j(Et_m) = E(j \circ t_m)$ it follows that $j(EX_1) = E(j \circ X_1)$.

Therefore $\|1/n \sum_{k=1}^n (j \circ X_k) - j(EX_1)\| \to 0$ a.e.

From the properties of j, it follows that $d_1(1/n \sum_{k=1}^n X_k, EX_1) \to 0$ a.e. (2)

Consider now the general case, i.e. $X_k: \Omega \to \mathscr{F}(\mathbb{R}^p)$. Since $\{\operatorname{co} X_k \mid k \geqslant 1\}$ are IID, and $E\|\operatorname{supp}(\operatorname{co} X_1)\| \leqslant E\|\operatorname{supp} X_1\| < \infty$, it follows from part (1) that

$$\left(\sum_{k=1}^{n} \operatorname{co} X_k\right)\Big/ n \xrightarrow{d_1} E\left(\operatorname{co} X_1\right).$$

The Shapley–Folkman lemma (see Arrow & Hahn 1971) gives;

$$d\left[L_\alpha\left(\frac{1}{n}\sum_{k=1}^{n} X_k\right), L_\alpha\left(\frac{1}{n}\sum_{k=1}^{n} \operatorname{co} X_k\right)\right] \leqslant \frac{\sqrt{p}}{n} \max_{1 \leqslant k \leqslant n} \|L_\alpha X_k\|,$$

for every $\alpha > 0$. This implies immediately

$$d_\infty\left(\frac{1}{n}\sum_{k=1}^{n} X_k, \frac{1}{n}\sum_{k=1}^{n} \operatorname{co} X_k\right) \leqslant \frac{\sqrt{p}}{n} \max_{1 \leqslant k \leqslant n} \|\operatorname{supp} X_k\|.$$

Since $(1/n)\max_{1\leqslant k\leqslant n} \|\operatorname{supp} X_k\| \to 0$ a.e. (see, for example, Chow & Teicher 1978, p. 122), it follows that $d_1((1/n)\sum_{k=1}^{n} X_k, (1/n)\sum_{k=1}^{n} \operatorname{co} X_k) \to 0$ a.e. Finally, from the triangle inequality, it follows that $d_1[(1/n)\sum_{k=1}^{n} X_k, E(\operatorname{co} X_1)] \to 0$ a.e. and the proof is complete.

Note. From the above proof it seems that we could prove the SLLN with respect to d_∞, at least for $\mathscr{F}_c(\mathbb{R}^p)$-valued fuzzy random variables. This is not the case since $(\mathscr{F}_c(\mathbb{R}^p), d_\infty)$ is not separable. This follows from Proposition 3.3 (the same example works).

6. The embedding theorem

Let $S^{p-1} = \{x \in \mathbb{R}^p \mid \|x\| = 1\}$ denote the unit sphere in $\mathbb{R}^p$. It is well known that $\mathscr{K}_c(\mathbb{R}^p)$ can be embedded isometrically into $C(S^{p-1})$ (the Banach space of continuous functions on S^{p-1}). This result goes back to Minkowski (see, for example, Artstein & Vitale 1975).

Such an embedding is realized via the support function of a compact convex set. More precisely, if $K \in \mathscr{K}_c(\mathbb{R}^p)$, its support function is defined by

$$s_K(x) = \sup_{a \in K} \langle x, a\rangle, \quad x \in S^{p-1}.$$

In the case of fuzzy sets, more restrictions should be imposed in order to achieve a similar embedding.

Consider the space $\mathscr{F}_L(\mathbb{R}^p)$ of fuzzy sets $u \in \mathscr{F}(\mathbb{R}^p)$ such that the map $\alpha \to L_\alpha u$ is Lipschitz. More specifically, $u \in \mathscr{F}_L(\mathbb{R}^p)$ if there exists a constant $M > 0$, such that

$$d(L_\alpha u, L_\beta u) \leqslant M|\alpha - \beta|$$

for every $\alpha, \beta \in (0, 1]$.

Define the space $\mathscr{F}_{CL}(\mathbb{R}^p) = \mathscr{F}_L(\mathbb{R}^p) \cap \mathscr{F}_c(\mathbb{R}^p)$. The next theorem gives the desired embedding.

Theorem 6.1 *There exists a function*

$$j: \mathscr{F}_{CL}(\mathbb{R}^p) \to C([0,1] \times S^{p-1})$$

such that:

(*a*) *j is an isometry* (*i.e.* $\|j(u)-j(v)\|_\infty = d_\infty(u,v)$),

(*b*) $j(u+v) = j(u)+j(v)$,

(*c*) $j(\lambda u) = \lambda j(u), \lambda \geqslant 0$.

Proof. Define $j(u) = s_u$, where

$$s_u(\alpha, x) = \begin{cases} s_{L_\alpha u}(x), & \text{if} \quad \alpha > 0 \\ s_{\text{supp}\, u}(x), & \text{if} \quad \alpha = 0. \end{cases}$$

The function s_u is Lipschitz on $[0,1]\times S^{p-1}$:

$$\begin{aligned} |s_u(\alpha,x)-s_u(\beta,y)| &\leqslant |s_{L_\alpha u}(x)-s_{L_\beta u}(x)|+|s_{L_\beta u}(x)-s_{L_\beta u}(y)| \\ &\leqslant \|s_{L_\alpha u}-s_{L_\beta u}\|_\infty + \|L_\beta u\|\,\|x-y\| \\ &\leqslant \|u\|_L(|\alpha-\beta|+\|x-y\|), \end{aligned} \tag{6.1}$$

where
$$\|u\|_L = \sup_{\alpha\neq\beta} \frac{d(L_\alpha u, L_\beta u)}{|\alpha-\beta|} + \sup_{\alpha>0} \|L_\alpha u\|$$

and we have used the fact that $K \in \mathscr{K}_c(\mathbb{R}^p) \to s_K$ is an isometry, as well as the fact that s_K satisfies a Lipschitz condition with constant $\|K\|$.

Since any two norms in $\mathbb{R}^{p+1}$ are equivalent, inequality (6.1) becomes

$$|s_u(\alpha,x)-s_u(\beta,y)| \leqslant q\|u\|_L(|\alpha-\beta|^2+\|x-y\|^2)^{\frac{1}{2}}, \tag{6.2}$$

where q is a constant.

Thus $s_u = j(u) \in C([0,1]\times S^{p-1})$.

It is now easy to show that j is an isometry, i.e.

$$\|j(u)-j(v)\|_\infty = \sup_{\alpha,\, x} |s_u(\alpha,x)-s_v(\alpha,x)| = d_\infty(u,v)$$

as well as properties (*b*) and (*c*), and the proof is completed.

7. Central limit theorem

The CLT for random sets was first given, in a particular case, by Cressie (1979). The general theorem first appeared in Weil (1982) and, independently, in Giné *et al.* (1982); Vitale (1981); Trader & Eddy (1981); and Puri & Ralescu (1985). Lyashenko (1979) even considered the non IID case, although in a metric different from the Hausdorff metric.

We will consider here fuzzy random variables with values in the space $(\mathscr{F}_L(\mathbb{R}^p), d_\infty)$. Our main result is as follows.

Theorem 7.1. $X_1, X_2, \ldots : \Omega \to \mathscr{F}_L(\mathbb{R}^p)$ *be independent and identically distributed fuzzy random variables, satisfying*:

(*a*) $E\|\text{supp}\, X_1\|^2 < \infty$,

(*b*) $E\left[\sup_{\alpha\neq\beta} \frac{d(L_\alpha X_1, L_\beta X_1)}{|\alpha-\beta|}\right]^2 < \infty$.

Then there exists a Gaussian random element Z in $C([0,1]\times S^{p-1})$, *such that* $\sqrt{n}\, d_\infty(X_1+X_2+\ldots+X_n/n, E(\text{co}\, X_1)) \to \|Z\|_\infty$ *weakly*.

Proof. (1) Consider first the convex case, i.e. $X_n : \Omega \to \mathscr{F}_{CL}(\mathbb{R}^p)$.

Let $j: \mathscr{F}_{CL}(\mathbb{R}^p) \to C([0,1] \times S^{p-1})$ be the isometry given by Theorem 6.1, and consider $Y_n = j(X_n)$ as random elements in $C([0,1] \times S^{p-1})$.

Observe that inequality (6.2) in the proof of Theorem 6.1 can be rewritten as

$$|Y_1(s) - Y_1(t)| \leqslant q\|X_1\|_L \|s-t\| \tag{7.1}$$

for $s, t \in [0,1] \times S^{p-1}$. Also note that hypotheses (a) and (b) imply $E\|X_1\|_L^2 < \infty$.

Next, note that the minimal number of closed balls of radius $\leqslant \epsilon$ which cover S^{p-1} is of the order $c\epsilon^{1-p}$ for some constant c.

Then, the minimal number of such balls which cover the cylinder $[0,1] \times S^{p-1}$ is $N(\epsilon) = c\epsilon^{-p}$. If $H(\epsilon) = \lg N(\epsilon)$ denotes the metric entropy of $[0,1] \times S^{p-1}$ (corresponding to the Euclidean distance), it is easy to check that

$$\int_0^1 H^{\frac{1}{2}}(\epsilon)\,\mathrm{d}\epsilon < \infty. \tag{7.2}$$

Conditions (7.1) and (7.2) show that the central limit theorem of Jain & Marcus (1975) can be applied. Thus,

$$\sqrt{n}\left[\sum_{i=1}^{n} j(X_i)/n - Ej(X_1)\right] \to Z \quad \text{weakly}, \tag{7.3}$$

where Z is Gaussian in $C([0,1] \times S^{p-1})$.

Since $E(s_f) = s_{Ef}$ for any random set $f: \Omega \to \mathscr{K}_c(\mathbb{R}^p)$, it is not difficult to show that $Ej(X_1) = j(EX_1)$.

This fact, together with (7.3), the continuous mapping theorem, and the fact that j is an isometry, implies

$$\sqrt{n}\, d_\infty\left(\sum_{i=1}^{n} X_i/n, EX_1\right) \to \|Z\|_\infty \quad \text{weakly}.$$

(2) Assume now that $X_n : \Omega \to \mathscr{F}_L(\mathbb{R}^p)$, and consider the fuzzy random variables $\operatorname{co} X_n$ (see §5). The Shapley–Folkman lemma (see Arrow & Hahn 1971) gives

$$d_\infty\left(\sum_{i=1}^{n} X_i, \operatorname{co} \sum_{i=1}^{n} X_i\right) \leqslant \sqrt{p} \max_{1 \leqslant i \leqslant n} \|\operatorname{supp} X_i\|. \tag{7.4}$$

Therefore, we can write:

$$\sqrt{n}\, d_\infty\left(\sum_{i=1}^{n} X_i/n, E(\operatorname{co} X_1)\right)$$
$$\leqslant \sqrt{n}\, d_\infty\left(\sum_{i=1}^{n} X_i/n, \sum_{i=1}^{n} \operatorname{co} X_i/n\right) + \sqrt{n}\, d_\infty\left(\sum_{i=1}^{n} \operatorname{co} X_i/n, E(\operatorname{co} X_1)\right).$$

Note that if $u \in \mathscr{F}_L(\mathbb{R}^p)$, then $\|\operatorname{co} u\|_L \leqslant \|u\|_L$ so $E\|\operatorname{co} X_1\|_L^2 < \infty$ and the second term of the above inequality converges weakly to $\|Z\|_\infty$ by part (1) of the proof. For the first term, we use (7.4), so that

$$\sqrt{n}\, d_\infty\left(\sum_{i=1}^{n} X_i/n, \sum_{i=1}^{n} \operatorname{co} X_i/n\right), \leqslant \sqrt{p} \max_{1 \leqslant i \leqslant n} \|\operatorname{supp} X_i\|/\sqrt{n}.$$

But $E\|\mathrm{supp}\, X_1\|^2 < \infty$ implies that $\max_{1\leq i\leq n} \|\mathrm{supp}\, X_i\|/\sqrt{n} \to 0$ weakly (see, for example, Breiman 1968, p. 188). It follows, finally, that

$$\sqrt{n}\, d_\infty\left(\sum_{i=1}^{n} X_i/n, E(\mathrm{co}\, X_1)\right) \to \|Z\|_\infty$$

weakly, concluding the proof.

8. Concluding remarks

One of the main difficulties in proving the SLLN for fuzzy random variables was the fact that the space $(\mathscr{F}(\mathbb{R}^p), d_\infty)$ is not separable. It would be possible to overcome this, by assuming that $\{X_k \mid k \geq 1\}$ are IID fuzzy random variables which are separably-valued a.e. Such a hypothesis, however, does not seem to be natural in the context of the SLLN.

We also mention that it is possible to define a metric in $\mathscr{F}(\mathbb{R}^p)$, more general than d_1, by the formula

$$d_r(u, v) = \left[\int_0^1 d^r(L_\alpha u, L_\alpha v)\, \mathrm{d}\alpha\right]^{1/r}, \quad r \geq 1.$$

The SLLN holds with respect to d_r as well but for the clarity of the exposition we have chosen the metric d_1.

The central limit theorem proved in §7 works for fuzzy random variables with values in $\mathscr{F}_L(\mathbb{R}^p)$. From the practical point of view, it might be difficult to check the Lipschitz condition for the function $\alpha \to L_\alpha u$.

In the following, we state some sufficient conditions which imply this Lipschitz condition.

(1) Let $u \in \mathscr{F}(\mathbb{R}^p)$. If there exists a constant $M > 0$ such that $|u(x) - u(y)| \geq M\|x - y\|$ for $x, y \in \mathrm{supp}\, u$, then $u \in \mathscr{F}_L(\mathbb{R}^p)$.

This statement can be proved by using properties of the Hausdorff distance and compactness arguments. It gives examples of fuzzy sets in $\mathscr{F}_L(\mathbb{R}^p)$ but it is quite restrictive, since u satisfying the above condition must be a one-to-one function.

(2) Let $u \in \mathscr{F}(\mathbb{R}^p)$. If

$$\min_{1\leq j\leq p} \inf\left\{\left|\frac{\partial u}{\partial x_j}(x)\right| : x \in \mathrm{supp}\, u \backslash L_1 u\right\} > 0$$

then $u \in \mathscr{F}_L(\mathbb{R}^p)$. Here the infimum is taken over all points in $\mathrm{supp}\, u \backslash L_1 u$ where the derivatives of u exist.

In the one dimensional case (i.e. $\mathscr{F}(\mathbb{R})$), this condition states that if the derivative of u (where it exists) is bounded away from zero on $\mathrm{supp}\, u \backslash L_1 u$, then $u \in \mathscr{F}_L(\mathbb{R})$.

M. L. P. and D. A. R. were supported by the National Science Foundation Grant IST-7918468, and the Office of Naval Research Contract N00014-85-K-0648.

References

Arrow, K. J. & Hahn, F. H. 1971 *General competitive analysis*. Holden-Day.

Artstein, Z. & Vitale, R. A. 1975 A strong law of large numbers for random compact sets. *Ann. Prob.* **3**, 879–882.

Aumann, R. J. 1965 Integrals of set-valued functions. *J. math. Anal. Appl.* **12**, 1–22.

Billingsley, P. 1968 *Convergence of probability measures*. Wiley.

Breiman, L. 1968 *Probability*. Addison-Wesley.

Chow, Y. S. & Teicher, H. 1978 *Probability theory*. Springer.

Cressie, N. 1979 A central limit theorem for random sets. *Z. Wahrscheinlichkeitstheor. Verw. Geb.* **49**, 37–47.

Debreu, G. 1967 Integration of correspondences. In *Proc. Fifth Berkeley Symp. math. statist. probability*, vol. II, part 1, 351–372. University of California Press (1967).

Féron, R. 1979 Ensembles aléatoires flous dont la fonction d'appartenance prend ses valeurs dans un treillis distributif fermé. In *Publ. Econométriques*, vol. XII, Fasc. 1, 81–118.

Giné, E., Hahn, M. G. & Zinn, J. 1982 Limit theorems for random sets: an application of probability in Banach space results. In *Proc. Fourth Int. Conf. on Probability in Banach Space* (In the press.)

Jain, N. C. & Marcus, M. B. 1975 Central limit theorem for $C(S)$-valued random variables. *J. Funct. Anal.* **19**, 216–231.

Lowen, R. 1980 Convex fuzzy sets. *Fuzzy Sets and Systems* **3**, 291–310.

Lyashenko, N. N. 1979 On limit theorems for sums of independent compact random subsets in the Euclidean space. *Zap. Nauch. Semin., LOMI* **85**, 113–128. (In Russian.)

Matheron, G. 1975 *Random sets and integral geometry*. Wiley.

Negoita, C. V. & Ralescu, D. A. 1975 *Applications of fuzzy sets to systems analysis*. Wiley.

Puri, M. L. & Ralescu, D. A. 1983*a* Strong law of large numbers for Banach space valued random sets. *Ann. Prob.* **11**, 222–224.

Puri, M. L. & Ralescu, D. A. 1983*b* Differentials of fuzzy functions. *J. math. Anal. Appl.* **91**, 552–558.

Puri, M. L. & Ralescu, D. A. 1985 Limit theorems for random compact sets in Banach space. *Math. Proc. Camb. Phil. Soc.* **97**, 151–158.

Puri, M. L. & Ralescu, D. A. 1986 Fuzzy random variables. *J. math. Anal. Appl.* **114**, 151–158.

Rådström, H. 1952 An embedding theorem for spaces of convex sets. *Proc. Am. math. Soc.* **3**, 165–169.

Taylor, R. L. 1978 *Stochastic convergence of weighted sums of random elements in linear spaces*. (*Lecture Notes in Mathematics*, no. 672.) Springer.

Trader, D. A. & Eddy, W. F. 1981 A central limit theorem for the Minkowski sum of random sets. Preprint, Carnegie-Mellon University.

Vitale, R. A. 1981 A central limit theorem for random convex sets. Preprint. Claremont Graduate School, Claremont, California.

Weil, W. 1982 An application of the central limit theorem for Banach space-valued random variables to the theory of random sets. *Z. Wahrscheinlichkeitstheor. Verw. Geb.* **60**, 203–208.

Zadeh, L. A. 1965 Fuzzy sets. *Inf. Control* **8**, 338–353.

Reprinted by permission of the publisher,
© 1986 by The Royal Society

CONVERGENCE THEOREM FOR FUZZY MARTINGALES

MADAN L. PURI* and DAN A. RALESCU†

Department of Mathematics, Indiana University, Bloomington, Indiana 47405
Department of Mathematics, University of Cincinnati, Cincinnati, Ohio 45221

We study fuzzy set-valued measures in a Banach space and their relationships to fuzzy random variables. Our main result is a convergence theorem for fuzzy martingales. Our tool is a Radon–Nikodym type theorem for fuzzy measures which are absolutely continuous with respect to probability measures. © 1991 Academic Press, Inc.

1. INTRODUCTION

In recent years the concept of random set which has its roots in the field of geometric probability (see Kendall [11]) has been formalized as an extension of random variables and random vectors (see Matheron [14]).

Various limit theorems have been proved for sequences of independent compact random sets. First, Artstein and Vitale [3] established a strong law of large numbers for i.i.d. random sets in $\mathbb{R}^n$. Later Weil [23] proved the central limit theorem in the same setting. In a Banach space setting, limit theorems for random sets are given by Puri and Ralescu [16], Giné, Hahn, and Zinn [8], Puri and Ralescu [18], and Artstein and Hansen [2].

It is important to realize that the theory of random sets is intimately related to the somewhat more general concept of multivalued function; in

* Research supported by the Office of Naval Research Contracts N00014-85-K-0648 and N00014-91-J-1020.

† Research supported by the National Science Foundation Grant.

particular, the integral of a multifunction (see Aumann [4] and Debreu [5]) is used to define the expected value of a random set.

The dual concept of a set-valued measure was also studied quite extensively; in particular the Radon–Nikodym theorem for sch measures was initiated by Debreu and Schmeidler [6]. More general results for set-valued measures in $\mathbb{R}^n$ can be found in Artstein [1].

Hiai [9] and Hiai and Umegaki [10] have studied set-valued measures in a Banach space, including the concepts of conditional expectation and set-valued martingale. Various limit results for this type of martingales are also given in the last two papers just quoted.

The concept of a fuzzy random variable was introduced by Puri and Ralescu [20] as a model to represent relationships between the outcomes of a random experiment and inexact data. By inexactness here we mean non-statistical inexactness, due to the subjectivity and imprecision of human knowledge rather than to the occurrence of random events. In particular, the concept of Gaussian fuzzy random variable was defined and a characterization was provided by Puri and Ralescu [19]. Limit theorems which generalize the results for random sets, are obtained by Klement, Puri, and Ralescu [12].

In this paper we continue this study and, more specifically, we focus on the concept of fuzzy-valued measure, the Radon–Nikodym theorem in this setting, and the concept of conditional expectation (of a fuzzy random variable). Finally, we introduce the notion of fuzzy martingale and prove a convergence theorem which generalizes the result for set-valued martingales of Hiai [9]. This generalization is twofold: first our variables take values in a space which includes the space of compact convex sets, and second, the Banach space $\mathscr{X}$ is not necessarily finite dimensional.

The motivation for the theory that we develop here comes from the need to give a rigorous framework for the problem of inexact measurement. In many instances of practical interest, the measurement process is imprecise, or not totally accurate, and so the values of the measure of various objects are only known within some tolerances. It is important to note that set-valued measures found their use in statistics, in particular in Bayesian analysis (see DeRobertis and Hartingan [7]), and a strong law of large numbers with respect to a set-valued probability measure was given by Puri and Ralescu [17]. The importance of these concepts and techniques in Bayesian analysis comes from the fact that prior information is often vague, imprecise, or not totally reliable.

This paper is organized as follows: in Section 2 we give the essential definitions and results pertinent to fuzzy random variables and their expected value. In Section 3 we discuss fuzzy measures whose values are upper semicontinuous. In Section 4 we introduce the conditional expectation of a fuzzy random variable, by using a Radon–Nikodym theorem for fuzzy set-valued measures. The latter result generalizes the Rieffel [21] theorem for vector measures, as well as the Radon–Nikodym theorem for set-valued measures of Hiai [9]. Finally, in Section 5 we define fuzzy martingales and prove a convergence theorem which extends the corresponding result of Hiai [9].

2. Fuzzy Random Variables

In the remainder of this paper, $\mathscr{X}$ will denote a (real) separable Banach space with norm $\| \ \|$ and with dual $\mathscr{X}^*$. We will also constantly use the notations

$$\mathscr{K}(\mathscr{X}) = \{K \subseteq \mathscr{X} \mid K \text{ nonempty, compact}\}$$

$$\mathscr{K}_c(\mathscr{X}) = \{K \subseteq \mathscr{X} \mid K \text{ nonempty, compact, and convex}\}.$$

A linear structure in $\mathscr{K}(\mathscr{X})$ is defined via the operations

$$K + L = \{a + b \mid a \in K, b \in L\}$$

$$\lambda K = \{\lambda a \mid a \in K\}$$

for K, $L \in \mathscr{K}(\mathscr{X})$, $\lambda \in \mathbb{R}$. However, note that $(\mathscr{K}(\mathscr{X}), +, \cdot)$ is *not* a vector space, since $K + (-K) \neq \{0\}$, unless K is a singleton.

The topology in $\mathscr{K}(\mathscr{X})$ is introduced through the Hausdorff metric

$$d(K, L) = \max\{\sup_{a \in K} \inf_{b \in L} \|a - b\|, \sup_{b \in L} \inf_{a \in K} \|a - b\|\}.$$

It is well known (see, e.g., Debreu [5]) that $\mathscr{K}(\mathscr{X})$ is a complete and separable metric space, and that $\mathscr{K}_c(\mathscr{X})$ is a closed subspace.

Denote the norm of a set $K \in \mathscr{K}(\mathscr{X})$ by $\|K\|$; defined as

$$\|K\| = d(K, \{0\}) = \sup_{a \in K} \|a\|.$$

If $(\Omega, \mathscr{A}, P)$ is a probability space, a *random set* is defined as a Borel measurable function $f: \Omega \to (\mathscr{K}(\mathscr{X}), d)$. The expected value Ef (see Aumann [4] and Debreu [5]) is defined by

$$Ef = \{E\phi \mid \phi: \Omega \to \mathscr{X}, E\|\phi\| < \infty, \phi(\omega) \in f(\omega) \text{ a.e.}\}.$$

Note that $E\|f\| < \infty$ implies the existence of $Ef \in \mathscr{K}(\mathscr{X})$. Furthermore, if the probability measure P is nonatomic, then $E \|f\| < \infty$ implies $Ef \in \mathscr{K}_c(\mathscr{X})$.

A *fuzzy set* in $\mathscr{X}$ is a function $u : \mathscr{X} \to [0, 1]$ (see Zadeh [24]). Denote by $L_\alpha u = \{x \in \mathscr{X} \mid u(x) \geqslant \alpha\}$ for $0 \leqslant \alpha \leqslant 1$, the α-level set of u. Further, denote by supp u the support of u (i.e., the closure of $\{x \in \mathscr{X} \mid u(x) > 0\}$).

An extension of $\mathscr{K}(\mathscr{X})$ is obtained by defining the space $\mathscr{F}(\mathscr{X})$ of all fuzzy sets $u : \mathscr{X} \to [0, 1]$ with the properties

(a) u is upper semicontinuous

(b) supp u is compact

(c) $\{x \in \mathscr{X} \mid u(x) = 1\} \neq \varnothing$.

The space $\mathscr{F}_c(\mathscr{X})$ contains all fuzzy sets in $\mathscr{F}(\mathscr{X})$ which are also fuzzy convex (see Zadeh [24] ($u : \mathscr{X} \to [0, 1]$ is *fuzzy convex* if $u(\lambda x + (1 - \lambda) y) \geqslant \min [u(x), u(y)]$, for all $x, y \in \mathscr{X}$, $\lambda \in [0, 1]$).

It is clear that $K \in \mathscr{K}(\mathscr{X})$ implies $\chi_K \in \mathscr{F}(\mathscr{X})$, while $K \in \mathscr{K}_c(\mathscr{X})$ implies $\chi_K \in \mathscr{F}_c(\mathscr{X})$ (where χ_K is the characteristic function of the set K).

A linear structure in $\mathscr{F}(\mathscr{X})$ is defined by the operations

$$(u + v)(x) = \sup_{y + z = x} \min [u(y), v(z)]$$

$$(\lambda u)(x) = \begin{cases} u(x/\lambda), & \text{if } \lambda \neq 0 \\ \chi_{\{0\}}(x), & \text{if } \lambda = 0, \end{cases}$$

where $u, v \in \mathscr{F}(\mathscr{X})$ and $\lambda \in \mathbb{R}$.

The Hausdorff metric can be extended to $\mathscr{F}(\mathscr{X})$ in various ways; two of the possible metrics are

$$d_1(u, v) = \int_0^1 d(L_\alpha u, L_\alpha v)\, d\alpha$$

$$d_\infty(u, v) = \sup_{0 < \alpha \leqslant 1} d(L_\alpha u, L_\alpha v).$$

For reasons which will become clear later, we will use in this paper the metric d_∞. Note that $(\mathscr{F}(\mathscr{X}), d_\infty)$ is complete (see Puri and Ralescu [18, 19]) but it is not separable (see Klement, Puri, and Ralescu [12]).

The norm $\| u \|$ of a fuzzy set $u \in \mathscr{F}(\mathscr{X})$ is defined by

$$\| u \| = d_\infty(u, \chi_{\{0\}}) = \sup_{\alpha > 0} \| L_\alpha u \|$$

and it is obvious that $K \in \mathscr{K}(\mathscr{X})$ implies $\| \chi_K \| = \| K \|$.

A *fuzzy random variable* (*f. r. v.*, for short) is a Borel measurable function $X : \Omega \to (\mathscr{F}(\mathscr{X}), d_\infty)$. Quite often we will consider f.r.v.'s whose values are in $\mathscr{F}_c(\mathscr{X})$.

Note that if X is a f.r.v. then supp X is a random set (i.e., Borel measurable).

If $E \|\text{supp } X\| < \infty$, the *expected value* EX is defined as the unique fuzzy set which satisfies the property

$$L_\alpha(EX) = E(L_\alpha X), \qquad 0 < \alpha \leqslant 1.$$

It also follows that $E \|\text{supp } X\| < \infty$ implies $EX \in \mathscr{F}(\mathscr{X})$ (for this and more details on the construction of EX, see Puri and Ralescu [20]).

The following inequality will be useful later: if X is a f.r.v. with $E \|\text{supp } X\| < \infty$, then

$$\|EX\| \leqslant E \|\text{supp } X\|. \tag{2.1}$$

The concept of normality for a special class of f.r.v.'s was defined in Puri and Ralescu [19], while a strong law of large numbers and a central limit theorem for independent and identically distributed f.r.v.'s were proved in Klement, Puri, and Ralescu [12]. Kruse [13] gave a law of large numbers using a different concept of expected value.

3. Fuzzy Measures

In this section we define fuzzy set-valued measures and indicate some of their properties that are of interest to us. Note that we use the term fuzzy measure in a sense different from Sugeno [22] (the latter concept is that of a nonadditive, real-valued set-function).

A series $\sum_{j=1}^{\infty} u_j$ of fuzzy sets $u_j \in \mathscr{F}(\mathscr{X})$ is called *unconditionally convergent* if $\sum_{j=1}^{\infty} \|u_j\| < \infty$.

Let Ω be a set and let $\mathscr{A}$ be a σ-algebra of subsets of Ω.

Definition 3.1. A *fuzzy set-valued measure* (or *fuzzy measure*, for short) is a set-function $\mu : \mathscr{A} \to \mathscr{F}(\mathscr{X})$ with the properties

(a) $\mu(\varnothing) = \chi_{\{0\}}$

(b) $\mu(\bigcup_{j=1}^{\infty} A_j) = \sum_{j=1}^{\infty} \mu(A_j)$, for every sequence $\{A_j\}_{j \geqslant 1}$, $A_j \in \mathscr{A}$, such that $A_j \cap A_k = \varnothing$ for $j \neq k$.

Note that in part (b) of the above definition, it is implicit that the series in the right-hand side is unconditionally convergent.

This concept of fuzzy measure extends that of set-valued measure (see Debreu and Schmeidler [6] and Artstein [1]).

If $\mu : \mathscr{A} \to \mathscr{F}(\mathscr{X})$ is a fuzzy measure, and if $0 < \alpha \leqslant 1$, we can define a set-valued measure $\mu_\alpha : \mathscr{A} \to \mathscr{K}(\mathscr{X})$ by

$$\mu_\alpha(A) = L_\alpha[\mu(A)], \qquad A \in \mathscr{A}.$$

Another set-valued measure that can be associated to a fuzzy measure μ is

$$(\text{supp } \mu)(A) = \text{supp } [\mu(A)], \qquad A \in \mathscr{A},$$

where supp in the right-hand side denotes the support of the corresponding fuzzy set.

The fact that these definitions make sense is a consequence of the following result.

PROPOSITION 3.1. (1) *μ_α is a set-valued measure.*

(2) supp *μ is a set-valued measure.*

Proof. (1) All we have to show is

$$L_\alpha\left(\sum_{j=1}^{\infty} u_j\right) = \sum_{j=1}^{\infty} L_\alpha u_j, \qquad \forall 0 < \alpha \leqslant 1, \tag{3.1}$$

where $\sum_{j=1}^{\infty} u_j$ is unconditionally convergent.

Since $\sum_{j=1}^{\infty} \|u_j\| < \infty$ and since $(\mathscr{F}(\mathscr{X}), d_\infty)$ is complete, it follows that $\lim_{k\to\infty} d_\infty(\sum_{j=1}^{k} u_j, u) = 0$, for some $u \in \mathscr{F}(\mathscr{X})$.

Fix an $\alpha \in (0, 1]$; then $d(L_\alpha(\sum_{j=1}^{k} u_j), L_\alpha u) \to 0$ as $k \to \infty$. Since $L_\alpha(\sum_{j=1}^{k} u_j) = \sum_{j=1}^{k} L_\alpha u_j$, the result follows.

(2) Here we have to show that $\text{supp}(\sum_{j=1}^{\infty} u_j) = \sum_{j=1}^{\infty} \text{supp } u_j$ (the fact that supp μ assumes values in $\mathscr{K}(\mathscr{X})$ is clear).

But $\text{supp}(\sum_{j=1}^{\infty} u_j) = \text{closure}[\bigcup_{\alpha>0} L_\alpha(\sum_j u_j)] = \text{closure}[\bigcup_{\alpha>0} (\sum_j L_\alpha u_j)]$ by using (3.1).

The fact that $\alpha \leqslant \beta \Rightarrow L_\alpha u \supseteq L_\beta u$ (for $u \in \mathscr{F}(\mathscr{X})$) implies $\bigcup_{\alpha>0} (\sum_j L_\alpha u_j) = \sum_{j=1}^{\infty} (\bigcup_{\alpha>0} L_\alpha u_j)$. The proof now follows by using simple properties of the closure.

It is also possible to associate to a fuzzy measure μ, a positive measure $\|\mu\|$—its *total variation.* It is defined by

$$\|\mu\|(A) = \sup \sum_{j=1}^{\infty} \|\mu(A_j)\|, \qquad A \in \mathscr{A},$$

where the supremum is taken over all partitions $\{A_j\}_{j\geqslant 1}$ of A.

The relationships between μ, supp μ, and the family $(\mu_\alpha)_\alpha$ are given in

PROPOSITION 3.2. *The following formulas hold*

$$\|\mu\| = \|\text{supp } \mu\| = \sup_{\alpha>0} \|\mu_\alpha\|.$$

Proof. Note that $\|\text{supp}\, u\| = \|u\|$ for $u \in \mathscr{F}(\mathscr{X})$. So we get

$$\|\mu\|(A) = \sup \sum_{j=1}^{\infty} \|\mu(A_j)\| = \sup \sum_{j=1}^{\infty} \|\text{supp}\, \mu(A_j)\| = \|\text{supp}\, \mu\|(A)$$

which gives the first formula in the statement.

Next, observe that the fact that L_α is nonincreasing in α implies

$$\sup_{\alpha>0} \left(\sum_{j=1}^{\infty} \|L_\alpha u_j\| \right) = \sum_{j=1}^{\infty} \sup_{\alpha>0} \|L_\alpha u_j\|,$$

where $u_j \in \mathscr{F}(\mathscr{X})$. Therefore we can write

$$\begin{aligned} \|\mu\|(A) &= \sup \sum_j \|\mu(A_j)\| = \sup \sum_j (\sup_{\alpha>0} \|\mu_\alpha(A_j)\|) \\ &= \sup \sup_{\alpha>0} \left(\sum_j \|\mu_\alpha(A_j)\| \right) = \sup_{\alpha>0} \sup \left(\sum_j \|\mu_\alpha(A_j)\| \right) \\ &= \sup_{\alpha>0} \|\mu_\alpha\|(A). \end{aligned}$$

4. Conditional Expectation

In order to define the conditional expectation of a fuzzy random variable, we have to use a Randon–Nikodym theorem for fuzzy measures. Such a result, which we discuss next, is obtained as a generalization of the Radon–Nikodym theorem for set-valued measures (see Debreu and Schmeidler [6] and, more recently, Hiai [9]). The fact that this is not a straightforward generalization will be clear from the sequel.

Although it is possible to consider a more general framework, we will limit our discussion to fuzzy measures with values in $\mathscr{F}_c(\mathscr{X})$.

Clearly, if $X : \Omega \to \mathscr{F}_c(\mathscr{X})$ is a f.r.v. with $E\,\|\text{supp}\, X\| < \infty$, we can define μ_X by

$$\mu_X(A) = \int_A X\, dP = E(X\chi_A), \qquad A \in \mathscr{A}.$$

The fact that μ_X is a fuzzy measure follows from the Lebesgue dominated type of convergence theorem for f.r.v.'s (see Puri and Ralescu [19]). Also μ_X is *absolutely continuous* with respect to P, denoted as $\mu_X \ll P$, in the following sense

$$P(A) = 0 \Rightarrow \mu_X(A) = \chi_{\{0\}}.$$

The Radon–Nikodym theorem is concerned with the converse problem: Given a fuzzy measure μ with $\mu \ll P$, does there exist an f.r.v. X such that $\mu = \mu_X$?

Clearly, for an affirmative answer, additional hypotheses are necessary. The precise statement, giving necessary and sufficient conditions, is as follows.

THEOREM 4.1. *Let $\mu : \mathcal{A} \to \mathcal{F}_c(\mathcal{X})$ be a fuzzy measure. There exists a fuzzy random variable $X : \Omega \to \mathcal{F}_c(\mathcal{X})$ with $E\,\|\text{supp}\, X\| < \infty$ and such that*

$$\mu(A) = \int_A X\, dP, \qquad A \in \mathcal{A} \tag{4.1}$$

if and only if the following conditions are satisfied

(i) $\mu \ll P$

(ii) $\|\text{supp}\,\mu\|(\Omega) < \infty$

(iii) $\forall A \in \mathcal{A},\ P(A) > 0 \Rightarrow \exists B \subset A,\ P(B) > 0$, *and* $\exists M > 0$ *such that*

$$C \subset B,\ P(C) > 0 \Rightarrow \frac{1}{P(C)}\, \|\mu(C)\| \leqslant M.$$

Proof. (1) We first show that μ given by (4.1) satisfies (i)–(iii). Clearly (i) is satisfied. For (ii) we have

$$\begin{aligned} \|\text{supp}\,\mu\|(\Omega) &= \sup_{\alpha > 0} \|\mu_\alpha\|(\Omega) \\ &= \sup_{\alpha > 0} E\, \|L_\alpha X\| = E\,(\sup_{\alpha > 0} \|L_\alpha X\|) = E\,\|\text{supp}\, X\| < \infty \end{aligned}$$

(where we have used Proposition 3.2 and Hiai [9, Proposition 4.1]).

To proce (iii), consider the positive measure

$$\xi(A) = \int_A \|\text{supp}\, X\|\, dP, \qquad A \in \mathcal{A}$$

It is obvious that (iii) is satisfied with ξ in place of μ. Inequality (2.1) implies

$$\|E(X\chi_A)\| \leqslant E(\|\text{supp}\, X\|\, \chi_A)$$

and this, in turn, implies (iii).

(2) The fact that (i)–(iii) imply the existence of the Radon–Nikodym derivative X and the representation (4.1) for μ, is obtained through the

following construction. Fix $\alpha \in (0, 1]$ and consider the set-valued measure μ_α (Section 3). It is clear that $\mu_\alpha \ll P$, $\|\mu_\alpha\|(\Omega) < \infty$, and that μ_α satisfies the equivalent of (iii). By applying Hiai [9, Theorem 5.2] we get a random set $X_\alpha : \Omega \to K_c(\mathscr{X})$ with $E\|X_\alpha\| < \infty$ and such that

$$\mu_\alpha(A) = \int_A X_\alpha dP, \qquad A \in \mathscr{A}.$$

Consider now the family of random sets $\{X_\alpha \mid 0 < \alpha \leqslant 1\}$. Since $\alpha \leqslant \beta$ implies $\mu_\alpha(A) \supseteq \mu_\beta(A)$ for each $A \in \mathscr{A}$, it follows that $X_\alpha \supseteq X_\beta$ a.e.

If $\{\alpha_n\}_{n \geqslant 1}$ is increasing with $\alpha_n \to \alpha$, we similarly get $X_\alpha = \bigcap_{n=1}^{\infty} X_{\alpha_n}$ a.e. (this is a consequence of the fact that $\mu_\alpha(A) = \bigcap_{n=1}^{\infty} \mu_{\alpha_n}(A)$ for each $A \in \mathscr{A}$).

Application of the representation theorem for fuzzy sets in Negoita and Ralescu [15, Theorem 5, p. 29] provides a (unique) fuzzy random variable $X : \Omega \to \mathscr{F}_c(\mathscr{X})$, with $L_\alpha[X(\omega)] = X_\alpha(\omega)$ a.e. for every $\alpha \in (0, 1]$. Thus

$$L_\alpha\left(\int_A X\, dP\right) = \int_A (L_\alpha X)\, dP = \int_A X_\alpha dP = \mu_\alpha(A) = L_\alpha[\mu(A)]$$

for every α. Finally this implies

$$\mu(A) = \int_A X\, dP, \qquad A \in \mathscr{A}$$

and also $E\|\operatorname{supp} X\| \leqslant \|\operatorname{supp} \mu\|(\Omega) < \infty$ which completes the proof.

We are now ready to define the conditional expectation of f.r.v. X with respect to a sub-σ-algebra $\mathscr{B} \subseteq \mathscr{A}$.

Consider $X : \Omega \to \mathscr{F}_c(\mathscr{X})$ with $E\|\operatorname{supp} X\| < \infty$.

DEFINITION 4.1. The *conditional expectation* of X with respect to $\mathscr{B}$ is the f.r.v. $E(X \mid \mathscr{B})$ with the following properties

(a) $E(X \mid \mathscr{B})$ is $\mathscr{B}$-measurable

(b) $\int_B E(X \mid \mathscr{B})\, dP = \int_B X\, dP$ for every $B \in \mathscr{B}$.

It is not clear at this point whether $E(X \mid \mathscr{B})$ exists. This can be obtained by a construction similar to the classical one, since we have now available Theorem 4.1. We briefly sketch this construction below.

Consider the set-function $\nu : \mathscr{B} \to \mathscr{F}_c(\mathscr{X})$,

$$\nu(B) = \int_B X\, dP, \qquad B \in \mathscr{B}. \tag{4.2}$$

Clearly ν is a fuzzy measure which satisfies hypotheses (i) through (iii) of Theorem 4.1. Thus there exists an f.r.v., denoted by $E(X\mid\mathscr{B})$ (which is implicitly $\mathscr{B}$-measurable, so (a) is satisfied), such that

$$\nu(B)=\int_B E(X\mid\mathscr{B})\,dP, \qquad B\in\mathscr{B}. \tag{4.3}$$

From (4.2) and (4.3) we get

$$\int_B E(X\mid\mathscr{B})\,dP=\int_B X\,dP, \qquad B\in\mathscr{B}$$

which shows that (b) is also satisfied.

One of the properties of this conditional expectation is stated next.

PROPOSITION 4.1. *If X is an f.r.v. with $E\,\|\operatorname{supp} X\|<\infty$, then*

$$L_\alpha[E(X\mid\mathscr{B})]=E(L_\alpha X\mid\mathscr{B})$$

for every $\alpha>0$.

Proof. It suffices to show

$$\int_B L_\alpha[E(X\mid\mathscr{B})]\,dP=\int_B L_\alpha X\,dP$$

for every $B\in\mathscr{B}$. We have

$$\int_B L_\alpha[E(X\mid\mathscr{B})]\,dP=L_\alpha\left(\int_B E(X\mid\mathscr{B})\,dP\right)=L_\alpha\left(\int_B X\,dP\right)=\int_B L_\alpha X\,dP$$

according to our definition of the expected value of an f.r.v.

5. CONVERGENCE OF FUZZY MARTINGALES

We first define the concept of fuzzy martingale. Our main result is a convergence theorem which is a generalization of the convergence theorem for set-valued martingales (see Hiai [9, Theorem 6.3]) in two ways. First, we consider f.r.v.'s which take values in $\mathscr{F}_c(\mathscr{X})$ and this includes, in particular, $\mathscr{K}_c(\mathscr{X})$-valued martingales. Second, the result of Hiai [9] is only proved for the case when $\mathscr{X}$ is finite dimensional. We drop the restriction of $\dim\mathscr{X}<\infty$; its place is taken by a less restrictive assumption on the range of the fuzzy random variables that we consider.

First, start with a sequence $\{X_n\}_{n\geqslant 1}$ of f.r.v.'s, $X_n:\Omega\to\mathscr{F}_c(\mathscr{X})$, and a sequence $\{\mathscr{B}_n\}_n$ of sub-σ-algebras of $\mathscr{A}$, such that $\mathscr{B}_n\subset\mathscr{B}_{n+1}$, $n\geqslant 1$.

DEFINITION 5.1. The sequence $\{X_n, \mathscr{B}_n\}_n$ of f.r.v.'s and σ-algebras is called a *fuzzy martingale* if we have, for each $n \geq 1$:

(a) X_n is $\mathscr{B}_n$-measurable and $E\,\|\text{supp}\, X_n\| < \infty$

(b) $E(X_{n+1} \mid \mathscr{B}_n) = X_n$.

If property (b) is replaced by

(b′) $E(X_{n+1} \mid \mathscr{B}_n) \geq X_n$ $(E(X_{n+1} \mid \mathscr{B}_n) \leq X_n)$

then $\{X_n, \mathscr{B}_n\}_{n \geq 1}$ is called a *fuzzy submartingale* (*supermartingale*), respectively.

In order to prove our convergence theorem, we need to restrict the values of X_n to a subspace of $\mathscr{F}_c(\mathscr{X})$.

Define $\mathscr{F}_{cL}(\mathscr{X})$ as the space of fuzzy sets $u \in \mathscr{F}_c(\mathscr{X})$ with the property that the function $\alpha \to L_\alpha u$ is Lipschitz (with respect to the Hausdorff distance); i.e., there exists a constant $C > 0$, such that

$$d(L_\alpha u, L_\beta u) \leq C\,|\alpha - \beta|$$

for every $\alpha, \beta \in (0, 1]$.

The space $\mathscr{F}_{cL}(\mathscr{X})$ can be embedded isometrically into a Banach space, as we will see in a moment.

Denote by ball $\mathscr{X}^* = \{x^* \in X^* \mid \|x^*\| \leq 1\}$ the unit ball of the dual $\mathscr{X}^*$. It is well known that ball $\mathscr{X}^*$ with the weak*-topology is compact and metrizable. Denote by ρ a metric which induces the weak*-topology on ball $\mathscr{X}^*$.

Also let $C([0, 1] \times \text{ball}\, \mathscr{X}^*)$ denote the Banach space of functions $\phi : [0, 1] \times \text{ball}\, \mathscr{X}^* \to \mathbb{R}$ which are continuous with respect to the product of the topology in $\mathbb{R}$ and the weak*-topology.

The precise embedding is given in

THEOREM 5.1. *There exists a function*

$$j : \mathscr{F}_{cL}(\mathscr{X}) \to C([0, 1] \times \text{ball}\, \mathscr{X}^*)$$

such that

(i) *j is an isometry (i.e., $\|j(u) - j(v)\| = d_\infty(u, v)$)*

(ii) *$j(u + v) = j(u) + j(v)$*

(iii) *$j(\lambda u) = \lambda j(u)$, $\lambda \geq 0$.*

Proof. This can be found in Puri and Ralescu [19] (see also Klement, Puri, and Ralescu [12]).

To be more specific, the embedding j is defined by $j(u) = s_u$ where

$$s_u(\alpha, x^*) = \begin{cases} s_{L_\alpha u}(x^*), & \text{if } \alpha > 0 \\ s_{\text{supp } u^{(x^*)}}, & \text{if } \alpha = 0. \end{cases} \tag{5.1}$$

Note that we use here the *support function* s_K of a set $K \in \mathscr{K}_c(\mathscr{X})$; it is given by

$$s_K(x^*) = \sup_{a \in K} \langle x^*, a \rangle, \qquad x^* \in \mathscr{X}^*.$$

Some more notations are in order before we state our main result. Let $u \in \mathscr{F}_c(\mathscr{X})$ be fixed. For $v \in \mathscr{F}_c(\mathscr{X})$ we denote by

$$\| v \|_u = \inf\{c \geq 0 \mid v \leq cu\}$$

and we are interested in fuzzy sets v such that $\| v \|_u < \infty$.

We also denote by

$$\| L \|_K = \inf\{c \geq 0 \mid L \subseteq cK\},$$

where $K, L \in \mathscr{K}_c(\mathscr{X})$.

The following inequality will be useful

$$\| L_\alpha v \|_{L_\alpha u} \leq \| v \|_u \tag{5.2}$$

for every $\alpha > 0$, where $u, v \in \mathscr{F}_c(\mathscr{X})$.

Finally, for every $v \in \mathscr{F}_{cL}(\mathscr{X})$, denote by $\| v \|_{u,L}$ the quantity

$$\| v \|_{u,L} = \| v \|_u + \sup_{\alpha \neq \beta} \frac{d(L_\alpha v, L_\beta v)}{|\alpha - \beta|}.$$

We can now state our convergence theorem.

THEOREM 5.2. *Let $(X_n, \mathscr{B}_n)$ be a fuzzy submartingale (supermartingale), where X_n assumes values in $\mathscr{F}_{cL}(\mathscr{X})$. Suppose that $u \in \mathscr{F}_{cL}(\mathscr{X})$ is fixed and that*

$$\sup_{n \geq 1} E(\| X_n \|_{u,L}) < \infty.$$

Then there exists an f.r.v. X with $E \|\text{supp } X\| < \infty$ and such that

$$\lim_{n \to \infty} d_\infty(X_n, X) = 0 \qquad a.e.$$

Proof. We prove the result, without loss in generality, for the case of fuzzy submartingales, i.e., $E(X_{n+1} \mid \mathscr{B}_n) \geq X_n$ for $n \geq 1$.

Consider first the positive random variables $\| X_n \|_{u,L}$. We have

$$\int_B \| X_n \|_{u,L} \, dP \leqslant \int_B \| E(X_{n+1} | \mathscr{B}_n) \|_{u,L} \, dP \leqslant \int_B \| X_{n+1} \|_{u,L} \, dP$$

for $B \in \mathscr{B}_n$, $n \geqslant 1$.

Thus $(\| X_n \|_{u,L}, \mathscr{B}_n)$ is a submartingale and, since $\sup_{n \geqslant 1} E \| X_n \|_{u,L} < \infty$, it follows that $\| x_n \|_{u,L} \to \eta$ a.e., where η is a positive random variable with $E\eta < \infty$.

Fix now $\alpha \in [0, 1]$ and $x^* \in \text{ball}\, \mathscr{X}^*$ and consider $s_{X_n}(\alpha, x^*)$ (given by (5.1)) as real-valued random variables. We have

$$\int_B s_{X_n}(\alpha, x^*) \, dP = s_{\int_B X_n dP}(\alpha, x^*) \leqslant s_{\int_B X_{n+1} dP}(\alpha, x^*) = \int_B s_{X_{n+1}}(\alpha, x^*) \, dP.$$

The first and last equalities follow from Puri and Ralescu [19, Lemma 3.1] while the inequality is true since $(X_n, \mathscr{B}_n)$ is a fuzzy submartingale.

Thus $(s_{X_n}(\alpha, x^*), \mathscr{B}_n)$ is a submartingale. Moreover, we have

$$\begin{aligned} &\sup_{n \geqslant 1} E | s_{X_n}(\alpha, x^*) | \\ &\qquad \leqslant \| x^* \| \sup_{n \geqslant 1} E \| X_n \| \\ &\qquad \leqslant \| x^* \| \, \| u \| \sup_{n \geqslant 1} E \| X_n \|_u \leqslant \| x^* \| \, \| u \| \sup_{n \geqslant 1} E \| X_n \|_{u,L} < \infty. \end{aligned}$$

Therefore $s_{X_n}(\alpha, x^*)$ cnverges a.e. (for fixed α and x^*). Recall that ρ is the distance which induces the weak*-topology on ball $\mathscr{X}^*$. Define another distance ρ_0 on ball $\mathscr{X}^*$ by

$$\rho_0(x^*, y^*) = \sup_{a \in \text{supp}\, u} |\langle x^* - y^*, a \rangle|$$

and note that ρ_0 is continuous in the weak*-topology. Take now $\varepsilon > 0$; there exists $\delta > 0$ such that $\rho(x^*, y^*) < \delta$ implies $\rho_0(x^*, y^*) < \varepsilon$.

Compactness of $[0, 1]$ and of (ball $\mathscr{X}^*$, wk^*) implies the existence of $\alpha_1, ..., \alpha_r \in [0, 1]$ and $a_1^*, ..., a_r^* \in \text{ball}\ \mathscr{X}^*$ such that $[0, 1] \subseteq \bigcup_{j=1}^r \{\alpha \,|\, |\alpha - \alpha_j| < \varepsilon\}$, ball $\mathscr{X}^* \subseteq \bigcup_{j=1}^r \{x^* \,|\, \rho(x^*, a_j^*) < \delta\}$.

We have the following inequalities, for $v, w \in \mathscr{F}_{cL}(\mathscr{X})$:

$$\begin{aligned} &| s_v(\alpha, x^*) - s_w(\alpha, x^*) | \\ &\qquad \leqslant | s_v(\alpha, x^*) - s_v(\alpha_j, a_j^*) | + | s_w(\alpha, x^*) - s_w(\alpha_j, a_j^*) | \\ &\qquad\quad + | s_v(\alpha_j, a_j^*) - s_w(\alpha_j, a_j^*) | \end{aligned} \tag{5.3}$$

$$\begin{aligned}
|s_v(\alpha, x^*) - s_v(\alpha_j, a_j^*)| &\leqslant |s_{L_\alpha v}(x^*) - s_{L_\alpha v}(a_j^*)| + |s_{L_\alpha v}(a_j^*) - s_{L_{\alpha_j} v}(a_j^*)| \\
&\leqslant d(L_\alpha v, L_{\alpha_j} v) + \| L_\alpha v \|_{L_\alpha u} \cdot \sup_{a \in L_\alpha u} |\langle x^* - a_j^*, a\rangle| \\
&\leqslant \| v \|_{u,L} |\alpha - \alpha_j| + \| v \|_u \cdot \sup_{a \in \operatorname{supp} u} |\langle x^* - a_j^*, a\rangle|. \qquad (5.4)
\end{aligned}$$

For inequality (5.4) we have used (5.2) and the fact that $v \in \mathscr{F}_{cL}(\mathscr{X})$. Thus

$$|s_v(\alpha, x^*) - s_v(\alpha_j, a_j^*)| \leqslant \| v \|_{u,L} [|\alpha - \alpha_j| + \rho_0(x^*, a_j^*)]. \qquad (5.5)$$

Similarly, we get

$$|s_w(\alpha, x^*) - s_w(\alpha_j, a_j^*)| \leqslant \| w \|_{u,L} [|\alpha - \alpha_j| + \rho_0(x^*, a_j^*)]. \qquad (5.6)$$

Now for α, x^* choose j such that

$$|\alpha - \alpha_j| < \varepsilon, \qquad \rho(x^*, a_j^*) < \delta.$$

Then $\rho_0(x^*, a_j^*) < \varepsilon$ and (5.5), (5.6) together with (5.3) give

$$|s_v(\alpha, x^*) - s_w(\alpha, x^*)| \leqslant 2\varepsilon[\| v \|_{u,L} + \| w \|_{u,L}] + |s_v(\alpha_j, a_j^*) - s_w(\alpha_j, a_j^*)|.$$

Taking the supremum over α, x^*, and taking into account the isometric property (i) in Theorem 5.1, we get

$$d_\infty(v, w) \leqslant 2\varepsilon[\| v \|_{u,L} + \| w \|_{u,L}] + \max_{1 \leqslant j \leqslant r} |s_v(\alpha_j, a_j^*) - s_w(\alpha_j, a_j^*)|.$$

Substituting X_m, X_n for v, w, respectively, we obtain

$$d_\infty(X_m, X_n) \leqslant 2\varepsilon(\| X_m \|_{u,L} + \| X_n \|_{u,L}) + \max_{1 \leqslant j \leqslant r} |s_{X_m}(\alpha_j, a_j^*) - s_{X_n}(\alpha_j, a_j^*)|.$$

Since $\| X_n \|_{u,L}$ and $s_{X_n}(\alpha_j, a_j^*)$, $j = 1, \ldots, r$, converge a.e. it follows that the sequence $\{X_n\}_{n \geqslant 1}$ is a.e. Cauchy.

Completeness if $(\mathscr{F}(\mathscr{X}), d_\infty)$ implies that there exists an f.r.v. X with values in $\mathscr{F}_{cL}(\mathscr{X})$ such that $d_\infty(X_n, X) \to 0$ a.e.

Finally, to show the integrability of X, note that $\| X \|_{u,L} = \eta$ a.e. so $E \| X \|_{u,L} < \infty$. This obviously implies $E \| X \|_u < \infty$, which implies, further that $E \| \operatorname{supp} X \| \leqslant \| u \| \, E \| X \|_u < \infty$. The proof is completed.

Note. As a particular case of Theorem 5.2, assume that X_n takes values in $\mathscr{K}_c(\mathscr{X})$. Then $(X_n, \mathscr{B}_n)$ is a set-valued submartingale (or supermartingale). It is clear that $\| X_n \|_K$ is finite a.e. (where K is a fixed compact convex set) only if X_n assumes values a.e. in $\bigcup_{p=1}^{\infty} pK$. However, the range of

X_n does not have to be finite dimensional. Our result takes the following form.

COROLLARY 5.1. *If* $(X_n, \mathscr{B}_n)$ *is a set-valued submartingale (supermartingale) with* X_n *assuming values a.e. in* $\bigcup_{p=1}^{\infty} pK$ *(where* $K \in \mathscr{K}_c(\mathscr{X})$ *is fixed) and if* $\sup_{n \geq 1} E \|X_n\|_K < \infty$, *then there exists a random set* X *with* $E\|X\| < \infty$ *and such that* $\lim_{n \to \infty} d(X_n, X) = 0$ *a.e.*

This result generalizes Theorem 6.3 of Hiai [9] where $\mathscr{X}$ was considered finite dimensional.

REFERENCES

1. Z. ARTSTEIN, Set-valued measure, *Trans. Amer. Math. Soc.* **165** (1972), 103–125.
2. Z. ARTSTEIN AND J. C. HANSEN, Convexification in limit laws of random sets in Banach spaces, *Ann. Probab.* **13** (1985), 307–309.
3. Z. ARTSTEIN AND R. A. VITALE, A strong law of large numbers for random compact sets, *Ann. Probab.* **3** (1975), 879–882.
4. R. J. AUMANN, Integrals of set-valued functions, *J. Math. Anal. Appl.* **12** (1965), 1–12.
5. G. DEBREU, Integration of correspondences, *in* "Proceedings, Fifth Berkeley Symp. Math. Statist. Prob.," Vol. 2, pp. 351–372, 1966.
6. G. DEBREU AND D. SCHMEIDLER, The Radon–Nikodym derivative of a correspondence, *in* "Proceedings, Sixth Berkeley Symp. Math. Statist. Prob.," Vol. 2, pp. 41–56, 1972.
7. L. DE ROBERTIS AND J. A. HARTIGAN, Bayesian inference using intervals of measures, *Ann. Statist.* **9** (1981), 235–244.
8. E. GINÉ, M. G. HAHN, AND J. ZINN, Limit theorems for random sets: An application of probability in Banach space results, *in* "Proceedings, Fourth Int. Conf. on Prob. in Banach Space, Oberwolfach," pp. 112–135, Lecture Notes in Math., Vol. 990, Springer-Verlag, New York/Berlin, 1983.
9. F. HIAI, Radon–Nikodym theorems for set-valued measures, *J. Multivariate Anal.* **8** (1978), 96–118.
10. F. HIAI AND H. UMEGAKI, Integrals, conditional expectations, and martingales of multivalued functions, *J. Multivariate Anal.* **7** (1977), 149–182.
11. D. G. KENDALL, Foundations of a theory of random sets, *in* "Stochastic Geometry" (E. F. Harding and D. G. Kendall, Eds.), Wiley, New York, 1974.
12. E. P. KLEMENT, M. L. PURI, AND D. A. RALESCU, Limit theorems for fuzzy random variables, *Proc. Roy. Soc. London Ser. A* **407** (1986), 171–182.
13. R. KRUSE, The strong law of large numbers for fuzzy random variables, *Inform. Sci.* **28** (1982), 233–241.
14. G. MATHERON, "Randon Sets and Integral Geometry," Wiley, New York, 1975.
15. C. V. NEGOITA AND D. A. RALESCU, "Application of Fuzzy Sets to Systems Analysis," Wiley, New York, 1975.
16. M. L. PURI AND D. A. RALESCU, Strong law of large numbers for Banach space valued random sets, *Ann. Probab.* **11** (1983), 222–224.
17. M. L. PURI AND D. A. RALESCU, Strong law of large numbers with respect to a set-valued probability measure, *Ann. Probab.* **11** (1983), 1051–1054.
18. M. L. PURI AND D. A. RALESCU, Limit theorems for random compact sets in Banach space, *Math. Proc. Cambridge Philos. Soc.* **97** (1985), 151–158.

19. M. L. Puri and D. A. Ralescu, The concept of normality for fuzzy random variables, *Ann. Probab.* **13** (1985), 1373–1379.
20. M. L. Puri and D. A. Ralescu, Fuzzy random variables, *J. Math. Anal. Appl.* **114** (1986), 402–422.
21. M. A. Rieffel, The Radon–Nikodym theorem for the Bochner integral, *Trans. Amer. Math. Soc.* **131** (1968), 466–487.
22. M. Sugeno, Fuzzy measures and fuzzy integrals—A survey, *in* "Fuzzy Automata and Decision Processes" (M. M. Gupta, G. N. Saridis, and B. R. Gaines, Eds.), pp. 89–102, North-Holland, Amsterdam, 1977.
23. W. Weil, An application of the central limit theorem for Banach space-valued random variables to the theory of random sets, *Z. Wahrsch. Verw. Gebiete* **60** (1982), 203–208.
24. L. A. Zadeh, Fuzzy sets, *Inform. and Control* **8** (1965), 338–353.

Reprinted by permission of the publisher,
© 1991 by Academic Press

LIMIT THEOREMS FOR FUZZY RANDOM VARIABLES

F. N. PROSKE and M. L. PURI

Department of Mathematics, Ulm University, D-89069 Ulm, Germany
Department of Mathematics, University Indiana, Bloomington, IN 47405, U.S.A.

ABSTRACT

In this article a survey of limit theorems for Borel measurable nonseparably valued random elements in the case of fuzzy random variables is given, comprising strong laws of large numbers and central limit theorems.

1. INTRODUCTION

In this survey paper a strong law of large numbers (SLLN) and a central limit theorem (CLT) for (pairwise) independent and identically distributed ((p.)i.i.d.) fuzzy random variables with compact α-level sets are presented, extending similar results for random sets in a natural way (see *e.g.* Artstein and Vitale (1975), Giné, Hahn and Zinn (1983), Puri and Ralescu (1985a) and Weil (1982)).

The concept of a fuzzy random variable (frv) was introduced by Puri and Ralescu (1986) as a model to describe random experiments whose outcomes are inexact data. By inexactness here we understand non-statistical inexactness, due to the subjectivity and imprecision of human knowledge rather than to a random mechanism. This kind of inexactness can be *e.g.* found in form of linguistic expressions like "warm" or "cold", which can be represented by fuzzy sets.

This paper is organized as follows: In section 2 we record some preliminaries pertinent to random sets and fuzzy random variables, being essential for the main results, a SLLN respectively a CLT for (p.)i.i.d. frv's, which are proven in section 3 respectively section 4. Here we choose the framework of separably Banach space valued fuzzy random variables.

Various SLLN's for frv's are already established. Such a result in the case of i.i.d. fuzzy random variables was obtained by Klement, Puri and Ralescu (1986), but they made use of two different metrics: The Borel measurability of a frv is defined via the one metric, while the convergence is induced by the other. For its proof they employ embedding theorems and results from probability theory on Banach spaces. A stronger convergence result for p.i.i.d. frv's was attained by Kruse (1982) in the case of $\mathbb{R}$, by exploiting its total order. The SLLN here, stated for i.i.d. frv's on separable Banach spaces (see Proske and Puri (1999a)), is a stronger convergence result than that of Klement, Puri and Ralescu, i.e. Borel measurability and convergence are only based upon one metric, which generalizes the Hausdorff distance. In handling the non-separability of involved metric spaces of fuzzy sets our main approach relies on associating linear isometrically each convex frv with a stochastic process in $l^\infty(T)$. By investigating its limit behavior, a Glivenko-Cantelli problem is solved (see Hoffmann-Jørgensen (1985), Talagrand (1987) and van der Vaart and Wellner (1996)).

As for the CLT for frv's, Klement, Puri and Ralescu (1986) (see also Puri and Ralescu (1985b)) could prove such a result for i.i.d. frv's with fuzzy set values, whose compact α-level sets fulfill a Lipschitz condition with respect to the Hausdorff distance. From the practical point of view, this means a rather strong restriction of frv's to a subclass of compact fuzzy sets. Also in this case, their main tools in proving the CLT are based on Banach space techniques. The CLT, presented here for i.i.d. Banach space valued frv's (see Proske and Puri (1999b)), is a generalization of the latter authors' result, because no Lipschitz condition is needed. In order to face the non-separability of the corresponding metric spaces of fuzzy sets, the same embedding as in the SLLN case is used, *e.g.* a Donsker problem is set (see van der Vaart and Wellner (1996)). In solving it, we verify a certain finiteness condition of entropy with bracketing for the resulting empirical process. Furthermore a SLLN for p.i.i.d. $\mathbb{R}^p$-valued frv's is also given here, which can be proved by a method similar to the proof in Proske and Puri (1999c).

First, Artstein and Vitale (1975) proved a SLLN for i.i.d. compact random sets in $\mathbb{R}^p$ (i.e. random sets, whose values are compact subsets of $\mathbb{R}^p$). Weil could derive a CLT for the same setting. Later on these results were extended by Giné, Hahn and Zinn (1983) and, Puri and Ralescu (1985a) to the case of separable Banach spaces.

The results here are natural generalizations of all previous SLLN's and CLT's for random sets.

In the sequel the underlying probability space $(\Omega, \mathfrak{I}, P)$ is assumed to be complete.

2. PRELIMINARIES

In the remainder of this paper. let $(X, \|\cdot\|)$ be a real separable Banach space. Denote by $\mathcal{K}(X)$ $(co\mathcal{K}(X))$ the class of all nonempty (convex and) compact subsets of X. A linear Abelian semigroup in $\mathcal{K}(X)$ $(co\mathcal{K}(X))$ is defined via the following operations:

$$A + B := \{a + b \mid a \in A, b \in B\},$$
$$\lambda A := \{\lambda a \mid a \in A\}$$

for $A, B \in \mathcal{K}(X)$ $(co\mathcal{K}(X))$, $\lambda \in \mathbb{R}$. Note that $\mathcal{K}(X)$ $(co\mathcal{K}(X))$ together with the operations above doesn't form a vector space, since $(A - B) + B$ needs not equal A. $\mathcal{K}(X)$ is metrizable by the well-known Hausdorff distance d_H, given by

$$d_H : \mathcal{K}(X) \times \mathcal{K}(X) \longrightarrow [0, \infty);$$
$$(A, B) \mapsto \max\{\sup_{a\in A} \inf_{b\in B} \|a - b\|, \sup_{b\in B} \inf_{a\in A} \|a - b\|\}.$$

$(\mathcal{K}(X), d_H)$ and $(co\mathcal{K}(X), d_H)$ are then complete separable metric spaces (see Puri and Ralescu (1985a)). Let us also define

$$\|A\| = d_H(A, \{0\}) = \sup\{\|a\| \mid a \in A\}$$

for $A \in \mathcal{K}(X)$. A *fuzzy set* of X is a function $u : X \longrightarrow [0, 1]$. Denote by $L_\alpha u := \{x \in X \mid u(x) \geqslant \alpha\}, 0 < \alpha \leqslant 1$ the *α-level set* of u. Further, by $\operatorname{supp} u$ we understand the closure of $\{x \in X \mid u(x) > 0\}$ in X. An extension of $\mathcal{K}(X)$ is obtained by defining the space $F_c(X)$ of all fuzzy sets $u : X \longrightarrow [0, 1]$ with the properties (i) $L_\alpha u$ is compact for all $0 < \alpha \leqslant 1$; (ii) $\operatorname{supp} u$ is compact; (iii) $\{x \in X \mid u(x) = 1\} \neq \emptyset$.

In the sequel we set $L_0 u = \operatorname{supp} u$. The space $U_c(X)$ consists of all $u \in F_c(X)$ with convex α-level sets. Then, $A \in \mathcal{K}(X)$ $(co\mathcal{K}(X))$ implies that the characteristic function χ_A is in $F_c(X)$ $(U_c(X))$. We denote the closure of the convex hull of $u \in F_c(X)$ by $\overline{co}u \in F_c(X)$, determined by its convex compact α-level sets $L_\alpha \overline{co}u := \overline{co}(L_\alpha u)$, $0 < \alpha \leqslant 1$ (see Klement, Puri and Ralescu (1986)) (where $\overline{co}A$ is the closure of the convex hull of a subset $A \subseteq X$ in X).

Zadeh's Extension Principle (1965) supplies a linear structure in $F_c(X)$ $(U_c(X))$ by the operations

$$(u + v)(x) := \sup_{y+z=x} \min(u(y), v(z))$$

In the sequel the underlying probability space $(\Omega, \mathfrak{F}, P)$ is assumed to be complete.

2. PRELIMINARIES

In the remainder of this paper, let $(X, \|\cdot\|)$ be a real separable Banach space. Denote by $\mathcal{K}(X)$ $(co\mathcal{K}(X))$ the class of all nonempty (convex and) compact subsets of X. A linear Abelian semigroup in $\mathcal{K}(X)$ $(co\mathcal{K}(X))$ is defined via the following operations:

$$A + B := \{a + b \mid a \in A, b \in B\},$$
$$\lambda A := \{\lambda a \mid a \in A\}$$

for $A, B \in \mathcal{K}(X)$ $(co\mathcal{K}(X))$, $\lambda \in \mathbb{R}$. Note that $\mathcal{K}(X)$ $(co\mathcal{K}(X))$ together with the operations above doesn't form a vector space, since $(A - B) + B$ needs not equal A. $\mathcal{K}(X)$ is metrizable by the well-known Hausdorff distance d_H, given by

$$d_H : \mathcal{K}(X) \times \mathcal{K}(X) \longrightarrow [0, \infty);$$
$$(A, B) \mapsto \max\{\sup_{a\in A}\inf_{b\in B} \|a - b\|, \sup_{b\in B}\inf_{a\in A} \|a - b\|\}.$$

$(\mathcal{K}(X), d_H)$ and $(co\mathcal{K}(X), d_H)$ are then complete separable metric spaces (see Puri and Ralescu (1985a)). Let us also define

$$\|A\| = d_H(A, \{0\}) = \sup\{\|a\| \mid a \in A\}$$

for $A \in \mathcal{K}(X)$. A *fuzzy set* of X is a function $u : X \longrightarrow [0, 1]$. Denote by $L_\alpha u := \{x \in X \mid u(x) \geqslant \alpha\}, 0 < \alpha \leqslant 1$ the *α-level set* of u. Further, by $\operatorname{supp} u$ we understand the closure of $\{x \in X \mid u(x) > 0\}$ in X. An extension of $\mathcal{K}(X)$ is obtained by defining the space $F_c(X)$ of all fuzzy sets $u : X \longrightarrow [0, 1]$ with the properties (i) $L_\alpha u$ is compact for all $0 < \alpha \leqslant 1$; (ii) $\operatorname{supp} u$ is compact; (iii) $\{x \in X \mid u(x) = 1\} \neq \emptyset$.

In the sequel we set $L_0 u = \operatorname{supp} u$. The space $U_c(X)$ consists of all $u \in F_c(X)$ with convex α-level sets. Then, $A \in \mathcal{K}(X)$ $(co\mathcal{K}(X))$ implies that the characteristic function χ_A is in $F_c(X)$ $(U_c(X))$. We denote the closure of the convex hull of $u \in F_c(X)$ by $\overline{co}u \in F_c(X)$, determined by its convex compact α-level sets $L_\alpha \overline{co}u := \overline{co}(L_\alpha u)$, $0 < \alpha \leqslant 1$ (see Klement, Puri and Ralescu (1986)) (where $\overline{co}A$ is the closure of the convex hull of a subset $A \subseteq X$ in X).

Zadeh's Extension Principle (1965) supplies a linear structure in $F_c(X)$ $(U_c(X))$ by the operations

$$(u + v)(x) := \sup_{y+z=x} \min(u(y), v(z))$$

$$(\lambda u)(x) := \begin{cases} u(\lambda^{-1}x), & \text{if } \lambda \neq 0 \\ \chi_{\{0\}}(x), & \text{if } \lambda = 0 \end{cases}$$

for $u, v \in F_c(X), \lambda \in \mathbb{R}$. Note that $L_\alpha(u+v) = L_\alpha u + L_\alpha v$, $L_\alpha(\lambda u) = \lambda L_\alpha u$ and $\overline{co}(L_\alpha u + L_\alpha v) = \overline{co}L_\alpha u + \overline{co}L_\alpha v$, $\overline{co}(\lambda L_\alpha u) = \lambda \overline{co} L_\alpha u$ for all $u, v \in F_c(X)$, $0 < \alpha \leqslant 1$ (see Dunford and Schwartz (1958), Puri and Ralescu (1983), (1986)). As in the case of $\mathcal{K}(X)$ ($co\mathcal{K}(X)$) the operations above induce a linear Abelian semigroup in $F_c(X)$ ($U_c(X)$).

$F_c(X)$ can be endowed with the metric d_∞ defined by

$$d_\infty(u, v) = \sup_{0<\alpha\leqslant 1} d_H(L_\alpha u, L_\alpha v)$$

for $u, v \in F_c(X)$, generalizing the Hausdorff distance d_H to the case of $F_c(X)$ in the sense that the mapping $\pi : \mathcal{K}(X) \longrightarrow F_c(X); A \mapsto \chi_A$ is a (continuous) semigroup monomorphism (i.e. π is injective and $\pi(A+B) = \pi(A) + \pi(B)$ holds for all $A, B \in \mathcal{K}(X)$). $(F_c(X), d_\infty)((U_c(X), d_\infty))$ is a complete, but in contrast to $(\mathcal{K}(X), d_H)$ $((co\mathcal{K}(X), d_H))$ a nonseparable metric space (see Puri and Ralescu (1986)).

A *fuzzy random variable* (*frv* for short) is a Borel measurable function $\xi : (\Omega, \Im, P) \longrightarrow (F_c(X), d_\infty)$. As one can see, frv's generalize *compact random sets* A (i.e. Borel measurable functions $A : \Omega \longrightarrow (\mathcal{K}(X), d_H)$) by identification with the characteristic function χ_A. Note that, if $\xi : \Omega \longrightarrow (F_c(X), d_\infty)$ is a frv, then $L_\alpha \xi, \overline{co}(L_\alpha \xi)$, $0 < \alpha \leqslant 1$ and suppξ, defined in an obvious way, are (convex) compact random sets. Besides this, $\overline{co}\xi$ is a $U_c(X)$-valued frv.

Let $\xi : \Omega \longrightarrow (F_c(X), d_\infty)$ be a frv with $E\|\xi\|_\infty < \infty$ ($\|\xi(\omega)\|_\infty := \sup_{\alpha>0} \|L_\alpha \xi(\omega)\|$). Then the *expected value of* ξ (see Puri and Ralescu (1986)) is the unique fuzzy set $E\xi \in U_c(X)$ such that the α-level sets of $E\xi$ are the Aumann expectations of the convex compact random sets $\overline{co}(L_\alpha \xi)$ (see for the definition of the Aumann expectation Aubin and Frankowska (1990), Debreu (1966)), i.e. $L_\alpha u = E\overline{co}(L_\alpha \xi)$ for all $0 < \alpha \leqslant 1$.

A crucial step in the proof of our main results is based on an embedding, which involves the representation of convex compact sets as elements of a Banach space of continuous functions. The next theorem stems in this form from Giné, Hahn and Zinn (1983), but goes back to Hörmander (1954):

THEOREM 1. *The function* $j : (co\mathcal{K}(X), d_H) \longrightarrow (C(\text{ball } X^*, d_*), \|\cdot\|_\infty)$ ($\|\cdot\|_\infty$ *is the supremum norm*), *given by* $A \longmapsto (f \mapsto \sup_{a \in A} f(a))$ *satisfies*

(i) j *is an isometry (i.e.* $d_H(A, B) = \|j(A) - j(B)\|_\infty$*)*.

(ii) $j(A+B) = j(A) + j(B)$, $j(\lambda A) = \lambda j(A)$ $(\lambda \geqslant 0)$.

d_* denotes a metric inducing the weak* topology of ball X^*, where ball X^* is the norm-closed unit ball of the dual of X.

A central idea in deriving a SLLN and a CLT for not necessarily convex frv's is to reduce first the study of frv's to the convex case by convexifying the underlying variables and then to drop the convexity restriction. by applying inequalities of the type, contained in the next two theorems.

In view of the first result, let us mention that a Banach space is of type $p > 1$, if there exists a constant $M > 0$ such that $E\|\sum_{i=1}^{n}\xi_i\|^p \leqslant M\sum_{i=1}^{n}E\|\xi_i\|^p$ for every finite sequence of zero-mean independent X-valued random elements $(\xi_i)_{i=1}^{n}$.

Every Hilbert space is of type 2, the spaces L^p with $1 < p < \infty$ are of type $\min(p, 2)$, whereas $C[0, 1]$ and L^∞ are not of type p for any $p > 1$ (they are of type 1).

The following two results allow to show, to which extent addition of sets is "convexifying":

THEOREM 2 (Puri and Ralescu (1985a)). *Let X be a Banach space of type $p > 1$. Then there exists a constant $M > 0$ such that*

$$d_H(A_1 + \ldots + A_n, \overline{co}(A_1 + \ldots + A_n)) \leqslant M^{1/p}(r^p(A_1) + \ldots + r^p(A_n))$$

for all $A_1, \ldots, A_n \in \mathcal{K}(X)$, where $r(A)$ denotes the inner radius of $A \in \mathcal{K}(X)$, given by

$$r(A) = \sup_{a\in\overline{co}A} \inf\{R \geqslant 0 \mid \exists a_1, \ldots, a_s \in A, a \in co\{a_1, \ldots, a_s\} \text{ and } \|a - a_i\| \leqslant R\}.$$

In the finite dimensional case even more is valid (see Arrow and Hahn (1971)):

THEOREM 3 (Shapley-Folkman-Starr). *If $\dim X < \infty$, we obtain for all $A_1, \ldots, A_n \in \mathcal{K}(X)$ and $n \in \mathbb{N}$:*

$$d_H(A_1 + \ldots + A_n, \overline{co}(A_1 + \ldots + A_n)) \leqslant \sqrt{p} \max_{1\leqslant i\leqslant n} \|A_i\|.$$

3. STRONG LAW OF LARGE NUMBERS FOR FUZZY RANDOM VARIABLES

The proof of the main result of this section (Theorem 4) requires the notion of a *Blackwell σ-algebra*: A σ-algebra $\Im$ on a set Ω is called so, if $\varphi(\Omega)$ is an analytic subset of $\mathbb{R}$ for all real $\Im$-measurable functions φ on Ω (see Hoffmann-Jørgensen (1985)).

First, we state a Kolmogorov type of strong law of large numbers for convex frv's:

THEOREM 4. *Let $(X, \|.\|)$ be a separable Banach space. If $\xi_i : \Omega \longrightarrow (U_c(X), d_\infty)$ is an i.i.d. sequence of frv's on the probability space $(\Omega, \Im, P)$*

with $E\|\xi_1\|_\infty < \infty$ *then*

$$d_\infty\left(\frac{1}{n}\sum_{i=1}^{n}\xi_i, E\xi_1\right) \longrightarrow 0 \; a.e.$$

Proof. We intend to apply a result of J. Hoffmann-Jørgensen (1985) (Corollary 4.6., p. 349) to the stochastic process

$$g : (U_c(X), d_\infty) \times T \longrightarrow \mathbb{R}$$

with index set $T = ([0,1] \times \text{ball}X^*, |.| + d_*)$ given by

$$(u, (\alpha, f)) \mapsto \begin{cases} \sup_{a \in L_\alpha u} f(a), & \text{if } \alpha > 0 \\ \sup_{a \in \text{supp} u} f(a), & \text{if } \alpha = 0. \end{cases}$$

Therefore, we check the assumptions of the result, what is accomplished in several steps:

1. Let (T, d) be a compact metric space. Then $\mathcal{B}a(T) :=$ {smallest σ-algebra on T, such that all continuous $f : T \longrightarrow \mathbb{R}$ are measurable} is a Blackwell σ-algebra on T (see Hoffmann-Jørgensen (1985)). Let V be an open set in (T, d), then $f : T \longrightarrow \mathbb{R}; x \mapsto \exp(-d(V^c, x))$ is continuous on (T, d) with $f^{-1}[\mathbb{R} - \{1\}] = V$. Hence $\mathcal{B}(T) = \mathcal{B}a(T)$ for the Borel σ-algebra on T. Thus $\mathcal{B}(T)$ is a Blackwell σ-algebra on T.

2. In this step we gather some facts, being due to the limit behavior of α-level sets. These are used in the next step to verify the existence of certain limits of the process g. For that purpose, pick up $s \in U_c(X)$ and define $S_\alpha = \{x \in X \mid s(x) > \alpha\}$ for $0 \leqslant \alpha < 1$. It is possible then to show the existence of the right limits and the left continuity of the α-level sets of s with respect to the Hausdorff distance (see Klement, Puri and Ralescu (1986)), i.e.

$$\lim_{\substack{\alpha \longrightarrow \alpha_0 \\ \alpha > 0}} d_H(L_\alpha s, S_{\alpha_0}) = 0 \; (\alpha_0 < 1) \text{ and } \lim_{\substack{\alpha \longrightarrow \alpha_0 \\ \alpha \leqslant \alpha_0}} d_H(L_\alpha s, L_{\alpha_0} s) = 0. \qquad (1)$$

From the last assertion (1) and Theorem 1 we immediately obtain

$$\begin{aligned} &\lim_{\substack{\alpha \longrightarrow \alpha_0 \\ \alpha \leqslant \alpha_0}} |\sup f(L_\alpha s) - \sup f(L_{\alpha_0} s)| \\ &\leqslant \lim_{\substack{\alpha \longrightarrow \alpha_0 \\ \alpha \leqslant \alpha_0}} \sup_{f \in \text{ball}X^*} |\sup f(L_\alpha s) - \sup f(L_{\alpha_0} s)| \\ &= \lim_{\substack{\alpha \longrightarrow \alpha_0 \\ \alpha \leqslant \alpha_0}} d_H(L_\alpha s, L_{\alpha_0} s) = 0 \text{ for all } f \in \text{ball}X^*. \end{aligned} \qquad (2)$$

Bearing in mind the monotony of α-level sets and that $\cup_{\alpha > \alpha_0} L_\alpha s = S_{\alpha_0}$ is valid, one realizes easily that

$$\lim_{\substack{\alpha \longrightarrow \alpha_0 \\ \alpha > \alpha_0}} \sup f(L_\alpha s) = \sup f(S_{\alpha_0}). \qquad (3)$$

Furthermore, let $(\alpha_n)_{n\geqslant 1} \subseteq (\alpha_0, 1]$ with $\alpha_n \longrightarrow \alpha_0$. Since $d_H(L_{\alpha_n}s, L_{\alpha_m}s) \leqslant d_H(L_{\alpha_n}s, S_{\alpha_0}) + d_H(S_{\alpha_0}, L_{\alpha_m}s)$ holds for all m, n, we find that $((f \mapsto \sup_{a\in L_{\alpha_n}s} f(a)))_{n\geqslant 1}$ is a Cauchy sequence in $C(\text{ball}X^*, d_*)$. Assertion (3) implies then

$$(f \mapsto \sup_{a\in S_{\alpha_0}} f(a)) \in C(\text{ball}X^*, d_*). \tag{4}$$

3. On the one hand we can follow now from (1) and Theorem 1

$$\begin{aligned}&\Big|\sup_{a\in L_\alpha s} f(a) - \sup_{a\in L_{\alpha_0}s} f(a)\Big| \\ &\quad\leqslant d_H(L_\alpha s, L_{\alpha_0}s) + \Big|\sup_{a\in L_{\alpha_0}s} f(a) - \sup_{a\in L_{\alpha_0}s} f_0(a)\Big| \underset{\substack{(\alpha,f)\longrightarrow(\alpha_0,f_0)\\ \alpha>\alpha_0}}{\longrightarrow} 0\end{aligned}$$

for all $\alpha \leqslant \alpha_0$ and $f, f_0 \in \text{ball}X^*$.

On the other hand we conclude from (1), (3) and (4) that

$$\begin{aligned}\Big|\sup_{a\in L_\alpha s} f(a) - \sup_{a\in S_{\alpha_0}} f_0(a)\Big| &= \lim_{\substack{\beta\longrightarrow\alpha_0\\ \beta>\alpha_0}} \Big|\sup_{a\in L_\alpha s} f(a) - \sup_{a\in L_\beta s} f(a)\Big| \\ &\leqslant \lim_{\substack{\beta\longrightarrow\alpha_0\\ \beta>\alpha_0}} \left(d_H(L_\alpha s, S_{\alpha_0}) + d_H(S_{\alpha_0}, L_\beta s) + \Big|\sup_{a\in L_\beta s} f(a) - \sup_{a\in L_\beta s} f_0(a)\Big|\right) \\ &= d_H(L_\alpha s, S_{\alpha_0}) + \Big|\sup_{a\in S_{\alpha_0}} f(a) - \sup_{a\in S_{\alpha_0}} f_0(a)\Big| \underset{\substack{(\alpha,f)\longrightarrow(\alpha_0,f_0)\\ \alpha>\alpha_0}}{\longrightarrow} 0\end{aligned}$$

for all $\alpha > \alpha_0$ and $f, f_0 \in \text{ball}X^*$.

4. Finally, we summarize all properties of the process g, sufficient for the applicability of the above mentioned result: Let $\mathcal{A} = \mathcal{B}(U_c(X))$. Then the process g is $\mathcal{A} \otimes \mathcal{B}(T), \mathcal{B}(\mathbb{R})$-measurable, because it is left continuous in α and continuous in f for the argument $(\alpha, f) \in T$. The topological basis of T is contained in $\mathcal{B}(T)$.

Further, let $\mu = \xi_1 P$ and $(\alpha_0, f_0) \in T$. Theorem 1 supplies the representation $d_\infty(u, u_0) = \sup_{\alpha\geqslant 0} \sup_{f\in\text{ball}X^*} |\sup_{a\in L_\alpha u} f(a) - \sup_{a\in L_\alpha u_0} f(a)|$ for all $u, u_0 \in U_c(X)$. Hence, the mapping $(s \mapsto g(s, (\alpha_0, f_0)))$ is continuous. This implies that $(s \longmapsto g(s, (\alpha, f)))$ is μ-measurable.

Besides this we obtain for the outer integral $\int^*$

$$\int^* \sup_{(\alpha,f)\in T} |g(s, (\alpha, f))|\mu(ds) = \int \|.\|_\infty d\mu = \int_\Omega \|\xi_1\|_\infty dP < \infty.$$

Let $t = (\alpha_0, f_0) \in T$. Then $A_t := \{[0, \alpha_0] \times \text{ball}X^*, (\alpha_0, 1] \times \text{ball}X^*\} = \{A_1, A_2\}$ is a finite cover of T with $A_t \subseteq \mathcal{B}(T)$ and $t \in \text{cl}(A_1\cap T)\cap\text{cl}(A_2\cap T)$.

Thus, by step 3 we see that

$$\mu^*\left(\left\{s \mid \lim_{\substack{(\alpha, f) \xrightarrow[d]{} (\alpha_0, f_0) \\ (\alpha, f) \in A_1}} g(s, (\alpha, f)) \text{ exists}\right\}\right) = \mu(U_c(X)) = 1$$

as well as

$$\mu^*\left(\left\{s \mid \lim_{\substack{(\alpha, f) \xrightarrow[d]{} (\alpha_0, f_0) \\ (\alpha, f) \in A_2}} g(s, (\alpha, f)) \text{ exists}\right\}\right) = \mu(U_c(X)) = 1.$$

Taking into consideration that $E \sup_{a \in A} f(a) = \sup_{a \in EA} f(a)$ holds for the Aumann expectation EA of compact random sets A and $f \in X^*$ (see Aubin and Frankowska (1990)), the result of Hoffmann-Jørgensen (1985) (Corollary 4.6, p. 349) provides then

$$d_\infty\left(\frac{1}{n}\sum_{i=1}^{n} \xi_i, E\xi_1\right) = \sup_{(\alpha, f) \in T}\left|\frac{1}{n}\sum_{i=1}^{n} g(\xi_i, (\alpha, f)) - g(E\xi_1, (\alpha, f))\right|$$

$$= \sup_{(\alpha, f) \in T}\left|\frac{1}{n}\sum_{i=1}^{n} g(\xi_i, (\alpha, f)) - Eg(\xi_1, (\alpha, f))\right| \longrightarrow 0 \text{ a.e.}$$

concluding the proof. □

We use the last result and Theorem 2 to prove a version of the SLLN for $F_c(X)$-valued frv's in the case of separable Banach spaces of type $p > 1$:

COROLLARY 5. *Let $(X, \|.\|)$ be separable Banach space of type $p > 1$ and let $\xi_i : \Omega \longrightarrow (F_c(X), d_\infty)$ be an i.i.d. sequence of frv's on the probability space $(\Omega, \Im, P)$ with $E\|\xi_1\|_\infty < \infty$. Then the strong law is valid, i.e.*

$$d_\infty\left(\frac{1}{n}\sum_{i=1}^{n} \xi_i, E\xi_1\right) \longrightarrow 0 \text{ a.e.}$$

Proof. By Theorem 2 we know there exists a constant $M > 0$ such that

$$d_\infty\left(\frac{1}{n}\sum_{i=1}^{n} \xi_i, \frac{1}{n}\sum_{i=1}^{n} \overline{co}\xi_i\right) \leqslant M^{1/p}\left(\frac{1}{n^{1/1/p}}\sum_{i=1}^{n} \|\xi_i\|_\infty^p\right)^{1/p}.$$

The last term tends to zero a.e. as $n \longrightarrow \infty$ by virtue of a result of Marcinkiewicz and Zygmund (see Chow and Teicher (1988)). The rest of the proof follows from the triangle inequality and Theorem 4. □

The finite dimensional case yields:

COROLLARY 6. *Let $\xi_i : \Omega \longrightarrow (F_c(\mathbb{R}^p), d_\infty)$ be an i.i.d. sequence of frv's with existing first moment then the SLLN holds.*

4. CENTRAL LIMIT THEOREM FOR FUZZY RANDOM VARIABLES

For preparing the proof of the CLT for Banach space valued frv's let us give some another definitions: If (T, d) is a metric space, the *ε-entropy* of T, denoted by $N(\varepsilon, T, d)$ $(\varepsilon > 0)$, is the minimal number of balls of radius $< \varepsilon$, with centers in T, which cover T. The *Minkowski functional* $\|\cdot\|_K$ of a convex, compact, symmetric subset K of X is given by $\|A\|_K = \inf\{\lambda > 0 \mid A \subseteq \lambda K\}$ $(\inf\emptyset := \infty)$. We also aim at employing the theory of empirical processes (see van der Vaart and Wellner (1996)) to investigate frv's. For that purpose we will refer to the term of a *bracketing number* to measure the "size" of certain function classes: Let $(\mathcal{F}, \|\cdot\|)$ be a subset of a normed space of real functions $f : \chi \longrightarrow \mathbb{R}$ on a set χ. Then the *bracket* $[l, u]$ for real functions l, u is the set of all functions $f : \chi \longrightarrow \mathbb{R}$, satisfying $l \leqslant f \leqslant u$. An *ε-bracket* is a bracket $[l, u]$ with $\|l - u\| < \varepsilon$. Then the *bracketing number* $N_{[\,]}(\varepsilon, \mathcal{F}, \|\cdot\|)$ is the minimal number of ε-brackets covering $\mathcal{F}$. Note, that in the definition of the bracketing number the functions l and u are assumed to have finite norms, but not to belong to $\mathcal{F}$ itself.

THEOREM 7 (CLT for frv's). *Let $(X, \|\cdot\|)$ be a separable Banach space over $\mathbb{R}$ and $\xi_i : \Omega \longrightarrow (U_c(X), d_\infty)$ an i.i.d. sequence of frv's with*

(i) There exists a convex, compact, symmetric $K \subseteq X$ such that

$0 < E\|\mathrm{supp}\xi_1\|_K^2 < \infty$ *and*

$\int_0^1 (\log N(\varepsilon, ball X^*, d_K))^{1/2} \lambda^1(d\varepsilon) < \infty$, *where d_K is the semimetric on $ball X^*$ with $d_K(f_1, f_2) = \sup_{a \in K} |f_1(a) - f_2(a)|$.*

(ii) $\int_0^1 (\log N(\varepsilon^2, K, \|\cdot\|))^{1/2} \lambda^1(d\varepsilon) < \infty$.

Then a CLT in the following sense is valid:

$$\sqrt{n}\, d_\infty\left(\sum_{i=1}^{n} \xi_i / n, E\xi_1\right) \longrightarrow \|\underline{Y}\|_\infty \text{ weakly}$$

for a tight Gaussian process $\underline{Y}$ in $l^\infty([0, 1] \times \mathrm{ball}X^*)$.

Remark. Theorem 7 would be more convenient if conditions (i) and (ii) could be directly expressed in terms of K and X, without incorporating the dual of X. It is somehow quite plain that the "size" of $\mathrm{ball}X^*$ with respect to the metric d_K is connected with the "size" of K for $\|\cdot\|$. Generally, it is not so easy to determine the relation between the entropy of $(K, \|\cdot\|)$ and $(\mathrm{ball}X^*, d_K)$. But one can show for example that for an ellipsoid K in a Hilbert space H the metric entropy of $(K, \|\cdot\|)$ equals that of $(\mathrm{ball}H^*, d_K)$ (see for details Giné, Hahn and Zinn (1983)).

Proof (of Theorem 7). To prove this result we want to apply a CLT for empirical processes (see Theorem 2.11.23, p. 221 in van der Vaart and

Wellner (1996)), what comes down to bound above bracketing numbers in an appropriate way for the function class $\mathcal{F} := \{f_t \mid t \in T\}$, consisting of $\mathcal{B}(U_c(X))$, $\mathcal{B}(\mathbb{R})$−measurable functions $f_t : (U_c(X), d_\infty) \longrightarrow \mathbb{R}: u \longmapsto \sup_{a \in L_\alpha u} f(a)$ for $t = (\alpha, f) \in T := [0, 1] \times \text{ball}X^*$. In the sequel we adopt the notions in van der Vaart and Wellner (1996) and define $F = \|\cdot\|_\infty$ and $\mathcal{P}$ as the distribution of ξ_1 under P. The proof is accomplished step by step:

1. The existence of the second moment of $\|\text{supp}\xi_1\|_K$ implies the existence of that of $\|\xi_1\|_\infty$. Then one is able to prove easily, that there exists a unique fuzzy set $E(\|\xi_1\|_\infty \xi_1) \in U_c(X)$, such that $L_\alpha E(\|\xi_1\|_\infty \xi_1) = E(\|\xi_1\|_\infty L_\alpha \xi_1)$ for all $0 < \alpha \leqslant 1$. Let us equip then the interval $[0, 1]$ with the pseudodistance

$$\begin{aligned}&\rho_1 : [0, 1] \times [0, 1] \longrightarrow [0, \infty);\\ &(\alpha_1, \alpha_2) \mapsto d_H^{1/2}(L_{\alpha_1} E(\|\xi_1\|_\infty \xi_1), L_{\alpha_2} E(\|\xi_1\|_\infty \xi_1)).\end{aligned}$$

Taking into account that the α-level sets of a fuzzy set $u \in F_c(X)$ are non-increasing and that the mapping $(\alpha \mapsto L_\alpha u)$ on the interval $[0, 1]$ is left continuous in $(0, 1]$ as well as right continuous in 0 with respect to the Hausdorff distance (see Klement, Puri and Ralescu (1986)), one verifies that the ρ_1-balls with radius $< \varepsilon$ and centers $0 \leqslant \alpha_0 \leqslant 1$, $U_\varepsilon^{(\rho_1)}(\alpha_0)$, are non-degenerate subintervals of $[0, 1]$.

2. For our convenience we denote $f_t(\cdot) = f((\alpha, \widetilde{f}), \cdot)$ for $t = (\alpha, \widetilde{f})$. By using the definition of the Minkowski functional $\|\cdot\|_K$, we can find that

$$|f((\alpha_2, f_1), \xi_1) - f((\alpha_2, f_2), \xi_1)| \leqslant \|\text{supp}\,\xi_1\|_K d_K(f_1, f_2) \text{ holds for all}$$

$(\alpha_2, f_1), (\alpha_2, f_2) \in T$.

This leads us to following inequality

$$\begin{aligned}A &:= |f((\alpha_1, f_1), \xi_1) - f((\alpha_2, f_2), \xi_1)|^2\\ &\leqslant 2(|f((\alpha_1, f_1), \xi_1) - f((\alpha_2, f_1), \xi_1)|^2 + \|\text{supp}\,\xi_1\|_K^2 d_K^2(f_1, f_2))\end{aligned}$$

for $(\alpha_1, f_1), (\alpha_2, f_2) \in T$.

We proceed to bound above and obtain

$$\begin{aligned}\|A\|_{\mathcal{P},2} &= (E|f((\alpha_1, f_1), \xi_1) - f((\alpha_2, f_2), \xi_1)|^2)^{1/2}\\ &\leqslant \sqrt{2}((\sup_{\widetilde{f} \in \text{ball}X^*} E|f((\alpha_1, \widetilde{f}), \xi_1) - f((\alpha_2, \widetilde{f}), \xi_1)|^2)^{1/2}\\ &\quad + (E\|\text{supp}\xi_1\|_K^2)^{1/2} d_K(f_1, f_2))\\ &\leqslant D((\sup_{\widetilde{f} \in \text{ball}X^*} E(\|\xi_1\|_\infty |f((\alpha_1, \widetilde{f}), \xi_1) - f((\alpha_2, \widetilde{f}), \xi_1)|))^{1/2}\\ &\quad + d_K(f_1, f_2))\\ &= D((\sup_{\widetilde{f} \in \text{ball}X^*} E\big|f((\alpha_1, \widetilde{f}), \|\xi_1\|_\infty \xi_1) - f((\alpha_2, \widetilde{f}), \|\xi_1\|_\infty \xi_1)\big|)^{1/2}\end{aligned}$$

$$+ d_K(f_1, f_2))$$
$$= D\rho((\alpha_1, f_1), (\alpha_2, f_2)) \text{ for all } (\alpha_1, f_1), (\alpha_2, f_2) \in T$$

with the pseudodistance $\rho = \rho_1 + \rho_2$ on T, where $\rho_2 = d_K$.

The last relation follows from Theorem 4 and the fact, that $E \sup_{a \in A} f(a) = \sup_{a \in EA} f(a)$ ($f \in X^*$) is valid for the Aumann expectation of compact random sets A (see Aubin and Frankowska (1990)).

3. We intend to construct a finite δ−net for the pseudodistance ρ_1.

For that purpose, let us define: $\alpha_0 := 1$

$\alpha_1 := \inf\{\alpha \in [0, 1] \mid \alpha \leqslant \alpha_0, \rho_1(\alpha, \alpha_0) < \delta\}$.

If $\alpha_1 = 0$, cut short the procedure, otherwise define:

$\alpha_2 := \inf\{\alpha \in [0, 1] \mid \alpha \leqslant \alpha_1, \rho_1(\alpha, \alpha_1) < \delta\}$

continuing by induction.

Suppose now, that the procedure doesn't stop.

The left-continuity of $(\alpha \mapsto \rho_1(\alpha, \alpha_{n-1}))$ on $(0, 1]$ for all n implies then $\rho_1(\alpha_n, \alpha_{n-1}) \geqslant \delta$ for all n. Because of the compactness of $L_{\alpha_n} E\xi_1$ and the continuity of $(x \mapsto \inf_{y \in L_{\alpha_{n-1}} E\xi_1} \|x - y\|)$ for all n there exists a sequence $(a_n)_{n \in \mathbb{N}}$ in $\operatorname{supp} E\xi_1$ such that $\|a_i - a_j\| \geqslant \delta^2$ for $i \neq j$. But this is inconsistent with the compactness of $\operatorname{supp} E\xi_1$.

We can summarize that there exists a natural number n such that $\alpha_0 = 1$, $\alpha_n = 0$, $\alpha_m < \alpha_{m-1}$ and $\rho_1(\alpha, \alpha_{m-1}) < \delta$ on $(\alpha_m, \alpha_{m-1}]$ for all $m = 1, \ldots, n$. Thus $(\alpha_m)_{m=0}^n$ is a δ−net for ρ_1. Since the identity map $id : (\text{ball} X^*, d_*) \longrightarrow (\text{ball} X^*, d_K)$ is continuous, we conclude in addition that there is a finite δ−net for ρ.

4. From step 3 we know that there exists a finite ρ-δ-net $(S_i^{(\delta)})_{i=1}^n$. Since $U_\delta^{(\rho)}(S_i^{(\delta)})$ is a subset of $U_\delta^{(\rho_1)}(\alpha_i^{(\delta)}) \times U_\delta^{(\rho_2)}(f_i^{(\delta)})$ for $S_i^{(\delta)} = (\alpha_i^{(\delta)}, f_i^{(\delta)})$, $i = 1, \ldots, n$, we get

$$C := \sup_{\substack{(\alpha, \tilde{f}) \in \\ U_\delta^{(\rho_1)}(S_i^{(\delta)})}} \left| f((\alpha, \tilde{f}), \xi_1) - f((\alpha_i^{(\delta)}, f_i^{(\delta)}), \xi_1) \right|^2$$
$$\leqslant 2(\underbrace{\sup_{\alpha \in U_\delta^{(\rho_1)}(\alpha_i^{(\delta)})} \left| f((\alpha, f_i^{(\delta)}), \xi_1) - f((\alpha_i^{(\delta)}, f_i^{(\delta)}), \xi_1) \right|^2}_{:= f(\delta)}$$
$$+ \|\operatorname{supp} \xi_1\|_K^2 \sup_{\tilde{f} \in U_\delta^{(\rho_2)}(f_i^{(\delta)})} d_K^2(\tilde{f}, f_i^{(\delta)})).$$

Because of step 1 and monotony reasons there exist sequences $\underline{\alpha}_n \searrow$ and $\overline{\alpha}_n \nearrow$ independent of $\omega \in \Omega$ and from $U_\delta^{(\rho_1)}(\alpha_i^{(\delta)})$ such that

$$f(\delta) = \max\{\lim_n \overbrace{|f((\underline{\alpha}_n, f_i^{(\delta)}), \xi_1) - f((\alpha_i^{(\delta)}, f_i^{(\delta)}), \xi_1)|^2}^{=:C_1^{(n)}}, \lim_n \underbrace{|f((\overline{\alpha}_n, f_i^{(\delta)}), \xi_1) - f((\alpha_i^{(\delta)}, f_i^{(\delta)}), \xi_1)|^2}_{=:C_2^{(n)}}\}.$$

Therefore, using the monotone convergence theorem and step 2, we find

$$\begin{aligned} Ef(\delta) &= \lim_n E\max\{C_1^{(n)}, C_2^{(n)}\} \\ &\leqslant \lim_n (EC_1^{(n)} + EC_2^{(n)}) \\ &\leqslant \lim_n D(\rho_1^2(\underline{\alpha}_n, \alpha_i^{(\delta)}) + \rho_1^2(\overline{\alpha}_n, \alpha_i^{(\delta)})) \\ &\leqslant \widetilde{D}\delta^2. \end{aligned}$$

Since the continuity of the expressions $\inf_{t\in A} f_t(\cdot)$ and $\sup_{t\in A} f_t(\cdot)$ for $A \subseteq T$ ensures the measurability of C, we can write then

$$(EC)^{1/2} \leqslant \sqrt{2}((Ef(\delta))^{1/2} + (E\|\mathrm{supp}\xi_1\|_K^2)^{1/2}\delta) \leqslant D_0\delta.$$

5. First, we want to find a finite number of ε-brackets, which cover the function class $\mathcal{F} = \{f_t \mid t \in T\}$. In doing so, let us define for $t_0 \in T$ and $\delta > 0$ $f_{\delta,t_0}^u := \sup_{t\in U_\delta^{(\rho)}(t_0)} f_t(\cdot)$ and $f_{\delta,t_0}^l := \inf_{t\in U_\delta^{(\rho)}(t_0)} f_t(\cdot)$. As for the same reasons, mentioned at the end of step 4, f_{δ,t_0}^u and f_{δ,t_0}^l are measurable functions. Because of step 3 there exists a finite ρ-δ-net, $S(\delta) \subseteq T$. We see then that for all $t \in T$ there is a $s \in S(\delta)$ such that for all $v \in U_c(X)$, $f_{\delta,s}^l(v) \leqslant f_t(v) \leqslant f_{\delta,s}^u(v)$ holds.

Furthermore, using the last inequality of step 4, we realize that

$$\begin{aligned} \|f_{\delta,s}^u - f_{\delta,s}^l\|_{\mathcal{P},2} &= (E(f_{\delta,s}^u(\xi_1) - f_{\delta,s}^l(\xi_1))^2)^{1/2} \\ &= (E(\sup_{t_1,t_2\in U_\delta^{(\rho)}(s)} |f_{t_1}(\xi_1) - f_{t_2}(\xi_1)|^2))^{1/2} \\ &\leqslant \sqrt{2}(E(\sup_{t\in U_\delta^{(\rho)}(s)} |f_t(\xi_1) - f_s(\xi_1)|^2))^{1/2} \\ &\leqslant D\delta \text{ for all } s \in S(\delta) \text{ and a constant } D. \end{aligned}$$

This enables us to give an upper bound for bracketing numbers by means of entropy numbers:

$$N_{[]}(\varepsilon, \mathcal{F}, L_2(\mathcal{P})) \leqslant N(\varepsilon/D, T, \rho).$$

Without loss of generality, we set $D = 1$ and follow further

$$\int_0^1 (\log N_{[]}(\varepsilon, \mathcal{F}, L_2(\mathcal{P})))^{1/2}\lambda^1(d\varepsilon)$$
$$\leqslant \int_0^1 (\log N(\varepsilon, T, \rho))^{1/2}\lambda^1(d\varepsilon)$$
$$\leqslant \int_0^1 (\log N(\varepsilon/2, [0,1], \rho_1))^{1/2}\lambda^1(d\varepsilon)$$
$$+ \int_0^1 (\log N(\varepsilon/2, \operatorname{ball} X^*, d_K))^{1/2}\lambda^1(d\varepsilon).$$

6. Let $(\alpha_i)_{i=1}^n$ a finite δ-net for ρ_1 with the corresponding $(a_i)_{i=1}^n \subseteq \operatorname{supp} E\xi_1 \subseteq (E\|\operatorname{supp}\xi_1\|_K)K$, constructed as in step 3.
We claim that $n-1 \leqslant N(\delta^2/3, (E\|\operatorname{supp}\xi_1\|_K)K, \|\cdot\|) =: m$.
Let's say, $n-1 > m$. Take $\{U_{\delta^2/3}(x_i) \mid i = 1, \ldots, m\}$, a minimal $\delta^2/3$-covering of $(E\|\operatorname{supp}\xi_1\|_K)K$ with respect to $\|\cdot\|$. We can find then pairwise disjoint sets $I_1, \ldots, I_m$ such that $\cup_{i=1}^m U_{\delta^2/3}(x_i) = \cup_{i=1}^m I_i$ and $\operatorname{diam} I_i \leqslant 2/3\delta^2$ for all i. Then there must be natural numbers $i_0, i \neq j$ with $i, j \in \{1, \ldots, n-1\}$ and $i_0 \in \{1, \ldots, m\}$ such that $a_i, a_j \in I_{i_0}$. On the other hand, by construction, $\|a_i - a_j\| \geqslant \delta^2$ is valid for all $i \neq j$. But this leads to a contradiction.
Hence

$$\int_0^1 (\log N(\varepsilon, [0,1], \rho_1))^{1/2}\lambda^1(d\varepsilon)$$
$$\leqslant \int_0^1 (\log 2N(\varepsilon^2/3, (E\|\operatorname{supp}\xi_1\|_K)K, \|\cdot\|))^{1/2}\lambda^1(d\varepsilon)$$
$$= \int_0^1 (\log 2N(\varepsilon^2/3E\|\operatorname{supp}\xi_1\|_K), K, \|\cdot\|))^{1/2}\lambda^1(d\varepsilon).$$

7. Finally, by the last two steps, we proved the following integral condition:

$$\int_0^1 (\log N_{[]}(\varepsilon, \mathcal{F}, L_2(\mathcal{P})))^{1/2}\lambda^1(d\varepsilon) < \infty.$$

Further, we see that

$$\mathcal{P}^* F^2 = E\|\xi_1\|_\infty^2 = O(1) \text{ as well as}$$
$$\mathcal{P}^* F^2 1_{\{F > \eta\sqrt{n}\}} =$$
$$\mathcal{P}^* F^2 1_{\{\|\xi_1\|_\infty > \eta\sqrt{n}\}} \longrightarrow 0 \text{ for all } \eta > 0.$$

In addition, the last inequality of step 3 provides

$$\begin{aligned}\sup_{\rho(s,t)<\delta_n} \mathcal{P}(f_s - f_t)^2 &= \sup_{\rho(s,t)<\delta_n} E|f_s(\xi_1) - f_t(\xi_1)|^2 \\ &\leqslant \sup_{\rho(s,t)<\delta_n} D\rho^2(s,t) \\ &\leqslant D\delta_n^2 \longrightarrow 0 \text{ for all } \delta_n \downarrow 0.\end{aligned}$$

With these verifications we are now ready to apply a CLT result for empirical processes (see Theorem 2.11.23, p. 221 in van der Vaart and Wellner (1996)). Together with Theorem 1 and the continuous mapping theorem we obtain the desired result:

$$\begin{aligned}&\sqrt{n}d_\infty\left(\sum_{i=1}^n \xi_i/n,\, E\xi_1\right) \\ &= \sup_{t\in T}\left|n^{-1/2}\left(\sum_{i=1}^n f_t(\xi_1) - Ef_t(\xi_i)\right)\right| \longrightarrow \|\underline{Y}\|_\infty \text{ weakly}\end{aligned}$$

for a tight Gaussian process $\underline{Y}$ in $l^\infty([0,1]\times \text{ball}X^*)$.

□

The finite dimensional case yields then:

COROLLARY 8. *If $\xi_i : \Omega \to (F_c(\mathbb{R}^p), d_\infty)$ is an i.i.d. sequence of frv's with $E\|\xi_1\|_\infty^2 < \infty$ then*

$$\sqrt{n}d_\infty\left(\sum_{i=1}^n \xi_i/n,\, E\xi_1\right) \longrightarrow \|\underline{Y}\|_\infty \text{ weakly}$$

for a tight Gaussian process $\underline{Y}$ in $l^\infty([0,1]\times \text{ball}\mathbb{R}^p)$.

Proof. Let us first consider the i.i.d. frv's $\overline{co}\xi_i : \Omega \longrightarrow (U_c(\mathbb{R}^p), d_\infty)$. We choose $K = \text{ball}\,\mathbb{R}^p$. Since $\text{supp}\overline{co}\xi_1 \subseteq \|\xi_1\|_\infty \text{ball}\mathbb{R}^p$ we see that $E\|\text{supp}\overline{co}\xi_1\|_K^2 < \infty$. The spaces $(\text{ball}\mathbb{R}^p, \|\cdot\|)$ and $(\text{ball}(\mathbb{R}^p)^*, d_K)$ are isometrical to each other. Therefore we get $N(\delta, \text{ball}\mathbb{R}^p, \|\cdot\|) = N(\delta, \text{ball}(\mathbb{R}^p)^*, d_K)$ for all $\delta > 0$. Taking into consideration that

$$\int_0^1 (\log N(\varepsilon^2, K, \|\cdot\|))^{1/2}\lambda^1(d\varepsilon) \leqslant \int_0^1 (\sqrt{p}(\log(4\sqrt{p}/\varepsilon^2))^{1/2}\lambda^1(d\varepsilon) < \infty,$$

we follow the existence of the integrals in Theorem 7. Thus it implies

$$\sqrt{n}d_\infty\left(\sum_{i=1}^n \overline{co}\xi_i/n,\, E\xi_1\right) \longrightarrow \|\underline{Y}\|_\infty \text{ weakly}$$

for a tight Gaussian process $\underline{Y}$ in $l^\infty([0,1]\times \text{ball}\mathbb{R}^p)$.

Further, the Shapley-Folkman-Starr theorem gives

$$\sqrt{n}d_\infty\left(\sum_{i=1}^{n}\xi_i/n, \sum_{i=1}^{n}\overline{co}\xi_i/n\right) \leqslant \sqrt{p}\max_{1\leqslant i\leqslant n}\|\xi_i\|_\infty/\sqrt{n}.$$

We can write then

$$\sqrt{n}d_\infty\left(\sum_{i=1}^{n}\xi_i/n, E\xi_1\right) \leqslant \sqrt{p}\max_{1\leqslant i\leqslant n}\|\xi_i\|_\infty/\sqrt{n}+\sqrt{n}d_\infty\left(\sum_{i=1}^{n}\overline{co}\xi_i/n. E\xi_1\right).$$

The first term of the above inequality converges weakly to zero (see, for example, p. 188 in Breiman (1968)). whereas the second to $\|\underline{Y}\|_\infty$, concluding the proof. □

The technique of proof in Theorem 7 can be even used to derive an Etemadi's type of SLLN for frv's (see for more details Proske and Puri (1999c)). That means that Corollary 6 is also true in the p.i.i.d. case (see Etemadi (1981)):

THEOREM 9. *Let $\xi_i : \Omega \longrightarrow (F_c(\mathbb{R}^p), d_\infty)$ be an p.i.i.d. sequence of frv's with $E\|\xi_1\|_\infty < \infty$ then*

$$d_\infty\left(\frac{1}{n}\sum_{i=1}^{n}\xi_i, E\xi_1\right) \longrightarrow 0 \ a.e.$$

5. CONCLUSION

As for the SLLN for compact frv's (Theorem 5) it seems to be quite natural to conjecture that it holds also in the general separable Banach space setting. The CLT for $U_c(X)$-valued frv's (Theorem 7) is analogous to that of Giné, Hahn and Zinn (1983) for random convex compact sets in the infinite dimensional case. But comparatively Theorem 7 requires in addition another integral condition. It would be interesting to check out possible relations between these two integrals. Regarding the finite dimensional case a natural generalization of the result of Weil (1982) for random compact sets was attained. Further, it seems to be that the assumptions of Theorem 7 are sufficient to state a law of iterated logarithm for i.i.d. $\mathbb{R}^p$-valued frv's.

REFERENCES

Arrow, K. J. and Hahn, F. H. (1971). *General Competitive Analysis*. Holden-Day.

Artstein, Z. and Vitale, R. A. (1975). A strong law of large numbers for random compact sets. *Ann. Prob.* **3**, 879–882.

Aubin. J. P. and Frankowska. H. (1990). *Set-valued Analysis. Systems & Control: Foundations & Applications*. Birkhäuser.

Breiman, L. (1968). *Probability*. Addison-Wesley.

Chow, Y. S. and Teicher, H. (1988). *Probability Theory*. Second Ed. Springer.

Debreu, G. (1966). Integration of correspondences. *Proc. Fifth Berkeley Symp. Math. Statist. Prob.*, Vol. 2, pp. 351–372. University of California Press.

Dunford, N. and Schwartz, J. T. (1958). *Linear Operators*, Part I. Interscience.

Etemadi, N. (1981). An elementary proof of the strong law of large numbers. *Z. Wahrscheinlichkeitstheor. Verw. Geb.* 55, 119–122.

Giné, E., Hahn. G. and Zinn. J. (1983). Limit theorems for random sets. *Springer Lecture Notes in Math.* **990**, 112–135.

Hörmander, L. (1954). Sur la fonction d'appui des ensembles convexe dans un espace localement convexe. *Ark. Mat.* **3**, 181–186.

Hoffmann-Jørgensen, J. (1985). The law of large numbers for non-measurable and non-separable random elements. *Colloque en l'honneur de L. Schwartz*, Astérisque Vol. 131, pp. 299-356, Hermann, Paris.

Klement, E. P., Puri, M. L. and Ralescu, D. A. (1986). Limit theorems for fuzzy random variables. *Proc. R. Soc. Lond.* **A407**, 171–182.

Kruse, R. (1982). The strong law of large numbers for fuzzy random variables. Inform. Sci. **28**, 233–241.

Proske, F. N. and Puri, M. L. (1999a). Strong law of large numbers for Banach space valued fuzzy random variables. To appear.

Proske, F. N. and Puri, M. L. (1999b). Central limit theorem for Banach space valued fuzzy random variables. To appear.

Proske, F. N. and Puri, M. L. (1999c). An application of the strong law of large numbers for empirical processes for generalized random sets from the viewpoint of empirical processes. To appear.

Puri, M. L. and Ralescu, D. A. (1983). Differentials of fuzzy functions. *J. Math. Anal. Appl.* **91**, 552–558.

Puri, M. L. and Ralescu, D. A. (1985a). Limit theorems for random compact sets in Banach spaces. *Math. Proc. Cam. Phil. Soc.* **97**, 151–158.

Puri, M. L. and Ralescu, D. A. (1985b). The concept of normality for fuzzy random variables. *Ann. Prob.* **13**, 1373–1379.

Puri, M. L. and Ralescu, D. A. (1986). Fuzzy random variables. *J. Math. Anal. Appl.* **114**, 409–422.

Talagrand, M. (1987). The Glivenko-Cantelli problem. *Ann. Prob.* **15**, 837–870.

van der Vaart, A. W. and Wellner, J. A. (1996). *Weak Convergence and Empirical Processes*. Springer.

Weil, W. (1982). An application of the central limit theorem for Banach space valued random variables to the theory of random sets. *Z. Wahrscheinlichkeitstheor. Verw. Geb.* **60**, 203–208.

Zadeh, L. A. (1965). Fuzzy Sets. *Information and Control* **8**, 338–353.

Reprinted from
Asymptotics in Statistics and Probability:
Papers in Honor of George Gregory Roussas,
edited by M.L. Puri,
© 2000 by VSP BV

CENTRAL LIMIT THEOREM FOR BANACH SPACE VALUED FUZZY RANDOM VARIABLES

FRANK N. PROSKE and MADAN L. PURI

(Communicated by Claudia M. Neuhauser)

ABSTRACT. In this paper we prove a central limit theorem for Borel measurable nonseparably valued random elements in the case of Banach space valued fuzzy random variables.

1. INTRODUCTION

In this paper we plan to derive a central limit theorem (CLT) for independent and identically distributed fuzzy random variables with compact $\alpha-$level sets, extending similar results for random sets (see Giné, Hahn and Zinn [6], Weil [14]). Here we are concerned with the concept of a fuzzy random variable (frv) introduced by Puri and Ralescu [12].

Fuzzy random variables are suitable for representing random mechanisms with vague outcomes of non-statistical nature. Consider for example an opinion poll where people are randomly chosen and asked about the weather. The resulting data of this random experiment arise from the imprecision of human knowledge, expressed by opinions like "warm" or "cold" which can be described by fuzzy sets rather than by a single real number or subsets of real numbers.

In section 2 we discuss some facts about frv's and other preliminaries which are needed in section 3 to present our main result, a CLT for separably Banach space valued frv's, independent and identically distributed. Klement, Puri and Ralescu [8] proved such a result for frv's with fuzzy set values whose compact α-level sets fulfill a Lipschitz condition with respect to the Hausdorff distance. From the practical point of view, this means a rather strong restriction of frv's to a subclass of compact fuzzy sets. Their main tools in proving the CLT are based on embedding theorems and results from probability theory on Banach spaces.

The CLT, given here, is a generalization of the latter authors' result, because no Lipschitz condition is needed. In order to then face the non-separability of involved metric spaces of fuzzy sets, our main approach is to identify linear isometrically each frv with convex and compact α-level sets with an empirical process in $l^{\infty}(T)$ to solve a Donsker problem (see van der Vaart and Wellner [13]).

Weil [14] proved a CLT for R^p-valued compact random sets (i.e. random sets whose values are compact subsets of R^p). Later on these results were extended by Giné, Hahn and Zinn [6] and Puri and Ralescu [11] to the case of separable Banach

Received by the editors May 20, 2000 and, in revised form, November 22, 2000.
2000 *Mathematics Subject Classification.* Primary 60F05; Secondary 46B09.

spaces. Our result is a natural extension of all previous CLT's for random sets, because no Lipschitz condition is imposed.

In the sequel, we confine ourselves to complete probability spaces $(\Omega, \Im, P)$.

2. Preliminaries

Let $(X, \|\cdot\|)$ be a real separable Banach space.

A *fuzzy set* of X is a function $u : X \longrightarrow [0,1]$. For each fuzzy set u the *α-level set* is defined by $L_\alpha u := \{x \in X \mid u(x) \geq \alpha\}, 0 < \alpha \leq 1$. By $\operatorname{supp} u$ we understand the closure of $\{x \in X \mid u(x) > 0\}$ in X. By $F_c(X)$, we denote the collection of those fuzzy sets $u : X \longrightarrow [0,1]$ which have the following properties: (i) $L_\alpha u$ is compact for all $0 < \alpha \leq 1$; (ii) $\operatorname{supp} u$ is compact; (iii) $\{x \in X \mid u(x) = 1\} \neq \emptyset$.

In the sequel, we set $L_0 u = \operatorname{supp} u$. Zadeh's Extension principle [15] supplies a linear structure in $F_c(X)$ by the operations

$$(u+v)(x) := \sup_{y+z=x} \min(u(y), v(z)),$$

$$(\lambda u)(x) := \begin{cases} u(\lambda^{-1}x), & \text{if } \lambda \neq 0, \\ \chi_{\{0\}}(x), & \text{if } \lambda = 0, \end{cases}$$

$$\text{for } u, v \in F_c(X),\ \lambda \in R.$$

For later use, we define $U_c(X)$ as the collection of all $u \in F_c(X)$ with convex α-level sets and denote the closure of the convex hull of $u \in F_c(X)$ by $\overline{co}u \in F_c(X)$, determined by its convex compact α-level sets $L_\alpha \overline{co}u := \overline{co}(L_\alpha u)$, $0 < \alpha \leq 1$ (see Klement, Puri and Ralescu [8]), where $\overline{co}A$ is the closure of the convex hull of a subset $A \subseteq X$ in X.

Denote by $\mathcal{K}(X)$ ($co\mathcal{K}(X)$) the class of all non-empty (convex and) compact subsets of X. The space $\mathcal{K}(X)$ ($co\mathcal{K}(X)$) together with the following operations forms a linear Abelian semigroup:

$$A + B := \{a + b \mid a \in A, b \in B\}, \lambda A := \{\lambda a \mid a \in A\}$$
$$\text{for } A, B \in \mathcal{K}(X)(co\mathcal{K}(X)), \lambda \in R.$$

Then, it is easy to check that (see Dunford and Schwartz [5], Puri and Ralescu [10, 12])

$$L_\alpha(u+v) = L_\alpha u + L_\alpha v, L_\alpha(\lambda u) = \lambda L_\alpha u, \quad \lambda > 0,$$

and

$$\overline{co}(L_\alpha u + L_\alpha v) = \overline{co}L_\alpha u + \overline{co}L_\alpha v, \overline{co}(\lambda L_\alpha u) = \lambda \overline{co} L_\alpha u, \quad \lambda > 0,$$

for all $u, v \in F_c(X)$, $0 < \alpha \leq 1$. Also $F_c(X)$ and $U_c(X)$ are closed under the above operations and form like $\mathcal{K}(X)$ or $co\mathcal{K}(X)$ a linear Abelian semigroup.

$\mathcal{K}(X)$ is metrizable by the well-known Hausdorff distance d_H, given by

$$d_H : \mathcal{K}(X) \times \mathcal{K}(X) \longrightarrow [0, \infty); (A, B) \mapsto \max\{\sup_{a \in A} \inf_{b \in B} \|a - b\|, \sup_{b \in B} \inf_{a \in A} \|a - b\|\}.$$

$(\mathcal{K}(X), d_H)$ and $(co\mathcal{K}(X), d_H)$ are then complete separable metric spaces (see Puri and Ralescu [11]). By identifying each $A \in \mathcal{K}(X)$ with its characteristic function $\chi_A \in F_c(X)$, we can generalize the Hausdorff distance to the case of $F_c(X)$, getting a continuous semigroup monomorphism:

Proposition 1 (see Puri and Ralescu [12]). *$(F_c(X), d_\infty)$ $((U_c(X), d_\infty))$ is a complete, but in contrast to $(\mathcal{K}(X), d_H)$ $((co\mathcal{K}(X), d_H))$, a nonseparable metric space, endowed with the metric d_∞ defined by $d_\infty(u,v) = \sup_{0<a\leq 1} d_H(L_\alpha u, L_\alpha v)$ for $u, v \in F_c(X)$.*

Let $(\Omega, \Im, P)$ be a probability space. A *fuzzy random variable* (frv) is a Borel measurable function $\xi : \Omega \longrightarrow (F_c(X), d_\infty)$. As one can see, frv's generalize *compact random sets A* (i.e. Borel measurable functions $A : \Omega \longrightarrow (\mathcal{K}(X), d_H)$) by identification with the characteristic function χ_A.

Proposition 2 (see Klement, Puri and Ralescu [8]). *Let $\xi : \Omega \longrightarrow (F_c(X), d_\infty)$ be an frv. Then:*
(i) *$\overline{co}\xi : \Omega \longrightarrow (U_c(X), d_\infty); \omega \mapsto \overline{co}\xi(\omega)$ is an frv.*
(ii) *$L_\alpha\xi : \Omega \longrightarrow (\mathcal{K}(X), d_H); \omega \mapsto L_\alpha\xi(\omega)$ and $\overline{co}(L_\alpha\xi) : \Omega \longrightarrow (co\mathcal{K}(X), d_H); \omega \longmapsto \overline{co}(L_\alpha\xi(\omega))$ are (convex) compact random sets for all $0 < \alpha \leq 1$.*
(iii) *$supp\,\xi : \Omega \longrightarrow (\mathcal{K}(X), d_H); \omega \mapsto L_0\xi(\omega)$ is a compact random set.*

Proposition 3 (see Puri and Ralescu [12]). *Let $\xi : \Omega \longrightarrow (F_c(X), d_\infty)$ be an frv with $E\,\|\xi\|_\infty < \infty$ $(\|\xi(\omega)\|_\infty := \sup_{\alpha>0} d_H(\{0\}, L_\alpha\xi(\omega)))$. Then there exists a unique fuzzy set $u \in U_c(X)$ such that the α-level sets of u are the Aumann expectations of the convex compact random sets $\overline{co}(L_\alpha\xi)$ (see for definition Aubin, Frankowska [2], Debreu [4]), i.e. $L_\alpha u = E\overline{co}(L_\alpha\xi)$ for all $0 < \alpha \leq 1$.*

If $\xi : \Omega \longrightarrow (F_c(X), d_\infty)$ is an frv with $E\,\|\xi\|_\infty < \infty$, then, by the last result, the *expected value* $E\xi$ is the unique convex fuzzy set satisfying the property $L_\alpha E\xi = E\overline{co}(L_\alpha\xi)$, $0 < \alpha \leq 1$.

One of our main tools in proving the CLT is a result for embedding convex compact sets into the space of continuous functions which stems in this form from Giné, Hahn and Zinn [6], but is originally due to Hörmander [7]:

Theorem 4. *The function $j : (co\mathcal{K}(X), d_H) \longrightarrow (C(ball X^*, d_*), \|\cdot\|_\infty)$ ($\|\cdot\|_\infty$ is the supremum norm), given by $A \longmapsto (f \mapsto \sup_{a\in A} f(a))$, satisfies:*
(i) *j is an isometry (i.e. $d_H(A,B) = \|j(A) - j(B)\|_\infty$).*
(ii) *$j(A+B) = j(A) + j(B)$, $j(\lambda A) = \lambda j(A)$ $(\lambda \geq 0)$.*

d_* denotes a metric inducing the weak* topology of ball X^*, where ball X^* is the norm-closed unit ball of the dual of X.

A crucial idea in deriving a CLT for $F_c(R^p)$-valued frv's (see Corollary 6.1) is to first reduce the study of frv's to the convex case by convexifying the underlying variables and then to get rid of the convexity restriction by applying the following inequality:

Theorem 5 (Shapley-Folkman-Starr). *For all $A_1,\ldots,A_n \in \mathcal{K}(R^p)$ and $n \in \mathbb{N}$*

$$d_H(A_1 + \ldots + A_n, \overline{co}(A_1 + \ldots + A_n)) \leq \sqrt{p} \max_{1\leq i\leq n} \|A_i\| \; (\|A\| := d_H(\{0\}, A)).$$

We wish to employ the theory of empirical processes (see van der Vaart and Wellner [13]) to investigate frv's. To this end, we need the term of a *bracketing number* to measure the "size" of certain function classes: Let $(\mathcal{F}, \|\cdot\|)$ be a subset of a normed space of real functions $f : \chi \longrightarrow R$ on a set χ. Then the *bracket* $[l, u]$ for real functions l, u is the set of all functions $f : \chi \longrightarrow R$ satisfying $l \leq f \leq u$.

An *ε-bracket* is a bracket $[l, u]$ with $\|l - u\| < \varepsilon$. Then the *bracketing number* $N_{[\ ]}(\varepsilon, \mathcal{F}, \|\cdot\|)$ is the minimal number of ε−brackets covering $\mathcal{F}$. Note that in the definition of the bracketing number, the functions l and u are assumed to have finite norms, but not to belong to $\mathcal{F}$ itself.

3. Central limit theorem for fuzzy random variables

In order to prepare for our main result, let us state two more definitions: If (T, d) is a metric space, the *ε-entropy* of T, denoted by $N(\varepsilon, T, d)$ $(\varepsilon > 0)$, is the minimal number of balls of radius $< \varepsilon$, with centres in T which cover T. The *Minkowski functional* $\|\cdot\|_K$ of a convex, compact, symmetric subset K of X is given by $\|A\|_K = \inf\{\lambda > 0 \mid A \subseteq \lambda K\}$ $(\inf \emptyset := \infty)$.

Theorem 6 (CLT for frv's). *Let $(X, \|\cdot\|)$ be a separable Banach space over R, and $\xi_i : \Omega \longrightarrow (U_c(X), d_\infty)$ an i.i.d. sequence of frv's such that:*

(i) *there exists a convex, compact, symmetric $K \subseteq X$ such that $0 < E\,\|supp\,\xi_1\|_K^2 < \infty$ and $\int_0^1 (\log N(\varepsilon, ball\,X^*, d_K))^{1/2} \lambda^1(d\varepsilon) < \infty$, where d_K is the semimetric on $ball\,X^*$ with $d_K(f_1, f_2) = \sup_{a\in K} |f_1(a) - f_2(a)|$, and*

(ii) $\int_0^1 (\log N(\varepsilon^2, K, \|\cdot\|))^{1/2} \lambda^1(d\varepsilon) < \infty$.

Then, a CLT in the following sense is valid:

$$\sqrt{n} d_\infty\Big(\sum_{i=1}^{n} \xi_i/n, E\xi_1\Big) \longrightarrow \|\underline{Y}\|_\infty \quad \textit{weakly}$$

for a tight Gaussian process $\underline{Y}$ in $l^\infty([0,1] \times ball\,X^*)$.

Proof. For proving this result we aim at applying a CLT for empirical processes (see Theorem 2.11.23, p. 221 in van der Vaart and Wellner [13]), which comes down to bounding above bracketing numbers in an appropriate way for the function class $\mathcal{F} := \{f_t \mid t \in T\}$, consisting of $\mathcal{B}(U_c(X))$, $\mathcal{B}(R)$-measurable functions $f_t : (U_c(X), d_\infty) \longrightarrow R; u \longmapsto \sup_{a\in L_\alpha u} f(a)$ for $t = (\alpha, f) \in T := [0,1] \times \text{ball}\,X^*$. In the sequel we adopt the notions in van der Vaart and Wellner [13] and define $F = \|\cdot\|_\infty$. The proof is accomplished step-by-step as follows:

1. The existence of the second moment of $\|\text{supp}\xi_1\|_K$ implies the existence of $\|\xi_1\|_\infty$. Then, one can easily prove that there exists a unique fuzzy set $E(\|\xi_1\|_\infty \xi_1) \in U_c(X)$ such that $L_\alpha E(\|\xi_1\|_\infty \xi_1) = E(\|\xi_1\|_\infty L_\alpha \xi_1)$ for all $0 < \alpha \le 1$. Let us equip the interval $[0,1]$ with the pseudodistance

$$\rho_1 : [0,1] \times [0,1] \longrightarrow [0,\infty); (\alpha_1, \alpha_2) \mapsto d_H^{1/2}(L_{\alpha_1} E(\|\xi_1\|_\infty \xi_1), L_{\alpha_2} E(\|\xi_1\|_\infty \xi_1)).$$

Taking into account that the α-level sets of a fuzzy set $u \in F_c(X)$ are non-increasing and that the mapping $(\alpha \mapsto L_\alpha u)$ on the interval $[0,1]$ is left continuous in $(0,1]$ as well as right continuous at 0 with respect to the Hausdorff distance (see Klement, Puri and Ralescu [8]), one sees that the ρ_1-balls with radius $< \varepsilon$ and centres $0 \le \alpha_0 \le 1$, $U_\varepsilon^{(\rho_1)}(\alpha_0)$ are non-degenerated subintervals of $[0,1]$.

2. For convenience, we denote $f_t(\cdot) = f((\alpha, \widetilde{f}), \cdot)$ for $t = (\alpha, \widetilde{f})$. Using the definition of the Minkowski functional $\|\cdot\|_K$, we find that

$$|f((\alpha_2, f_1), \xi_1) - f((\alpha_2, f_2), \xi_1)| \le \|\text{supp}\xi_1\|_K\, d_K(f_1, f_2)$$

holds for all (α_2, f_1),$(\alpha_2, f_2) \in T$.

This leads us to following inequality:

$$\begin{aligned} A &:= |f((\alpha_1, f_1), \xi_1) - f((\alpha_2, f_2), \xi_1)|^2 \\ &\leq 2(|f((\alpha_1, f_1), \xi_1) - f((\alpha_2, f_1), \xi_1)|^2 + \|\text{supp}\xi_1\|_K^2 \, d_K^2(f_1, f_2)) \end{aligned}$$

for (α_1, f_1),$(\alpha_2, f_2) \in T$.

We proceed to bound the above and obtain

$$\begin{aligned} \|A\|_{\mathcal{P},2} &= (E\,|f((\alpha_1, f_1), \xi_1) - f((\alpha_2, f_2), \xi_1)|^2)^{1/2} \\ &\leq \sqrt{2}((\sup_{\widetilde{f} \in \text{ball} X^*} E\left|f((\alpha_1, \widetilde{f}), \xi_1) - f((\alpha_2, \widetilde{f}), \xi_1)\right|^2)^{1/2} \\ &\quad +(E\,\|\text{supp}\xi_1\|_K^2)^{1/2} d_K(f_1, f_2)) \\ &\leq K((\sup_{\widetilde{f} \in \text{ball} X^*} E(\|\xi_1\|_\infty \left|f((\alpha_1, \widetilde{f}), \xi_1) - f((\alpha_2, \widetilde{f}), \xi_1)\right|))^{1/2} \\ &\quad +d_K(f_1, f_2)) \\ &= K((\sup_{\widetilde{f} \in \text{ball} X^*} E\left|f((\alpha_1, \widetilde{f}), \|\xi_1\|_\infty \xi_1) - f((\alpha_2, \widetilde{f}), \|\xi_1\|_\infty \xi_1)\right|)^{1/2} \\ &\quad +d_K(f_1, f_2)) \\ &= K\rho((\alpha_1, f_1), (\alpha_2, f_2)) \text{ for all } (\alpha_1, f_1), (\alpha_2, f_2) \in T \end{aligned}$$

with the pseudodistance $\rho = \rho_1 + \rho_2$ on T, where $\rho_2 = d_K$.

The last relation follows from Theorem 4 and the fact that $E \sup_{a \in A} f(a) = \sup_{a \in EA} f(a)$ $(f \in X^*)$ is valid for the Aumann expectation of compact random sets A (see Aubin and Frankowska [2]).

3. We intend to construct a finite $\delta-$net for the pseudodistance ρ_1.

For that purpose, let us define:

$\alpha_0 := 1$.

$\alpha_1 := \inf\{\alpha \in [0, 1] \mid \alpha \leq \alpha_0, \rho_1(\alpha, \alpha_0) < \delta\}$.

If $\alpha_1 = 0$, cut short the procedure, otherwise define:

$\alpha_2 := \inf\{\alpha \in [0, 1] \mid \alpha \leq \alpha_1, \rho_1(\alpha, \alpha_1) < \delta\}$

.

.

.

continue by induction.

Suppose now that the procedure doesn't stop.

The left-continuity of $(\alpha \mapsto \rho_1(\alpha, \alpha_{n-1}))$ on $(0, 1]$ for all n implies $\rho_1(\alpha_n, \alpha_{n-1}) \geq \delta$ for all n. Because of the compactness of $L_{\alpha_n} E(\|\xi_1\|_\infty \xi_1)$ and the continuity of $(x \mapsto \inf_{y \in L_{\alpha_{n-1}} E(\|\xi_1\|_\infty \xi_1)} \|x - y\|)$ for all n, there exists a sequence $(a_n)_{n \in \mathbb{N}}$ in supp$E(\|\xi_1\|_\infty \xi_1)$ such that $\|a_i - a_j\| \geq \delta^2$ for $i \neq j$. But this is inconsistent with the compactness of supp $E(\|\xi_1\|_\infty \xi_1)$.

To summarize, there exists a natural number n such that $\alpha_0 = 1$, $\alpha_n = 0$, $\alpha_m < \alpha_{m-1}$ and $\rho_1(\alpha, \alpha_{m-1}) < \delta$ on $(\alpha_m, \alpha_{m-1}]$ for all $m = 1, ..., n$. Thus $(\alpha_m)_{m=0}^n$ is a δ-net for ρ_1. Since the identity map $id : (\text{ball} X^*, d_*) \longrightarrow (\text{ball} X^*, d_K)$ is continuous, we conclude in addition that there is a finite δ-net for ρ.

4. From step 3 we know that there exists a finite ρ-δ-net $(S_i^{(\delta)})_{i=1}^n$. Since $U_\delta^{(\rho)}(S_i^{(\delta)})$ is a subset of $U_\delta^{(\rho_1)}(\alpha_i{}^{(\delta)}) \times U_\delta^{(\rho_2)}(f_i^{(\delta)})$ for $S_i^{(\delta)} = (\alpha_i^{(\delta)}, f_i^{(\delta)})$, $i = 1, ..., n$, we get

$$C := \sup_{\substack{(\alpha,\widetilde{f})\in \\ U_\delta^{(\rho)}(S_i^{(\delta)})}} \left| f((\alpha,\widetilde{f}),\xi_1) - f((\alpha_i^{(\delta)}, f_i^{(\delta)}),\xi_1)\right|^2$$

$$\leq 2(\underbrace{\sup_{\alpha\in U_\delta^{(\rho_1)}(\alpha_i^{(\delta)})} \left| f((\alpha, f_i^{(\delta)}),\xi_1) - f((\alpha_i^{(\delta)}, f_i^{(\delta)}),\xi_1)\right|^2}_{:=f(\delta)}$$

$$+ \|\operatorname{supp}\xi_1\|_K^2 \sup_{\widetilde{f}\in U_\delta^{(\rho_2)}(f_i^{(\delta)})} d_K^2(\widetilde{f}, f_i^{(\delta)})).$$

Because of step 1 and monotonicity reasons there exist sequences $\underline{\alpha}_n \searrow$ and $\overline{\alpha}_n \nearrow$ independent of $\omega \in \Omega$ and $U_\delta^{(\rho_1)}(\alpha_i^{(\delta)})$ such that

$$f(\delta) = \max\left\{\lim_n C_1^{(n)}, \lim_n C_2^{(n)}\right\},$$

where $C_1^{(n)} = \left| f((\underline{\alpha}_n, f_i^{(\delta)}),\xi_1) - f((\alpha_i^{(\delta)}, f_i^{(\delta)}),\xi_1)\right|^2$, and

$$C_2^{(n)} = \left| f((\overline{\alpha}_n, f_i^{(\delta)}),\xi_1) - f((\alpha_i^{(\delta)}, f_i^{(\delta)}),\xi_1)\right|^2.$$

Using the monotone convergence theorem and step 2, we find

$$\begin{aligned} Ef(\delta) &= \lim_n E\max\{C_1^{(n)}, C_2^{(n)}\} \\ &\leq \lim_n (EC_1^{(n)} + EC_2^{(n)}) \\ &\leq \lim_n K(\rho_1^2(\underline{\alpha}_n, \alpha_i^{(\delta)}) + \rho_1^2(\overline{\alpha}_n, \alpha_i^{(\delta)})) \\ &\leq \widetilde{K}\delta^2. \end{aligned}$$

Since the continuity of the expressions $\inf_{t\in A} f_t(\cdot)$ and $\sup_{t\in A} f_t(\cdot)$ for $A \subseteq T$ ensures the measurability of C, we can write

$$(EC)^{1/2} \leq \sqrt{2}((Ef(\delta))^{1/2} + (E\|\operatorname{supp}\xi_1\|_K^2)^{1/2}\delta) \leq K_0\delta.$$

5. First, we want to find a finite number of ε-brackets which cover the function class $\mathcal{F} = \{f_t \mid t \in T\}$. In doing so, let us define for $t_0 \in T$ and $\delta > 0$ $f_{\delta,t_0}^u := \sup_{t\in U_\delta^{(\rho)}(t_0)} f_t(\cdot)$ and $f_{\delta,t_0}^l := \inf_{t\in U_\delta^{(\rho)}(t_0)} f_t(\cdot)$. For the reasons mentioned at the end of step 4, f_{δ,t_0}^u and f_{δ,t_0}^l are measurable functions. Because of step 3 there exists a finite ρ-δ-net, $S(\delta) \subseteq T$. We then see that for all $t \in T$ there is a $s \in S(\delta)$ such that for all $v \in U_c(X)$, $f_{\delta,s}^l(v) \leq f_t(v) \leq f_{\delta,s}^u(v)$ holds.

Furthermore, using the last inequality of step 4, we note that

$$\begin{aligned} \|f_{\delta,s}^u - f_{\delta,s}^l\|_{\mathcal{P},2} &= (E(f_{\delta,s}^u(\xi_1) - f_{\delta,s}^l(\xi_1))^2)^{1/2} \\ &= (E(\sup_{t_1,t_2\in U_\delta^{(\rho)}(s)} |f_{t_1}(\xi_1) - f_{t_2}(\xi_1)|^2))^{1/2} \\ &\leq \sqrt{2}(E(\sup_{t\in U_\delta^{(\rho)}(s)} |f_t(\xi_1) - f_s(\xi_1)|^2))^{1/2} \\ &\leq K\delta \text{ for all } s \in S(\delta) \text{ and a constant } K. \end{aligned}$$

This enables us to give an upper bound for bracketing numbers by means of entropy numbers:

$$N_{[\,]}(\varepsilon, \mathcal{F}, L_2(\mathcal{P})) \leq N(\varepsilon/K, T, \rho).$$

Without loss of generality we set $K = 1$ and then

$$\begin{aligned}
&\int_0^1 (\log N_{[\,]}(\varepsilon, \mathcal{F}, L_2(\mathcal{P})))^{1/2} \lambda^1(d\varepsilon) \\
\leq\ &\int_0^1 (\log N(\varepsilon, T, \rho))^{1/2} \lambda^1(d\varepsilon) \\
\leq\ &\int_0^1 (\log N(\varepsilon/2, [0,1], \rho_1))^{1/2} \lambda^1(d\varepsilon) \\
&+ \int_0^1 (\log N(\varepsilon/2, \mathrm{ball} X^*, d_K))^{1/2} \lambda^1(d\varepsilon).
\end{aligned}$$

6. Let $(\alpha_i)_{i=1}^n$ be a finite δ-net for ρ_1 with the corresponding

$$(a_i)_{i=1}^n \subseteq \operatorname{supp} E(\|\xi_1\|_\infty \xi_1) \subseteq (E\,\|\xi_1\|_\infty \|\operatorname{supp} \xi_1\|_K) K,$$

constructed as in step 3.

We claim that $n - 1 \leq N(\delta^2/3, (E\,\|\xi_1\|_\infty \|\operatorname{supp}\xi_1\|_K)K, \|\cdot\|) =: m$.

Let us say $n-1 > m$. Take $\{U_{\delta^2/2}(x_i) \mid i = 1, ..., m\}$, a minimal $\delta^2/3$-covering of $(E\,\|\xi_1\|_\infty \|\operatorname{supp}\xi_1\|_K)K$ with respect to $\|\cdot\|$. We can then find pairwise disjoint sets $I_1, ..., I_m$ such that $\bigcup_{i=1}^m U_{\delta^2/3}(x_i) = \bigcup_{i=1}^m I_i$ and $\operatorname{diam} I_i \leq 2/3\delta^2$ for all i. Then there must be natural numbers i_0, $i \neq j$, with $i, j \in \{1, ..., n-1\}$ and $i_0 \in \{1, ..., m\}$ such that $a_i, a_j \in I_{i_0}$. On the other hand, by construction, $\|a_i - a_j\| \geq \delta^2$ is valid for all $i \neq j$. But this leads to a contradiction.

Hence

$$\begin{aligned}
&\int_0^1 (\log N(\varepsilon, [0,1], \rho_1))^{1/2} \lambda^1(d\varepsilon) \\
\leq\ &\int_0^1 (\log 2N(\varepsilon^2/3, (E\,\|\xi_1\|_\infty \|\operatorname{supp}\xi_1\|_K)K, \|\cdot\|))^{1/2} \lambda^1(d\varepsilon) \\
=\ &\int_0^1 (\log 2N(\varepsilon^2/3E(\|\xi_1\|_\infty \|\operatorname{supp}\xi_1\|_K), K, \|\cdot\|))^{1/2} \lambda^1(d\varepsilon).
\end{aligned}$$

7. Finally, by the last two steps, we prove the following integral condition:

$$\int_0^1 (\log N_{[\,]}(\varepsilon, \mathcal{F}, L_2(\mathcal{P})))^{1/2} \lambda^1(d\varepsilon) < \infty.$$

Further, we see that

$$\begin{aligned}
\mathcal{P}^* F^2 &= E\,\|\xi_1\|_\infty^2 = O(1) \text{ as well as} \\
\mathcal{P}^* F^2 1_{\{F > \eta\sqrt{n}\}} &= \mathcal{P}^* F^2 1_{\{\|\xi_1\|_\infty > \eta\sqrt{n}\}} \longrightarrow 0 \text{ for all } \eta > 0.
\end{aligned}$$

In addition, the last inequality of step 3 provides

$$\begin{aligned}
\sup_{\rho(s,t)<\delta_n} \mathcal{P}(f_s - f_t)^2 &= \sup_{\rho(s,t)<\delta_n} E\,|f_s(\xi_1) - f_t(\xi_1)|^2 \\
&\leq \sup_{\rho(s,t)<\delta_n} K\rho^2(s,t) \\
&\leq K\delta_n^2 \longrightarrow 0 \text{ for all } \delta_n \downarrow 0.
\end{aligned}$$

With these verifications we are now ready to apply a CLT result for empirical processes (see Theorem 2.11.23, p. 221 in van der Vaart and Wellner [13]). Together with Theorem 4 and the continuous mapping theorem we obtain the desired result:

$$\begin{aligned}
&\sqrt{n} d_\infty \left(\sum_{i=1}^{n} \xi_i / n, E\xi_1 \right) \\
= &\sup_{t \in T} \left| n^{-1/2} \left(\sum_{i=1}^{n} E f_t(\xi_1) - f_t(\xi_i) \right) \right| \longrightarrow \|\underline{Y}\|_\infty \text{ weakly}
\end{aligned}$$

for a tight Gaussian process $\underline{Y}$ in $l^\infty([0,1] \times \text{ball} X^*)$.

□

The finite dimensional case then yields:

Corollary 6.1. *If $\xi_i : \Omega \longrightarrow (F_c(R^p), d_\infty)$ is an i.i.d. sequence of frv's with $E \|\xi_1\|_\infty^2 < \infty$, then*

$$\sqrt{n} d_\infty (\sum_{i=1}^{n} \xi_i / n, E\xi_1) \longrightarrow \|\underline{Y}\|_\infty$$

weakly for a tight Gaussian process $\underline{Y}$ in $l^\infty([0,1] \times \text{ball}\, R^p)$.

Proof. Let us first consider the i.i.d. frv's $\overline{co}\xi_i : \Omega \longrightarrow (U_c(R^p), d_\infty)$. We choose $K = \text{ball}\, R^p$. Since $\text{supp}\, \overline{co}\xi_1 \subseteq \|\xi_1\|_\infty \text{ball}\, R^p$ we see that $E \|\text{supp}\, \overline{co}\xi_1\|_K^2 < \infty$. The spaces $(\text{ball}\, R^p, \|\cdot\|)$ and $(\text{ball}\, (R^p)^*, d_K)$ are isometrical to each other. Therefore we get $N(\delta, \text{ball}\, R^p, \|\cdot\|) = N(\delta, \text{ball}\, (R^p)^*, d_K)$ for all $\delta > 0$. Taking into consideration that

$$\int_0^1 (\log N(\varepsilon^2, K, \|\cdot\|))^{1/2} \lambda^1(d\varepsilon) \leq \int_0^1 \sqrt{p} (\log(4\sqrt{p}/\varepsilon^2))^{1/2} \lambda^1(d\varepsilon) < \infty,$$

we follow the existence of the integrals in Theorem 6. Thus it implies that

$$\sqrt{n} d_\infty (\sum_{i=1}^{n} \overline{co}\xi_i / n, E\xi_1) \longrightarrow \|\underline{Y}\|_\infty$$

weakly for a tight Gaussian process $\underline{Y}$ in $l^\infty([0,1] \times \text{ball}\, R^p)$.

Further, the Shapley-Folkman-Starr theorem gives

$$\sqrt{n} d_\infty (\sum_{i=1}^{n} \xi_i / n, \sum_{i=1}^{n} \overline{co}\xi_i / n) \leq \sqrt{p} \max_{1 \leq i \leq n} \|\xi_i\|_\infty / \sqrt{n}.$$

We can then write

$$\begin{aligned}
&\sqrt{n} d_\infty (\sum_{i=1}^{n} \xi_i / n, E\xi_1) \\
&\qquad \leq \sqrt{p} \max_{1 \leq i \leq n} \|\xi_i\|_\infty / \sqrt{n} + \sqrt{n} d_\infty (\sum_{i=1}^{n} \overline{co}\xi_i / n, E\xi_1).
\end{aligned}$$

The first term of the above inequality converges weakly to zero (see, for example, p. 188 in Breiman [3]), whereas the second converges to $\|\underline{Y}\|_\infty$, concluding the proof. □

4. Conclusion

The CLT for $U_c(X)$-valued frv's (Theorem 6) is analogous to that of Giné, Hahn and Zinn [6] for random convex compact sets in separable Banach spaces. But comparatively Theorem 6 requires in addition another integral condition. It would be worth checking out possible relations between these two integrals. Regarding the finite dimensional case we attained a natural generalization of the result of Weil [14] for random compact sets.

References

[1] Z. Artstein and R.A. Vitale, *A strong law of large numbers for random compact sets,* Ann. Prob. **3** (1975), 879-882. MR **52**:6825

[2] J.P. Aubin and H. Frankowska, *Set-valued Analysis. Systems and Control: Foundations and Applications*, Birkhäuser, 1990. MR **91d**:49001

[3] L. Breiman, *Probability.* Addison-Wesley, 1968. MR **37**:4841

[4] G. Debreu, *Integration of correspondences,* Proc. Fifth Berkeley Symp. Math Statist. Prob. **2** (1966), 351-372, University of California Press. MR **37**:3835

[5] N. Dunford and J.T. Schwartz, *Linear Operators, Part I.* Interscience, 1958. MR **22**:8302

[6] E. Giné, M.G. Hahn and J. Zinn, *Limit theorems for random sets,* Springer Lecture Notes in Math. **990** (1983), 112-135. MR **85d**:60019

[7] L. Hörmander, *Sur la fonction d'appui des ensembles convexe dans un espace localement convexe,* Ark. Mat. **3** (1954), 181-186. MR **16**:831e

[8] E.P. Klement, M.L. Puri and D.A. Ralescu, *Limit theorems for fuzzy random variables,* Proc. Royal Soc. Lond. **A 407** (1986), 171-182. MR **88b**:60092

[9] F.N. Proske, *Limit theorems for fuzzy random variables,* Ph.D. Thesis, Ulm University, Germany, 1997.

[10] M.L. Puri and D.A. Ralescu, *Differentials of fuzzy functions,* J. Math. Anal. Appl. **91** (1983), 552-558. MR **85c**:26014

[11] M.L. Puri and D.A. Ralescu, *Limit theorems for random compact sets in Banach spaces,* Math. Proc. Cam. Phil. Soc. **97** (1985), 151-158. MR **86c**:60024

[12] M.L. Puri and D.A. Ralescu, *Fuzzy random variables,* J. Math. Anal. Appl. **114** (1986), 409-422. MR **87f**:03159

[13] A.W. van der Vaart and J.A. Wellner, *Weak convergence and empirical processes,* Springer, 1996. MR **97g**:60035

[14] W. Weil, *An application of the central limit theorem for Banach space valued random variables to the theory of random sets,* Z. Wahrscheinlichkeitstheor. Verw. Geb. **60** (1982), 203-208. MR **83h**:60010

[15] L.A. Zadeh, *Fuzzy Sets,* Information and Control **8** (1965), 338-353. MR **36**:2509

Abt. Math. III, Universität Ulm, 89069 Ulm, Germany

E-mail address: `frproske@metronet.de`

Current address: Department of Mathematics, University of Oslo, 1053 Blindern, 0316 Oslo, Norway

E-mail address: `proske@math.uio.no`

Department of Mathematics, Indiana University, Bloomington, Indiana 47405

E-mail address: `puri@indiana.edu`

Reprinted by permission of
The American Mathematical Society,
© 2001 by The American Mathematical Society

STRONG LAW OF LARGE NUMBERS FOR BANACH SPACE VALUED FUZZY RANDOM VARIABLES

FRANK N. PROSKE[1] and MADAN L. PURI[2,3]

Received May 1, 2001

In this paper we prove a strong law of large numbers for Borel measurable nonseparably valued random elements in the case of Banach space valued fuzzy random variables.

KEY WORDS: Banach space; strong law; fuzzy random variable.

1. INTRODUCTION

In this paper we aim at extending the strong law of large numbers (SLLN) for random sets (see e.g., Artstein and Vitale,[1] Giné *et al.*,[6] Puri and Ralescu[13]) to the concept of a fuzzy random variable (frv) introduced by Puri and Ralescu.[14]

Fuzzy random variables are used for describing random phenomena whose outcomes are not sharp (e.g., a single real number) and rather due to subjectivity of human knowledge, e.g., compounded by vague linguistic terms which can be represented by fuzzy sets.

Section 2 is devoted to the foundations of frv's derived from facts about random sets, and further results and definitions, necessary to prove the main result, a SLLN for convex frv's which is gone ahead with in Section 3. Here we choose the framework of independent and identically distributed frv's on separable Banach spaces.

In literature one can already find SLLN's for frv's: Klement, Puri and Ralescu[9] proved such a result by using two different metrics: The Borel measurability of a frv is defined via one metric whereas the convergence is

[1] Mathematical Institute, Ulm University, Germany.
[2] Department of Mathematics, Indiana University, Bloomington, Indiana 47405-7106.
[3] To whom correspondence should be addressed. E-mail: puri@indiana.edu

induced by the other. Their approach relies on embedding theorems and results from probability theory on Banach spaces. A stronger convergence result for pairwise independent and identically distributed frv's could be shown by Kruse[10] in the case of R, utilizing its total order.

In handling the nonseparability of involved metric spaces, the main basis of our result is to identify linear isometrically each convex frv with a stochastic process in $l^\infty(T)$, i.e., we consider a Glivenko–Cantelli problem (see Hoffmann-Jørgensen,[8] Talagrand[15] and, van der Vaart and Wellner[16]).

Artstein and Vitale[1] proved a SLLN for R^p-valued random sets with respect to the Hausdorff distance (i.e., random sets whose values are compact subsets of R^p). Later on their results was extended by Puri and Ralescu[13] and by Giné, Hahn and Zinn[6] to the case of separable Banach spaces.

Our result is a natural generalization of all previous SLLN's for random sets, because Borel measurability and convergence is only based upon one metric which generalizes the Hausdorff distance.

In the sequel, a probability space $(\Omega, \mathfrak{I}, P)$ is supposed to be complete.

2. PRELIMINARIES

Let $(X, \|\cdot\|)$ be a real separable Banach space.

A *fuzzy set* of X is a function $u\colon X \to [0, 1]$. For each fuzzy set u, the α-*level set* is defined by $L_\alpha u := \{x \in X \mid u(x) \geqslant \alpha\}$, $0 < \alpha \leqslant 1$. By $\operatorname{supp} u$, we understand the closure of $\{x \in X \mid u(x) > 0\}$ in X. We denote by $F_c(X)$, the collection of those fuzzy sets $u\colon X \to [0, 1]$ with the following properties:

(i) $L_\alpha u$ is compact for all $0 < \alpha \leqslant 1$;

(ii) $\operatorname{supp} u$ is compact;

(iii) $\{x \in X \mid u(x) = 1\} \neq \emptyset$.

In the sequel, we set $L_0 u = \operatorname{supp} u$. Zadeh's Extension principle[17] supplies a linear structure in $F_c(X)$ by the operations

$$(u+v)(x) := \sup_{y+z=x} \min(u(y), v(z))$$

$$(\lambda u)(x) := \begin{cases} u(\lambda^{-1}x), & \text{if } \lambda \neq 0 \\ \chi_{\{0\}}(x), & \text{if } \lambda = 0 \end{cases} \qquad \text{for} \quad u, v \in F_c(X), \quad \lambda \in R$$

Denote by $U_c(X)$, the collection of all $u \in F_c(X)$ with convex α-level sets and denote the closure of the convex hull of $u \in F_c(X)$ by $\overline{\text{co}}\, u \in F_c(X)$, determined by its convex compact α-level sets $L_\alpha\, \overline{\text{co}}\, u := \overline{\text{co}}(L_\alpha u)$, $0 < \alpha \leqslant 1$

(see Klement, Puri, and Ralescu[9]), where $\overline{\mathrm{co}}\, A$ is the closure of the convex hull of a subset $A \subseteq X$ in X.

Also denote by $\mathscr{K}(X)$ (co $\mathscr{K}(X)$), the class of all nonempty (convex and) compact subsets of X. The space $\mathscr{K}(X)$ (co $\mathscr{K}(X)$) together with the following operations forms a linear abelian semigroup:

$$A+B := \{a+b \mid a \in A, b \in B\}, \qquad \lambda A := \{\lambda a \mid a \in A\}$$
$$\text{for} \quad A, B \in \mathscr{K}(X)\ (\mathrm{co}\ \mathscr{K}(X)), \quad \lambda \in R$$

Then, the following properties can be shown, which are needed later (see Dunford and Schwartz,[5] and Puri and Ralescu[12]):

$$L_\alpha(u+v) = L_\alpha u + L_\alpha v, \qquad L_\alpha(\lambda u) = \lambda L_\alpha u$$

and

$$\overline{\mathrm{co}}(L_\alpha u + L_\alpha v) = \overline{\mathrm{co}}\, L_\alpha u + \overline{\mathrm{co}}_\alpha\, v, \qquad \overline{\mathrm{co}}(\lambda L_\alpha u) = \lambda\, \overline{\mathrm{co}}\, L_\alpha u$$

for all $u, v \in F_c(X)$, $0 < \alpha \leqslant 1$. Note that $F_c(X)$ and $U_c(X)$ are closed under the operations above and form like $\mathscr{K}(X)$ or co $\mathscr{K}(X)$ a linear abelian semigroup.

$\mathscr{K}(X)$ is metrizable by the well-known Hausdorff distance d_H, given by

$$d_H\colon \mathscr{K}(X) \times \mathscr{K}(X) \to [0, \infty)$$
$$(A, B) \mapsto \max\{\sup_{a \in A} \inf_{b \in B} \|a-b\|, \sup_{b \in B} \inf_{a \in A} \|a-b\|\}$$

$(\mathscr{K}(X), d_H)$ and (co $\mathscr{K}(X), d_H)$ are then complete separable metric spaces (see Puri and Ralescu[13]). By identifying each $A \in \mathscr{K}(X)$ with its characteristic function $\chi_A \in F_c(X)$ we can generalize the Hausdorff distance to the case of $F_c(X)$, getting a continuous semigroup monomorphism:

Proposition 1 (see Puri and Ralescu[14]). $(F_c(X), d_\infty)$ $((U_c(X), d_\infty))$ is a complete, but in contrast to $(\mathscr{K}(X), d_H)$ ((co $\mathscr{K}(X), d_H$)) a nonseparable metric space, endowed with the metric d_∞ defined by $d_\infty(u, v) = \sup_{0 < a \leqslant 1} d_H(L_\alpha u, L_\alpha v)$ for $u, v \in F_c(X)$.

Let $(\Omega, \mathfrak{I}, P)$ be a probability space. A *fuzzy random variable* (frv) is a Borel measurable function $\xi\colon \Omega \to (F_c(X), d_\infty)$. As one can see, frv's generalize *compact random sets* A (i.e., Borel measurable functions $A\colon \Omega \to (\mathscr{K}(X), d_H)$) by identification with the characteristic function χ_A.

Proposition 2 (see Klement, Puri, and Ralescu[9]). Let $\xi: \Omega \to (F_c(X), d_\infty)$ be a frv. Then

(i) $\overline{\text{co}}\, \xi: \Omega \to (U_c(X), d_\infty); \omega \mapsto \overline{\text{co}}\, \xi(\omega)$ is a frv.

(ii) $L_\alpha \xi: \Omega \to (\mathscr{K}(X), d_H); \omega \mapsto L_\alpha \xi(\omega)$ and $\overline{\text{co}}(L_\alpha \xi): \Omega \to (\text{co}\, \mathscr{K}(X), d_H); \omega \mapsto \overline{\text{co}}(L_\alpha \xi(\omega))$ are (convex) compact random sets for all $0 < \alpha \leqslant 1$.

(iii) $\text{supp}\, \xi: \Omega \to (\mathscr{K}(X), d_H); \omega \mapsto L_0 \xi(\omega)$ is a compact random set.

Proposition 3 (see Puri and Ralescu[14]). Let $\xi: \Omega \to (F_c(X), d_\infty)$ be a frv with $E\, \|\xi\|_\infty < \infty$ $(\|\xi(\omega)\|_\infty := \sup_{\alpha > 0} d_H(\{0\}), L_\alpha \xi(\omega)))$. Then, there exists a unique fuzzy set $u \in U_c(X)$ such that the α-level of u are the Aumann expectations of the convex compact random sets $\overline{\text{co}}(L_\alpha \xi)$ (see for definition, Aubin and Frankowska,[2] and Debreu[4]), i.e., $L_\alpha u = E\, \overline{\text{co}}(L_\alpha \xi)$ for all $0 < \alpha \leqslant 1$.

If $\xi: \Omega \to (F_c(X), d_\infty)$ is a frv with $E\, \|\xi\|_\infty < \infty$ then by the last result, the *expected value* $E\xi$ is the unique convex fuzzy set, satisfying the property $L_\alpha E\xi = E\, \overline{\text{co}}(L_\alpha \xi)$, $0 < \alpha \leqslant 1$.

One of our main tools for proving the SLLN is based on an embedding of convex, compact sets into the space of continuous functions, which stems in this form from Giné, Hahn, and Zinn,[6] but is originally due to Hörmander:[7]

Theorem 4. The function $j: (\text{co}\, \mathscr{K}(X), d_H) \to (C(\text{ball}\, X^*, d_*), \|\cdot\|_\infty)$ ($\|\cdot\|$ is the supremum norm), given by $A \mapsto (f \mapsto \sup_{a \in A} f(a))$ satisfies the following

(i) j is an isometry (i.e., $d_H(A, B) = \|j(A) - j(B)\|_\infty$).

(ii) $J(A+B) = j(A) + j(B)$, $j(\lambda A) = \lambda j(A)$ $(\lambda \geqslant 0)$.

d_* denotes a metric inducing the weak* topology of ball X^*, where ball X^* is the norm-closed unit ball of the dual of X.

Let us mention that a Banach space is of type $p > 1$, if there is a $C > 0$ such that $E\, \|\sum_{i=1}^n \xi_i\|^p \leqslant C \sum_{i=1}^n E\, \|\xi_i\|^p$ for every finite sequence of zero-mean independent X-valued random elements $(\xi_i)_{i=1}^n$. A central idea in proving a SLLN for $F_c(X)$-valued frv's in the case of Banach spaces of type $p > 1$ is to reduce first the study of frv's to the convex case, by convexifying the underlying variables, and then to drop the convexity assumption, by applying following inequality:

Proposition 5 (Puri and Ralescu[13]). Let X be a Banach space of type $p > 1$. Then there exists a constant $K > 0$ such that

$d_H(A_1+\cdots+A_n, \overline{\text{co}}(A_1+\cdots+A_n)) \leqslant K^{1/p}(r^p(A_1)+\cdots+r^p(A_n))$ for all $A_1,\ldots, A_n \in \mathscr{K}(X)$, where $r(A)$ denotes the inner radius of $A \in \mathscr{K}(X)$, given by $r(A) = \sup_{a \in \overline{\text{co}}\, A} \inf\{R \geqslant 0 \mid \exists a_1,\ldots, a_s \in A, a \in \text{co}\{a_1,\ldots, a_s\}$ and $\|a-a_i\|\}$.

Finally, for preparing the proof of our main result, we refer to the notion of a *Blackwell σ-algebra*. A σ-algebra $\mathfrak{J}$ on a set Ω is called a Blackwell σ-algebra, if $\varphi(\Omega)$ is an analytic subset of R for all real $\mathfrak{J}$-measurable functions φ on Ω (see Hoffmann-Jørgensen[8]).

3. STRONG LAW OF LARGE NUMBERS FOR FUZZY RANDOM VARIABLES

First, we state a Kolmogorov type of strong law of large numbers for convex frv's:

Theorem 6. Let $(X, \|\cdot\|)$ be a separable Banach space. If $\xi_i: \Omega \to (U_c(X), d_\infty)$ is an i.i.d. sequence of frv's on the probability space $(\Omega, \mathfrak{J}, P)$ with $E\,\|\xi_1\|_\infty < \infty$, then

$$d_\infty\left(\frac{1}{n}\sum_{i=1}^{n} \xi_i, E\xi_1\right) \to 0 \quad \text{a.e.}$$

Proof. We intend to apply a result of J. Hoffmann-Jørgensen[8] (Corollary 4.6, p. 349) to the stochastic process

$$g: (U_c(X), d_\infty) \times T \to R$$

with index set $T = ([0, 1] \times \text{ball } X^*, |\cdot| + d_*)$ given by

$$(u, (\alpha, f)) \mapsto \begin{cases} \sup_{a \in L_\alpha u} f(a), & \text{if } \alpha > 0 \\ \sup_{a \in \text{supp } u} f(a), & \text{if } \alpha = 0 \end{cases}$$

To this end, we check the assumptions of the result which is accomplished in the following steps:

1. Let (T, d) be a compact metric space. Then $\mathscr{B}a(T) := \{$smallest σ-algebra on T, such that all continuous $f: T \to R$ are measurable$\}$ is a Blackwell σ-algebra on T (see Hoffmann-Jørgensen[8]). Let V be an open set in (T, d), then $f: T \to R$; $x \mapsto \exp(-d(V^c, x)$ is continuous on (T, d) with $f^{-1}[R - \{1\}] = V$. Hence $\mathscr{B}(T) = \mathscr{B}a(T)$ for the Borel σ-algebra on T. Thus $B(T)$ is a Blackwell σ-algebra on T.

2. In this step we gather some facts due to the limit behaviour of α-level sets. These are used in the next step to verify the existence of certain limits of the process g. For that purpose, pick up $s \in U_c(X)$ and define $S_\alpha = \{x \in X \mid s(x) > a\}$ for $0 \leqslant \alpha < 1$. It is possible then to show the existence of the right limits and the left continuity of the α-level sets of s with respect to the Hausdorff distance (see Klement, Puri, and Ralescu[9]), i.e.,

$$\lim_{\substack{\alpha \to \alpha_0 \\ \alpha > 0}} d_H(L_\alpha s, S_{\alpha_0}) = 0 \ (\alpha_0 < 1) \qquad \text{and} \qquad \lim_{\substack{\alpha \to \alpha_0 \\ \alpha \leqslant \alpha_0}} d_H(L_\alpha s, L_{\alpha_0} s) = 0 \tag{1}$$

From the last assertion (1) and Theorem 4 we immediately obtain

$$\begin{aligned} &\lim_{\substack{\alpha \to \alpha_0 \\ \alpha \leqslant \alpha_0}} |\sup f(L_\alpha s) - \sup f(L_{\alpha_0} s)| \\ &\quad\leqslant \lim_{\substack{\alpha \to \alpha_0 \\ \alpha \leqslant \alpha_0}} \sup_{f \in \text{ball}\, X^*} |\sup f(L_\alpha s) - \sup f(L_{\alpha_0} s)| \\ &\quad= \lim_{\substack{\alpha \to \alpha_0 \\ \alpha \leqslant \alpha_0}} d_H(L_\alpha s, L_{\alpha_0} s) = 0 \qquad \text{for all} \quad f \in \text{ball}\, X^* \end{aligned} \tag{2}$$

Bearing in mind the monotonicity of α-level sets and that $\bigcup_{\alpha > \alpha_0} L_\alpha s = S_{\alpha_0}$ is valid, one notices easily that

$$\lim_{\substack{\alpha \to \alpha_0 \\ \alpha > \alpha_0}} \sup f(L_\alpha s) = \sup f(S_{\alpha_0}) \tag{3}$$

Furthermore, let $(\alpha_n)_{n \geqslant 1} \subseteq (\alpha_0, 1]$ with $\alpha_n \to \alpha_0$. Since $d_H(L_{\alpha_n} s, L_{\alpha_m} s) \leqslant d_H(L_{\alpha_n} s, S_{\alpha_0}) + d_H(S_{\alpha_0}, L_{\alpha_m} s)$ holds for all m, n, we find that $((f \mapsto \sup_{a \in L_{\alpha_n} s} f(a)))_{n \geqslant 1}$ is a Cauchy sequence in $C(\text{ball}\, X^*, d_*)$. Assertion (3) then implies

$$(f \mapsto \sup_{a \in S_{\alpha_0}} f(a)) \in C(\text{ball}\, X^*, d_*) \tag{4}$$

3. On the one hand, it follows from (1) and Theorem 4 that

$$\begin{aligned} &|\sup_{a \in L_\alpha s} f(a) - \sup_{a \in L_{\alpha_0} s} f(a)| \\ &\quad\leqslant d_H(L_\alpha s, L_{\alpha_0} s) + |\sup_{a \in L_{\alpha_0} s} f(a) - \sup_{a \in L_{\alpha_0} s} f_0(a)| \xrightarrow[\alpha \leqslant \alpha_0]{(\alpha, f) \to (\alpha_0, f_0)} 0 \\ &\qquad \text{for all} \quad \alpha \leqslant \alpha_0 \qquad \text{and} \qquad f, f_0 \in \text{ball}\, X^* \end{aligned}$$

On the other hand, we conclude from (1), (3), and (4) that

$$\Big|\sup_{a \in L_\alpha s} f(a) - \sup_{a \in S_{\alpha_0}} f_0(a)\Big|$$

$$= \lim_{\substack{\beta \to \alpha_0 \\ \beta > \alpha_0}} \Big|\sup_{a \in L_\alpha s} f(a) - \sup_{a \in L_\beta s} f(a)\Big|$$

$$\leqslant \lim_{\substack{\beta \to \alpha_0 \\ \beta > \alpha_0}} \Big(d_H(L_\alpha s, S_{\alpha_0}) + d_H(S_{\alpha_0}, L_\beta s) + \Big|\sup_{a \in L_\beta s} f(a) - \sup_{a \in L_\beta s} f_0(a)\Big|\Big)$$

$$= d_H(L_\alpha s, S_{\alpha_0}) + \Big|\sup_{a \in S_{\alpha_0}} f(a) - \sup_{a \in S_{\alpha_0}} f_0(a)\Big| \xrightarrow[\substack{(\alpha, f) \to (\alpha_0, f_0) \\ \alpha > \alpha_0}]{} 0$$

$$\text{for all} \quad \alpha > \alpha_0 \quad \text{and} \quad f, f_0 \in \text{ball } X^*$$

4. Finally, we summarize all properties of the process g, sufficient for the applicability of the above mentioned result: Let $\mathscr{A} = \mathscr{B}(U_c(X))$. Then the process g is $\mathscr{A} \otimes \mathscr{B}(T)$, $\mathscr{B}(R)$-measurable, because it is left continuous in α and continuous in f for the argument $(\alpha, f) \in T$. The topological basis of T is contained in $\mathscr{B}(T)$.

Further, let $\mu = \xi_1 P$ and $(\alpha_0, f_0) \in T$. Theorem 4 supplies the representation $d_\infty(u, u_0) = \sup_{\alpha \geqslant 0} \sup_{f \in \text{ball } X^*} |\sup_{a \in L_\alpha u} f(a) - \sup_{a \in L_\alpha u_0} f(a)|$ for all u, $u_0 \in U_c(X)$. Hence, the mapping $(s \mapsto g(s, (\alpha_0, f_0)))$ is continuous. This implies that $(s \mapsto g(s, (\alpha, f)))$ is μ-measurable.

Besides this we obtain $\int^* \sup_{(\alpha, f) \in T} |g(s, (\alpha, f))| \, \mu(ds) = \int \|\cdot\|_\infty \, d\mu = \int_\Omega \|\xi_1\|_\infty \, dP < \infty$.

Let $t = (\alpha_0, f_0) \in T$. Then $A_t := \{[0, \alpha_0] \times \text{ball } X^*, (\alpha_0, 1) \times \text{ball } X^*\} = \{A_1, A_2\}$ is a finite cover of T with $A_t \subseteq \mathscr{B}(T)$ and $t \in \text{cl}(A_1 \cap T) \cap \text{cl}(A_2 \cap T)$. Thus, by step 3 we see that

$$\mu^*\Big(\Big\{s \,\Big|\, \lim_{\substack{(\alpha, f) \xrightarrow[d]{} (\alpha_0, f_0) \\ (\alpha, f) \in A_1}} g(s, (\alpha, f)) \text{ exists}\Big\}\Big) = \mu(U_c(X)) = 1 \quad \text{as well as}$$

$$\mu^*\Big(\Big\{s \,\Big|\, \lim_{\substack{(\alpha, f) \xrightarrow[d]{} (\alpha_0, f_0) \\ (\alpha, f) \in A_2}} g(s, (\alpha, f)) \text{ exists}\Big\}\Big) = \mu(U_c(X)) = 1$$

Taking into consideration that $E \sup_{a \in A} f(a) = \sup_{a \in EA} f(a)$ holds for the Aumann expectation EA of compact random sets A and $f \in X^*$ (see Aubin and Frankowska[(2)]), the result of Hoffmann-Jørgensen[(8)] (Corollary 4.6, p. 349) provides then

$$d_\infty\left(\frac{1}{n}\sum_{i=1}^{n}\xi_i, E\xi_1\right) = \sup_{(\alpha, f)\in T}\left|\frac{1}{n}\sum_{i=1}^{n} g(\xi_i, (\alpha, f)) - g(E\xi_1, (\alpha, f))\right|$$

$$= \sup_{(\alpha, f)\in T}\left|\frac{1}{n}\sum_{i=1}^{n} g(\xi_i, (\alpha, f)) - Eg(\xi_1, (\alpha, f))\right| \to 0 \quad \text{a.e.}$$

concluding the proof. □

We use the last result and Proposition 5 to derive a version of the SLLN for $F_c(X)$-valued frv's in the case of separable Banach spaces of type $p > 1$:

Corollary 1. Let $(X, \|\cdot\|)$ be separable Banach space of type $p > 1$ and let $\xi_i: \Omega \to (F_c(X), d_\infty)$ be an i.i.d. sequence of frv's on the probability space $(\Omega, \Im, P)$ with $E\,\|\xi_1\|_\infty < \infty$. Then the strong law is valid, i.e.,

$$d_\infty\left(\frac{1}{n}\sum_{i=1}^{n}\xi_i, E\xi_1\right) \to 0 \quad \text{a.e.}$$

Proof. By Proposition 5 we know there exists a constant $K > 0$ such that

$$d_\infty\left(\frac{1}{n}\sum_{i=1}^{n}\xi_i, \frac{1}{n}\sum_{i=1}^{n}\overline{\text{co}}\,\xi_i\right) \leqslant K^{1/p}\left(\frac{1}{n^{1/1/p}}\sum_{i=1}^{n}\|\xi_i\|_\infty^p\right)^{1/p}$$

The last term tends to zero a.e. as $n \to \infty$ in virtue of a result of Marcinkiewicz–Zygmund.[3] The rest of the proof follows from the triangle inequality and Theorem 6. □

The finite dimensional case yields:

Corollary 2. Let $\xi_i: \Omega \to (F_c(R^p), d_\infty)$ be an i.i.d. sequence of frv's with finite first moment. Then the SLLN holds.

4. CONCLUSION

To the best of our knowledge, our main result is the first version of the SLLN, extending the result of Klement, Puri, and Ralescu[9] to the natural case of the d_∞-metric. In the context of Glivenko–Cantelli problems there are some other possibilities of attaining such a result, e.g., another approach is the verification of the finiteness condition of the entropy with bracketing for the process g (see for details van der Vaart and Wellner,[16] and Proske[11]). Besides this we conjecture that Corollary 1 is also true in the general case of separable Banach spaces because it seems to be quite natural.

REFERENCES

1. Artstein, Z., and Vitale, R. A. (1975). A strong law of large number for random compact sets. *Ann. Prob.* **3**, 879–882.
2. Aubin, J. P., and Rankowska, H. (1990). *Set-Valued Analysis. Systems & Control: Foundations & Applications*, Birkhäuser.
3. Chow, Y. S., and Teicher, H. (1978). *Probability Theory*, Birkhäuser.
4. Debreu, G. (1966). *Integration of Correspondences*, Proc. Fifth Berkeley Symp. Math. Statist. Prob., Vol. 2, pp. 351–372, University of California Press.
5. Dunford, N., and Schwartz, J. T. (1958). *Linear Operators, Part I*, Interscience.
6. Giné, E., Hahn, G., and Zinn, J. (1983). Limit theorems for random sets. *Springer Lecture Notes in Math.* **990**, 112–135.
7. Hörmander, L. (1954). Sur la fonction d'appui des ensembles convexe dans un espace localement convexe. *Ark. Mat.* **3**, 181–186.
8. Hoffmann-Jørgensen, J. (1985). *The Law of Large Numbers for Nonmeasurable and Non-Separable Random Elements*, Colloque en l'honneur de L. Schwartz, pp. 299–356, Astérisque, Vol. 131, Hermann, Paris.
9. Klement, E. P., Puri, M. L. and Ralescu, D. A. (1986). Limit theorems for fuzzy random variables. *Proc. R. Soc. Lond. A* **407**, 171–182.
10. Kruse, R. (1982). The strong law of large numbers for fuzzy random variables. *Inform. Sci.* **28**, 233–241.
11. Proske, F. N. (1997). Limit theorems for fuzzy random variables. Ph.D. Thesis, Ulm University, Germany.
12. Puri, M. L. and Ralescu, D. A. (1983). Differentials of fuzzy functions. *J. Math. Anal. Appl.* **91**, 552–558.
13. Puri, M. L., and Ralescu, D. A. (1985). Limit theorems for random compact sets in Banach spaces. *Math. Proc. Cam. Phil. Soc.* **97**, 151–158.
14. Puri, M. L., and Ralescu, D. A. (1986). Fuzzy random variables. *J. Math. Anal. Appl.* **114**, 409–422.
15. Talagrand, M. (1987). The Glivenko–Cantelli problem. *Ann. Prob.* **15**, 837–870.
16. van der Vaart, A. W. and Wellner, J. A. (1996). *Weak Convergence and Empirical Processes*, Springer.
17. Zadeh, L. A. (1965). Fuzzy Sets. *Information and Control* **8**, 338–353.

Reprinted by permission of the publisher,
© 2002 by Kluwer Academic/Plenum Publisher

Part III

Miscellaneous Topics

LOCAL MAXIMA OF THE SAMPLE FUNCTIONS OF THE N-PARAMETER BESSEL PROCESS

MADAN L. PURI* and LANH TAT TRAN**

University of Washington, Seattle, U.S.A.
Indiana University, Bloomington, U.S.A.

Received 25 August 1978
Revised 14 March 1979

In this paper we show that almost every sample function of the N-parameter Bessel process associated with the N-parameter Wiener process has a local maximum. In addition some properties related to the local maxima are investigated.

sample functions	Wiener process
Bessel process	local maxima

1. Introduction and preliminaries

Let $W^{(N)}$ be the N-parameter Wiener process, that is a real-valued separable Gaussian process with zero means and covariance $\prod_{i=1}^{N}(s_i \wedge t_i)$ where $s=\langle s_i\rangle$, $t=\langle t_i\rangle$, $s_i \geqslant 0$, $t_i \geqslant 0$, $i=1,\ldots,N$. Then $W^{(N,d)}$ is to be the process with values in the d-dimensional Euclidean space R^d such that each component is an N-parameter Wiener process, the components being independent. Write $W=W^{(N,d)}$ for simplicity, and denote the ith component of W by W^i. Define the N-parameter Bessel process associated with W by

$$B_t=\left[\sum_{i=1}^{d}(W_t^i)^2\right]^{1/2}. \tag{1}$$

It is shown that almost every sample function of B_t has a local maximum. Furthermore, some properties related to the local maxima of B_t are investigated.

As in Orey and Pruitt [5][1], our parameter space is R_+^N, that is the set of $t \in R^N$ with all components nonnegative. When dealing with a point t in the parameter space we sometimes write $t=\langle t_1,\ldots,t_N\rangle$ or simple $\langle t_i\rangle$. In case all $t_i=0$, we write $t=\langle 0\rangle$. For $s=\langle s_i\rangle$ and $t=\langle t_i\rangle$ with $s_i \leqslant t_i$, the interval $\mathbf{X}_{i=1}^{N}[s_i,t_i]$ is denoted by $\Delta(s,t)$, and by

* Research supported by the Air Force Office of Scientific Research AFSC, USAF, under Grant No. AFOSR76-2927. Some of this research was done at the University of Washington during my leave of absence from Indiana University. I wish to express my sincere thanks to Professor Donovan Thompson and his colleagues for their kind hospitality at the Department of Biostatistics in the University of Washington, Seattle.

** Research supported by National Science Foundation Grant No. MCS76-00951.

[1] Wherever possible, we shall use the notation of Orey and Pruitt [5].

$\Delta(t)$ in case $s=\langle 0\rangle$. Denote by $S(s, t)$, the symmetric difference of $\Delta(s)$ and $\Delta(t)$. Then it is easy to check that if $s, t \in R_+^N$, the variance of $W^i(t)-W^i(s)$ is $|S(s, t)|$ where $|\cdot|$ denotes the N-dimensional Lebesgue measure. Furthermore, W has continuous sample functions and independent increments. We denote the increment of W over $\Delta(s, t)$ by $W(\Delta(s, t))$. For further information on W, the reader is referred to [1, 4–13].

Local maxima of the sample functions of the two-parameter Wiener process have been studied by Tran [8]. However, due to the complex sample function behavior of the N-parameter Bessel process and the complicated geometrical structures in higher dimensions, Tran's method does not provide a simple generalization to the problem considered in this paper. A much more involved argument is needed here.

Throughout the paper, we will assume that a set of probability zero has been deleted from the probability space so that all sample functions of B_t are continuous.

Definition 1. The sample function $B(\cdot, \omega)$ has a local maximum at s if there exists an open set 0 containing s such that $0 \subset R_+^N$ and $B(t, \omega) \leqslant B(s, \omega)$ for all $t \in 0$.

We shall need the Orey–Pruitt analogue of the familiar zero-one law. Let $\mathscr{C}_n$ be the class of time intervals in R_+^N with vertices of the form $\langle k_i 2^{-n}\rangle$, k_i nonnegative integers, and having all sides of equal length, and for $n>0$ each member of $\mathscr{C}_n$ is to be a subcube of one in $\mathscr{C}_0$. Let $\mathscr{C}_\infty = \bigcup_{n=0}^\infty \mathscr{C}_n$, and $\mathscr{F}_n = \mathscr{B}(W(\Delta), \Delta \in \mathscr{C}_n)$, $\mathscr{F}_\infty = \bigvee_{n=0}^\infty \mathscr{F}_n$. Thus $\mathscr{F}_n$ is the Borel field generated by the indicated class of random variables and $\mathscr{F}_\infty$ is the smallest Borel field including all $\mathscr{F}_n$. For a subset D of R_+^N, we put $\mathscr{C}_n(D) = \{\Delta \in \mathscr{C}_n : \Delta \subseteq D\}$, $\mathscr{F}_n(D) = \mathscr{B}\{W(\Delta) : \Delta \in \mathscr{C}_n(D)\}$, $\mathscr{F}_\infty(D) = \bigvee_{n=0}^\infty \mathscr{F}_n(D)$ then we have the following lemma.

Lemma 1 ([5]). *Let $D_m \subseteq R_+^N$, $m=1, 2, \ldots$, with $D_m \downarrow \emptyset$. If $A \in \mathscr{F}_\infty(D_m)$ for every m, then $P(A)=\{0, 1\}$.*

Lemma 2. *Let ϕ be a nonnegative, nondecreasing, continuous function defined for large arguments. Then for almost all ω there is an $\varepsilon(\omega)$ such that for all intervals $\Delta(s, t)$ with $\Delta(s, t) \subset \Delta(\langle 1\rangle)$ and $|\Delta(s, t)| < \varepsilon(\omega)$,*

$$|W(\Delta(s, t))| < |\Delta(s, t)|^{1/2} \phi(|\Delta(s, t)|^{-1})$$

if and only if

$$\int_0^\infty (\log \xi)^{3N+d/2-2}\, \mathrm{e}^{-\phi^2(\xi)/2}\, \mathrm{d}\xi$$

converges.

For the proof of Lemma 2, see [5, p. 147].

2. Local maxima

In this section we prove the main theorem dealing with the existence of the local maxima of the sample functions of the Bessel process B_t.

Theorem 1. *For almost all sample functions of the Bessel process B_t defined in* (1), *there exists a local maximum.*

Proof. Let s be the center of the unit interval U, and let $C_n \subset U$ be a cube with center at s, sides parallel to the coordinate axes and equal to a_n. Let u^n and v^n be the smallest and the largest vertex of C_n, i.e. closest and farthest from the origin $\langle 0 \rangle$. Pick C_n with $\min(u_1^n, \ldots, u_N^n) > \frac{1}{4}$.

Consider two points s^{nk} and v^{nk} of R_+^N determined by $s_k^{nk} = \frac{1}{2}$, $v_k^{nk} = v_n^k$, $s_j^{nk} = v_j^{nk} = u_j^n$ for $j \neq k$ where $1 \leq j \leq N$.

Define

$$A_{ni} = \bigcap_{k=1}^{N} [W^i(s^{nk}) - W^i(u) > 2a_n^{1/2},\ W^i(s^{nk}) - W^i(v^{nk}) > 2a_n^{1/2}],$$

$$B_{ni} = \bigcap_{k=1}^{N} [W^i(s^{nk}) - W^i(u) < -2a_n^{1/2},\ W^i(s^{nk}) - W^i(v^{nk}) < -2a_n^{1/2}],$$

$$C_{ni} = [\inf_{t \in C_n} W_t^i \geq 0], \qquad E_{ni} = [\sup_{t \in C_n} W_t^i \leq 0],$$

$$F_{ni} = [\sup_{s,t \in U} |W^i(\Delta(s, t))| < (2^{N-1} - 1)^{-1} a_n^{1/2} : |s_i - t_i| \leq a_n,$$
$$|s_j - t_j| \leq a_n \text{ for some } i, j \text{ with } i \neq j].$$

The variables $W^i(s^{nk}) - W^i(u)$, $W^i(s^{nk}) - W^i(v^{nk})$ are normally distributed with mean 0 and variances greater than $4^{-N+1/2} a_n$. Thus $P(A_{ni}) > \beta$ for some constant β.

Let $\{a_n\}$ be a sequence of positive numbers with $a_n \downarrow 0$ and let D_n be the interior of $S(u^n, v^n)$. Clearly $D_n \downarrow \emptyset$ as $a_n \downarrow \emptyset$. Observe that the event $[A_{ni}$ infinitely often$] \in \mathscr{F}_\infty(D_n)$. Thus, from Lemma 1, it follows that

$$\mathbf{P}[A_{ni} \text{ infinitely often}] = 1.$$

Analogously,

$$\mathbf{P}[B_{ni} \text{ infinitely often}] = 1.$$

Let $\eta > 0$. Then by Lemma 2, or by the continuity of the sample functions of W,

$$\mathbf{P}\left[\bigcap_{n=n_0}^{\infty} C_{n_i} \mid W_s^i > \eta\right] \to 1 \quad \text{as } n_0 \to \infty.$$

Furthermore,

$$\mathbf{P}[A_{ni}C_{ni} \text{ infinitely often}] \geqslant \mathbf{P}\left[(A_{ni} \text{ infinitely often}) \bigcap_{n=n_0}^{\infty} C_{ni}\right]$$

$$\geqslant \mathbf{P}\left[\bigcap_{n=n_0}^{\infty} C_{ni}\right] \geqslant \mathbf{P}\left[\bigcap_{n=n_0}^{\infty} C_{ni}[W_s^i > \eta]\right].$$

Since $\mathbf{P}[W_s^i > \eta]$ converges to $\frac{1}{2}$ as $\eta \to 0$, by picking η small enough and then n_0 large enough, the probability of the last event can be made as close to $\frac{1}{2}$ as desired.

It is now clear that

$$\mathbf{P}[A_{ni}C_{ni} \cup B_{ni}E_{ni}) \text{ i.o.}] = 1.$$

Also, by Lemma 2 and by the independence of the components of W

$$\mathbf{P}\left[\bigcap_{i=1}^{d} (A_{ni}C_{ni}F_{ni} \cup B_{ni}E_{ni}F_{ni}) \text{ i.o.}\right] = 1. \tag{2}$$

We claim that

$$\left[\bigcap_{i=1}^{d} (A_{ni}C_{ni}F_{ni} \cup B_{ni}E_{ni}F_{ni})\right] \subset [\sup_{t \in C_n^0} B_t > \sup_{t \in \partial C_n} B_t]. \tag{3}$$

Observe that $[\bigcap_{i=1}^{d} (A_{ni}C_{ni}F_{ni} \cup B_{ni}E_{ni}F_{ni})]$ can be written as the union of 2^d events. Each of these events is the intersection of d sets with the ith set $(1 \leqslant i \leqslant d)$ being either $A_{ni}C_{ni}F_{ni}$ of $B_{ni}E_{ni}F_{ni}$. To complete the proof of (3), it is sufficient to show that every one of these 2^d events is a subset of $[\sup_{t \in C_n^0} B_t > \sup_{t \in \partial C_n} B_t]$. Without loss of generality we will show

$$\left[\bigcap_{i=1}^{d} A_{ni}C_{ni}F_{ni}\right] \subset [\sup_{t \in C_n^0} B_t > \sup_{t \in \partial C_n} B_t]. \tag{4}$$

A slight variation of the proof along the same lines can be applied to show that the remaining $2^d - 1$ events are also subsets of $[\sup_{t \in C_n^0} B_t > \sup_{t \in \partial C_n} B_t]$.

Let $t \in \partial C_n$ where ∂C_n is the boundary of C_n, and let

$$p^t = \langle \sigma(t_1), \ldots, \sigma(t_N) \rangle$$

where

$$\sigma(t_i) = \begin{cases} t_i, & \text{if } u_i^n < t_i < v_i^n, \\ s_i, & \text{otherwise.} \end{cases}$$

Observe that p^t lies in the interior of C_n.

Let

$$H_n = \bigcap_{t \in \partial C_n} \left\{ \bigcap_{i=1}^{d} [W^i(p^t) - W^i(t) > a_n^{1/2}] \right\}.$$

We will complete the proof of (4) by showing

$$\left[\bigcap_{i=1}^{d} A_{ni}C_{ni}F_{ni}\right] \subset \left[H_n \bigcap_{i=1}^{d} C_{ni}\right], \tag{5}$$

$$\left[H_n \bigcap_{i=1}^{d} C_{ni}\right] \subset [\sup_{t\in C_n^0} B_t > \sup_{t\in\partial C_n} B_t]. \tag{6}$$

Observe that for any $1 \leqslant i \leqslant d$,

$$\begin{aligned} W^i(p^t) - W^i(t) &= W^i(\langle\sigma(t_1), t_2, \ldots, t_N\rangle) - W^i(\langle t_1, t_2, \ldots, t_N\rangle) \\ &+ W^i(\langle\sigma(t_1), \sigma(t_2), t_3, \ldots, t_N\rangle) - W^i(\langle t_1), t_2, \ldots, t_N\rangle) \\ &+ \cdots + W^i(\langle\sigma(t_1), \sigma(t_2), \ldots, \sigma(t_n)\rangle) - W^i(\langle\sigma(t_1), \ldots, \sigma(t_{N-1}), t_N\rangle). \end{aligned} \tag{7}$$

Consider the random variable

$$\begin{aligned} &W^i(\langle\sigma(t_1), \ldots, \sigma(t_{j-1}), \sigma(t_j), t_{j+1}, \ldots, t_N\rangle) \\ &\quad - W^i(\langle\sigma(t_1), \ldots, \sigma(t_{j-1}), t_j, t_{j+1}, \ldots, t_N\rangle). \end{aligned} \tag{8}$$

The variance of this variable is equal to

$$(u_1^n + e_1^n) \cdots (u_{j-1}^n + e_{j-1}^n)|\sigma(t_j) - t_j|(u_{j+1}^n + e_{j+1}^n) \cdots (u_N^n + e_N^n) \tag{9}$$

where $0 \leqslant e_i^n \leqslant a_n$. Observe that (9) can be expanded into a sum of 2^{N-1} terms. At this point, it is easy to see that (8) is equal to

$$\begin{aligned} &W^i(\langle u_1^n, \ldots, u_{j-1}^n, \sigma(t_j), u_{j+1}^n, \ldots, u_N^n\rangle) \\ &\quad - W^i(\langle u_1^n, \ldots, u_{j-1}^n, t_j, u_{j+1}^n, \ldots, u_N^n\rangle) + L^i \end{aligned}$$

where L^i can be decomposed into no more than $2^{N-1} - 1$ random variables, such that each random variable is the increment of W^i over an interval in U with at least two sides smaller than or equal to a_n.

The random variables in (8) is zero if $\sigma(t_j) = t_j$. Assume $\sigma(t_j) \neq t_j$ and let $\omega \in [\bigcap_{i=1}^d A_{ni}C_{ni}F_{ni}]$. Then $\omega \in F_{ni}$ for each i. Hence

$$|L^i(\omega)| < (2^{N-1} - 1)(2^{N-1} - 1)^{-1} a_n^{1/2} = a_n^{1/2}.$$

Now, $\sigma(t_j) = \frac{1}{2}$ and $t_j = u_j^n$ or v_j^n since $\sigma(t_j) \neq t_j$. Therefore

$$\begin{aligned} &W^i(\langle u_1^n, \ldots, u_{j-1}^n, \sigma(t_j), u_{j+1}^n, \ldots, u_N^n\rangle, \omega) \\ &\quad - W^i(\langle u_1^n, \ldots, u_{j-1}^n, t_j, u_{j+1}^n, \ldots, u_N^n\rangle, \omega) \\ &= \begin{cases} W^i(s^{nj}, \omega) - W^i(u, \omega), & \text{if } t_j = u_j^n, \\ W^i(s^{nj}, \omega) - W^i(v^{nj}, \omega), & \text{if } t_j = v_j^n, \end{cases} \end{aligned}$$

which is greater than $2a_n^{1/2}$ since $\omega \in A_{ni}$. We obtain

$$W^i(\langle\sigma(t_1), \ldots, \sigma(t_{j-1}), \sigma(t_j), t_{j+1}, \ldots, t_N\rangle, \omega) \\ - W^i(\langle\sigma(t_1), \ldots, \sigma(t_{j-1}), t_j, t_{j+1}, \ldots, t_N\rangle, \omega) \\ > 2a_n^{1/2} - a_n^{1/2} = a_n^{1/2}. \tag{10}$$

By (7) and (10) we have

$$W^i(p^t, \omega) - W^i(t, \omega) > [\# \text{ of } j\text{'s}, 1 \leq j \leq N, \text{ with } \sigma(t_j) \neq t_j] a_n^{1/2}.$$

If $t \in \partial C_n$, t_j is equal to u_j^n or v_j^n and $\sigma(t_j) = \frac{1}{2}$ for some $1 \leq j \leq N$. Therefore $\sigma(t_j) \neq t_j$ for some j and $[\# \text{ of } j\text{'s}, 1 \leq j \leq N, \text{ with } \sigma(t_j) \neq t_j] \geq 1$. Hence for $\omega \in A_{ni}$, $W^i(p^t, \omega) - W^i(t, \omega) > a_n^{1/2}$ for all $t \in \partial C_n$ and for all i.

We have shown $\omega \in H_n$. Furthermore $\omega \in \bigcap_{i=1}^d C_{ni}$ since $\omega \in [\bigcap_{i=1}^d A_{ni} C_{ni} F_{ni}]$. The proof of (5) is completed.

Let us turn to the proof of (6). We argue by contradiction. Let $\omega \in [H_n \bigcap_{i=1}^d C_{ni}]$ and assume $\omega \notin [\sup_{t \in C_n^0} B_t > \sup_{t \in \partial C_n} B_t]$. Then $\sup_{t \in \partial C_n} B(t, \omega) \geq \sup_{t \in C_n^0} B(t, \omega)$. Now, $B(t, \omega)$ is a continuous function of t on ∂C_n. Hence for some $l \in \partial C_n$,

$$B(l, \omega) = \sup_{t \in \partial C_n} B(t, \omega) \geq \sup_{t \in C_n^0} B(t, \omega) \geq B(p^l, \omega)$$

since p^l is an interior point of C_n.

We obtain $(W^i(l, \omega))^2 \geq (W^i(p^l, \omega))^2$ for some $1 \leq i \leq d$ since

$$\left[\sum_{i=1}^d (W^i(l, \omega))^2\right]^{1/2} \geq \left[\sum_{i=1}^d (W^i(p^l, \omega))^2\right]^{1/2}.$$

Since $\omega \in [\inf W_t^i \geq 0]$, both $W^i(l, \omega)$ and $W^i(p^l, \omega)$ are nonnegative. Therefore $W^i(l, \omega) \geq W^i(p^l, \omega)$ for some $1 \leq i \leq d$, contradicting the fact that $W^i(p^l, \omega) - W^i(l, \omega) > a_n^{1/2}$.

The proof of (6) is completed. From (2) and (3), we conclude that with probability one, the event $[\sup_{t \in C_n^0} B_t > \sup_{t \in \partial C_n} B_t]$ occurs infinitely often.

Pick $\omega \in [\sup_{t \in C_n^0} B_t > \sup_{t \in \partial C_n} B_t]$. Since $B(t, \omega)$ is a continuous function of t on C_n, there exists an $l \in C_n$ such that $B(l, \omega) = \sup_{t \in C_n} B(t, \omega)$. If $l \in \partial C_n$, then

$$\sup_{t \in \partial C_n} B(t, \omega) = B(l, \omega) = \sup_{t \in C_n} B(t, \omega) \geq \sup_{t \in C_n^0} B(t, \omega),$$

contradicting the fact that $\omega \in [\sup_{t \in C_n^0} B_t > \sup_{t \in \partial C_n} B_t]$. Finally, C_n^0 can be chosen to be the open set $0 \subset R_+^N$ mentioned in Definition 1.

The proof is now completed. Recall that s was picked to be the center of U. Actually, s can be chosen to be any point in U^0. Therefore, for almost all sample functions of B_t, the set of local maxima is dense in R_+^N.

We shall now investigate some properties of the local maxima of B_t.

Definition 2. The sample function $B(\cdot, \omega)$ has a strict local maximum at s if there exists an open set 0 containing s such that $0 \subset R_+^N$ and $B(t, \omega) < B(s, \omega)$ for all $t \in 0$.

We have the following theorem.

Theorem 2. *For almost every sample function of $\{B_t, t \in R_+^N\}$, all the local maxima are strict and the set of local maxima is countable.*

Proof. Let I and J be two disjoint, closed intervals in the interior of R_+^N. We claim that

$$\mathbf{P}\{\sup_{t \in I} B_t = \sup_{t \in J} B_t\} = 0. \tag{11}$$

Let $I = \Delta(u, v)$, $J = \Delta(s, t)$. Denote the complements of $\Delta(t)$ and $\Delta(v)$ by $[\Delta(t)]'$ and $[\Delta(v)]'$. Since I and J are disjoint intervals, it is clear that

$$[\Delta(u) \cap [\Delta(t)]'] \cup [\Delta(s) \cap [\Delta(v)]']$$

contains a nondegenerate interval, i.e., an interval with a positive N-dimensional Lebesgue measure. Let I' be any such interval, and without loss of generality assume that

$$I' \subset \Delta(u) \cap [\Delta(t)]'.$$

Consider now

$$\mathbf{P}[\sup_{t \in I} B_t = \sup_{t \in J} B_t] = \mathbf{P}\left[\sup_{t \in I}\left[(W_t^1)^2 + \sum_{i=2}^{d}(W_t^i)^2\right]^{1/2}\right.$$
$$\left. = \sup_{t \in J}\left[\sum_{i=1}^{d}(W_t^i)^2\right]^{1/2}\right]. \tag{12}$$

Let $W^1(I')$ be the increment of W^1 over I'. Since W^1 has independent increments, for $t \in I$, we can write

$$W_t^1 = W_t^1 - W^1(I') + W^1(I')$$

such that $W^1(I')$ is independent of $W_t^1 - W^1(I')$ for all $t \in I$. Also, $W^1(I')$ is independent of $\sup_{t \in J}[\sum_{i=1}^d (W_t^i)^2]$ since $I' \subset [\Delta(t)]'$.

Let $X = W^1(I')$, $Y_t = W_t^1 - W^1(I')$. Then the σ-fields generated by X and $(Y_t, W_t^2, \ldots, W_t^d, t \in I, W_t^1, \ldots, W_t^d, t \in J)$ are independent. Let $(\Omega, \mathscr{F}, P)$ be the original probability space and let $(\Omega_1, \mathscr{F}_1, P_1)$ and $(\Omega_2, \mathscr{F}_2, P_2)$ be two identical copies of $(\Omega, \mathscr{F}, P)$. Define

$$S = \left[(\omega_1, \omega_2) \in \Omega_1 \times \Omega_2 : \sup_{t \in I}\left\{(X(\omega_1) + Y_t(\omega_2))^2 + \sum_{i=2}^{d}(W_t^i(\omega_2))^2\right\}^{1/2}\right.$$
$$\left. = \sup_{t \in J}\left\{\sum_{i=1}^{d}(W_t^i(\omega_2))^2\right\}^{1/2}\right].$$

Let χ_S denote the indicator function of S. Using independence and Fubini's theorem and following an argument similar to that of [3, p. 37], we obtain that (12) is equal to

$$\int_{\Omega_2}\int_{\Omega_1} \chi_S(\omega_1, \omega_2) P_1(\mathrm{d}\omega_1) P_2(\mathrm{d}\omega_2) =$$
$$= \int_{\Omega_2} P_1\left[\sup_{t\in I}\left\{(X + Y_t(\omega_2))^2 + \sum_{i=2}^{d} (W_t^i(\omega_2))^2\right\}^{1/2}\right.$$
$$\left. = \sup_{t\in J}\left\{\sum_{i=1}^{d} (W_t^i(\omega_2))^2\right\}^{1/2}\right] P_2(\mathrm{d}\omega_2).$$

For an arbitrary fixed ω_2, consider

$$P_1\left[\sup_{t\in I}\left\{(X + Y_t(\omega_2))^2 + \sum_{i=2}^{d} (W_t^i(\omega_2))^2\right\}^{1/2} = \sup_{t\in J}\left\{\sum_{i=1}^{d} (W_t^i(\omega_2))^2\right\}^{1/2}\right]. \tag{13}$$

We shall now show that (13) equals zero for a fixed ω_2. Consider the function $f(x)$, defined by

$$f(x) = \sup_{t\in I}\left\{(x + Y_t(\omega_2))^2 + \sum_{i=2}^{d} (W_t^i(\omega_2))^2\right\}^{1/2}.$$

Note that $f(x)$ equals the supremum of the distance from the origin of the set D_x in R^d defined by

$$D_x = \{z \in R^d : z_1 = Y_t(\omega_2) + x,\ z_2 = W_t^2(\omega_2), \ldots, z_d = W_t^d(\omega_2) \text{ for some } t \in I\}.$$

It is now easy to see that as x varies from $-\infty$ to $+\infty$, the set D_x is translated along a vector parallel to a coordinate axis and so $f(x)$ decreases and then increases as x goes from $-\infty$ to $+\infty$. For a fixed ω_2, (13) equals

$$P_1\left[X = f^{-1}\left(\sup_{t\in J} \sum_{i=1}^{d} (W_t^i(\omega_2))^2\right)\right] \tag{14}$$

where f^{-1} is the inverse of f. It is clear that there are at most 2 values of $f^{-1}(\sup_{t\in J} \sum_{i=1}^{d} (W_t^i(\omega_2))^2)$, and since X is normal random variable, (14) equals 0. Thus, for each fixed ω_2, (13) equals zero. The proof of (11) follows by integrating (13) over the probability space.

Consider the set $\bigcap[\sup_{t\in I} B_t \neq \sup_{t\in J} B_t]$ where the intersection is taken over all intervals I and J with rational least and largest vertices, i.e., the coordinates of u, v, s, t are all rational.

We claim that $\bigcap[\sup_{t\in I} B_t \neq \sup_{t\in J} B_t]$ is contained in the set of ω such that all local maxima of $B(\cdot, \omega)$ are strict. Let $\omega \in \bigcap[\sup_{t\in I} B_t \neq \sup_{t\in J} B_t]$ and suppose $B(\cdot, \omega)$ has a local maximum at s. Then there exists an open set $0 \subset R_+^N$ with $s \in 0$ and $B(t, \omega) \leq B(s, \omega)$ for all $t \in 0$. If this local maximum is not strict, then there exists an $l \in 0$ with $l \neq s$ such that $B(l, \omega) = B(s, \omega)$. Let I, J be two disjoint, closed intervals with rational least and largest vertices such that $l \in I \subset 0$ and $s \in J \subset 0$. Then

$$\sup_{t \in I} B(t, \omega) = B(l, \omega) = B(s, \omega) = \sup_{t \in J} B(t, \omega),$$

contradicting the fact that $\omega \in \bigcap [\sup_{t \in I} B_t \neq \sup_{t \in J} B_t]$. Clearly $\mathbf{P}[\bigcap (\sup_{t \in I} B_t \neq \sup_{t \in J} B_t)] = 1$.

Countability of the set of local maxima is a consequence of the following lemma.

Lemma 3. *Let f be a continuous, real valued function of R_+^N with all local maxima strict. Then f has countably many local maxima.*

This lemma is a straightforward generalization of the univariate case, the proof of which can be found in [3, p. 38].

Acknowledgment

The authors would like to express their sincere thanks to Professor J. Gani and the referee for the critical examination of the original draft. Their detailed comments and suggestions for improvements are gratefully acknowledged.

References

[1] N.N. Chentsov, Wiener random fields depending on several parameters, Dokl. Akad. Nauk SSSR 106 (1956) 607–609.

[2] J. Delporte, Fonctions aléatoires de deux variables presque surement a echantillons continus sur un domaine rectangulaire borne, Z. Wahrscheinlichkeitstheorie und Verw. Gebiete 6 (1966) 181–205.

[3] D. Freedman, Brownian Motion and Diffusion (Holden-Day, San Francisco, CA, 1971).

[4] T. Kitagawa, Analysis of variance applied to function spaces, Mem. Fac. Sci. Kyushu Univ. Ser. A 6 (1951) 41–53.

[5] S. Orey and W.E. Pruitt, Sample functions of the *N*-parameter Wiener process, Ann. Probability 1 (1973) 138–163.

[6] C. Park, A generalized Paley–Wiener–Zygmund integral and its applications, Proc. Amer. Math. Soc. 23 (1969) 388–400.

[7] W.J. Park, A multiparameter Gaussian process, Ann. Math. Statist. 41 (1970) 1582–1595.

[8] L.T. Tran, Local maxima of the sample functions of the two-parameter Wiener process, Proc. Amer. Math. Soc. 58 (1976) 250–253.

[9] L.T. Tran, The Hausdorff dimension of the *N*-parameter Wiener process, Ann. Probability 5 (1977) 235–242.

[10] J. Yeh, Wiener measure in a space of functions of two variables, Trans. Amer. Math. Soc. 95 (1960) 443–450.

[11] J. Yeh, Cameron–Martin translation theorems in the Wiener space of functions of two variables, Trans. Amer. Math. Soc. 107 (1963) 409–420.

[12] J. Yeh, Orthogonal developments of functionals and related theorems in the Wiener space of functions of two variables, Pacific J. Math. 13 (1963) 1427–1436.

[13] G. Zimmerman, Some sample function properties of the two-parameter Gaussian process, Ann. Math. Statist. 43 (1972) 1235–1246.

Reprinted by permission of the publisher,
© 1979 by Elsevier Science

REALIZATION OF l_p BY SPACES OF RANDOM VARIABLES

MADAN LAL PURI† and SHEY SHIUNG SHEU
Indiana University, Bloomington, U.S.A.

AMS 1970 *subject classification:* Primary 60B99, Secondary 46A15.

KEY WORDS AND PHRASES: Metric linear space, linear homeomorphism, symmetric stable laws, domain of normal attraction, Khinchine inequality.

Let $L^0(\Omega, \mathscr{A}, P)$ be the space of equivalent classes of random variables defined on a probability space $(\Omega, \mathscr{A}, P)$. Let $\mathscr{H}$ be the closed subspace of $L^0(\Omega, \mathscr{A}, P)$ spanned by a sequence of i.i.d. (independent and identically distributed) random variables having the symmetric nondegenerate law F. It is proved that $\mathscr{H}$ is linearly homeomorphic to l_p for $0<p\leqq 2$ if F belongs to the domain of normal attraction of symmetric stable law with exponent p.

1. INTRODUCTION

Let E be a linear metric space. According to Mushtari (1969), E is said to be *realized* by a space of random variables if E is isomorphic (linearly homeomorphic) to a subspace of $L^0(\Omega, \mathscr{A}, P)$. As usual, two random variables are equivalent if they agree with each other with probability one and the topology of $L^0(\Omega, \mathscr{A}, P)$ is the topology of convergence in probability. Spaces like l_p, $L^p(0, 1)$, $0<p\leqq 2$ have such realization but not spaces like l_p, $p>2$, c_0; see e.g. Mushtari (1969), Schilder (1970), and Schwartz (1969). The standard way to realize l_p, $0<p\leqq 2$ is defined as follows:

Take a sequence of i.i.d. random variables $\{X_n\}$ with the same

†Work supported by the Air Force Office of Scientific Research, AFSC, USAF, under Grant No. AFOSR 71-2009 and AFOSR 76-2927.

Reproduction in part or in whole permitted for any purpose of the U.S. Government.

symmetric stable law of exponent p; i.e. $E\,e^{itX_1}=e^{-|t|^p}$. Define

$$T((a_n))=\sum a_n X_n, \quad (a_n)\in l_p. \tag{1.1}$$

It is easy to see that T is an isomorphism. Naturally, one would ask what other laws will do the same job. More precisely, given a sequence $\{X_n\}$ of i.i.d. random variables with the same symmetric, nondegenerate law F, when is T defined as in (1.1) an isomorphism? We will show that for $p=2$, the necessary and sufficient condition is $\int x^2\,dF(x)<\infty$, and for $0<p<2$, a sufficient condition is that F belongs to the domain of normal attraction of symmetric stable law with exponent p. Of course, when T is an isomorphism, the closed subspace generated by $\{X_n\}$ is isomorphic to l_p.

When E is a separable Banach space and there exists a positive definite, real-valued function φ on E such that

$$\varphi(x)\to\varphi(0)=1\Leftrightarrow\|x\|=0, \tag{1.2}$$

we will show that E can be realized by a space of random variables. These spaces are important because some theory pertaining to central limit theorems can be extended on them [see LeCam (1969)]. For other related topics, see Kuelbs (1973) and Mushtari (1973).

2. PRELIMINARIES

We state the main results in this section.

PROPOSITION 1 *Let E be a linear metric space. The following two statements are equivalent.*

a) *There exists a positive definite function φ on E such that for every sequence $\{x_n\}$ in E,*

$$\varphi(tx_n)\to\varphi(0)=1, \qquad \forall t \quad real \quad \Leftrightarrow x_n\to 0. \tag{2.1}$$

b) *There exists an isomorphism T,*

$$T: E\to L^0(\Omega, \mathscr{A}, P)$$

for some $L^0(\Omega, \mathscr{A}, P)$.

Remark 1 Condition (1.2) is stronger than (2.1). Therefore, a Banach space satisfying (1.2) can be realized by a space of random variables.

THEOREM 1 *Let $\{X_n, n \geqq 1\}$ be a sequence of* i.i.d. *random variables in $L^0(\Omega, \mathscr{A}, P)$. Assume that the distribution function F of X_1, is symmetric and nondegenerate. Then, the following are equivalent.*

a) $$\int x^2\,dF(x) < \infty.$$

b) *The map T defined by*

$$T((a_n)) = \sum a_n X_n, \qquad (a_n) \in l_2 \tag{2.2}$$

is an isomorphism.

c) *F belongs to the domain of normal attraction of the standard normal law.*

Remark 2 Recall that a distribution function G belongs to the domain of normal attraction of the stable law $V(x)$ with exponent α if $\exists\, a>0$, $\exists\, A_n$, $\ni$

$$\lim_{n\to\infty} P\left(a^{-1}n^{-1/\alpha}\sum_{i=1}^{n} X_i - A_n < x\right) = V(x) \qquad \forall\, x, \tag{2.3}$$

where X_i, $i \geqq 1$, are i.i.d. random variables having the same law G.

Remark 3 If we drop the assumption of equi-distribution in Theorem 1, then the map T defined as in (2.2) will be an isomorphism if the distribution of X_n is symmetric for every n, and

$$E(X_n^2) = 1 \qquad \forall\, n \tag{2.4}$$

and

$$\exists\, \delta > 0 \ni P(|X_n| \geqq \delta) \geqq \delta \qquad \forall\, n. \tag{2.5}$$

COROLLARY 1 *(Khinchine Inequality). Let $\gamma_n(t)$ be the Rademacher functions. Then, $\forall\, r>0, \exists\, A_r, B_r \ni$*

$$A_r\left(\sum a_n^2\right)^{1/2} \leqq \left(\int_0^1 \left|\sum a_n\gamma_n(t)\right|^r dt\right)^{1/r} \leqq B_r\left(\sum a_n^2\right)^{1/2} \qquad \forall\, (a_n) \in l_2. \tag{2.6}$$

THEOREM 2 *Under the assumptions of Theorem 1, if F belongs to the domain of normal attraction of symmetric stable law with exponent p, $0<p<2$, then the map T defined by*

$$T((a_n)) = \sum a_n X_n, \qquad (a_n) \in l_p, \tag{2.7}$$

is an isomorphism.

3. PROOFS

We will give here the proofs of the statements in Section 2.

(A) *The proof of Proposition* 1. First we show that (a)⇒(b). Since φ is positive definite, continuous, and $\varphi(0)=1$, it induces a probability measure P on the algebraic dual E^a of E, and

$$P(x)=\int_{E^a} e^{ix(y)}P(dy). \tag{3.1}$$

Define the map $T:E\to L^0(E^a,P)$ by

$$T(x)=x(\cdot). \tag{3.2}$$

Obviously, T is linear. T is one-to-one because if $T(x)=T(x')$, then $T(t(x-x'))=0$ a.e. P, $\forall t$ real. By (3.1)

$$\varphi(t(x-x'))=\varphi(0)=1, \qquad \forall t \text{ real.}$$

According to (a)

$$x-x'=0, \qquad \text{i.e.} \quad x=x'.$$

Next, T is bi-continuous, because if $x_n\to 0$, then $\varphi(tx_n)\to 1$, $\forall x$ real. Hence,

$$\int_{E^a} e^{itx_n(y)}P(dy)\to 1, \qquad \forall t.$$

By Lévy,

$$x_n(\cdot)\to 0 \quad \text{in probability,}$$
$$\text{i.e.} \quad T(x_n)\to 0.$$

Conversely, if $T(x_n)\to 0$, then

$$\varphi(tx_n)=\int_{E^a} e^{itx_n(y)}P(dy)\to\varphi(0)=1, \qquad \forall t \text{ real.}$$

By (a) $x_n\to 0$.

We now show that (b)⇒(a). Define $\varphi(x)=E\,e^{iT(x)}$. Apparently, φ is positive-definite. Also, $x_n\to 0\Rightarrow tx_n\to 0\Rightarrow\varphi(tx_n)\to\varphi(0)=1$, $\forall t$ real. Conversely, if $\varphi(tx_n)\to\varphi(0)=1$, $\forall t$ real, then $E\,e^{itT(x_n)}\to 1$, $\forall t$. Hence, $T(x_n)\to 0$ in probability $\Rightarrow x_n\to 0$.

To prove Theorem 1, we need some lemmas. First, recall that a family of random variables $\{X_\gamma\}_{\gamma\in\Gamma}$ is *tight* if $\forall\varepsilon>0$, $\exists K(\varepsilon)$ finite $\ni\sup_{\gamma\in\Gamma}P(|X_\gamma|\geqq K)\leqq\varepsilon$. If X_n, $n\geqq 1$ are independent, symmetric random variables, then $\{\sum_{i=1}^n a_iX_i\}$ is tight if and only if $\sum_{i=1}^n a_iX_i\to$a.s.

LEMMA 3.1 *Suppose X_n, $n \geqq 1$, are independent and have symmetric non-degenerate laws. Let $\mathscr{H}$ be the closed subspace spanned by $\{X_n\}$. Then, $\forall Z \in \mathscr{H}$, $\exists$ a sequence $\{a_n\}$ of real numbers $\ni$*

$$Z = \sum a_n X_n. \tag{3.3}$$

Proof Let

$$\mathscr{Y} = \{\sum a_n X_n : a_n = 0 \text{ except for finitely many } n\}.$$

Then, $\mathscr{Y}$ is a dense subspace of $\mathscr{H}$.

For every integer m, define a map T^m on $\mathscr{Y}$ by

$$T_m(\sum a_n X_n) = a_m \tag{3.4}$$

Obviously, T_m is well defined and linear.

Now denote by f_n, the characteristic function of X_n. If

$$Z_k = \sum_n a_n^{(k)} X_n \in \mathscr{Y}$$

such that $Z_k \xrightarrow{p} 0$, as $k \to \infty$, then

$$\prod_n f_n(a_n^{(k)} t) \to 1, \qquad \forall t.$$

Since $|f_n(a_n^{(k)} t)| \leqq 1, \qquad \forall n, \quad \forall t,$

$$f_n(a_n^{(k)} t) \to 1, \qquad \text{as} \quad k \to \infty,$$

which implies $a_n^{(k)} \to 0$, since $f_n \not\equiv 1$. Hence T_m is continuous in $\mathscr{Y}$ for every m. Now extend T_m on $\mathscr{H}$, and let $Z \in \mathscr{H}$.

Consider

$$Z - \sum_{k=1}^{n} T_k(Z) X_k.$$

$Z - \sum_{k=1}^{n} T_k(Z) X_k$ is independent of X_k, $k = 1, \ldots, n$, when $Z \in \mathscr{Y}$. Passing to the limit, we see that

$$Z - \sum_{k=1}^{n} T_k(Z) X_k$$

is independent of X_k, $k = 1, \ldots, n$, when $Z \in \mathscr{H}$. Since

$$Z = \left(Z - \sum_{k=1}^{n} T_k(Z) X_k\right) + \sum_{k=1}^{n} T_k(Z) X_k,$$

by independence, $\forall \varepsilon>0, \exists \delta>0, \ni |t| \leqq \delta$,

$$\left| E \exp\left[i\left(\sum_{k=1}^{n} T_k(Z) X_k \right) t \right] \right| \geqq |E \exp[itZ]| \geqq 1-\varepsilon$$

i.e., $\forall n$,

$$1-\left| E \exp\left[i\left(\sum_{k=1}^{n} T_k(Z) X_k \right) t \right] \right| \leqq \varepsilon, \qquad \text{if} \quad |t| \leqq \delta.$$

By Lévy,

$$\left\{ \sum_{k=1}^{n} T_k(Z) X_k \right\} \qquad \text{is tight}$$

$$\Rightarrow \sum_{k=1}^{n} T_k(Z) X_k \to \text{a.s.}$$

Since $Z-\sum_{k=1}^{\infty} T_k(Z) X_k$ is measurable over the tail σ-field of $\{X_n, n \geqq 1\}$, it follows that $Z-\sum_{k=1}^{\infty} T_k(Z) X_k = a$ for some constant a.

By symmetry of X_n, $n \geqq 1$, $a=0$. The proof follows by letting $a_n = T_n(Z)$.

LEMMA 3.2 *Let X_i, $i \geqq 1$, be* i.i.d. *with the (same) symmetric, non-degenerate law F. Then*

$$\sum a_n X_n \to \text{a.s.} \Rightarrow \sum a_n^2 < \infty. \tag{3.5}$$

Proof According to Kolmogorov,

$$\sum a_n X_n \to \text{a.s.} \Rightarrow \sum_n P\left(|X_n| \geqq \frac{d}{|a_n|} \right) < \infty, \qquad \forall d>0, \tag{3.6}$$

and

$$\sum_n \int_{|x| \leqq \frac{1}{|a_n|}} a_n^2 x^2 F(dx) < \infty.$$

Claim: $a_n \to 0$.

Suppose the claim is not true. Then $\exists \delta>0, \ni$

$$|a_n| \geqq \delta$$

for infinitely many n. Obviously, one can choose d so small that

$$P\left(|X_n| \geqq \frac{d}{|a_n|} \right) \geqq \tfrac{1}{2} \quad \text{for those} \quad n \ni |a_n| \geqq \delta,$$

Since F is non-degenerate.

Then,

$$\sum_n P\left(|X_n| \geqq \frac{d}{|a_n|}\right) = \infty$$

which contradicts to (3.6). Since $a_n \to 0$, $\exists\, c > 0$, and $\exists\, n_0 \ni n \geqq n_0$

$$\int_{|x| \leqq \frac{1}{|a_n|}} x^2\, dF \geqq \tfrac{1}{2} c.$$

Therefore,

$$\tfrac{1}{2} c \left(\sum_{n=n_0}^{\infty} a_n^2 \right) \leqq \sum_{n=n_0}^{\infty} \int_{|a_n x| \leqq 1} a_n^2 x^2\, dF(x) < \infty.$$

Therefore,

$$\sum a_n^2 < \infty.$$

LEMMA 3.3 *Given* $b_n \uparrow \infty$, $\exists\, a_n \downarrow 0 \ni$

$$\sum a_n < \infty,$$

but

$$\sum a_n b_n = \infty.$$

Proof Choose n_k so that $\forall\, n \geqq n_k$,

$$b_n \geqq k^2$$

and

$$n_k - n_{k-1} \geqq n_{k-1} - n_{k-2}, \qquad \forall\, k.$$

Define

$$a_n = \frac{1}{k^2(n_{k+1} - n_k)} \quad \text{if} \quad n_k \leqq n < n_{k+1}.$$

Clearly, $a_n \downarrow 0$.

Now,

$$\sum a_n = \sum_k \sum_{n_k \leqq n < n_{k+1}} \frac{1}{k^2(n_{k+1}-n_k)} = \sum_k \frac{1}{k^2} < \infty.$$

But,

$$\sum a_n b_n = \sum_k \sum_{n_k \leqq n < n_{k+1}} \frac{b_n}{k^2(n_{k+1}-n_k)} \geqq \sum_k \sum_{n_k \leqq n < n_{k+1}} \frac{1}{(n_{k+1}-n_k)} = \sum_k 1 = \infty.$$

LEMMA 3.4 *Suppose* X_n, $n \geqq 1$, *are* i.i.d. *with the same symmetric law* F. *If* $\sum a_n X_n \to$ a.s. $\forall (a_n) \in l_2$, *then*

$$\int x^2 \, dF(x) < \infty. \tag{3.8}$$

Proof Since $\sum a_n X_n \to$ a.s. $\forall (a_n) \in l_2$,

$$\sum_n a_n^2 \int_{|a_n x| \leq 1} x^2 \, dF < \infty, \qquad \forall (a_n) \in l_2. \tag{3.9}$$

Suppose (3.8) fails, i.e., $\int x^2 \, dF(x) = \infty$. Let

$$b_n = \int_{-\sqrt{n}}^{\sqrt{n}} x^2 \, dF(x).$$

Then $b_n \uparrow \infty$. By Lemma 3.3, $\exists \, a_n^2 \downarrow 0 \ni$

$$\sum a_n^2 < \infty, \sum a_n^2 b_n = \infty.$$

But

$$a_n^2 \downarrow 0, \sum a_n^2 < \infty \Rightarrow a_n^2 = o\left(\frac{1}{n}\right).$$

Therefore, $\exists \, n_0 \ni$

$$\sum_{n=n_0}^{\infty} a_n^2 \int_{-\frac{1}{|a_n|}}^{\frac{1}{|a_n|}} x^2 \, dF \geqq \sum_{n=n_0}^{\infty} a_n^2 \int_{-\sqrt{n}}^{\sqrt{n}} x^2 \, dF = \sum_{n=n_0}^{\infty} a_n^2 b_n = \infty$$

i.e.,

$$\sum_n a_n^2 \int_{|xa_n| \leqq 1} x^2\, dF = \infty, \quad \sum a_n^2 < \infty,$$

which contradicts to (3.9).

(B) *The proof of Theorem* 1 That (a)⇔(c) is well known. See Gnedenko and Kolmogorov [1], page 181.

To prove (a)⇒(b), observe that T is $1-1$, linear, continuous. To show T is an isomorphism, it is sufficient to show

$$\mathscr{H} \equiv \{\sum a_n X_n : (a_n) \in l_2\}$$

is closed and that is an immediate consequence of Lemma 3.1 and Lemma 3.2. Conversely, to prove (b)⇒(a), use Lemma 3.4.

(C) *The proof of Remark* 3 It is enough to show that Lemma 3.2 holds if X_n, $n \geqq 1$, are independent and symmetric and satisfy (2.4) and (2.5). That is a consequence of a theorem proved by Marcinkiewicz and Zygmund [4].

(D) *The proof of Corollary* 1 (2.6) is equivalent to that l_2 is isomorphic to a closed subspace of $L^r(0, 1)$, $r > 0$, where the isomorphism T is defined by

$$T((a_n)) = \sum a_n \gamma_n(t), \qquad (a_n) \in l_2, \tag{3.10}$$

$\{\gamma_n(t)\}$ being the Rademacher functions.

Since $\{\gamma_n(t)\}$are i.i.d. with probability $\frac{1}{2}$ each at 1 and -1, by Theorem 1,

$$\mathscr{W} = \{\sum a_n \gamma_n(t) : (a_n) \in l_2\}$$

is closed in $L^0(0, 1)$.

But,

$$\begin{aligned} E \exp(\sum a_n \gamma_n(t)) &= \prod_n E \exp(a_n \gamma_n(t)) \\ &= \prod_n \left(\frac{e^{a_n} + e^{-a_n}}{2} \right) \\ &\leqq \prod_n e^{a_n^2} = \exp(\sum a_n^2) < \infty. \end{aligned}$$

Therefore,

$$\mathscr{W} \subset L^r(0, 1), \qquad \forall r > 0.$$

The topology of L^0 is weaker than that of L^r.

Therefore,

$$\mathscr{W} \text{ is closed in } L^r(0, 1), \qquad r > 0.$$

Since T^{-1} is continuous, T is an isomorphism.

(E) *The proof of Theorem* 2 By assumption, F belongs to the domain of normal attraction of symmetric stable law of exponent P. According to [1], page 181.

$$\lim_{x \to \infty} x^P(F(-x) + 1 - F(x)) = c, \qquad 0 < c < \infty. \tag{3.11}$$

Consider

$$\int_{-x}^{x} y^2 \, dF(y) = 2 \int_0^x y^2 \, dF(y)$$

$$= -2y^2(1 - F(y)) \Big|_0^x + 4 \int_0^x y(1 - F(y)) \, dy$$

Applying (3.11), we have

$$\lim_{x \to \infty} x^{P-2} \int_{-x}^{x} y^2 \, dF(y) = \frac{p}{2-p} c.$$

Therefore, by (3.11),

$$\sum_n P(|a_n X_n| \geqq 1) = \sum_n P\left(|X_n| \geqq \frac{1}{|a_n|}\right)$$

$$\leqq K(\sum |a_n|^p) < \infty, \quad \text{some constant } K$$

and

$$\sum a_n^2 \int_{|xa_n| \leqq 1} x^2 \, dF(x) \leqq K'(\sum a_n^2 |a_n|^{p-2})$$

$$\leqq K'(\sum |a_n|^p) < \infty,$$

some constant K'.

By Kolmogorov, T is well defined. Clearly, T is $1-1$ and linear. By Banach–Steinhaus, T is continuous. To finish the proof, we need only to show

$$\mathcal{M}=\{\sum a_n X_n:(a_n)\in l_p\}$$

is closed in $L^0(\Omega, \mathscr{A}, P)$.

Let $\bar{\mathcal{M}}$ be the closure of $\mathcal{M}$ and $Z\in\bar{\mathcal{M}}$. Then, $Z=\sum c_n X_n$, some $\{c_n\}$ such that $c_n\to 0$, by Lemma 3.1 and Lemma 3.2. But,

$$\infty>\sum_n P(|c_n X_n|\geqq 1)$$

$$\geqq K''\left(\sum_n |c_n|^p\right),$$

some constant $K''\neq 0$, by (3.11).

Therefore

$$\sum|c_n|^p<\infty \quad \text{implies} \quad Z\in\mathcal{M}$$

i.e., $\mathcal{M}$ is closed.

References

[1] B. Gnedenko and A. Kolmogorov, *Limit Distributions for Sums of Independent Random Variables*, Addison-Wesley, 1954.

[2] J. Kuelbs, Fourier analysis on linear metric spaces. *Tran. Amer. Math. Soc.* **181** (1973), 293–311.

[3] L. LeCam, Remarques sur le théoreme limite central dans les espaces localement convexes. *Les Probabilités sur les Structures Algébriques, Colloques Internationaux du Centre National de la Recherche Scientifique*, France, 1969.

[4] J. Marcinkiewicz and A. Zygmund, Sur les fonctions indépendantes. *Fund. Math.* **29** (1937), 60–90.

[5] D. Mushtari, On a problem of Laurent Schwartz and on realization of l_p spaces by spaces of random variables, *Theory Prob. Applications* **14** (1969), 699–701.

[6] D. Mushtari, Certain general questions of the theory of probability measures in linear spaces, *Theory Prob. Applications* **18** (1973), 64–75.

[7] M. Schilder, Some structure theorems for the symmetric stable laws, *Ann. Math. Stat.* **41** (1970), 412–421.

[8] L. Schwartz, Probabilités cylindriques et applications radonifiantes, *C.R. Acad. Sci. Paris* Ser A, **268** (1969), 646–648.

Reprinted by permission of the publisher,
© 1979 by Taylor and Francis Ltd.
http://www.tandf.co.uk

SHORTED OPERATORS AND GENERALIZED INVERSES OF MATRICES

SUJIT KUMAR MITRA and MADAN L. PURI

Indiana University, Bloomington, Indiana and Indian Statistical Institute New Delhi, India
Indiana University, Bloomington, Indiana

1. INTRODUCTION

The shorted operator in finite dimensional linear spaces was defined by Anderson [1] in terms of an explicit matrix construction. It was shown to be the solution to a maximization problem. In a subsequent paper by Anderson and Trapp [3], the same maximization problem was used to furnish the definition, and the definition extends itself to an arbitrary Hilbert space (not necessarily finite dimensional).[1] The concepts of a minimum seminorm g-inverse and a semi-least-squares inverse of a complex matrix were introduced and studied by Rao and Mitra [8]. In the present paper we obtain two explicit representations for the shorted operator (in finite dimensional linear spaces) in terms of these inverses. These representations lead to some new theorems on the shorted operators and to new proofs of old theorems. The shorting operation could also be viewed as a hermitian order preserving map from $\mathcal{C} \times \mathcal{C}$ onto $\mathcal{C}$, where $\mathcal{C}$ is the cone of nonnegative definite matrices and $\mathcal{C} \times \mathcal{C}$ is the cartesian product of $\mathcal{C}$ with itself. The operation is not commutative but it satisfies associativity.

*Work supported by the Air Force Office of Scientific Research, AFSC USAF, Grant No. AFOSR 76-2927, and National Science Foundation Grant No. MCS 76-00951. Reproduction in whole or in part permitted for any purpose of the United States Government.

[1]The referee has very kindly pointed out that this alternative definition is actually due to Krein [4].

We use the following notation and definitions. For a positive integer n, $\mathcal{E}^n$ is the linear space of complex n-tuples. Column vector representations of vectors in $\mathcal{E}^n$ will be denoted by lower case letters such as y, v, etc. $\mathcal{C}^{m\times n}$ represents the linear space of complex matrices of order $m\times n$. $\mathcal{C}_n$ represents the cone of nonnegative definite (n.n.d.) matrices in $\mathcal{C}^{n\times n}$. Matrices are denoted by capital letters such as A, B, M, etc. For a matrix A, $\mathfrak{M}(A)$ denotes its column span, $\mathfrak{N}(A)$ its null space and A^* its complex conjugate transpose. If $B,C\in\mathcal{C}_n$, we write $B\geqslant C$ if $B-C\in\mathcal{C}_n$.

DEFINITION. If $\mathcal{S}$ is a subspace of $\mathcal{E}^n$ and $B\in\mathcal{C}_n$, the shorted matrix $\mathcal{S}(B)$ is the unique matrix in $\mathcal{C}_n$ such that

$$\mathfrak{M}[\mathcal{S}(B)]\subset\mathcal{S}, \tag{1.1i}$$

$$B\geqslant\mathcal{S}(B), \tag{1.1ii}$$

$$\text{if } C\in\mathcal{C}_n, B\geqslant C \quad\text{and}\quad \mathfrak{M}(C)\subset\mathcal{S}, \tag{1.1iii}$$

then $\mathcal{S}(B)\geqslant C$. The existence of $\mathcal{S}(B)$ was established by Anderson and Trapp [3].

For a complex matrix $A\in\mathcal{C}^{m\times n}$, the matrix $G\in\mathcal{C}^{n\times m}$ is a g-inverse of A if $x=Gy$ is a solution of the consistent equation $Ax=y$ $\forall y\in\mathfrak{M}(A)$. We represent such an inverse by A^- and the entire class by $\{A^-\}$. G is a g-inverse of A iff

$$AGA = A. \tag{1.2}$$

For a vector $x\in\mathcal{E}^n$ we consider the seminorm of x defined by

$$\|x\|_n = (x^*Nx)^{1/2},$$

where N is a given matrix in $\mathcal{C}_n$. G is a minimum N seminorm g-inverse of A if $G\in\{A^-\}$ and $\forall y\in\mathfrak{M}(A)$, $Au=y\Rightarrow$

$$\|Gy\|_n\leqslant\|u\|_n.$$

We represent such a G by $A^-_{m(N)}$ and the class by $\{A^-_{m(N)}\}$. $G\in\{A^-_{m(N)}\}$ iff

$$AGA = A, \qquad (NGA)^* = NGA. \tag{1.3}$$

Let the seminorm of vector $y \in \mathcal{E}^m$ be similarly defined in terms of a given $M \in \mathcal{C}_m$. $x_0 \in \mathcal{E}^n$ is called an M-semi-least-squares solution of the possibly inconsistent equation $Ax = y$ if $\forall x \in \mathcal{E}^n$

$$\|Ax_0 - y\|_m \leqslant \|Ax - y\|_m. \tag{1.4}$$

G is an M-semi-least-squares inverse of A if $\forall y \in \mathcal{E}^m$, Gy is an M-semi-least-squares solution of $Ax = y$. We represent such a G by $A_{l(M)}$ and the class by $\{A_{l(M)}\}$. $G \in \{A_{l(M)}\}$ iff

$$A^*MAG = A^*M, \tag{1.5}$$

or equivalently

$$MAGA = MA, \qquad (MAG)^* = MAG. \tag{1.6}$$

$x_0 \in \mathcal{E}^n$ is a minimum N seminorm M semi-least-squares solution of $Ax = y$ if $\forall x \in \mathcal{E}^n$, either

$$\|Ax_0 - y\|_m < \|Ax - y\|_m \tag{1.7}$$

or

$$\|Ax_0 - y\|_m = \|Ax - y\|_m \quad \text{and} \quad \|x_0\|_n \leqslant \|x\|_n. \tag{1.8}$$

G is a minimum N seminorm M-semi-least-squares inverse of A if $\forall y \in \mathcal{E}^m$, Gy is a minimum N seminorm M-semi-least-squares solution of $Ax = y$. We represent such a G by A_{MN}, the class by $\{A_{MN}\}$. $G \in \{A_{MN}\}$ iff

$$A^*MAG = A^*M, \qquad \mathfrak{M}(NG) \subset \mathfrak{M}(A^*MA), \tag{1.9}$$

or equivalently

$$\begin{gathered} MAGA = MA, \qquad (MAG)^* = MAG; \\ NGAG = NG, \qquad (NGA)^* = NGA; \\ \mathfrak{M}(NG) \subset \mathfrak{M}(A^*MA). \end{gathered} \tag{1.10}$$

The results on minimum seminorm and semi-least-squares inverses listed in the section are proved in [8], and those on minimum seminorm semileast squares inverse in [7].

2. THE SHORTED OPERATOR IN TERMS OF A MINIMUM SEMINORM g-INVERSE

We show that the matrix NGA as considered in (1.3) is indeed the shorted operator $\mathcal{S}(N)$, where $\mathcal{S}=\mathcal{M}(A^*)$. Before we do this let us first establish the uniqueness and nonnegativity of NGA.

THEOREM 2.1. *If* $G\in\{A^-_{m(N)}\}$,

(a) *NGA is unique with respect to choice of G in this class;*
(b) $NGA\in\mathcal{C}_n$, $(N-NGA)\in\mathcal{C}_n$;
(c) *$\mathcal{M}(N-NGA)$ is virtually disjoint with $\mathcal{M}(A^*)$;*
(d) *NGA and $N-NGA$ are disjoint matrices* [5], *or equivalently*

$$\mathcal{M}(N)=\mathcal{M}(NGA)\oplus\mathcal{M}(N-NGA);$$

(e) $\mathcal{M}(NGA)=\mathcal{M}(N)\cap\mathcal{M}(A^*)$.

Proof.

(a): If $G_1,G_2\in\{A^-_{m(N)}\}$,

$$[(G_1-G_2)A]^*N(G_1-G_2)A=N(G_1-G_2)A(G_1-G_2)A=0.$$

Since $N\in\mathcal{C}_n$, this implies $N(G_1-G_2)A=0$.

(b): Since $NGA=(GA)^*NGA$, $N-NGA=(I-GA)^*N(I-GA)$ and $N\in\mathcal{C}_n$, (b) is established.

(c): Let $A^*u=(N-NGA)v$ be a vector in $\mathcal{M}(A^*)\cap\mathcal{M}(N-NGA)$, then $A^*u=A^*G^*A^*u=A^*G^*(N-NGA)v=0$. This proves (c).

(d): follows from (c), since $NGA=A^*G^*N\Rightarrow\mathcal{M}(NGA)\subset\mathcal{M}(A^*)$.

(e): (d) implies that if $x\in\mathcal{M}(N)\cap\mathcal{M}(A^*)$, it can be uniquely expressed as $x=x_1+x_2$ such that $x_1\in\mathcal{M}(NGA)$, $x_2\in\mathcal{M}(N-NGA)$. However, $x_2=x-x_1\in\mathcal{M}(A^*)$. Using (c), therefore, we have $x_2=0$ and $x=x_1\in\mathcal{M}(NGA)$. This shows $\mathcal{M}(N)\cap\mathcal{M}(A^*)\subset\mathcal{M}(NGA)$. That $\mathcal{M}(NGA)\subset\mathcal{M}(N)\cap\mathcal{M}(A^*)$ is trivial. This concludes the proof of (e) and of Theorem 2.1. ■

The unique matrix $NA^-_{m(N)}A$ will henceforth be denoted by $I(N,A)$. $P_0=A^*(A^-_{m(N)})^*$ is a projector onto $\mathcal{M}(A^*)$. If $P=A^*G^*$ is also a projector onto $\mathcal{M}(A^*)$, the minimum seminorm property of $A^-_{m(N)}$ implies

$$P_0NP_0^*\leqslant PNP^*.$$

Let $\mathcal{P}$ be the class of all matrices which are projectors onto $\mathfrak{M}(A^*)$. We have the following theorem:[2]

THEOREM 2.2.

$$I(N,A) = \min_{P \in \mathcal{P}} PNP^*.$$

The following theorem identifies $I(N,A)$ as a shorted operator.

THEOREM 2.3. *If* $\mathcal{S} = \mathfrak{M}(A^*)$, $I(N,A)$ *is the shorted operator* $\mathcal{S}(N)$.

Proof. We recall the definition (1.1). The condition of (1.1i) follows from Theorem 2.1(e), (1.1ii) from Theorem 2.1(b). Also, if $C \in \mathcal{C}_n$, $\mathfrak{M}(C) \subset \mathcal{S}$ and $N \geqslant C$, then

$$x^* I(N,A)x = x^* A^* (A^-_{m(N)})^* N A^-_{m(N)} Ax$$

$$\geqslant x^* A^* (A^-_{m(N)})^* C A^-_{m(N)} Ax = x^* Cx.$$

Thus the condition (1.1iii) is also satisfied. This completes proof of Theorem 2.3. ∎

THEOREM 2.4.

(a) *If* P_N *is the orthogonal projector onto* $\mathfrak{M}(N)$ *under the inner product* $(x,z) = z^* x$, *then* $I(P_N,A)$ *is the orthogonal projector onto* $\mathfrak{M}(N) \cap \mathfrak{M}(A^*)$.
(b) *Every g-inverse of N is a g-inverse of* $I(N,A)$.
(c)

$$\{I(N,A)\}^+ = I(P_N,A) N^- I(P_N,A).$$

(d) *If x is an eigenvector of N corresponding to the eigenvalue* λ *and* $x \in \mathfrak{M}(A^*)$, *then x is also an eigenvector of* $I(N,A)$ *corresponding to the same eigenvalue* λ.

Proof.

(a): is given as Corollary 6 in [3].
(b): is a consequence of the fact that

$$I(N,A)N^- I(N,A) = A^*(A^-_{m(N)})^* NN^- NA^-_{m(N)} A$$

$$= A^*(A^-_{m(N)})^* NA^-_{m(N)} A = I(N,A).$$

[2]Since in Theorem 2.3, $I(N,A)$ is shown to be the shorted operator $\mathcal{S}(N)$, Theorem 2.2 is same as Theorem 5 of [3]. However, the approaches are different.

(c): follows from (a), (b) and Note 8, p. 52 of Rao and Mitra [8].
(d):

$$Nx = \lambda x \quad \Rightarrow \quad I(N,A)x = A^*(A^-_{m(N)})^* Nx$$

$$= \lambda A^*(A^-_{m(N)})^* x$$

$$= \lambda x \qquad \text{if} \quad x \in \mathfrak{M}(A^*).$$

Note that the condition $x \in \mathfrak{M}(A^*)$ is superfluous if $\lambda = 0$. This establishes (d), and the proof of Theorem 2.4 is concluded. ■

NOTE 1. A general solution to a g-inverse of $I(N,A)$ is given by $N^- + X$ where N^- is an arbitrary g-inverse of N and X is an arbitrary solution of the equation $AXA^* = 0$. Since $N^- \in \{I(N,A)^-\}$, a general solution to $I(N,A)^-$ is given by $N^- + X_0$, where X_0 is an arbitrary solution to the equation $I(N, A)X_0 I(N,A) = 0$. In view of Theorem 2.1(b) and (c), any such X_0 can always be expressed as $X_0 = X_1 + X_2$, where $NX_1N = 0$ and $AX_2A^* = 0$. This establishes the above claim.

The following two theorems are concerned with square matrices.

THEOREM 2.5. *If NA is hermitian,*

$$\mathfrak{M}[N - I(N,A)] = \mathfrak{M}(N) \cap \mathfrak{N}(A^*).$$

Proof.

$$NA = A^*N \quad \Rightarrow \quad A^*[N - NA^-_{m(N)}A] = NA - NAA^-_{m(N)}A = 0.$$

Dimensionality considerations show

$$\mathfrak{M}[N - I(N,A)] = \mathfrak{M}(N) \cap \mathfrak{N}(A^*).$$ ■

THEOREM 2.6. *If $A, N \in \mathcal{C}_n$, then*

(a) $P[I(N,A), I(A,N)] = P(N,A)$, *where $P(\ ,0)$ represents parallel sum* [2, 6, 8];
(b) $[I(A,N)]^-$ *and* $[I(N,A)]^-$ *are arbitrary g-inverses, then*

$$[I(A,N)]^- + [I(N,A)]^- = N^- + A^-$$

for suitable choice of N^- and A^-.

Proof.

(a):

$$P(N,A) = N(N+A)^{-}A = I(N,A)(N+A)^{-}A + [N-I(N,A)](N+A)^{-}A$$

$$= I(N,A)[I(N,A)+A]^{-}A$$

$$= P[I(N,A),A],$$

noting that since $N-I(N,A)$ and $I(N,A)+A$ are disjoint matrices by Lemma 2.7 of Mitra [5], every g-inverse of $N+A$ is a g-inverse of $I(N,A)+A$. Also $[N-I(N,A)](N+A)^{-}[I(N,A)+A]=0 \Rightarrow [N-I(N,A)](N+A)^{-}A=0$. A similar argument shows $P[I(N,A),A]=P[I(N,A),I(A,N)]$.

(b) follows from Theorem 10.18(d) of Rao and Mitra [8] or alternatively from Note 1 following Theorem 2.4.

3. THE SHORTED OPERATOR IN TERMS OF SEMI-LEAST-SQUARES INVERSE

We obtain results here which are similar to those in Sec. 2.

THEOREM 3.1. *If* $G \in \{A_{l(M)}\}$,

(a) *MAG is unique with respect to choice of G in this class,*

(b) $MAG \in \mathcal{C}_m$, $M-MAG \in \mathcal{C}_m$,

(c) $\mathfrak{M}(MAG)$ *is virtually disjoint with* $\mathfrak{N}(A^*)$

(d) *MAG and* $M-MAG$ *are disjoint matrices, or equivalently*

$$\mathfrak{M}(M) = \mathfrak{M}(MAG) \oplus \mathfrak{M}(M-MAG),$$

(e) $\mathfrak{M}(M-MAG) = \mathfrak{M}(M) \cap \mathfrak{N}(A^*)$.

Proof.

(a): If $G_1, G_2 \in \{A_{l(M)}\}$,

$$[A(G_1-G_2)]^*MA(G_1-G_2) = (G_1-G_2)^*A^*MA(G_1-G_2)$$

$$= (G_1-G_2)^*(A^*M-A^*M) = 0.$$

Since $M \in \mathcal{C}_m$, this implies $MAG_1 = MAG_2$.

(b): Since $MAG=G^*A^*MAG$, $M-MAG=(I-AG)^*M(I-AG)$ and $M\in\mathcal{C}_m$, (b) is established.

(c): Let $MAGu$ be a vector in $\mathfrak{M}(MAG)\cap\mathfrak{N}(A^*)$: then

$$A^*Mu = A^*MAGu = 0$$

$$\Rightarrow \quad MAGu = G^*A^*Mu = 0.$$

This proves (c).

(d): $A^*(M-MAG)=0 \Rightarrow \mathfrak{M}(M-MAG)\subset\mathfrak{N}(A^*)$. Hence (d) follows from (c).

(e): Clearly $\mathfrak{M}(M-MAG)\subset\mathfrak{M}(M)\cap\mathfrak{N}(A^*)$. Let $y\in\mathfrak{M}(M)\cap\mathfrak{N}(A^*)$. Using (d), we write $y=y_1+y_2$ such that $y_1\in\mathfrak{M}(MAG)$, $y_2\in\mathfrak{M}(M-MAG)$. Observe that $y_1=y-y_2\in\mathfrak{N}(A^*)$. In view of (c), this implies $y_1=0$, $y=y_2\in\mathfrak{M}(M-MAG)$. Thus $\mathfrak{M}(M)\cap\mathfrak{N}(A^*)\subset\mathfrak{M}(M-MAG)$ and (e) is established. ■

The unique matrix $M-MAA_{l(M)}$ will henceforth be denoted by $J(M,A)$. $Q_0=AA_{l(M)}$ is a matrix in $\mathcal{C}^{m\times m}$ with column span contained in $\mathfrak{M}(A)$. If $Q=AG$ is any other such matrix, the semi-least-squares property of $A_{l(M)}$ implies

$$(I-Q_0)^*M(I-Q_0) \leqslant (I-Q)^*M(I-Q).$$

Let $\mathcal{Q}$ be the class of all matrices of all matrices in $\mathcal{C}^{m\times m}$ with column span contained in $\mathfrak{M}(A)$. We have thus the following theorem.

THEOREM 3.2.

$$J(M,A) = \min_{Q\in\mathcal{Q}} (I-Q)^*M(I-Q).$$

THEOREM 3.3. *If* $\mathcal{S}=\mathfrak{N}(A^*)$, *then* $J(M,A)$ *is the shorted operator* $\mathcal{S}(M)$.

Proof. Observe that the condition (1.1i) follows from Theorem 3.1(e), and (1.1ii) from Theorem 3.1(b). Also if $C\in\mathcal{C}_m$, $\mathcal{C}(C)\subset\mathcal{S}$ and $M\geqslant C$, then

$$\begin{aligned} y^*J(M,A)y &= y^*(I-AA_{l(M)})^*M(I-AA_{l(M)})y \\ &\geqslant y^*(I-AA_{l(M)})^*C(I-AA_{l(M)})y \\ &= y^*Cy. \end{aligned}$$

Thus the condition (1.1iii) is also satisfied, and the proof of Theorem 3.3 is concluded. ■

THEOREM 3.4.

(a) *If* P_M *is the orthogonal projector onto* $\mathfrak{M}(M)$ *under the inner product* $(y,w)=w^*y$, *then* $J(P_M,A)$ *is the orthogonal projector onto* $\mathfrak{M}(M)\cap \mathfrak{N}(A^*)$.

(b) *Every g-inverse of M is a g-inverse of* $J(M,A)$.

(c) $\{J(M,A)\}^+ = J(P_M,A)M^-J(P_M,A)$.

(d) *If* y *is an eigenvector of* M *corresponding to the eigenvalue* λ *and* $y\in\mathfrak{N}(A^*)$, *then* y *is an eigenvector of* $J(M,A)$ *corresponding to the same eigenvalue.*

Proof. The proof is similar to that of Theorem 2.4 and is therefore omitted. ■

NOTE 2. A general solution to a g-inverse of $J(M,A)$ is given by $M^- + AY + W^*A^*$ where M^- is an arbitrary g-inverse of M, and $Y,W\in \mathcal{C}^{n\times m}$ but are otherwise arbitrary. This claim is established in the same way as Note 1 following Theorem 2.4.

The next two theorems are concerned with square matrices.

THEOREM 3.5. *If MA is hermitian,*

$$\mathfrak{M}[J(M,A)] = \mathfrak{M}(M)\cap\mathfrak{M}(A^*).$$

Proof. The proof is trivial and is therefore omitted. ■

THEOREM 3.6. *If* $A,M\in\mathcal{C}_m$,

(a) $J(M,A)J(A,M)=J(M,A)I(M,A)=J(M,A)I(A,M)= P[J(M,A),J(A,M)]=P[J(M,A),I(M,A)]=P[J(M,A),I(A,M)]=0$

(b) $[M-J(M,A)][A-J(A,M)]=MA$.

4. THE MINIMUM SEMINORM SEMI-LEAST-SQUARES INVERSE

Theorem 4.1 gives an alternative definition of generalized Moore-Penrose inverses (Mitra and Rao [7]).

THEOREM 4.1. *The following statements are equivalent*:

(a) $G \in \{A_{MN}\}$,
(b) $M - MAG = J(M,A)$, $NGA = I(N, A^*MA)$, $\operatorname{rank} NGA = \operatorname{rank} NG$.

Proof.

(a) $\Rightarrow$ (b): Since $\{A_{MN}\} \subset \{A_{l(M)}\}$, the first part of (b) follows from the definition of $J(M,A)$. Also (1.10)$\Rightarrow$

$$NGA = A^*G^*NGA,$$

$$N - NGA = (I - GA)^*N(I - GA).$$

Hence $NGA \in \mathcal{C}_n$ and $N \geq NGA$. Clearly $\mathfrak{M}(NGA) \subset \mathfrak{M}(A^*MA)$. Also if $C \in \mathcal{C}_n$, $N \geq C$ and $\mathfrak{M}(C) \subset \mathfrak{M}(A^*MA)$,

$$x^*NGAx = x^*A^*G^*NGAx \geq x^*A^*G^*CGAx = x^*Cx,$$

since $A^*MAGA = A^*MA$. Hence NGA satisfies the conditions (1.1i), ii, iii) of (1.1), and it is seen that $NGA = I(N, A^*MA)$. The last part of (b) is trivial.

(b) $\Rightarrow$ (a): From Theorem 3.1(e) it is seen that $M - MAG = J(M,A) \Rightarrow G \in \{A_{l(M)}\}$. Also $NGA = I(N, A^*MA) \Rightarrow \mathfrak{M}(NGA) \subset \mathfrak{M}(A^*MA)$, and the rank condition implies $\mathfrak{M}(NG) = \mathfrak{M}(NGA) \subset \mathfrak{M}(A^*MA)$. Thus G satisfies (1.9) and hence $G \in \{A_{MN}\}$. ∎

It was noted in Mitra and Rao [7] that the conditions

$$A^*MAG = A^*M, \qquad G^*NGA = G^*N \tag{4.1}$$

do not determine a minimum N seminorm M-semi-least-squares inverse of A. A matrix G which satisfies (4.1) is denoted by the symbol $A_{(MN)}$. The following theorem gives an explicit description of the class $\{A_{(MN)}\}$.

THEOREM 4.2. $G \in \{A_{(MN)}\}$ *iff* $G \in \{A_{M_0 N}\}$ *for some* $M_0 = M + M_1$, *where* $M_1 \in \mathcal{C}_m$ *and is such that* A^*MA *and* A^*M_1A *are disjoint matrices* (*that is, the row spans of* A^*MA *and* A^*M_1A *are virtually disjoint and so are the column spans*).

Proof.

"*If*" *part*: A^*MA and A^*M_1A are disjoint
$\Leftrightarrow \operatorname{rank}(A^*M_0A) = \operatorname{rank}(A^*MA) + \operatorname{rank}(A^*M_1A)$ (by Lemma 2.1 of Mitra [5])
$\Leftrightarrow \operatorname{rank}(A^*M_0) = \operatorname{rank}(A^*M) + \operatorname{rank}(A^*M_1)$

$\Leftrightarrow A^*M$ and A^*M_1 are disjoint matrices. Hence $G \in \{A_{M_0N}\}$
$\Rightarrow A^*M_0AG = A^*M_0,\ G^*NGA = G^*N$
$\Rightarrow A^*MAG + A^*M_1AG = A^*M + A^*M_1,\ G^*NGA = G^*N$
$\Rightarrow A^*MAG = A^*M,\ G^*NGA = G^*N$
$\Rightarrow G \in \{A_{(MN)}\}$.

"Only if" part: Conversely, let $G \in \{A_{(MN)}\}$. Put $A^*M_1A = NGA - I(NGA, A^*MA)$, $A^*M_2A = I(NGA, A^*MA)$ and note, using Theorem 2.1(d), that A^*M_1A and A^*M_2A are disjoint matrices. Hence

$$NGAG = NG \quad \Rightarrow \quad A^*M_1AG = A^*M_1.$$

Further, by construction, A^*M_1A is disjoint with A^*MA. Hence $G \in \{A_{(MN)}\}$
$\Leftrightarrow A^*MAG = A^*M,\ G^*NGA = G^*N$
$\Rightarrow A^*(M+M_1)AG = A^*(M+M_1),\ G^*NGA = G^*N$
$\Rightarrow G \in \{A_{M_0N}\}$, since

$$\begin{aligned}\mathfrak{M}(NG) = \mathfrak{M}(NGA) &= \mathfrak{M}(A^*M_1A) \oplus \mathfrak{M}(A^*M_2A)\\ &\subset \mathfrak{M}(A^*M_1A) + \mathfrak{M}(A^*MA) = \mathfrak{M}(A^*M_0A).\end{aligned}$$ ■

THEOREM 4.3. $\{A_{(MN)}\} = \{A_{MN}\}$ *iff*

$$\mathfrak{M}(N) \cap \mathfrak{M}(A^*) = \mathfrak{M}(N) \cap \mathfrak{M}(A^*MA). \tag{4.2}$$

Proof. The "if" part is trivial. For the "only if" part observe that if (4.2) is untrue, $\mathfrak{M}(N) \cap \mathfrak{M}(A^*MA)$ is a proper subset of $\mathfrak{M}(N) \cap \mathfrak{M}(A^*)$. Consider the decomposition

$$I(N,A) = A^*M_1A + A^*M_2A$$

as in the proof of the "only if" part of Theorem 4.2, where $A^*M_2A = I[I(N,A), A^*MA]$. Put $M_0 = M + M_1$, and observe that $G = A_{M_0N} \in \{A_{(MN)}\}$. However,

$$\begin{aligned}\mathfrak{M}(NGA) = \mathfrak{M}(N) \cap \mathfrak{M}(A^*M_0A) &= \mathfrak{M}(N) \cap \mathfrak{M}[I(N,A)]\\ &= \mathfrak{M}(N) \cap \mathfrak{M}(A^*) \not\subset \mathfrak{M}(A^*MA).\end{aligned}$$

Hence $G \notin \{A_{MN}\}$. Thus negation of (4.2) implies negation of (1.9). This completes the proof of Theorem 4.3. ■

The authors would like to express their sincere thanks to the referee for his prompt and thorough reading of the manuscript and many valuable comments and suggestions.

REFERENCES

1 W. N. Anderson, Jr., Shorted operators, *SIAM J. Appl. Math.* 20:520–525 (1971).
2 W. N. Anderson, Jr. and R. J. Duffin, Series and parallel addition of matrices, *J. Math. Anal. Appl.* 11:576–594 (1963).
3 W. N. Anderson, Jr. and G. Trapp, Shorted operators II, *SIAM J. Appl. Math.* 28:60–71 (1975).
4 M. G. Krein, The theory of self-adjoint extensions of semi-bounded Hermitian operators and its applications, *Mat. Sbornik N.S.* 20(62):431–495 and 21(63):365–404 (1947).
5 S. K. Mitra, Fixed rank solutions of linear matrix equations, *Sankhyā*, Ser. A, 35:387–392 (1972).
6 S. K. Mitra and Madan L. Puri, On parallel sum and difference of matrices, *J. Math. Anal. Appl.* 44:92–97 (1973).
7 S. K. Mitra and C. R. Rao, Projections under seminorms and generalized Moore-Penrose inverses, *Linear Algebra Appl.* 9:155–167 (1974).
8 C. R. Rao and S. K. Mitra, *Generalized Inverse of Matrices and Its Applications*, Wiley, New York, 1971.

Received 9 March 1978

Reprinted by permission of the publisher,
© 1979 by Elsevier Science

COMPLEX PLANAR SPLINES*

GERHARD OPFER and MADAN L. PURI

Department of Mathematics, Indiana University, Bloomington, Indiana 47405
University Hamburg, Institut für Angewandte Mathematik, Bundesstrasse 55, D-2000 Hamburg 13, West Germany
Department of Mathematics, Indiana University, Bloomington, Indiana 47405

In contradistinction to the known theory on complex splines which are defined on the boundary of a region in $\mathbb{C}$, we define complex planar splines on a region itself as a complex valued continuous function which is defined piecewise on suitable meshes of that region. The main idea is to use nonholomorphic functions as pieces, since holomorphic pieces would lead to just one holomorphic function on the whole region. Some of the techniques used are available from the theory of finite elements. But we also consider new aspects, namely, mapping properties of a complex planar spline v and the difference $f-v$, where f is, in general, a holomorphic function. For triangular meshes, rectangular and parallelogrammatic meshes, and meshes on circular sectors, explicit expressions are provided; also properties of the newly introduced complex planar splines are studied.

1. Introduction

In this paper we are concerned with the approximation of complex valued functions by functions which we would like to call complex *planar* splines.

In the current literature (see e.g., Ahlberg [1], Ahlberg *et al.* [2–4], Atteia [7], and Schoenberg [12, 13]), complex splines are defined on the boundary of a given region and are then extended into the interior by Cauchy's integral

* Research supported by the Office of Naval Research under Contract N00014-77-C-0659. The U.S. Government's right to retain a nonexclusive royalty-free license in and to the copyright covering this paper, for governmental purposes, is acknowledged.

formula. However, this extension process is not easy to execute numerically. Therefore, we offer another approach, which in spirit originates from the theory of finite elements. (For full information on this subject, see the references in Schwarz [15].) We subdivide a given region into *meshes* and define a complex valued function on that region piecewise. The functions defined on each individual mesh will be called *elements*. The vertices of the occurring meshes will be called *grid points* and the set of all grid points a *grid*.

The minimum requirement we impose is the continuity of that piecewise defined function. Any such piecewise defined complex function which is continuous will be called a *complex planar spline*.

The continuity already has a very serious implication. If we try to define a complex planar spline by holomorphic elements like polynomials, then by the well known identity theorem (e.g., Diederich and Remmert [9, p. 132, Theorem 60]) all the elements represent just one holomorphic function. The consequence is that it makes no sense to work with holomorphic elements. Therefore, we have to use nonholomorphic elements.

Very simple nonholomorphic functions are polynomials in the complex variable z and its complex conjugate $\bar{z}$. These functions have the form

$$p(z, \bar{z}) = \sum_{j,k=0}^{n} a_{jk} z^j \bar{z}^k, \qquad a_{jk} \in \mathbb{C}. \tag{1.1}$$

The number

$$\partial p = \max_{j,k=0,1,\ldots,n} \{j + k : a_{j,k} \neq 0\} \tag{1.2}$$

will be called the *degree* of p. If $\partial p = 1$ we shall say that p is *linear*; if $\partial p = 2$ we shall say that p is *quadratic*; and if $\partial p = 3, 4, 5$ we shall use the words *cubic*, *quartic*, and *quintic*, respectively.

If we use the representation

$$z = x + iy, \qquad \bar{z} = x - iy \tag{1.3}$$

then

$$x = \frac{1}{2}(z + \bar{z}); \qquad y = \frac{1}{2i}(z - \bar{z}), \tag{1.4}$$

which means that a function in the real variables x and y can be transformed into a function depending on the complex variables z and $\bar{z}$, and vice versa.

Thus, a function $u = u(z, \bar{z})$ may also be regarded as a function in x and y. If u is continuously differentiable with respect to x and y, then we have

$$u_x = u_z + u_{\bar{z}}; \qquad u_y = i(u_z - u_{\bar{z}}), \tag{1.5}$$

$$u_z = \frac{u_x - iu_y}{2}; \qquad u_{\bar{z}} = \frac{u_x + iu_y}{2}. \tag{1.6}$$

If u is twice continuously differentiable, then

$$\Delta u = u_{xx} + u_{yy} = 4u_{z\bar{z}}. \tag{1.7}$$

In the sequel we treat triangular, rectangular and parallelogrammatic meshes, and meshes on circular sectors. The elements defined on those meshes will be as simple as possible. Besides the aspects known from the theory of finite elements, (for instance, interpolating properties, computational aspects, and an error analysis), there are new aspects which can be summarized by the term *mapping properties* of the newly defined complex planar splines. These new aspects concern the following questions, among others:

(I) How close are complex planar splines to comformality?

(II) Are complex planar splines quasiconformal?

(III) Are complex planar splines open mappings?

(IV) Is the boundary maximum principle valid for complex planar splines?

In some cases these questions apply also to the difference $\varepsilon = f - v$ between a certain function f and a complex planar spline v.

Since the interpolating formulae for the complex case appear different from the corresponding formulae for the real case, we believe that it is reasonable to state these formulae here. We shall see that the complex interpolation and L_2-approximation problem reduce to two real problems such that no new error analysis is needed. But the situation changes, for instance, for the uniform approximation problem.

Besides the approximation of functions by complex planar splines, still another application seems possible, namely, solving complex differential equations like Beltrami's equation

$$f_{\bar{z}} = \mu f_z + \nu \overline{f_z} \tag{1.8}$$

without splitting them into real and imaginary parts. Equation (1.8) was recently treated numerically by Weisel [16] for the case $\mu = 0$ by solving the corresponding real system with finite element techniques. A systematic treatment of (1.8) can be found in Wendland [17]. Furthermore, for the conformal mapping problem there also exists an investigation by Bosshard [8] on the use of finite elements, again by treating corresponding real

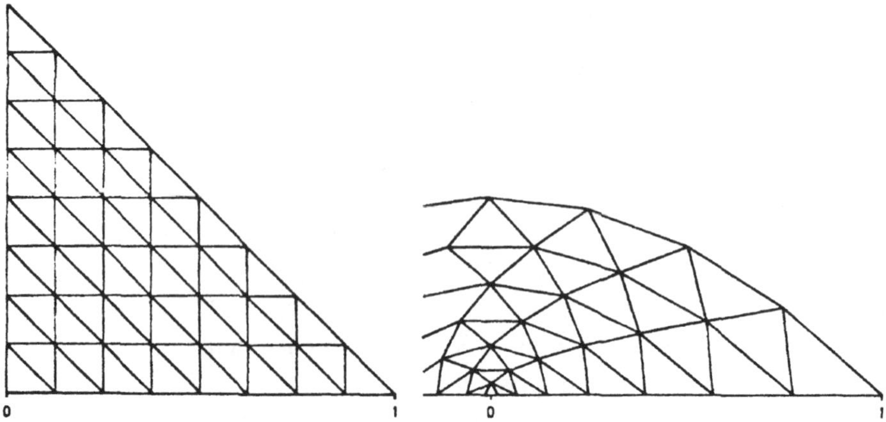

FIG. 1. Complex planar spline interpolating z^2, domain and range.

problems. Also in this case a direct attack seems possible, which then avoids the computation of the conjugate function to the computed real part of the mapping.

Still another application is the automatic construction of meshes with desired behavior. In many cases, for instance, it is desirable to construct meshes which concentrate at a certain point. An example is shown in Fig. 1, where the range of an interpolating complex planar spline is sketched. These applications provide motivation for our study of complex planar splines and their basic properties.

2. LINEAR COMPLEX PLANAR SPLINES ON TRIANGULAR MESHES

A *triangle* Δ is defined by its three vertices P_1, P_2, P_3 which are supposed to be three pairwise different complex numbers not located on a straight line. The three edges of Δ will be designated by $\overline{P_1P_2}$, $\overline{P_2P_3}$, $\overline{P_3P_1}$, where the order of the indices is irrelevant.

We first investigate one single element on a triangle Δ. We will use a linear element of the form

$$p(z, \bar{z}) = a + bz + c\bar{z}; \qquad a, b, c \in \mathbb{C}. \tag{2.1}$$

For simplicity of notation we shall write $p(z)$ instead of $p(z, \bar{z})$.

It is clear that the interpolation problem $p(P_j) = \zeta_j$, $j = 1, 2, 3$, has a unique solution for any three complex numbers $\zeta_1, \zeta_2, \zeta_3$. This solution is given by

$$a = (\zeta_1(P_2\overline{P_3} - \overline{P_2}P_3) + \zeta_2(\overline{P_1}P_3 - P_1\overline{P_3})$$
$$+ \zeta_3(P_1\overline{P_2} - \overline{P_1}P_2))/\delta, \tag{2.2}$$
$$b = (\zeta_1(\overline{P_2} - \overline{P_3}) + \zeta_2(\overline{P_3} - \overline{P_1}) + \zeta_3(\overline{P_1} - \overline{P_2}))/\delta, \tag{2.3}$$
$$c = (\zeta_1(P_3 - P_2) + \zeta_2(P_1 - P_3) + \zeta_3(P_2 - P_1))/\delta, \tag{2.4}$$

where

$$\delta = 2i \operatorname{Im}(P_1\overline{P_2} + P_2\overline{P_3} + P_3\overline{P_1}). \tag{2.5}$$

(Note that $\operatorname{Re}(z)$ and $\operatorname{Im}(z)$ mean the real and imaginary parts of z, respectively.)

It is easy to see that any three points $z_1, z_2, z_3 \in \mathbb{C}$ form a triangle if and only if

$$\operatorname{Im}(z_1\overline{z_2} + z_2\overline{z_3} + z_3\overline{z_1}) \neq 0 \tag{2.6}$$

and this property is invariant under translation.

For the *standard triangle* $P_1 = 0$, $P_2 = 1$, $P_3 = i$ the above formulae reduce to

$$a = \zeta_1, \tag{2.7}$$
$$b = \tfrac{1}{2}(\zeta_1(i-1) + \zeta_2 - i\zeta_3), \tag{2.8}$$
$$c = \tfrac{1}{2}(-\zeta_1(i+1) + \zeta_2 + i\zeta_3), \tag{2.9}$$

with

$$\delta = -2i. \tag{2.10}$$

For any given triangle with vertices P_1, P_2, P_3 we can construct three special elements p_j, $j = 1, 2, 3$, by solving $p_j(P_k) = \delta_{jk}$, $j, k = 1, 2, 3$, where δ_{jk} is the common Kronecker symbol. These elements are usually called *form elements*. They have the property that the general interpolating element p with $p(P_j) = \zeta_j$, $j = 1, 2, 3$, can be written in the simple form

$$p = \sum_{j=1}^{3} \zeta_j p_j. \tag{2.11}$$

Their importance lies in the fact that they can be used to construct a basis for the linear space of all complex planar splines, as will be seen later.

Definition 2.1. A linear element $p = a + bz + c\bar{z}$ for any $a, b, c \in \mathbb{C}$ is called *degenerate* if $|b| = |c|$.

Let R be a region in $\mathbb{C}$ and $f: R \to \mathbb{C}$ a mapping which has continuous partial derivatives f_z and $f_{\bar{z}}$. If f is quasiconformal, the numbers

$$d(z) = \frac{|f_z(z)| + |f_{\bar{z}}(z)|}{|f_z(z)| - |f_{\bar{z}}(z)|} \tag{2.12}$$

and

$$\mu(z) = \frac{f_{\bar{z}}(z)}{f_z(z)} \tag{2.13}$$

are called the *dilatation quotient* and *complex dilatation* at the point z, respectively (Lehto and Virtanen [10, p. 52, 191]). We always have $d(z) \geqslant 1$. In case $d(z) = 1$, z could be called a conformal point. If $d(z) = 1$ for all z, then f is conformal.

The following theorem gives some properties of a linear element p.

THEOREM 2.1. *A linear element* $p(z) = a + bz + c\bar{z}$, $a, b, c \in \mathbb{C}$ *has the following properties*:

1. *If* $z_1, z_2 \in \mathbb{C}$, *then*

$$p((1-\lambda) z_1 + \lambda z_2) = (1-\lambda) p(z_1) + \lambda p(z_2) \qquad \text{for all } \lambda \in \mathbb{R}. \tag{2.14}$$

2. *The range of* p *applied to a triangle is again a triangle if and only if* p *is nondegenerate.*

3. p *is an orientation preserving homeomorphism if and only if* $|b| > |c|$.

4. p *is an open mapping if and only if* p *is nondegenerate.*

5. p *is quasiconformal if and only if* $|b| > |c|$. *In that case its dilatation is a constant given by*

$$d = \frac{|b| + |c|}{|b| - |c|} \tag{2.15}$$

and the complex dilatation is a constant given by

$$\mu = \frac{c}{b}. \tag{2.16}$$

(6) *If we apply* p *to an angle* τ, $0 < \tau < \pi$, *the angle* τ *is distorted by the angle*

$$\tilde{\tau} = \arctan \frac{|c|^2 + \mathrm{Re}(b\bar{c}) - \mathrm{Im}(b\bar{c}) \tan \tau}{\mathrm{Im}(b\bar{c}) + \frac{1}{2}(|c|^2 - |b|^2) \tan \tau - \frac{1}{2}|b + c|^2 \cot \tau}. \tag{2.17}$$

Proof. Properties 1–5 are immediate. The proof of property 6 follows by standard computations. ∎

We now come to another important application of Theorem 2.1.

DEFINITION 2.2. Two triangles Δ_1, Δ_2 are called *neighbors* or *neighboring* if they share a common edge.

THEOREM 2.2. *Let Δ_1, Δ_2 be two neighboring triangles with the common edge $\overline{P_1 P_2}$ (see Fig. 2). Further let p_j be a linear element in Δ_j, $j = 1, 2$. If*

$$p_1(P_j) = p_2(P_j), \qquad j = 1, 2, \tag{2.18}$$

then $p_1(z) = p_2(z)$ for all $z \in \overline{P_1 P_2}$, and consequently

$$\begin{aligned} p(z) &= p_1(z) \qquad \text{for} \quad z \in \Delta_1 \\ &= p_2(z) \qquad \text{for} \quad z \in \Delta_2 \end{aligned} \tag{2.19}$$

is continuous on $\Delta_1 \cup \Delta_2$.

Proof. Follows directly from Theorem 2.1 (property 1). ∎

Now if we subdivide a region into triangular meshes and define a linear element in each mesh we obtain a complex planar spline if we impose condition (2.18) for each pair of neighboring triangles. In the triangulation, however, we do not allow that a vertex of any triangle is interior to any other edge. More specifically the triangulation has to be *proper* (Prenter [11, p. 127]).

The form elements introduced earlier are used to construct a basis for the linear space V of all complex planar splines. Let R be a region in $\mathbb{C}$, subdivided into finitely many triangles and $P_1, P_2, \ldots, P_N$ its grid points. A complex planar spline associated with the grid point P_j and defined by

$$\begin{aligned} v_j(z) &= 1 \qquad \text{for} \quad z = P_j \\ &= 0 \qquad \text{for} \quad z \in \{P_1, P_2, \ldots, P_N\} - \{P_j\}, \quad j = 1, \ldots, N, \end{aligned}$$

will be called a (*global*) *form function*. It can be constructed piecewise from the (local) form elements already known. Assume that $T_1, T_2, \ldots, T_{k_j}$ are the

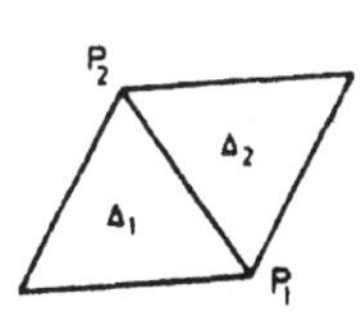

FIG. 2. Neighboring triangles.

triangles which have P_j as a common vertex. Then combine those form elements defined in $T_1, T_2, \dots, T_{k_j}$ which have the value one at P_j and zero at the two other vertices with the zero elements on all other triangles. Clearly $V = \langle v_1, v_2, \dots, v_N \rangle$.

A complex planar spline in general does not represent an open mapping even if all elements are not degenerate. To see this we define a complex planar spline on the square

$$Q = \{z: 0 \leqslant \operatorname{Re} z \leqslant 1, 0 \leqslant \operatorname{Im} z \leqslant 1\}, \tag{2.20}$$

which we divide into two triangles,

$$\Delta_1 = \{z \in Q: \operatorname{Re} z + \operatorname{Im} z \leqslant 1\}, \tag{2.21}$$

$$\Delta_2 = \{z \in Q: \operatorname{Re} z + \operatorname{Im} z \geqslant 1\}, \tag{2.22}$$

and define

$$\begin{aligned} v &= z && \text{for} \quad z \in \Delta_1 \\ &= 1 + i - i\bar{z} && \text{for} \quad z \in \Delta_2, \end{aligned} \tag{2.23}$$

which is then a complex planar spline on Q.

However, the range of Q under v is Δ_1 such that the range of any open set in Q containing parts of the diagonal of Q is not open.

But we have the following lemma (where the proof is obvious).

LEMMA 2.1. *If a complex planar spline represents a univalent function it is an open mapping.*

A consequence of this lemma is the following.

THEOREM 2.3. *Let v be a complex planar spline which interpolates a univalent, holomorphic mapping f on the grid points of a sufficiently fine triangular grid which is inside of the domain of definition of f. Then v is univalent and open.*

Proof. Since f is conformal, a small triangle is mapped such that the images of the three vertices form a triangle. Therefore, the interpolating complex planar spline maps the triangular grid onto another triangular grid in a univalent way. The result follows from Lemma 2.1. ∎

LEMMA 2.2. *Let f be a holomorphic function in a region R and g a holomorphic function in $\bar{R} = \{z: \bar{z} \in R\}$. Define a function h on R by $h(z) =$*

$f(z)+g(\bar{z})$, $z \in R$. If h is not constant, then $|h|$ does not admit a maximum in R.

Proof. It is sufficient to show that $|h|^2$ has no maximum in R. If Δ represents the Laplace operator as defined in (1.7) we obtain

$$\Delta |h|^2 = 4\, |h|^2_{z\bar{z}} = 4(|f_z|^2 + |g_{\bar{z}}|^2) \geqslant 0.$$

But this implies that $|h|^2$ is subharmonic. Since h is not constant, $|h|^2$ does not admit a maximum in R (Ahlfors [6, p. 245]). ∎

Particularly a nonconstant linear element p itself has the property that it admits no maximum in the interior of the triangle in which it is defined. This is a stronger property than property 4 of Theorem 2.1.

Clearly, a complex planar spline is not subharmonic in general. As an example take an interpolating spline which is 1 at one interior grid point and 0 at the other grid points.

3. Quadratic Complex Planar Splines on Rectangular and Parallelogram Meshes

First we study splines on rectangles whose sides are parallel to the axes. Such a rectangle □ (called *quabla* in physics) is defined by its four vertices P_1, P_2, P_3, P_4, which are to be understood as pairwise distinct complex numbers with $\operatorname{Im} P_1 = \operatorname{Im} P_2$, $\operatorname{Im} P_3 = \operatorname{Im} P_4$, $\operatorname{Re} P_1 = \operatorname{Re} P_4$, $\operatorname{Re} P_2 = \operatorname{Re} P_3$ arranged in positive orientation.

We first investigate one single element on a rectangle □. A quadratic element of the form

$$p(z) = a + bz + c\bar{z} + d(z^2 - \bar{z}^2), \qquad a, b, c, d \in \mathbb{C}, \tag{3.1}$$

will be used. That the use of this element is reasonable will be seen in Theorem 3.1. The interpolation problem $p(P_j) = \zeta_j$, $j = 1, 2, 3, 4$, again has a unique solution for any four complex numbers $\zeta_1, \zeta_2, \zeta_3, \zeta_4$:

$$a = (1/\delta)((P_3^2 - \bar{P}_3^2)\,\zeta_1 - (P_4^2 - \bar{P}_4^2)\,\zeta_2 + (P_1^2 - \bar{P}_1^2)\,\zeta_3 - (P_2^2 - \bar{P}_2^2)\,\zeta_4), \tag{3.2}$$

$$b = (2/\delta)(-P_3\zeta_1 + P_4\zeta_2 - P_1\zeta_3 + P_2\zeta_4), \tag{3.3}$$

$$c = (2/\delta)(\bar{P}_3\zeta_1 - \bar{P}_4\zeta_2 + \bar{P}_1\zeta_3 - \bar{P}_2\zeta_4), \tag{3.4}$$

$$d = (1/\delta)(\zeta_1 - \zeta_2 + \zeta_3 - \zeta_4), \tag{3.5}$$

where

$$\delta = 4(P_2 - P_1)(P_3 - P_2). \tag{3.6}$$

In the case of quadratic complex planar splines on rectangular meshes, we have the following analogue of Theorem 2.1.

THEOREM 3.1. *Let P, Q any two distinct points of $\mathbb{C}$ and p a quadratic element as defined in* (3.1) *with $d \neq 0$.*
Then

$$p((1-\lambda)P + \lambda Q) = (1-\lambda)\, p(P) + \lambda p(Q) \qquad \text{for all } \lambda \in [0, 1] \tag{3.7}$$

if and only if the straight line through P and Q is parallel to the x- or y-axis.

Proof. Set $z_\lambda = (1-\lambda)P + \lambda Q$. Then because of Theorem 2.1 and $d \neq 0$, (3.7) is equivalent to

$$z_\lambda^2 - \bar{z}_\lambda^2 = (1-\lambda)(P^2 - \bar{P}^2) + \lambda(Q^2 - \bar{Q}^2) \qquad \text{for all } \lambda \in [0, 1]. \tag{3.8}$$

Let $P = x + iy$ and $Q = u + iv$. After routine computations we deduce from (3.8) that $x(v-y) = u(v-y)$, which yields the assertion. ∎

DEFINITION 3.1. Two rectangles will be called *neighbors* or *neighboring* if they share a common edge (Fig. 3).

THEOREM 3.2. *Let $\square_1, \square_2$ be two neighboring rectangles whose sides are parallel to the axes, with the common edge $\overline{P_1P_2}$* (*Fig.* 3). *Let p_j be a quadratic element of the form* (3.1) *defined on $\square_j$, $j = 1, 2$. If*

$$p_1(P_j) = p_2(P_j), \qquad j = 1, 2, \tag{3.9}$$

then

$$p_1(z) = p_2(z) \qquad \text{for all } z \in \overline{P_1P_2}.$$

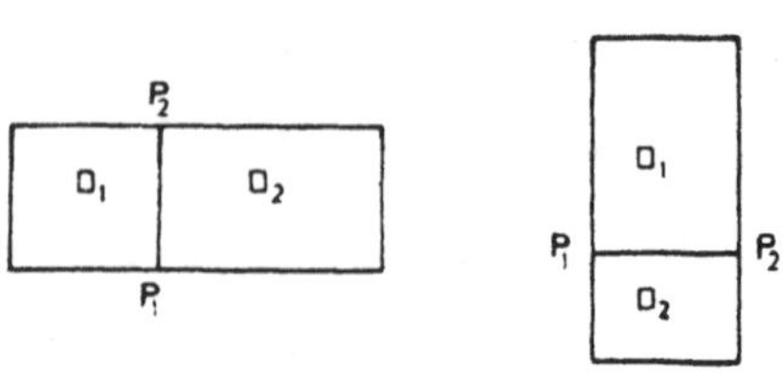

FIG. 3. Neighboring rectangles.

Consequently

$$\begin{aligned} p(z) &= p_1(z) \quad \text{for} \quad z \in \Box_1 \\ &= p_2(z) \quad \text{for} \quad z \in \Box_2 \end{aligned} \tag{3.10}$$

is continuous on $\Box_1 \cup \Box_2$.

Proof. Follows directly from Theorem 3.1. ∎

Now if we subdivide a region into rectangular meshes where the sides of the rectangles have to be parallel to the axes then we obtain a complex planar spline if we impose condition (3.9) for each pair of neighboring rectangles.

THEOREM 3.3. *Let R be any region in $\mathbb{C}$, f a holomorphic function in R and p a quadratic element as defined in* (3.1). *If $f-p$ is not constant, then $|f-p|$ does not admit a maximum in R.*

Proof. Since $f-p$ is a sum of a holomorphic function in z and a holomorphic function in $\bar{z}$, Lemma 2.2 applies and yields the desired result. ∎

This theorem applies particularly to p itself.

More information about the quadratic element p as defined in (3.1) can be deduced from its Jacobi determinant

$$J = |p_z|^2 - |p_{\bar{z}}|^2 \tag{3.11}$$

(see Lehto and Virtanen [10, p. 136]), which reads, in the present case,

$$\begin{aligned} J(z) &= |b + 2dz|^2 - |c - 2d\bar{z}|^2 \\ &= |b|^2 - |c|^2 + 2(d\bar{b} + \bar{d}c)z + 2(\bar{d}b + d\bar{c})\bar{z}. \end{aligned} \tag{3.12}$$

The set

$$H = \{z : J(z) = 0\} \tag{3.13}$$

is either a straight line or a point in $\mathbb{C}$.

DEFINITION 3.2. The quadratic element $p(z) = a + bz + c\bar{z} + d(z^2 - \bar{z}^2)$, $a, b, c, d, z \in \mathbb{C}$, is called *degenerate* if

$$\sigma = d\bar{b} + \bar{d}c = 0. \tag{3.14}$$

Now, H of (3.13) is a straight line in $\mathbb{C}$ if and only if p is not degenerate. If p is degenerate then (3.14) implies $d = 0$ or $|b| = |c|$. If p is not degenerate then the straight line H forms the angle α with the real axis, which is given by

$$\tan \alpha = \frac{\operatorname{Re}(d\bar{b} + \bar{d}c)}{\operatorname{Im}(d\bar{b} + \bar{d}c)}, \qquad 0 \leqslant \alpha < \pi. \tag{3.15}$$

Let us assume now that p is not degenerate. If the domain of definition of p is any compact set S in $\mathbb{C}$ which is located in the half plane

$$H^+ = \{z: J(z) > 0\} \tag{3.16}$$

then p is locally an orientation preserving homeomorphism in S which then is also quasiconformal in S since its dilatation quotient (see 2.12) is bounded.

In order to find out whether p is a global homeomorphism we study the solutions of

$$p(z_2) - p(z_1) = 0. \tag{3.17}$$

If we use the abbreviations

$$x = z_2 - z_1, \qquad y = z_2 + z_1 \tag{3.18}$$

Eq. (3.17) reads

$$p(z_2) - p(z_1) = x(b + dy) + \bar{x}(c - d\bar{y}) = 0. \tag{3.19}$$

From this it follows that

$$d\overline{(p(z_2) - p(z_1))} + \bar{d}(p(z_2) - p(z_1)) = x\bar{\sigma} + \bar{x}\sigma = 0, \tag{3.20}$$

where σ was already introduced in (3.14). If we use this equation to eliminate $\bar{x}$ from (3.19) we obtain

$$y\sigma + \bar{y}\bar{\sigma} + |b|^2 - |c|^2 = 0 \tag{3.21}$$

in case $x \neq 0$. Let $J(z_j) > 0$, $j = 1, 2$. Then from (3.12) by forming $J(z_1) + J(z_2)$ it follows that

$$2(|b|^2 - |c|^2) + 2\sigma y + 2\bar{\sigma}\bar{y} > 0, \tag{3.22}$$

which contradicts (3.21).

To summarize: If p is not degenerate then it is univalent in both half planes H^+ and $H^- = \{z: J(z) < 0\}$.

Since the domain of definition of p is a rectangle R whose sides are parallel to the axes, one can find out whether $R \subset H^+$ just by inserting a suitable vertex into J. To explain that we distinguish four cases according to the special location of H^+ in $\mathbb{C}$.

Case 1. $\pi/2 \leqslant \alpha < \pi$ and H^+ is a right half plane in $\mathbb{C}$.

Case 2. $\alpha = 0$ and H^+ is an upper half plane in $\mathbb{C}$, or $0 < \alpha < \pi/2$ and H^+ is a left half plane in $\mathbb{C}$.

Case 3. $\pi/2 \leqslant \alpha < \pi$ and H^+ is a left half plane in $\mathbb{C}$.

Case 4. $\alpha = 0$ and H^+ is a lower half plane in $\mathbb{C}$, or $0 < \alpha < \pi/2$ and H^+ is a right half plane in $\mathbb{C}$.

The words *left, right, lower, upper* half plane are used in the ordinary sense.

THEOREM 3.4. *Let R be a rectangle whose sides are parallel to the axes. Call P_1 the lower left, P_2 the lower right, P_3 the upper right and P_4 the upper left vertex of R. Further let p be a nondegenerate quadratic element (as defined in (3.1)) on R. Then the mapping p is an orientation preserving homeomorphism and quasiconformal on R if and only if $J(P_j) > 0$, where j is determined by the case number j to which H^+ belongs, $j \in \{1, 2, 3, 4\}$.*

Proof. If we are in case j, $j \in \{1, 2, 3, 4\}$, then $J(P_j) > 0$ is equivalent to $R \subset H^+$. ∎

If for mnemonic reasons one would like to give this theorem a name, then *four corner theorem* seems to be very suitable, since all four corners of the rectangle R are involved.

Under the assumptions stated p will be an orientation *inverting* homeomorphism on R if and only if $J(P_k) < 0$, where $k = (j + 2) \bmod 4$ and j is determined as before.

If a rectangle is subdivided in this way into $m \cdot n$ little rectangles, then there are $m \cdot n$ elements having $p = 4m \cdot n$ parameters altogether. Further there are $s = 3mn - m - n - 1$ continuity conditions leaving $p - s = (m + 1)(n + 1)$ parameters free, where $(m + 1)(n + 1)$ is also the number of grid points.

Now we can adjust the element (3.1) to the case where the rectangle has any position in the plane. If one of the edges of a given rectangle forms the angle α with the x-axis then instead of (3.1) one must use

$$p(z) = a + bz + c\bar{z} + d(z^2 - e^{4i\alpha}\bar{z}^2). \tag{3.23}$$

Since a parallelogram can be mapped by a linear transformation l of the type (2.1) onto a rectangle whose sides are parallel to the axes one can also work with parallelograms.

If $\alpha_1, \alpha_2 \in \mathbb{C}$, with $\alpha_1 \neq \lambda\alpha_2$ for any $\lambda \in \mathbb{R}$, describe the directions of the parallelogram grid, which means that the two angles β_j, $j = 1, 2$, with the real axis are given by $\tan \beta_j = \operatorname{Re} \alpha_j / \operatorname{Im} \alpha_j$, $j = 1, 2$, then the aforementioned linear transformation l yields an element of the form

$$p(z) = a + bz + c\bar{z} + d(\alpha_1 z + \overline{\alpha_1 z})(\alpha_2 z + \overline{\alpha_2 z}), \quad a, b, c, d \in \mathbb{C}. \quad (3.24)$$

4. Splines on Disks and Circular Sectors

Let S be a circular sector. If we divide the radius into k subintervals and the opening angle into $l - 1$ subintervals, we obtain, in total, $k(l-1)$ meshes of two types, which we would call *rectangular* and *triangular* meshes, respectively. The triangular meshes contain the origin O of S, whereas the rectangular meshes do not contain the origin (see Fig. 4).

It is straightforward to use polar coordinates with respect to the origin O of S in this situation. Therefore for rectangular meshes we use an element of the form

$$p(z) = p(re^{i\phi}) = a + br + c\phi + dr\phi, \qquad a, b, c, d \in \mathbb{C}. \quad (4.1)$$

Let P_1, P_2, P_3, P_4 be the vertices of one specific rectangular mesh in positive orientation such that $\phi_1 = \arg P_1 = \arg P_2$, $\phi_2 = \arg P_3 = \arg P_4$, $r_1 = |P_1| = |P_4|$, $r_2 = |P_2| = |P_3|$ (see Fig. 5).

Then the solution of the interpolation problem

$$p(P_j) = \zeta_j, \qquad \zeta_j \in \mathbb{C}, \quad j = 1, 2, 3, 4, \quad (4.2)$$

is given by

$$a = (r_2\phi_2\zeta_1 - r_1\phi_2\zeta_2 + r_1\phi_1\zeta_3 - r_2\phi_1\zeta_4)/\delta, \quad (4.3)$$

$$b = (-\phi_2\zeta_1 + \phi_2\zeta_2 - \phi_1\zeta_3 + \phi_1\zeta_4)/\delta, \quad (4.4)$$

$$c = (-r_2\zeta_1 + r_1\zeta_2 - r_1\zeta_3 + r_2\zeta_4)/\delta, \quad (4.5)$$

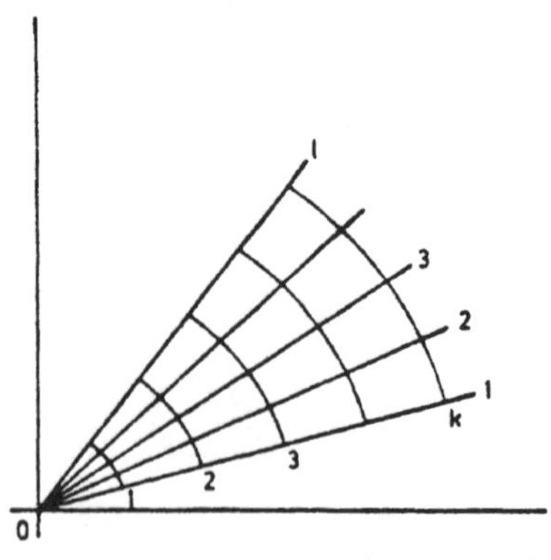

Fig. 4. Sector S subdivided into rectangular and triangular meshes.

and

$$d = (\zeta_1 - \zeta_2 + \zeta_3 - \zeta_4)/\delta, \tag{4.6}$$

where

$$\delta = 1/(r_2 - r_1)(\phi_2 - \phi_1). \tag{4.7}$$

Now assume that O, P, Q are the vertices of a triangular mesh in positive orientation, where O is the origin of the sector S (see Fig. 5). In order to find out what type of element to use, we study the interpolation problem for a rectangular mesh, where $P_1 \to 0$, $P_4 \to 0$ and $\zeta_1 = \zeta_4$, such that $|P_1| = |P_4|$ and $\arg P_1 = \arg P_2$, $\arg P_4 = \arg P_3$.

After some computation we find $c \to 0$ in (4.1). This means that we have to use an element of the form

$$p(z) = \alpha + \beta r + \gamma r\phi, \qquad \alpha, \beta, \gamma \in \mathbb{C}. \tag{4.8}$$

If we assume that P and Q have the polar coordinates (r_1, ϕ_1) and (r_2, ϕ_2), respectively, then the interpolation problem

$$p(O) = \zeta_1, \qquad p(P) = \zeta_2, \qquad p(Q) = \zeta_3, \qquad \zeta_1, \zeta_2, \zeta_3 \in \mathbb{C}, \tag{4.9}$$

has the solution

$$\alpha = \zeta_1, \tag{4.10}$$

$$\beta = (-(\phi_2 - \phi_1)\,\zeta_1 + \phi_2\zeta_2 - \phi_1\zeta_3)/\delta, \tag{4.11}$$

and

$$\gamma = (-\zeta_2 + \zeta_3)/\delta, \tag{4.12}$$

where

$$\delta = 1/r_1(\phi_2 - \phi_1). \tag{4.13}$$

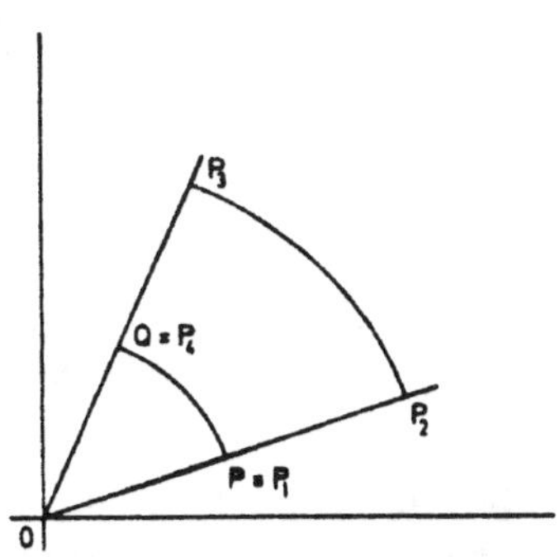

FIG. 5. Triangular and rectangular mesh of a circular sector.

THEOREM 4.1. *Let S be a circular sector subdivided into meshes as described above and m_j, $j = 1, 2$ two neighboring meshes sharing the common edge $E = \overline{P_1 P_2}$. Let p_j be an element defined on m_j, where p_j has the form (4.1) if m_j is a rectangular mesh or the form (4.8) if m_j is a triangular mesh, $j = 1, 2$.*

If

$$p_1(P_j) = p_2(P_j), \qquad j = 1, 2, \tag{4.14}$$

then

$$p_1(z) = p_2(z) \qquad \text{for all } z \in E. \tag{4.15}$$

Consequently,

$$\begin{aligned} p(z) &= p_1(z) \qquad \text{for} \quad z \in m_1 \\ &= p_2(z) \qquad \text{for} \quad z \in m_2 \end{aligned} \tag{4.16}$$

is continuous on $m_1 \cup m_2$.

Proof. The elements (4.1) on rectangular meshes as well as the elements (4.8) on triangular meshes are linear on the edges of their respective domain of definition. ∎

The element p introduced in (4.1) has the same form as the element p defined in (3.1) and used for ordinary rectangles.

In order to see this, one has only to identify $\operatorname{Re} z$ with r and $\operatorname{Im} z$ with ϕ. The consequence is that no particular analysis is required besides that for elements on ordinary rectangles.

We end this section with the computation of the number of free parameters in a complex planar spline on a circular sector.

The $k(l-1)$ meshes distribute in $(k-1)(l-1)$ rectangular and $l-1$ triangular meshes. In order to make a piecewise defined function continuous we have to impose $l(3k-1)-4k$ conditions. The number of parameters is $(l-1)(4k-1)$, leaving $lk+1$ parameters free. This is also the number of grid points.

5. Least-Squares Approximation and Interpolation with Complex Planar Splines

If we want to approximate a complex valued function by complex planar splines of a certain type, we end up with minimizing a real valued functional defined in $\mathbb{C}^N$, $N \in \mathbb{N}$.

To treat such a problem it is not necessary to rewrite it in real form. Let us assume that we have to handle the problem

$$\phi(a) = \min, \qquad \phi\colon \mathbb{C}^N \to \mathbb{R}, \tag{5.1}$$

where ϕ may also explicitly depend on $\bar{a}$. If ϕ has continuous partial derivatives with respect to all components of a, then

$$\frac{\partial \phi(\hat{a})}{\partial a_j} = 0, \qquad j = 1, 2, \ldots, N, \quad a = (a_1, a_2, \ldots, a_n), \tag{5.2}$$

is a necessary condition for $\hat{a}$ to be a minimum of ϕ. This follows immediately from (1.5) and (1.6).

If $g, h\colon \mathbb{C}^N \to \mathbb{C}$ are complex valued functions possessing continuous partial derivatives with respect to a and $\bar{a}$, then

$$(\bar{g})_a = \overline{g_{\bar{a}}}, \qquad (\bar{g})_{\bar{a}} = \overline{g_a}, \tag{5.3}$$

$$(gh)_a = g_a h + \overline{g h_a}, \tag{5.4}$$

$$\frac{\partial}{\partial a} |g|^2 = \frac{\partial}{\partial a}(g\bar{g}) = g_a \bar{g} + \overline{g g_{\bar{a}}} = \overline{\frac{\partial}{\partial \bar{a}} |g|^2}, \tag{5.5}$$

and if g does not depend on $\bar{a}$ explicitly, then

$$\frac{\partial}{\partial a} |g|^2 = g_a \bar{g}. \tag{5.6}$$

The least-squares problem can be treated along the lines of Schultz [14, Chap. 6] as follows.

Let R be a compact set in $\mathbb{C}$ subdividable into meshes of the discussed form, $f \in L_2(R)$ and V the linear space of all complex planar splines, where $V = \langle v_1, v_2, \ldots, v_N \rangle$ and the v_j are the global form functions defined earlier, $j = 1, 2, \ldots, N$. It should be noticed that the form functions are real by definition.

The problem here is to minimize

$$\phi(a) = \int_R \left| f(z) - \sum_{j=1}^{N} a_j v_j \right|^2 dx\,dy, \qquad a = (a_1, a_2, \ldots, a_N) \in \mathbb{C}^N. \tag{5.7}$$

Using (5.2) to (5.7), we obtain

$$\frac{\partial}{\partial a_j} \phi(a) = \int_R \left(\sum_{k=1}^{N} \bar{a}_k v_j v_k - v_j \bar{f} \right) dx\,dy = 0,$$

which reduces to the linear system

$$Ca = r, \tag{5.8}$$

where

$$C = (c_{jk}) = \int_R v_j v_k \, dx\, dy, \qquad j, k = 1, 2, \ldots, N, \tag{5.9}$$

$$r = (r_j) = \int_R f v_j \, dx\, dy, \qquad j = 1, 2, \ldots, N. \tag{5.10}$$

Since C is a real matrix, system (5.8) can be partitioned into the two real systems,

$$C \operatorname{Re} a = \operatorname{Re} r, \qquad C \operatorname{Im} a = \operatorname{Im} r, \tag{5.11}$$

where $\operatorname{Re} a$ is the vector of the real parts of a; analogous meanings apply to $\operatorname{Re} r$, $\operatorname{Im} a$, and $\operatorname{Im} r$.

Clearly (5.8) has a unique solution since ϕ is strictly convex.

Let $\hat{v}$ be the best least-squares approximation of f. The error analysis can be directly taken from the real case (e.g., Schultz [14, Chap. 6]) since

$$\begin{aligned} \|f - \hat{v}\| &= \|\operatorname{Re}(f - \hat{v}) + i \operatorname{Im}(f - \hat{v})\| \\ &\leqslant \|\operatorname{Re} f - \operatorname{Re} \hat{v}\| + \|\operatorname{Im} f - \operatorname{Im} \hat{v}\|. \end{aligned} \tag{5.12}$$

This means that the order of convergence is the same as that in the real case, but the convergence constants have to be doubled, provided, of course, that $\operatorname{Re} f$ and $\operatorname{Im} f$ are of the same smoothness.

An interpolation problem

$$v(z_k) = f(z_k), \qquad k = 1, 2, \ldots, K, \tag{5.13}$$

where v is a complex planar spline, f is a given function, and z_k, $k = 1, 2, \ldots, K$, are given grid points, may also be partitioned into two real problems by splitting Eq. (5.13) into real and imaginary parts. The above remarks on the error analysis therefore also apply here.

6. A Numerical Example

We subdivide the standard triangle 0, 1, i in the usual way by dividing its two smaller sides into $1/h = 2^k$, $k = 0, 1, \ldots, 5$, pieces of equal length and divided the resulting little squares of side length h diagonally by parallels through the hypotenuse of the standard triangle. We obtained a complex

TABLE I

Spline Interpolating the Exponential Function on Standard Triangle

h	e_h	c_h	d_h	$\tilde{c}_h$
1	0.3730	1.54	2.229	1.45
1/2	0.1285	1.79	1.451	1.19
1/4	0.0372	1.90	1.198	1.09
1/8	0.00996	1.95	1.093	1.05
1/16	0.00257	1.98	1.045	1.03
1/32	0.000653		1.022	

planar spline v_h by interpolating the exponential function at the grid points of that triangle.

In Table I we list the computed values $e_h = \|\exp - v_h\|_\infty$, the corresponding numerical convergence order c_h of e_h, the maximal dilatation quotient d_h over all meshes (compare Formula (2.15)), and the corresponding convergence order $\tilde{c}_h$ of d_h.

The fact that e_h approaches 0 with order 2 is of course known (Schultz [14, Chap. 2]). According to our computation the maximal dilatation approaches one with order one. The number $d_h - 1$ could be called *deviation* from conformality. To the best of our knowledge, neither numerical values for the dilatation d_h nor theoretical investigations on the behavior of d_h as $h \to 0$ appear in literature.

Acknowledgments

The assistance of Mr. Grothkopf and Mr. Zieher for preparing the figures and the table is acknowledged.

References

1. J. H. Ahlberg, Splines in the complex plane, *in* "Approximations with Special Emphasis on Spline Functions" (I. J. Schoenberg, Ed.), pp. 1–27, Academic Press, New York/London, 1969.
2. J. H. Ahlberg, E. N. Nilson, and J. L. Walsh, "The Theory of Splines and Their Applications," Academic Press, New York/London, 1967.
3. J. H. Ahlberg, E. N. Nilson, and J. L. Walsh, Complex cubic splines, *Trans. Amer. Math. Soc.* **129** (1967), 391–413.
4. J. H. Ahlberg, E. N. Nilson, and J. L. Walsh, Properties of analytic splines. I. Complex polynomial splines, *J. Math. Anal. Appl.* **27** (1969), 262–278.
5. J. H. Ahlberg, E. N. Nilson, and J. L. Walsh, Complex polynomial splines on the unit circle, *J. Math. Anal. Appl.* **33** (1971), 234–257.

6. L. Ahlfors, "Complex Analysis," 3rd ed., McGraw–Hill, New York, 1979.
7. M. Atteia, Fonctions ⟪spline⟫ dans le champ complexe, *C.R. Acad. Sci. Paris* **273** (1971), 678–681.
8. F. Bosshard, "Die Konstruktion konformer Abbildungen mit der Methode der finiten Elemente," Dissertation, Universität Zürich, 1980.
9. K. Diederich and R. Remmert, "Funktionentheorie I," Springer-Verlag, Berlin/Heidelberg/New York, 1972.
10. O. Lehto and K. I. Virtanen, "Quasikonforme Abbildungen," Springer-Verlag, Berlin/Heidelberg/New York, 1965.
11. P. Prenter, "Splines and Variational Methods," Wiley, New York, 1975.
12. I. J. Schoenberg, On polynomial spline functions on the circle. I. Their interpolatory properties, *in* "Proceedings, Conference on Constructive Theory of Functions, Budapest, 1969" (G. Alexits and S. B. Stechkin, Eds.), pp. 403–418, Akadémiai Kiadó, Budapest, 1972.
13. I. J. Schoenberg, On polynomial spline functions on the circle. II. Monosplines and quadrature formulae, *in* "Proceedings, Conference on Constructive Theory of Functions, Budapest, 1969" (G. Alexits and S. B. Stechkin, Eds.), pp. 419–433, Akadémiai Kiadó, Budapest, 1972.
14. M. H. Schultz, "Spline Analysis," Prentice–Hall, Englewood Cliffs, N.J., 1973.
15. H. R. Schwarz, "Methode der finiten Elemente," Teubner, Stuttgart, 1980.
16. J. Weisel, Numerische Ermitlung quasikonformer Abbildungen mit finiten Elementen, *Numer. Math.* **35** (1980), 201–222.
17. W. L. Wendland, "Elliptic Systems in the Plane," Pitman, San Francisco/Melbourne, 1979.

Reprinted by permission of the publisher,
© 1981 by Academic Press

SHORTED MATRICES — AN EXTENDED CONCEPT AND SOME APPLICATIONS

SUJIT KUMAR MITRA* and MADAN L. PURI
Indiana University, Bloomington, Indiana, Indian Statistical Institute, New Delhi, India
Indiana University, Bloomington, Indiana

ABSTRACT

The shorting of an operator, hitherto considered by Krein [11] and by Anderson and Trapp [3] only for positive operators, is extended to rectangular matrices and square matrices not necessarily hermitian nonnegative definite. Some applications of the shorted matrix in mathematical statistics are discussed.

1. INTRODUCTION

Anderson and Duffin [2] have introduced the concept of "parallel sum" of a pair of matrices and have deduced interesting and important properties of this operation when the matrices concerned are nonnegative definite. They were led to this concept from a parallel connection of "resistors" through a vectorial generalization of Kirchhoff's and Ohm's laws in which resistors become nonnegative definite linear operators. The concept of parallel sum was extended and its elegance further demonstrated by Rao and Mitra [20],

*Research supported by National Science Foundation Grant No. MCS76-00951 at Indiana University.

†Research supported by the Air Force Office of Scientific Research, AFSC, USAF under Grant No. AFOSR 76-2927. Reproduction in whole or in part permitted for any purpose of the United States Government.

who showed that most of the properties proved by Anderson and Duffin [2] are indeed true for a much wider class of pairs of matrices, designated by these authors as "parallel summable." Similar extensions of the concept of "hybrid sum," introduced by Duffin and Trapp [9] in analogy with a hybrid connection of resistors, were made by Mitra and Trapp [18]. The object of this paper is to offer a comparable extension of the notion of a shorted operator studied by Anderson [1],[1] by Anderson and Trapp [3], and by the present authors [15]. The key point in this development is a theorem of Anderson and Trapp which exhibits the shorted n.n.d. matrix as the limit of a sequence of parallel sum matrices. We may mention that even though in this paper we confine ourselves exclusively to complex vector spaces and matrices, Theorem 4.1 and most of the discussion in Section 5, with obvious modifications, remain valid in more general fields. We also describe some applications of the shorted operator in mathematical statistics.

2. PRELIMINARIES

We use the following notations and definitions. For a positive integer n, $\mathcal{E}^n$ is the linear space of complex n-tuples. Column vector representations of vectors in $\mathcal{E}^n$ are denoted by lowercase letters such as y, v etc. $\mathcal{C}^{m\times n}$ represents the linear space of complex matrices of order $m\times n$. $\mathcal{C}_n$ represents the cone of hermitian nonnegative definite (n.n.d.) matrices in $\mathcal{C}^{n\times n}$. Matrices are denoted by capital letters such as A, B, M etc. For a matrix A, $\mathfrak{M}(A)$ denotes its column span and $\mathfrak{N}(A)$ its null space. A^-, the generalized inverse of A, is defined by the equation $AA^-A=A$ [20]. Two subspaces of a linear space are said to be virtually disjoint if they have only the null vector in common. Matrices A and B in $\mathcal{C}^{m\times n}$ are said to be disjoint if their column spans are virtually disjoint and so are their row spans [13]. For matrices $A, B\in\mathcal{C}_n$ we write $A\geqslant B$ if $A-B\in\mathcal{C}_n$.

Before we proceed any further let us record here some known results on the parallel sum.

Definition 2.1. Matrices A and B in $\mathcal{C}^{m\times n}$ are said to be parallel summable (p.s. [20]) if $A(A+B)^-B$ is invariant under the choice of the generalized inverse $(A+B)^-$. If A and B are p.s., $P(A, B)=A(A+B)^-B$ is called the parallel sum of A and B.

The following theorem is proved in [20].

---

[1] See also Krein [11].

THEOREM 2.1. *A and B are p.s. iff*

$$\mathfrak{M}(A) \subset \mathfrak{M}(A+B), \qquad \mathfrak{M}(A^*) \subset \mathfrak{M}(A^*+B^*),$$

or equivalently

$$\mathfrak{M}(B) \subset \mathfrak{M}(A+B), \qquad \mathfrak{M}(B^*) \subset \mathfrak{M}(A^*+B^*).$$

Theorem 2.2 lists certain known properties of the parallel sum [2, 14, 20].

THEOREM 2.2. *If A and B are p.s. matrices in $\mathcal{C}^{m\times n}$, then*

(a) $P(A,B)=P(B,A)$,
(b) A^* *and* B^* *are also p.s. and* $P(A^*,B^*)=[P(A,B)]^*$,
(c) $P(A,B)$ *is n.n.d. when* $m=n$ *and* A, B *are n.n.d.*,
(d) *for* C *of rank* m *in* $\mathcal{C}^{p\times m}$, CA *and* CB *are p.s. and* $P(CA,CB)=CP(A,B)$,
(e) $\{[P(A,B)]^-\}=\{A^-+B^-\}$,
(f) $\mathfrak{M}[P(A,B)]=\mathfrak{M}(A)\cap\mathfrak{M}(B)$,
(g) *if* P_* *is the orthogonal projector onto* $\mathfrak{M}(A^*)\cap\mathfrak{M}(B^*)$ *and* P *is the orthogonal projector onto* $\mathfrak{M}(A)\cap\mathfrak{M}(B)$, *then* $[P(A,B)]^+=P_*(A^-+B^-)P$,
(h) $P[P(A,B),C]=P[A,P(B,C)]$ *if all the parallel sum operations are defined*,
(i) *if* P_A *and* P_B *are the orthogonal projectors onto* $\mathfrak{M}(A)$ *and* $\mathfrak{M}(B)$, *respectively, then the orthogonal projector onto* $\mathfrak{M}(A)\cap\mathfrak{M}(B)$ *is given by* $P=2P(P_A,P_B)$.

DEFINITION 2.2. If $\mathcal{S}$ is a subspace of $\mathcal{E}^m$ and $A\in\mathcal{C}_m$, the shorted matrix $\mathcal{S}(A)$ is the unique matrix in $\mathcal{C}_m$ such that

$$\mathfrak{M}[\mathcal{S}(A)]\subset\mathcal{S},$$

$$A \geqslant \mathcal{S}(A).$$

If $C\in\mathcal{C}_m$, $A\geqslant C$, and $\mathfrak{M}(C)\subset\mathcal{S}$, then $\mathcal{S}(A)\geqslant C$.

The existence of $\mathcal{S}(A)$ was established by Anderson and Trapp [3].

THEOREM 2.3. *Let* $A,B\in\mathcal{C}_m$ *and* $\mathfrak{M}(B)=\mathcal{S}$; *then*

$$\mathcal{S}(A)=\lim_{\lambda\to 0}\frac{1}{\lambda}P(\lambda A,B).$$

Theorem 2.3 was proved by Anderson and Trapp for the special case where B is the orthogonal projector [3, Theorem 12]. The general case could be proved on the same lines. See, for example, Ando [4], which uses a similar limit to present a Lebesgue type decomposition of positive operators. Alternatively, a direct proof could be constructed using simultaneous diagonalization of the pair A, B of n.n.d. matrices (see e.g. [20, Theorem 6.2.3]).

3. THE SHORTED MATRIX—AN EXTENDED CONCEPT

In this section, as in Theorem 2.3, we shall define the generalized shorted matrix as the limit, as $\lambda \to 0$, of a sequence of matrices depending on the symbol λ. It will however be clear in Theorems 3.3 and 4.1, and more so in Section 5, that the computation of the shorted matrix does not depend on the symbolic computations with the symbol λ.

Let A, B be a pair of matrices in $\mathcal{C}^{m\times n}$, and for some nonnull complex number c, let A and cB be p.s. and

$$\lim_{\lambda\to 0} A(\lambda A+B)^{-}B=C \tag{3.1}$$

exist and be finite. Theorem 3.1 gives certain properties of the matrix C when it exists.

THEOREM 3.1.

(a) *We have*

$$C=\lim_{\lambda\to 0} B(\lambda A+B)^{-}A. \tag{3.2}$$

(b) $\mathfrak{M}(C)\subset \mathfrak{M}(A)\cap \mathfrak{M}(B)$, *and* $\mathfrak{M}(C^*)\subset \mathfrak{M}(A^*)\cap \mathfrak{M}(B^*)$.

(c) $\mathfrak{M}(A-C)$ *is virtually disjoint with* $\mathfrak{M}(B)$, *and so is* $\mathfrak{M}(A^*-C^*)$ *with* $\mathfrak{M}(B^*)$

(d) C *and* $A-C$ *are disjoint matrices* [13], *that is,*

$$\mathfrak{M}(A)=\mathfrak{M}(C)\oplus\mathfrak{M}(A-C),$$

$$\mathfrak{M}(A^*)=\mathfrak{M}(C^*)\oplus\mathfrak{M}(A^*-C^*). \tag{3.3}$$

(e) *Furthermore*

$$\mathfrak{M}(C)=\mathfrak{M}(A)\cap\mathfrak{M}(B),$$

$$\mathfrak{M}(C^*)=\mathfrak{M}(A^*)\cap\mathfrak{M}(B^*). \tag{3.4}$$

(f) *Let E be any matrix such that* $\mathfrak{M}(E)\subset\mathfrak{M}(B)$, $\mathfrak{M}(E^*)\subset\mathfrak{M}(B^*)$; *then*

$$\operatorname{rank}(A-E)\geqslant\operatorname{rank}(A-C),$$

the sign of equality holding if and only if $E=C$.

Proof. (a): If A and cB are p.s. for some nonnull c, then λA and B are p.s. for each λ sufficiently small. Hence

$$C=\lim_{\lambda\to 0}\frac{1}{\lambda}P(\lambda A, B).$$

Since $P(\lambda A, B)=P(B,\lambda A)$ by Theorem 2.2(a), (3.2) follows.

(b): Consider a typical vector $C\mathbf{x}$ in $\mathfrak{M}(C)$:

$$C\mathbf{x}=\lim_{\lambda\to 0}\mathbf{y}_\lambda,$$

where $\mathbf{y}_\lambda=A(\lambda A+B)^-B\mathbf{x}\in\mathfrak{M}(A)$. Since $\mathfrak{M}(A)$ is closed,

$$C\mathbf{x}=\lim_{\lambda\to 0}\mathbf{y}_\lambda\in\mathfrak{M}(A).$$

Similarly, using (3.2) we have $C\mathbf{x}\in\mathfrak{M}(B)$ and hence $C\mathbf{x}\in\mathfrak{M}(A)\cap\mathfrak{M}(B)$. The other part of (b) is established in a like manner.

(c): Let $(A-C)\mathbf{u}=B\mathbf{v}$ be a vector in $\mathfrak{M}(A-C)\cap\mathfrak{M}(B)$. Then $A\mathbf{u}=C\mathbf{u}+B\mathbf{v}$. Hence

$$C\mathbf{u}=\lim_{\lambda\to 0}B(\lambda A+B)^-A\mathbf{u}=\lim_{\lambda\to 0}B(\lambda A+B)^-(C\mathbf{u}+B\mathbf{v})$$

$$\Rightarrow\quad\lim_{\lambda\to 0}B(\lambda A+B)^-B\mathbf{v}=\lim_{\lambda\to 0}\lambda A(\lambda A+B)^-C\mathbf{u}.\quad(3.5)$$

Since $C=BK$ for some matrix K in $\mathcal{C}^{n\times n}$, the R.H.S. of (3.5) is seen to be equal to the null vector, while

$$B=B(\lambda A+B)^-(\lambda A+B)$$

for each λ sufficiently small implies, on taking limits of both sides, that

$$B=\lim_{\lambda\to 0}B(\lambda A+B)^-B.$$

Hence $B\mathbf{v}=\mathbf{0}$ and the first part of (c) is established. The proof of the second part is similar.

(d): (d) is a simple consequence of (c).

(e): Let $\mathbf{x}\in\mathfrak{M}(A)\cap\mathfrak{M}(B)$. Using (3.3), we write

$$\mathbf{x}=\mathbf{x}_1+\mathbf{x}_2,$$

where $\mathbf{x}_1\in\mathfrak{M}(C)$ and $\mathbf{x}_2\in\mathfrak{M}(A-C)$. Observe that $\mathbf{x}_2=\mathbf{x}-\mathbf{x}_1\in\mathfrak{M}(B)$ and hence $\in\mathfrak{M}(B)\cap\mathfrak{M}(A-C)$. This implies $\mathbf{x}_2=\mathbf{0}$ and $\mathbf{x}=\mathbf{x}_1\in\mathfrak{M}(C)$. This, in view of (b), establishes the first part of (e). The other part is similarly deduced.

(f): In view of (c), the expression

$$A-E=(C-E)+(A-C).$$

exhibits $A-E$ as the sum of two disjoint matrices $C-E$ and $A-C$. Hence

$$\operatorname{rank}(A-E)=\operatorname{rank}(C-E)+\operatorname{rank}(A-C).$$

This concludes the proof of (f) and of Theorem 3.1. ■

The matrix C will henceforth be called the matrix A shorted by the matrix B and denoted by $S(A|B)$. Since $S(A|B)=S(A|dB)$ for each nonnull complex number d, we shall hereafter assume without any loss of generality that the matrix B is such that A and B are p.s. Theorem 3.2 gives some more properties of the shorted matrix.

THEOREM 3.2. *Let $S(A|B)$ exist. The following holds:*

(a) *$S(A^*|B^*)$ also exists and $S(A^*|B^*)=[S(A|B)]^*$,*

(b) *if $K\in\mathcal{C}^{p\times m}$ and $\operatorname{rank} K=m$, then $S(KA|KB)$ exists and $S(KA|KB)=KS(A|B)$,*

(c) *if $m=n$, A is n.n.d., and further*

$$\mathfrak{M}(B)\cap\mathfrak{M}(A)=\mathfrak{M}(B^*)\cap\mathfrak{M}(A), \tag{3.6}$$

then $S(A|B)$ is n.n.d.

Proof. Proofs of (a) and (b) are straightforward and are omitted.

(c): Observe that by part (a) if $C=S(A|B)$, then $C^*=S(A|B^*)$. Further, if (3.6) holds, both C and C^* have identical row and column spans. Application of Theorem 3.1(f) now shows $C=C^*$.

Choose and fix a hermitian g-inverse A^- of A. If for some $\mathbf{x}$, $\mathbf{x}^*C\mathbf{x}<0$, then

$$\mathbf{x}^*CA^-AA^-C\mathbf{x}=\mathbf{x}^*C\mathbf{x}<0.$$

The equality follows from the fact that A^-AA^- is a g-inverse of A and, A being the sum of disjoint matrices C and $A-C$, every g-inverse of A is a g-inverse of C. This contradicts the assumption that A is n.n.d., and establishes the claim in (c). ■

That (c) is not true in general can be seen from the following counterexample. Let I and H be the identity matrix and an idempotent matrix in $\mathcal{C}^{m\times m}$. It is not difficult to see that $S(I|H)$ exists and is equal to H. H need not be hermitian, and it is conceivable that a vector $\mathbf{x}$ might exist such that $\mathbf{x}^*H\mathbf{x}<0$. Consider the idempotent matrix

$$H=\begin{pmatrix} 0 & 1 \\ 0 & 1 \end{pmatrix}$$

for such a counterexample.

Remarks. A matrix $A\in\mathcal{C}^{m\times m}$ is said to be almost positive definite (a.p.d.) [8, 12] if $\forall\mathbf{x}\in\mathcal{E}^m$, $\operatorname{Re}(\mathbf{x}^*A\mathbf{x})\geqslant 0$ and $\mathbf{x}^*A\mathbf{x}=0\Rightarrow A\mathbf{x}=0$. Unlike n.n.d. matrices, an a.p.d. matrix need not be hermitian. Similarly to Theorem 3.2(c), we can prove that if A is a.p.d., $S(A|B)$ exists, and (3.6) holds, then $S(A|B)$ is a.p.d.

This can be proved as follows. Since A is a.p.d., it follows as in [12, Theorem 2] that A is an EP matrix [that is, $\mathfrak{M}(A)=\mathfrak{M}(A^*)$]. Equations (3.4) and (3.6) therefore imply that $C=S(A|B)$ is EP. Further, as in Corollaries 2 and 3 of [12], A^+ is seen to be a.p.d. Since C is EP,

$$C[I-A^+C]=0 \quad\Rightarrow\quad C^*[I-A^+C]=0 \quad\Rightarrow\quad C^*=C^*A^+C.$$

If $\operatorname{Re}(\mathbf{x}^*C\mathbf{x})<0$, then $\operatorname{Re}(\mathbf{x}^*C^*\mathbf{x})=\operatorname{Re}(\mathbf{x}^*C^*A^+C\mathbf{x})<0$, which contradicts almost positive definiteness of A^+. Also,

$$\mathbf{x}^*C\mathbf{x}=0 \quad\Rightarrow\quad A^+C\mathbf{x}=0 \quad\Rightarrow\quad C\mathbf{x}=0.$$

A matrix $A\in\mathcal{C}^{m\times m}$ is said to be positive semidefinite (p.s.d.) if $\forall\mathbf{x}\in\mathcal{E}^m$, $\operatorname{Re}(\mathbf{x}^*A\mathbf{x})\geqslant 0$. When A is p.s.d. and (3.6) holds, $S(A|B)$ if it exists is also p.s.d. This can be proved on similar lines.

Let the matrices A and B be p.s. and $A+B$ be of rank r. Consider a rank factorization of $A+B$:

$$A+B=LR,$$

and the representations

$$A=LDR, \qquad B=L(I-D)R \tag{3.7}$$

implied by parallel summability, where D and $I-D$ are square matrices in $\mathcal{C}^{r\times r}$. It is easily seen that in any such representation, the matrix D (and naturally $I-D$) is uniquely determined up to a similarity transformation. Theorem 3.3 gives a necessary and sufficient condition for $S(A|B)$ to exist.

THEOREM 3.3. *Let A and B be p.s. Then $S(A|B)$ exists iff $I-D$ is of Drazin index* 1, *that is*,

$$\operatorname{rank}(I-D)^2=\operatorname{rank}(I-D), \tag{3.8}$$

or equivalently,

$$\operatorname{rank} B(A+B)^- B=\operatorname{rank} B. \tag{3.9}$$

Proof. Without any loss of generality we assume that D is already in Jordan canonical form, and write D as the sum of two disjoint matrices D_1 and D_2, each of order $r\times r$, where D_1 is identical with D everywhere including all its diagonal Jordan blocks, except for the Jordan block corresponding to the eigenvalue 1, if any, which is replaced by a null matrix; and $D_2=D-D_1$. The above assumption (3.8) implies that $I-D$ and $\lambda D_1+(I-D)$ are disjoint with D_2. Hence

$$D_2[\lambda D_1+(I-D)+\lambda D_2]^-[\lambda D_1+(I-D)]=0$$

$$\Rightarrow \quad D_2[\lambda D_1+(I-D)+\lambda D_2]^-(I-D)=0,$$

and a g-inverse of $\lambda D_1+(I-D)+\lambda D_2$ is also a g-inverse of $\lambda D_1+(I-D)$.

Thus

$$
\begin{aligned}
A(\lambda A+B)^{-}B &= LD(\lambda D+I-D)^{-}(I-D)R \\
&= L(D_1+D_2)[\lambda D_1+(I-D)+\lambda D_2]^{-}(I-D)R \\
&= LD_1[\lambda D_1+(I-D)+\lambda D_2]^{-}(I-D)R \\
&= LD_1[\lambda D_1+(I-D)]^{-}(I-D)R \\
&= LD_1R-\lambda LD_1[\lambda D_1+(I-D)]^{-}D_1R.
\end{aligned}
$$

Taking the limit as $\lambda\to 0$, we have

$$\lim_{\lambda\to 0} A(\lambda A+B)^{-}B=LD_1R.$$

This concludes the proof of the "if" part.

Assume now that (3.1) holds, and consider the representation

$$A+B=(B+C)+(A-C),$$

where $B+C$ and $A-C$ are disjoint matrices. Since A and B are p.s., $B+C$ has the same row and column spans as that of B. Consider rank factorizations of $B+C$ and $A-C$:

$$
\begin{aligned}
B+C &= L_1R_1, \\
A-C &= L_2R_2,
\end{aligned}
$$

leading to the rank factorization $A+B=LR$, where

$$L=\begin{pmatrix} L_1 & L_2 \end{pmatrix} \quad \text{and} \quad R=\begin{pmatrix} R_1 \\ R_2 \end{pmatrix}.$$

If $B=L_1FR_1$, clearly F is nonsingular. Hence $I-D=\mathrm{diag}(F,0)$ is of Drazin index 1. The last part of Theorem 3.3 is trivial. This concludes the proof of Theorem 3.3. ∎

THEOREM 3.4. *If* $S(A|B)$ *is defined, then* $S[A|S(A|B)]$ *is also defined and*

$$S[A|S(A|B)] = S(A|B) \tag{3.10}$$

Proof. The proof is fairly straightforward and is omitted. ■

THEOREM 3.5. *A general solution to a g-inverse of* $S(A|B)$ *is* $A^- + X_b$, *where* A^- *is an arbitrary g-inverse of* A, *and* X_b *is any arbitrary solution of the homogeneous equation*

$$BX_bB = 0. \tag{3.11}$$

Proof. Every g-inverse of A is a g-inverse of $S(A|B)$. This we have already noted while proving Theorem 3.2. Consider now a matrix $A^- + X_b$ as determined above:

$$S(A|B)(A^- + X_b)S(A|B) = S(A|B) + S(A|B)X_bS(A|B) = S(A|B)$$

on account of Theorem 3.1(b). This shows

$$\{A^- + X_b\} \subset \{[S(A|B)]^-\}.$$

Now choose and fix a g-inverse G of A. A general solution to a g-inverse of $S(A|B)$ is $G + X_s$, where X_s is a general solution to the homogeneous equation

$$S(A|B)X_sS(A|B) = 0.$$

Theorem 3.1(e) implies that any such matrix X_s can be written as $X_s = X_a + X_b$ where X_a and X_b satisfy respectively the equations

$$AX_aA = 0, \qquad BX_bB = 0.$$

The matrix $G + X_a \in \{A^-\}$. Hence

$$\{[S(A|B)]^-\} \subset \{A^- + X_b\}.$$

Theorem 3.5 is thus proved. ■

THEOREM 3.6.

(a) *If* $S(A|B)$ *and* $S(B|A)$ *are both defined, then* $S(A|B)$ *and* $S(B|A)$ *are p.s. and*

$$P[S(A|B),S(B|A)]=P(A,B). \tag{3.12}$$

(b) *Furthermore,*

$$\{[S(A|B)]^{-}+[S(B|A)]^{-}\}=\{A^{-}+B^{-}\}. \tag{3.13}$$

Proof. (a): From Theorem 3.3 it is seen that $S(B|A)$ will exist iff rank $D^2=$rank D. Let $(I-D)_1$ denote a matrix which is identical with $I-D$ everywhere except for the diagonal block corresponding to the zero eigenvalue of D, which is replaced by a null matrix. As in the proof of Theorem 3.3, it is seen that

$$S(B|A)=\lim_{\lambda\to 0}A(\lambda B+A)^{-}B=L(I-D)_1R.$$

When both $S(A|B)$ and $S(B|A)$ exist, it is seen that D_1 and $(I-D)_1$, which are both block diagonal, have nonnull diagonal blocks at identical positions, each such nonnull pair adding up to an identity matrix of the same order as that of the diagonal block concerned. This shows that $S(A|B)$ and $S(B|A)$ are parallel summable and

$$\begin{aligned}P[S(A|B),S(B|A)]&=LD_1(I-D)_1R=LD(I-D)R\\&=P(A,B).\end{aligned}$$

(b): Clearly $\{A^{-}+B^{-}\}\subset\{[S(A|B)]^{-}+[S(B|A)]^{-}\}$. Conversely, by Theorem 3.5, $[S(A|B)]^{-}$ can be written as G_a+X_b where $G_a\in\{A^{-}\}$ and $BX_bB=0$. Similarly $[S(B|A)]^{-}$ can be written as G_b+X_a where $G_b\in\{B^{-}\}$ and $AX_aA=0$. Hence

$$\begin{aligned}[S(A|B)]^{-}+[S(B|A)]^{-}&=G_a+X_b+G_b+X_a\\&=(G_a+X_a)+(G_b+X_b).\end{aligned}$$

This shows $\{[S(A|B)]^{-}+[S(B|A)]^{-}\}\subset\{A^{-}+B^{-}\}$, which concludes the proof of part (b) and of Theorem 3.6. ∎

THEOREM 3.7.

$$S[P(A,B)|C]=P[S(A|C),B]=P[S(B|C),A] \tag{3.14}$$

when the parallel sum and shorted matrices involved are defined.

Proof. Using Theorem 3.5 and Theorem 2.2(e),

$$\{(S[P(A,B)|C])^{-}\}=\{A^{-}+B^{-}+X_c\},$$

$$\{(P[S(A|C),B])^{-}\}=\{A^{-}+X_c+B^{-}\},$$

$$\{(P[S(B|C),A])^{-}\}=\{B^{-}+X_c+A^{-}\}.$$

This shows the three matrices in (3.14) have identical general solutions for a g-inverse. Since a matrix is uniquely determined by its class of g-inverses [20, Theorem 2.4.2], (3.14) is established. ■

The following theorem can also be proved using a similar argument. We omit the proof.

THEOREM 3.8.

$$\begin{aligned} S[S(A|B)|C]&=S[A|S(B|C)]=S[A|S(C|B)] \\ &=S[A|P(B,C)] \end{aligned} \tag{3.15}$$

when the parallel sum and shorted matrices involved are defined.

4. ANOTHER APPROACH

In view of Theorem 3.1(f), one is tempted to put forward the following definition of the shorted matrix, imitating Anderson and Trapp's definition given in Section 2.

Let A be a given matrix in $\mathcal{C}^{m\times n}$ and $\mathcal{S},\mathcal{T}$ be given subspaces in $\mathcal{E}^m,\mathcal{E}^n$ respectively. The shorted matrix $S(A|\mathcal{S},\mathcal{T})$ is a matrix C in $\mathcal{C}^{m\times n}$ such that

(a) We have

$$\mathcal{M}(C)\subset\mathcal{S},\qquad \mathcal{M}(C^*)\subset\mathcal{T}, \tag{4.1}$$

(b) if $E\in\mathcal{C}^{m\times n}$, $\mathfrak{M}(E)\subset\mathcal{S}$, and $\mathfrak{M}(E^*)\subset\mathcal{T}$, then

$$\operatorname{rank}(A-E)\geqslant\operatorname{rank}(A-C). \tag{4.2}$$

This definition however does not always lead to a unique answer. Consider for example the matrix

$$A=\begin{pmatrix}1&0&0\\1&1&1\\0&1&0\end{pmatrix},$$

and let

$$\mathcal{S}=\mathcal{T}=\mathfrak{M}\begin{pmatrix}1&0\\0&1\\0&0\end{pmatrix}.$$

It is seen that for arbitrary scalars a and b the matrix

$$\begin{pmatrix}1&0&0\\a&b&0\\0&0&0\end{pmatrix}$$

satisfies the conditions required of the matrix C in the above definition. The following theorem gives necessary and sufficient conditions on the triplet $(A,\mathcal{S},\mathcal{T})$ so that $\mathrm{S}(A|\mathcal{S},\mathcal{T})$ may exist uniquely.

Let $\mathcal{S}$ and $\mathcal{T}$ be column spans of matrices L_1 and R_1^* respectively, and let the columns of L_2 and R_2^* span respectively complementary subspaces of $\mathcal{S}$ in $\mathcal{E}^m$ and of $\mathcal{T}$ in $\mathcal{E}^n$. We assume that L_1, L_2, R_1^*, R_2^*, are of full column rank. Let us write

$$A=\begin{pmatrix}L_1 & L_2\end{pmatrix}\begin{pmatrix}W_{11}&W_{12}\\W_{21}&W_{22}\end{pmatrix}\begin{pmatrix}R_1\\R_2\end{pmatrix}. \tag{4.3}$$

THEOREM 4.1. *The shorted matrix* $\mathrm{S}(A|\mathcal{S},\mathcal{T})$ *exists and is unique iff*

$$\mathfrak{M}(W_{21})\subset\mathfrak{M}(W_{22}),\qquad \mathfrak{M}(W_{12}^*)\subset\mathfrak{M}(W_{22}^*). \tag{4.4}$$

When (4.4) *is satisfied,*

$$\mathrm{S}(A|\mathcal{S},\mathcal{T})=L_1(W_{11}-W_{12}W_{22}^-W_{21})R_1.$$

Proof. The required conditions are seen to be independent of the specific choice of matrices L_1, L_2, R_1, and R_2 in the sense that if the conditions are met for one choice, they would also be met for an alternative choice.

Assume now that (4.4) holds, and write

$$\begin{aligned}
A &= L_1W_{11}R_1 + L_1W_{12}R_2 + L_2W_{21}R_1 + L_2W_{22}R_2 \\
&= L_1(W_{11} - W_{12}W_{22}^{-}W_{21})R_1 + (L_1W_{12} + L_2W_{22})(R_2 + W_{22}^{-}W_{21}R_1) \\
&= L_1(W_{11} - W_{12}W_{22}^{-}W_{21})R_1 + (L_1W_{12}W_{22}^{-} + L_2)(W_{21}R_1 + W_{22}R_2) \\
&= A_1 + A_2, \qquad \text{say.}
\end{aligned} \tag{4.5}$$

Clearly $\mathfrak{M}(A_1)\subset\mathcal{S}$, $\mathfrak{M}(A_1^*)\subset\mathcal{T}$. We shall show that $\mathfrak{M}(A_2)$ is virtually disjoint with $\mathcal{S}$, and $\mathfrak{M}(A_2^*)$ with $\mathcal{T}$. Let

$$\begin{aligned}
A_2\mathbf{x} &= (L_1W_{12} + L_2W_{22})(R_2 + W_{22}^{-}W_{21}R_1)\mathbf{x} \\
&= (L_1W_{12} + L_2W_{22})\mathbf{y}
\end{aligned}$$

be a vector in $\mathfrak{M}(A_2)\cap\mathfrak{M}(L_1)$. Then

$$\begin{aligned}
L_2W_{22}\mathbf{y} = \mathbf{0} \quad &\Rightarrow \quad W_{22}\mathbf{y} = \mathbf{0} \\
&\Rightarrow \quad W_{12}\mathbf{y} = \mathbf{0} \qquad \text{on account of (4.4)} \\
&\Rightarrow \quad A_2\mathbf{x} = \mathbf{0}.
\end{aligned}$$

That $\mathfrak{M}(A_2^*)$ is virtually disjoint with $\mathcal{T}$ is similarly established. That the matrix A_1 satisfies also the condition (4.2) required of the matrix C and is the unique matrix to do so follows as in the proof of Theorem 3.1(f).

Conversely, suppose that there exists a unique matrix C satisfying (4.1) and (4.2). This implies that $\mathfrak{M}(A-C)$ is virtually disjoint with $\mathcal{S}$, and $\mathfrak{M}(A^*-C^*)$ with $\mathcal{T}$. Let $C=L_{11}R_{11}$, $A-C=L_{21}R_{21}$ be rank factorizations of C and $A-C$ respectively. Let the columns of $L_1=(L_{11}\ L_{12})$ provide a basis for $\mathcal{S}$, and those of $R_1^*=(R_{11}^*\ R_{12}^*)$ a basis for $\mathcal{T}$. Further, let the columns of $(L_1\ L_{21}\ L_{22})$ form a basis of $\mathcal{E}^m$, and those of $(R_1^*\ R_{21}^*\ R_{22}^*)$ a basis for $\mathcal{E}^n$. Let us write L_2 for the matrix $(L_{21}\ L_{22})$ and R_2 for the matrix $(R_{21}\ R_{22})$. Clearly by construction the matrices C and $A-C$ are of the form $C=L_1W_{11}R_1$, $A-C=L_2W_{22}R_2$, which shows that in a representation of the type (4.3), $W_{12}=0$, $W_{21}=0$, and the condition (4.4) is trivially satisfied. ■

In [18], Mitra and Trapp defined the generalized shorted operator as the strong hybrid sum of A with the null matrix. Theorem 4.1 is closely related to this definition. The reader is also referred to Theorem 2 in Carlson [6], which considers decompositions of the matrix A with a somewhat different emphasis.

When A is a square matrix (that is, $m=n$ and $\mathcal{S}=\mathcal{T}$), the condition (4.4) is seen to be equivalent to the condition that A is $\mathcal{S}^{\perp}$-complementable (Ando [5]), where $\mathcal{S}^{\perp}$ denotes the orthogonal complement of $\mathcal{S}$. Further,

$$\mathrm{S}(A|\mathcal{S},\mathcal{S})=A_{/\mathcal{S}^{\perp}},$$

the generalized Schur complement. The verification is fairly straightforward. For the case $m\neq n$, the notion could be extended as follows: Let $\mathfrak{M},\mathfrak{N}$ be given subspaces in $\mathcal{E}^m,\mathcal{E}^n$ respectively and following Ando's notation let $I_{\mathfrak{M}}$ and $I_{\mathfrak{N}}$ denote respectively the orthogonal projectors onto $\mathfrak{M}$ and $\mathfrak{N}$ under the respective dot products.

Definition. The matrix A is said to be $\mathfrak{M},\mathfrak{N}$-complementable if there exist matrices $M_l\in\mathcal{C}^{m\times m}$, $N_r\in\mathcal{C}^{n\times n}$ such that

$$M_l I_{\mathfrak{M}}=M_l,\qquad I_{\mathfrak{N}}N_r=N_r, \tag{4.6}$$

$$I_{\mathfrak{M}}AN_r=I_{\mathfrak{M}}A,\qquad M_lAI_{\mathfrak{N}}=AI_{\mathfrak{N}}. \tag{4.7}$$

When (4.6) and (4.7) are satisfied, we have

$$M_lA=M_lAN_r=AN_r. \tag{4.8}$$

M_lAN_r, which clearly depends only on $\mathfrak{M}$, $\mathfrak{N}$, and A, is called the Schur compression of A and denoted by $A_{\mathfrak{M},\mathfrak{N}}$. The matrix $A-M_lAN_r$ is called the generalized Schur complement of A and denoted by $A_{/\mathfrak{M},\mathfrak{N}}$.

Theorem 4.2. *Let A be $\mathfrak{M},\mathfrak{N}$-complementable. Then*

$$A_{/\mathfrak{M},\mathfrak{N}}=\mathrm{S}(A|\mathfrak{M}^{\perp},\mathfrak{N}^{\perp}). \tag{4.9}$$

Proof. Observe that

$$I_{\mathfrak{M}}(A-M_lAN_r)=I_{\mathfrak{M}}(A-AN_r)=0,$$

$$(A-M_lAN_r)I_{\mathfrak{N}}=(A-M_lA)I_{\mathfrak{N}}=0.$$

These show that

$$\mathcal{M}(A_{/\mathcal{M},\mathcal{N}}) \subset \mathcal{M}^{\perp}, \qquad \mathcal{M}[(A_{/\mathcal{M},\mathcal{N}})^*] \subset \mathcal{N}^{\perp}.$$

Further, if

$$M_l A N_r a = A N_r a \in \mathcal{M}(A_{\mathcal{M},\mathcal{N}}) \cap \mathcal{M}^{\perp},$$

$AN_ra = M_l I_{\mathcal{M}} AN_r a = 0$. This shows $\mathcal{M}(A_{\mathcal{M},\mathcal{N}})$ is virtually disjoint with $\mathcal{M}^{\perp}$. Similarly $\mathcal{M}[(A_{\mathcal{M},\mathcal{N}})^*]$ is seen to be virtually disjoint with $\mathcal{N}^{\perp}$. Hence $A_{/\mathcal{M},\mathcal{N}}$ is seen to be the unique matrix satisfying (4.1) and (4.2) with $\mathcal{S} = \mathcal{M}^{\perp}, \mathcal{T} = \mathcal{N}^{\perp}$. ■

For an application of Theorem 4.1 consider the following problem. Let $A, B \in \mathcal{C}^{m\times n}$, and Λ denote the matrix

$$\Lambda = \begin{pmatrix} A & A \\ A & A+B \end{pmatrix}.$$

Put

$$\mathcal{S} = \mathcal{M}\begin{pmatrix} I_m \\ 0 \end{pmatrix}, \qquad \mathcal{T} = \mathcal{M}\begin{pmatrix} I_n \\ 0 \end{pmatrix}.$$

By Theorem 4.1, $S(\Lambda|\mathcal{S},\mathcal{T})$ exists uniquely iff $\mathcal{M}(A) \subset \mathcal{M}(A+B)$, $\mathcal{M}(A^*) \subset \mathcal{M}(A^*+B^*)$, that is, if A and B are p.s. Further, if this condition is satisfied,

$$S(\Lambda|\mathcal{S},\mathcal{T}) = \begin{pmatrix} A - A(A+B)^{-}A & 0 \\ 0 & 0 \end{pmatrix} = \begin{pmatrix} P(A,B) & 0 \\ 0 & 0 \end{pmatrix}. \qquad (4.10)$$

Anderson and Trapp [3] have used (4.10) to define the parallel sum of hermitian n.n.d. matrices through the concept of a shorted operator.

5. COMPUTATION OF THE SHORTED MATRIX

Let $A \in \mathcal{C}^{m\times n}$, $X \in \mathcal{C}^{m\times p}$, $Y \in \mathcal{C}^{q\times n}$. Let F denote the matrix $\begin{pmatrix} A & X \\ Y & 0 \end{pmatrix}$, where 0 is the null matrix in $\mathcal{C}^{q\times p}$. The following result can easily be established. We omit the proof.

LEMMA 5.1. *Let $\mathcal{S}=\mathfrak{M}(X)$, $\mathcal{T}=\mathfrak{M}(Y^*)$. Then the condition* (4.4) *is equivalent to each of the following conditions*:

$$\operatorname{rank} F=\operatorname{rank}(A \quad X)+\operatorname{rank} Y=\operatorname{rank}\begin{pmatrix} A \\ Y \end{pmatrix}+\operatorname{rank} X, \tag{5.1}$$

$$\mathfrak{M}\begin{pmatrix} A \\ 0 \end{pmatrix}\subset\mathfrak{M}(F), \qquad \mathfrak{M}\begin{pmatrix} A^* \\ 0 \end{pmatrix}\subset\mathfrak{M}(F^*). \tag{5.2}$$

Let

$$G=\begin{pmatrix} C_1 & C_2 \\ C_3 & -C_4 \end{pmatrix}\in\{F^-\}.$$

LEMMA 5.2. *If the condition* (5.1) *or the equivalent* (5.2) *holds, then*

(i) $XC_3X=X$, $YC_2Y=Y$,
(ii) $YC_1X=0$, $AC_1X=0$, $YC_1A=0$,
(iii) $AC_2Y=XC_3A=XC_4Y$,
(iv) $AC_1AC_1A = AC_1A$, $\operatorname{Tr} AC_1 = \operatorname{rank}(A \;\; X)-\operatorname{rank} X = \operatorname{rank}\begin{pmatrix} A \\ Y \end{pmatrix} -$ $\operatorname{rank} Y$,
(v) $\begin{pmatrix} C_1 \\ C_3 \end{pmatrix}\in\{(A \;\; X)^-\}$, $(C_1 \;\; C_2)\in\left\{\begin{pmatrix} A \\ Y \end{pmatrix}^-\right\}$.

Proof. Lemma 5.2 except for the second part of (iv) follows from the following equations:

$$FGF=F, \tag{5.3a}$$

$$FG\begin{pmatrix} A \\ 0 \end{pmatrix}=\begin{pmatrix} A \\ 0 \end{pmatrix}, \tag{5.3b}$$

$$(A \quad 0)GF=(A \quad 0), \tag{5.3c}$$

the last two equations being consequences of (5.2).

Since $G\in\{F^-\}$, $C_3\in\{X^-\}$, $C_2\in\{Y^-\}$,

$$\operatorname{rank} F=\operatorname{rank} FG=\operatorname{Tr} FG=\operatorname{Tr}(AC_1+XC_3)+\operatorname{Tr} YC_2$$

$$=\operatorname{Tr} AC_1+\operatorname{rank} X+\operatorname{rank} Y.$$

The second part of (iv) therefore follows from (5.1). ■

THEOREM 5.1. *Let* $\mathcal{S}=\mathcal{M}(X)$, $\mathcal{T}=\mathcal{M}(Y^*)$. *If the condition* (5.1) *or equivalently* (5.2) *holds, the matrix*

$$XC_4Y=AC_2Y=XC_3A \tag{5.4}$$

is the unique shorted matrix $S(A|\mathcal{S},\mathcal{T})$.

Proof. (5.3b) $\Rightarrow$

$$AC_1A+XC_3A=A. \tag{5.5}$$

Since $A-XC_3A=AC_1A$, it suffices to show that

$$\mathcal{M}(AC_1A)\cap\mathcal{M}(X)=\{0\}, \tag{5.6a}$$

$$\mathcal{M}[(AC_1A)^*]\cap\mathcal{M}(Y^*)=\{0\}. \tag{5.6b}$$

If $AC_1Aa=Xb\in\mathcal{M}(AC_1A)\cap\mathcal{M}(X)$, then

$$AC_1Aa=AC_1AC_1Aa=AC_1Xb=0 \quad\Rightarrow\quad (5.6a).$$

Equation (5.6b) is similarly established. The rest of the proof of Theorem 5.1 is similar to the proof of the "if" part of Theorem 4.1. ∎

REMARK 1. Lemma 5.2 for the special case where A is a real symmetric n.n.d. matrix and $Y=X'$ was proved by Rao [19]. That C_2 in such a case is a minimum A seminorm g-inverse of Y was shown in [20, Corollary 1, p. 47]. Using Theorem 2.1 of Mitra and Puri [15], it is seen that $AC_2Y=\mathcal{S}(A)$, where $\mathcal{S}=\mathcal{M}(X)$. It is remarkable that in the general case, the same formula also provides the shorted matrix $S(A|\mathcal{S},\mathcal{T})$, when in no conceivable way can C_2 be interpreted a minimum A seminorm g-inverse of Y.

REMARK 2. The formula (5.4) appears to be more direct and therefore simpler to compute than the expressions given in Theorems 3.3 and 4.1, as it does not require the rank factorization of $A+B$ and determination of the matrix D in (3.7), or the identification of complementary subspaces of $\mathcal{S}$ and $\mathcal{T}$ and determination of the W_{ij} matrices in (4.3).

Similar to [15, Theorem 2.4(a)] we have the following theorem.

THEOREM 5.2. *If* $m=n$ *and the matrix* A *is idempotent, so also is the unique shorted matrix* $S(A|\mathcal{S},\mathcal{T})$ *in Theorem* 5.1.

Proof. $(XC_3A)^2 = XC_3A^2C_2Y = XC_3AC_2Y = XC_3XC_3A = XC_3A.$ ■

THEOREM 5.3. *Let the matrices A and B in $\mathcal{C}^{m\times n}$ be p.s., and in addition let* (3.8) *or equivalently* (3.9) *hold. Then the matrix*

$$F = \begin{pmatrix} A & B \\ B & 0 \end{pmatrix}$$

satisfies the condition (5.1), *and if*

$$G = \begin{pmatrix} C_1 & C_2 \\ C_3 & -C_4 \end{pmatrix} \in \{F^-\},$$

then $AC_2B = S(A|B)$.

Proof.

$$\operatorname{rank} F = \operatorname{rank}\begin{pmatrix} A+B & B \\ B & 0 \end{pmatrix} = \operatorname{rank}\begin{pmatrix} A+B & B \\ 0 & -B(A+B)^-B \end{pmatrix},$$

since $\mathfrak{M}(B^*) \subset \mathfrak{M}(A^* + B^*)$. The R.H.S. is further equal to

$$\begin{aligned} \operatorname{rank}(A+B) + \operatorname{rank}\left[B(A+B)^-B\right] &= \operatorname{rank}(A \quad B) + \operatorname{rank} B \\ &= \operatorname{rank}\begin{pmatrix} A \\ B \end{pmatrix} + \operatorname{rank} B. \end{aligned}$$

The second part of Theorem 5.3 follows from Theorem 5.1. ■

6. SOME APPLICATIONS OF THE SHORTED MATRIX

A. *Recovery of Interblock Information in Incomplete Block Experiments* [7, 10]

Consider a pair of consistent linear equations

$$Ax = a, \tag{6.1}$$

$$Bx = b \tag{6.2}$$

and the combined equation

$$(A+B)x = a+b. \tag{6.3}$$

We shall assume that $\mathfrak{M}(A) \subset \mathfrak{M}(A+B)$, so that the equations (6.3) may be consistent whenever (6.1) and (6.2) are so. This condition is satisfied for example when A and B are p.s. matrices. The linear function p^*x assumes a unique value for every solution x of (6.1) iff

$$p \in \mathfrak{M}(A^*). \tag{6.4}$$

Among such linear functions we are interested in identifying those for which substitution of a solution of (6.3) or of (6.1) leads to identical answers. Such problems crop up in the theory of recovery of interblock information in incomplete block experiments, where (6.1) and (6.2) are respectively the normal equations for deriving intra- and interblock estimates and (6.3) is the normal equation for deriving combined intra-interblock estimates. When $S(A|B)$ exists, a neat answer is provided by Theorem 6.1.

THEOREM 6.1. *If* $S(A|B)=C$ *exists, then*

$$p^*(A+B)^-(a+b)=p^*A^-a \qquad \forall a \in \mathfrak{M}(A),\, b \in \mathfrak{M}(B) \tag{6.5}$$

iff

$$p \in \mathfrak{M}(A^*-C^*). \tag{6.6}$$

Proof. "If" part: We note first that $\mathfrak{M}(A^*-C^*) \subset \mathfrak{M}(A^*)$. Let x_0 satisfy (6.1). Then

$$\begin{aligned}(A-C)(A+B)^-(a+b)\\ &=(A-C)(B+C+A-C)^-[(A+B)x_0+b-Bx_0]\\ &=(A-C)(B+C+A-C)^-(A+B)x_0,\end{aligned}$$

since $B+C$ and $A-C$ are disjoint matrices and $b-Bx_0 \in \mathfrak{M}(B)=\mathfrak{M}(B+C)$. The R.H.S. further simplifies to

$$(A-C)(A+B)^-(A+B)x_0=(A-C)x_0,$$

since $\mathfrak{M}(A^*-C^*) \subset \mathfrak{M}(A^*) \subset \mathfrak{M}(A^*+B^*)$.

"Only if" part:

$$(6.5) \quad \Rightarrow \quad p^*(A+B)^- b=0 \quad \forall b\in\mathfrak{M}(B)$$

$$\Rightarrow \quad p^*(A+B)^- B=0; \tag{6.7}$$

$$(6.4) \quad \Rightarrow \quad p^*=\lambda^* A \quad \text{for some } \lambda\in\mathcal{E}^m. \tag{6.8}$$

Substituting (6.8) in (6.7), we have

$$\lambda^* P(A, B)=0. \tag{6.9}$$

Since $\mathfrak{M}[P(A, B)]=\mathfrak{M}[S(A|B)]$ and $A^-\in\{C^-\}$, it follows that

$$(6.9) \quad \Rightarrow \quad \lambda^* C=0$$

$$\Rightarrow \quad \lambda^*=\mu^*[I-CA^-] \quad \text{for some } \mu\in\mathcal{E}^m. \tag{6.10}$$

Hence $p^*=\lambda^* A=\mu^*[I-CA^-]A=\mu^*(A-C) \Leftrightarrow$ (6.6). This concludes the proof of Theorem 6.1. ∎

b. Test of Linear Hypotheses in Linear Models

Let the random vector $\mathbf{Y}\sim N_n(X\beta, \sigma^2 I)$, where β, an m-tuple, is an unknown parameter vector, and $\sigma^2>0$ is also an unknown parameter. X is a known matrix.

Consider a hypothesis

$$H_0: H\beta = h. \tag{6.11}$$

We shall assume that the equation (6.11) is consistent, as otherwise the hypothesis could be rejected without any formal statistical test. It was shown in [16] that when $H\beta$ is not estimable, only the estimable part of (6.11) could be tested. To be more precise, let K be a matrix such that

$$\mathfrak{M}(H'K')=\mathfrak{M}(H')\cap\mathfrak{M}(X')$$

where $'$ on a matrix indicates its transpose. Then one can only test if

$$KH\beta=Kh, \tag{6.12}$$

and deviations from (6.11) that do not result in deviations from (6.12) will go undetected. In the same paper the authors suggested an expression for K. One could alternatively use

$$K=X'X(H'H+X'X)^{-}H'$$

in view of Theorem 2.2(f). We shall however recommend

$$K=CH^{-}_{m(C)},$$

where $C=X'X$ is the matrix of normal equations that provide least squares estimates for the parameters β, and $H^{-}_{m(C)}$ is a minimum C seminorm g-inverse of H.

Observe that

$$KH=\mathcal{S}(C)$$

is the shorted matrix C where $\mathcal{S}=\mathcal{M}(H')$ [15]. The shorted matrix is symmetric and

$$\mathcal{M}(KH)=\mathcal{M}(H'K')=\mathcal{M}(H')\cap\mathcal{M}(C)=\mathcal{M}(H')\cap\mathcal{M}(X')$$

Further, the dispersion matrix $D(KH\hat{\beta})$ of the BLUE of $KH\beta$ is given by

$$D(KH\hat{\beta})=[\mathcal{S}(C)]C^{-}[\mathcal{S}(C)]\sigma^2$$
$$=\mathcal{S}(C)\sigma^2.$$

Note that C^{-}/σ^2 is a g-inverse of this dispersion matrix. If $\mathbf{u}=\mathcal{S}(C)\hat{\boldsymbol{\beta}}-\mathbf{g}$, where $\mathbf{g}=CH^{-}_{m(C)}h$, we have the following simple formula for computing the expression which appears in the numerator of the variance ratio test.

We note in this connection that under H_0,

$$\mathbf{u}'C^{-}\mathbf{u}/\sigma^2\sim x^2_v,$$

where x^2_v is a chi squared random variable with degrees of freedom $v=$ rank $\mathcal{S}(C)=\operatorname{Tr}C^{-}\mathcal{S}(C)$. Since any routine least squares analysis of the data would provide C, C^{-}, and $C^{-}C$, the suggested use of the shorted matrix is more attractive than alternatives proposed earlier. It requires fewer additional matrix inversions, and as a by-product gives $H^{-}_{m(C)}$, which can be used to test the consistency of the equation (6.11).

The authors thank Professor David H. Carlson and the referee for drawing their attention to the papers [5] and [4] respectively by Ando.

REFERENCES

1 W. N. Anderson Jr., Shorted operators, *SIAM J. Appl. Math.* 20:520–525 (1971).

2 W. N. Anderson, Jr., and R. J. Duffin, Series and parallel addition of matrices, *J. Math. Anal. Appl.* 11:576–594 (1969).

3 W. N. Anderson, Jr., and G. Trapp, Shorted operators II, *SIAM J. Appl. Math.* 28:60–71 (1975).

4 T. Ando, Lebesgue type decomposition of positive operators, *Acta Sci. Math. (Szeged)* 38:253–260 (1976).

5 T. Ando, Generalized Schur complements, *Linear Algebra Appl.* 27:173–186 (1979).

6 D. Carlson, Matrix decompositions involving the Schur complement, *SIAM J. Appl. Math.* 28:577–587 (1975).

7 W. G. Cochran and G. M. Cox, *Experimental Designs*, 2nd ed., Wiley, New York, 1957.

8 R. J. Duffin and T. D. Morley, Almost definite operators and electromechanical systems, *SIAM J. Appl. Math.* 35:21–30 (1978).

9 R. J. Duffin and G. E. Trapp, Hybrid addition of matrices—network theory concept, *Applicable Anal.* 2:241–254 (1972).

10 O. Kempthorne, *The Design and Analysis of Experiments*, Wiley, New York, 1952.

11 M. G. Krein, The theory of self-adjoint extensions of semibounded Hermitian operators and its applications, *Mat. Sb.* 20(62):431–495, 21(63):365–404 (1947).

12 T. O. Lewis and T. G. Newman, Pseudoinverses of positive semidefinite matrices, *SIAM J. Appl. Math.* 10:701–703 (1968).

13 S. K. Mitra, Fixed rank solutions of linear matrix equations, *Sankhyā Ser. A* 35:387–392 (1972).

14 S. K. Mitra and M. L. Puri, On parallel sum and difference of matrices, *J. Math. Anal. Appl.* 44:92–97 (1973).

15 S. K. Mitra and M. L. Puri, Shorted operators and generalized inverses of matrices, *Linear Algebra Appl.* 25:45–56 (1979).

16 S. K. Mitra and C. R. Rao, Some results in estimation and tests of linear hypothesis under the Gauss-Markoff models, *Sankhyā Ser. A* 30:281–290 (1968).

17 S. K. Mitra and C. R. Rao, Projections under seminorms and generalized Moore-Penrose inverses, *Linear Algebra Appl.* 4:155–167 (1974).

18 S. K. Mitra and G. E. Trapp, On hybrid addition of matrices, *Linear Algebra Appl.* 10:19–35 (1975).

19 C. R. Rao, Unified theory of linear estimation, *Sankhyā Ser. A.* 33:371–394 (1971).

20 C. R. Rao and S. K. Mitra, *Generalized Inverse of Matrices and Its Applications*, Wiley, New York, 1971.

Reprinted by permission of the publisher,
© 1982 by Elsevier Science

THE FUNDAMENTAL BORDERED MATRIX OF LINEAR ESTIMATION AND THE DUFFIN–MORLEY GENERAL LINEAR ELECTROMECHANICAL SYSTEMS

SUJIT KUMAR MITRA† and MADAN L. PURI‡

Indian Statistical Institute, New Delhi, India, Indiana University, Bloomington, U.S.A.

(*Received June 24, 1981, in revised form, May 17, 1982*)

AMS (MOS)'s: Primary: 15A09 Secondary: 15A99

A complex matrix V of order $n \times n$ is said to be almost definite if for a complex n-tuple x, $x^*Vx = 0 \Rightarrow Vx = 0$. Interrelations between the blocks of a generalized inverse of the complex matrix

$$F = \begin{pmatrix} V & X \\ X^* & 0 \end{pmatrix}$$

are obtained when V is an almost definite matrix. These results are used to study the transfer impedance in the Duffin–Morley general linear electromechanical system.

1. INTRODUCTION

Consider the Gauss Markov model $(Y, X\beta, \sigma^2 V)$ with a possibly singular covariance matrix $\sigma^2 V$. The matrix

$$F = \begin{pmatrix} V & X \\ X' & 0 \end{pmatrix}$$

†Research supported by National Science Foundation Grant No. MCS 76-00951 at Indiana University.

‡Research supported by the Air Force Office of Scientific Research AFSC, USAF under Grant No. AFOSR 76-2927. Reproduction in whole or in part permitted for any purpose of the United States Government.

has been termed the fundamental bordered matrix of linear estimation by Hall and Meyer [7] on account of its role in the IPM (Inverse Partitioned Matrix) method of statistical analysis proposed by Rao [14]. The matrix X here is a real matrix of order $n\times m$ and V a real symmetric nonnegative definite (n.n.d.) matrix of order $n\times n$. Let

$$G=\begin{pmatrix} C_1 & C_2 \\ C_3 & -C_4 \end{pmatrix}$$

be a generalized inverse (g-inverse) of F. Khatri [6] and later Rao [14] have established interesting interconnections between the matrices C_1, C_2, C_3 and C_4. We shall extend their results as follows: We shall allow the matrices V and X to be complex and require the matrix V to be almost definite (a.d.) instead of being hermitian nonnegative definite (h.n.n.d.). These results will then be used to derive in an alternative way the explicit expression for the transfer impedance matrix in the Duffin–Morley general linear electromechanical system and to deduce additional properties of the same.

We use the following notations and definitions. For a positive integer n, $\mathscr{E}^n$ is the linear space of complex n-tuples. Column vector representations of vectors in $\mathscr{E}^n$ are denoted by lower case letters such as y, v etc. $\mathscr{C}^{m\times n}$ represents the linear space of complex matrices of order $m\times n$. Matrices are denoted by capital letters such as A, B, M etc. For a matrix A, A^* denotes its complex conjugate transpose, $\mathscr{M}(A)$ its column span, and $\mathscr{N}(A)$ its null space. A^-, the generalized inverse of A is defined by the equation $AA^-A=A$ [15]. A^+, the Moore Penrose inverse of A is the unique matrix G satisfying the following four conditions: $AGA=A$, $GAG=G$, $(AG)^*=AG$, $(GA)^*=GA$ [15]. Two subspaces of a linear space are said to be virtually disjoint if they have only the null vector in common.

A matrix $V\in\mathscr{C}^{n\times n}$ is said to be almost definite (a.d. [3]) if for $x\in\mathscr{E}^n$, $x^*Vx=0\Rightarrow Vx=0$. V is positive semidefinite (p.s.d. [3, 7]) if $\operatorname{Re}(x^*Vx)\geqq 0$, $x\in\mathscr{E}^n$. A matrix which is both almost definite and positive semidefinite is termed almost positive definite (a.p.d. [3, 7]). We introduce here the notion of a quasipositive definite matrix. A p.s.d. V is said to be quasi positive definite (q.p.d.) if in addition $\operatorname{Re}(x^*Vx)=0\Rightarrow Vx=0$. A q.p.d. matrix is always a.p.d. but the converse is not necessarily true. Other notations and definitions are introduced as and when they are required.

2. SOME PRELIMINARY RESULTS

Matrices V and W of the same order are said to be parallel summable (p.s. [15], see also [13]) if $V(V+W)^-W$ is invariant under choice of the g-

inverse of $(V+W)$. When this condition is satisfied, the matrix $P(V, W) = V(V+W)^- W$ is called the parallel sum of V and W.

Let the matrices V and W of the same order be conformably partitioned as

$$V=\begin{pmatrix} V_{11} & V_{12} \\ V_{21} & V_{22} \end{pmatrix}, \quad W=\begin{pmatrix} W_{11} & W_{12} \\ W_{21} & W_{22} \end{pmatrix}$$

where $V_{11}, W_{11} \in \mathscr{C}^{p\times q}$, V and W are said to be (p, q) hybrid summable (h.s. [12], see also [13]) if the matrices

$$H_{11}=V_{11}(V_{11}+W_{11})^- W_{11}$$

$$H_{12}=V_{11}(V_{11}+W_{11})^-(W_{12}-V_{12})+V_{12}$$

$$H_{21}=(W_{21}-V_{21})(V_{11}+W_{11})^- V_{11}+V_{21}$$

$$H_{22}=V_{22}+W_{22}-(W_{21}-V_{21})(V_{11}+W_{11})^-(W_{12}-V_{12})$$

are invariant under the choice of the g-inverse of $(V_{11}+W_{11})$. When this condition is satisfied, the matrix

$$H(V, W; p, q)=\begin{pmatrix} H_{11} & H_{12} \\ H_{21} & H_{22} \end{pmatrix}$$

is called the (p, q) hybrid sum of V and W. We shall abbreviate $H(V, W; p, q)$ by $H(V, W)$ when p and q are otherwise clear from the context.

Let $V \in \mathscr{C}^{m\times n}$ and $\mathscr{S}, \mathscr{T}$ be given subspaces of $\mathscr{E}^m$ and $\mathscr{E}^n$ respectively. The shorted matrix $S(V|\mathscr{S}, \mathscr{T})$ (see [11]) is a matrix $C \in \mathscr{C}^{m\times n}$ such that

$$\mathscr{M}(C)\subset\mathscr{S}, \ \mathscr{M}(C^*)\subset\mathscr{T} \tag{2.1}$$

and if $E \in \mathscr{C}^{m\times n}$ and E satisfies (2.1) then

$$\operatorname{Rank}(V-E)\geqq\operatorname{Rank}(V-C) \tag{2.2}$$

We shall however reserve the nomenclature and the notation only for those situations where the matrix C is uniquely determined by conditions (2.1) and (2.2) (Theorem 4.1 [11]).

Some of the properties of a.d., a.p.d. and q.p.d. matrices are listed below. Not all will be needed in this paper.

LEMMA 2.1 *An a.d. matrix V is also an EP matrix (that is V and V^* have identical column spans or equivalently identical null spaces).*

Proof Since $\dim[\mathcal{N}(V)]=\dim[\mathcal{N}(V^*)]$ if $\mathcal{N}(V)\neq\mathcal{N}(V^*)$ there exists $x\in\mathcal{N}(V^*)$ such that $x\notin\mathcal{N}(V)$. Clearly

$$x^*Vx=0$$

but $Vx\neq 0$ which is a contradiction.

Lemma 2.1 is essentially equivalent to Lemma 1 of [3] though the proof is different. Lemma 2.2 is also due to Duffin and Morley [3].

LEMMA 2.2 *If V* is *a.d. (a.p.d./q.p.d.) so is V^*.*

Proof Trivial and therefore omitted.

LEMMA 2.3 *If V is a.d. (a.p.d./q.p.d.) and V_1 is a principal minor of V, then so is V_1.*

Proof Trivial and therefore omitted.

LEMMA 2.4 *If*

$$V=\begin{pmatrix} V_{11} & V_{12} \\ V_{21} & V_{22} \end{pmatrix} \tag{2.3}$$

is a.d. where $V_{11}\in\mathscr{C}^{m\times m}$, then $\mathscr{M}(V_{12})\subset\mathscr{M}(V_{11})$, $\mathscr{M}(V_{21}^*)\subset\mathscr{M}(V_{11}^*)$.

Proof We show $\mathcal{N}(V_{11})\subset\mathcal{N}(V_{21})$. Let $x_1\in\mathcal{N}(V_{11})$. Then $x_1^*V_{11}x_1=0\Rightarrow x^*Vx=0$ where $x^*=(x_1^*,0^*)$. This implies $Vx=0\Rightarrow V_{11}x_1=0$, $V_{21}x_1=0$. Thus $x_1\in\mathcal{N}(V_{21})$. This establishes the second part of Lemma 2.4. The first part follows from this and Lemma 2.2.

LEMMA 2.5 *If V is a.d. (a.p.d./q.p.d.) so is V^+.*

Proof Lemma 2.5 can be proved in a way similar to that of Corollary 3 in [7]. We omit the proof.

LEMMA 2.6 *Let $A=B\oplus C$ be a.d. (a.p.d./q.p.d.) and B be EP, then B is a.d. (a.p.d./q.p.d.).*

Proof Let $G\in\{A^-\}$. Then $G^*\in\{(A^*)^-\}$ and since B^* and C^* are disjoint matrices and $A^*=B^*+C^*$, $G^*\in\{(B^*)^-\}$. Also since B is EP, $(B^*G^*-I)B^*=0\Rightarrow(B^*G^*-I)\,B=0\Rightarrow B^*G^*B=B$. Hence

$$x^*Bx=0\Rightarrow x^*B^*G^*AGBx=x^*B^*G^*Bx=x^*Bx=0\Rightarrow AGBx=Bx=0.$$

Hence B is a.d. The a.p.d./q.p.d. part is easily established.

Lemma 2.6 is not true without the EP assumption on B. Consider for example

$$A=\begin{pmatrix}1 & 0\\ 1 & 1\end{pmatrix},\quad B=\begin{pmatrix}0 & 0\\ 1 & 1\end{pmatrix},\quad C=\begin{pmatrix}1 & 0\\ 0 & 0\end{pmatrix}$$

and check that B is not a.d. even though A and C are a.p.d.

For a matrix $V\in\mathscr{C}^{n\times n}$, let sym V denote the hermitian matrix $(V+V^*)/2$. The following lemma is easily established. We omit the proof.

LEMMA 2.7 *If V is EP, then*

$$\mathscr{N}(V)\subset\mathscr{N}(\text{sym } V) \tag{2.4}$$

Further, equality holds in (2.4) *if V is q.p.d.*

LEMMA 2.8 *V is q.p.d. iff sym V is hermitian p.s.d. and*

$$\text{Rank}\,(sym\ V)=\text{Rank}\ V. \tag{2.5}$$

Proof The "only if" part follows from Lemma 2.7. For the "if" part observe that if sym V is hermitian p.s.d., then

$$\text{Re}\,(x^*Vx)=x^*(\text{sym } V)x\geqq 0.$$

Hence V is p.s.d. As in the proof of Theorem 2 [7] it follows that V is EP. (2.4) and (2.5) now imply that

$$\mathscr{N}(\text{sym } V)=\mathscr{N}(V)$$

Hence $\text{Re}\,(x^*Vx)=0\Rightarrow x^*(\text{sym } V)x=0\Rightarrow(\text{sym } V)x=0\Rightarrow Vx=0\Rightarrow V$ is q.p.d.

We notice that the condition (2.5) need not hold for an a.p.d. matrix. Consider for example the matrix

$$V=\begin{pmatrix}1+i & i\\ i & i\end{pmatrix}$$

where $i=\sqrt{-1}$. V is a.p.d. Check that:

$$\frac{V+V^*}{2}=\begin{pmatrix}1 & 0\\ 0 & 0\end{pmatrix}.$$

The same example incidentally shows that a pair of a.p.d. matrices need not be parallel summable. It is easily verified that V and V^* fail to satisfy condition (10.1.29) of [15] for parallel summability. Note that even though V and V^* are a.p.d. $\mathrm{diag}(V, V^*)$ is not even a.d.

The following lemma due to Duffin and Morley (Lemma 4 [3]) gives an interesting characterization of a.d. matrices. Ando (Theorem 1 [2]) gives an alternative proof of the same proposition.

LEMMA 2.9 (Duffin–Morley) *If V is a.d., $\exists$ a nonnull complex number α such that αV is a.p.d.*

Lemma 2.10 is equivalent to Theorems 1 and 2 of [3].

LEMMA 2.10 *The following two statements are equivalent*

a) *V is a.d.*

b)
$$\operatorname{Rank}\begin{pmatrix} V & X \\ X^* & 0 \end{pmatrix} = \operatorname{Rank}[V \,\vdots\, X] + \operatorname{Rank} X$$
$$= \operatorname{Rank}[V^* \,\vdots\, X] + \operatorname{Rank} X$$
$$= \dim \mathscr{M}(V) + 2\dim \mathscr{M}(X) - \dim \mathscr{M}(V) \cap \mathscr{M}(X)$$

for every matrix X with same number of rows as in V.

Proof It is seen that (b) is tantamount to claiming that

$$\mathscr{M}\begin{pmatrix} V \\ X^* \end{pmatrix} \quad \text{and} \quad \mathscr{M}\begin{pmatrix} X \\ 0 \end{pmatrix}$$

are virtually disjoint. If (b) is not true and $X \in \mathscr{C}^{n\times m}$, $\exists a \in \mathscr{E}^n$, $b \in \mathscr{E}^m$ such that

$$Va = Xb \neq 0$$
$$X^*a = 0 \tag{2.6}$$

(2.6)$\Rightarrow a^*Va = 0$ which contradicts (a). Hence (a)$\Rightarrow$(b). Conversely if (a) is not true $\exists a \in \mathscr{E}^n$ such that $Va \neq 0$ but $a^*Va = 0$. Choose $X = Va$ and verify that

$$\begin{pmatrix} V \\ X^* \end{pmatrix} a = \begin{pmatrix} X \\ 0 \end{pmatrix}$$

Hence for this X, (b) is not true. Thus (b)$\Rightarrow$(a).

LEMMA 2.11 *A pair of q.p.d. matrices of the same order are p.s. and the parallel sum is q.p.d.*

Proof Let $V, W \in \mathscr{C}^{n \times n}$ and be q.p.d. Then sym V and sym W are hermitian p.s.d. and by Lemmas 2.1 and 2.8

$$\mathscr{M}(V) = \mathscr{M}(V^*) = \mathscr{M}(\text{sym } V).$$

$$\mathscr{M}(W) = \mathscr{M}(W^*) = \mathscr{M}(\text{sym } W).$$

Also since $V + W$ is q.p.d.

$$\mathscr{M}(V + W) = \mathscr{M}(V^* + W^*) = \mathscr{M}[\text{sym}(V + W)].$$

The matrices sym V and sym W being hermitian p.s.d. it is easily seen that

$$\mathscr{M}(\text{sym } V) \subset \mathscr{M}[\text{sym}(V + W)]$$

Hence $\mathscr{M}(V) \subset \mathscr{M}(V + W)$, $\mathscr{M}(V^*) \subset \mathscr{M}(V^* + W^*)$ and condition (10.1.29) of [15] for parallel summability is seen to be satisfied. Observe that $P = P(V, W)$ is EP on account of Theorem 10.1.8(b) and (e) of [15] and the fact that both V and W are EP. Also Theorem 10.1.8(d) implies that $V^+ + W^+$ is a g-inverse of P. Hence

$$P = P(V^+ + W^+)P \Rightarrow P^* = P^*(V^+ + W^+)P$$

From Lemma 2.5 it is seen that $V^+ + W^+$ is q.p.d. This shows P^* and hence P is q.p.d.

LEMMA 2.12 *Let $V, W \in \mathscr{C}^{n \times n}$ and be q.p.d. Let p be an integer $< n$. Then V, W are (p, p) h.s. and the hybrid sum is q.p.d.*

Proof The proof of the first part is similar to that in Lemma 2.11. We defer the proof of the second part to the end of the next section.

Interestingly enough, unlike the hermitian case, the property of an a.d., p.s.d., a.p.d. or a q.p.d. matrix is not preserved under the Kronecker product. Thus the matrix diag$(1, i)$ is a.p.d. However its Kronecker product with itself, diag$(1, i, i, -1)$ is neither a.d. nor p.s.d. Also if I_n denotes the identity matrix in $C^{n \times n}$, the matrix $(1 + 2i)I_n$ is q.p.d. However its Kronecker product with itself $(-3 + 4i)I_{n^2}$ is not even p.s.d.

3. THE GENERAL LINEAR ELECTROMECHANICAL SYSTEMS

Before we proceed to a discussion of the general linear electromechanical system let us consider the following extension of the theorems of Khatri [6] and Rao [14].

THEOREM 3.1 *Let $X \in \mathscr{C}^{n \times m}$ and $V \in \mathscr{C}^{n \times n}$ and V be a.d. Let*

$$G=\begin{pmatrix} C_1 & C_2 \\ C_3 & -C_4 \end{pmatrix}, \quad F=\begin{pmatrix} V & X \\ X^* & 0 \end{pmatrix} \quad \text{and} \quad G \in \{F^-\} \tag{3.1}$$

Then

a) $XC_3X = X$, $XC_2^*X = X$

b) $X^*C_1X = 0$, $VC_1^*X = 0$, $VC_1X = 0$

c) The matrices VC_2X^*, XC_3V and XC_4X^* are unique with respect to the choice of the g-inverse in (3.1) *and*

$$VC_2X^* = XC_3V = XC_4X^*$$

d) VC_2X^* *is the shorted matrix* $S(V|\mathscr{S}, \mathscr{S})$, *where* $\mathscr{S} = \mathscr{M}(X)$.

Proof $FGF = F$ gives

$$VC_1V + XC_3V + VC_2X^* - XC_4X^* = V \tag{3.2}$$

$$VC_1X + XC_3X = X \tag{3.3}$$

$$X^*C_1V + X^*C_2X^* = X^* \tag{3.4}$$

$$X^*C_1X = 0 \tag{3.5}$$

(a) and (b) follow from these equations. Rest of the proof follows from Lemma 2.10 and Theorem 2.1 of [8].

It will be seen that Theorem 3.1 (a), (b) and (c) or Theorem 2.1 excluding the part (biii) of [8] are not just routine extensions of the corresponding results of Khatri or Rao. The proofs given by these authors are very much dependent on the fact that V is real symmetric n.n.d. while the proof of Theorem 2.1 of [8] capitalizes on the rank additivity condition (see Lemma 2.10 (b)). The structure of VC_2X^* reminds us of the representation of the shorted matrix given in [10] for the special case where V is hermitian n.n.d..

Remark 1 For an a.d. matrix $V \in \mathscr{C}^{n \times n}$ and a matrix $A \in \mathscr{C}^{m \times n}$, let us introduce a subclass of g-inverses of A through the equations

$$AGA = A \tag{3.6}$$

$$VGA = A^*G^*VGA \tag{3.7}$$

or equivalently

$$\mathcal{M}(VGA) \subset \mathcal{M}(A^*) \tag{3.8}$$

Let A^- be an arbitrary but fixed g-inverse of A and U a solution of the equation

$$(I-A^-A)^*V(I-A^-A)UA=(I-A^-A)^*VA^-A \tag{3.9}$$

then $A^- -(I-A^-A)U=G_0$ satisfies (3.6) and (3.7). It will be seen that Eq. (3.9) is consistent whenever V is a.d.

Let G_1 and G_2 be two distinct members of the above subclass. Put $VG_1A=A^*K_1$, $VG_2A=A^*K_2$. Then

$$A^*(G_1^*-G_2^*)V(G_1-G_2)A=A^*(G_1^*-G_2^*)A^*(K_1-K_2)=0$$

$$\Rightarrow V(G_1-G_2)A=0 \quad \text{since } V \text{ is a.d.}$$

$$\Rightarrow VG_1A=VG_2A.$$

The discussion in the proof of Theorem 3.1 will convince us that this unique matrix VGA is indeed the shorted matrix $S(V|\mathcal{S},\mathcal{S})$ where $\mathcal{S}=\mathcal{M}(A^*)$.

In terms of a particular solution G_0 of (3.6) and (3.7 or 3.8) a general solution is given by

$$G_0+(I-G_0A)T+W(I-AG_0)$$

where W is arbitrary and T is an arbitrary solution of the homogeneous equation

$$V(I-G_0A)T=0.$$

When V is hermitian n.n.d., (3.6) and (3.7 or 3.8) determine the class of minimum V seminorm g-inverses of A.

Remark 2 Let V and X be respectively the matrices

$$\begin{pmatrix} 1 & 0 \\ 1 & 1 \end{pmatrix} \quad \text{and} \quad \begin{pmatrix} 1 \\ 2 \end{pmatrix}.$$

It is seen that the matrix $V-VC_2X^*$ is not q.p.d. even though V is q.p.d. In fact $V-VC_2X^*$ is not even a.d. This shows that the Krein–Anderson–Trapp definition of the shorted operator (reproduced in [10]) cannot be extended literally for q.p.d. matrices (that is replacing the hermitian n.n.d. matrix by a q.p.d. matrix wherever it occurs). The same example also shows that $XC_3V \neq XC_2^*V$. Hence

$$\begin{pmatrix} C_1 & C_3^* \\ C_2^* & -C_4 \end{pmatrix}$$

is not another choice of a g-inverse of F as is true when V is hermitian n.n.d. [14]. In fact here F is invertible and $C_3 \neq C_2^*$.

A general linear electromechanical system is defined by a pair of equations

$$Ax = b$$

$$Zx - A^*v = 0 \tag{3.9}$$

where $b \in \mathscr{M}(A)$ and Z is a.d. For an interpretation of these equations in terms of Kirchhoff's and Ohm's laws as applicable for an electrical network, the reader is referred to the original paper of Duffin and Morley [3]. The matrix A is termed the structure operator and Z the constitutive operator of the system.

Duffin and Morley have shown that these equations are consistent and for a solution $\binom{x}{v}$ of this system, v is unique if one requires that $v \in \mathscr{M}(A)$. Further there is a unique matrix $\psi(Z)$ with $\mathscr{M}[\psi(Z)] \subset \mathscr{M}(A)$, $\mathscr{M}[\psi^*(Z)] \subset \mathscr{M}(A)$ such that $\psi(Z)b$ is the unique v as stipulated above. The matrix $\psi(Z)$ is called the transfer impedance.

Let $\begin{pmatrix} C_1 & C_2 \\ C_3 & -C_4 \end{pmatrix}$ be a g-inverse of $K = \begin{pmatrix} Z & A^* \\ A & 0 \end{pmatrix}$.

A solution of (3.9) is given by $x = C_2 b$ and $v = C_4 b$ and if b is nonnull every solution is obtainable this way through a g-inverse of K.

By Theorem 3.1, $A^*C_4A = ZC_2A$. If C_4 is required to satisfy the condition $\mathscr{M}(C_4) \subset \mathscr{M}(A)$, $\mathscr{M}(C_4^*) \subset \mathscr{M}(A)$, we have

$$\begin{aligned} C_4 = AA^+C_4(A^+)^*A^* &= (A^+)^*A^*C_4AA^+ \\ &= (A^+)^*(ZC_2A)A^+ \end{aligned} \tag{3.10}$$

which is essentially the expression for $\psi(Z)$ given in Theorem 7 [3].

We note that C_2AA^+ is another choice of C_2 and in fact of a g-inverse of A satisfying the following conditions similar to that of a generalized Moore Penrose inverse

$$AGA = A$$

$$GAG = G$$

$$(AG)^* = AG$$

$$\mathcal{M}(ZGA) \subset \mathcal{M}(A^*)$$

Using A_{IZ}^+ to denote such a g-inverse of A and noting that

$$A^*C_2^*ZC_2A = ZC_2A$$

from (3.10) we are led to the following alternative expression for C_4 and hence that of $\psi(Z)$

$$C_4 = (A_{IZ}^+)^*ZA_{IZ}^+ \tag{3.11}$$

Note that the right side of (3.11) is unique even though A_{IZ}^+ may not be so.

Also proceeding as in the unified least squares approach to a Gauss Markov model with a singular covariance matrix initiated in [16] we are led to yet another expression for C_4 that is $\psi(Z)$:

$$C_4 = [A(Z + \bar{\alpha}A^*A)^- A^*]^+ - \bar{\alpha}AA^+ \tag{3.12}$$

where α is a complex number such that αZ is a.p.d. (Lemma 2.9). Note that $Z + \bar{\alpha}A^*A$ is a.d. and $\mathcal{M}(A^*) \subset \mathcal{M}(Z + \bar{\alpha}A^*A)$. From the expressions (3.10) and (3.11) it is seen that $\psi(Z)$ is a.d. and when Z is a.p.d./q.p.d., $\psi(Z)$ is so.

Since for the shorted matrix ZC_2A, $\mathcal{M}(ZC_2A) = \mathcal{M}(Z) \cap \mathcal{M}(A^*)$ (Theorem 3.1(e) [11]).

$$\mathcal{M}[\psi(Z)] = \mathcal{M}[(A^+)^*Z] \cap \mathcal{M}[(A^+)^*A^*] = \mathcal{M}[(A^+)^*Z]$$

since $\mathcal{M}[(A^+)^*Z] \subset \mathcal{M}[(A^+)^*] = \mathcal{M}[(A^+)^*A^*]$. We have thus arrived at the following result.

THEOREM 3.2 *The range (column span) of $\psi(Z)$ and also of $[\psi(Z)]^*$ is*

$$\mathcal{M}[(A^+)^*Z]. \tag{3.13}$$

Observe that $(A^+)^*Z = (A^*)^+Z$ is the best approximate solution in the sense of Penrose of the matrix equation

$$A^*Y = Z$$

We shall now prove

THEOREM 3.3 *AZ^-A^* is a g-inverse of $\psi(Z)$.*

Proof $\psi(Z)AZ^-A^*\psi(Z)=(A^+)^*(ZC_2A)A^+AZ^-A^*(A^+)^*(A^*C_3Z)A^+$

$$=(A^+)^*(ZC_2A)Z^-(A^*C_3Z)A^+$$

$$=(A^+)^*A^*C_3ZZ^-ZC_2A(A^+)$$

$$=(A^+)^*A^*C_3ZC_2A(A^+)$$

$$=(A^+)^*ZC_2AC_2A(A^+)=(A^+)^*ZC_2A(A^+)=\psi(Z)$$

Equation (3.12) is actually a special case of a more general result we present below.

THEOREM 3.4 *If Z and $Z+A^*UA$ are a.d., then*

$$\psi(Z+A^*UA)=\psi(Z)+P_AUP_A$$

where $P_A=AA^+$ is the orthogonal projector onto $\mathscr{M}(A)$.

Proof Note that

$$K_U=\begin{pmatrix} Z+A^*UA & A^* \\ A & 0 \end{pmatrix}=K\begin{pmatrix} I & 0 \\ UA & I \end{pmatrix}.$$

Theorem 3.4 is a simple consequence of the fact that if

$$\begin{pmatrix} C_1 & C_2 \\ C_3 & -C_4 \end{pmatrix}\in\{K^-\},$$

then

$$\begin{pmatrix} I & 0 \\ -UA & I \end{pmatrix}\begin{pmatrix} C_1 & C_2 \\ C_3 & -C_4 \end{pmatrix}\in\{K_U^-].$$

THEOREM 3.5 *Let T and W be complex matrices such that $\mathscr{M}(T)=\mathscr{N}(A^*C_3)$, $\mathscr{M}(W)=\mathscr{N}(A^*C_2^*)$ and both Z* and *$Z+TUW^*$ be a.d., then*

$$\psi(Z+TUW^*)=\psi(Z)$$

Proof Put $Z_0=Z+TUW^*$ and let

$$\begin{pmatrix} C_1^0 & C_2^0 \\ C_3^0 & C_4^0 \end{pmatrix}\in\{K_0^-\}\quad\text{where}\quad K_0=\begin{pmatrix} Z_0 & A^* \\ A & 0 \end{pmatrix}$$

and

$$\begin{pmatrix} C_1 & C_2 \\ C_3 & C_4 \end{pmatrix} \in \{K^-\}.$$

Since $Z_0C_2A = ZC_2A$, $G = C_2$ satisfies (3.6) and (3.7) with V replaced by Z_0. Hence $Z_0C_2A = Z_0C_2^0A$. Similarly it is seen that $A^*C_3Z_0 = A^*C_3^0Z_0$. The following matrix obtained through block transplanting

$$\begin{pmatrix} C_1^0 & C_2 \\ C_3 & -C_4 \end{pmatrix}$$

is thus seen to belong to $\{K_0^-\}$. This proves Theorem 3.5.

It was shown by Duffin and Morley [3] that using the transfer impedance one could present the theories of parallel addition, hybrid addition, etc. in a unified manner. Thus let R, S be q.p.d. matrices in $\mathscr{C}^{n\times n}$. Put $Z = \operatorname{diag}(R, S)$, which is q.p.d. and $A = (I_n \vdots I_n)$, each I_n an identity matrix in $\mathscr{C}^{n\times n}$. Then

$$\psi(Z) = P(R, S)$$

Also if R and S are conformably partitioned as

$$R = \begin{pmatrix} R_{11} & R_{12} \\ R_{21} & R_{22} \end{pmatrix}, \quad S = \begin{pmatrix} S_{11} & S_{12} \\ S_{21} & S_{22} \end{pmatrix}$$

where $R_{11}, S_{11} \in \mathscr{C}^{p\times p}$, and

$$A = \begin{pmatrix} I_p & 0 & I_p & 0 \\ 0 & I_{n-p} & 0 & 0 \\ 0 & I_{n-p} & 0 & -I_{n-p} \end{pmatrix} \in \mathscr{C}^{2n-p\times 2n}$$

Then $H(R, S)$, the (p, p) hybrid sum of R and S, is the leading principal minor of order $n\times n$ of $\psi(Z)$. Since $\psi(Z)$ is q.p.d., $H(R, S)$ is also q.p.d. by Lemma 2.3. This will conclude the proof of Lemma 2.11. The structure operators A, used by us are somewhat different from those in [3].

4. CONCLUDING REMAKS

Consider the linear system (3.9) without insisting, however, that the matrix

Z be a.d. We ask ourselves the following question: When does such a system determine a unique transfer impedance $\psi(Z)$ subject of course to the usual stipulations that $\mathcal{M}[\psi(Z)] \subset \mathcal{M}(A)$, $\mathcal{M}[(\psi(Z))^*] \subset \mathcal{M}(A)$?

Let

$$G = \begin{pmatrix} C_1 & C_2 \\ C_3 & -C_4 \end{pmatrix} \in \{K^-\}.$$

Looking at equations like (3.2)–(3.5), which will ultimately determine the C_i matrices ($i=1(1)4$), it is seen that C_4 explicitly occurs only in one equation and that also through A^*C_4A. The above question could therefore be rephrased as follows. When does the system (3.9) uniquely determine

$$A^*C_4A = -(0 \vdots A^*)G\begin{pmatrix} 0 \\ A \end{pmatrix}$$

independent of the choice of G in $\{K^-\}$. Since the matrix $\binom{0}{A}$ is nonnull, using standard results we have the following necessary and sufficient conditions

$$\mathcal{M}\begin{pmatrix} 0 \\ A \end{pmatrix} \subset \mathcal{M}(K), \ \mathcal{M}\begin{pmatrix} 0 \\ A \end{pmatrix} \subset \mathcal{M}(K^*).$$

See, for example Ex. 14 ([15], p. 43).

Let us consider the following conditions

$$c1 \ \mathcal{M}\begin{pmatrix} 0 \\ A \end{pmatrix} \subset \mathcal{M}(K) \qquad r1 \ \mathcal{M}\begin{pmatrix} 0 \\ A \end{pmatrix} \subset \mathcal{M}(K^*)$$

$$c2 \ \mathcal{M}\begin{pmatrix} Z \\ 0 \end{pmatrix} \subset \mathcal{M}(K) \qquad r2 \ \mathcal{M}\begin{pmatrix} Z^* \\ 0 \end{pmatrix} \subset \mathcal{M}(K^*).$$

It is easily seen that the following are equivalent necessary and sufficient conditions

1) any one of $c1$ or $c2$ coupled with any one $r1$ or $r2$

or

2) $\text{Rank } K = \text{Rank}\,(Z \vdots A^*) + \text{Rank } A = \text{Rank}\begin{pmatrix} Z \\ A \end{pmatrix} + \text{Rank } A.$

The properties of the transfer impedance stated earlier in this paper remain true under these conditions.

It was pointed out by Duffin and Morley [3] that the unique transfer impedance exists for every structure operator A (with same number of rows as in Z) iff the constitutive operator Z is a.d. We refer to Lemma 2.10 in this connection.

One could in fact in a purely formal manner consider an extension of the linear system (3.12) itself using the pair of equations.

$$Ax=b$$

$$Zx-Bv=0 \tag{4.1}$$

where $b\in\mathcal{M}(A)$ and under proper further assumptions extend results stated earlier in this paper. Equations (4.1) are consistent for every such b if

$$\mathcal{M}\begin{pmatrix}0\\A\end{pmatrix}\subset\mathcal{M}(K) \tag{4.2}$$

where

$$K=\begin{pmatrix}Z & B\\A & 0\end{pmatrix}$$

When condition (4.2) is satisfied we may agree to call (A,B) the pair of structure operators and Z the constitutive operator of the system. We may then ask under what further conditions the system will determine a unique transfer impedance $\psi(Z)$ subject of course to the stipulations that $\mathcal{M}[\psi(Z)]\subset\mathcal{M}(B^*)$, $\mathcal{M}[(\psi(Z))^*]\subset\mathcal{M}(A)$ so that $v=\psi(Z)b$. Following a similar argument as above, it is seen that unless precisely one of A or B is a null matrix for $\psi(Z)$ to be uniquely determined it is necessary and sufficient that in addition to (4.2) the following holds.

$$\mathcal{M}\begin{pmatrix}0\\B^*\end{pmatrix}\subset\mathcal{M}(K^*) \tag{4.3}$$

(4.2) and (4.3) together are equivalent to

$$\text{Rank}\,K=\text{Rank}\,(Z\;\;B)+\text{Rank}\,A=\text{Rank}\begin{pmatrix}Z\\A\end{pmatrix}+\text{Rank}\,B. \tag{4.4}$$

A system $(Z;A,B)$ obeying (4.4) will be called a conforming system. We shall prove Theorem 4.1. Let $\otimes$ denote the Kronecker product.

THEOREM 4.1 *If the systems* $(Z;A,B)$ *and* $(V;X,Y)$ *are conforming so is the system* $(Z\otimes V;\ A\otimes X,\ B\otimes Y)$ *and further*

$$\psi(Z\otimes V)=\psi(Z)\otimes\psi(V). \tag{4.5}$$

Conversely if the system $(Z\otimes V;\ A\otimes X,\ B\otimes Y)$ *is conforming and* $\mathcal{M}(Z)\cap\mathcal{M}(B)$, $\mathcal{M}(Z^*)\cap\mathcal{M}(A^*)$, $\mathcal{M}(V)\cap\mathcal{M}(Y)$ *and* $\mathcal{M}(V^*)\cap\mathcal{M}(X^*)$ *are each of positive dimensions (or equivalently* $\psi(Z\otimes V)$ *is nonnull), then the systems* $(Z{:}\,A,B)$ *and* $(V;X,Y)$ *are also conforming.*

Proof (first part) It is a simple matter to check that if $(Z;A,B)$ and $(V;X,\ Y)$ obey conditions (4.2) and (4.3) then so does $(Z\otimes V;\ A\otimes X,\ B\otimes Y)$. Let $(Z;A,B)$ and $(V;X,Y)$ be confirming and

$$\begin{pmatrix} C_1 & C_2\\ C_3 & -C_4\end{pmatrix}\in\left\{\begin{pmatrix} Z & B\\ A & 0\end{pmatrix}^-\right\},\ \begin{pmatrix} D_1 & D_2\\ D_3 & -D_4\end{pmatrix}\in\left\{\begin{pmatrix} V & Y\\ X & 0\end{pmatrix}^-\right\}.$$

Then from Lemma 5.2 [11] it is seen that

$$BC_3B=B, AC_2A=A, BC_3Z=ZC_2A=BC_4A.$$

$$YD_3Y=Y, XD_2X=X,\ YD_3V=VD_2X=YD_4X.$$

Hence

$$B\otimes Y[C_3\otimes D_3]B\otimes Y=B\otimes Y$$

$$A\otimes X[C_2\otimes D_2]A\otimes X=A\otimes X$$

$$B\otimes Y[C_3\otimes D_3][Z\otimes V][C_2\otimes D_2]A\otimes X=Z\otimes V[C_2\otimes D_2]A\otimes X$$

$$=B\otimes Y[C_3\otimes D_3]Z\otimes V. \tag{4.6}$$

Let R denote the matrix $Z\otimes V-[Z\otimes V][C_2\otimes D_2]A\otimes X$. If $[B\otimes Y]a=Rb$ is a vector common to $\mathcal{M}(R)$ and $\mathcal{M}[B\otimes Y]$, then using (4.6) it is seen that

$$[B\otimes Y]a=[B\otimes Y][C_3\otimes D_3][B\otimes Y]a=[B\otimes Y][C_3\otimes D_3]Rb=0$$

It is similarly seen that $\mathcal{M}(R^*)$ and $\mathcal{M}([A\times X]^*)$ are also virtually disjoint. Hence arguing as in the proof of Theorem 2.1 [8], we conclude that $[Z\otimes V][C_2\otimes D_2]A\otimes X$ is the unique shorted matrix $S(Z\otimes V|\mathcal{M}(B\otimes Y),\ \mathcal{M}(A^*\otimes X^*))$ (4.5) follows from an argument similar to that used in the derivation of (3.13) since the Moore Penrose inverse of the Kronecker product is the Kronecker product of the respective Moore Penrose inverses.

Proof (*converse part*) Here one shows that under the stated assumptions if $(Z; A, B)$ and/or $(V; X, Y)$ are nonconforming then so is $(Z \otimes V;\, A \otimes X,\, B \otimes Y)$.
While proving Theorem 4.1 we have virtually proved the following corollary.

COROLLARY 4.1 *If the shorted matrices* $S(Z \mid \mathscr{M}(B), \mathscr{M}(A^*))$ *and* $S(V \mid \mathscr{M}(Y), \mathscr{M}(X^*))$ *are uniquely determined then so is* $S(Z \otimes V \mid \mathscr{M}(B \otimes Y), \mathscr{M}(A^* \otimes X^*))$ *and further*

$$S(Z \otimes V \mid \mathscr{M}(B \otimes Y), \mathscr{M}(A^* \otimes X^*))$$
$$= S(Z \mid \mathscr{M}(B), \mathscr{M}(A^*)) \otimes S(V \mid \mathscr{M}(Y), \mathscr{M}(X^*)). \tag{4.6}$$

Conversely if $S(Z \otimes V \mid \mathscr{M}(B \otimes Y), \mathscr{M}(A^* \otimes X^*))$ *is uniquely determined and is nonnull then so are* $S(Z \mid \mathscr{M}(B), \mathscr{M}(A^*))$ *and* $S(V \mid \mathscr{M}(Y), \mathscr{M}(X^*))$.

Acknowledgement

The authors express their sincere thanks to Professor R. J. Duffin for valuable suggestions.

References

[1] W. N. Anderson, Jr. and R. J. Duffin, Series and parallel addition of matrices, *J. Math. Anal. Appl.* **11** (1969), 576–594.
[2] T. Ando, Generalized Schur complements, *Linear Algebra Appl.* **27** (1979), 173–186.
[3] R. J. Duffin and T. D. Morley, Almost definite operators and electromechanical systems, *SIAM J. Appl. Math.* **35** (1978), 21–30.
[4] R. J. Duffin and G. E. Trapp, Hybrid addition of matrices—a network theory concept, *Applicable Anal.* **2** (1972), 241–254.
[5] F. J. Hall and C. D. Meyer, Jr., Generalized inverses of the fundamental bordered matrix used in linear estimation, *Sankhyā* Ser A, **37** (1975), 428–438.
[6] C. G. Khatri, Some results for the singular multivariate regression models, *Sankhyā*, Ser A, **30** (1968), 268–280.
[7] T. O. Lewis and T. G. Newman, Pseudoinverses of positive semidefinite matrices, *SIAM J. Appl. Math.* **10** (1968), 701–703.
[8] S. K. Mitra, Simultaneous diagonalization of rectangular matrices, *Linear Algebra Appl.* **47** (1982), 139–150.
[9] S. K. Mitra and M. L. Puri, On parallel sum and difference of matrices, *J. Math. Anal. Appl.* **44** (1973), 92–97.

[10] S. K. Mitra and M. L. Puri, Shorted operators and generalized inverses of matrices, *Linear Algebra Appl.* **25** (1979) 45–56.

[11] S. K. Mitra and M. L. Puri, Shorted matrices—an extended concept and some applications, *Linear Algebra Appl.* **42** (1982) 57–79.

[12] S. K. Mitra and G. E. Trapp, On hybrid addition of matrices, *Linear Algebra Appl.* **10** (1975), 19–35.

[13] T. D. Morley, Parallel summation, Maxwell's Principle and the infimum of projections, *J. Math. Anal. Appl.* **70** (1979), 33–41.

[14] C. R. Rao, Unified theory of linear estimation, *Sankhyā*, Ser A. **33** (1971), 371–394.

[15] C. R. Rao and S. K. Mitra, *Generalized inverse of matrices and its applications*, Wiley, New York, (1971).

[16] C. R. Rao and S. K. Mitra, Further contributions to the theory of generalized inverse of matrices and its applications, *Sankhyā*, Ser A. **33** (1971), 289–300.

Reprinted by permission of the publisher,
© 1983 by Taylor and Francis Ltd.
http://www.tandf.co.uk

COMPLEX CHEBYSHEV POLYNOMIALS AND GENERALIZATIONS WITH AN APPLICATION TO THE OPTIMAL CHOICE OF INTERPOLATING KNOTS IN COMPLEX PLANAR SPLINES*

GERHARD OPFER and MADAN L. PURI

Universität Hamburg, Institut für Angewandte Mathematik, Bundesstraße 55, D-2000 Hamburg 13, West Germany

Department of Mathematics, Indiana University, Bloomington, Indiana 47405, U.S.A.

This paper contains a brief account on complex planar splines which are complex valued functions defined piecewise on a grid. For noncontinuous (so called nonconforming) splines the problem of the placement of knots at which these splines are required to be continuous is investigated. It is shown that this problem reduces to finding complex Chebyshev polynomials under the additional requirement that the zeros of the polynomials are on the boundary of the corresponding domains. It is proved that the zeros of a generalized Chebyshev polynomial are in the convex hull of the domain on which the Chebyshev polynomials are defined. Some open problems are stated. A numerical and graphical display for the optimal location of three and six points on certain triangles is provided.

1. Introduction

To motivate this study, a brief account of the so-called complex planar splines is given. These splines were introduced by Opfer and Puri [15] and were further studied by Opfer and Schober [16]. The problem of the optimal

* Research supported by the Office of Naval Research under Contract N00014-77-C-0659. The U.S. Government's right to retain a nonexclusive royalty-free license in and to the copyright covering this paper, for governmental purposes, is acknowledged.

choice of interpolating knots is closely related to the concept of nonconforming complex planar splines.

Let $f: D \subset \mathbb{C} \to \mathbb{C}$ be a continuous mapping, where D is the closure of some bounded region in $\mathbb{C}$. If f is complicated, then it is desirable to find an approximation for f which is easier to compute than f itself. Since f is intrinsically connected to its domain of definition D, D in general needs to be approximated by a simpler set, too.

One approach to this problem is to replace D by a union of suitable simple configuration like triangles or rectangles and to approximate f piecewise by simple *elements*.

Since triangles play a special role when dividing up a given two-dimensional set like D, the study here will be restricted to triangular grids.

It is explained in Opfer and Puri [15] that the continuity requirement which is imposed upon the complex planar splines has the consequence that holomorphic elements like polynomials cannot be used reasonably.

However, if one requires continuity of the approximation of f only in the interior points and in certain selected points (knots) of the boundary of the triangles, one obtains approximations which (in analogy to the real case) should be called *nonconforming* complex planar splines.

The advantage of relaxing the continuity requirement is that one can use holomorphic functions like complex polynomials as elements.

In this connection the question of placing the knots in a certain optimal fashion when using polynomials on triangles is of much interest and is the subject of study in this paper. In a forthcoming paper by Opfer and Werner some further properties of nonconforming elements will be exhibited.

To illustrate the problem suppose we take a triangular grid and define a quadratic complex polynomial in each triangle of that grid. Suppose further that we require that the resulting function is either continuous in all grid-points (i.e., vertices of the triangles) or continuous in all the midpoints of the edges. The question is which of the two requirements is favorable.

To illustrate further we interpolate $f(z) = \exp(z)$ at the vertices (case 1) or at the midpoints (case 2) of the triangle

$$T = \{(x, y) \in \mathbb{R}^2 : |y| \leqslant x \leqslant \sqrt{0,5}\} \tag{1.1}$$

by a complex polynomial of degree two and suppose we denote the resulting polynomials by p_1 and p_2 respectively. We now compare the uniform errors on T and we obtain after some numerical calculations that

$$\|f - p_1\|_\infty = 0.0984, \qquad \|f - p_2\|_\infty = 0.1106 \tag{1.2}$$

which indicates that case 1 is preferable to case 2.

2. Complex Chebyshev Polynomials and Generalizations

Let Π_n denote the set of all polynomials with complex coefficients and degree at most $n \in \mathbb{N}^0 = \{0, 1, \ldots\}$. Let $\Pi_n^{\mathbb{R}}$ denote the subset of Π_n consisting of all the polynomials with real coefficients.

If $p \in \Pi_n$ interpolates a given function $f: D \to \mathbb{C}$ at points $z_0, z_1, \ldots, z_n \in D$, where D is a compact set in $\mathbb{C}$, then the error takes the form

$$r(z) = f(z) - p(z) = f[z_0, z_1, \ldots, z_n, z] \prod_{j=0}^{n} (z - z_j) \tag{2.1}$$

(Davis [7, p. 64]) regardless of the assumptions on f since expression (2.1) can be derived by purely algebraic means. Here the expression $f[z_0, z_1, \ldots, z_n, z]$ is defined in terms of divided differences. Implicity we always assume that D contains sufficiently many points; in this context at least $n+2$.

However, the assumptions imposed on f play an important role in deriving equivalent expressions for $f[z_0, z_1, \ldots, z_n, z]$ or estimates for $|f[z_0, z_1, \ldots, z_n, z]|$.

For a selection of these expressions and estimates see Davis [7, pp. 67–69] or Abramowitz and Stegun [1, Chap. 25].

For any $z_0, z_1, \ldots, z_n, z \in \mathbb{C}$ we set

$$\zeta = (z_0, z_1, \ldots, z_n) \tag{2.2}$$

and

$$\omega(z, \zeta) = \prod_{j=0}^{n} (z - z_j).$$

A uniformly best choice of $z_0, z_1, \ldots, z_n$ in (2.1) which is independent of f is apparently made if we require that

$$\|\omega(\cdot, \zeta)\|_\infty = \max_{z \in D} |\omega(z, \zeta)| \tag{2.3}$$

is as small as possible by appropriate choice of ζ. Since

$$\omega(z, \zeta) = z^{n+1} + p(z), \qquad p \in \Pi_n, \quad n \in \mathbb{N}^0, \tag{2.4}$$

the problem of minimizing $\|\omega(\cdot, \zeta)\|_\infty$ may be regarded as a linear complex approximation problem. Π_n is a Haar space of finite dimension $n+1$ which implies that the stated approximation problem is uniquely soluble (cf. Meinardus [14, p. 16]). It also implies that the zeros of the best $\omega(z, \zeta)$ are unique. For each fixed $n \in \mathbb{N}^0$, the polynomial

$$t_{n+1}(z) = z^{n+1} + \hat{p}(z), \qquad z \in D, \quad \hat{p} \in \Pi_n \tag{2.5}$$

defined by

$$\|t_{n+1}\|_\infty \leqslant \|z^{n+1}+p\|_\infty \qquad \text{for all } p \in \Pi_n \tag{2.6}$$

wil be called the *Chebyshev polynomial* (or for short T-polynomial) of degree $n+1$ with respect to D, and the corresponding zeros the *Chebyshev points* of D. In addition, we define

$$t_0(z)=1 \qquad \text{for all } z \in D. \tag{2.7}$$

The ordinary T-polynomials usually called T_n (cf. Rivlin [17]) are obtained by letting $D=[-1, 1]$. However note that usually the normalization $T_n = t_n/\|t_n\|_\infty$ implying $\|T_n\|_\infty = 1$ is used.

Interestingly the ordinary T-polynomials for $D=[-1, 1]$ are at the same time the T-polynomials for all confocal ellipses with foci -1 and $+1$. This was already observed by Faber [8].

Since the T-polynomials are essentially uneffected by the linear transformation of D,

$$w(z)=az+b, \qquad z\in D, \quad a\neq 0, \quad a, b\in \mathbb{C}, \tag{2.8}$$

one can compute the nth degree T-polynomial of $w(D)$ if the nth degree T-polynomial of D is known. We note that the inverse of w in (2.8) is given by $z(w)=(w-b)/a$.

To be precise let

$$t_n(z)=z^n+\sum_{j=0}^{n-1} \alpha_j z^j \tag{2.9}$$

be the nth T-polynomial of any compact nonempty set $D\subset \mathbb{C}$ and $\tilde{D}=w(D)$. Then

$$\begin{aligned}\tilde{t}_n(w) &= a^n t_n\left(\frac{w-b}{a}\right)\\ &= (w-b)^n+\sum_{j=0}^{n-1} \alpha_j a^{n-j}(w-b)^j\end{aligned} \tag{2.10}$$

is the T-polynomial of degree n with respect to $\tilde{D}$. We also note that T-polynomials with respect to D are in $\Pi_n^{\mathbb{R}}$, $n\in\mathbb{N}$, provided D is symmetric with respect to the real axis (cf. Meinardus [14, p. 28]).

EXAMPLE 2.1. Let E be any ellipse in the plane with foci $f_1, f_2 \in \mathbb{C}$, $f_1\neq f_2$. Then

$$w(z)=[(f_2-f_1)z+(f_2+f_1)]/2$$

maps $D=[-1,1]$ onto the segment $[f_1,f_2]$ in $\mathbb{C}$. Let

$$t_n(z)=z^n+\alpha_{n-2}z^{n-2}+\alpha_{n-4}z^{n-4}+\cdots+\alpha_{n-2[n/2]}z^{n-2[n/2]} \tag{2.11}$$

be the (ordinary) T-polynomial for $D=[-1,1]$ where the coefficients α_{n-2j}, $j=1,2,\ldots,[n/2]$ are given up to a factor 2^{n-1} in Abramowitz and Stegun [1, see 22.3.6].

Explicit numbers for all $n\leqslant 20$ are given by Luke [13, p. 458]. Then the T-polynomials of E are given explicitly by

$$t_n(z)=\left(z-\frac{f_2+f_1}{2}\right)^n+\sum_{j=1}^{[n/2]}\alpha_{n-2j}\left(\frac{f_2-f_1}{2}\right)^{2j}\left(z-\frac{f_2+f_1}{2}\right)^{n-2j}. \tag{2.12}$$

It should be noticed that for ellipses E the polynomials given in (2.12) are at the same time also the so called Faber polynomials of E (cf. Gaier [10, p. 47]).

EXAMPLE 2.2. Define

$$D=\{z\in\mathbb{C}:|z-2|\leqslant 1\}\cup\{z\in\mathbb{C}:|z+2|\leqslant 1\}. \tag{2.13}$$

In this case the first degree T-polynomial with respect to D is

$$t_1(z)=z,\qquad z\in D, \tag{2.14}$$

and the only zero of t_1 is at $z_0=0\notin D$. However

$$t_2(z)=z^2-5,\qquad z\in D \tag{2.15}$$

has its zeros $z_0=-\sqrt{5}$, $z_1=+\sqrt{5}$ in D. It may be remarked that applications do occur where the approximation problems are treated on unconnected sets (cf. de Boor and Rice [6] and Fuchs [9]). A slightly more general problem than stated so far is to find a vector $\zeta=(z_0,z_1,\ldots,z_n)\in\mathbb{C}^{n+1}$ of knots such that for a given vector $\nu=(\nu_0,\nu_1,\ldots,\nu_n)\in\mathbb{N}^{n+1}$ of natural numbers the uniform norm of a polynomial ω_ν defined by

$$\omega_\nu(z,\zeta)=\prod_{j=0}^{n}(z-z_j)^{\nu_j},\qquad z\in D, \tag{2.16}$$

is minimal. This allows for multiple knots in the mentioned interpolation problem.

The problem of minimizing the norm of ω_ν by appropriate choice of the knots ζ was recently treated for the real case by Bojanov [4] under the additional requirement that the knots are pairwise distinct. Related problems for the real case were investigated by Boyanov [5], Barrar and Loeb [2], and Barrar, Loeb and Werner [3].

A polynomial ω_ν with the described minimal property will be called *generalized Chebyshev polynomial* or generalized T-polynomial.

By the following example it is shown that in the complex case we cannot require the knots to be pairwise distinct, or in other words the requirement of distinct knots would imply that a generalized T-polynomial will not necessarily exist.

EXAMPLE 2.3. The uniform norm of the polynomial ω defined by

$$\omega(z, \zeta) = (z - z_0)(z - z_1), \qquad z \in D = \{z : |z| \leqslant 1\}$$

is minimal if and only if $z_0 = z_1 = 0$. The proof is elementary and omitted.

THEOREM 2.1. *Let $\nu = (\nu_0, \nu_1, ..., \nu_n) \in \mathbb{N}^{n+1}$ be given. Then there exists a vector $\hat{\zeta} = (\hat{z}_0, \hat{z}_1, ..., \hat{z}_n) \in \mathbb{C}^{n+1}$ such that*

$$\|\omega_\nu(\cdot, \hat{\zeta})\|_\infty \leqslant \|\omega_\nu(\cdot, \zeta)\|_\infty \qquad \textit{for all } \zeta \in \mathbb{C}^{n+1}.$$

If H is the convex hull of D in $\mathbb{C}$, then $\hat{z}_j \in H$ for all $j = 0, 1, ..., n$.

Proof. Let $\Theta = (0, 0, ..., 0) \in \mathbb{C}^{n+1}$. The set

$$K = \{\zeta \in \mathbb{C}^{n+1} : \|\omega_\nu(\cdot, \zeta)\|_\infty \leqslant \|\omega_\nu(\cdot, \Theta)\|_\infty\}$$

is nonempty and compact in $\mathbb{C}^{n+1}$ and $\|\omega_\nu(\cdot, \zeta)\|_\infty$ as a function of $\zeta \in \mathbb{C}^{n+1}$ is continuous on K. Therefore a vector $\hat{\zeta}$ which minimizes $\|\omega_\nu(\cdot, \zeta)\|_\infty$ exists. Thus the existence of generalized T-polynomials is established. To show that all $\hat{z}_j \in H$, $j = 0, 1, ..., n$, we assume the contrary, namely, that one of the $\hat{z}_j$, say $\hat{z}_0$ is outside of H. Since H is compact and convex there is a uniquely determined point $u \in H$ with

$$0 < |\hat{z}_0 - u| \leqslant |\hat{z}_0 - z| \qquad \text{for all } z \in H.$$

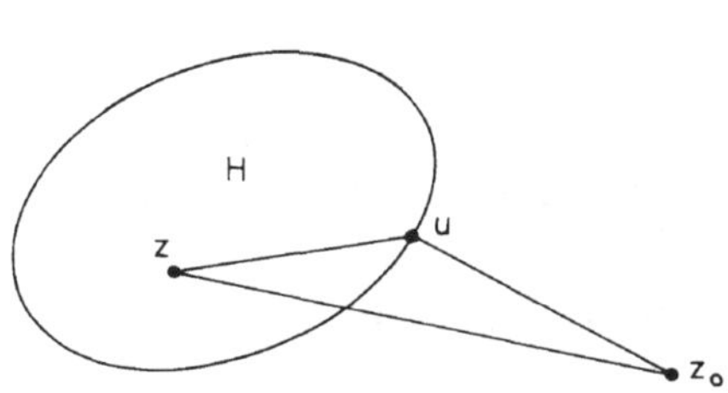

FIG. 1. Zero z_0 outside of H.

In addition, since H can be separated strictly from $\{\hat{z}_0\}$ (see Fig. 1) we have

$$|z-u| < |z-\hat{z}_0| \qquad \text{for all } z \in H. \tag{2.17}$$

Let

$$\omega_\nu(z, \zeta) = \prod_{j=0}^{n} (z-\hat{z}_j)^{\nu_j},$$

then there is a $\hat{\hat{z}} \in D$ with

$$|\hat{\hat{z}}-u|^{\nu_0} \prod_{j=1}^{n} |\hat{\hat{z}}-z_j|^{\nu_j} = \max_{z\in D} \left\{ |z-u|^{\nu_0} \prod_{j=1}^{n} |z-\hat{z}_j|^{\nu_j} \right\}$$

$$\geqslant \max_{z\in D} \prod_{j=0}^{n} |z-\hat{z}_j|^{\nu_j} \geqslant \prod_{j=0}^{n} |z-\hat{z}_j|^{\nu_j} \qquad \text{for all } z \in D. \tag{2.18}$$

By letting $z=\hat{\hat{z}}$ in (2.18) we obtain

$$|\hat{\hat{z}}-u|^{\nu_0} \prod_{j=1}^{n} |\hat{\hat{z}}-\hat{z}_j|^{\nu_j} \geqslant |\hat{\hat{z}}-\hat{z}_0|^{\nu_0} \prod_{j=1}^{n} |\hat{\hat{z}}-\hat{z}_j|^{\nu_j}$$

which contradicts (2.17) in case $\prod_{j=1}^{n} |\hat{\hat{z}}-\hat{z}_j|^{\nu_j} > 0$. If, however, $\prod_{j=1}^{n} |\hat{\hat{z}}-\hat{z}|^{\nu_j} = 0$, then ω_ν vanishes identically, which implies $D \subset \{z_0, z_1, \ldots, z_n\}$, a case which we have excluded.

If all ν_j, $j=0, 1, \ldots, n$, are the same, then the problem of finding the corresponding generalized T-polynomial may be regarded as finding a best approximation of z^{n+1} by polynomials of degree n or less. This problem has a unique solution. However, in general the unicity problem is open here.

It is easy to see that $t_n(z) = z^n$ is the nth T-polynomial with respect to all disks centered at zero. An even sharper result, given by Geiger and Opfer [7], says that $t_n(z) = z^n$ is the nth T-polynomial with respect to any sector.

$$S^\alpha = \{z \in \mathbb{C} : |z| \leqslant 1, |\arg z| \leqslant \alpha\}, \tag{2.19}$$

provided $\alpha \geqslant n\pi/(n+1)$. In example 2.2 we have also observed that $t_1(z) = z$ is the T-polynomial for the unconnected set D defined in (2.13). Also the ordinary T-polynomials are T-polynomials on a whole family of sets (see Example 2.1).

On the other hand, Rivlin [18] established that any monic polynomial (i.e., the coefficient of the highest power is one) is *the* T-polynomial for a certain set D.

It seems therefore natural to pose the following:

Problem 2.1. Let $t_n(z) = z^n + p(z)$, $p \in \Pi_{n-1}$, $n \in \mathbb{N}$ be any given polynomial.

(A) Characterize (or determine explicitly) those compact sets $D \subset \mathbb{C}$ for which t_n is the nth T-polynomial with respect to D.

(B) Problem A with additional restriction that D is convex.

(C) Consider problems A and B for the special case when $p \in \Pi_{n-1}^{\mathbb{R}}$.

As mentioned earlier a compact set D which is symmetric with respect to the real axis yields real T-polynomials. To see that the converse is not true, assume that $t_n(z) = z^n$ is the T-polynomial with respect to a compact set D which is not a disk centered at zero. Examples of such a set D are the sectors S^α introduced in (2.19) for $\alpha \in [\pi n/(n+1), \pi]$ (cf. Geiger and Opfer [11]).

An application of formula (2.10) shows that $t_n(z) = z^n$ is also the T-polynomial for $e^{i\phi}D$ for all angles $\phi \in [0, 2\pi]$. Since D is not a disk, the sets $e^{i\phi}D$ cannot be symmetric with respect to the real axis for all ϕ.

Definition 2.1. Let $t_n(z) = z^n + p(z)$, $p \in \Pi_{n-1}$, $n \in \mathbb{N}$, be a monic polynomial. All compact sets $D \subset \mathbb{C}$ which have the property that t_n is the T-polynomial with respect to D will be called *Chebyshev sets* with respect to t_n. The family of all Chebyshev sets with respect to t_n will be called *Chebyshev cluster* with respect to t_n.

It seems even difficult to determine the Chebyshev cluster $\mathscr{C}$ with respect to $t_1(z) = z$. Clearly all compact sets D which are symmetric with respect to the real and imaginary axes belong to $\mathscr{C}$. Also all sectors $S^\alpha \in \mathscr{C}$ for $\alpha \in (\pi/2, \pi)$. It may be noted that S^α is not convex for $\alpha \in (\pi/2, \pi)$. If $D \in \mathscr{C}$, then $a \cdot D \in \mathscr{C}$ for all $a \in \mathbb{C}$, $a \neq 0$.

Let $f\colon D \subset \mathbb{C} \to \mathbb{C}$ be any continuous function. Then

$$E_f = \{z \in D : |f(z)| = \|f\|_\infty\} \tag{2.20}$$

will be called the *set of extremal points of f*. The number and distribution of extremal points is studied by Grothkopf and Opfer [12] for T-polynomials on sector S^α.

3. Optimal Knots on Triangles

There are two aspects of choosing knots. One is the purely interpolating aspect to reduce the error in (2.1) by an appropriate choice of the zeros of $\omega(z, \zeta)$ as defined in (2.2).

The other aspect is relevant if we want to use the knots for constructing splines. In that case we have to impose the additional side condition that the knots are on the boundary of the corresponding meshes.

If, for example, the underlying set D is a triangle, then Theorem 2.1 implies that all zeros of the corresponding T-polynomial with respect to D

are in D. However, the zeros will usually not be on the boundary ∂D of D as can be seen in the following:

EXAMPLE 3.1. Let D be any triangle. The first degree T-polynomial $t_1(z) = z - a$ is obtained either by letting a be the center of the circumscribed circle about D (in case D is acute, i.e., all angles are less or equal to $\pi/2$), or by letting a be the midpoint of the longest side of D (in case D is obtuse). It may also be noted that in the acute case t_1 has three extremal points, namely, the vertices of D, whereas in the obtuse case there are only two extremal points, namely, the endpoints of the longest side of D.

If we want to construct splines on triangular grids as explained earlier, then the knots must be on the edges of the corresponding triangles which imposes additional side conditions. Usually the desired number of knots is a multiple of three since we want to place the same number of knots on each edge of the triangles.

Let us now study the problem for the triangles

$$D = \text{convex hull}\{0, e^{i\alpha}, e^{-i\alpha}\}, \qquad \alpha \in [0, \pi/2]. \tag{3.1}$$

Let $t_1(z) = z - a$ be the first T-polynomial with respect to D^α. Then

$$\begin{aligned} a &= 1/(2\cos\alpha) &&\text{for} \quad 0° \leqslant \alpha \leqslant 45° \\ &= \cos\alpha &&\text{for} \quad 45° \leqslant \alpha \leqslant 90°, \end{aligned} \tag{3.2}$$

$$\begin{aligned} \|T_1\|_\infty &= 1/(2\cos\alpha) &&\text{for} \quad 0° \leqslant \alpha \leqslant 45° \\ &= \sin\alpha &&\text{for} \quad 45° \leqslant \alpha \leqslant 90°. \end{aligned} \tag{3.3}$$

It may be noted that the coefficient a as a function of α is not differentiable at $\alpha = 45°$, whereas $\|T_1\|_\infty$ is differentiable in $(0, \pi/2)$ with respect to α.

If we call t_n^α the nth T-polynomial with respect to D^α for $\alpha \in [0, \pi/2]$, then t_n^α may be regarded as a homotopy from t_n^0 to $t_n^{\pi/2}$. The T-polynomial t_n^0 is the usual T-polynomial (up to normalization) on $[0, 1]$ and $t_n^{\pi/2}$ is the T-polynomial with respect to the segment $[-i, i]$ which can easily be determined from the ordinary T-polynomials on $[-1, 1]$ just by changing all coefficients into their absolute values. This follows from (2.10).

Thus the number of extremal points of t_n will start with $n+1$ for $\alpha = 0$ and will end with the same number for $\alpha = \pi/2$. Even for a small n explicit expressions for the nth T-polynomials t_n with respect to α are difficult to obtain. The additional requirement that the zeros of t_n should be on ∂D^α makes the problem even more complicated.

Thus we have made an attempt to find the optimal knots for the interesting cases $n = 3$ and $n = 6$ by numerical means. The T-polynomials

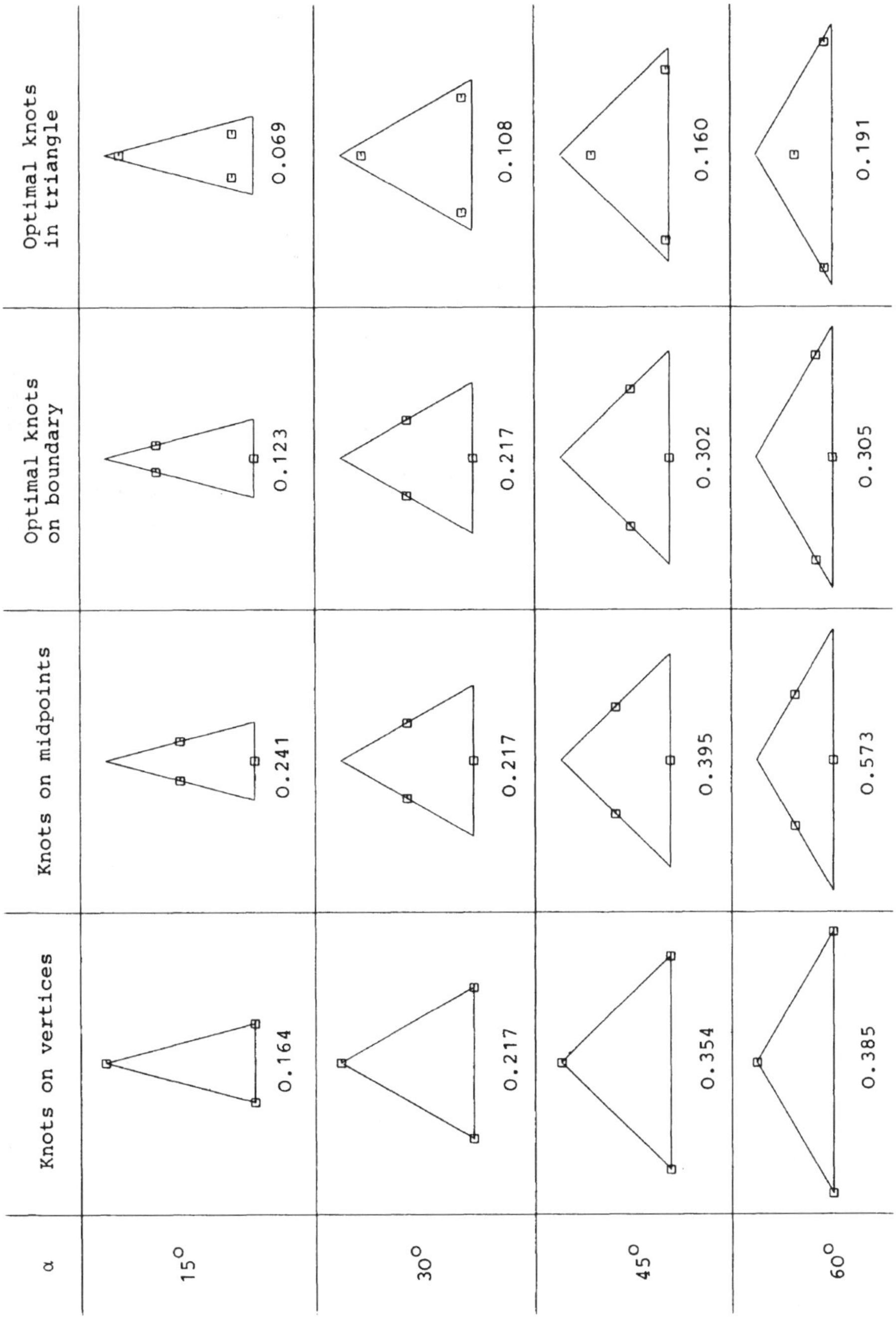

FIG. 2. Location of knots z_1, z_2, z_3 in $T_3(z) = (z - z_1)(z - z_2)(z - z_3)$ and $\|T_3\|_\infty$ for different cases.

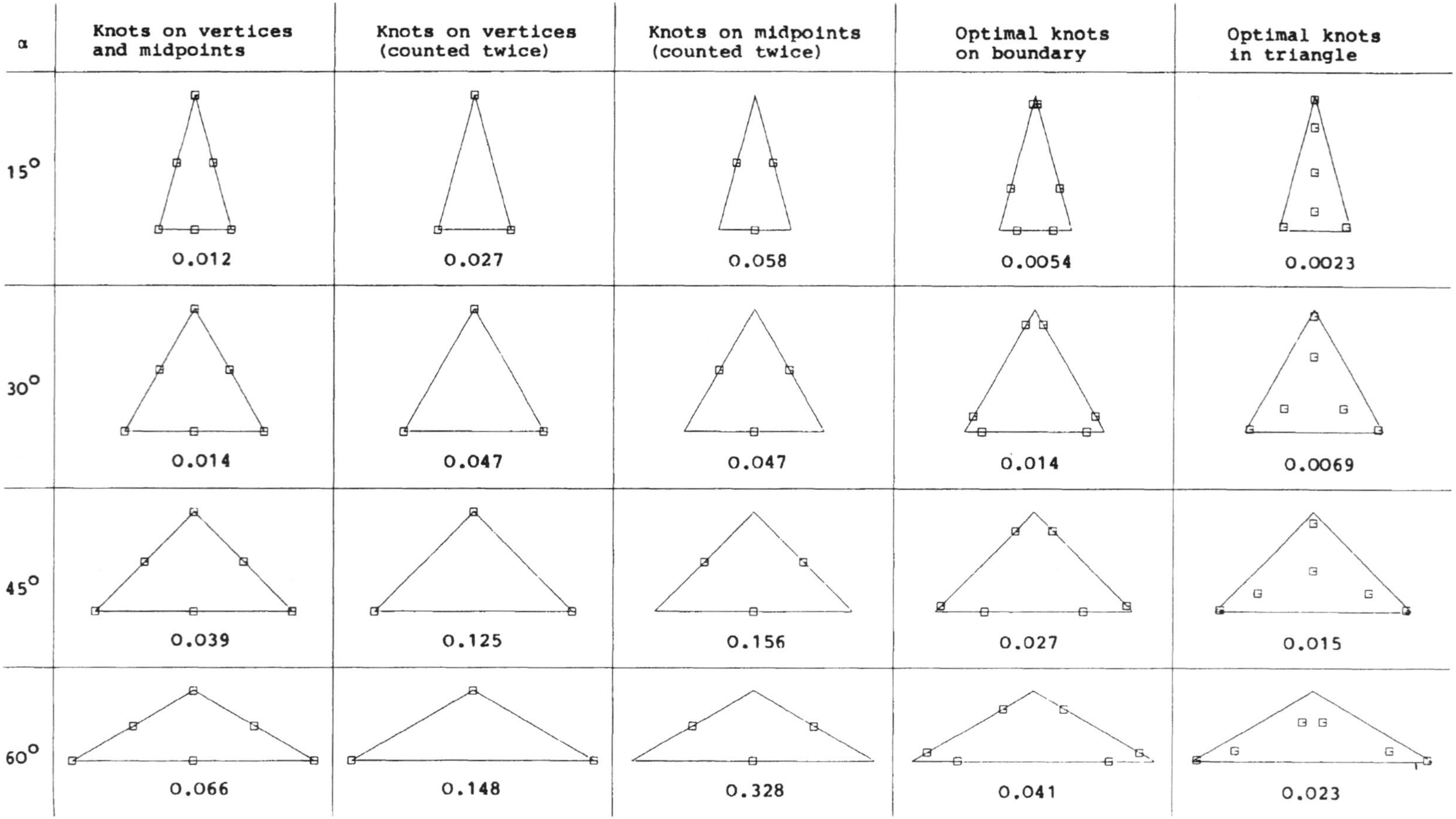

FIG. 3. Location of knots $z_1, z_2, \ldots, z_6$ in $T_6(z) = \prod_{j=1}^{6} (z - z_j)$ and $\| T_6 \|_\infty$ for different cases.

were computed by a method which is described in Grothkopf and Opfer [12]. Optimal knots on the boundary ∂D (which are not necessarily unique) were computed by a direct search method.

We selected the values $\alpha = 15°, 30°, 45°, 60°$, including the equilateral and the rectangular isosceles triangle. The results are displayed in Figs. 2 and 3.

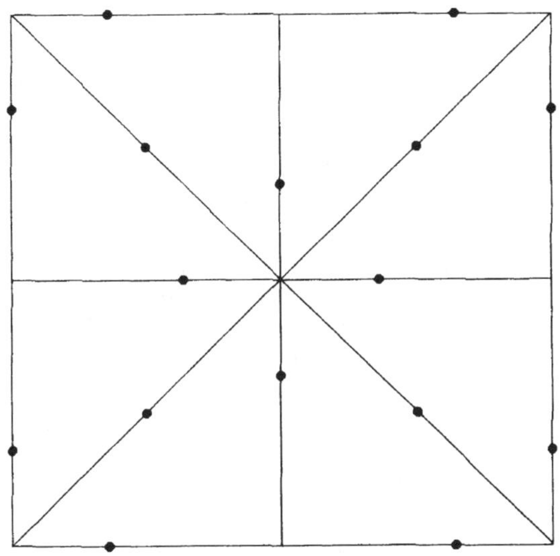

FIG. 4. "Union Jack" triangulation, optimal knots coincide on common edges of neighboring triangles

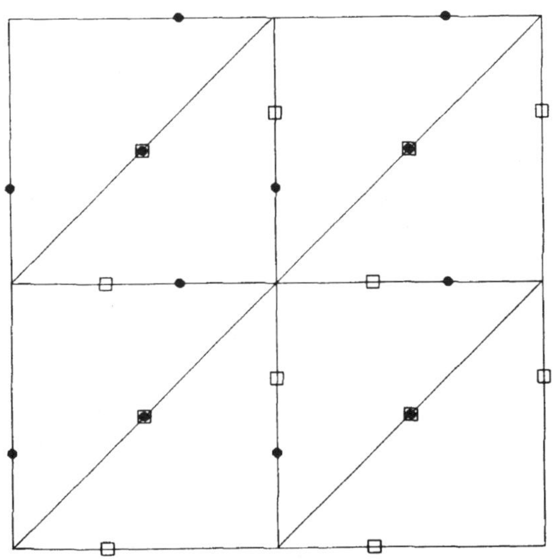

FIG. 5. Triangulation with noncoinciding optimal knots.

If one wants to use the optimal knots in the grids, then the so called "Union Jack" triangulation of a rectangular grid of congruent rectangles is well suited (see Fig. 4).

On the other hand, an unsymmetric triangulation of the same rectangular grid cannot be used since, on a common edge of two triangles, the optimal knots will, in general, not coincide (see Fig. 5).

References

1. M. Abramowitz and I. A. Stegun (eds.), "Handbook of Mathematical Functions," National Bureau of Standards, Applied Mathematics Series 55, U.S. Government Printing Office, Washington, 1972.
2. R. B. Barrar and H. L. Loeb, On Monosplines with odd multiplicity of least norm, J. Analyse Math. **33** (1978), 12–38.
3. R. B. Barrar, H. L. Loeb, and H. Werner, On the uniqueness of the best uniform extended totally positive monospline, J. Approx. Theory **28** (1980), 20–29.
4. B. D. Bojanov, A generalization of Chebyshev polynomials, J. Approx. Theory **26** (1979), 293–300.
5. B. D. Boyanov, Extremal problems in a set of polynomials with fixed multiplicities of zeros, *C. R. Acad. Bulgare Sci.* **31** (1978), 377–380.
6. C. de Boor and J. R. Rice, Extremal polynomials with application to Richardson iteration for indefinite linear systems, preprint, 1980.
7. P. J. Davis, "Interpolation and approximation," Blaisdell, New York/Toronto/London, 1963.
8. G. Faber, Über Tschebysheffsche Polynome, *J. Reine Angew. Math.* **150** (1920), 79–106.
9. W. H. J. Fuchs, On Chebyshev approximation on several disjoint intervals, *in* "Complex Approximation, Proceedings, Quebec, 1978" (B. Aupetit, Ed.), Progress in Mathematics Vol. 4, pp. 67–74 Birkhäuser, Basel/Boston/Stuttgart.
10. D. Gaier, "Vorlesungen über Approximation im Komplexen," Birkhäuser, Basel/Boston/Stuttgart, 1980.
11. C. Geiger and G. Opfer, Complex Chebyshev polynomials on circular sectors, *J. Approx. Theory* **24** (1978), 93–118.
12. U. Grothkopf and G. Opfer, Complex Chebyshev polynomials on circular sectors with degree six or less, *Math. Comp.*, in press.
13. Y. L. Luke, "Mathematical Functions and their Approximations," Academic Press, New York/San Francisco/London, 1975.
14. G. Meinardus, "Approximation of Functions: Theory and Numerical Methods," Springer-Verlag Berlin/Heidelberg/New York, 1967.
15. G. Opfer and M. L. Puri, planar splines, *J. Approx. Theory* **31** (1981), 383–402.
16. G. Opfer and G. Schober, On convergence and quasiregularity of interpolating complex planar splines, *Math. Z.* **180** (1982), 469–481.
17. T. J. Rivlin, "The Chebyshev Polynomials," Wiley, New York/London/Sydney/Toronto, 1974.
18. T. J. Rivlin, Best uniform approximation by polynomials in the complex plane, *in* "Approximation Theory III" (E. W. Cheney, Ed.), pp. 75–86, Academic Press, New York/London/Toronto/Sidney/San Francisco.

Printed by the St. Catherine Press Ltd., Tempelhof 41, Bruges, Belgium
Reprinted by permission of the publisher,
© 1983 by Academic Press

STRONG SOLUTIONS OF STOCHASTIC DIFFERENTIAL EQUATIONS FOR MULTIPARAMETER PROCESSES

MARKUS DOZZI† and MADAN L. PURI‡

Department of Mathematical Statistics, University of Berne, Sidlerstr. 5 3012 Berne, Switzerland
Department of Mathematics, Swain Hall East, Indiana University, Bloomington, Indiana 47405, USA

(*Accepted for publication August 28, 1985*)

We consider the stochastic differential equation (SDE) $X_t = V_t + \int_{]0,t]} f(s,\omega,X.(\omega))\,dZ_s(\omega)$, where V and Z are vector valued process indexed by $t \in \mathbb{R}^p_+$. The assumptions we make on Z and on the increasing process controlling Z are satisfied by certain classes of square integrable martingales, by processes of finite variation and by mixtures of these types of processes. Existence, uniqueness and the possibility of explosions of a strong solution X are investigated under Lipschitz conditions on f. A well-known sufficient condition for non-explosion is shown to work also in the multiparameter case and stability of X under perturbation of V, f and Z is proved. Finally more special SDE without Lipschitz conditions are considered, including a class of SDE of the Tsirel'son type.

KEY WORDS: Stochastic differential equations, multiparameter processes, explosions, stability.

---

†Research supported by the Natural Sciences and Engineering Research Council of Canada, during the visit at the Department of Mathematics of the University of British Columbia and by the Swiss National Foundation for Scientific Research (grant nr. 82.185.0.84) during the visit at the Laboratoire de Probabilité, Paris VI.

‡Research supported by the Office of Naval Research Contract N00014-85-K-0648.

1. INTRODUCTION AND NOTATION

We consider the stochastic differential equation (SDE)

$$X_t(\omega) = V_t(\omega) + \int_{]0,t]} f(s,\omega,X_{\cdot}(\omega))\, dZ_s(\omega) \tag{1.1}$$

where V and Z are vector-valued processes indexed by $t \in \mathbb{R}^p_+$ and f is matrix-valued; $f(s,\omega,X_{\cdot}(\omega))$ may depend on the past of X strictly before s and on ω explicitely. The assumptions we make on Z are weak enough to include square integrable martingales, processes of finite variation and certain mixtures of these types of processes (i.e. processes which are martingales in certain components and processes of finite variation in the others) in the case where the increasing process controling Z in the sense of (2.1) satisfies the hypothesis of proposition 3.1. In fact (1.1) is general enough to include the SDE considered in [5], [14], [18], [26], [29] and [32]. Nevertheless (1.1) looks similar to SDE for one-parameter processes and the results on the existence and uniqueness of strong solutions of (1.1) are similar to those known for one-parameter processes. One main difference is the notion of stopping a process and, associated to it, the possibility of localization of the stochastic integral when the integrands are not necessarily bounded. Let us notice that localization is easier when we assume a Lipschitz condition for f.

After a review of the notion of stopping boundaries and of stochastic integrals dominated by an increasing process in Section 2, we consider in Section 3 the existence and uniqueness of a solution of (1.1) on compact intervals under a uniform and a local Lipschitz condition on f. We prove that in the latter case explosions are possible in points on a stopping boundary. In Section 4 we show that a well-known sufficient condition for non-explosion works also for p-parameter processes and prove stability of the solution of (1.1) under perturbation of V, f and Z. Finally Section 5 is devoted to the study of more special SDE without Lipschitz conditions on f. In particular, following [27] rather closely, we give a class of SDE of the Tsirel'son-type without strong solutions.

A summary of these results appeared in [11]. We would like to thank Professors R. Cairoli, J. Walsh and M. Yor for fruitful discussions concerning this paper.

NOTATION The notation established for two-parameter processes (see [24]) can easily be extended to the p-parameter case. In particular we suppose given a stochastic basis $(\Omega, \mathscr{F}, (\mathscr{F}_t; t \in \mathbb{R}^p_+), P)$, where the filtration is supposed to be complete ($\mathscr{F}_0$ contains all $A \in \mathscr{F}$ with $P(A)=0$), increasing ($t \leqq t' \Rightarrow \mathscr{F}_t \subseteq \mathscr{F}_{t'}$) and right-continuous ($\cap\{\mathscr{F}_{t'}; t<t'\} = \mathscr{F}_t$ for all $t \in \mathbb{R}^p_+$). Here $t \leqq t'$ (resp. $t<t'$) is understood component-wise, i.e. $t_i \leqq t'_i$ (resp. $t_i < t'_i$) for $i=i,\dots,p$. For $T \subseteq \{1,\dots,p\}$ let $\mathscr{F}^T_t = \bigvee\{\mathscr{F}_{t'}; t'_i \leqq t_i \text{ for } i \in T\}$. We assume moreover that the following conditional independence holds: For all bounded random variables c, all $t \in \mathbb{R}^p_+$ and $T \subset T' \subseteq \{1,\dots,p\}$ we have $E[c \mid \mathscr{F}^{T'}_t] = E[E[c \mid \mathscr{F}^T_t] \mid \mathscr{F}^{T' \div T}_t]$. For $p=2$ this is the well-known condition (F4) of [6].

Let us denote by $C(\mathbb{R}^p_+, \mathbb{R}^m)$ (resp. $D(\mathbb{R}^p_+, \mathbb{R}^m)$) the class of continuous functions (resp. right-continuous functions with limits in all the orthants) defined on $\mathbb{R}^p_+$ with values in $\mathbb{R}^m$. The limits of the orthants are understood in the following sense: Let $\sigma = (\sigma_1, \dots, \sigma_p)$, where the σ_i are $+$ or $-$ signs. For fixed $t \in \mathbb{R}^p_+$ let $Q_\sigma = \{t'; t'_i < t_i$ if $\sigma_i = -$ and $t'_i \geqq t_i$ if $\sigma_i = +\}$. Then $x \in D(\mathbb{R}^p_+, \mathbb{R}^m)$ means that for each $t \in \mathbb{R}^p_+$ all the limits $\lim_{t' \to t, t' \in Q_\sigma, t' \neq t} X_{t'}$ exist and that x_t equals to the limit in the quadrant with $\sigma_i = +$ for all i. A stochastic process $X = (X_t; t \in \mathbb{R}^p_+)$ with values in $\mathbb{R}^m$ will be called right-continuous with limits in all the orthants if $X_.(\omega) \in D(\mathbb{R}^p_+, \mathbb{R}^m)$ P-a.s.

A real-valued right-continuous process $A = (A_t; t \in \mathbb{R}^p_+)$ is said to be increasing if $A_t = 0$ P-a.s. for $t \in [0,\infty[\div]0,\infty[$ and $A(]t,t']) \geqq 0$ for all intervals $]t,t'] = \Pi^p_{i=1}]t_i, t'_i]$. We note that the increment of $x \in D(\mathbb{R}^p_+, \mathbb{R}^m)$ on the interval $]t^1, t^2]$ can be given by

$$x(]t^1,t^2]) = \sum_{i(1),\dots,i(p)=1}^{2} (-1)^{\sum_{j=1}^{p} i(j)} x(t_1^{i(1)},\dots,t_p^{i(p)}).$$

We notice that the hypothesis of conditional independence of the filtration $(\mathscr{F}_t; t \in \mathbb{R}^p_+)$ stated above is satisfied if the following condition holds for every $T \subseteq \{1,\dots,p]$: Every $\mathscr{F}^T$-martingale M (with $|T|$-dimensional index, $|T|$ denoting the number of elements of T) is a strong $\mathscr{F}^T$-martingale i.e. $E[M(]t,t']) \mid \bigvee_{i \in T} \mathscr{F}^{\{i\}}_t] = 0$ for all intervals $]t,t'] \subset {}^{|T|}_{+}$. For $p=2$ this is the condition (F5) of [35].

2. PRELIMINARIES

2.1. Stopping boundaries

Let us first consider a notion which is an immediate generalization of the notion of stopping lines for two-parameter processes [3], [7], [21], [24]. Let $A \subseteq \mathbb{R}_+^p \times \Omega$ a progressively measurable set and let $]A, \infty[(\omega) := \cup_{t \in A(\omega)}]t, \infty[$. Then $]A, \infty[$ is predictable. Let $[A, \infty[(\omega) := \overline{]A, \infty[(\omega)}$ and let $D(A) := [A, \infty[\div]A, \infty[$. We say that a progressively measurable set B is a *stopping boundary* if $B = D(B)$. Let $[0, B] :=]B, \infty[^c$. Then for every stopping boundary B we have $\{\omega; [0, t] \subseteq [0, B]\} \in \mathscr{F}_t$ for all $t \in \mathbb{R}_+^p$. Let $\{B^k; k \in \mathbb{N}\}$ be a sequence of stopping boundaries B^k. Then $\sup_k B^k := D(\cap_{k \in \mathbb{N}}]B^k, \infty[)$ and $\inf_k B^k := D(\cup_{k \in \mathbb{N}}]B^k, \infty[)$ are stopping boundaries. Moreover every stopping boundary B can be approximated "from above" by *simple stopping boundaries* in the following sense: let $D_k = \{l 2^{-k};\ l = (l_1, \ldots, l_p\} \in \mathbb{N}^p, l_i \leqq 2^{2k}, i = 1, \ldots, p\}$ and let $B^k = D(D_k \cap]B, \infty[)$. Then $\{B^k; k \in \mathbb{N}\}$ is said to be a sequence of simple stopping boundaries and $B = \inf_k B^k$.

2.2. Stochastic integrals

In the proof for the existence and uniqueness of a solution of (1.1) we shall assume that Z is an $\mathbb{R}^n$-valued right-continuous process with limits in all the orthants and that the stochastic integral of an *adapted simple process* $Y_t(\omega) = \sum_{k=1}^r a_k(\omega) I_{]u^k, v^k]}(t)$ (where a_k is $\mathscr{F}_{u^k}$-measurable) is controlled by an increasing process in the following sense:

(H_1) There exists an adapted increasing process A such that for every simple $(m \times n)$-matrix valued process Y and every simple stopping boundary B

$$E\left[\sup_{t \in [0, B]} \left| \int_{]0, t]} Y_s \, dZ_s \right|^2\right] \leqq E\left[c(A, B) \int_{]0, B]} \left| Y_s \right|^2 dA_s\right], \tag{2.1}$$

where $|\cdot|$ on the left (resp. right) side of (2.1) is some norm in $\mathbb{R}^m$ (resp. $\mathbb{R}^m \times \mathbb{R}^n$) and where $c(A, B)$ is a random variable, depending on A and B such that $[0, B] \subset [0, B']$ implies $c(A, B) \leqq c(A, B')$ and $\lim_k c(A, B^k) = c(A, \inf_k B^k)$ for every decreasing sequence $\{B^k; k \in \mathbb{N}\}$ of simple stopping boundaries.

Remark We notice that the integral on the left side of (2.1) equals $\sum_{k=1}^{r} a_k Z(]u^k \wedge t, v^k \wedge t])$ and that the hypotheses made in (H_1) imply that (2.1) holds for all stopping boundaries. Moreover the results of Allain [1], Bakry [4] and Hürzeler [15] imply that the stochastic integral on the left side of (2.1) exists for all predictable processes Y such that the right side of (2.1) is finite.

Let us now mention classes of stochastic processes Z which satisfy (H_1) on a compact interval $[0, t^0]$. We restrict ourselves to classes of real-valued processes and observe that if $Z^1, \ldots, Z^n$ (real-valued) satisfy (H_1) on $[0, t^0]$ with respect to increasing processes $A^1, \ldots, A^n$, then $Z = (Z^1, \ldots, Z^n)$ satisfies (H_1) on $[0, t^0]$ with respect to $\sum_{i=1}^{n} A^i$ and with $|\cdot|^2 = \sum_{j=1}^{m} (\cdot)^2$ on the left side of (2.1) and $|\cdot|^2 = \sum_{i=1}^{n} \sum_{j=1}^{m} (\cdot)^2$ on the right side of (2.1).

1) Adapted processes of square integrable variation. Choose $A = \operatorname{var} Z$ and $c(A, B) = \int_{]0, B]} d(\operatorname{var} Z)_t$.

2) Square integrable right-continuous martingales. By the maximal inequality [6] we have

$$E\left[\sup_{t \in [0, B]} \left| \int_{]0, t]} Y_s \, dZ_s \right|^2\right] \leq E\left[\sup_{t \in [0, t^0]} \left| \int_{]0, r]} I_{[0, B]} Y_s \, dZ_s \right|^2\right]$$

$$\leqq 2^{2p} E\left[\int_{]0, B]} \left| Y_s \right|^2 d\langle Z \rangle_s\right], \tag{2.2}$$

where $\langle Z \rangle$ is the (unique) predictable increasing process associated to Z. Notice that certain multiple integrals [6], [22], [31], [33] are included in this case: For $p = 2$ let $Z_t = \int_{]0, t]^2} \psi_{s, s'} \, dM_s \, dM_{s'}$, where M is a right-continuous strong martingale such that $E[M_{t^0}^4] < \infty$ and where ψ is given as in [6, p. 131]. Then Z is a square integrable, right-continuous martingale with associated increasing process $\langle Z \rangle_t = \int_{]0, t]^2} \psi_{s, s'}^2 \, d\langle M \rangle_s \, d\langle M \rangle_{s'}$. Notice also, that right-continuous, square integrable martingales admit limits in all the quadrants [2], [25].

3) Mixed stochastic integrals [6], [16], [22], [31], [34]. Consider again $p = 2$ and let

$$Z_t = \int_{]0, t]^2} \psi_{s, s'} \, d\mu_s \, dM_{s'} \text{ and } Z'_t = \int_{]0, t]^2} \psi'_{s, s'} \, dM_s \, d\mu_{s'}, \tag{2.3}$$

where μ is a process of bounded variation, M is given as in 2) and ψ and ψ' satisfy the assumptions of [31, p. 577]. By Theorems 2.5 and 3.1 of [31] there exists a constant c such that

$$E\left[\sup_{t\in[0,B]}\left|\int_{]0,t]} Y_s\,dZ_s\right|^2\right]$$

$$\leq cE\left[\sup_{t\in[0,t^0]}\left|\int_{]0,t]^2} I_{[0,B]}(s\vee s')Y_{s\vee s'}\psi_{s,s'}\,d\mu_s\,dM_{s'}\right|^2\right]$$

$$\leq cE\left[\int_{[0,t^0]^2} I_{[0,B]}(s\vee s')Y^2_{s\vee s'}\psi^2_{s,s'}\,d(\operatorname{var}\mu)_s\,d\langle M\rangle_{s'}\right]. \tag{2.4}$$

Therefore Z satisfies (H_1) under each of the following conditions: (a) ψ is bounded, (b) $\psi_{s,s'}=\psi_{s\vee s'}$. In case (b) (2.4) is majorized by $cE[\int_{]0,B]}Y_s^2\,dA_s]$, where $A_t=\int_{]0,t]}\psi_s^2\,d_{s_1}(\operatorname{var}\mu)_s\,d_{s_2}\langle M\rangle_s$.

4) McShane-integrators [19], [9]. An adapted, right-continuous process with limits in all the orthants is called a McShane integrator if there exist continuous increasing functions G^i $(i=1,\ldots,p)$, defined on $\mathbb{R}_+$ with $G^i(0)=0$ and such that $G:=\prod_{i=1}^p G^i$ satisfies the following condition:

$$\max\{|E[Z(]t,t'])|\mathscr{F}_t^*]|,E[Z(]t,t'])^2|\mathscr{F}_t^*]\}\leq G(]t,t'])$$

for all intervals $]t,t']\subset[0,t^0]$. Here $\mathscr{F}_t^*=\bigvee_{i=1}^p\mathscr{F}_t^{\{i\}}$. Notice that under the condition of conditional independence of $(\mathscr{F}_t;t\in\mathbb{R}_+^p)$, Z can be decomposed into the sum of a square integrable strong martingale and an adapted process of square integrable variation ([10] for $p=2$).

2.3. Approximation and localization of stochastic integrals

Let us show that the approximation by simple processes is uniform on compact intervals and that the stochastic inegral $\int_{]0,.]}Y\,dZ$ is itself right-continuous with limits in all the orthants.

PROPOSITION 2.5 *Let Z be an adapted right-continuous process with limits in all the orthants that satisfies (H_1) on an interval $[0,t^0]$. Let Y be a predictable process such that $E[c(A,t^0)\int_{]0,t^0]}|Y_s|^2\,dA_s]<\infty$ and* *let* $\{Y^k; k\in\mathbb{N}\}$ *be a sequence of simple processes such that* $E[c(A,t^0)\int_{]0,t^0]}|Y_s-Y_s^k|^2\,dA_s]\to 0$ *as* $k\uparrow\infty$. *Then* P-a.s. *the paths of the process* $X_t^k := \int_{]0,t]} Y_s^k\,dZ_s$ *converge uniformly on* $[0,t^0]$ *to the paths of a right-continuous modification with limits in all the orthants of* $X_t := \int_{]0,t]} Y_s\,dZ_s$.

Proof By choosing a subsequence we may suppose $E[c(A,t^0)\int_{]0,t^0]}|Y_s-Y_s^k|^2\,dA_s]<8^{-k}$. Let $B^k := D(\{(t,\omega);\ |X_t^{k+1}-X_t^k|^2>2^{-k}\}\cap[0,t^0])$. Then B^k is a (simple) stopping boundary and we have by (2.1)

$$E\left[\sup_{t\in[0,t^0]}\left|\int_{]0,t]} I_{[0,B^k]}(s)\,(Y_s^{k+1}-Y_s^k)\,dZ_s\right|^2\right]\leqq 2.8^{-k}. \tag{2.6}$$

Moreover the left side of (2.6) is at least $2^{-k}P(\{\omega;[0,B^k]\subset[0,t^0]\})$. This implies $P(\limsup_{k\uparrow\infty}\{\omega;[0,B^k]\subset[0,t^0]\})=0$, i.e. for almost all ω there exists $k_0(\omega)$ such that for $k>k_0(\omega)$ we have $|X_t^k-X_t^{k+1}|\leqq 2^{-k}$ for all $t\in[0,t^0[$. Therefore P-a.s. $\{X_t^k(\omega);k\in\mathbb{N}\}$ is a Cauchy sequence in $\mathbb{R}^m$, which converges uniformly on $[0,t^0[$ to a function $\hat{X}_t(\omega)$. Since X_t^k converges in L^2 to $X_t, \hat{X}_t$ is a modification of X with right-continuous trajectories and limits in all the orthants. □

Definition 2.7 An adapted process Y is said to be *locally bounded on* $[0,t^0]$ if there exists a sequence $\{B^k;k\in\mathbb{N}\}$ of stopping boundaries B^k such that $\lim_{k\to\infty}P([0,B^k]\subset[0,t^0])=0$ and $|YI_{[0,B^k]}|\leqq k$ P-a.s. $\{B^k;k\in\mathbb{N}\}$ will be called a *localizing sequence for Y*.

The following proposition shows that the stochastic integral of locally bounded processes exists.

Proposition 2.8 *Let Z be a right-continuous process with limits in all the orthants that satifies* (H_1) *on* $[0,t^0]$ *and let Y be predictable and locally bounded on* $[0,t^0]$ *with localizing sequence* $\{B^k;k\in\mathbb{N}\}$. *Let* $X_t^k=\int_{]0,t]}Y_sI_{[0,B^k]}(s)\,dZ_s$. *Then,* P-a.s., $X_t^k(\omega)=X_t^{k+l}(\omega)$ *for all* $l\in\mathbb{N}$ *and* $t\in[0,B^k(\omega)]$.

Proof $E[\sup_{t\in[0,B^k]}|X_t^k-X_t^{k+l}|^2]=0$ by (2.1) and the remarks following (2.1) □

2.4. Lipschitz conditions

Let us formulate the condition on f in (1.1). We assume that f as an application defined on $[0, t^0] \times \Omega \times D(\mathbb{R}^p_+, \mathbb{R}^m)$ with values in the set of $m \times n$-matrices is predictable with respect to the basis $(\Omega \times D(\mathbb{R}^p_+, \mathbb{R}^m), (\mathscr{F}_t \otimes \mathscr{D}([0,t], \mathbb{R}^m)_{t \in \mathbb{R}^p_+})$, where $\mathscr{D}([0,t], \mathbb{R}^m)$ is the σ-algebra generated by the sets $\{x \in D(\mathbb{R}^p_+, \mathbb{R}^m); x_s \in B\}$ for $s \leqq t$ and Borel sets $B \subseteq \mathbb{R}^m$.

LEMMA 2.9 *For every $\mathbb{R}^m$-valued right-continuous process X with limits in all the orthants the process $Y_t(\omega) := f(t, \omega, X_{.}(\omega))$ is predictable with respect to the basis $(\Omega, (\mathscr{F}_t)_{t \in \mathbb{R}^p_+})$. Moreover $Y_t(\omega)$ depends only on the values of $X_s(\omega)$ for $s < t$.* □

The proof is easily adapted from Proposition 6.4 of [23] and is therefore omitted here. We consider now a uniform and a local Lipschitz property of f.

(H_2^u) There exists an adapted, increasing and left-continuous process L (i.e. $L_t = \lim_{t' < t, t' \to t, t' \neq t} L_{t'}$) such that

$$|f(t, \omega, x) - f(t, \omega, x')|^2 \leqq L_t(\omega) \sup_{s<t} |x_s - x'_s|^2 \tag{2.10}$$

for all $t \in [0, t^0]$ and $x, x' \in D(\mathbb{R}^p_+, \mathbb{R}^m)$, where $|\cdot|$ on the left (resp. right) side of (2.10) is the same as $|\cdot|$ on the right (resp. left) side of (2.1).

(H_2^l) For every $b > 0$ there exists an adapted, increasing and left-continuous process L^b such that

$$|f(t, \omega, x) - f(t, \omega, x')|^2 \leqq (L^b)_t(\omega) \sup_{s<t} |x_s - x'_s|^2 \tag{2.11}$$

for all $t \in [0, t^0]$ and $x, x' \in \{y \in D(\mathbb{R}^p_+, \mathbb{R}^m); \sup_{s<t} |y_s|^2 \leqq b\}$, where the norms in (2.11) are the same as in (2.10).

(H_3) $f(t, \cdot, 0)$ is locally bounded.

3. EXISTENCE, UNIQUENESS AND EXPLOSIONS OF STRONG SOLUTIONS

In this section we consider the existence and uniqueness of strong solutions of (1.1) under Lipschitz conditions on f. Just as in the one-parameter case (see [8], [17], [23] for example) the proof of the existence and uniquness of a solution of (1.1) reduces to the fixed-point theorem in a suitably chosen complete metric space and requires that $[0, t^0]$ can be decomposed into (random) subintervals on which $\int L\,dA < 1$. Whereas in the one-parameter case such a decomposition can be constructed explicitely, we could give such a decomposition in specific cases only, which, however, are general enough to cover the SDE investigated so far for multiparameter processes. In the proof for the existence of a solution of (1.1) we shall therefore assume in (H_4^u) below that such a decomposition exists locally (i.e. between the "big" discontinuities of A which are, by a result of [20], contained in a countable number of stopping boundaries) and indicate how to construct this decomposition in the cases we have in mind.

(H_4^u) For each $r \in \mathbb{N}$ there exist increasing sequences $\{B_r^{k,l};\ k \in \mathbb{N}\}$ $(l \in \mathbb{N})$ of stopping boundaries $B_r^{k,l}$ with $[0, B_r^{k,l}] \subseteq [0, t^0]$ such that

a) $[0, B_r^{k,l}] \subseteq [0, B_r^{k',l+1}]$ for all $k, k', l \in \mathbb{N}$;

b) for all $l \in \mathbb{N}$ $c(A, B_r^{k,l}) \int_{]B_r^{k-1,l}, B_r^{k,l}]} L_s\, dA_s \leqq \frac{1}{2}$ P-a.s. for all $k \in \mathbb{N}$, where A and c are given in (H_1) and where $B_r^{0,1} = [0, \infty[\div]0, \infty[$; $B_r^{0,l} = \sup_k B_r^{k,l-1}$;

c) $\lim\limits_{r \uparrow \infty} P\left(\overline{\bigcup_l \bigcup_k]B_r^{k-1,l}, B_r^{k,l}]} \subset [0, t^0]\right) = 0.$

(H_4^l) For each $b > 0$ there exist families of increasing sequences $\{B_r^{b,k,l}; k \in \mathbb{N}\}$ $(l, r \in \mathbb{N})$ of stopping boundaries $B_r^{b,k,l}$ with $[0, B_r^{b,k,l}] \subseteq [0, t^0]$ such that (a)–(c) in (H_4^u) are satisfied for L^b (defined in (H_2^l)) and $B_r^{b,k,l}$ instead of L and $B_r^{k,l}$.

PROPOSITION 3.1 *Suppose that A and c in (H_1) are given by $A_t = \int_{]0,t]} U_s \prod_{i=1}^p dA_{s_i}^i$ and $c(A, t) = \prod_{i=1}^p c^i(A^i, t_i)$, where $(U_t : t \in \mathbb{R}_+^p)$ is non-negative, predictable and locally bounded on $[0, t^0]$ and where $(A_{t_i}^i; t_i \geqq 0)$ and $(c^i(A^i; t_i); t_i \geqq 0)$ are one-parameter increasing processes, predictable with respect to $(\mathscr{F}_t; t_j = 0$ for $j \neq i)$. Then (H_4^u) holds.*

Proof Let $r \in \mathbb{N}$ and let $(T_i^{l_i}; l_i \in \mathbb{N})$ be the times of the (predictable) jumps of A^i of size $\geqq \frac{1}{2}(2r^2)^{-1/p}$. Let us consider the following stopping times on $]T_i^{l_i}, T_i^{l_i+1}]$: $S_i^{0,l_i} = T_i^{l_i}$, $S_i^{k,l_i} = \inf\{t_i > S_i^{k-1,l_i}; c^i(A^i, t_i) A^i(]S_i^{k-1,l_i}, t_i]) > \frac{1}{2}(2r)^{-1/p}\} \wedge T_i^{l_i+1}$ $(k \in \mathbb{N})$. Let $\bar{T}_i^r = \inf\{t_i; c^i(A^i, t_i) \geqq r^{1/p}\}$. Let $(B_r; r \in \mathbb{N})$ be a sequence of stopping boundaries B_r

such that $L_t U_t I_{[0,B_r]}(t) \leqq r$. Together with stopping times announcing the $T_i^{l_i}$ $(l_i \in \mathbb{N})$ and the $\bar{T}_i^r$, the $S_i^{k,l_i}(i=1,\ldots,p)$ generate a grid of $[0,t^0]$ that allows to construct sequences $\{B_r^{k,l}; k \in \mathbb{N}\}$ $(l \in \mathbb{N})$ of stopping boundaries $B_r^{k,l}$ such that $\bar{B}_r^{k,l} := B_r^{k,l} \wedge B_r$ satisfy (a) and (b) of (H_4^u). Since $P(\bigcup_k \bigcup_l [0, \bar{B}_r^{k,l}] \subset [0,t^0]) \to 0$ $(r \uparrow \infty)$, also (c) is satisfied. □

Let us now look at the restrictions that proposition 3.1 imposes on the examples considered in Section 2.2 If $Z_t = \int_{]0,t]^2} \psi_{s,s'} \, dM_s \, dM_{s'}$, suppose that $\langle M \rangle_t = A^1_{t_1} A^2_{t_2}$, where A^1 and A^2 satisfy the hypotheses of proposition 3.1 and suppose that ψ is (a) bounded or (b) $\psi_{s,s'} = \psi_{s \vee s}$ and ψ is locally bounded. Then $\langle Z \rangle$ satisfies 3.1. This is the case for $(J_Z)_t := \int_{]0,t]^2} 1_{s \wedge s'} \, dM_s dM_{s'}$ in particular. In the same way Z given by (2.3) satisfies 3.1 if M and ψ (or ψ') are as above and μ is a product of one-parameter processes. Let us mention that $Z_t := M^1_{t_1} M^2_{t_2}$ satisfies 3.1 for $(M^i_{t_i}; \mathscr{F}^i_{t_i}; t_i > 0)$ $(i=1,\ 2)$ independent square integrable (one-parameter) martingales. In fact Z is an $\mathscr{F}^1_{t_1} \otimes \mathscr{F}^2_{t_2}$-martingale with increasing process $\langle Z \rangle_t = \langle M^1 \rangle_{t_1} \langle M^2 \rangle_{t_2}$. Finally McShane integrators satisfy 3.1 because the controlling process G is deterministic and of product type.

Remark (3.2) Let $i \in \{1,\ldots,p\}$ and suppose that Z admits $(\mathscr{F}_t^{\{i\}}; t_i \geqq 0)$-predictable integrands. This is the case for right-continuous, square integrable strong martingales, for example. Then stopping boundaries satisfying (H_4^u) with respect to the filtrations $\mathscr{F}_t^{\{i\}}$ can be constructed as follows:
Consider (for $i=1$)

$$T_1^0 = 0,\ T_1^k = \inf\left\{t_1 > T_1^{k-1}; c(A,(t_1,t_2^0,\ldots,t_p^0)) \int_{]T_1^{k-1},t_1]^{(1)}} L_s \, dA_s > \frac{1}{2}\right\},$$

where

$$]T_1^{k-1},t_1]^{(1)} =]T_1^{k-1},t_1] \times \prod_{i=2}^{p} [0,t_i^0].$$

Then $\{T_1^k\} \times \prod_{i=2}^p [0,t_i^0]$ $(k \in \mathbb{N})$ are stopping boundaries and can be used to construct $B_r^{k,l}$ in (H_4^u) in a similar way as in Proposition 3.1.

Let us now give a first result of existence and uniqueness of a strong solution of (1.1). Notice that the conditional independence property of the filtration is not used in the proof of the following two theorems.

THEOREM 3.3 *Let* V (resp. Z) *a* $\mathbb{R}^m$-(resp. $\mathbb{R}^n$-) *valued, adapted, right-continuous processes with limits in all the orthants such that* Z *satisfies* (H_1) *on* $[0, t^0]$. *Let* f *be an* $(m \times n)$*-matrix-valued predictable process defined on* $\mathbb{R}^p_+ \times \Omega \times D(\mathbb{R}^p_+, \mathbb{R}^m)$ *such that* (H^u_2) *and* (H_3) *is satisfied. Suppose finally that* (H^u_4) *holds. Then there exists an* $\mathbb{R}^m$-*valued, adapted, right-continuous process* X *with limits in all the orthants, unique up to P-equivalence, such that (1.1) holds on* $[0, t^0]$.

Proof 1) Let us first notice, that we can suppose $V \equiv 0$, since X is a solution of (1.1) iff $X - V$ is a solution of $Y_t(\omega) = \int_{]0,t]} \bar{f}(s, \omega, Y_{\cdot}(\omega))\, dZ_s(\omega)$, where $\bar{f}(s, \omega, Y_{\cdot}(\omega)) := f(s, \omega, (Y+V)_{\cdot}(\omega))$. Notice that $\bar{f}$ satisfies (H^u_2) with respect to the same increasing process L as f.

2) Uniqueness up to P-equivalence: Suppose that X and X' are two solutions of (1.1) on $[0, \bar{B}]$ and $[0, \bar{B}']$, respectively, both with the properties mentioned in the theorem. Let us show that $X = X'$ P-a.s. on $[0, \bar{B}] \cap [0, \bar{B}']$. Let $\bar{B}^{k,l}_r := \bar{B} \wedge \bar{B}' \wedge B^{k,l}_r$ and $\hat{B} = D(\{(t, \omega); X_t \neq X'_t\})$. We assume that $[0, \bar{B}^{k,l}_r] \subset [0, \hat{B}]$ and show that $[0, \bar{B}^{k+1,l}_r] \subset [0, \hat{B}]$. Fix r and l and let $\bar{J}^k =]\bar{B}^{k,l}_r, \bar{B}^{k+1,l}_r]$. Then by Lemma 2.9

$$\|X - X'\|^2_{(k)} := E\left[\sup_{t \in \bar{J}^k} |X_t - X'_t|^2\right] \leqq E[c(A, \bar{B}^{k+1,l}_r) \int_{\bar{J}^k} L_t$$

$$\sup_{s<t} |X_s - X'_s|^2\, dA_s] \leqq \tfrac{1}{2} \|X - X'\|^2_{(k)}.$$

Therefore P-a.s. $X'_t(\omega) = X_t(\omega)$ for $t \in \bar{J}^k(\omega)$. By (c) of (H^u_4) P-a.s. $X'_t = X_t$ for $t \in [0, \bar{B}] \cap [0, \bar{B}']$.

3) Successive construction of the solution: Fix r, let $l = 1$ and write B^k instead of $B^{k,1}_r$. Part 4) below will imply that the solution exists on $[0, B^1]$. Let $b_1 > 0$ and let ${}^1B = D(\{t, (\omega); |X_t(\omega)| > b_1\})$. Consider ${}^1B^k := B^k \wedge {}^1B$. Then X is bounded by b_1 on $[0, {}^1B^1[$ and part 4) below will imply that the solution exists on $[0, {}^1B^2]$. Let ${}^2B = D(\{(t, \omega); |X_t(\omega)| > b_2\})$ for some $b_2 \geqq b_1$. Then X will be bounded by b_2 on $[0, {}^2B^2[$, where ${}^2B^2 = B^2 \wedge {}^2B$, and so on. Since the solution X is

right-continuous and has left limits in all the orthants, we can choose b_k such that $P([0, {}^kB^k] \neq [0, {}^{k-1}B^k]) < \varepsilon 2^{-k}$. (Let ${}^0B^1 := B^1$.) Therefore $P(\cup_k [0, {}^kB^k] \neq \cup_k [0, B^k]) < \varepsilon$ for any $\varepsilon > 0$. Therefore the solution of (1.1) will exist on $\cup_k [0, B_r^{k,l}]$ P-a.s.

4) We assume that the solution X exists on $[0, {}^kB^k]$ and is bounded by $b_k > 1$ on $[0, {}^kB^k[$ and show that the solution exists on $[0, {}^kB^{k+1}]$. Let $J^k =]{}^kB^k, {}^kB^{k+1}]$ and let

$$D^k := \left\{ Y; Y_\cdot \in D(\mathbb{R}_+^p, \mathbb{R}^m), YI_{[0,{}^kB^k]} = XI_{[0,{}^kB^k]} \text{ P-a.s.}, \right.$$

$$\left. \|Y\|_{(k)} := \left\{ E\left[\sup_{t \in J^k} |Y_t|^2 \right] \right\}^{1/2} < \infty \right\}.$$

Moreover, for $Y \in D^k$ and $t \in [0, t^0]$ let

$$(LY)_t = \int_{]0,t]} f(s, \cdot, Y_\cdot)\, dZ_s. \tag{3.4}$$

Notice that the stochastic integral on the right side of (3.4) is well defined by Proposition 2.8 since $f(s, \omega, Y_\cdot(\omega))$ is locally bounded on $[0, t^0]$ by (H_2^u) and (H_3). Moreover LY can be chosen right-continuous with limits in all the orthants by Proposition 2.5. In order to conclude by the fixed point theorem that there exists a (unique) element in D^k that is a solution of (1.1) on $[0, {}^kB^{k+1}]$ we have to show that (a) $\|LY\|_{(k)} < \infty$ and (b) $\|LY - LY'\|_{(k)} \leqq d\|Y - Y'\|_{(k)}$ for all $Y, Y' \in D^k$, where $d < 1$.

a) $(LY)_t = (XI_{[0,{}^kB^k]})_t + \left(\int_{]0,\cdot]} f(s, \cdot, Y_\cdot)\, dZ_s I_{J^k} \right)_t$

$$\|LY\|_{(k)}^2 \leqq 2 \left\| \int_{]0,\cdot]} [f(s,\cdot,Y_\cdot) - f(s,\cdot,0)]\, dZ_s \right\|_{(k)}^2 + 2 \left\| \int_{]0,\cdot]} f(s,\cdot,0)\, dZ_s \right\|_{(k)}^2$$

$$\leqq 4E\left[c(A, {}^kB^{k+1}) \int_{]0,{}^kB^{k+1}]} \left(L_t \sup_{s<t} \left| Y_s \right|^2 + |f(t,\cdot,0)|^2 \right) dA_s \right]$$

$$\leqq 4(b_k^2 + \|Y\|_{(k)}^2) E[c(A, {}^kB^{k+1}) \int_{]0,{}^kB^{k+1}]} (L_s + |f(s,\cdot,0)|^2)\, dA_s]. \tag{3.5}$$

Since L and $f(\cdot,\cdot,0)$ are locally bounded we may suppose $L_t+|f(t,\cdot,0)|^2 \leqq 2r$ for the r chosen in part 3) of the proof. This implies $\|LY\|_{(k)}<\infty$.

$$\text{b) } \|LY-LY'\|^2_{(k)}=E\left[\sup_{t\in J^k}\left|\int_{]0,t]}(f(s,\cdot,Y_\cdot)-f(s,\cdot,Y'_\cdot))\,dZ_s\right|^2\right]$$

$$\leqq E\left[c(A,{}^kB^{k+1})\int_{J^k}L_s\sup_{s<t}|Y_s-Y'_s|^2\,dA_s\right]\leqq \tfrac{1}{2}\|Y-Y'\|^2_{(k)}.$$

5) Extension of the solution from $[0,\sup_k B^k[$ to $[0,\sup_k B^k]$. Let Y be adapted, right-continuous and have limits in all the orthants such that $(YI_{[0,\sup_k B^k[})_t=(XI_{[0,\sup_k B^k[})_t$. For $t\in\sup_k B^k$

$$\int_{]0,t]} f(s,\cdot,Y_\cdot)\,dZ_s=\int_{]0,t]} f(s,\cdot,X_\cdot)\,dZ_s \text{ (P-a.s.)}$$

by Lemma 2.9. We define $X_t:=\int_{]0,t]}f(s,\cdot,X_\cdot)\,dZ_s$.

6) The procedure described in parts 3)–5) can be applied to prove the existence of the solution of (1.1) on $\bigcup_{l\in\mathbb{N}}\bigcup_{k\in\mathbb{N}}]B_r^{k-1,l},B_r^{k,l}]$ for all $r\in\mathbb{N}$. By (c) in (H_4^u) this defines the solution of (1.1) on $[0,t^0]$ P-a.s. □

Let us now replace the hypotheses (H_2^u) and (H_4^u) by (H_2^l) and (H_4^l) and show that a unique solution still exists but that explosions occur on a stopping boundary if the solution cannot be defined on the whole interval $[0,t^0]$.

THEOREM 3.6 *Let V, Z and f in (1.1) satisfy the same hypotheses as in Theorem (3.3), but with (H_2^u) and (H_4^u) replaced by (H_2^l) and (H_4^l). Then there exists a stopping boundary B and an $\mathbb{R}^m$-valued, adapted, right-continuous process X with limits in all the orthants, unique up to P-equivalence, defined on $[0,B[$ such that*

i) (1.1) holds for every $t\in[0,B(\omega)[$ and $\sup_{t\in[0,B(\omega)[}|X_t|=\infty$ P-a.s. *on $\{\omega;[0,B(\omega)]\subset[0,t^0]\}$;*

ii) $[0,B']\subseteq[0,B]$ P-a.s. *for all stopping boundaries B' with property (i).*

Proof As in the proof of the previous theorem we can assume $V \equiv 0$. Also the uniqueness of a solution of (1.1) is proved similarly, with L and $B_r^{k,l}$ replaced by L^b and $B_r^{b,k,l}$.

1) Construction of B: For $x \in D(\mathbb{R}_+^p, \mathbb{R}^m)$ let $x_.^b = x_.(1 \wedge b/|x_.|)$ and let $f^b(s, \omega, x_.) = f(s, \omega, x_.^b)$. As in [23, p. 79] one shows that f^b is predictable and satisfies (2.10) with $4L^b$ instead of L. Applying part (4) of the proof of the previous theorem to

$$X_t' = \int_{]0,t]} f^b(s, \omega, X')\, dZ_s \tag{1.1^{b}}$$

for $b_k = b$ $(k \in \mathbb{N})$ and for $B^{b,k}$ instead of B^k, one obtains stopping boundaries ${}^kB^{b,k}$ and the solution X' of (1.1^{b}) on $[0, {}^kB^{b,k}]$. Since X' is bounded by b on $[0, {}^kB^{b,k}[$, it is also solution of (1.1) on $[0, {}^kB^{b,k}]$ by Lemma 2.9. Let $B_r^{b,1} := \sup_k {}^kB_r^{b,k,1}$, apply the same procedure to $l > 1$ and let $B^b := \sup_l B_r^{b,l}$. Then the solution of (1.1) is known on $[0, B^b]$ and bounded by b on $[0, B^b[$ P-a.s. Let $B := \sup_{b \in \mathbb{N}} B^b$.

2) Explosions on B: Denote again by X the solution of (1.1) on $[0, B[$ constructed in 1). Let $B^n := D(\{(t, \omega); |X_t(\omega)| > n\} \cap [0, t^0])$. Suppose on the contrary that with probability $\varepsilon > 0$ there are no explosions on B i.e. $P([0, B \wedge B^n] = [0, B] \subset [0, t^0]) = \varepsilon$ for some $n \in \mathbb{N}$. Choose b in part (1) so that $P(\sup_{t \in [0, B \wedge B^n]} |X_t| > b/2) < \varepsilon/2$. Then by part (4) in the proof of the previous theorem, by part (1) above and by the right continuity of X we have $P([0, B^b] \not\subseteq [0, B \wedge B^n]) \geqq P([0, B \wedge B^n] \subset [0, t^0]) - \varepsilon/2$. Therefore $P([0, B^b] \not\subseteq [0, B] \subset [0, t^0] \geqq \varepsilon/2 > 0$ what contradicts the definition of B. Therefore $\sup_{t \in [0, B[} |X_t| = \infty$ P-a.s. on $[0, B] \neq [0, t^0]$. This proves (i) of Theorem 3.6; (ii) follows from the construction of B.

Remark The sequence $(B^n; n \in \mathbb{N})$ defined in part 2) of the proof above announces the explosions of X in the following sense: The explosions of $X_.(\omega)$ on $B(\omega)$ are contained in $\{t \in B(\omega); t \notin [0, B^n(\omega)]$ for all $n\}$. If X has continuous trajectories $\{B^n; n \in \mathbb{N}\}$ announces B strictly (in the sense of [3]).

Let $(W_t; t \in \mathbb{R}_+^2)$ be the (two-parameter) Brownian sheet, i.e. the centered Gaussian process with covariance function $EW_tW_{t'} = (t_1 \wedge t_1')(t_2 \wedge t_2')$ and let J_W be the associated product measure [6, p.

147]. Then the following corollary follow immediately from Proposition 3.1. and Theorem 3.3.

COROLLARY 3.7 *Let $(V_t; t \in \mathbb{R}^2_+)$ be an adapted, continuous, $\mathbb{R}^m$-valued process. Let $f=(f^{i,j}; i=1,\dots,m; j=1,\dots,5)$ be adapted, measurable and satisfy (H^u_2) and (H_3). Then the system of SDE in $X=(X^i; i=1,\dots,m)$*

$$X^i_t = V^i_t + \int_{]0,t]} [f^{i,1}(s,\cdot,X_.)\,dW_s + f^{i,2}(s,\cdot,X_.)\,d(J_W)_s$$

$$+ f^{i,3}(s,\cdot,X_.)\,d_{s_1}W_s\,ds_2$$

$$+ f^{i,4}(s,\cdot,X_.)\,ds_1\,d_{s_2}W_s + f^{i,5}(s,\cdot,X_.)\,ds] \tag{3.8}$$

has an adapted, continuous solution that is unique up to P-equivalence.

4. CONDITIONS FOR NON-EXPLOSION AND STABILITY OF SOLUTIONS

In this section we restrict ourselves to the case where the increasing process A in (H_1) satisfies the conditions of Proposition 3.1. First we show that a well-known condition for non-explosion works also in the multiparameter case. We follow [23, chapter 7] rather closely.

THEOREM 4.1 *Suppose that the conditions of Theorem 3.6. are satisfied such that the increasing process A in (H_1) satisfies the conditions of Proposition 3.1. We suppose moreover that there exists $d>0$ such that $|f(t,\omega,x_.)|^2 \leqq d(1+\sup_{s<t}|x_s|^2)$ for all $x \in D(\mathbb{R}^p_+,\mathbb{R}^m)$. Then there exists an adapted right-continuous process with limits in all the orthants that is a solution of (1.1) on $[0,t^0]$.*

Proof Again we may suppose $V \equiv 0$ in (1.1). Let $\bar{T}^r_i = \inf\{t_i; c^i(A^i,t_i) \bigvee A^i_{t_i} > r^{1/p}\}$ and let $\tilde{B}_r$ be given by $[0,\tilde{B}_r] = [0, B\wedge B_r] \cap \prod_{i=1}^p [0,\bar{T}^r_i]$, where B is given in Theorem 3.6 and B_r is a stopping boundary such that $U_t I_{[0,B_r]}(t) \leqq r$. We have to show that $\sup_{t\in[0,\tilde{B}_r[}|X_t|^2 < \infty$ P-a.s. for all $r \in \mathbb{N}$. This will imply that $[0,B]=[0,t^0]$, i.e. there exists a solution of (1.1) on $[0,t^0[$ that can be extended to $[0,t^0]$ by setting $X_t := \int_{]0,t]} f(s,\omega,X_.(\omega))\,dZ_s(\omega)$ for $t\in[0,t^0]\div[0,t^0[$.

Let now $(B^l; l\in\mathbb{N})$ be a sequence of stopping boundaries containing all the discontinuities of A generated by the jumps of A^i of size $>\frac{1}{2}(2r^2 d)^{-1/p}$, i.e. $]B^l, B^{l+1}]$ are intervals, say $\prod_{i=1}^{p}]T_i^{l_i}, T_i^{l_i+1}]$ where $T_i^{l_i}(l_i\in\mathbb{N})$ are the times of the jumps of A^i of size $>\frac{1}{2}(2r^2 d)^{-1/p}$. Let $T_i^0=0$, $T_i^{0,l_i}=T_i^{l_i}(l_i\geqq 0)$ and

$$T_i^{k,l_i}=\inf\{t_i>T_i^{k-1,l_i}; c^i(A^i,t_i)A^i(]T_i^{k,l_i},t_i])>\tfrac{1}{2}(2rd)^{-1/p}\}\wedge T_i^{l_i+1}.$$

Let $B^{0,l}=B^l$ and let $(B^{k,l}; k\in\mathbb{N})$ $(l\geqq 0)$ be sequences of stopping boundaries, exhausting the grid generated by the T_i^{k,l_i} $(k,l_i\geqq 0, i=1,\dots,p)$. Let furthermore $(\bar{B}_{r,n}; n\in\mathbb{N})$ be a sequence of stopping boundaries announcing $\prod_{i=1}^{p}[0,\bar{T}_i^r]$, let $\hat{B}^n=D(\{(t,\omega); |X_t|>n\})$, and let $(B^{l,n}; n\in\mathbb{N})$ be sequence announcing B^l $(l\in\mathbb{N})$. Consider $B_{r,n}^{k,l}=B^{k,l}\wedge\bar{B}_{r,n}\wedge\hat{B}^n\wedge B^{l+1,n}$ and $C_r^{k,l}=\cup_n[0,B_{r,n}^{k,l}]$. Then we have

$$c(A,B_{r,n}^{k,l})\int_{]B_{r,n}^{k-1,l},B_{r,n}^{k,l}]} dA_s \leqq (2d)^{-1}$$

and

$$\begin{aligned} E\Big[\sup_{t\in C_r^{k,l}}|X_t|^2\Big] &= \lim_{n\uparrow\infty} E\Big[\sup_{t\in[0,B_{r,n}^{k,l}]}|X_t|^2\Big] \\ &\leqq \lim_{n\uparrow\infty} E\Big[c(A,B_{r,n}^{k,l})\int_{[0,B_{r,n}^{k,l}]}|f(t,\cdot,X_.)|^2\,dA_t\Big] \\ &\leqq K+dE\Big[c(A,C_r^{k,l})\int_{C_r^{k,l}}\sup_{s<t}|X_s|^2\,dA_t\Big], \end{aligned} \tag{4.2}$$

where $K=dE[c(A,t^0)A_{t^0}]$. Then, denoting the left side of (4.2) by $x_r^{k,l}$ we have $x_r^{k,l}\leqq K+dr^3x_r^{k-1,l}+\frac{1}{2}x_r^{k,l}$. Therefore $x_r^{k,l}\leqq 2K+2dr^3x_r^{k-1,l}$ and $x_r^{k,l}\leqq 2K\sum_{j=0}^{k-1}(2dr^3)^j+(2dr^3)^k x_r^{0,l}$. Since the number of jumps of A^i of size $>\frac{1}{2}(2r^2d)^{-1/p}$ are bounded on $[0,\tilde{B}^r[$, $\overline{C_r^{k,l}}=[0,\tilde{B}^r]$ P-a.s. after a finite number (depending on d, r, p only) of steps. Therefore $E[\sup_{[0,\tilde{B}^r[}|X_t|^2]<\infty$ for all $r\in\mathbb{N}$. $\square$

Let us now proceed to the stability of the solution of (1.1). Let V and V^k (resp. Z and Z^k) ($k \in \mathbb{N}$) be $\mathbb{R}^m$-(resp. $\mathbb{R}^n$-)valued, adapted, right-continuous processes with limits in all the orthants and suppose that Z and Z^k satisfy (H_1) on $[0, t^0]$ and that the increasing processes $(\bar{A}^k; k \in \mathbb{N})$ controling $Z - Z^k$ in (H_1) satisfy the conditions of Proposition 3.1. Let $\bar{A}_t^k = \int_{]0,t]} \bar{U}_s^k \prod_{i=1}^p d\bar{A}_{s_i}^{k,i}$. Suppose that the processes $\bar{U}^k$ ($k \in \mathbb{N}$) are uniformly locally bounded (i.e. $\sup_{k \in \mathbb{N}} \bar{U}_t^k I_{]0,B_r]}(t) \leqq r$ for a localizing sequence $(B_r; r \in \mathbb{N})$). Let f and f^k be $(m \times n)$-matrix-valued predictable processes that satisfy (H_2^u) and (H_3) with the same increasing process L. Therefore solutions X of (1.1) and X^k of

$$X_t^k = V_t^k + \int_{]0,t]} f^k(s, \omega, X^k_{\cdot}(\omega))\, dZ_s^k(\omega) \tag{1.1^k}$$

exist on $[0, t^0]$ and are unique up to P-equivalence by Theorem 3.3. The perturbation is measured by the following quantities, that are assumed to be finite:

$$d_1^k = E\left[\sup_{t<t^0} |f(t,\cdot,X_{\cdot}) - f^k(t,\cdot,X_{\cdot})|^2\right], \quad d_2^k = E\left[\sup_{t<t^0} |V_t - V_t^k|^2\right],$$
$$d_3 = E\left[\sup_{t<t^0} |f(t,\cdot,X_{\cdot})|^2\right]. \tag{4.3}$$

THEOREM 4.4 *Suppose that, for $i = 1, \ldots, p$, $c^i(\bar{A}^{k,i}, t_i^0)\bar{A}_0^{k,i}$ converges to zero in probability as $k \uparrow \infty$, and suppose that $\lim_{k\uparrow\infty} d_1^k = \lim_{k\uparrow\infty} d_2^k = 0$. Then, under the assumptions made above, $\sup_{t<t^0} |X_t - X_t^k|^2$ converges to zero in probability as $k \uparrow \infty$.*

Proof Let $\bar{B}_r$ be given by $[0, \bar{B}_r] = \prod_{i=1}^p [0, \bar{T}_i^r]$, where $\bar{T}_i^r = \inf\{t_i; c^i(A^i, t_i) \vee A_{t_i}^i > r^{1/p}\} \wedge t_i^0$. Let $\varepsilon > 0$ and let $\bar{B}_\varepsilon^k$ be defined for $\bar{A}^k$ and ε in the same way as $\bar{B}_r$ for A and r. Let B_r be a stopping boundary such that $\sup_{k\in\mathbb{N}}(\bar{U}_t^k + L_t) I_{[0,B_r]}(t) \leqq r$. By assumption, for $\delta > 0$ given, there exists $r_0 = r_0(\delta)$ such that for $r \geqq r_0$ we have $P([0, \bar{B}_\varepsilon^k \wedge \bar{B}_r \wedge B_r] \subset [0, t^0]) < \delta$ for any $\varepsilon > 0$ and $k \geqq k_0(\delta, \varepsilon)$. Let us write

$$X_t - X_t^k = V_t - V_t^k + \int_{]0,t]} (f(s,\cdot,X_{\cdot}) - f^k(s,\cdot,X_{\cdot}))\, dZ_s$$

$$+\int_{]0,t]} (f^k(s,\cdot,X_\cdot)-f^k(s,\cdot,X^k_\cdot))\,dZ_s$$

$$+\int_{]0,t]} (f^k(s,\cdot,X^k_\cdot)-f^k(s,\cdot,X_\cdot))\,d(Z_s-Z^k_s)$$

$$+\int_{]0,t]} (f^k(s,\cdot,X_\cdot)-f(s,\cdot,X_\cdot))\,d(Z_s-Z^k_s)$$

$$+\int_{]0,t]} f(s,\cdot,X_\cdot)\,d(Z_s-Z^k_s).$$

Notice that $\bar{B}^k_{\varepsilon,r}:=\bar{B}^k_\varepsilon\wedge\bar{B}_r$ is predictable and can therefore be announced by a sequence of stopping boundaries $(\bar{B}^{k,n}_{\varepsilon,r}; n\in\mathbb{N})$. For each n and each stopping boundary B such that $[0,B]\subseteq[0,\bar{B}^{k,n}_{\varepsilon,r}\wedge B_r]$ we have

$$E\left[\sup_{t\in[0,B]}|X_t-X^k_t|^2\right]\leqq 15(d^k_2+d^k_1 r^3+d^k_1\varepsilon^2 r+d_3\varepsilon^2 r)$$

$$+15r(r+\varepsilon)E\left[\int_{]0,B]}\sup_{s<t}|X_s-X^k_s|^2\prod_{i=1}^{p} d(A^i+\bar{A}^{k,i})_s\right]$$

Let $K=15(d^k_2+d^k_1 r^3+d^k_1\varepsilon^2 r+d_3\varepsilon^2 r)$ and $d=15r(r+\varepsilon)$ to see that we are in a similar situation as in (4.2) of the proof of the previous theorem. By the same arguments as there we conclude that, for each r fixed, $E[\sup_{t\in[0,\bar{B}^{k,\varepsilon}_r[}|X_t-X^k_t|^2]$ converges to zero as $k\uparrow\infty$ and $\varepsilon\downarrow 0$. Therefore, for $\delta>0$ given, we can choose r and ε such that we have simultaneously $P([0,\bar{B}^{k,\varepsilon}_r]\subset[0,t^0])<\delta$ and $E[\sup_{t\in[0,\bar{B}^{k,\varepsilon}_r[}|X_t-X^k_t|^2]<\delta$ for $k>k_0(\delta,\varepsilon)$. This implies the assertion of the theorem. □

5. SPECIAL SDE WITHOUT LIPSCHITZ CONDITIONS

The following (simple) example of a SDE shows that Lipschitz conditions are necessary for uniqueness of solutions. Consider $p=2$ and $Y_t=\int_{]0,t]}Y^a_s\,ds$ $(0<a<1)$. In addition to the trivial solution $Y=0$, $Y_t=((1-a)^2 t_1 t_2)^{1/(1-a)}$ is a solution of the equation. If $(W_t; t\in\mathbb{R}^2_+)$ is a two-parameter Brownian sheet, the equation $X_t=W_t+\int_{]0,t]}(X-$

$W)^a_s\,ds$ reduces to the preceding equation by $Y=X-W$ and has therefore the solutions $X^1=W$ and $X^2=Y+W$.

The next example is a two-parameter analogue of a well-known SDE of Tsirel'son [28]. It shows that (similarly as in the one-parameter case) Lipschitz conditions are even necessary for the existence of strong solutions. Consider

$$X_t=W_t+\int_{]0,t]} f(s,X_.)\,ds, \quad t\in[0,1]^2, \tag{5.1}$$

where, for $w\in C(\mathbb{R}^2_+,\mathbb{R})$,

$$\begin{aligned} f(t,w)&=w(\Delta_{k+1,l+1})\,|\Delta_{k+1,l+1}|^{-1} \ (\text{mod } 1)\\ &\qquad \text{if } t\in\Delta(k,l):=[2^{-k},2^{-(k-1)}[\times[2^{-l},2^{-(l-1)}[\quad k,l=1,2,\dots\\ &=0 \ \text{ if } t\notin[0,1[\end{aligned} \tag{5.2}$$

Suppose that there exists a solution adapted to $\mathscr{W}_t:=\sigma(W_s;s\leqq t)$, $t\in[0,\infty[^2$. Writing $\theta(x)$ for $x(\text{mod } 1)$ we have

$$\begin{aligned} X_{k,l}=W_{k,l}+\int_{\Delta_{k,l}} f(s,X_.)\,ds&=W(\Delta_{k,l})\\ &+\theta(X(\Delta_{k+1,l+1})\,|\Delta_{k+1,l+1}|^{-1})\,|\Delta_{k,l}|. \end{aligned}$$

Therefore, if we set $n_{k,l}=X(\Delta_{k,l})\,|\Delta_{k,l}|^{-1}$ and $\xi_{k,l}=W(\Delta_{k,l})\,|\Delta_{k,l}|^{-1}$, we have $\eta_{k,l}=\xi_{k,l}+\theta(\eta_{k+1,l+1})$. This implies $\exp(2\pi i\eta_{k,l})=\exp(2\pi i\xi_{k,l})\exp(2\pi i\eta_{k+1,l+1})$. Therefore, for all $j\geqq 0$ we have $\exp(2\pi i\eta_{k,l})=\prod_{j=0}^m \exp(2\pi i\xi_{k+j,l+j})\exp(2\pi i\eta_{k+m+1,l+m+1})$. Since $\xi_{k,l}$ and $\sigma(\xi_{k+j,l+j};j\geqq 1)\vee\sigma(\eta_{k+j,l+j};j\geqq 1)$ are independent

$$\begin{aligned} E[\exp(2\pi i\eta_{k,l})]&=\textstyle\prod_{j=0}^m E[\exp(2\pi i\xi_{k+j,l+j})]\,E[\exp(2\pi i\eta_{k+m+1,l+m+1})]\\ &\leqq\exp\left(-2\pi^2\sum_{j=0}^m|\Delta_{k+j,l+j}|^{-1}\right)\to 0\ (m\uparrow\infty). \end{aligned}$$

Therefore writing

$\sigma_m^{k,l}$ for $\sigma(W(]t,t']);\ (2^{-k-m},2^{-l-m})\leqq t<t'\leqq(2^{-k+1},2^{-l+1}))$,

$$E[\exp(2\pi i\eta_{k,l})|\sigma_m^{k,l}]=\prod_{j=0}^{m}\exp(2\pi i\xi_{k+j,l+j})$$

$$\times E[\exp(2\pi i\eta_{k+m+1,l+m+1})]=0. \tag{5.3}$$

It follows that

$$\exp(2\pi i\eta_{k,l})=E[\exp(2\pi i\eta_{k,l})|\sigma(W_s;s\leqq(2^{-k+1},2^{-l+1})]$$

$$=\lim_{m\uparrow\infty}E[\exp(2\pi i\eta_{k,l})|\sigma_m^{k,l}]=0.$$

This implies that (5.1) has no solution that is adapted to $\mathscr{W}_t, t\in[0,1]^2$. With this fact one can construct a class of SDE for two-parameter processes without solutions adapted to W. We follow [27, p. 152] and refer to [13] for the Girsanov theorem for two-parameter processes. In particular, we suppose that a stochastic basis satisfying the conditions listed in the introduction is given on $C(\mathbb{R}_+^2,\mathbb{R})$ and that it carries an ($\mathscr{F}_t$-)Brownian sheet $(B_t; t\in\mathbb{R}_+^2)$ and an ($\mathscr{F}_t$-adapted) solution Y of

$$Y_t=B_t+\int_{]0,t]}h(s,Y_\cdot)\,ds,\quad t\in[0,1]^2, \tag{5.4}$$

where, for $w\in C(\mathbb{R}_+^2,\mathbb{R})$, $h(t,w_\cdot):=f(t,w_\cdot)+g(t,w_\cdot-\int_{]0,\cdot]}f(s,w_\cdot)\,ds$, f is given by (5.2) and g is supposed to be ($\mathscr{F}_t$)-adapted, uniformly bounded and such that $g(t,w)=0$ for $t\notin[0,1]^2$. Let $\mathscr{Y}_t=\sigma(Y_s; s\leqq t)$ and $\mathscr{B}_t=\sigma(B_s; s\leqq t)$.

PROPOSITION 5.5 *Under the assumptions made above* $\mathscr{B}_t\subsetneqq\mathscr{Y}_t (t\in[0,1]^2)$.

Proof. Let $W_t=Y_t-\int_{]0,t]}f(s,Y_\cdot)\,ds$. Then by (5.4) $W_t=B_t+\int_{]0,t]}g(s,W_\cdot)\,ds$. Therefore, for $t\in[0,1]^2$, $\mathscr{B}_t\subseteq\mathscr{W}_t:=\sigma(W_s; s\leqq t)$. Let

$$\Lambda_t=-\int_{]0,t]}g(s,W_\cdot)\,dB_s-\tfrac{1}{2}\int_{]0,t]}g^2(s,W_\cdot)\,ds.$$

Then, by the results of [13, p. 8–67], W is an $(\mathscr{F}_t, Q)$-Brownian sheet, where $Q = \exp \Lambda_1 \cdot P$. By definition of W, Y is a solution of (5.1) and therefore $\mathscr{W}_t \subset \mathscr{Y}_t$. This implies $\mathscr{B}_t \subset \mathscr{Y}_t$ $(t \in [0,1]^2)$. $\square$

Let us finally consider the SDE $X_t = B_t + \int_{]0,t]} f(s, X_.)\, ds$ for more general functions f. Suppose that f is measurable and adapted to the σ-field generated by the two-parameter Brownian sheet $(B_t; t \in \mathbb{R}_+^2)$ and suppose that f is bounded. Then this SDE has a weak solution (i.e. a solution-measure) that can be constructed as follows: let $\bar{\Lambda}_t = -\int_{]0,t]} f(s, B_.)\, dB_s - \frac{1}{2}\int_{]0,t]} f^2(s, B_.)\, ds$, and let $Q = \exp \bar{\Lambda}_{t0} \cdot P$, where P is the Wiener measure associated with B. By the results of [13, p. 8–67] there exists a Q-Brownian motion $\bar{B}$ such that $\bar{B}_t = B_t - \int_{]0,t]} f(s, B_.)\, ds$. Therefore B is a solution under Q of the SDE we consider.

We also notice that, under the assumptions made above for f, the SDE $X_t = B_t + \int_{]0,t]} f(s, X_.)\, ds$ has at most one solution that is adapted to the filtration generated by B. In fact, if X^1 and X^2 are two strong solutions of this SDE, we consider again $\tilde{\Lambda}_t^i = -\int_{]0,t]} f(s, X_.^i)\, dB_s - \frac{1}{2}\int_{]0,t]} f^2(s, X_.^i)\, ds$ $(i = 1, 2)$. As in the proof of the Proposition 5.5 we conclude that X^i is a Q^i-Brownian sheet, where $Q^i = \exp \tilde{\Lambda}_{t0}^i \cdot P$. Therefore the measure defined by $\nu^i = P((X^i)^{-1})$ is equivalent to the Wiener measure. By a result of I. V. Girsanov (see for example [12, p. 270] we conclude that $X^1 = X^2$ P-a.s.

References

[1] M. F. Allain, Semimartingales indexées par une partie de $\mathbb{R}^d$ et formule d'Itô. Cas continu. *Z. Wahrscheinlichkeitstheorie verw. Geb.*, **65** (1984), 421–444.

[2] F. Bakry, Limites quadrantales des martingales. Processus aléatoires à deux indices, edited by H. Korezlioglu, G. Mazziotto and J. Szpirglas, *Lecture Notes in Math.*, **863** (1981), 40–49.

[3] D. Bakry, Théorèmes de section et de projection pour les processus à deux indices. *Z. Wahrscheinlichkeitstheorie verw. Geb.*, **55** (1981), 55–71.

[4] D. Bakry, Semimartingales à deux indices. Seminaire de Probabilitiés, XVI, *Lecture Notes Math.*, **920** (1981), 355–369.

[5] R. Cairoli, Sur une équation différentielle stochastique, *C.R. Acad. Sci. Paris*, **274** (1972), 1739–1742.

[6] R. Cairoli and J. B. Walsh, Stochastic integrals in the plane, *Acta Math.*, **134** (1975), 111–183.

[7] R. Cairoli and J. B. Walsh, Régions d'arrêt, localisations et prolongements de martingales, *Z. Wahrscheinlichkeitstheorie verw. Geb.*, **44** (1978), 279–306.

[8] C. Doléans-Dade and P. A. Meyer, Equations différentielles stochastiques. Seminaire de Probabilitiés XI, *Lecture Notes in Math.*, **511** (1976), 376–382.

[9] M. Dozzi, Intégrales stochastiques pour processus à paramètre multidimensionnel: formule de Stokes et indépendance de chemin, *C. R. Acad. Sci. Paris*, **289** (1979), 445–448.

[10] M. Dozzi, On the decomposition and integration of two-parameter stochastic processes. Processus aléatoires à deux indices, edited by H. Korezlioglu, G. Mazziotto and J. Szpirglas, *Lecture Notes in Math.*, **863** (1981), 162–171.

[11] M. Dozzi and M. L. Puri, Solutions fortes des équations différentielles stochastiques pour des processus à indice multidimensionnel, *C. R. Acad. Sci. Paris,* **301** (1985), 845–848.

[12] I. I. Gihman and A. V. Skorohod, *The theory of stochastic processes III.* Springer Verlag, Berlin, 1979.

[13] X. Guyon and B. Prum, Semimartingales à indice dans $\mathbb{R}^2$. Thèse d'Etat. Université de Paris Sud, 1980.

[14] B. Hajek, Stochastic equations of hyperbolic type and a two-parameter Stratonovich calculus. *Ann Prob.*, **10** (1982), 451–463.

[15] H. Huerzeler, Stochastic integration on partially ordered sets, *J. Multivariate Anal.*, **17** (1985), 279–303.

[16] P. Imkeller, Itô's formula for continuous (N, d)-Processes, *Z. Wahrscheinlichkeitstheorie verw. Geb.*, **65** (1984), 535–562.

[17] J. Jacod, Calcul stochastique et problèmes de martingales, *Lecture Notes in Math.*, **714** (1979).

[18] H. Korezlioglu, P. Lefort and G. Mazziotto, Une propriété markovienne et diffusions associées. Processus aléatoires à deux indices, edited by H. Korezlioglu, G. Mazziotto and J. Szpirglas, *Lecture Notes in Math.*, **863** (1981), 245–274.

[19] E. McShane, Stochastic differential equations, *J. Mult. Anal.*, **5** (1975), 121–177.

[20] G. Mazziotto and J. Szpirglas, Sur les discontinuitès d'un processus cad-lag à deux indices. Processus aléatoires à deux indices, edited by H. Korezlioglu, G. Mazziotto and J. Szpirglas, *Lecture Notes in Math.,* **863** (1981), 84–90.

[21] E. Merzbach, Stopping for two-dimensional stochastic processes, *Stochastic processes Appl.*, **10** (1980), 49–63.

[22] E. Merzbach and M. Zakai, Bimeasures and measures induced by planar stochastic integrators. Preprint (1984).

[23] M. Metivier and J. Pellaumail, *Stochastic integration*, Academic Press, New York, 1980.

[24] P. A. Meyer, Théorie élémentaire des processus à deux indices. Processus aléatoires à deux indices, edited by H. Korezlioglu, G. Mazziotto and J. Szpirglas., *Lecture Notes in Math.,* **863** (1981), 1–39.

[25] A. Millet and L. Sucheston, On regularity of multiparameter amarts and martingales, *Z. Wahrscheinlichkeitstheorie verw. Geb.*, **56** (1981), 21–45.

[26] J. Reid, Estimate on moments of the solutions to stochastic differential equations in the plane, *Ann. Prob.*, **11** (1983), 656–668.

[27] D. W. Stroock and M. Yor, On extremal solutions of martingale problems, *Ann. Scient. Ec. Norm. Sup.*, **13** (1980), 95–164.

[28] B. S. Tsirel'son, An example of a stochastic differential equation having no strong solution, *Theory Prob. Appl.*, **20** (1975), 416–418.

[29] C. Tudor, Stochastic integral equations in the plane, *Rev. Roum. Math. Pures et Appl.*, Tome XXVI (1981), 507–538.

[30] C. Tudor and M. Tudor, On approximation in quadratic mean for the solutions of two-parameter stochastic differential equations in Hilbert space, *Analele Universitatii Bucuresti*, XXXII (1983), 73–88.

[31] E. Wong and M. Zakai, Weak martingales and stochastic integrals in the plane, *Ann. Prob.*, **4** (1976), 570–586.

[32] J. Yeh, Existence of strong solutions for stochastic integral equations in the plane, *Pac. J. Math.*, **97** (1981), 217–247.

[33] M. Yor, Représentation de martingales de carré intégrable relative aux processus de Wiener et de Poisson à n paramètres, *Z. Wahrscheinlichkeitstheorie verw. Geb.*, **35** (1976), 121–129.

[34] M. Zakai, Some remarks on integration with respect to weak martingales. Processus aléatories à deux indices, edited by H. Korezlioglu, G. Mazziotto and J. Szpirglas, *Lecture Notes in Math.*, **863** (1981), 149–161.

[35] G. Zbăganu, and X. W. Zhuang, Two-parameter filtrations with respect to which all martingales are strong. *Z. Wahrscheinlichkeitstheorie verw. Geb.*, **61** (1982), 437–452.

Reprinted by permission of the publisher,
© 1986 by Taylor and Francis Ltd.
http://www.tandf.co.uk

A NEW SEMIGROUP TECHNIQUE IN POISSON APPROXIMATION

P. DEHEUVELS, D. PFEIFER and MADAN L. PURI

ABSTRACT. We present a unified and self-contained approach to Poisson approximation problems for independent Bernoulli summands with respect to several metrics by a general semigroup technique, expanding and completing earlier work on this subject by the first two authors [4], [5], [6].

1. Introduction

We consider the approximation of the distribution of $S_n = \sum_{k=1}^{n} X_k$ where $X_1, \ldots, X_n$ are independent Bernoulli random variables with success probabilities $p_k = P(X_k = 1) = 1 - P(X_k = 0)$ by suitable Poisson distributions. As a measure of deviation, it is convenient to consider probability metrics over a set $\mathcal{M}$ of probability measures defined on a measurable space $(\Omega, \mathcal{A})$, for example

(1.1) $d(P,Q) = \sup_{A \in \mathcal{A}} |P(A) - Q(A)|$ (total variation)

(1.2) $d_0(P,Q) = \sup_{x \in \mathbf{R}} |F_P(x) - F_Q(x)|$ (uniform or Kolmogorov metric)

AMS 1980 *subject classifications.* Primary 60F05, 62E20; Secondary 47D05, 20M30.

Key words and phrases. Poisson approximation, operator semigroups, Kolmogorov metric, Fortet-Mourier-Wasserstein metric, minimal metrics.

(1.3) $d_1(P,Q) = \inf\{E_L|X-Y| \mid \mathcal{L}(X,Y) = L,\ \mathcal{L}(X) = P,\ \mathcal{L}(Y) = Q\}$
(Fortet-Mourier or Wasserstein metric).

Here F_P denotes the cumulative distribution function of the measure P, E_L means expectation with respect to the measure L, and $\mathcal{L}(X,Y)$ denotes the distribution of (X,Y). Note that the definitions of d_0 and d_1 make sense only in the case $(\Omega,\mathcal{A}) = (\mathbf{R},\mathcal{B})$ where $\mathcal{B}$ is the σ-field of Borel sets. Moreover, d and d_1 are special cases of minimal metrics d_ρ obtained via

(1.4) $d_\rho = \inf\{\rho(\mathcal{L}(X,Y)) \mid \mathcal{L}(X) = P,\ \mathcal{L}(Y) = Q\}$

where ρ is a probability metric in the sense of Zolotarev ([21], [22]; see also Deheuvels, Karr, Pfeifer and Serfling [7]).

Until very recently, only estimates and expansions for $d(\mathcal{L}(S_n),\mathcal{P}(\mu))$ and $d_0(\mathcal{L}(S_n),\mathcal{P}(\mu))$ (where $\mathcal{P}(\mu)$ denotes a Poisson distribution with expectation $\mu > 0$) have been studied in the literature, of which the following ones are of specific importance:

(1.5) $d(\mathcal{L}(S_n),\mathcal{P}(\sum_{k=1}^n p_k)) \le \sum_{k=1}^n p_k^2$ (Le Cam [12])

(1.6) $d(\mathcal{L}(S_n),\mathcal{P}(\sum_{k=1}^n \nu_k)) \le \frac{1}{2}\sum_{k=1}^n \nu_k^2$ (Serfling [18], [19])
where $\nu_k = -\log(1-p_k),\ 1 \le k \le n$,

(1.7) $d(\mathcal{L}(S_n),\mathcal{P}(\sum_{k=1}^n p_k)) \le (1 - exp(-\sum_{k=1}^n p_k))\theta_n$ (Barbour and Hall [2])
where $\theta_n = \sum_{k=1}^n p_k^2 / \sum_{k=1}^n p_k$,

(1.8) $d_0(\mathcal{L}(S_n),\mathcal{P}(\sum_{k=1}^n p_k)) \le \frac{1}{2}\sum_{k=1}^n p_k^2$ (Daley; cf. Serfling [19])

(1.9) $d_0(\mathcal{L}(S_n),\mathcal{P}(\sum_{k=1}^n \nu_k)) \le \frac{1}{2}\sum_{k=1}^n \nu_k^2$ (Serfling [19])

(1.10) $d_0(\mathcal{L}(S_n),\mathcal{P}(\sum_{k=1}^n p_k)) \le \frac{\pi}{4}(\sum_{k=1}^n p_k^2/(1-p_k))/(\sum_{k=1}^n p_k(1-p_k))$
(Hipp [9])

(1.11) $d_0(\mathcal{L}(S_n),\mathcal{P}(\sum_{k=1}^n p_k)) \le (\frac{1}{2} + \sqrt{\frac{\pi}{8}})\theta_n/(1-\sqrt{\theta_n})$ (Shorgin [20]).

Although most of these authors consider $\mu = \sum_{k=1}^n p_k = E(S_n)$ as the favorable choice for an approximation, it is noteworthy that for $n = 1$, Serfling's [18] choice for d is actually the best possible, i.e.

(1.12) $d(\mathcal{L}(S_1), \mathcal{P}(\nu_1)) = \min_\mu d(\mathcal{L}(S_1), \mathcal{P}(\mu)) = 1 - (1+\nu_1)e^{-\nu_1} \le \frac{1}{2}\nu_1^2$

whenever $p_1 \le 1 - 1/e$.

Although it was not explicitly stated in Serfling's paper [19], it is easy to see that a corresponding optimization procedure for d_0 would yield

(1.13) $\min_\mu d_0(\mathcal{L}(S_1), \mathcal{P}(\mu)) = d_0(\mathcal{L}(S_1), \mathcal{P}(\overline{p})) = 1 - (1+\overline{p})e^{-\overline{p}}$

where $\overline{p}$ is the unique positive solution of the equation

(1.14) $p_1 = 2 - 2e^{-\overline{p}} - \overline{p}e^{-\overline{p}}$.

Some further analysis shows in fact that

(1.15) $p_1 < p_1 + \frac{1}{6}p_1^3 - \frac{1}{12}p_1^4 \le \overline{p} \le p_1 + \frac{1}{6}p_1^3 < \nu_1$

which in turn gives rise to the estimate

(1.16) $d_0(\mathcal{L}(S_n), \mathcal{P}(\sum_{k=1}^n \overline{p}_k)) \le \frac{1}{2}\sum_{k=1}^n p_k^2(1 - \frac{1}{3}p_k + \frac{1}{2}p_k^2 + \frac{1}{36}p_k^4)$

where $\overline{p}_k$ is the corresponding solution of (1.14) when p_1 is replaced by p_k. Note that when for example $max(p_1, \dots, p_n) \le \frac{1}{2}$, then the right hand side of (1.6) is actually smaller than the right hand side of (1.8).

Similarly, it can easily be shown that the optimal $\mu = \widetilde{p}$ for minimizing $d_1(\mathcal{L}(S_1), \mathcal{P}(\mu))$ with respect to μ is given by

(1.17) $\widetilde{p} = \min(\nu_1, \log 2)$

with a corresponding distance given by

(1.18) $d_1(\mathcal{L}(S_1), \mathcal{P}(\overline{p})) = \begin{cases} \nu_1 - p_1, & p_1 \le \dfrac{1}{2} \\ \log 2 - 1 + p_1, & p_1 > \dfrac{1}{2} \end{cases} = \Delta(p_1)$, say.

Let (X_k, Y_k) be a coupling such that $E|X_k - Y_k| = d_1(\mathcal{L}(X_k), \mathcal{L}(Y_k)) = d_1(\mathcal{L}(X_k), \mathcal{P}(\widetilde{p}_k))$, $1 \le k \le n$ (which always exists). Then

$$\begin{aligned} d_1(\mathcal{L}(S_n), \mathcal{L}(\sum_{k=1}^n Y_k)) &\le E|\sum_{k=1}^n X_k - \sum_{k=1}^n Y_k| \le \sum_{k=1}^n E|X_k - Y_k| \\ &\le \sum_{k=1}^n d_1(\mathcal{L}(X_k), \mathcal{L}(Y_k)), \end{aligned}$$

or

$$(1.19)\quad d_1(\mathcal{L}(S_n), \mathcal{P}(\sum_{k=1}^{n} \widetilde{p}_k)) \le \sum_{k=1}^{n} \Delta(p_k).$$

Note that by (1.18), $\Delta(p_k)$ is always larger than $\frac{1}{2}p_k^2$, but bounded by $\frac{1}{2}p_k^2/(1-p_k)$, for $p_k < 1$.

The last examples show that we cannot hope for an explicit closed solution for the problem of the best choice for the Poisson parameter, not even for commonly used distance measures. However, it was possible to approach the problem at least asymptotically (cf. Deheuvels and Pfeifer [4], [5], [6]) by a successively refined application of convolution semigroups on different Banach spaces, in the spirit of Le Cam's [12] paper which at the same time also allowed for asymptotic expansions of the metrics under consideration with essentially arbitrary choices of μ. Unfortunately, the 'telescoping technique' used there gave only results in the range where $\sum_{k=1}^{n} p_k^2 = o(1)$ or $\sum_{k=1}^{n} p_k^2 = 0(1)$ and $\sum_{k=1}^{n} p_k$ unbounded, whereas some of the above estimates show that Poisson convergence with respect to d or d_0 takes place as long as $\theta_n = \sum_{k=1}^{n} p_k^2 / \sum_{k=1}^{n} p_k = o(1)$ which is a weaker condition. In the present paper we present a new approach in the semigroup setting which covers also this case, and at the same time provides in a unified and self-contained fashion, upper and lower bounds for the metrics under consideration, which are sometimes considerably sharper than all the competitors outlined in (1.5) to (1.11). A particular advantage of this technique is also the fact that we do not need any longer different methods for different metrics, such as characteristic functions (Shorgin [20], Hipp [9]), couplings (Serfling [18], [19]) or other specific approaches such as Stein's method (Barbour and Hall [2], Barbour [1]), and that we are also able to present at least nearly optimal solutions to the problem of the best parameter choice for the approximating Poisson distribution.

2. The Semigroup Setting

Let $\mathcal{X}$ denote either the Banach space ℓ^1 of summable sequences $f = (f(0), f(1), \ldots)$ or the Banach space ℓ^∞ of all absolutely bounded sequences. Define the shift operator $B: \mathcal{X} \to \mathcal{X}$ by

$$(2.1)\quad Bf = (0, f), \text{ i.e. } Bf(n) = \begin{cases} 0 & \text{if } n = 0 \\ f(n-1) & \text{if } n \ge 1. \end{cases}$$

Actually B corresponds to a convolution of the unit mass ε_1 concentrated at 1 with f (cf. Deheuvels and Pfeifer [4], [5]). Then, the operator $A = B - I$ (I denotes the identity) generates the Poisson convolution semigroup

$$(2.2)\quad e^{tA}f = e^{-t}e^{tB}f = e^{-t}\sum_{k=0}^{\infty} \tfrac{t^k}{k!}B^k f = \mathcal{P}(t) * f,\ f \in \mathcal{X}$$

where $*$ means convolution. Similarly,

$$(2.3)\quad (I+p_kA)f = ((1-p_k)I+p_kB)f = \mathcal{L}(X_k)^*f,\ f\in\mathcal{X}.$$

Hence, by independence,

$$(2.4)\quad \sum_{k=1}^{n}(I+p_kA)f = \mathcal{L}(S_n)^*f,\ f\in\mathcal{X}.$$

Then, we have the following relationship between such semigroups and the metrics outlined in the introduction.

Lemma 2.1. *Let* $\|\cdot\|_{(\mathcal{X},\mathcal{X})}$ *denote the corresponding operator norm. Then*

$$(2.5)\quad d(\mathcal{L}(S_n),\mathcal{P}(\mu)) = \tfrac{1}{2}\|e^{\mu A}-\prod_{k=1}^{N}(I+p_kA)\|_{(\ell^1,\ell^1)}$$

$$(2.6)\quad d_1(\mathcal{L}(S_n),\mathcal{P}(\mu)) = \|\{e^{\mu A}-\prod_{k=1}^{N}(I+p_kA)\}A^{-1}\|_{(\ell^1,\ell^1)}$$

Proof. Relation (2.5) is essentially given in Deheuvels and Pfeifer [4]; see also Deheuvels and Pfeifer [6]. To prove (2.6), observe that with $f_0=(1,0,0,\cdots)$ we have $\|f_0\|_{\ell^1}=1$, $A^{-1}f_0=(1,1,1\ldots)$, and $d_1(\mathcal{L}(S_n),\mathcal{P}(\mu)) = \|\{e^{\mu A}-\prod_{k=1}^{N}(I+p_kA)\}(A^{-1}f_0)\|_{\ell^1}$ (Deheuvels and Pfeifer [5]) from which the assertion follows. (Note that $-A^{-1}$ is also called the potential operator associated with the semigroup.)

Unfortunately, the corresponding d_0-metric cannot be expressed directly by an operator norm; however, it is valid that

$$(2.7)\quad d_0(\mathcal{L}(S_n),\mathcal{P}(\mu)) = \|\{e^{\mu A}-\prod_{k=1}^{N}(I+p_kA)\}(A^{-1}f_0)\|_{\ell^\infty}$$

(Deheuvels and Pfeifer [5]), which will also be sufficient for our purposes.

The following theorem is a key result for a unified treatment of estimates and expansions for the expressions above.

Theorem 2.1. *Let* $\lambda=\sum_{k=1}^{n}p_k$, $\lambda_m=\sum_{k=1}^{n}p_k^m$, $m\geq 2$. *Then, for an arbitrary operator semigroup* $\{e^{tA},\ t\geq 0\}$ *with bounded generator* A,

$$(2.8)\quad e^{\lambda A}-\prod_{k=1}^{n}(I+p_kA) = \tfrac{1}{2}\lambda_2A^2e^{\lambda A}-\sum_{k=3}^{\infty}a_k(-A)^ke^{\lambda A}$$

where a_k *is recursively defined by*

(2.9) $a_k = -\frac{1}{k}(\lambda_k + \sum_{i=2}^{k-2} a_i\lambda_{k-i}),\ k \geq 3,$ *and* $a_2 = -\frac{1}{2}\lambda_2.$

The convergence in (2.8) is absolute and to be understood in the uniform operator topology. Likewise

(2.10) $\{e^{\lambda A} - \prod_{k=1}^{n}(I + p_k A)\}A^{-1} = \frac{1}{2}\lambda_2 A e^A + \sum_{k=3}^{\infty} a_k(-A)^{k-1}e^{\lambda A}.$

Proof. By Lemma 3 of Shorgin [20], it follows that

(2.11) $\prod_{k=1}^{n}(1 + p_k z)e^{-p_k z} = 1 - \frac{1}{2}\lambda_2 z^2 + \sum_{k=3}^{\infty} a_k(-z)^k,$

where the right hand side converges absolutely for every real z since by Lemma 5 of Shorgin [20], we have

(2.12) $|a_k| \leq (e\lambda_2/k)^{k/2} \leq |2z|^{-k}$ for sufficiently large k $(z \neq 0)$.

Thus, we have

(2.13) $\prod_{k=1}^{n}(1 + p_k A)e^{-p_k A} = 1 - \frac{1}{2}\lambda_2 A^2 + \sum_{k=3}^{\infty} a_k(-A)^k$

where the right hand side coverges absolutely, from which (2.8) follows by application of $e^{\lambda A}$. (2.10) is proved similarly.

It should be pointed out that Theorem 2.1 also allows for a full expansion related to the Charlier-Poisson expansion in Barbour [1] since e.g. for the metric d, we have

(2.14) $\| \sum_{j=2}^{\infty} a_j(-A)^j e^{\lambda A}\|_{(\ell^1,\ell^1)} = \sum_{k=0}^{\infty} e^{-\lambda}\frac{\lambda^k}{k!}|\sum_{j=2}^{\infty}(-1)^j a_j \sum_{i=0}^{j}(-1)^i\binom{j}{i}\frac{k_{(i)}}{\lambda^i}|$

where $k_{(i)} = \prod_{m=0}^{i-1}(k-m)$; similar ideas lead to the expansions for the metrics d_0 and d_1 even in the more general setting of this paper.

We should like to mention that Theorem 1 also applies to the more general Banach space setting used in connection with minimal metrics described in Deheuvels, Karr, Pfeifer and Serfling [7]. It was shown there that such metrics can be estimated by

(2.15) $d_\alpha(\mathcal{L}(S_n), \mathcal{P}(\mu)) = \sum_{k=0}^{\infty} \alpha_k |P(S_n = k) - \mathcal{P}(\mu; k)|$

where $\{\alpha_k\}$ is a non-negative sequence which does not grow faster than at a geometric rate, i.e. $\sup_{k\geq 1}(\alpha_{k+1}/\alpha_k) = M < \infty$. Note that d_α is itself a metric for all such α, and that d_α can be represented as

(2.16) $d_\alpha(\mathcal{L}(S_n), \mathcal{P}(\mu)) = \|\{e^{\mu A} - \prod_{k=1}^{n}(I + p_k A)\} f_0\|_{\ell^1_\alpha}$

where ℓ^1_α is the Banach space of all α-summable sequences with norm

(2.17) $\|f\|_{\ell^1_\alpha} = \sum_{k=0}^{\infty} \alpha_k |f(k)|.$

Then $\|A\|_{(\ell^1_\alpha),(\ell^1_\alpha)} \leq 1 + M$, and (2.8) also holds for this case. Note that d_α-metrics have also been considered earlier by Johnson and Simons [10], however, for the case of equal p_k's only. The next result will provide a tool for estimating some distances in the case of $\mu = \lambda$.

Lemma 2.2. *Suppose there exists a constant $K > 0$ such that*

$$\|Ae^{\lambda A}\|_{(\mathcal{X},\mathcal{X})} \leq \sqrt{K/e\lambda}. \tag{2.18}$$

Then

$$\|e^{\lambda A} - \prod_{k=1}^{n}(I + p_k A)\|_{(\mathcal{X},\mathcal{X})} \leq \frac{K\theta}{1 - \sqrt{K\theta}} \tag{2.19}$$

provided $\theta = \theta_n < \frac{1}{k}$.

Proof. By (2.18),

$$\|A^k e^{\lambda A}\|_{(\mathcal{X},\mathcal{X})} \leq \|Ae^{(\lambda/k)}A\|^k_{(\mathcal{X},\mathcal{X})} \leq \sqrt{Kk/e\lambda}^{\,k}, \tag{2.20}$$

hence by (2.8) and (2.12),

$$\|e^{\lambda A} - \prod_{k=1}^{n}(I + p_k A)\|_{(\mathcal{X},\mathcal{X})} \leq \sum_{k=2}^{\infty} |a_k| \; \|A^k e^{\lambda A}\|_{(\mathcal{X},\mathcal{X})} \leq \sum_{k=2}^{\infty} \sqrt{K\theta}^{\,k}$$

which proves the result.

Note that in a similar way we obtain

$$|\{e^{\lambda A} = \prod_{k=1}^{n}(I + p_k A)\}A^{-1}\|_{(\ell^1,\ell^1)} \leq \sqrt{e\lambda}\frac{K\theta}{1 - \sqrt{K\theta}}, \quad \theta < \frac{1}{K}. \tag{2.21}$$

As has been shown in Deheuvels and Pfeifer [4], [5] we have

$$\|Ae^{\lambda A}\|_{(\ell^1,\ell^1)} = 2e^{-\lambda}\frac{\lambda^{[\lambda]}}{[\lambda]!} \leq \sqrt{2/e\lambda} \tag{2.22}$$

with equality reached for $\lambda = \frac{1}{2}$, hence we may choose $K = 2$ in Lemma 2.2. Here $[\lambda]$ denotes the integral part of λ. Thus we obtain:

Corollary 2.1.

$$d(\mathcal{L}(S_n),\mathcal{P}(\lambda)) \le \frac{\theta}{1-\sqrt{2\theta}},\ \theta < \frac{1}{2}, \tag{2.23}$$

$$d_1(\mathcal{L}(S_n),\mathcal{P}(\lambda)) \le \frac{4\sqrt{\lambda}\theta}{1-\sqrt{2\theta}},\ \theta < \frac{1}{2}. \tag{2.24}$$

Better estimations are obtained if the leading term $\frac{\lambda_2}{2}\|A^2e^{\lambda A}\|_{(\mathcal{X},\mathcal{X})}$ is considered separately; since it is completely known in closed form (see Deheuvels and Pfeifer [4], [5]) one can easily derive upper and lower bounds by estimating the remainder terms in the series expansion (2.8). By application of some of Shorgin's [20] auxiliary results in the case of d_0, one obtains, for instance,

Corollary 2.2.

$$|d(\mathcal{L}(S_n),\mathcal{P}(\lambda)) - \frac{\lambda_2}{2}\, e^{-\lambda}\{\frac{\lambda^{a-1}(a-\lambda)}{a!} + \frac{\lambda^{b-1}(\lambda-b)}{b!}\}| \le \frac{(2\theta)^{3/2}}{2(1-\sqrt{2\theta})}, \tag{2.25}$$

$$\theta < 1/2,$$

where $a = [\lambda + 1/2 + \sqrt{\lambda + 1/4}],\ b = [\lambda + 1/2 - \sqrt{\lambda+1/4}]$;

$$|d_o(\mathcal{L}(S_n),\mathcal{P}(\lambda)) - \frac{\lambda_2}{2}\, e^{-\lambda}\max\{\frac{\lambda^{a-1}(a-\lambda)}{a!}, \frac{\lambda^{b-1}(\lambda-b)}{b!}\}| \le \frac{C\theta^{3/2}}{1-\sqrt{\theta}}, \tag{2.26}$$

$$\theta < 1;$$

$$|d_1(\mathcal{L}(S_n),\mathcal{P}(\lambda)) - \lambda_2\, e^{-\lambda}\frac{\lambda^{[\lambda]}}{[\lambda]!}| \le \frac{2\sqrt{\lambda}(2\theta)^{3/2}}{1-\sqrt{2\theta}},\ \theta < 1/2. \tag{2.27}$$

Here we have used $C = \frac{1}{2} + \sqrt{\frac{\pi}{8}} < 1.13$.

This shows clearly that the condition $\theta \to 0$ is necessary and sufficient for Poisson convergence with $\mu = \lambda$ for the metrics d and d_o, and $\sqrt{\lambda}\theta \to 0$ for the metric d_1. The bounds obtained in Corollary 2.2 are usually much sharper than all the competitors from (1.5) to (1.11). This will be demonstrated in the next section.

In order to attack the problem of the (nearly) best Poisson parameter not only in an asymptotic setting as in Deheuvels and Pfeifer ([4], [5]) we proceed as follows. Instead of using the parameter $\mu = \lambda$ we make an adjustment of the form

$$\mu = \lambda + \delta\lambda_2 \ \text{for some}\ 0 \le \delta \le 1/2. \tag{2.28}$$

Then

$$e^{\mu A} - \prod_{k=1}^{n}(I + p_k A) = \frac{\lambda_2}{2}(A^2 + 2\delta A)e^{\lambda A} + \frac{\delta^2\lambda_2^2}{2}A^2 e^{\lambda A} + \sum_{k=3}^{\infty}\{\frac{(\delta\lambda_2)^k}{k!} - a_k(-1)^k\}A^k e^{\lambda A} \tag{2.29}$$

which is obvious from $e^{\mu A} - e^{\lambda A} = e^{\lambda A}(e^{\delta\lambda_2 A} - I)$ and the Taylor expansion for the second factor. Since the leading term in the above expansion will be the dominating one in general, a suboptimal solution is obtained by choosing δ such that $\|(A^2 + 2\delta A)e^{\lambda A}\|_{(\mathcal{X},\mathcal{X})}$ is minimal, paralleling the selection procedure in the asymptotic case. Since also the last norm terms can explicitly be given, the determination of δ is mainly a numerical problem. For these we shall restate Example 3.1 in Deheuvels and Pfeifer [5] giving the minimizing δ for each case. In what follows, $L(\mu)$ denotes the resulting leading term for the metric obtained from (2.29) and Lemma 2.1.

Lemma 2.3. *I. For the metric* d, *we have*

$$\delta = 1/2,\ L(\mu) = \frac{1}{2}\lambda_2 e^{-\lambda} \text{ if } 0 < \lambda < 1, \tag{2.30}$$

$$\delta = 1/2,\ L(\mu) = \frac{1}{2}\lambda_2\lambda e^{-\lambda} \text{ if } 1 < \lambda < \sqrt{2}, \tag{2.31}$$

$$\delta = \frac{1}{2} - \frac{3}{2\lambda}\cdot\frac{2-\lambda}{3-\lambda},\ L(\mu) = \frac{1}{2}\lambda_2 e^{-\lambda}\{\lambda + (1 - \frac{\lambda^2}{2})\frac{3}{\lambda}\cdot\frac{2-\lambda}{3-\lambda}\}, \tag{2.32}$$

$$\delta = 0,\ L(\mu) = \frac{1}{2}\lambda_2 e^{-\lambda}\{\frac{\lambda^2}{2} + (1 - \frac{\lambda^3}{6})\} \text{ if } \sqrt[3]{6} < \lambda \le \sqrt{2}. \tag{2.33}$$

II. For the metric d_0, *we have*

$$\delta = \frac{1}{2} - \frac{1}{2(1+\lambda)},\ L(\mu) = \frac{\lambda_2 e^{-\lambda}}{2(1+\lambda)} \text{ if } 0 < \lambda \le \sqrt{3} - 1, \tag{2.34}$$

$$\delta = \frac{1}{2} - \frac{\lambda}{2+\lambda^2},\ L(\mu) = \frac{\lambda_2\lambda e^{-\lambda}}{(2+\lambda^2)} \text{ if } \sqrt{3} - 1 < \lambda \le 1. \tag{2.35}$$

III. For the metric d_1, *we have*

$$\delta = \frac{1}{2},\ L(\mu) = \lambda_2/2 \text{ if } 0 < \lambda < \log 2, \tag{2.36}$$

$$\delta = 0,\ L(\mu) = \lambda_2 e^{-\lambda} \text{ if } \log 2 < \lambda \le 1, \tag{2.37}$$

$$\delta = \frac{1}{2} - \frac{1}{2\lambda},\ L(\mu) = \frac{1}{2}\lambda_2\{1 - \frac{1}{\lambda}(1 - 2e^{-\lambda})\} \text{ if } 1 < \lambda \le \alpha \tag{2.38}$$

where α *is the positive root of the equation* $2e^{-\alpha}(1+\alpha) = 1$,

$$\delta = 0,\ L(\mu) = \lambda_2\lambda e^{-\lambda} \text{ if } \alpha < \lambda \le 2. \tag{2.39}$$

If $\delta > 0$ *is an arbitrary choice, and if again* $L(\mu)$ *denotes the resulting leading term, then (2.29) leads to the following:*

Corollary 2.3.

$$(2.40)\qquad |d(\mathcal{L}(S_n),\mathcal{P}(\mu))-L(\mu)| \le \frac{\delta^2\lambda_2^2}{2e\lambda}+\frac{1}{2}\Big|\frac{\delta^3\lambda_2^3}{6}-\frac{\lambda_3}{3}\Big|\Big(\frac{6}{e\lambda}\Big)^{3/2} + \frac{2\theta^2}{\sqrt{1-2\theta}}+8\frac{\delta^4\lambda_2^4}{e^2\lambda^2}e^{2\delta\lambda_2},\ \theta<1/2;$$

$$(2.41)\qquad |d_o(\mathcal{L}(S_n),\mathcal{P}(\mu))-L(\mu)| \le \frac{\delta^2\lambda_2^2}{2e\lambda}+\Big|\frac{\delta^3\lambda_2^3}{6}-\frac{\lambda_3}{3}\Big|\Big(\frac{3}{2\lambda}\Big)^{3/2} + \frac{C\theta^2}{1-\sqrt{\theta}}+\frac{4\delta^4\lambda_2^4}{e^2\lambda^2}e^{2\delta\lambda_2},\ \theta<1;$$

$$(2.42)\qquad |d_1(\mathcal{L}(S_n),\mathcal{P}(\mu))-L(\mu)| \le \frac{\delta^2\lambda_2^2}{\sqrt{2}e\lambda}+\Big|\frac{\delta^3\lambda_2^3}{6}-\frac{\lambda_3}{3}\Big|\frac{4}{e\lambda} + \sqrt{e\lambda_2}\frac{(2\theta)^{3/2}}{\sqrt{1-2\theta}}+\delta^4\lambda_2^4\sqrt{6/e}\,e^{2\delta\lambda_2},\quad \theta<1/2.$$

In a similar manner, estimates and expansions can be obtained for the d_α-metrics which for lack of space we omit here.

Of course, evaluations of the above distances can be made up to an arbitrary precision using more terms from the full expansion given in Theorem 2.1, and a corresponding remainder treatment as in Lemma 2.2 or Corollary 2.3, together with (2.29). The advantage of the one-term procedure, however, relies on the fact that the expressions involved are much more easy to compute.

3. Applications and Numerical Comparisons

In order to show the power of the approach introduced above, we shall briefly investigate a problem stemming from the theory of extremes. Let $\{Z_n, n\ge 1\}$ be a sequence of independent and identically distributed random variables with continuous distribution function F. Set $Z_{(n)}=\max(Z_1,\dots,Z_n)$, and

$$(3.1)\qquad X_n=\begin{cases}1 & \text{if } Z_{(n)}>Z_{(n-1)}\\ 0 & \text{otherwise}\end{cases}$$

for $n\ge 2$ (by convention $X_1\equiv 1$).

We say that $Z_{(n)}$ is a *record value* of $\{Z_n, n\ge 1\}$ if for the corresponding n we have $X_n=1$. By a well known result of Rényi [16] we know that $\{X_n\}$ is an independent sequence with $p_k=\frac{1}{k}$, $k\in\mathbf{N}$. Hence $S_n(m)=\sum_{k=n+1}^{nm}X_k$, the number of record values between observation $n+1$ and nm, is approximately Poisson-distributed with $E(S_n(m)) = \sum_{k=n+1}^{nm} \frac{1}{k} \sim \log \frac{n \cdot m}{n} = \log m$, for large n. The following table provides a comparison between the different expansions and estimates outlined above, for the Poisson approximation of $\mathcal{L}(S_n(m))$.

Figure 1

Metric d, $n = 10$, $m = 5$ $\log m = 1.6094$

μ	δ	$L(\mu)$	lower bound	upper bound	authors
		.0124	.0020	.0228	Deheuvels, Pfeifer, Puri
1.5702	0	--	.0015	.0380	Barbour and Hall
		--	--	.0754	LeCam
		.0124	.0017	.0229	Deheuvels, Pfeifer, Puri
1.6094	.5199	--	--	.0398	Serfling
1.5863	.2129	.0092	--	.0196	Deheuvels, Pfeifer, Puri

Figure 2

Metric d_o, $n = 10$, $m = 2$ $\log m = .6931$

μ	δ	$L(\mu)$	lower bound	upper bound	authors
		.0119	.0038	.0200	Deheuvels, Pfeifer, Puri
.6688	0	--	--	.0232	Daley
		--	--	.0631	Hipp
		--	--	.1060	Shorgin
		.0119	.0037	.0201	Deheuvels, Pfeifer, Puri
.6931	.5247	--	--	.0246	Serfling
.6780	.2004	.0071	.0011	.0142	Deheuvels, Pfeifer, Puri

Note that in last line of each figure, δ is chosen to minimize the corresponding leading term from (2.29).

We should like to point out that with the former semigroup approach developed in Deheuvels and Pfeifer ([4], [5], [6]), no reasonable upper and lower bounds are to be obtained for this case, for the metrics d_0 and d_1.

Further situations in which Poisson approximation problems of such kind (i.e. with varying success probabilities p_k) occur are to be found in Ross [17] (average-case analysis for the simplex method in Linear Programming), Kemp [11] (average-case analysis of search algorithms), Gastwirth and Bhattacharya [8] (chain letter systems, multilevel marketing systems), Bruss [3] (secretary problems), Nevzorov [13] (generalized record problems), Pfeifer [14] (extremal processes), and Pfeifer [15] (best choice problems, non-homogeneous extremal processes). The results developed in this paper are especially suited to provide a unified treatment of the approximation problems considered in these papers with a strong emphasis on numerical tractability.

Acknowledgements

This paper was written while the second author was visiting the Department of Mathematics at Indiana University, Bloomington. He is grateful to the Deutsche Forschungsgemeinschaft for financial support by the Heisenberg Research Foundation.

We are also indebted to Dr. H. J. Witte for pointing out and correcting an error in an earlier version of the paper.

References

[1] Barbour, A. D. (1987), *Asymptotic expansions in the Poisson limit theorem*, Ann. Prob. **15**, 748-766.

[2] Barbour, A. D. and Hall, P. (1984), *On the rate of Poisson convergence*, Math. Proc. Cambridge Philos. Soc. **95**, 473-480.

[3] Bruss, F. T. (1984), *Patterns of relative maxima in random permutations*, Ann. Soc. Sci. Bruxelles **98 (I)**, 19-28.

[4] Deheuvels, P. and Pfeifer, D. (1986 a), *A semigroup approach to Poisson approximation*, Ann. Prob. **14**, 663-676.

[5] Deheuvels, P. and Pfeifer, D. (1986 b), *Operator semigroups and Poisson convergence in selected metrics*, Semigroup Forum **34**, 203-224.

[6] Deheuvels, P. and Pfeifer, D. (1987), *Semigroups and Poisson approximation. In: New Perspectives in Theoretical and Applied Statistics*, M. L. Puri, J. P. Vilaplana, W. Wertz (eds.), John Wiley, N.Y., 439-448.

[7] Deheuvels, P., Karr, A. F., Pfeifer, D. and Serfling, R. J. (1988), *Poisson approximations in selected metrics by coupling and semigroup methods with applications.* To appear in: J. Stat. Planning Inf.

[8] Gastwirth, J. L. and Bhattacharya, P. K. (1984), *Probability models of pyramid or chain letter systems demonstrating that their promotional claims are unreliable*, Operat. Res. **32**, 527-536.

[9] Hipp, C. (1985), *Approximation of aggregate claims distributions by compound Poisson distributions.* Insurance: Math. and Econ. **4**, 227-232.

[10] Johnson, N. L. and Simons, G. (1971), *On the convergence of binomial to Poisson distributions*, Ann. Math. Stat. **42**, 1735-1736.

[11] Kemp, R. (1984), *Fundamentals of the Average Case Analysis of Particular Algorithms*, Wiley-Teubner, N.Y.

[12] Le Cam, L. (1960), *An approximation theorem for the Poisson binomial distribution*, Pacific J. Math. **10**, 1181-1197.

[13] Nevzorov, V. B. (1986), *Two characterizations using records. In: Stability Problems for Stochastic Models*, Lecture Notes in Math. **1233**, Springer, Berlin, 79-85.

[14] Pfeifer, D. (1986), *Extremal processes, record times and strong approximation*, Pub. Inst. Stat. Univ. Paris **31**, 47-65.

[15] Pfeifer, D. (1988), *Extremal processes, secretary problems and the 1/e-law.* To appear in: J. Appl. Prob.

[16] Rényi, A. (1962), *Théorie des éléments saillants d'une suite d'observations*, Coll. Comb. Meth. Prob. Th. **104-115**, Mathematisk Institut, Aarhus Universitet, Denmark.

[17] Ross, S. M. (1982), *A simple heuristic approach to simplex efficiency*, Eur. J. Oper. Res. **9**, 344-346.

[18] Serfling, R. J. (1975), *A general Poisson approximation theorem*, Ann. Prob. **3**, 726-731.

[19] Serfling, R. J. (1978), *Some elementary results on Poisson approximation in a sequence of Bernoulli trials*, SIAM Rev. **20**, 567-579.

[20] Shorgin, S. Y. (1977), *Approximation of a generalized binomial distribution*, Th. Prob. Appl. **22**, 846-850.

[21] Zolotarev, V. M. (1976), *Metric distances in spaces of random variables and their distributions*, Math. U.S.S.R. Sbornik **30**, 3, 373-401.

[22] Zolotarev, V. M. (1984), *Probability metrics*, Th. Prob. Appl. **28**, 278-302.

Reprinted by permission of the publisher,
© 1989 by Springer-Verlag GmbH & Co. KG

INFORMATION

H. L. RESNIKOFF and M. L. PURI[1]

Abstract--This paper examines the mutual relationships of computers, information, and uncertainty at the fundamental level of *measurement*, which is the interface between science and experience, and *calculation*, which we think of as the interface between measurement and mind. Axioms for the measurement of information gain are presented and used to show how the information gained from a calculation can be explicitly evaluated.

- - - - -

"An addition to knowledge is won
at the expense of an addition to ignorance."
--*Arthur S. Eddington*[2]

Immediately upon its invention, the digital computer was recognized as an entirely new kind of machine whose major influence would be on information-intensive activities--the life of the mind--rather than on energy-intensive activities, which had been the case for all previous machine inventions. In a prescient analysis in 1948 of the potential effect of computers on the social and economic structure,

---

[1]Research of Madan L. Puri was supported by the Office of Naval Research Contract N000-14-85-K-0648.

[2]Eddington (1928, p. 229).

Norbert Wiener (1961, pp. 27-28) drew attention to the distinctive and unprecedented role of this machine at the level of mind:

> Long before Nagasaki and the public awareness of the atomic bomb, it had occurred to me that we were here in the presence of another social potentiality of unheard of importance for good and for evil.... For one thing, it makes the metaphorical dominance of the machine, as imagined by Samuel Butler, a most immediate and non-metaphorical problem. It gives the human race a new and most effective collection of mechanical slaves to perform its labor. Such mechanical labor has most of the economic properties of slave labor, although, unlike slave labor, it does not involve the direct demoralizing effects of human cruelty. However any labor that accepts the conditions of competition with slave labor accepts the conditions of slave labor, and is essentially slave labor. The key word of this statement is *competition.*
>
> Perhaps I may clarify the historical background of the present situation if I say that the first industrial revolution, the revolution of the "dark satanic mills", was the devaluation of the human arm by the competition of machinery. There is no rate of pay at which a United States pick-and-shovel laborer can live which is low enough to compete with the work of a steam shovel as an excavator. The modern industrial revolution is similarly to devalue the human brain, at least in its simpler and more routine decisions. Of course, just as the skilled carpenter, the skilled mechanic, the skilled dressmaker have in some degree survived the first industrial revolution, so the skilled scientist and the skilled administrator may survive the second. However, taking the second revolution as accomplished, the average human being of mediocre attainments or less has nothing to sell that it is worth anyone's money to buy.

The intervening 40 years have done nothing to dull the edge of Wiener's implicit call for a reorganization of society. The digital computer now dominates the scene in psychology and biology as the principal metaphor for the operation of the brain (see Resnikoff, 1987a; 1987b; 1989),[3] and some physicists and computer scientists have argued that the physical universe itself can best be understood and modeled as an enormous digital computer whose evolution represents the working out of a program the source and purpose of which remain unknown.

Thus it is not surprising that the word *information* has taken on special meanings in all phases of everyday life as well as in science and philosophy: we speak of and hear about *information age, information glut, information management, information overload, information society, information technology, information theory*, and *information workers* without end. These modifications of vocabulary merely reflect the now universally recognized truth that information is a fundamental constituent of the world that interacts with the more traditional and immediately physical constituents such as the mass and charge of material bodies, and space and time as the stage on which natural philosophy plays (see Resnikoff 1986; 1987b; 1989). Nevertheless, philosophers and scientists no less than economists still have great difficulty in identifying exactly what information is, how it interacts with the other constituents of physical or societal reality, how to quantify it in the national as well as the natural accounts, and even how to recognize it.

In this paper we will examine the curious role of measurement and computation in the Janus-like interplay of information and uncertainty at the fundamental level of *measurement*, which is the interface between science and experience, and *calculation*, which we think of as the interface between measurement and mind.

---

[3]This metaphor has been implicitly adopted by art historians in their analysis of the role of the psychophysics of vision. The standard presentation is Gombrich (1960).

I. DIGITAL COMPUTERS AND MEASUREMENT

A digital computer can only deal with dyadic rational numbers, that is, with numbers of the form

$$r = \pm \frac{m}{2^n}$$

where m and n are non-negative integers. Moreover, any operation that a digital computer can perform can be thought of, or encoded as, a finite sequence of bits--of zeros and ones--and hence as a particular binary expansion of a dyadic rational number. Thus the entire realm of digital computers is limited to the ring **D** of dyadic rationals.[4] Moreover, every reading produced by a measuring instrument is a rational number[5] and, if the instrument is constrained so that the bits in the binary expansion of a measured value can be read off one by one, then the instrument only produces dyadic rationals. Thus it appears that direct observation of the physical world can only produce numbers of exactly the sort that a digital computer can use.[6]

---

[4]A *ring* is a mathematical structure in which addition, subtraction and multiplication can be performed. If division by every non-zero element is also possible, mathematicians call the structure a *field*. The reader will note that the reciprocal of a dyadic rational is not in general a dyadic rational, whence the numbers known to a computer do not form a field.

[5]This argument is not strictly correct. The computer can represent an integer by its finite binary expansion, which is a dyadic rational number. Therefore, an arbitrary rational number can be identified with the ordered pair of dyadic rational integers that are the numerator and denominator of the rational number in reduced form. Similarly algebraic numbers such as $\sqrt{2}$, $\sqrt{17}$, etc., can be coded by finite sequences of integers, so they can also be expressed exactly in a digital computer. A measuring apparatus, however, cannot usually be arranged to measure the numerator and denominator of a rational number separately.

[6]A rational number is one of the form m/n where m and n are integers and $n \neq 0$. The set of all rational numbers, denoted **Q**, is the smallest infinite field and the ring **D** is contained within it.

Until recently, when one spoke of using mathematics to model natural phenomena, one meant the calculus. The ideas that underlie the calculus were systematically introduced in the late seventeenth century and the principal properties of derivatives and integrals were obtained by means of informal arguments that employed infinitesimal quantities. Leibniz recognized that the introduction of infinitesimals implied the existence of ideal infinitely small and infinitely large numbers that share the properties of the ordinary real numbers and, although mathematicians were unable to justify the use of these ideal numbers--indeed, the real numbers themselves were not fully understood until the late nineteenth century--infinitesimals were freely used in research and in mathematics education, as they still tend to be used in engineering and other branches of applied mathematics today. As Abraham Robinson (1966, p. 2) observed, "The idea of the infinitesimally small seems to appeal to our intuition." It was not until 1960 that the notion of infinitesimals was put on a sound logical basis by Robinson's introduction of non-standard analysis, which relies on the modern theory of models in mathematical logic to introduce a structure that extends the field **R** of real numbers to include an infinite hierarchy of infinite and infinitesimal numbers in a way that insures the validity of first order predicates in the extended system; cf. Robinson (1966).

The kinds of numbers that seem to be needed for the theoretical mathematics of the calculus are evidently very different from the numbers that are produced by direct observation of the physical world: measurements are not unrestricted real numbers which, like the transcendental π, require infinitely many bits to specify their binary expansion, and they certainly are not one of Leibniz's and Robinson's infinitesimals. We appear to be faced with two unpleasant alternatives: either real numbers (and possibly infinitesimals) occur in the world but cannot be measured exactly (which introduces an inherent uncertainty in measurement and in all deductions in which measurement plays a part), or measurements reveal all that there truly is to know (so that our mental models, dependent upon the real numbers and perhaps even upon infinitesimals, are overdetermined: they contain far too many constraints to accurately reflect the phenomena).

These remarks recall Kronecker's century-old criticism of the "completed infinite," that nothing could be said to have mathematical existence unless it could actually be constructed in terms of a finite number of positive integers. In his view, the rational numbers exist since they can be represented as the ratio of two integers but transcendental numbers like π do not exist, since they require infinitely many fractions or operations for their representation. Kronecker is thus the mathematician in whom a digital computer can believe. He proposed a program to "arithmetize" mathematics and eliminate from it all "non-constructive" concepts.[7] It is indeed time to reexamine the relationship between mathematics and natural philosophy in the new light provided by the theory of computability and in terms of the information that calculations and physical measurements can actually supply.

II. UNCERTAINTY AND NATURAL PHILOSOPHY

During the past 100 years philosophy and science have conspired to teach man the limits of reason and observation. The set-theoretic paradoxes of Cantor and the logicians, Einstein's ([1905] 1955) elimination of absolute time and space, Heisenberg's (1927) introduction of uncertainty as the indispensable ingredient in the microworld, and Gödel's ([1931] 1986) proof that mathematics and logic are not strong enough to prove all that is true, limit what can be known. It may seem paradoxical that this period of lowered expectations for what *can be* known has coincided with the period of greatest expansion and precision of what *is* known in mathematics and science. Evidently

---

[7]"And if I can't do this," he said, "it will be done by those who come after me!" (Reid, 1976, p. 26). Indeed, beginning in 1907, L.E.J. Brouwer put forward a program of "constructivist mathematics" which made little headway among practicing mathematicians, who were concerned that Brouwer's views would unnaturally and unnecessarily limit the development of mathematics. More recently, Bishop (1967) continued the attempt to "purge [mathematics] completely of its idealistic content." Hear Bishop (1967, p. 2):

> Mathematics belongs to man, not to God. We are not interested in the properties of the positive integers that have no descriptive meaning for finite man. When a man proves a positive integer to exist, he should know how to find it. If God has mathematics of his own that needs to be done, let him do it himself.

it is better to now one's limits than to thrash about in frenzied but unfulfillable expectation.

It was during this period that probability and statistics entered scientific thinking in a fundamental way. The concepts of probability and information first influenced each other in the second half of the nineteenth century in the work of Maxwell (1908) and Boltzmann (1909) on statistical thermodynamics, where entropy gain was associated with loss of information. Boltzmann's distribution for a thermodynamical system in equilibrium defines an entropy measure which is equivalent to the information measure introduced by Shannon (Shannon and Weaver, 1949) three-quarters of a century later in the context of electrical communication theory.

At the macroscopic level there are circumstances in which it is possible--in principle, at least--to isolate the information gained from a single measurement without prior knowledge of, or reference to, probability distributions. Where quantum phenomena are concerned, the outcome of a measurement is conditioned by the initial state of the system which, if it is not pure--not an "eigenstate"--will be represented by a state vector in Hilbert space, say

$$\psi = \Sigma \, c_a \, \psi_a,$$

where the ψ_a are a system of eigenstates and the complex coefficients c_a define a probability distribution $\{p_a\}$ by the formula

$$p_a = |\, c_a \,|^2.$$

The number p_a is the probability that measurement of the quantum system will show it to be in the state labelled a. It is the essence of quantum phenomena that systems cannot long remain in pure states, so measurements cannot be sharp (see von Neumann, 1955, for a profound treatment of this question). Shannon's measure of information associated with the discrete probability distribution $\{p_a\}$ is

$$I(\{p_\alpha\}) = -\sum_\alpha p_\alpha \lg p_\alpha,$$

where 'lg' denotes the base-2 logarithm and information I is measured in bits. The minus sign is introduced to insure that I is non-negative. If the state of a quantum system prior to an observation is

$$\psi_\alpha^{\text{initial}} = \Sigma\, c_\alpha^{\text{initial}}\, \psi_\alpha$$

and its state subsequent to measurement is

$$\psi_\alpha^{\text{final}} = \Sigma\, c_\alpha^{\text{final}}\, \psi_\alpha\,,$$

then the information gain is

$$\Delta I = I(\{\, | c_\alpha^{\text{initial}} |^2\}) - I(\{\, | c_\alpha^{\text{final}} |^2\}).$$

Quantum mechanics asserts that a measurement forces a physical system into a pure state, which immediately begins to decay into a superposition of states. Thus the "final" state, by which we understand the pure state in which the physical system finds itself as an immediate consequence of the measurement, is one for which exactly one of the coefficients c_α^{final} is not zero, whence $\psi^{\text{final}} = \psi_\beta$ for some state β. It follows from the definition of I that $I(\{\, | c_\alpha^{\text{final}} |^2\}) = 0$. Therefore the gain in information is measured by the quantity,

$$\Delta I = I(\{\, | c_\alpha^{\text{initial}} |^2\}) = -\Sigma\, | c_\alpha^{\text{initial}} |^2 \lg (\, | c_\alpha^{\text{initial}} |^2),$$

which is non-negative. This shows that a quantum mechanical measurement will produce an increase in information as long as the initial state is not pure.

Uncertainty is the one part of life in the quantum world of which we have no longer any doubt (Heisenberg, 1927), but it is still not generally realized that quantum uncertainty is inherited from the mathematical model that is used to express the physical phenomena. So-called "conjugate" physical quantities, such as position and momentum, or time and energy, correspond to linear functionals that are Fourier transform pairs. It is a mathematical fact that a function (or functional) $x \mapsto f(x)$ and its Fourier transform $y \mapsto \hat{f}(y)$ satisfy the uncertainty inequality (see Resnikoff, 1989, pp. 132-39):

$$\Delta x \quad \Delta y \geq \frac{1}{4\pi}.$$

Here the physical interpretation of x and y will be that of conjugate dynamical variables. If a measurement *increases* information about one quantum mechanical dynamical variable by *decreasing* the uncertainty concerning is magnitude, then the uncertainty principle requires that the measurement *increase* the uncertainty about the magnitude of the conjugate dynamical variable and thereby produce a corresponding *loss* of information.

III. AXIOMS FOR MEASURES OF INFORMATION GAIN

The goal of a calculation is to determine the value of some number within certain bounds of accuracy. The information gained from the calculation is therefore measured by the difference between the information about the number that was known before the calculation began, and the information known about the number after the calculation is completed. Since all that the calculation can produce is some finite sequence of bits of the binary representation of the number,[8] the gain in information must be measured by the new bits that the calculation produces. Indeed, as Wiener (1961) may have been the first to observe, the measure of information gained is just the number of new bits that the calculation produces. This raises an important problem, for it means that it is impossible to know the exact numerical value as a "real number" of any physical measurement because a real number requires infinitely many bits for its complete specification. Since the measurement of a bit, or the calculation of a bit, requires energy, the only way in which a numerical value could be determined exactly from observations would be if the energy E_n required to observe the nth bit in the binary expansion of a number were to decrease sufficiently rapidly with increasing n that the infinite series ΣE_n (which represents the total energy required to observe the number) converged to a finite value. But the consequences of such an assumption are contrary to experience. Even if the observed numerical quantity were a rational number--even if it were the number 1--there would be no way to verify this by direct physical measurement or observation of the bits of the number's binary expansion. This is a type of "uncertainty principle" which seems to be more fundamental and universal than the principle of Heisenberg; indeed, the latter may be a consequence of it.

At this point we must take stock of where we are and where we want to go. Our goal is to quantify the information gained from a calculation. Shannon's measure of information will not do for this purpose since there is no natural way to associate a probability distribution with a calculation. But *ad hoc* introduction of a new measure of information will not do either. So let us return to first principles and attempt to axiomatize the properties of a reasonable measure of information gain, and then *derive* the quantitative form that information gain must assume. We follow the discussion in Resnikoff (1989, pp. 25-32).

A prototypical physical measurement consists of determining the length of a rod by aligning it alongside a ruler. Let the true length of the rod be denoted by l, and suppose that the ruler is the real number line subdivided by marks in the ordinary way. The process of

---

[8]For convenience, and without loss of generality, we may suppose that all numbers are represented in base 2 positional notation, i.e., as a sequence of "bits".

measurement will produce not the true length l but some finite sequence of bits that can be interpreted as the initial (i.e. most significant) bits in the binary expansion of some real number. The finite number of bits produced by the measurement exactly define the binary expansion of some dyadic rational number. Successive refinement of the measurement (with the aid of a magnifying lens, perhaps) will lead to increasingly accurate dyadic rational approximations to l, but we cannot infer anything at all about the as yet unmeasured bits.

This process of successive refinement should be thought of as selection of successively smaller intervals of the real line that contain the number l. That is, a measurement of the length of the rod produces a pair of (dyadic rational) numbers x_1 and y_1 that are the coordinates of the interval containing l, such that $x_1 < l < y_1$. A second measurement that refines the first might produce the pair of numbers x_2 and y_2 such that $x_1 < x_2 < l < y_2 < y_1$. Our task is to determine the gain in information provided by the pair of measurements.

Since we are assuming that the gain in information depends only on the measured intervals in which l is contained, we may express the gain in information as a function of the endpoint coordinates of the pair of intervals:

$$\Delta I = \Delta I(x_1, y_1, x_2, y_2).$$

The coordinates of the endpoints of the measurement intervals depend on the unit of measurement of the ruler (the numbers x_1, y_1, x_2, y_2 will be different if centimeters or inches are marked) and on where the zero-point of the ruler is placed alongside the rod, but the gain in information is surely independent of both. This implies that the function ΔI that measures the gain in information must be invariant under the transformations that correspond to scale changes, i.e., $t \mapsto at$, where $a > 0$, and under transformations that correspond to a rigid motion which shifts the position of the zero point of the scale without changing the distance between any pair of marks, i.e., $t \mapsto t + b$. The most general combination of these two types of transformation has the form

$$t \mapsto at + b$$

with $a > 0$ and an arbitrary real number, b. Let us write this transformation as $t \mapsto \gamma_{a,b}(t)$. The collection of these transformations constitutes the *affine group*, and our first axiom states that the function ΔI that measures the information gain is invariant--i.e. unchanged--by the action of affine transformations:

Axiom 1: For any dyadic rational numbers, w, x, y, z and any affine transformation γ, the gain in information satisfies the equation

$$\Delta I(\gamma(w),\gamma(x),\gamma(y),\gamma(z)) = \Delta I(w,x,y,z).$$

Suppose that scientists in two laboratories are racing to be the first to measure the length l of the unicorn's horn. Both laboratories have found that $x_1 < l < y_1$. Proceeding with caution, Laboratory A then found $x_1 < x_2 < l < y_2 < y_1$. Emboldened by this result, they improved their measurement technique and obtained the refinement $x_2 < x_3 < l < y_3 < y_2$. Laboratory B, off to a late start, threw caution to the winds and, in a single astonishing experiment, found the remarkable estimate $x_1 < x_3 < l < y_3 < y_1$. What is the relationship of the information gains reported by Laboratory A to the information gain realized by Laboratory B, and what does it mean to the harried director of the National Science Foundation, who is preparing the budget request for the following year?

Clearly, the gain in information obtained by Laboratory A in its first experiment is $\Delta I_A(x_1,y_1,x_2,y_2)$, and in the second, $\Delta I_A(x_2,y_2,x_3,y_3)$, for a total gain of $\Delta I_A(x_1,y_1,x_2,y_2) + \Delta I_A(x_2,y_2,x_3,y_3)$, whereas the gain reported by Laboratory B is $\Delta I_A(x_1,y_1,x_3,y_3)$. Were the total gain in information reported by Laboratory A greater than the gain reported by Laboratory B, then investigations that proceed by a sequence of small incremental improvements would be preferable, but if the information gain reported by Laboratory B were greater, then a research policy of fewer but more significant experiments would be indicated. Experience supports reason when it tells us that both procedures produce the same gain in information. This is the content of our second axiom:

Axiom 2:

$$\Delta I(x_1,y_1,x_3,y_3) = \Delta I(x_1,y_1,x_2,y_2) + \Delta I(x_2,y_2,x_3,y_3).$$

These two axioms imply that the gain in information $\Delta I(x_1,y_1,x_2,y_2)$ is a function of the *ratio* (i.e. not of the variables separately) $(x_1-y_1)/(x_2-y_2)$, that is,

as

$$\Delta I(x_1,y_1,x_2,y_2) = g\left(\frac{x_1 - y_1}{x_2 - y_2}\right),$$

for some as yet unknown function g satisfying the relation, $g(uv) = g(u) + g(v)$, for positive numbers u and v.

It is not hard to show that g is the logarithm function.[9] If the unit of information is the bit, then

$$\Delta I(x_1,y_1,x_2,y_2) = lg\left(\frac{x_1 - y_1}{x_2 - y_2}\right).$$

One important consequence of this formula is that more than one measurement must be made in order to gain information. A single measurement lacks a standard against which information gain can be evaluated.[10]

The essential nature of comparison has been intuitively understood for a long time. Circa 1630 Constantijn Huygens, father of

---

[9]Relatively weak technical mathematical assumptions, such as continuity of g, are required to prove this result. Without any additional assumptions one can prove that g is completely determined by the numbers $\{g(p_k): 1 \le k < \infty\}$, where the p_k run through the prime numbers, and where the values $g(p_k)$ can be chosen arbitrarily. Thus g is a kind of "generalized" logarithm.

[10]Resnikof. (1989) shows how Shannon's formula for information appears as the measure of the average gain in information produced by a collection of measurements.

the scientist and secretary to the first stadholder of the Dutch Republic, wrote in his *Autobiography* (see S. Alpers, 1983, p. 18):

> ... the estimation which we commonly make of things is variable, untrustworthy, and fatuous insofar as we believe that we can eliminate every comparison and can discern any great difference in size merely by the evidence of our senses. Let us in short be aware that it is impossible to call anything "little" or "large" except by comparison. And then, as a result, let us firmly establish the proposition that the multiplying of bodies...is infinite; once we accept this as a fundamental rule then no body, even the most minute, may be so greatly magnified by lenses without there being reason to assert that it can be magnified more by other lenses, and then by still others, and so on endlessly.

Leone Battista Alberti (see S. Alpers, 1983, p. 22), writing circa 1450 on painting and sculpture, had already acknowledged the relativism of measurement:

> [Properties]...which philosophers term "accidents" because they may or may not be present in things,--all these are such as to be known only by comparison.

He continued with the perceptive gloss that

> [a]s man is best known of all things to man, perhaps Protagoras, in saying that man is the scale and measure of all things, meant that accidents in all things are duly compared to and known by the accidents in man.

IV. MENTAL INFORMATION GAIN

Thus far we have considered the information gained from an *observation* or a *measurement*. It has been implicitly assumed that something "out there" is being observed or measured, so theories and models of external reality play some role in what can and cannot be measured. It is instructive to turn our attention to the internal world of the mind to consider the information that is gained from mental activity. By far the simplest and most clear-cut case concerns mathematical mental activity, which is precisely defined and easily described. So let us consider the information gained from a *computation*. Clearly it makes no difference whether the computation is performed entirely in a mind, or is assisted by paper and pencil, or even whether it be performed by a machine.

The kind of idealized computation we have in mind has as its goal the determination of the bits in the binary expansion of some number, let us say

$$x = \sum_{n=n_0}^{\infty} \frac{x_n}{2^n},$$

by applying some prescription, or "algorithm", that generates a finite number of bits--i.e. a better approximation--each time it is applied. In a finite time and with only a finite amount of energy at our disposal, we will only be able to determine a finite number of the bits x_n in this expression. At the end, we will not know the exact value of x, but we will have gained information about it.

This difficult problem has not yet been treated in full generality, but there is a simple although relatively representative case which exhibits the essentials. Suppose that the calculation consists of summing the terms of an infinite series that represents some number. A concrete example is Gregory's famous series for π:

$$\frac{\pi}{4} = 1 - \frac{1}{3} + \frac{1}{5} - \ldots + (-1)^k \frac{1}{2k+1} \ldots .$$

A *measurement* will consist of explicit numerical calculation of the binary expansion of the partial sums

$$s_n = 1 - \frac{1}{3} + \frac{1}{5} - \ldots + (-1)^n \frac{1}{2n+1} = \sum_{k=0}^{n} (-1)^k \frac{1}{2k+1}.$$

Gregory's series is a so-called *alternating series*, and this implies that the partial sum determines an interval which contains the exact value of the series according to the inequalities

$$s_n - \left| \frac{(-1)^{n+1}}{2n+3} \right| < \frac{\pi}{4} < s_n + \left| \frac{(-1)^{n+1}}{2n+3} \right|,$$

which hold for any n. If we think of the left and right sides of this chain of inequalities as the endpoints of an interval that correspond to a "measurement" of $\pi/4$, then we can evaluate the gain in information $\Delta I(s_n, s_{n+1})$ that is provided by successive measurements which correspond to the calculation of successive partial sums s_n and s_{n+1}; we find

$$\Delta I(s_n, s_{n+1}) = \lg \left| \frac{s_{n+1} - s_n}{s_{n+2} - s_{n+1}} \right|.$$

This formula can be dramatically simplified because the difference of successive partial sums is

$$| s_{n+1} - s_n | = \frac{1}{2n+3}.$$

Therefore

$$\Delta I(s_n, s_{n+1}) = \lg \left(\frac{2n+5}{2n+3} \right) = \lg \left(1 + \frac{2}{2n+3} \right).$$

Each "turn of the crank" produces less information--the nth "turn" about $\frac{1}{2n+3}$ bit--so, when Gregory's series is used to calculate π, the cost per bit of information increases the greater the required accuracy.

This formula also expresses the total number of bits of information I_n that are provided by the partial sum s_n: it is just the expression

$$I_n = \sum_{k=0}^{n} \Delta I(s_k, s_{k+1}) = \lg\left(\frac{2n+5}{3}\right).$$

Hence, 1.5 million terms of the series will yield only about 20 bits of accuracy--about 6 decimal places.

Gregory's series provides a concrete example of how the finitary nature of measurement and calculation appear to impose limits on internally generated mental information as well as on knowledge of the world derived from physical experience. Should we become discouraged by these realities, then is it time to recall Kronecker's famous words and take fresh heart:

> "Die ganzen Zahlen hat der liebe Gott gemacht, alles andere ist Menschenwerk."[11]

REFERENCES

Alpers, S. (1983), *The Art of Describing*, Chicago: The University of Chicago Press.

Bishop, E. (1967), *Foundations of Constructive Analysis*, New York: McGraw-Hill Book Company.

[11] "God made the natural numbers, all else is the work of man" (Reid, 1976, p. 26).

Boltzmann, L. (1909), *Abhandlungen*, Vol. I, Leipzig: Barth.

Eddington, A.S. (1928), *The Nature of the Physical World*, Cambridge: Cambridge University Press.

Einstein, A. (1955), *The Meaning of Relativity*, Princeton: Princeton University Press, 5th Ed.

Gödel, K. (1986), *Collected Works*, Vol. I, S. Feferman et al., eds. Oxford and New York: Oxford University Press.

Gombrich, E.H. (1960), *Art and Illusion*, A.W. Mellon Lectures in the Fine Arts 1956, Princeton: Princeton University Press.

Heisenberg, W. (1927), "Über den anschaulichen Inhalt der quantentheoretischen Kinematik and Mechanik," *Zeitschrift für Physik*, 43, pp. 172-98.

Maxwell, J.C. (1908), *Theory of Heat*, London: Longmans, Green & Co.

Neumann, J. von (1955), *Mathematical Foundations of Quantum Mechanics*, Princeton: Princeton University Press (translation of the original German edition).

Reid, C. (1976), *Hilbert--Courant*, New York and Heidelberg: Springer-Verlag.

Resnikoff, H.L. (1985), "The New Science of Information," in Jane Chance and R.O. Wells, Jr., eds., *Mapping the Cosmos*, Houston: Rice University Press, pp. 129-50.

Resnikoff, H.L. (1986), "Information Technology," in Robert M. Hayes, ed., *Universities, Information Technology and Academic Libraries*, Norwood: Ablex Publishing Corporation, pp. 77-133.

Resnikoff, H.L. (1987a), "The Duality Between Noise and Aliasing," SPIE Vol. 758: *Image Understanding and the Man-Machine Interface*, Bellingham, WA: International Society for Optical Engineering, pp. 31-38.

Resnikoff, H.L. (1987b), "Concurrent Computation and Models of Biological Information Processing," in Manfred Kochen and Harold M. Hastings, eds., *Advances in Cognitive Science*, AAAS Selected Symposium N. 104, Boulder: Westview Press, pp. 78-99.

Resnikoff, H.L. (1989), *The Illusion of Reality*, New York and Heidelberg: Springer-Verlag.

Robinson, A. (1966), *Non-standard Analysis*, Amsterdam: North-Holland Publishing Company.

Shannon, C. and Weaver, W. (1949), *Mathematical Theory of Communication*, Urbana: University of Illinois Press.

Wiener, N. (1961), *Cybernetics: or Control and Communication in the Animal and the Machine*, Cambridge, MA: The MIT Press, 2nd Ed.

Reprinted from
Acting Under Uncertainty: Multidisciplinary Conceptions
(1990), pages 451-469, edited by George von
Furstenberg, by permission of the publisher,
© 1990 by Kluwer Academic/Plenum Publisher

A LOCAL ALGORITHM FOR CONSTRUCTING NON-NEGATIVE CUBIC SPLINES

BERND FISCHER, GERHARD OPFER and MADAN L. PURI

Institut für Angewandte Mathematik, University of Hamburg, Bundesstrasse 55, D-2000 Hamburg 13, West-Germany

Department of Mathematics, Indiana University, Bloomington, Indiana 47405

An interpolating spline which interpolates positive function values is not necessarily positive for all arguments; a typical example of this occurs in the spline interpolation of the normal density function of probability theory. We describe an algorithm which produces non-negative interpolating splines provided the function to be interpolated is positive. We start with the interpolating natural spline and redefine it in those intervals for which the spline has negative values; we do this by adding few extra knots to those intervals. First we show that for any choice of extra knots it is possible to construct a spline which is non-negative. We call such a spline feasible. Second we show that within the set of feasible splines there is a best spline, in the sense that the functional which reflects the strain energy is minimized. Finally, by using a necessary optimality criterion we obtain an *optimal* spline which minimizes the strain energy also with respect to the free knots. The resulting formulae give explicit results and are very simple. Our numerical results show that the final step of selecting the interior knots appropriately gives very satisfactory results in a very efficient manner. The algorithm can be applied also to produce splines which stay below upper or stay above lower constant bounds. It is also very useful as a starting procedure for finding the globally best spline. © 1991 Academic Press, Inc.

1. Introduction

Shape-preserving approximation is nowadays understood as an approximation which preserves relevant properties of the underlying function. These properties may be non-negativity (positivity), monotonicity

* Research supported by the Office of Naval Research Contract N00014-85-K-0648.

(strict monotonicity), boundedness of the first derivative, etc.; in certain cases the goal is simply to produce "visually pleasing" approximations (cf. Carlson [6]).

Not so long ago several of these types of approximations were broadly referred to as monotone approximations (cf. Lorentz and Zeller [21]). The concept and subject of monotone approximation, a major trend in approximation theory, were introduced by Shisha [29] (see also [20] in this connection). In particular preservation of (ordinary) convexity has attracted several authors. One of the early papers is by Hornung [19]. More recent results are by Burmeister, Hess, and J. W. Schmidt in several papers [5, 16, 25, 26]. Convergence rates for monotone spline interpolation were given by Utreras [31].

We are concerned here with the preservation of non-negativity by using interpolating cubic splines. One immediate application is a class of problems arising in non-parametric statistics. Let $X_1, X_2, ..., X_n$ be the order statistics from a random sample of independent and identically distributed random variables arising from a population with unknown continuous density function f. Consider the problem of obtaining an estimate $\hat{f}$ of f, based on the sample and on some simple assumptions regarding the degree of smoothness of f. The first attempt at a solution of this problem was made by Boneva, Kendall, and Stefanov [3]. Using an area-matching condition on the histogram, they constructed a quadratic spline approximation to f. This work was extended by Schoenberg [27], who used the area-matching condition to construct cubic and quintic spline approximations to the cumulative distribution function F. In all of these applications, the constraint $f \geqslant 0$ was ignored.

More recently de Montricher, Tapia, and Thompson [14] developed the theory of a maximum penalized likelihood estimator of f in which the constraint $f \geqslant 0$ was incorporated in the analysis. However, there was no numerical implementation.

Another important statistical application, where the non-negativity constraint has been ignored, is the construction of smoothing spline approximations to the spectral density of a second order stationary random process. Cogburn and Davis [7] developed the theory of periodic smoothing L-splines, which in a particular case yielded a polynomial spline approximation to the spectral density . Wahba [32] developed a smoothing periodic polynomial spline approximation to the logarithm of the spectral density, and she has also done considerable research on the practical problem of estimating the optimal value of the Lagrange multiplier. De Figueiredo and Thompson [11] constructed smoothing cubic spline approximations to the spectral density. All of the above authors started with the periodogram of the random process which was then smoothed in some manner.

Another area in which non-negative polynomial splines are applicable is in the construction of quadrature rules with non-negative weights. Although no use of splines was made, Davis [8] gave a constructive proof of a result by Tchakaloff [30] on the existence of non-negative quadrature formulas. Applications of shape-preserving interpolation were made by Hill and Passow [17].

The rapid development of digital microprocessors has increased the area of feasibility for digital signal processing. Communications engineers are becoming aware of the potentialities of the use of splines, as evidenced, for example, in the work of Polge and Bhagavan [23], in which polynomial spline approximations to various signal waveforms were constructed. Many functions considered in this subject, such as the signal envelope and the ambiguity function, are non-negative functions [24].

There have been two main attempts on this problem in the literature. One concept was followed by de Boor and Daniel [10]. It consisted of representing a spline by non-negative linear combinations of B-splines. Naturally such a spline is non-negative.

The other concept was to use quadratic Bernstein-polynomials. In this area there have been many investigations, both theoretical and numerical, including those of Akima [2], Briggs and Rubel [4], Deimel, Doss, Fornuaro, McAllister, and Roulier [12], Deimel, McAllister, and Roulier [13] and Schumaker [28].

The latter concept has been also applied in CAD (see [15]).

The present authors use an entirely different approach to the construction of non-negative interpolating cubic splines. They consider first the natural spline which satisfies only the obvious constraint of passing through the data points.

In the second step in each subinterval between two successive knots in which this spline assumes negative values, the relevant portion of the spline is replaced by a non-negative cubic spline. This is carried out by the addition of extra knots, the location of which is variable and depends only on the values of the original spline and of its derivatives at the end points of the relevant subintervals. Hence, the fixing up procedure is entirely local in nature. In a first step one can determine a best non-negative spline, in the sense that the strain energy integral in each such sub-interval is minimized under the assumption that the extra knots are fixed.

In the last step the knots are also varied, again with the goal of minimizing the strain energy integral which require to solve a nonlinear (but not too difficult) equation. Here we make use of a characterization of such splines as given in Opfer and Oberle [22].

As a by-product we obtain an algorithm which can be used to produce splines which stay locally (i.e., between two consecutive knots) below or above a given constant.

The results can be extended to splines of arbitrary degree without difficulties.

2. Preliminaries

We present here a few well known results on cubic splines that we shall require. This material is treated well in [1].

We consider an interval $[a, b]$, $a<b$, and subdivide it by a mesh of points Δ, where

$$\Delta : a = x_1 < x_2 < \cdots < x_n = b.$$

An associated set of ordinates $y = \{y_i \mid i = 1, 2, ..., n\}$ is prescribed. A cubic spline is a function $S_\Delta(y; \cdot)$ which is continuous together with its first and second derivatives on $[a, b]$, and which coincides with a cubic polynomial in each subinterval $[x_{j-1}, x_j]$, $j = 2, 3, ..., n$, and satisfies the interpolation conditions $S_\Delta(y; x_j) = y_j$, $j = 1, 2, ..., n$. The points $\{x_j \mid j = 1, 2, ..., n\}$ are termed the *knots* of the spline. Moreover, a *natural* cubic spline is one satisfying the boundary conditions $S''_\Delta(y; x_1) = S''_\Delta(y; x_n) = 0$. The existence and uniqueness of an interpolating natural cubic spline is well established. Moreover, we denote by $W_2^{(2)}$ the Sobolev space of real valued functions on $[a, b]$ with absolutely continuous first derivative and square integrable second derivative. The first result that we require is a theorem due to Holladay [18].

THEOREM 2.1. *Let $\Delta : a = x_1 = x_2 < \cdots < x_n = b$ and $y = \{y_i \mid i = 1, 2, ..., n\}$ be given. Then of all functions f in $W_2^{(2)}$ such that $f(x_i) = y_i$, the natural cubic spline $S_\Delta(y; \cdot)$ minimizes*

$$\int_a^b f''^2(x)\, dx$$

uniquely.

The second result that we shall require is termed the first integral relation or simply Theorem of Pythagoras (cf. de Boor [9, p. 66]):

$$\int_a^b f''^2(x)\, dx - \int_a^b S''^2_\Delta(y; x)\, dx = \int_a^b |f''(x) - S''_\Delta(y; x)|^2\, dx \qquad (2.1)$$

for all $f \in W_2^{(2)}$ which passes through the data points. And clearly (2.1) is valid if in addition $f \geqslant 0$.

3. Local Argument for the Construction of Non-negative Splines

In this paper we consider the obvious constraint that the prescribed vector y is positive; i.e., $y_i > 0$, $i = 1, 2, ..., n$. In view of Holladay's theorem, the minimizer of (2.1) is the natural cubic spline $S_\Delta(y; \cdot)$, but this does not in general satisfy the non-negativity constraint. In the sequel, for brevity, we abbreviate by $S_\Delta(y; \cdot)$ the unconstrained interpolating spline, since all splines under consideration pass necessarily through the data points.

To search for a non-negative function f in the subset of $W_2^{(2)}$ consisting of all non-negative cubic splines (passing through the data points) that minimizes the semi-norm $\int_a^b f''^2(x)\,dx$ would require the repeated computation of natural splines on finer grids than the given Δ.

If $S_\Delta(y; \cdot) \geqslant 0$, $f = S_\Delta(y; \cdot)$ minimizes $\int_b^a f''^2(x)\,dx$ and then our problem is solved.

Therefore we consider the case in which $S_\Delta(y; \cdot)$ has two distinct zero crossings in just one interval, say $[x_j, x_{j+1}]$. Also we set $R_j = [a, x_j] \cup [x_{j+1}, b]$. In view of the first integral relation, we have

$$\int_a^b f''^2(x)\,dx - \int_a^b S_\Delta''^2(y; x)\,dx$$

$$= \int_{R_j} |f''(x) - S_\Delta''(y; x)|^2\,dx + \int_{x_j}^{x_{j+1}} |f''(x) - S_\Delta''(y; x)|^2\,dx. \quad (3.1)$$

A consequence of the first integral relation is that the minimum of $\int_a^b |f''(x) - S_\Delta''(y; x)|^2\,dx$ can be computed by minimizing $\int_a^b f''^2(x)\,dx$ alone.

In order to obtain a local minimizing procedure which depends only on the interval $[x_j, x_{j+1}]$ we set $f(x) = S_\Delta(y; x)$ for all $x \in R_j$. Since in this case $\int_{R_j} |f''(x) - S_\Delta''(y; x)|^2\,dx = 0$, it remains to minimize

$$\int_{x_j}^{x_{j+1}} f''^2(x)\,dx. \quad (3.2)$$

We have put $f(x) = S_\Delta(y; x)$ for all $x \in R_j$. In order that the resulting spline is still in C^2, the minimizer of (3.2) has to be restricted to the conditions

$$\begin{aligned} f(x) &\geqslant 0 \qquad \text{for all} \quad x \in [x_j, x_{j+1}], \\ f^{(k)}(x_l) &= S_\Delta^{(k)}(y; x_l); \qquad k = 0, 1, 2; \quad l = j, j+1. \end{aligned} \quad (3.3)$$

Moreover, we see that, regardless of the number of intervals in $[a, b]$ in which $S_\Delta(y; \cdot)$ possesses negative values, the above considerations may be applied separately to each such interval and result in each case in minimizing (3.2) subject to (3.3). We emphasize also that we have not established as yet the existence of such an non-negative spline. We have, however, shown that if such a cubic spline exists, then it is determined in each interval solely by the boundary conditions on $S_\Delta(y; \cdot)$ in that interval.

Thus we seek a non-negative cubic spline f in $[x_j, x_{j+1}]$ which matches $S_\Delta(y; \cdot)$, $S'_\Delta(y; \cdot)$, and $S''_\Delta(y; \cdot)$ at the endpoints and minimizes (3.2). It is plausible that such an f, if it exists at all, should have the x-axis as a tangent at some interior $\hat{x} \in]x_j, x_{j+1}[$ and $f''(\hat{x}) \geqslant 0$.

It should be emphasized that our local algorithm suggested will not solve the global problem of minimizing $\int_b^a f''(x)^2\,dx$ subject to $f \geqslant 0$. However, it will produce—as we will see from the examples—a visually pleasing non-negative spline with little expense.

4. Construction of Feasible and Best Cubic Splines

Let j be an index such that the natural cubic spline $S_\Delta(y; \cdot)$ has negative values in $I_j = [x_j, x_{j+1}]$. Since the second derivatives $S''_\Delta(y; x_j)$ and $S''_\Delta(y; x_{j+1})$ are already uniquely determined by the four values $S_\Delta(y; x_j)$, $S'_\Delta(y; x_j)$, $S_\Delta(y; x_{j+1})$, $S'_\Delta(y; x_{j+1})$ we cannot expect to find a non-negative spline in I_j which matches all six boundary conditions mentioned in (3.3) and minimizes (3.2) as well.

Let us simplify our notation by calling the above interval only $I = [u, v]$. The spline $S_\Delta(y; \cdot)$ restricted to I is a cubic polynomial and will be called p in the sequel of this section.

Two typical polynomials which are positive at the endpoints of I and negative somewhere in the interior are

$$p_1 = 1 - 7x + 4x^2 + 4x^3, \qquad x \in [0, 1], \tag{4.1a}$$

$$p_2 = 2 - 24x - 75x^2 + 50x^3, \qquad x \in [0, 1]. \tag{4.1b}$$

These polynomials will serve as test cases later on. The graphs are shown in Fig. 1. The relevant zeros of p_1 are $x_l = 0.15977540$, $x_r = 0.79886759$, and those of p_2 are $x_l = 0.61705073$, $x_r = 0.95110062$.

The problem of the preceding section can now be reformulated as finding a function $f \in W_2^2[u, v]$ with

$$\int_u^v f''^2(x)\,dx = \min \tag{4.2}$$

subject to

$$f(x) \geqslant 0 \qquad \text{for all} \quad x \in I = [u, v], \tag{4.3}$$

$$f(u) = \sigma_0, \qquad f'(u) = \sigma'_0, \qquad f(v) = \sigma_1, \qquad f'(v) = \sigma'_1, \tag{4.4}$$

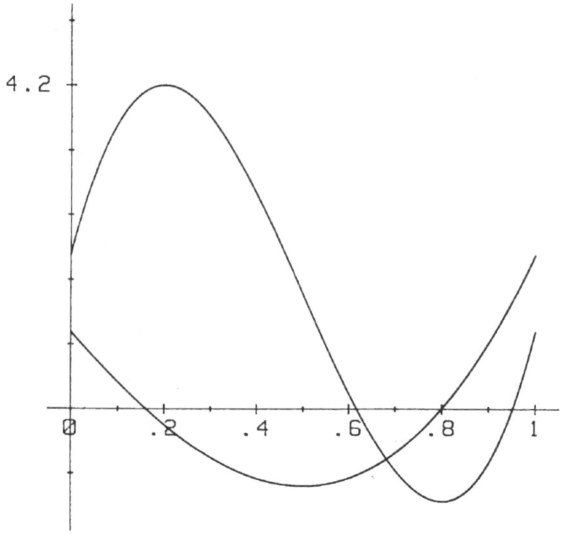

FIG. 1. Two test polynomials with positive endpoints and negative interior points.

where the four numbers σ are given by

$$\sigma_0 = p(u) > 0, \qquad \sigma_0' = p'(u), \qquad \sigma_1 = p(v) > 0, \qquad \sigma_1' = p'(v), \quad (4.5)$$

for a cubic polynomial p for which $p(x) < 0$ for some $x \in I$. Thus p has precisely two zeros $x_l < x_r$ in I. A necessary condition which is crucial in the construction of non-negative cubic spline which solves the above problem can be derived by variational techniques or by methods used in optimal control theory. The main information is contained in the following

THEOREM 4.1. *Let f solve the problem* (4.2) *to* (4.5). *Then f is a cubic* C^2-*spline with precisely one or two additional* (*simple*) *knots in the interior of I at which f vanishes. If there are two interior knots then f vanishes identically between these knots. Furthermore there are no zeros of f apart from or between the knots.*

Proof. Compare Opfer and Oberle [22]. ∎

A cubic spline with precisely one or two knots in the interior of I is called *feasible* if it matches the constraints (4.3) to (4.5). It is *optimal* it if solves the problem (4.2) to (4.5).

THEOREM 4.2. *The set of feasible splines is not empty.*

Proof. Let B be a cubic B-spline with vanishing values and vanishing first derivatives at the endpoints of I and with exactly one additional knot in the interior of I. For all $\alpha \in \mathbb{R}$, the spline

$$s = p + \alpha B \tag{4.6}$$

matches the boundary conditions (4.4) and (4.5). If α is sufficiently large it follows that $s(x) \geqslant 0$ for all $x \in I$. A similar construction is possible if we do have precisely two interior knots. ∎

COROLLARY 4.1. *There is a feasible spline s for which there exists a point* (called a contact point) $\hat{x} \in]u, v[$ *such that* $s(\hat{x}) = s'(\hat{x}) = 0$, $s''(\hat{x}) \geqslant 0$.

Proof. If s is a feasible spline such that $s(\hat{x}) = 0$ for some interior point $\hat{x}$, then necessarily $s'(\hat{x}) = 0$ and $s''(\hat{x}) \geqslant 0$ since $\hat{x}$ is a local minimum of s. By varying α in (4.6) we can produce splines s with positive and with negative minima. Therefore there must be an α such that the corresponding s has minimum value zero. Since $s(u) = \sigma_0 > 0$, $s(v) = \sigma_1 > 0$ this minimum must be attained at an interior point $\hat{x}$. ∎

A feasible spline with a contact point $\hat{x}$ as described in Corollary 4.1 can be constructed easily. The non-negativity condition of s reads

$$s(x) = p(x) + \alpha B(x) \geqslant 0 \qquad \text{for all } x \in I.$$

Since $B(x) > 0$ for all $x \in]u, v[$ we must require that

$$\alpha \geqslant -p(x)/B(x) \qquad \text{for all} \quad x \in]u, v[.$$

Therefore the smallest α which defines a feasible spline is

$$\hat{\alpha} = \max_{x \in [x_l, x_r]} \{-p(x)/B(x)\}, \tag{4.7}$$

where

$$p(x_l) = p(x_r) = 0 \qquad \text{and} \qquad u < x_l < x_r < v. \tag{4.8}$$

Let $\hat{\alpha} = -p(\hat{x})/B(\hat{x})$; then clearly $s(\hat{x}) = p(\hat{x}) + \hat{\alpha}B(\hat{x}) = 0$ and $s(x) \geqslant 0$ for all $x \in I$.

For the two polynomials p_1, p_2 of (4.1) the corresponding curves $-p/B$ are sketched in Fig. 2.

For p_1 we obtain $\hat{\alpha} = 2.42814970$ and for p_2 we obtain $\hat{\alpha} = 18.9384921$ where the additional knot is taken at 0.3.

A feasible spline $\hat{s}$ is called a *best* spline if for fixed interior knot(s) it approximates the given polynomial best, i.e.,

$$\| p'' - \hat{s}'' \|_2 \leqslant \| p'' - s'' \|_2 \qquad \text{for all feasible } s, \tag{4.9}$$

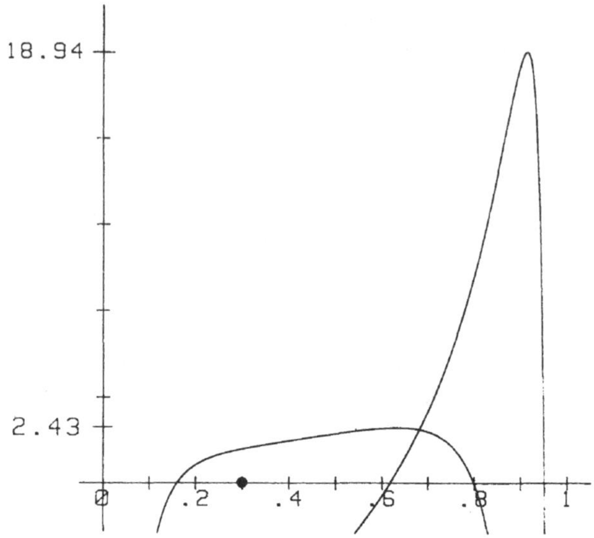

FIG. 2. Quotients $-p/B$ for the two test polynomials $p=p_1$ and $p=p_2$.

where $\| \ \|_2$ is the usual 2-norm. If we use $s=p+\alpha B$ from (4.6) then (4.9) reads $\hat{\alpha}\,\|B''\|_2 \leqslant \alpha\,\|B''\|_2$ for all α which define feasible splines. Since we have already computed the smallest α which defines a feasible spline, namely $\hat{\alpha}$ of (4.7), this $\hat{\alpha}$ also defines a best spline in the sense of (4.9).

In view of Theorem 4.1, splines with precisely two interior knots also must be taken into account. If we choose two interior knots, say $u<\xi_l<\xi_r<v$ arbitrarily, then we cannot expect to find a feasible C^2-spline which vanishes between the two given knots, since this would require finding two cubic polynomials each defined by five conditions. We will call a feasible C^2-spline with two interior knots which vanishes between these two knots a spline with a *contact line* (sometimes also called spline with *boundary arc*). It should be remarked here that by using a representation of the form $s=p+\alpha_1 B_1+\alpha_2 B_2$ it is not so easy to construct a feasible spline with two interior knots.

It should be observed that the evaluation of $\|B''\|_2$, if at all necessary, can be carried out exactly by applying Simpson's rule in the subintervals defined by the knots.

5. Optimal Knots Selection

Let us consider the same setting as before. The only difference now is that the interior knots are regarded as free. As measure of approximation we still use the semi-norm $\|p''-s''\|_2$. We have to distinguish between two

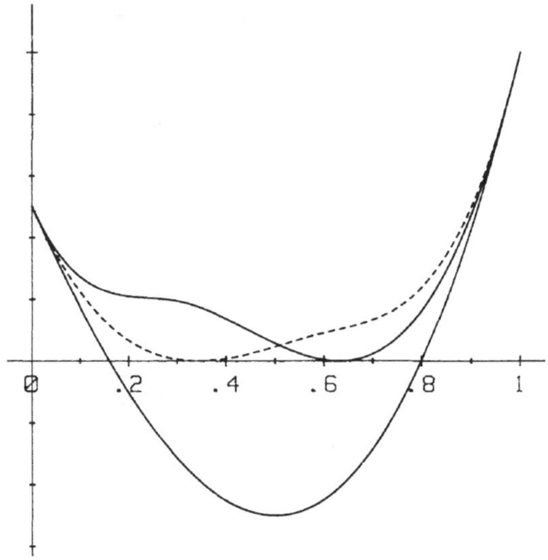

FIG. 3. Best splines for test polynomial p_1.

cases according to the number of interior knots. In Figs. 3 and 4 we present best splines for different placements of one additional knot which demonstrate that it is worthwhile to think about the placement of the interior knot(s).

A. *Case of Two Interior Knots*

The case of two interior knots $u < \xi_l < \xi_r < v$ turns out to be simpler. Therefore we treat this case first. For this case we apply Theorem 4.1 and

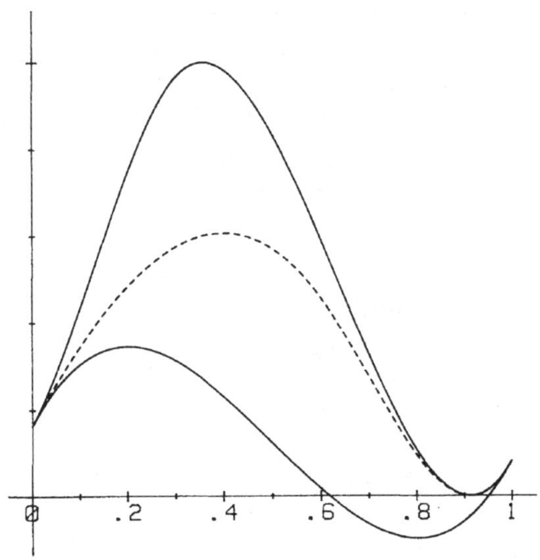

FIG. 4. Best splines for test polynomial p_2.

take the spline s in a form which is more convenient for our purpose, namely

$$s(x)=\begin{cases} s_l(x)=a_l(\xi_l-x)^3 & \text{for} \quad x\in[u,\xi_l],\\ 0 & \text{for} \quad x\in]\xi_l,\xi_r[,\\ s_r(x)=a_r(x-\xi_r)^3 & \text{for} \quad x\in[\xi_r,v].\end{cases} \tag{5.1}$$

This representation makes s automatically a C^2-spline. We have to satisfy the conditions (4.4), (4.5) which read explicitly

$$s(u)=\sigma_0=a_l(\xi_l-u)^3, \qquad s'(u)=\sigma_0'=-3a_l(\xi_l-u)^2, \tag{5.2a}$$

$$s(v)=\sigma_1=a_r(v-\xi_r)^3, \qquad s'(v)=\sigma_1'=3a_r(v-\xi_r)^2. \tag{5.2b}$$

The first two equations are equations for the unknowns a_l, ξ_l. The last two equations are equations for the unknowns a_r, ξ_r. These equations can be easily solved. And the solution is

$$\xi_l=u-\frac{3\sigma_0}{\sigma_0'}, \qquad a_l=-\frac{\sigma_0'^3}{27\sigma_0^2}, \qquad \xi_r=v-\frac{3\sigma_1}{\sigma_1'}, \qquad a_r=\frac{\sigma_1'^3}{27\sigma_1^2}. \tag{5.3}$$

Since we must have $u<\xi_l<\xi_r<v$, we obtain the requirements

$$\sigma_0'<0, \qquad \sigma_1'>0, \qquad v-u>3\left(\frac{\sigma_1}{\sigma_1'}-\frac{\sigma_0}{\sigma_0'}\right). \tag{5.4}$$

The first two conditions imply that the coefficients a_l, $a_r>0$ with the consequence that $s(x)\geqslant 0$ for all $x\in[u,v]$. They also say that p is convex on $[u,v]$. However, convexity alone does not imply the existence of a contact line. For, replacing p by $p+\text{const}$ does not change the convexity. But the last condition of (5.4) will not hold if the constant is only large enough. We turn now to the case of one interior knot.

B. *Case of One Interior Knot*

We use a similar representation as in the previous case, namely

$$s(x)=\begin{cases} s_l(x)=b_l(\xi-x)^2+a_l(\xi-x)^3 & \text{for} \quad x\in[u,\xi[,\\ s_r(x)=b_l(x-\xi)^2+a_r(x-\xi)^3 & \text{for} \quad x\in[\xi,v],\end{cases} \tag{5.5}$$

where ξ is the unknown knot. The representation (5.5) again makes s a C^2-spline. Since ξ is a contact point we must have $s''(\xi)/2=b_l\geqslant 0$. The spline must meet the four conditions (4.4). Therefore we have four equations for the unknowns a_l, a_r, b_l, ξ. From these four equations one can easily eliminate the quantities a_l, a_r and one is left with two equations for b_l, ξ which can be put into the form

$$b_l(\xi-u)^2\,(v-\xi)^2=(v-\xi)^2\,(3\sigma_0+\sigma_0'(\xi-u)), \tag{5.6a}$$

$$b_l(\xi-u)^2\,(v-\xi)^2=(\xi-u)^2\,(3\sigma_1-\sigma_1'(v-\xi)). \tag{5.6b}$$

Since $b_l\geqslant 0$ we must have

$$3\sigma_0+\sigma_0'(\xi-u)\geqslant 0,\quad 3\sigma_1-\sigma_1'(v-\xi)\geqslant 0 \tag{5.7}$$

in order that the above equations have a solution. Every solution is then a zero of the cubic polynomial

$$g(\xi)=(v-\xi)^2\,(3\sigma_0+\sigma_0'(\xi-u))-(\xi-u)^2\,(3\sigma_1-\sigma_1'(v-\xi)). \tag{5.8}$$

This polynomial always has at least one zero in $]u,v[$ since $g(u)>0$ and $g(v)<0$.

In dependence on the signs of σ_0' and σ_1' the conditions (5.7) are given in Table I.

From Table I we see that conditions (5.7) are valid exactly if conditions (5.4) are not valid.

If the given problem (4.2) to (4.5) would have two different solutions, any convex combination of these solutions would also be a solution. Thus we would have an infinite number of solutions. Our conditions, however, allow only finitely many solutions (namely at most three, corresponding to the zeros of g defined in (5.8)). Therefore only one solution is possible.

We summarize.

THEOREM 5.1. *Let p be a polynomial on $[u,v]$ with properties mentioned in* (4.5). *If p satisfies conditions* (5.4) *the optimal spline s is defined by two interior knots $u<\xi_l<\xi_r<v$ and it is given by* (5.1) *and* (5.3). *If the conditions*

TABLE I

Conditions (5.7) in Dependence on the Signs of the σ''s

Sign σ_0'	Sign σ_1'	Conditions (5.7)
+	+	$u\leqslant v-3\sigma_1/\sigma_1'\leqslant\xi\leqslant v$
+	−	Cannot occur
−	+	$v-3\sigma_1/\sigma_1'\leqslant\xi\leqslant u-3\sigma_0/\sigma_0'$
−	−	$\xi\leqslant u-3\sigma_0/\sigma_0'\leqslant v$

(5.4) *are not satisfied, then the optimal spline s is defined by precisely one interior knot $u < \xi < v$, and s is defined by* (5.5), (5.6), (5.8).

Proof. By the above considerations. ∎

We apply the above theorem to the two test polynomials p_1, p_2 defined in (4.1a), (4.1b), respectively. For p_1 we obtain

$$\sigma_0 = 1, \qquad \sigma_0' = -7, \qquad \sigma_1 = 2, \qquad \sigma_1' = 13$$

and condition (5.4) is valid. Thus the resulting spline s has a *contact line* and is defined in (5.1), (5.3) with

$$\xi_l = 3/7 = 0.4286, \qquad a_l = 7^3/27 = 12.7037, \tag{5.9a}$$

$$\xi_r = 7/13 = 0.5385, \qquad a_r = 13^3/108 = 20.3426, \tag{5.9b}$$

$$\|p'' - s''\|_2 = 13.88044188. \tag{5.9c}$$

From the last condition in (5.4) we deduce that we could replace p_1 with $p_1 + \text{const}$ without violating this condition as long as const $< 1/6$. For p_2 we have

$$\sigma_0 = 2, \qquad \sigma_0' = 24, \qquad \sigma_1 = 1, \qquad \sigma_1' = 24$$

and condition (5.4) is not valid. Thus the solution has only a *contact point* and is obtained from (5.8), (5.6), (5.5):

$$\xi = (2 + 3\sqrt{2})/7 = 0.89180581, \tag{5.10a}$$

$$\begin{aligned} b_l = 2340.968830, \qquad a_l = -2622.156167, \qquad a_r = -20{,}847.16817, \\ \hat{\alpha} = 4.05250462, \qquad \|p'' - s''\|_2 = 45.1936094, \end{aligned} \tag{5.10b}$$

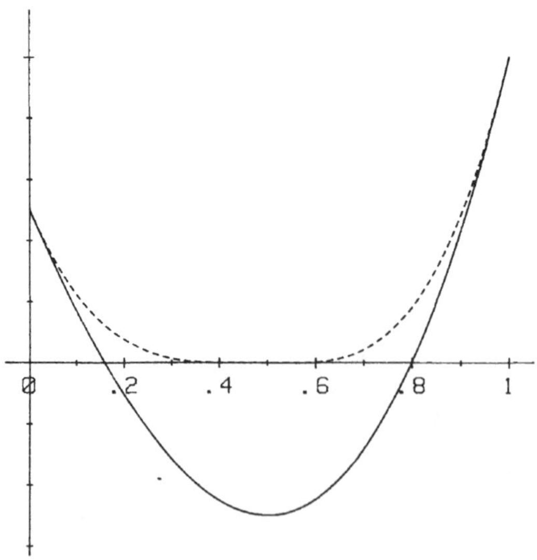

FIG. 5. Optimal spline for test polynomial p_1.

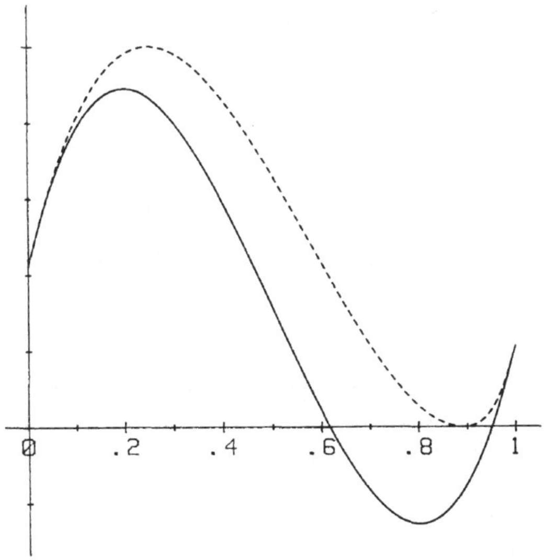

FIG. 6. Optimal spline for test polynomial p_2.

where $\hat{\alpha}$ refers to the former representation (4.6). The optimal splines are presented in Figs. 5 and 6.

C. *General Remarks*

The procedure described is very simple and the results are very satisfactory in many cases. The procedure may serve also another purpose. When computing non-negative splines according to a global strategy (cf. [22]) the start selection of knots turns out to be crucial, since these knots serve as initial values for a Newton-iteration. From our own numerical experience it is apparent that the optimal knots as described in this section are in general very useful as starting values for the above mentioned Newton-iteration, cf. Dauner and Reinsch.

ACKNOWLEDGMENT

The authors express their sincere thanks to Professor Oved Shisha for bringing to their attention the work of Shisha [29] and Lorentz [20].

REFERENCES

1. J. H. AHLBERG, E. N. NILSON, AND J. L. WALSH, "The Theory of Splines and Their Applications," Academic Press, New York/London, 1967.
2. H. AKIMA, A new method of interpolation and smooth curve fitting based on local procedures, *J. Assoc. Comput. Mach.* **17** (1970), 589–602.

3. L. I. Boneva, D. Kendall, and I. Stefanov, Spline transformations: Three new diagnostic aids for the statistical data-analyst, *J. Roy. Statist. Soc. Ser. B* **33** (1971), 1–70.
4. J. Briggs and L. A. Rubel, Interpolation by non-negative polynomials, *J. Approx. Theory* **30** (1980), 160–168.
5. W. Burmeister, W. Hess, and J. W. Schmidt, Convex spline interpolants with minimal curvature, *Computing* **35** (1985), 219–229.
6. R. E. Carlson, Shape preserving interpolation, *in* "Algorithms for Approximation: Proceedings of IMA Conference on Algorithms for the Approximation of Functions and Data, Shrivenham, England, 1985" (J. C. Mason and M. G. Cox, Eds.), pp. 97–114, Clarendon Press, Oxford, 1987.
7. R. Cogburn and H. T. Davis, Periodic splines and spectral estimation, *Ann. Statist.* **2** (1974), 1108–1126.
7′. H. Dauner and C. H. Reinsch, An analysis of two algorithms for shape-preserving cubic spline interpolation. *IMA Journal of Numerical Analysis* **9** (1989), 299–314.
8. P. J. Davis, A construction of nonnegative approximate quadratures. *Math. Comp.* **21** (1967), 578–582.
9. C. de Boor, A practical guide to splines, *in* "Applied Mathematical Sciences," Vol. 27, Springer-Verlag, New York/Heidelberg, 1978.
10. C. de Boor and J. W. Daniel, Splines with nonnegative *B*-spline coefficients, *Math. Comput.* **28** (1974), 565–568.
11. R. J. P. de Figueiredo and J. R. Thompson, Power spectral density estimation by spline smoothing in the frequence domain, *in* "Proceedings, Third Symposium on Nonlinear Estimation Theory and Its Application, San Diego, California, September 1972" (H. Sorenson, Ed.), pp. 50–57, Western Periodical Company, North Hollywood, CA.
12. L. E. Deimel, L. L. Doss, R. J. Fornuaro, D. F. McAllister, and J. A. Roulier, Application of shape-preserving spline interpolation to interactive editing of photogrammetric data, *Proc. SIGGRAPH 78* **12** (1978), 93–99.
13. L. E. Deimel, D. F. McAllister, and J. A. Roulier, Smooth curve fitting with shape-preserving osculatory, quadratic splines, *in* "Proceedings, 11th Annual Conference on the Interface between Statistics and Computer Science, 1978," pp. 343–347.
14. G. F. de Montricher, R. A. Tapia, and J. R. Thompson, Nonparametric maximum likelihood estimation of probability densities by penalty function methods, *Ann. Statist.* **3** (1975), 1329–1348.
15. W. J. Gordon and R. F. Riesenfeld, Bernstein–Bezier methods for the computer-aided design of freeform curves and surfaces, *J. Assoc. Comput. Mach.* **21** (1974), 293–310.
16. W. Hess and J. W. Schmidt, Convexity preserving interpolation with exponential splines, *Computing* **36** (1986), 335–342.
17. D. R. Hill and E. Passow, Quadrature formulae based on shape preserving interpolation, *in* "Approximation Theory and Applications (Proc. Workshop, Technion-Israel Inst. Tech., Haifa, 1980)," pp. 283–289, Academic Press, New York/London, 1981.
18. J. C. Holladay, A smoothest curvature approximation, *Math. Tables Aids Comput.* **11** (1957), 233–243.
19. U. Hornung, Numerische Berechnung monotoner und konvexer Spline-Interpolierender, *Z. Angew. Math. Mech.* **59** (1979), T64–T65.
20. G. G. Lorentz, Monotone approximation, *in* "Inequalities—Proceedings of the Third Symposium on Inequalities, University of California, Los Angeles, September 1–9, 1968" (O. Shisha, Ed.), pp. 201–215, Academic Press, New York, 1972.
21. G. G. Lorentz and K. L. Zeller, Monotone approximation by algebraic polynomials, *Trans. Amer. Math. Soc.* **149** (1970), 1–18.
22. G. Opfer and H. J. Oberle, The derivation of cubic splines with obstacles by methods of optimization and optimal control, *Numer. Math.* **52** (1988), 17–31.

23. R. J. POLGE AND B. K. BHAGAVAN, Fourier transform and ambiguity function of piecewise polynomial functions. *IEEE Trans. Aerospace Electron. Systems* **13** (1977), 403–408.
24. W. SCHEMPP, Analog radar signal design and digital signal processing—A Heisenberg nilpotent Lie group approach, *in* "Lie Methods in Optics," pp. 1–27, Lecture Notes in Physics, Vol. 250, Springer-Verlag, Berlin/Heidelberg/New York/Tokyo, 1986.
25. J. W. SCHMIDT, Convex interval interpolation with cubic splines, *BIT* **26** (1986), 377–387.
26. J. W. SCHMIDT AND W. HESS, Schwach verkoppelte Ungleichungssysteme und konvexe Spline-Interpolation, *Elem. Math.* **39** (1984), 85–95.
27. I. J. SCHOENBERG, "Splines and Histograms," with an appendix by Carl de Boor, Math. Research Center Report No. 1273, University of Wisconsin, Madison, 1972.
28. L. L. SCHUMAKER, On shape-preserving quadratic spline interpolation, *SIAM J. Numer. Anal.* **20** (1983), 854–864.
29. O. SHISHA, Monotone approximation, *Pacific J. Math.* **15** (1965), 667–671.
30. V. TCHAKALOFF, Formules de cubatures mécaniques à coefficients non négatifs, *Bull. Sci. Math.* **81** (1957), 123–134.
31. F. I. UTRERAS, Convergence rates for monotone cubic spline interpolation, *J. Approx. Theory* **36** (1982), 86–90.
32. G. WAHBA, Smoothing noisy data with spline functions, *Numer. Math.* **24** (1975), 383–393.

Reprinted by permission of the publisher,
© 1991 by Academic Press

THE TIME SPENT BY THE WIENER PROCESS IN A NARROW TUBE BEFORE LEAVING A WIDE TUBE

ANTONIA FÖLDES and MADAN L. PURI

ABSTRACT. The almost sure behavior of the time spent by the Wiener process in a "small interval" before first leaving a "bigger interval" is investigated.

1. INTRODUCTION

In this paper we investigate the almost sure behavior of the time spent by the Wiener process in a "small" interval, before first leaving a "bigger interval". Let $W(t)$ $(t \geq 0)$ be an ordinary Wiener process, and define

$$m_b = \inf\{s \geq 0, \ W(s) \notin [-b, b]\}.$$

Moreover, let

$$\mu_x(\alpha, b) = \mu_x\{0 \leq s; \ |W(s)| \leq \alpha b, \ s \leq m_b\}$$

be the Lebesgue measure of the time spent by $W(s)$ in $[-\alpha b, \alpha b]$ until m_b, where the lowercase subscript x indicates that $W(0) = x$. We will prove the following theorems.

Theorem 1. *For any* $0 < \alpha \leq 1$ *and for any fixed* x,

$$\overline{\lim_{b \to \infty}} \frac{\mu_x(\alpha, b) c_0^2(\alpha)}{2\alpha^2 b^2 \log\log b} = 1 \tag{1.1}$$

where $c_0(\alpha)$ *is the unique root of the equation*

$$u \tan u = \alpha/(1-\alpha)$$

in the interval $(0, \pi/2]$.

Remark 1. Observe that the case $\alpha = 1$ states that

$$\overline{\lim_{b \to \infty}} \frac{m_b \cdot \pi^2}{8b^2 \log\log b} = 1,$$

which is a natural reformulation of Chung's law of iterated logarithm for the maximum of the Wiener process.

Received by the editors December 12, 1990 and, in revised form, June 18, 1991.

1991 *Mathematics Subject Classification.* Primary 60J65, 60G17.

Key words and phrases. Wiener process, almost sure behavior.

Research of Antonia Földes is partially supported by the Hungarian National Foundation for Scientific Research Grant No. 1905 and a PSC CUNY grant #662349; the research of Madan L. Puri is supported by the Office of Naval Research Contracts N00014-85-K-0648 and N00014-91-J-1020.

Theorem 2. *For any* $0 < \alpha < 1$ *and for any fixed* x,

$$\lim_{b\to\infty} \frac{\mu_x(\alpha, b)}{\alpha^2 b^2} 2\log\log b = 1. \tag{1.2}$$

Remark 2. Theorem 2 can be proved (just like Theorem 1) based on the asymptotic behaviour of the Laplace transform of $\mu_x(\alpha, b)$ and then applying the Erdös-Rényi extension of the Borel-Cantelli lemma, however, a much simpler proof reveals that it can be deduced from the law of iterated logarithm.

2. Proof of the results

In our first lemma, we give the Laplace-Stieltjes transform of

$$F_x(t) = P(\mu_x(\alpha, b) < t)$$

where the lowercase subscript x means again that $W(0) = x$.

Lemma 1. *Let, for* $0 < \alpha \le 1$,

$$H(x) = H(x, \lambda) = E(\exp\{-\lambda\mu_x(\alpha, b)\})$$

and denote $q = \alpha b$. *Then*

$$H(x) = \begin{cases} \dfrac{\cosh(q\sqrt{2\lambda}) - (x+q)\sqrt{2\lambda}\sinh(q\sqrt{2\lambda})}{\cosh(q\sqrt{2\lambda}) + (b-q)\sqrt{2\lambda}\sinh(q\sqrt{2\lambda})} & \text{if } -b \le x \le -q, \\[2ex] \dfrac{\cosh(x\sqrt{2\lambda})}{\cosh(q\sqrt{2\lambda}) + (b-q)\sqrt{2\lambda}\sinh(q\sqrt{2\lambda})} & \text{if } -q \le x \le q, \\[2ex] \dfrac{\cosh(q\sqrt{2\lambda}) + (x-q)\sqrt{2\lambda}\sinh(q\sqrt{2\lambda})}{\cosh(q\sqrt{2\lambda}) + (b-q)\sqrt{2\lambda}\sinh(q\sqrt{2\lambda})} & \text{if } q \le x \le b. \end{cases} \tag{2.1}$$

Remark 3. Observe the symmetry of $H(x)$, $H(x) = H(-x)$, a fact that will be used frequently without mentioning it.

Remark 4. The special cases $x = 0$ and $x = q$ are treated in [7, Theorem 4.2.16].

Remark 5. The method needed to compute the Laplace-Stieltjes transform is very general and goes back to Kac [6] (see also [7, Theorem 4.2.11, p. 77; 1; 5]).

Theorem. *Let* $f(x)$, $x \in [a, b]$, *be a piecewise continuous function and let* $A(t) = \int_0^t f(W(s))\,ds$. *Then the function*

$$H(x) = E_x(\exp\{-A(m_b)\})$$

(*where* $E_x(\ )$ *means expectation under the condition that* $W(0) = x$) *is the unique continuous solution of the problem*

$$\begin{gathered} \tfrac{1}{2}H''(x) - f(x)H(x) = 0, \\ H(-b+0) = H(b-0) = 1, \qquad x \in (-b, b). \end{gathered} \tag{2.2}$$

Proof of Lemma 1. We apply the above theorem for

$$f(x) = \begin{cases} \lambda & \text{if } |x| \le q, \\ 0 & \text{otherwise}. \end{cases}$$

Then we get the general solution of (2.2) in the following form:

$$H(x) = H(x, \lambda) = \begin{cases} vx + z & \text{if } -b < x \leq -q, \\ \gamma \exp(\sqrt{2\lambda} x) + \beta \exp(-\sqrt{2\lambda} x) & \text{if } |x| \leq q, \\ lx + r & \text{if } q \leq x \leq b. \end{cases}$$

Using the continuity of $H(x)$ and $H'(x)$ and the boundary conditions in (2.2), we get a system of equations for v, z, γ, β, l, and r. The unique solution of this system produces (2.1). □

Consequently the ordinary Laplace transform of $F_x(t)$ is $H(x, \lambda)/\lambda$.

To investigate the asymptotic behavior of $F_x(t)$, when $t \to \infty$, we have to study $H(x, \lambda)/\lambda$ when $\lambda \to 0$.

The following lemma is most probabily well known; however, we give a brief sketch of its proof.

Lemma 2. *The function*

$$U(\lambda) = \cosh(q\sqrt{2\lambda}) + (b - q)\sqrt{2\lambda} \sinh(q\sqrt{2\lambda})$$

(*where* $q = \alpha b$, $0 < \alpha < 1$) *has roots only on the negative real axis and among them the root of the smallest absolute value is*

$$\lambda_0 = -c_0^2(\alpha)/2\alpha^2 b^2$$

where $c_0(\alpha)$ *is the unique root of* $u \tan u = \alpha/(1 - \alpha)$ *in the interval* $0 < u \leq \pi/2$.

Proof. To find the roots on the negative real axis, let $2\lambda = -z^2$ where $z > 0$. Applying the well-known formulas $\sinh(ix) = i \sin x$ and $\cosh ix = \cos x$, our equation $U(\lambda) = 0$ transforms into

$$zq \tan(zq) = \frac{q}{b - q} \qquad \left(= \frac{\alpha}{1 - \alpha}\right).$$

Now let $u = zq$ (where $z > 0$, $q > 0$, hence $u > 0$). Then we want the positive solutions of

$$u \tan u = \alpha/(1 - \alpha). \tag{2.3}$$

Equation (2.3) has one root in each of the intervals $(k\pi, k\pi + \pi/2]$ $(k \geq 0)$. Denote the smallest one by $c_0(\alpha)$, $0 < c_0(\alpha) \leq \pi/2$. Then $\lambda_0 = -c_0^2(\alpha)/2q^2$ is the negative real root of $U(\lambda)$ of smallest absolute value.

To see that all roots of $U(\lambda)$ are on the negative real axis, first observe that $U(\lambda) > 0$ when $\lambda > 0$ and real, so we have only to show that no root of the form

$$q\sqrt{2\lambda} = A + Bi, \qquad A \neq 0, \ B \neq 0 \text{ real}, \tag{2.4}$$

exists. Arguing indirectly, if there were a root of the form (2.4), then from $U(\lambda) = 0$ we would get

$$(A + Bi) \tanh(A + Bi) = -q/(b - q).$$

Elementary calculations lead to the point that $(A + Bi)[e^{4A} - 1 + i(2e^A \sin 2B)]$ has to be real, which is the same as (being $A \neq 0$, $B \neq 0$)

$$B(e^{4A} - 1) + 2Ae^{2A} \sin 2B = 0,$$

implying that

$$\frac{\sin 2B}{2B} = -\frac{\sinh 2A}{2A}. \tag{2.5}$$

But for all $x \neq 0$, $|\frac{\sin x}{x}| < 1$ and $|\frac{\sinh x}{x}| > 1$, which proves that (2.5) cannot be true for any real $A \neq 0$, $B \neq 0$. □

Based on Lemma 2, using standard methods (see, e.g., [3]) we get the following result:

Let $x = \delta b$, $|\delta| < 1$ and $\widetilde{F}_x(t) = P(\frac{\mu_x(\alpha, b)}{b^2} < t)$.

Lemma 3. *For $t \to \infty$, $0 < \alpha < 1$,*

$$1 - \widetilde{F}_x(t) \sim c_1(\alpha) \cdot c_2(\alpha, x) \exp\left\{-\frac{c_0^2(\alpha)t}{2\alpha^2}\right\} \tag{2.6}$$

where $c_0(\alpha)$ (as before) is the unique root of $u \tan u = \alpha/(1-\alpha)$ in $(0, \pi/2)$,

$$c_1(\alpha) = \frac{2}{1/(1-\alpha) + c_0^2(\alpha)(1-\alpha)/\alpha}, \tag{2.7}$$

and

$$c_2(\alpha, x) = \begin{cases} (1+\delta)/(1-\alpha) & \text{if } -1 < \delta \leq -\alpha, \\ \cos\left(c_0(\alpha)\frac{\delta}{\alpha}\right) / \cos c_0(\alpha) & \text{if } |\delta| \leq \alpha, \\ (1-\delta)/(1-\alpha) & \text{if } \alpha \leq \delta \leq 1 \end{cases}$$

in the case that $\alpha = 1$,

$$1 - \widetilde{F}_x(t) \sim \frac{4}{\pi} \cos\left(\frac{\delta\pi}{2}\right) \exp\left\{-\frac{\pi^2 t}{8}\right\} \quad \text{as } t \to \infty.$$

(Here $\sim$ means that the ratio of the two sides $\to 1$ as $t \to \infty$.)

Remark 7. Observe that $0 < c_2(\alpha, x) \leq 1/\cos c_0(\alpha)$.

Proof of Theorem 1. Convergent part. Let $b_k = \theta^k$ where $\theta > 1$ will be specified later on. For any arbitrary $\varepsilon > 0$, define

$$A_x(k) = \left\{\mu_x(\alpha, b_k) < \frac{2\alpha^2 b_{k-1}^2 \log\log b_{k-1}}{c_0^2(\alpha)}(1+\varepsilon)\right\}. \tag{2.8}$$

x being fixed, let k be big enough such that $x \in (-b_k, b_k)$. Applying Lemma 3

$$\begin{aligned} P(A_x(k)) &\leq 2c_1(\alpha) \exp\left\{-\frac{b_{k-1}^2}{b_k^2} \log\log b_{k-1}(1+\varepsilon)\right\} \\ &\leq 2c_1(\alpha) \exp\left\{-\frac{1+\varepsilon}{\theta^2}(\log(k-1) + \log\log\theta)\right\} \\ &\leq 2c_1^*(\alpha) \exp\left\{-\frac{1+\varepsilon}{\theta^2} \log(k-1)\right\} \\ &\leq 2c_1^*(\alpha) \frac{1}{(k-1)^{(1+\varepsilon)/\theta^2}}, \end{aligned} \tag{2.9}$$

where $c_1^*(\alpha)$ is a constant depending on α, θ, ε.

Now choose θ to be so small that

$$\theta^2 < 1+\varepsilon. \tag{2.10}$$

Then

$$\sum_{k=1}^{\infty} P(A_x(k)) < \infty. \tag{2.11}$$

Consequently, for $k > k_0(\omega)$

$$\mu_x(\alpha, b_k) \le (1+\varepsilon)2\alpha^2 b_{k-1}^2 \log\log b_{k-1}/c_0^2(\alpha).$$

Let $b_{k-1} \le b < b_k$. Then, obviously

$$\begin{aligned}\mu_x(\alpha, b) \le \mu_x(\alpha, b_k) &< (1+\varepsilon)\frac{2\alpha^2 b_{k-1}^2 \log\log b_{k-1}}{c_0^2(\alpha)}\\ &\le (1+\varepsilon)\frac{2\alpha^2 b^2 \log\log b}{c_0^2(\alpha)},\end{aligned}$$

which proves the convergent part of the theorem.

Divergent part. Let $b_k = \theta^k$, where $\theta > 1$ will again be specified later on. For an arbitrary $\varepsilon > 0$, define

$$A_x(k) = \{\mu_x(\alpha, b_k) > u_k\}, \qquad k = 1, 2, \ldots, \tag{2.12}$$

where

$$u_k = (1-\varepsilon)2\alpha^2 b_k^2 \log\log b_k/c_0^2(\alpha).$$

Moreover, let

$$\overline{A}_x(k+1) = \{\mu_x(\alpha, [b_k, b_{k+1})) > u_{k+1}\} \tag{2.13}$$

where

$$\mu_x(\alpha, [b_k, b_{k+1})) = \mu_x(s; m_{b_k} \le s < m_{b_{k+1}}, |W(s)| \le \alpha b_{k+1}). \tag{2.14}$$

Now choose $\theta > 1$ so large that $\alpha\theta > 1$. This implies that $\alpha b_{k+1} > b_k$. Then, as in the first part, x being fixed, for k big enough for $k > k_0$, $|x| < b_k$, and therefore

$$P(\overline{A}_x(k+1)) = P(\mu_x(\alpha, [b_k, b_{k+1})) > u_{k+1}) = P(\mu_{b_k}(\alpha, b_{k+1}) > u_{k+1}). \tag{2.15}$$

Observe that in (2.15) the symmetry of $\mu_{b_i}(\alpha, b_j)$ plays an important role. Namely, $\mu_{b_i}(\alpha, b_j)$ and $\mu_{-b_i}(\alpha, b_j)$ have the same distribution. According to Lemma 3, if $0 < \alpha < 1$, we have (for k big enough)

(2.16)

$$\begin{aligned}P(\mu_{b_k}(\alpha, b_{k+1}) > u_{k+1}) &\ge \tfrac{1}{2}c_1(\alpha)(\cos(c_0(\alpha)))^{-1}\exp\{-(1-\varepsilon)\log\log b_{k+1}\}\\ &= \tfrac{1}{2}c_1(\alpha)(\cos(c_0(\alpha)))^{-1}/((k+1)\log\theta)^{1-\varepsilon}.\end{aligned}$$

In the case that $\alpha = 1$, (2.16) is also valid except that for the constant term before the exponential. Observe that the events $\overline{A}_x(k)$ are independent for $k > k_0$. (2.16) implies, via the Borel-Cantelli lemma, that $P(\overline{A}_x(k) \text{ i.o.}) = 1$. But obviously

$$\overline{A}_x(k) \subset A_x(k) = \{\mu_x(\alpha, b_k) > u_k\},$$

which implies the second half of our theorem. □

Proof of Theorem 2. Let $W(t)$ be a standard Wiener process with $W(0) = x$, and recall that

$$m_b = \inf\{t \geq 0,\ \ W(t) \notin [-b, b]\}. \tag{2.17}$$

First we prove that

$$\liminf_{b\to\infty} \frac{2\mu_x(\alpha, b)\log\log b}{\alpha^2 b^2} \geq 1 \quad \text{a.s.} \tag{2.18}$$

Observe that $\mu_x(\alpha, b) \geq m_{\alpha\cdot b}$ (if $b \geq x$, which we might assume). Hence it is enough to prove (as $0 < \alpha < 1$ is fixed, and $b \to \infty$) that

$$\liminf_{b\to\infty} \frac{2m_b \log\log b}{b^2} \geq 1. \tag{2.19}$$

We will show that (2.19) follows from the law of iterated logarithm. Denote $M(t) = \sup_{0\leq s\leq t} |W(s)|$. Then, for any $\varepsilon > 0$ and t large enough

$$M(t) \leq ((2+\varepsilon)t \log\log t)^{1/2}. \tag{2.20}$$

This clearly implies that again for t large enough

$$M^2(t)/\log\log M(t) \leq (2+2\varepsilon)t. \tag{2.21}$$

Now apply (2.21) for $t = m_b$ to get

$$b^2/\log\log b \leq (2+2\varepsilon)m_b. \tag{2.22}$$

$\varepsilon > 0$ being arbitrary, this clearly implies (2.18).

Turning to the converse part of the theorem, first observe that for any $\varepsilon > 0$,

$$m_b \leq b^2/((2-\varepsilon)\log\log b) \tag{2.23}$$

if b is large enough. To get (2.23), start with the converse part of (2.20),

$$M(t) \geq ((2-\varepsilon)t\log\log t)^{1/2}, \tag{2.24}$$

and apply it for m_b. Now choose a $0 < \delta < 1 - \alpha$. Clearly, one can find an increasing (random) sequence $\{c_n\}$ with the property $c_{n+1} \geq c_n/\alpha$ such that

$$m_{c_n} \leq c_n^2/(2-\varepsilon)\log\log c_n \tag{2.25}$$

holds and m_{c_n} is a stopping time for every n. Choose $b_n = c_n/(\alpha+\delta)$. Then we have

$$m_{(\alpha+\delta)b_n} \leq \frac{(\alpha+\delta)^2 b_n^2}{(2-2\varepsilon)\log\log b_n} \tag{2.26}$$

if n is large enough.

Define for $n = 1, 2, \ldots$ the following two sequences of events:

$$E_n = \left\{\mu_x(\alpha, b_n) \leq \frac{(\alpha+\delta)^2 b_n^2}{(2-2\varepsilon)\log\log b_n}\right\}$$

and

$$F_n = \{|W_x(s)| \geq \alpha b_n \text{ for every } s \in [m_{(\alpha+\delta)b_n}, m_{b_n}]\}.$$

Observe first that $F_n \subset E_n$, as if F_n occurs then

$$\mu_x(\alpha, b_n) \le \mu_x(\alpha, (\alpha+\delta)b_n) \le m_{(\alpha+\delta)b_n},$$

implying E_n by (2.26).

Moreover, $P(F_n) = \delta/(1-\alpha)$ independently of n. Finally, the condition $c_{n+1} \ge c_n/\alpha$ and the strong Markov property also imply the independence of the events $\{F_n\}$. Hence infinitely many of the events $\{F_n\}$ occur with probability 1 and the same is true for $\{E_n\}$. □

3. A similar open problem

In this section we propose a problem of similar nature. Let us formulate the question at first for the simple symmetric random walk case, which seems to be more natural.

Let $X_1, X_2, \dots, X_n, \dots$ be a sequence of i.i.d. r.v. with $P(X_i = +1) = P(X_i = -1) = \frac{1}{2}$. Define the simple symmetric random walk as $S_n = X_1 + X_2 + \cdots + X_n$, $n = 1, 2, \dots$. Moreover let

$$\begin{aligned} \rho_1 &= \inf\{n;\ n > 0,\ S_n = 0\} \\ &\ \vdots \\ \rho_k &= \inf\{n;\ n > \rho_{k-1},\ S_n = 0\} \\ &\ \vdots \end{aligned}$$

be the endpoints of the consecutive excursions of S_n.

What can we say about

$$\nu(\alpha, k) = \#\{n;\ 0 < n \le \rho_k,\ |S_n| \le \alpha k\} \tag{3.1}$$

where α might be a fixed constant or could go to infinity?

Thus we are asking the behaviour of the amount of time spent by the simple symmetric random walk in an appropriate tube until completing k excursions.

The corresponding problem for the Wiener case goes as follows:

Let us consider a standard Wiener process $W(t)$, $W(0) = 0$, and denote its local time at x by $L(x, t)$. Define

$$T_r = \inf\{s,\ s > 0,\ L(0, s) \ge r\}, \tag{3.2}$$

the so-called inverse local time process. Then we might be interested in the following problem: How much time is spent by the Wiener process in a symmetric tube of size αr in the (random) time interval $0 < t < T_r$. (Here $0 < \alpha < 1$, α fixed, $\alpha \to 1$, $\alpha \to \infty$ cases are all meaningful questions.)

Define

$$\gamma(r, \alpha) = \gamma\{y,\ 0 < y \le T_r,\ |W(y)| < \alpha r\} \tag{3.3}$$

where γ is the ordinary Lebesgue measure. Then

$$D(x) = \int_0^\infty e^{-\lambda r} E_x(e^{-s\gamma(r, \alpha)})\, dr \tag{3.4}$$

can be evaluated based on Kac's method just as in our original problem. After some tedious computations one arrives at

$$E_0\left(\exp\left\{-s\frac{\gamma(r\,,\,\alpha)}{r^2}\right\}\right)=\exp\{-\sqrt{2s}\tanh(\alpha\sqrt{2s})\}\,. \tag{3.5}$$

Now the investigation of the asymptotic behaviour of this Laplace transform would probably give us some insight into the behaviour of $\gamma(r\,,\,\alpha)$ and would lead to similar results to the ones discussed in this paper.

We propose to investigate the Wiener case problem in which (3.5) might be helpful.

Acknowledgment

The authors are indebted to the unknown refree whose suggestions significantly improved the quality of this paper.

References

1. A. N. Borodin, *On the distribution of functionals of Brownian local time*, LOMI preprint E-4-35, U.S.S.R. Academy of Sciences, Steklov Mathematical Institute, Leningrad Department, 1985.
2. K. L. Chung, *On the maximum partial sums of sequences of independent random variables*, Trans. Amer. Math. Soc. **64** (1948), 205–233.
3. G. Doetsch, *Introduction to the theory and application of the Laplace transformation*, Springer-Verlag, New York, Heidelberg, and Berlin, 1970.
4. I. S. Gradshteyn and I. M. Ryzhik, *Table of integrals, series and products*, Academic Press, New York, 1980.
5. R. Z. Hasminskiĭ, *Probability distributions for functionals of the trajectories of diffusion type random processes*, Dokl. Akad. Nauk SSSR **100** (1985), 22–25. (Russian)
6. M. Kac, *On some connections between probability theory and differential and integral equations*, Proc. Second Berkeley Sympos. Math. Statist. and Prob. 1950, Univ. of California Press, Berkeley and Los Angeles, pp. 189–215.
7. F. B. Knight, *Essentials of Brownian motion and diffusion*, Amer. Math. Soc., Providence, RI, 1981.
8. A. Rényi, *Probability theory*, North-Holland, Amsterdam, 1970.

Department of Mathematics, CUNY, College of Staten Island, Staten Island, New York 10301

Department of Mathematics, Indiana University, Bloomington, Indiana 47405

Reprinted by permission of
The American Mathematical Society,
© 1993 by The American Mathematical Society

MAXIMUM LIKELIHOOD ESTIMATION FOR STATIONARY POINT PROCESSES

MADAN L. PURI† and PHAM D. TUAN‡

†*Department of Mathematics, Indiana University, Bloomington, IN 47405*

‡*Mathématiques Appliquées, Université de Grenoble, 38041 Grenoble Cedex, France*

Communicated by Herbert Robbins, September 27, 1985

ABSTRACT **In this paper we derive the log likelihood function for point processes in terms of their stochastic intensities by using the martingale approach. For practical purposes we work with an approximate log likelihood function that is shown to possess the usual asymptotic properties of a log likelihood function. The resulting estimates are strongly consistent and asymptotically normal (under some regularity conditions). As a by-product, a strong law of large numbers and a central limit theorem for martingales in continuous times are derived.**

The Log Likelihood Function for Point Processes

We shall be concerned with multitype point processes, that is, point events of r different types randomly occurring along the real line. This process can be described by a multivariate counting process $N(t) = \{N_1(t), \ldots, N_r(t)\}$, $t \in R$, defined on some probability space $(\Omega, \mathcal{A}, P)$. Here $N_j(t) - N_j(s)$, $t > s$ denotes the number of point events of type j that occur in $(s, t]$. By convention $N_j(0) = 0$. We shall suppose that at each t, at most one event, regardless of its type, can occur. Let $\mathcal{A}_t$, $t \geq 0$ be an increasing family of sub-σ fields such that $N(t)$ is $\mathcal{A}_t$-measurable, $t > 0$. Then there is for each $j = 1, \ldots, r$ a natural increasing process $\Lambda_j(t)$, called the compensator of $N_j(t)$, relative to $(\mathcal{A}_t, P)$ such that $m_j(t) = N_j(t) - \Lambda_j(t)$, $t \geq 0$ is an $\mathcal{A}_t$-local martingale. Here we shall be interested only in the case in which the measure $d\Lambda_j(t)dP$ admits a density $\lambda_j(t)$ with respect to $dtdP$. The processes $\lambda_j(t)$ can be chosen to be $\mathcal{A}_t$-predictable; that is, $(t, \omega) \to \lambda_j(t, \omega)$ is measurable with respect to the σ-field generated by all the λ_t-adapted processes with left continuous sample paths and is called the stochastic intensity of the $N(t)$ process. Intuitively, $\lambda_j(t)$ can be interpreted as

$$\lim_{\Delta t \downarrow 0} P\{\Delta N_j(t) = 1 | \mathcal{A}_t\}/\Delta t, \qquad [1]$$

where $\Delta N(t) = N(t + \Delta t) - N(t)$. Indeed, if τ_n^j is the time of occurrence of the nth event of type j after the origin, then $m_j(t \wedge \tau_n^j)$, $t \geq 0$ is a martingale and so for $s > t$, $A \in \mathcal{A}_t$,

$$E[1_A\{N_j(s \wedge \tau_n^j) - N_j(t \wedge \tau_n^j)\}] = E\left\{1_A \int_{t\wedge\tau_n^j}^{s\wedge\tau_n^j} d\Lambda_j(u)\right\}$$
$$= \int_t^s E[1_A 1_{\{u \leq \tau_n^j\}} \lambda_j(u)]du. \qquad [2]$$

Heuristically, if $s = t + \Delta t$ and Δt is small, then the left-hand side of Eq. **2** is approximately $P(A \cap \{t \leq \tau_n^j\} \cap \{\Delta N(t) = 1\})$ and the right-hand side of Eq. **2** is approximately $E[1_A 1_{\{t \leq \tau_n^j\}} \lambda_j(t)]\Delta t$. Since $\{\tau_n^j < t\} \in \mathcal{A}_t$, we obtain the interpretation **1**. Historically Eq. **1** was proposed as the definition of the stochastic intensity. This definition requires the existence of the limit in Eq. **1** and is equivalent to our definition only under some regularity conditions (1).

If $\mathcal{A}_t = \bar{\mathcal{A}}_t$ represents the sub σ-field generated by $N(s)$, $0 \leq s \leq t$, then the stochastic intensity, which we now write as $\bar{\lambda}_j(t)$, completely defines the probability distribution of the process (2, 3). Let π be the probability measure under which the $N_j(t)$ are independent Poisson processes with unit rate. Observe that the random measure $\{j\} \times (s, t) \to \int_s^t \bar{\lambda}_j(u)du$ on $\{1, \ldots, r\} \times R^+$ is precisely the dual predictable projection of the random measure $\{j\} \times (s, t) \to N_j(t) - N_j(s)$ in the sense of Jacod (2). By a result of this paper, P is absolutely continuous with respect to π on any $\bar{\mathcal{A}}_T$, $T > 0$, with density

$$\frac{dP}{d\pi} = \exp\left(\int_0^T \sum_{j=1}^r \{\log \bar{\lambda}_j(t)dN(t) - \bar{\lambda}_j(t)dt\}\right).$$

Now let $\{P_\theta, \theta \in \Theta\}$ be a family of probability distributions on $(\Omega, \mathcal{A})$ and let $\bar{\lambda}_{\theta,j}(t)$ be the corresponding stochastic intensity of the $N_j(t)$ process. Then our result shows that the log likelihood function corresponding to an interval of observation $[0, T]$ is, up to an additive constant,

$$\bar{L}_T(\theta) = \int_0^T \sum_{j=1}^r \{\log \bar{\lambda}_{\theta,j}(t)dN_j(t) - \bar{\lambda}_{\theta,j}(t)dt\}. \qquad [3]$$

Remark: The multitype process is a special case of the marked point process when the space of the marks is just $\{1, \ldots, r\}$. It is not necessary to suppose the existence of the stochastic intensity to write down the log likelihood function for the marked point process; it suffices that the measure $d\bar{\Lambda}_{\theta,j}(t)dP_\theta$ be absolutely continuous with respect to some measure $d\nu(t)dP_\theta$, where $\bar{\Lambda}_{\theta,j}(t)$ is the natural increasing process of $N_j(t)$ relative to $\bar{\mathcal{A}}_t$ and to P_θ. The special case we considered is convenient for further development concerning the asymptotic properties of the log likelihood function.

The Approximate Log Likelihood Function

The process we consider is defined on the whole line although only an observation on $[0, T]$ is available. Denote by $\lambda_{\theta,j}(t)$ the stochastic intensity of the $N_j(t)$ process relative to P_θ and the sub σ-fields $\mathcal{A}_t$, $t \in R$ generated by $N(s)$, $-\infty < s < t$. Models of point processes are usually described in terms of $\lambda_\theta(t) = (\lambda_{\theta,1}(t), \ldots, \lambda_{\theta,r}(t))$. For example, the self-exciting process introduced by Hawkes (4) and also considered by Hawkes and Oakes (5) can be defined by

$$\lambda_\theta(t) = \alpha_\theta + \int_{-\infty}^t g_\theta(t - s)dN(s), \qquad [4]$$

where α_θ is a constant vector and $g_\theta(\cdot)$ is some appropriate matrix function. Thus, it is desirable to obtain the log likelihood function in terms of $\lambda_\theta(t)$. We are led to the problem of computing $\bar{\lambda}_\theta(t)$ in terms of $\lambda_{\theta,j}(t)$. Now, from the interpretation **1** of $\lambda_{\theta,j}(t)$ one can expect that $\bar{\lambda}_{\theta,j}(t) = E_\theta$

The publication costs of this article were defrayed in part by page charge payment. This article must therefore be hereby marked "*advertisement*" in accordance with 18 U.S.C. §1734 solely to indicate this fact.

$\{\lambda_{\theta,j}(t)|\tilde{\mathscr{A}}_t\}$ in case $\lambda_{\theta,j}(t)$ is integrable. The following theorem and its corollary are of the type often called *innovation theorems* (see, e.g., p. 372 of ref. 6).

THEOREM 1. *Let τ_n^j be the time of occurrence of the nth event of type j after the origin. Then, for almost all t, the random variable $1_{\{t\le\tau_n^j\}}\tilde{\lambda}_{\theta,j}(t)$ is P_θ-integrable and $1_{\{t\le\tau_n^j\}}\tilde{\lambda}_{\theta,j}(t) = E_\theta\, 1_{\{t\le\tau_n^j\}}\lambda_{\theta,j}(t)|\tilde{\mathscr{A}}_t]$ almost surely.*

Proof: From Eq. 2, we have for all $s > 0$,

$$n \ge \int_0^s E_\theta[1_{\{t\le\tau_n^j\}}\lambda_{\theta,j}(t)]dt$$

and hence the function

$$(t, \omega) \to X_n(t, \omega) = 1_{\{t\le\tau_n^j\}}(\omega)\lambda_{\theta,j}(\omega)$$

belongs to $L^1(R^+ \times \Omega, \mathscr{B}_R + \otimes \mathscr{A}, dtdP_\theta)$. We show that there exists an $\mathscr{A}_t$-predictable process $\tilde{X}_n(t)$ with $\tilde{X}_n(t) = E\{X_n(t)|\tilde{\mathscr{A}}_t\}$ for almost all t, almost surely. Indeed, there exists a sequence of simple functions of the form

$$X_{n,k}(t, \omega) = \sum_{m=1}^{m_k} Z_m^{(k)}(\omega)\, 1_{(t_{m-1}^{(k)}, t_m^{(k)}]}(t), \qquad k = 1, 2, \ldots,$$

which converges in L^1 to X_n. For $t_{m-1}^{(k)} < t \le t_m^{(k)}$, set $\tilde{X}_{n,k}(t)$ to be a version of $E_\theta\{Z_m^{(k)}|\tilde{\mathscr{A}}_t\}$, such that the process $\tilde{X}_{n,k}(t)$ has left continuous sample paths and is $\tilde{\mathscr{A}}_t$-adapted. This is possible because of the martingale property of the $E_\theta\{Z_m^{(k)}|\tilde{\mathscr{A}}_t\}$, $t_{m-1}^{(k)} < t \le t_m^{(k)}$, and hence $\tilde{X}_{n,k}(t)$ is $\tilde{\mathscr{A}}_t$-predictable.

From $|\tilde{X}_{n,k}(t) - E_\theta\{X_n(t)|\tilde{\mathscr{A}}_t\}| \le E_\theta\{|X_{n,k}(t) - X_n(t)||\tilde{\mathscr{A}}_t\}$ and the fact that $X_{n,k} \to X_n$ in L^1 as $k \to \infty$, we note that the sequence $\tilde{X}_{n,k}$, $k \ge 1$ is a Cauchy sequence in L^1 and converges to some $\tilde{\mathscr{A}}_t$-predictable process $\tilde{X}_n$ that equals $E_\theta\{X_n(t)|\tilde{\mathscr{A}}_t\}$ almost surely for almost all t.

From the definition of $X_n(t)$ and the fact that $\{t \le \tau_n^j\} \in \tilde{\mathscr{A}}_t$, we have, almost surely, $1_{\{t\le\tau_n^j\}}\, E_\theta\{X_m(t)|\tilde{\mathscr{A}}_t\} = E_\theta\{X_n(t)|\tilde{\mathscr{A}}_t\}$ for $m \ge n$ and hence there is an $\tilde{\mathscr{A}}_t$-predictable process $\tilde{X}(t)$ such that

$$1_{\{\tau_n^j>t\}}\, \tilde{X}(t) = \tilde{X}_n(t) = E_\theta[1_{\{\tau_n^j>t\}}\, \lambda_{\theta,j}(t)|\tilde{\mathscr{A}}_t]$$

for almost all t, almost surely.

We now show that $\tilde{X}(t)$ is $\tilde{\lambda}_{\theta,j}(t)$. For this, let $A \in \tilde{\mathscr{A}}_t$. Then the right-hand side of Eq. 2 equals

$$E_\theta\left[\int_t^s 1_A X_n(u)du\right] = E_\theta\left\{1_A \int_t^{s\wedge\tau_n^j} \tilde{X}(u)du\right\}.$$

By Eq. 2 the process $N(t) - \int_0^t\tilde{X}(u)du$, $t \ge 0$ is a local martingale, which gives the desired result.

COROLLARY. *Suppose that $\lambda_{\theta,j}(t)$ is integrable for almost all t. Then $\tilde{\lambda}_{\theta,j}(t) = E_\theta\{\lambda_{\theta,j}(t)|\tilde{\mathscr{A}}_t\}$ almost surely, for almost all t.*

Although the above result provides a means of computing $\tilde{\lambda}_\theta(t)$ in terms of $\lambda_\theta(t)$, the actual computation is not easy. Instead, we approximate $\lambda_\theta(t)$ by some $\hat{\lambda}_\theta(t)$ that depends only on $N(s)$, $0 \le s \le t$. The approximate log likelihood function is then

$$L_T(\theta) = \int_0^T \left\{\sum_{j=1}^r \log \hat{\lambda}_{\theta,j}(t)dN_j(t) - \hat{\lambda}_{\theta,j}(t)dt\right\}. \qquad [5]$$

Since $T^{-1}\tilde{L}_T(\theta)$ depends essentially on the values of $\tilde{\lambda}_\theta(t)$ for large t, if T is large, we would expect L_T to be a good approximation to $\tilde{L}_T$ for large T if $\hat{\lambda}_\theta(t)$ is a good approximation to $\tilde{\lambda}_\theta(t)$ for large t. By the *Corollary* of *Theorem 1*, and assuming that $N(t)$ is a process with stationary increments, we have that $E\|\tilde{\lambda}_\theta(t) - \lambda_\theta(t)\| \to 0$ as $t \to \infty$, which suggests that L_T is a good approximation to $\tilde{L}_T$ for large T if $\hat{\lambda}_\theta(t)$ is a good approximation to $\lambda_\theta(t)$ for large t.

To illustrate our point, consider the self-exciting process when $\lambda_\theta(t)$ is given by Eq. 3. In order that $\lambda_{\theta,j}(t) \ge 0$ for all t, we shall assume that $\alpha_{\theta,jl} \ge 0$, $g_{\theta,jl}(t) \ge 0$ for all t. If the $g_{\theta,jl}$ are integrable and $N(t)$ is of stationary increments with $EN_j(1) = \mu_{\theta,j} < \infty$, then the integral in Eq. 3 exists almost surely. Indeed

$$E_\theta\left\{\int_{-\infty}^t g_{\theta,jl}(t - s)dN_l(s)\right\} \le \mu_{\theta,l}\int_{-\infty}^{\infty} g_{\theta,jl}(t)dt < +\infty,$$

and hence the integral in the above left-hand side is finite almost surely. As an approximation of $\lambda_\theta(t)$, one might consider

$$\hat{\lambda}_\theta(t) = \alpha_\theta + \int_0^t g_\theta(t - s)dN(s),$$

which is evidently $\tilde{\mathscr{A}}_t$-measurable and would be a good approximation to $\lambda_\theta(t)$ for large t if $g_\theta(t) \to 0$ sufficiently fast.

Asymptotic Properties of the Approximate Log Likelihood

From now on, we shall suppose that the $N(t)$ process is of stationary increments and metrically transitive in the sense of Doob (p. 510 of ref. 7). Let $\bar{\Omega}$ be the space of $\bar{\omega} = (\bar{\omega}_1, \ldots, \bar{\omega}_r)$, where the $\bar{\omega}_i$ are nondecreasing integer-valued functions on $(-\infty, \infty)$, $\bar{\mathscr{A}}$ is the σ-field generated by the projections π_{st}: $\bar{\omega} \to \bar{\omega}(t) - \bar{\omega}(s)$, $s < t$, $\bar{P}_\theta$ is the restriction to $\bar{\mathscr{A}}$ of the image of P_θ by the application $N{:}\omega \to N(\cdot)$, and T_h is the shift operator $(T_h\bar{\omega})(t) = \bar{\omega}(t + h)$. The stationarity means that T_h conserves the probability $\bar{P}_\theta$; that is, $\bar{P}_\theta(T_h^{-1}A) = \bar{P}(A)$, $\forall\, A \in \bar{\mathscr{A}}$, and metric transitivity means that the invariant sets, which are those sets $A \in \bar{\mathscr{A}}$ with $T_h^{-1}A = A$, a.s., have probabilities zero or one. Now, from the fact that $N(t) - N(s) = \pi_{st} \circ N$ and that T_h conserves the probability $\bar{P}_\theta$, one can show that $\lambda_\theta(t) = \bar{\lambda}_\theta(t) \circ N$ with $\bar{\lambda}_\theta(t + h) = \bar{\lambda}_\theta(t) \circ T_h$. Hence if $\lambda_\theta(t)$ is integrable, then

$$\lim_{T\to\infty} \frac{1}{T}\int_0^T \lambda_\theta(t)dt = E\{\lambda_\theta(0)\}, \text{ almost surely.}$$

We assume that $\theta \subset R^k$ and $\lambda_\theta(t)$, $\hat{\lambda}_\theta(t)$ are twice continuously differentiable with respect to θ, almost surely. We shall use the notation $x_\theta^{(1)}$, $x_\theta^{(2)}$ to denote the vector and the matrix of first and second derivatives of x_θ with respect to θ. Let $\psi(t)$ be a function of the $\lambda_{\theta,j}(t)$, $\lambda_{\theta,j}^{(i)}(t)$, $i = 1, 2$ such that $\psi(t)$ is integrable. Then, by the same argument as above

$$\frac{1}{T}\int_0^T \psi(t)dt \to E\psi(0), \qquad \text{a.s. } (T \to \infty). \qquad [6]$$

In this section, we are interested in the limiting behavior of $L_T(\theta)$, $L_T^{(i)}(\theta^*)$, $i = 1, 2$ as $T \to \infty$, θ^* being the true value of θ. We have that $dm_j = dN_j(t) - \lambda_j(t)dt$ when $\theta = \theta^*$, and putting $\hat{\phi}_{\theta,j}(t) = \log \hat{\lambda}_{\theta,j}(t)$, it follows that

$$L_T(\theta) = \sum_{j=1}^r \left[\int_0^T \hat{\phi}_{\theta,j}(t)dm_j(t) + \int_0^T \{\hat{\phi}_{\theta,j}(t)\lambda_j(t) - \hat{\lambda}_j(t)\}dt\right] \qquad [7]$$

$$L_T^{(1)}(\theta) = \sum_{j=1}^{r} \left[\int_0^T \hat{\phi}_{\theta,j}^{(1)}(t)\{\lambda_j(t) - \hat{\lambda}_{\theta,j}(t)\}dt + \int_0^T \hat{\phi}_{\theta,j}^{(1)}(t)dm_j(t) \right] \quad [8]$$

and

$$L_T^{(2)}(\theta) = \sum_{j=1}^{r} \left[\int_0^T \hat{\phi}_{\theta,j}^{(2)}(t)\{\lambda_j(t) - \hat{\lambda}_{\theta,j}(t)\}dt + \int_0^T \hat{\phi}_{\theta,j}^{(2)}(t)dm_j(t) - \int_0^T \hat{\phi}_{\theta,j}^{(1)}(t)\hat{\phi}_{\theta,j}^{(1)}(t)'\hat{\phi}_{\theta,j}(t)dt \right]. \quad [9]$$

To use the result **6** we replace $\hat{\lambda}_{\theta,j}(t)$, $\hat{\phi}_{\theta,j}(t)$, . . . by $\lambda_{\theta,j}(t)$, $\phi_{\theta,j}(t) = \log \lambda_{\theta,j}(t)$, ... The following result can easily be proven and is quite useful.

LEMMA 1. *Let* ε_t, x_t *be such that* $\varepsilon_t/x_t \to 0$ *as* $t \to \infty$. *Also let* $\int_0^T|\varepsilon_t|d\mu(t) < \infty$ *and* $\mu(T)^{-1}\int_0^T|x_t|d\mu(t)$ *be bounded where* μ *is a nondecreasing function with* $\mu(\infty) = \infty$. *Then* $\mu(T)^{-1}\int_0^T \varepsilon_t d\mu(t) \to 0$ *as* $T \to \infty$.

To obtain convergence results for the stochastic integrals with respect to $dm_j(t)$ in Eqs. **7** and **8**, we will need the following result, which is also of independent interest.

LEMMA 2. *Let* M_t, $t \geq 0$ *be a locally square integrable martingale with continuous natural increasing process* $\langle M \rangle_t$. *Let* g_t, $t > 0$ *be a nondecreasing left continuous function with* $g_\infty = \infty$, *such that* $\langle M \rangle_t = 0(g_t^{2-\varepsilon})$, $\varepsilon > 0$, *almost surely as* $t \to \infty$. *Then* $M_T/g_T \to 0$ *almost surely as* $T \to \infty$.

Proof: Let $a > 0$ be such that $g_a > 0$. Let c be an arbitrary positive constant. Define the stopping time $\tau = \text{Inf}\,\{t: \int_a^t g_s^{-2}d\langle M \rangle_s \geq c\}$. Set $\tilde{M}_t = M_{t\wedge\tau}$, and $Z_t = \int_a^t g_s^{-1}d\tilde{M}_s = \int_a^{t\wedge\tau} g_s^{-1}dM_s$. The process Z_t, $t \geq a$ is a martingale with a natural increasing process $\langle Z \rangle_t = \int_a^{t\wedge\tau} g_s^{-2}d\langle M \rangle_s \leq c$, $t \geq a$. Since $EZ_t^2 \leq c$, $t \geq a$, we know (pp. 354 and 316 of ref. 7) that almost surely Z_{a+}, Z_{t-}, Z_{t+}, $t \geq a$, Z_∞ exist and Z_t is bounded on any finite interval. Set $t_i^{(k)} = a + (T - a)i/k$ and write

$$g_T Z_T = \sum_{i=1}^{k} [\{g_{t_i^{(k)}} - g_{t_{i-1}^{(k)}}\}Z_{t_i^{(k)}} + g_{t_{i-1}^{(k)}}\{Z_{t_i^{(k)}} - Z_{t_{i-1}^{(k)}}\}]$$

$$= \int_0^T Z_t^{(k)}dg_t + \int_a^T g_t^{(k)}dZ_t,$$

where $Z_t^{(k)} = Z_{t_i^{(k)}}$, $g_t^{(k)} = g_{t_{i-1}^{(k)}}$ for $t_{i-1}^{(k)} \leq t < t_i^{(k)}$.

By the Lebesgue-dominated convergence theorem and the property of stochastic integral, almost surely as $k \to \infty$

$$\int_a^{T-} Z_t^{(k)}dg_t \to \int_a^{T-} Z_{t+}dg_t$$

and

$$\int_a^T g_t^{(k)}dZ_t \to \int_a^T g_t dZ_t = \int_a^T d\tilde{M}_t = \tilde{M}_T - \tilde{M}_a.$$

Therefore,

$$g_T^{-1}(\tilde{M}_T - \tilde{M}_a) = Z_T - (1 - g_a g_T^{-1})Z_\infty - g_T^{-1}\int_a^{T-}(Z_{t+} - Z_\infty)dg_t.$$

Since $g_T \uparrow \infty$ as $T \to \infty$, from *Lemma 1*, we obtain $g_T^{-1}M_T^a \to 0$ almost surely as $T \to \infty$.

Now, we have $\tilde{M}_t = M_t$ for all t on the set $\{\tau = \infty\} = \{\int_a^\infty g_{s-}^2 d\langle M \rangle_s \leq c\}$. Since c is arbitrary, we obtain $g_T^{-1}M_T \to 0$ almost surely as $T \to \infty$ on the set $\{\int_a^\infty g_s^{-2}d\langle M \rangle_s < \infty\}$. By assumption $\langle M \rangle_s \leq \text{const}\, g_s^{2-\varepsilon}$, for all $s \geq b$, for some $b \geq a$, almost surely. So,

$$\int_b^\infty g_s^{-2}d\langle M \rangle_s < \infty; \text{ also } \int_a^\infty g_s^{-2}d\langle M \rangle_s < \infty.$$

The proof follows.

THEOREM 2. *Suppose that*

(i) $P\{\lambda_{\theta,j}(t) \geq c\} = 0$ *for some* $c > 0$ (*c is independent of* t *by stationarity*) *and* $\hat{\lambda}_{\theta,j}(t) - \lambda_{\theta,j}(t)$, $\hat{\lambda}_{\theta,j}^{(i)}(t) - \lambda_{\theta,j}^{(i)}(t)$, $i = 1, 2$ *tend to zero almost surely as* $t \to \infty$.

(ii) $\lambda_{\theta,j}(t)$ *is integrable, and* $\phi_{\theta,j}(t)$ *and the elements of* $\phi_{\theta,j}^{(1)}(t)$, $\phi_{\theta,j}^{(2)}(t)$ *are square integrable with respect to the measure* $\lambda_j(t)dP$.

Then, almost surely as $T \to \infty$,

$$T^{-1}L_T(\theta) \to E\left[\sum_{j=1}^{r} \phi_{\theta,j}(0)\lambda_j(0) - \lambda_{\theta,j}(0)\right] = \Lambda(\theta)$$

$$T^{-1}L_T^{(1)}(\theta^*) \to 0$$

$$T^{-1}L_T^{(2)}(\theta^*) \to -E\left\{\sum_{j=1}^{r} \phi_j^{(1)}(0)\phi_j^{(1)}(0)'\phi_j(0)\right\} = -J.$$

Proof: Consider the right-hand side of Eq. **7**. By (*i*),

$$\hat{\phi}_{\theta,j}(t)\lambda_j(t) - \hat{\lambda}_{\theta,j}(t) = \phi_{\theta,j}(t)\lambda_{\theta,j}(t) - \lambda_{\theta,j}(t) + o(t)$$

where in this proof $o(t)$ denotes a quantity tending to zero almost surely as $t \to \infty$. By *Lemma 1* and the fact that $\phi_{\theta,j}(t)\lambda_j(t)$ is integrable, we obtain from Eq. **6** that T^{-1} times the last term of the extreme right-hand side of Eq. **7** tends to $\Lambda(\theta)$ almost surely as $T \to \infty$. On the other hand, $m_j(t)$ is a martingale with natural increasing process given by $d\langle m_j \rangle_t = \lambda_j(t)dt$. Therefore, $M_{j,t} = \int_0^t \hat{\phi}_{\theta,j}(t)dm_j(t)$ is also a martingale with natural increasing process $\langle M_j \rangle_t = \int_0^t \hat{\phi}_{\theta,j}^2(t)\lambda_j(t)dt$. If $\langle M_j \rangle_t = 0(t)$ almost surely as $t \to \infty$, then, by *Lemma 2*, $M_{j,T}/T \to 0$ and hence $T^{-1}L_T(\theta) \to \Lambda(\theta)$ almost surely as $T \to \infty$. Since from (*i*)

$$\hat{\phi}_{\theta,j}^2(t)\lambda_j(t) = \phi_{\theta,j}^2(t)\lambda_j(t) + \phi_{\theta,j}(t)\lambda_j(t)o(t) + \lambda_j(t)o(t)$$

and so $M_{j,T}/T \to E\{\phi_{\theta,j}^2(0)\lambda_j(0)\}$ almost surely as $T \to \infty$, we obtain the result.

The proofs for the convergence of $T^{-1}L_T^{(1)}(\theta^*)$ and $T^{-1}L_T^{(2)}(\theta^*)$ follow analogously.

Consider now the asymptotic distribution of $T^{-1/2}L_T(\theta^*)$. From Eq. **8**, we expect that this is asymptotically distributed like

$$Z_T = T^{-1/2}\int_0^T \sum_{j=1}^{r} \phi_j^{(1)}(t)dm_j(t). \quad [10]$$

The asymptotic distribution of Z_T can be obtained from the following result, which is of independent interest.

LEMMA 3. *Let* M_t, $\mathscr{A}_t$, $t \geq 0$ *be a square integrable martingale with natural increasing process* $\langle M_t \rangle$ *satisfying* $d\langle M \rangle_t = X_t dt$. *Suppose that there is a semigroup of shift operator* T_s, $s \geq 0$, *conserving the probability and metrically transitive, such that* $T_h^{-1}\mathscr{A}_t = \mathscr{A}_{t+h}$, $(M_t - M_s) \circ T_h = M_{t+h} - M_{s+h}$. *Then as* $T \to \infty$, $T^{-1/2}M_T$ *is asymptotically normal with mean zero and variance* EX_0.

Proof: Let n_T be integers such that $\Delta_T = T/n_T \to 0$ as $T \to \infty$. Set $Y_{T,j} = T^{-1/2}(M_{j\Delta T} - M_{(j-1)\Delta T})$. Then, as $T \to \infty$, $T^{-1/2}M_T = T^{-1/2}M_0 + \sum_{j=1}^{n_T} Y_{T,j}$ is asymptotically normal with mean zero and variance EX_0, provided

$$(i)\ \sum_{j=1}^{n_T} E\{Y_{T,j}^2|\mathcal{A}_{(j-1)\Delta T}\} \to EX_0 \text{ in probability}$$

$$(ii)\ \sum_{j=1}^{n_T} E[Y_{T,j}^2 1_{\{Y_{T,j}^2>\varepsilon\}}] \to 0,\ \varepsilon > 0.$$

To verify *i*, observe that the sum in *i* equals

$$\frac{1}{T}\int_0^T X_t^{(T)}dt = \langle M\rangle_T/T + \frac{1}{T}\int_0^T (X_t^{(T)} - X_t)dt,$$

where $X_t^{(T)} = E(X_t|\mathcal{A}_{j\Delta T})$ for $j\Delta_T \le t < (j+1)\Delta_T$. Clearly $X_{t+h} = X_t \circ T_h$ and $E(X_{t+h}|\mathcal{A}_{s+h}) = E(X_t|\mathcal{A}_s) \circ T_h$. Therefore, as $\Delta_T \to 0$,

$$\operatorname*{Sup}_{0\le t\le T} E|X_t^{(T)} - X_t| \le \operatorname*{Sup}_{0\le u\le \Delta T} E|E(X_1|\mathcal{A}_{1-u}) - X_1| \to 0.$$

Note that X_t is $\mathcal{A}_t$-measurable because this process is predictable. Thus, the sum in *i* differs from $\langle M\rangle_T/T$ by a term tending to 0 in the mean and since $\langle M\rangle_T/T \to EX_0$ by the metric transitivity of T_h, we obtain the result.

To verify *ii*, observe that the $Y_{T,j}, j = 1, \ldots, n_T$ have the same distribution. Putting $\xi(t) = M_t^2 - M_0$, the result will follow by verifying that

$$n_T[E(Y_{T,1}^2\ 1_{\{Y_{T,1}^2>\varepsilon\}}] = \Delta_T^{-1}E[\xi(\Delta_T)1_{\{\xi(\Delta_T)>\varepsilon T\}}]$$

tends to zero as $T \to \infty$.

We now show that the difference between $T^{1/2}L_T(\theta^*)$ and Z_T, defined by Eqs. **7** and **10**, respectively, tends to zero in probability as $T \to \infty$.

LEMMA 4. *Let* $M_t, t \ge 0$ *be a locally integrable martingale with continuous natural increasing process* $\langle M\rangle_t$. *Let* $h_t \ge 0$. *If* $\langle M\rangle_T/h_T^2 \to 0$ *in probability as* $T \to \infty$, *then so does* M_T/h_T.

Proof: Let $\varepsilon > 0$. Define the stopping time σ_t to be the value of s such that $\langle M\rangle_t = \varepsilon h_t^2$. Then as $T \to \infty$, $P\{M_{T\wedge\sigma_T} \ne M_T\} \to 0$. On the other hand, $E(M_{T\wedge\sigma_T}^2) = E\langle M\rangle_{T\wedge\sigma_T} \le E\langle M\rangle_{\sigma_T} = \varepsilon h_T^2$ and, hence, by the Chebyshev inequality,

$$P\{|M_T/h_T| > \delta\} \le P\{M_{T\wedge\sigma_T} \ne M_T\} + \varepsilon/\delta^2,$$

from which the result follows.

THEOREM 3. *Suppose the conditions of* Theorem 2 *are satisfied and also*

$$(i)\ T^{-1/2}\int_0^T \hat{\phi}_j^{(1)}(t)\{\hat{\lambda}_j(t) - \lambda_j(t)\}dt \to 0$$

in probability as $T \to \infty$. *Then* $T^{-1/2}L_T(\theta^*)$ *is asymptotically normal with mean zero and covariance matrix* J *given in* Theorem 2.

Proof: Let α be a vector and Z_T be given by Eq. **10**. Then $M_T = T^{1/2}\alpha' Z_T, T \ge 0$ is a martingale with natural increasing process

$$\langle M\rangle_t = \int_0^t \sum_{j=1}^r \{\alpha'\phi_j^{(1)}(s)\}^2\lambda_j(s)ds.$$

By *Lemma 3*, $\alpha' Z_T$ is asymptotically normal with mean zero and variance $\alpha' J\alpha$. On the other hand, by *i* and *Lemma 4* with $dM_t = \Sigma_j\alpha'\{\hat{\phi}_j^{(1)}(t) - \phi_j^{(1)}(t)\}dm_j(t)$, $h_t = \sqrt{t}$, we see that $T^{-1/2}L_T^{(1)}(\theta^*) - Z_T \to 0$ in probability as $T \to \infty$. The result follows.

Asymptotic Properties of the Maximum Likelihood Estimator

We now study the asymptotic properties of the estimate θ_T, which maximizes L_T in Θ. For convenient reference, we state the general result.

THEOREM 4. 1. *Let* Λ_T *be a random function on* $\theta \subset R^k$, *satisfying*

(i) $\Lambda_T(\theta) \to \Lambda(\theta)$ *almost surely as* $T \to \infty$, *with* Λ *being continuous, admitting a unique maximum* θ^*,

(ii) *For any* $\theta \ne \theta^*$,

$$\lim_{T\to\infty}\operatorname{Sup}\ \operatorname*{Sup}_{\theta'\in U(\theta)} \{\Lambda_T(\theta') - \Lambda_T(\theta)\} \to 0\ a.s.$$

as the neighborhood $U(\theta)$ *of* θ *shrinks to* θ.

Then any θ_T *realizing the maximum of* Λ_T *on a compact* C *of* Θ *containing* θ^*, *converges almost surely to* θ^* *as* $T \to \infty$.

2. *Suppose that* θ^* *is an interior point of* Θ, *and* Λ *admits continuous first and second derivatives with respect to* θ, *denoted by the vector* $\Lambda_T^{(1)}$ *and the matrix* $\Lambda_T^{(2)}$, *satisfying*

(iii) *As* $T \to \infty$, $\Lambda_T^{(2)}(\theta^*) \to -J$ *in probability, and* $\sqrt{T}\,\Lambda_T^{(1)}(\theta)$ *is asymptotically normal with mean zero and covariance matrix* J.

(iv) *For every* $\varepsilon > 0$,

$$\liminf_{T\to\infty} P\{|\Lambda_T^{(2)}(\theta) - \Lambda_T^{(2)}(\theta^*)|\} < \varepsilon \qquad \forall\ \theta \in U(\theta^*)$$

increases to 1 *as the neighborhood* $U(\theta^*)$ *of* θ^* *shrinks to* θ^*.

Then $\hat{\theta}_T$ *is asymptotically normal with mean* θ^* *and covariance matrix* $T^{-1}J^{-1}$. *Moreover, if* θ_T *is* $T^{1/2}$*-consistent, that is the distributions of* $\sqrt{T}(\theta_T - \theta^*)$ *are tight, then* $\hat{\theta}_T - \{\theta_T - \Lambda_T^{(2)}(\theta_T)^{-1}\Lambda_T^{(1)}(\theta_T)\} \to 0$ *in probability as* $T \to \infty$.

We apply the above results with $\Lambda_T = T^{-1}L_T$. By *Theorem 2*, condition *i* is satisfied except for the continuity of $\Lambda(\theta)$, which we shall assume. To see that θ^* realizes the maximum of $\Lambda(\theta)$, write

$$\Lambda(\theta) - \Lambda(\theta^*) = \sum_{j=1}^r [E\log\{\lambda_{\theta,j}(0)\lambda_j^{-1}(0)\}\lambda_j(0) + \lambda_j(0) - \lambda_{\theta,j}(0)]$$

and note that $\log x \le x - 1$ with equality if and only if $x = 1$, we get $\Lambda(\theta) \le \Lambda(\theta^*)$ with equality if and only if $\lambda_{\theta,j}(0) = \lambda_j(0)$ almost surely, implying, by stationarity that $\lambda_{\theta,j}(t) = \lambda_j(t)$ almost surely. If the parametrization is such that for $\theta' \ne \theta$, $\lambda_{\theta,j}(t)$ is not equal to $\lambda_{\theta',j}(t)$ for all t almost surely, then θ^* is the unique maximum of Λ.

Theorems 2 and *3* show that condition *iii* is satisfied. So all we need is to verify conditions *ii* and *iv*. This would require rather strong assumptions on $\hat{\lambda}_\theta(t)$ and $\lambda_\theta(t)$. A sufficient set of assumptions is as follows.

A0: *For any compact* C *of* Θ, *there is a* $c > 0$ *such that* $\lambda_{\theta,j}(t) \ge c$ *almost surely for all* t, *all* $\theta \in C$.

A1: *For any compact* C *of* Θ

$$\lim_{t\to\infty}\operatorname*{Sup}_{\theta\in C} |\hat{\lambda}_{\theta,j}(t) - \lambda_{\theta,j}(t)| = 0\ a.s.,$$

$$E\left\{\operatorname*{Sup}_{\theta\in C} \phi_{\theta,j}^2(t)\right\}\lambda_j(t) < \infty.$$

A2: *For some compact neighborhood* U *of* θ^*,

$$\lim_{t\to\infty}\operatorname*{Sup}_{\theta\in U} \|\hat{\lambda}_{\theta,j}^{(i)}(t) - \lambda_{\theta,j}^{(i)}(t)\| = 0\ a.s., \qquad i = 1, 2,$$

$$E\left[\left\{\underset{\theta\in U}{Sup}\ \|\phi^{(2)}_{\theta,j}(t)\|^2\right\}\lambda_j(t)\right] < \infty,$$

$$E\left\{\underset{\theta\in U}{Sup}\ \|\lambda^{(2)}_{\theta,j}(t)\|\right\} < \infty.$$

THEOREM 5. *Under assumptions* A0 *and* A1, *condition* ii *of* Theorem 4 *is satisfied, and under assumptions* A0 *and* A1, *condition* iv *of* Theorem 4 *is satisfied.*

Proof: Let $U(\theta)$ be a compact neighborhood of θ. Then from Eq. **7**,

$$\underset{\theta'\in U(\theta)}{\text{Sup}}\ T^{-1}L_T(\theta') \leq T^{-1}\sum_{j=1}^{r}\left[\int_0^T\left\{\underset{\theta'\in U(\theta)}{\text{Sup}}\ \hat{\phi}_{\theta',j}(t)\right\}dN_j(t) - \int_0^T\left\{\underset{\theta'\in U(\theta)}{\text{Inf}}\ \hat{\lambda}_{\theta,j}(t)\right\}dt\right].$$

By the same argument as in the proof of *Theorem 2*, the above right-hand side is seen to converge almost surely as $T \to \infty$ to

$$\sum_{j=1}^{r}\left[E\left[\left\{\underset{\theta'\in U(\theta)}{\text{Sup}}\ \phi_{\theta',j}(t)\right\}\lambda_j(t)\right] - E\left\{\underset{\theta'\in U(\theta)}{\text{Inf}}\ \lambda_{\theta,j}(t)\right\}\right].$$

By the monotone convergence theorem, as $U(\theta)$ shrinks to θ the above expression converges to $\Lambda(\theta)$ and hence

$$\lim_{T\to\infty}\sup\left\{\underset{\theta'\in U(\theta)}{\text{Sup}}\ T^{-1}L_T(\theta')\right\} \to \Lambda(\theta)\ a.s.,$$

which gives the result.

Verification of condition *iv* follows essentially analogously.

We are highly obliged to Prof. R. N. Bhattacharya for the very careful and critical examination of the first draft of this paper. His extremely valuable comments and helpful suggestions are gratefully acknowledged. The research of M.L.P. was supported in part by Office of Naval Research Contract N00014-85-K-0648.

1. Jacobsen, M. (1982) *Statistical Analysis of Counting Process*, Lecture Notes in Statistics (Springer, New York), Vol. 12, p. 63.
2. Jacod, J. (1975) *Z. Wahrscheinlichkeitstheor. Verw. Geb.* **31**, 235–253.
3. Lipster, R. S. & Shiryayev, A. N. (1978) *Statistics of Random Processes II* (Springer, Berlin), p. 252.
4. Hawkes, A. G. (1971) *Biometrika* **58**, 83–90.
5. Hawkes, A. G. & Oakes, D. (1974) *J. Appl. Probab.* **11**, 493–503.
6. Brémaud, P. & Jacod, J. (1977) *Adv. Appl. Probab.* **9**, 369–416.
7. Doob, J. L. (1953) *Stochastic Processes* (Wiley, New York).

Reprinted from
Proceedings of The National Academy of Sciences,
USA, Volume 83 (1986), pages 541-545

ON TIME-REVERSIBILITY AND THE UNIQUENESS OF MOVING AVERAGE REPRESENTATIONS FOR NON-GAUSSIAN STATIONARY TIME SERIES

MARC HALLIN, CLAUDE LEFEVRE and MADAN L. PURI

Institut de Statistique, Université Libre de Bruxelles, B-1050 Bruxelles, Belgium
Department of Mathematics, Indiana University, Bloomington, Indiana 47405, U.S.A.

SUMMARY

A recent result by Findley (1986) on the uniqueness of moving average representations for non-Gaussian time series is shown to establish a conjecture by Weiss (1975) on the time-reversibility of general linear processes.

Some key words: ARMA process; Gaussian process; General linear process; Reversible.

A stationary discrete-time process $\{X_t; t = \ldots, -1, 0, 1, \ldots\}$ is time-reversible if, for any $k = 1, 2, \ldots$ and any k-tuple $t_1 < \ldots < t_k$, the joint distribution of $(X_{t_1}, \ldots, X_{t_k})$ is the same as that of $(X_{t_k}, \ldots, X_{t_1})$. Time-reversibility for finite-order autoregressive-moving average, ARMA, processes was shown to be a typically Gaussian property by Weiss (1975), who establishes that the only time-reversible non-Gaussian ARMA processes are pure moving average processes

$$X_t = \varepsilon_t + \sum_{j=1}^{q} b_j \varepsilon_{t-j}, \tag{1}$$

where $\{\varepsilon_t\}$ is an independent white noise process, with symmetric or skew-symmetric sets of coefficients, i.e.

$$b_q = -1, \quad b_{q-i} = -b_i \quad (i = 1, \ldots, q-1) \tag{2}$$

or

$$b_q = 1, \quad b_{q-i} = b_i \quad (i = 1, \ldots, q-1). \tag{3}$$

Weiss concludes his paper by conjecturing that some extended form of his result holds for general linear processes, i.e. processes of the form

$$X_t = \sum_{j=-\infty}^{\infty} c_j \varepsilon_{t-j}. \tag{4}$$

This open problem was implicitly solved in an apparently unrelated paper by Findley (1986), where a non-Gaussian process is shown to admit at most one mean square convergent representation of the form (4), ignoring multiplications by scale factors and shifts in the time origin of the process $\{\varepsilon_t\}$. To be specific, the following time-reversibility result can be deduced from Findley's (1986) theorem.

PROPOSITION. *Let $\{X_t\}$ be a stationary process of the form* (4), *where $\{\varepsilon_t\}$ is a sequence of independent and identically distributed random variables with mean zero and finite moments of all orders. Assume that the corresponding spectral density is positive almost everywhere. If $\{X_t\}$ is time-reversible, then either $\{\varepsilon_t\}$ is a Gaussian white noise, or there exists a k and $s \in \{0, 1\}$ such that $c_{2k+j} = (-1)^s c_{2k-j}$.*

Proof. If $\{X_t\}$ is time-reversible, then it admits an alternative representation

$$X_t = \sum_{j=-\infty}^{\infty} c_j \varepsilon'_{t+j} = \sum_{j=-\infty}^{\infty} c_{-j} \varepsilon'_{t-j}, \tag{5}$$

where $\{\varepsilon'_t\}$ and $\{\varepsilon_t\}$ are two copies of the same white noise process. By Findley's result, if $\{\varepsilon_t\}$ is not Gaussian, (5) has to result from (4) by means of a positive or negative change of scale and/or a shift in the time origin. Since $\Sigma\, c_j^2$ is proportional to var (X_t), the only scale change left is a change of sign. The sequence $(\ldots, c_{i+1}, c_i, c_{i-1}, \ldots)$ thus can be obtained from the original sequence $(\ldots, c_{j-1}, c_j, c_{j+1}, \ldots)$ by a shift in the time origin or by a shift and change of all signs. Accordingly, both sequences either have to be symmetric of the form $(\ldots, z, y, x, y, z, \ldots)$ or $(\ldots, y, x, x, y, \ldots)$ or skew-symmetric of the form $(\ldots, -y, -x, 0, x, y, \ldots)$ or $(\ldots, -y, -x, x, y, \ldots)$.

Clearly, these are infinite order generalizations of (3) and (2) respectively, and the above proposition therefore constitutes a general linear process-extension of Weiss's time-reversibility result. □

References

FINDLEY, D. F. (1986). The uniqueness of moving average representations with independent and identically distributed random variables for non-Gaussian stationary time series. *Biometrika* **73**, 520-1.

WEISS, G. (1975). Time-reversibility of linear stochastic processes. *J. Appl. Prob.* **12**, 831-6.

[*Received May* 1987. *Revised September* 1987]

Reprinted by permission of the Biometrika Trustees,
© 1988 by Oxford University Press

LOCALLY ASYMPTOTICALLY RANK-BASED PROCEDURES FOR TESTING AUTOREGRESSIVE MOVING AVERAGE DEPENDENCE

MARC HALLIN† and MADAN L. PURI‡

†*Institut de Statistique, Université Libre de Bruxelles, Campus Plaine C.P. 210, B-1050 Brussels, Belgium*

‡*Department of Mathematics, Indiana University, Bloomington, IN 47405*

Communicated by Herbert Robbins, November 24, 1987

ABSTRACT **The problem of testing a given autoregressive moving average (ARMA) model (in which the density of the generating white noise is unspecified) against other ARMA models is considered. A distribution-free asymptotically most powerful test, based on a *generalized linear serial rank statistic*, is provided against contiguous ARMA alternatives with specified coefficients. In the case in which the ARMA model in the alternative has unspecified coefficients, the asymptotic sufficiency (in the sense of Hájek) of a finite-dimensional vector of rank statistics is established. This asymptotic sufficiency is used to derive an asymptotically maximin most powerful test, based on a *generalized quadratic serial rank statistic*. The asymptotically maximin optimal test statistic can be interpreted as a rank-based, weighted version of the classical Box–Pierce portmanteau statistic, to which it reduces, in some particular problems, under gaussian assumptions.**

Section 1. Rank Tests for Time Series Analysis

Time series analysis is one of the areas in statistics whose development has been most impressive during the past two decades, and its importance for practical applications, ranging from economics to engineering, has been widely recognized. However, in spite of the growing interest in the subject, and in spite of the fact that the need for robust or rank-based procedures has been emphasized by many authors, little has been done to introduce such procedures (especially rank-based ones) in time series context.

A step toward a systematic treatment of this subject was taken in three papers by Hallin *et al.* (1, 2, 8), in which asymptotically optimal and maximin-optimal rank tests (in the Pitman–Noether sense) were derived for the problem of testing randomness against alternatives of autoregressive moving average (ARMA) dependence. Two classes of *serial rank statistics* were introduced for that purpose.

The first one is the class of linear serial rank statistics, of order p, of the form

$$S^{(n)} = (n-p)^{-1} \sum_{t=p+1}^{n} a^{(n)}(R_t^{(n)}, R_{t-1}^{(n)}, \ldots, R_{t-p}^{(n)}), \quad [1]$$

where $R_t^{(n)}$ denotes the rank of $X_t^{(n)}$ in some observed series $\mathbf{X}^{(n)} = (X_1^{(n)}, \ldots, X_n^{(n)})$ of length n, and $a^{(n)}(\ldots)$ is some *score function*. Of special interest is the particular linear serial rank statistic—i.e., the *rank autocorrelation* $r_{i;f}^{(n)}$ (for definition, see Eq. **9**) *of order* i (= 1, . . ., $n-2$) *associated with density* $f(\cdot)$ that enjoys (cf. ref. 2) most of the asymptotic properties of the usual sample autocorrelations $r_i^{(n)}$ for gaussian series.

The second class of rank statistics considered (in ref. 2) is that of *quadratic serial rank statistics*, which are quadratic forms of linear serial rank statistics. Of particular interest are the *rank portmanteau statistics* $\Sigma_{i=1}^{K} (r_{i;f}^{(n)})^2$ that provide a rank version of the *classical portmanteau statistics* $\Sigma_{i=1}^{K} (r_i^{(n)})^2$ of Box and Pierce (3).

Section 2. Testing an ARMA Model Against Another ARMA Model: Asymptotically Most Powerful Test

The problem of testing a given ARMA model (with unspecified density function) against other ARMA models is important because of its implications in the various *identification* and *validation* steps that are part of any time series model-building procedure.

Denote by $H^{(n)}(\mathbf{A}, \mathbf{B}; \cdot)$ the null hypothesis under which the observed series $\mathbf{X}^{(n)}$ is generated by the ARMA (p_1, q_1) model

$$\begin{aligned} A(L)X_t &= X_t - A_1X_{t-1} - \ldots - A_{p_1}X_{t-p_1} \\ &= \varepsilon_t + B_1\varepsilon_{t-1} + \ldots + B_{q_1}\varepsilon_{t-q_1} \\ &= B(L)\varepsilon_t, \qquad t \in \mathbb{Z}, \end{aligned} \quad [2]$$

where $\{\varepsilon_t\}$ is an independent white-noise process with unspecified density function f [L denotes the lag operator, and $A(L) = 1 - A_1L - \ldots - A_{p_1}L^{p_1}$, $B(L) = 1 + B_1L + \ldots + B_{q_1}L^{q_1}$ are difference operators].

We assume the following conditions.

(*i*) $A(L)$ and $B(L)$ are of orders p_1 and q_1, respectively (i.e., $A_{p_1} \neq 0 \neq B_{q_1}$).

(*ii*) The polynomials $A(z)$ and $B(z)$, $z \in \mathbb{C}$ have no common roots.

(*iii*) The polynomials $A(z)$ and $B(z)$, $z \in \mathbb{C}$ have all their roots outside the unit circle.

(*iv*) ε_t has finite moments up to the sixth order. $E(\varepsilon_t) = 0$ and $E(\varepsilon_t^2) = \sigma^2$ (unspecified).

(*v*) $f(x)$ is (a.e.) derivable [denote by $f'(x)$ its derivative] and *absolutely continuous on finite intervals* and has finite Fisher's information $I(f): 0 < I(f) = E[(f'(\varepsilon_t)/f(\varepsilon_t))^2] < \infty$.

(*vi*) Denote by $F(x)$ the cumulative distribution function associated with $f(x)$, and let $F^{-1}(u) = \inf\{x | F(x) \geq u\}$, $0 < u < 1$. Define $\phi(x)$ as $-f'(F^{-1}(F(x)))/f(F^{-1}(F(x)))$. Assume that $\phi(x)$ is a.e. derivable and that its derivative $\phi'(x)$ is (a.e.) lipschitzian and square integrable: $|\phi'(x) - \phi'(y)| < K|x-y|$, and $\int_0^1 \phi'^2(F^{-1}(u))du < \infty$. Note that $\phi(x)$ is a.e. equal to $-f'(x)/f(x)$.

Conditions *iv*–*vi* are assumed to hold throughout this paper. Some results (mainly those involving *f-rank autocorrelations*) require an additional assumption on f, as follows:

(*vii*) f is strongly unimodal [cf. page 15 of Hájek and Šidák (4)].

The publication costs of this article were defrayed in part by page charge payment. This article must therefore be hereby marked "*advertisement*" in accordance with 18 U.S.C. §1734 solely to indicate this fact.

Abbreviation: ARMA, autoregressive moving average.

The alternative hypotheses of interest here are those under which $\mathbf{X}^{(n)}$ is an observed series from some other ARMA model $\alpha(L)X_t = \beta(L)\varepsilon_t$ distinct from Eq. **2**. To investigate locally optimal procedures, we shall consider sequences of alternatives that are contiguous to the null hypotheses. Let

$$\begin{aligned}\alpha^{(n)}(L)X_t &= X_t - \alpha_1^{(n)}X_{t-1} - \ldots - \alpha_{p_2}^{(n)}X_{t-p_2}\\ &= \varepsilon_t + \beta_1^{(n)}\varepsilon_{t-1} + \ldots + \beta_{q_2}^{(n)}\varepsilon_{t-q_2}\\ &= \beta^{(n)}(L)\varepsilon_t \end{aligned} \qquad [3]$$

be a sequence of ARMA (p_2, q_2) models, with $p_2 \geq p_1$, $q_2 \geq q_1$, $\alpha_{p_2} \neq 0 \neq \beta_{q_2}$, and

$$\alpha_i^{(n)} = \begin{cases} A_i + n^{-1/2}\gamma_i, & 1 \leq i \leq p_1 \\ n^{-1/2}\gamma_i, & p_1 < i \leq p_2 \end{cases},$$

$$\beta_i^{(n)} = \begin{cases} B_i + n^{-1/2}\delta_i, & 1 \leq i \leq q_1 \\ n^{-1/2}\delta_i, & q_1 < i \leq q_2 \end{cases}. \qquad [4]$$

Furthermore, assume that the polynomials $\alpha^{(n)}(z)$ and $\beta^{(n)}(z)$, $z \in \mathbb{C}$ have no root that is common for any value of $n \in \mathbb{N}$. Denote by $\boldsymbol{\gamma}$ and $\boldsymbol{\delta}$ the vectors $(\gamma_1, \ldots, \gamma_{p_2})$ and $(\delta_1, \ldots, \delta_{q_2})$, respectively, and by $K^{(n)}(\mathbf{A}, \mathbf{B}; \boldsymbol{\gamma}, \boldsymbol{\delta}; f)$ the sequence of ARMA (p_2, q_2) alternatives (Eq. **3**) corresponding to specified $\boldsymbol{\gamma}$, $\boldsymbol{\delta}$, and f.

Denoting by $\{Z_t\}$ the filtered process $(A(L)/B(L))X_t$ (i.e., the exact residuals with respect to Eq. **2**), we note that, under the null hypothesis, $\{Z_t\}$ is a white-noise process and, under the alternative hypothesis, $K^{(n)}(\mathbf{A}, \mathbf{B}; \boldsymbol{\gamma}, \boldsymbol{\delta}; f)$ (see Eqs. **3** and **4**), $\{Z_t\}$ is generated by a general linear model of the form

$$Z_t - n^{-1/2}\sum_{i=1}^{\infty} a_i Z_{t-i} = \varepsilon_t + n^{-1/2}\sum_{i=1}^{\infty} b_i\varepsilon_{t-i}, \qquad t \in \mathbb{Z}, \qquad [5]$$

where

$$a_i = \sum_{j=1}^{\min(p_2, i+p_1-1)} \gamma_j G_{i-j} \quad \text{and}$$

$$b_i = \sum_{j=1}^{\min(q_2, i+q_1-1)} \delta_j H_{i-j}, \qquad i = 1, 2, \ldots. \qquad [6]$$

Here G_u and H_u, $u \in \mathbb{Z}$ are the *Green's functions* of the operators $A(L)$ and $B(L)$, respectively, ($p_1 > 0$ and $q_1 > 0$):

$$[A(L)]^{-1} = 1 + \sum_{u=1}^{\infty} G_u L^u \quad \text{and}$$

$$[B(L)]^{-1} = 1 + \sum_{u=1}^{\infty} H_u L^u, \qquad [7]$$

with

$$\sum_{u=1}^{\infty} |G_u| < \infty \quad \text{and} \quad \sum_{u=1}^{\infty} |H_u| < \infty, \qquad [8]$$

which imply $\Sigma_{i=1}^{\infty} |a_i| < \infty$ and $\Sigma_{i=1}^{\infty} |b_i| < \infty$.

Let $R_t^{(n)}$ be the rank of Z_t among $(Z_1, \ldots, Z_n)$, and define the rank autocorrelation coefficient associated with density f (or f-rank autocorrelation) of order i as

$$r_{i;f}^{(n)} = \left[(n-i)^{-1}\sum_{t=i+1}^{n} \phi\left(F^{-1}\left(\frac{R_t^{(n)}}{n+1}\right)\right) \times F^{-1}\left(\frac{R_{t-i}^{(n)}}{n+1}\right) - m^{(n)}\right] \Big/ \sigma^{(n)}, \qquad [9]$$

where

$$m^{(n)} = (n(n-1))^{-1} \sum\sum_{1 \leq i_1 \neq i_2 \leq n} \phi\left(F^{-1}\left(\frac{i_1}{n+1}\right)\right) F^{-1}\left(\frac{i_2}{n+1}\right) \qquad [10]$$

and

$$\begin{aligned}[\sigma^{(n)}]^2 = {} & \frac{1}{n(n-1)} \sum\sum_{1 \leq i_1 \neq i_2 \leq n} \left[\phi\left(F^{-1}\left(\frac{i_1}{n+1}\right)\right) F^{-1}\left(\frac{i_2}{n+1}\right)\right]^2 \\ & + 2\,\frac{(n-2i)}{(n-i)} \cdot \frac{1}{n(n-1)(n-2)} \sum\sum\sum_{i \leq i_1 \neq i_2 \neq i_3 \leq n} \left[\phi\left(F^{-1}\left(\frac{i_1}{n+1}\right)\right)\right. \\ & \left.\phi\left(F^{-1}\left(\frac{i_2}{n+1}\right)\right) F^{-1}\left(\frac{i_2}{n+1}\right) F^{-1}\left(\frac{i_3}{n+1}\right)\right] \\ & + \frac{n^2 - n(2i+3) + i^2 + 5i}{n(n-1)(n-2)(n-3)(n-i)} \sum\sum\sum\sum_{1 \leq i_1 \neq i_2 \neq i_3 \neq i_4 \leq n} \\ & \left[\phi\left(F^{-1}\left(\frac{i_1}{n+1}\right)\right)\phi\left(F^{-1}\left(\frac{i_2}{n+1}\right)\right) F^{-1}\left(\frac{i_3}{n+1}\right) F^{-1}\left(\frac{i_4}{n+1}\right)\right] \\ & - (n-i)(m^{(n)})^2. \end{aligned} \qquad [11]$$

Then

PROPOSITION 2.1. *The asymptotically most powerful test for* $\mathrm{H}^{(n)}(\mathrm{A}, \mathrm{B}; \cdot)$ *against* $\mathrm{K}^{(n)}(\mathbf{A}, \mathrm{B}; \boldsymbol{\gamma}, \boldsymbol{\delta}; \mathrm{f})$ *at given level* α *rejects if*

$$\mathrm{n}^{1/2}\sum_{i=1}^{n-2} (a_i + b_i) r_{i;f}^{(n)} > \left[\sum_{i=1}^{n-2} (a_i + b_i)^2\right]^{1/2} k_{1-\alpha},$$

where $k_{1-\alpha}$ *denotes the* $(1 - \alpha)$*-quantile of the standard normal distribution.*

The proof of this *Proposition* is given in ref. 5.

Section 3. Asymptotic Sufficiency of *f*-Rank Autocorrelations

Consider the general problem of testing $H^{(n)}$ against $K^{(n)}$ where $H^{(n)} = \{l_0^{(n)}\}$ and $K^{(n)} = \{l_1^{(n)}\}$ are families of densities for some observed $\mathbf{X}^{(n)}$ defined over a sample space $(\mathcal{X}^{(n)}, \mathcal{A}^{(n)})$. Recall that the *envelope power function* $\beta(\alpha, H^{(n)}, K^{(n)})$ for this problem is defined as

$$\beta(\alpha, H^{(n)}, K^{(n)}) = \sup_{\psi^{(n)}} \inf_{l_1^{(n)} \in K^{(n)}} E_{l_1^{(n)}}[\psi^{(n)}], \qquad [12]$$

where the supremum is taken over all the tests $\psi^{(n)}$ satisfying, for fixed $\alpha \in [0, 1]$,

$$E_{l_0^{(n)}}[\psi^{(n)}] \leq \alpha, \qquad l_0^{(n)} \in H^{(n)}. \qquad [13]$$

Let $T^{(n)}$ be some statistic from $(\mathcal{X}^{(n)}, \mathcal{A}^{(n)})$ to some range space $(\mathcal{T}^{(n)}, \mathcal{B}^{(n)})$. Define the *envelope power function* $\beta_T(\alpha, H^{(n)}, K^{(n)})$ *associated with* $T^{(n)}$ by taking the supremum, in Eq. **12**, over the set of all $T^{(n)}$-measurable tests $\psi^{(n)}$ satisfying Eq. **13**. If we denote by $H_T^{(n)}$ and $K_T^{(n)}$ the families of densities [on $(\mathcal{T}^{(n)}, \mathcal{B}^{(n)})$] induced from $H^{(n)}$ and $K^{(n)}$ by $T^{(n)}$, then clearly

$$\beta_T(\alpha, H^{(n)}, K^{(n)}) = \beta(\alpha, H_T^{(n)}, K_T^{(n)}). \qquad [14]$$

Obviously, we also have

$$\beta(\alpha, H^{(n)}, K^{(n)}) \geq \beta_T(\alpha, H^{(n)}, K^{(n)}). \qquad [15]$$

If, however, an equality in Eq. **15** is achieved for every $\alpha \in [0, 1]$ [this happens, e.g., if $T^{(n)}$ is sufficient for the structure

$(\mathscr{X}^{(n)}, \mathscr{A}^{(n)}, H^{(n)} \cup K^{(n)})]$, then $T^{(n)}$ can be considered, in some sense, as *sufficient for the problem of testing* $H^{(n)}$ *against* $K^{(n)}$.

Accordingly [see LeCam (6) and theorem VII of Hájek and Šidák (4)] we say that $T^{(n)}$ is asymptotically sufficient for the problem of testing $H^{(n)}$ against $K^{(n)}$ if

$$\lim_{n\to\infty} [\beta(\alpha, H^{(n)}, K^{(n)}) - \beta_T(\alpha, H^{(n)}, K^{(n)})] = 0, \qquad \alpha \in [0, 1]. \quad \textbf{[16]}$$

An important consequence is that if $T^{(n)}$ is asymptotically sufficient for $H^{(n)}$ against $K^{(n)}$, then any asymptotically maximin most powerful test $\psi^{(n)*}$ for $H_T^{(n)}$ against $K_T^{(n)}$ is also asymptotically maximin most powerful for $H^{(n)}$ against $K^{(n)}$.

Let us consider now the problem of testing $H^{(n)}(\mathbf{A}, \mathbf{B}; \cdot)$ against *unspecified* ARMA alternatives. Denote by

$$K^{(n)}(\mathbf{A}, \mathbf{B}; \mathbb{R}^{p_2}, \mathbb{R}^{q_2}; f) = U\{K^{(n)}(\mathbf{A}, \mathbf{B}; \boldsymbol{\gamma}, \boldsymbol{\delta}; f) | \boldsymbol{\gamma} \in \mathbb{R}^{p_2}, \boldsymbol{\delta} \in \mathbb{R}^{q_2}\}, \quad \textbf{[17]}$$

the alternative under which the observed series $\mathbf{X}^{(n)}$ is generated by some ARMA model (Eqs. **3** and **4**, where $\boldsymbol{\gamma}$ and $\boldsymbol{\delta}$ take on unspecified values, although being of specified dimensions p_2 and q_2. Let also $d \in \mathbb{R}^+$, and denote by $\mathscr{D}(d)$ the set of values of $(\boldsymbol{\gamma}, \boldsymbol{\delta}) \in \mathbb{R}^{p_2} \times \mathbb{R}^{q_2}$ such that $\|\mathbf{a} + \mathbf{b}\| \geq d$, with **a** and **b** given in Eq. **6**. In order to obtain asymptotically maximin most powerful tests, we need to consider subhypotheses of the form

$$K^{(n)}(\mathbf{A}, \mathbf{B}; \mathbb{R}^{p_2}, \mathbb{R}^{q_2}; f|d) = U\{K^{(n)}(\mathbf{A}, \mathbf{B}; \boldsymbol{\gamma}, \boldsymbol{\delta}; f) | (\boldsymbol{\gamma}, \boldsymbol{\delta}) \in \mathscr{D}(d)\}. \quad \textbf{[18]}$$

We are now able to characterize a finite-dimensional rank-based statistic that is asymptotically sufficient (in the sense defined above) for testing a given ARMA model against another, unspecified, one.

PROPOSITION 3.1. *Let* $\{\psi_t^{(1)}, \ldots, \psi^{(p_1+q_1)}\}$ *denote an arbitrary fundamental system of solutions of the homogeneous equation*

$$\mathrm{A(L)B(L)}\psi_t = 0, \qquad t \in \mathbb{Z}. \quad \textbf{[19]}$$

Writing π *for* $max(p_2 - p_1, q_2 - q_1)$, *consider the* $max(p_1 + q_2, p_2 + q_1)$*-tuple of rank statistics* [*assume* $n \geq max(p_2 - p_1 + 3, q_2 - q_1 + 3, p_1 + q_1 + 2)$]

$$\mathbf{T}_{\psi;f}^{(n)} = \left(r_{1;f}^{(n)}, \ldots, r_{\pi;f}^{(n)}, \sum_{i=\pi+1}^{n-2} \psi_i^{(1)} r_{i;f}^{(n)}, \ldots, \sum_{i=\pi+1}^{n-2} \psi_i^{(p_1+q_1)} r_{i;f}^{(n)}\right)'. \quad \textbf{[20]}$$

Then, for any $d \in \mathbb{R}^+$, $\mathbf{T}_{\psi;f}^{(n)}$ *is asymptotically sufficient for the problem of testing* $H^{(n)}(\mathbf{A}, \mathbf{B}; \cdot)$ *against Eq.* **18.**

The proof mainly consists in establishing that the sufficient conditions for asymptotic sufficiency of theorem VII 1.1 of Hájek and Šidák (4) are satisfied. Let $l_{0;f}^{(n)}(\mathbf{X}^{(n)}) = \Pi_{t=1}^{n} f(Z_t^{(n)})$. Then we have to show that there exist $\mathbf{T}_{\psi;f}^{(n)}$-measurable variables $h_{\boldsymbol{\gamma},\boldsymbol{\delta}}^{(n)}$, such that the functions

$$\bar{l}_{\boldsymbol{\gamma},\boldsymbol{\delta}}^{(n)} = l_{0;f}^{(n)} h_{\boldsymbol{\gamma},\boldsymbol{\delta}}^{(n)} \quad \textbf{[21]}$$

are densities for all $(\boldsymbol{\gamma}, \boldsymbol{\delta}) \in \mathscr{D}(d)$ and n (n sufficiently large) and satisfy

$$\lim_{n\to\infty} \sup_{\mathscr{D}(d)} \|l_{\boldsymbol{\gamma},\boldsymbol{\delta}}^{(n)} - \bar{l}_{\boldsymbol{\gamma},\boldsymbol{\delta}}^{(n)}\|_{L_1} = 0, \quad \textbf{[22]}$$

where $\|p - q\|_{L_1}$ is the L_1-distance $\int |p - q| d\mu$ for densities p and q defined with respect to the σ-finite measure μ [for details, see Hallin and Puri (5)].

Asymptotic Distribution of $\mathbf{T}_{\Psi;f}^{(n)}$. We conclude this section by establishing the joint asymptotic normality of any asymptotically sufficient vector $\mathbf{T}_{\Psi;f}^{(n)}$.

PROPOSITION 3.2. $n^{1/2}\mathbf{T}_{\Psi;f}^{(n)}$ *is asymptotically normal, with mean* 0 *under* $H^{(n)}(\mathbf{A}, \mathbf{B}; \cdot)$ *mean* $((a_1 + b_1), \ldots, (a_\pi + b_\pi), \Sigma_{i=\pi+1}^{\infty} \psi_i^{(1)}(a_i + b_i), \ldots, \Sigma_{i=\pi+1}^{\infty} \psi_i^{(p_1+q_1)}(a_i + b_i))[\sigma^2(I(f)]^{1/2}$ *under* $K^{(n)}(\mathbf{A}, \mathbf{B}; \boldsymbol{\gamma}, \boldsymbol{\delta}; f)$, *and full-rank covariance matrix*

$$\mathbf{W}_\Psi^2 = \begin{pmatrix} I_{\pi\times\pi} & \vdots & 0 \\ \cdots & \vdots & \cdots \\ 0 & \vdots & W_\Psi^2 \end{pmatrix}$$

under both $H^{(n)}(\mathbf{A}, \mathbf{B}; \cdot)$ *and* $K^{(n)}(\mathbf{A}, \mathbf{B}; \boldsymbol{\gamma}, \boldsymbol{\delta}; f)$, *where* $W_\Psi^2 = (W_{\psi;kl})$ *and* $W_{\psi;kl} = \Sigma_{i=\pi+1}^{\infty} \psi_i^{(k)} \psi_i^{(l)} < \infty$; $k, l = 1, \ldots, p_1 + q_1$.

Proof: Let $\kappa > \pi$, and consider an arbitrary linear combination of $\log L^{(n)}(\mathbf{Z}^{(n)})$, $n^{1/2} r_{1;f}^{(n)}, \ldots, n^{1/2} r_{\pi;f}^{(n)}$, $n^{1/2} \Sigma_{i=\pi+1}^{\kappa} \psi_i^{(1)} r_{i;f}^{(n)}, \ldots$ and $n^{1/2} \Sigma_{i=\pi+1}^{\kappa} \psi_i^{(p_1+q_1)} r_{i;f}^{(n)}$. [Here $\log L^{(n)}(\mathbf{Z}^{(n)})$ denotes the log likelihood ratio for testing $H^{(n)}(\mathbf{A}, \mathbf{B}; f)$ against $K^{(n)}(\mathbf{A}, \mathbf{B}; \boldsymbol{\gamma}, \boldsymbol{\delta}; f)$]. The asymptotic normality under the null hypothesis of this linear combination can easily be established along the lines of proposition 3.1 of Hallin *et al.* (2), and then using theorem 7.7.1 of Anderson (7), we obtain the asymptotic normality of this linear combination in the case when $\kappa = \infty$. *Proposition 3.2* then follows from LeCam's third lemma (6).

The nonsingularity of W_Ψ^2, hence that of $\mathbf{W}_\Psi^2$, results from the fact that the $\Psi_t^{(j)}$ values are linearly independent solutions of Eq. **19**.

Section 4. Testing an ARMA Model Against Another ARMA Model: Asymptotically Maximin Most Powerful Tests

Generalized Quadratic Serial Rank Statistics. Quadratic serial rank statistics, providing rank-based versions of the portmanteau statistic of Box and Pierce (3) were introduced in Hallin *et al.* (2, 8) to palliate the poor overall performances of linear serial rank statistics in the problem of testing randomness against unspecified ARMA alternatives. The asymptotic (central and noncentral) χ^2 distributions of such quadratic statistics (when computed from the ranks of the filtered process $\{Z_t\}$) can easily be obtained under $H^{(n)}(\mathbf{A}, \mathbf{B}; \cdot)$ and $K^{(n)}(\mathbf{A}, \mathbf{B}; \boldsymbol{\gamma}, \boldsymbol{\delta}; f)$.

Since, however, the asymptotically most powerful test for $H^{(n)}(\mathbf{A}, \mathbf{B}; \cdot)$ against $K^{(n)}(\mathbf{A}, \mathbf{B}; \boldsymbol{\gamma}, \boldsymbol{\delta}; f)$ cannot, in general, be found within the class of linear serial rank tests, quadratic serial rank statistics cannot be expected to provide optimal tests against unspecified values of $\boldsymbol{\gamma}$ and $\boldsymbol{\delta}$ (unless, e.g., $p_1 = 0 = q_1$). Consequently a more general type of quadratic rank statistics has to be considered. To this end, let $\mathbf{S}^{(n)} - \mathbf{m}^{(n)} = (S_1^{(n)} - m_1^{(n)}, \ldots, S_l^{(n)} - m_l^{(n)})'$ be a column vector of generalized linear serial rank statistics centered about their means [under $H^{(n)}(\mathbf{A}, \mathbf{B}; \cdot)$]. Assuming that $n^{1/2}(\mathbf{S}^{(n)} - \mathbf{m}^{(n)})$ has an asymptotic covariance matrix $\mathbf{V}^2$ of full rank, we define a *generalized quadratic serial rank statistic* as a statistic of the form $Q^{(n)} = n(\mathbf{S}^{(n)} - \mathbf{m}^{(n)})'\mathbf{V}^{-2}(\mathbf{S}^{(n)} - \mathbf{m}^{(n)})$.

Since we have shown that asymptotically sufficient statistics $\mathbf{T}_{\Psi;f}^{(n)}$ exist for testing $H^{(n)}(\mathbf{A}, \mathbf{B}; \cdot)$ against $K^{(n)}(\mathbf{A}, \mathbf{B}; \boldsymbol{\gamma}, \boldsymbol{\delta}; f)$, it is intuitively quite natural to consider generalized quadratic serial rank statistics of the form

$$Q_{A,B;f}^{(n)*} = n\, \mathbf{T}_{\Psi;f}^{(n)\prime} \mathbf{W}_\Psi^{-2} \mathbf{T}_{\Psi;f}^{(n)}. \quad \textbf{[23]}$$

Actually, we shall show that $Q^{(n)*}_{A,B;f}$ provides the asymptotically maximin most powerful test for our problem. We first establish an important property that $Q^{(n)*}_{A,B;f}$, unlike $\mathbf{T}^{(n)}_{\Psi;f}$, does not depend on the particular fundamental system $\{\Psi^{(i)}_t\}$ adopted.

PROPOSITION 4.1. *The quadratic statistic* $Q^{(n)*}_{A,B;f}$ *does not depend on the fundamental system* $\{\Psi^{(i)}_t\}$.

Proof: Let $\{\theta^{(1)}_t, \ldots, \theta^{(p_1+q_1)}_t\}$ and $\{\Psi^{(1)}_t, \ldots, \Psi^{(p_1+q_1)}_t\}$ denote two arbitrary fundamental systems for Eq. **19**. Then $(\theta^{(1)}_t, \ldots, \theta^{(p_1+q_1)}_t) = (\Psi^{(1)}_t, \ldots, \Psi^{(p_1+q_1)}_t)\mathbf{K}'$, $t \in \mathbb{Z}$, where $\mathbf{K}$ is some nonsingular constant matrix of dimension $(p_1 + q_1) \times (p_1 + q_1)$. Hence

$$\mathbf{T}^{(n)}_{\theta;f} = \begin{pmatrix} I & 0 \\ 0 & \mathbf{K} \end{pmatrix} \mathbf{T}^{(n)}_{\Psi;f}, \quad \text{and} \quad W^2_\theta = \mathbf{K} W^2_\Psi \mathbf{K}'.$$

Thus

$$\begin{aligned} &\mathbf{T}^{(n)'}_{\theta;f} \mathbf{W}^{-2}_\theta \mathbf{T}^{(n)}_{\theta;f} \\ &= \mathbf{T}^{(n)'}_{\Psi;f} \begin{pmatrix} I & 0 \\ 0 & \mathbf{K} \end{pmatrix} \begin{pmatrix} I & 0 \\ 0 & \mathbf{K}^{-1} W^{-2}_\Psi \mathbf{K}'^{-1} \end{pmatrix} \begin{pmatrix} 1 & 0 \\ 0 & \mathbf{K} \end{pmatrix} \mathbf{T}^{(n)}_{\Psi;f} \\ &= \mathbf{T}^{(n)}_{\Psi;f} \mathbf{W}^{-2}_\Psi \mathbf{T}^{(n)}_{\Psi;f}. \end{aligned}$$

Let us illustrate the above proposition by giving an example of a particular formulation of $Q^{(n)*}_{A,B}$.

Example 4.1: Denote by $\lambda_{(1)}, \ldots, \lambda_{(p_1+q_1)}$ the roots of $A(z^{-1})B(z^{-1}) = 0$, $z \in \mathbb{C}$, and assume that they all are distinct (hence of multiplicity one—also recall that they all lie inside the unit circle). Then a fundamental system of solutions of Eq. **19** is provided by $\{\Psi^{(j)}_t = \lambda_j^{t-(\pi+1)}, t \in \mathbb{Z}; j = 1, \ldots, p_1 + q_1\}$. $Q^{(n)*}_{A,B}$ is thus a quadratic form involving the rank autocorrelations of orders 1 through π and geometrically weighted sums of rank autocorrelations of orders $\pi + 1$ through $n - 2$. W^2_Ψ is of the form $(W_{\Psi;kl}) = ((1 - \lambda_k\lambda_l)^{-1})$.

We now give the asymptotic distribution of $Q^{(n)*}_{A,B;f}$.

PROPOSITION 4.2. *Under* $\mathrm{H}^{(n)}(\mathbf{A}, \mathbf{B}; \cdot)$, $Q^{(n)*}_{A,B;f}$ *is asymptotically* χ^2 *with* $max(p_1 + q_2, p_2 + q_1)$ *degrees of freedom. Under* $\mathrm{K}^{(n)}(\mathbf{A}, \mathbf{B}; \boldsymbol{\gamma}, \boldsymbol{\delta}; f)$, *it is asymptotically noncentral* χ^2, *with the same degrees of freedom and with noncentrality parameter*

$$\lambda^*_f(\boldsymbol{\gamma}, \boldsymbol{\delta}) = \tfrac{1}{2}\|\mathbf{a} + \mathbf{b}\|^2 \sigma^2 I(f), \qquad [24]$$

where a *and* b *are given by* Eq. **6**.

Proof: The proof follows straightforwardly from *Proposition 4.1*. Note that by letting $\mathbf{k}' = ((a_1 + b_1), \ldots, (a_\pi + b_\pi), k_1, \ldots, k_{p_1+q_1})$, the asymptotic mean of $n^{1/2}\mathbf{T}^{(n)}_{\Psi;f}$ under $K^{(n)}(A, B; \boldsymbol{\gamma}, \boldsymbol{\delta}; f)$ takes the form $\mathbf{W}^2_\Psi \mathbf{k}[\sigma^2 I(f)]^{1/2}$. On the other hand,

$$\mathbf{k}'\mathbf{W}^2_\Psi\mathbf{k} = \|\mathbf{a} + \mathbf{b}\|^2,$$

yielding the noncentrality parameter (Eq. **24**).

Asymptotically Maximin Most Powerful Test for Testing an ARMA Model Against Unspecified ARMA Alternatives. Let us define an asymptotically maximin most powerful test for $H^{(n)}(\mathbf{A}, \mathbf{B}; \cdot)$ against an unspecified ARMA (p_2, q_2) alternative as a test that is asymptotically maximin most powerful against any alternative of the form $K^{(n)}(\mathbf{A}, \mathbf{B}; \mathbb{R}^{p_2}, \mathbb{R}^{q_2}; f|d)$ (cf. Eq. **18**), $d \in \mathbb{R}^+$. We then have the following result.

PROPOSITION 4.3. *An asymptotically maximin most powerful test for* $\mathrm{H}^{(n)}(\mathbf{A}, \mathbf{B}; \cdot)$ *against an unspecified ARMA* (p_2, q_2) *alternative is provided by*

$$\psi^{(n)*}_\alpha(\mathbf{A}, \mathbf{B}; p_2, q_2; f) = 1 \quad \text{if} \quad Q^{(n)*}_{A,B;f} > \chi^2_{max(p_1+q_2,p_2+q_1);1-\alpha}. \qquad [25]$$

Accordingly, the envelope power function $\beta(\alpha, \mathrm{H}^{(n)}(\mathbf{A}, \mathbf{B}; \cdot), \mathrm{K}^{(n)}(\mathbf{A}, \mathbf{B}; \mathbb{R}^{p_2}, \mathbb{R}^{q_2}; f|d)$ *converges, as* $n \to \infty$, *to* $1 - F_{max(p_1+q_2,p_2+q_1)}(\chi^2_{max(p_1+q_2,p_2+q_1);1-\alpha}; \tfrac{1}{2}d^2\sigma^2 I(f))$, *where* $\chi^2_{m;1-\alpha}$ *denotes the* $(1 - \alpha)$*-quantile of the* χ^2 *distribution with* m *degrees of freedom and* $F_m(\cdot; \lambda)$, *the distribution function of the noncentral* χ^2 *with* m *degrees of freedom and noncentrality parameter* λ.

COROLLARY 4.4. *The rank portmanteau statistic* $n\sum^p_{i=1}(r^{(n)}_{i;f})^2$ [*cf. Hallin* et al. (2)] *is asymptotically maximin most powerful against unspecified ARMA* (p_2, q_2) *alternatives if and only if* $p_1 = q_1 = 0$ *and* $p = max(p_2, q_2)$ (*i.e., for testing randomness against serial dependence of order* p).

Proof: We know from *Section 3* that $\mathbf{T}^{(n)}_{\Psi;f}$ is asymptotically sufficient for testing $H^{(n)}(\mathbf{A}, \mathbf{B}; \cdot)$ against Eq. **18**. $n^{1/2}\mathbf{W}^{-1}_\Psi\mathbf{T}^{(n)}_{\Psi;f}$ is thus also asymptotically sufficient, since $\mathbf{W}^2_\Psi$ is a full-rank covariance matrix.

Under $K^{(n)}(A, B: \boldsymbol{\gamma}, \boldsymbol{\delta}; f)$, the asymptotic distribution of $n^{1/2}\mathbf{W}^{-1}_\Psi\mathbf{T}^{(n)}_{\Psi;f}$ is normal, with mean $\boldsymbol{\mu}(\boldsymbol{\gamma}, \boldsymbol{\delta}) = \mathbf{W}_\Psi\mathbf{k}[\sigma^2 I(f)]^{1/2}$ and identity covariance matrix.

Let us show that when $(\boldsymbol{\gamma}, \boldsymbol{\delta})$ describes $\mathbb{R}^{p_2} \times \mathbb{R}^{q_2}$, $\mathbf{k}$ and thus $\boldsymbol{\mu}(\boldsymbol{\gamma}, \boldsymbol{\delta})$ describes $\mathbb{R}^{\max(p_1+q_2,p_2+q_1)}$. Any value of $(a_1, \ldots, a_{p_2-p_1}) \in \mathbb{R}^{p_2-p_1}$ can be reached by choosing appropriate values of $\gamma_1, \ldots, \gamma_{q_2-q_1}$, and a_t, $t \geq p_2 - p_1 + 1$ can be made equal to any solution of $A(L)\Psi_t = 0$ by adjusting the remaining p_1 components $\gamma_{p_2-p_1+1}, \ldots, \gamma_{p_2}$ of $\boldsymbol{\gamma}$. A similar result holds for $\mathbf{b}$ and $\boldsymbol{\delta}$. Furthermore, any solution of $A(L)B(L)\Psi_t = 0$, $t \geq \pi + 1$, can be decomposed into a solution a_t of $A(L)\Psi_t = 0$ plus a solution b_t of $B(L)\Psi_t = 0$. This follows indeed from the fact that $A(z)$ and $B(z)$ have no common root; a fundamental system for $A(L)B(L)\Psi_t = 0$ can thus be formed by juxtaposing a fundamental system $\{\Psi^{(1)}_t, \ldots, \Psi^{(p_1)}_t\}$ of $A(L)\Psi_t = 0$ with a fundamental system $\{\Psi^{(p_1+1)}, \ldots, \Psi^{(p_1+q_1)}\}$ of $B(L)\Psi_t = 0$.

Now, under the condition $(\boldsymbol{\gamma}, \boldsymbol{\delta}) \in \mathcal{D}(d)$, i.e., $\|\mathbf{a} + \mathbf{b}\|^2 \geq d^2$, the values of $\boldsymbol{\mu}(\boldsymbol{\gamma}, \boldsymbol{\delta})$ are restricted to those satisfying $\boldsymbol{\mu}'(\boldsymbol{\gamma}, \boldsymbol{\delta})\boldsymbol{\mu}(\boldsymbol{\gamma}, \boldsymbol{\delta}') = \mathbf{k}'\mathbf{W}^2_\Psi\mathbf{k}\sigma^2 I(f) \geq d^2\sigma^2 I(f)$ (cf. *Proposition 4.2*). Thus the family of asymptotic distributions of $n^{1/2}\mathbf{W}^{-1}_\Psi T^{(n)}_{\Psi;f}$ under $K^{(n)}(\mathbf{A}, \mathbf{B}; \mathbb{R}^{p_2}, \mathbb{R}^{q_2}; f|d)$ is the set of all normal distributions with mean $\boldsymbol{\mu}$ and identity covariance matrix such that $\boldsymbol{\mu}'\boldsymbol{\mu} \geq d^2\sigma^2 I(f)$, whereas under $H^{(n)}(\mathbf{A}, \mathbf{B}; \cdot)$ the asymptotic distribution of $n\mathbf{W}^{-1}_\Psi\mathbf{T}^{(n)}_{\Psi;f}$ is normal with mean $\mathbf{0}$ and identity covariance matrix.

It then follows from an invariance argument and the Hunt–Stein theorem that the maximin most powerful test for this problem is based on the test statistic $n\mathbf{T}^{(n)}_{\Psi;f}\mathbf{W}^{-2}_\Psi\mathbf{T}^{(n)}_{\Psi;f}$—which is precisely $Q^{(n)*}_{A,B;f}$. Because of the asymptotic sufficiency of $\mathbf{T}^{(n)}_{\Psi;f}$, this test is also maximin most powerful for $H^{(n)}(\mathbf{A}, \mathbf{B}; \cdot)$ against any $K^{(n)}(\mathbf{A}, \mathbf{B}; \mathbb{R}^{p_2}, \mathbb{R}^{q_2}; f|d)$, $d \in \mathbb{R}^+$.

Corollary 4.4 gives the particular form of the asymptotically maximin most powerful statistic $Q^{(n)*}_{A,B;f}$ in the case where $p_1 = q_1 = 0$ and $\pi = \max(p_2, q_2)$. This is the only case where $Q^{(n)*}_{A,B;f}$ turns out to be a quadratic serial rank statistic in the sense of Hallin *et al.* (2).

Remark 4.1: Denoting by $Q^{(n)}_f = n\sum^\pi_{i=1}(r^{(n)}_{i;f})^2$ the rank portmanteau statistic of order $\pi = \max(p_2 - p_1, q_2 - q_1)$, maximin optimal quadratic statistics $Q^{(n)*}_{A,B;f}$ can be decomposed into $Q^{(n)*}_{A,B;f} = Q^{(n)}_f + Q^{(n)}_{A,B;f}$. Because $Q^{(n)}_f$ depends only on $r^{(n)}_{1;f}, \ldots, r^{(n)}_{\pi;f}$ whereas $Q^{(n)}_{A,B;f}$ depends only on $r^{(n)}_{\pi+1;f}, r^{(n)}_{\pi+2;f}, \ldots$, $Q^{(n)}_f$ and $Q^{(n)}_{A,B;f}$ are asymptotically independent χ^2-distributed statistics [under $H^{(n)}(\mathbf{A}, \mathbf{B}; \cdot)$]. $Q^{(n)}_f$ accounts for those models in the alternative that differ from the tested model because of the existence of higher order terms (autoregressive terms of orders larger than p_1 and/or moving average terms of orders larger than q_1). It contributes for $\max(p_2 - p_1, q_2 - q_1)$ degrees of freedom, the maximal number of such possible higher order terms. $Q^{(n)}_{A,B;f}$, which has the form of a *weighted rank portmanteau statistic* accounts for those models in the alternative that differ from the tested model because their autoregressive coefficients of orders 1 through p_1 and moving average coefficients of orders 1 through q_1 differ from the tested ones (i.e., the A_i and B_i values). It contributes for $(p_1 + q_1)$ degrees of freedom, one for each tested coefficient $A_1, \ldots, A_{p_1}, B_1, \ldots, B_{q_1}$. The consideration of this decomposition in the case when $H^{(n)}(\mathbf{A}, \mathbf{B}; \cdot)$ has been rejected may provide a useful insight into the reasons why rejection occurred and thus suggest alternative model specifications.

Remark 4.2: The noncentrality parameter $\lambda_f^*(\gamma, \delta)$ of the asymptotic distribution of $Q_{A,B;f}^{(n)*}$ under alternatives of the form $K^{(n)}(A, B; \gamma, \delta; f)$ also can be decomposed into $\frac{1}{2}\Sigma_{i=1}^{\pi}(a_i + b_i)^2\sigma^2 I(f)$ (contribution of the "unweighted part" $Q_f^{(n)}$) and $\frac{1}{2}\Sigma_{i=\pi+1}^{\infty}(a_i + b_i)^2\sigma^2 I(f)$ (contribution of the "weighted part" $Q_{A,B;f}^{(n)}$). A study of the relation between these two quantities and the coefficients of the tested model (Eq. **2**) provides an interesting insight into the importance of the respective contributions of the "weighted" and "unweighted" parts to the asymptotic power of Eq. **25**. As a general rule, the contribution of the unweighted part does not depend on $A(L)$ and $B(L)$, whereas the contribution of the weighted part is an increasing function of the "closeness to the unit circle" of the roots of $A(z)B(z) = 0$, $z \in \mathbb{C}$. To show this, let us consider a simple example.

Example 4.2: Consider the problem of testing the $AR(1)$ model $X_t - \rho X_{t-1} = \varepsilon_t$, $t \in \mathbb{Z}$, when $|\rho| < 1$, against ARMA (2, 1) alternatives. Here $\pi = 1$, and

$$a_i = \begin{cases} \gamma_1 & \text{if} \quad i = 1 \\ \gamma_1\rho^{i-1} + \gamma_2\rho^{i-2} & \text{if} \quad i \geq 2 \end{cases}$$

and

$$b_i = \begin{cases} \delta_1 & \text{if} \quad i = 1 \\ 0 & \text{if} \quad i \geq 2 \end{cases}$$

if an alternative of the form $K^{(n)}(\mathbf{A}, \mathbf{B}; \gamma = (\gamma_1, \gamma_2), \delta = (\delta_1, 0); f)$ is considered. The noncentrality parameter (Eq. **24**) accordingly decomposes into $(\gamma_1 + \delta_1)^2$ (contribution of the unweighted $Q_f^{(n)}$), which does not depend on ρ, and $\Sigma_{i=2}^{\infty}(\gamma_1\rho^{i-1} + \gamma_2\rho^{i-2})^2 = (\gamma_1\rho + \gamma_2)^2(1/(1 - \rho^2))$ (contribution of the weighted $Q_{A,B;f}^{(n)}$). Clearly, this latter contribution is approximately γ_2^2 if $|\rho| \simeq 0$—and the optimal test (Eq. **25**) is then approximately equivalent to the test based on an unweighted rank portmanteau statistic of order two; if $|\rho| \simeq 1$, then the contribution of $Q_{A,B;f}^{(n)}$ can be arbitrarily large. The absence of such a weighted part in the classical portmanteau statistic could possibly be responsible for the somewhat disappointing performance of the usual Box–Pierce test.

Section 5. Exact and Approximate Residuals

All the results in the sections above are based on the ranks $R_t^{(n)}$ of the exact residuals

$$Z_t = [A(L)/B(L)]X_t^{(n)}, \qquad t = 1, \ldots, n. \qquad \textbf{[26]}$$

Clearly, even in the case of purely autoregressive processes ($q_1 = 0$), these exact residuals cannot be all computed from the observations. Replacing them with the *approximate residuals* [obtained by putting $X_t^{(n)} = 0$, $t \leq 0$ in Eq. **26**] does not affect any of the asymptotic results given here [cf. Hallin and Puri (5)].

The research of M.L.P. was supported by Office of Naval Research Contract N00014-85-K-0648.

1. Hallin, M., Ingenbleek, J.-F. & Puri, M. L. (1985) *Ann. Stat.* **13,** 1156–1181.
2. Hallin, M., Ingenbleek, J.-F. & Puri, M. L. (1985) *C.R. Seances Acad. Sci. Ser. A* **301,** 935–938.
3. Box, G. E. P. & Pierce, D. A. (1970) *J. Am. Stat. Assoc.* **65,** 1509–1526.
4. Hájek, J. & Šidák, Z. (1967) *Theory of Rank Tests* (Academic, New York).
5. Hallin, M. & Puri, M. L. (1988) *Ann. Stat.*, in press.
6. LeCam, L. (1960) *Univ. Calif. Publ. Stat.* **3,** 27–98.
7. Anderson, T. W. (1971) *The Statistical Analysis of Time Series* (Wiley, New York).
8. Hallin, M., Ingenbleek, J.-F. & Puri, M. L. (1987) *J. Time Ser. Anal.* **8,** 409–424.

Reprinted from
Proceedings of The National Academy of Sciences,
USA, Volume 85 (1988), pages 2031-2035

ON SOME LIMIT LAWS FOR PERTURBED EMPIRICAL DISTRIBUTION FUNCTIONS

MANFRED DENKER[a] and MADAN L. PURI[b,1]

[a] *Universität Göttingen, Göttingen, Germany*

[b] *Indiana University, Swain Hall East, Bloomington, IN 47405, USA*

Abstract

In this note, we establish the convergence properties for a broad class of random variables of the form $S_n = \int F_n(T_n - s)\nu_n(\mathrm{d}s)$ where T_n is some random variable, F_n is an empirical distribution function based on an independent sample of size n, and ν_n is some measure.

Keywords: Asymptotic normality; Perturbed empirical distribution functions

1. Introduction and main results

Let $\{X_i, i \geqslant 1\}$ be a sequence of independent real-valued random variables with common continuous distribution function F. We denote by F^{-1}, the left continuous inverse of F, i.e.

$$F^{-1}(x) = \inf\{y \in \mathbb{R}: F(y) \geqslant x\}.$$

In case F is differentiable, F' will denote the derivative of F. Let $\{\nu_n, n \geqslant 1\}$ be a sequence of finite measures on $\mathbb{R}$ such that ν_n is weakly convergent to some finite measure ν on $\mathbb{R}$, i.e. $\lim_{n\to\infty} \int f \mathrm{d}\nu_n = \int f \mathrm{d}\nu$ for all bounded continuous functions f on $\mathbb{R}$. Note that ν_n and ν need not be probability measures. Let $\{T_n, n \geqslant 1\}$ be a sequence of random variables (not necessarily independent of $\{X_i, i \geqslant 1\}$), and define

$$S_n = \int_{-\infty}^{\infty} F_n(T_n - s)\nu_n(\mathrm{d}s), \tag{1.1}$$

where F_n is the empirical cumulative distribution function based on $X_1, \ldots, X_n$, i.e.

$$F_n(x) = n^{-1} \sum_{i=1}^{n} u(x - X_i), \tag{1.2}$$

where $u(x) = 1$ if $x \geqslant 0$, and $= 0$ otherwise.

[1] Research supported by the Office of Naval Research Contract N00014-91-J-1020 and Deutsche Forschungsgemeinschaft.

Throughout the paper we shall use the notation $\to$ to denote convergence when $n \to \infty$. The type of convergence will always be indicated. Finally we denote the identity function by I.

Our main results are the following theorems.

Theorem 1.1. *If $\xi \in \mathbb{R}$ is a real constant such that $T_n \to \xi$ a.s. (in probability, respectively), then $S_n \to \theta = \int F(\xi - s)\nu(\mathrm{d}s)$ a.s. (in probability, respectively).*

It will be clear from the proof that this theorem extends to the case when F_n is replaced by an arbitrary random function converging a.s. to F (in the topology of pointwise convergence).

Theorem 1.2. *Assume that the function F is continuously differentiable, and that there exist a real number ξ and a random variable (B, Z) with values in $C([0,1]) \times \mathbb{R}$ such that*

$$(n^{1/2}(F_n \circ F^{-1} - I), a_n(T_n - \xi)) \to (B, Z) \quad \textit{weakly in } \mathscr{D}([0,1]) \times \mathbb{R} \tag{1.3}$$

for some sequence $\{a_n > 0, n \geqslant 1\}$ of real constants.

Let $\lim_{n\to\infty} n^{1/2}/a_n = c \in [0, \infty]$. Then

$$\min(a_n, n^{1/2})\left(S_n - \int F(\xi - t)\nu_n(\mathrm{d}t)\right)$$

$$\to \int\left[\min\left(\frac{1}{c}, 1\right)B(F(\xi - t)) + \min(c, 1)F'(\xi - t)Z\right]\nu(\mathrm{d}t) \quad \textit{weakly.} \tag{1.4}$$

Remarks. We note two special cases of Theorem 1.2:

(a) If $\lim_{n\to\infty} a_n/n^{1/2} = 0$, then

$$a_n\left(S_n - \int F(\xi - t)\nu_n(\mathrm{d}t)\right) \to \int F'(\xi - t)\nu(\mathrm{d}t)Z \quad \text{weakly.} \tag{1.5}$$

(b) If $\lim_{n\to\infty} n^{1/2}/a_n = 0$, then

$$n^{1/2}\left(S_n - \int F(\xi - t)\nu_n(\mathrm{d}t)\right) \to B(F(\xi - t))\nu(\mathrm{d}t) \quad \text{weakly.} \tag{1.6}$$

In particular, since B must be the standard Brownian bridge, $n^{1/2}(S_n - \int F(\xi - t)\nu(\mathrm{d}t))$ is asymptotically normal with mean 0, and variance

$$\sigma^2 = 2\iint_{\{s<t\}} F(\xi - t)(1 - F(\xi - s))\nu(\mathrm{d}s)\nu(\mathrm{d}t) + \iint_{\{s=t\}} F(\xi - t)(1 - F(\xi - s))\nu(\mathrm{d}s)\nu(\mathrm{d}t). \tag{1.7}$$

Theorem 1.3. *Let F be continuously differentiable, and assume that there exist a real number ξ and a $C([0,1]^2) \times C([0,1])$-valued process $(K(s,t), W(s'))$, $(0 \leqslant s', s, t \leqslant 1)$ such that*

$$\left(\frac{[ns]}{n^{1/2}}(F_{[ns]}(F^{-1}(t)) - t), \frac{[ns]}{n^{1/2}}(T_{[ns']} - \xi)\right) \to (K(s,t), W(s')) \tag{1.8}$$

weakly in $D([0,1]^2)\times D([0,1])$. *Then*

$$\frac{[nu]}{n^{1/2}}\left(S_{[nu]}-\int F(\xi-t)\nu_n(\mathrm{d}t)\right)$$

$$\to\int[K(F(\xi-t),u)+F'(\xi-t)W(u)]\nu(\mathrm{d}t),\quad 0\leqslant u\leqslant 1, \tag{1.9}$$

weakly. ($[nu]$ *means the integral part of* nu).

2. Applications

Before giving the proofs of the above theorems, we provide some applications. We may mention that the first example is a generalization of the main theorem of Puri and Ralescu (1986). The other two examples provide new applications of Theorem 1.2.

Example 2.1. Let T_n be a U-statistic based on $X,\ldots,X_n$ with symmetric kernel $h:\mathbb{R}^m\to\mathbb{R}$, $m\leqslant n$, i.e.

$$T_n=\binom{n}{m}^{-1}\sum_C h(X_{i_1},\ldots,X_{i_m}), \tag{2.1}$$

where C denotes the set of all $\binom{n}{m}$ subsets $\{i_1,\ldots,i_m\}$ of cardinality m. Let $Eh^2(X_1,\ldots,X_m)<\infty$, and let $E(T_n)=Eh(X_1,\ldots,X_m)=\xi$. Then it is well known (see, e.g. Serfling, 1980) that $T_n\to\xi$ a.s. as $n\to\infty$, and if h is non-degenerate, then $n^{1/2}(T_n-\xi)\xrightarrow{\mathscr{D}}\mathrm{N}(0,\tau^2)$ where

$$\tau^2=m^2\int\left(\int\cdots\int h(x_1,\ldots,x_m)\prod_{i=2}^m F(\mathrm{d}x_i)-\xi\right)^2F(\mathrm{d}x_1). \tag{2.2}$$

Moreover, letting $Z_i=n^{1/2}(F_n(F^{-1}(t_i))-t_i-F_n(F^{-1}(t_{i-1})+t_{i-1})$, $1\leqslant i\leqslant r$ where $t_0=0<t_1<t_2<\cdots<t_r\leqslant 1$, it can be shown by standard arguments (following Billingsley, 1968, Theorem 16.4) that $(Z_1,\ldots,Z_r,n^{1/2}(T_n-\xi))$ has asymptotically (as $n\to\infty$) a multinormal distribution with mean vector $\mathbf{0}$, and covariance matrix $\Sigma=((\sigma_{ij}))_{1\leqslant i,j\leqslant r+1}$ where

$$\sigma_{ij}=\begin{cases}(t_i-t_{i-1})(1-t_i+t_{i-1}) & \text{if } 1\leqslant i=j\leqslant r,\\ \tau^2 & \text{if } i=j=r+1,\\ (t_i-t_{i-1})(t_j-t_{j-1}) & \text{if } 1\leqslant i\neq j\leqslant r,\\ mE(h^*(X_1)I[F^{-1}(t_{i-1})\leqslant X_1\leqslant F^{-1}(t_i)]) & \text{if } i<j=r+1,\end{cases} \tag{2.3}$$

where $h^*(x)=\int\cdots\int h(x,x_2,\ldots,x_m)\prod_{i=2}^m F(\mathrm{d}x_i)-\xi$.

It follows that $(n^{1/2}(F_n(F^{-1}(t)-t),n^{1/2}(T_n-\xi))$ converges weakly in $\mathscr{D}([0,1])\times\mathbb{R}$ to some random variable Y. The first coordinate of Y is clearly the Brownian bridge B and the second one of Y is a normal variable Z with mean 0 and variance τ^2. The covariances are given by Σ. From Theorem 1.2, it follows that

$$n^{1/2}\left(S_n-\int F(\xi-t)\nu(\mathrm{d}t)\right)$$

has an asymptotic normal distribution given by the distribution of

$$\int\left[B(F(\xi-t))+cF'(\xi-t)Z\right]\nu(\mathrm{d}t).$$

In particular, if ν is the unit mass at 0, then $n^{1/2}(S_n - F(\xi))$ converges weakly to the normal variable with mean 0 and variance σ^2, where

$$\sigma^2 = E(B(F(\xi)) + F'(\xi)Z)^2 = F(\xi)(1 - F(\xi)) + (F'(\xi))^2\tau^2 + 2mF'(\xi)E(h^*(X_1)I_{[X_1 \leqslant \xi]}).$$

For special statistical applications of this result, see Puri and Ralescu (1986).

Example 2.2. Let T_n be a U-statistic with $E(h^2(X_1, \ldots, X_m)) < \infty$. Then, for some $k \geqslant 2$, $n^{k/2}(T_n - \xi)$ has asymptotically (as $n \to \infty$) some non-normal limit distribution, provided the kernel h is degenerate. From Theorem 1.2, it is easily seen that $n^{1/2}(S_n - \int F(\xi - t)\nu(dt))$ has a limiting normal distribution with mean 0 and variance σ^2 given by (1.7). In particular, if ν is the unit mass at 0, then $\sigma^2 = F(\xi)[1 - F(\xi)]$.

Example 2.3. Let the random variable Z in Theorem 1.2 have some non-normal distribution, and let $a_n = n^{1/2}$. Then, from Theorem 1.2, $n^{1/2}(S_n - \int F(\xi - t)\nu(dt)) \to \int [B(F(\xi - t)) + Z \cdot F'(\xi - t)]\nu(dt)$. Suppose T_n is independent of the X_i's. If ν is the unit mass at 0, then the distribution of $B(F(\xi)) + Z \cdot F'(\xi)$ is the convolution of the normal distribution with the distribution of $Z \cdot F'(\xi)$.

3. Proofs of the theorems

3.1. Proof of Theorem 1.1.

We show the a.s. convergence first. Let $\Omega_0 = \{T_n \to \xi\} \cap \{F_n \to F \text{ uniformly}\}$ and let $\omega \in \Omega_0$ be fixed. Then f_n defined by

$$f_n(s) = F_n(T_n(\omega) - s, \omega)$$

converges continuously to f where $f(s) = F(\xi - s)$. Applying Rubin's theorem (see Billingsley, 1968, Theorem 5.5) and by boundedness, $\int f_n \, d\nu_n \to \int f d\nu$. This proves convergence on Ω_0.

The convergence in probability can be deduced from the a.s. convergence, since if the assertion were not true, there is a subsequence so that no subsubsequence converges in probability. However, for some subsubsequence (k_n), $T_{k_n} \to \xi$ a.s., hence also $S_{k_n} \to \theta$ a.s., a contradiction.

3.2. Proof of Theorem 1.2

We first prove the theorem in the case $n^{1/2}/a_n \leqslant 1$. Note that by Billingsley (1968, Theorem 4.4),

$$(n^{1/2}(F_n \circ F^{-1} - I), T_n, a_n(T_n - \xi)) \to (B, \xi, Z)$$

weakly as random elements in $\mathscr{D}([0,1]) \times \mathbb{R}^2$. For each $n \geqslant 1$, define $\phi_n : \mathscr{D}([0,1]) \times \mathbb{R}^2 \to \mathbb{R}$ by

$$\phi_n(x, u, v) = \int \left\{ x(F(u - t)) + n^{1/2}\left[F\left(\frac{v}{a_n} + \xi - t\right) - F(\xi - t)\right]\right\} \nu_n(dt). \tag{3.1}$$

Let $\phi : \mathscr{D}([0,1]) \times \mathbb{R}^2 \to \mathbb{R}$ be defined by

$$\phi(x, u, v) = \int [x(F(u - t)) + cvF'(\xi - t)]\nu(dt). \tag{3.2}$$

By the assumption on F, if $v_n \to v$ and $t_n \to t$, then

$$n^{1/2}\left[F\left(\frac{v_n}{a_n} + \xi - t_n\right) - F(\xi - t_n)\right] = \frac{n^{1/2}}{a_n} \frac{F(v_n/a_n + \xi - t_n) - F(\xi - t_n)}{v_n/a_n} v_n \to cF'(\xi - t)v. \tag{3.3}$$

Thus, following Billingsley (1968, Theorem 5.5), for $v_n \to v$,

$$\int n^{1/2}\left[F\left(\frac{v_n}{a_n}+\xi-t\right)-F(\xi-t)\right]v_n(\mathrm{d}t)\to c\int F'(\xi-t)\,v\nu(\mathrm{d}t). \tag{3.4}$$

Similarly, for $x_n \to x$ and $u_n \to u$,

$$\int x_n(F(u_n-t))\,\nu_n(\mathrm{d}t)\to\int x(F(u-t))\nu(\mathrm{d}t).$$

It follows that, for $x_n \to x$, $u_n \to u$ and $v_n \to v$,

$$\phi_n(x_n, u_n, v_n) \to \phi(x, u, v).$$

Next observe that $F_n(F^{-1}(F(t))) = F_n(t)$ a.s. Hence, using Billingsley (1968) again,

$$\phi_n(n^{1/2}(F_n\circ F^{-1}-I);\, T_n;\, a_n(T_n-\xi)) = n^{1/2}\int [F_n(T_n-t)-F(\xi-t)]\,\nu_n(\mathrm{d}t)$$

$$\to \phi(B,\xi,Z)=\int [B(F(\xi-t))+cF'(\xi-t)]\,\nu(\mathrm{d}t).$$

This shows the first part when $n^{1/2}$ is asymptotically smaller than a_n.

To prove the remaining case, replace (3.1) by

$$\phi_n(x,u,v)=\int\left\{\frac{a_n}{n^{1/2}}x(F(u-t))+a_n\left[F\left(\frac{v}{a_n}+\xi-t\right)-F(\xi-t)\right]\right\}\nu_n(\mathrm{d}t).$$

and (3.2) by

$$\phi(x,u,v)=\int\left[\frac{1}{c}x(F(u-t))+F'(\xi-t)v\right]\nu(\mathrm{d}t)$$

and proceed as in the first part.

The proof of Theorem 1.3 follows analogously.

Acknowledgement

The proof of Theorem 1.1 was indicated by the referee, which is more elegant than our original one. We thank the referee for providing this proof, and express our sincere appreciation for his very constructive and helpful suggestions.

References

Billingsley, P. (1968), *Convergence of Probability Measures* (Wiley, New York).

Puri, M.L. and S.S. Ralescu (1986), Central limit theorem for perturbed empirical distribution functions evaluated at a random point, *J. Multivariate Anal.* **19**, 273–279.

Serfling, R.J. (1980), *Approximation Theorems of Mathematical Statistics* (Wiley, New York).

Reprinted by permission of the publisher,
© 1994 by Elsevier Science

Part IV

Appendices A, B and C

Appendix A

Volume 1

Nonparametric Methods in Statistics and Related Topics

Contents

Part IV. MISCELLANEOUS TOPICS

Part V. APPENDICES

Appendix B

Volume 2

Probability Theory and Extreme Value Theory

Contents

Part II. LIMIT THEOREMS (DEPENDENT CASE)

Part III. EXTREME VALUE THEORY

Part IV. APPENDICES

Appendix C

The Publications of Madan Lal Puri

BOOKS AND MONOGRAPHS

1970

[1] Nonparametric Techniques in Statistical Inference. Cambridge University Press, England, 623 pages. (Editor).

1971

[2] Nonparametric Methods in Multivariate Analysis. John Wiley and Sons, New York, 440 pages. Co-author: P.K. Sen. Revised Reprint by Krieger Publication Company, Florida (1993).

1975

[3] Statistical Inference and Related Topics. Academic Press, New York, 352 pages. (Editor).

[4] Stochastic Processes and Related Topics. Academic Press, New York, 315 pages. (Editor).

1982

[5] Nonparametric Statistical Inference, Vol I. North Holland Publishing Company, Amsterdam, 463 pages. Co-editors: B.V. Gnedenko (Moscow State University, Moscow, U.S.S.R.) and Istvan Vincze (Hungarian Academy of Sciences, Budapest, Hungary).

[6] Nonparametric Statistical Inference, Vol. II. North Holland Publishing Company, Amsterdam, 446 pages. Co-editors: B.V. Gnedenko (Moscow State University, Moscow, U.S.S.R.) and Istvan Vincze (Hungarian Academy of Sciences, Budapest, Hungary).

1985

[7] Nonparametric Methods in General Linear Models. John Wiley and Sons, New York, 399 pages. Co-author: P.K. Sen.

1987

[8] New Perspectives in Theoretical and Applied Statistics. John Wiley and Sons, New York, 544 pages. Co-editors: J.P. Vilaplana (University of Bilbao, Spain) and Wolfgang Wertz (Technische Universität, Vienna, Austria).

[9] Mathematical Statistics and Probability Theory: Volume A, Theoretical Aspects. D. Reidel Publication Company, Holland, 326 pages. Co-editors: Pal Révész (Hungarian Academy of Sciences, Budapest, Hungary) and Wolfgang Wertz (Technische Universität, Vienna, Austria).

1993

[10] Statistical Sciences and Data Analysis: Proceedings of the Third Pacific Area Statistical Conference. VSP International Science Publishers, The Netherlands, 570 pages. Co-editors: Kameo Matusita and Takesi Hayakawa.

1994

[11] Recent Advances in Statistics and Probability: Proceedings of the 4th International Meeting of Statistics in the Basque Country, Spain. VSP International Science Publishers, Utrecht, The Netherlands, 458 pages. Co-editor: J.P. Vilaplana.

2000

[12] Asymptotics in Statistics and Probability: Papers in Honor of George Gregory Roussas. VSP International Science Publishers, The Netherlands, 429 pages. (Editor).

RESEARCH PAPERS

1964

[13] *Asymptotic efficiency of a class of c-sample tests.* The Annals of Mathematical Statistics, 35, 102-121.

1965

[14] *On the estimation of contrasts in linear models.* The Annals of Mathematical Statistics, 36, 198-202. Co-author: S. Bhuchongkul.

[15] *Some distribution-free k-sample rank tests of homogeneity against ordered alternatives.* Communications on Pure and Applied Mathematics, 18, 51-63.

[16] *On the combination of independent two-sample tests of a general class.* Review of the International Statistical Institute, 33, 229-241.

[17] *On some tests of homogeneity of variances.* Annals of the Institute of Statistical Mathematics, 17, 323-330.

[18] *On a class of testing procedures in linear models.* The Journal of Indian Statistical Association, 3, 1-8.

1966

[19] *On a class of multivariate multi-sample rank-order tests.* Sankhyā Series A, 28, 353-376. Co-author: P.K. Sen.

1967

[20] *Multi-sample analogues of some one-sample tests.* The Annals of Mathematical Statistics, 38, 523-549. Co-author: K.L. Mehra.

[21] *On the theory of rank-order tests for location in the multivariate one-sample problem.* The Annals of Mathematical Statistics, 38, 1216-1228. Co-author: P.K. Sen.

[22] *On robust estimation in incomplete block designs.* The Annals of Mathematical Statistics, 38, 1587-1591. Co-author: P.K. Sen.

[23] *On some optimum nonparametric procedures in two-way layouts.* Journal of the American Statistical Association, 62, 1214-1229. Co-author: P.K. Sen.

[24] *Combining independent one-sample tests of significance.* Annals of the Institute of Statistical Mathematics, 19, 285-300.

[25] *On some rank tests for homogeneity of scale parameters.* Trabajos de Estadistica, 28, 353-376. Co-author: P.S. Puri.

1968

[26] *On Chernoff-Savage tests for ordered alternatives in randomized blocks.* The Annals of Mathematical Statistics, 39, 967-972. Co-author: P.K. Sen.

[27] *Nonparametric confidence regions for some multivariate location problems.* Journal of the American Statistical Association, 62, 1373-1378. Co-author: P.K. Sen.

[28] *Selection procedures based on ranks: Scale parameter case.* Sankhyā, Series A, 30, 291-302. Co-author: P.S. Puri.

[29] *On a class of rank-order estimates of contrasts in MANOVA.* Sankhyā, Series A, 30, 31-36. Co-author: P.K. Sen.

[30] *On a class of multivariate multisample rank order tests II. Tests for homogeneity of dispersion matrices.* Sankhyā, Series A, 30, 1-22. Co-author: P.K. Sen.

[31] *Multi-sample scale problem: Unknown location parameter.* Annals of the Institute of Statistical Mathematics, 22, 99-106.

1969

[32] *Multiple decision procedures based on ranks for certain problems in analysis of variance.* The Annals of Mathematical Statistics, 40, 619-632. Co-author: P.S. Puri.

[33] *A class of rank order tests for a general linear hypothesis.* The Annals of Mathematical Statistics, 40, 1325-1343. Co-author: P.K. Sen.

[34] *Analysis of covariance based on general rank scores.* The Annals of Mathematical Statistics, 40, 610-618. Co-author: P.K. Sen.

[35] *Rank order tests for multivariate paired comparisons.* The Annals of Mathematical Statistics, 40, 2101-2117. Co-author: H. Shane.

[36] *On a class of rank-order tests for the identity of two multiple regression surfaces.* Zeitschrift für Wahrscheinlichkeitstheorie und verw. Geb., 12, 1-8. Co-author: P.K. Sen.

[37] *On the asymptotic normality of one-sample rank order test statistics.* Teoria Veroyatnostey i ee Primeneniya, (also in Theory of Probability and its Applications, Translation by SIAM) 14, 167-173. Co-author: P.K. Sen.

[38] *The Van Elteren W Test and non-null hypotheses.* Review of the International Statistical Institute, 37, 166-175.

[39] *On the asymptotic theory of rank order tests for experiments involving paired comparisons.* Annals of the Institute of Statistical Mathematics, 20, 163-173. Co-author: P.K. Sen.

[40] *On robust nonparametric estimation in some multivariate linear models.* Proceedings of the Second International Symposium on Multivariate Analysis, Dayton, Ohio, 33-52, Academic Press. Co-author: P.K. Sen.

[41] *Structure of distribution-free multivariate paired comparison tests.* Proceedings of the 37th Session of the International Statistical Institute, 320-322.

1970

[42] *Asymptotic theory of likelihood ratio and rank order tests in some multivariate linear models.* The Annals of Mathematical Statistics, 41, 87-100. Co-author: P.K. Sen.

[43] *On the estimation of location parameters in the multivariate one-sample and two-sample problems.* Metrika, 16, 58-73. Co-author: P.K. Sen.

[44] *On a class of rank-order tests for independence in multivariate distributions.* Sankhyā, Series A, 32, 271-298. Co-author: P.K. Sen.

[45] *Statistical inference based on incomplete blocks designs.* Nonparametric Techniques in Statistical Inference, Cambridge University Press, 131-153. Co-author: H. Shane.

1972

[46] *On the robustness of rank order tests and estimates in the generalized multivariate one-sample location problem.* Zeitschrift für Wahrscheinlichkeitstheorie und Verw. Geb., 22, 226-241-author: P.K. Sen.

[47] *On some selection procedures in two-way layouts.* Zeitschrift für Wahrscheinlichkeitstheorie und Verw. Geb., 22, 242-250. Co-author: P.K. Sen.

[48] *Some aspects of nonparametric inference.* Review of the International Statistical Institute, 40, 298-327.

[49] *Rank tests for some linear hypotheses in paired comparisons designs.* Sankhyā, Series A, 34, 257-264.

1973

[50] *On parallel sum and difference of matrices.* Journal of Mathematical Analysis and Applications, 44, 92-97. Co-author: S.K. Mitra.

[51] *A note on asymptotic distribution-free tests for subhypotheses in multiple linear regression.* The Annals of Statistics, 1, 553-556. Co-author: P.K. Sen.

1974

[52] *Joint asymptotic multinormality for a class of rank order statistics in multivariate paired comparisons.* Journal of Multivariate Analysis, 4, 88-105. Co-author: C.T. Russell.

[53] *The asymptotic behavior of the minimax risk for multiple decision problems.* Sankhyā, Series A, 36, 1-12. Co-author: O. Krafft.

[54] *Asymptotic distribution of rank statistics for experiments involving incomplete block designs.* Annals of the Institute of Statistical Mathematics, 26, 421-446. Co-author: N.L. Wykoff.

[55] *Asymptotic results for rank statistics in some multivariate problems*. Bulletin of the International Statistical Institute, 35(2), 285-293.

1975

[56] *Order of normal approximation for rank test statistics distribution*. Annals of Probability, 3, 526-533. Co-author: J. Jureckova.

[57] *Distribution-free approaches to general linear models*. Proceedings of the International Symposium in Statistical Design and Linear Models, Fort Collins, Colorado; North Holland Publishing Company, 459-474. Co-author: P.K. Sen.

[58] *Miscellaneous problems of rank order theory*. Proceedings of the International Conference on Asymptotic Methods of Statistics and Their Applications in Physical and Social Sciences, Charles University, Prague, Czechoslovakia, 273-311.

1976

[59] *Augmenting Shapiro-Wilk Test for normality*. Contributions to Applied Statistics, Birkhauser Verlag, 129-139. Co-author: C.R. Rao.

[60] *On linear combination of order statistics*. Essays in Probability and Statistics, 433-449. Co-author: F. Eicker.

[61] *Remainder term estimate in the distribution of rank statistics under alternatives*. Proceedings of the Statistics Days Conference, 136-142. Co-author: H. Bergström.

1977

[62] *Convergence and remainder terms in linear rank statistics*. Annals of Statistics, 5, 671-680. Co-author: H. Bergström.

[63] *Asymptotically distribution-free aligned rank order tests for composite hypotheses for general linear models*. Zeitschrift für Wahrscheinlichkeitstheorie und verw. Geb., 39 (1977), 175-186. Co-author: P.K. Sen.

[64] *Problems of association for bivariate circular data and new test of independence*. Proceedings of the Fourth International Symposium on Multivariate Analysis, North Holland Publishing Company, 513-522. Co-author: J.S. Rao.

1978

[65] *A spherical correlation coefficient robust against scale*. Biometrika, 65, 391-395. Co-author: K.V. Mardia.

1979

[66] *Local maxima of the sample functions of the N-parameter Bessel process. Stochastic Processes and Their Applications, 9, 137-145. Co-author: L.T. Tran.*

[67] *Analysis of central place theory.* Bulletin of the International Statistical Institute, 47(2), 93-110. Co-authors: R. Edwards and K.V. Mardia.

[68] *Realization of ℓ_p by spaces of random variables.* Applicable Analysis, 8, 337-347. Co-author: S.S. Sheu.

[69] *Shorted operators and generalized inverses of matrices.* Linear Algebra and its Applications, 25, 45-56. Co-author: S.K. Mitra.

[70] *Rank order estimators in the case of grouped data.* Sankhyā, Series B, 41, 239-259. Co-author: A.R. Padmanabhan.

[71] *Asymptotic behavior of stochastic linear rank statistics.* Bulletin of the International Statistical Institute, 47(4), 292-297. Co-author: N.S. Rajaram.

[72] *A simple test for goodness-of-fit based on spacing with some efficiency comparisons.* Contributions to Statistics — Jaroslav Hájek Memorial Volume, Academia Prague, Czechoslovak Academy of Sciences, 197-209. Co-authors: J.S. Rao and Y. Yoon.

[73] *Sample size, parameter rates and contiguity — the i.i.d. case.* Communication in Statistics — Theory and Methods, A8 (1), 71-83. Co-authors: M.G. Akritas and G.G. Roussas.

[74] *Matrices G satisfying simultaneous equations $A * MAG = A * M$ and $G * NGA = G * N$.* Journal of Indian Statistical Association, 17, 103-108. Co-authors: C.G. Khatri and S.K. Mitra.

1980

[75] *Empirical distribution functions and functions of order statistics for mixing random variables.* Journal of Multivariate Analysis, 10, 405-425. Co-author: L.T. Tran.

[76] *Asymptotic normality and convergence rates of linear rank statistics under alternatives.* Banach Center Publications, 6, 267-277. Co-author: N.S. Rajaram.

1981

[77] *Invariance principles for rank statistics for testing independence*. Contributions to Probability Theory. Academic Press, New York, 267-282. Co-author: L.T. Tran.

[78] *A note on predicting the results of chess championship.* Behavioral Science, 26, 85-87. Co-author: Robert Bartoszynski.

[79] *Complex Planar Splines*. Journal of Approximation Theory, 31, 383-402. Co-author: Gerhard Opfer.

[80] *Some remarks on strategy in playing tennis*. Behavioral Science, 26, 379-387. Co-author: Robert Bartoszynski.

[81] *Différentielle d'une fonction floue*. C.R. Acad. Sc. Paris, 293, Série I, 237-239. Co-author: Dan A. Ralescu.

[82] *A nonparametric test of ordered alternatives in the case of skewed data, with a biomedical application*. Statistics and Related Topics, North Holland Publishing Company, 279-283. Co-authors: A.R. Padmanabhan and A.K. Md. Ehsanes Saleh.

1982

[83] *Shorted matrices – an extended concept and some applications*. Linear Algebra and its Applications, 42, 57-79. Co-author: S.K. Mitra.

[84] *On estimating intersubject variability of choice probabilities under observability constraints*. Journal of Mathematical Psychology, 25, 175-184. Co-author: Robert Bartoszynski.

[85] *Integration on fuzzy sets*. Advances in Applied Mathematics, 3, 430-434. Co-author: Dan Ralescu.

[86] *A possibility measure is not a fuzzy measure*. Fuzzy Sets and Systems, 7, 311-313. Co-author: Dan A. Ralescu.

[87] *On the degeneration of the variance in the asymptotic normality of signed rank statistics*. Statistics and Probability: Essays in Honor of C.R. Rao. North Holland Publishing Company, 591-608. Co-author: Stefan Ralescu.

[88] *Stochastic integrals and rank statistics*. Nonparametric Statistical Inference, Volume II, North Holland Publishing Company, 729-747. Co-author: N.S. Rajaram.

[89] *The asymptotic distribution theory of one sample signed rank statistic.* Statistical Decision Theory and Related Topics III, Volume 2, Academic Press, New York, 213-232. Co-author: Stefan S. Ralescu.

1983

[90] *Theory of nonparametric statistics for rounded-off data with applications.* Mathematische Operations-Forschung und Statistik, Series Statistics, 14, 301-349. Co-author: A.R. Padmanabhan.

[91] *Strong law of large numbers for Banach space valued random sets.* Annals of Probability, 11, 222-224. Co-author: Dan A. Ralescu.

[92] *On the order of magnitude of cumulants of von Mises functionals and related statistics.* Annals of Probability, 11, 346-354. Co-author: R.N. Bhattacharya.

[93] *Strong law of large numbers with respect to a set-valued probability measure.* Annals of Probability, 11, 1051-1054. Co-author: Dan A. Ralescu.

[94] *Differentials of fuzzy functions.* Journal of Mathematical Analysis and Applications, 91, 552-558. Co-author: Dan A. Ralescu.

[95] *Complex Chebyshev polynomials and generalizations with an application to the optimal choice of interpolating knots in complex planar splines.* Journal of Approximation Theory, 37, 89-101. Co-author: Gerhard Opfer.

[96] *The fundamental bordered matrix of linear estimation and the Duffin–Morley general linear electromechanical systems.* Applicable Analysis, 14, 241-258. Co-author: S.K. Mitra.

[97] *Improving upon the best invariant estimator in multivariate location problems.* The Australian Journal of Statistics, 25, 453-462. Co-author: Dan A. Ralescu.

1984

[98] *On Berry-Esséen rates, a law of the iterated logarithm and an invariance principle for the proportion of the sample below the sample mean.* Journal of Multivariate Analysis, 14, 231-247. Co-author: Stefan S. Ralescu.

[99] *Centering of signed rank statistics with continuous score generating function.* Teoria Veroyatnostei i ee Primeneniya (also in Theory of Probability and Its Applications, Translation by SIAM), 29, 580-584. Co-author: Stefan S. Ralescu.

[100] *Law of large numbers and central limit theorem for fuzzy random variables.* Cybernetics and Systems Research, 2, 525-529. Co-authors: E.P. Klement and D.A. Ralescu.

[101] *The Hausdorff α-dimensional measures of the level sets and the graph of the N-parameter Wiener process.* Metrika, 31, 275-283. Co-author: L.T. Tran.

[102] *Convergence of generalized inverses with applications to asymptotic hypothesis testing.* Sankhyā, Series A, 46, 277-286. Co-authors: Carl T. Russell and Thomas Mathew.

[103] *Rank order tests for the parallelism of several regression surfaces.* Journal of Statistical Planning and Inference, 10, 43-57. Co-author: Ching-Yuan Chiang.

[104] *Edgeworth expansions for signed linear rank statistics with regression constants.* Journal of Statistical Planning and Inference, 10, 137-149. Co-author: Munsup Seoh.

[105] *Edgeworth expansions for signed linear rank statistics under near location alternatives.* Journal of Statistical Planning and Inference, 10, 289-309. Co-author: Munsup Seoh.

[106] *Asymptotic normality of the lengths of a class of nonparametric confidence intervals for regression parameters.* Canadian Journal of Statistics, 12, 217-228. Co-author: T.J. Wu.

[107] *Linear minimax estimators for estimating a function of the parameter.* The Australian Journal of Statistics, 26, 277-283. Co-authors: Dan A. Ralescu and Stefan S. Ralescu.

[108] *Rank procedures for testing subhypotheses in linear regression.* Annals of the Institute of Statistical Mathematics, 36, 35-50. Co-author: Ching-Yuan Chiang.

[109] *Locally most powerful rank tests for the two-sample problem with censored samples.* Journal of Organizational Behavior and Statistics, 1, 189-196. Co-author: A.K. MD. Ehsanes Saleh.

1985

[110] *Linear serial rank tests for randomness against ARMA alternatives.* Annals of Statistics, 13, 1156-1181. Co-authors: Marc Hallin and Jean-François Ingenbleek.

[111] *Cramér-type large deviations for generalized rank statistics.* Annals of Probability, 13, 115-125. Co-authors: Stefan S. Ralescu and Munsup Seoh.

[112] *The concept of normality for fuzzy random variables.* Annals of Probability, 13, 1373-1379. Co-author: Dan A. Ralescu.

[113] *Limit theorems for random compact sets in Banach space.* Mathematical Proceedings of the Cambridge Philosophical Society, 97, 151-158. Co-author: Dan A. Ralescu.

[114] *On the rate of convergence in the central limit theorem for signed rank statistics.* Advances in Applied Mathematics, 6, 23-51. Co-author: Stefan S. Ralescu.

[115] *Asymptotic distributions of likelihood ratio criteria for testing latent roots and latent vectors of a covariance matrix under an elliptic population.* Biometrika, 72, 331-338. Co-author: T. Hayakawa.

[116] *Gaussian approximations of signed linear rank statistics process.* Journal of Statistical Planning and Inference, 11, 277-312. Co-author: T.J. Wu.

[117] *A sharpening of the remainder term in the higher dimensional central limit theorem for multilinear rank statistics.* Journal of Multivariate Analysis, 17, 148-167. Co-authors: Manfred Denker and Uwe Rösler.

[118] *Asymptotic expansions of the distributions of some test statistics.* Annals of the Institute of Statistical Mathematics, 37, 95-108. Co-author: Takesi Hayakawa.

[119] *On the rate of convergence to asymptotic normality for generalized linear rank statistics.* Annals of the Institute of Statistical Mathematics, 37, 51-69. Co-author: Munsup Seoh.

[120] *Berry-Esséen theorems for signed linear rank statistics with regression constants.* Limit Theorems in Probability Statistics, (P. Révész, Editor). North Holland Publishing Company, 875-905. Co-author: Munsup Seoh.

[121] *Central limit theorem for linear rank statistics process.* Statistical Theory and Data Analysis, (K. Matusita, Editor). North Holland Publishing Company, 103-145. Co-author: Mark A. Carlson.

[122] *On the Cramér-Fréchet-Rao inequality for translation parameter in the case of finite support.* Mathematische Operationsforschung und Statistik, Series Statistics, 16, 495-506. Co-author: I. Vincze.

[123] *Tests de rangs linéaires pour une hypothèse de bruit blanc.* C.R. Acad. Sc. Paris, 301, Série I, 49-52. Co-authors: Marc Hallin and Jean-François Ingenbleek.

[124] *Solutions fortes des équations différentielles stochastiques pour des processus à indice multidimensionnel.* C.R. Acad. Sc. Paris, 301, Série I, 845-848. Co-author: Markus Dozzi.

[125] *Tests de rangs quadratiques pour une hypothèse de bruit blanc.* C.R. Acad. Sc. Paris, 301, Série I, 935-938. Co-authors: Marc Hallin and Jean-François Ingenbleek.

1986

[126] *Maximum likelihood estimation for stationary point processes.* Proceedings of the National Academy of Sciences, U.S.A., 83, 541-545. Co-author: Pham D. Tuan.

[127] *Fuzzy random variables.* Journal of Mathematical Analysis and Applications, 114, 409-422. Co-author: Dan A. Ralescu.

[128] *The order of normal approximation for signed linear rank statistics.* Teoria Veroyatnostei i ee Primeneniya (also in Theory of Probability and Its Applications, Translation by SIAM), 31, 156-163. Co-author: Tiee-Jian Wu.

[129] *Gaussian random sets in Banach space.* Teoria Veroyatnostei i ee Primeneniya (also in Theory of Probability and Its Applications, Translation by SIAM), 31, 598-601. Co-authors: Dan A. Ralescu and Stefan S. Ralescu.

[130] *Strong solutions of stochastic differential equations for multiparameter processes.* Stochastics, 17, 19-41. Co-author: Markus Dozzi.

[131] *Central limit theorem for perturbed empirical distribution functions evaluated at a random point.* Journal of Multivariate Analysis, 19, 273-279. Co-author: Stefan S. Ralescu.

[132] *Limit theorems for fuzzy random variables.* Proceedings of the Royal Society of London, 407, 171-182. Co-authors: E.P. Klement and D.A. Ralescu.

[133] *Bhattacharyya bound of variances of unbiased estimators in nonregular cases.* Annals of the Institute of Statistical Mathematics, 39, 35-44. Co-authors: M. Akahira and K. Takeuchi.

[134] *Almost sure linearity for signed rank statistics in the non-i.i.d. case.* Acta Mathematica Hungarica, 48, 273-284. Co-author: Stefan S. Ralescu.

[135] *Les tests de rangs dans l'analyse des séries chronologiques.* Cahiers du C.E.R.O., 28, 41-55. Co-authors: Marc Hallin and Jean-François Ingenbleek.

[136] *Tests de rangs localement optimaux pour une hypothèse de bruit blanc multivarié.* C.R. Academy of Sciences, Paris, 303, Série I, 901-904. Co-authors: Marc Hallin and Jean-François Ingenbleek.

1987

[137] *Linear and quadratic serial rank tests for randomness against serial dependence.* Journal of Time Series Analysis, 8, 409-424. Co-authors: Marc Hallin and Jean-François Ingenbleek.

[138] *Asymptotic normality of a class of nonparametric statistics.* Econometric Theory, 3, 313-347. Co-author: Munsup Seoh.

[139] *Linear rank statistics, bounded operators and weak convergence.* New Perspectives in Theoretical and Applied Statistics, John Wiley and Sons, New York, 171-206. Co-authors: Manfred Denker and Gerhard Keller.

[140] *Locally asymptotically optimal tests for randomness.* New Perspectives in Theoretical and Applied Statistics, John Wiley and Sons, New York, 87-99. Co-author: Marc Hallin.

[141] *Berry-Esséen theorems for signed rank statistics under near location alternatives.* Studia Scientiarum Mathematicarum Hungarica, 20, 197-211. Co-author: Munsup Seoh.

[142] *Tests of subhypotheses in linear regression based on rank order estimates.* Studia Scientiarum Mathematicarum Hungarica, 20, 237-247. Co-author: Ching-Yuan Chiang.

[143] *Limit theorems for random central order statistics.* Adaptive Statistical Procedures and Related Topics, IMS Lecture Notes and Monograph Series, 8, 447-475. Co-author: Stefan Ralescu.

[144] *On admissible estimation in exponential families with imprecise information.* In Statistical Decision Theory and Related Topics IV, (Editors: Shanti S. Gupta and James O. Berger), Springer Verlag, 403-408. Co-author: Stefan Ralescu.

[145] *Convergence faible de la statistique serielle linéaire de rang en condition de dépendance avec applications aux séries chronologiques et processus de Markov.* C.R. Academy of Sciences, Paris, 304, Série I, 583-586. Co-author: Michel Harel.

[146] *Convergence faible du processus empirique multidimensionnel corrigé en condition de mélange.* C.R. Academy of Sciences, Paris, 305, Série I, 93-95. Co-author: Michel Harel.

[147] *Weak convergence of weighted multivariate empirical processes under mixing conditions.* Proceedings of the Sixth Pannonian Symposium on Mathematical Statistics, Volume A: Mathematical Statistics and Probability (Editors: M.L. Puri, P. Révész and W. Wertz), Reidel Publishing Company, Holland, 121-141. Co-author: Michel Harel.

1988

[148] *Locally asymptotically rank-based procedures for testing autoregressive moving average dependence.* Proceedings of the National Academy of Sciences, U.S.A., 85, 2031-2035. Co-author: Marc Hallin.

[149] *Optimal rank-based procedures for time series analysis: testing an ARMA model against other ARMA models.* Annals of Statistics, 16, 402-432. Co-author: Marc Hallin.

[150] *On time-reversibility and the uniqueness of moving average representations for non-Gaussian stationary time series.* Biometrika, 75, 170-171. Co-authors: Marc Hallin and Claude Lefevre.

[151] *Asymptotic behavior of multi-response permutation procedures.* Advances in Applied Mathematics, 9, 200-210. Co-author: Manfred Denker.

[152] *Adaptive nonparametric procedures and applications.* Journal of the Royal Statistical Society, Series C (Applied Statistics), 37, 205-218. Co-authors: N.J. Hill and A.R. Padmanabhan.

[153] *Asymptotic properties of linear functions of order statistics.* Journal of Statistical Planning and Inference, 18, 203-223. Co-author: Cheun-Der Lea.

[154] *An invariance principle for processes indexed by two parameters and some statistical applications.* Probability and Mathematical Statistics, 9, 25-76. Co-author: Manfred Denker.

[155] *Asymptotic properties of perturbed empirical distribution functions evaluated at a random point.* Journal of Statistical Planning and Inference, 19, 201-215. Co-author: Cheun-Der Lea.

[156] *L'espace $\tilde{D}_k$ et la convergence faible du processus empirique indexé par rectangles en condition de mélange.* C.R. Academy of Sciences, Paris, 306, Série I, 207-210. Co-author: Michel Harel.

[157] *On locally asymptotically maximin tests for ARMA processes.* In Statistical Theory and Data Analysis II (Editor: K. Matusita), 495-499. North Holland Publishing Company. Co-author: Marc Hallin.

[158] *Comportement limite de la U-statistique, de la V-statistique et d'une statistique de rang à un échantillon pour des processus absolument réguliers non stationnaires.* C.R. Academy of Sciences, Paris, 306, Série I, 625-628. Co-author: Michel Harel.

[159] *Tests de rangs signés localement optimaux pour une hypothèse de dépendance ARMA.* C.R. Academy of Sciences, Paris, 307, Série I, 355-358. Co-author: Marc Hallin.

[160] *Convergence faible de la statistique sérielle linéaire de rang avec des fonctions de scores et des constantes de régression non bornées en condition de mélange.* C.R. Academy of Sciences, Paris, 307, Série I, 617-620. Co-author: Michel Harel.

1989

[161] *Asymptotic expansions for sums of nonidentically distributed Bernoulli random variables.* Journal of Multivariate Analysis, 28, 282-303. Co-authors: Paul Deheuvels and Stefan Ralescu.

[162] *Asymptotically most powerful rank tests for multivariate randomness against serial dependence.* Journal of Multivariate Analysis, 30, 34-71. Co-authors: Marc Hallin and Jean François Ingenbleek.

[163] *Central limit theorems under alternatives for a broad class of nonparametric statistics.* Journal of Statistical Planning and Inference, 22, 271-294. Co-author: Munsup Seoh.

[164] *Limiting behavior of U-statistics, V-statistics and one sample rank order statistics for nonstationary absolutely regular processes.* Journal of Multivariate Analysis, 30, 181-204. Co-author: Michel Harel.

[165] *On the rate of convergence in normal approximation and large deviation probabilities for a class of statistics.* Teoria Veroyatnostei i ee Primeneniya (also in Theory of Probability and its Applications, Translation by SIAM), 33, 735-750. Co-author: Munsup Seoh.

[166] *A new semigroup technique in Poisson approximation.* Semigroup Forum, 38, 189-201. Co-authors: Paul Deheuvels and Dietmar Pfeifer.

[167] *Weak convergence of the U-statistic and weak invariance of the one sample rank order statistic for Markov processes and ARMA models.* Journal of Multivariate Analysis, 31, 258-265. Co-author: Michel Harel.

[168] *Convergence faible de la U-statistique generalisée pour des processes nonstationnaires absolument réguliers.* C.R. Academy of Sciences, Paris, 309, Série I, 135-138. Co-author: Michel Harel.

1990

[169] *Weak convergence of serial rank statistics under dependence with applications in time series and Markov processes.* Annals of Probability, 18, 1361-1387. Co-author: Michel Harel.

[170] *Weak invariance of generalized U-statistics for non-stationary absolutely regular processes.* Stochastic Processes and Their Applications, 34, 341-360. Co-author: Michel Harel.

[171] *The space $\tilde{D}_k$ and weak convergence for the rectangle-indexed processes under mixing.* Advances in Applied Mathematics, 11, 443-474. Co-author: Michel Harel.

[172] *Asymptotic normality of L-statistics based on decomposible time series.* Journal of Multivariate Analysis, 35, 260-275. Co-authors: K.C. Chanda and F.H. Ruymgaart.

[173] *A strong invariance principle concerning J-upper order statistics for stationary m-dependent sequences.* Journal of Statistical Planning and Inference, 25, 43-51. Co-author: George Haiman.

[174] *Weak convergence of the serial linear rank statistics with unbounded scores and regression constants under mixing conditions.* Journal of Statistical Planning and Inference, 25, 163-186. Co-author: Michel Harel.

[175] *Information.* In: Acting Under Uncertainty: Multidisciplinary Conceptions (Editor: George von Furstenberg). Kluwer Academic Publishers, Boston/Dordrecht/Lancaster, 451-469. Co-author: Howard L. Resnikoff.

[176] *Measure of information and contiguity.* Statistics and Probability Letters, 9, 223-228. Co-author: Istvan Vincze.

[177] *Information and Mathematical Statistics.* Proceedings of the Fourth Prague Symposium on Asymptotic Statistics, Charles University, Prague (Editors: P. Mandl and M. Hušková) August 29, 1988 September 2, 1988, 447-456. Co-author: Istvan Vincze.

1991

[178] *A local algorithm for constructing nonnegative cubic splines.* Journal of Approximation Theory, 64, 1-16. Co-authors: Bernd Fischer and Gerhard Opfer.

[179] *Time series analysis via rank order theory: signed-rank tests for ARMA models.* Journal of Multivariate Analysis, 39, 1-29. Co-author: Marc Hallin.

[180] *Convergence theorem for fuzzy martingales.* Journal of Mathematical Analysis and Applications, 160, 107-122. Co-author: Dan Ralescu.

[181] *Limiting behavior of one sample rank-order statistics with unbounded scores for nonstationary absolutely regular processes.* Journal of Statistical Planning and Inference, 27, 1-23. Co-author: Michel Harel.

[182] *Weak invariance of the multidimensional rank statistic for nonstationary absolutely regular processes.* Journal of Multivariate Analysis, 36, 204-221. Co-author: Michel Harel.

[183] *On Hilbert space valued U-statistics.* Teoria Veroyatnostei i ee Primeneniya (also in Theory of Probability and Its Applications, Translation by SIAM), 36, 604-605. Co-author: V.V. Sazonov.

[184] *Théorème central limite de l'estimateur de densité et de l'erreur quadratique intégrée pour des variables aléatoires nonstationnaires.* C.R. Academy of Sciences, Paris, 312, Série 1, No. 1, 145-148. Co-author: Michel Harel.

[185] *Invariance faible de la statistique de rang multidimensionnelle. Applications.* C.R. Academy of Sciences, Paris, 312, Série 1, No. 4, 349-352. Co-author: Michel Harel.

1992

[186] *Rank tests for time series analysis.* In: New Directions in Time Series Analysis (the IMA Volume in Mathematics and Its Applications) 45, 111-153. (Editors: David Brillinger, Emanuel Parzen, Murray Rosenblatt et al.). Springer-Verlag, New York/Heidelberg/Berlin. Co-author: Marc Hallin.

[187] *Rank-based estimation of ratio of scale parameters and applications.* Journal of Statistical Planning and Inference, 31, 23-49. Co-author: A.R. Padmanabhan.

[188] *The Neyman-Pearson probability ratio and information.* In: Data Analysis and Statistical Inference, Festschrift in Honor of Friedhelm Eicher (Editors: Siegfried Schach and Götz Trenkler), 53-64. Verlag Josef Eul, Bergisch Gladbach, Köln. Co-author: Istvan Vincze.

[189] *Asymptotic normality of two-sample linear rank statistics under U-statistic structure.* Journal of Statistical Planning and Inference, 32, 89-110. Co-author: Manfred Denker.

[190] *Some asymptotic results for a broad class of nonparametric statistics.* Journal of Statistical Planning and Inference, 32, 165-196. Co-author: Marc Hallin.

1993

[191] *A strong invariance principle concerning the Jth upper order statistics for stationary Gaussian sequences.* Annals of Probability, 21, 86-135. Co-author: George Haiman.

[192] *The time spent by the Wiener process in a narrow tube before leaving a wide tube.* Proceedings of the American Mathematical Society, 117, 529-536. Co-author: Antonia Földes.

[193] *Weak convergence of the simple linear rank statistic under mixing conditions in the nonstationary case.* Teoria Veroyatnostei i ee Primeneniya (also in Theory of Probability and Its Applications, Translation by SIAM) 38, 579-599. Co-author: Michel Harel.

[194] *Nonparametric prediction for random fields.* Stochastic Processes and Their Applications, 48, 139-156. Co-author: Frits Ruymgaart.

[195] *Asymptotic behavior of L-statistics for a large class of time series.* Annals of the Institute of Statistical Mathematics, 45, 687-701. Co-author: Frits H. Ruymgaart.

[196] *On the central limit theorem in Hilbert space with application to U-statistics.* In: Statistical Sciences and Data Analysis: Proceedings of the Third Pacific Area Statistical Conference. VSP International Science Publishers, Holland. (Editors: Kameo Matusita, Madan L. Puri and Takesi Hayakawa), 407-413. Co-author: V.V. Sazonov.

[197] *Asymptotics of the perturbed sample quantile for a sequence of m-dependent stationary random process.* In: Statistical Sciences and Data Analysis: Proceedings of the Third Pacific Area Statistical Conference. VSP International Science Publishers, Holland. (Editors: Kameo Matusita, Madan L. Puri and Takesi Hayakawa), 415-426. Co-author: Shan Sun.

1994

[198] *Weak convergence of weighted empirical U-statistics processes for dependent random variables.* Journal of Multivariate Analysis, 48, 297-314. Co-authors: Colm Art O'Cinneide and Michel Harel.

[199] *Aligned rank tests for linear models with autocorrelated error terms.* Journal of Multivariate Analysis, 50, 175-237. Co-author: Marc Hallin.

[200] *Law of the iterated logarithm for perturbed empirical distribution functions evaluated at a random point for nonstationary random variables.* Journal of Theoretical Probability, 7, 831-855. Co-author: Michel Harel.

[201] *Berry–Esséen rate in asymptotic normality for perturbed sample quantiles.* Metrika, 41, 83-98. Co-author: Munsup Seoh.

[202] *On some limit laws for aligned perturbed empirical distribution functions.* Statistics and Probability Letters, 21, 317-321. Co-author: Manfred Denker.

[203] *Change curves in the presence of dependent noise.* In: Change-Point Problems, a volume in the IMS Lecture Notes and Monograph Series, Volume 23, 242-254. Co-author: Frits H. Ruymgaart.

1995

[204] *Nonparametric methods for stratified two-sample designs with application to multiclinic trials.* Journal of the American Statistical Association, 90, 1004-1014. Co-authors: Edgar Brunner and Shan Sun.

[205] *Extremes of Markov sequences.* Journal of Statistical Planning and Inference, 45, 185-201. Co-authors: George Haiman and Maxime Kiki.

[206] *A multivariate Wald-Wolfowitz rank test against serial dependence.* The Canadian Journal of Statistics, 23, 55-65. Co-author: Marc Hallin.

[207] *Higher order asymptotic theory for normalizing transformations.* Annals of the Institute of Statistical Mathematics, 17, 581-600. Co-author: Masanobu Taniguchi.

[208] *Asymptotic expansions in statistics: a review of methods and applications.* In: Advances in Econometrics and Quantitative Economics: Essays in honor of C.R. Rao. Blackwell Publishers, Oxford, United Kingdom. (Editors: G.S. Maddala, Peter C.B. Phillips and T.N. Srinavasan), 88-122. Co-author: Rabi N. Bhattacharya.

[209] *Comportement asymptotique des estimateurs de la fonction de regression pour des variables aléatoires dependantes et nonstationnaires.* C.R. Academy of Sciences, Paris, 320, Série I, 75-78. Co-author: Michel Harel.

1996

[210] *Nonparametric approach for non-Gaussian vector stationary processes.* Journal of Multivariate Analysis, 56, 259-283. Co-authors: Masanobu Taniguchi and Masao Kondo.

[211] *Valid Edgeworth expansions of M-estimators in regression models with weakly dependent residuals.* Econometric Theory, 12, 331-346. Co-author: Masanobu Taniguchi.

[212] *Conditional U-statistics for dependent random variables.* Journal of Multivariate Analysis, 57, 84-100. Co-author: Michel Harel.

[213] *Asymptotic normality of nearest neighbor regression function estimates based on nonstationary dependent observations.* Special Issue on Multivariate Statistical Analysis in Honor of Professor Minoru Siotani, American Journal of Mathematical and Management Sciences, 15, 255-290. Co-author: Michel Harel.

[214] *Weak convergence of sequences of first passage processes and applications.* Stochastic Processes and Their Applications, 62, 327-345. Co-author: Stefan Ralescu.

[215] *Characterization of weak convergence for perturbed empirical and quantile processes under φ-mixing.* Journal of Statistical Planning and Inference, 53, 285-295. Co-authors: H.J.A. Degenhardt, Shan Sun and Martien van Zuijlen.

[216] *Nonparametric density estimators based on nonstationary absolutely regular random sequences.* Journal of Applied Mathematics and Stochastic Analysis, 9, 233-254. Co-author: Michel Harel.

[217] *Nonparametric methods in design and analysis of experiments.* In: Handbook of Statistics, volume 13, Design of Experiments. (Editors: S. Ghosh and C.R. Rao) North Holland, Amsterdam, Netherlands, 631-703. Co-author: Edgar Brunner.

1997

[218] *Normal approximation of U-statistics in Hilbert spaces.* Teoria Veroyatnostei i ee Primeneniya (also in Theory of Probability and its Applications, Translation by SIAM), 41, 481-504. Co-authors: Yu.V. Borovskich and V.V. Sazonov.

1998

[219] *Autoregressive quantiles and related rank scores processes for generalized random coefficient autoregressive processes.* Journal of Statistical Planning and Inference, 68, 271-294. Co-author: Michel Harel.

[220] *Records and 2-block designs of 1-dependent stationary sequences under local dependence.* Annales de l'institut Henri Poincaré, 34, 481-503. Co-authors: George Haiman, Nelly Mayeur and Valery Nevzorov.

1999

[221] *Conditional empirical processes defined by nonstationary absolutely regular sequences.* Journal of Multivariate Analysis, 70, 250-285. Co-author: Michel Harel.

[222] *Rank score tests in factorial designs with repeated measures.* Journal of Multivariate Analysis, 70, 286-317. Co-author: Edgar Brunner with Ullrich Munzel.

[223] *Robust one-way ANOVA under possibly non-regular conditions.* Biometrical Journal, 41, 321-339. Co-authors: G.J. Babu and A.R. Padmanabhan.

2000

[224] *The limiting density of unit root test statistics: A unifying technique.* The Journal of Time Series Analysis, 21, 249-260. Co-authors: Mithat Gönen, Frits H. Ruymgaart and Martien C.A. van Zuijlen.

[225] *An estimation method in time series errors-in-variables models.* Journal of the Japan Statistical Society, 30, 75-87. Co-authors: Narihisa Tsuga and Masanobu Taniguchi.

[226] *Limit theorems for fuzzy random variables.* In: Asymptotics in Statistics and Probability: Papers in Honor of George Gregory Roussas. VSP International Science Publishers, The Netherlands, 331-346. Co-author: Frank Proske.

2001

[227] *Nonparametric methods in factorial designs.* Statistical Papers (Springer-Verlag journal), 42, 1-52. Co-author: Edgar Brunner.

[228] *Point and interval estimators for nonparametric treatment effects in designs with repeated measures.* In: Data Analysis from Statistical Foundations: Festschrift in Honor of Donald A.S. Fraser. (Editor: A.K.E. Saleh) Nova Science Publishers, Inc., New York, 167-178. Co-author: Edgar Brunner.

[229] *U-statistiques conditionnelles universellement consistantes pour des modèles de Markov cachés.* Comptes rendus de l'Académie des Sciences, Paris, t.333, Série I/Mathématique, 953-956. Co-author: Michel Harel.

2002

[230] *A class of rank score tests in factorial designs.* Journal of Statistical Planning and Inference, 103, 329-359. Co-author: Edgar Brunner.

[231] *Central limit theorem for Banach space valued fuzzy random variables.* Proceedings of the American Mathematical Society 130, 1493-1501. Co-author: Frank Proske.

[232] *Strong law of large numbers for Banach space valued fuzzy random variables.* Journal of Theoretical Probability 15, 543-552. Co-author: Frank Proske.

[233] *Weighted rank statistics in factorial designs with fixed effects.* Statistica Neerlandica, 56, 179-194. Co-authors: Edgar Brunner and Sebastian Domhof.

[234] *Asymptotics of tests for a unit root in autoregression.* Journal of Statistical Planning and Inference (To appear). Co-authors: Kenji Sakiyama and Masanobu Taniguchi.

[235] *The multivariate nonparametric Behrens-Fisher problem.* Journal of Statistical Planning and Inference (To appear). Co-authors: Edgar Brunner and Ullrich Munzel.

[236] *LAN based asymptotic theory for time series.* In: Proceedings of the Fourth International Triennial Calcutta Symposium on Probability and Statistics, December 26-28, 2000. (Editors: Arijit Chaudhuri and Malay Ghosh), Calcutta Statistical Association Bulletin, 52, 143-170. Co-author: Masanobu Taniguchi.

2003

[237] *A strong law of large numbers for the generalized random sets from the viewpoint of empirical processes.* Proceedings of the American Mathematical Society. (To appear.) Co-author: Frank Proske.